Ernst Heinrich Hirschel · Horst Prem · Gero Madelung

Aeronautical Research in Germany

Springer
Berlin
Heidelberg
New York
Hong Kong
London
Milan
Paris
Tokyo

Ernst Heinrich Hirschel · Horst Prem · Gero Madelung

Aeronautical Research in Germany

From Lilienthal until Today

With 557 Figures and 24 Tables

Springer

Professor Dr. Ernst Heinrich Hirschel
Herzog-Heinrich-Weg 6
85604 Zorneding
Germany

Horst Prem
Prinz-Otto-Str. 9
85521 Ottobrunn
Germany

Professor Gero Madelung
Bonselsstr. 10
81925 München
Germany

First published in the German language under the title "Luftfahrtforschung in Germany" by Bernard & Graefe Verlag, Bonn, Germany, 2001

ISBN 3-540-40645-X Springer-Verlag Berlin Heidelberg New York

Library of Congress Cataloging-in-Publication Data
A catalog record for this book is available from Library of Congress.
Bibliographic information published by Die Deutsche Bibliothek.
Die Deutsche Bibliothek lists this publication in the Deutsche Nationalbibliografie;
detailed bibliographic data is available in the Internet at <http:/dnb.ddb.de>.

Springer-Verlag is a part of Springer Science+Business Media

springeronline.com

Printed in Germany

Production Editor: gaby maas, Berlin
Cover Design: deblik Berlin
Typesetting: medio Technologies AG, Berlin

Printed on acid-free paper 62/3020/gm 5 4 3 2 1 0

Preface

In May 2001 the book "Luftfahrtforschung in Deutschland" was brought out by the publishing house Bernard & Graefe, Bonn, Germany. Now, at the end of the year 2003, its English version "Aeronautical Research in Germany – from Lilienthal until Today" is published by Springer Verlag.

Although the decision to publish this English edition came somewhat late, it was possible to have the book appear at the end of 2003, because several enthusiastic people worked on the translation, and because the Springer Verlag was very helpful.

The text of this edition follows with only a few exceptions the original German text. For technical reasons, parts of the original text and also some figures have been omitted. On the other hand some text has been added if an update was deemed desirable. Apart from this, the text in general is unchanged that from the years 2000 and 2001.

We have tried, despite the number of translators, to achieve a consistent English text. This regards especially the nomenclature and the technical language, where all effort was made to follow the established international custom, i.e., the Anglo-American based "aerospace language". Where ambiguities existed with the translation of names of German institutions, the full German name is given in parentheses. The reader should also keep in mind, that the term "Staat" (state) in German can mean the "Reich", i.e., the "Empire", or, after 1945, the "Federal Republic of Germany" or one of the federal states. The meaning of the word "state" hence hopefully is clear in the respective context.

Another problem were the many references given in the text. In the German edition it was a rule to cite as many relevant primary and secondary literature as deemed necessary. We decided to keep these references also in this English edition. The reader will hence find a large number of German references without a translation of the title. However, if a English translation of an item of literature is available, the data of both the German and the English edition are given.

Finally we wish to acknowledge the substantial sponsoring of the book by the European Aeronautic Defense and Space Company (EADS), and furthermore by the Deutsches Zentrum für Luft- und Raumfahrt (DLR), to whom we are most grateful. We wish to thank the Springer-Verlag for publishing this book, and especially *Thomas Ditzinger* and *Gaby Maas* for their big help, advise and efforts. Big thanks also merit our translators, some of them also being the original contributors. First mentioned is *Egon Stanewsky*, who bore the brunt of the work, and then *Hans Fabian, Claudia*

Grant, Andreas Hafer, Xafer Hafer, Peter Hamel, Dietrich Hanke, Heinz Hoheisel, Heather King, Anke Kovar, Holger Poese, Hansgeorg Riedel, and finally *Stephan Eelman,* who reworked many of the figures. Last not least we would like to express our special thanks to *Kyrill von Gersdorff* and *Klaus Peters,* the former and the present coordinator and co-publisher of the book series "Die deutsche Luftfahrt" ("German Aeronautics") of Bernard & Graefe, who effectively helped us in every way also with this English edition of our book.

The book is appearing at the end of the year 2003, the 100th Anniversary of Powered Flight. We wish to dedicate it to the memories of *Otto Lilienthal, Wilbur* and *Orville Wright,* and of all of those who pursued – and still pursue – the dream of human flight.

Munich,
October 2003

Ernst Heinrich Hirschel
Horst Prem
Gero Madelung

Foreword of the German Edition by L. Bölkow (†)

In May 1980, Volume 1 of the book series "German Aeronautics" was presented at the Deutsches Museum (German Science and Technology Museum at Munich). It was dedicated to "Kurt Tank – Designer and Test Pilot at Focke Wulf". In the meantime, almost all important personalities, who have decisively influenced German Aeronautics, have been honored by their own volume within this book series.

Now, 21 years later, the volume on "Aeronautical Research in Germany" is published as Volume 30 presenting a summary of contributions, which Germany made to aeronautical research, a field which was always characterized by worldwide cooperation. The situation regarding the exchange of thoughts and information with foreign countries was not even affected by the politically enforced interruption of 12 years. The present volume explains the inner relations between the aeronautical research areas, which become explicit also through corresponding exhibits at the Deutsches Museum. It appreciates the researcher, who, often in the shadow of designers, managers and pilots, creates the prerequisites for success.

It was the particular desire of the then publisher and co-initiator of the book series, the late *Dr. Theodor Benecke,* to present the aeronautical research as it encompasses research at universities, research establishments and industry. The ones, who have now completed this work, know how demanding this has been. The secret of success lay in the willingness of more than 30 scientists and engineers to cooperate in establishing this comprehensive survey.

German aircraft engineering meant from the beginnings of flight until today a cooperation of practitioners, experimentalists and personalities searching for the physical laws underlying the phenomena observed and, herewith, also searching for the possibility of a repeatable and predictable use of these effects. Dependent on talent was, during the entire period, one as experimentalist and the other as calculating truth-seeker the provider of impulses.

Military customers as well as the civil market for aeronautical products have, with their demands, set the focal points in applied research. All crucial aerodynamic and gas dynamic basics, as, for instance, the effect of laminarity in the boundary layer as well as the swept wing, have already been worked out before World War Two. In addition, one must not forget that also basic computational methods for thin sheet metal under static and dynamic loads had already been formulated.

Foreword of the German Edition

Improvements in the area of materials with regard to the intensive utilization of fiber-reinforced materials and especially the triumphant advance of electronics in the area of simulation and flight control have, together with gas-turbine aircraft propulsion, revolutionized aeronautics after the war. Added must be sensor technology, which made automatic flight possible. Aircraft engineering and astronautics grow together in conceiving air-breathing space transport systems. New on-board energy systems, as, for instance, fuel cells, but also new, light electric propulsion systems, will sustain the pace-setting function of aeronautics for other industry branches.

Courage to really attack new aircraft configurations with large transport capacities is needed because the conventional aircraft design, which has only changed marginally since the Junkers F 13 (1919), is at its end. Within the worldwide research exchange – especially within the narrower European framework – the German contribution will continue to have its appropriate significance.

Dipl.-Ing. Dr.-Ing. E. h. mult.
Ludwig Bölkow (1912–2003)

Foreword of the German Edition by W. Kröll

In the present book, over one hundred years of German Aeronautical Research are depicted. It is fascinating to observe what great momentum for aeronautics has resulted from this research. The German Aerospace Center (DLR) – and her predecessor organizations – was and is the central element, positioned between the universities, which are more oriented towards fundamental research, and the industrial experimental development. The effective cooperation between universities, large research establishments and industry is today, more than ever, decisive for the competitiveness of industrialized nations.

Aeronautical research is still an important technology motor for the national economy. Knowledge, results and impulses of aeronautical research, the latter still attracting excellent university graduates, find in numerous ways entry into other branches. The "Mathematization-wave" of more or less empirical disciplines, initiated about one hundred years ago at Göttingen has – with *Ludwig Prandtl* as a leading figure – given fluid mechanics and aerodynamics, hence flight technology, decisive impulses. The second "Mathematization-wave", driven by the development of the information technologies, also includes now the highly complex, multi-disciplinary design problems associated with aircraft engineering, with large challenges for aeronautical research.

In view of the rapidly increasing air-traffic volume, the critical research areas of the upcoming ten to twenty years will, most of all, be concerned with safety aspects, the reduction of pollutant emissions and means of noise reduction. Besides research involving the air vehicle proper, optimizing the infrastructure of airspace, the runway and the airport, which are steadily becoming more congested will, in addition, gain crucial importance.

The research activities of all involved are, since almost 50 years and to an increasing degree, being carried out within an international framework. This means for us, of course, primarily, within Europe. DLR has for a long time actively supported a consolidation of European aerospace research. The foundation of the Association of the European Aeronautical Research Establishments (EREA) was co-initiated by us. Concerning large research facilities, there exists for a long time a European cooperation, which is steadily being extended. Generally, there is, however, a European aeronautical research policy – on the level of the European Union – needed, which harmonizes the activities of the large research establishments, university research and industry further and in a stronger way.

The aeronautical research centers must follow the step of the consolidated European aerospace industry to remain competitive on a global scale. This requires also, besides the engagement of the research centers themselves, political initiatives to overcome the strive for national autonomy and the national "preservation of assets". Should it be possible to "Europeanize" in the foreseeable future the traditional and fruitful cooperation of German industry, research establishments and universities – which is appreciated in this book - we may expect from this new alliance pace-setting impulses for the future of European and global air transport.

Prof. Dr. rer. nat. Walter Kröll
then Head, Board of Directors, DLR

Authors' Preface of the German Edition

This book is Volume 30 of the Series "German Aeronautics" of the publishing house Bernard & Graefe where all areas of German aeronautical technology, from the very beginnings to the newest developments in air vehicles within an international framework, are being depicted. It was the desire of the publisher and the publishing house to present with Volume 30 a portrayal of aeronautical research in Germany from the beginnings until today, giving a summarizing survey over the various research activities at universities, state-sponsored research establishments and industry.

Aeronautical research is part of the research within the engineering sciences and thus predominantly "applied" research. It has, in so far, a special status in the general research landscape since – for various reasons – it was carried out very early, i.e., already at the beginning of the second decade of the 20th Century, not only at universities, but also at state-sponsored research institutions. Research was also always, and will remain, part of the industrial activity. The reason for this is the fact that aircraft engineering, with many participating disciplines – including new developments, such as the information technology – still makes, and will make again, in cooperation with institutional research, great technical progress. In the process, a change in the paradigm of product philosophy can be observed for some time, namely, from the "technology-oriented" design characterized by "faster, higher, farther" which was for a long time the trade mark of aircraft engineering, to "cost-oriented" and "environment-conscious" design which is guided by the market and social acceptance. This applies stronger to civil than to military aircraft designs.

Which were – and are – the special features of the German aeronautical research? This question implies that there were, at least, such special features: Features in comparison to the research within other fields, but in particular also in comparison to research within other nations. There was, as in all fields, always a "Give and Take" between nations. Germany has, however, for some time, in many disciplines of aeronautical research occupied a pioneering role with, on the one hand, outstanding research personalities, such as *Ludwig Prandtl*, and, on the other hand, with just as outstanding business personalities, such as *Hugo Junkers*, who was also an eminent researcher. This, too, shall be made visible with this book.

This book is a result of a joint effort lasting more than three years because the authorship of a single person would have been inconceivable considering the width of the field. About three dozen people – active or already retired – from German universities, research establishments and industry have, as scientists, engineers and historians, provided their contributions – as authors of individual sections, as scientific advisers or as authors of the book.

The book addresses a broad readership: the expert, the interested layman, the scientist, the engineer and the student. It is not a textbook but it provides orientation and also addresses detail problems, their background and solution approaches. Importance was attached to a comprehensive literature survey as well as a subject index. Short biographies of outstanding research personalities shall ease the access to the topic "German Aeronautical Research".

There is a tendency to associate the topic "Aeronautical Research", hence also the topic "Aircraft Engineering", most of all in Germany, with "war research". However, aeronautical research is also just subject to the ambivalence of human actions. "*The limits in the technique of aircraft construction lie not in the technique itself but in its application by humans*" remarked *Ludwig Bölkow* 1990 with respect to this topic. It is indisputable that the airplane, as the airship, was already very early for all great powers, including Germany, also a tool of war. The "also" must be emphasized since it has to be shown that the motivation of many aeronautical researchers, also in industry, was initially simply the opening-up of flying and then the possible transport of people and goods by air. Today, the latter is realized to an extent, which has rendered flying as taken for granted to a degree that the entire scientific, technical, industrial and infrastructural background is hardly being perceived. Thus has the German airline Lufthansa transported in 2000 a total of 48 million passengers with the entire world air traffic amounting to about 1.68 billion.

Now to the contents of the book. Corresponding to the character of the series, the book does not claim to be a scientific publication, and it will also not give a complete portrayal of German aeronautical research in all its details. The book places main emphasis on "high-light" presentations trying to follow the most important lines of research chronographically, also naming the most important persons involved. A subjective view of the contributors can, in the process, not be ruled out. Considering the wealth of topics in aeronautical research, one had – also to limit the size of this volume – to omit a number of areas. Thus, the book does not contain a detailed presentation concerning research with respect to ballooning, airships, aviation medicine, and also not concerning astronautical research where many relations to aeronautics (space planes) exist. Concerning illustrations, it was attempted to present the whole scope of the research. The aircraft realized, which is, so to speak, the "end-product" of the aeronautical research, was in the process given a large amount of space. Concerning literature, also the other volumes of the book series "German Aeronautics", where details of the German aeronautical technology are comprehensively covered, were quoted (partly only with the number of the volume).

In the forty contributions to the present book, the individual way of writing and the form of expression of the contributing authors were largely retained. Repetitions were accepted if thereby an improvement in the understanding was attained. German aeronautical research has in many areas commenced with *Otto Lilienthal* and solely for that reason he and his work is, and this also holds for others, quoted more than once.

Munich
March 2001

Ernst Heinrich Hischel
Horst Prem
Gero Madelung

Authors' Acknowledgements of the German Edition

The authors, contributors, coordinators and the publishing house like to thank the archives of the Deutsches Museum, especially *Messrs. Dr. W. Füssl* and *A. Limmer* and *Ms. C. Hennet,* the archives of DLR, especially *Ms. Dr. K. Weingarten* and *Mr. H. Fütterer,* as well as the archives of Dasa/EADS for the provision of material including illustrations.

We thank as well the many expert colleagues and contemporary witnesses, especially *Messrs. Dipl.-Ing. Dr. E. h. mult. L. Bölkow (†), H. Deinhardt, Prof. Dr. K. Doetsch (†), Dr. H. Grieb, Dr. H. Hagen (†), K.-H. Mewes, Prof. Dr. H.-G. Münzberg (†), H. Zacher (†),* who have contributed by suggestions, discussions, critical advice, contributions and illustrations to the realization of this book. Thanks are also due to *Alke Prem* for proofreading the extensive manuscript and for the many suggestions concerning the linguistic form of this complex matter.

The readiness to cooperate, regardless of all technical boundaries, with the objective of a comprehensive popular-scientific presentation of German aeronautical research was the prerequisite for the success. The fact that so many agreed to help shows that great and general interest in the topic of the book exists. We hope that we have done justice to this interest.

Munich
March 2001

Ernst Heinrich Hischel
Horst Prem
Gero Madelung

Contents

Contents

Introduction

Part 1

Aeronautical Research and Technology

Ernst Heinrich Hirschel

Definition, Contents, Background

This volume of the book series "German Aeronautics" is dedicated to German aeronautical research, whose beginnings are here set to coincide with the work and flights of *Otto Lilienthal*. The deliberations are predominantly related to the airplane. The airship as well as astronautics are only treated marginally, in case of the latter considering aspects of the so called space planes which today, like the American "Space Shuttle", ascend to earth orbit by rocket propulsion and return glider-like. They will, in some decades, most likely employ air-breathing engines so that by then today's sharp differentiation will have been lifted.

Aeronautical research is, as part of the research within the engineering sciences, predominantly "applied research". Within the engineering sciences, aeronautical technology is a relatively young discipline compared to the older "sisters", such as, for instance, bridge- and fortress construction, mining or metallurgic engineering. It is also within the field of transportation technology the youngest discipline, following shipping and railroads, but of the same age as automobile technology. Astronautics, still younger, can probably not yet be considered a transportation-technology discipline.

The terms "aeronautical technology", "aeronautical science" and "aeronautical research" cover an interdisciplinary complex. Aeronautics, respectively air traffic, is, in the engineering sciences, considered as a "system" where the airplane itself is only one element, although the dominant one. The element "airplane", itself seen as "system", consists of subsystems, typically on the first level of the airframe, the propulsion system and the equipment. Concerning the "technique" of the subsystems (the term "technique" is today increasingly replaced by "technology"), there are often several disciplines involved. These disciplines contribute, respectively, to the design of the subsystems, such as, for instance, aerodynamics to the design of the overall aircraft, the airframe and the propulsion system. Electronics, or rather "Avionics", is part of almost all subsystems. The design of the overall aircraft, i.e., the functional integration of the contributions of all disciplines, is, finally, a discipline of its own that owes its origin to the work of the pioneers of aircraft engineering.

This brief reflection shows that the spectrum regarding the content of aeronautical research is very manifold with difficult, but also very fascinating questions. Although the human dream of "flying" has come true, it was only with the aid of complex techniques which were created with great idealistic personal and financial effort, and which are being further developed, a process whose end cannot be foreseen.

What is research? This question aims at the conceptual definition of research as well as its assignment to universities, research establishments and industry. Of course, this whole complex is today defined throughout since the roles as well as the motives of the participants, whether financing or executing, are for legal – also with respect to taxation law –, organizational and also for guarantee reasons different. Research, particularly in the natural sciences and in engineering, requires large financial resources since investments in staff, equipment, technical devices and buildings must be made, and all must be maintained, frequently over decades. Nevertheless, the definition changes in the daily usage of the language and also alters its contents with time. In the present section, the above question will, therefore, not be answered by pure definition, but it will be attempted, with particular regard for the topic, to give an introduction to the question, also with respect to the historical development. From the literature concerning the total complex, only a few references are quoted [1–3].

One can observe that in the ranking of social values, especially in Germany, research will intuitively be awarded primacy over business, i.e., the industry. Yet, the distribution of the roles of research, development and production is especially in aeronautics – with aircraft construction as the main element – very fluid. The aeronautical industry is a branch with special features. These are: a long technology lead time with great risks, long product-definition and product-development times, small series at large unit costs and a limited correction potential for product deficits due to the small series. The resulting large and long commitment of capital, but particularly also the military business component, have led, always and everywhere, to a strong governmental engagement, which manifested itself not only in development, production and sales support but most of all in the provision and maintenance of specific, government-sponsored research institutions which underlines this fluid character.

The highlights in the development of the natural sciences were set by a few great scholars. The separation of research and technology, advocated by the purists, never existed, particularly not during and since the natural-sciences revolution in the 16th and 17th century which is characterized by names such as *Kopernikus, Bacon, Galilei, Kepler, Descartes, Newton, Leibniz* and others [4]. The history of technology is, however, up to the present time full of examples of a rather tense relation between science and application. In an American study of 1985, it was attempted to shed light on the problem of the interrelation between science and technology [5]. Accordingly, technology is, other than the natural sciences, under the constraint to face the question of purpose, to foresee one's own possible stagnation or even failure and to plan countermeasures. Furthermore, technological development means not only to invent individual devices and procedures, but also to achieve within large systems a component-development level as uniform as possible and their perfect interaction. Basic knowledge has in both instances rather the character of a large reservoir and not the one of a source. Different from the generation of basic knowledge, is the development of complex technical systems – e.g., a transport aircraft – characterized by great financial, personnel and organizational expenses. It must, however, be stated that this also holds for large research establishments, for instance, as associated with particle physics.

Research in aeronautics, whoever is performing it, is as generator of specific knowledge the basis for the performance of industry. Before an explanation of this fact will be provided, some definitions shall be given. Following the "Rationalization Committee of German Economics" [2], a registered association, ("Rationalisierungs-Kuratorium der Deutschen Wirtschaft, RKW"), the concept "Research" embraces:

- **Basic Research**, i.e., research work aiming at the generation of new scientific or technical knowledge without being initially or predominantly oriented towards the goal of a practical applicability.
- **Applied Research**, i.e., research work aiming at the generation of new scientific or technical knowledge with a predominantly specified practical objective or application.
- **Experimental Development,** i.e., the utilization of scientific knowledge, including theoretical and experimental methods, to come to new or considerably improved material, equipment, procedures and system performances, which, after a successful "technology" demonstration and verification, finally lead to a new product.[1]

Therefore, certain activities of the industry are *research* if prior to a project or a product-realization program, respectively, individual physical-technical phenomena and their interrelation within a possible product ("system-aspect") are being investigat-

[1] The term "Experimental Development" is in so far misleading as it contradicts in the scientific domain linguistic usage, which distinguishes between "theoretical" and "experimental" research (and development), with the latter being "facility-related" research.

ed and researched. The *experimental development*, in aircraft engineering, more aptly called *development by experiment*, yields as visible result of a publicly financed project or a (self-financed) "Research and Development (R&D)" project, besides research reports, also research "hardware" up to experimental and demonstration aircraft. In the previous volumes of the series "German Aeronautics", as elsewhere [6], it is repeatedly shown that the experimental development, predominantly conducted by industry, is, indeed, also research, namely, systematic creative work to extend scientific and technical knowledge and its utilization with the objective to find new possibilities of application [7]. After 1945, it was up to the Americans to verbally express this fact by assigning names to their experimental and demonstration aircraft (X-aircraft and Y-aircraft, respectively) [8]. It is noted that "Research and Development", as used in

Between these examples of aeronautical research lie 100 years!

In the summer of 1891, *Otto Lilienthal* took up his flight tests in the Krielower mountains near Derwitz located west of Potsdam. From a take-off height of 5–6 m, he managed with his Flying Apparatus No. 3 glides of about 20–25 m distance. *Otto Lilienthal* carried out his investigations and tests personally; he built his flight apparatus with the assistance of a few helpers.

Aeronautical research today: an example is the German-American Enhanced-Fighter-Maneuverability (EFM) research program with the experimental aircraft X-31 with participants on the German side being MBB/EADS (Ottobrunn and Manching), the Institute of Flight Mechanics of DLR and the Test Center WTD 61 (Manching), on the American side Rockwell and NASA (Dryden).

the USA, encompasses both "Applied Research" and "Experimental Development".

But how is *development* with regard to the primary activity of industry, i.e., the realization of a product, defined? The RKW-handbook [2] distinguishes here between *constructive development* and *routine-development.* The former follows the experimental development and still contains certain risks. In case of the latter, success is more or less guaranteed due to the application of approved standards and procedures since the necessary development routine is by definition given.

Before we deal with the protagonists and the characteristics of aeronautical research, we shall introduce the concept of the economic cycle. A presentation in the form of cycles simplifies, however, the situation strongly and also does not show the dependence of the products on the preceding and accompanying research, which is actually of interest here. On the other hand can it be used to give an overview of developments and events over a long period of time. But it also leads to the S-curve concept, which is today an important economic forecast and consultation tool [9].

The theory of cultural cycles, with its origins in the 16th century and its rather questionable prominence due to *Oswald Sprengler's* "The Decline of the West", tries to establish an analogy between cultures, understood as organisms, and living beings with birth, youth/growth, maturity and decline/death. Something similar was tried by the Russian economist *N.D. Kondratieff* for the medium- to long-term economic development by means of the now called Kondratieff-waves or cycles [10].

These cycles, with an apparent wavelength of 40–50 years, represent in their ascending part the life cycle of "growth-bearers" from the – new – basic theory, via the concept, the proof of feasibility and the development up to the introduction of the new product. The descending part is here more related to the (economic) situation of research and the product development since the product possibly holds its position for a period of several Kondratieff-waves, as, for instance, the railroad system. That a technical product and its use is taken for granted has manifested itself some years ago in a funny way by the polemic statement against new nuclear plants: "We do not need nuclear plants, we get our electricity from the wall socket". As one may observe, important cycles in aeronautical technology can quite well be assigned to the general Kondratieff-cycles.

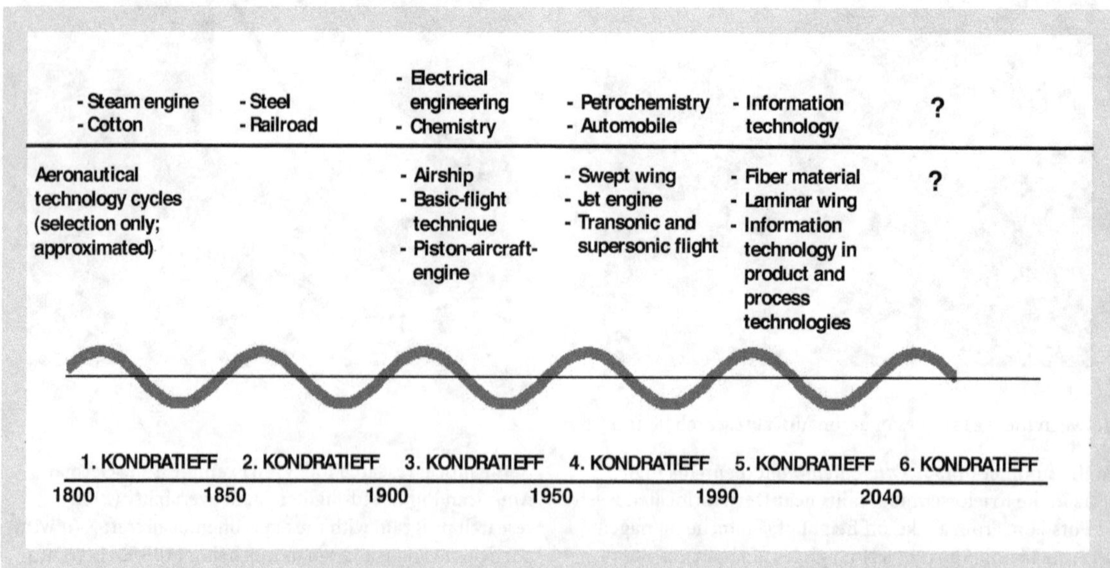

Technology-determined economic cycles (schematic) due to *N.D. Kondratieff* and approximate assignment of aeronautical-technology cycles.

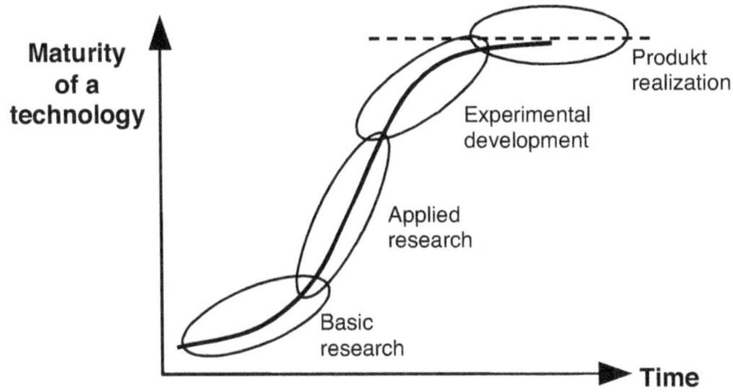

The S-curve concept as schematic equivalent of the ascending Kondratieff-wave.

The ascending part of a Kondratieff-cycle can schematically be visualized by the S-curve concept, which goes back to the Italian mathematician *Vito Volterra* (1860–1940) and was extended by *Cesare Marchetti* [11]. The "maturation" of a technology happens in the stages mentioned above, namely, "basic research", "applied research" and "experimental development".

This leads finally into the product realization with, at first, the "constructive" and then the "routine development".

The maturity of a technology is also a qualitative measure for the accumulated research and development effort if one replaces "time" on the abscissa by this effort. This then shows the known predicament, namely, that in the late phases of the technological development great effort is possibly required to achieve small improvements. This can, however, for economical, political, and increasingly also for ecological reasons be quite justified. Examples are here the "fleet-consumption" of fuel in the case of aircraft or also of automobiles. Some percent in individual reductions will, by means of the large lever "number of vehicles", grow to a magnitude, which makes a vast and long research effort acceptable.

Now, who are the protagonists of German aeronautical research? Today, they are schematically combined in three groups, following viewpoints of sponsorship [12]; but history (see Part 2) also played a role:

- **Aeronautical research at universities** which is carried out at special Technical Universities and regular universities, respectively, within different disciplines in the classical close association of research and teaching. There exist universities with aerospace faculties but also with relevant institutes/chairs within other faculties.
- **Non-university aeronautical research** at specialized *research establishments*, which are, as a rule, publicly, i.e., today by the Federal Government and the States, institutionally funded. They are largely aligned to the industrial activities. Scientific research and investigations in conjunction with special test facilities are priority tasks. The "German Aerospace Center, DLR" is today, as "collecting-organization" of almost all former research institutions, the actual German research establishment for aeronautics (see Part 2).
- **Industrial aeronautical research** is carried out by industry, including medium-sized enterprises, generally linked to a self-financed or – today – to a federal-government- or a state-sponsored or commissioned program or project. Research establishments, but also universities, will be integrated, with the test facilities of the research establishments playing a special role. The aeronautical industry of Germany operates, for instance, aside from a very few exceptions, no large wind tunnels. Organizationally, research, including the experimental development, within industry takes place in groups which may be formed on an ad-hoc basis or which are subsets of organizational units (e.g., centers of merit, departments) of a company.

The roles of these three groups can briefly be described as follows:

- **Basic research** is predominantly carried out at universities. If it is associated with large expenditures in equipment and personnel, it is also performed by research establishments and, for special reasons (competitive constraints, secrecy), also by industry.
- **Applied research** is predominantly performed at research establishments, but partly also at universities and industry. Stronger than in the case of basic research, there may exist research cooperations also including industry.
- The **experimental development** is mainly carried out by industry, but again, also at research establishments and universities. Research cooperations are being aimed at. The proof of success of the experimental development will usually be furnished by technology demonstrators before a product realization is decided upon.
- The **product realization** with the four phases[2] "definition (conception, pre-design, design)", "development", "manufacturing", "support" finally is the task of industry, although there are exceptions.

The role distribution with its overlaps is also reflected by the S-curve concept. Generally, it holds that the aeronautical technology is today again on the ascending part of a Kondratieff-wave, or, to be more exact, on a "sub-wave". This applies stronger to military than to civil aircraft due to the still large technology thrust, and even stronger to space planes; it is for the various disciplines involved in aircraft engineering more or less distinct. The information technologies, which are, due to the innovations taking place just now, indeed, only at the beginning of their own Kondratieff-wave, will have large effects on product technologies, but also on many process technologies (design methods, manufacturing techniques, etc.) in aircraft engineering. While today in case of the product-related operating costs of aircraft savings potentials of 10–15 per cent in aerodynamics (drag reduction), up to 25 per cent in propulsion technology (reduction in fuel consumption) and about 5 per cent in construction and materials (weight reduction) are seen [13], there are no robust estimates of further savings potentials due to the new product and process technologies resulting from the further development of the information technologies. But conversely it holds that for reasons of competitiveness nobody can evade this development.

[2] It shall be remarked that the division into these four phases is not used everywhere like this. Consideration for the European partners, the English-dominated language within the aeronautical industry, etc., have, up to now, not led to a common linguistic rule.

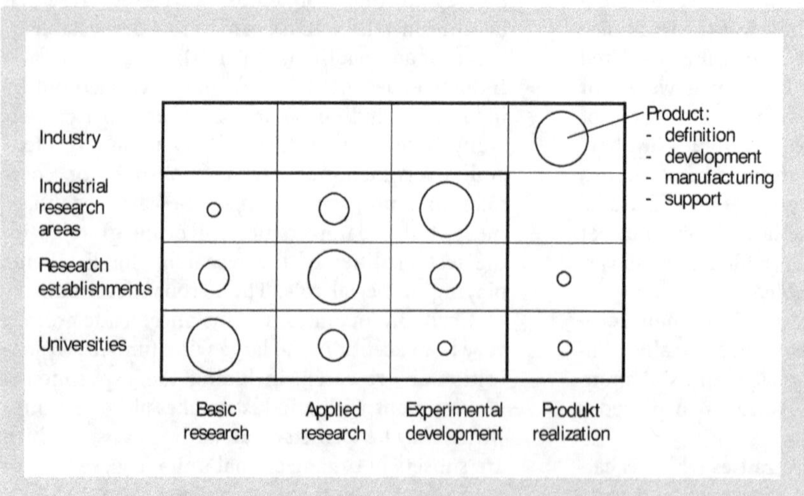

Schematic presentation of the role distribution between universities, research establishments and industry.

Aeronautical Research and Technology During the More Recent German History

In the second part, the political, institutional and industrial environment of German aeronautical research from its beginnings up to today will be depicted. In order to give the reader some guidance for this part of the book, a brief introductory overview over the aeronautical research and development during periods of the more recent German history will be given.

Adolf Baeumker distinguishes in his contribution [14] to the *Benecke*-commemorative-publication [1] five periods in the history of the development of the aeronautical technology up to 1945:

1. The discoverer or pioneer period 1900–1914; it has to be remarked here that the development of the aeronautical technology shifted after *Lilienthal's* death at first to the USA (*Wright Brothers*) then to France (*Clement Ader, Louis Bleriot,* and others [6]).
2. First combat operations of German aircraft, collapse, destruction, 1914–1918.
3. German air-traffic development, glider flight without motor, 1923–1932.
4. Large-scale German air armament, 1933–1939.
5. Aerial warfare, collapse, second destruction, 1940–1945.

German aeronautics had to accept two major interruptions in its activities:

1. From 1919–1923 largely for the entire aeronautical technology, for the military one even up to 1935.
2. The time from 1945 to about 1955; thereafter most allied injunctions were lifted.

A division of the time after 1955/56 into periods presents itself, but shall, for the present, not be made here. The detailed presentation of the entire development will be provided in Part 2. The five periods from 1900–1945 and the time after 1955/56 have, however, to be sketched here individually as an introduction, again following mainly *A. Baeumker* [14].

1900 to 1914

During the pioneer period, the activated public opinion was the decisive factor. Germany reached with the aid of the "Zeppelin Foundation" the technical leadership regarding dirigibles. In aircraft engineering, France was ahead until the "German National Flight Donation" ("Deutsche Nationalflugspende") 1912/13 also allowed top performances here. The public interest led to the foundation of associations related to flight sciences. The interest of the military in the airplane became so great that the army aviation was founded. In 1912, aeronautics in Germany was thus militarized, after this had already happened in the USA and in France. The first large aeronautical research organization, the "German Test Establishment of Aeronautics" ("Deutsche Versuchsanstalt für Luftfahrt, DVL") was founded in Berlin-Adlershof still within the same year. In 1904, *Ludwig Prandtl* was appointed to the chair for Technical Physics at the University of Göttingen. The first chair for aeronautical technology was set up in 1911 at Stuttgart. The industrial build-up was not yet specifically supported by the government, but the flight sciences started to develop, and so did the play between the free scientific interest and the interests of government as well as industry [15].

1914 to 1918

Following the outbreak of the World War One, aeronautical technology was entirely oriented towards war-related tasks. It was, at the same time, strongly supported. The aircraft prevailed over the airship. Many revolutionary results in aeronautical technology and research were produced up to the end of the war in November 1918. The airplane had become a military weapon.

1918 to 1932

The loss of World War One brought with the drastic regulations of the Versailles Treaty the end of German military aviation and hence the end of the corresponding research. However, with the experience gained from military-aircraft and aircraft-engine engineering, Germany could successfully enter into commercial-aircraft manufacturing. This was up to the late twenties most of all accomplished by private enterprises with the aid of public sponsorship. The support of aeronautical engineering did, however, not correspond to the economic importance of aeronautics: Funds for the establishment of modern research facilities were not provided to a sufficient degree. The research establishments, the DVL and the "Aerodynamic Test Establishment Göttingen" ("Aerodynamische Versuchsanstalt Göttingen, AVA") played, nevertheless, an important role.

An omission in the ban of German aeronautics by the Versailles Treaty could be utilized in 1920 at the universities to consider aeronautical research and technology related to gliding. "Academic Flying Groups, Akaflieg" were formed, who used the research potential at the universities.

The worldwide economic crisis 1929–1932 finally caused great damage to aeronautical research and the industry since it led to far-reaching limitations in all activities.

1933 to 1939

All aspects of the aeronautical technology were, from the beginning, very strongly supported by the Third Reich. In the time period between 1933 and 1936, equipment and facilities of the aeronautical research organizations were extended. Teaching at the universities was intensified. Important and internationally acknowledged research activities commenced. Germany became, as already in the twenties, once more almost the Mecca of aeronautical research. German aeronautical research, although strongly government-supported, kept its rather substantial independence and remained, initially, also open to an international exchange. (The strong scientific interaction with foreign nations, the back and forth of impulses, hypotheses, theories and ways of technical realizations is frequently being underestimated so that today several nations consider themselves the cradle of aeronautical technology.)

Despite the technically positive assessment of the aeronautical research during this period, one must not ignore that it was, indeed, part of a large-scale armament. The League of Nations had failed with its disarmament intentions and Germany tried, with the plans of her rulers in the background, to fill the gap which lasted by now for fifteen years. Aeronautical industry and research were, consequently, built up with a number of supporting measures – amalgamation of the technical reporting, foundation of the Lilienthal society, foundation of the German Academy of Aeronautical Research, installation of national research-professorships – all to be able to create a powerful air force, even if, according to *Alfred Baeumker*, the latter was more seen as support for the army and navy, respectively, and less as an individual arm of the services, as became internationally the rule after World War Two.

1940 to 1945

The grand time of German aeronautical research ended with the beginning of the World War Two. Under the impression of the success of the German "blitz", research was initially limited to subjects, which became effective immediately. This changed as a longer duration of the war had to be accepted. Simultaneously, new propulsion techniques (turbo-jet and rocket propulsion) as well as wing/fuselage design principles (swept wing, high-speed airfoil, area rule) matured so that with jet- and rocket-propelled aircraft a great qualitative leap forward could be envisaged. All participants oriented their research towards that leap and the barriers between basic research, applied research and experimental development became permeable.

Research gained during the second half of the war, remarkably, again a certain autonomy. Even if research did not succeed to completely overcome its structural weaknesses, it must be noticed that German aeronautics, with research as a partner, achieved considerable product-related success, with aircraft, which penetrated to the transonic regime, with new propulsion technologies, radar technologies, et cetera.

With the end of World War Two, German aeronautical research succumbed almost completely, the aeronautical industry totally. Only in the mid-fifties, with membership in NATO, was the German Federal Republic granted permission to carry out her own aeronautical development and research. The non-university aeronautical research in Germany after 1955 was, as the industry, initially characterized by the totally new beginning after a ten-year abstinence. The promotion of aeronautics by the Federal Government was initially justified by flanking measures related to the German defense contribution within the framework of NATO, later by the need to utilize the arisen production capacities also for civil purposes [16]. The aeronautical industry produced aircraft initially under license and soon found its way back to an innovative aeronautical technology.

The research institutions in Western Germany, to a large degree revived or re-established, passed, as did industry, through a consolidation process which ended in 1969 with the formation of the unified organization "German Aerospace Research and Test Establishment" ("Deutsche Forschungs- and Versuchsansalt für Luft- und Raumfahrt, DFVLR"). Constraints in the public budget and market constraints, the end of the Cold War, the disappearance of the Eastern Block and, finally, the globalization of the aerospace business led to a further consolidation, also in industry, which reached its culmination with the foundation of the German Aerospace Corporation (Dasa AG) in 1989. The period after 1945 is, on the other hand, also characterized by a strong internationalization, but mainly Europeanization, of the aeronautical industry and aeronautical research. After decades of cooperation in military and civil aircraft engineering, a European Aerospace Corporation started to become apparent by the end of the nineties. But the universities, research establishments and industry also fitted themselves stronger and stronger into the European research framework with the European Union playing an increasingly important role.

In the German Democratic Republic, the situation was different. Free enterprise with independent research was not granted anyway. Nevertheless, as early as 1954, the first German jet-powered passenger aircraft was built as a prototype (first flight in 1959). However, this interesting East-German development had, for political but also for economic reasons, to be abandoned. With it, also aeronautical research in the DDR was essentially terminated. After the reunification in 1990, the remaining support industry merged into Dasa (Elbe-Aircraft-Plant).

An important aspect of the aeronautical technology in Germany must not remain unmentioned: the transfer of knowledge to other branches (also see Part 4). Since the beginning of the twentieth century until about 1985, the aeronautical technologies were for some branches of the German industry technically of formative influence, which is, however, today of secondary importance. This has various reasons. Germany possibly lacks the political understanding, as it exists in the USA and in some West-European countries, that aeronautical engineering may be an important economic factor, but is definitely an important "technology motor" that should be supported accordingly. It is certainly no longer a growth industry, such as today, for example, the information technologies, but its technological potential is not nearly exhausted and can, therefore, continue to act as technological pace setter for other branches. In Germany, however, foreign and trade political considerations towards the Allied Powers of World War Two also may have had an influence, and are still playing a role. In concluding, the following may be stated: aeronautical research is today characterized by

- the still ascending part of its present Kondratieff-(sub-) cycle,
- the high pressure on industry to rationalize,
- the declining self-financed research and development (R&D) share of industry,
- the intensification of the cooperation between universities, research establishments and industry which is urgently needed to meet the competition Europe – USA,
- the introduction of research results into specific aeronautical programs which shall lead to success in the market,
- international sponsorship rules of the European Union and the World Trade Organization (WTO).

"Publicity" in Aeronautical Research: Publications and Patents

In concluding this part, some aspects of publicity related to research shall be discussed. Research, too, lives by communication. Research has its great successes if it is unrestrictedly open, even if there have been, still are, and always will be times and segments with restricted public dissemination. Within aeronautical research, the "experimental development" is, as mentioned initially, predominantly a matter of industry, which may have restricting effects on its publicity.

Aside from such restrictions in publicity due to reasons of competitiveness or security policy, there

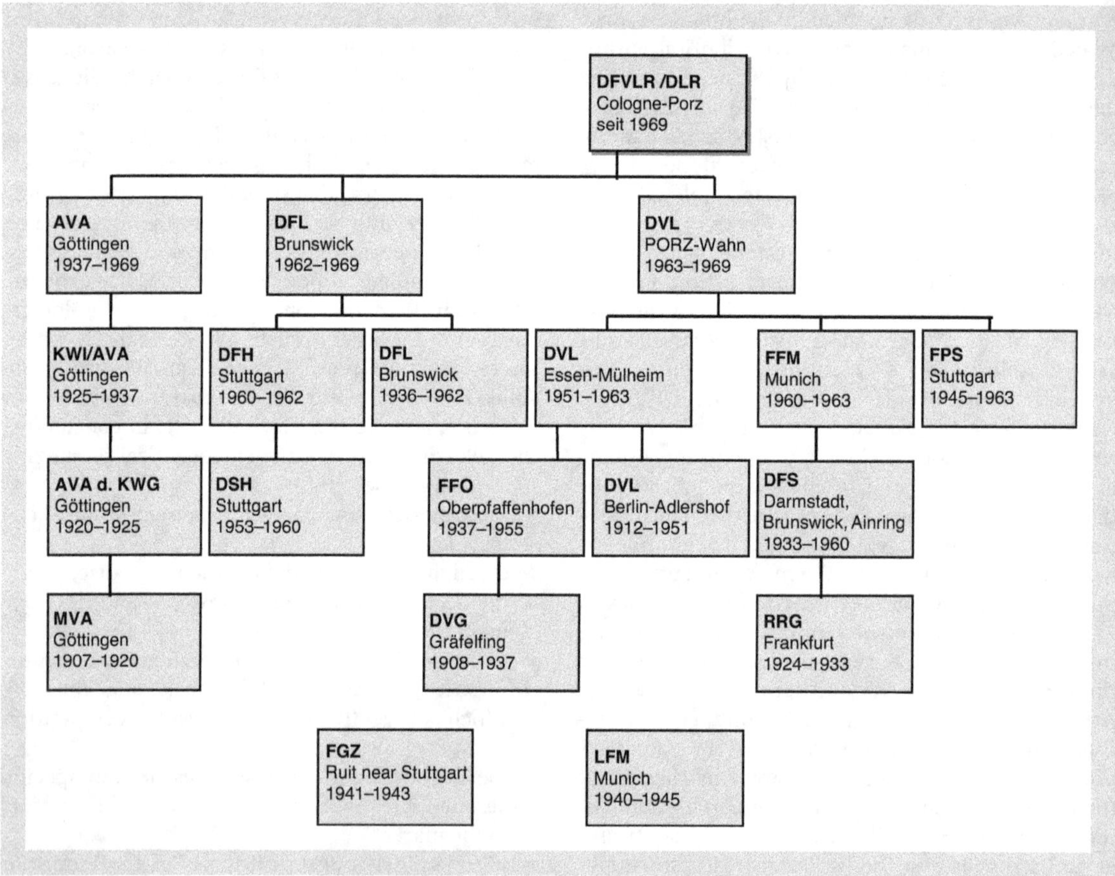

The German Aerospace Research and Test Establishment (Deutsche Forschungs- and Versuchsanstalt für Luft- und Raumfahrt, DFVLR) and its predecessor organizations from 1907–1969, according to *K. Weingarten* (for details see Part 2).

exist two typical forms of communication in aeronautical research, but not only there:

- the publication of results in form of lectures, contributions to journals, and books which also serves simultaneously to secure the intellectual authorship for a research result,
- the patent which guarantees the inventor (also research may lead to inventions) on account of the government time-limited rights for the economic utilization of his invention.

The publication as well as the patent is a means of communication within an "innovation system", which could not exist without them. However, it is important to note that the publication is mainly the communication means of the university researcher and the researcher at a research organization, while the patent predominantly represents the means of communication of the researcher at industry.

Although there are these two forms of communication in research, obviously only the patent has today for the general public a high rating. The number of patents of a nation, compared to the number of patents of other nations, is periodically taken up by the media and commented; it has apparently become the only yardstick for the industrial and economic ranking of a country.

In what follows, the "publicity" of aeronautical research in Germany from the beginnings up to now shall, therefore, briefly be sketched considering the two typical forms of communication. Hereby it shall be shown that both have their value and both are, in the end, indispensable elements of an innovation system (other aspects of "publicity", such as scientific journalism, technical museums, especially the Deutsches Museum in Munich, etc., will not be dealt with here).

The pioneers of flight were frequently not primarily scientists and also not engineers but simply visionaries, who wanted to enable humans to fly. Their success was based on the respective level of technical and/or scientific knowledge, which they deepened and extended by experimental and theoretical means. They were convinced of their ideas and if they were successful in their experiments, or thought to be successful in the future, they generally tried to secure the rights on their invention by patents. To formulate an idea in a patent application is, however, one thing, but to prove the feasibility is another, which can frequently be observed afterwards.

Many German, but also foreign researchers have given to the emerging aeronautics, either directly or indirectly, important contributions and impulses. But their names can generally not be found among the patent applicants. However, they are found in the scientific literature, in conference publications and journals. Emanating from Göttingen, *Prandtl's* workplace, the "Scientific Society of Flight Technology" ("Wissenschaftliche Gesellschaft für Flugtechnik") was founded in 1912, becoming in 1914 the "Scientific Society of Aeronautics" ("Wissenschaftliche Gesellschaft für Luftfahrt, WGL"), today's "German Society of Aeronautics and Astronautics" ("Deutsche Gesellschaft für Luft- und Raumfahrt, DGLR"). One of the main objectives was always the promotion of aeronautics, later of aeronautics and astronautics, by *publications* [17].

The patent is, as was already stated, the communication means of the industry researcher. The scientist as applicant for a patent is rather the exception. However, well-known scientists were enlisted as experts as early as 1912 if controversial questions had to be settled or the scientific elucidation of a patent demand was necessary.

After 1933, the research and development results were, due to the plans of the new rulers, increasingly subject to government control, leading frequently to secrete patents and publications restricted by the State. Between 1933 and 1938 were, on the other

DEUTSCHES REICH

AUSGEGEBEN AM 14. APRIL 1921

REICHSPATENTAMT

PATENTSCHRIFT

№ 303328

KLASSE 77h GRUPPE 5

Hugo Junkers in Dessau.

Kühleranordnung für Luft- und andere Fahrzeuge.

Zusatz Patent 299799.

Cover of a German patent document of 1921.

hand, time and again important international symposia held in Germany, where the scientific exchange of ideas was cultivated. With the beginning of the war in 1939, these activities were, however, rigorously stopped, something that was actually the case for all powers being at war.

The secret patents of the time period between 1933 and 1945 served the same purpose and had the same form as the general German "Reichs"(State)-patent, but they were not accessible to everybody. They are today a goldmine for the one interested in history since during that time a great number of inventions were made which initiated and advanced the substitution of conventional aircraft by jet- and rocket-propelled aircraft with new aerodynamic shapes.

The Federal Republic of Germany resumed in 1948 the pre-war patent tradition. But after 1955, the situation with regard to the consideration and assessment of a patent again changed completely. Since then, patents reflect, as mentioned above, more and more the economic significance of a nation and they are today increasingly regarded as an indicator of her technical und economic assets.

The publication means have broadened in an unexpected way. Here, a real globalization has taken place. Aeronautical congresses "wander" around the world. The Internet is increasingly becoming a means of communication. The German "Journal of Flight Sciences and Space Research" ("Zeitschrift für Flugwissenschaften and Weltraumforschung, ZFW"), the classical publication organ of DGLR, is since 1997 combined with "La Recherche Aérospatiale", the corresponding French professional journal, to the journal "Aerospace Science and Technology, AST". Other European countries are joining. Missing is a European Aerospace Congress like the yearly Aerospace Sciences Meeting of the American Institute of Aeronautics and Astronautics (AIAA). But also representing increasingly a deficiency is the lack of summarizing and synthesizing monographs, which ease, for the student or the job novice, the access to the respective knowledge, which is constantly becoming more strongly divided and more highly specialized.

However, the "publicity" of research, not only in aeronautics, runs more and more into another problem. The transfer of industrial research abroad and the "slimming-down" of the development departments of industry in favor of "Shareholder-value" concepts hinders increasingly the interplay between research at universities, research establishments and industry. The elimination of qualified personnel within companies because it is "only" doing research, whose results cannot be realized on a short-term basis, is detrimental to the communication within the national or even the European innovation system. The companies need highly qualified experts, who are able to judge research results, who can communicate with and are also able to affect research. On the other hand occurs here, with the industry engagement for the "management of knowledge", again a change of heart which accepts that research must also always be seen on a medium-to long-term basis and that one cannot abscond an innovation system without penalty. This does not rule out that innovation systems, in part or as a whole, must frequently be checked.

Innovations can, however, other than knowledge, not be "managed". One may create favorable conditions: personal and financial freedom, provision of personnel and equipment, etc., and later the innovation-chain analogous to the food chain within the animal realm. But all this does not guarantee innovations. Great (basic) innovations are reserved to only a few people: "The spirit wanders wherever – and whenever – it wants".

The Contents of the Book

After having introduced definitions and background, and supplementary topics of aeronautical research, the further contents of the book shall now briefly be sketched.

The subsequent second part is devoted to the presentation of the political, institutional and industrial environment of German aeronautical research from the beginnings to the end of the twentieth century, also considering the situation and development within the former German Democratic Republic during the period from 1945–1961, the year when aircraft engineering and research ended there.

The third part, as the core of the book, introduces, in individual contributions, outstanding developments and results of aeronautical research in Germany, initially from the beginnings up to the end of World War Two, then from after World War Two to the present time. System aspects as well as all

ZEITSCHRIFT FÜR FLUGWISSENSCHAFTEN

S 7473 E

ORGAN DER
DEUTSCHEN GESELLSCHAFT FÜR
LUFT- UND RAUMFAHRT E.V.
UND DER
DEUTSCHEN FORSCHUNGS-
UND VERSUCHSANSTALT FÜR
LUFT- UND RAUMFAHRT E.V.

Unter Mitwirkung von F. Bollenrath, Aachen – W. Bunk, Porz-Wahn – G. Czerwenka, München – Ph. Hartl, Oberpfaffenhofen – O. Heer, Frankfurt/Main – R. Lüst, Garching – O. Lutz, Braunschweig – K. Marguerre, Darmstadt – H. G. Müller, Oberpfaffenhofen – H. G. Munzberg, München – W. Oldekop, Braunschweig – K. Oswatitsch, Wien und Aachen – A. W. Quick, Aachen – E. Rößger, Berlin – S. Ruff, Bonn-Bad Godesberg – I. Sänger-Bredt, Stuttgart – H. Schlichting, Braunschweig und Göttingen – J. Weissinger, Karlsruhe

herausgegeben von H. Blenk, Braunschweig

Schriftleitung: W. Schulz, DGLR, 33 Braunschweig, Flughafen

VERLAG FRIEDR. VIEWEG & SOHN GMBH 33 BRAUNSCHWEIG BURGPLATZ 1

INHALT DES HEFTES

Hauptaufsätze

W. GEISSLER: Berechnung der Potentialströmung um rotationssymmetrische Rümpfe, Ringprofile und Triebwerkseinläufe 457–462

M. TANNER: Experimentelle Untersuchung des Widerstands und Auftriebs von Hügeln mit stumpfer Hinterkante bei kompressibler Strömung 462–470

E. H. HIRSCHEL: The Influence of the Accommodation Coefficients on the Flow Variables in the Viscous Interaction Region of a Hypersonic Slip-Flow Boundary Layer 470–475

O. P. NAGPAL: Optimum Deboost from Hyperbolic Orbit 476–479

D. RADAJ: Vergleichende Betrachtung zu parametrischen Kerbspannungsuntersuchungen mit finiten Elementen und mit komplexen Spannungsfunktionen 479–484

Nachrichten 485–487

Fritz Leitke 90 Jahre. – Persönliches. – DGLR-Symposium „Tragflügel-Aerodynamik bei schallnahen Strömungen". – 13. Otto-Lilienthal-Vorlesung in Paris. – GAMM-Tagung 1973. – 16. Ludwig-Prandtl-Gedächtnis-Vorlesung in München. – Jahrestagung 1973 der Deutschen Gesellschaft für Zerstörungsfreie Prüfverfahren. – AGARD Avionics Panel in München. – Handbook of Time Code Formats.

Kurzberichte aus der Forschung 487–489

Deutsche Luft- und Raumfahrt, FB 72-60 bis 72-70. – MPI für Strömungsforschung, Bericht 10/1972. – Forschungsinstitut für Anthropotechnik, Anthropotechnische Mitteilung 2/72. – Verkehrswissenschaftliches Institut an der Universität Stuttgart, Bericht 6 (1972).

Fortsetzung 2. Umschlagseite

Z. FLUGWISS. 20. JAHRGANG DEZEMBER 1972 HEFT 12

The Journal of Flight Sciences (ZFW) has been published since 1952 and merged 1997 into the European Journal "Aerospace Science and Technology, AST".

essential disciplines will be treated. The activities of German engineers and scientists abroad in the years after 1945 till about 1955 will be traced. As in Part 2, short biographies of outstanding research personalities will be given here.

The fourth part describes the importance of German aeronautical research internationally and for the national economy. It also provides a summary and an outlook onto prospective developments in aeronautics and aeronautical technologies within the next years.

The book concludes with an appendix containing milestones, conversions, a list of abbreviations, indices, short biographies of the authors, contributors and advisers, and, finally, the picture sources.

References

1 M. Schröder (Ed.): Management in Wissenschaft und Technik, Festschrift für Theodor Benecke zum 60. Geburtstag, DFVLR, Porz-Wahn, 1971

2 H.H. Moll, J.J. Warnecke (Eds.): RKW-Handbuch Forschung, Entwicklung, Konstruktion (F+E), Erich Schmidt Verlag, Berlin, 1976

3 E. Zahn (Ed.): Handbuch Technologiemanagement, Schäffer/Poeschel-Verlag, Stuttgart, 1995

4 S.F. Mason: A History of the Sciences. Main Currents of Scientific Thought. Collier, New York, 1977

5 G. Wise: Science and Technology, Osiris, 2nd Ser., 1, 1985, pp. 229–246, see also W. Kaiser: Forschung und Technik in Deutschland nach 1945, D.M., Bonn/München, 1995, p. 76

6 L. Bölkow (Ed.), Ein Jahrhundert. Flugzeuge, Geschichte und Technik des Fliegens, VDI-Verlag, Düsseldorf, 1990

7 Stifterverband für die deutsche Wissenschaft: allgemeine Richtlinien für statistische Übersichten in Forschung und experimenteller Entwicklung, Frascati-Handbuch II, Arbeitsschrift 1971C, Essen, 1971

8 J. Miller: The X-Planes X-1 to X-29. Speciality Press Publication and Wholesalers, Inc., Marine on St.Croix, Maine, USA, 1983

9 Th. Modis, Prediction. Societies' Tell-tale Signature Reveals the Past and Forecasts the Future, Simon and Schuster, New York, 1992

10 N.D. Kondratieff, Die langen Wellen der Konjunktur, Archiv für Sozialwissenschaft und Sozialpolitik, Vol. 56, 1926

11 C. Marchetti: A Forecasting Model for Research and Innovation Activities in Selected Areas: a Support for Strategic Choices, International Course on Research and Innovation Management, organized by UNIDO and ICS, Venice, September 1991

12 H.A. Hertrich, Wirtschaftliche Rahmenbedingungen und Schwerpunkte der BMFT-Förderung der Luftfahrtforschung in den achtziger Jahren, BMFT, October 1982

13 H. Ambos, Technologiemanagement in der Luft- und Raumfahrtindustrie, Handbuch Technologiemanagement (E. Zahn, Ed.), Schäffer/Poeschel-Verlag, Stuttgart, 1995, pp. 861–880

14 A. Baeumker, Von der Führung der deutschen Luftfahrttechnik im ersten halben Jahrhundert 1900–1945, Man-

agement in Wissenschaft und Technik, Festschrift für Theodor Benecke zum 60. Geburtstag (M. Schröder, Ed.), DFVLR, Porz-Wahn, 1971, pp. 19–43

15 H. Trischler, Luft- und Raumfahrtforschung in Deutschland 1900–1970, Politische Geschichte einer Wissenschaft, Campus Verlag, Frankfurt/New York, 1992

16 G. Madelung: 40 Jahre Flugzeugbau in der Bundesrepublik Deutschland, DGLR-Jahrbuch 1995, Vol. III, 1995

17 W. Schulz: Von der WGL zur DGLR, aus sechs Jahrzehnten Tätigkeit für die Luftfahrt, Management in Wissenschaft und Technik, Festschrift für Theodor Benecke zum 60. Geburtstag (M. Schröder, Ed.), DFVLR, Porz-Wahn, 1971, pp. 43–71

Part 2

The Political, Institutional and Industrial Environment of German Aeronautical Research

This part of the book

This part of the book summarizes briefly the environment of German aeronautical research from its beginnings up to the end of the twentieth century. The *political environment*, in form of the varying German government and state structures during the time of the Empire up to the Federal Republic of Germany, represents the, more or less, globally controlling and financing environment. The *institutional* environment concerns the research institutions, the universities, but also the scientific societies. The *industrial* environment, finally, includes the airframe as well as the engine manufacturers and the suppliers, which are, on the one hand, users of research results, but, on the other hand, also carry out research with emphasis, however, on the experimental development, as already pointed out in Part 1.

The structure of the present part does not strictly correspond to the periods of aeronautical research as sketched following *Baeumker* in Part 1. Here the environment during the time of the Empire, i.e., up to 1918, is depicted followed by the environment during the Weimar Republic and the time of the Third Reich. Concluding, the situation from 1945 until today is described.

In presenting the historical environment of German aeronautical research, which can only roughly be sketched, the original sources are, due to lack of space, not always quoted. Generally, the common literature is cited initially, which is also more easily accessible to the reader, such as, for example, historical publications and source books of German aeronautical research and industry [1–6], descriptions of the aeronautical research establishments [7–10] and investigations concerning the history of aeronautics and its technical disciplines [11, 12].

German aeronautical research – and its environment – is strongly determined by individual personalities, as is all research. The people, who were initially more influential than the environment – who have actually created the environment – are being included in this presentation. They must however, inevitably stand back with respect to their own individual history in order to limit the size of the book. Their résumés, as far as they are not considered here, may be taken from the literature. It is also impossible to consider the developments in Europe and the non-European countries, which took place at the same time. This is done if and to the extent they were of strong influence on German aeronautical research.

Aeronautical Research Comes into Being During the Time of the Empire

HANS FABIAN

From the Beginnings to Lilienthal

Going back to the roots of the German word "Forschung", i.e. "Research", one finds in the Old-High-German language "forscon" as "asking for". Therefore it is justified to date the beginnings of German aeronautical research to *Lilienthal*. He was the first, who enquired scientifically why a bird flies. This shall not belittle the preceding work, on the contrary. It is, with its early experiments, important. Its failure is due to the fact that systematic "research" was missing and that just "trying" was not sufficient. Some of these early tests shall be described exemplarily to show how far back experimental trials in Germany reached.

To fly like a bird is an ancient dream of mankind, which is reflected in myths, fairytales and legends. However, it is known that not until the late Middle Ages tests were conducted, which seriously served to move through the air. Not by chance became in Germany Nuremberg, the town of the famous watchmakers and mechanics, the scene of the commencement of such tests. The first flight attempt within the German-speaking regions, reported in 1612, was performed around 1500 by the Nuremburg Kantor (master of the church choir) *Senecio* [13], also called *Lobsinger* or *Vorsinger* (singer of praise and singer leading the congregation in song); he is reported to have broken his arms and legs attempting to sail with wings through the air. It was *Hans Hautsch* (1595–1670), a famous Nuremberg mechanic and compass-smith, who was said to have, about one hundred years later, carried out flight tests; however, nothing is imparted in the book of *J.J. Becher* [14] about the nature and the results of these tests.

A fresco in the library hall of the monastery Schussenried shows the flying father *Kaspar Mohr* (1575–1625), of whom *C.F. Burgwein* [15] reports that he must have been a very skillful universal genius. He is said to also have conducted flight tests. His superiors forbade him, however, to fly down, with wings made of goose feathers, from the third floor of the monastery into the garden. His apparatus was taken away and any further tests forbidden. *Lilienthal* had many other predecessors, but they all failed because of the lack of knowledge concerning the nature of the lift-generating air forces. Remarkable is the *handwritten treatise of Melchior Bauer* [16]. The flight device, which is described in the treatise and depicted by seven design drawings, shows the high technical understanding of the author.

After the invention of the brothers *Montgolfier*, who showed with their first balloon ascend on November 21, 1783 the possibility that man could really lift himself up into the air, activities shifted soon towards the maneuverability of the balloons. The entire nineteenth century is characterized by unsuccessful tests in that regard. So it is not surprising that parallel further efforts were invested to rise, with a flight apparatus, like a bird into the air.

Thus, two "lines of testing" emerged: flying according to the principle "lighter-than-air" and flying according to the principle "heavier-than-air". The former, i.e., flying with balloons, and its further development towards the maneuverable airship was one direction, which was considered to be successful since proof had been given that man could rise into the air with the help of heated air or light gas. The latter, which attempted to imitate the flight of birds, had the reputation to pursue an impossible approach; it was, therefore, regarded as a folly. Nevertheless, there were always men, who saw in the imitation of the flight of birds the more promising direction, i.e., giving the principle "heavier-than-air" the greater chances of success. It shall here already be noted that, subsequently, the tests according to this principle proceeded in two main directions. The one that did not succeed tried to imitate the flapping of bird wings to apparently generate lift, but essentially to produce thrust (flapping-wing flying machines). The other one pursued the "rigid" shape flying apparatus (fixed-wing flying machine), leaving for the time being the question of thrust open, although the propeller was known since the middle of the nineteenth century at the latest. That birds could glide with their wings and bodies in a "rigid" position was certainly an old observation.

The early flight tests were not only confined to the German-speaking regions. In all of Europe, many people dealt with the question of flight, among them also reputable scientists. Most of all the latter discovered, step by step, the most important fundamentals of flight. Already *Galilei* (1564–1642) recognized the dependence of drag on the speed and size of an object; however, he did not carry out any measurements to that effect. The first quantitative measurements of drag go back to the French physicist *Edme Mariotte* (1620–1684). On the other hand, the attempt of *Isaac Newton* to derive the laws of drag theoretically led to incorrect formulations, which served primarily the opponents of the "heavier-than-air" flight as arguments to establish its impossibility.

Daniel Bernoulli (1700–1782) discovered that the static pressure of a liquid decreases with increasing velocity and increases as the velocity decreases. This fact, formulated in the equation carrying his name, made it possible to explain the decrease in pressure on the surface of cambered airfoils and hence the lift.

Sir George Cayley (1773–1857) was the first to recognize the basic principles of a successful "heavier-than-air" flight, namely, the necessity of lift, the realization of an appropriate propulsion, and the need for sufficient flight-stability and controllability of the flight apparatus [17]. An additional condition was, of course, the demand for minimum weight, i.e., light-weight construction. *Cayley* can also be considered the first scientist, who knew how to utilize the knowledge existing at his time for the technique of flying. He was first to subdivide the forces acting on a wing into the components "lift" and "drag" and also first to build airplane models, which, as fixed-wing gliders, could already in 1809 have led to the same conclusions as *Lilienthal* arrived at considerably later. In his publication "On Aerial Navigation", the principles of longitudinal, directional and lateral stability were treated for the first time. He was in contact with many European scientists, with whom he tried, although in vain, to found an "Aeronautical Society".

The best-known tests concerned with flapping-wing flight within the German-speaking regions were the ones of *Degen* and *Berblinger* in the 19th Century. The Viennese watchmaker *Jakob Degen* [18] (1761–1848) developed a flying machine with two flapping wings of 6.7 m (22 ft) span and a wing area of 12 m^2 (129 sq ft). The wings were equipped with about 3,500 valves (flaps) each, which opened on the upbeat and closed on the downbeat, i.e., a principle other than that involved in the flight of birds. That way, he obtained lift corresponding to about 43.7 kg (96 lbs). But his physical strength was insufficient to lift himself up into the air. He demonstrated his apparatus to scholars and journalists at the Viennese University in 1807. An imitator of *Degen*, the tailor *Albrecht Ludwig Berblinger* of Ulm, became much more famous by legend and the mockery of his contemporaries as "The Tailor of Ulm" [19], but he also failed.

During the 19th century, it was gradually realized that the "heavier-than-air" flight could only be accomplished with the aid of appropriately powerful propulsion. The invention of the steam engine led, commencing in the middle of the 19th century, to many suggestions of German inventors. In 1867, for example, *Joseph M. Kaufmann* [20] built the first German steam-engine-powered flying machine, which he also exhibited at the London Crystal Palace where it attracted the greatest attention. However, it never left the ground.

Otto Lilienthal (1848–1896).

Otto Lilienthal, born at Anklam (near the Baltic Sea) on May 23, 1848. From his earliest childhood on, he wants to realize human flight following the example of the birds.

1867 after a traineeship at the engineering company Louis Schwarzkopff, he started in October with his studies at the "Royal Vocational Academy" ("Königliche Gewerbeakademie") at Berlin.

1872 as a young engineer, he joins the engineering company Carl Hoppe.

1873 design of flapping-wing apparatus.

1881 independent businessman as manufacturer of steam boilers; the coiled-pipe boiler becomes the standard product of the "Mechanical Engineering Company Otto Lilienthal" ("Maschinenfabrik Otto Lilienthal"). Reflections on a "steam-jet wheel" ("Dampfstrahlrad") led into the direction of early steam turbines to be utilized for the propulsion of ships. Numerous patents resulted.

1889 a summary of all his tests regarding bird flight is published as a book: "Bird Flight as the Basis of the Art of Flying" ("Der Vogelflug als Grundlage der Fliegekunst"), a contribution to the technology of flight on the basis of the numerous tests carried out by *Otto* and his brother Gustav *Lilienthal.*

1896 he dies after a flight accident on August 10 in Berlin-Lichterfelde. The causes for this accident are still unresolved. (Problems of stability, gust problems, fracture of a wing rib, or a combination of all.)

Inventor, Researcher and Businessman Otto Lilienthal

Modern aeronautical research starts with *Otto Lilienthal. Otto Lilienthal* was born on May 23, 1848, his brother *Gustav* on October 9, 1849 at Anklam. The interest in flying was awakened in both boys while reading the book of travels of the aeronaut *Count Zambeccari,* who met his death in 1812 during a balloon journey. They also watched the many birds in the surrounding area, which was rich on meadows, making one of their first and most important observations watching young storks, which they approached by sneaking carefully in their direction, although with the wind. Nevertheless, since the scent of storks is very low, they came very close and noticed that the storks always hopped a little way towards them before they ascended. The young men concluded that lifting up against the wind must be easier than with the wind, an amazing deduction for boys of their age.

At the age of 14, they set out to build their first wings, however, they failed. Although *Otto* was the more active of the brothers and worked later more engineer-like, he always emphasized that his brother *Gustav* had a great share in their work. *Otto Lilienthal* left high school in 1864 and moved to the provincial vocational school at Potsdam, where he graduated after two years with the best examination ever taken there. Already at that time, *Otto Lilienthal* took up his flight-technical experiments on a scientific basis, the latter being his greatest merit.

He was the first who carried out systematic measurements of the drag of (fixed) wings, initially with the wings mounted vertically. In the process, he noticed that the formulas given in the technical handbooks of the time must have been derived mostly by theoretical considerations. They could not be confirmed by the measurements. His measurements, starting in 1866, continued, partly with longer interruptions, until 1889 and produced, among other things, the important result that a slightly cambered surface, similar to a bird's wing, gives the highest lift at the lowest drag.

In October 1867, *Otto Lilienthal* began his studies at the Vocational Academy Berlin, which later became the Institute of Technology (TH) Charlottenburg, today's Technical University (TU) Berlin. His brother *Gustav* enrolled 1869 at the Civil Engineering Academy. In 1868, the brothers constructed a flapping-wing apparatus, which they suspended from the gable of a barn, succeeding to lift a total weight of 80 kg (90 lbs) with this device using a counterweight of 40 kg (45 lbs). But *Otto* found that the physical effort necessary to do so for a longer period of time was too great. He carried out specific measurements related to the energy needed and discovered that the forward movement is at least as important to flight as the wing stroke.

Otto Lilienthal finished his studies at the beginning of the war in 1870 and joined, on July 22, 1870, the Prussian "Garde-Füsilier-Regiment" ("Infantry-Guards Regiment"). He observed, as did *Count Zeppelin*, the French military endeavor to employ balloons; both came, however, to different conclusions. In *Lilienthal*, it reinforced the opinion that bird-flight must be exemplary.

After returning from the war, work related to the systematic investigation of flight diminished due to the necessity to earn his cost of living, but it was not abandoned. In 1874, first systematic aerodynamic tests commenced. Here *Lilienthal* made the decision to construct a rotating experimental apparatus since other methods seemed too expensive. After all, he bore all the expenses.

With a rotating device, driven by dropping weights and called "measurement apparatus", *Otto Lilienthal* carried out extensive measurements on plane and cambered plates as well as on airfoils during the years 1871–1873.

In the spring of 1886, the brothers *Lilienthal* became members of the "German Association for the Promotion of Airship Aviation" ("Deutscher Verein zur Förderung der Luftschiffahrt"), founded in 1881, and soon *Otto Lilienthal* gave his first lecture to the association concerned with light engines and their possible use in aviation. His first of three lectures on bird flight took place on October 29, 1888, the subject being "The power consumption associated with bird flight and its effect on the viability of free flight".

In 1889, he published his book "Bird Flight as the Basis of the Art of Flying" ("Der Vogelflug als Grundlage der Fliegekunst") [21]. The results of the investigations and the test equipment employed are described in the book in a very careful and generally understandable way. It can be considered as the first publication of scientific work in aeronautics, where the work was performed according to modern aspects. Within the book, Figures 30 and 33–43 are of particular interest since they show how *Lilienthal* intuitively, based on observations and experimental results, depicts flow forms, known today as laminar and turbulent flows, as well as a favorable airfoil shape. His work made him well-known among the experts concerned with flight in England, France and, especially, in the USA.

Otto Lilienthal started on July 1, 1890, based on the knowledge described in his book, his practical tests with cambered wings. Setbacks did not discourage him and in order to obtain a favorable base for further flight tests, he had his own "flight-hill" raised near his house, protruding about 15 m (49 ft) above the ground; the hill was topped with a shed for the storage of the flight apparatus. Later, *Otto Lilienthal* moved his flight tests to the Rhinow Mountains near Berlin, firstly, because of the more favorable wind conditions, secondly, to be able to pursue his tests without being unduly disturbed. In the time period between 1890 and his death on August 10, 1896, he carried out numerous glider flights with distances up to 250 m (820 ft), developing 18 gliders in different versions, namely, gliders with rudder, biplanes and a rudimentary flapping-wing machine. For the latter, he had someone develop an engine which ran on carbonic acid, had a weight of about 20 kg (44 lbs) and a power output of about 1.5 kW (2 hp). He settled for this design, because he thought that an aircraft with propeller would have needed a much stronger engine and such engines were not available at that time.

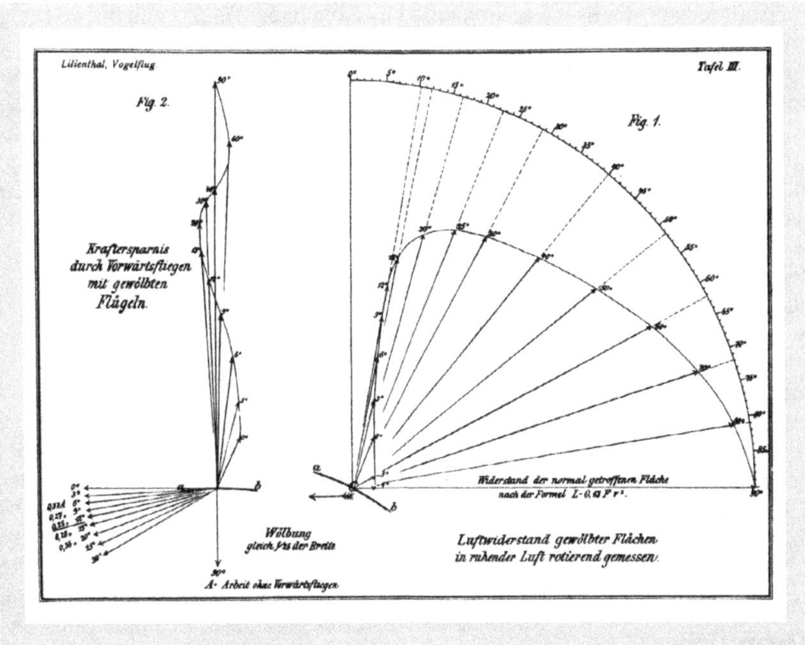

The measurement of lift and drag with the rotating rig led *Otto Lilienthal* to the so-called "polar diagram", which is still used today to depict lift and drag coefficients.

Otto Lilienthal with his large biplane after a take-off from the "Flight-hill" ("Fliegeberg") at Berlin-Lichterfelde (1895).

Lilienthal's influence on the subsequent development of the airplane is undisputed. Well-known became, most of all, his "students" *Octave Chanute, A.M. Herring* and *Samuel Pierpont Langley* in the USA, *Percy S. Pilcher* in England, and *Ferdinand Ferber* in France, who adopted and passed on his findings and his approach to the subject. *Langley*, at the time secretary of the Smithsonian Institution in Washington, visited *Lilienthal* in August of 1895. In the fall of 1895, *Nikolai Jegerowitsch Joukowsky*, the leading theoretician of flight in Russia, also visited *Lilienthal*. The visit proceeded extremely friendly and, afterwards, *Joukowsky* took along a "Standard-Glider" ("Normal-Segelapparat") to Russia, which is still displayed at the Joukowsky-Museum near Moscow.

Otto Lilienthal's fatal crash on August 10, 1896 struck everybody interested in aviation very hard. Subsequently, all work concerned with the problem of flight ceased in Germany for a longer period of time.

First Flight Engines as Propulsive Power Sources

Werner von Siemens recognized that the lightweight engine is a prerequisite for a flight vehicle to be controlled by man. An important part of aeronautical research is, therefore, closely linked to engine design and construction. The prerequisite was the invention of the four-stroke gas-engine by *Nicolaus August Otto* (1832–1891). It was the basis for further developments by *Gottlieb Daimler* (1834–1900), *Wilhelm Maybach* (1846–1929) and *Carl Friedrich Benz* (1844–1929), who created, by increasing the rotational speed (number of revolutions per minute, rpm), engines light enough to serve as propulsive devices for air vehicles. One must also not forget *Robert Bosch* (1861–1942) and his co-worker *Gottlob Honold*, who provided with the high-voltage magneto-ignition scheme the prerequisite for making the ignition process in an internal combustion engine so safe that it could be operated in flight vehicles.

Hermann Wölfert installed in 1895, for the first time, a four-stroke *Daimler* engine with a performance 5.9 kW (8 hp) into an airship. After initially satisfying tests, the flight on June 12, 1897 ended disastrously; after about 5 min flight time, a mighty flame erupted from the engine nacelle and the airship caught fire, accompanied by a big bang. *Wölfert* and his mechanic, *Knabe*, could only be recovered dead. The distance 2 m (6.6 ft) between nacelle and the body of the airship was, presumably, insufficient and led to the ignition of hydrogen escaping from the body. The brothers *Wright* also produced in 1901 their 17.7 kW (24 hp) flight engine on the basis of the engines of *Daimler* and *Benz*.

The Controllable Airship – The Dirigible

The development of the dirigible [22] is a part of early aeronautical research. Here, *Count Zeppelin* was, above all, the outstanding personality [23]. The controllable airship, as a logical advanced development of the "free" balloon, was, not least, promoted by the interest of the military in balloons. Already in 1784, the Academy of Lyon organized a competition concerning the subject "About the control of balloons", which drew, in the course of one year, 96 submissions. In spite of many quite naive proposals, it became clear that adequate propulsion was needed to achieve an effective control.

An important early approach to controllability is the one of *David Schwarz* (1850–1897). The timber-merchant from Agram (today Zagreb, Croatia) recognized that only a body, in itself rigid, could achieve satisfactory flight characteristics. He, therefore, developed an airship, whose planking was to consist of thin sheet aluminum. For that purpose, he negotiated in August 1892 a contract with *Carl Berg* (1851–1906), who was, at that time, the only manufacturer in Germany able to produce and process aluminum in greater quantities, for the delivery of aluminum for the construction of his controllable airship. The airship consisted of a frame covered with long sheets of aluminum of 0.18–0.20 mm (0.0071–0.0079 in) thickness. The sheets were folded and riveted airtight. A gondola (nacelle) was suspended from the main body by four grated girders. An 18.5 kW (16 hp) Daimler-engine powered the two propellers. Unfortunately, the ascent on November 3, 1897 failed after a successful lift-off, since stabilizer and control surfaces, which could have compensated the failure of one of the propellers, were not provided for.

The first all-metal airship, "Schwarz II", with a Daimler gasoline engine was built according to proposals of *David Schwarz* and with the support of *Carl Berg*. The first and only ascent took place in Berlin on November 3, 1897.

Count Zeppelin had, as an observer during the American Civil War on the side of the Union and during the siege of Paris, become acquainted, on the one hand, with the merits of balloons, on the other hand, with their limited possibilities. When, in 1890, he retired from active military duty, he devoted himself completely to his ideas related to the construction of airships. Here, he saw himself less as an inventor than as the driving force behind the realization of his ideas.

Since 1891, *Count Zeppelin* intensified the pursuit of his ideas and also solicited the military, because he realized that greater resources would be needed. He consulted the Daimler factory since he considered the engines as the key to success. And he engaged the engineer *Theodor Kober* (1865–1930) of the Augsburg Balloon Factory, with whom he drew up the plans which led in 1895 to the patent document "Controllable air-train with several lifting bodies in series" ("Lenkbarer Luftfahrzug mit mehreren hintereinander angeordneten Tragkörpern"). The apparatus was categorized by the patent office under Class 77 "Sports". *Rudolf von Tschudi* (1851–1925), commander of the Prussian Airship Battalion and delegated by the Prussian military to evaluate the apparatus, justly considered it to be unsuitable with regard to maneuverability and aerodynamics, because it consisted of at least three sections, which where not rigidly coupled. On February 6, 1896, *Count Zeppelin* gave a lecture to the Württemberg Section of the "Association of German Engineers" ("Verein Deutscher Ingenieure, VDI") entitled "Designs of controllable air vehicles" ("Entwürfe für lenkbare Luftfahrzeuge"). This lecture, which was held before 435, in part technically highly educated participants, was insofar important as it led the executive committee of the VDI, after lengthy internal discussions, in December 1896 to appeal to the public, and especially to the German industry, to support *Count Zeppelin* financially.

In his lecture, *Count Zeppelin* obviously attached great importance to the depiction of high structural stability at minimum weight. Regarding speed measurements, he referred to data of the French airship "La France", which was built by *Charles Renard* and *Arthur Krebs* and presented to the public on August 9, 1884. Powered by a 6.3 kW (8.5 hp) electric motor, the airship flew a circle of about 7.5 km (4.7 mi) length and landed at its take-off point. It had the shape of a pointed cigar, was 50.42 m (165.4 ft) long with its largest diameter being 8.4 m (27.6 ft) and a volume 1,864 m^3 (85,866 ft^3). On this flight, the airship reached an average speed of 21.6 km/h (13.4 mi/h).

As choice material, *Count* Zeppelin named aluminum of the Carl Berg Company, which was alloyed with 6–10 per cent copper and nickel additives and had, at equal weight, a higher buckling strength than steel. This led to a first contact with the aluminum producer *Carl Berg*, who pointed out, among other things, a new and more resistant aluminum alloy. *Count Zeppelin* had tests of the material be performed at the "Material Test Laboratory" ("Materialprüfungsanstalt") of the Institute of Technology (TH) Stuttgart. He proceeded similarly with strength tests on molded parts and joints. With his approach he followed a

Ferdinand Graf von Zeppelin (1838–1917).

Count Ferdinand von Zeppelin, born in Constance on July 8, 1838.

1853 he takes up an army officer's career.
1863 he was given leave for a stay in the USA.
1866 and 1870 he becomes well known by spectacular patrol missions.
1869 he marries the Baltic baroness *Isabella von Wolff.*
1879 birth of the only daughter, who supported him in later years actively as a secretary.
1884 promotion to colonel.
1885 delegate of the King of Württemberg in Berlin.
1887 special ambassador of Württemberg in Berlin and delegate of his state to the Council of Ministers (Bundesrat).
1890 promotion to lieutenant general and discharge from the military service; devotion to the airship projects which made him famous.
1907 following intensive discussions with his co-worker *Kober*, he recognizes the importance of aircraft engineering next to airship engineering.
1912 he contributes financially to the "Aircraft Engineering Company Friedrichshafen" ("Flugzeugbau Friedrichshafen"), founded by *Kober*, which in
1917/18, together with three licensees, delivered about 850 multi-engine aircraft (61 per cent of the German production).
1914 he commissions *Claudius Dornier* with the development of giant flying boats (seaplanes) and establishes, for that purpose, the "Seaplane Department Dornier" in Seemoos. He, furthermore, founds, together with *Robert Bosch*, the "Experimental Works Gotha-East" ("Versuchsbau Gotha-Ost"), which moves in
1916 as "Zeppelin Company, Ltd" ("Zeppelin-Werke, GmbH") to Staaken. Cooperators, consultants, and providers of ideas are, at this time, *Helmuth Hirth, Alexander Baumann* and *Adolf Rohrbach.*
1917 he dies, after surgery, in Berlin on March 8, and is buried in Stuttgart.

scientifically based test philosophy that was later also applied in the newly developing aircraft industry.

In the spring of 1897, *Count Zeppelin* started to organize in Stuttgart a joint-stock "Society for the promotion of airship aviation", where the Schwarz heirs participated. The protocol of the foundation was drawn up on June 28, 1898. Among the subscribers of shares were *Carl Berg, Gottlieb Daimler, Max von Eyth, Philip Holzmann, Carl von Linde* and *Friedrich Voith. Count Zeppelin* himself had to subscribe 441 shares at 1,000 Marks each in order to obtain the needed capital of 800.000 Marks.

The construction of the first airship, consisting of a rigid aluminum frame made up of stringers and 16 ribs, commenced in 1898; integrated into the frame were 16 gas cells with a total volume of 11,000 m^3 (388,685 ft^3). Two engines of 11.8 kW (16 hp) each, powered via cardan shafts four propellers located laterally near the longitudinal axis. Control was accomplished by a small rudder and, in lieu of an elevator, by a 100 kg (221 lb) trimming weight, which could be moved horizontally on a steel cable suspended under the airship. The airship was of cylindrical shape with pointed ends; it had a diameter of 11.7 m (38.4 ft) and a length of 128 m (420 ft). The outer skin consisted of impregnated cotton fabric.

The first take-off of the Zeppelin-airship LZ 1 occurred on July 2, 1900 from a raft anchored on Lake Constance outside Friedrichshafen.

On July 2, 1900, LZ 1 performed first test flights over Lake Constance, reaching a speed of 31 km/h (19 mi/h), the highest speed reached by an airship up to that time. The test flight was well controlled and elaborately recorded. However, after two more test flights, the "Society for the Promotion of Airship Aviation" had to go into liquidation in February of 1901 due to lack of funding.

But *Count Zeppelin* did not give up and accomplished that the King of Württemberg approved a lottery, which brought in about 125,000 Marks, and that, in addition, Prussia contributed 50,000 Marks. The Daimler Company provided the engines free of charge and also *Carl Berg* participated by the provision of aluminum parts. In 1905, the construction of the LZ 2 could commence incorporating significant improvements especially with regard to the rigidity of the stringers, which, following a suggestion of *Zeppelin*, were designed with a triangular cross section. The engines now had 125 kW (170 hp) giving the airship, besides higher speed, a better maneuverability. On January 17, 1906, LZ 2 performed its only flight. It had to make a forced landing due to control problems and was, on the ground, so heavily damaged by a storm that it had to be scrapped. The parts still functional, among them the engines, were utilized for the construction of the LZ 3, which made its first flight already on October 9, 1906.

The year 1906 is insofar of importance to aeronautical research, as, at the instigation of Emperor *Wilhelm II*, the "Society for the Study of Motor-Airships" ("Motorluftschiff-Studiengesellschaft") was founded and endowed with a capital of one Million Marks. It supported scientific research and experimental aviation. *Richard von Kehler* (1866–1943), at the time the president of the Aero-club of Germany, became its first secretary. *Von Kehler* had in 1894, together with *Georg von Tschudi* (1862–1928), visited *Otto Lilienthal* and had performed a flight test with his glider. With his support, the above society pushed the development of the non-rigid airship, following ideas of *August von Parseval*. The ideas were based on experiences made with the "Kite-balloon" ("Drachenballon") of *Parseval* and *Sigsfeld*. The so-called "tout airship" ("Prall-Luftschiff") obtained its stiff outer shape by means of a (air) fan. Lift was generated by a number of small balloons (ballonets) filled with gas; the engine powered a propeller consisting of canvas, which adapted its rigid form due to the rotation of the propeller. The essential disadvantage was that after a pressure loss in one of the small balloons, the rigid form disappeared and the airship buckled.

Count Zeppelin was, incidentally, despite of his belief in the value of his idea, open-minded enough not to underestimate the importance of the airplane. Already between 1899 and 1902, he had his co-worker *Heinrich Rüb* design a "paddlewheel aircraft", which, however, did not go into operation. In 1907, he commissioned *Theodor Kober*, who had already worked out and calculated the design of LZ 1, with the construction of an aircraft. The installations at Manzell were placed at his disposal. There, *Kober* founded in 1912 the "Aircraft-Factory Friedrichshafen, Ltd." ("Flugzeugbau Friedrichshafen GmbH"), which, in the course of the World War One, became a large enterprise with about 3,000 employees.

Powered Flight

The fatal flight accident of *Otto Lilienthal* motivated the *Wright Brothers* in the US to seriously devote themselves to the problem of flight [24]. *Octave Chanute*, at that time secretary of the "Aerodynamic Association" in Washington, had bought a Lilienthal-glider and shipped it to the US. Here, he tested it together with his students *Herring* and *Avery*. As members of the association, the *Wright Brothers* became acquainted with this apparatus. The message of the death of *Lilienthal* affected them so much, that they started to familiarize themselves with the problem of flight, studying all attainable books in order to explore the cause of the crash. In October 1900, they started with practical tests. They carryed out, with self-constructed gliders of a Lilienthal design, in the sand dunes of Kitty Hawk, North Carolina, more than 1,000 flights within a time period of two years. When *Chanute* heard about these tests, which were going on very quietly, he visited the *Wright Brothers* and supported them with his advice. They flew with their bodies in a horizontal position in order to reduce drag, and they contrived wing twist to achieve a better roll-control than possible by shifting the body.

Their approach is, in many respects, similar to the one of *Lilienthal*. At first they tested their flight vehicle gliding, then they developed a 9.8 kW (12 hp) engine since the available automobile engines were too heavy, and, in addition, a propeller with an efficiency of 66 per cent. On December 17, 1903, it finally happened: the first flight took off, self-propelled, from a launching rack. On the occasion, a distance of 50 m (164 ft) was covered in 12 sec.

Did the *Wright Brothers* achieve the first powered flight, or was it on August 14, 1901 *Gustav Weißkopf*, who had emigrated in 1895 from Germany to the US? This question, again being discussed today, shall and cannot be pursued here [25, 26].

In Europe especially the studies performed in France advanced powered flight. Well-known is *Clement Ader* (1825–1911), who built in 1889 a flapping-wing airplane with a span of 14 m (46 ft) and a 14.7 kW (20 hp) steam engine, which powered a propeller. He tried to obtain money from the French military administration since his own assets were nearly exhausted. The name AVION I for this airplane has led in the French language to the expression "*avion*" for airplane, still used today. Since this and subsequent flight vehicles were modeled after bats, they had a totally insufficient controlability. This led at the demonstration of AVION III in front of a military commission in October 1899, to the situation that a gust pressed one wing of the airplane to the ground after being in the air for a short period of time. The apparatus was heavily damaged und the commission passed negative judgment. This led to the termination of the financial support.

After the first flights of the *Wright Brothers*, the idea of powered flight again gained momentum in Europe. The fame of having performed the first powered flight in Europe is claimed by three personalities: *Traian Vuia*, who performed on March 18, 1906 a first "jump" of 12 m (39 ft) at a height of 1 m (3.28 ft) with an airplane of his own design, whose wings were modeled after the Lilienthal-gliders. *Jacob Christian Ellehammer* from Denmark had performed a "jump" of 40 m (131 ft) on September 12, 1906. But the first real flight was accomplished by the Brazilian *Alberto Santos-Dumont* (1873–1932), who made his first "jumps" in Paris in early September 1906 with a flight vehicle that had the form of a box-kite and an Antoinette-engine of the Frenchman *Levasseur*. Thereafter, he asked the "Sports Witnesses" of the Paris Aero-club for a meeting on October 23, 1906, in order to apply for the prize donated by the distinguished Parisian patent-attorney *Ernest Archdeacon*. Here, *Santos-Dumont* accomplished a flight over a distance of about 50–60 m (164–197 ft). In November 1906, he improved his performance with a flight of up to 220 m (722 ft) at a height of 6 m (20 ft) lasting 21 sec.

While in France people were seriously concerned with the further development of aircraft – among others *Farman*, *Voisin*, *Delagrange* – the interest in Germany stagnated. However, the year 1908 brought a change when, on September 21, *Wilbur Wright* succeeded to break all records up to that time held by *Farman* and *Delagrange*. He increased the record in distance to 66.6 km (41 mi) and the record in flight duration to 1 h 31 min and 24.5 sec, reaching a speed of 60 km/h (37 mi/h). The news about these accomplishments revived the interest in Germany.

Friedrich Ahlborn (1858–1937).

Friedrich Ahlborn, born 1858

1882 graduates from the University of Göttingen with a PhD in zoology. After a short period as an assistant, he becomes a teacher at the scientifically oriented high school in Hamburg (Realgymnasium des Johanneums in Hamburg). In the eighties, he began his investigations concerning the flight tools of flying fish; later, analyses of the flight of birds followed.

1903 he receives a professorship. He tries to utilize his zoological observations for the construction of artificial wings. His studies of the flight movements of the "Zanonia macrocarpa" seed become important. Concerning the influence of his findings on the "Etrich-pigeon" ("Etrich-Taube"), the reader is referred to the book of *Peter Supf* "Book of the German Flight History" ("Buch der deutschen Fluggeschichte"), pp 221–224. *Ahlborn* conducted, furthermore, experiments by moving his models through standing water and visualizing the induced three-dimensional movement of the water by means of the shavings of oak and maple wood. For surface-flow visualizations, the seed of "Bärlapp" (club-moss, lycopodium) proved to be especially well suited.

1916 he becomes head of the "Hydrodynamic Research Institute of the (military) Aircraft Department" ("Hydrodynamische Versuchsanstalt der Flugzeugmeisterei") of the DVL in Berlin-Adlershof.

1918 in Hamburg he observes and studies the physics of the free atmosphere, conducting experiments to gain insight into the large circulation phenomena in the atmosphere.

1937 he dies in Hamburg.

In France, the *Wright Brothers* competed in demonstrating the airworthiness of their flight vehicles. The flight of *Blériots* across the English Channel on July 25, 1909 attracted the attention of the world. The airplane had become a device to be taken seriously. The great flight enthusiasm, existing in Germany up to that time mainly for the work of *Count Zeppelin*, finally turned to the aircraft.

One of the early scientists concerned with aeronautics was *Friedrich Ahlborn*, a senior teacher at a scientifically oriented high school in Hamburg (Realgymnasium des Johanneums in Hamburg), who was concerned with the investigation of drag-generating processes and streamlines in air and water. He was a co-founder of the "Hamburg Association of Airship Aviation". A very important stimulus for German aircraft engineering was his paper "On the stability of flight vehicles" ("Über die Stabilität der Flugapparate"), published in 1897 in the "Treatises of the Natural Sciences Association" ("Abhandlungen des Naturwissenschaftlichen Vereins"). He had discovered that in nature the seeds of certain plants are particularly stable during flight, and that especially the seeds of the tropical plant "Zanonia macrocarpa" had excellent flight characteristics; he had requested and obtained these seeds from a botanical garden on Java.

His findings were used by the Austrian *Igo Etrich*. In 1903, he built his first glider whose form was based on the "Zanonia" seed, performing glides with a distance of up to one kilometer. *Etrich* had recognized that the flight stability of the "Zanonia" seed resulted from air flowing off the wing tips keeping the wing in a position parallel to the freestream. In 1905, he received an Austrian patent on this wing shape, as well as on the propeller for a bigger flight apparatus with a wingspan of 12 m (39 ft).

Based on the glider, *Etrich* developed the well-known "Etrich-pigeon" ("Etrich-Taube"). On May 17, 1910, his co-worker and pilot *Karl Illner* carried out the first Austrian cross-country flight of 40 km (25 mi) between Vienna-Neustadt and Vienna at an altitude of 300 m (948 ft) employing an advanced version of the "Etrich-pigeon". Of crucial importance was here a new 44/51.5 kW (60/70 hp) engine, which had only a third of the weight of the automobile engine used before. The new engine had been designed by

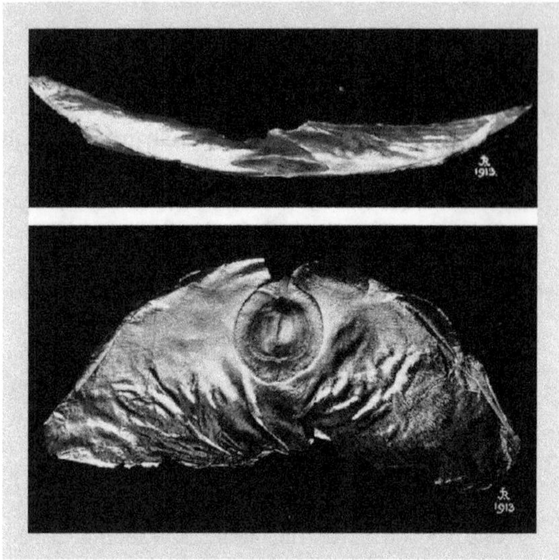

Friedrich Ahlborn performed in 1903 investigations concerning the flight characteristics of the "Zanonia macrocarpa" seed initiating the special wing form of the "Pigeon-airplanes".

Ferdinand Porsche, at that time an employee at the Daimler Company. The "Zanonia» wing form, together with a rudder and an elevator, gave the airplane the pigeon-like form and a very high natural stability. Thus, the "Etrich-pigeon" became an ideal plane for fligth training and was until about 1913 the most frequently flown vehicle.

The shape of the wing could, due to *Ahlborn's* earlier publication, in Germany not be protected by a patent. As a result, other "Pigeons" evolved, besides the ones produced under license by Etrich-Rumpler, such as the "Gotha-pigeon", the "Hansa-pigeon", and the "Jatho-steel-pigeon", all being of similar wing shape as the "Rumpler-pigeon". *Etrich* founded in 1912 in Liebau, Silesia (now Poland), the "Etrich-Airplane-Company" ("Etrich-Flieger-Werke"), whose design office was since the spring of 1914 headed by *Ernst Heinkel*. The transfer of the company to Brandenburg and its expansion, now becoming the "Brandenburg Aircraft Company, Ltd" ("Brandenburgische Flugzeugwerke GmbH"), made it one of the largest aircraft companies in Germany during World War One.

Karl Jatho (1873–1933) designed already in 1897 a tri-plane with 48 m² (517 ft²) wing surface, which he equipped with a rubber-tire landing gear. But he succeeded only in 1903 to equip the plane with an 8.8 kW (12 hp) engine of 64 kg (144 lbs) weight. With this flight vehicle, he performed on August 18, 1903 his first "jump" of 18 m (59 ft) distance at a height of three quarters of a meter (2.5 ft). He later succeeded in making "jumps" (flights) of up to 60 m (197 ft). Although he thus performed with his powered vehicle flights prior to the flights of the *Wright Brothers*, he did not succeed in carrying out a real flight before they did, since his engine did not have enough power. Following the tri-plane and the biplane, he considered the monoplane. In 1911 he created the first monoplane made of steel-tubes, which he fitted with a 36.8 kW (30 hp) water-cooled Argus-engine; with this vehicle, he reached a height of 26 m (85 ft) and flew a distance of 2.1 km (1.3 mi). In 1912 he installed a 73.5 kW (100 hp) engine into his apparatus winning the prize for the first flight circling Hanover. His steel-tube monoplanes were of such high stability that one could even perform nosedives. In 1914 he founded the "Hanover Aircraft Company" ("Hannoversche

A Rumpler "Pigeon" (Rumpler "Taube") exhibiting the typical wing shape of the "Zanonia" seed.

Flugzeugwerke"). However, since his company did not receive any contracts from the Army Administration, he had to relinquish it.

Aluminum, the Material of Aeronautics

During the early years of aeronautics, mostly natural materials as light as possible have been used: all kinds of wood, including balsa, willow-twigs and similarly deformable materials. As cover material served cotton and other fabrics. However, it was soon attempted to use metal since strength and deformability were important. Here, aluminum soon gained a special role. Aluminum is a light metal, discovered in 1827 by the German chemist *Friedrich Woehler*, with a specific weight of 2.7 g/cm^3 (0.1 lb/in^3) and thus three times lighter than iron. At the time of its discovery as expensive as precious metal, declined the price from decade to decade. In 1855 the cost of one kilogram was 1,000 Marks dropping to just 2 Marks in 1900. Initially, it was due to its softness not very suitable. Only after discovering that the addition of magnesium, manganese and copper resulted in hardened light-metal alloys did it become the most important material in aircraft construction, after having already been proven effective in airship design.

Carl Berg, owner of a copper foundry in Eveking/Westphalia, was, as already mentioned, the one who could produce forgeable aluminum alloys in sufficient quantity. His personal engagement for the use of aluminum in the construction of airships has considerably contributed to its success also in the design of aircraft.

The Beginnings of Institutionalized Aeronautical Research – the Roots of Scientific Aerodynamics

Considering the above description of the beginnings of aeronautics, it is apparent that, with few exceptions, as, for instance, *Otto Lilienthal*, it were enterprising men, who, utilizing the contemporary knowledge in natural sciences available, constructed their flying machines with a lot of courage and imagination rather than trying to explore the fundamentals. In the process, some succeeded in finding solutions to the multifarious flight-technical problems intuitively rather than through a scientific recognition of the underlying physical relations. This also holds for the years following *Lilienthal* and the *Wright Brothers* until just prior to the beginning of World War One.

The early history of the institutionalized and scientifically pursued aeronautical research in Germany is coined by a few personalities, among whom the professor of mathematics at the University of Göttingen, *Felix Klein* (1849–1925), played an outstanding role. He was the one who already in 1893, under the impression of a study trip to Chicago, advocated the idea to add to the mathematical-natural-sciences lectures at the University of Göttingen the subject of "technical physics". *Henry Theodor von Böttinger*, at the time director of the "Paint Factories" ("Farbenfabriken") in Elberfeld and son-in-law of the founder of the company, *Friedrich Bayer*, actively supported this idea. *Friedrich Althoff*, a director at the Prussian "Ministry of Religious, Educational and Medical Affairs" ("Ministerium der geistlichen, Unterrichts- und Medizinalangelegenheiten") was the public servant who helped to realize this.

In 1897, after overcoming considerable resistance, and with the aid of a larger donation from *von Böttinger* and from two other donors from industry, a "technical department" could be established within the Institute of Physics of the University of Göttingen, whose management was coupled with an associate professorship. The first head was *Richard Mollier*, who stayed until the autumn of 1897, followed by *Eugene Meyer* who held the position until 1900. In the wake of the foundation of this department, the "Association for the Promotion of Applied Physics of Göttingen" ("Göttinger Vereinigung zur Förderung der angewandten Physik, GV") constituted itself on February 26, 1898.

Felix Klein showed very early an interest in aeronautics. Already in 1902, the volume of the Encyclopedia of Mathematical Sciences relating to mechanics, edited by *Klein*, contained a chapter on aerodynamics written by *Sebastian Finsterwalder* [27]. This chapter may be considered as the first relevant article in a major German scientific treatise. In 1904, *Felix Klein* arranged for *Ludwig Prandtl* (1875–1953) to be appointed associate professor at the chair of Technical Physics at the University of Göttingen. The subsequent developments in aerodynamics and later the

Felix Klein (1849–1925).

Felix Klein, born in Düsseldorf 1849, becomes at the age of 17 assistant to the geodetic surveyor *Pluecker* in Bonn, and at the age of 23, i.e., in

1872 he is offered the professorship for mathematics at the university of Erlangen.

1875 he moves to the University of Munich and from there in

1880 to the University of Leipzig, where he expresses his views on the problem of "the relation of modern mathematics to applications".

1886 he moves to Göttingen to assume the chair of mathematics, and he publicly expresses his opinion "on the plan to establish a physical-technical institute at the University of Göttingen".
He starts a personnel policy specifically designed to realize his plans and calls in

1895 *Hilbert* to Göttingen. He gives, together with *Sommerfeld*, in

1895/96 lectures on the theory of gyroscopes, from which evolves the four-volume treatise "On the theory of gyroscopes" ("Über die Theorie des Kreisels").

1898 he comments on the relation "University and Technical University (or Institute of Technology)" ("Universität und Technische Hochschule"). He sets up the "Association for the Promotion of Applied Physics of Göttingen" ("Göttinger Vereinigung zur Förderung der angewandten Physik, GV").

1902 he calls *Minkowski* and in

1904 *Runge* and *Prandtl*, two important partners in the realization of his ideas regarding applied mechanics, to Göttingen. In the wake of this endeavor, institutions arise, which *Blenk* calls the nucleus of aeronautical research and technologies. These are:

1906 the "Powered-airship Study Association" ("Motorluftschiff-Studiengesellschaft, M.ST.G.") and in

1907 the "Model-test Establishment" ("Modellversuchsanstalt"), as predecessor of the later "Aerodynamic Test Establishment Göttingen" ("Aerodynamische Versuchsanstalt Göttingen, AVA"). Up to

1908 four additional institutes are established, namely, the "Institute of Applied Mathematics", the "Institute of Applied Mechanics", the "Institute of Applied Electricity" and the "Institute of Geophysics". In

1912 follows the foundation of the "Scientific Aeronautical Society" ("Wissenschaftliche Gesellschaft für Luftfahrt") as predecessor of today's DGLR, and finally in

1925 the foundation of the "Emperor-Wilhelm Institute for Fluid Research" ("Kaiser-Wilhelm-Institut für Strömungsforschung").
In addition, he publishes the "Encyclopedia of the Mathematical Sciences Including their Applications" ("Enzyklopaedie mathematischer Wissenschaften mit Einschluß ihrer Anwendungen".

1925 he dies in Göttingen.

development of the "Aerodynamic Test Establishment Göttingen" ("Aerodynamische Versuchsanstalt Göttingen, AVA") are inseparably associated with the name of *Prandtl*.

Ludwig Prandtl studied at the Institute of Technology (TH) Munich and received in 1900 his PhD from the University of Munich, since the TU did not have the right to bestow a PhD. During his occupation as an engineer at the "Machine Company Nuremberg" ("Maschinenfabrik Nürnberg"), he encountered during the construction of a suction system for the removal of wood shavings the problem that the flow on the wall of a conical air duct separated, which could not be explained with the fluid mechanical laws known at that time. This was the starting point of his occupation with fluid mechanics. In 1901, *Prandtl* became associate professor at the "Institute of Technology (TH) Hanover". During that time, he developed the theory of fluids with very low friction, i.e., the boundary layer theory.

At the same time as *Prandtl*, *Carl Runge* received in 1904 a call to the first ever chair for applied mathe-

Ludwig Prandtl (1875–1953) visualized in 1903 in his simple water tunnel, whose flow was generated by a hand-driven paddle wheel, fluid boundary layers, separations and vortices. A copy of the historic water tunnel can be seen at the "Deutsches Museum" in Munich, the original is exhibited at DLR Göttingen.

Ludwig Prandtl, born in Freising near Munich on February 4, 1875.

1898 completes his studies of engineering at the Institute of Technology (TH) Munich.

1900 PhD from the University of Munich.

1901 Professor of Mechanics at the TH Hanover.

1904 establishes boundary-layer theory. He receives a call from *Felix Klein* to join the chair for Technical Physics at the University of Göttingen.

1905 foundation of the "Association for the Promotion of Applied Physics and Mathematics of Göttingen" ("Göttinger Vereinigung zur Förderung der angewandten Physik und Mathematik").

1907 planning and foundation of the "Powered-airship-model Test Establishment" ("Motorluftschiffmodell-Versuchsanstalt"), which very soon became the "Model Test Establishment Göttingen" ("Modell-Versuchsanstalt Göttingen").

1908 he constructs the first wind tunnel of "Göttingen-type design" ("Göttinger Bauart") for the investigation of drag.

1909 he is assigned the first German lectureship for aeronautics at the University of Göttingen; several memoranda on German aeronautical research evolve.

1910 he becomes scientific head of the "Journal of Flight-techniques and Powered-airship Aviation" ("Zeitschrift für Flugtechnik und Motorluftschifffahrt").

1911 *Prandtl*, together with *Hergesell*, prepares an expertise concerning the organization and technical equipment of a "State Test Establishment for Airship Aviation" ("Reichsversuchsanstalt für Luftschiffahrt"), which leads in

1912 to the foundation of the "German Test Establishment of Aeronautics» ("Deutsche Versuchsanstalt für Luftfahrt, DVL").

1913 the senate of the "Emperor-Wilhelm Society" ("Kaiser-Wilhelm Gesellschaft, KWG") decides, following a proposal by *Prandtl*, to establish the "Emperor-Wilhelm Institute of Aerodynamics and Hydrodynamics" ("Kaiser-Wilhelm-Institut für Aerodynamik und Hydrodynamic") in Göttingen.

1917 the second wind tunnel is constructed and the M.St.G. (see *Felix Klein* 1906) is renamed "Model Test Establishment for Aerodynamics" ("Modellversuchsanstalt für Aerodynamik, MVA"), becoming in

1920 the "Aerodynamic Test Establishment" ("Aerodynamische Versuchsanstalt, AVA") within the KWG.

1918 formulation of the "Theory of the lifting wing (lifting-line theory)".

1922 foundation of the "German Engineering Sciences Association" ("Deutsche Ingenieurwissenschaftliche Vereinigung") in Leipzig, which later develops into the "Society of Applied Mathematics and Mechanics" ("Gesellschaft für angewandte Mathematik und Mechanik, GAMM"), with *Prandtl* becoming first secretary.

1925 he in addition, becomes head of the "Emperior-Wilhelm Institute of Fluid Research" ("Kaiser-Wilhelm Institut für Strömungsforschung").

1933 he becomes chairman of the "Committee of Fluid Mechanics of the Aeronautical Research Association" ("Ausschuß für Strömungsforschung der Vereinigung für Luftfahrtforschung, VLF"), which in

1936 is renamed "Lilienthal Society for Aeronautical Research" ("Lilienthal-Gesellschaft für Luftfahrtforschung"). He becomes a member of the "German Academy of Aeronautical Research" ("Deutsche Akademie der Luftfahrtforschung").

1937 the AVA (see 1920) becomes an independent registered association with eight institutes and *Prandtl* as its director.

1939 he takes over as head of the associated group "Meteorological Fluid Research" ("Meteorologische Strömungsforschung") of the German Aero Lloyd.

1942 he is put in charge of the "Aviation Research Command" ("Forschungsführung") at the "Ministry of Aeronautics" ("Reichsluftfahrtministerium, RLM").

1945 presentation of the results of aeronautical research in the "Göttingen Monographs".

1947 retirement.

1953 dies in Göttingen on August 15.

matics at a German university, the University of Göttingen. *Runge*, in friendship closely liaised to the younger *Prandtl*, originally took care that *Prandtl* came to Hanover. In the autumn of 1905, "technical physics" and "applied mathematics" were combined in Göttingen within an "Institute of Applied Mathematics and Mechanics", where *Runge* and *Prandtl*, together with *Klein*, realized their envisaged study courses. Both institutes remained, however, administratively separated. *Prandtl's* interest in fluid mechanics was awakened, and he started to devote himself to the problems within this field.

Theodore von Kármán (1881–1961) came in October 1906 from Budapest as an assistant to this institute, receiving in 1908 his PhD with a topic from the theory of elasticity. In 1909, he attained his "venia legendi" (right to lecture at a university), and, as lecturer, disburdened *Prandtl* with respect to his lecturer's obligations.

In July of 1906, the "Powered-airship Study Association, Ltd." ("Motorluftschiff-Studiengesellschaft GmbH") was founded with major participation of the "Ministerial Department Head" ("Ministerialdirektor") *Althoff* and endowed with a capital of one Million Marks. The management was entrusted to two experienced airship pilots (aeronauts), *Richard von Kehler* and, after January 1, 1907, *August von Parseval*. The objective of the association was: "*the promotion of the technique and application of aviation, especially of airships. ...*". In September of the same year, a technical committee was elected which established four groups, with one speaker each, viz., the meteorological group, the dynamics group, the design and construction group, and the machinery-group.

Felix Klein as speaker of the dynamics group asked *Prandtl* to consider which task this group should take on. *Prandtl* recommended a model test establishment, which could conduct drag measurements as well as pressure and velocity distribution measurements on airships. In the process, the construction of a wind tunnel, analogous to test installations in a ship-model-towing facility, was considered. Following *Prandtl's* application, the establishment of a provisional model test establishment was approved on November 9, 1907. This date is generally considered as the date of the foundation of the organization which later became the "Aerodynamic Test Establishment" ("Aerodynamische Versuchsanstalt, AVA").

In 1907, one of the test facilities was the first ever closed-circuit wind tunnel [28], a forerunner of the so-called "Göttingen-type" wind tunnels. Due to many delays, especially during the procurement of the tunnel drive system and the instrumentation, and during the commissioning phase, where emphasis was placed on an essentially vorticity-free air stream, the organization could take up full test activities only in early 1910. The test establishment, called "Model Test Establishment of the Powered-airship Study Association" ("Modellversuchsanstalt der Motorluftschiff-Studiengesellschaft"), for short "Model Test Establishment Göttingen" ("Modell-Versuchsanstalt Göttingen"), was at that time something very new. It may be considered the cradle of scientific institutes in aeronautical research.

Prandtl was also actively involved in the foundation of the "Lower Saxony Association of Airship Aviation" ("Niedersächsischer Verein für Luftschiffahrt") seated in Göttingen. In 1909 he obtained the license to pilot free-balloons. In April 1908, he undertook, together with *von Parseval*, on behalf of the "Powered-airship Study Association" an excursion to Paris to study

Theodore von Kármán (1881–1963).

Theodore von Kármán, born in Budapest on May 11, 1881, *Prandtl's* student

1908 PhD from the University of Göttingen with a thesis on the buckling-strength of bars.

1911 conducts studies on vortex formations, later called Kármán vortex-street (Kármán'sche Wirbelstrasse)

1912 Professor of mechanics and aerodynamics at the Institute of Technology (TH) Aachen, where he founded the Institute of Aerodynamics; advisor of *Junkers and Zeppelin.*

1917 helicopter investigations in Hungary.

1930 Director of the Guggenheim Aeronautical Laboratories of the California Institute of Technology (Caltech), Pasadena, California. He becomes one of the leading scientists in the USA in the fields of applied mathematics and mechanics, as well as on issues of strength and elasticity.

1939 consultant, Army Air Corps and US Air Force. During that time, booster rockets for aircraft were developed. A corresponding working group resulted later in the Aerojet-General Corporation. On a small rocket test range near the Caltech campus the Jet Propulsion Laboratory (JPL) develops under von *Kármán's* supervision during World War Two.

1944 he is charged with the organization of a scientific advisory staff, which later required his presence in Washington, DC. He was, therefore, in

1949 given leave of absence from his duties at Caltech. He proposes an advisory group to NATO and becomes in

1951 director of the "Advisory Group for Aeronautical Research and Development, AGARD" of NATO.

1963 he dies in Aachen on May 7.

developments in France. After their return, both began with first flight-mechanical studies on stability. Here, *von Parseval* pursued the idea to improve the stability of an aircraft by making the wings flexible, which, however, turned out to be the wrong approach.

The flights of *Wilbur Wright* on August 8, 1908, which attracted appreciable attention due to the elegant turns, demonstrated the importance of control about the longitudinal axis of an aircraft, which received little attention by other flight pioneers. *Parseval* initiated a design, which was to surpass the controllability, or maneuverability, of *Wright's* airplane. Although he failed due to insufficient engine power, it shows that the "Powered-airship Study Association" now turned towards the airplane.

Also the "Association for the Promotion of Applied Physics of Göttingen" ("Göttinger Vereinigung zur Förderung der angewandten Physik") discussed the problem of flight and tried to introduce "Aeronautical Sciences" as a study subject at the University of Göttingen. On January 11, 1909 *Prandtl* was requested to cover, in addition to his already existing subjects, the complete field of aeronautics in lectures and exercises. The news of the installation of a professorship for aeronautics in Göttingen, the first worldwide, was spread by newspapers and journals.

Ludwig Prandtl took up his new lecture series in the summer semester of 1909 with a two-hour weekly lecture entitled "Scientific Fundamentals of Airship Aviation" to 31 students. The lectures covered the areas of aerostatics, aerodynamics, theory of propellers, stability and controllability of air vehicles.

A further step towards scientific aeronautical research was the foundation of the "Journal of Flight-techniques and Powered-airship Aviation" ("Zeitschrift für Flugtechnik und Motorluftschiffahrt, ZFM"), which was published since January 31, 1910 by the publishing house R. Oldenbourg in Munich and Berlin. The publisher was *Ansbert Vorreiter*, the scientific editor *Prandtl.*

Promotion of Flight and Foundation of the First Aircraft Companies

At about the same time as the scientific research on flight commenced, the public interest changed towards the flight vehicle "heavier-than-air"[29]. Especially

Alfred Hildebrandt (1870–1949) was a protagonist of this idea. He organized the first "flight days (air shows)" in Germany, which became quite important to the early engineers and technicians concerned with flight. As a young officer, he supervised the first ascends of kites for meteorological purposes, thus coming into contact with airship aviation. Visiting *Lilienthal* in 1893 enthused him for aviation. In 1896, he founded, together with his older regimental fellow, *Moedebeck*, the "Upper-Rhine Association for Airship Aviation" ("Oberrheinischer Verein für Luftschiffahrt") also taking charge of the organization. In October 1906, he privately traveled to the US to get to the bottom of the rumors and reports concerning the flights of the *Wright Brothers*. After many discussions with eyewitnesses, he came to the conclusion that, in Germany, one should seriously consider the construction of airplanes.

In 1907 *Hildebrandt* suffered during a balloon voyage an accident resulting in severe bone fractures, which forces him to resign from the services. He started studies in mathematics and natural sciences. In the years between 1907 and 1909, *Hildebrandt* performed extensive educational work, which made him an important pioneer in paving the way for the airplane. He traveled to France to observe the flights of *Wright* in Le Mans, and he witnessed the first "circle flight" in Europe, performed by *Henri Farman* in January of 1908. The "Tourist Association" of Kiel, headed by *Baron von Moltke*, decided, based on *Hildebrandt's* reports, to organize on June 28, 1908 an aviation day combined with a (prize) competition. *Hildebrandt* was in charge of the flight demonstration, but only the Dane *Jacob Christian Ellehammer* showed up, delivering a "jump" (flight) of about 47 m (154 ft) at a height of about two meters (6.6 ft), a performance far behind that achieved in other countries.

In 1909, *Hildebrandt* arranged for the publishing house Scherl to invite the French aviator *Armand Zipfel* to participate in an air show, to be held from January 29 until February 4, promoted by the Berlin newspaper "Local Gazette" ("Lokalanzeiger"). However, *Zipfel* was able to perform, with his self-made biplane of the "Voisin"-type, only on two days flights across the Tempelhof field of more than 1,000 m (3,300 ft) at heights between two and three meters (6.6 and 9.8 ft). In September 1909, *Orville Wright* and *Hubert Latham* conducted their air show at the Tempelhof field. On this occasion, *Wright* also took along passengers, among them *Alfred Hildebrandt*, the latter's first pupil, *Engelhard*, and, on September 9, also Mrs. *Hildebrandt*, who thus became the first German woman ever to fly as a passenger. *Hubert Latham* started on September 27, 1909 his show flights. On September 27, he took off at Tempelhof, circled the Tempelhof field once, and landed, after 14 min and 31 sec at the airfield Johannisthal, 10 km (6.25 mi) away from Tempelhof. This was the first cross-country flight in Germany.

Orville Wright demonstrated in 1908 his biplane in various European countries, thus boosting the development of airplanes in France and Germany. The company "Wright Flight Machines" ("Wright-Flugmaschinen"), located in Berlin-Johannisthal, produced this airplane in license.

During the same year, further events important to aeronautics occurred, such as, for example, the first "International Airship Aviation Exhibition" ("Internationale Luftschiffahrt-Ausstellung, ILA") in Frankfurt (Main) from July to October, organized by *Georg von Tschudi* (1862–1928). On the occasion, powered-airplane and glider competitions were held for the first time in Germany. Besides airships, a great number of airplanes were exhibited. However, also basic scientific lectures were given by *August von Parseval*, the privy councilors *Hergesell* and *Aßmann*, as well as by the distinguished professors *Ahlborn*, *Prandtl*, *Pütter* and *Süring*. The only German who took to the air during the exhibition, was *August Euler*, who was also able to win a prize for the longest flight of the day, a flight of 4 min and 54 sec duration.

At about the same time, the airfield of Johannisthal near Berlin was built. It was inaugurated on September 26, 1908 without a celebration, but with a "flight competition of the foremost aviators of the world",

Hugo Hergesell (1859–1938).

Hugo Hergesell, born in Bromberg (now Poland) on May 29, 1859. Already during his studies he tended towards meteorology; he headed until 1914 the "Imperial Meteorological State Institute" ("Kaiserliche Meteorologische Landesanstalt") in Strasbourg (now France).

1896 during the International Congress of Meteorologists in Paris, the "International Committee for Scientific Airship Aviation" ("Internationale Kommission für Wissenschaftliche Luftschiffahrt") was founded and *Hergesell* is charged with its first presidency.

1897 he suggests the ascent of kites equipped with instruments, which in

1900 was realized over Lake Constance. Initiated by the airship officer *Hermann Moedebeck*, conditions of the upper atmosphere were mapped through simultaneous ascents of kites and balloons at different locations of the globe.

1904 the measurements were continued over the Mediterranean Sea. In July/August, measurements with registering balloons were performed in the trade wind zones of the Northern Atlantic. Up to World War One, the data were exchanged every two years at international congresses. *Hergesell* was in close contact with *Count Zeppelin*, being responsible for weather consultation-services during the German reliability flights in

1911 and 1912 in the Upper-Rhine area, as well as in

1913 and 1914 for the "Prince Heinrich-Flights" ("Prinz-Heinrich-Flüge").

1915 as a director at the Prussian Observatory Lindenberg (county of Beeskow), he managed the German air-weather service. Since the mid-twenties he was, in addition, a honorary professor at the University of Berlin and at the Institute of Technology (TH) Charlottenburg.

1927 until 1935, he was once more president of the "International Committee for Scientific Airship Aviation"

1938 he dies in Berlin on June 6.

At the first "International Airship Aviation Exhibition" ("Internationale Luftschiffahrt-Ausstellung, ILA"), held in Frankfurt (Main) from July until October 1909, balloons and airships still dominated. But also some flight vehicles were exhibited and a flying competition was arranged. Exhibition and flight event found great interest with industrialists, scientists and the general public.

so the official announcement. On the second day, the sensation was the arrival of *Latham*; up to the end of the air show on October 3, foreign pilots won all available prizes. This clearly demonstrated especially the superiority of the French, awakening German national ambition.

Of the many early aeronautical companies, one has, in conjunction with aeronautical research, to name above all those, that successfully mastered the early stages due to aerodynamically and constructively carefully designed flight vehicles, and which later systematically adapted their designs according to the growing knowledge. The oldest of these German companies was already established in 1908 in Griesheim, near Darmstadt, by *August Euler* (1868–1957). The great-grandson of the famous mathematician *Leonard Euler* acquired, as the first German, in 1908 the patents for the Voisin-biplane in order to set up a flight school. He trained himself initially, similar to *Lilienthal*, by glides with a box-kite glider of *Chanute* design. When the French Ministry of War prevented the implementation of the license contract with *Voisin*, he designed an airplane nearly half the weight of the design of the brothers *Charles* and *Gabriel Voisin*. Based on his own calculations, he improved it so much, that it must be considered his own design. *Helmuth Hirth*, who became later well known, was at that time his employee.

On August 20, 1909, *August Euler* succeeded as the first German to perform a flight of 20 sec under official supervision; on December 31, 1909, he also passed as the first German the officially specified international pilot's examination, by flying twice, two closed circles each time, around the training ground in Griesheim put at his disposal. *Euler* received on February 1, 1910 the German Pilot's License No. 1, issued in six languages.

A second German airplane company, also founded in 1908 – and becoming well known later – was the one of *Edmund Rumpler* (1872–1940). Since he started his professional career with the "Daimler Motor Company" ("Daimler-Motorengesellschaft") in Marienfelde, now part of Berlin, it was natural that he first dealt with aircraft engines and, in the process, designed his own 4-cylinder water-cooled engine which he called "Aeolus". He built already in 1908 in his oxy-acetylene-welding (autogenously-welding) shop in Lichtenberg for the *Focke Brothers* of Bremen the "Duck" ("Ente") which they had jointly designed. When

Edmund Rumpler (1872 –1940) first dealt with aircraft engines developing his "Aeolus-engine". After entering a license contract with *Igo Etrich*, he built the successful "Rumpler-Dove" ("Rumpler-Taube") in larger quantities.

Alfred Hildebrandt brought the merits of the "Etrich-Dove" ("Etrich-Taube") to his attention, he entered in June 1910 into a license contract with *Igo Etrich*. After some quarrels concerning the patent situation in Germany and after design changes, among them the installation of his Aeolus-engine, became the airplane, from then on only known under the name "Rumpler-Dove" ("Rumpler-Taube"), one of the best-known flight vehicles of the pre-war era.

Hans Grade (1879–1946), from his youth on enthusiastic about flying and technically interested, studied after graduating from high school mechanical engineering at the Institute of Technology (TH) Charlottenburg, now the Technical University (TU) Berlin, and founded in 1905, at the age of 26, the "Grade-Engine Company" ("Grade-Motorenwerke") in Magdeburg, where fast motorbikes were built. In 1907, he built his first airplane, a tri-plane, equipped with one of his own engines. He won with a monoplane, built thereafter and also equipped with one of his own engines, in October of 1909 the "Lanz-prize of the Air" ("Lanz-Preis der Lüfte"). The requirements included that all parts of the airplane had to be built in Germany, that the designer as well as the pilot were German, and that a "horizontal eight" with a length of about 2.5 km (1.6 mi) had to be flown. He used the prize money of 40,000 Marks to set up an aircraft plant in Borkheide near Beelitz not far from Berlin.

Another early designer of airplanes was *Hermann Dorner* (1882–1963), who already as a child wanted to become a pilot and build airplanes. He studied nautical engineering. Since the subject of aeronautical engineering was not (yet) offered, this appeared to him the obvious choice. In 1906, he received his degree as a nautical engineer from the Institute of Technology (TH) Charlottenburg. Already in 1907, he embarked on the construction of a glider, to whose nose an engine could be attached. Scientifically very thorough by nature and education, his striving was to achieve "maximum efficiency". On June 11, 1910 he won with a monoplane of his own design and equipped with his own engine the 3rd Lanz-prize; in August, he won the first "passenger-prize" presented at an air show in Johannisthal. With the money, he founded his own "Dorner-aircraft Company" ("Dorner-Flugzeug-Gesellschaft").

Also founded during that time was the "Wright Flight Machines Company, Ltd." ("Flugmaschinen Wright GmbH") in Berlin-Reinickendorf with the objective to exploit the Wright patents. The managing director was *von Kehler*. The company built about 100 airplanes before it was dissolved in 1913, following a patent litigation, which it lost.

Walter Huth, a former first lieutenant, who resigned from the services due to a leg ailment, bought in France an Antoinette-monoplane and a Farman-biplane and founded the "Albatros-Company" ("Albatros-Werke") with headquarters in Johannisthal; the company later introduced its own design and built in 1911 a series of seven biplanes ordered by the Prussian army administration.

Jacob Goedecker, an engineer who studied in Aachen and Charlottenburg nautical and mechanical engineering, founded in 1909 the "Flying-Machine Company J. Goedecker" ("Flugmaschinen-Werke J. Goedecker") in Mainz-Gonsenheim. He had recognized as the best and most stable wing form the one of the seagull and had already built glider models with cranked wings at the time he worked for *Hugo Junkers* at the "Mechanics Laboratory" in Aachen. In 1909, he designed a monoplane with a 12-meter-span (39 ft) cantilever wing, which could be rotated about its lateral axis (pitched) and its vertical center axis (yawed) with respect to the fuselage. This first airplane without any wire-bracings was not successful because the elevator did not meet expectations.

In addition to these first companies, which persevered for a longer period of time, there were others that soon disappeared. The great number of "home-built" air vehicles must be mentioned, too. In the context of the emergence of aeronautical research, all have to be seen as an essential part of the experimental exploration of flight.

In engine manufacturing, at that time a domain of the French, where the engines "Gnôme" and "Antoinette" excelled, new companies in Germany were especially concerned with the construction of light aircraft engines. They also tried to reduce the gap to the French

Hans Grade won in 1909 with his monoplane "dragonfly" ("Libelle"), powered by a Grade-engine, the "Lanz-prize of the Air" ("Lanz-Preis der Lüfte").

manufacturers. In 1910, the "Aircraft-Engine Construction Company, Ltd." ("Luftfahrzeug-Motorenbau GmbH") was founded in Bissingen with a 40 per cent participation of the "Airship-Construction Company Zeppelin, Ltd." ("Luftschiffbau-Zeppelin GmbH"). *Karl Maybach*, the son of *Wilhelm Maybach*, became the technical manager. Further startups included the "Argus Engine Company, Ltd." ("Argus-Motoren-Gesellschaft mbH"), Berlin-Reinickendorf, the "NAG New Automobil Company, Ltd." ("NAG Neue Automobil-Gesellschaft mbH"), Berlin-Oberschöneweide, and the "Hermann Haake Engine Engineering Company" ("Hermann Haacke Motorenbau"), Berlin-Johannisthal.

The years between 1910 and 1912 can be characterized as the time of closing the gap with respect to France. This was promoted by a great number of competitive events. In 1911 there were eight events and in 1912 even eleven, but the zenith of public interest had been exceeded. Flying as a sport and for entertainment did not earn enough to form a secure financial basis. Although *Heinrich von Stephan*, the postmaster general of the "North-German Federation" ("Norddeutscher Bund"), had already in 1874 pointed out the likely importance of the transport of mail by air, only a first air-mail route from Bork to Brück opened on February 18, 1912. However, it was a purely private enterprise, which *Hans Grade* operated for some time before it was closed down again. On June 10 of the same year, the first air-mail service was put into operation on behalf of the "German Postal Service" ("Reichspost"); the airship "Swabia" ("Schwaben") and *Euler's* aircraft "Yellow Dog" ("Gelber Hund") served as mail carriers.

Aeronautical Research Obtains its Structure

The year 1912 marks an important turning point in the history of aeronautics in Germany. The great interest of the French military shown at the aeronautical exhibition in December 1911 in Paris, and the importance of the airplane as a means of reconnaissance, demonstrated at the "Imperial Maneuvers" ("Kaisermanöver") of 1911, changed the public mind to take the new technique seriously. *Prince Heinrich*, stimulated by his flight instructor *Euler* appointed a "State Committee" ("Reichskomitee"), which included leading representatives of science, trade and industry. It called for a "National Flight Donation" ("Nationalflugspende") aiming at acquiring a leading position in this new technology. Within six months, more than seven million Marks were raised.

In the area of research, two important events were subsequently to be registered: On April 3, 1912 opened under the patronage of *Prince Heinrich* the "General Aeronautical Exhibition" ("Allgemeine Luftfahrtausstellung, ALA") in the "Great Exhibition Halls" at the Zoological Garden in Berlin. On the same day, the "Scientific Flight-technical Society" ("Wissenschaftliche Gesellschaft für Flugtechnik, WGF") was founded at a meeting of the "Upper Chamber of the Prussian Parliament" ("Herrenhaus") under the honorary chairmanship of *Prince Heinrich*. The society became later the "Scientific Aeronautical Society" ("Wissenschaftliche Gesellschaft für Luftfahrt, WGL").

The initiation of the society is a result of the correspondence between *August Euler* and *Ludwig Prandtl*, which was started by *Euler* on the occasion of *Prandtl* being awarded the lectureship for aeronautics. The subject was heightened by a non-public lecture of *Euler* to the general assembly of the "Association for the Promotion of Applied Physics of Göttingen" ("Göttinger Vereinigung zur Förderung der angewandten Physik, GV") during the ILA ("International Airship Aviation Exhibition") 1911. *Euler* reported his very new flight experiences to the association. As a result, a first meeting between "Theory" and "Praxis" was arranged during the time between November 3 and November 5, 1911: *Prandtl* had invited representatives of science and technology to a joint event, which can be considered as the first aeronautical congress in Germany. Nine professional lectures, addressing aerodynamic and flight-mechanical questions, were presented. *Euler* and some of his pupils performed flight demonstrations, which had a scientific rather than a sporting character. In concluding it was suggested that the GV take the necessary steps to create a central organization, which was to take care of existing and newly found contacts.

In the letter of February 20, 1912 to the participants in the above congress, the foundation of a society was suggested which was to bring together technical and scientific aeronautical and flight-technical representatives. *Von Böttinger*, *Klein* and *Prandtl* signed this letter. Until the beginning of March, about 40 positive statements were returned so that the foundation was secured. On April 4, 1912, the day after the foundation of the WGF, *von Böttinger*, *von Parseval* and *Prandtl* were elected to the executive board of the society, and the Journal ZFM, published since 1910, was determined to be the official organ of the society.

Almost parallel in time occurred the foundation of the "German Test Establishment of Aeronautics" ("Deutsche Versuchsanstalt für Luftfahrt, DVL"). Its early history commenced with a suggestion of *Count Zeppelin* on September 28, 1909 at the board meeting of the "Deutsches Museum" in Munich, where he proposed the creation of a central test establishment analogous to the ones just established in England and France. These ideas were taken up by the "German Reichstag" (Parliament) in December. There was also a positive response by important associations, societies, and clubs. At the "German Office of the Interior" ("Reichsamt des Innern"), the councilor *Albert* and the head of the ministerial department, *Lewald*, were instructed to check possible procedures. They had two expert opinions worked out.

One, entitled the "The Project of a State Test Establishment for Aeronautics" ("Das Projekt einer Reichsversuchsanstalt für Luftfahrt"), was drawn up by *Hugo Hergesell*, head of the Meteorological Institute in Strasbourg, and *Prandtl* as head of the "Powered-Airship Test Establishment". The second one entitled "Concerning the Need, the Tasks and the Arrangement of a State Institution for the Promotion of Airship Aviation" ("Über Bedürfnis, Aufgaben und Gestaltung einer staatlichen Anstalt zur Förderung der Luftschiffahrt"), was drawn up by *Friedrich Bendemann*, head of the "Office of Flight Technology" ("Geschäftsstelle für Flugtechnik") in Lindenberg, Mark Brandenburg. This office was launched in 1908 by the "Jubilee Foundation of German Industry" ("Jubiläumsstiftung der deutschen Industrie"). It established a laboratory, where propeller tests and first aerodynamic force measurements on a tethered airplane, exposed to natural winds, were performed.

Both expertises supported such an institution, but not as a state organization in order to allow a greater freedom of research. This led on April 20, 1912 to the foundation of a corresponding registered association. Property near Adlershof was leased from the "Real Estate Corporation at the Airport Johannisthal-Adlershof" ("Terrain-Aktien-Gesellschaft am Flugplatz Johannisthal-Adlershof"). *Bendemann* was appointed director.

The first big task of the DVL was the organization of the "Competition for the best German aero-engine", which became known as the "Emperor's Prize" ("Kaiserpreis"), since it was announced, by cabinet order, coincident with the birthday of Emperor *Wilhelm II*, on January 27, 1912. The results of the competition were to be available one year later at the latest. The competition was initiated due to the fact that Germany trailed behind France in aircraft-engine construction. Five engine test rigs, including instrumentation, had to be erected as quickly as possible. A total of 68 engines by 26 competitors registered; 17 engines from eight competitors were admitted to the main trials, which were carried out in the time period October to December 1912, so that the results were available in time for a decision.

In addition to other test installations at the DVL, an open-circuit wind tunnel (Type Eiffel) with a drive power of 25.7 kW (35 hp), an open-jet diameter of 0.76 m (2.5 ft), and a flow velocity of 50 m/s (164 ft/s), was planned; its construction was completed in 1913.

The year 1912 is also in other respects remarkable. *Hugo Junkers* appeared for the first time as aircraft designer. His mode of operation and his developments will be described in more detail in Part 3.

Another engineer, who became famous later on, entered in 1910 the field of aeronautical research. It was *Claude Dornier*, who joined in November as an engineer the newly established test department of the Zeppelin Company. One of his tasks was to measure velocity, drag, and rudder forces of airships during flight, to determine the radius of curvature of the flight path, and to determine the loads acting on the airframe. To do so, the necessary measurement and control devices had to be developed. Later, around 1912, he worked on propellers, constructing one of metal, and extensively worked on the corresponding theory. *Count Zeppelin*, who recognized the capabilities of *Dornier* early on, soon gave him his own department, the "Department Do", which can be considered the nucleus of the later *Dornier* enterprise.

At the turn of the year 1912/1913, essential features of the future structure of German aeronautical research were thus established; they were, without any specific order:

- first professorships (chairs) at universities,
- independent scientific/technical institutions,
- a dedicated scientific society, and

Friedrich Bendemann (1874–1923).

Friedrich Bendemann, born in Berlin on January 19, 1874

1901 after finishing his studies at the Institute of Technology (TH) Berlin, a three-months stay in the USA followed; subsequently he became assistant at the TH Berlin.

1905 design engineer at the Machine Laboratory of the TH and PhD in the (special) field of steam flows ("Bendemann's Formula").

1908 the "Jubilee Foundation of German Industry" ("Jubiläumsstiftung der deutschen Industrie) engages him as head of the "Test Establishment for Propellers" ("Versuchsanstalt für Luftschrauben"), erected on the grounds of the "Aeronautical Observatory" in Lindenberg (County of Beeskow).

1911 on November 4, he presents his proposals "On the Benefits of Scientific Work for Aeronautics and Flight Technology" in Göttingen. Shortly thereafter, he started his preparations for the creation of the "German Test Establishment of Aeronautics" ("Deutsche Versuchsanstalt für Luftfahrt, DVL").

1912 he becomes the first head of DVL. At the same time, he is charged with the organization of the first "Emperor's-Prize" competition for the best German aero-engine.

1914 he was during the early war at the very front line. He was highly decorated as, for instance, with the Iron Cross, I. Class.

1915 he returns to the "Technical Service of the Flying Corps" ("Technischer Dienst der Fliegertruppe") and the management of DVL; he learns to fly. After the end of war, it is his responsibility to adjust DVL to peacetime conditions and to secure its future.

1919 due to his untiring work for more than ten years, with only a few days off, he is forced to take a longer vacation to recover. His resolve to abandon technical work matures. He changes to the "State Ministry of Transportation" ("Reichs-Verkehrsministerium").

1923 he dies in Berlin on December 14.

The first wind tunnel of DVL in Adlershof went into operation in 1913. With a drive power of 25.7 kW (35 hp), the open-jet (test section) with a diameter of 0.76 m (2.5 ft) reached a maximum flow velocity of 50 m/s (164 ft/s).

- institutes in close cooperation with industry and with development-oriented company departments.

This structure principally still exists today, despite of all the changes that occurred in the course of time. Although universities and research institutions were supposed to predominantly address fundamental phenomena, they were, from the beginning, involved in the solution of technical tasks of the arising industry. On the other hand, basic investigations were also carried out by industry.

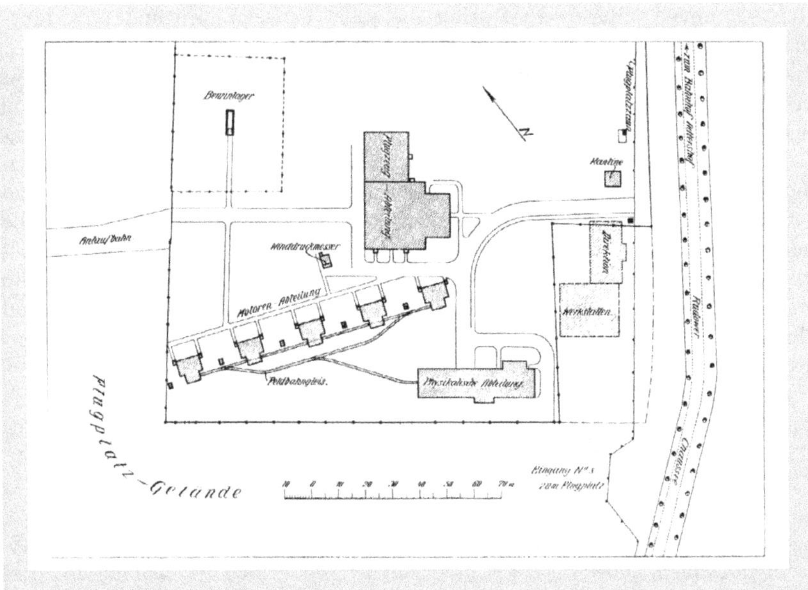

Plan of the first construction stage of DVL (1912) in Adlershof with the buildings for the aircraft department, the engine test rigs, the department of physics, and for management, workshops and the cafeteria.

The first aero-engine test rigs in Adlershof were erected in 1912 on the occasion of the "Emperor's Prize" ("Kaiserpreis") competition; the facilities have been repeatedly extended.

Aeronautical Research During the Time Period 1912 to 1914

In 1909, the worldwide first chair for Aeronautics was set up in Göttingen. Appointed to the chair was the already very renowned Professor *Prandtl*. This then was the beginning of the universities seriously concerning themselves, in research and teaching, with the problems of aeronautics, which also strengthened the reputation of the new technology, which was still in its infancy.

This first chair at a university had, of course, a "knock-on" effect for other German universities. In 1910, *Robert Bosch* had donated over one million Marks to the Institute of Technology (TH) Stuttgart. This enabled the TH Stuttgart to establish, among other things, a chair for "Flight Technology" ("Flugtechnik"), to which *Alexander Baumann* (later also designer of giant aircraft) was appointed in 1911.

The "Institute of Aerodynamics" ("Institut für Aerodynamik") of the Institute of Technology (TH) Aachen, founded in 1910/11, was, for a short period without a director, when *Hans Reissner* moved to Berlin in 1912. In 1913, *Theodore von Kármán* became the successor of *Hans Reissner*. At that time, the subject of "Mechanics" was firmly extended to include aerodynamics and the Institute of Aerodynamics became a permanent installation.

The Institute of Technology (TH) Darmstadt installed in 1913 a chair for aeronautics, which was held by *Karl Eberhard*, a former senior engineer at the Aircraft-Station in Doeberitz. The appointment of *Reissner* in 1912 to the chair for mechanics at the Institute of Technology (TH) Charlottenburg did not yet lead directly to an independent professorship in flight sciences, although special courses, later in cooperation with the DVL, were offered. Also at the other Institutes of Technology (THs), mostly specialized lectureships arose in conjunction with lectures in physics, lectures related to internal combustion-engines, meteorology, et cetera.

These developments were, as in Göttingen, above all promoted by regional aeronautical associations, which accordingly exerted their influence on the university administrators. There also existed a lively exchange of ideas between the university chairs and aircraft companies and their designers since it was, after all, in the interest of both sides that theory was followed by corresponding practical trials, respectively that envisaged designs were, as far as possible, supported or reinforced by theory.

The "Model Test Establishment Göttingen" ("Modell-Versuchsanstalt Göttingen") became operational in 1910 and started out with research tasks formulated by *Prandtl*, viz., the investigation of drag and pressure distributions on bodies of balloons (performed by *Georg Fuhrmann*), and the measurement of the forces acting on flat and cambered plates (performed by *Otto Föppl*). In addition, praxis-related tests, commissioned by external customers, were carried out, such as measurements on an airplane model provided by *August von Parseval*, on airship models of the "Airship-Engineering Company Zeppelin, Ltd." ("Luftschiffbau-Zeppelin GmbH"), calibrations of Pitot probes of the Siemens & Halske company, and, already, wind tunnel measurements of drag on bodies arranged in series to clarify questions related to railroad trains. The results were published in the Journal ZFM ("Journal of Flight-technology and (Powered-airship) Aviation") as "Reports of the Model Test Establishment Göttingen" ("Mitteilungen aus der Göttinger Modellversuchsanstalt").

During the years between 1910 and 1914, nineteen such reports were published. *Otto Foeppl* left the organization in 1911. Albert Betz, who, up to his retirement, stayed attached to the institution and its successor organizations, replaced him, beginning in September; he had a strong formative influence on their development. *Carl Wieselsberger* joined the organization in 1912. After the "Society for the Study of Motor-Airships" had served its purpose, it was in October 1913 transferred into the ownership of the University of Göttingen and officially renamed "Model Test Installation of the University of Göttingen" ("Modellversuchsanstalt der Universität Göttingen, MVG").

At the same time another important decision concerning Göttingen was made. On January 11, 1911 about 200 business women and men gathered to found a society, under the protectorate of the German Emperor and King of Prussia, which was to promote the sciences, especially by establishing and maintaining corresponding institutes. The "Emperor-Wilhelm-Society for the Promotion of the Sciences" ("Kaiser-Wilhelm-Gesellschaft zur Förderung der Wissenschaften, KWG") was thus established. The KWG, since 1948 known as "Max-Planck-Society for the Promotion of Science, MPG", had the well-known theologian and scientist *Adolf Harnack* as its first president.

Felix Klein had already during the founding phase of this society considered the installation of an "Emperor-Wilhelm-Institute for Aerodynamics" ("Kaiser-Wilhelm-Institut für Aerodynamik") and had, accordingly, approached *von Böttinger*. *Klein* asked *Prandtl* to prepare a cost estimate for this endeavor. These considerations were initially on a collision course with the foundation of DVL already discussed. However, after lengthy negotiations, in whose course especially *von Böttinger* supported the interests of Göttingen, it was finally, on June 18, 1913, unanimously decided by the KWG to establish an "Emperor-Wilhelm-Institute for Aerodynamics and Hydrodynamics" ("Kaiser-Wilhelm-Institut für Aerodynamik und Hydrodynamik") in Göttingen. A delay in the financing, caused by the Ministry of Finance, had the consequence that, due to the outbreak of World War One, a temporary interruption in these activities occurred.

The structure of DVL, whose build up had in the meantime also been initiated, envisaged, corresponding to the recommendations of the experts, three scientific departments:

Albert Betz (1885–1968).

Albert Betz, born in Schweinfurt on December 25, 1885.

1905 he studies mechanical engineering at the Institute of Technology (TH) Munich and has, thereafter, his practical training at the "Germanian Dockyard" ("Germania-Werft") in Kiel.

1906 he studies nautical engineering at the Institute of Technology (TH) Berlin and, after some time as an assistant, he moves in

1911 to the "Model Test Establishment of the Powered-airship Study Association" ("Modellversuchsanstalt, MVA der Motorluftschiff-Studiengesellschaft, MStG") to join *Prandtl*; the organization was affiliated with the Institute of Applied Mechanics of the University of Göttingen.

1912 heads the propeller test facility of the University of Göttingen.

1914 he is drafted into military service, but then, at the instigation of *Prandtl*, is discharged to continue his tests and the set up of the MVA.

1915 performance of propeller tests for the military.

1916 design and construction of the new wind tunnel with a 2.25 meter (7.4 ft) diameter test section.

1919 the MVA (see 1911) is incorporated as "Aerodynamic Test Establishment" ("Aerodynamische Versuchsanstalt, AVA") into the "Emperor-Wilhelm-Society" ("Kaiser-Wilhelm-Gesellschaft, KWG").

1924 assistant director of the AVA; performs fundamental research related to axial compressors.

1930 lifetime employment as "Scientific member and assistant director of the Emperor-Wilhelm Institute of Fluid Research associated with the AVA" ("Wissenschaftliches Mitglied und stellvertretender Direktor am Kaiser-Wilhelm-Institut für Strömungsforschung verbunden mit der AVA").

1937 he becomes director and head of the AVA (now a registered association) within the KWG (see 1919).

1945 publisher and author of the "Göttingen monographs" "Progress in German Aeronautical Research since 1939" ("Fortschritte der deutschen Luftfahrtforschung seit 1939").

1947 successor of *Prandtl* as director of the "Max-Planck-Institute of Fluid Research" ("Max-Planck-Institut, MPI, für Strömungsforschung") in Göttingen.

1953 *Betz* is elected director of the AVA at the first general meeting after World War Two on May 28, 1953. He held this position until his death.

1954 honorary member of the "Scientific Aeronautical Society" ("Wissenschaftliche Gesellschaft für Luftfahrt, WGL").

1957 he retires.

1958 recipient of the Ludwig-Prandtl-Ring.

1968 he dies in Göttingen on April 16.

- **Engine department**
  Activities started with the work for the "Emperor's Prize" competition. Management was assigned to *E. Seppeler*; however, *Bendemann* continued to be personally responsible for the competition.
- **Aircraft department**
  Management was, on January 1, 1913, assigned to *Wilhelm Hoff*, who became later the head of DVL (1920–1936). He devoted himself mainly to constructional and strength problems, developed viable test methods to prove the strength of airframes and provided the basics for calculating wing-spars and plywood-fuselage constructions.
- **Physics department**
  Georg Fuhrmann, who had obtained his PhD from *Prandtl* in March 1912, and who had become the first co-worker of *Prandtl* at the "Model Test Establishment" ("Modell-Versuchsanstalt"), joined DVL on October 7 and became on February 1, 1913 head of this department. He built the wind tunnel of DVL mentioned above, which was mainly utilized for the development of instrumentation, partic-

Wilhelm Hoff (1893–1945).

Wilhelm Hoff, born in Strasbourg in 1893.

1910 he acquires a pilot's license for free-balloons and in

1912 the pilot's license for airplanes.

1913 since January 1 manager of the newly established aircraft department of the "German Test Establishment of Aeronautics" ("Deutsche Versuchsanstalt für Luftfahrt, DVL") in Berlin-Adlershof. He receives his PhD from the Institute of Technology (TH) Aachen with a thesis related to aircraft engineering.

1920 since January 1 until 1936 head of DVL. He promotes the glider-flight movement in Germany as head of the technical commission for the annual glider competitions at the Wasserkuppe (Rhön-mountains).

1923 he is appointed to the chair of Aeronautics at the Institute of Technology (TH) Berlin. During this time, he is publisher of the "Ring-books of Aeronautical Technology" ("Ringbücher der Luftfahrttechnik"). Later he becomes a lifetime member of the "German Academy of Aeronautical Research" ("Deutsche Akademie der Luftfahrtforschung") and a member of the Strasbourg Scientific Society at the University of Frankfurt, a member of the "Prussian Academy of Civil Engineering" ("Preußische Akademie des Bauwesens") in Berlin, and a member of the Senate of the "Lilienthal-society of Aeronautical Research" ("Lilienthalgesellschaft für Luftfahrtforschung") until it is dissolved in 1945.

1924 *Hoff* is successful in winning the first long-term contracts from the government for DVL.

1927 he is successful in introducing into the Reich's budget, for the first time ever, the titles "Permanent Research Funds" ("laufende Forschungsmittel") and "One-time Research Funds" ("einmalige Forschungsmittel") under special consideration of DVL. The DVL was thus officially accepted as an institution of independent research.

1945 he commits, together with his wife, suicide when the Russians march into Berlin.

ularly for viable airspeed indicators. In addition, tests were carried out on a wind-pressure gauge, Type Giessen, which DVL had found on its premises and which was subsequently given to DVL.

The "Scientific Head Office" ("Wissenschaftliche Hauptstelle"), originally not planned, was established by *Bendemann* and taken over by *Everling* on February 1, 1914. It was concerned with the set up of a central library, but also with basic research in the field of flight mechanics. The small test installation of the "Jubilee Foundation of German Industry" ("Jubiläumsstiftung der deutschen Industrie") at Lindenberg, previously managed by *Bendemann*, was incorporated into DVL in May 1913 and its equipment, especially the propeller test rig with a drive power of 162 kW (220 hp), moved to Adlershof.

At the same time, an "Air-radio Research Laboratory" ("Flugfunk-Forschungseinrichtung"), which was initially not directly linked with aviation, was founded; it became, however, an important asset and is today part of the DLR, the successor organization of DVL (see also Part 1). The laboratory is closely linked to the name *Max Friedrich Wilhelm Dieckmann* (1882–1960).

Dieckmann received in 1907 his PhD in physics with Professor *Braun* in Strasbourg and went in the autumn of the same year as assistant to the Institute of Technology (TH) Munich. In 1906 he leased in Graefelfing near Munich a meadow, where he started, in a small wooden cabin, his "air electricity" measurements.

This led to the privately owned "Radiotelegraphic and Air-electricity Test-station Gräfelfing" ("Drahtlostelegraphische und Luftelektrische Versuchsstation Graefelfing, DVG"). The years until the outbreak of World War One were devoted, in addition to the set-up of practical-training courses, to investigations for the Zeppelin airships. In the course of the latter,

Max Dieckmann (1882–1960).

Max Dieckmann, born in Hermannsacker near Stolberg in the Harz-mountains on July 5, 1882.

He studies in Göttingen, Munich, and Strasbourg with main emphasis on the exploration of phenomena related to radio communication

1907 receives his PhD with Professor *Braun* in Strasbourg with a thesis related to high-frequency physics. In October, he accepts a position as research assistant at the Institute of Technology (TH) Munich.

1908 he leases a meadow in Graefelfing (near Munich) and sets up his "Radiotelegraphic and Air-electricity Test-station Gräfelfing" ("Drahtlostelegraphische und Luftelektrische Versuchsstation Graefelfing, DVG") in a small wooden cabin.

1910 he becomes "First Assistant" ("Assistent I. Ordnung").

1911 Habilitation (qualification as university lecturer) with work related to a subject on the borderline between radiotelegraphy and air electricity. Until the outbreak of World War One, effort is dedicated to the development of practical-training possibilities in Graefelfing and research for the Zeppelin aviation.

1914 he enters as a (war) volunteer the "Reserve-unit for Communication Munich" ("Nachrichtenersatzabteilung München").

1916 director of the scientific department of the "Prussian Inspection of the Airship Corps" ("Preussische Inspektion der Luftschiffertruppen").

1917 head of the tube-laboratory of the "Aviation-communications-test-department" ("Flieger-Nachrichten-Versuchsabteilung") in Doeberitz.

1920 he once more receives a lectureship in Munich.

1925 he develops, together with his students, a television set utilizing Braun's tubes, which is shown in operation at the "Traffic Exhibition" in Munich.

1927 he is appointed by the "Ministry of Transportation" ("Reichsverkehrsministerium") to the "Technical Advisory Board" of DVL.

193 his lectures now include the following subjects: radiotelegraphic and telephonic communication, praxis-oriented training in high-frequency physics, aerial navigation, and air-electricity.

1936 full professor for "Aeronautical Radio Communication" at the Institute of Technology (TH) Munich; establishment of the "Research Institute of Aeronautical Radio Communication" in Oberpfaffenhofen, where he becomes a director.

1939 full member of the "German Academy of Aeronautical Research" ("Deutsche Akademie der Luftfahrtforschung"); during this time basic studies related to microwave techniques are being conducted.

1943 he receives the "German Flight-navigation Prize" ("Deutscher Flugnavigationspreis").

1947 under a private contract, he becomes aconsultant to the Wright-Patterson Air Force Base, but after one year he returns back to Germany due to illness.

1958 retires in Munich.

1960 he dies in Munich on July 28.

he sent in May 1912, for the first time in aviation history, telegrams from the airship "Victoria Luise" to private recipients, and, nearly at the same time, conducted first radio communications over a distance of 200 km (124 mi) between the airships "Victoria Luise" and "Baden-Baden".

The Turning Point: World War One 1914 to 1918

The outbreak of World War One on August 1, 1914 strongly changed the structure and organization of research. The military requirements received total priority [30]. Sportsmanship and scientific interest turned into a deadly serious matter. Military aviation was in Germany hardly two years old, the airplanes were initially unarmed, and their number much less than that of the enemies. The aeronautical industry had to be built up to increase its production capacity. New, improved, and mission-adapted types of aircraft had to be designed, tested and manufactured. Research was aligned to these requirements.

The surge of national enthusiasm at the beginning of the war led, among other things, to the situation that, with one exception, all staff of the "Model Test Establishment" ("Modellversuchsanstalt, MVG") and of the Institute of Applied Mechanics, including their PhD-candidates, had left for war so that the activities of the MVG had to be terminated. During late summer of 1914, the department "Do" of the "Airship-Construction Company Zeppelin, Ltd." ("Luftschiffbau-Zeppelin GmbH") in Friedrichshafen commenced preparations for the design and construction of the giant seaplane Rs I, and *Claude Dornier* asked the MVG, weather it wanted to perform the model tests. *Prandtl* replied that all his mechanics were still present, but that the assistants had left for war. If, however, the task would not be too voluminous, he would perform it himself. The results of the measurements within this contract were presented on December 20, 1914 in form of a so-called "Resultant Diagram" ("Resultantenzeichnung"). In this type of diagram, the resultant force acting on the model is plotted according to magnitude, direction and location for various angles of attack. Further measurements on modified models were commissioned so that the investigation lasted until spring of 1915.

Further contracts of increasing volume and from other customers, such as DVL and the "Ministry of the Navy" ("Marineministerium") followed. This enabled *Prandtl* during the first half of 1915 to have his research assistants *Albert Betz* and *Carl Wieselsberger* exempted from military service, and, in addition, to win over the newly graduated "Master of Engineering" ("Diplom-Ingenieur") of the Institute of Technology (TH) Hanover, *Max M. Munk*, as "military-exempt" co-worker.

The plans for the built up of an "Emperor-Wilhelm Institute of Aerodynamics and Hydrodynamics" ("Kaiser-Wilhelm-Institut für Aero- und Hydrodynamik") in Göttingen, due to the outbreak of the war initially postponed, were taken up again on April 26, 1915. They were in revised form presented as memorandum "Build up of the Model-test Establishment Göttingen to an adequate aerodynamic research institution for army and navy". The suggestion was taken up by the "Ministry of War" ("Kriegsministerium"), and, at a meeting on Mai 11 and 12, 1915, it was decided to immediately commence the built up of the "Model Test Establishment Göttingen". Envisaged was, of course, to only establish an aerodynamic institute with a large wind tunnel.

Concerning the international environment, it is important to mention that on March 3, 1915 the Congress of the still neutral United States decided to establish the "National Advisory Committee for Aeronautics, NACA". This great aeronautical research establishment was in 1958 converted in today's "National Aeronautics and Space Administration, NASA".

In June 1915 it came to an agreement concerning the creation of an aerodynamic test establishment and its incorporation into the, still to be established, "Emperor-Wilhelm-Institute of Aerodynamics", as it reads in the document, which can be considered as the founding charter. For the institution, the designation "Model Test Establishment for Aerodynamics" ("Modellversuchsanstalt für Aerodynamik, MVA") was selected. After initial difficulties during construction, partly caused by war-related material shortages, started the trial runs for the new wind tunnel. On March 19, 1917, the first practical test, viz., the measurement of drag on a fuselage-model, was conducted. The instrumentation of the wind tunnel consisted of a simple balance for drag measurements and a provisional three-component balance, partly in-house made, which was later extended to a six-components balance. Until September 1918, both facilities, i.e., the already existing and the new wind tunnel, were in operation; only in late September of 1918 was the old wind tunnel closed and transferred to the new organization. Due to the difficulties existing at that time, the wind tunnel did not become operational again before the end of war.

Starting in 1915, the MVA was increasingly in demand by the aircraft industry. The large wind tunnel became only fully operational by the end of 1917. One of the initial difficulties resulted from the fact that even results of the investigations which were of general scientific interest could no longer be published in technical journals, because they were supposed to serve predominantly military purposes. In order to remedy the insufficient exchange of experience between companies, a "Scientific Consulting Agency for Aviation"

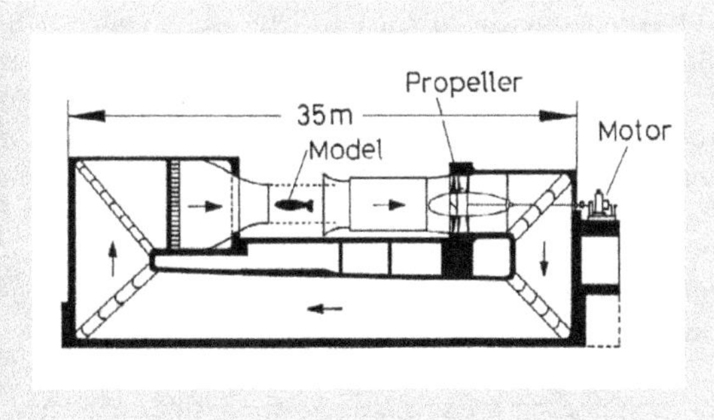

The first wind tunnel, Type "Göttingen", of the "Model Test Establishment" ("Modell-Versuchsanstalt, MVA") in Göttingen with a test section diameter of 4.6 m (15 ft) and a flow velocity of 58 m/s (190 ft/s) was built in 1917 on property next to the Böttingerstrasse. The drive-power was 295 kW (400 hp). The wind tunnel, designed by *Ludwig Prandtl*, had an open test section and a closed-loop return. (Propeller = Fan)

("Wissenschaftliche Auskunftei für Flugwesen, WAF") was installed in December, whose manager became *Wilhelm Hoff*, and whose membership consisted of almost all companies responsible for military aviation. The WAF published under the title "Technical Reports" rather irregularly the research and test results classified as "Secret" with the circulation being 1,000 copies. Volume I, which appeared in October 1917, contained the reports since March 1917. A total of three volumes with a total of about 1,880 pages were published; the results were presented, praxis-related, with many diagrams, nomograms (means of solving calculations by drawing), and simple formulas.

An already existing "Committee for Units and Denotations" ("Ausschuß für Einheiten und Formelgrößen") defined the units and symbols, which had to be used in the technical reports. Many of the symbols defined at that time were for a long time used in the German specialized literature on aerodynamics and flight mechanics. At the MVA, the previously used terms were replaced starting in 1917. Also new terms were coined, such as, e.g., "Leitwerk" ("empennage") for "Seitensteuerung" ("rudder control") *and* "Höhensteuerung" ("elevator control"), in order to avoid a mix-up with the term "Control" ("Steuer") relating to the control elements located inside the cockpit.

Let us now turn to the destiny of DVL, the second large research establishment in Germany. It was cought-up by the war during the first phase of its build up. Still in April 1914, DVL had signed a contract with the newly established "Transport-technologies Board of Examiners" ("Verkehrstechnische Prüfungskommission") of the army administration, which provided for DVL to take over scientific tests and preparatory work for the flight corps. For that purpose, it made the unused part of its grounds available. Still during the same summer, huts were erected and occupied by a military "Test Company" ("Versuchskompanie"). A large hangar, adjacent to the DVL hangar for aircraft tests, was completed just at the outbreak of the war.

The leading employees *Bendemann, Everling, Fuhrmann, Hoff, Madelung, Seppeler*, and others, mostly officers in the reserve, immediately entered, at the outbreak of the war, military service. Consequently, scientific work was terminated here too. The military took over the buildings and facilities of DVL and used them for non-research purposes, for example, as a flight school, repair yard, a depot for spare parts, and as a scrap yard. Only in 1915 was the military department "Test Station and (maintenance) Yard of the Flight Corps" ("Prüfanstalt und Werft der Fliegertruppe") under the command of *Felix Wagenführ* established; it was later renamed into "Aircraft Establishment of the Army» ("Flugzeugmeisterei des Heeres, FLZ"). *Wagenführ* took care that the leading personnel of DVL, and other qualified engineers and scientists, were detailed to his organization. *Bendemann, Everling, Hoff* and *Madelung* thus returned; *Fuhrmann* had already been killed on September 4, 1914, and *Seppeler* stayed with his field-radio detachment.

After his return, *Bendemann* dedicated himself to the project of a towing-track test installation which was, however, already in 1916 terminated as military-irrelevant. In the rooms set up for this project, engine test rigs were developed and built together with the appropriate instrumentation. On October 1, 1916, the DVL was also instated as a test station for the propellers of the army aircraft. The existing 162 kW (220 hp) propeller test rig was replaced in 1917 by one with a power of 294 kW (400 hp) in order to meet the requirements of increasing engine performances.

The Eiffel-type wind tunnel, closed-down when *Fuhrmann* left DVL, was put back into operation in 1917 and utilized for the evaluation and calibration of velocity probes and instrumentation. With an

A towing-track with a tower-cart was installed at DVL in 1915 for the measurement of forces on airplanes. Since the needed reinforcement of the tracks of the military railway Zossen-Jüterbog was, due to low military priority, not authorized, this project had to be abandoned in 1916.

attainable velocity of 50 m/s, i.e., 180 km/h, (164 ft/s or 112 mi/h), the tunnel satisfied the requirements of the time.

The war also brought large changes to industry. Previously financially weak companies, which could not afford to do any research, now received larger funds from the military. These could also be used, although dedicated to meet the military requirement to improve aircraft, to finance fundamental investigations generally needed in aeronautics. Urgently needed was new knowledge concerning the reduction of drag and the increase of lift of bodies exposed to the flow, about higher structural integrity, engine performance, et cetera.

Since the existing research institutions could by far not accommodate all the required work, research tasks and tests were also performed by industry. *Dornier* began in December of 1914 his work for the giant airplane FS I, later Rs I, which was designed as seaplane with a wingspan of 43.5 m (143 ft). Experience gained from the light-metal constructions at the Zeppelin Company was to be transferred. Aerodynamics were treated in cooperation with *Prandtl*, and towing tests with the boat-shaped fuselage were carried out at the "Royal Test Establishment for Hydraulic and Naval Engineering" ("Königliche Versuchsanstalt für Wasserbau und Schiffbau") in Berlin, especially to correctly place a spoiler (a device to induce separation), which had, in 1912, been introduced by the American *Glenn Curtiss*. Questions concerning the wing structure, material selections, shapes for increased strength, etc., were accompanied by tests in in-house test facilities. In the design, large amounts of light metal were used, above all for the frame, parts of the boat planking and the fairings. Although the Rs I did not succeed to break water during the first take-off, it was the predecessor of later very successful seaplanes.

Claude Dornier also considered small "landplanes" (as counterpart of seaplanes). His designs are characterized by monocoque fuselages with smooth, load carrying planking. In 1918, the first completely cantilever biplane was created with the wings also equipped with a smooth, load carrying sheet-metal skin. This single-seat fighter plane was no more brought into action.

Hugo Junkers extended the tasks of his private test establishment to aerodynamic investigations initiated by his occupation with the concept of the "thick wing". He had noticed that the MVA too was only concerned with thin airfoils, although the measurements of drag on bodies, like spheres, cylinders, disks and airships, had shown that, for the same cross-sectional area, drag was very different, i.e., shape must be of decisive influence. In order to investigate this problem, an Eiffel-type wind tunnel was provided to the "Test Organization Prof. Junkers" and installed at the "Frankenburg" near Aachen. The tunnel went into operation on May 1, 1914. In addition to aerodynamical research, *Junkers* also charged his collaborators with tasks related to the development and construction of a thick metallic wing, initially considering steel sheet metal as material.

The objective of this development, viz., the realization of a cantilever monoplane, was reached in November 1915 with the completion of the J 1 aircraft. The flight tests commenced in January 1916 in Doeberitz, where the aircraft, on the first attempt, reached a velocity of 170 km/h (106 mi/h). But its climb rate was too low to be able to compete with other fighters. In 1915, the "Research Organization Prof. Junkers" was established within the company, now residing in Dessau, taking over the main part of the work previously performed at the test organization in Aachen. However, the wind tunnel in Aachen remained active, and a second tunnel went into operation in February 1916 in Dessau. The management of the research institution in Dessau was assigned to *Otto Mader*.

Hugo Junkers (1859–1935) as engineer, university professor and industrialist – largely based on his own research – has decisively affected modern aircraft design, especially with regard to all-metal constructions.

Influenced by the experiences made with the airplane of *Reissner* and the knowledge gained in conjunction with the Dornier Seaplane Rs II, *Junkers* used for the construction of the J 3, light-metal, however, not in a flat-sheet design. For stability reasons, the load-bearing outer skin should consist of corrugated sheet-metal. In 1914, *Junker's* research organization also embarked on the development of high-power aircraft engines to be used to propel "giant airplanes".

In addition to the principal innovations, which *Junkers* and *Dornier* introduced with their developments into the research domain, there were many small steps performed within the respective companies, which led to a rapid increase of knowledge in all fields associated with aviation. At the outbreak of World War One, there existed 51 companies offering aircraft; 20 of these companies were allowed to train pilots and ten of them built aircraft engines. During the war, the number of aircraft companies was reduced to 36 due to the demand for mass-production; of the 36, many were pure manufacturing plants and component suppliers for other companies. The number of companies involved in the construction of engines increased to 21.

Hugo Junkers had in 1913 an Eiffel-type wind tunnel built next to the Frankenburg (near Aachen), where he lived. With a 74 kW (100 hp) electric motor, flow velocities of up to 50 m/s (164 ft) were generated.

The Difficult Situation of Aeronautical Research and the Aeronautical Industry During the Weimar Republic, 1919–1932

Hans Fabian

The, for Germany, bitter ending of the war was to be anticipated since the summer of 1918 after the successful offensives of the Allies. On September 29, 1918 the supreme command felt compelled to request an immediate cease-fire. Thus, the war was lost. On November 9, the revolution occurred, the Emperor resigned, and a new government, under *Friedrich Ebert*, was formed. On November 11, the cease-fire agreement was signed. On January 19, 1919, a National Assembly was elected, which convened on February 6,

The giant aircraft Staaken R VI had four 190 kW (260 hp) Daimler engines in a tandem arrangement with push-pull propellers. These R-aircraft were, towards the end of World War One, deployed as long-range bombers on the western front. With a wingspan of over 42 m (138 ft), they were the largest airplanes of their time.

The two-seated all-metal reconnaissance airplane Junkers J 10, a further developed J 3, which in May 1918 took-off for its first flight, was in many respects the predecessor of the Junkers civil aircraft F 13.

1919 in Weimar and accepted the "Weimar Constitution for the Reich" ("Weimarer Reichsverfassung") on July 31, 1919; it became effective on August 14. On June 28, 1919, the "Versailles Treaty" was signed; it was ratified on January 10, 1920. This peace treaty, dictated by the Allies, had severe consequences for the aeronautical industry, hence also strongly affected aeronautical research [2].

The Transition to Peace

For the management of the two research establishments, MVA and DVL, the foremost task was to secure the existence of their institutions and to compensate the funding up to now received from the military. *Ludwig Prandtl* of MVA consulted *Henry Theodor von Böttinger* already in late October and considered the foundation of a registered organization analogous to the legal status of DVL. It was also considered to stronger involve industry. A corresponding request of *von Böttinger* was answered by the "General War Department of the Ministry of War" ("Allgemeines Kriegs-Department des Kriegsministeriums") on December 14, 1918: *"Before the size of the future peace-time air force has not been decided upon, it cannot be judged to what extent the Model Test Establishment MVA will be needed"*.

This forced immediately restrictions in research since, in the meantime, government contracts as well as orders from the industry had strongly declined. The number of employees was, at the end of November, still 42; however, this number had decreased to 30 by the end of 1918. The head of the workshop at the time, *H.Naumann*, tried to obtain contracts from the industry in order to keep a skeleton-crew of skilled workers.

The head of DVL, *Friedrich Bendemann*, judged the situation less pessimistically since he assumed that his well-equipped research establishment was well qualified to also perform tasks other the ones strictly related to aeronautics.

The new government, functioning as the "Council of the People's Delegates" ("Rat der Volksbeauftragten"), quickly recognized the importance of civil aviation. Already on November 26, 1918, it issued the "Decree concerning the preliminary regulation of airship aviation" ("Verordnung betreffend die vorläufige Regelung der Luftschiffahrt"), which was followed, on December 4, 1918, by a decree concerning the foundation of a "State (Reichs) Aviation Office" ("Reichsluftamt, RLA"). Management was assigned to *August Euler*, who, in the process, was appointed undersecretary (Unterstaatssekretär) of the government. The main tasks of this office were stated in the protocol of December 9, 1918 as being: the licensing of aircraft, air-transport enterprises, and pilots and participation in legal matters.

There was, during the war, a rapid development of the airplane from a device of limited use, except for sporting purposes, to a robust military combat instrument with an already remarkable operational range and a relatively high payload capacity in the case of bombers. As a consequence, the possibilities of air traffic were partly overestimated, at least in some circles, and many of the existing aircraft were considered suitable for air transportation. Here, one must add that in Germany, at the time of the cease-fire, somewhat more than 21,000 aircraft existed. *Euler* warned against this interpretation in his memorandum of January 9, 1919, where he stated, among other things: *"During the war, means available to aeronautics multiplied so that we now have a situation of over-development compared to the peacetime utilization possibilities...."* and *"The aircraft companies will again become research and test "workshops", and in the creation of new types of aircraft for air transportation one must see the means for the economic build up of German aeronautics, later to be internationally competitive....".*

Already on December 18, 1918, the, at that time, existing "Air Vehicle Construction Company Schütte-Lanz" ("Luftfahrzeugbau Schütte-Lanz") inaugurated, with some airplanes of the "German Aircraft Company" ("Deutsche Flugzeugwerke, DFW") and some airplanes of "Albatros", an on-demand taxi-flight service, which had, however, to be suspended soon since the airfield located in Zeesen near Berlin was too difficult to reach.

Walther Rathenau, who became later the general manager of the AEG, was one of the first to consider the idea of long-distance flight across the Atlantic. Still during the war, on December 13, 1917, the AEG founded a study company, which received on January 8, 1919 as "German Air-shipping Company, Ltd." ("Deutsche Luftreederei GmbH, DLR") a temporary license to perform flights for publicity purposes during the time period between January 10 and the end of January 1919. On January 28, 1919, the temporary character was lifted, and DLR became the first, officially approved airline in Germany. Its emblem, a stylized crane, is today the emblem of German Airlines "Lufthansa", the successor organization. On February 5, 1919, a regular flight service between Berlin and Weimar was established, which served, first of all, to transport mail and newspapers to the National Assembly, which convened in Weimar. Aircraft employed were converted military vehicles of the type AEG J II, a former double-seated infantry-support plane, and the LVG C V, a double-seated multipurpose plane. Both aircraft were biplanes.

Junkers too established in March of 1919 an air service, viz., between Dessau and Weimar, utilizing a civil version of the J 10, a double-seated low-wing monoplane of all-metal design, transporting, besides passengers, also mail. During the latter operation, a hood covered the second seat.

Junkers suggested to his engineers, immediately after the cease-fire in November 1918, the design and construction of a civil transport aircraft. Originally, a further developed J 10 was envisaged, but it turned out that many new ideas could not be realized that way. Therefore, a new design was initiated in January of 1919. For this aircraft, systematic airfoil data were, for the first time, available. These data had already been obtained a year earlier, on the instigation of *Junkers*, by the engineer *Franz Wilhelm Schmitz* in test series of wings of equal span, equal contour, and equal center-line camber, but with different thicknesses. This allowed the airplane, designated F 13, to be equipped with an aerodynamically highly effective wing. The F 13 made its first flight on July 25, 1919 with *Emil Monz* at the controls. On September 13, the aircraft set a world altitude record reaching an altitude of 6,750 m (22,145 ft) with eight passengers in an 86-minute climb. As engine served the 185 kW (250 hp) BMW-engine IIIa. Although this world record was not recognized, since Germany was not a member of the FAI, i.e., the "Fédération Aeronautique Internationale" ("International Aeronautical Federation"), the aircraft became rapidly world renown. The F 13 was designed as a (the first) pure passenger plane with a capacity of four passengers and two crewmembers. A total of 322 aircraft of this type were built and predominantly sold to foreign countries, where they were in service of up to 20 years.

Of the time period 1919/20, further drafts of *Junkers* related to sea and land planes exist. They envisaged up to 60 passengers and a range of up to 1,900 km (1,181 mi); they were very advanced for their time, and all were based on "thick" airfoils with favorable lift-to-drag ratios. The projects of larger commercial aircraft were based on two assumptions, which, however, turned out to be erroneous: firstly, it was assumed that there would be no restrictions concerning the manufacturing of civil aircraft, and, secondly, one had expected a higher traffic volume than later materialized. In the spring of 1920, *Junkers* projected the four-engine wide-body aircraft JG I with an envisaged payload of 1,200 kg (2,700 lbs), nine passengers, three crewmembers, and a range of 2,500 km (1,554 mi) at an average altitude of 3,000 m (9,840 ft).

Besides many attempts to convert existing giant airplanes into passenger aircraft or to redesign these aircraft, respectively, the "Zeppelin Company Staaken" projected a very advanced transport aircraft for 12–18 passengers. The designer was *Adolf A. Rohrbach*, who not only envisaged a very thorough aerodynamic layout of the four-engine high-wing monoplane, designated E4/20, but also, for the first time, an appreciable increase in wing loading compared to the one common up to now. The aircraft, with a take-off weight of 8,500 kg (19,100 lbs) and an overall engine power of 735 kW (1,000 hp), was designed for a wing loading of 80 kg/m^2 (55 lbs/sq ft). The construction of the aircraft commenced in May 1919 and already in April 1919 first model tests were scheduled in the large MVA wind tunnel. In addition to the usual tests, also measurements with a ground plate were performed in order to determine velocity and glide path during the landing-approach of the aircraft.

These tests, which, with three phases, lasted until September 1919, constituted a rather important source of funding for the MVA. Public funds were extremely reduced. Only the GV and the KWG contributed with their subscription fees to the maintenance of the MVA. The MVA supported itself by accepting contracts for the workshop, which, of course, was rather irrelevant to aerodynamic research. Added were later non-aeronautical but paid-for fluid-mechanical tasks, such as, e.g., measurements of the drag of railway cars, wind-pressure measurements on tall steel constructions and on electrical cables exposed to the wind.

The revenues from these tasks enabled the MVA to also perform scientific research. *Prandtl* reports

The Junkers F 13 was the first all-metal commercial aircraft, which was, immediately after the cease-fire, developed by *Hugo Junkers* in cooperation with *Otto Reuter* and *Franz Wilhelm Schmitz*. The first flight occurred on July 26, 1919. The second F 13, "Anneliese", started operation in 1920 at "Danzig Air Transportation" ("Danziger Luftverkehr").

Adolf Rohrbach designed during the last year of the war, 1918, an all-metal bomber, which served, after the cease-fire, as base for the development of a four-engine passenger airplane with the designation Staaken E4/20. The first flight took place on September 30, 1920. By order of the allied control commission, the airplane had to be scrapped during the same year.

that, in addition to the 60 measurements for external customers, 95 in-house sponsored measurements were conducted. Personnel had reduced to 17 employees by the end of 1919. At DVL, the number of employees rose in 1918 to 117 and fell back to 43 after the end of war. The first post-war assignments consisted in consulting the newly installed "State Aviation Office" ("Reichsluftamt, RLA") with respect to licensing and the examination of aircraft, and the preparation of legal bills relating to civil aviation. DVL was, in addition, charged with the performance of type-tests, which, as a consequence, were already in the spring of 1919 followed by the first regular orders. In addition to individual vehicles, which were the result of remodeling former military aircraft, also the first German civil aircraft, the Junkers F 13, was type-tested here. The DVL also received, in trust, in 1919 the machinery and test facilities, as well as some hangars, of the "Aircraft Establishment, FLZ". The material assets were considered as part-compensation for the occupation of DVL installations during the war.

Friedrich Bendemann moved at the end of 1919 to the "State (Reichs) Office for Matters of Air and Motor Vehicles" ("Reichsamt für Luft-und Kraftfahrzeugwesen"). *Wilhelm Hoff* took over as head of the DVL.

The Versailles Treaty, Restrictions and Inflation, 1920 to 1923

During this time, when, despite of all the difficulties, hopeful new starts were still possible, the peace conditions, fixed in the "Versailles Treaty", were felt as a particularly hard and catastrophic blow. On July 16, 1919, the Reichs government announced the acceptance of the treaty and published its contents. For aviation, the following articles were of importance:

- **Article 198:** The armed forces of Germany are not allowed to have an air force either as part of the army or as part of the navy.

This meant that the research establishments could no longer expect any funding from the military.

The second important statement was:

- **Article 201:** During a six-months period after the present treaty has become effective, the manufacture and import of air vehicles and parts thereof, as well as the manufacture of aircraft engines or parts thereof, is prohibited for the entire German territory.

Not only military aviation was prohibited, but bans and restrictions were also imposed on civil aviation; these were later reinforced to a degree that partly even surpassed the initial peace conditions, all to completely cut-off any kind of activity.

Between March 1919 and May 1920, about 600 civil airplanes received their licenses, which were, however, revoked when the "Versailles Treaty" took effect on January 10, 1920 and corresponding German laws were issued. One had to obtain new licenses, which enabled the "Inter-allied Aviation Control Commission" ("Interallierte Luftfahrt-Überwachungs-Kommission, ILÜK") to register all aircraft. The commission agreed, however, to accept the aircraft Junkers F 13, Fokker F II, a cantilever high-wing monoplane for four passengers, the Sablatnig P I and P III, designed for four and six passengers, respectively, and the sports plane Sablatnig KE 1 as civil aircraft. Later decreed aircraft demolitions, possible due to supplementary manufacturing bans, led, by order of the ILÜK, among other things, also to the destruction of the Rorbach monoplane E4/20 on November 21, 1922.

It is, therefore, not surprising that, anticipating the restrictions, the RLA contemplated the continuance of the aeronautical research institutions. This led to the idea, to combine the different civil and military institutions, scattered State-wide, also for cost reasons, to a "State Test Establishment" ("Reichsversuchsanstalt") [2]. The airfield at Lake-Müritz near Rechlin (state of Mecklenburg), which was owned by the treasury, was selected as location.

When *Ludwig Prandtl* and *Hugo Andres Krüss*, department head at the "Ministry of Science, Art and Education" ("Ministerium für Wissenschaft, Kunst und Volksbildung"), learned about these plans during the annual meeting of the DVL on April 12, 1919, they intervened, and *Euler* abandoned these plans.

The question of how the financing of the MVA could be continued remained open. Neither the "State (Reichs) Ministries" nor the individual states considered themselves in a position to contribute to the financing. The still existing "Emperor-Wilhelm-Society" ("Kaiser-Wilhelm-Gesellschaft") was also not in a position to fund the MVA over and above the obligations assumed earlier. The idea of a registered association, already once before considered by *Prandtl*, was again contemplated. After difficult pre-negotiations, and after the necessary seven members were found, the foundation of a registered association was achieved on December 3, 1919 [8]. *Mardersteig* and *Georg von Tschudi* drew up the statutes and articles in the presence of *Prandtl* and *Friedrich Glum*, an executive member of the KWG. In January 1920, the constituent assembly convened. The organization was given the name "Association for the Promotion of the Aerodynamic Test Establishment Göttingen" ("Verein zur Förderung der aerodynamischen Versuchsanstalt Göttingen"). *Felix Kasinger*, director of the "Federation of German Aviation Industrialists", became its chairman; his deputy was *Georg von Tschudi*.

Following a first advertising campaign on behalf of the association in December of 1919, mainly addressing industry and aviation circles, 14 applications for membership were received. Thus, a contractual agreement between the present association, the "Emperor-Wilhelm-Society" ("Kaiser-Wilhelm-Gesellschaft, KWG") and the "Association for the Promotion of Applied Physics and Mathematics of Göttingen" ("Göttinger Vereinigung zur Förderung der angewandten Physik und Mathematik") could be reached. *Friedrich Glum* drew up statutes for the test establishment, which designated that the organization was to be incorporated into the KWG, however, not as an "Emperor-Wilhelm-Institute, KWI" in the usual sense, but in the sense that the KWG *"supports the organization in accordance with the funds to be provided by the "Association for the Promotion of the Aerodynamic Test Establishment Göttingen""*.

Early 1920, at the age of 72, *von Böttinger* resigned his function as chairman of the administrative committee of the MVA. As his successor, he had won the former "State (Reichs) Minister" ("Staatsminister") at the "Ministry of Education and the Arts", *Friedrich Schmidt-Ott*, who was presently second vice president of the KWG. He took over office in a difficult time since financing had to be secured and adjusted to the ever-increasing prices.

The new statutes were approved by the KWG on May 11, 1920, and the organization attained the name "Aerodynamic Test Establishment in Göttingen" ("Aerodynamische Versuchsanstalt in Göttingen"), abbreviated AVA. The preliminary Board of Trustees of the MVA was replaced by a final one, which consisted of nine members. The Senate of the KWG elected to this board *von Böttinger*, *Schmidt-Ott*, and *Glum*, as well as, from the "State Ministry of Finance" ("Reichsfinanzministerium"), *E. Wagner*, and councilor *Kruess* from the "Prussian Ministry of Science, Art and Education". *Bendemann* was a member as the representative of the "State Office for Matters of Air and Motor Vehicles" ("Reichsamt für Luft- und Kraftfahrzeugwesen"), *Richard Gradenwitz* as industrialist and member of the board of directors of the DVL, and, finally, *von Kehler* and *Listemann* as representatives of the aviation industry. After the sudden death of *von Boettinger* on June 9, 1920, *Max Planck*, as representative of the KWG, was elected to the board in November 1920.

Despite the reduction in personnel to 15 employees, a significant increase in the research activities occurred in 1920. Although in-house research dropped to 42 projects, i.e., less than half of the previous high, externally financed projects rose to about 175, i.e., they tripled. Of those, 91 measurements concerned aircraft investigations, especially for the Zeppelin plant in Staaken, and tests with a model of the airship LZ 120 of the "Airship-Engineering Company Zeppelin" ("Luftschiffbau Zeppelin") in Friedrichshafen.

The currency depreciation, worsening since the end of 1920 and leading to the "great inflation" in 1923, confronted aeronautical research not only with the common financial difficulties, but also with the problem that, although the increasing contracts from the aviation industry assured the continued existence of the organization, at the same time, and for that reason, scientific work was more and more reduced, particularly so when, in 1922, the first relaxation of the allied construction ban was announced.

Prandtl received during the summer of 1920 a call to the chair for mechanics at the Institute of Technology (TH) Munich, previously held by his father-in-law *August Foeppl*. The future of the AVA had thus become questionable, so that, most of all *Felix Klein* and *Schmidt-Ott* tried to keep *Prandtl* in Göttingen. *Prandtl* attained, on November 21, 1921, the full professorship for technical physics and aeronautics at the University of Göttingen, replacing the associate professorship for technical physics he had held up to now. He was, in addition, in October 1922 appointed a member of the KWG for the time he held this new position; he did not have to pay membership fees, a privilege that was, according to the statutes of the KWG, reserved for the directors of the "Emperor-Wilhelm-Institutes". The "tug-of-war" concerning the stay of *Prandtl* was finally decided in December 1923 in favour of Göttingen, after, as a concession to *Prandtl*, permission was granted to establish an institute of hydrodynamics.

The situation at the DVL was similarly difficult. Early in 1920, a change in the management of DVL occurred with the departure of *Bendemann* and the assignment of *Hoff*. In addition, a change of the statutes was decided upon to make the organization more flexible. During the inflation, DVL lost almost all of its assets except for some small material possessions. But it also began to secure its existence and to initiate a gradual increase in personnel by acquiring test contracts.

The Alternative to Powered Fight: the Sailplane

Unpowered flight was, during the war, largely pushed into the background. After the war, the prohibitions of the Versailles Treaty concerned engine-powered aircraft, so it was natural that aviation enthusiasts leaned increasingly towards non-powered flight [31]. Even during the war, there were still some steadfast individuals. Here, especially *Oskar Ursinus* must be named since he created in 1908 with the journal "Aviation Sports" ("Flugsport") the means that would decisively contribute to the later stormy development of glider flight. An important aviation-related group was made up of high-school graduates from Darmstadt, who, under the impression of the ILA, teamed up to form the "Aviation Sports Association Darmstadt" ("Flugsportvereinigung Darmstadt, FSV"). They designed and constructed their own flight vehicles, and they can pride themselves for having "discovered" the "Wasserkuppe", a mountain in the "Rhön-mountains" – still today very famous for gliding – already prior to the war when exploring the region for a suitable area for gliding. Here, *Hans Gutermuth*, a high-school pupil from Darmstadt, covered already on July 22, 1912 with the biplane FSV X a distance of 838 m (2,750 ft) in 112 sec. *Friedrich Harth* and *Willy Messerschmitt* had performed glider flights in the Rhön-mountains since 1914. With the glider S 6, which had direct lift control, they succeeded in 1916 to soar for 3.5 min while attaining 15 m (49 ft) height above the take-off point.

In July of 1920, a first competition was held in the Rhön-mountains, initiated by an announcement of the – at the time – students at the Institute of Technology (TH) Dresden, *Wolfgang Klemperer* and *Erich Meyer*, in the journal "Aviation Sports". *Klemperer*, who, in the meantime, had moved to Aachen to become a research assistant to *von Kármán*, attained with his glider "Black Devil" ("Schwatze Düwel") spectacular results: he covered a distance of 1,830 m (6,004 ft) in 2 min and 22 sec. The sailplane, which was, based on *Junker's* ideas, designed as a monoplane with thick wings, was built by the "Flight-sciences Association Aachen" ("Flugwissenschaftliche Vereinigung Aachen"). For the first time, take-off was accomplished with the aid of a "rubber-band" catapult invented by the same *Klemperer*.

Prandtl had participated as an observer in the first Rhön-mountain competition; with his lectures and publications, he contributed to the acceptance of aviation-related sports as "scientifically serious". As design objectives he had defined: large aspect ratio wings, high camber for higher lift and low drag coefficients, hence reduced sink rates, in order to be able to perform "a non-powered flight without a loss in altitude". These requirements were largely met by the H 1 "Vampire" of the (later) "Academic Flying Group Hanover" ("Akaflieg Hannover"), which appeared towards the end of the second competition in the Rhön-mountains in 1921; Georg Madelung had designed the glider. With this glider, very successful flights were accomplished at once with *Arthur Martens*, for instance, reaching a distance of up to 7.5 km (4.7 mi) and a flight-duration of 15 min 40 sec.

The "Vampire" had been slightly improved in 1922, and *Martens* employed, for the first time, a tactic suggested by *Walter Georgii*, viz., to utilize the upwind near a slope, flying ahead of the slope large "figures of eight". This way he managed, for the first time, on

For the first Rhön-mountain competition in 1920, *Wolfgang Klemperer* at the "Flight-sciences Association Aachen" ("Flugwissenschaftliche Vereinigung Aachen") designed and built the sailplane "Black Devil" ("Schwatze Düwel") with a wingspan of only 9.5 m (31 ft) and a lift-to-drag ratio of 10.

A great success was the in 1922 completed glider "Vampyr", a design of *Georg Madelung* in cooperation with *Walter Blume*, *Fritz Hentzen* and *Arthur Martens*, all member of the "Academic Flying Group Hanover" ("Akademische Fliegergruppe Hannover").

August 18, 1922 a flight of 1 h, 6 min duration while attaining 108 m (354 ft) height above the take-off point. On August 19, *Fritz Hentzen* managed a flight of 2 h, 10 min duration, and on August 24 he even stayed in the air for 3 h, 6 min. Also other pilots, utilizing different sailplanes, stayed long in the air. *Georgii* later called the August 18, 1922 "the real birthday of soaring". The news from the "Wasserkuppe" was considered a world sensation. Gliding was at once taken seriously.

Air Traffic Stimulates Aeronautical Research

In Germany, air traffic commenced already in 1910 with the airship LZ 6, owned by the worldwide first air-shipping company, the "German (Airship) Aviation Corporation" ("Deutsche Luftschiffahrts AG, Delag") founded in 1909. All were tourist voyages, but the demand was, in spite of the high prices, considerable so that four additional airships were ordered bearing the names "Deutschland", "Viktoria Luise", "Hansa" and "Sachsen" ("Saxony"). Until World Ware One, these airships transported 37,250 passengers on 1,600 cruises traveling a total of 150,000 km (93,200 mi). Airplanes sometimes also took passengers along; however, this was mainly out of courtesy and cannot really be considered as passenger transport in any sense.

As already mentioned, the "German Air-shipping Company, Ltd." ("Deutsche Luftreederei GmbH, DLR") was founded still during the war on December 13, 1917, by the AEG, then headed by *Walther Rathenau*, as a study society for air transportation. It was believed that in civil aviation one still had the possibility to operate aircraft. Accordingly, many aviation companies were founded immediately after the war, which mainly utilized former military aircraft after appropriate reconstructions. In addition, the first purely civil aircraft were developed and put into operation; here, the Junkers F 13 played a superior role.

In the beginning, one mainly flew by sight, and only between Mai and October, and with the aid of a compass, which played, however, only a minor role. *Hans Baur*, becoming later a well-known flight captain with Lufthansa, reports, among other things, that the first Junkers F 13 had the compass installed behind the pilot; this, indeed, shows the importance one attached to this instrument at that time.

Night flights commenced in 1924, mainly transporting mail; at the same time, tests started to extend flight operations beyond October into the winter. These events were to be considered by aeronautical researchers as an encouragement to not only address structures and aerodynamics but also flight mechanics and controls.

1924: The New Beginnings of Aeronautical Research

The currency reform and the introduction of the new currency, the "Rentenmark", in November 1923 gave the research institutions again a reliable financial basis, which turned out to have a positive effect on their further development. With the aid of 100,000 Rentenmarks, donated by *Walter Hoehne*, an industrialist involved in railway engineering, and the same amount provided by the government, it was possible to commence construction of the "Institute of Hydrodynamics" in Göttingen early 1924; in the spring of 1925, the institute was ready for occupancy.

Already in his earlier memoranda, *Prandtl* had divided fluid-mechanical research into four "Problem Areas":

1. Fluid flows with low viscosity and constant volume (hydrodynamics).
2. Fluid flows of high compressibility (gas dynamics).
3. Fluid flows close to their evaporation point (flow cavitation).
4. Fluid flows with very high viscosity (creeping flows, e.g., bearing lubrication).

The laws of fluid flow in hydrodynamics are, within certain limits, the same for water and air, i.e., the equipment of the AVA with its wind tunnels was very well suited for research in Area 1 above. It was, therefore, important to also create viable research facilities for the Problem Areas 2–4. Supplementing the existing wind tunnels, two steel vessels with a volume of 10 m^3 (353 ft^3) each were procured. They could be evacuated or pressurized, allowing, by either inflow or outflow into or from the vessels, respectively, to generate very high flow velocities, thus allowing measurements in the high subsonic speed range or at supersonic velocities (gas dynamics). In addition, it was possible to simultaneously fill one tank with water and the second one with compressed air so that the water could be ejected at very high velocities; the installation could thus be used for experiments with instantaneously evaporating flows.

The new "Department of Fluid Research" was headed by *Jakob Ackeret*; theoretical work started with *Walter Tollmien*, who had taken up his research work when still a student with *Prandtl*. In 1925, the department was joined, among others, by *Adolf Busemann*, *Oskar Tietjens* and *Johann Nikuradse*, names, which later gave the institution its world-wide reputation.

The new installations were supposed to exclusively serve research, while the already existing parts of the AVA, whose management was taken over by *Albert Betz* as deputy director, should still be able to perform tests under external contracts. In order to have this distribution of tasks also reflected in the name of the organization, it was given, on the occasion of the inauguration of the new building on July 16, 1925, the name "Emperor-Wilhelm-Institute of Fluid Research associated with the Aerodynamic Test Establishment Göttingen" ("Kaiser-Wilhelm-Institut für Strömungsforschung, verbunden mit der Aerodynamischen Versuchsanstalt in Göttingen"), for short "KWI associated with AVA".

During the years 1924 and 1925, the AVA obtained increasingly more research contracts. The "State Ministry of Transportation" promoted research projects and also partly financed research contracts on behalf of the airframe manufacturers. The latter had, at this time, already proceeded to, more often than not, carefully investigate their designs with the aid of wind-tunnel model tests. Since the six-component balance, constructed during World War One, was no longer fully operational, it was initially tried to get by with supplementary modifications.

After the inflation, also the DVL started again a steady development [7]. A fire during the night between August 5 and August 6, 1925 destroyed part of the engine-department building and the wind tunnel erected by *Fuhrmann* prior to the war; this caused severe disturbances in the progression of work. During 1926, DVL was totally reorganized and after March 31, 1927, the following situation prevailed: the overall management was in the hands of *Wilhelm Hoff*, science and technology was the responsibility of *Georg Madelung*, and administration and finances were supervised by *Ottfried von Dewitz*. Directly attached to the executive board was the administrative secretariat, the scientific secretariat, the sales office, and the office of high-altitude flight (closed in 1928). Added later were the office of standards (1928), the department responsible for new technical constructions (1928), and the office for the investigation of aviation accidents (1932). The DVL enclosed the following departments:

- Administration,
- Operations,
- Statics,
- Engines,
- Physics (closed in 1928),
- Aerial Photography and Navigation,
- Flight,
- Inspection,
- Aerodynamics,
- Materials,
- Radio Communication and Electricity.

At DVL, emphasis was, among others, placed on the following activities:

- **Development of load assumptions, design and inspection regulations for all aviation related equipment, and performance of the inspections**
  The extensive material, taken over from the "Aircraft Establishment of the Army" ("Flugzeugmeisterei") after the war, was evaluated providing a good basis for the above tasks. The inspection and

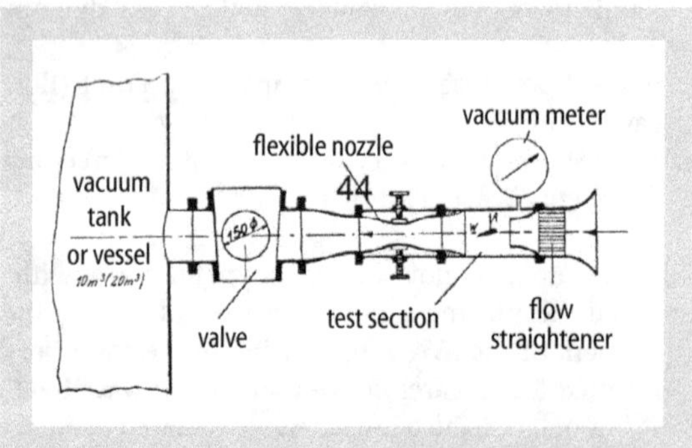

Test set up and layout of the intermittent high-velocity wind tunnel of the AVA in the configuration of 1928. The flow is from right to left, i.e., air flows into the big vacuum tank. The cross-section of the test section is 6 cm × 7 cm (15 in x 18 in).

Georg Madelung (1889–1972).

design regulations were defined in negotiations between DVL and the parties concerned. Initially, each department had to attend to the inspections relevant to its area of responsibility. However, due to the increasing workload, a special inspection department was created during the reorganization in 1926, which later had to install branch offices in many parts of Germany and had to move construction supervision to the manufacturers' plants.

- **Participation in the development of special standards**
  During the years 1929 and 1930, the "German Aircraft Committee" ("Deutsche Luftfahrzeug-Ausschuß, DLA") was installed, which served as a neutral authority for the specification of design and inspection regulations; employees of DVL participated here as members and experts. The offices of the DLA and the "Committee for Special Aviation Standards" ("Fachnormenausschusses für Luftfahrt"), founded in 1925, merged within DVL. This organizational unit soon contacted corresponding international boards of standards leading to close cooperations.
- **Basic research and performance of investigations related to the solution of current problems in aircraft engineering, especially also related to safety**
  Immediately after the inflation, scientific research was resumed and decisively supported by increasing financial contributions by the "State Ministry of Transportation". Already in 1925, research constituted the major part of the activities of DVL. Research was mainly related to fundamental aspects of aircraft and engine design, but also,

Georg Hans Madelung, born in Rostock in 1889.

1908 Second Lieutenant of the reserve; he studied at the Institutes of Technology (TH) Karlsruhe, Hanover, and Berlin.

1910 joins the "Association of German Aviation Engineers" ("Verein Deutscher Flugtechniker").

1912 founding member of the "Scientific Society for Flight Technology" ("Wissenschaftliche Gesellschaft für Flugtechnik"), a predecessor of today's DGLR.

1914 joins DVL.

1915 conducts, together with *Wilhelm Hoff*, type-tests for all airplanes of the time; he is author of the first German design specifications for airplanes.

1919 master's degree at the Institute of Technology (TH) Berlin-Charlottenburg. At Junkers' he is responsible for the strength of the "F 13" multi-celled wings, planked with corrugated sheet-metal.

1921 receives his PhD; he designs the sailplane "Vampyr" with a cantilever single-spar wing.

1922 on the recommendation of *Ludwig Prandtl*, he goes to *Glenn Martin* in Cleveland, Ohio. Metal airframe design is being introduced.

1925 head of the aircraft department of DVL in Berlin-Adlershof.

1926 scientific director and member of the board of directors of DVL; associate professor at the Institute of Technology (TH) Berlin-Charlottenburg.

1929 as full professor of Aviation Technology at the Institute of Technology (TH) Stuttgart, he builds up the "Flight-technical Institute Stuttgart" ("Flugtechnisches Institut Stuttgart, FIST") and in 1940 the "Research Establishment Count Zeppelin" ("Forschungsanstalt Graf Zeppelin, FGZ").

1946 he joins the "Naval Medical Research Institute" in Bethesda, Md., USA, studying the effect of accelerations on the human body.

1954 returns to Stuttgart and establishes the "Research Association Count Zeppelin" ("Arbeits- und Forschungsgemeinschaft Graf Zeppelin"); he advises the government (military) office "Blank" in matters related to the reconstruction of the German Armed Services.

1972 he dies in Uffing at the Staffelsee (upper Bavaria) on August 17.

Walter Georgii (1888–1968).

increasingly, to the areas of equipment and navigation, as reflected in the organizational chart of 1926. However, the inadequate equipment of the DVL with experimental facilities, a result of unresolved property rights in Adlershof and the strong recession at the end of the twenties/beginning of the thirties, led to a negative effect on the development, and only in 1931/32 new constructions commenced.

- **Publications**
  In 1927, the publication series "Aeronautical Research" ("Luftfahrt-Forschung") was, together with the AVA and the Institute of Aerodynamics in Aachen, established by the publishing house "Oldenburg" for the dessimination of scientific reports. The WGL and the Institute of Aerodynamics in Hanover joined this series. Since 1929, DVL again published a "Literature Review", where approximately 200 journals were evaluated; of these were two thirds English and American, some were French and Italian, the rest German.

Soaring, too, became more and more attractive to research. Although it started, and is continuing, as a sports discipline, the various flight vehicles showed that only scientifically gained knowledge was able to contribute decisively to a further development, but that, on the other hand, gliding represented an ideal instrument for the experimental verification of new developments. Of course, interest had already been demonstrated by the attendance of various university professors at the Rhön-mountain competitions, and it is, therefore, not surprising that there was the desire to bring science and technology, on the one hand, and sports, on the other hand, closer together in a more strictly organized way.

On August 31, 1924, the "Rhön-Rossitten Society, RRG", with Rossitten at the shores of the Baltic Sea

---

Walter Georgii, born in Meiningen on August 12, 1888; he studied physics, mathematics and geography at the University of Leipzig

1913 he received his PhD and, after habilitation, was appointed lecturer for flight meteorology at the University of Frankfurt.

1924 he joins the "German Naval Observatory" ("Deutsche Seewarte") in Hamburg.

1926 he becomes full professor and receives the newly installed chair of Flight-meteorology at the Institute of Technology (TH) Darmstadt.

1930 he founds the "International Study Committee for unpowered Flight" ("Internationale Studienkommission für den motorlosen Flug, ISTUS").

1937 foundation of the "Research Establishment for Gliding" ("Forschungsanstalt für Segelflug, DFS"), which originates within his institute in Darmstadt; he initially heads the flight meteorology department and later assumes the overall management of the organization. During the war, the research establishment was moved to Brunswick and Ainring, Bavaria.

1939 head of the German gliding expedition in the Libyan Desert.

1943 executive member of the "Aviation Research Command" ("Forschungsführung der Luftfahrt").

1946 consulting engineer for the "Technical Services of the Air Ministry in Paris" ("Service Technique du Ministère de L 'Air-Paris").

1948 professor of aero-physics and director of the research department of the University Mendoza, Argentine, and foundation of the "International Scientific and Technical Organization of Gliding" ("Organisation Scientifique et Technique Internationale de Vol à Voile, OSTIV") as successor organization of the ISTUS.

1953 the general assembly of the DFS decides to resume activities in Munich.

1955 after returning from Argentine, he sets up the "Institute of Airspace Research" ("Institut für Flugraumforschung") in Munich, which he heads until 1961.

1968 he dies in Munich on July 24.

---

being another center for soaring, was founded at the "Wasserkuppe". Within the society, the two existing important centers for gliding combined forces, in order to jointly advance new developments. *Karl Kotzenberg*, head of an important commercial enterprise, who had already financially supported *Ursinus* during the organization of the 1. Rhön-mountains competition, became the chairman. On April 1, 1925, a research institute was affiliated to the RRG which was initially headed by *Wilhelm Schlink*, at that time professor of technical mechanics in Darmstadt. On October 1, the "Martens Flight School" ("Martens-Fliegerschule") joined the above institute; head of the school became *Fritz Stamer*. A flight-technical department was assigned to *Alexander Lippisch*. Starting in 1926, *Walter Georgii*, professor of meteorology in Darmstadt, became head of the institute, and the following structure of responsibilities evolved:

- Exploration of the atmospheric conditions for soaring *(Georgii)*.
- Design of efficient sailplanes, including the preparation of drawings, to enable aviation groups to construct these sailplanes *(Lippisch)*.
- Systematic glider-flight training and continuous development of the method of self-training *(Stamer)*.

A New Beginning for the Aviation Industry

The total ban of aircraft manufacturing had hit the industry particularly hard, partly leading to the evasion of the ban by shifting production to formerly neutral countries; alternatively, design work, which was not prohibited, was performed on behalf of third parties. These activities not only served to retain a minimum in company assets but also to preserve relevant knowledge. On April 14, 1922, the construction ban was partially lifted; on that occasion, so called "Definitions" ("Begriffsbestimmungen") were issued by the "Ambassador Conference of the Allies", in order to be able to distinguish between civil and the still prohibited military aircraft design. These definitions were as follows:

- An aircraft able to fly without pilot will be considered as a means of war.
- Aircraft carrying weapons or being armored will be regarded as means of war.
- Single-seat aircraft with an engine power of more than 44 kW (60 hp) will be considered as military aircraft.
- The absolute ceiling of German aircraft, utilizing full engine power, may not exceed 4,000 m (13,100 ft); otherwise they will be considered as bombers.
- The speed at an altitude of 2,000 m (6,500 ft) and full throttle will be limited to 170 km/h (106 mi/h).
- The payload, including crew and equipment, may not exceed 600 kg (1,395 lbs).
- The quantity of fuel allowed onboard, is limited according to a formula.
- The engines may not be equipped with superchargers.
- The number of spare parts for engines is limited.

The German (Reichs) government at the time, as a consequence, advised neighboring countries that commercial aircraft of foreign carriers, which did not adhere to these restrictions and landed on German territory, had to be considered a means of war and hence had to be confiscated. In addition, German airspace could not be entered by aircraft, which were, according to the allied definitions, military aircraft. No one had anticipated such a reaction by the Germans; it led to heavy discussions among the Allies and, particularly, with the neutral states.

An interesting commentary concerning the consequences of the restrictions listed above, and the additional demand on German aircraft not to enter the airspace above the occupied Rhineland, was given by the American journalist *F. Britten Austin*, who visited Europe in 1924, being especially interested in air traffic matters. He writes, among other things: "*The consequence of the restrictions imposed on German aircraft was the necessity to built commercial aircraft, which were to be superior to the ones of other nations, with excellent aerodynamic features but restricted engine performance. German aircraft companies established, in addition, branches in all of Europe. Junkers went to Sweden, Rohrbach and Caspar to Denmark, Dornier took up production in Italy and Heinkel in Sweden – just to name a few*".

The allied "Definitions" listed above were replaced in 1925 by the "Paris Agreements", which allowed, among other things, the manufacturing of "aircraft

Claude Dornier (1884–1969) has made, as engineer and later as businessman, with his successful early all-metal seaplanes the first transoceanic air traffic possible.

with the characteristics of modern fighters" for competition and record-setting purposes. This soon led to the invitation to compete in the "German Seaplane Competition 1926" ("Deutscher Seeflugwettbewerb 1926"). Besides technical tests with climbs and speed flights and the determination of fuel consumptions, also a flight over a distance of 4,000 km (2,486 mi) over the North and Baltic Seas was required. *Heinkel* built for this competition the He 5 and won.

A further German design which made history was the "Whale" Seaplane ("Wal" Flugboot) of *Claude Dornier*. The patented basic concepts of the tandem arrangement of the engines and the stub planes as elements of stabilization on the water, already developed during the war, were perfectly implemented on this aircraft. They resulted in an unusually seaworthy aircraft, which on November 6, 1922 executed its first flight. During the following year, on May 21, 1925, two of these aircraft took off in conjunction with the North-Pole-Expedition of *Amundsen*. They did not reach the North Pole, but, nevertheless, a position corresponding to 87°43' north and 10°37' west. One seaplane had to be left behind in the pack-ice; with the second, the return of the expedition succeeded.

The first civil versions of the "Whale" went to South America, where they started to open up this continent for air transportation. In 1926, *Ramon Franco* succeeded with a "Dornier-Whale" to cross the South Atlantic for the first time. All "Whale" seaplanes built up to now were manufactured at the Italian plant Marina di Pisa as a result of the restrictive "Definitions" listed above. Only the "Paris Agreements", initialed on May 5, 1926, freed the German aircraft industry of the restrictions in civil aircraft engineering.

With the new freedom, the time had also come to realize the many existing ideas without the danger of an early termination. How difficult civil aircraft engineering was at that time is shown by the story of the development of Junkers' tri-motor aircraft G 24, the predecessor of the Ju 52 [32]. *Junkers* applied to the so-called "Guarantee Committee" ("Garantiekommission") for a license for the tri-motor aircraft G 23 with Junkers engines L 2 of 143 kW (195 hp) each. On February 8, 1925, *Junkers* was informed that the tri-motor G 23a L aircraft could not be approved with three engines; however, corresponding pre-designs with one and two engines, respectively, were approved. It is not possible to go here into any more detail. However, the records of the time show that a G 23 really never existed and that, from the very beginning, the aircraft in question was the later G 24, which was licensed in Switzerland and had appropriately powerful engines.

It was, however, now possible to actively pursue the long harbored ideas of large aircraft for transatlantic flights and flights with many passengers – promoted by the "State Ministry of Transportation" – without any restrictions. Dornier produced the seaplane Do X with 12 engines in a tandem of always two arrangement. It was designed for about 70 passengers, who should have all the comfort of a luxury liner; but also the utilization by the navy was considered [33]. Its first flight occurred on July 12, 1929. Although the demonstration flights, following the trials, were successful, the economic crisis at the beginning of the thirties did not allow the aircraft to become a commercial success. But the "seaplane" was, ultimately, a misconception; the future belonged to the "landplane".

During the same time period, *Junkers* had developed his "Air Giant", the G 38, as a landplane. The aircraft was based on the original patented thick-wing concept, had observation decks in the wings, and was designed for 34 passengers and a crew of seven. The engines could be reached via a passageway, thus inspected and maintained during flight. Starting in May 1931, the first G 38 participated in the flight service of Lufthansa (German Airlines), a second aircraft was delivered on July 1, 1932, and both aircraft served on long-distance routes within Europe, proving extremely successful. However, the number of passengers was not sufficient to employ more aircraft

The seaplane Dornier "Whale" ("Wal"), a high-wing monoplane with all-metal strut-supported stub planes for stabilization, flew for the first time in November 1922 in Pisa – at a time when the construction ban of the Versailles Treaty was still in effect.

The Junkers G 24, the first three-engine all-metal passenger aircraft, utilized in air transportation since 1925, was the predecessor of the later so successful and renowned Ju 52/3m.

The first flight of the famous all-metal "tri-motor" passenger aircraft Junkers Ju 52/3m was performed in 1931. The second Ju 52/3m was, during the same year, employed by the Lufthansa to test the route between Munich and Rome. The Lufthansa fleet consisted in the years 1938 and 1939 to 85 per cent of this type of aircraft.

of this type; mass tourism had not yet emerged [34]. The successor of the G 24, the Ju 52/3m, built since 1931 and equipped with twelve seats, became, on the other hand, a commercial aircraft operating worldwide.

Of interest in conjunction with aeronautical research is a special development at Junkers. In 1928, DVL charged Junkers with the development of an aircraft for high-altitude research. It was designated

The four-engine all-metal passenger aircraft Junkers G 38, based on early patents of *Hugo Junkers* related to a flying-wing aircraft, was a "giant" aircraft with cabins for 34 passengers. The engines were, together with the cargo, totally integrated into the wings.

Ju 49. The development lasted three years. The Ju 49 was a cantilever, low-wing monoplane with corrugated-sheet metal planking. The pressure-proof, double-skin, temperature-insulated high-altitude chamber for a two-member crew was integrated into the fuselage as an independent unit. The engine was a 588 kW (800 hp) Junkers L 88a, which drove a four-blade propeller with a diameter of 5.6 m (18 ft); this required a high-legged landing gear. The first flight was performed on October 2, 1931. The flight tests lasted nearly two years until September 1933, when the aircraft was taken over by the DVL. With the Ju 49, initially altitudes of over 9,000 m (29,500 ft) were attained which were in 1935 raised to 13,000 m (42,700 ft).

Another pioneering development, which, above all, has to be attributed to applied aerodynamics, are the high-speed aircraft built by *Ernst Heinkel*. In his biography, *Heinkel* describes how fascinated he was, when visiting the air-race for the "Schneider-trophy" in Venice in 1927, by the British racing aircraft. The "Supermarine" monoplane had, all through, an aerodynamically-smooth, low-drag design, so that one could feel its high speed by just looking at the aircraft. *Heinkel* was lucky that, in 1931, he was able to employ the twin-brothers *Siegfried* and *Walter Günter*, two congenial designers, who initiated, most of all, the development towards aerodynamically designed aircraft shapes. Born on December 8, 1899 in Keula/Thüringen (Thuringia), they were flight enthusiasts since *Blériot's* flight in 1909; as 16-year-old boys, they already developed their own propeller theory. Both studied mechanical engineering at the Institute of Technology (TH) Hanover with emphasis on the topics aircraft design, aircraft statics, and aerodynamics, the latter with *Arthur Pröll*. *Walter Günter* designed, together with his fellow students *Walter Mertens* and

Ernst Heinkel (1888–1958) considered very early, as engineer and businessman, "high-speed flight" and has, by promoting research to demonstrate jet propulsion, considerably influenced this technical development.

Siegfried Günter (1899–1969) has, as highly gifted engineer, decisively influenced the design of high-speed aircraft.

W. Meyer-Kassel, the glider H 6. Already in 1924, *Siegfried Günter* had pointed out the importance of a thorough aerodynamic aircraft design; he built the first "Buzzing Wind" ("Sausewind") airplane for the "German Roundtrip Flight 1925" ("Deutscher Rundflug 1925") competition while working at the "Bäumer-Aero, Ltd." company in Hamburg. The aircraft was an aerodynamically refined low-wing monoplane with wings and empennage, for the first time, of elliptic design according to *Prandtl's* wing theory published in 1918. After joining *Heinkel*, *Walter Günter* was soon charged with the development of a low-speed and a high-speed wind tunnel.

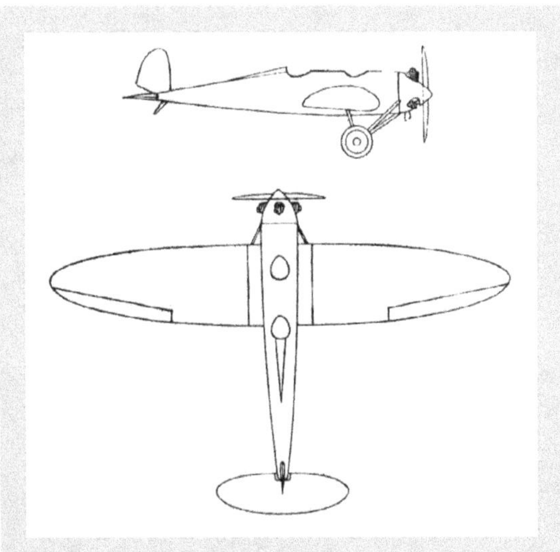

For the "Rhön Light-aircraft Competition of 1924", the brothers *Siegfried* and *Walter Günter* – at that time still students – developed at the "Bäumer-Aero, Ltd." company in Hamburg the sports plane "Buzzing Wind" ("Sausewind"). It attracted attention due to its excellent aerodynamic shape. Here the wings had, for the first time, an elliptical plan form, a design that found its continuation in various subsequent Heinkel aircraft.

The He 64 of *Heinkel* was, in 1932, the first aircraft that was given – though still constructed of wood – an unusually slim, aerodynamically smooth form. With a 112 kW (150 hp) engine, its pilot won the "European Roundtrip Flight" competition over a distance of 7,500 km (4,660 mi) by a wide margin. In 1932, the high-speed aircraft He 70 was produced as answer to the American Lockheed "Orion" of 1931, which was, at a cruising speed of 260 km/h (162 mi/h) and with retractable landing gear, considerably faster than other commercial aircraft which reached about 200 km/h (124 mi/h). The detail-design of the He 70 commenced on July 1, 1932 and the first flight occurred on December 1, 1932. The aircraft reached in horizontal flight a speed of 362 km/h (225 mi/h) and was, with a cruising speed of 323 km/h (200 mi/h), the leading high-speed aircraft worldwide. It carried four passengers, or 500 kg (1,124 lbs) of mail, had a two-men crew, and had blind- and night-flight capabilities.

The He 70 was mainly designed by *Siegfried Günter*. It had a sandwiched monocoque fuselage

The high-speed transport aircraft Heinkel He 70 – "Blitz" ("Lightning") – was built in response to the American Lockheed "Orion". The He 70 was designed by the *Guenter* brothers and had wooden wings of elliptical plan form, a fuselage with a smooth metal skin, glycol cooling of the engine, and a retractable landing gear. The cruising speed was about 320 km/h (199 mi/h).

Test section and three-component balance of the small wind tunnel of DVL at Adlershof completed in July 1932. The drive power was 200 kW (150 hp), the free-jet diameter 1.2 m (3.9 ft) and the flow velocity 65 m/s (213 ft/s).

of plywood covered with duralumin, and, for the first time for Germany, a retractable landing gear, which *Walter Günter* had developed. The *Günter Brothers* also introduced the elliptical wing plan-form, characteristic of the later Heinkel-aircraft. With this type of aircraft, an officially recognized speed record of 357.427 km/h (222.142 mi/h) was established over a distance of 100 km (62 mi) with payloads of 500 kg (1,124 lbs) and 1,000 kg (2,248 lbs), respectively. For comparison, one has to consider the absolute speed record of 1933 of 417.06 km/h (259.20 mi/h) held by an American racing-aircraft especially designed for that purpose. Lufthansa named the He 70 the "Heinkel-Blitz" ("Heinkel-Lightning") and established with this type of aircraft a "blitz" air-route network between Berlin, Hamburg, Cologne and Frankfurt (Main) with travel times, indeed, compatible to today's.

Until the End of the Weimar Republic 1932

At the end of 1923, the inflation in Germany had been overcome; six years later, the maelstrom of the World Economic Crisis, triggered by collapsing stock prices at the New York stock exchange on October 24, 1929, gripped the Weimar Republic. This also was to have a strong influence on German aeronautical industry and aeronautical research, which, as mentioned above, had just entered relatively calm waters, though being dependend on governmental support [2].

After, in 1919, efforts to convert DVL into a government organization had failed, DVL had at the beginning of the twenties considerable financing problems since the membership contributions dropped. These problems were finally overcome, but the government was now almost the sole financer. The situation was similar at the AVA, whose revenues were not sufficient to cover its expenses. Here too was governmental support required. Due to the strong dependence of aeronautical research on government support – the situation in industry was similar – the desire of the financial sponsor, the "State Ministry of Transportation, RVM", for the coordination of development plans and activities grew.

A (coordinating) organization like the American NACA did not exist in Germany, and the "Scientific Aeronautical Society, WGL" was not able to carry out the task of coordination. Plans of the RVM to trans-

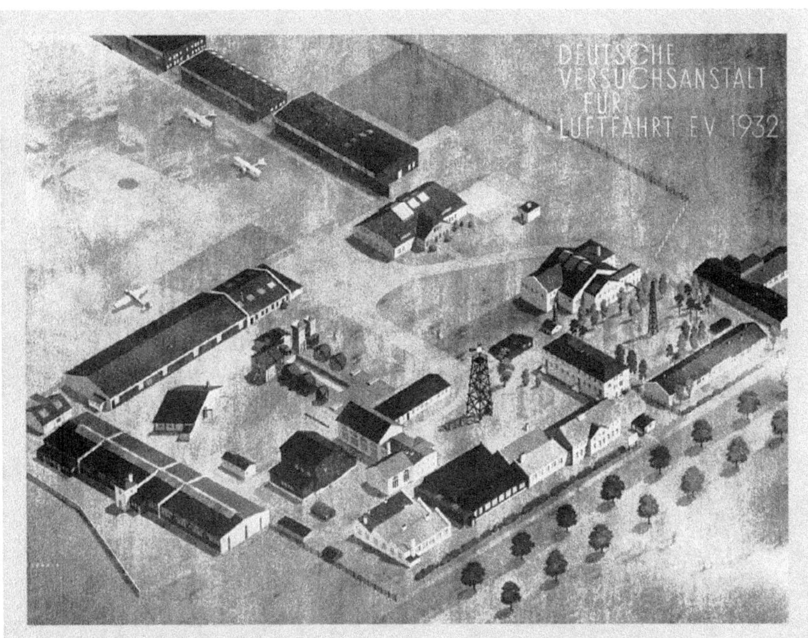

The DVL installations in Adlershof at beginning of 1932.

fer coordination to DVL were blocked in May 1928 by the AVA and the universities involved. *Ludwig Prandtl* and *Theodore von Kármán* made, however, a counterproposal, which envisaged a body of leading researchers in this field to take over the coordination. This way also the scientific autonomy of the research establishments and the university institutes would be preserved.

During the year preceding the world economic crisis, the "German Aeronautical Research Council" ("Deutscher Forschungsrat für Luftfahrt") was founded, thus creating an institution, which, although without legal form and statute, in a non-bureaucratic and effective way provided the desired coordination of the research establishments, adjusted research objectives, improved the flow of information between industry and research, and also, officially, advised the RVM. The research council, which generally met its self-appointed tasks, finally became a sub-committee of the "Aeronautical Committee" within the "Emergency Association of the German Sciences" ("Notgemeinschaft der Deutschen Wissenschaft"), founded in 1920, which, beginning in 1930 and lasting until 1945, was called "German Research Association" ("Deutsche Forschungsgemeinschaft, DFG").

The status reached at the end of the twenties was considerably disrupted by the world economic crisis. The government budget, showing already in 1928/29 a deficit, went into a deep crisis, which had severe negative effects on the aeronautical industry as also on aeronautical research. The cancellation of government subsidies due to the rigorous austerity policy of the government led in industry to a number of mergers and bankruptcies [35]. Industry had, at the beginning of 1933, only about 4,000 employees left. German aviation industry fell back behind foreign industries, qualitatively as well as quantitatively.

Aeronautical research suffered twofold: by the budget cutbacks, and, in addition, due to the fact that industry contracts for testing ceased. The extension plans of the research establishments had to be canceled; this was initially particularly hard for the DVL, whose facilities were in urgent need for renewal. The government subsidies for DVL dropped, indeed, considerably, but with various measures DVL could, at least, keep almost all of its personnel [7]. The AVA ran into a severe crisis, which was overcome only with considerable difficulties [2].

The High Rating of Aeronautical Research During the Third Reich

Ernst Heinrich Hirschel

Consequences of the Power Seizure for Aeronautical Research

The "Seizure of Power" on January 30, 1933 was immediately felt in the area of aeronautical research as well as in aeronautical industry. *Hermann Göring* became "State Secretary without Portfolio" ("Reichsminister ohne Geschäftsbereich") and took over the position of a "State Commissioner for Air Transportation" ("Reichskommissar für den Luftverkehr"). The office "State Commissioner's Department for Aviation" ("Reichskommissariat für Luftfahrt") was founded on February 2, 1933 and preliminarily attached to the "State Ministry of Transportation, RVM". *Erhard Milch*, a board member of German Airlines (Lufthansa), took over as head of the air-transportation department of the RVM, being at the same time *Göring's* deputy.

With the decree concerning the "Protection of People and State", which was issued on February 28, 1933, the government created the legal prerequisite for the arbitrary prosecution of political opponents and for the removal of unwanted persons. This also held for members of the aeronautical research and aircraft industry community. The most prominent victim was *Hugo Junkers*, who, in 1936, was expropriated [32].

On April 29, 1933, the "State Commissioner's Department for Aviation" was converted to the "State Ministry of Aviation" ("Reichsluftfahrtministerium, RLM") with *Göring* as Secretary (Minister) and *Milch* as Undersecretary (Staatssekretär). The newly created RLM received a "Technical Office", from which later, in February 1939, the domain of the "Aircraft Master General" ("Generalluftzeugmeister") arose. The top position within the technical management was thus held by the military. Only the department "Aeronautical Research", one of three, later one of four departments of the Technical Office, responsible for armament, largely kept, at least initially, its civilian character. Head of the department was *Adolf Baeumker*, who previously was the "official in charge" ("Referent") for new technical developments in the areas of civil aeronautics and aeronautical research at the RVM. Still earlier, he had been a delegate, for matters of aeronautics, to the disarmament conference of the League of Nations.

On March 16, 1935, *Hitler* announced the re-establishment of the German air sovereignty. The restrictions imposed on the German aircraft industry by the Versailles Treaty were, unilaterally, lifted and the air force, since 1933 intensively being built up, was "exposed".

The developments related to the built up of an air force, which already commenced prior to 1933 – the "Secrete Flight Trials" in the Russian city of Lipezk even went back to 1924 [36] – were, after the power seizure, characterized by the extension of the industrial production capacities and the training of flight personnel. The aircraft companies were at the time of the power seizure relatively small; only the Junkers-plants could be considered a large-scale enterprise. The companies had not built any military aircraft since the end of World War One; however, the secrete trials in Russia had given them the opportunity to experiment with temporary aircraft types.

Adolf Baeumker (1891–1976).

Adolf Baeumker, born in Breslau on July 14, 1891.

1927 until then he was active in various departments of the Prussian Army and worked, after the end of World War One, at the "State Defense Ministry" ("Reichswehrministerium"), until he became the "official in charge" ("Referent") of aeronautical research and development at the "State Ministry of Transportation" ("Reichsverkehrsministerium"). *Baeumker* may be considered as one of the first "Science Managers" in Germany. He established the "Aeronautical Research Council" ("Forschungsrat für Luftfahrt"), initiated by *Prandtl* and *von Kármán*, to promote the exchange between research and industry.

1933 he becomes head of the aeronautical research office at the "State Commissioner's Department for Aviation" ("Reichskommissariat für Luftfahrt") and subsequently department head in charge of aeronautical research at the Technical Office of the "State Ministry of Aviation" ("Reichsluftfahrtministerium, RLM").

1935 he created the "Lilienthal-Society", which replaced the "Scientific Aeronautical Society" ("Wissenschaftliche Gesellschaft für Luftfahrt, WGL"), and he founded in

1936 the "German Academy for Aeronautical Research" ("Deutsche Akademie der Luftfahrtforschung"). During the same time, the DFL in Brunswick and the FFO in Oberpfaffenhofen were newly founded. When the head of the "Technical Office", *Ernst Udet*, committed suicide, *Baeumker*, in

1941 resigned from his position and the office was dissolved.

1942 the research department within the former "Technical Office" is substituted by the "Aviation Research Command" ("Forschungsführung"), headed by *Prandtl*; *Baeumker* was also a member. This body of experts, which directly reported to the "State Secretary of Aeronautics" ("Reichsminister der Luftfahrt"), *Göring*, and his undersecretary *Milch*, was also not able to secure more prestige for aeronautical research.

1945 he goes to the USA, where he worked for various military offices.

1958 he is transferred to the European Headquarter of the American Air Force (USAFE) in Wiesbaden to work for the cooperation in research and development between the Federal Republic of Germany and the USA. He advises in the sixties German and American head offices in Bonn, among other things, also to unite non-university German aerospace research.

1976 he dies in Bad Godesberg on March 4.

The built up of the aircraft industry was after 1933 pursued with great force. At the time of the power seizure, about 4,000 people were employed; in 1938, this number had risen to almost 300,000. The development of the aircraft industry during the time of the Third Reich – as the "environment" of aeronautical research – can here, for lack of space, not be redrawn, especially since the industry was highly divided and the companies were able to preserve their independence in principle. The reader is, therefore, referred to the literature, which illuminates various aspects of this subject [5, 11, 37, 2].

Aeronautical research also was subject of the plans of the government to create a powerful air force. This meant that, quantitatively and qualitatively, a strong expansion in personnel and equipment had to be realized. The instructions, given by *Göring* in March 1935 to aeronautical research, read that aeronautical research had in 1938 to attain the level of foreign nations and had, subsequently, to surpass these nations in all important areas.

Adolf Baeumker had, obviously from the very beginning, a concept for the realization of such an extension. It was, however, towards the end of the thirties subject to modifications, not the least due to the technical developments in aircraft engineering. Looking back, this concept presents itself briefly as follows [1]:

At the start of the first extension phase of aeronautical research, which lasted from early 1933 to early 1935, stood management measures to promote the "development of intellectual activities", and "material" measures. As already mentioned, in February 1933, the research department within the "Technical Office" of the (later) RLM, headed by *Baeumker*, was founded as an instrument to realize *Baeumker's* concept. It followed the formation of the "Center of Scientific Publications" ("Zentrale für wissenschaftliches Berichtswesen, ZWB"), the formation of the "Aeronautical Research Association" ("Vereinigung für Luftfahrtforschung, VLF"), and the preparation of an educational schedule to "introduce" engineers to industry, research and the administrative bodies. The material measures during the first phase concerned new facilities for DVL and AVA, but also for other relevant research institutions.

During the second extension phase, 1935–1938/39, the "Lilienthal Society of Aeronautical Research" and the "German Academy of Aeronautical Research" ("Lilienthal-Gesellschaft für Luftfahrtforschung" and "Deutsche Akademie der Luftfahrtforschung") were founded; furthermore, "State Research Professorships" ("Reichsforschungsprofessuren") were created. The material measures were related to the set up of large research facilities, especially the "German Research Establishment of Aeronautics" ("Deutsche Forschungsanstalt für Luftfahrt, DFL") near Brunswick and the "Air Radio Research Institute Oberpfaffenhofen" ("Flugfunkforschungsinstitut Oberpfaffenhofen, FFO") near Munich, and further institutions.

During the third and last extension phase, 1939–1945, the equipment of the research establishments was to be further improved. The rapid development of aircraft and propulsion technology led, in 1940, to the foundation of the "Aeronautical Research Establishment Munich" ("Luftfahrtforschungsanstalt München, LFM"), which became, however, up to 1945, only partly operational.

In the following sections, details of the institutional development of German aeronautical research during these three phases is sketched. At first, brief surveys concerning the "large" research establishments DVL, AVA and the newly founded DFL are presented.

DVL – the German Test Establishment of Aeronautics [7]

Within the framework of the overall extension of aeronautics, having commenced in 1933, greater financial means for further growth were available to DVL at its now secure location Adlershof. As a consequence, a new extension plan was drawn up. It envisaged the erection of further large test facilities, which was possible since in 1934 the construction site had been considerably enlarged. This also allowed a new space-wise arrangement of the organization.

On the northern campus, which corresponded to the original DVL grounds, the following facilities were located:

- Three wind tunnels with test-section dimensions of 1.2 m (3.9 ft) and 2.5 m (8.2 ft) diameter, and 5 m × 7 m (16.4 ft × 23 ft) and 6 m × 8 m (19.7 ft × 26 ft), respectively, cross-sectional areas, a free-spin test facility, and a high-speed wind tunnel,
- a propeller test facility with a test rig of 1,200 hp and a whirling rig of 4,000 hp (since December 1932 in operation),
- two engine test hangars, with 12 one-cylinder and multi-cylinder test stands each,
- one sound-proof and four open pendulum-frame engine test rigs,
- one altitude engine test stand.

Located on the southern campus, i.e., on the other side of the "Rudow High Road" ("Rudower Chaussee"), were the laboratories and offices of "Material Research", "Electro Physics", "Photography and Navigation", etc., as well as the "Scientific Secretariat" ("Wissenschaftliches Sekretariat") and the "Center of Scientific Publications" ("Zentrale für Wissenschaftliches Berichtswesen, ZWB").

On November 17, 1933, the general assembly of DVL agreed on a change in the statutes, which was approved on December 14 by the "State Minister of Aeronautics" ("Reichsminister der Luftfahrt"). The changes corresponded to the leadership demands of the government offices. The "State Minister of Aeronautics" appointed all members of the board of directors; he also appointed the chairman of the board from its members and two deputies. The board appointed two managing directors, who, at the same time, were the heads of the organization. The "Assem-

At a "Colonial Show" ("Kolonialschau"), which was held in February/March 1938 in the exhibition halls at the "Radio Tower" ("Funkturm") in Berlin, *Hanna Reitsch* performed, towards the end of the show, demonstration flights with the Focke-Wulf helicopter FW 61. During her demonstration, she flew forwards, backwards, laterally, and remained standing in the air for a prolonged period of time, all of that within the very limited space of the "German Exhibition Hall" ("Deutschlandhalle"). German and foreign experts were astonished and their attention was directed towards the many possibilities associated with a helicopter but also towards the great progress achieved in Germany during the thirties.

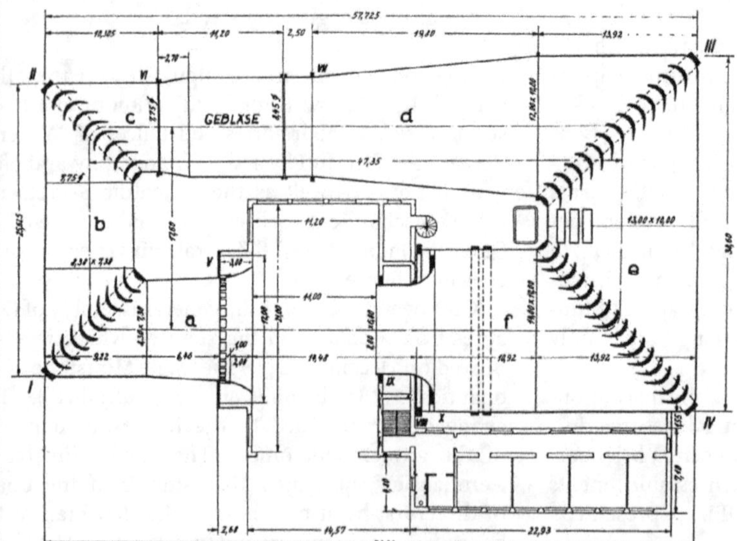

Preparatory work for the large wind tunnel in Adlershof started in 1931. Only in 1933, when sufficient funding was provided, construction could commence. The tunnel had a drive power of 2,800 kW (3,810 hp) and test-sections with cross-sections of 5 m × 7 m (16.4 ft × 23 ft) and 6 m × 8 m (19.7 ft × 26 ft), respectively. The tunnel was fully occupied until shortly before the end of World War Two. (Gebläse = Fan)

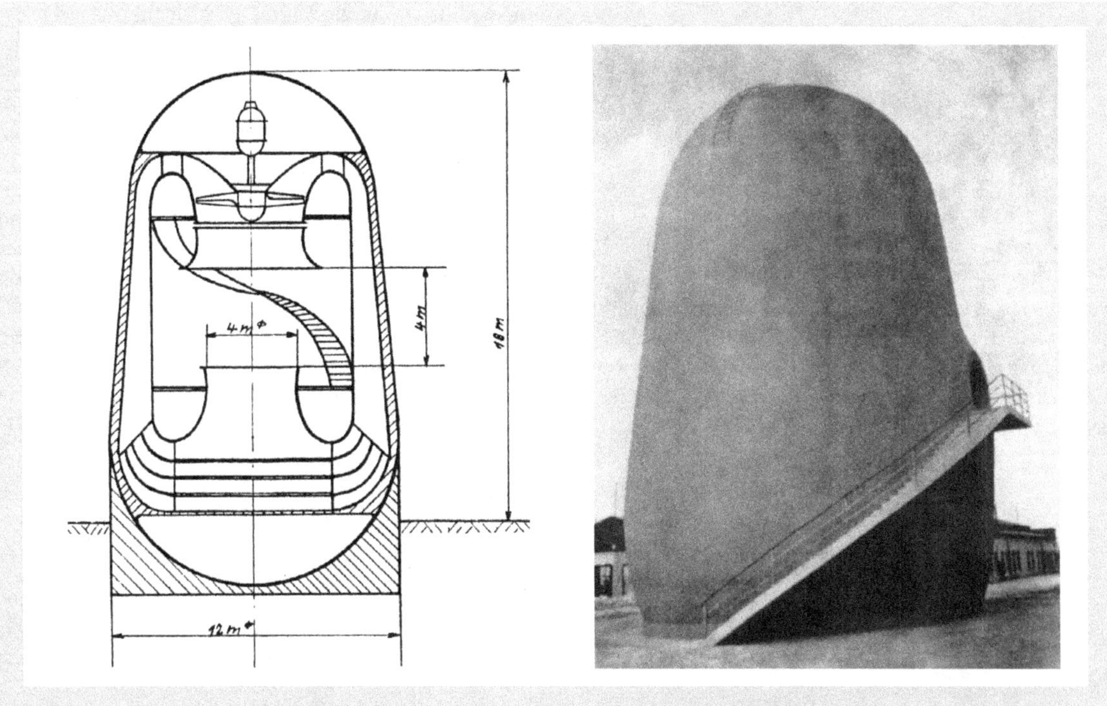

The free-spin wind tunnel of DVL was built during the years 1934/35. It looks from the outside like a giant egg. The empty concrete shell still exists. To investigate the spin behavior, the model floated freely in front of camera objectives. The ascending air stream had a diameter of 4 m (13 ft) and a velocity of 40 m/s (131 ft/s).

bly of the Department Heads of the Organization" ("Versammlung der Abteilungsleiter der Anstalt"), an organ of management participation, was abolished, and the departments were combined within "Main Sections", each with a corresponding section head.

Within the DVL management and the board of directors, positions changed several times. *Wilhelm Hoff*, who was with DVL since January 1, 1913, with interruptions only during World War One, and who took over as head of DVL on January 1, 1920, abandoned on March 31, 1936 his position as scientific head in order to completely devote himself to his lectureship at the Institute of Technology (TH) Berlin. *Friedrich Seewald* was on April 1, 1936 assigned the scientific leadership of DVL, first as *Hoff's* deputy and in October 1936 on his own behalf.

Essential changes also occurred in the subject areas of the DVL. On April 1, 1933, the "Testing Department" ("Prüfabteilung") was separated from DVL and attached to the newly formed "State Office of Air Traffic Control" ("Reichsamt für Flugsicherung") as "Air Vehicle Testing Board" ("Prüfstelle für Luftfahrzeuge"). Management remained the same, as did the location, Adlershof. This was to secure the continuous exchange of know-how. On April 1, 1934, the "Board of Standards" ("Normenstelle") and the "Management Office" ("Geschäftsstelle") of the "Committee of Standards for Aeronautics" ("Fachnormenausschuss Luftfahrt, FaLu") were transferred to the "State Association of the German Aeronautical Industry" ("Reichsverband der Deutschen Luftfahrtindustrie"). The "Examination Office of Aviation Accidents" ("Untersuchungsstelle für Luftfahrtunfälle") was in 1935 integrated into the RLM. Originally it was formed in February 1932 following order of the "State Ministry of Transportation" ("Reichsverkehrsministerium") by separation from the "Testing Department".

On August 1, 1933, the "Department of Junior Engineers" ("Abteilung für Ingenieurnachwuchs") was

The high-speed wind tunnel of DVL was commissioned in 1938; its drive power was 13,000 kW (17,700 hp). Velocities of up to almost 300 m/s (984 ft/s), corresponding to a Mach number of M = 0.95, were attained. The test section had a cross section of 2,7 m × 2.7 m (8.9 ft x 8.9 ft).

formed. It cooperated directly with the research department of the "Technical Office" of the RLM. Its purpose was to overcome the pronounced lack in technical-scientific junior personnel by corresponding promotional measures and the performance of work in the area of occupational planning.

Organizational changes in the area of aeronautical research during 1936 included the establishment of a central administrative office for the flight vehicles of all aeronautical research establishments including the "Flight-technical Academic Groups" ("Flugtechnische Fachgruppen") and corresponding "Working Groups". A "Pilot Training Center" was created which pursued the important task of training engineering pilots. It was attached to the "Flight Depart-

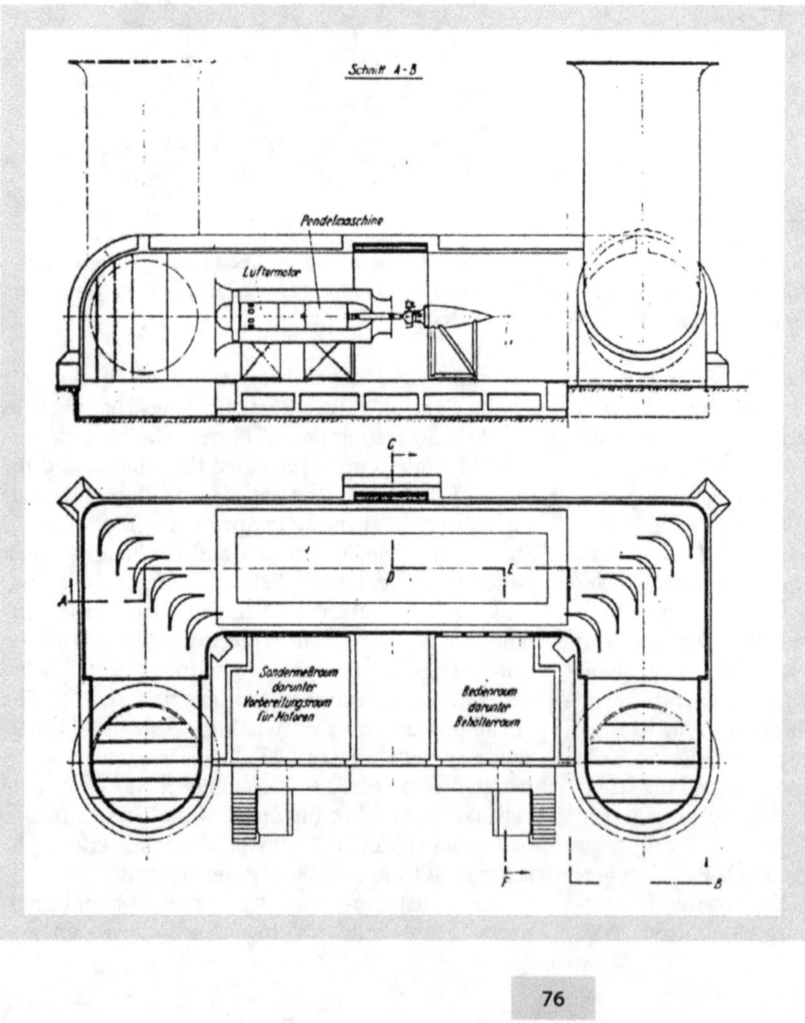

During 1933/35, a universally employable mufffled engine test stand was erected on the north campus of DVL, which allowed the investigation of flight engines with propellers of up to 5 m (16 ft) diameter, and the braking of engines with a pendulum-frame of up to 1,500 hp. For the investigation of air-cooled engines, a cooling-fan system was available. The two towers of reinforced concrete, 15 m (49 ft) high, also served as mufflers.

Friedrich Seewald (1895–1974).

Friedrich Seewald, born in Oberkirchen, Rhineland on January 15, 1895.

1925 on January 1, he joined the Physics Department of DVL in Berlin-Adlershof.

1926 he becomes head of the Aerodynamics Department.

1934 he becomes head of the "Main Section for Aircraft Engineering" ("Hauptgruppe Flugzeugbau") of DVL.

1936 on April 1, he took over as scientific head of DVL, after *Hoff* had left, and on August 1 he assumed the overall leadership, a position he held until July 21, 1941. *Seewald* remained, however, a member of the board of directors of DVL.

1941 he is appointed "Full Professor of Fluid Mechanics" at the Institute of Technology (TH) Aachen.

1942 since the establishment of the State Office "Aviation Research Command" ("Forschungsführung") on June 30 until the end of the war, he was *Prandtl's* deputy and executive member of the "Aviation Research Command".

1944 on July 1, the board of directors of DVL is converted into a "Supervisory Committee" ("Aufsichtsausschuss"); *Seewald* is appointed chairman, a position he keeps until September 29, 1955.

1955 on the same day, he is voted "Honorary Chairman" by the General Assembly of DVL.

1974 he dies in Aachen on February 4.

ment" ("Flugabteilung, Fluga"). In 1939, the center was moved to Brunswick due to the heavy load on the airfield Johannisthal-Adlershof, remaining, however, under the management of DVL.

Due to the expansion and the changes in tasks, the following organization of the DVL was realized in the years between 1933 and 1936. Management and overall supervision of DVL were in the hands of *Wilhelm Hoff* (until March 31, 1936) and *Friedrich Seewald* (starting on April 1, 1936). Directly attached to the board of directors were administration and financing with the administrative secretariat, the scientific secretariat, the Center of Scientific Publications (ZBW), the Department of Junior Engineers, and the "Examination Office of Aviation Accidents". DVL was divided into four Main Sections:

- **Main Section I, Aircraft Engineering** with the departments Aerodynamics, Statics and Materials, and the Hydrodynamics Department with the group for load determination.
- **Main Section II, Aircraft-engine Engineering** with the departments Carburator Motors and Engines (Vergasermotoren und Triebwerke), Fuel-injection Engines, Engine Mechanics, Thermodynamics, and the groups "Petroleum, Oils and Lubricants (POL-products)" and "test rigs".
- **Main Section III, Applied Flight** with the Flight Department and the departments of Electro-technique and Radio Communication, Aerial Photography, Terrestrial and Astronomical Navigation.
- **Main Section IV, Administration.**

In 1936, considerable changes occurred, since a new aeronautical research establishment, viz., the DFL had been founded in Brunswick-Völkenrode, which started operations in October. DVL participated in the set up of this organization not only in an advisory and preparatory way, but also by providing part of its scientific personnel to DFL, especially *Hermann Blenk*, who took over as head of the organization and head of the Institute of Aerodynamics, and *Adolf Busemann*, who was assigned manager of the Institute of Gas Dynamics.

Within DVL, a new organizational set up became effective after a corresponding change in the statutes on August 1, 1936. Appointed to the board of directors were *Friedrich Seewald* and *Ernst Oskar Müller*. On October 7, 1936 – the new organization was, in the meantime, already in effect – *Seewald* was appointed sole manager of the test establishment. Deputy managers and heads of the organization became *Günther*

Hermann Blenk (1901–1995).

Bock and *E.O. Müller*. The test establishment was now divided into four "Directorates":

- **Airframe Directorate** with the Institutes of Aerodynamics, Seaplane Engineering (Seeflugwesen), Flight Mechanics, Strength and Material Research.
- **Engine Directorate** with the Institutes of Engine Design, Motor Processes and Thermodynamics, Engine Mechanics, POL-products, Gas Dynamics, Turbo-machines (since 1938).

Wind-tunnel model of the rocket-powered interceptor Messerschmitt-Lippisch Me 163 inside the large wind tunnel of the DVL. The Me 163 was the first aircraft faster than 1,000 km/h (622 mi/h), which corresponds to a Mach number of M = 0.84.

Hermann Blenk, born in Bad Hersfeld on December 9, 1901.

1923 PhD with *Ludwig Prandtl* with a thesis concerning the extension of *Prandtl's* "lifting-line theory" to a "wing theory" applicable to low aspect ratio wings. He worked for twelve years successfully at DVL in Berlin-Adlershof, where he, in his last position, headed the Institute of Aerodynamics. In his work, emphasis was placed on questions of aircraft stresses, coupled torsion and bending oscillations of wings, rudder oscillations, wing oscillations of cantilever monoplanes, control surface buffeting, and many problems of flight mechanics.

1935 he is a founding member of the "German Research Establishment of Aeronautics" ("Deutsche Forschungsanstalt für Luftfahrt, DFL") in Brunswick, which he headed between 1936 and 1945, in addition to managing the Institute of Aerodynamics and teaching at the Institute of Technology (TH) Brunswick the subjects Aerodynamics and Flight Mechanics.

1952 renewal of DFL, whose president *Blenk* was between 1953 and 1958. He headed, in addition, the Institute of Flight Mechanics until April 1, 1971.

1959 merger of AVA, DFL and DVL within the "German Society of Flight Sciences" ("Deutsche Gesellschaft für Flugwissenschaften, DGF"), whose chairmanship he assumed from 1960 to 1965. Parallel to these activities, he published the "Journal of Flight Sciences" ("Zeitschrift für Flugwissenschaften") and the "Yearbook of the WGLR" ("Jahrbuch der WGLR"). (WGLR = Scientific Aerospace Society)

1962 he becomes an honorary member of the WGLR and bearer of the Ludwig-Prandtl-Ring.

1995 he dies in Brunswick.

- **Equipment Directorate** with the Institutes of Electro-physics, On-board Instrumentation and Navigation, Photography, Flight Medicine and Control Engineering, the latter with the group on terrestrial and astronomical navigation.
- **Administration Directorate** with the main workshop, hangar operations, research flight-operations, and the central administrative office for air vehicles.

Directly attached to the board of directors were, again, the general secretariat, the scientific secretariat,

View of the test facilities of the "German Test Establishment of Aeronautics, DVL" in Adlershof (North Campus). From left to right one observes part of the large wind tunnel and the free-spin wind tunnel, the assembly building, the cooling towers, and the sound-proof engine test facility (1939).

"Specific Scientific Tasks" ("Wissenschaftliche Sonderaufgaben"), the Department of Junior Engineers, and the Center of Scientific Publications (ZBW), which was, however, separated from DVL in 1940.

Beginning in 1942, the increasing severity of the war was strongly felt. Far-reaching organizational changes were planned, which were, however, due to the difficulties of the time, only with strong resistance, and also not completely, realized. On August 1, 1941 *Friedrich Seewald* left his position as head of the organization and entered, as full professor and Director of the Institute of Aerodynamics of the Institute of Technology (TH) Aachen, a new sphere of activity. According to the statutes, management of DVL now fell jointly to the two deputies *Bock* and *Müller*.

AVA – the Aerodynamic Test Establishment [9]

The organizational structure, introduced in 1925, was in effect until 1937. The construction of the large wind tunnel, in planning since 1928, could commence in 1934. Within the framework of the extension of the institution, the RLM demanded a new organizational form, namely, as a registered association with its own legal capacity. After lengthy negotiations, the reorganization was executed on April 1, 1937. The AVA was, in the process, separated from the "Emperor-Wilhelm Institute for Fluid Research" ("Kaiser-Wilhelm-Institut für Strömungsforschung"). Its new name was

> "Aerodynamic Test Establishment Göttingen, e.V. within the Emperor-Wilhelm-Society for the Promotion of the Sciences" ("Aerodynamische Versuchsanstalt Göttingen e.V. in der Kaiser-Wilhelm-Gesellschaft zur Förderung der Wissenschaften").

The new organizational structure was completed shortly before the war. *Ludwig Prandtl* was appointed as chairman of the board, the latter having seven honorary members. *Albert Betz*, also a member of the board, assumed the scientific management of the organization. Administrative management was in the hands of *Walter Engelbrecht*. The scientific activities of the organization were, since 1939 and remaining so during the war, carried out within eight institutes:

- **Institute of Wind-tunnel-related Research**, which operated the wind tunnels, among them the large tunnel with an elliptic cross section of 4.7 m × 7 m (15.4 ft × 23 ft).
- **Institute of Flight Research and Flight Engineering**, which investigated the flight characteristics

Günther Bock (1898–1970), on the right, presents the "Ludwig-Prandtl-Ring" to *Heinrich Focke.*

Günther Bock, born in Berlin on June 10, 1898.

1916 after graduating from high school (Gymnasium), he is drafted for military service.

1919 studies general mechanical engineering at the Institute of Technology (TH) Berlin.

1923 joins as "Master of Engineering" ("Diplomingenieur") the scientific secretariat of Junkers in Dessau.

1924 transfer to the "Research Organization Prof. Junkers" ("Forschungsanstalt Prof. Junkers").

1926 transfer to the "Junkers Aircraft Corporation" ("Junkers Flugzeugwerke, A.G.").

1931 appointed full professor at the Institute of Technology (TH) Danzig.

1933 appointed Chief Engineer at the RLM.

1934 appointed to the "Chair of Mechanics and Aerodynamics" at the Institute of Technology (TH) Aachen.

1935 becomes a member of the "Academy of Aeronautical Research" and chairman of the board of directors of the "Scientific Aeronautical Society, WGL".

1936 becomes director of the "German Test Establishment of Aeronautics, DVL" in Berlin-Adlershof in charge of the "Airframe Directorate".

1937 appointed full professor in the services of the State ("im Reichsdienst") and "Corresponding Member" ("korrespondierendes Mitglied") of the "German Academy of Aeronautical Research".

1941 assumes the scientific management of DVL Berlin.

1942 becomes a full member of the "German Academy of Aeronautical Research" and a member of the senate of the "Lilienthal-Society of Aeronautical Research".

1944 appointed member of the board of directors of DVL.

1945 he becomes head of the expert committee of the DFL in Brunswick, instituted by *Göring*, to evaluate and assess the long-range bomber projects submitted to the RLM. He voluntarily returns from Brunswick to Berlin to stay with his co-workers during the march of the Russians into Berlin. After the occupation of the DVL-facilities in Berlin-Adlershof by Soviet troupes, *Bock* is taken prisoner, shipped to the USSR. After an eight-months imprisonment in the "Lubjanka", he is, in July

1946 put to work as a scientist at the "Central Aero-hydrodynamics Institute, ZAGI" near Moscow.

1953 he is transferred to the "German Aeronautics Collective" ("Deutsches Luftfahrt-Kollektiv") in Sawjelowo, which is headed by *Baade.*

1954 he returns from the Soviet Union to Pirna. He uses a leave to visit his family in East Berlin to escape to the Federal Republic. On October 1, he is appointed "Full Professor of Aeronautical Technology" at the Institute of Technology (TH) Darmstadt. He becomes chairman of the "Technical Committee of the European Launcher Deployment Organization, ELDO", member of the "Senate and the Board of Trustees of the German Research Association" ("Senat und Kuratorium der Deutschen Forschungsgemeinschaft"), and chairman of the "Committee of Flight Research of the Scientific Aerospace Society" ("Kommission für Flugforschung bei der Wissenschaftlichen Gesellschaft für Luft- und Raumfahrt, WGLR").

1963 on his 65th birthday, he is made an honorary member of DVL. He dies in 1970.

of aircraft and helicopters and also attended to the "flow-control-by-suction aircraft" ("Absaugeflugzeug").

- **Institute of Icing Research,** operating a large icing tunnel. The main subject area was the wing icing problem and the effectiveness of chemical and warm-air de-icing.
- **Institute of Theoretical Aerodynamics** with the subject areas airfoil design, laminar airfoils, Pitot intakes (cooler intakes), et cetera.

- **Institute of Turbo Machinery**, which considered axial superchargers for high-altitude engines and axial compressors for jet engines.
- **Institute of Unsteady Processes** whose field of activities concerned the problems related to the interaction between aerodynamics and the elastic airframe (aeroelastics). Here, pioneering work, especially with respect to the phenomena flutter and buffeting, was performed.
- **Institute of High-speed Problems**, where, among other things, swept-wing aerodynamics and projectile aerodynamics were treated.
- **Institute of Instrumentation Development**, which considered electrical measuring instrumentation and auxiliary equipment as well as propulsion devices for model engines of wind tunnel models.

The organizational structure shows that the tasks of the AVA approached those of the DVL. While fundamental research remained the domain of the "Emperor-Wilhelm-Institutes", AVA worked almost exclusively on problems of the aeronautical industry.

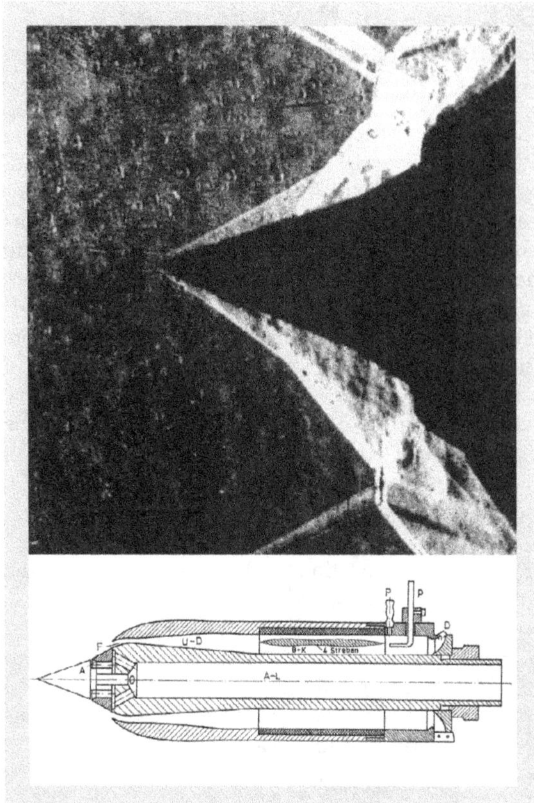

▲
At the end of the thirties, an oblique-shock diffuser for a ramjet engine, suggested by *Klaus Oswatitsch*, was investigated in the supersonic wind tunnel "B 2" of the AVA. The upper photograph shows a Schlieren picture of the intake section of the engine at supersonic speeds.

◄ The nozzle of the test section of the large wind tunnel "VI" of AVA with a cross section of 4.7 m × 7 m (15.4 ft × 23 ft) as of 1936.

DFL/LFA – the Newly Founded German Research Establishment of Aeronautics [10]

With the substantial expansions of the AVA in Göttingen and the DVL in Berlin-Adlershof, these organizations reached their space-wise limits. Since the foreign aeronautical research accomplishments were to be equaled in 1938 and subsequently surpassed, more research capacity was needed; therefore, it was decided to establish a new aeronautical research institution. But also the desire of decentralization played a role. In addition, the cooperation with the regional industry was to be supported. The driving force behind the expansion of German aeronautical research was still the head of the research department within the RLM, *Adolf Baeumker*.

In April 1935, *Göring* had given his permission. Concerning the location for the new organization, the following criteria applied:

- Central location within the State (Reich),
- ajar a medium-size city with a university or an institute of technology,
- best possible camouflage of the complete installation against ground and air observation,
- installation of its own airfield.

It soon became obvious that Brunswick met all these criteria very well. The city had very good traffic links, further improved by the just completed "Interstate Highway" ("Autobahn") Berlin-Ruhr Valley. The central-German aeronautical industry, e.g., Junkers in Dessau, was not far away. The Institute of Technology (TH) Brunswick had just set up a special aeronautical department and founded new institutes related to this subject area. The University of Göttingen with the KWI ("Emperor-Wilhelm-Institute") and the AVA were virtually in the neighborhood.

The new organization was to concentrate on fundamental research in the most important areas of aeronautical technology. In the process, areas only weakly treated by DVL and AVA, were to be preferred. The five envisaged institutes were to utilize one central airfield.

Anticipating the importance of the work envisaged for the organization to the military, the buildings were possibly to be arranged in wooded areas as protection from observations from the air, and the longitudinal axes of the buildings were principally to be aligned in a north-south direction in order to reduce the formation of shadows when the sun was at or near its zenith. The ridges of the roofs should, in addition, not exceed the treetops and the purpose of the facilities should not be recognizable from traffic routes or highways. It should, furthermore, be possible to hold congresses and meetings without the participants being able to observe the entire facility complex.

Groundwork started in October 1935 on premises south of the village of Völkenrode near Brunswick. Research work commenced in October of 1936 after the first buildings had been completed. On the campus in Völkenrode, 80 buildings were erected; eight buildings were constructed at the branch in Trauen near Fassberg, which had been established in 1937/38. DFL-personnel grow almost linearly, starting in 1936, reaching 1,500 employees, among them 150 scientists, by the end of the war.

The organizational form was the "Registered Association" ("eingetragener Verein, e. V."). The foundation assembly took place on February 10, 1936 in Berlin. *Adolf Baeumker* was elected chairman of the board, *Hermann Blenk* became his deputy and, at the same time, head of the entire organization. In 1938, DFL was renamed "Aeronautical Research Establishment Hermann Göring" ("Luftfahrtforschungsanstalt Hermann Göring, LFA").

The technical structure of DFL/LFA corresponded to the new structure just before given to the DVL. However, while the DVL consisted of thirteen institutes, DFL only set up five:

- **Institute of Aerodynamics** with five wind tunnels. One of the tunnels, with a test-section diameter of 8 m (26 ft), designed for the investigation of full-scale aircraft, was only constructed later.
- **Institute of Gas Dynamics** with, among others, a high-speed wind tunnel, which allowed to attain flow velocities up to three-times the speed of sound. The test rigs for rocket and jet engines represented such a danger potential that only a small test stand was erected at the main site, while the large facility was set up on the grounds of the Trauen branch.
- **Institute of Strength Properties (Institut für Festigkeit)**, especially with facilities for static tests on large components and complete aircraft.
- **Institute of Engine Research.**

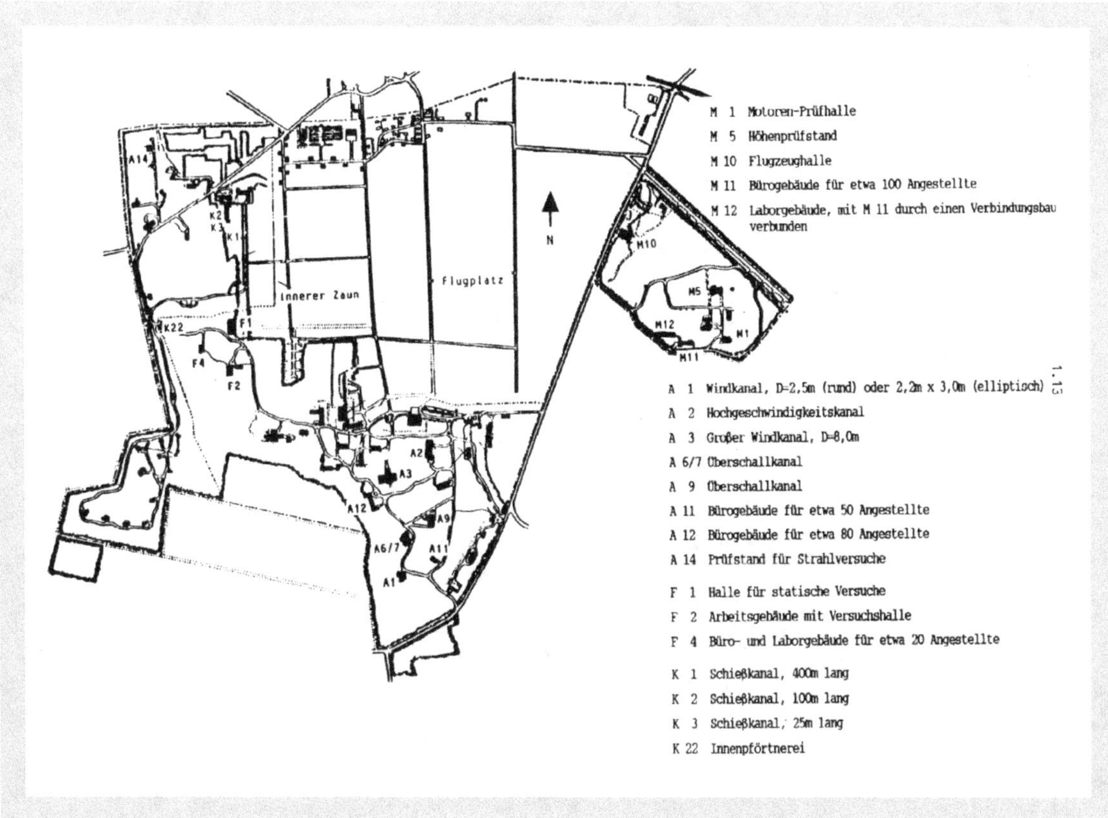

Plan view of the campus of the "German Research Establishment of Aeronautics" ("Deutsche Forschungsanstalt für Luftfahrt, DFL") in Brunswick-Völkenrode. The construction of the wind tunnel "A 1" started in 1935. Most of the construction work was completed shortly before the begin of World War Two.
M1: Engine test building; **M5:** high-altitude test stand; **M10:** aircraft hangar; **M11:** office building for 100 employees; **M12:** laboratory building, connected with M 11 by a building; **A1:** wind tunnel with D = 2.5 m (8.2 ft) circular, or 2.2 m × 3 m (7.2 ft × 9.8 ft) elliptic test section; **A2:** high-speed wind tunnel; **A3:** large wind tunnel with D = 8 m (26 ft) test section; **A6/7:** high-speed wind tunnel; **A9:** high-speed wind tunnel; **A11:** office building for about 50 employees; **A12:** office building for about 80 employees; **A14:** test rig for jet investigations; **F1:** hall for static tests; **F2:** work building with test hall; **F4:** office and laboratory building for about 20 employees; **K1:** ballistic range 400 m (1,312 ft) long; **K2:** ballistic range 100 m (328 ft) long; **K3:** ballistic range 25 m (82 ft) long; **K22:** internal gate house).

- **Institute of Kinematics**, respectively, weapons research, with a 400-meter (1,312-foot) long ballistic range, which could be evacuated, for the investigation of external ballistics.

Of importance is to note that at DVL an Institute of Flight Mechanics had been established, however, not at DFL. There was, obviously, within the Institute of Aerodynamics initially not even an independent department concerned with flight mechanics; such a department was only created when pressured by *Hermann Blenk*. This indicates that, at the time, "Flight Mechanics" had obviously the problem of being recognized as a technical-scientific discipline.

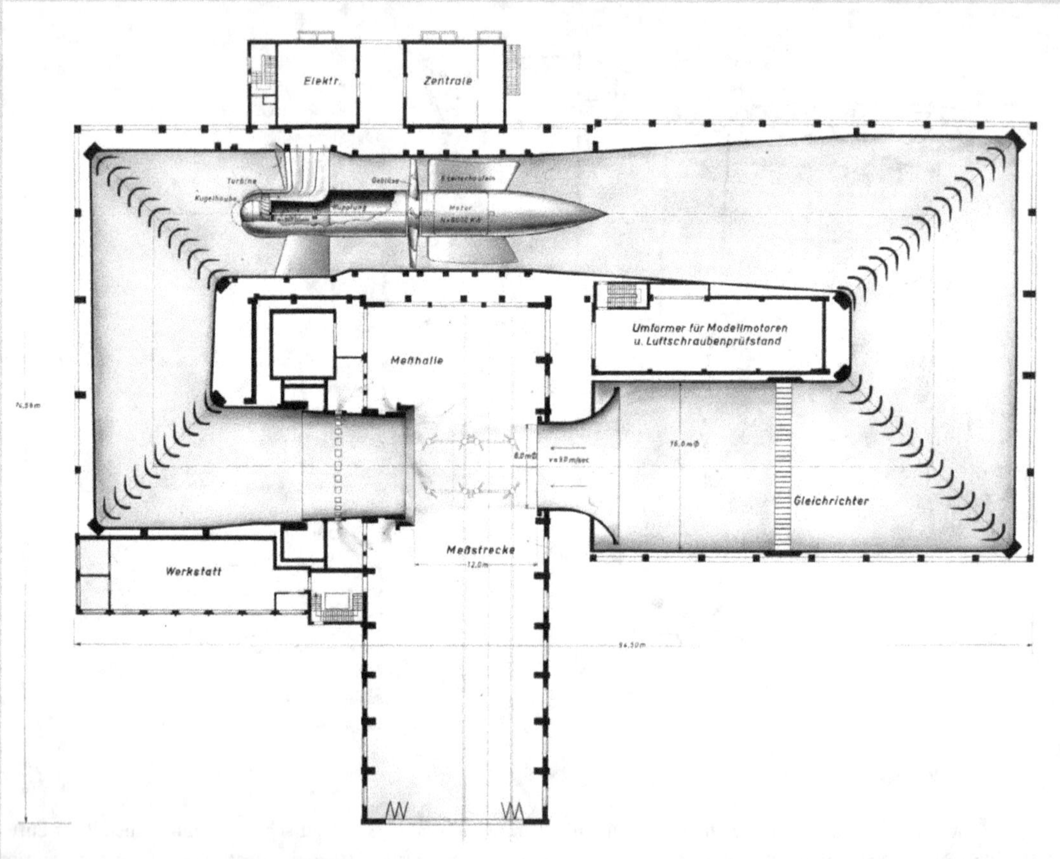

▲
The large wind tunnel of the DFL/LFA "A 3", also called "Hermann-Göring-Tunnel", was in 1940 put into operation. (Meßstrecke = test section, Gleichrichter = flow straightener)

◄ The large wind tunnel of DFL/LFA allowed testing of full-scale aircraft; the figure shows a Messerschmitt Me 109 E.

The large ram jet engine with a diameter of 1.5 m (4.9 ft), developed by *Eugen Sänger*, was tested in 1942 on top of a Do 217 as test carrier. Velocities of up to 720 km/h (447 mi/h) were reached during an inclined dive.

Scientific Societies – From the Scientific Aeronautical Society to the Lilienthal Society

The American National Advisory Committee for Aeronautics (NACA), founded in 1915, was, for the German aeronautical research, an exemplary institution. The objective of the NACA, since 1958 the NASA (National Aeronautics and Space Administration), was the "supervision and direction of the scientific study of the problems of flight". NACA was directly answerable to the President of the USA, and performed an effective coordination of the interests of research and industry within the "youthful" aeronautics. Such an institution would have also been needed in Germany, at the latest during the second half of the twenties, considering the expansion plans of the aeronautical research establishments and the flourishing (civil) aeronautical industry, although these processes were considerably dampened in 1929 as a result of the World Economic Crisis [35].

Such coordination was, indeed, lacking in Germany [2]. The "Scientific Aeronautical Society" ("Wissenschaftliche Gesellschaft für Luftfahrt, WGL") was not up to the task. Plans to assign this task to DVL did not meet the approval of the AVA, but also not of the universities and the institutes of technology. Finally, the "German Aeronautical Research Council" ("Deutscher Forschungsrat für Luftfahrt") was formed, which, as the leading committee in aeronautical research provided the necessary coordination for the "State Ministry of Transportation" ("Reichsverkehrsministerium").

The leadership demands of the Third Reich, concerning aeronautical research, went, however, further. The suggestions of *Adolf Baeumker* with regard to the reformation of aeronautical research, which he had presented in March 1933, envisaged a stronger governmental control. Already in April 1933, the "Aeronautical Research Association" ("Vereinigung für Luftfahrtforschung, VLF"), directly attached to the RLM, was founded. This event was preceded by the installation of the "Center of Scientific Publications concerning Aeronautical Research" ("Zentrale für wissenschaftliches Berichtswesen über Luftfahrtforschung, ZWB") in March 1933. One of its objectives was the preparation of periodical report summaries, which were to serve the exchange of information between research and industry. With these institutions one approached the "model" NACA rather closely. The WGL was, also by these measures, systematically deprived of its scientific advisory and coordinating functions. Subsequently, WGL also increasingly dried up financially, members left; finally, it had to surrender its field of activity to its competitor, the VLF. In May 1936, WGL dissolved.

The VLF had already before the break up of the WGL received orders to extend its activities. VLF was, especially, to take over representative tasks up to now attended to by the WGL. After renaming it "Lilienthal-Society of Aeronautical Research" ("Lilienthal-Gesellschaft für Luftfahrtforschung") and the enthronement on June 12, 1936 under the patronage of the "State Minister of Aeronautics", *Hermann Göring*, it was thus established as the successor organization of the WGL. It was headed by *Ludwig Prandtl*, *Adolf Baeumker* and *Carl Bosch*. The Lilienthal-Society organized before the war, for the last time in 1938, with its so called "general meetings" international congresses of the highest scientific quality, thus also being a representation of German aeronautical research during the Third Reich [2]. Especially the committee meetings held during the war must be mentioned, where representatives of research and industry exchanged ideas about all aspects of aeronautical research. The Lilienthal-Society-reports about these meetings, mostly secrete at that time, give a good overview of the status of research at that time, which influenced strongly the developments of industry.

The Academic Upgrading of Aeronautical Research and the Problem of Junior Researchers

The establishment of the Lilienthal-Society was one element of *Baeumker's* plans, with the aspirations of the rulers in the background, to attain, scientifically and socially, a top position for German aeronautical research. A further element was the "Academy of Aeronautical Research", which was, originally, envisaged to head the Lilienthal-Society [1]. An "Academy of Engineering Sciences" was at that time, and would be today too, a novelty. Plans of the "State Minister of Education" ("Reichsminister für Erziehung") to create a "State Academy of Research" had just foundered. After succeeding in convincing *Göring* of this plan, the "Academy of Aeronautical Research" was a settled matter. However, it became organizationally independent.

The academy was officially founded on July 24, 1936. It had 40 full and 45 associate members. *Göring* was its president, *Erhard Milch* vice-president, and *Baeumker* chancellor. In its work, the academy initially ruled out subjects of military relevance in order not to endanger its foreign contacts, which existed via its corresponding members. The start of the encroachment of the Third Reich on neighboring countries (October 1, 1938 march on the Sudetenland, "Agreement of Munich") marks the commencement of the academy to address also military subjects with, in part, very specific research contracts. Beginning in 1938, the overall situation made many foreign members leave the organization. Beginning in 1939, the academy was politically pressured to dismiss members who were not "State (Reichs) Germans".

The World Economic Crisis had led, beginning in 1929, to a reduction in the number of students. Studying aeronautical technology was particularly uninteresting since the industry, on the one hand, was, due to the Versailles Treaty, restricted to civil aircraft engineering, and, on the other hand, the State was forced to save rigorously thus drastically reducing funding [35]. This situation clearly dimmed the professional chances in aeronautical engineering.

After the "Power Seizure" it was soon recognized that the lack of junior engineers and scientists would constitute a serious obstacle in the build up of aeronautical research and industry. Several steps were taken to overcome this problem.

For the build up of the education in aeronautical technology, a master plan [1] was devised in the autumn of 1933, which was to ensure that the estimated requirement of yearly 200 "Masters of Engineering" ("Diplom-Ingenieure") and 800 "Technical-college Engineers" ("Fachschul-Ingenieure") was met. "Faculties of Aeronautical Technology" were installed at the "Institutes of Technology" ("Technische Hochschulen, TH") Berlin, Brunswick and Stuttgart, with others to be added later; also installed were a number of individual chairs and technical colleges. However, these measures only slowly took effect and the need for personnel in research and industry remained high for a long time.

Two further steps must be mentioned, though they too only slowly became effective. For both, the DVL "Department of Junior Engineers" ("Abteilung für Ingenieurnachwuchs") was responsible. The "Academic Flying Groups" ("Akademische Fliegergruppen"), dissolved after the "Power Seizure", were reestablished as "Flight-technical Study Groups" ("Flugtechnische Fachgruppen") or "Working Groups", respectively. Within the eleven "Study Groups" at institutes of technology (TH) and nine "Working Groups" at technical colleges, members were educated in "engineering flight" ("ingenieurmässiges Fliegen") and practical design. The second measure was the creation of advanced studies, which led, subsequent to the regular studies, to becoming a "Aircraft Master Builder" ("Flugbaumeister"). Also here, the possibility of an advanced pilot's training existed.

Analogous to the problem of junior engineers or scientists was, for the research establishments, the problem of winning over personnel qualified for leadership, which had to come from universities or industry. With the introduction of "State Research Professorships" ("Reichsforschungsprofessuren") an additional incentive was given 1936 to accept, but also keep, leading positions at the research institutions. *Baeumker* had already suggested this in 1935. Many qualms and objections were the reason that only in

December of 1937 the first full and associate professorships were bestowed on leading scientists of DVL, AVA and DFL.

The scientific community was generally positively inclined towards the Third Reich [38]. This was basically related to the overall situation at the end of the twenties/beginning of the thirties. It seemed to the scientists that the "Weimar Republic" was not able to see the necessity of a strong research, especially in aeronautics, and to promote it accordingly. This changed with the "Power Seizure", especially since – subsequently – there was no bureaucratic control, as one could have expected, but a self-accountable responsibility of the researchers according to functional and cooperative principles, respectively [39]. The installation of the "Academy of Aeronautical Research", the research professorships and other attractive offers to the scientists also contributed to the positive attitude.

Ideologically motivated actions of those in power, respectively of subordinate agencies, against Jewish or foreign colleagues generally met the resistance of management. Although the latter might have more or less only been motivated by the desire to keep up operating conditions of the respective organization, it was, nevertheless, effective and prevented a drain in personnel as had occurred in physics and mathematics. Painful, though, for the German scientific community was the departure of *Theodore von Kármán* in 1934 from the TH Aachen to the California Institute of Technology (CALTECH). Though he had already contacts with CALTECH since 1926, the developments within the Third Reich might have enforced his decision, respectively even forced him, to leave Germany for California [40].

Aeronautical Research During World War Two

The Third Reich commenced its expansion politics openly in 1938. For that purpose, armament was pushed and the armed forces built up since 1936. However, this process was in 1939, at the beginning of the war, not yet completed. *Göring's* desire, expressed in 1935, namely, that German aeronautical research should have caught up, and even surpassed, foreign, and especially American, accomplishments by 1938, had not been realized [2]. Still, aircraft engineering had reached a level in 1937 that allowed to attain, for the first time, the speed record for land planes (as compared to sea planes). This was accomplished with a "racing aircraft" – derived from a production aircraft, a Me 109 equipped with an increased-performance DB-601-engine. In 1940, the research department of the RLM developed a five-year plan, which comprised as main topics: aerodynamics of high-speed, high-altitude and long-range flight, high-performance reciprocating engines as well as jet and rocket engines, weapons technology, radio navigation and measurement techniques, and materials.

After the foundation of the DFL in Brunswick in 1936, one envisaged now an additional aeronautical research establishment in the south of Germany, viz., the "Aeronautical Research Establishment Munich" ("Luftfahrtforschungsanstalt München, LFM"). The organization was to take into account the progress achieved, in the meantime, in jet-engine technology. This had consequences with regard to facility planning: large high-speed (according to today's diction "transonic") wind tunnels, a transonic engine-test wind tunnel for original-size engines, test stands for jet engines and rocket motors, etc. were foreseen.

The LFM was founded in 1940 in Ottobrunn near Munich on part of the property now occupied by EADS. The originally envisaged "Institute of Biological-physiological Aeronautical Research" ("Institut für biologisch-physiologische Luftfahrtforschung") was no longer pursued. The construction of a high-speed wind tunnel with a maximum Mach number of M = 0.9 and a diameter of 8 m (26 ft) for jet engine, but also aerodynamic investigations, commenced during the war in the Ötztal (valley of the Ötz in Austria). In Ottobrunn, a pilot tunnel with a diameter of 3 m (9.8 ft) was completed shortly before the end of the war. The already partly erected "Ötztal-Wind Tunnel" was, in 1946, moved to Modane in the Savoy Alps (France), completed and put into operation by the ONERA (Office National d'Etudes et de Récherches Aéronautique). It stimulated the later Arnold Engineering Development Center (AEDC) of the American Air Force in Tullahoma, Tennessee [41].

Work at, and the build-up of the research establishments was, at the beginning of the war, strongly handicapped. For German aeronautical research, there were, in spite of early recommendations, at the beginning of the war no mobilization plans whatsoever. A

Aircraft engineering and aeronautical research had reached in Germany in the mid-thirties a well-recognized high standard. A good example is the four-seated civil aircraft Messerschmitt Me 108 "Typhoon" (first flight in 1934), an aerodynamically high-quality all-metal design with slats, landing flaps, variable-pitch propeller and retractable landing gear.

Willy Messerschmitt (1889–1978) was as an engineer and businessman trend-setting through his successful design of high-performance aircraft, optimized for manufacturing including the first jet-powered aircraft built in quantity production.

personnel register did also not exist. In 1940, aeronautical research lost half of its personnel, which was either drafted or put to work in industry.

As background for this development one has to see the "Blitz-war strategy" ("Blitzkriegstrategie") of the Third Reich, which was to avoid a lengthy war. Research was essentially controlled through funding. Long-term projects were, in favor of so-called special tasks, e.g., ad-hoc developments for the air force, to be abandoned or postponed. Secrecy was tightened and work plans were secret and not allowed to be communicated to other research institutions or industry. Also all relations with foreign researchers had to be broken off. Initially, the activities of the "Lilienthal-Society" and of the "Academy of Aeronautical Research" largely ceased.

After the campaign against France, the "Blitz" ended in 1940. The air battle for England failed in the autumn of 1940. In 1941/42, one had to anticipate a longer duration of the war. In the autumn of 1941, planning of a reorganization of research control and funding commenced [2]. *Baeumker* suggested in early 1942 to generally switch from the political and bureaucratic control of research to a professional/technical self-control. A research council was to be attached to the "Technical Office" of the RLM. On this occasion, the standardization of the statutes of the research establishments, postponed in 1939 due to the beginning of the war, was resumed.

Mid-1942, the foundation of the "Aviation Research Command of the State Minister of Aviation and Supreme Commander of the Air Force" ("Forschungsführung des Reichsministers der Luftfahrt und Oberbefehlshabers der Luftwaffe, FoFü") was announced. The authority of the FoFü was far reaching [3]. From the specification of work priorities and the management of funds to the reorganization of the respective committees of the research institutions, all was in the hands of the researchers themselves. A research council, as initially proposed by *Baeumker*, and favored by *Prandtl* as complement to the FoFü, was not intended. *Prandtl* had desired such a council, with representatives of research and industry, as an institutional link between FoFü and research. Indeed, a "State Research Council" ("Reichsforschungsrat, RFR") was founded in 1942, which was, however, to be of no consequence to aeronautical research. Although the "State Minister of Aviation" ("Reichsluftfahrtminister") *Göring* was the president of the RFR, he insisted that "his" Aer-

Several large wind tunnels, which belonged to the "Aeronautical Research Establishment Munich" ("Luftfahrtforschungsanstalt München, LFM"), were built, but not fully completed before the end of the war, in the Ötztal (Tyrol), where sufficient water-power was available. Almost all facilities, but especially the 8-meter (26-foot) high-speed wind tunnel, were re-erected in Modane in the French Alps and put into operation by ONERA (this figure).

onautical Research was directly subordinate to him, independent of the RFR.

Members of the FoFü were *Ludwig Prandtl*, head of the AVA, as chairman; *Friedrich Seewald*, former head of the DVL, since August 1942 director and full professor at the Institute of Aerodynamics of the Institute of Technology (TH) Aachen, as deputy of the chairman and executive member; *Adolf Baeumker* as representative of the RLM; *Walter Georgii*, head of the "German Research Institute of Gliding" ("Deutsches Forschungsinstitut für Segelflug, DFS").

With the foundation of the FoFü, the structural problems of aeronautical research were, nevertheless, not yet resolved. The coordination of the research plans of the individual institutions suffered from the fact that no government work programs were provided that reflected the needs of the air-armament efforts. The drafting of scientists continued and the personnel shortage increased. However, the FoFü successfully developed a control concept with representatives for the individual subject areas of aeronautical research, and with "foremen" ("Obleute"), who managed, in combination, closely related subject areas and provided the contact to development and trial endeavors. Their main task was to secure the transfer of research results into industry and their development departments. The FoFü finally succeeded with this control concept, at least partly, to bring some systematics into research and to guide research according to the war-relevance of the individual topics.

The landscape of the aeronautical research establishments was, during the war, very polymorph [2]. The "classical" organizations DVL and AVA were, as discussed above, in 1936 joined by the DFL, and in 1940, the set-up of the LFM commenced. Further important institutions are listed in the following:

- "Research Institute of Motor Vehicles and Motor-vehicle Engines" ("Forschungsinstitut für Kraftfahrwesen und Fahrzeugmotoren, FKFS"), founded in 1930 by *Wunibald Kamm* at the Institute of Technology (TH) Stuttgart, located in Ruit near Stuttgart. The FKFS included a "Department of Flight Engines"[42], which considered, most of all, the development of large engines. Towards the end of the thirties, an altitude test rig for aircraft engines was set up. It was the desire of the RLM to combine the FKFS with the "Flight-technical Institute" ("Flugtechnisches Institut, FIST") of *Georg Madelung* at the Institute of Technology (TH) Stuttgart, which, however, failed due to the objections of *Klamm*.

- "German Research Institute of Gliding" ("Deutsches Forschungsinstitut für Segelflug, DFS"), which emerged from the "Research Institute of the Rhön-Rossitten-Society" ("Forschungsinstitut der Rhön-Rositten-Gesellschaft, RRG") in Darmstadt; it was headed by *Walter Georgii* [43]. It was in 1939 transferred to Brunswick and later to Ainring near Freilassing and Bad Reichenhall. During the war, it included institutes of Physics of the Atmosphere, Flight Techniques, Flight Equipment, Flight Mechanics, Flight Testing, and Special Tasks. At the DFS, *Alexander Lippisch* developed the tailless rocket-powered aircraft, which, after his transfer in 1939 to the Messerschmitt Corporation, underwent in 1941 as Me 163 "Comet" [44] its flight tests at the "Flight-test Center" ("Erprobungsstelle") Peenemünde-West/Usedom as a rocket fighter-aircraft.
- "Air Radio Research Institute Oberpfaffenhofen" ("Flugfunk-Forschungsinstitut Oberpfaffenhofen, FFO"), which emerged in 1937 in Oberpfaffenhofen near Munich from the "Radiotelegraphic and Air-electricity Test-station Gräfelfing" ("Drahtlostelegraphische und Luftelektrische Versuchsstation Gräfelfing, DVG") of *Max Dieckmann* [7]. Its areas of work were high-frequency technology, radio measurement technique (today's radar-technique), radar-jamming, radar-signatures of aircraft determined in model tests, antennae development, et cetera. [45].
- "Research Establishment Count Zeppelin" ("Forschungsanstalt Graf Zeppelin, FGZ"), founded in 1941 by *Georg Madelung* in Ruit near Stuttgart. The FGZ did not, say, work on problems related to airship technology but was rather strongly aligned to military topics: ballistics of bombs, mobile (land) aircraft catapults (take-off support), parachutes and braking parachutes, underwater detonations, torpedo stabilization. FGZ and FIST were both headed by *Madelung*.

At the beginning of the forties, the RLM looked for new resources for aeronautical research in view of the lack of personnel at the research establishments and the reduced chances, with exceptions (see the LFM), for new facility projects. These presented themselves,

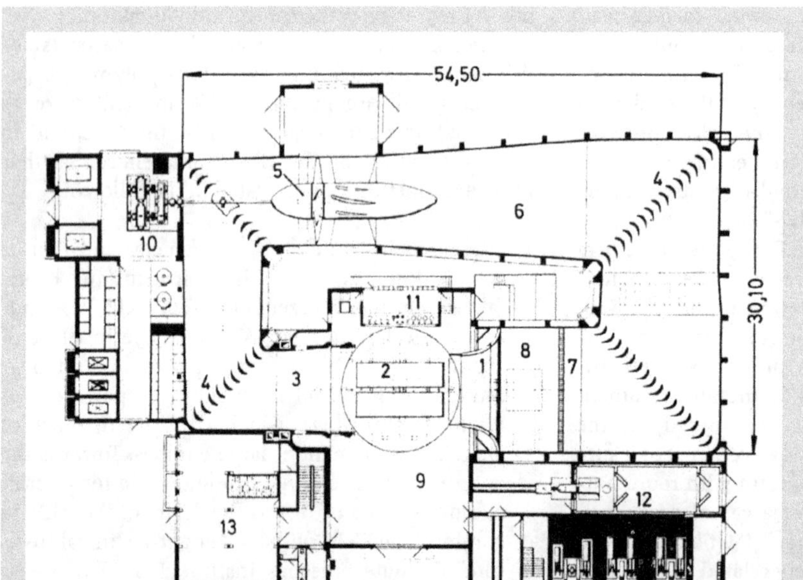

Plan view of the "motor-vehicle full-scale test field" ("Kraftwagen-Vollprüffeld"), with large wind tunnel and moving floor, of the "Research Institute of Motor Vehicles and Motor-vehicle Engines" ("Forschungsinstitut für Kraftfahrwesen und Fahrzeugmotoren, FKFS") in Stuttgart-Untertürkheim (1940). The wind tunnel was also employed for projects related to aeronautical research.

Employees of the Test Station Gräfelfing registered passing steamers on the Ammer-lake and measured their speed by means of the Doppler effect (radar principle).

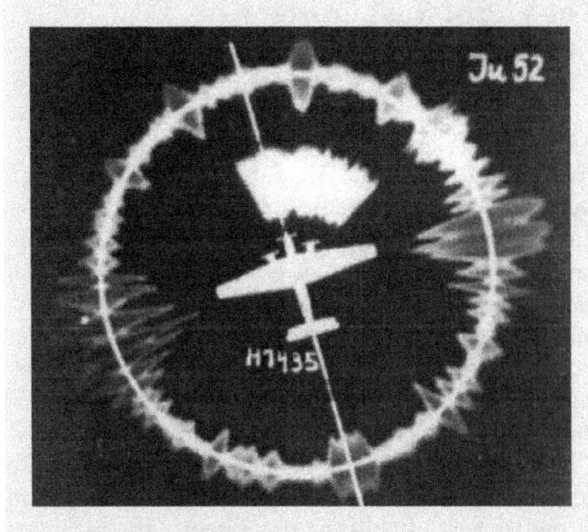

Radar reflections generated by a 1:10-scale model of a Ju 52 transport aircraft documented in 1940 at the Test Station Gräfelfing (from the DLR-news Issue 86/July 1997).

if only to a limited extent, in the annexed and occupied countries.

Already subsequent to the annexation of Bohemia and Mähren, the first step in that direction was made. The "Flight-technical Research Establishment Prague" ("Flugtechnische Versuchsanstalt Prag, FVP") was, in August 1939, given into the care of DVL [7]. The FVP worked with Czech personnel and under Czech management on projects defined by DVL.

The "Etablissement d'Expériences Techniques de Chalet Meudon des Ministére de l'Air", located near Paris, with the largest European wind tunnel (16 m × 8 m or 52 ft × 26 ft) [46] was in 1940 put under the management of DFL. As in the case of the FVP was the tunnel operated by French personnel and under French management. Besides developmental tasks for the German industry, research topics for DFL were treated.

The AVA attended since 1940 to the Dutch "Nationaal Luchtvaartlaboratorium, NLR" in Amsterdam. Here, even the board of directors, besides the management, remained in Dutch hands. A branch of the AVA, set up in Paris in 1941, employed the wind tunnel of the "Societé Française Hispano-Suiza", a French engine manufacturer, with a test section jet diameter of 5 m (16 ft), for its work [9].

The "Aeronautical Research Establishment Vienna" ("Luftfahrt-Forschungsanstalt Wien"), originally founded to develop jointly with the Messerschmitt Company a high-pressure gas-turbine engine, was governed by the LFM. *Alexander Lippisch* took over as head of this organization in 1943, after he had left Messerschmitt, to develop fundamentals for future "Delta"-configuration supersonic aircraft.

Efforts of the RLM to interlock aeronautical research at universities in Germany and the occupied countries with non-university research institutions in order to overcome the personnel shortages that way, had mixed success [2]. The successful cooperation in Berlin and Stuttgart was, similarly, strived for in Brunswick and Stuttgart. Attempts in Strasbourg, Prague and Posen, but also in Heidelberg (Aeronautical Research Heidelberg) brought, in the above sense, not the desired success or failed completely.

The Flight-Test Centers of the Air Force

Although the flight test centers for aircraft and aircraft equipment of the Air Force of the Third Reich were not research establishments in a strict sense, they are here briefly considered because of their proximity to the experimental development. Flight-test centers, or E-centers, which were in existence in Germany in this form since the beginning of World War One until the end of World War Two, were an important element between aircraft industry and their military customers [36]. While the "Certification Board" ("Prüfstelle") considered the operational safety of an aircraft, which then, possibly, led to the admission of the aircraft, established the E-center the technical (and, so to speak, also the military) viability of the aircraft with regard to the customer demands. The most important and largest Flight-test Centers during the Third Reich were [36]:

- **Test Center Rechlin, Mecklenburg,** with predecessor centers, since 1926 until 1945 concerned with testing aircraft, aircraft engines, propellers, jet engines, fuel and lubricants, parachutes, ejection seats.
- **Test Center (Sea) Travemünde, Sleswig-Holstein,** since 1928 until 1945 concerned with testing seaplanes, fighters, helicopters.
- **Test Center of the Air Force Tarnewitz, Mecklenburg,** since 1934 until 1945 concerned with testing aircraft weapons (armament) and rockets.
- **Trial and Test Center of the Air Force Peenemünde-West/Usedom, Pomerania,** since 1938 until 1945 concerned with testing aircraft and missiles with rocket propulsion, rockets for take-off assistance. heavy-duty parachutes.

Besides the above test centers, there were E-centers, also for winter trials, in France, Italy, Libya, Estonia, and, in addition, a number of test squadrons and test commandos.

The trials, as well as the type testing, took place with an "experimental specimen" ("Versuchsmuster" or "V-Muster") of the aircraft or with aircraft from the Zero-series, after the latter had already gone through company testing. In many instances, the trials, as also the type-testing, led to an optimization of the aircraft, if objections were, correspondingly, implemented by the manufacturer. However: "the Certification Board (Prüfstelle) favors regulations, while the E-centers are inclined to give guidelines in order not to obstruct progress" was formulated at a meeting of the "Lilienthal-Society" on "Flight Characteristics" in April 1943. The post-trials, and during large production series, the test of a new sub-series or of new armaments, e.g., the integration of new weapons systems, reflect this iterative and interactive process at that time.

The E-centers especially came into play in the case of invitations to tender – determining the degree of compliance – and when the "specimen aircraft" of competing companies had to be evaluated. Competitive flights or the comparative trials, respectively, were only performed in the case of fighter planes. Here, different approaches existed: the analytical method, which evaluated individual flight segments, such as climb, horizontal flight, banking, etc., and a method, where, as in air combat, one situation led to another, which were subsequently evaluated.

During regular trials, the flight performance and flight characteristics of an aircraft were determined by formal, but flexible approaches (see Part 3). This corresponds to today's "parameter identification", or largely to today's "system identification" of an aircraft [47]. For this purpose, a, for the time, advanced measuring technique was developed. At that time, one also already experienced the "Man-Machine" problem, as it is called today, which may be generated by adverse interactions – in both directions – between pilot and aircraft.

The flight-test centers were particularly important at the time of the Third Reich, but especially so during the war. This was due to the fact that, at that time, on the one hand large technology thrusts occurred, for instance, the introduction of the jet engine and new wing shapes. On the other hand, new aircraft developments – which, in part, had hardly left the experimental-development stage – of a great many competing aircraft manufacturers [37] crowded into the test centers, despite of or because of the war. There were certainly also during the World War One large technology thrusts and many competing companies, while today's situation is again completely different. Today, there are, at least within Europe, individual aircraft companies within countries, which evolved through a long consolidation process, and which are now in the process of merging into a European corporation. The technology thrusts are, however, still large, examples

The Messerschmitt fighter plane Bf/Me 109 was built in large quantities with a total production 35,000 aircraft. The version G-6 reached, at an altitude of 7,000 m (22,960 ft), a maximum speed of 630 km/h (390 mi/h), the Series K towards the end of the war even 680 km/h (420 mi/h).

The final stage in the development of the Fock-Wulf fighter plane Fw 190 was reached in 1944 with the high-altitude fighter plane Ta 152 H equipped with a pressurized cockpit. The Ta 152 H, developed by *Kurt Tank*, had a Jumo 213 E engine and reached a maximum speed of 720 km/h (447 mi/h) at an altitude of 10,500 m (34,450 ft).

being the step from the "Tornado" to the "Eurofighter" with the introduction of artificial-stability flight with the delta-canard configuration, or the advance of digital data acquisition and processing in avionics.

If you are a Visionary ...

... and do not think about the purpose,...you will accomplish something [in business]..., thus the author *Robert Musil* was advised in a conversation in 1914 by the industrialist and later politician and publicist *Walther Rathenau* [48]. How can one judge, from a scientific-technical standpoint, the results of aeronautical research achieved during the Third Reich, and what were the reasons – this shall be said in advance – for its success? These questions shall be discussed below, proceeding from the three categories of "research" introduced in the first part of the present book:

- Basic Research,
- Applied Research,
- Experimental Development.

Before, however, some words about the freedom of research, which is, like any freedom, on the one hand, a freedom "to", and on the other hand, a freedom "from". *Theodore von Kármán* formulated three criteria with regard to the freedom of science [49]:

- Freedom in the choice of topics within a field of research.
- Freedom in the choice of co-workers in the pursuit of solutions to a given problem.
- Freedom to carry out scientific discussions and to publish the results of the scientific work.

This is the freedom "to", the freedom "from" will be addressed later.

Basic Research

The "Mathematization" of empirical sciences was, at the beginning of the twentieth century, primarily introduced and advanced by *Felix Klein* in Göttingen. Within aerodynamics, this process had with *Ludwig Prandtl* its outstanding protagonist. "*He (Prandtl) was clearly the father of modern aerodynamics and a monumental figure in fluid dynamics. The impact of his work will reverberate for centuries to come.*", constitutes *John David Anderson* in his "History of Aerodynamics" [50]. *Prandtl* put down the fundamentals of theoretical/scientific aerodynamics during the first three decades of the twentieth century. He also worked successfully in other areas, relevant to aeronautics, such as in applied mechanics (elasticity, plasticity, rheology), in aeroelastics and meteorology.

In the years of the Third Reich his school of thought and working in Germany was the basis of fundamental research in aerodynamics, which played, at that time, together with lightweight construction and engine technology, the primary role (today, avionics has joined these three classical disciplines). Although *Prandtl's* main place of work, the AVA, was a non-university research establishment, his connections to the universities, also via his former students, were strong. Here, especially Aachen has to be mentioned [51]. Although fundamental research in aerodynamics, but also within other disciplines, was restricted during the war, the advancement towards high-speed flight has initiated work at research institutions and universities on a number of subjects, which partly only after the war, and frequently abroad, led to results.

Applied Research

Applied research was, in principle, the domain of the research establishments. They represented, personnel- and equipment-wise, the center of German aeronautical research. Applied research was finally, after the built-up of the classical organizations, like DVL and AVA, and the set up of new organizations, such as the DFL, and despite the shortages in personnel, which became more severe during the war, and despite the stalling of the construction of new facilities, qualitatively in an international comparison at a very high standard. Several investigative committees of the Allies confirmed this after the war. An indication of this situation was also provided by the fact that, after the war, a considerable number of scientists found, voluntarily or forced, a new field of activity at foreign research establishments. A certain deficiency only existed in high-frequency technology and measuring techniques.

Experimental Development

The experimental development, as a domain of the aircraft industry, including the propulsion and equipment industry, was, especially during the war, definitely foremost in the world, and has, directly or indirectly, lastingly determined the development of aeronautics (and astronautics) in the West as well as in the East. The owners of the aircraft companies were, at the beginning of the Third Reich, almost all designers with a high innovation potential [2], and their enterprises remained, up to 1945, also characteristically "inventor enterprises". This might have been problematic in cases where the role of the enterprise as a production company was impaired, which was, however, obviously the exception [5]. Also the industrial managers and their co-workers were, after the war, in demand in foreign countries and worked, forced or voluntarily, all over the world.

Now, what were the reasons? Were the criteria of *von Kármán* and the dictum of *Rathenau*, at least in the sense of the words, met? Yes, it was obviously to an adequate degree the case [39]. Aeronautics was, from the very beginning, promoted by the new rulers, even if these rulers, according to *Baeumker*, never completely understood the implications of an air war [52]. Important is, however, in this context the person of *Adolf Baeumker*. He has with his farsighted strategic approach set the course for the expansion of aeronautical research and succeeded in realizing his vision, with adaptability and ambition, even exploiting *Göring's* vanity, to a large degree [2]. Thereby, aeronautical research had gained the freedom "from", namely, from adverse political and financial constraints. Its capabilities were appreciated and desired, and they were supported all the way up to the experimental development. Self-mobilization commenced [38], also *Baeumker* was in that sense self-mobilized, and success was achieved soon.

Of course, aeronautical research was no island of the blessed. Fights for power and power conservation, influence and independence were carried out in research as well as in industry. One had to ward off undesirable political intervention, frequently with the aid of *Baeumker*, and there were, at all levels, personal tragedies; one may remember *Hugo Junkers* being forced out of his enterprise, and the suicide of *Ernst Udet* in November 1941. *Baeumker* himself entered in 1941 a lengthy health crisis.

Von Kármán's first and second criterion applied, of course, only to leading personnel at universities, research establishments and industry. As long as there is no incompetence, no lack of concepts, or conservatism within management, this is acceptable. Important is to run, conceptually, on two tracks (or stand on two legs), otherwise one will end up inside an ivory tower away from reality. Universities have to know the daily problems of industry so that they can adjust their teaching accordingly. At the same time, at least the longer ranging developments within the respective fields should be reflected in the teaching subjects and in fundamental research.

Research institutions have to know, and support to a certain degree, the daily business of industry, in order to be able to align the main task of the organization, i.e., the medium- and long-time applied research, in such a way that it may becomes effective later. Pure contract work or services should be organizationally separated. The industry must, finally, always also carry out sufficient research – besides the experimental development – to be able to communicate with the research institutions. If this is not at all, or only partly, the case, the entire system in the end will become second or even third rate.

How then was the interplay between research and industry during the Third Reich? Obviously not bad since the success of the industrial experimental development would not have been possible without the existing research environment. The structural or organizational problems of aeronautical research, which led in 1942 to the introduction of the FoFü, have already been addressed above.

The answer is different (from the one just given) if one asks for the importance of aeronautical research to the war efforts during the Third Reich. The accusations are aplenty: duplication of work, too many things done at once, lack of contact to industry and air force, et cetera. The legitimacy of these accusations, which are certainly partly valid – with faults on all sides – shall not be followed up here. A linearly deterministic relation between research and industrial implementation of the results is, however, always rather the exception [2]. It may in certain situations – for the industry almost always – be desirable. Such a relation should, however, be given if research, as applied research, is far advanced, otherwise there exists a fundamental error within the system, which immediately must be eliminated.

The fighter and bomber aircraft of the Allies, especially of the Americans, were during the second half of the war, as far as conventional aircraft were concerned, superior to the German ones[1]. This held already in the thirties for transport aircraft; only in 1937 was it possible to introduce with the Fw 200 "Condor" a superior airplane.

In the second half of the thirties began in Germany the substitution of the old aerodynamic aircraft shapes and piston-type engines by new aerodynamic shapes and jet engines. At the end of the war, Germany had a lead of several years [53a]. It is interesting to note that German research and developments were hardly noticed abroad. Especially in the USA was the window of perception essentially closed, which may be partly explained with a, in the short-term successful, conservatism and an exaggerated practical orientation.

Von Kármán complained resentfully about the fact that the USA missed, since 1938, the development of jet engines, which can be traced to a gross misjudgement by the experts [53]. The principle of the swept wing was investigated in the USA in the middle of the forties by *Robert T. Jones* [50], but only when the application of this principle by the Germans became

[1] The turbo charger, whose introduction in Germany was, due to a lack of high-temperature materials, not possible, played here an important role.

The world's first turbo-jet aircraft, the Heinkel He 178, took off on August 27, 1939 in Rostock with *Erich Warsitz* at the controls.

The Messerschmitt Me 262 (first flight on April 18, 1941) was the first military jet aircraft. It went into operation at the western front beginning in October 1944. The maximum speed was 870 km/h (540 mi/h) at an altitude of 10,000 m (32,800 ft).

A top-quality product of German aeronautical research and development was the long-range reconnaissance and bomber aircraft Arado Ar 234 (first flight in July 1943) with Jumo 004 turbo-jet engines. Until the end of World War Two, more than 200 planes were built. The maximum speed was 760 km/h (470 mi/h).

The rocket-powered fighter Messerschmitt Me 163B "Comet", a development of *Alexander Lippisch*, went into operation for the first time during the summer of 1944. Its maximum speed was 960 km/h (597 mi/h).

The turbo-jet engine BMW 003 was built in quantity production since the middle of January of 1944, some months later than the Jumo 004 engine. In Germany began, at this point, the era of jet engines and jet aircraft.

In many aspects progressive was the transport aircraft Focke-Wulf Fw 200 "Condor" (first flight in 1937) for 26 passengers. Until the begin of World War Two, 16 aircraft for civil use were built.

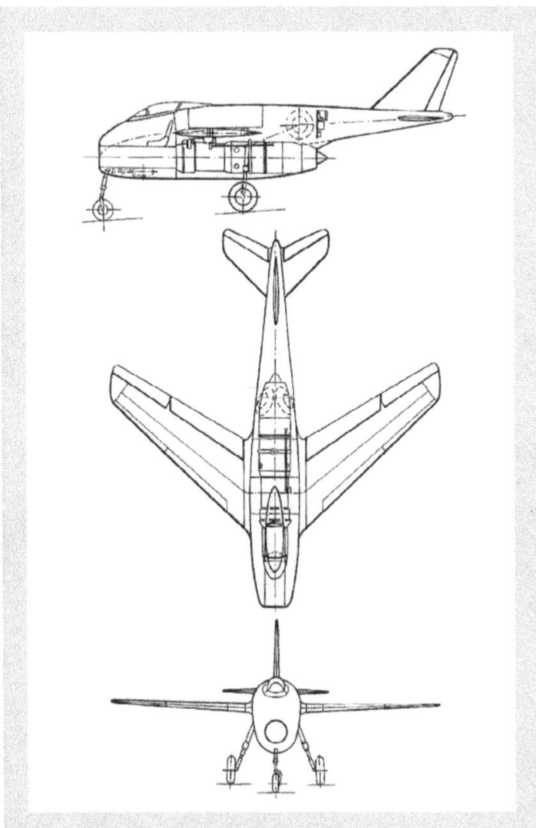

The Messerschmitt P 1101 was the first swept-wing fighter aircraft worldwide. The prototype was, at the end of World War Two, shortly before completion. The Americans discovered the aircraft on April 29, 1945 – 80 per cent complete but without the He S 011 engine – in Oberammergau, Bavaria.

known, led this to an application in American aircraft design. Apparently had in American aeronautical research, as well as in industry, contrary to German research, empirical experience priority over theory [50]. This was, for conventional-aircraft design, certainly one of the elements of success. The USA, however, assumed after 1945 very soon the lead in military and civil aircraft technology.

The Reconstruction of German Aeronautical Research after 1945

Ernst Heinrich Hirschel

Important Applications of Research Results in Post-War Aircraft Design

Before describing the environment of German aeronautical research after 1945, technical developments in aircraft design that have occurred during the second half of the twentieth century shall be briefly addressed. These developments are supported by basic research at universities, applied research at research institutions and the industrial experimental development, in total by research.

The international European interrelations played, for West German aircraft manufacturing, after 1955 an important role. All products involving quantity production and large financial volumes, from the Airbus-family to the fighter aircraft "Tornado" and "Eurofighter", were developed by European consortia. Though sole German developments existed, they had generally a hard time to prevail, except for some helicopter developments and, especially, high-performance glider developments.

If one asks for the most important technical developments in aircraft engineering during the second half of the twentieth century, one has, firstly, to note that many of these developments had their roots in the time period before 1945 and frequently in Germany – this will become visible in the detailed depictions in Part 3 – and that, secondly, aircraft engineering commenced in (West-) Germany only after she, in essence, regained her sovereignty in May of 1955. Although German engineers developed between 1945 and 1955 aircraft in many countries (see Part 3), there prevailed in the Federal Republic of Germany, due to the restrictions imposed by the Allies, after 1955 initially a deficit compared to the USA and the Soviet Union, but also compared to Great Britain, France, Sweden, Italy and the Netherlands.

Now, what are the most important technical developments and in what products were they utilized? Considering civil transport aircraft, these are the swept wing with a transonic (initially called supercritical) airfoil, the jet engine and the pressurized cabin [11]. These were finally also applied, in various combinations, in (West) Germany on the HFB 320 of the "Hamburg Aircraft Company" ("Hamburger Flugzeugbau, HFB") in the form of forward-swept wings, on the VFW 614 of the "United Flight-technical Company Bremen" ("Vereinigte Flugtechnische Werke Bremen, VFW") with the engines on top of the wings, and on the 152 of the "VEB Aircraft Company Dresden" ("VEB Flugzeugwerke Dresden") in East Germany. They are, today, a common characteristic of all aircraft of the Airbus Consortium, but also of the jet aircraft of Fairchild-Dornier.

The most important technical developments concerning military-aircraft are as well the swept wing, but also the "thin" wing, the jet engine with afterburner and the pressurized cockpit. In addition, there is the variable-sweep wing (Tornado), the delta wing (Eurofighter), the application of the area-rule – today also applied to transport aircraft, but generally not directly recognizable – the Vertical Take-off-and-Landing (VTOL) technology (in Germany not in operation), the unstable flight (Eurofighter), the flight at high angles of attack ("post-stall flight"), and the low-signature

Some products of the German aeronautical industry built in quantity production after 1955. (Note that the numbers in parentheses give the year of the first flight and the number of air vehicle built, respectively.)
From left top to bottom: HFB 320 (1964/48), Bölkow BO 105 (1967/1481), Airbus A 320 (1987/1481 until September of 2000), Dornier/Dassault "Alpha Jet" (1973/500+). *From right top to bottom:* VFW 614 (1971/12), Bölkow "Phoebus" (1964/255), Nord Aviation/Weser Flugzeugbau/HFB "Transall" (1963/204), Panavia "Tornado" (1974/992).

technology (radar, infrared) based on contouring and other measures. All these techniques were investigated in West Germany in several experimental-development programs with the corresponding aircraft such as: the VTOL [54] aircraft VJ 101 of the "Development Group South" ("Entwicklungsring Süd") and the VAK 191 of the "United Flight-technical Company" ("Vereinigte Flugtechnische Werke, VFW"), the CCV-F-104G demonstrator of Messerschmitt-Bölkow-Blohm (MBB), Ltd. (CCV: control-configured vehicle for artificial-stability flight), as well as the X-31 of MBB and the North American company Rockwell International (post-stall flight). Some were subsequently utililized [55, 56] on the "Alpha-jet" of Dornier and the French company Dassault-Breguet, on the "Tornado" of the tri-national (Germany, Great Britain, Italy) "PANAVIA Aircraft, Ltd.", and finally on the "European Fighter Aircraft (EFA, Fighter 90 Typhoon)" of the four-nation (Germany, Great Britain, Italy, Spain) "Eurofighter Fighter Aircraft, Ltd".

Concerning helicopter technologies, primarily pure (West) German developments in rotor technology (main and tail rotors) must be mentioned [57]. With the progress in fiber construction (glass fiber), MBB developed in the sixties a rotor with joint-free blade connections, which were then employed on the BO 105 and BK 117 helicopters. In the seventies and eighties followed the bearing-free rotor utilized on the BO 108 and the EC 135 of "Eurocopter", the German-French helicopter company which came out of MBB and Aerospatiale. Finally followed the fiber-elastomer-(rubber-like polymeric substance)-rotor, that was used, for instance, as no-joint rotor in the combat helicopter PAH 2 "Tiger".

Fiber-reinforced construction, which allows very large aspect ratios and extreme smooth surfaces, were the most important technical developments, which, together with laminar airfoils, led to sailplanes of an outstanding aerodynamic quality [31]. Germany has, with a whole series of manufacturers and academic flight groups, today almost a monopoly in high-performance glider engineering.

Last not least, the development in the information technologies (electronic data acquisition, storage, processing and transfer) during the second half of the twentieth century shall be mentioned; they, primarily, made modern aeronautics possible with such features as "fly-by-wire"-systems, the electronic cockpit, flight control and navigation systems, artificial-stability flight (at first only for military aircraft), simulation and training systems, et cetera. Information technologies play, today, also a large role in aircraft design (see the discussion of the "Virtual Product" in Part 4). The development of computer performance and capacity, but also of the numerical algorithms, will lead, within the definition and development processes, step by step, to faster and more accurate mathematical descriptions and optimizations of the physical characteristics and functions of the system "aircraft" and its components. This will lead to a reduction in development costs and risks at a simultaneous increase in product quality. Similar effects of the development of the information technologies can be observed in the automobile industry (crash tests etc.), the chemical industry (molecule-modeling), but also in meteorology (weather and climate predictions) and in other areas.

The development of the environment of German aeronautical research after 1945 shall now be sketched. Note that this is essentially, since 1961, the West German aeronautical research, which, increasingly took place within a European network.

The New Beginnings of Aeronautical Research Within the Divided Post-War Germany

The end of World War Two meant the end of German aeronautical research at universities, research institutions and industry. It is remarkable that this does not hold across the board and also not totally [2]. There were also differences between the two German states evolving later.

Warring parties are always very interested in the equipment of the enemy, its performances, its strengths and weaknesses. The own equipment has to be improved accordingly, tactical adjustments have to be made, the own research and development has possibly to be redirected fast and effectively. Military reconnaissance was in that sense on all sides highly developed. Lucid World-War-Two examples are given by *Eric Brown*, who during the war and thereafter has flown, investigated and assessed German aircraft, which had either to make forced landings in England or were highjacked by commando raids from the parts of France, which were occupied by Germany [58]. With the end of World War Two, all victors took the chance to thoroughly familiarize themselves with the knowledge and results of German aeronautical research [59], especially since the war in the Pacific against Japan had at that time not yet ended.

The Allies had already planned during the war various measures to get hold of persons, documents and equipment of importance to aeronautical technology. Different "teams" were formed. One of the teams, of more interest to the later Federal Republic of Germany, was the "Combined Intelligence Objectives Subcommittee, CIOS" of the Anglo-Americans, which had the task to collect information about military-relevant knowledge in the areas of aeronautical technology, rocket technology, etc., but also to localize and question German scientists and engineers [60]. (The "hunt" for German scientists and engineers led to voluntary and forced emigrations of considerable dimensions: between 1946 and 1950, approximately 500 people stayed in the USA, 2,500 to 3,000 in the USSR, 2,000 to 3,000 in France and roughly 200 to 800 in Great Britain [61]. The activities of the aeronautical scientists and engineers abroad after 1945 are briefly discussed in Part 3.) Already at the end of May 1945, the necessity of a better coordination of the various activities became obvious. The "Field Information Agency Technical, FIAT" was founded, into which the CIOS-teams were integrated.

The CIOS-teams advanced initially with the fighting troops to execute their mission all over Germany. Known is the story of *Theodore von Kármán* discovering at the DFL in Brunswick the activities concerned with the swept wing, which soon should become *the* characteristic of military combat aircraft and later also of civil aircraft [53].

The CIOS/FIAT-questioning also covered the company design teams as, e.g., the Focke-Wulf team, questioned in May 1945; the protocol of this interrogation is even today interesting reading for the expert [62]. Also of great significance was the directive to the German scientists to reveal their knowledge. Within the British occupation zone, the "Göttingen Monographs" [63] were written at the AVA in 1945/46 by order of the British Ministry of Supply, the "Brunswick Monographs" at DFL. Similar things happened in the Soviet occupation zone.

The Allies soon came to the conclusion that within the defeated Germany basic research should be continued with as little restriction as possible, whereas applied research should be strictly controlled. This course of action was initially favored primarily by Great Britain and the Soviet Union, but later also accepted by the USA and France, who first wanted to prohibit research all together. The Control Council Law No. 25 of April 29, 1946 then prohibited any research of a military nature. Prohibited was work related to applied aerodynamics, rocket and jet propulsion, positioning and remote control of missiles, et cetera. Institutions and installations of German aeronautical research, confiscated by the Allies after the war, were later returned into German custody step by step, unless they had been taken abroad as reparation.

In Göttingen, employees of the AVA and the "Emperor-Wilhelm Institute of Fluid Research" ("Kaiser-Wilhelm-Institute für Strömungsforschung") were initially prohibited to enter their work area; later all activities had to be stopped. However, this was annulled for the scientists who contributed to the monographs. The AVA was dissolved in 1947 in favor of an enlarged "Emperor-Wilhelm Institute of Fluid Research within the KWG" ("Kaiser-Wilhelm-Institut für Strömungsforschung in der KWG"), with the KGW transformed in 1948 into the "Max-Planck-Society" ("Max-Planck-Gesellschaft, MPG"). After the return of its assets in 1953, aeronautical research could be resumed within the AVA, now called "Aerodynamic Test Establishment within the Max-Planck-Society for the Promotion of Science" ("Aerodynamische Versuchsanstalt e.V. in der Max-Plank-Gesellschaft zur Förderung der Wissenschaft"). It was now headed by *Albert Betz* and remained a member of the MPG until 1968.

For the DVL the situation was different. Due to the advancing Soviet troops, the board of directors of DVL decided in February 1945 to terminate work in Berlin-Adlershof. A number of institutes had already been transferred, mostly to the southern part of Germany. In April 1945, the Red Army marched in. Leading staff and some employees, who had stayed behind, were either deported or found work elsewhere. The facilities in Adlershof, and within the Russian occupation zone, were either shipped to Russia or destroyed. Institutes, which had been transferred, for instance, to the French occupation zone, suffered a similar fate. Personnel and facilities were transported to France. Only within the American occupation zone, the institutions remained. They worked, partly until 1946, on behalf of Allied committees.

August Wilhelm Quick (1906–1982). Symbolic presentation of the key to the preliminary home of DVL at the airfield Essen-Mühlheim on April 10, 1957; *A. W. Quick* (right) and State Minister President *Steinhoff*.

August Wilhelm Quick, born in Oese, county of Bremervörde on January 25, 1906.

1926 starts his studies at the Institute of Technology (TH) Danzig and becomes there in
1929 research assistant at the chair of "Air-vehicle Engineering".
1932 changes to the "Junkers Aircraft and Engine Company" in Dessau.
1939 becomes institute head at DVL in Berlin-Adlershof.
1940 appointment as (civil servant) full professor in the services of the State (Reich).
1942 becomes deputy member of the board of directors of DVL and corresponding member of the "Academy of Aeronautical Research" in Berlin.
1946 changes to the "Société Nationale d'Etude et de Construction de Moteurs d' Aviation, SNECMA" in Paris.
1950 becomes head of the gas-turbine department.
1952 becomes head of the combustion department and, during the same year, receives his PhD from the Institute of Technology (TH) Darmstadt.
1954 appointed to the "Chair of Aeronautics" at the Institute of Technology (RWTH) Aachen. At the same time he concerns himself with the revival of DVL.
1956 becomes national delegate of the German Federal Republic at AGARD; he holds this position until 1960.
1958 it is decided to establish the DVL-center in Porz-Wahn (near Cologne/Bonn airport), today's location of the head office of DLR and one of its main centers. *Quick's* main concern was the incorporation of astronautical research into the field of activities of DVL.
1960 he gives the 24th Wright-Brothers-Memorial Lecture in Washington and Los Angeles.
1961 the "Institute of Astronautical Research" ("Institut für Raumfahrtforschung") is founded in Porz-Wahn.
1963 he receives the "Croix d'Officier de la Légion d'Honneur".
1975 he receives the "Ludwig-Prandtl-Ring".
1982 he dies in Aachen on April 27.

In 1947, the Office of the Military Government of the United States (OMGUS) confirmed that the DVL did not belong to the organizations dissolved by the Allied Control Council. However, aeronautical research was prohibited. During the following years, members of the "Supervisory Board" ("Aufsichtssausschuß") tried to free the assets of DVL. They tried to re-establish the DVL in some of the states of the Federal Republic of Germany, established in 1949. In the spring of 1951, changing the statutes and electing a new supervisory board finally secured the continuance of DVL in West Germany. Of the 110 members of DVL, mainly companies, only 34 participated in the vote.

The actual research work was taken up again in 1951 in so-called "Working Groups". Since 1946 various former members of DVL accepted chairs at the "Rhine-Westphalia Institute of Technology Aachen" ("Rheinisch-Westfälische Technische Hochschule Aachen, RWTH"). A part of the working groups were guests at the corresponding RWTH institutes, mostly with the same person heading the DVL working group and the corresponding institute. At the time, the following working groups existed:

- since 1951 "Aerodynamics" and "High-frequency" in Aachen,
- since 1952 "Flight Medicine" and "Material Research" in Bonn and Aachen, respectively,
- since 1953 "Thermodynamics" and "Fuel and Lubricants" in Essen-Mühlheim/Garmisch-Grainau, respectively Munich,
- since 1954 "Turbo Machinery" and "Machine Dynamics" in Aachen.

Beginning in 1954, the "Working Groups" were converted into institutes and new institutes were founded. The DVL was, at least initially, primarily located in Aachen and attained under the management of *August Wilhelm Quick* and *Hans Ebner* successively the form, which it, more or less, kept until it merged into the "German Aerospace Research and Test Establishment" ("Deutsche Forschungs- und Versuchsanstalt für Luft- und Raumfahrt, DFVLR") in April of 1969.

The DFL, from 1938–1945 named "Aeronautical Research Establishment Hermann Göring" ("Luftfahrtforschungsanstalt Hermann Göring, LFA"), largely terminated its work in Brunswick-Völkenrode in May 1945 [64]. Some employees participated in writing the monographs already mentioned, but some facilities also went back into operation. This lasted until 1947/48; then the installations were dismantled or destroyed and grounds reserved by the British military government for other utilizations.

The registered organization DFL, however, existed further. In the summer of 1952, first steps were undertaken to revive DFL (installation of an "emergency board" with *Hermann Blenk* and *Fritz H. Schmidt*). The General Assembly decided in July 1953 on new statutes and elected the governing bodies, so that DFL again became fully operational. Former members of DFL had after 1945, similar to the situation at DVL in Aachen, become full professors at the Institute of Technology (TH) Brunswick. They actively supported DFL. Finally, in September 1953, the prerequisites were given to resume the research activities. The six institutes were headed, in a personal union, with the corresponding institutes of the TH. However, the DFL no longer had the use of its grounds in Brunswick-Völkenrode. But since the TH-Brunswick was no longer interested in its property at the airport Brunswick-Waggum, DFL could resettle there.

Besides the DVL, AVA and DFL, which were the German aeronautical research institutions with the greatest tradition and visibility, there were others, which also re-established themselves in West Germany after 1945, such as the "German Research Establishment of Gliding" ("Deutsche Forschungsanstalt für Segelflug, DFS"), 1953 in Munich, the "Air Radio Research Establishment Oberpfaffenhofen" ("Flugfunk-Forschungsinstitut Oberpfaffenhofen, FFO"), 1954 also in Munich. New foundations were the "German Study-Association Helicopter" ("Deutsche Studiengemeinschaft Hubschrauber, DSH"), 1953 in Stuttgart and the "Research Establishment of Jet Propulsion Physics" ("Forschungsinstitut für Physik der Strahlantriebe, FPS"), 1954 also in Stuttgart, with the area of responsibility, so *Eugen Sänger*, "basic and applied research in the interim area of aeronautics and astronautics". All these institutions were later to merge into DFVLR.

An interesting situation arose at the West German Institutes of Technology and Universities, respectively, in that, after 1945, employees of the former research establishments were already very soon appointed professorships there. This applied particularly to the RWTH Aachen, where, since 1946, former State (Reichs) professors held full professorships, but it also applied to the TH-Brunswick (DFL members) and others. Obviously these appointments were related to the desire of the (later) West German federal states to secure personnel, within the context of their regional economic structural policy, for the re-implementation of aeronautical research institutions, particularly since the abolition of the research bans seemed, since the beginning of the fifties, to be near. The flight ban for gliders was lifted in 1951, which led to the re-establishment of the academic flying groups, which then became frequently engaged, with great success, in the experimental development of gliders.

Considering all the effort concerning the revival of West German aeronautical research, it takes no wonder that the desire rose to also restore the "publicity" of aeronautics research. The "Scientific Aeronautical Society, WGL" had dissolved itself in May 1936 in the course of the national-socialistic "Gleichschaltung" (elimination of organisations outside the party realm) [65]. In its place, the Lilienthal-Society had been established, but it was also dissolved at the end of the war. In April of 1952, the TH Brunswick seized the initiative. It invited to an aeronautical conference, whose preparation was carried out jointly with the "Association of German Engineers" ("Verein Deutscher Ingenieure, VDI"). On the occasion of this conference, the WGL, since 1967 "German Aerospace Society" ("Deutsche Gesellschaft für Luft- und Raumfahrt, DGLR"), was founded anew. Of the two initiators of the conference, *Hermann Blenk* and *Heinrich Koppe*, the former was elected president. (On the occasion of the 75th birthday of the society, *Peter Hamel* held, in 1987, an interview concerning the history of the DGLR with the contemporary witness *Hermann*

Eugen Sänger (1905–1964).

Blenk. Many historical details of the period between 1912 and 1987, not published in this form before, have been recorded during this interview [66].)

The missions of the WGL acccording to its bylaws were again the publishing of brochures, research reports, lectures, et cetera. In continuation of the tradition of the old WGL, the "Journal of Flight Sciences" ("Zeitschrift für Flugwissenschaften, ZFW") as well as the WGL-yearbooks were published as of 1952. Due to an initiative of the WGL, the "Central Office of Aeronautical Documentation and Information" ("Zentralstelle für Luftfahrtdokumentation und- information, ZLDI") was set up in 1954.

The WGL had already very early contacts with the "Advisory Group for Aeronautical Research and Development, AGARD" of NATO, founded on the initiative of *Theodore von Kármán* in 1952 [67], where the Federal Republic of Germany became a member in 1955. These contacts were, at that time, intensified. The WGL formed during the fifties, following the tradition of the old WGL, special committees to, among other things, also advise the Federal Ministries of Transportation and of Defense, respectively, with regard to the promotion of research.

This then led to the question of how to finance the slowly recovering aeronautical research activities in West Germany. A financing by the federal states, of which some wanted to re-establish DVL within their state, was, due to the high financial requirements, only possible in a limited way [2]. It became soon clear that the Federal Government had to bear the brunt of financing the re-establishment of the research organ-

Eugen Sänger, born in Preßnitz (Bohemia) on September 23, 1905

1923 he began his studies at the Institutes of Technology (TH) Graz and Vienna in civil engineering. After the rejection of a thesis-outline on "rocket flight technology", he received in

1930 his PhD with a conventional topic concerning airplane statics from the TH Vienna. He summarized his preceding considerations in the book "Rocket Flight Technology".

1937 for the DFL (Brunswick), he sets up the "Research Branch of Rocket Flight Technology" ("Raketenflugtechnische Forschungsstelle") in Trauen in the Luneburg heath.

1943 *Eugen Sänger* has to continue his research at the "German Research Establishment of Gliding" ("Deutsche Forschungsanstalt für Segelflug, DFS") in Ainring on instructions by the RLM. He summarizes the results in a report about "Rocket Propulsion for Long-range Bombers", written together with *Irene Bredt*, his later wife.

1945 he follows the invitation of the French government and demonstrates to the Arsénal de' l'Aéronautique and several companies the key tests for the preparation of the ram-jet test aircraft "Griffon".

1950 he publishes a manuscript "Concerning the Limits of Space Travel".

1951 due to the supra-national tasks in aeronautics, he commits himself to the implimentation of an "International Astronautical Federation", which is then founded in London.

1954 he founds, on invitation of the Federal State Baden-Wurttemberg, the "Research Establishment of Jet Propulsion Physics" ("Forschungsinstitut für Physik der Strahlantriebe") in Stuttgart and builds up the "Rocket Test Site Lampoldshausen". He receives a lectureship at the Institute of Technology (TH) Stuttgart and becomes in

1957 honorary professor.

1963 he follows the call of the Technical University Berlin to the first German chair of astronautics and completes in

1964 on February 9, a project proposal for a two-stage space transportation system for Junkers, one day before his fatal heart attack on February 10.

DER VORSTAND:	Prof. Dr. phil. HERMANN BLENK, Braunschweig Vorsitzender
	Prof. Dipl.-Ing. WALTER BLUME, Düsseldorf stellv. Vorsitzender
	Flughafendirektor WILLI SONDERMANN, Braunschweig stellv. Vorsitzender
DER VORSTANDSRAT:	Prof. Dr. phil. Dipl.-Ing. Dr.-Ing. E. h. ALBERT BETZ, Direktor des Max-Planck-Instituts für Strömungsforschung, Göttingen
	Prof. Dr.-Ing. PAUL BRENNER, Direktor der Vereinigten Leichtmetall-Werke G. m. b. H., Bonn
	Prof. Dipl.-Ing., Dr.-Ing. E. h. HENRICH FOCKE, São José dos Campos, Est. São Paulo (Brasilien)
	Dipl.-Ing. OTTO FUCHS, Geschäftsführer der Deutschen Versuchsanstalt für Luftfahrt E. V., München
	Prof. Dr.-Ing. E. h., Dr. phil. h. c. ERNST HEINKEL, Stuttgart
	Prof. Dr. phil. HEINRICH KOPPE, o. Prof. a. d. TH Braunschweig
	Prof. Dr.-Ing. KURT LÖHNER, o. Prof. a. d. TH Braunschweig
	Prof. Dipl.-Ing., Dr.-Ing. E. h. WILLY MESSERSCHMITT, München
	Prof. Dr.-Ing. AUG. WILH. QUICK, Decize (Nièvre)
	Prof. Dr. phil. HERMANN SCHLICHTING, o. Prof. a. d. TH Braunschweig
	Prof. Dr.-Ing., Dr. rer. nat. E. h. ERNST SCHMIDT, o. Prof. a. d. TH München
	Prof. Dr.-Ing. FRIEDRICH SEEWALD, o. Prof. a. d. TH Aachen
	Prof. Dr. phil. WALTER TOLLMIEN, o. Prof. a. d. Universität Göttingen
	Prof. Dr.-Ing. HERMANN WINTER, o. Prof. a. d. TH Braunschweig
	Dipl.-Ing. ERNST ZINDEL, Bad Homburg
DER GENERALSEKRETÄR:	Flugkapitän WILL ENNENBACH, Braunschweig

Board of directors and board council of the in 1952 newly established "Scientific Aeronautical Society" ("Wissenschaftliche Gesellschaft für Luftfahrt, WGL").

isations. The promotion of the university research was, and remained, a matter of the individual federal states. But here, too, evolved a certain centralism, especially since the change of the German constitution in 1969, whereby the federal government gained authority in the promotion of research.

The "German Research Association" ("Deutsche Forschungsgemeinschaft, DFG") resulted in 1951 in West Germany by merger of the "German Research Council" (founded in 1949) and the "Emergency Association of German Science" ("Notgemeinschaft der Deutschen Wissenschaft"), originally founded in 1920 and re-established in 1949. The DFG-committee of aeronautical research commented, for instance, in 1952 on the financial requirements and possible control mechanisms of aeronautical research [3], and played, generally, a strong role during its re-establishment. This is not surprising since the committee consisted of people affected. Initially, the committee members were all university teachers, who were opposed to centralized aeronautical research establishments.

Finally, one must ask, how the effort related to the reconstruction of West German aeronautical research during the first ten years after the war was to be justified. Was it to continue a glorious, though seemingly stained tradition, or was it simply to perform academic research in a foremost technological area? Aircraft engineering was prohibited in Germany after the war so that industrial needs were initially non-existant.

The first two of the above motivations certainly played a role for some of the persons involved. It is, however, important to note that, within the Federal

Republic, people very early started to specifically think about possible (military) aircraft developments [35]. This was related to the overall political situation, i.e., the beginning of the cold war [68]. The "North Atlantic Treaty Organization, NATO" was founded in 1949 to integrate the American forces with the West European ones for the defense against the Soviet Union. The Korean War, starting in 1950, then aroused the desire, particularly among the political representatives of the USA, that Germany participate in the defense of Western Europe with its own forces.

In conjunction with the subsequent considerations concerning the formation of a "European Defense Community, EDC", which failed in 1954 due to French objections, representatives of the old German aeronautical industry met in November of 1951 to discuss the reconstruction of their line of business. At that time, also the "Association for the Promotion of Aeronautics" ("Verband zur Förderung der Luftfahrt"), formerly "State Association of the Aeronautical Industry" ("Reichsverband der Luftfahrt-Industrie") founded in 1923, evolved anew. The organization was renamed in 1962 into "Federal Association of the German Aeronautical Industry" ("Bundesverband der Deutschen Luftfahrtindustrie"), in 1962 "Federal Association of the German Aerospace Industry" ("Bundesverband der Deutschen Luft- und Raumfahrtindustrie, BDLI"). However, there existed, at the time, not yet any specific ideas concerning the approach to the reconstruction of West German aircraft manufacturing. Nevertheless, the (military) office "Blank", which later evolved into the "Federal Ministry of Defense" ("Bundesverteidigungsministerium") already considered, even before Germany became sovereign, questions concerning the manufacturing of aircraft for the German forces, within the framework of NATO, under license. This was to be accomplished by a German aeronautical industry. There was, in addition, due to the success of civil air transportation – the German Lufthansa was founded anew in 1953 and went into operation in 1955 – hope to be able to resume (civil) transport-aircraft manufacturing. This then describes the climate surrounding the re-establishment of West German aeronautical research after the war outlined here.

The development was quite different within the east of Germany. After the American troops, at the beginning of July 1945, had pulled out of Saxony and Thuringia, which they had occupied [68], taking along relevant material on aeronautical research, the Soviet Union was in her occupation zone master of almost two thirds of the aeronautical activities of the defeated Third Reich [35]. Similar to the Western Allies, the Soviet Union also tried to acquire as much knowledge and as many results of German aeronautical research as possible [60]. Already in the autumn of 1944 corresponding plans had been worked out.

In research and development offices of former German airplane manufacturers, "special technical offices" were established, where, initially, material was collected and employees were questioned, who later also had, as had been the case for AVA and the DFL in West Germany, to prepare reports concerning their activities and their knowledge. While research activities in the West were continued on in a small scale, the Soviets continued on a large scale, but secret, German experimental developments at a number of locations [69]. These activities became finally known, and, since the Western Allies insisted on the compliance with the Control-council Law No. 25, issued in April of 1946, it was decided to transfer all activities to the Soviet Union. In October 1946, two- to three-thousand [61] aeronautical scientists and engineers from German industry (Askania, BMW, Heinkel, Junkers, Siebel, ...) and from research (DVL) were deported with their families to several different locations within the Soviet Union. All facilities and factories were either transferred or destroyed.

In the fall of 1949, the "German Democratic Republik, GDR" [68] was created in response to the foundation of the Federal Republic in May the same year. Since the scientists and engineers, deported to the Soviet Union, were, beginning in 1950, gradually being returned – they were no longer needed, however, also political pressure was imposed by the West – it was decided to have them work in the GDR in their original profession. This was also to prevent that old and new knowledge was transferred to the West.

The set up of the new aeronautical industry goes back to 1952 and was initially orientated towards military aircraft. After the national uprising on July 17, 1953, the military component was renounced for political reasons in favor of civil aircraft development. After the return of the elite of the aeronautical experts, who were still retained in the Soviet Union in 1950, the industrial activities started in 1954 with the center being in Dresden. The "Association of the People's Enterprise Aircraft Construction – VVB Aircraft

Brunolf Baade (1904–1969).

Construction" ("Vereinigung Volkseigener Betriebe Flugzeugbau – VVB Flugzeugbau") was founded in May 1958. Within the "VVB Aircraft Construction" *Brunolf Baade* was responsible for the overall coordination of research and development in industry and at the universities.

For aeronautical research, the "Research Center of the Aeronautical Industry" ("Forschungszentrum der Luftfahrtindustrie, FZL") was founded in 1955 in Dresden; the center was attached to the "Ministry of Mechanical Engineering" ("Ministerium für Maschinenbau"). This center, also known as "Factory 817" ("Werk 817"), consisted of six institutes and a "Central Office of Standardization". In addition, there were research facilities at industry and the ones of other ministries and the university institutes. Up to 1,500 people worked at the FZL.

The Universities (or Institutes of Technology) of the GDR established their first chair of aeronautics in Rostock in 1952 [70]. However, with the concentration of aircraft engineering and research in Dresden [69, 71], teaching of the subject "aircraft engineering" was transferred to the Institute of Technology (TH) Dresden, with the "Faculty of Aeronautics" ("Fakultät für Luftfahrtwesen") being established in 1956. The following aeronautics-related organizations or institutions were located in or in the vicinity of Dresden:

- "VVB Aircraft Construction",
- "Research Center of the Aeronautical Industry",
- "VEB Aircraft Company Dresden" ("Flugzeugwerke Dresden, FWD") for development and testing of aircraft and equipment,

Brunolf Baade, born in Berlin on March 15, 1904.

1926 graduation from the Institute of Technology (TH) Munich; he becomes a designer for special tasks at BMW.

1930 studies and engineering activities in the US aeronautical industry.

1936 head of type-testing and chief designer of the "Junkers Aircraft and Engine Company" ("Junkers Flugzeug- und Motorenwerke, JFM"); he has part-responsibility for the Ju 88, Ju 188, Ju 288 and the first bomber with jet engines mounted on (forward-) swept wings, the Ju 287.

1945 he is initially instructed by the Soviet Military Administration to reestablish the JFM and becomes head of the design and construction office OKB Nr. 1.

1946 further development of the Ju 287, resulting in the EF 131; documentation concerning the JFM (history) until 1945. On October 24, he is moved to Podberesje near Moscow and becomes chief designer of the "Junkers-team" working on order of the Soviet Union.

1952 he is involved in planning GDR aircraft and engine engineering.

1954 he returns to the GDR and becomes technical manager for the build-up of the aeronautical industry of the GDR. Simultaneously he becomes chief designer and director of the "Research Center" with the task to develop a four-jet engine commercial transport aircraft, 15, based on the Samaljot 150.

1956 he is assigned a chair at the TH Dresden.

1958 foundation of the "VEB Aircraft Company Dresden" and test of the four-jet engine commercial transport aircraft, now designated 152; project studies for the medium-range airplane 153 and the short-range aircraft 155.

1961 *Baade* becomes head of the "Institute of Lightweight Construction" at the TH-Dresden.

1969 he dies in Dresden.

in the close vicinity of Dresden:

- "VEB Development Company Pirna" ("VEB Entwicklungsbau Pirna") near Dresden for the development of engines,
- "VEB Apparatus Factory Lommatsch" ("VEB Apparatebau Lommatsch") in Meissen for the development and construction of gliders,

Take-off preparations for the 152 V1 in Dresden-Klotzsche for its maiden flight on December 4, 1958. The aircraft was equipped with four Mikulin-RD 9B jet engines.

and farther away from Dresden:

- "VEB Industrial Company Karl-Marx City" ("VEB Industriewerke Karl-Marx-Stadt") in Chemnitz for the construction of engines, propellers, landing gears…,
- "VEB Machine and Apparatus Company Schkeuditz" ("VEB Maschinen- und Apparatebau Schkeuditz") near Leipzig for aircraft repairs, manufacture of rigs, empennage construction…,
- "VEB Industrial Company Ludwigsfelde" ("VEB Industriewerke Ludwigsfelde") near Berlin for the quantity production of jet engines (and motor scooters).

A number of experimental developments related to commercial transport aircraft was carried out within the GDR, with the best-known aircraft being the Aircraft 152 (see the description in Part 3). Corresponding test facilities, low- and high-speed wind tunnels, etc., were provided as basic equipment. The engine development commenced in 1955/56 [72]. Two prototypes of the 152 were completed with first flights being performed in 1958 and 1960, respectively. The first prototype crashed in 1959. Glider design and construction, whose beginnings in the GDR go back to 1950 [69] was initially very successful. It could, however, before its discontinuation in 1962, not reach the international standing of its West German counterpart [31].

The end of aircraft engineering in the GDR came at the beginning of the sixties in the context of "Production Control" ("Produktionslenkung") within the Eastern Bloc. Following a resolution of the Central Committee of the SED (East German ruling party at the time) on February 28, 1961, all was abandoned, aircraft- and aero-engine research and development, glider construction (1962), even the chair of aeronautics in Dresden was dissolved in 1961. After 1961, low-key activities were continued with the repair of military aircraft. In 1991, the "VEB Elbe Aircraft Company" ("VEB Elbe Flugzeugwerke") in Dresden and the "VEB Repair Plant Ludwigsfelde" ("VEB Instandsetzungswerk Ludwigsfelde") were incorporated into Dasa [70].

The Development of Aeronautical Research in West Germany in View of the Atlantic Alliance 1955–1970

With the Paris Treaty, which became effective in May 1955, the Federal Republic of Germany attained her far-reaching sovereignty. At the same time, she committed herself to make a defense contribution within the context of the West European-American Alliance. The Federal Republic had already joined NATO on October 23, 1954, when the Paris Treaty was signed. The subsequent reconstruction of the German aeronautical industry was at first aligned to the set-up phases of the Federal Armed Forces [35, 56]: initial equipment within a five-year program 1956–1960, main equipment within the manufacturing-in-license-programs Lockheed F-104G (multi-purpose fighter "Starfighter") and Fiat G-91 (light fighter aircraft), and initiation of the follow-up generation.

Already the first large contracts were awarded to company consortia. This shows that, from the beginning, the government aimed at a concentration of the (old) companies. Already during the time of the "Office Blank" combined production capacities were considered a prerequisite for future contracts. It is also interesting to note that, in 1955, a continental European industry association was considered to ward off U.S. and British competitors.

During the concentration process, the old aircraft companies, which after 1945 had only been joined by "Bölkow Engineering" ("Ingenieurbüro Bölkow") in 1948 (1956: "Bölkow Developments, KG", a limited partnership company, 1958/1965: "Bölkow Ltd."), and the "Railway-car and Mechanical Engineering, Ltd." ("Waggon- und Maschinenbau, GmbH") in 1946, merged until the end of the sixties into the following enterprises [73, 35, 4, 6]:

- "Messerschmitt-Bölkow-Blohm (MBB), Ltd.", in 1969,
- "Central Association VFW-Fokker, Ltd." ("Zentralgesellschaft VFW-Fokker mbH") in 1969 (VFW = Vereinigte Flugtechnische Werke = United Flight-technologies Company),
- "Engine and Turbo-machinery Association, Ltd." ("Motoren- und Turbinen-Union, MTU, GmbH") in 1969.

During this time, only "Dornier, Ltd.", 1950, with the "Dornier Repair Yard" ("Dornier Reperaturwerft"), 1957, and the "Dornier Systems, Ltd.", 1962, as well as AEG-Telefunken had preserved their independence. The "Project Partnership German Airbus, Ltd." ("Projektgesellschaft Deutsche Airbus GmbH"), which had evolved from the "German Work Association Airbus" ("Deutsche Arbeitsgemeinschaft Airbus"), founded in December 1965, was incorporated into MBB in 1969.

In Part 3, details of the research activities of these companies and of the company mergers will be reported; also addressed will be the experimental-development programs, conducted with governmental support, related to the "Vertical Take-off and Landing, VSTOL" fighter aircraft VJ 101 and VAK 191 and the VSTOL transport aircraft Do 31.

Industrial aeronautical research had, from the very beginning, government support. In one of the later "Federal Reports on Research" ("Bundesberichte Forschung"), respectively the "Reports-on-Facts-Supplement to the Federal Reports" ("Faktenberichte zu den Bundesberichten"), which also relates to the research institutions, the following objectives were stated [74]:

"The Federal Government promotes aeronautical research to

Ludwig Bölkow, (1912–2003), has, as engineer and businessman, decisively influenced the development of modern technologies during the rebuilding of German aircraft manufacturing after World War Two.

- augment the competitiveness of the aeronautical industry so it may carry out international cooperations,
- secure the participation in advanced technological developments,
- ensure an adequate participation in civil aeronautics,
- gain the technological basis for a participation in covering the demand of the Air Force."

The so-called FFE-regulations (FFE: free research and development) of the federal government, issued at the beginning of the sixties, allowed companies to charge in the case of public contracts, at the time mainly to cover the demand of the armed forces, a certain percentage – varying from 2 per cent at the time to 3–4 per cent today – of their prime cost to maintain and extent their technological basis. This was, and still is, related to new technologies, methods and procedures, but under certain conditions also to experimental developments.

Individual technologies related to military aircraft engineering were supported within the ZTL-program (Future Technologies Air = Zukunftstechnologien Luft) from 1967/68 to about 1982. Within this framework, essential technologies were developed for the later Eurofighter within the sub-program "Component-related Experimental Program Air" ("Komponenten-Experimentalprogramm Luft, KEL") during the seventies and eighties. Such as, for instance, within the CCV experimental program (Configuration Controlled Vehicle CCV-104 G), the control-configured

The experimental development of vertical-take-off aircraft (in parentheses the year of the first flight). *From top to bottom*: EWR VJ 101 C (1963), Dornier Do 31 (1967), VKA 191 (1971).

At MBB, a Lockheed F-104G of the air force was converted to a CCV experimental aircraft.

design, the digital flight control system for the unstable airplane, carbon-fiber construction of the wing, et cetera. After 1982, support was provided within various "Research and Technology (R&T) Programs" ("Forschungs- und Technologie (FuT) Programmen").

Technologies concerning civil aircraft engineering were promoted since the early seventies. A first overall program "Aeronautical Research and Aeronautical Technology 1975–1978" was established in 1974/75 jointly by the Federal Ministries of "Research and Technology" ("Forschung und Technologie, BMFT"), "Transportation" ("Verkehr, BMV"), "Defense" ("Verteidigung, BMVg"), and "Economics" ("Wirtschaft, BMWi"). The BMFT Technical Program (1974–1982/85) contained the "Civil Components Program" ("Ziviles Komponentenprogramm, ZKP"), where subjects, such as "transonic wing sections for large transport aircraft", "components of large aircraft in hybrid construction", "novel wing designs for general-aviation aircraft", and "vibration control for helicopters", were sponsored. In addition, air-traffic technologies, test facilities (DNW = German-Dutch Wind Tunnel, ETW = European Transonic Wind Tunnel), and avionics technologies were promoted.

The aeronautical research establishments were subject to a similar concentration process as the manufacturers. Initially, the actual (re-) construction process had priority, which was, however, during the first half of the sixties essentially completed.

DVL built between 1955 and 1959 in Aachen two large test facilities, which were later transferred to Cologne Porz-Wahn. However, there were already plans in 1954 to provide a new home for DVL at the airport Essen-Mühlheim [7]. Accordingly, construction projects were initiated and also completed. However, it soon became clear that the prerequisites for the establishment of the organization at this location could not be met. Finally, a new location for DVL was found in the autumn of 1958 in Porz-Wahn, close to the airport Cologne-Bonn. Construction started in 1959 and continued until the middle of the sixties. The central administration settled there in 1964.

After the "Air Radio Research Institute Oberpfaffenhofen" ("Flugfunkforschungssinstitut Oberpfaffenhofen, FFO") had been incorporated into DVL, Oberpfaffenhofen near Munich became the second large DVL site. Added there was in 1963 the "German Research Establishment for Gliding" ("Deutsche Forschungsanstalt für Segelflug, DFS"), which had changed its name in 1960 into "Research Establishment of Flight Sciences Munich" ("Flugwissenschaftliche Forschungsanstalt München, FFM"). The build up in Oberpfaffenhofen also lasted until the middle of the sixties. In the early sixties, DVL consisted of 15 institutes and the "Airworthiness Authority" ("Luftfahrtprüfstelle"), distributed over the cities Porz-Wahn, Aachen, Essen-Mühlheim, Godesberg, Freiburg, Oberpfaffenhofen and Munich.

At the end of the fifties, beginning of the sixties, the authorities of the Federal Republic started to consider space technology within a European framework. Correspondingly, new institutes were founded beginning in 1961. The "Research Establishment of Jet Propulsion Physics" ("Forschungsinstitut für Physik der Strahlantriebe, FPS") was incorporated into DVL in 1962/63. On its campus in Lampoldshausen near Stuttgart, altitude test-cells for rockets were erected within the framework of an ELDO program.

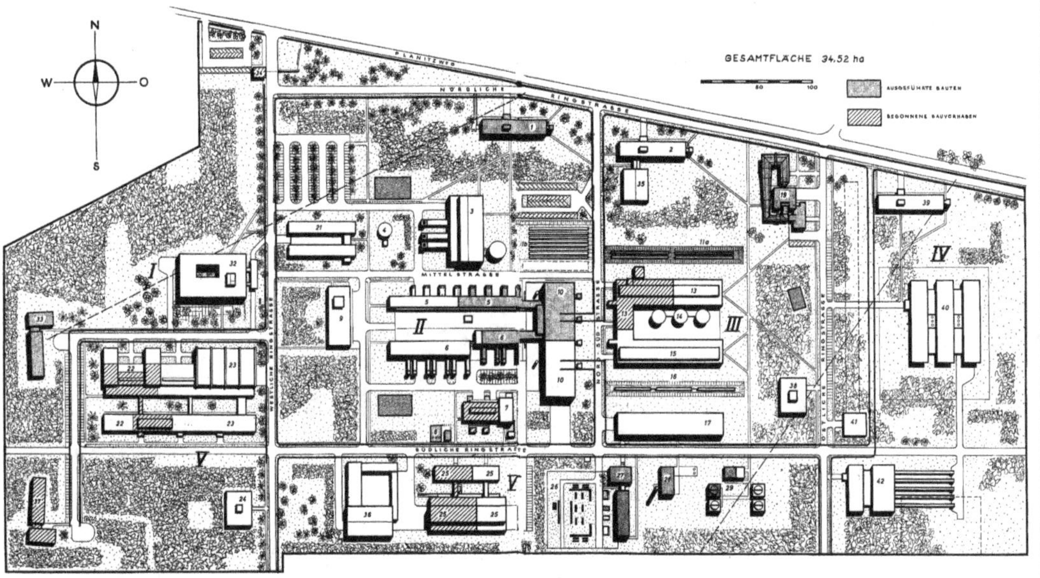

I	Central Building	V	Institute of Space Research
II	Institute of Propulsion Technologie	V	Institute of Strength and Materials Research
III	Institute of Applied Gas Dynamics	VI	Utility Department and Means of Operation

Plan view of 1962 related to the build up of the DVL facilities in Porz-Wahn.

The DVL operated already since 1961 under the name "German Aerospace Test Establishment" ("Deutsche Versuchsanstalt für Luft- und Raumfahrt"). In 1968, i.e., before the creation of the "German Aerospace Research and Test Establishment" ("Deutsche Forschungs- und Versuchsanstalt für Luft- und Raumfahrt, DFVLR"), DVL operated 19 institutes and various inter-institutional installations with a total of about 1,800 employees, about 28 per cent of them being scientists.

DFL expeditiously build-up its new location Brunswick-Waggum, starting in 1953. DFL was able to retain its off-site location Trauen, which had been established in 1937/38 under the management of *Eugen Sänger*, as "Aircraft Test Station Trauen" ("Flugzeugprüfstelle Trauen") in the south of the "Luneburg Heath" ("Lüneburger Heide") between Munster and Fassberg, for rocket-motor tests. In the years 1958 and 1964, funding was available to erect wind tunnel facilities in Waggum and engine and rocket test stands in Trauen [75]. The "German Study Association Helicopters" ("Deutsche Studiengemeinschaft Hubschrauber, DSH"), which had, in 1957, changed its name into "German Research Establishment of Helicopters and Vertical-flight Techniques" ("Deutsche Forschungsanstalt für Hubschrauber und Vertikalflugtechnik, DFH"), joined DFL in 1962 (officially, the merger took only place in 1963). Since the DFL in 1962 also had become active in the area of space technology, its name was changed to "German Aerospace Research Establishment" ("Deutsche Forschungsanstalt für Luft- und Raumfahrt"). In 1969, prior to the merger into DFVLR, it consisted of five institutes with about 1,030 employees, 25 per cent of them being scientists.

The AVA, since 1953 a member of the "Max-Planck-Society", created a number of new facilities on its premises in Göttingen. Many of the old employees initially stayed at the "Max-Planck Institute of Fluid Research", participating, however, in the reconstruction of the AVA. Following *Albert Betz*, who retired

The DVL Research Center Southwest was, in 1969, set up in the former buildings of the "Research Establishment of Jet Propulsion Physics" ("Forschungsinstitut für Physik der Strahlantriebe, FPS") in Stuttgart-Pfaffenwald.

Within the framework of the 1958-engine-program of DFL, an engine test facility with a "hot-water drive" ("Heißwasserantrieb") according to *Otto Frenzl*, was built in Trauen, becoming operational in 1968. It was primarily utilized to investigate ramjet engines. The cross-section of the test section was 70 cm × 70 cm (28 in × 28 in), the Mach number could be varied between M = 2 and 3.5.

in 1967, *Hermann Schlichting* took over as head of the AVA. *Schlichting* was, at the time, also head of the DFL Institute of Aerodynamics and the Institute of Fluid Mechanics of the Institute of Technology (TH) in Brunswick. In July 1965, also AVA took up space research and the department "Space Aerodynamics" was established. At the end of its independence in 1969, the AVA consisted of six departments with approximately 220 employees, a little more than a quarter of them being scientific employees [76].

The start of the concentration processes among the aeronautical research institutions actually goes back to the time of their re-establishment [2]. After the commencement of their reconstruction phases and scientific activities, the respective federal states were quickly financially overstrained. The federal government was to step in. The government, however, demanded an organizational streamlining, which was, of course, not necessarily in the interest of the research institutions. The demand for federal subsidies of 50–70 million German Marks (DM) respectively 25–35 million Euro per annum – this was, at the time, about the magnitude of the French government support for aeronautical research – made, nevertheless, certain concessions by the research establishments advisable. The "Presidential Council of the

Aeronautical Research Establishments" ("Präsidialrat der Luftfahrtforschungsanstalten"), a council without legal form founded in 1956, was the result of these concessions.

This did, however, not correspond to the ideas of the federal ministries involved. The "Ministry of Transportation" ("Verkehrsministerium, BMV") took over the lead. The "Knipferplan" of 1957, named for the ministry official in the BMV responsible, envisaged the transformation of the "Presidential Council" into a legal body, which could have been named "Ludwig-Prandtl-Society", and which was to be authorized to represent all research establishments [3]. Discussions led in May 1959 to the "German Society of Flight Sciences" ("Deutsche Gesellschaft für Flugwissenschaften, DGF") with *Hermann Blenk* as chairman. Its main task was to establish joint research and economic plans and forecasts for the member organizations (AVA, DFH, DFL, DFS, DVL), and to coordinate these with the sponsors, i.e., the participating federal states and the federal ministries, as well as the DFG ("German Research Association").

The financial problems were exacerbated by the necessity to create wind tunnel facilities, which, on the one hand, satisfied the requirements of research, and, on the other hand, the demands of the recovering aeronautical industry. After the war, most German wind tunnels "disappeared", either by destruction or by dismantling [46]. Furthermore, industry did not see itself able, financially and personnel-wise, to construct their own wind tunnels, and wanted therefore these to be set up and operated by the research establishments. There existed thereafter, indeed, only one smaller low-speed wind tunnel at Dornier in Immenstaad at Lake Constance (since 1958) and at the "United Flight-technologies Company Bremen" ("Vereinigte Flugtechnische Werke Bremen"), since 1970.

The WGL seized the initiative with the formation of a wind tunnel committee. This committee presented mid-1956 its wind tunnel memorandum. Approximately 90 million DM (approx. 45 Million Euro) were estimated to be required for the facilities envisioned for universities and research establishments. The industry formed its own committee, which supported, but also extended, the WGL-proposal. The final DGF-proposal envisaged 29 Million DM (approx. 14,5 million Euro) for the wind tunnel program, with an additional 26 million DM (approx. 13 million Euro) for the engine (facilities) program [2]. Since especially the large German wind tunnels did no exist anymore after the war, a large subsonic wind tunnel (GUK) was considered, which was primarily needed by the industry. After many discussions, which were also characterized by rivalry between the research institutions, this tunnel was erected during the second half of the seventies, jointly by the DFVLR and the NLR, as "German-Dutch Wind Tunnel" ("Deutsch-Niederländischer Windkanal, DNW"), at the Northeast Polder in the Netherlands [77].

Ludwig Prandtl (1875–1953), photograph from 1936. *J.D. Anderson* stated: *"He was clearly the father of modern aerodynamics".*

Despite of these activities and interactions, all sides involved criticized the DGF. The government criticized the lack of coordination, the researchers, in turn, missed an adequate improvement of their working conditions. The situation escalated and became completely muddled in the mid-sixties. The demands for government control on the one side and scientific autonomy on the other side strongly opposed each other [2]. In order to evade the demand for an

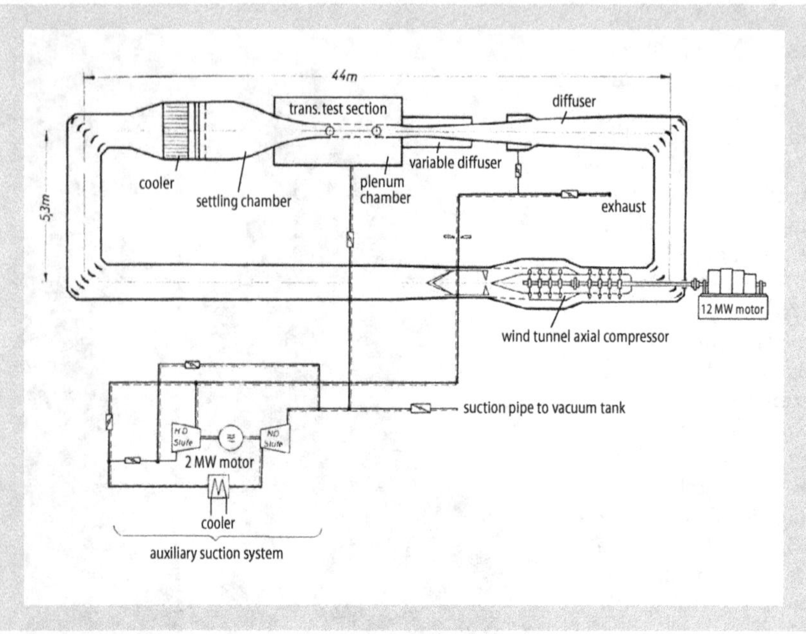

The transonic wind tunnel of the AVA in Göttingen has been in operation since 1963. The tunnel has been, to a large extend, utilized for research on behalf of the aeronautical industry. The cross-section of the test section is 1 m × 1 m (3.3 ft × 3.3 ft), the Mach number range is M = 0.3–2.25.

"Integrated Research Establishment" ("Einheitsgesellschaft") – the individual institutions had the wish to stay independent – the DGF tried to offset the critique of the government by more specific research plans. In May 1966, the first "coordinated" DGF research plan was presented. It was more than the loosely joined research plans of DVL, DFL and AVA so far.

The government pressure did not cease and, finally, in 1967, the course for the "Integrated Research Establishment" was set. The institutions AVA, DFL and DVL were dissolved in March and April, respectively, of 1969. On April 1, 1969, the "German Aerospace Research and Test Establishment ("Deutsche Forschungs- und Versuchsanstalt für Luft- und Raumfahrt, DFVLR, e. V.") with headquarters in Porz-Wahn, was founded. The board of directors consisted of *Volker Aschoff*, who came from the TH Aachen (full professor of communication engineering), *Theodor Benecke*, who was president of the "Federal Office of Defense Technology and Procurement" ("Bundesamt für Wehrtechnik und Beschaffung, BWB") and *Horst Niemeyer*, administrative board member of DVL.

The concentration process within the aeronautical research institutions did not have any direct repercussions on the overall environment of research. The WGL, but also the DFG, had voiced their opinions and exerted their influence [2]. The WGL was subject to its own process of change [65]. WGL reacted, during the early sixties, to the increasing German engagement in space technology with a corresponding expansion of its activities and a change of its name into "Scientific Society of Aeronautics and Astronautics" ("Wissenschaftliche Gesellschaft für Luft- und Raumfahrt, WGLR"). In December 1967, it merged with the "German Society of Rocket Technology and Astronautics" ("Deutsche Gesellschaft für Raketentechnik und Raumfahrt, DGRR") to form the "German Aerospace Society" ("Deutsche Gesellschaft für Luft- und Raumfahrt, DGLR").

Since the fifties, the DFG promoted aeronautical research primarily at universities including institutes of technology [78]. The focused financial support of aeronautical research amounted between 1953 and 1959 to 20 million DM (approx. 10 million Euro). Of these, about 4.5 million DM (approx. 2.3 million Euro) were supplied to the aeronautical research establishments for the acquisition of standard scientific equipment. Funds for aeronautical research totaling of 29.1 million DM (approx. 14.9 million Euro) were provided up to 1965. At the end of the sixties, the first "Collaborative Research Centers" ("Sonderforschungsbereiche, SFB") for aeronautics were established [79]:

Volker Aschoff (1907–1996).

- **SFB 58:** "Flight control", Institute of Technology (TH) Brunswick, 1968 until the early eighties.
- **SFB 85:** "Thermodynamics and fluid mechanical problems of aeronautical and space propulsion", Institute of Technology (TH) Stuttgart, 1968 to 1985.
- **SFB 83:** "Fluid mechanics and thermo-gas dynamics", Institute of Technology (RWTH) Aachen, 1970 to 1983.

The end of the sixties represents in so far a turning point that the research institutions were now united within the DFVLR (1969), the industry was, with few exceptions, united within MBB (also in 1969), and the Federal Government had officially defined its engagement for aero- and astronautics. (In 1970, the first annual report of the Federal Government on the aerospace industry "Basic program for the aeronautical industry 1970–1974" was issued.)

Volker Aschoff, born in Freiburg im Breisgau on June 14, 1907, studied Electrical Engineering in Bonn, Danzig and Karlsruhe

1936 received his PhD with *H. Backhaus* with a topic related to electro-acoustics. He then joined the central laboratory of AEG, and in

1940 took over the technical management of the "Torpedo Test Establishment" ("Torpedoversuchsanstalt") in Gotenhafen.

1945 after brief internment, he attends to development tasks at the W. Zeh Company in Freiburg.

1950 he is appointed to the chair of "Electrical Communications Engineering" ("Elektrische Nachrichtentechnik") at the RWTH Aachen.

1952/53 Dean of the "Department of Mechanical and Electrical Engineering" ("Maschinenwesen und Elektrotechnik").

1962/63 Dean of the "Department of Electrical Engineering".

1963 Chancellor (Rector) of the RWTH Aachen for two years, thereafter Vice-chancellor for another two years.

1967 he is appointed to the "Science Council" ("Wissenschaftsrat"), the "Advisory Council for Planning of the Minister of Education and the Arts of North-Rhine-Westphalia" ("Planungsbeirat des Kultusministers von Nordrhein-Westfalen") and to the "Rheinisch-Westphalian Academy of the Sciences" ("Rhenish-Westfälische Akademie der Wissenschaften"). He becomes, in addition, chairman of the "Expert Committee on Space Technology and Space Research of the Federal Minister of Education and Science" ("Fachausschuß Weltraumtechnik und Weltraumforschung beim Bundesminister für Bildung und Wissenschaft").

1969 he took over the chairmanship of the board of directors of the newly founded DFVLR, which combined DVL, DFL and AVA.

1996 he dies on July 14.

Consolidation of West German Aeronautical Research 1970–1990

At the beginning of the time period of approximately twenty years considered here (1970–1990), stood the consolidation of non-university aeronautical research - DFVLR had first to establish its structure and mode of operation [80] - and at the end stood the consolidation of the aerospace industry within the "German Aerospace Corporation (AG), Dasa". The statutes of the new DFVLR envisaged the following tasks:

- Research in the fields of aeronautics and astronautics.
- Participation in the planning and execution of projects.
- Erection and operation of large test facilities.

- Advanced education of junior scientists and engineers.
- Advising authorities of the Federal Republic.

This way, DFVLR stayed in the tradition of its predecessor organizations. Initially, its structure also reflected this tradition [81]:

- **Subject Area 1 "Fluid Mechanics"**, six institutes (Brunswick (1), Göttingen (3), Porz-Wahn (1), plus three university-attached off-site institutes in Aachen, Berlin and Freiburg).
- **Subject Area 2 " Flight Mechanics, Flight Control"**, four institutes (Brunswick (2), Stuttgart (1), Oberpfaffenhofen (1)).
- **Subject Area 3 "Strength, Materials, Structural Design"**, four institutes (Brunswick (1), Göttingen (1), Porz-Wahn (1), Stuttgart (1)).
- **Subject Area 4 "Propulsion, Energetics"**, nine institutes (Brunswick (3), Porz-Wahn (1), Stuttgart (4), plus one university-attached off-site institute in Munich).
- **Subject Area 5 "Air-Space Physics, Space Simulation, Flight Medicine"**, three institutes (Porz-Wahn (2), Oberpfaffenhofen (1)).

In addition, there were the "Technical Aeronautical Section", consisting of the Flight Departments in Brunswick and Oberpfaffenhofen, the "Technical Astronautical Section", with, among others, the rocket test facilities in Trauen and Lampoldshausen and the "Central Office of Aeronautical Documentation" ("Zentralstelle für Luftfahrtdokumentation, ZLDI") in Munich. The 3,115 employees (at the end of 1969) were spread over five Research Centers: Porz-Wahn (headquarters) with the off-site office Bad-Godesberg, Brunswick with the off-site installation Trauen, Göttingen, Oberpfaffenhofen, Stuttgart with the off-site installation Lampoldshausen and the four university-attached off-site institutes in Aachen, Berlin, Freiburg and Munich.

The budget of the DFVLR amounted in 1970 to about 380 million DM (approx. 190 million Euro), the targeted 9 to 1 split in the budget between the federal government and the federal states hosting institutes had not yet been reached. In order to meet the demand of the sponsors, i.e., the federal and state parliaments and ministries, for a more effective control of research topics and efficiency, but also of the expenditures, which was the very reason to create DFVLR, the "Main Department Science and Technology" ("Hauptabteilung Wissenschaft und Technik") was established in 1970 with *Werner Schulz* as head. Its department "Research Planning" had to present, annually, the "Research Plan" and in the following year the so called "Yearly Report" containing the "Target/Accomplishment"-comparison. The research-worthiness of a topic was judged by four groups of criteria: science-immanent criteria, research-political criteria, economic criteria and feasibility criteria.

The overall personnel-plan, which had for the combined organizations almost linearly risen from about 1,500 employees in 1960, envisaged for 1971 somewhat more than 3,500 positions. Since 1971, scientific employees were able to exert a formal influence on all matters related to research through the newly instituted "co-determination" of the "Institute Councils" ("Institutsräte"). Following a directive of the "Federal Ministry of Research" ("Bundesforschungsministerium, BMFT") at the time, a "Scientific-Technical Council" ("Wissenschaftlich-Technischer Rat") was assigned to the board of directors as "Management Participation Organ" ("Mitwirkungsorgan"), consisting of elected representatives of the institute heads and the institute councils.

With the implementation of an international office, DFVLR started during this year to supervise, on behalf of the "Federal Ministry of Education and Science" ("Bundesministerium für Bildung und Wissenschaft"), the cooperations in the area of aerospace research, which had been agreed upon between the Federal Republic and European as well as non-European countries. At the same time, the in-house concentration process commenced. The low-speed wind tunnels were (organizationally) taken out of their original institutes in Porz-Wahn, Brunswick and Göttingen and combined in the "Central Department Low-speed Wind Tunnels" ("Zentralabteilung Niedergeschwindigkeits-Windkanäle").

In 1972, the federal budget was only after a delay approved, which led to restrictions in the financial management of DFVLR. At the beginning of the year, DFVLR had taken over the "Society of Space Research" ("Gesellschaft für Weltraumforschung mbH, GFW"), which carried out, on behalf of the federal government, the project management in the case of astronautics-related research. But the overall personnel-plan remained as in 1971, leading to a redistri-

Hermann Jordan (1922–1998).

bution of personnel. However, in 1972/73, also the first era of the new DFVLR came to an end. *Volker Aschoff* retired in 1972; *Theodor Benecke* and *Horst Niemeyer* left the DFVLR board of directors in 1973.

The new board of directors, whose structure had been decided upon by the senate of the DFVLR in November 1974, consisted of the board chairman, *Hermann Jordan*, the technical board member, *Walter Luksch*, and the administrative board member, *Wolfgang Hasenclever*. In addition, there were local board members with seats in Porz-Wahn (*Alfred Heyser*), Göttingen (*Hermann Schlichting*) and Brunswick (*Fred Thomas*), respectively. The old subject-related structure was abandoned in favor of a structure related to "Research Areas" ("Forschungsbereichen"):

- "Flight mechanics and flight control",
- "Fluid mechanics",
- "Materials and structural design",
- "Communication technology and observation (reconnaissance)",
- "Energetics" (preliminary).

In 1975, the "Annual Report" ("Rechenschaftsbericht") was, for the first time, not arranged according to the subject-oriented structure, but according to "Focused Areas" ("Schwerpunkte"), programs and projects, which were aligned to the overall program of the federal government "Aeronautical Research and Technology 1975–1978", thus essentially directed towards a later industrial application. The seven "Focused Research Areas" were:

- "Transportation and communication systems",
- "Aeronautical technology",
- "Space (astronautical) technology",
- "Observation (reconnaissance) technology",
- "Other advanced technologies",
- "Program-preparation and supplementary research and development (R&D) activities",
- "Contractual services".

The number of employees, which had in 1971 and 1972 its maximum of 3,776, declined to 3,574 in 1974. The process of coordinating research objectives of the DFVLR with the ones of industry was continued so

Hermann Jordan, born in Munich in 1922, studies physics at the Technical University (TU) and the University of Munich. He is one of the first students, who, after the war, gets the opportunity to study in the USA, viz., at the University of Berkeley, California.

1951 he obtains his PhD in Munich with a topic related to quantum theory. He becomes a research assistant at the Institute of Technology (RWTH) Aachen and is appointed in

1958 director of the "Institute of Plasma Physics" of the "Nuclear Research Establishment" ("Kernforschungsanstalt") in Jülich.

1966–1969 he obtains a leave-of-absence to manage, as director, the "European Space Research Institute" in Frascati/Italy.

1973 he is appointed chairman of the board of DFVLR, the "German Aerospace Research and Test Establishment"

1978 he resigns as head of the "Institute of Plasma Physics" in Jülich and fully concentrates, with success, on the integration of the DFVLR predecessor organizations (DVL, DFL, AVA). Through a consistent appointment politic, he is able to bring a new generation of institute heads into DFVLR, which considerably accelerated the integration process. He favored very early joint international projects, which led, for instance, to the German Dutch Wind Tunnel (DNW) and the European Transonic Wind Tunnel (ETW).

1979 he prepared the cooperation between Germany and China related to aerospace research. He always tried to generate an environment, where matters could develop by themselves.

1987 he yields the management of the DFVLR to *Walter Kröll*.

1998 he dies on December 11 after a long illness.

that in 1984, following an initiative of *Fred Thomas* and *Horst Prem* (MBB), the long-term planning of the DFVLR and the FFE-planning of MBB could be harmonized. This finally led to the "Reference Concept Approach" ("Leitkonzeptansatz") of *Hans Ambos* (Dasa) and the cooperation contract between DLR (formally DFVLR), Dasa and Daimler-Benz Research of 1996.

Similar to the DFVLR focused research areas "Aeronautics and Space Technology" [82], existed at the universities and institutes of technology during this time period the following DFG sponsored "Collaborative Research Centers" relevant to aeronautical technologies [79]:

- **SFB 25:** "Vortical flows relevant to flight technology", RWTH Aachen, 1981–1994.
- **SFB 212:** "Air traffic safety", TH Brunswick, 1983–1995.
- **SFB 167:** "Highly loaded combustion chambers – stationary constant-pressure combustion", University of Karlsruhe, 1984 to 1998.
- **SFB 339:** "Blades and disks of gas turbines – material and component behavior", TU Berlin, 1988 until the mid-nineties.

The German "Hypersonic-Technology Program" [83] (1988–1995) attempted a synthesis between aeronautics and astronautics by developing technologies for an air-breathing first stage for a two-stage space plane. Following its "Reference Concept" ("Leitkonzept") "SÄNGER", four "Collaborative Research Centers" were created:

- **SFB 253:** "Fundamentals of the design of space planes", RWTH Aachen, since 1989.
- **SFB 255:** "Trans-atmospheric flight systems – fundamentals of aero-thermodynamics, propulsion and flight systems", TU Munich, since 1989.
- **SFB 257:** "Flight physics of hypersonic aircraft and space transports", TH Brunswick, 1990/1991, preliminary set up.
- **SFB 259:** "High-temperature problems of re-usable space transport systems", University of Stuttgart, since 1990.

In the early seventies, the course was set for the large German-Dutch Wind Tunnel DNW. Primarily the developments in civil aircraft engineering, which tended towards the flight at high subsonic or low transonic Mach numbers (supercritical or transonic wings), respectively, initiated, at that time, also the discussion concerning a large transonic wind tunnel, which was not included in the wind-tunnel program of the DGF of 1965. The need arose in Germany only with the plans for the Airbus (1970). The discussion was initially carried out within the framework of AGARD. The AGARD "Working Group on Large Wind Tunnels, LaWs", chaired by *Dietrich Küchemann*, presented in 1972 its recommendations. Within Europe, the "AEROTEST-group" (France, Germany, Great Britain and the Netherlands) formed to consider this topic. During the second half of the seventies it had been decided to build a so-called cryogenic wind tunnel, which operates at very low temperatures (-280 degrees Celsius up to ambient temperatures). The "European Transonic Wind Tunnel, ETW" [84] was finally erected during the early nineties in Porz-Wahn, in the immediate neighborhood of DFVLR, and started in 1994 its trial runs.

Because of the strong engagement of Germany in astronautics, the "German Agency of Aeronautical Affairs" ("Deutsche Agentur für Raumfahrtangelegenheiten, DARA") was founded, which later had its seat in Bonn-Oberkassel. DARA also took over "project-management tasks" (Projektträgeraufgaben) in space research and technology, which had previously been the domain of DFVLR. The DFVLR changed its name to "German Aerospace Research Establishment" ("Deutsche Forschungsanstalt für Luft- und Raumfahrt, DLR").

The DFVLR had in 1969 taken up its operation at a critical time [2]. The financial situation and priorities of the federal government were such that an increase in funding was not, as previously hoped for, to be expected. On the contrary, one had already at the beginning to consider the possible shutdown of institutes or even research centers. Nevertheless, restrictions were not as severe as that. Industry also stabilized only in the early seventies after the reduction of the federal budget during the economic crisis in 1966/67 and the resulting reduction in the support of industry [35]. In June 1970, the federal government issued its first yearly report concerning the "Aeronautical Industry". At the same time received the industry, with the "Basic Program for the German Aerospace Industry 1970–1974" ("Basisprogramm für die deutsche Luft- und Raumfahrtindustrie 1970–1974") the

The German-Dutch Wind Tunnel (DNW) with a maximum cross-section area of the test section of 9.5 m × 9.5 m (31.15 ft × 31.15 ft) started operations in 1980. The photograph shows a 1:10.6-scale model of the Airbus A 340 being investigated in a take-off position in the presence of a moving floor (1990).

Aerial view of the ETW-facility in Cologne-Porz. The photograph was taken in 1997.

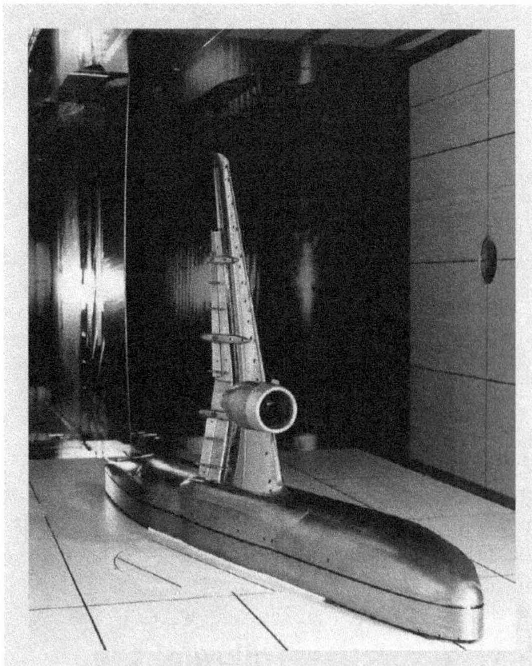

Half-model of an Airbus A 320 in the ETW.

urgently needed framework for planning the future. In December 1974, the federal government nominated a coordinator for the aerospace industry, since several federal ministries (BMV, BMVg, BMWi, BMFT, BMBF) were involved. His task was the continued consolidation of the industry and the tuning of a "Structural Concept".

After the first "concentration wave", which ended in 1968/1969, VFW and Fokker separated and the former merged with MBB. Now, the Dornier Group, MTU and AEG-Telefunken were still independent despite the pressure of the public costumer. In 1985, the Daimler-Benz Corporation took over the majority at Dornier and owned, at that time, 100 per cent of the shares of MTU. Finally, the "Deutsche Aerospace (Dasa) AG" was founded in May 1989, which united under its roof the companies Dornier, MBB, MTU and

DLR Research Center Cologne-Porz (in front right is the ETW). The photograph was taken in 1993.

TST (Telefunken Systems Technology, Ltd). The companies partly retained their status as "bodies corporate" ("Rechtspersonen"). Within the business areas

- "Aeronautics",
- "Astronautics (Space)",
- "Defense",
- "Propulsion",
- "Diversification",

almost 50 per cent of the employees were, in 1988/89, on military contracts. The increase of the share of non-military work was initially one of the most pressing tasks of Dasa.

New Perspectives Following the German Reunification

During the night between November 9 and November 10, 1989 the inner-German borders fell. In October 3, 1990, the GDR ceased to exist. These events mark stations on the way to the end of the Cold War; they belong to the symptoms of the break up of the USSR and the Eastern Block.

At the beginning of the nineties, the end of the Cold War brought the expected reduction in military business ("peace dividend"), and the global recession a corresponding drop in the civil business. But at the early nineties, also the dispute between the USA and Europe concerning the from an American viewpoint unfair Airbus loans of the European countries participating in Airbus, culminated [85].

After a long start-up period, which began around 1970, Airbus became during the eighties a serious competitor for major American aircraft manufacturers. This happened due to the strategy of Airbus to create an aircraft family concept and due to a number of technical innovations.

As long as Airbus was not a danger to US-products, nobody was disturbed. The American companies Boeing, McDonnell-Douglas, and up to the mid-eighties also Lockheed, dominated the world market including the European [86] with their jet aircraft. The initial success of European products (DeHavilland "Comet", BAC VC10, Breguet 763, Sud-Est "Caravelle") was quickly negated by the American ones (Boeing 707, Douglas DC-8). The reasons were that the USA after 1945 with the beginning of the Cold War had developed into the dominating military power in the West. Extensive development and manufacturing contracts for military aircraft (bombers, tankers, transport aircraft), accompanied by corresponding research efforts of NACA (since 1958 NASA) and other research institutions, were, in the fifties, the technological foundation for the civil transport aircraft, which were quickly to succeed all over the world, except for the East Block countries.

In that context, the Airbus support based on public credits could be justified – one was, nevertheless, not doing anything much different from what the Americans were doing in the fifties – but it was interpreted as illegal subsidies by the American competitors. When in 1990 it was appealed to GATT ("General Agreement on Tariffs and Trade", since 1995 the "World Trade Organization, WTO"), GATT supported the American viewpoint. Finally, however, bilateral agreement between the US and Europe was reached in 1992, which set for the European governments upper limits with regard to the direct support and for the Americans upper limits for the indirect support [85].

Airbus adapted its design processes [87], and the USA did not feel impeded to consolidate her aerospace industry, which represented one of her largest economic assets [88]. The neo-mercantile attitude of the USA caused a company concentration and thereby a globalization of aerospace [89]. To the merger of the (American) companies to today's corporations Boeing, Lockheed Martin, Northrop-Grumman and Raytheon Hughes, as well as the specific research programs, has Europe not yet found the right answer.

The concentration of the American companies had, however, still another reason. Already at the onset of the nineties, attention in the USA was directed to something that sometimes is in the danger to be overloooked at the higher management levels of industry and politics, namely, the knowledge, the ability and experience of the company personnel related to all production processes, but especially related to product definition and development of aircraft. This of course holds essentially also for military aircraft manufacturing, which was, and still is, dependent on government contracts.

An American aeronautical engineer, who entered his professional life in 1950, worked in an industry that designed and built, during his professional career, about 80 new aircraft. If he started in 1960, it were still 40, and, starting in 1990, it were only four new aircraft, a number that was soon to reduce to two [90]. Now, how could engineers in the future gain the necessary range of expertise and experience and how was the knowledge transfer to junior professionals to take place? The answer was: by means of experimental and demonstration programs. In view of the enormous technical complexity of modern military aircraft, and the correspondingly high unit costs, this would have been associated with exorbitant costs considering the many companies existing at the time. The consequence was that mergers became unavoidable just to solve this "personnel problem".

The American authorities made it unmistakably clear that they wanted a concentration. During the jokingly called "Last Supper", the directives, which led to the reduction of 32 companies to now four, were given. Over a million jobs disappeared, but, on the other hand, the needed advantages of size ("economies of scale") and a product multitude with an enormous synergetic potential (civil-military, aeronautics-astronautics) was gained [89].

The controversy about Airbus loans and the recession during the first half of the nineties, the subsequent American neo-mercantile attitude and the globalization of the aerospace industry during the second half of the nineties, had, partly, a dramatic effect on the German aeronautical industry and aeronautical research.

In 1992, aircraft orders were, for the first time, higher for Airbus than for Boeing. The worldwide recession in conjunction with the aftermath of the deregulation of American air traffic at the end of the seventies [86], led, however, in 1992/93 to massive cancellations of orders for commercial aircraft. The value of the dollar dropped so low that, despite the (official) exchange-rate protection [85], revenues from aircraft sales, which also declined, fell through the floor.

Dasa reacted in various ways to this development [91]. With the "Dolores"-program (**Do**llar **lo**w **res**cue), the number of employees was reduced from 80,000 to 45,000 between 1993 and 1998. During the years 1993/94 people worked reduced hours. The Dasa Aeronautics Division started in September 1992 to set up a so-called "Structured Development Concept" ("Entwicklungsstrukturkonzept"), which was to soften the impact of the "Dolores" related reduction in personnel, but was also to align the division to the prevailing boundary conditions, which had changed with the beginning of the nineties.

In mid-1993, the possibility of a new national aeronautical research program could be foreseen (from 1988 to 1995 predominantly hypersonic technologies were sponsored), which was essentially to be funded by the "Federal Ministry of Education and Research, BMFT", with a participation of the "Federal Ministry of Economics, BMWi", which, prior to the bilateral agreement, was the provider of the loans, and the integration of the "Eight-column"-research and technology program of the "Federal Ministry of Defense, BMVg". The latter was adopted in 1992 and ran until 1999; since then, it is being continued as military technology program "Air Vehicles" ("Luftfahrzeuge").

Dasa and the supply industry had to accept the fact that the aeronautical research program was not just a compensation for the discontinued loans, but a pre-competitive research and development program, GATT-conform and in agreement with the promotion directives of the European Union, and, in addition, that the companies had to participate with 50 per cent in the funding of the respective programs. Consequently, a forward strategy was developed. With the definition of "Reference Concepts", "Focused Technology Areas" ("Technologieschwerpunkte") were created (1993). These became in 1994, with a change in the meaning of the word, research and development projects, which were to align fundamental scientific and technological work to marketable products by an early cooperation between research and industry [92].

Already with the specification of the aeronautical research program in the summer of 1994, Dasa pursued the incorporation of the DLR and of relevant universities into the program. In February 1995, *Dieter Volk* organized the first of the aeronautical colloquia, which supported a closer cooperation between research and industry, and, at the same time, served to structure the cooperation within the aeronautical research program.

Dasa operated since October 1994 under the name "Daimler-Benz Aerospace". The Dutch aircraft manufacturer Fokker, of whom Dasa had acquired 51 per cent of the shares in April 1993, went bankrupt in March of 1996. Dornier Aeronautics was passed on to the American company Fairchild in June of 1996. In 1990, the engine manufacturing company BMW Rolls-Royce (BRR, today RR-Germany) was founded with its (later) seat in Dahlewitz near Berlin. (At the end of 1999, BMW left this company.) Fairchild-Dornier and RR-Germany are today the only aeronautical companies in Germany remaining independent of Dasa.

The aeronautical research program (Program framework "Aeronautical research and technology") of the federal government ran from 1995 to 1998 on a promotional (funding) volume of 600 Million DM (approx. 300 Million Euro). It was conceived around several subject groups and also included the DLR and several universities, but also had econnections to the "Collaborative Research Centers" and "Priority Programs" of the DFG [93]. Federal-state funding of aeronautical research within official programs was only provided in Bavaria.

The recession 1992/93 had other consequences, too. One was that the number of new university students, not only in aerospace engineering, dropped drastically, similarly to the situation during the world economic crisis seventy years ago. Industry and research are still negatively affected even today. During the years 2000/2001, the number of new graduates reached with 40–50 per cent of the long-term average a minimum. A further consequence of the recession was the fact that the DFG reduced the initial salary of young scientists in sponsored projects by half due to the large number of job-seeking graduates. This was later, with a large phase shift, repealed. During the last years it was therefore difficult to attract young researchers.

Despite this problem, some new "Collaborative Research Centers" were established [79]:

- **SFB 340:** "Flow control and flow-structure interaction on wings", RWTH Aachen, since 1997.
- **SFB 409:** "Adaptive structures in aircraft engineering and light-weight designs", University Stuttgart, since 1997.
- **SFB 1664:** "Control of complex shear flows", Technical University (TU) Berlin, since 1998.

DLR incorporated in 1997 the "German Space Agency" ("Deutsche Raumfahrt-Agentur, DARA"), only founded in 1989, which was as project-management agency also responsible for the aeronautical research program. DLR changed its name to "German Aerospace Center" ("Deutsches Zentrum für Luft- und Raumfahrt") as already mentioned, it sees itself as a "Research Enterprise" [94].

The DGLR merged early 1993 with the "Hermann-Oberth-Society" ("Hermann-Oberth-Gesellschaft, HOG"), the "Society of Space Research and Astro-

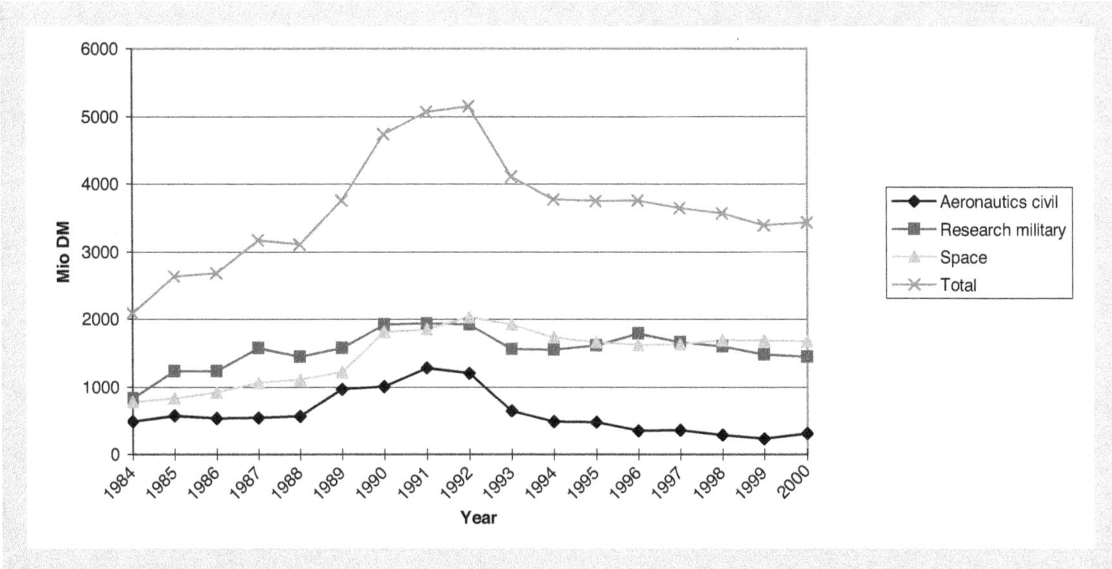

Payments of the federal government for aerospace research between 1984 and 2000. (Mio DM = Million Deutsch Marks; 1 DM ≈ 0,50 Euro)

nautics" ("Gesellschaft für Weltraumforschung und Raumfahrt, GWR") and the "Professional Aeronautical Association" ("Fachverband Luftfahrt, FVLF"), founded in 1964 in the GDR as member of the "Technical-Engineering Association" ("Ingenieur-Technischer Verband, KDT"). The resulting scientific-technical association was renamed "German Society of Aeronautics and Astronautics – Lilienthal-Oberth" ("Deutsche Gesellschaft für Luft- und Raumfahrt – Lilienthal-Oberth e.V.").

Has now the aeronautical industry finally become a "normal" industry: highly technological, capital intensive (a consequence of the "de-nationalization" of the companies is that the financing of (civil) product developments has now to take place via the capital markets), and global in its structures, processes and markets [89]? There is no definite answer. As long as military and civil aeronautics are both part of a comprehensive political strategy, like in the US, one cannot consider it a "normal" industry.

Germany, as Europe's economically major technology site (2000), seems to have even today difficulties with aeronautics. The public enthusiasm for the field is still large. This is evidenced by the corresponding university study courses, which still draw a large number of excellent students, by the interest of young people in a research job at DLR or an occupation in the aircraft industry, but also by the visits to flight shows, "open-door" events of research and industry, and sporting events. On the government side, it is different. A continuous and sustainable promotion of aeronautics and aeronautical research, as in the USA and by the West European partner states, is barely visible. Of course, there was, up until the recent past, the necessity of political and economic considerations for the former allied powers [4], which should have, however, today been replaced by other points of view. Even if today the information technology, gene technology and others are considered the "Growth-industries", the aeronautical industry, with its product-, process-, and systems-integration-technologies and the resulting high demands on research, is still a technological pace maker per se, which should receive appropriate political attention. This is especially true with regard to the continued pursuit of European concentration, which requires economically and technically strong partners, a requirement that is met by Germany.

Europe has to merge her aerospace companies [95]. After Daimler-Benz Aerospace became first Daimler-Chrysler Aerospace, the road to a European company was consequently entered with the establishment of the "European Aeronautic Defense and Space

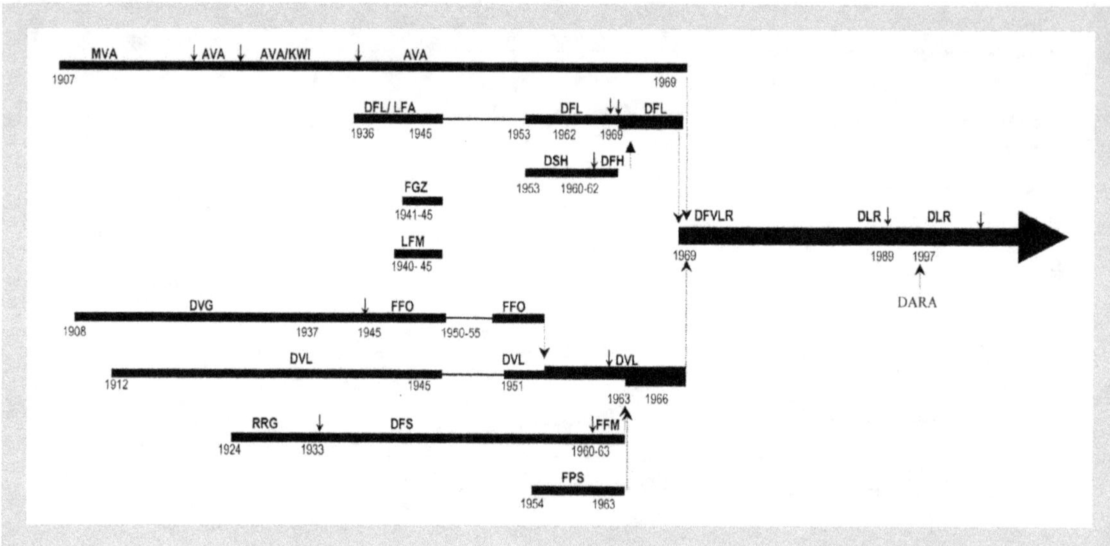

Concentration process within German institutional aeronautical research – the road to the "German Aerospace Center" ("Deutsches Zentrum für Luft- und Raumfahrt, DLR") according to K. Weingarten.

Company, EADS". Important is an overall European approach, which brings research and industry continuously and lastingly together, and which secures, with experimental and demonstration programs as well as with an adequate civil/military "dual-use strategy", the future of this line of business. In Germany, a first important step was undertaken by DFG and universities, the DLR and the German aerospace industry in January of 1999 with a memorandum concerning the future cooperation in the area of aerospace research.

Growing International Engagement of German Aeronautical Research

After World War Two and the formation of the Western/Eastern blocks during the Cold War, and the subsequent sovereignty of the Federal Republic in 1954/55, West German aeronautical research and industry were soon accepted by and integrated into the western "Community". The West German aeronautical industry was already in 1955 invited to join the "Association Internationale des Constructeurs de Martériel Aéronautique, AICMA", the later AECMA ("Association Européenne des Constructeurs de Matériale Aérospatiale"). The East German aeronautical industry and research was, however, after a successful start soon abolished.

The landscape of aeronautical research changed fundamentally after 1945, compared to that existing prior to the war. One finds within research (research establishments, universities, but also industry) since the seventies a growing interlocking of the activities of the western European countries. This is similar to the situation in civil and military aircraft manufacturing proper, which takes place today within international consortia (which are on the way to a European company).

At the very beginning of this development were certainly the initiatives of *Theodore von Kármán*, who early on strove for a membership of the federal republic within AGARD, which realized in 1955 [67]. (In 1997, AGARD combined with the "Defense Research Group, DRG", to the "Research and Technology Organization, RTO" [67].) For researchers from West Germany, the "Fluid Dynamics Panel, FDP" was initially the most important field of action [96]. They were also from the very beginning fully involved in AGARD-Symposia, "Working Groups", which considered specific problems of a scientific-technical nature, in "consultant and exchange" programs, et cetera. In 1956, the

Theodore von Kármán (center) with the first two German delegates to AGARD, *August Wilhelm Quick* (left) and *Theodor Benecke*, 1959 in Aachen.

"Training Center for Experimental Aerodynamics", the later "Von Kármán Institute, VKI", was founded in Rhode-Saint-Gènese near Brussels. German students already participated in the first courses. Since 1959/60, the German republic also belonged, with other NATO countries, to the "Sponsors" of the VKI. *Von Kármán* made another important contribution to the advance of international cooperation by the foundation of the "International Council of the Aeronautical Sciences, ICAS", which organized in 1958 in Madrid its first congress.

An important role in interlocking especially the European research activities was played by *Dietrich Küchemann*. He founded in 1964, together with *George F. Batchelor*, the "European Mechanics-(EUROMECH-) Colloquia", small informal gatherings with invited participants, which are dedicated to problems of fluid mechanics and structural mechanics. At the beginning of the seventies, he initiated the "European Program on Viscous Flows, EUROVISC" and the corresponding program "on Hypersonic Flows, EUROHYP". In both programs, administered by the aerospace research establishments, yearly surveys concerning the investigations at universities, research establishments and industry were given in the form of brief reports, working groups were formed, and workshops organized.

Following the start of the Airbus A 300 development at the end of sixties, the scientific-technical exchange within Western Europe reached a new dimension with the foundation of the "Group for Aeronautical Research and Technology in Europe, GARTEUR". English, French, and German representatives of the ministries and research institutions involved, founded GARTEUR in 1973. After the Netherlands in

AGARD publishes since 1952 one of its main document series, the "AGARDographs".

Dietrich Küchemann (1911–1976).

Dietrich Küchemann, born in Göttingen in 1911, studied in Munich and Göttingen.

1936 PhD with *Prandtl* in Göttingen with the topic "Stability of a flat-plate gaseous flow with a linear velocity distribution." He transfers to the AVA and joins the Theoretical Gasdynamics Department, whose head he becomes in

1938 together with his colleague *Johanna Weber*, he develops the basic principles of propulsion aerodynamics of jet aircraft.

1946 he transfers to the RAE in Farnborough; his wife, *Johanna Weber*, follows him later.

1953 their book "Aerodynamics of Propulsion" is published.

1946–1956 development of methods for the computation of flows about swept wings of arbitrary aspect ratio at subsonic speeds; RAE-standard-method.

1956–1966 research related to supersonic aircraft at M = 1.2 with swept wings and M = 2.0 with slender delta wings ("Concorde"). He is decisively involved in defining the configuration of the "Concorde". He engages himself since

1964 successfully for the scientific cooperation within Europe and proposes test facilities for the correct Mach number und Reynolds number simulation at transonic speeds.

1966–1971 he heads the Department of Aerodynamics at RAE.

1970 he receives the Ludwig-Prandtl-Ring

1971 he becomes chairman of the AGARD Fluid Dynamics Panel and gives lectures on the "Aerodynamic Design of Aircraft".

1981 joined the group, GARTEUR was given in 1981 an official status by a Memorandum of Understanding of the respective governments. Later, further European countries joined. GARTEUR carries out research activities with members of the research establishments and the (civil) aeronautical industry, as well as university representatives, within "Action Groups" in the field of aerodynamics, flight mechanics, structures, and propulsion, but also helicopter topics are considered.

The military pendant to "GARTEUR" is "EUCLID (European Cooperation for Long Term Defense)", initiated by the "IEPG (Independent European Program Group)" of NATO, since 1992, the "Western European Armament Group, WEAG". EUCLID coordinates various military aeronautical research activities, for instance, related to aerodynamics and materials. A specific German-American exchange exists since the seventies within the framework of "Data-Exchange Agreements, DEAs" of the ministries (departments) of defense, where industry, research establishments and universities are participating.

The European Union started during the second half of the eighties to consider questions related to the support of pre-competitive engineering technologies for civil aircraft. In the "EUROMART (European Cooperative Measures for Aeronautical Research and Technology)" study, representatives of ten European aircraft companies worked out, on behalf of the EU in the time period February 1987 to February 1988, a proposal for a research and technology program [97]. The pilot phase of a corresponding program commenced in 1989 within the second Framework Program of the EU imbedded in BRITE-EURAM ("Basic Research in Industrial Technologies for Europe – European Research in Advanced Materials)". Subsequently, no independent programs were established and the support of aeronautical research and technology remained within the framework programs that followed (5th Framework Program 1998–2002) embedded in BRITE-EURAM [98]. Topics relevant to aeronautics are also being sponsored within other EU-programs, for instance, topics related to information technology within ESPRIT ("European Strategic Program for Research and Development in Information Technologies"). The EU endeavours, in addition, to establish a closer coordination of its sponsored topics with the ones of "EUREKA (European Research Coordination Agency)". EUREKA supports – without financial contributions – the "border-crossing" cooperation of European industry and research establishments in research and technology endeavours for civil purposes, where, however, aeronautics is not a central topic.

Triggered by the "Airbus-loans controversy" with the USA, but also due to the general situation during the first half of the nineties, the European aerospace research establishments:

- Centro Italiano Ricerche Aerospaziali (CIRA), Italy,
- Defense Research Agency (DRA), the former Royal Aircraft Establishment (RAE), today the Defense Evaluation Research Agency (DERA), Great Britain,
- German Aerospace Center (Deutsches Zentrum für Luft- und Raumfahrt, DLR), Germany,
- Flygtekniska Försöksanstalten (FFA), Sweden,
- Instituto Nacional de Técnica Aerospacial (INTA), Spain,
- Nationaal Lucht- en Ruimtevaartlaboratorium (NLR), the Netherlands,
- Office National d`Etudes et de Récherches Aérospatiales (ONERA), France

established the "Association of European Research Establishments in Aeronautics, AEREA".

In 1999 the organization was registered in Brussels as legal entity in the sense of an "Association of Interests", as, for instance, the AECMA. On this occasion, the acronym was changed to EREA and registered as legal trademark. The tasks of EREA are: the coordination of research projects, the exchange of personnel, the mutual provision of research facilities, the joined acquisition of research contracts. Regarding the research facilities, DLR and NLR led the way by combining, in 1996, the operation of various wind tunnels under the auspices of DNW (in the legal form of a foundation). ONERA is likely to join with some of its wind tunnels.

There was also in the case of the European professional aerospace associations the formation of an alliance. In 1992/93, the eight societies founded the "Confederation of European Aerospace Societies, CEAS". Specialist committees of CEAS organize conferences, short courses, et cetera. A large "European Aerospace Sciences Conference", analogous to the "Aerospace Sciences Meeting" of the "American Institute of Aeronautics and Astronautics,AIAA", is, considering the existing close interlock between industry and research within Europe, long overdue and still missing.

An important step, also long overdue, was made in 1997 on the initiative of *Fred Thomas*, by DLR, together with the DGLR, and ONERA by merging their Journals "ZFW – Zeitschrift für Flugwissenschaften und Weltraumforschung" and "La Recherche Aerospatiale" to the international Journal "Aerospace Science and Technology, AST". Since mid-1999, the publication of the AST-Journal has been joined by the research establishments of Italy (CIRA), the Netherlands (NLR), Sweden (FFA) and Spain (INTA), so that a centralized European Aeronautical Journal is almost reality.

References

1 A. Baeumker, Ein Beitrag zur Geschichte der Führung der deutschen Luftfahrttechnik im ersten halben Jahrhundert, 1900–1945, Heft XXXIV der Schriftenreihe "Langfristiges Planen der Forschung und Entwicklung", Bad Godesberg, 1971

2 H. Trischler, Luft- und Raumfahrtforschung in Deutschland 1900–1970, Politische Geschichte einer Wissenschaft, Campus Verlag, Frankfurt/New York, 1992

3 H. Trischler, Dokumente zur Geschichte der Luft- und Raumfahrtforschung in Deutschland 1900–1970, DLR-Mitteilung 92–08, 1992

4 C. M. Andres, Die bundesdeutsche Luft- und Raumfahrtindustrie 1945–1970,Ein Industriebereich im Spannungsfeld von Politik, Wirtschaft und Militär, Peter Lang, Frankfurt, 1996

5 L. Budraß, Flugzeugindustrie und Luftrüstung in Deutschland 1918–1945, Droste Verlag, Düsseldorf, 1998.

6 U. Kirchner, Geschichte des bundesdeutschen Verkehrsflugzeugbaus – Der lange Weg zum Airbus, Campus Verlag, Frankfurt/New York, 1998

7 P. Bruders (Eds.), Beiträge zur Geschichte der Deutschen Versuchsanstalt für Luft- und Raumfahrt e.V., Festschrift aus Anlaß des 50jährigen Bestehens der DVL im April 1962, Berlin, 1962

8 J. C. Rotta, Die Aerodynamische Versuchsanstalt in Göttingen, ein Werk Ludwig Prandtls, Ihre Geschichte von den Anfängen bis 1925, Vandenhoek & Ruprecht, Göttingen, 1990

9 W. Wuest, Sie zähmten den Sturm, Verlag Göttinger Tageblatt, Göttingen, 2nd Ed., 1991

10 H. Schubert (Schriftleitung), Luftfahrtforschung, Luftfahrtindustrie und Luftfahrtwissenschaft in Braunschweig, Textbeiträge einer Vortragsveranstaltung der DGLR – Fachgruppe 12, "Geschichte der Luft- und Raumfahrt", DLR, Braunschweig, 1990

11 L. Bölkow (Herausgeber), Ein Jahrhundert Flugzeuge, Geschichte und Technik des Fliegen, VDI-Verlag, Düsseldorf, 1990

12 Th. von Kármán, Aerodynamics: Selected Topics in the Light of their Historical Development, Cornell University Press, Ithaka, New York, 1954

13 J. E. Burggravius, Achilles Redivivus, seu panoplia physico-vulcania, Amsterdam, 1612

14 J. J. Becher, Närrische Weisheit und weise Narrheit, Frankfurt, 1682

15 C. F. Burgwein, Die Kunst zu fliegen nach der Art der Vögel, Basel, 1784

16 Die Flugzeughandschrift des Melchior Bauer, Greiz, Faksimiledruck des Thüringer Staatsarchivs, Heft 1, 1922

17 G. Cayley, On Aerial Navigation, in Nicholsons Journey of Philosophy Bd. XXIV und XXV, London, 1809/10

18 Jakob Degens Flugmaschine, Zeitung für die elegante Welt, 30. August 1808

19 B. Berblinger, Das Denkmal für den Schneider von Ulm, Chemnitzer Tageblatt, 8. August 1933

20 Article about Kaufmann in: Leipziger Illustrierte Zeitung, 5. September 1868

21 O. Lilienthal, Der Vogelflug als Grundlage der Fliegekunst, Nachdruck Dr. Martin Sändig oHG, Wiesbaden, 1965. (English edition: Otto Lilienthal, Birdflight as the Basis of Aviatation: A Contribution toward a System of Aviation, Motorbooks Intl./MBI Publishing Company, St. Paul, MN, USA, 2001)

22 P. Meyer, Luftschiffe, Bernard & Graefe Verlag, Bonn, 1992

23 H. G. Knäusel, LZ 1, der erste Zeppelin, Geschichte einer Idee, Kirschbaum Verlag, Bonn, 1985

24 W. Rathjen, Historische Entwicklung des Flugzeuges im Überblick, Ein Jahrhundert Flugzeuge, Geschichte und Technik des Fliegens (L. Bölkow, Herausgeber), VDI-Verlag, Düsseldorf, 1990, S. 8–51

25 A. Wüst (Herausgeber), Gustaf Weißkopf: "Ich flog vor Lilienthal", Erster Motorflug 14. August 1901, Fritz Majer und Sohn, Leutershausen, 1999

26 W. Heinzerling, Ohne echte Beweise zum ersten Motorflieger der Welt hochstilisiert?, VDI Nachrichten, Nr. 40, 8.10.1999

27 S. Finsterwalder, AËRODYNAMIK, Enzyklopädie der mathematischen Wissenschaften IV 2, Leipzig, 1902

28 L. Prandtl, Die Bedeutung von Modellversuchen für die Luftschifffahrt und Flugtechnik und die Einrichtungen für solche Versuche in Göttingen, Z. VDI 53, 1909, S. 1711–1719

29 W. Hackenberger, Die alten Adler, Pioniere der deutschen Luftfahrt, J.F. Lehmanns Verlag, München, 1960

30 H. Trischler, Die neue Räumlichkeit des Krieges: Wissenschaft und Technik im Ersten Weltkrieg, Berichte zur Wissenschaftsgeschichte 19, 1996, S. 95–103

31 G. Brinkmann, H. Zacher, Die Evolution der Segelflugzeuge, Bernard & Graefe Verlag, Bonn, 2. Auflage 1999

32 W. Wagner, Hugo Junkers, Pionier der Luftfahrt – seine Flugzeuge, Bernard & Graefe Verlag, Bonn, 1996

33 J. Roeder, Giganten am Himmel, Rudolf Leuthold Verlag, München, 1982

34 K.-D. Seifert, Der deutsche Luftverkehr 1926–1945 – auf dem Weg zum Weltverkehr, Bernard & Graefe Verlag, Bonn, 1999

35 A. L. Schmitz, Die Luftfahrtindustrie, Ein Jahrhundert Flugzeuge, Geschichte und Technik des Fliegens (L. Bölkow, Herausgeber), VDI-Verlag, Düsseldorf, 1990, S. 544–575

36 H. Beauvais, K. Kössler, M. Mayer, C. Regel, Flugerprobungsstellen bis 1945, Johannisthal, Lipezk, Rechlin, Travemünde, Tarnewitz, Peenemünde-West, Bernard & Graefe Verlag, Bonn, 1998. (English edition: H. Beauvais, K. Kössler, M. Mayer, C. Regel, German Secret Flight Centres, Midland Publishing (Ian Allen Group), Hinckey, UK, 2003)

37 K.H. Kens, H.J. Nowarra, Die deutschen Flugzeuge 1933–1945, J. F. Lehmann, München, 5. Auflage, 1977

38 H. Trischler, Self-Mobilization or Resistance? Aeronautical Research and National Socialism, Science and Technology under National Socialism (M. Walker, M. Renneberg, editors), Cambridge, 1994

39 G. Wissmann, Luftfahrtforschung und Luftfahrttechnik in Deutschland 1933–1945, DGLR Jahrbuch II, Bonn, 1993, S. 739–756

40 U. Kalkmann, Von Kármán zwischen Aachen und Pasadena, "...der ganzen Hochschule zu Weltruhm verholfen ...", Abhandlungen aus dem Aerodynamischen Institut der Rheinisch-Westfälischen Technischen Hochschule Aachen, Heft 32, 1996, S. 1–8

41 E. Thiel, Vom Ötztal nach Modane, Aus der Geschichte des großen Hochgeschwindigkeits-Windkanals "Bauvorhaben 101" der Luftfahrtforschungsanstalt München (LFM), später S1 MA der ONERA, DGLR-Jahrbuch 1986, Teil II, S. 773–795, also: E. Thiel, Windkanal Ötztal

als Grundidee und Anregung für die Planung des US-Luftwaffenzentrums AEDC Tullahoma, DGLR-FG12, 1987
42 K. von Gersdorff, K. Grasmann, H. Schubert, Flugmotoren und Strahltriebwerke, Bernard & Graefe Verlag, Bonn, 3. Auflage, 1995. (English edition: K. von Gersdorff, K. Grasmann, H. Schubert, German Aviation Power Plants, the Beginnings - Aircraft Engines of WW II - Postwar Developments, SAE Aerospace International, Warrendale, PA, USA, 2003/2004)
43 W. Georgii, Forschen und Fliegen, ein Lebensbericht, Tübingen, 1976
44 H.J. Ebert, J.B. Kaiser, K. Peters, Willi Messerschmitt - Pionier der Luftfahrt und des Leichtbaues, Bernard & Graefe Verlag, Bonn, 1992. (English edition: H.J. Ebert, J.B. Kaiser, K. Peters, Willi Messerschmitt - Pioneer of Aviation Design, Schiffer Publishing Ltd., Atglen, PA, USA, 2000)
45 F. Trenkle, Bordfunkgeräte - Vom Funkensender zum Bordradar, Bernard & Graefe Verlag, Koblenz, 1986
46 W. Heinzerling, Windkanäle, Ein Jahrhundert Flugzeuge, Geschichte und Technik des Fliegens (L. Bölkow, Herausgeber), VDI -Verlag, Düsseldorf, 1990, S. 304–333
47 The Challenge of Flight Research, A Historical Account and Technical Guide, DLR-Institut für Flugmechanik, DLR-IB 111-99/02, Brunswick, 1999
48 W. Berghahn, Robert Musil in Selbstzeugnissen und Bilddokumenten, rowohlts monographien, Rowohlt, Reinbeck b. Hamburg, 1963
49 Th. von Kármán, Freiheit und Organisation in der Luftfahrtforschung, Heft 57 der Arbeitsgemeinschaft für Forschung des Landes Nordrhein-Westfalen, Köln und Opladen, 1959, S. 7–16
50 J.D. Anderson, jr., A History of Aerodynamics and its Impact on Flying Machines, Cambridge University Press, New York, 1997
51 E. Krause, U. Kalkmann, Prandtls Schüler in Aachen, in: Ludwig Prandtl, ein Führer in der Strömungslehre, biographische Artikel zum Werk Prandtls (G.E.A. Meier, Herausgeber), Vieweg, Braunschweig/Wiesbaden, 2000, S. 243–276
52 A. Baeumker, Von der Führung der deutschen Luftfahrttechnik im ersten halben Jahrhundert 1900–1945, Management in Wissenschaft und Technik, Festschrift für Theodor Benecke zum 60.Geburtstag (M. Schröder, Herausgeber), DFVLR, Porz-Wahn, 1971, S. 19–43
53 Th. von Kármán (with Lee Edson), The Wind and Beyond: Theodore von Kármán, Pioneer in Aviation and Pathfinder in Space, Little, Brown and Company, Boston, Mass., USA, 1967
53a W. Wagner, Die ersten Strahlflugzeuge der Welt, Bernard & Gräfe Verlag, Koblenz, 1989. (English edition: W. Wagner, The First Jet Aircraft, Schiffer Military/Aviation History, Atglen, PA, USA, 1998.
54 O.E. Pabst, Kurzstarter und Senkrechtstarter, Bernard & Graefe Verlag, Koblenz, 1984
55 R. Kosin, Die Entwicklung der deutschen Jagdflugzeuge, Bernard & Graefe Verlag, Koblenz, 1983. (English edition: R. Kosin, The German Fighters since 1915, Conway Maritime Press, London, 1988)
56 R. Cescotti, Kampfflugzeuge und Aufklärer, Bernard & Graefe Verlag, Koblenz, 1989. (English edition: R. Cescotti, Bombers and Reconnaissance Aircraft 1939 to the Present, Schiffer Publishing LTd., Atglen, PA, USA, 2001)
57 K. von Gersdorff, K. Knobling, Hubschrauber und Tragschrauber, Bernard & Graefe Verlag, Bonn, 3. Aufl., 1999.
58 E. Brown, Wings of the Luftwaffe, Flying German Aircraft of the World War Two, Jane's Publishing Company, London/Sydney, 1977
59 M. Judt, B. Ciesla (editors), Technology Transfer out of Germany after 1945, Harwood Academic Publishers, Amsterdam, 1996
60 J. Gimbel, Science, Technology and Reparation, Exploitation and Plunder in Postwar Germany, Stanford University Press, 1990
61 B. Ciesla, "Intellektuelle Reparationen" der SBZ an die alliierten Siegermächte? Begriffsgeschichte, Diskussionsaspekte und ein Fallbeispiel - Die Deutsche Flugzeugindustrie 1945–1946, Wirtschaftliche Folgelasten des Krieges in der SBZ/DDR, C. Buchheim (Herausgeber), Nomos Verlagsgesellschaft, Baden-Baden, 1995, S. 79–109
62 V. W. Petersen, G. B. Stephens, P. Cherney, Focke-Wulf Designing Offices and General Management Bad Eilsen; Appendix XVIII: Notes on Discussion with the Staff of Focke-Wulf Ltd. by R. Smelt, CIOS Item No. 25, File No. XXVI-6, May 1945
63 A. Betz (Herausgeber), Monographien über Fortschritte der deutschen Luftfahrtforschung seit 1939, im Auftrage des britischen Ministry of Supply bei der AVA Göttingen verfasst, 1945/46
64 R. Ulken, Neuaufbau der DFL in Personalunion mit der Technischen Hochschule Braunschweig, DGLR FG 12, Braunschweig, 1990
65 W. Schulz, Von der WGL zur DGLR, aus sechs Jahrzehnten Tätigkeit für die Luftfahrt, Management in Wissenschaft und Technik, Festschrift für Theodor Benecke zum 60. Geburtstag, (M. Schroeder, Ed.), DFVLR, Porz-Wahn, 1971, S. 43–71

66 P. Hamel, Nachwuchs fördern und Tradition pflegen: ein Interview mit Prof.(em.) Dr. phil. Hermann Blenk, Luft- und Raumfahrt, DGLR, 3/87, 1987, S. 18–21
67 J. A. van der Bliek (Editor), AGARD –The History 1952 – 1997, The NATO Research and Technology Organization, RTO, Paris, 1999
68 C. Kleßmann, Die doppelte Staatsgründung, Deutsche Geschichte 1945–1955,Vandenhoeck & Ruprecht, Göttingen, fünfte Auflage, 1991
69 J. Michels, J.Werner, Luftfahrt Ost, 1945–1990, Bernard & Graefe Verlag, Bonn, 1994
70 L. Brehmer (Herausgeber), Luftfahrt in Sachsen, ein historischer Abriss, UniMedia, Leipzig, 1998
71 H.-D. Tack, personal communication, Berlin, 1998
72 K.-H. Meves, Pirna 014 – Flugtriebwerke der DDR, Entwicklung, Erprobung und Bau von Strahltriebwerken und Propellerturbinen, Aviatic Verlag, Oberhaching, 1997
73 Zeit im Flug 1994, Eine Chronologie der Geschichte der Daimler-Benz Aerospace AG, München, 1995
74 Faktenbericht zum Bundesbericht Forschung 1990, BMFT, Bonn, 1990
75 H. Dissen, DLR-Standort Braunschweig 1969–1997, Daten, Fakten, Ergebnisse, Trends, Interne Schrift, DLR Braunschweig, 1997
76 F. Riegels (Schriftleitung), Die Aerodynamische Versuchsanstalt Göttingen von 1945 bis 1969, AVA, Göttingen, 1969.
77 M. Seidel (editor), Construction 1976–1980, Design, Manufacturing, Calibration of the German-Dutch Wind Tunnel (DNW), DNW, 1982
78 K. Zierold, Forschungsförderung in drei Epochen, Deutsche Forschungsgemeinschaft, Geschichte – Arbeitsweise – Kommentar, Franz Steiner Verlag, Wiesbaden, 1968
79 W. Lachenmeier, personal communication, Bonn, 2000
80 V. Aschoff, Über die Aufgabe und die Organisation der DFVLR – Zwei Jahre Einheitsgesellschaft der deutschen Luft- und Raumfahrtforschung, Management in Wissenschaft und Technik, Festschrift für Theodor Benecke zum 60. Geburtstag, (M. Schroeder, Herausgeber), DFVLR, Porz-Wahn, 1971, S. 115–146
81 DFVLR, Jahresbericht 1969, Porz-Wahn, 1970
82 30 Jahre Deutsches Zentrum für Luft- und Raumfahrt 1969–1999, DLR Presse- und Öffentlichkeitsarbeit, Köln, 2000
83 Förderkonzept Hyperschalltechnologie, Bundesministerium für Forschung und Technologie, Bonn, 1988
84 J. A. van der Bliek, ETW, a European Resource for the World of Aeronautics, The History of ETW in the Context of European Aeronautical Research and Development Cooperation, ETW, Cologne, 1996
85 P. K. Lawrence, Europe vs. America: Strategic Trade in Civil Aeronautics, Strategic Issues in European Aerospace, P. K. Lawrence, D. L. Braddon (editors), Ashgate, Aldershot, 1999, pp. 27–61
86 H.-L. Dienel, P. Lyth (editors), Flying the Flag, European Commercial Air Transport since 1945, Macmillan Press, Houndmills, 1998
87 D. Schmitt, New Perspective on Aeronautical R&T for Europe, Strategic Issues in European Aerospace, P.K. Lawrence, D.L. Braddon (editors), Ashgate, Aldershot, 1999, pp. 213–227
88 H. Prem, Integration and Consolidation in European Military and Civil Aerospace: an Urgent Task, Strategic Issues in European Aerospace, P.K. Lawrence, D.L. Braddon (editors), Ashgate, Aldershot, 1999, pp. 133–142
89 K. Hayward, The World Aerospace Industry: from Internationalism to Globalization, Strategic Issues in European Aerospace, P. K. Lawrence, D. L. Braddon (editors), Ashgate, Aldershot, 1999, pp. 3–14
90 J. A. Drezner, G. K. Smith, L. E. Horgan, C. Rogers, R. Schmidt, Maintaining Future Military Aircraft Design Capability, RAND, R-4199-AF, 1992
91 Die DaimlerChrysler Aerospace AG im Rückblick, Dasa Presse und Information, München, 2000
92 H. Ambos, Geschäftsfeldstrategie und Technologieplanung der Dasa, München, 1995
93 H. Diehl, Cooperation Project Aeronautical Research, New Results in Numerical and Experimental Fluid Mechanics, H. Körner, R. Hilbig (editors), Notes on Numerical Fluid Mechanics, Vol. 60, Vieweg, Brunswick/ Wiesbaden, 1997, pp. 1–6
94 Das Forschungsunternehmen DLR – Ziele und Strategien, DLR Vorstand, Köln, 1999
95 Die Europäische Luft- und Raumfahrtindustrie – Antworten auf globale Herausforderungen, Europäische Kommission, Brüssel, 1997
96 F. Wattendorf, AGARD – German Cooperation, Management in Wissenschaft und Technik, Festschrift für Theodor Benecke zum 60. Geburtstag, (M. Schroeder, Herausgeber), DFVLR, Porz-Wahn, 1971, S. 80–87
97 EUROMART Study Report, Executive Summary, 1988.
98 Aeronautics Related Research, Synopsis of Current Projects Selected under the 1995 Calls for Proposal, European Commission, Brussels/Luxemburg, 1996

Results of German Aeronautical Research

Part 3

From the Beginning to the End of World War Two

CONCEPT AND CONFIGURATION

"The scientist explains what already exists
The engineer creates, what he wants to be".
Theodore von Kármán

Visions and First "Flight Vehicles" – the Beginnings are Promoted Nationally

DIETER SCHMITT, HORST PREM

The age of flying with objects heavier than air starts in Germany with Otto Lilienthal (1848–1896). After numerous theoretical studies [1] and practical experiments, he was able to glide in the summer of 1891, for the very first time, a distance of 30 m with his flight apparatus at moderate wind speeds. Other nations consider the beginning of aircraft engineering coincident with the first powered flight of the Wright Brothers. There are many "firsts" in flight technology, depending on which "first" we want to honor. Before Lilienthal there were many trials with small unmanned flight models of different kinds and, of course, experiments with large-scale models and people as payload or pilots, respectively, but none of them reached the goal they had hoped for. The overall designation was, at that time, "Airship Aviation" and was divided into Aeronautics (Balloon technology) and Aviation (Flight technology).

A comprehensive overview of the different approaches to realize man's old dream of flying was given by Hoernes [2] and by Supf [3].

At this point the question is posed, which influence science had on the different activities related to flight at that time and before? Was science actively stimulating or carefully cautious?

There are publications of the 17th century, which are already explaining that the power of human muscles alone is not enough to become airborne. Also the great German physicist *Hermann von Helmholtz* wrote [4] in 1873: "… *nature already reached with vultures the limit, with muscles as work-performing organs and under the condition of a suitable diet, which can be attained for the size of a creature, which wants to become airborne and stay, for a longer period of time, up in the air. … Under these circumstances it is hardly realistic, that a human would become airborne and stay there, even with the most adept wing-like mechanism, which he would have to move with his own muscles power.*"

There was initially not too much help to be expected from science in terms of supporting the establishment of the basic principles of flight technology. But science and research also depend on stimulations and visions from the outside or from "above".

It is remarkable that before *Lilienthal's* gliding experiments and even after, there was hardly any interest in the development of flight technical apparatus. However, there were already very early some remarkable visionaries like *Zeidler*, who, in his publication [5] of 1710 the "The Flying Traveler", developed considerable fantasies not only concerning the benefits but also the damages associated with the "art of the airship and the art of flying":

"Concerning benefits, not only private people would have the advantages of flight instead of fatiguing travels by water and land. Firstly, they would be faster, secondly wouldn't have to be afraid of enemies, thieves and wild animals; thirdly, they wouldn't fatigue; fourthly, they would not throw nor be thrown off while driving or riding; fifthly, they would be in a better position to give way; sixthly, costs would be minimized, which otherwise would be spent on horses, carriages and ships, also on provisions and eighthly, they would easily escape their enemies. Whole countries would gain incredible profit, wealth and reputation. Ninthly, mail could be sent at highest speed; tenthly, trading would be easier and much more comfortable than by car or ship. Eleventhly, whole nations could evade winter and fly to warmer climates like storks and swallows. Twelfthly, hostile fortifications could be forced by fire and taken from above. Thirteenthly, "descendants" (and here it is already thought of parachute jumpers) could be brought into hostile countries without endangering them. And as fourteenth, one could take in Turkey and France and the whole world. Zeidler also makes interesting points about the dangers of the "art of flying."

This means that there existed already very early many ideas and visions concerning the possible benefits of flight technology. But there are only very few instances that a sovereign charged his scholars with the investigation and development of the basic principles of flight technology. Interesting is also the indication that all flight- and war-relevant drawings and ideas of *Leonardo da Vinci* were, for a long time, locked away in the *Biblioteca Ambrosiana* in Milan before *Napoleon I* carried them off to Paris. This treasure of technical drafts and sketches was not published before the end of the 19th century [3]. This way, science and research gained access to and knowledge of these early ideas of the genius *Leonardo da Vinci* only 350 years later.

At the end of the 19th century (around 1890), science and also possible beneficiaries of flight technology were convinced that a flight vehicle heavier than air operated by muscle power was impossible. The higher we have to esteem the performance of *Otto Lilienthal*, who only had his belief in the art of flying. He endured all the mockery of science and learned people, financed his experiments himself, and just held on to his idea. Science and research stood on the sidelines.

Interesting in this regard is a remark of *Werner von Siemens*, who writes in his memories [6]: *"Undoubtedly we can produce flight vehicles modeled after flying animals if the basic condition is met, which is that we have machines as light and strong as the propulsive muscles of flying animals and with a fuel consumption no bigger than theirs. If this kind of machine is invented, any skilful mechanic can build such a flight vehicle. But the inventors always start at the wrong end and invent machines without having the power to propel these machines."*

Werner von Siemens certainly made here an important remark, which underlines the importance and significance of a light and efficient engine for propulsion. But he misjudged the importance of aerodynamics and flight mechanics if he thinks that any skilful mechanic could build such a flight vehicle.

The importance of *Otto Lilienthal* for the development of flight technology is emphasized by the fact that he was the first, who systematically analyzed the flight of birds, recognized the cambered wing and the importance of cambered airfoils and provided this knowledge to his competitors by publishing all results in his book [1]. It is only surprising that *Lilienthal* had very few pupils in Germany but found the more interest and recognition in France, England and America. Also the *Wright Brothers* refer to the encouraging results of *Lilienthal's* glides.

Lilienthal, an excellent observer of nature, titled Chapter 40 of his book "Calculation of flight work". Besides determining the form of the wings of white and black storks, which he kept in his garden for experimental purposes, he also develops a model for the calculation of the work needed for the "up and down flapping" of wings (as he calls it) by dividing the wing in the spanwise direction into four sections with different angles of attack. Wing twist during flapping is given in detail in the sketch. The flight work in normal flight and the minimum velocity needed for the stork to glide is calculated assuming the drag values of a flat plate at different angles of attack (neglecting the "induced angle of attack").

Today, discretization techniques for aerodynamic wing design are used analogously. Although they are considerably more refined and backed up by vortex models (*Prandtl's* lifting wing theory), hence more precise in terms of the acting forces lift and drag, *Lilienthal* obtained with his model results, which had the correct order of magnitude and formed the

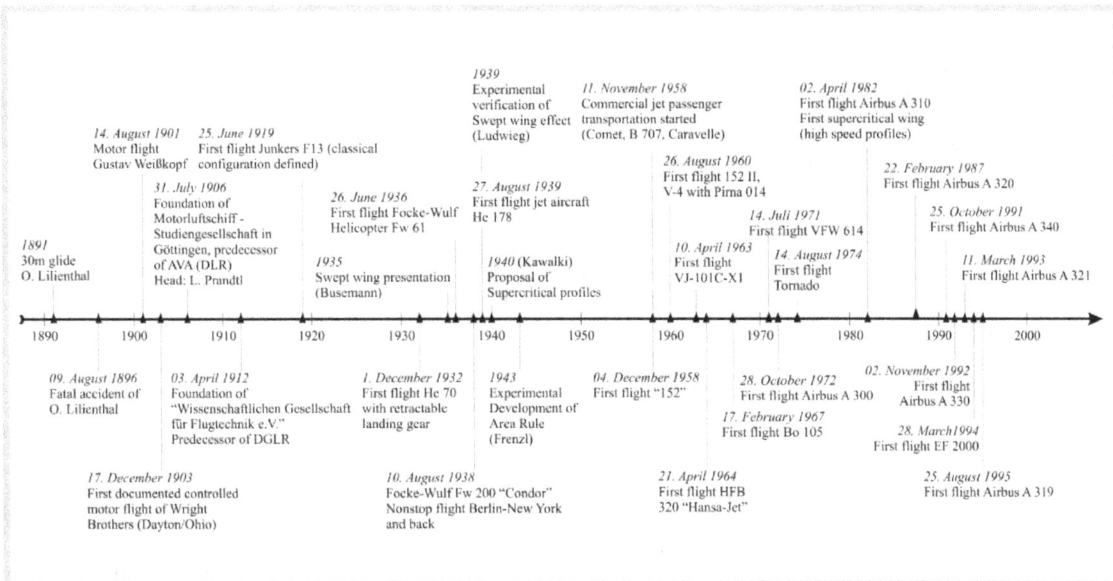

Selected dates of German aeronautical history.

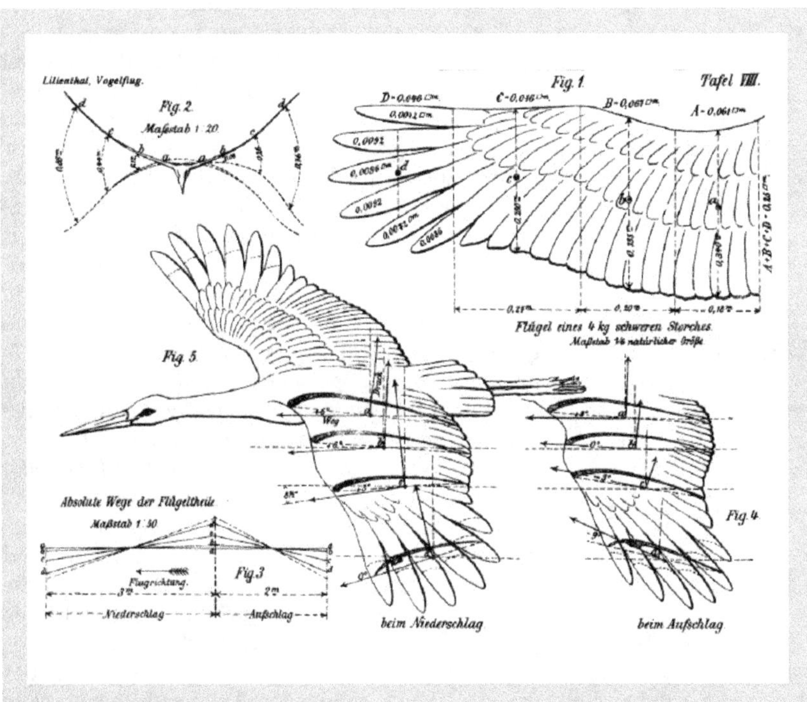

Lilienthal's observations concerning the flight of storks.

Otto Lilienthal in Derwitz (1891).

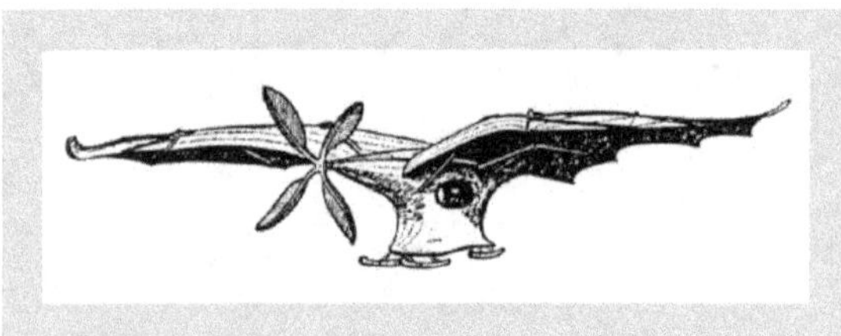

Ader Eole (1890).

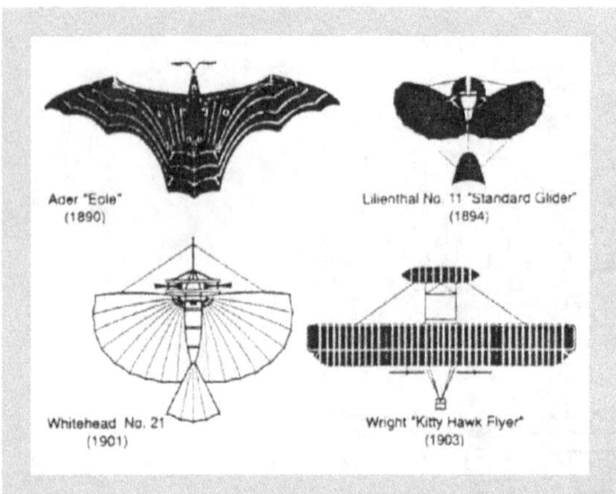

Comparison of flight vehicle configurations (same scale).

basis for the design of the first technical flight vehicles.

The observation of birds encouraged early designers over and over again to copy the flapping of wings. This complex, and technically hard to realize movement certainly led to a number of detours before also *Lilienthal* found to his glider, as predecessor of today's sailplanes [7], and others finally replaced the energy provision per wing flapping by motor-powered propellers.

Following *Lilienthal* there were various engineers all over the world, who, based on *Lilienthals* findings, continued to experiment and came up with numerous other flight vehicle configurations.

Flight-stability problems were considered to account for *Lilienthal's* accident, and this led to different concepts and configurations. Subsequently, different designs originated such as the "lattice kite" of *Chanute*, a multiplane glider, biplane gliders, and many other shapes, all modeled after different kinds of birds. Aim of these experiments was to gain experience with flight devices as *Hoernes* [2] quotes the Franco-American *Chanute*: *"I do not know how many more experiments I will conduct, all in a similar way. I paid for all of them and they all were performed only for the benefit of knowledge and regardless of pecuniary gains. To me it also seems impossible that a really perfect flight machine will be build soon; too many experiments are still needed. Here, I wrote down all my investigations and tests so that maybe another researcher finds what he needs in order to succeed."*

From this one can observe the competition, arising from the researchers' thirst for knowledge and sportive ambition, to be the first to develop a viable machine and to prove its capability to fly and to be controllable. Different concepts, which were considered at the time, are displayed in the figure. Besides the kite shape, common today, nearly all possible configurations have been investigated at one time or another.

Of course, *Gustav Weißkopf* must be mentioned, a German from Leutershausen near Ansbach, who immigrated to the USA and had built there a flight vehicle. In the "Illustrated Aeronautical Announcements" (Journal No. 4 of 1901) he wrote the following article:

"My machine is essentially modeled after a bird, is 4.8 m in length, 0.9 m high and has a maximum width of the body (fuselage) of 0.75 m. This body rests with four wheels on the ground. The diameter of the wheels is 1 m. A 10-horsepower machine drives the front wheels, whereas the back wheels run freely.

On either side is a wing reinforced with bamboo sticks and covered with silk. The span is 10.8 m and the surface area of the wings is 50 m^2. The wings are highly concave on the lower side and have no slack parts. At the height of the wings, a 20 horsepower double-expansion machine is mounted laterally within the body, which moves two counter-rotating propellers with 700 revolutions per minute. Stability is sustained by an apparatus engaging automatically. As fuel calcium carbide or acetylene gas, respectively, is used. The motor weighs 0.9 kg per horsepower and is a miracle regarding the compendious design. I made two test runs with my machine. At both runs, the apparatus landed without any damage. During the first test, a load of 100 kg was taken on board so that the total weight was about 226.5 kg. When the motor worked, the apparatus taxied approximately 30 m, became airborne and flew roughly 1.5 min. During the second test (one hour later), I took out the load and climbed in myself. I will never forget the feeling I had. The success was the same as during the first test. The duration of the flight was 1.5 min and the distance covered in the air was 1,270 m."

On August 14, 1901, nearly two years before the *Wright Brothers*, he is said to have performed the first powered flight near Bridgeport, Connecticut. There is no photo-documentary of *Weißkopf's* flight by a third party, so there is no hard proof whether he really flew.

Reproductions of *Weißkopf's* "flight apparatus 21", as it was called, flew in December 1986 in the United States of America and on September 12, 1997 in Manching in Germany and proved his aircraft concept's ability to fly. The small museum in Leutershausen, honoring *Gustav Weißkopf*, documents his thoughts concerning his configuration and the carbide test-engines.

Also worth mentioning of that early phase of aviation is *Clement Ader's* flight apparatus "Eole" and *Langley's* "Aerodrome". These were more the scientific approaches to find laws concerning lift and drag, controllability, stability and propulsion in order to devise a flyable and controllable configuration. The concepts of kite-like planes with a separation of lift, generated by wings, and propulsion provided by a propeller (motor) succeeded in the end.

Scientists as we know them today, were not present during this first phase of trial and error. Major progress was gained by engineers, who relied on tests and experiments and who wanted to find rules and to make progress, respectively. They were, of course, driven by their ambition to be "The First". Through open discussion and the exchange of experience, by test-flying other configurations, and driven by the generally huge enthusiasm for "personal aerobatics", important findings and progress were made rapidly and in a rather short time. But it is interesting to note that only the engineers and inventors finally succeeded, who, from the beginning, incorporated man as pilot into a configuration, and who were designers, pilots and financiers in one. No military financed engineering group was charged to conduct investigations. Science did not believe in the viability of human flight. Only individual pioneers like *Lilienthal*, and the successful powered flight of the *Wright Brothers* produced the break-through of flight technology.

There were no successors in Germany after *Otto Lilienthal's* death to continue his work. The breakthrough was accomplished by the *Wright Brothers* in the USA, who performed the first documented controlled flight with their "Kitty Hawk" on December 17, 1903.

The French captain *Ferber* was perhaps the first to recognize the importance of aircraft for military as well as civil use. In Germany, it was not before 1908 that captain *Alfred Hildebrandt* made public the importance of the successful flights of the *Wright Brothers* and also pointed out the military importance of this technology. In the following three years, several initiatives then developed, which promoted the further development of flight technology. Now, also science became interested in flight technology.

On April 3, 1912, the "Scientific Society of Flight Technology" ("Wissenschaftliche Gesellschaft für Flugtechnik, WGF") was founded in Berlin. Honorary chairman of the society was His Royal Highness *Prince Heinrich of Prussia* (Heinrich von Preußen). One of

The Whitehead-airplane No. 21 (1901) of *Gustav Weißkopf.*

the three chairmen of the society was *Ludwig Prandtl.* Now, government and science in Germany became interested in flight technology and a period of very successful scientific work commenced.

The Kaiser emphasized his interest in the Scientific Society by telegraphing to the honorary chairman: *"... to inform the members of the society that they are still assured my deepest interest in their efforts with respect to defending the country and industry."*

Around 1910 then, an atmosphere of awakening developed in Germany, which also had an effect on the enthusiasm and support for the new subject aeronautics. Besides the foundation of the scientific society, there was an appeal for a national flight donation, which accumulated the enormous amount of 7.5 million gold marks. Arisen was now the shared interest of government, industry and science to support aeronautical technology, and succeeded with German efficiency. The table lists the various performance records achieved between 1910 and 1914. In 1914, before the outbreak of World War One, Germany held all records except for the speed record [8].

Within only four years, the altitude record was improved by a factor of 2.5, the velocity was nearly doubled, and the endurance and distance records were nearly tripled. Then World War One commenced and configurations changed. Initially the biplane, with front engine and propeller and rear empennage dominated.

	Year	Country	Aircraft	Engine		Result
Height	1910	F	Bleriot	50 PS	Gnôme	3100 m
	1912	F	Morane	80 PS	Gnôme	5610 m
	1914	D	D.F.W.	100 PS	Mercedes	7850 m
Velocity	1910	USA	Bleriot	100 PS	Gnôme	109 km/h
	1912	USA	Deperdussin	100 PS	Gnôme	170 km/h
	1913[a]	F	Deperdussin	160 PS	Rhône	204 km/h
Duration	1910	F	H. Farman	50 PS	Gnôme	8 h 12 min
	1912	F	H. Farman	70 PS	Renault	13 h 18 min
	1914	D	Albatros	100 PS	Mercedes	24 h 12 min
Range	1910	F	H. Farman	70 PS	Renault	585 km
	1912	F	H. Farman	70 PS	Renault	1017 km
	1914	D	Albatros	100 PS	Mercedes	1900 km

Source: J. P. Roeder, Evolution of the Art of Flying since Lilienthal, DGLR-Congress 1991, Berlin, September 10–13, 1991)
[a] No improvement in 1914

The performance records show the dominance of France at the beginnings of powered flight. (For the conversion of metric units into english units see the appendix).

Alfred Hildebrandt (1870–1949).

Alfred Hildebrandt, born in 1870
1888 he enters a military career.
1893 he is enthused for aeronautics by *Lilienthal.*
1896 he founds, together with captain *Moedebeck*, the "Upper-Rhine Association of Aeronautics" ("Oberrheinischer Verein für Luftfahrt").
1897 he is finally ordered to the "Airship-aviator Department".
1901 he manages the first Baltic-sea crossing in a balloon cruising between Berlin and Sweden. He becomes
1905 a teacher at the Airship-aviator School of the Airship-aviator Battalions in Berlin and takes up
1907 studies to supplement his practical knowledge. He also undertakes a study trip to the *Wright Brothers* and brings the knowledge about their flight apparatus to Germany. Vocally and in writing he repeatedly points out the effort, which other nations invest into aviation.
1914 an article of his appears, where he predicts parachute jumps from aircraft for reconnaissance purposes and shooting from aircraft. During World War One he is a member of the flight corps, initially as an observer and later as a squadron and division commander. Both predictions become true. When military aviation was banned after the Versailles-treaty, he became effective as supporter of German aviation.
1934 he becomes "Reichs-Expert-Adviser of Aeronautics" ("Reichsfachberater der Luftfahrt") within the "Reichs-Association of the German Press" ("Reichsverband der Deutschen Presse").
1940 he returns to the air force.
1949 he dies in Oberkochen (Württemberg) on February 24.

The first aerodynamic results of research, such as slightly thicker airfoils, more streamlined fuselage geometries, and profiled bracings prevailed in practical applications. The military leadership realized only during the World War One that aircraft can be armed, can carry bombs to be dropped by hand leading to great confusions on the ground. But the aircraft does still not decide wars. The technical development was not yet far enough advanced for the military to recognize the importance of air superiority, which would have given aircraft a dominating role. On the German side, roughly 300 aircraft were in service [8].

During this time it was mainly *Hugo Junkers*, who gave decisive momentum to aircraft engineering and determined with his investigations the aircraft configuration, which is common up to today. Therefore we pay in the following section special attention to this scientist, designer and entrepreneur. A detailed description of his contributions to aircraft engineering is given in Volume 24 of the book series "Die deutsche Luftfahrt"[9].

Junkers Aircraft as a Result of Applied Research

PETER PLETSCHACHER, BERND JUNKERS

From today's point of view it is astonishing that Hugo Junkers had already exceeded the age of 50, was a successful businessman and a scientist of great reputation, before he turned up as a designer of aircraft. This distinguished him from most of the other aeronautical pioneers, who mainly as pilots or engineers tried to solve empirically the problems of flight technology. Junkers, on the other hand, worked analytically, decomposing complex problems into single fields and finding new solutions on a scientific basis. This approach already distinguished him in his early days as an entrepreneur and this is also the reason for his later success in aircraft and engines manufacturing.

Hugo Junkers graduated in 1883 at the age of 24 from the Institute of Technology (TH) Aachen as "Governmental Mechanical-engineering Supervisor" ("Regierungs-Maschinenbauführer"). On the suggestion of *Adolf Slaby*, who was his professor in Berlin,

Wilhelm von Oechelhaeuser called him to Dessau in 1888 and charged him with the development of a large high-performance gas engine. In 1890, they founded jointly the "Gas-engine Test Station von Oechelhaeuser and Junkers" ("Versuchsstation für Gasmotoren von Oechelhaeuser and Junkers"), where, after two years of intensive work, in 1892 a new two-stroke opposed-cylinder engine was developed, which may be considered the great-grandfather of the later Junkers heavy-oil and flight-diesel engines.

For his first gas engine *Junkers* had already to solve some difficult problems. Under his supervision, a reliable ignition system was developed. *Adolph Nägel* of Dresden described *Junkers'* approach to work in a way that holds for all of the pioneer's activities: *"Junkers decided from the very beginning to explore experimentally to what limits in supercharging one could go while still maintaining the efficiency of the gas engine (...) Initially, he isolates, based on the wealth of experience, the basic problems, which have to be solved individually before one can start with the synthesis of the practically useful machine. For the solution of the individual problems, he develops experimental devices or machines of surprising originality and simplicity."*

After a short period of cooperation with *von Oechelhaeuser, Hugo Junkers* founds, in 1892, in Dessau his first company under the name "Hugo Junkers Civil-Engineer".

There, he started to exploit his patent for the "calorimeter", a measurement device to determine the caloric contents of gases. One of the first commercial products was, besides the calorimeter, the gas geyser, which was derived from the calorimeter. This device was since 1895 industrially produced by the new company "Junkers & Co., Ico", followed by many other products related to "gas-thermo-technology". The profits of this successful operating enterprise provided the economic foundation for *Hugo Junkers'* test and research institutions where he could conduct his pioneering and fundamental research free of financial constraints.

The instant success *Junkers* and his partners had in the area of aircraft engineering is certainly based on the wealth of experience gathered within decades in the field of applied research, especially in fluid mechanics and material sciences as well as in the area of manufacturing technology.

His career as a university lecturer began in September 1896 with his appointment as professor of thermodynamics and as head of the "Mechanical-engineering Laboratory" at the TH Aachen. Only some weeks later, he founded in Aachen the "Test Organization Professor Junkers" and further institutions followed within the next years. So *Junkers* consequently operated, apart from his enterprise Ico in Dessau and his teaching activities in Aachen, his own private test organizations. Essentially funded by the profits of Ico, he was able to work independently on his self-posed scientific and technical problems.

It is not exactly identifiable at what time *Hugo Junkers* became interested in aeronautical technology. Certain is, that he already gathered literature about topics related to airship aviation during his university studies. In 1885, he wrote down the following questions concerning aeronautical technology:

"1. *Under what conditions of "Mechanics" can a body be kept floating in the air?*
2. *What kind of shape needs a body to develop at a certain velocity a lifting force?*
3. *What are suitable propulsion systems for such an airship?"*

Astonishing are, in discussing these questions, *Junkers'* thoughts concerning propulsion, which he wants to realize by *"sucking in or ejecting air"*: *"One could propel an airship forward by the reaction (force) due to an outpouring body (gas et cetera)"*. He thinks about a

As proposed by *Hugo Junkers, Hans Reissner* used for his "Ente" ("duck") a wing of slightly corrugated sheet metal, which, together with other structural parts, was manufactured in 1912 at the Ico-workshop in Dessau.

Hans Jakob Reissner (1874–1967).

machine *"working with high pressure and compressing a gas at low pressure"*. These ideas are still utopian at the time when his first theoretical and practical tests commenced, but they show how far the creative thinking of *Hugo Junkers* was ahead of its time. Also many years later, he was occupied with questions of unconventional propulsion means far away from the propeller principle common at that time.

Hugo Junkers' first practical activities in aeronautics were inspired by *Hans Reissner*, who in 1906 had been appointed professor of technical mechanics at the TH Aachen. Since then, he conducted experiments with a Voisin-biplane, for which he could raise *Junkers'* interest. They had a lively exchange of thoughts and it became obvious that *Junkers* intensively thought about improvements. On January 15, 1908 he wrote to *Reissner*, among other things, about "*one's own comprehensive tests with corresponding wing geometries*" which one should conduct. *Reissner* now decided to design his own aircraft, which should, as advised by *Junkers*, be equipped with corrugated metal wings. The production of the aluminum surfaces caused problems since none of the metal-processing companies addressed saw fit to camber previously corrugated sheet metal, i.e., to realize a cambered wing shape. This difficult manufacturing task was finally solved at Ico. The "Reissner duck" was tested in 1912 and for *Junkers* this meant the first step into manufacturing sheet metal in aircraft engineering. *Junkers* himself describes his first direct consideration of questions of aeronautical technology in a presentation at the WGL in 1920 as following: "*Although the problem of flight interested me from youth on, I neither found, in earlier years, the time nor the opportunity to concern myself with this topic. Only in 1909, when my colleague Reissner asked me in Aachen to jointly work on the problems of flying machines, there seemed to emerge favorable prospects for an efficient cooperation in this field (...). After all, I could participate in the application of metal in the field of flight technology.*"

Hans Jakob Reissner, born in Berlin on January 18, 1874.

1894 studies civil engineering at the Institute of Technology (TH) Berlin.

1897 stays in the USA as civil engineer.

1898 studies physics at the University of Berlin.

1902 receives his PhD with a thesis related to framework oscillations in collaboration with *Count Zeppelin*.

1903 he stays, on a scholarship from the TH Berlin, a second time in the USA considering the topic of "steel high-rise construction" (structural engineering).

1904 associate professor at the chair of "Structural Statics" ("Baustatik") at the TH Berlin.

1906 full professor at the chair of mechanics at the TH Aachen. He turns to aeronautics and founds, together with *Junkers*, the aerodynamic laboratory at this chair.

1912 founding member of the "Scientific Society of Flight Technology".

1913 he designs the "Reissner-duck"; he follows a call to the TH Berlin. Years of collaboration with the "Zeppelin Aircraft Company Staaken", with work on strength and variable propellers of giant aircraft, follow.

1923 founding member of the GAMM and deputy chairman of the society.

1926 he heads the Berlin-section of GAMM; he works out aircraft registration rules at the "German Air Vehicle Committee of the RVM" and in

1930 he becomes chairman of this committee.

1936 due to his Jewish origin, he is prematurely retired as full professor; he becomes a consultant to the Argus Engine Company.

1938 he immigrates to the USA and receives a professorship at the Illinois Institute of Technology in Chicago.

1944 he follows a call to the Polytechnic Institute of Brooklyn and pursues his teaching and consulting obligations.

1967 he dies in his new home country.

The cooperation with *Reissner* has brought *Junkers* to aircraft engineering. Between the two teachers a lively and fruitful exchange of ideas and suggestions developed. Among them was also a short exchange concerning the possibilities of a helicopter. *Junkers* wrote to *Reissner* on April 21, 1911: *"In my opinion should the design and realization of a helicopter (propeller flyer) be tackled with highest priority". Reissner's* message, that also at velocities of up to 300 m/s (984 ft/s) at the rotor one could expect a ratio of drag to lift of 1:10, interested him vividly, and he asked *Reissner* for literature and results of preceding tests. *Reissner* informed *Junkers* about helicopter tests of *Breguet* and *Cornu*, who were able *"to lift considerable loads into the air"*, but the tests then failed because of problems with power transmission and stability. *Reissner* also considered the efficiency of the rotor and pointed out that close to 300 m/s *"elastic bow waves start to develop"*. For that reason, the outer tip and hence *"the uncomfortably high velocities could easily be sacrificed". Junkers* on his part contemplated the weight, the propulsion system, the number of revolutions, and the reduction gear. The considerations concerning the *"helicopter or propeller flyer"* were apparently no longer pursued but they show that both scientists, again and again, tackled new technical questions.

Meanwhile, *Hugo Junkers* had written down in December of 1909 his first ideas concerning a thick, hollow wing, which led to the patent, issued on February 1, 1910, *"Glider with hollow bodies to accept non-lifting parts"*. It reads in the patent write-up that, in the interest of keeping the engine power as low as possible, *"to make the ratio of lift to drag as large as possible". Hugo Junkers* wanted to optimize the lift to drag ratio by stowing all non-lifting parts, including the propulsion system, the crew, and the payload within the hollow body designed wing. This concept was later named the *"flying wing patent" ("Nurflügelpatent")*, but actually important was the recognition that a thick airfoil produces, at about the same drag, a much higher lift than the cambered surfaces modeled after the wings of birds. This patent showed, in addition, for the first time an ideal integration of the wing, the propulsion system and the payload compartments. *Junkers* knew, of course, how far ahead of the technical possibilities he was with his ideas, because he wrote already in 1920: *"Probably such an ideal aircraft will never entirely become reality, but in my opinion will the further development of aircraft engineering move into that direction so that we will, in the foreseeable future, come very close to this ideal."* With his large-scale aircraft G 38, he made in 1929, indeed, an essential step in that direction.

Voisin-Delagrange III (1908), test object of *Reissner* and *Junkers*.

On December 28, 1911, *Hugo Junkers* as well as *Reissner* participated in a meeting, whose minutes also discuss the fundamentally different airfoil shapes, which are called "jackets" ("Ummantelungen"): *"To produce lift, the "jacket" has to have a fish- or club-like cross sectional shape, whose imaginary center line between the upper and the lower bounds is curved upwards."* Airfoils with concave, straight or convex lower sides are sketched. Since the accommodation of people necessitates large wing chords with a corresponding airfoil thickness, it is pointed out in the minutes *"that it is perhaps not necessary to put all non-lifting parts of an air vehicle into special, lift-generating jackets or hollow wings. In such cases it is sufficient to provide for these parts jackets with the lowest possible drag"*. It thus became clear that for *Junkers* not the flying wing was important but the principle that a thick wing produces less drag and requires less propulsive power or allows higher velocities, respectively. *Junkers* did not claim the entire credit for the idea of the thick wing; he himself wrote later: *"I would hardly have had the courage to tackle such thick wings with the prospect of success if there had not been already some highly interesting results of Göttingen and Eiffel's tests showing that the drag of a body in air is not primarily depending on the cross sectional area perpendicular to the flow."*

For the new research area "Aerodynamics", *Junkers* had first to apply for funding at the governmental university administration. It is interesting to note his arguments justifying the application of January 23, 1910: *"The importance that (airship) aviation*

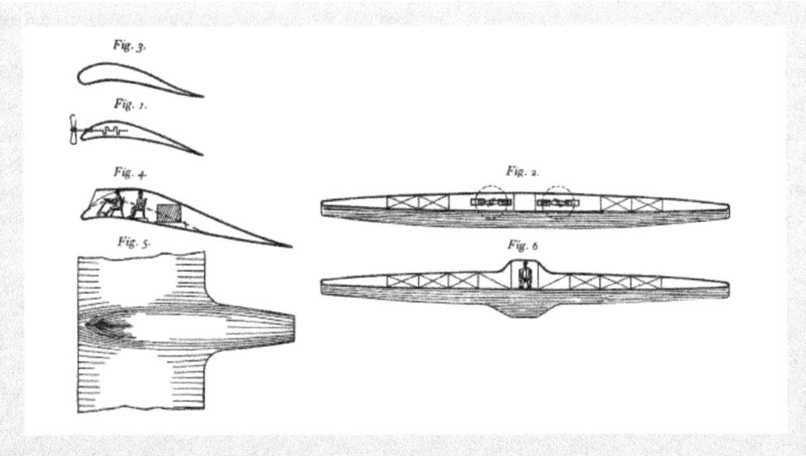

The so called "Flying Wing Patent" ("Nurflügelpatent"), issued to *Hugo Junkers* on February 1, 1910 by the "Government (Reichs) Patent Office", had the title "Glider with hollow bodies to accept non-lifting parts".

and flight technology will have in the future, as traffic, cultural and military tools, has more and more forged ahead. It will have to become a part of the engineering work. A task of the universities and institutes of technology will therefore be to introduce students of the respective departments to the disciplines, which have this youngest child of technology as subject, and provide them with the corresponding scientific tools." Junkers also pointed out that the theoretical and structural lectures had to be supplemented by experimental activities, which were of great importance not only to flight technology but also to general mechanical engineering technology; as examples, he mentioned the movement of air, mass flow measurements and static and dynamic investigations of the flow conditions in channels.

In an attachment to his proposal, *Junkers* also pointed to the number of already existing research institutes at other locations. At many universities, experimental aerodynamics had become a focus of interest due to the progress in flight technology: *"Within and outside of Germany, universities compete ambitiously to establish for this new branch of technology a location for the cultivation of research and teaching by creating professorships and by providing special equipment for laboratories. For instance, in Paris, Vienna and in Göttingen special laboratories have been established."* This also held for Danzig and other German universities and institutes of technology. *Junkers* also mentioned corresponding initiatives of the State: *"The German Reichstag has shown its interest in this modern research topic by suggesting the establishment of a special (airship) aviation academy to offer engineers the opportunity to gain their expert knowledge quicker and more comprehensively as otherwise possible at universities or institutes of technology."* Though authorities had correctly pointed out, as *Junkers* remarks, that the time for the establishment of such an academy is not favorable, since the experience in (airship) aviation had not yet been sufficiently consolidated and that also the audience would probably be limited. This educational task would actually be up to the German universities.

In 1911, *Junkers* was appointed head of the Institute of Aerodynamics of the TH Aachen, initially together with *Reissner*. Due to his application, a small wind tunnel was, in addition to a rotating-arm apparatus of 10 m (32.8 ft) diameter, built at the TH Aachen, which *Junkers* now used for systematic measurements. Initially, he tested models with equal drag and found that large, aerodynamically shaped bodies produced as much drag as plates, spheres or rods with considerably smaller cross sections.

Since *Junkers* was not satisfied with the possibilities of this small wind tunnel, he built his own wind tunnel at the "Test Organization Professor Junkers" ("Versuchsanstalt Prof. Junkers") in Aachen, which achieved with a 74 kW electric motor and a cross section area of 2.5 m^2 (26.9 ft^2) velocities of up to 50 m/s (164 ft/s). In 1916, another wind tunnel was built in Dessau, which already had a closed-circuit return. *Junkers* was already aware of the problem that the results of the measurements were affected by wind-tunnel turbulence. This was something he possibly wanted to avoid in case of his tunnel in Aachen, so he let the air flow directly from the atmosphere into the tunnel (bell-mouth). To determine the influence of turbulence, he conducted measurements during calm days, which were then compared with measurements during windy days.

During the extensive test sequences that followed, he adopted the principle that was characteristic for all his work: "*I try to find a new position and do not start from what already exists and has been proven successful, in order not to enter a trodden path right at the start*". Thus he started his measurements not on airfoils, but on a sphere followed by increasingly more slender ellipsoids of the same cross section. That way he proved that the cross section does not have a large influence and "*that thick cross sections are not only permissible but, within certain limits, are even better than thin cross sections (that is without considering the advantages due to the omission of the wing braces)*".

Junkers was not only concerned with the airfoil shape in his tests but also with the optimum wing aspect ratio. Initially, he experimented with small aspect ratios since he wanted to combine a small-span wing with favorable lift and drag values. Only through his measurements, he recognized the advantage of larger aspect ratios. The next step in his tests concerned the determination of the best camber of the airfoil skeleton (center) line. Without knowledge of *Lilienthal's* experiments, he reached the fundamental conclusion: "*It was shown that the degree of camber of this line was extraordinarily important for the amount of lift generated. It is an important result of my investigations that not so much the individual form of the pressure or suction side of the wing is important but the shape of the skeleton (center) line.*"

Junkers did not find the most favorable drag on the thinnest airfoil, as was generally assumed at that time, but on airfoils with average thickness ratios of 1:5–1:10. It was found that, at equal camber of the centerline, more favorable drag values were also found for airfoils with a convex lower surface. Systematically measured were, in addition, the aerodynamic moment and the movement of the center of pressure location. Here, too, was the thicker wing superior to the thinner one. *Junkers* summarized the results of his airfoil measurements as follows: "*Thus, the demand for the accommodation of drag-generating parts inside the wing surprisingly meets the demand for an aerodynamically advantageous airfoil. At the same time, another requirement on wings becomes much easier to be realized: sufficient strength at the lowest possible weight. This follows from a consideration of forces acting on the wing.*"

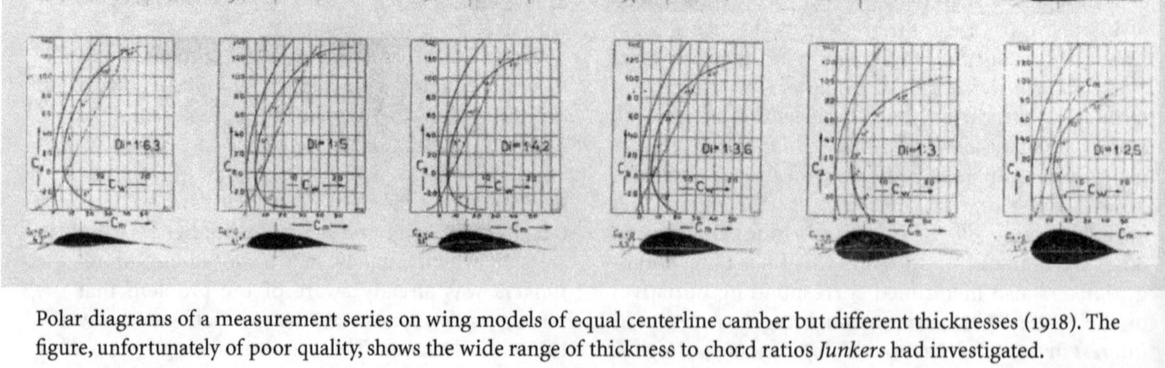

Polar diagrams of a measurement series on wing models of equal centerline camber but different thicknesses (1918). The figure, unfortunately of poor quality, shows the wide range of thickness to chord ratios *Junkers* had investigated.

Model of a cantilever monoplane which was investigated in 1916 in *Junkers'* wind tunnel in Aachen. At that time, this model represented a very progressive, far advanced configuration.

The problem of structural height was, at that time, fairly well solved by the biplane, where the upper and lower surfaces were joined by struts and braces to a very rigid frame. This is, however, negatively compensated by aerodynamic disadvantages. *Junkers* wanted to investigate experimentally, and without prejudice, the question whether biplanes or monoplanes were the best solution. After extensive biplane measurements he came to the conclusion "*that the interference between the two wings of the biplane is aerodynamically always an unfavorable one regardless of the arrangement of the wings*". *Junkers* summarized these findings in four points as follows:

1. The aspect ratio should be as large as possible.
2. The camber of the centerline is controlling the aerodynamic properties of a wing.
3. A thick airfoil, including the generally avoided convex lower side, may also provide good results.
4. The multi-wing plane is inferior to the monoplane.

Parallel to the aerodynamic research, *Junkers* and his partners were also concerned with very unusual propulsion-related questions. The development of gas and oil motors for stationary applications, but also for locomotives and ships was already considerably advanced, since *Junkers* here too had worked intensively on the solution of individual problems, e.g., on combustion conditions or heat transfer at the walls of the combustion chamber. This applied basic research was later to become the basis for the success in aero-engine design and manufacturing.

On August 14, 1912 a rather unusual patent was applied for. A patent for an "*ejector for the production of a gas flow for the propulsion of aircraft and other installations*" that should discharge gaseous jets or gas layers downstream of the leading edge over the wing surface to produce simultaneously lift and forward thrust. The invention refers to an apparatus with a certain kind of self-supporting compression and combustion, controlled by flap-valves, which was to proceed initially intermittently, but after sufficient heating of the walls of the combustion chamber steadily. The procedure was supposed "*to provide stationary lift for flight vehicles as well as to allow the take-off and landing without horizontal velocity*". In other words, *Junkers* already thought at that time about vertical take-off aircraft, which did not operate according to the helicopter principle. This is astonishing at a time when aircraft did not require very long runways anyway.

It seems that in Dessau already some elementary tests concerning propulsion without propellers were carried out, because in a memo of January 30, 1913 to *Otto Mader*, *Junkers* talks of a "*jet apparatus*" or "*jet propeller*" as propulsion for air vehicles, which was especially suitable at high velocities: "*The conversion of combustion heat into work by means of engine, crankshaft and propeller is very intricate. A jet propeller would here, at a sufficiently high efficiency, be advantageous because of its simplicity, especially with regard to safety and the ease of operation*". In the attached sketches, a "*jet apparatus with intermittent jet impingement*" is depicted, which has a certain similarity to the later intermittently-operating ram jets. This ejector could, as it says in the patent description, also be utilized for a certain kind of reaction propeller – a principle, which was very much later actually applied, although in a different form, for helicopter rotors with reactive propulsion.

In a meeting on August 22, 1914, where, besides *Hugo Junkers*, also *Otto Mader*, *Ludwig Wagenseil*, and *Hans Steudel* participated, it was already said that, *"at the high velocity of aircraft"*, take-off and landing were associated with difficulties and therefore means had to be found to ease these procedures. Here, the following proposals were made:

1. Wings of variable size.
2. Generation of an upward acceleration by pushing the aircraft off the ground (e.g., by stored spring force).
3. Utilization of energy from the propulsion unit to lift the aircraft.

It was discussed that also for an aircraft at rest lift could be generated if the normally horizontal airflow of the propeller would be deflected downwards, e.g., by means of deflector vanes. The perfect solution was seen in using the entire airflow of the propeller to generate the lifting force by deflecting the airflow downwards by 90°, either as an unbounded stream by Venetian-blinds-like flaps or in a closed channel. It was recognized that the necessary lifting force was only to be attained by a considerable increase in the accelerated amount of air. A means to accomplish this would be a nozzle with an integrated propeller. Such a shrouded propeller was, indeed, tested in the Aachen wind tunnel in 1914.

Of course, it was also referred to the possibility to utilize large propellers as *"lift-generating devices"*, but their application was said to have encountered large difficulties. The ideal was identified as an aircraft where the entire *"incoming air"* would be utilized, under addition of energy, not only for the generation of lift but also for propelling the aircraft forward. Even if the technical means were not sufficiently advanced to realize these ideas at that time, one has, also from today's perspective, to admire the far-sightedness of these concepts.

During the years 1913 and 1914, several tests, initiated by *Junkers* and supervised by *Otto Mader*, were performed in Dessau on nozzles, which did not lead to a propulsion concept, but were utilized in the development of a nozzle-cooler (ducted cooler) for engines. This very low resistance cooler was later realized and very successfully applied also by other aircraft manufacturers.

It is interesting to note how *Theodore von Kármán*, who taught at the same time at the TH Aachen and later provided essential impulses to aeronautical research not only in Germany but also since 1930 especially for some decades in the United States, judged the work of *Hugo Junkers*. That aeronautical technology was at all taught in Aachen is thanks to this man *"the great teacher, industrialist, and scien-*

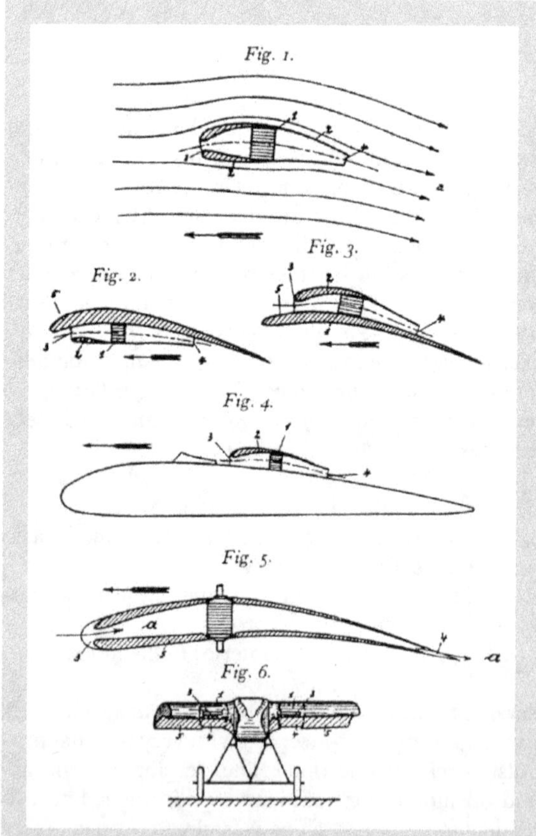

State (Reichs) Patent No. 303 328 of March 14, 1917, as supplement to Patent No. 299 799 of January 15, 1915 with the title: "Cooler arrangement for aircraft and other vehicles". With this proposed ducted cooler the lift to drag ratio was to be improved.

In 1918, a Junkers all-metal fighter aircraft J 9 was equipped with a belly radiator underneath the fuselage instead of the cooler in front of the engine.

tist, who is one of Germany's greatest aircraft designers, and who was to become one of my most important and most interesting clients". According to *von Kármán, Junkers* tried for years to overcome the gap between the theoretical science of thermodynamics and applied technology. *Von Kármán* called the concept of the thick wing a *"great idea"*. This, in the meantime generally accepted idea, was, at that time, considered revolutionary: *"A look at the history of aviation shows that these early ideas were the basis for the development of the American multi-engine airplane"*. *Von Kármán* was convinced *"that Junkers was the first of the great aircraft designers, who really understood the idea of circulation developed by Joukowsky. He realized that one could gain as much lift as necessary, independent of the thickness of the wing, as long as the wing was correctly designed."*

How much *von Kármán* was inspired by *Junkers* showed his work at the TH Aachen: *"Completely in the spirit of Junkers, the first thing I did in my new position, after choosing my staff, was to try to improve the laboratory equipment and the wind tunnel."*

Von Kármán admired the "*amazing and radical ideas*" of *Junkers*, who initiated *von Kármán* into his plans to build a metal airplane. *Junkers*, on the other hand, recognized the talent of the younger colleague and asked for his cooperation: *"One of my problems would be solved if you could develop an exact mathematical method to design any wing that would meet our requirements."* When *von Kármán* initially calculated an airfoil shape with the aid of *Joukowsky's* airfoil theory, the only viable procedure available at that time, the result was, in *von Kármán's* own words, a wing with a *"strangely thin and pointed trailing edge"*. *Junkers* was skeptical and feared that this wing would be deformed in the air and hence destabilize the aircraft. *"He had an instinct for these things, and he was right"*, confessed *von Kármán*, who, as a consequence of this disappointment, developed, together with *Erich Trefftz*, a new airfoil theory, which could be applied to a multitude of useful shapes. The airfoil of the Junkers J 1 aircraft was later developed with the aid of this theory; the J 1 was the first cantilever low-wing metal aircraft. It will be addressed again below. *Von Kármán* assessed this pioneering aircraft as follows: *"This type became the model for military and civil transport aircraft in Germany until the start of the jet-aircraft era; as a matter of fact, one may say that it directly opened the age of the modern propeller aircraft."*

Also during the years when *Junkers* was totally focused as businessman on his company in Dessau, *von Kármán* remained closely connected with the German aeronautical pioneer and accepted his offer, to be available as consultant for aerodynamics. *Von Kármán* highly appreciated *Junkers* not only for his *"wonderful imagination"* but also for his quality *"that he belonged to those very few businessmen, who always seemed to have time for you"*. In that he compared him to *Agnelli*, the founder of the Fiat company: *"It was the belief of both, Junkers and Agnelli, that a good entrepreneur was not supposed to work on a nine to five basis, but should arrange his time to be free to think and to make long-range plans."*

Due to *von Kármán's* profound knowledge of *Junkers'* aircraft designs, he was called to consult during patent conflicts in the USA. *Junkers* threatened *Henry Ford* with legal action concerning the latter's new tri-

The worldwide first cantilever all-metal monoplane Junkers J 1 performed its first flight on December 12, 1915. The aircraft consisted of a steel-tube frame planked with thin sheet iron.

motor aircraft, since *Ford* planned to build the wings of corrugated sheet metal. It was impossible to change the design to a flat-sheet construction. *Ford* decided, on advice of *von Kármán*, to pay the requested license fee. Also after 1933, when the National Socialists had already expelled *Junkers* from his companies, did *von Kármán* defend *Junkers'* patent rights against several American companies. For this, he was heavily criticized in the USA, but he defended his opinion: *"After thoroughly considering the situation, I had to come to the conclusion that it is not a political matter but a technical question"*. His expertise concluded that several American companies violated *Junkers'* patent rights. *Robert Gross*, at that time president of the Lockheed Aircraft Corporation, supported him: *"If von Kármán thinks a patent is violated, we will provide compensation. This is all about science and facts and not about politics."*

The systematic aerodynamic experiments had revealed the considerable advantages of the thick, brace-free wing, and when *Junkers* started to build such a wing he had to deal with suitable materials. He saw the deficiencies of wooden constructions, almost exclusively common at the time, but had, on the other hand, already long-time experience with processing metal in the form of thin sheets in his factory in Dessau. He therefore decided to use metal as material for his aircraft. This was in complete contradiction to the idea of light-weight construction common at the time.

Junkers argued that, although the specific weight was indeed high, the required cross sections needed for strength reasons were small, so that the performance per unit weight would come close to or were even superior to the performance of wood. If one could overcome the mechanical difficulties, *Junkers* believed, the advantages of metal would become fully effective.

Junkers also knew the prejudices generally associated with sheet-metal construction: *"Sheet metal is something the mechanical engineer looks at with a certain disregard"*. But he saw the advantage in the many possibilities to vary the design and in manufacturing, which wood could not provide. The advantages of all-metal constructions in aircraft engineering should lead to increased safety and durability.

However, in Dessau many more tests were necessary before the design and construction of the first all-metal aircraft J 1 could be started. *Junkers* wanted, for weight-reasons, from the very beginning to use the outer skin to carry loads, in order to be able to utilize all material to take up bending and torsion moments. At first, only sheet iron with a tensile strength of 25 kg/mm^2 (36,000 psi), as it was used in transformers for its magnetic properties, was available. Problems related to processing and bonding resulted from the small wall thickness, and *Junkers* decided on spot-welding for which already a certain experience existed from activities at the Dessau factory. The wings of the J 1 were put together with this welding technique.

An unsolved problem remained initially the buckling strength under pressure. *Junkers* conducted intensive tests with sheet metal, which were reinforced on the inside with corrugated lamellae. The tests concerning buckling strength were carried out, for time and cost reasons, in the in-house workshop in Dessau, and since there were no strength-testing machines

A wing rib of the Junker J1 with thin welded sheet-iron strips of 0.1–0.2 mm thickness.

for material tests available, it was necessary to rely on very simple but effective methods. Thus, the first tests with corrugated plates were performed on a milling machine, where the support was displaced for compressing and expanding the test specimen, and its change in length was determined from the displacement of the support. Also a press was used for strength testing.

After having determined the most important strength coefficients of the corrugated plate design, *Junkers* built various test wings of different lengths, which were, prior to the construction of the first test aircraft, tested for their ultimate-strength. The upper and lower sides of the wing boxes, reinforced on the inside by lamellae, were supported internally by struts. Since the pilot's seat was located in the center of gravity, and a continuous, one-piece spar was not possible, the introduction of the forces acting on the wing into the fuselage required special constructive solutions. The forces, concentrated within the wing skin, i.e., along one line, had to be focused in single points to introduce them into the fuselage. "*That was probably the most difficult problem I ever had to solve as an engineer*", *Junkers* confessed.

The actual manufacturing of the J1 took, due to the very thorough scientific preparation, only about four months, and the first flight took place on December 12, 1915. The aircraft had a span of 12.95 m (42.5 ft), a wing area of 24 m^2 (258 ft^2), a length of 8.62 m (28 ft), and a net mass of about 1,010 kg (2,227 lbs). With the

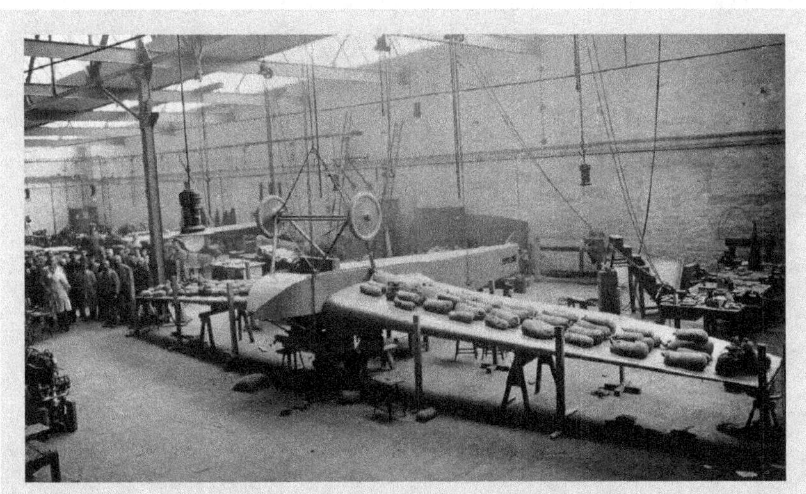

Load tests with sand bags on the wings of the inverted J1.

120-hp Daimler engine, the J1 reached a velocity of 170 km/h (105.7 mph). With this aircraft *Junkers* had realized, after years of basic scientific research, the first all-metal cantilever monoplane, which was to become not only the start of the successful Junkers aircraft but also a milestone in aircraft engineering per se.

Junkers was, however, not satisfied with the result. Although the J1 had achieved an acceptable velocity due to its good aerodynamic efficiency, the climb performance did not meet the expectations. *Junkers* paraphrased the problem as the following: *"For an aircraft, which is supposed to climb very steeply, the lifting work is, most of all, essential and the lifting work depends on its weight"*. Now, an aircraft was to be addressed that had less weight per horsepower. Therefore the steel used so far for the J1 was no longer adequate and another material had to be found, namely light metal as already used in airship construction. But *Junkers* was also not content with the structural design of the J1, and he abandoned the concept of the skin as load-carrying primary structure. He started out by considering that the contribution of the load carrying primary structure to the total wing weight could be increased, if the exploitation of the primary structure was increased: *"I have therefore decided to separate the cover (skin) and the load-carrying structures to be able to choose for the latter a more favorable shape."* The wing upper surface was only supposed to take up the local loads and torsion. He concluded that tubes were most suitable to take up pressure forces, and he now started the completely altered design of a tubular-frame wing.

At first, *Junkers* had to become thoroughly acquainted with the new material "duralumin". The following investigations were concerned with the strength of the material after various heat treatments, the fatigue phenomena, corrosion, and with possibilities to connect different duralumin parts. It became clear that duralumin, under varying loads, did not have any fatigue characteristics worse than steel. Unfavorable was, however, that it was poorly weldable, as *Junkers* confessed. *"So far we have not succeeded to improve the significant deterioration of the material during welding, although welding is generally possible"*. That was the reason to convert to riveted joints, which had, however, to be investigated with respect to their durability also in the presence of intensive engine vibrations. Here too, *Junkers* progressed very thoroughly and vibrated a horizontal-tail unit with the aid of an electric motor, with the number of impulses corresponding exactly to the ones of the aircraft engine. The result was that even after 1,000 h of testing, no loosening of the rivets could be observed.

Again, more test wings with ever increasing span were built and subjected to strength tests. Several tubular spars were the basis of the frame design; a special, patented splice, a result of lengthy tests, was used for the connection of two tubes. The spars were connected with riveted diagonal braces to form girders and over this frame the covers of corrugated sheet metal were riveted. This corrugated sheet metal design now became the standard construction principle with *Junkers*. It combined bending stiffness and low weight with an acceptable aerodynamic surface (friction) drag, at least up to velocities common until the mid-

Load test with 42 people on the fourth test wing of the Junkers J3, made of duralumin, on October 2, 1916.

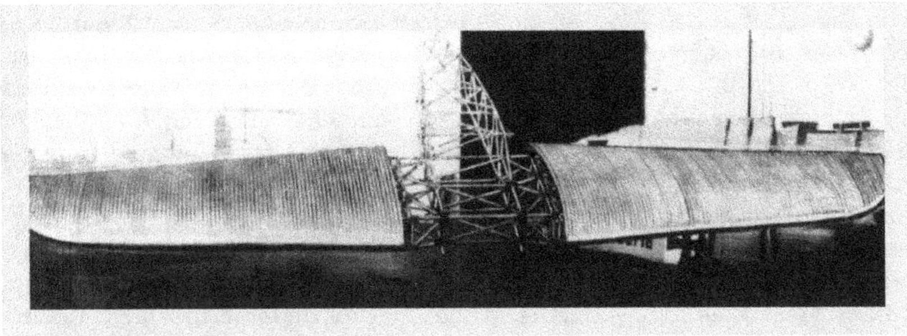

Already the unfinished Junkers J3 contained all essential elements of the later Junkers low-wing aircraft. These were especially fuselage and wings made of duralumin tubing and planked with corrugated sheet duralumin, and the connection of the wings via spherical bolts.

thirties. Wind tunnel tests showed that the drag of the wings hardly increased due to the wavy wall, which enlarged the surface by approximately 15 per cent.

For the various aircraft parts, the most favorable division and height of the corrugation had to be found, and here, too, many experiments were performed. It was not simple to solve the question of how the corrugated sheets could be joined among each other and with flat metal plates.

The new duralumin wings could be built with a weight of approx. 6 kg/m^2 (1.25 lbs/ft^2), whereas the iron wings of the J1 had a weight of 12 kg/m^2 (2.5 lbs/ft^2). The heavily armored aircraft J4, which *Junkers* had to build on request of the army administration as sesquiplane, was the first to follow this new design and construction. The aircraft was statically over-determined and was therefore able to negotiate spar damages without a loss in flight safety. With the J4, which proved itself in squadron service, *Junkers* managed the entry into the quantity production of metal aircraft.

On his own initiative, *Junkers* started in 1917 after a series of preliminary studies with the development of the light single-seat fighter J7, since he was strongly convinced that with a metal monoplane he could successfully compete with the dominant "wood-and-fabrics" biplane with respect to weight and maneuverability. The J7 was also the first Junkers aircraft where the wings were located below the center of gravity. Many experts expressed their fears that this would lead to unacceptable problems with regard to flight stability. But tests showed that these doubts were unfounded. The J7 turned out to be a very efficient fighter with extraordinary flight characteristics. *Junkers* thought that the greatest advantage of this type of low-wing design was the safety since in a crash the wing would hit the ground first and absorb a large part of the shock energy.

Despite of its constructive advantages, the low-wing aircraft design was criticized over and over again. Apart from the flight stability considerations already discussed, the main point of discussion was whether the high-wing or the low-wing aircraft should be preferred for aerodynamic reasons. Wind tunnel tests at Junkers' as well as in Göttingen initial-

Following an invitation to tender by the "Inspection of the Air Corps", for an infantry aircraft, *Hugo Junkers* was forced to build a sesquiplane, which he realized, however, as an all-metal construction with strut-supported wings. The J4 was built until the end of World War One in a series of more than 200 aircraft.

ly showed that a wing above the fuselage gives a better lift-to-drag ratio. However, later tests also took into account the effect of the propeller wash and this led to the result that additional lift is generated when the propeller wash flows over the wing. Especially aircraft with high engine performance and a correspondingly strong propeller wash tend to compensate the aerodynamic disadvantage of the low-wing aircraft.

During World War One, *Junkers* worked on numerous projects for various applications. Most of all for large-scale aircraft *Junkers* saw the advantages of his concept. He was convinced that the characteristics of light-weight design, viz., the transmission of small forces over large distances, would be the more effective the larger the aircraft would become. After the end of the war, *Junkers* planned a family-program of commercial aircraft:

- a large-scale aircraft with 740–1,000 hp,
- a medium-size transport aircraft with 160–185 hp,
- a small transport aircraft with 50–70 hp.

The first project started was, in early 1919, under the supervision of *Otto Reuter*, the 6-seat transport aircraft J 13 (later designation F 13) with a 160/185 hp engine. In the minutes of meetings, one finds notes like *"Security comes first, also in case of emergency landings (...) Good possibility of orientation, pilot's unobstructed view to both sides for landing (...). Good interior decoration is recommended. Dampen noise"*. It was possible to create inside the F 13 a large area for the comfortable accommodation of the passengers. The F 13, as the first cantilever all-metal transport aircraft, also became a great success for *Junkers* in civil aircraft engineering. The first flight on June 25, 1919 already showed, despite the relatively low performance of the 160 hp Daimler engine, the large potential of the aircraft, which achieved a speed of 165 km/h (102 mph). Only a few months later, an F 13 with the 185 hp BMW IIIa engine set, on September 13, 1919, a world record, when it climbed with eight people on board within 86 min to an altitude of 6,750 m (22,146 ft). Remarkable was one detail of the F 13, which was realized again on large aircraft only decades later: A small trim tank in the rear part of the aircraft was used, by pumping fuel into this tank, to shift the center of gravity during flight downstream thus reducing the aerodynamic trim drag.

Quickly, this robust, high-performance aircraft found a wide distribution, not only in Europe but also within extreme climate zones from the tropical Columbia to arctic regions of Canada. Within a period of ten years, various derivatives were developed and a total of 360 planes manufactured. After the F 13, a whole family of transport aircraft followed, from the W 33/W 34, via the G 24 and G 31, to the large-scale aircraft G 38, which attracted in 1929 as the world's largest landplane considerable attention. Already in 1928, *Junkers* was able to roll out his 1,000th aircraft in Dessau. In 1931, the Ju 52 took off for its first flight; the aircraft became in the thirties a symbol for safety and reliability in aviation. *Junkers* had become the leading manufacturer of transport aircrafts.

But *Hugo Junkers* also made an important contribution to pressurized cabins. In 1930, he realized the first 2-men pressurized chamber inside his high-altitude aircraft Ju 49, suitable for stratospheric flight.

Hugo Junkers turned out to be an ingenious designer, who, with his investigations, ideas and considerations, decisively advanced aeronautical engineering, and who initiated important new developments.

The almost market-dominating lead, gained in very few years, was only possible because *Junkers* had established the wide scientific basis needed for this development. He described in 1920 the necessity of pre-competitive research in the following words: *"The cornerstone of all of Junkers' enterprises is the systematic search for important assessments of and solutions*

Final assembly of the first F 13 in Dessau. The photograph was taken in June 1919.

The successful Junkers transport aircraft F 13, derived from the design of the J 13 of January 1919, was able to take-off on June 25, 1919 in Dessau for its first flight. A total of 360 Junkers F 13 aircraft was built.

"Junkers Air Traffic" operated the Junkers tri-motor transport aircraft G 24 since 1925 on numerous European main routes.

The high point of Junkers' transport aircraft development was the Ju 52/3m, which was built in a series of more than 5,000 aircraft; some oldtimers are still flying today (first flight 1932).

For the large-scale aircraft G38, whose first specimen was completed in 1929, *Hugo Junkers* was able to realize part of his ideas from the "Flying-wing Patent".

With the high-altitude test aircraft J 49 experience was gained with pressurized cabins and high-altitude flight engines, which benefited later developments.

to new problems. This is, most of all, what guaranties a permanent lead and superiority with respect to competitors, and allows, with respect to licensees, to keep things in one's own hands."

From the very beginning, *Junkers* followed in his scientific-technical research the principle that all new developments had to have an economic objective. He did not believe in inventions for their own sake, but in solving a technical problem, each single step should be such that it contained the economic basis for the next step. The economic objective had to have priority over the technical task. Thus research in *Junkers'* sense included, from the very beginning, not only the design but also all production-related questions. Contrary to the mass production of less demanding products, as they were at that time mainly produced in the USA, *Junkers* saw the importance of the technical developments in Germany with respect to world economics mainly in research-intensive pioneering work for innovative products.

The J 13, built by *Junkers* in 1919, already exhibits all essential characteristic features of today's transport aircraft such as cantilever wings in low-wing arrangement, end-to-end passenger cabin, and vertical and horizontal stabilizers. Only the pressurized cabin and the retractable landing gear were missing.

The Development of the Modern Passenger Aircraft

Dieter Schmidt, Horst Prem

A presentation at the Royal Aeronautical Society is concluded by *Junkers* as follows: *"There can be no doubt about the enormous boom aviation will still have, if one considers the development potential of aircraft with regard to design and application. Aviation, in extent and importance, will certainly catch up with shipping and the railway."*

Hugo Junkers had, as before him *Count Zeppelin*, a vision of air transportation and suggested a transcontinental air-transport system. But in the twenties, there was no great demand for air-transport services, neither by passengers nor by mail services. The acceptance as well as the possibilities of civil air transportation had to be developed slowly through a further increase in the reliability of the technical devices and by educating potential customers. Licensing procedures were in Germany developed and implemented by DVL.

But an important impulse came from the USA. In 1925 it was decided to withdraw airmail services from the governmental monopoly of the "Post Office" and award them to private enterprises under competitive conditions in order to reach cost reductions. This was a first impulse for civil aviation to realize with their products shorter delivery times. The second impulse came in 1926 with the "Air Commerce Act". Here, flight navigation, licensing, and acceptance tests for pilots and air vehicles as well as the investigation of aviation accidents were placed under government control. Not the least the first Atlantic crossing by *Charles Lindbergh* in 1927, which triggered great enthusiasm, changed people's attitude towards air transportation. Imported Fokker DVII aircraft were employed and operated successfully before the Boeing "Monomail Model 200" (first flight on May 6, 1930) appeared in the sky. It was an aircraft that, with its retractable landing gear and Townend-ring engine fairings, was already aerodynamically well designed. The comparable German He 70 followed 19 months later. But it lacked economic success. Further steps in the development of flight vehicles were needed to gain faster, larger, and more reliable aircraft with lower failure rates, thus being able to decrease the seat-mileage costs.

A further impulse in the military as well as in the civil domain was expected from the "High-altitude Aircraft". The following two essential advantages were envisaged:

- Improvement of comfort by flying above cloud level.
- Increase in velocity at a simultaneously reduced fuel consumption.

It was expected that the flight at higher altitudes above the "weather" would, in the higher atmosphere, result in a reduction in friction drag and in the unfavorable turbulence. Thus, the flight speed could be increased, provided sufficiently powerful engines were available. Another advantage arose from the desire to design the aircraft for higher altitudes and higher flight speeds since this would lead to an increase in wing area, hence to a reduction in wing loading. This led, at the same time, to improved take-off and landing velocities and a better-balanced configuration. Now, the airframe as well as the engine had to be modified to accommodate the flight at higher altitudes.

In the report "Stratospheric Flight and High-altitude Engines" [10], the problem area is summarized: To realize stratospheric flight, a sufficient amount of oxygen must be provided for the pilot and the passengers; for the engines, a sufficient air flow at an adequate charge pressure has to be supplied. The corresponding investigations and research activities, especially related to the improvement of the flight engines, were in Germany forcefully pursued since the early thirties. Special altitude test chambers were developed to be able to conduct the necessary basic investigations concerning air supply and the utilization of the exhaust energy. Alternative solutions to the pressure chamber, such as pressurized suits for the crew, had been developed and tested for high-altitude record flights, but were found to be unsuited for an extended use, since they strongly restricted the freedom of movement. Problems were initially caused by the cockpit windows due to fogging, icing and pressure loads and by the necessity to lead control cables and rods through the pressure-cabin walls to the engine and the control surfaces. But in the already quoted report "Stratospheric Flight and High-altitude Engines" it says *"that the main problems occurring due to the flight at altitudes between 6,000 and 10,000 m can be considered as solved"* [10]. In America, the consolidation of aviation began, with governmental support, in 1930. Thirty existing airlines were – under slight pressure from "above" – combined to the "Big Four": Eastern Air Transport, American Airways, Trans World Airlines and United Airlines.

Mail was now only an additional payload on the airplane. The profit was to come from passenger transportation. The establishment of the big airlines created a demand for economic aircraft. Following the "Vega" of Lockheed, which was on the market since 1927, offering seating space for five to eight passengers and having a cruise speed of 220 km/h (137 mph), Boeing developed, based on the "Monomail", the two-engine B-247 with ten seats; the aircraft was operated by United Airlines since 1933, and two aircraft were also delivered to Lufthansa in April 1934. The inner American competition then led to the inception of the Douglas aircraft. When TWA wanted to purchase the B-247, Boeing thankfully rejected due to an exclusive contract with United covering the American territory. The DC-1, which was put into operation at TWA in 1933, was quickly modified to meet market demands and subsequently further developed. In 1934, the DC-2 was rolled out, which showed, at that time, already 33 per cent lower costs per seat-mile than the Ford "tri-motor". The performance increase was, among other things, achieved through improved aerodynamics (the lift-to-drag ratio was 14:1). The next step in the development was the enlargement of the cabin. One more seat in a fuselage cross section resulted in an increased capacity of 50 per cent. In addition, an extension of the wing increased the aspect ratio from 7.7–9.1, reducing the wing loading down to 120 kg/m^2 (577 lbs/ft^2). The performance parameters of the DC-3, thus developed, were not decisively improved, but the direct operating costs (DOC) declined by approximately 25 per cent compared to the predecessor model DC-2.

Here, all development-related steps of successful aircraft engineering were already executed. Approximately 11,000 DC-3 aircraft were built, many of them as military derivatives.

It was recognized in the mid-thirties, in the USA as well as in Europe, that a modern long-range aircraft with pressurized cabin had to be equipped with four engines. The bomber Boeing B-17 "Flying Fortress", with a growth potential towards a pressurized cabin, was the world's first four-engine aircraft; it made its maiden flight on July 28, 1935. The Boeing 307 was derived from the B-17. It had a pressurized cabin holding 33 passengers and performed its first flight on December 31, 1938.

The development in the USA towards four-engine transport aircraft, with a potential for stratospheric flight, was carefully watched by *Kurt Tank* at "Focke-Wulf Aircraft Engineering, Ltd." ("Focke-Wulf Flugzeugbau GmbH"). He convinced *Rudolf Stüssel* of Lufthansa of the concept of the four-engine long-range aircraft "Condor", which envisaged a capacity of 25–30 passengers [11]. *Kurt Tank* and *Wilhelm Bansemir* realized the "Condor", with the type-designation "Focke-Wulf 200 (Fw 200)", within 12 months and 11 days. Its maiden flight was on July 27, 1937. One month later, on August 28, 1937, the slightly larger Ju 90 took-off for its maiden flight, so that Lufthansa

The Focke-Wulf Fw 200 "Condor" (D-ACON) on August 11, 1938 in New York following the flight Berlin – New York.

Selected aircraft types in chronological order of their maiden flights.

Type		First considerations	First flight	Entry into service	First customer	Total production
Heinkel He 70	D	1931	Nov. 1932	28.2.1933	Lufthansa	310[a]
Boeing 247	USA	Beginning 1932	8.2.1933	30.3.1933	United	75[b]
Douglas DC-1	USA	Aug./Sept. 1932	1.7.1933	Dez. 1933	TWA	1
Lockheed L.10	USA	Sept. 1932	23.2.1934	11.8.1934	Northwest	148
Douglas DC-2	USA	Middle 1933	11.5.1934	18.5.1934	TWA	220
Heinkel He 111 (zivil)	D	1934	März 1935	Middle 1936	Lufthansa	12
Fiat G 18/G 18V	I	1934	18.3.1935	1936	Avio Linea Italiana	9
Junkers Ju 86 (zivil)	D	1934	April 1935	Middle 1936	Lufthansa	40
Boeing B-17 (mil.)	USA	Aug. 1934	28.7.1935	Jan. 1937	US-Army	12.731
Bloch 220/221	F	1935	Dez. 1935	Dez. 1937	Air France	17
Douglas DC-3	USA	Beginning 1935	22.12.1935	25.6.1936	American	til Dec. 1941 approx. 800[c]
Focke-Wulf Fw 200	D	End 1935/Begin. 1936	27.7.1937	Juli 1938	Lufthansa	16
Junkers Ju 90	D	End 1935/Begin. 1936	28.8.1937	Juli 1938	Lufthansa	20
Armstr.-Whitworth A.W.27	GB	End 1934	24.1.1938	Sept. 1938	Imperial Airways	14
Boeing 307	USA	Middle 1936	31.12.1938	8.7.1940	TWA	10
Douglas DC-4 (C-54)	USA	Middle 1939	17.2.1942	Begin. 1943	US-Army	1.242[d]
				7.3.1946	American	

[a] 296 went to the German Air force;
[b] since August 1934 two aircraft were with Lufthansa;
[c] approximately 350 airplanes went to the US-Army. The total production in the USA was more than 11,000; after the war, a large number of aircraft was modified for civil use;
[d] including about 250 C-54 modified after the war for civil use as well as 79 DC-4, which were manufactured since 1945/46.

Transport aircraft Junkers Ju 90 (1937).

was able to meet with these two aircraft of 26 and 40 seats, respectively, the transport requirements on the continental routes.

A comparison of the configurations B-307, Ju 90 and FW 200 with the DC-4, which appeared in 1942, shows for the latter two a wing aspect ratio of more than 9. It was only at the end of the thirties, with a delay of more than 15 years with respect to sailplane engineering, that in powered-aircraft engineering *Prandtl's* finding, that the aerodynamic efficiency could be improved by increasing the aspect ratio of the wing, asserted itself. The B-307 did not yet have the aerodynamically optimized shape, but was already equipped with the comfort of a pressurized cabin. The DC-4

Four-engine long-range aircraft.

Manufacturer Type	Focke-Wulf Fw 200 »Condor«	Junkers Ju 90	Boeing 307 »Stratoliner«	Douglas DC-4
Maximum take-off mass [kg]	14.600	21.600	19.050	33.110
Engines [hp]	4 x 720	4 x 950	4 x 1.187	4 x 1.470
Power loading [kg/hp]	5,07	7,35	4,01	5,63
Crew,	3 + 1	3 + 1	5	5
Passengers	25 + 1	38 + 2	33–38	30
Span [m]	32,84	35,00	32,70	35,80
Wing area [m²]	118,00	184,00	210,00	135,00
Aspect ratio	9,14	6,67	5,09	9,49
Wing loading [kg/m²]	118,64	119,00	90,80	245,26
Maximum speed [km/h]	365	355	396	430
Cruise speed [km/h]	335	325	355	390
Maximum range [km]	1.450	1.500	3.824	5.300
First flight	27.07.1937	28.08.1937	31.12.1938	17.02.1942

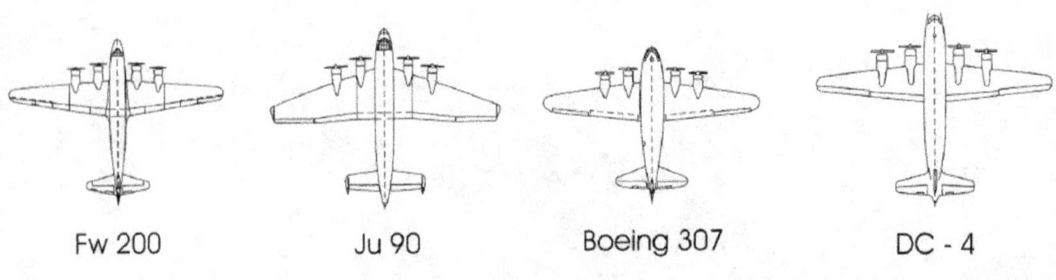

Comparison of four-engine long-range aircraft configurations.

with its optimized aerodynamics could, on the contrary, only in its advanced version DC-6 (first flight on February 2, 1946), similar to the Lockheed "Constellation" (first flight on January 9, 1943) be equipped with a pressurized cabin and thus offer stratospheric flight above the weather. But back to the Fw 200 [11]:

"The Fw 200 then mainly made headlines by its smooth record flight Berlin-New York and back in August of 1938. A total of 16 civil and 263 military Fw 200 are said to have been delivered. The first military aircraft versions were modified civil aircraft. If there had not been a war, the "Condor" would easily have become an export hit."

The concept of a Fw 300, basically an improved Fw 200, with pressurized cabin, optimized wing loading and aspect ratio had been completed, but this long-haul aircraft was not realized anymore [12]. The customer determined the objectives of research and development [13], which were at the end of the thirties definitely of military nature. But with the develop-

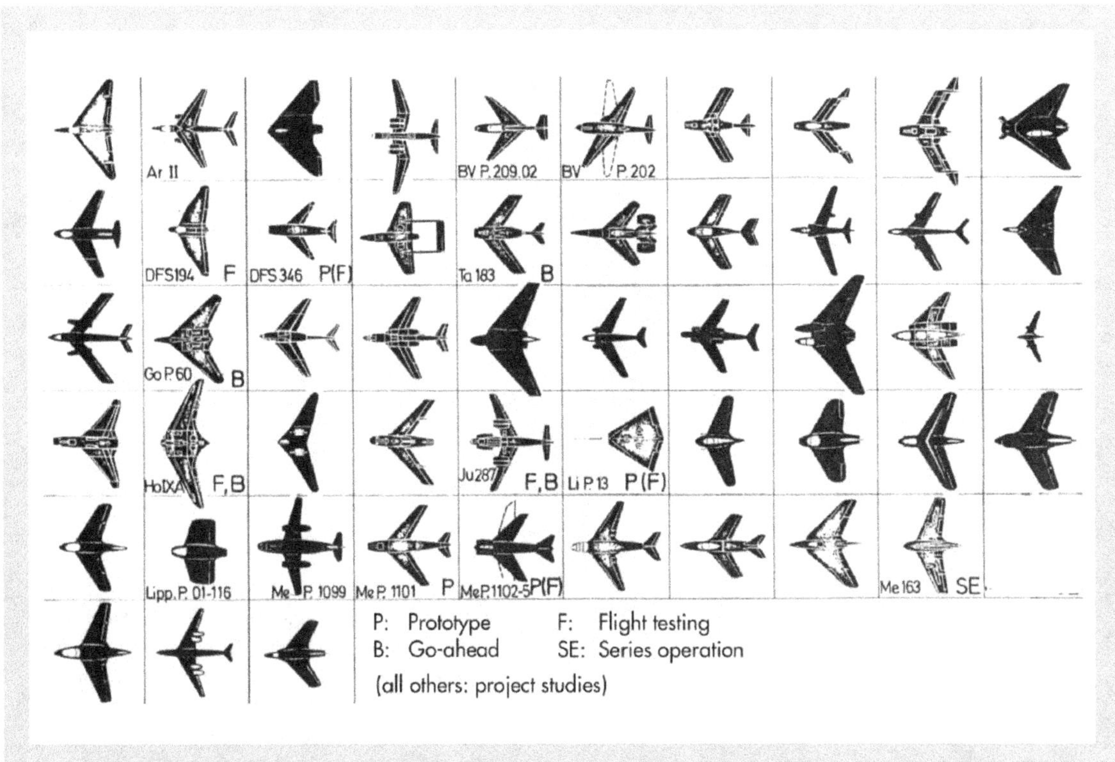

Examples of German jet-aircraft with swept wings up to 1945 (according to *W. Heinzerling*).

ment of jet engines and the research on swept wings et cetera for military aircraft also the technological basis for radically new future long-haul aircraft was provided.

The measurements on sweptback wings for the Messerschmitt AG [14, 15] at the AVA, carried out by *H. Ludwieg*, were inspired by a conversation between *Ludwig Bölkow* and *Albert Betz*, to finally use the Busemann-effect. This led to a radical change in aircraft configurations. The overview of the swept-wing jet-aircraft projects up to 1945, given in the preceding figure, shows the progression of this research in Germany. In this respect see also the recent paper of *P. Hamel* [16].

A further specific example concerns the wind tunnel measurements at the DFL in Brunswick since 1944 on a model of the Ju 287 with a forward-swept wing, from which four jet engines, mounted in two double nacelles, were suspended. The development teams of this project were brought to the Soviet Union in 1946, where they designed the first long-range bomber of the Soviet Union before they were gradually released to the German Democratic Republic (GDR), in the early fifties.

After the German wind tunnel results [17, 18] became known, Boeing adopted similar configurations within the framework of the development of the bomber B-47, which was realized with sweptback wings. The B-47 and B-52 were the predecessor developments for the application of the sweptback wing with attached jet engines in civil aviation. Since the mid-fifties, the B-707 was the standard configuration for long-range flights. With this type of aircraft, which was first utilized in large numbers as military tanker KC-135, the triumphant advance of Boeing in the passenger aircraft market began.

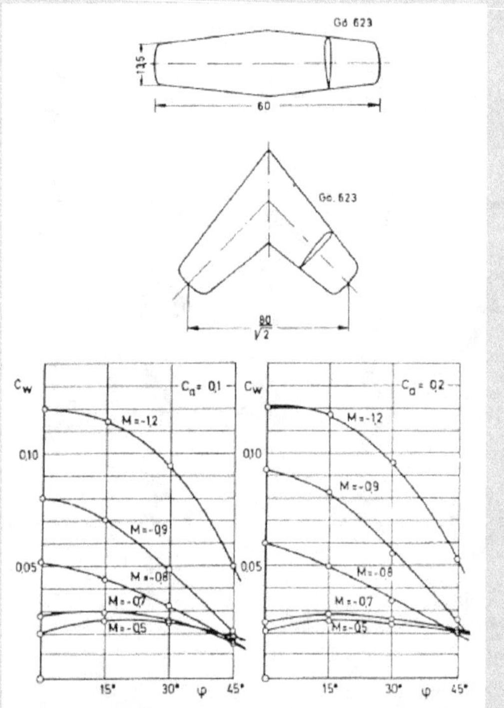

Hubert Ludwieg presented, for the first time, the results of the wind tunnel tests on swept wings carried out at the AVA in September 1940 at a meeting of the Lilienthal Society in Göttingen. This presentation of the sweep effect originated with *Ludwig Bölkow* (C_W is the drag coefficient, C_a the lift coefficient, and φ the sweep angle).

Model of the Junkers Ju 287 with forward-swept wings and four engines attached to the wing in the Brunswick high-speed wind tunnel (1944).

Highlights in the Evolution of Sailplanes until 1945*

HANS ZACHER

Two events, made possible by consequent research, mark special highlights: In 1921 the introduction of the cantilever single-spar wing with a large aspect ratio, torsion-nose and a thick airfoil and, since 1957, the utilization of fiber-reinforced plastics and monocoque designs in combination with the use of advanced laminar-type airfoils. In between there were lesser, but still significant developments, such as, since 1927 the use of the variometer and since 1934 the installation of airbrakes, without whom modern sailplanes are no more conceivable. Examples for these steps in the development are the "Vampyr", the HKS 3, the Ka 6 and the "Phönix", which are exhibited at the "Deutsches Museum" in Munich. Further prominent examples can be found at the Branch of the "Deutsches Museum" in Oberschleissheim and at the "German Sailplane Museum" ("Deutsches Segelflugmuseum") on the "Wasserkuppe" in the Rhön-mountains.

After the success of the powered flights of the brothers *Orville* and *Wilbur Wright*, the art of gliding was nearly forgotten. There were only a few, who tried to continue *Lilienthal's* work [1, 7, 19, 20], such as, for instance, *Erich Offermann* and *August Euler.* Exceptional was also the "Flight-sports Association Darmstadt" ("Flugsport-Vereinigung Darmstadt, FSV") [21], an association of high-school students, who successfully launched more than ten of their own designs and discovered in 1911 the "Wasserkuppe" in the Rhön mountains as an especially suitable area for gliding; there, they performed remarkable flights in 1912, among others, one of 838 m (2,749 ft) distance in 112 sec with their braced biplane glider FSV X.

After World War One, when powered flight was prohibited in Germany, *Oskar Ursinus* and others announced in the Journal "Flight Sports" ("Flugsport") in 1920 the first Rhön Glider Competition. A few useable and many useless gliders in many shapes, sometimes bizarre, showed up and skidded, mostly hopped, and sometimes even flew. Among the latter was the FVA 1 "Black Devil" ("Schwatze Düvel") of *Wolfgang Klemperer* [22, 23] from the TH Aachen, a cantilever low wing glider, which covered, after a rubber-cord start the longest distance 1,830 m (6,000 ft) in 142 sec. *Klemperer* got with the similar successor model FVA 2 "Blue Mouse" on August 30, 1921 the distance world record of 5 km (3.1 mi).

At the "Wasserkuppe" gathered also World War One pilots now studying at institutes of technology, universities and engineering schools [24]. They founded, first at Aachen, Berlin, Darmstadt and Hanover, "Academic Flying Groups" ("Akademische Fliegergruppen, Akaflieg") and flight-scientific associations to promote scientifically, in the spirit of *Ludwig Prandtl*, soaring and gliding flights and to improve sailplanes with respect to performance and safety.

* See Vol. 19 of the Series German Aeronautics "Die Evolution der Segelflugzeuge" ("The Evolution of Sailplanes")

Braced biplane glider FSV X (1912).

Cantilever low-wing sailplane FVA 2 "Blue Mouse" ("Blaue Maus"), 1921, which is similar to the "Black Devil" ("Schwatze Düvel"), 1920.

Ludwig Prandtl [25] had published after the Rhön competition the theoretical basis underlying a low sinking speed and a good glide angle: low drag coefficients, a large aspect ratio and high lift coefficients. Already at the second Rhön competition in 1921, the "Academic Flying Group Hanover" introduced, towards the end of the competition, a cantilever high-wing glider that showed what a sailplane should really look like. The sailplane H 1 "Vampyr" has been called the "Grandfather of all future gliders". Its shape and way of construction became exemplary for contemporary and future designers. The design was by *Georg Madelung*. In 1919, he had used *"Bredt's Theorem"* in his "master's thesis" ("Diplomarbeit") at the Institute of Technology (TH) Charlottenburg to compute a wing by "*considering the spar together with the forward part of the wing (the nose) as a closed pipe*"; this theorem had, for a long time, been disregarded. *Rudolph Bredt* had published a paper in the ZVDI Journal in 1896 [26] with the title "Critical Remarks on Torsional Elasticity" where he showed that a hollow body could easily be calculated "*if the wall thickness is small relative to the external dimensions*". His computing scheme is for hollow bodies of arbitrary shape but with essentially smooth (continuous) boundaries sufficiently accurate for practical purposes. The results of his thesis enabled *Georg Madelung*, possibly following a suggestion by *Arthur Pröll*, to develop a sailplane with a cantilever high wing with a thick airfoil and only one spar and a torsion nose. The fuselage was closed but still angular. The members of the "Flying Group Hanover", *Walter Blume*, *Arthur Martens* and *Fritz Hentzen*, carried out the design. The "Vampyr" was built at the "Wagon Factory Hanover" ("Hannoversche Waggonfabrik, HAWA") under the supervision of *Hermann Dorner*. Following the advice of *Albert Betz*, the 17 per cent thick airfoil "Göttingen 482" was utilized, which allowed a high spar and a voluminous torsional nose, together providing large bending and torsional stiffness. An additional aerodynamic advantage was provided by the fact that the plywood-planked nose, extending up to one third of the wing chord, did not sag between the wing ribs, as had been the case for nearly all previous sailplanes. In 1922, *Arthur Martens* and *Fritz Hentzen* flew with the "Vampyr" in the up-winds of the Rhön-mountain slopes successively more than one, then two and three hours. The transition from "gliding to soaring" had been achieved.

With a glide ratio of about 16 and a sinking speed of 0.8 m/s (2.6 ft/s), a span of 12.6 m (41 ft), only limited due to insufficient transportation capacity for the one-piece cantilever wing, and an aspect ratio of 10 was the "Vampyr" the model for all further famous sailplanes. It should be noted that the glide ratio and sinking speed were predicted – for the first time – by calculations based on wind tunnel measurements in Göttingen.

In 1923, the "Grandfather of all Sailplanes" was succeeded by the "Father of all Sailplanes": the D 9 "Konsul" was developed by the "Academic Flying Group Darmstadt" (*Albert Botsch, Fritz Hoppe* and

The Grandfather of all later sailplanes "Vampyr" equipped with a cantilever one-spar wing with torsion nose (1921).

In August 1922, *Fritz Hentzen* (picture) and *Arthur Martens* performed flights of up to several hours with the "Vampyr" in the up-winds of the Wasserkuppe in the Rhön mountains.

Rudolf Spies) and built in cooperation with the "Railroad-Demands Corporation" ("Bahnbedarf AG") in Darmstadt. Special features of the "Konsul" were: a large aspect ratio of 16.1, a large span of 18.7 m (61 ft), a divided single-spar wing with a torsion nose, a "Göttingen 535" airfoil, a differential aileron, a streamlined fuselage with an elliptical cross section, a large vertical tail, and a central-skid landing gear [27]. Within only three years since the beginning of the Rhön competitions, the sailplane had already developed into a form essentially valid still today. Single-spar designs with torsion nose and light-weight construction soon became also common in powered-aircraft designs, again permitted in Germany; examples are the Klemm L25, Messerschmitt M17 and M23, to name a few. *Bredt's* theory thus gained 25 years after its publication its full importance, first for sailplanes, then for powered aircraft, making stiff, solid and light cantilever wings with low drag possible.

The Rhön sailplane competitions were the events, which brought together scientists, designers and pilots, who directly exchanged their knowledge and stimulated further developments. The Wasserkuppe was their test range.

Soon the sailplanes became so efficient that only sometimes flight characteristics and, because of the good glide ratios, landing difficulties were criticized. Already in 1923, *Ludwig Prandtl* had suggested, considering the "Konsul", to equip high-quality sailplanes with air brakes or similar devices.

During these first four years, the academic flying groups developed, via the FVA1 "Black Devil" ("Schwatze Düwel") in 1920, the FVA2 "Blue Mouse" ("Blaue Maus") in 1921, the H1 "Vampyr" in 1921 and the D9 "Konsul" in 1923, the "ultimate sailplane" [28]. Today's sailplanes still have all their essential characteristics: cantilever streamlined wing with single spar, torsion nose and thick airfoil, streamlined fuselage of nearly elliptical cross section and a differential aileron. Their efficiency was demonstrated by flights of up to three hours, utilizing the up-winds of mountain slopes, distance flights of 19 km (12 mi), even against the wind, and elevations above take-off of 350 m (1,150 ft). This all, incidentally, with airplanes built during the time of the inflation 1922/23 under difficult conditions with almost no money.

Already in 1922, the still young "Academic Flying Groups" ("Akafliegs") had founded an association of interests ("Idaflieg") [29, 30] with the objective of providing mutual advice in theory and application, mutual support and representing their external interests. The existing groups were joined during the following years by groups from Brunswick, Breslau, Cöthen, Dresden, Munich and Stuttgart as well as by some purely sportive groups, which did no scientific-technical work and did not design or construct their own sailplanes. In order to spread gliding still further, the "Rhön-Rossitten-Society" ("Rhön-Rossitten-Gesellschaft, RRG") was founded in 1924, and in 1925 an

The "Father" of all sailplanes, the D 9 "Konsul" (1923), with large aspect-ratio wing and streamlined fuselage.

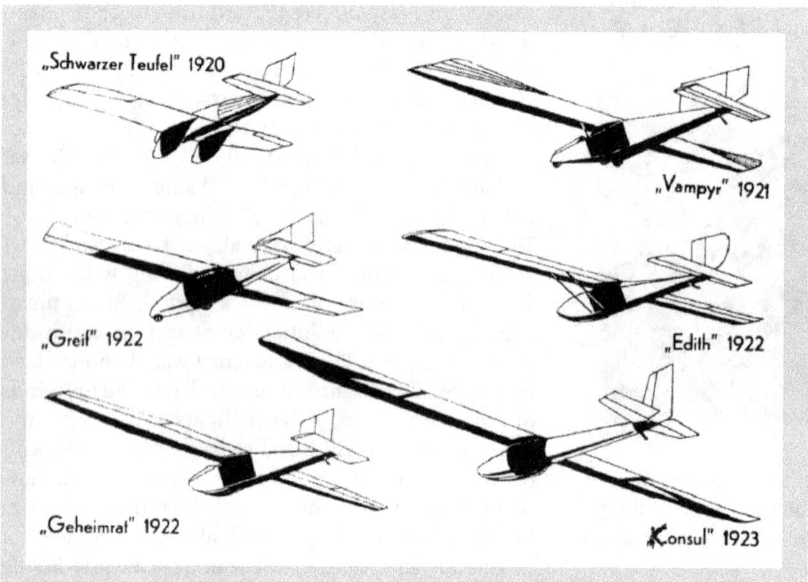

The sailplane development during the early twenties (according to *Elze-Jacobs*).

associated research institute was established with the following tasks:

- Exploration of the atmospheric conditions for gliding.
- Design of efficient sailplanes including the furnishment of drawings to allow other flying groups to subsequently rebuild these sailplanes.
- Systematic flight training in gliding and further development of the "Self-training Method" (which did not prove successful in the long run and was seldom applied by the "Akafliegs" and other groups).

Sports, technology and science complemented each other in gliding in a unique way: the propelling force "upslope wind" was discovered and tested by pilots; new knowledge was gained for meteorologists (e.g., about the movements of thunderstorm clouds and leeward waves) and for aerodynamicists and designers (e.g., about gust-load assumptions and flight characteristics).

Especially the very active "Akafliegs" were the ones, who, through research and the application of new knowledge, designed exemplary high-performance sailplanes, which always provided new impetus to the "Sailplane Movement". Here, especially the "Darmstadt School" ("Darmstädter Schule") [27] should be mentioned, which provided until the mid-thirties many competition sailplanes, such as, for instance, the D 15 "West-Prussia" ("Westpreußen") type reproductions with spans of 14–16 m (55–52 ft), elliptical wing plan forms, and good flight characteristics.

The interplay between man, machine and meteorology has brought to non-powered flight during the first decade of its existence important successes and high appreciation. In flights during thunderstorms and in thermal-lift flights distances of 15 km (9.3 mi), altitudes of 2,000 m (6,560 ft) and durations of 15 h were reached.

Since glider pilots and researchers of other nations wanted to learn from the German pioneers and to cooperate with them, the "International Study Committee of Gliding" ("Internationale Studienkommission für Segelflug, ISTUS") was founded in 1931 with *Walter Georgii* as chairman; its tasks were similar to the ones of the research institute of the RRG.

At the beginning of the thirties, gliding had abandoned rubber-cord-supported take-offs from mountain slopes. The development and introduction of the winch launch and the launch by aircraft towing allowed gliders to also become airborne in plain country so that gliding experienced a wider spreading, and the professional manufacturing of gliders for training and high performance flights increased.

A large variety of planes were built. *Alexander Lippisch*, for instance, introduced with the "Fafnir" in 1930 the cranked wing (seagull-type wing), which was copied many times and was until about 1960 in vogue. It was not the alleged better directional stability but rather the aesthetics of the seagull wing, which made the glider so attractive. At the same time others saw progress in the large 30-meter (98 ft) span of *August Kupper's* Ku 4 "Austria" as well as in the small D 28 "Windspiel" ("Greyhound") of the "Akaflieg" Darmstadt with a span of only 12 m (39 ft) and a mass of 54 kg (119 lb). Another independent and successful way was pursued by the "Akaflieg" Munich with their

The "Fafnir" with its seagull-type (cranked) wing in 1930.

D 28, the light "Windspiel" ("Greyhound") in 1933.

steel-tube fuselage and their in-house designed airfoils; their glider Mü 13 was later build in series.

At the end of the twenties, *Alexander Lippisch* built and tested at the RRG tailless gliders ("Stork", "Delta"), which were later improved and finally lead to the development of the rocket-powered aircraft Me 163. He himself admitted that these sailplanes were not very well suited for gliding. *Igo Etrich* and *Franz Wels* had already in 1906, following suggestions of *Friedrich Ahlborn* [31], built and flown a "flying wing" glider. In 1921, several tailless "World Gliders" ("Weltensegler") of *Friedrich Wenk* and the "Charlotte" of the "Akaflieg" Berlin flew at the Wasserkuppe, however, the performances and characteristics of these vehicles were never very satisfying. They had some advantages but many disadvantages, which were not easily remedied, as the young aircraft designer *Willy Messerschmitt* found out the hard way. Also the brothers *Reimar* and *Walter Horten* have built and tested in the thirties a number of interesting "flying-wing" gliders in Germany and after the war in Argentine. Only in recent times has the "Akaflieg" Brunswick with its tailless SB 13 shown, following extensive research and test work, where the limits of this (flying-wing) development lie.

At the beginning of the thirties, the name of the RRG ("Rhön-Rossitten-Gesellschaft") was changed to "German Research Establishment of Gliding" ("Deutsche Forschungsanstalt für Segelflug, DFS"); this organization now also designed under the supervision of *Hans Jacobs* [32, 33] cantilever high-performance sailplanes (e.g., "Rhönsperber" ("Sparrowhawk"), "Reiher" ("Heron"), "Weihe" ("Harrier"), "Meise" ("Tit")), which were built by industry in series and essentially met the growing demand. Also smaller companies developed new glider types.

When the "National-Socialist Party, NS-Party" came into power, the "NS-Flight-Sports Organizations" ("NS-Flugsport Organisationen") declared that "*in order to limit the number of types, new designs are undesirable*". This was a severe set-back for the advanced individual designs of the "Akafliegs" and for the groups themselves; the latter were supposed to be dissolved or "brought into line". They were, however, taken over by DVL (*Otto Fuchs*, "Department of Junior Engineers") and supported as scientific-technical "Flight Technical Expert Groups", respectively "Working Groups" ("Flugtechnische Fachgruppen, FFG", respectively "Arbeitsgemeinschaften, FAG"). These groups were allowed to continue work, to do research and to build new sailplanes (and also powered planes) [34] without being influenced by the NS-party.

Specifically, these groups were involved in the development of glued fittings (Esslingen E 3), two Olympic Sailplanes (Berlin B 8 and Munich Mü 17), a self-launching motor glider (Chemnitz C 10), prone positions for the pilot (Stuttgart fs 17), gliders for measurements, with cambered and other flaps. Particularly outstanding was in 1938 the "Akaflieg" Darmstadt D 30 "Cirrus", a high-performance sailplane with a span of 20 m (66 ft) and a for that time unusually large aspect ratio of 33.6; furthermore, with a dihedral that could be adjusted during flight, with cambered flaps and airbrakes, and with various vertical control surfaces. *Hans Zacher* carried out an extensive measurement

The D 30 "Cirrus" with variable dihedral, cambered flaps and airbrakes was used for performance and flight-characteristics measurements (1938).

program with the D 30 to determine performance and flight characteristics [35]; this program still serves as model to the "Idaflieg Comparison Flights" [29].

In 1934, *Hans Jacobs* (DFS) started the exploration and development of spoilers and airbrakes for sailplanes. At that time, flights within clouds were common and permitted. Thus, it was easily possible that pilots, unskilled in blind-flying, exceeded the permitted maximum speed, which often resulted in wing fracture. It was not possible to increase wing strength since this would mean too much additional weight for a low-drag light sailplane. Hence *Hans Jacobs* decided to develop an in-flight extendable flap system for both the upper and lower wing surfaces, which

Data of trend-setting and exceptional German sailplanes from the beginnings until 1941.

Sailplane type	Designer/ manufacturer	Year	Span [m]	Surface area [m²]	Aspect ratio	Net-mass [kg]	Gross-mass [kg]	Wing loading [kg/m²]	Sinking speed [m/s]	Glide ratio	Remarks
Normalsegelapparat	Lilienthal	1894	6,7	13,0	3,5	ca. 20	ca. 90	6,9			Monoplane, braced, hang glider
FSV X	Flugsportverein Darmstadt	1912	10,0	22,4	–	44	104	4,7			Biplane, pilot seated, braced
FVA 1 Schwatze Düwel	Flugwiss. Vereinigung Aachen	1920	9,5	15,0	6,0	62	137	9,1			Low wing, cantilever, multiple spars
H 1 Vampyr	Akaflieg Hannover	1921	12,6	16,0	10,0	120	195	12,2	0,80	ca. 16	Cantilever wing with one spar and torsion nose, angular fuselage
D 9 Konsul	Akaflieg Darmstadt	1923	18,7	21,8	16,1	200	280	12,8	0,70	ca. 21	Large aspect ratio, streamlined fuselage, elliptical wing shape, differential aileron
Wien	RRG	1929	19,1	18,6	20,0	168	248	13,3	0,71	ca. 22	Tapered wing, strut-braced
Fafnir	RRG	1930	19,0	18,6	20,0	200	315	16,9	0,76	24	Seagull-shaped cantilever wing, tapered
Ku 4 Austria	Kupper, Segelflugzeugbau Kassel	1931	30,0	35,0	25,7	392	482	13,8	0,55	26	Cambered flaps, twin vertical control surfaces, very large and tall
D 28 Windspiel	Akaflieg Darmstadt	1933	12,0	11,4	12,6	54	129	11,3	0,66	23,5	Cambered flaps, differential aileron and rudder, very light, one-piece wing
Mü 13 D	Akaflieg München	1936	16,0	16,2	15,8	170	270	16,7	0,60	28	Steel-tube fuselage, series production
Reiher	DFS	1937	19,0	19,4	18,6	238	315	16,4	0,5	33	Small series, contour-integrated cockpit
D 30 Cirrus	Akaflieg Darmstadt	1938	20,1	12,0	33,6	190	288	24,0	0,55	38	Cambered flaps and airbrakes, very large aspect ratio, variable dihedral
Olympia Meise	DFS	1939	15,0	15,0	15,0	160	252	17,0	0,67	25	15-meter-class/Olymia, series production
Ho IV	Gebrüder Horten	1941	20,0	21,1	19,0	230	349	16,5		33	Flying wing, prone pilot position

For further data see: G. Brinkmann, H. Zacher, Evolution der Segelflugzeuge, 2. Ed. Bernhard & Graefe Verlag, Bonn, 1999.

could be used to increase form as well as induced drag. With these "airbrakes" the terminal dive speed, being at that time for advanced gliders without flaps 400–500 km/h (249–311 mph), could be limited to 200 km/h (124 mph).

The DFS "dive-brakes" were later improved at the Schempp-Hirth Company by *Ulrich* and *Wolfgang Hütter*; they are today known as SH-flaps. They are not only used as dive-brakes but also, as suggested by *Prandtl*, to adjust the glide angle during the landing approach, allowing today, for instance, a reduction in the glide ratio from 60 to 6.

Flight characteristics were also improved by replacing the highly tapered wings, introduced by the "Wasserkuppe-School" for weight-saving reasons – which were, however, susceptible to wing-tip stall – by more elliptical wing planforms. Further performance improvements were achieved by the use of "laminar-type" airfoils, or partly also by thin NACA-type airfoils, by higher aspect ratios, as in the case of the "Austria" and, thereafter, in the case of the D30, and by closed and integrated cockpit enclosures. Retractable landing gears, various types of cambered flaps, well-shaped fuselages, and considerable detail work, allowed, in the late thirties, the attainment of glide ratios of nearly 40 and rates of descent as low as 0.5 m/s (1.6 ft/s).

With the beginning of the war in September of 1939 almost all activities ceased. Until then, distances of up to 750 km (466 mi), flight durations of up to 56 h and altitudes of 7,000 m (23,000 ft) above the take-off point were achieved in non-motorized flights. Finally, *Erich Klöckner* reached in 1940 an absolute altitude of 11,410 m (37,400 ft) during a "Föhn-(lee)-wave" flight.

The DFS moved from Darmstadt to Ainring near Salzburg, took over additional responsibilities and was enlarged. The "Akafliegs" were able to organize two more "Comparison Flights" in Prien/Chiemsee in 1941 and in Darmstadt in 1943, and to conduct measurement flights, e.g., with the D30. Later they were assigned special tasks, as, for instance, the construction of the manned "Delta-wing Aircraft DM 1" of *Alexander Lippisch*.

The table contains data of a selected trend-setting and extraordinary group of German sailplanes up to 1941.

References

1 O. Lilienthal, Der Vogelflug als Grundlage der Fliegekunst – Ein Beitrag zur Systematik der Flugtechnik, R. Gaertners Verlagsbuchhandlung, Berlin, 1889. (English edition: Otto Lilienthal, Birdflight as the Basis of Aviation: A Contribution toward a System of Aviation, Motorbooks Intl./MBI Publishing Company, St. Paul, MN, USA, 2001)

2 H. Hoernes, Luftschiffahrt der Gegenwart, A. Hartlebens Verlag, 1903

3 P. Supf, Das Buch der deutschen Fluggeschichte, Drei Brunnen Verlag, Stuttgart, 1956

4 H. von Helmholtz, Populärwissenschaftliche Vorträge, Braunschweig, 1865/1876

5 J.G. Zeidler, Fliegender Wandersmann oder philosophische Untersuchungen der Fliegekunst, Halle, 1710

6 W. von Siemens, Lebenserinnerungen, Berlin, 1892

7 W. Heinzerling, H. Trischler (Hrsg.), Otto Lilienthal, Flugpionier, Ingenieur, Unternehmer, Dokumente und Objekte, Deutsches Museum, München, 1991

8 H. Junkers, Grundsätze technisch-wirtschaftlicher Forschung, entwickelt aus ihren Zielen nach eigenen Erfahrungen, Vortrag im Haus der Technik in Essen am 11.10.1932

9 W. Wagner, Hugo Junkers Pionier der Luftfahrt – seine Flugzeuge, Die deutsche Luftfahrt, Bernard & Graefe Verlag, Bonn, 1996

10 W. von der Nüll, Stratosphärenflug und Höhentriebwerk, Luftwissen 1943, Band 10, Nr. 8

11 H. Bansemir, J.P. Roeder, Die Fw 200 „Condor" – ein Meilenstein der Entwicklung von Verkehrsflugzeugen, DGLR-Tagung, Bremen, 5–8. 10. 1998

12 K. Tank, Langstreckenflugzeuge, Luftwissen, 1944, Bd. 11, Nr. 9, pp. 239–242

13 J.P. Roeder, Evolution of the Art of Flying since Lilienthal, An Airplane Company's View, DGLR Kongress 1991, Berlin

14 W. Heinzerling, Die Geschichte des Pfeilflügels (2): Die projektmässige Anwendung des Pfeilflügels, Vortrag-Nr. 81-071, DGLR- Jahrestagung 1981, Aachen, 11.–14.Mai 1981

15 H. Ludwieg, Die Geschichte des Pfeilflügels (1): Die Entwicklung des Pfeilflügels in Deutschland bis 1945, Vortrag-Nr. 81-070, DGLR-Jahrestagung 1981, Aachen, 11.–14. Mai 1981

16 P. Hamel, The Birth of Sweepback – Related Research at LFA-Germany, AIAA Atmospheric Flight Mechanics Conference, Austin, Texas, August 12, 2003

17 L. Bölkow, Mit dem Pfeilflügel zum Hochgeschwindigkeitsflug, in: 50 Jahre Turbostrahlflug, DGLR-Bericht 89-05, Vorträge DGLR-Symposium 26./27.10.1989, München
18 G.S. Schairer, On the Early Large Swept Wing Aircraft, in: 50 Jahre Turbostrahlflug, DGLR-Bericht 89-05, Vorträge DGLR-Symposium 26./27.10.1989, Munich
19 G. Halle, Otto Lilienthal und seine Flugzeugkonstruktionen, VDI-Verlag, München und Düsseldorf, 1962
20 K. Kopfermann, Otto Lilienthal – Über meine Flugversuche, VDI-Verlag, Düsseldorf, 1987/88
21 H. Watzinger, Gedenke des Anfangs, Flugversuche der Darmstädter Jugend, Liebig Verlag, Darmstadt 1973
22 W. von Langsdorff, Das Segelflugzeug, V.J.F. Lehmann, München 1931
23 W. Klemperer, Theorie des Segelfluges, Abhandlungen aus dem Aerodynamischen Institut der TH Aachen, Berlin, 1927
24 H. Voigtländer, Der Student und die deutsche Luftfahrt, V. Klasing, Der Jungflieger,1929
25 L. Prandtl, Bemerkungen über den Segelflug, ZFM 1921
26 R. Bredt, Kritische Bemerkungen zur Drehungselastizität, ZVDI 1896
27 H. Zacher, Studenten forschen, bauen und fliegen, Verlag Akaflieg Darmstadt 1981
28 H. Zacher, The Shape of High Performance Sailplane Technical Development, in OSTIV, The World's Sailplanes, Vol. II, 1963
29 Idaflieg, 75 Jahre Idaflieg: Studenten forschen, bauen und fliegen, Idaflieg, Aachen, 1997
30 G. Brinkmann, H. Zacher, Die Evolution der Segelflugzeuge, Bernard & Graefe Verlag, Bonn, 1992
31 F. Ahlborn, Zur Mechanik des Vogelfluges, Hamburg, 1896
32 H. Jacobs, Die Entwicklung der Leistungssegelflugzeuge, in: W. Hirth, Handbuch des Segelfliegens, Franckh'sche Verlagsbuchhandlung, Stuttgart, 1938
33 G. Brütting, Die berühmtesten Segelflugzeuge, Motorbuch Verlag, Stuttgart, 1970
34 G. Brinkmann, K. von Gersdorff, W. Schwipps, Sport- und Reiseflugzeuge, Bernard & Graefe Verlag, Bonn, 1995
35 H. Zacher, Ergebnisse der Leistungsmessung und Flugeigenschaftsprüfung des Segelflugzeuges D 30 "Cirrus", DVL – Mittl. FFG Folge 6, 1944

AERODYNAMICS – THE KEY TO SUCCESSFUL FLIGHT

Derivation of the Aerodynamic Fundamentals

Ernst Heinrich Hirschel

Mathematical Description of Aerodynamic Phenomena

Businessmen and implementers of the aircraft idea like Hugo Junkers cannot be envisaged without the application-oriented basic research as background, as performed by Ludwig Prandtl in Göttingen. Ludwig Prandtl is called the "Father of Aerodynamics" [1, 2], which does, however, not allow the conclusion that Germany is the sole "Fatherland of Aerodynamics". Prandtl (1875–1953) is in line with other, mostly European, researchers and engineers, who considered basic questions of human flight. He has, however, in a unique way, discovered and formed essential elements of aerodynamics as they are known today. His inquisitive personality, which utilized theory and experiment "problem-oriented", found at the University of Göttingen, which he joined in 1904, an unique combination of science and technology, inspired by the mathematician Felix Klein (1849–1925). This was of decisive importance for the development of aerodynamics.

The "Mathematization" of the sciences, up to then essentially treated empirically, is certainly the most important characteristic of the Göttingen era at the time of the departing 19th/arriving 20th century. "One must have hypotheses and theories to organize ones findings..." was formulated by Georg Christoph Lichtenberg [3], a physicist at the University of Göttingen, one hundred years ago. The "Mathematization", i.e., the description of physical phenomena and of the functioning of technical devices with the aid of mathematical formulations, allows to understand these and to predict their behavior.

In the case of aircraft aerodynamics one may easily recognize how mathematical modeling permits, on the one hand, the understanding of the relationships of phenomena, and, on the other hand, allows finding the optimum aerodynamic form, for instance, of the wing. It is, of course, not mathematics by itself, but its handling by people with intuition and the gift "to see things in context", which was strongly developed by Ludwig Prandtl. Today, theory and computational methods play an ever-increasing role besides wind tunnel simulations, since the demands on modern aircraft, but also the development risks, cannot be met without them.

Here, we will sketch the development of aerodynamics in Germany until the end of World War Two. All crucial discoveries occurred within this time period (about 1880–1945). They do not only concern aircraft aerodynamics, but also concern fluid machinery including aircraft engines, automobile aerodynamics and wind-energy aerodynamics. The most important developments, which were made by research – but were also immediately applied in aircraft engineering – will subsequently be presented within three categories: "Lift, Drag, Lifting-wing Theory", "Boundary Layers and Boundary-layer Physics" and "Compressibility Effects in Aerodynamics". For an in-depth consideration of these three categories, the reader is referred to original literature but also to summarizing presentations [4–10].

Lift, Drag and Lifting-Wing Theory

Lift and drag of an air vehicle are its fundamental characteristics which are addressed by aerodynamics. For steady horizontal flight it holds that the lift generated by the wing has to compensate the weight of the air vehicle including the payload and the propulsion system. Drag, opposing the aircraft movement through the atmosphere, has to be overcome by propulsion, i.e., engine performance, (or by gliding, which is associated with a loss in altitude).

The discovery that a surface "inclined" at a small angle with respect to the flight direction generates lift and the distinction between lift and drag are today attributed to the Englishman *George Cayley* [11]. Essential aerodynamic design elements were discovered by *Alphonse Pénaud* [12] (among others, the horizontal tail to achieve longitudinal stability), and *Otto Lilienthal* (among others, polar diagrams and the fact that a longitudinally cambered wing generates more lift than a plane wing). Since at that time, except for Newton's Theory, which was obviously at fault, there was no theoretical possibility to make quantitative statements concerning air-vehicle lift and drag, experimental methods were common. Here, a decisive break-through was achieved through the development of the wind tunnel by the Englishman *Francis Herbert Wenham* [5] in 1871; in the wind tunnel, a model of the air vehicle is exposed to an artificially generated air stream.

The theory of lift and drag had initially to struggle with two paradoxes. *Isaac Newton* (1642–1727) [13] had determined analytically that the force acting on a plate inclined at an angle α with respect to the flow is proportional to $\sin^2\alpha$. Lift is accordingly proportional to $\sin^2\alpha \cos\alpha$, drag to $\sin^3\alpha$. As during the 19th century empirical data became increasingly available, Newton's law was shown to be obviously wrong since the lift of a wing is, within a large range of angles of attack, directly proportional to α itself. *Jean-Baptist d'Alembert* (1717–1783) [14] had, on the other hand, analytically determined that forces, except for friction forces, do not act on a body moving through air. This result thus excludes the generation of lift (and drag) by an inclined plate, which was later shown by *George Cayley's* results to be a paradox. It was already named a paradox by *d'Alembert.*

The consequences of these paradoxes are thoroughly explained elsewhere [5]. Still in 1903, *Hermann Hoernes* stated that "*to determine the laws of drag, the experimental approach should exclusively be employed*" and mentions in that regard the new results of *Friedrich Ritter von Lössl* [15]. However, during that time fundamental features of today's theory of lift and drag were already established. *John William Strutt* (1842–1919), third *Baron Rayleigh (Lord Rayleigh)*, investigated the Magnus-effect on tennis balls and found in 1878 [16] that in an ideal fluid the superposition of the flow past a cylinder with a circulation (rotation of the cylinder) resulted in a lift force normal to the freestream flow direction, thus invalidating *d'Alembert's* paradox. The formulation of the mathematical model, which relates lift of an airfoil to the circulation (lifting vortex) about this airfoil, is connected with the names of the German mathematician *Wilhelm Martin Kutta* (1867–1944), the Russian mathematician and physicist *Nikolai E. Joukowsky* (1847–1921) and the English engineer *Frederick W. Lanchester* (1878–1946). The latter went in his approach already beyond the two-dimensional airfoil problem considering the three-dimensional problem of the wing of finite span. A fourth researcher has to be mentioned, viz., the German physicist *Hermann von Helmholtz* (1821–1894), who provided with his vortex theorems [17] fundamentals concerning circulation theory and especially the theory of the finite wing, which merits highest regard.

The decisive element for circulation theory was the (today) so-called Kutta-Joukowsky condition: on an airfoil with a sharp trailing edge the flow always leaves the airfoil at the trailing edge, i.e., the flow cannot reach the aft stagnation point, located for positive angles of attack on the upper surface, which would be required to meet *d'Alembert's* paradox of force-free conditions. The consequence is a finite force, namely lift, acting on the airfoil. The Kutta-Joukowsky condition is exactly the compulsory condition determining circulation and hence lift [10]. The circulation theory holds for airfoils with attached flow, i.e., for angles of attack below the one for maximum lift.

Although *Lanchester* had a large part in the development of the theory of the finite-span lifting wing, the wing theory was formulated by *Prandtl* [18]. For large aspect ratio wings (span large compared to the average chord) he introduced the "lifting (vortex) line" ("bound vortex") analogous to the "lifting vortex" in the case of airfoils. Two free vortices extend from the wing tips back, hence the name "horseshoe vortex", and meet at infinity the "starting vortex". With this mathematical model, *Prandtl* generalized circulation theory, meeting the requirements of *Helmholtz's* vortex theorems, thus establishing the basis for modern wing theory.

Göttingen was subsequently the center of extensive research efforts especially considering the fundamental questions concerning lift and drag of a wing; the latter since *Prandtl's* wing theory also addresses drag of a finite-span wing. The free vortices, which are not present on an airfoil (they are located spanwise at infinity, hence are not of importance) generate drag, the "induced drag", which may represent a large part of the total drag. Induced drag can also be interpreted as lost kinetic energy bound in the free vortices, or as drag due to an induced angle of attack [10].

Based on wing theory, two questions were subsequently addressed:

a) How should an airfoil be shaped and
b) how should one lay out the wing planform (in a wider sense also considering spanwise airfoil variations) to achieve an optimum design with respect to lift, drag, pitching moment, et cetera?

During the years 1923–1927, systematic wind tunnel investigations were performed on "Joukowsky" airfoils at the "Aerodynamic Test Establishment Göttingen" ("Göttinger Airfoil Systematics" [19]). These investigations were mainly thought to check airfoil theories. More application-oriented were the in the

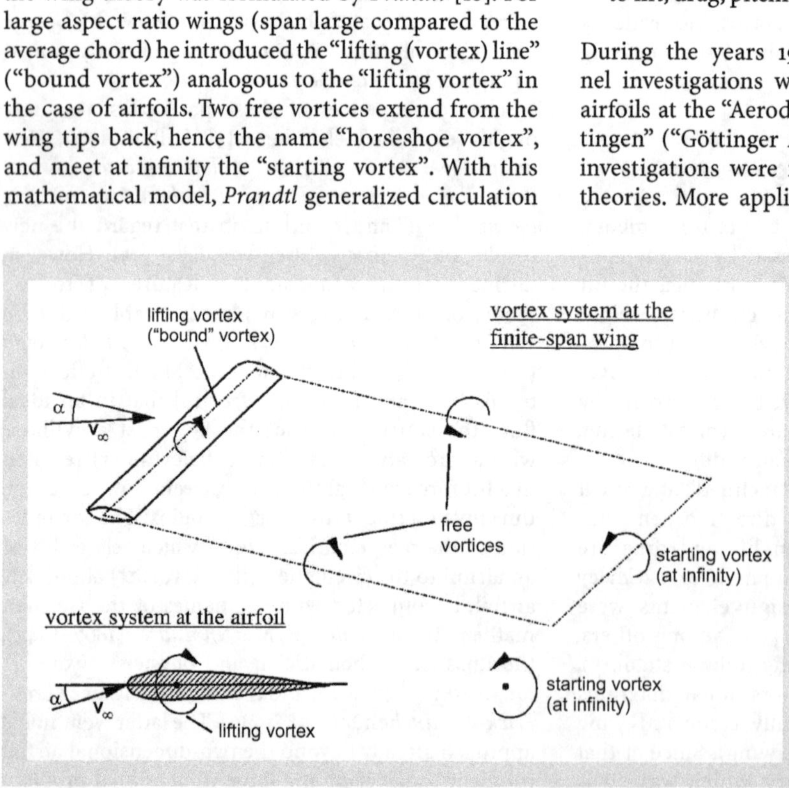

Vortex systems of the airfoil and the finite aspect-ratio wing.

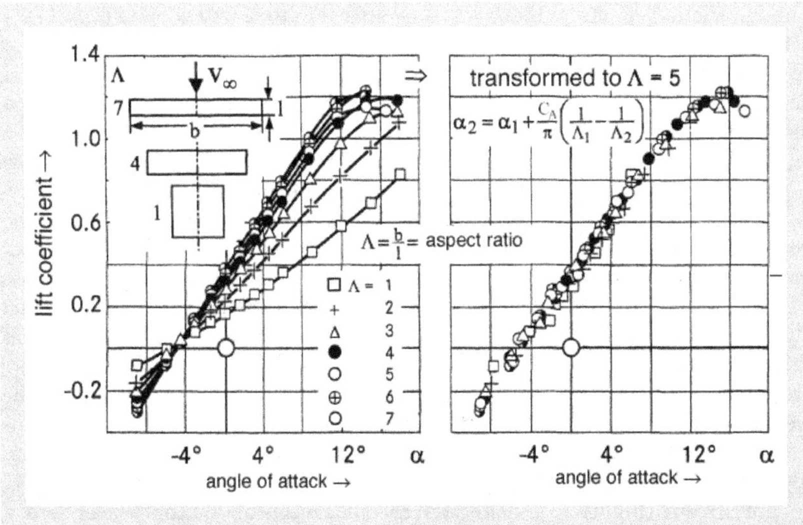

Measured polars for a rectangular wing with various aspect ratios (left) and transformation to the aspect ratio $\Lambda = 5$ (right) by means of the transformation equation of the lifting-line theory.

USA developed NACA airfoil systematics. Concerning airfoil theory, there existed two main directions: the theory of "conformal mapping" and the "singularity theory". In case of the former, the flow past an airfoil is traced back to the flow, say, past a circular cylinder by means of conformal mapping. In case of the latter, singularities (sources, sinks and vortices) are arranged in the airfoil. Both are able to treat practical problems including airfoils with deflected flaps.

Prandtl's wing theory provides by means of *Prandtl's* integral equations either the wing geometry (spanwise chord and angle of attack distributions) for a given circulation distribution (first main task of wing theory), or the circulation distribution for a given wing geometry (second main task). An important statement, namely, that the induced drag of a wing assumes a minimum if the spanwise circulation distribution is elliptic, was already made in 1918 by *Max Munk* [20]. This result was later reflected in the design of several subsonic aircraft with wings of an elliptical planform (two half-ellipses), since the elliptical circulation distribution is exactly realized for a straight zero-twist wing of that planform.

Prandtl's wing theory is a lifting-line theory for large aspect ratio wings. The standard computational method for unswept wings according to this theory is the method of *Hans Multhopp* [21]. For wings with lesser aspect ratios, it is expedient to work with a wing-sheet vortex model. *Hermann Blenk* [22] and the Englishman *Hermann Glauert* [23], who spread *Prandtl's* wing theory within the English-speaking countries, developed the corresponding theory.

Further developments, initially limited to incompressible flow, concerned the effect of wing thickness, more general wing forms and nonlinear effects. There were, in addition, the problems related to fuselage aerodynamics (at first initiated by airship technology), to wing-fuselage configurations, to the effect of the propeller jet (propwash), to stabilizer and control surfaces (rudder and flaps). These problems are extensively covered in the standard book of *H. Schlichting/E. Truckenbrodt* [10]. Mentioned must be the problems associated with small-aspect-ratio wings (highly swept wings, delta wings). For these *R.T. Jones* published in 1946 his theory [24], which is complementary to *Prandtl's* approach. A similar approach was already taken by *Max Munk* in 1923 to calculate the flow about airships [25].

Boundary Layers and Boundary Layer Physics

Many aerodynamic flows may for computational purposes considered to be essentially inviscid, which explains the success of wing theory, which is essentially a potential theory, i.e., it assumes inviscid (frictionless) and irrotational flow. Though friction effects are also present here, but hidden, e.g., in the Kutta-Joukowsky condition and in the vortex singularities. Friction effects in aerodynamics are explicitly only present within a small layer adjacent to the surface of a body subjected to the flow, i.e., within the so called boundary layer, and at locations where the boundary layer leaves the surface, either at the trailing edge (wake or vortex sheet downstream of an airfoil) or upstream of the trailing edge (separated flow, vortex formation).

Boundary layers are generally very thin; its principal features are depicted in the figure below. An observer on the surface of an aircraft would, depending on his position, register in the direction normal to the surface of the aircraft within a few millimeters or centimeters a decrease in the tangential velocity from the outer inviscid flow velocity (freestream (flight) velocity or higher) to zero velocity at the aircraft surface. Relative to the inviscid outer flow, this is a very large decrease in velocity within the boundary layer due to the adhesion of the flow at the surface, which leads to the friction drag. The latter may constitute a large part of the total drag of an aircraft and is, next to the induced drag, the second large drag contributor. The question whether the air at the flow conditions of aerodynamics adheres to the surface (no-slip condition) was only settled in 1904 by *Ludwig Prandtl* [26]. Up to that time, a slip of the flow along the surface was considered possible [27], as is actually only the case for highly rarefied flow (vacuum technique, hypersonic aerothermodynamics at high altitudes).

When the boundary layer leaves the rearward part of a body, a third drag contribution occurs, the friction-induced pressure or form drag. Due to the displacement effect of the boundary layer (finite boundary layer thickness δ) the inviscid streamlines are not able to close at the end of the body. This means that the recompression back to the total pressure does not occur thus resulting in a drag force. This shows that *d'Alembert's* paradox is a result of neglecting the boundary layer effect not known at that time. This effect is, however, partly accounted for in the Kutta-Joukowsky condition.

Although certain aspects of boundary layer theory, as it is understood today, were already known, it was *Ludwig Prandtl*, at the time still in Hanover, who published in 1904 the epoch-making fundamentals of boundary-layer theory [26]. He derived with the aid of order-of-magnitude considerations from the Navier-Stokes equations (at the time a not generally accepted approach) the "boundary layer equation" and showed, among other things, how the boundary layer and the outer inviscid flow interact. As long as

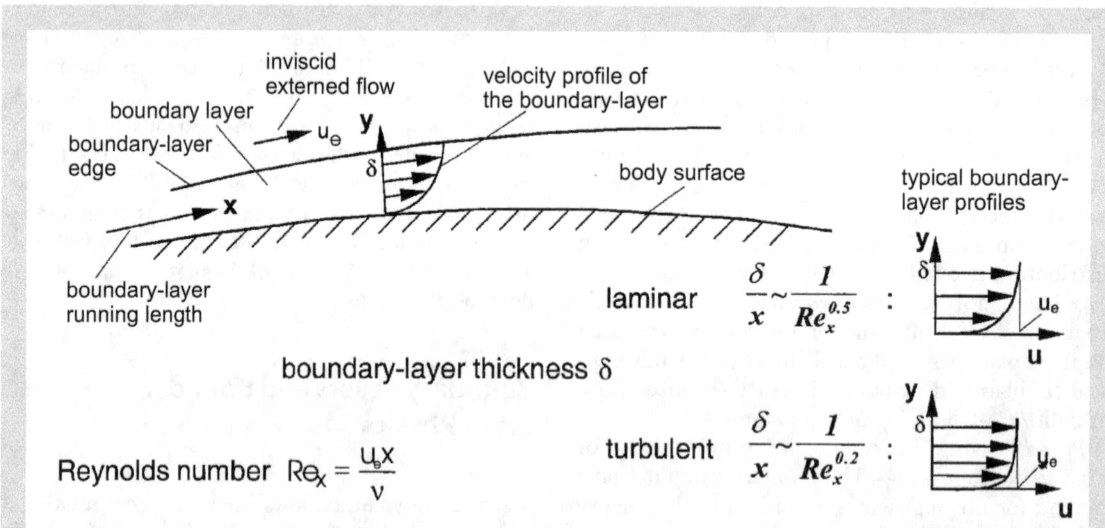

Fundamental boundary-layer characteristics.

Hermannn Schlichting (1908–1982).

Hermann Schlichting, born in Balje at the Elbe river, County of Stade on September 9, 1908.

1930 receives his PhD with *Ludwig Prandtl* in Göttingen.
1935 head of the wind tunnel department at Dornier in Friedrichshafen.
1938 he is assigned the chair of fluid mechanics at the Institute of Technology (TH) Brunswick
1941 corresponding member of the "German Academy of Aeronautical Research".
1943 full member of the "Scientific Society Brunswick" ("Braunschweigische Wissenschaftliche Gesellschaft").
1951 his standard book "Boundary Layer Theory" is published.
1953 in addition to his teaching obligations at the TH Brunswick, he takes over as head of the DFL.
1955 he becomes a fellow of the "Institute of Aeronautical Sciences, IAS".
1957 in addition to the above positions, he succeeds *A. Betz* as director of the AVA Göttingen.
1959 he publishes together with *E. Truckenbrodt* the two-volume book "Aircraft Aerodynamics" ("Aerodynamik des Flugzeuges". Through his scientific publications (more than 100) and his books, he becomes renown far beyond Germany.
1960 member of the board of the "German Society of Flight Sciences" ("Deutsche Gesellschaft für Flugwissenschaften, DGF").
1964 corresponding member of the "International Academy of Astronautics" in Paris.
1968 honorary doctor of the Technical University (TU) Munich.
1969 after the merger of the individual aerospace research and test establishments to form DFVLR, he becomes head of the "Research Center AVA Göttingen"; he receives the Ludwig-Prandtl-Ring of the DGLR.
1982 he dies in Göttingen on June 15.

the boundary layer is very thin, as, for instance, at high Reynolds numbers, its effect on the inviscid flow past the body, hence on the pressure distribution on the body surface, is very weak (weak interaction). This is the main reason why circulation theory, and in a wider sense wing theory, can be successful. The inviscid flow imposes its pressure field on the boundary layer, which is of paramount importance for the characteristic features of the boundary layer equation and the mathematical approaches to its solution.

The first solution to *Prandtl's* boundary layer equation was given by *Heinrich Blasius* [28] for the flow over a flat plate, the simplest boundary layer case. He reduced the boundary layer equation by means of an affinity transformation to ordinary differential equations whose solution provided results in excellent agreement with experimental data of *Johann Nikuradse* [29].

A different approach was suggested by *Theodore von Kármán*, viz., the reduction of the boundary layer equation to an ordinary differential equation, – the momentum equation of boundary layer theory – by the introduction of integral quantities in the direction normal to the body surface (displacement thickness, momentum thickness). *Karl Pohlhausen* provided a solution to the momentum equation with the aid of boundary-layer-profile families. With the Kármán-Pohlhausen-method and its further developments (*Alfred Walz* and others), one was able to also compute boundary layers with accelerating (negative) or decelerating (adverse) pressure gradients as long as adequate profile families could be provided. For boundary layers in compressible flow, where convective and diffusive heat transport as well as compression and dissipative work must be described, relations between velocity and temperature fields were initially given for special cases (*A. Busemann*, 1940). Thereafter, *W. Hantsche* and *H. Wendt* (1940) provided, almost concurrent with *L. Crocco* (1941) in Italy, relations for the computation of laminar compressible boundary layers with different wall-temperature boundary conditions, viscosity laws and Prandtl numbers [29].

In the preceding deliberations it was tacitly assumed that the boundary layers in question were laminar. In the aerodynamic reality there are always also turbulent boundary layers. If the flight Reynolds number is high enough, the boundary layer on the flight vehicle is predominantly turbulent. Turbulence is an extremely multifaceted phenomenon. In the case of aerodynamics, it can as "boundary-layer turbulence" be characterized as follows: it transports through mixing high-momentum fluid from the edge of the boundary layer towards the body surface. This results in comparison to a laminar profile in the proximity of the wall in a fuller (turbulent) profile. This is, due to its higher momentum close to the wall, able to negotiate a larger pressure increase without separating than the laminar profile. For a turbulent boundary layer, the friction-induced pressure drag of, say, a wing is therefore generally smaller than in the presence of a laminar boundary layer. However, friction drag is due to the fuller turbulent profile higher than in the case of a laminar one. It is hence a question of the operating range of a wing that shall be optimized, and thus of the aerodynamic design of it surface, whether a turbulent boundary layer may be considered beneficial or detrimental. For typical slender airfoils the type of pressure distribution is the main criterion.

The concept of turbulence is generally being related to the English physicist *Osborne Reynolds* (1842–1912), who performed pioneering work in this field [30]. It was *Arnold Sommerfeld* (1868–1952), who named the parameter that gives the ratio of inertial force to friction forces within a flow "Reynolds number (Re)". The first successful approach to render turbulent boundary layers accessible to computations was suggested by *Ludwig Prandtl* in 1925 based on the *Boussinesq* approach [31]. *Prandtl's* mixing-length approach combines the momentum transport due to the turbulent fluctuations with macroscopic characteristics of the time-averaged boundary layer profile, thereby providing an expression for the apparent turbulent viscosity. This is a function of the distance from the wall, containing the mixing length still to be determined. With his similarity hypothesis, *Theodore von Kármán* worked out an approach to present the mixing length as local function [32].

For the prediction of friction drag of a flat plate in turbulent flow, a computational tool was finally provided by *Prandtl's* and *Schlichting's* drag law [29] ("$1/7^{th}$–power-law"), still important today for practical applications, which allows a quick estimate of friction drag as long as the flow past a configuration is predominantly two-dimensional and incompressible and large pressure gradients are absent. In the years up to 1945, turbulent boundary layer theory was also in Germany further developed considering turbulent boundary layers exposed to negative and positive pressure gradients (*E. Gruschwitz*), the effect of wall roughness (*Fritz Schultz-Grunow, Johann Nikuradse*), et cetera [29].

As long as turbulence is a topic of research it is also the origin of turbulence. Systematic studies were already carried out by *Osborne Reynolds* and then, after 1880, especially by *Lord Rayleigh* in England. The transition from laminar to turbulent flow was, however, already observed earlier, e.g., by *Gotthilf Heinrich Hagen* (1797–1884) [33]. *Ludwig Prandtl* started in 1920 to study the origin of turbulence [34]. *Oskar Tietjens* calculated in 1922 solutions to the inviscid stability equation (Rayleigh-equation) and found first indications [35] that velocity profiles with an inflection point, for instance, due to an adverse pressure gradient, are unstable (inflection-point instability). It was thereafter *Wilhelm Tollmien*, who established in 1929 with his work [36] the foundation of modern stability theory. He provided a solution of the Orr-Sommerfeld-equation, which completely describes the linear stability problem. With *Tollmien's* theory, initially considerably extended by *Hermann Schlichting*, therefore today referred to as "Tollmien-Schlichting stability theory", it was possible to answer many, up to that time, still open questions, among others ones related also to the problem of the inflection-point instability. After that, it took, by the way, more than a decade before the Americans *H.L. Dryden, G.B. Schubauer* and *H.K. Skramstad* succeeded in confirming the theory experimentally [37], but this only after they had removed all wind tunnel disturbances, which had masked the stability effects predicted by *Tollmien* and *Schlichting*.

After the breakthrough achieved with *Tollmien's* work, the influence of various effects on the stability behavior of the boundary layer were investigated. *J. Pretsch* showed in 1941 that pressure gradients could implicitly be accounted for via the form of the boundary layer profile. *Hermann Schlichting* and *A. Ulrich* performed stability computations for families of

Wilhelm Tollmien (1900–1968).

Wilhelm Tollmien, born in Berlin on January 30, 1900.

1924 after studying mathematics, physics and chemistry in Berlin and Göttingen, he receives his PhD with *Prandtl* and becomes a scientific assistant at the "KWI of Fluid Research" ("Kaiser-Wilhelm-Institut für Strömungsforschung".

1931 Research Fellow at the Guggenheim Aeronautical Laboratory of the California Institute of Technology in Pasadena with *von Kármán*.

1933 return to the "KWI of Fluid Research" in Göttingen.

1936 habilitation and appointment as lecturer at the University Göttingen.

1937 full professor of technical mechanics at the Institute of Technology (TH) Dresden.

1945 cooperation in compiling the "Göttingen Monographs".

1946 he joins the Royal Aircraft Establishment in Farnborough.

1947 he becomes *Prandtl's* successor at the chair of applied mechanics of the University Göttingen and director of one of the three KW-Institutes of Fluid Research.

1952 he becomes permanent member of the "Organization Committee of the International Mechanics Congresses".

1957 director of the three "Max-Planck-Institutes of Fluid Research" in Göttingen.

1962 honorary doctor of the Technical University (TU) Clausthal.

1968 he dies in Göttingen on November 25.

boundary layer profiles (1940). This way stability theory could easily be applied to given bodies, say airfoils, after the laminar boundary layer development had been calculated based on an inviscid pressure distribution.

With the aid of the inflection-point criterion one can show that a downstream directed flow-accelerating pressure gradient "stabilizes" the boundary layer. The effect of a decelerating pressure gradient can be compensated by removing "retarded" boundary layer material by means of suction through, say, a porous surface. The effect of the stabilization by an accelerating pressure gradient, today one speaks of passive laminarization, was investigated by *H. Doetsch*, who published in 1939 first experimental results. The so called "laminar dip" ("Laminardelle"), i.e., a pronounced reduction of drag in parts of the polar, occurs, corresponding to the type of pressure distribution present, only at small angles of attack of the airfoil. While in the USA and the USSR during World War Two fighter aircraft with "laminar airfoils" were developed and put into operation, are these airfoils today mostly used for sailplanes. They are at moderate flight speeds an excellent means to visibly improve the aerodynamic quality (lift to drag ratio) of a wing. In Germany, this technology was obviously first applied to the prototype of the Me 309 (first flight in July 1942), which was, however, never produced in quantity.

Concerning active laminarization by suction (also demonstrable with the aid of the inflection-point criterion), *K. Busemann* and *H. Münz* published in 1942 a stability computation, after *H. Holstein* had demonstrated this effect experimentally in 1940. The "Göttingen suction aircraft", already employed by *Oskar Schrenk* at the beginning of the thirties for tests related to the increase of maximum lift by applying suction, was due to changing priorities not used to study boundary layer laminarization. Active laminarization has today again a high rating due to its possible application on long-range transport aircraft.

It is possible to demonstrate with the aid of the inflection-point criterion that, in air, cooling of the boundary layer, i.e., heat transfer to the wall, has a stabilizing effect. The opposite is true in fluids (water) since here viscosity reduces with increasing temperature. *W. Linke* observed already in 1942 experimentally that heating an air-flow boundary layer has a destabilizing effect. It seems, however, that this topic has not been treated in Germany to any large extent up to 1945. A similar situation existed in the case of the effect of compressibility on boundary-layer stability, a topic that was treated by *Dietrich Küchemann* in his PhD thesis (Göttingen 1938).

Theoretical investigations concerning the effect of a convex or concave wall curvature had initially shown that curvature has only a small effect on stability. It was *Henry Goertler*, who proved in 1940 theoretically that on concave walls a centrifugal instability exists. Stability problems in three-dimensional boundary layers were obviously not treated in Germany prior to 1945. The effect of roughness on stability and transition behavior was also largely disregarded.

In concluding it can be stated that since 1929, starting with *Tollmien's* work, the stability theory of boundary layers has gone through an appreciable, also for practical problems important development. Since then, it can be predicted where in the boundary layer primary instability occurs, and how one can understand it. The subsequent transition to turbulent conditions, hence the prediction of the location where the boundary layer becomes fully turbulent, is even today not amenable to a complete theoretical description. Empirical and semi-empirical transition criteria and models constitute the state of research, although many details of the transition process are already known from experimental, analytical and numerical ("direct numerical simulation, DNS") investigations.

Compressibility Effects in Aerodynamics

The initial developments in aerodynamics, described above, were related to low flight speeds, respectively freestream velocities. Since under these conditions the density of the air in the flow past a body does not, or only marginally, change, one speaks of incompressible flow. Density changes with the square of the velocity, or more exactly of the Mach number, the latter being the ratio of the flight speed, respectively flow velocity, to the speed of sound. Up to flight Mach numbers of M = 0.2 or 0.3, corresponding to a flight speed of 250–370 km/h (155–230 mph) at a temperature of 15 °C, compressibility effects may therefore in general, but still dependent on the shape of the body, completely be neglected (incompressible subsonic speeds). At flight Mach numbers of about M = 0.5–0.8, it is necessary to account, in a simple way, for compressibility (compressible subsonic flow). In both instances one speaks of subsonic flight, respectively subsonic aerodynamics.

At higher Mach numbers, a particular phenomenon occurs, namely, an abrupt increase in density after the Mach number within the flow field of a body locally has exceeded the speed of sound (M = 1). Along with this abrupt increase of density goes an abrupt decrease of velocity back to subsonic speed. This phenomen happens in a very thin sheet which is called a shock wave (compression shock). It is associated with a further type of drag, viz., wave drag. It may also result, due to its interaction with the boundary layer developing on the body, in an increase of the friction-induced pressure drag. The phenomenon of the shock wave is related to the finite propagation velocity of pressure disturbances, i.e., the speed of sound. The latter is close to the ground about a = 330 m/s or 1,200 km/h (1,080 ft/s or 746 mph). If a body flies below the speed of sound, i.e., at subsonic speeds M < 1, pressure disturbances precede the body and "announce" that the body is coming so that the air can gradually adjust to the body. Is a body faster than the propagation of the pressure disturbances, M > 1, the air can only adjust immediately in front of the body and then abruptly by means of a shock wave. Dependent on the shape of the shock wave, a new Mach number distribution evolves downstream of the shock. (Note that only a shock normal to the flow direction results in subsonic flow behind the shock.)

The phenomenon of the propagation of pressure or sound waves, respectively, was already investigated by *Isaac Newton* in relation to appearances in conjunction with artillery fire. The German mathematician *G.F. Bernhard Riemann* (1826–1866) was the first to consider the shock relations. In doing so he made a mistake, which was, however, remedied later by the English engineer *William John Maquorn Rankine* (1820–1872) and the French ballistician *Pierre Henry Hugoniot* (1851–1887). The (today) so called Rankine-Hugoniot relations couple the conditions upstream and downstream of a shock wave (which may normally be considered as a field discontinuity).

The aerodynamics at higher Mach numbers, at the beginning of the twentieth century developing first and very rapidly in Göttingen, is called gas dynamics but also aerothermodynamics if very high Mach numbers occur, that are accompanied by high-temperature effects.

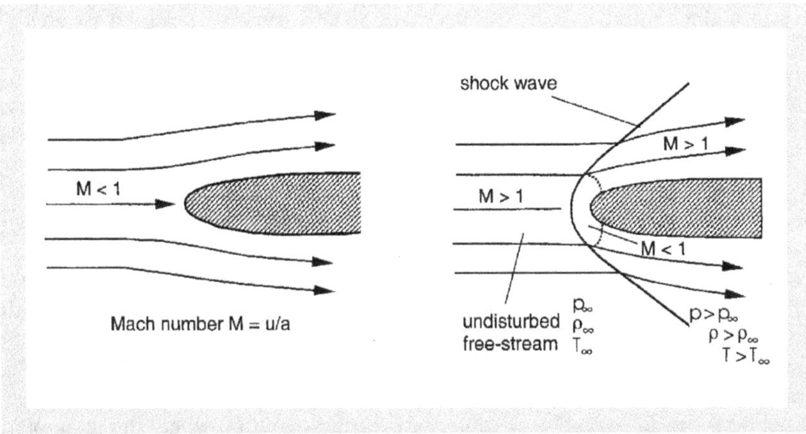

A body in subsonic (left) and supersonic flow.

At the end of the nineteenth century, important impulses were given by the Austrian physicist and philosopher *Ernst Mach* (1838–1916), who, for the first time with the aid of the Schlieren method, introduced by *August Töpler* in 1864, visualized shock waves on projectiles [38]. *Jakob Ackeret* (1898–1981) named in his honor the ratio of the flight speed to the speed of sound "Mach number (M)". Perhaps the most important impulse was given by the Swiss engineer *Carl G.P. de Laval* (1845–1913), who exhibited in 1893 in Chicago a one-stage steam turbine with convergent-divergent nozzles generating the jets impinging on the turbine blades therewith opening a completely new performance category. The discovery of *de Laval*, the "Laval-nozzle", permits, due to its form, to achieve almost any supersonic speed dependening on the imposed pressure ratio.

Stimulated by this discovery, the Hungarian-born engineer *Aurel Boleslav Stodola* (1859–1942) made the first systematic investigations on supersonic nozzles in 1903 at the "Swiss Institute of Technology, ETH" in Zurich. This, in turn, stimulated *Ludwig Prandtl* to build in 1905, already in Göttingen, a M = 1.5 supersonic nozzle for the investigation of steam-turbine-related flow problems. *Ludwig Prandtl* and *Theodor Meyer* performed systematic investigations with this nozzle, and in 1908 the latter received his PhD with a thesis [39], which already contained a description of the "Prandtl-Meyer expansion", in a certain sense the counterpart of the (compression) shock.

The interest in compressibility effects in aircraft aerodynamics proper was ignited at the beginning of the twenties by contemporary propellers – the airplanes themselves still flew very slowly – whose tip velocities already approached the speed of sound. *Ludwig Prandtl* stated in 1922 that the lift of a wing at subsonic speeds increases with

$$1/\sqrt{1-M_\infty^2}\,.$$

M_∞ is here the Mach number of the undisturbed freestream. Rigorous proof was given in 1928 by *Hermann Glauert* on the basis of linearized theory [40] ("Prandtl-Glauert rule"). *Jakob Ackeret* published in 1925 the corresponding corrective relation [41] for the supersonic range with

$$1/\sqrt{M_\infty^2-1}\,.$$

With these investigations, and later additions and extensions by different authors, the fundamentals for the further treatment of wings in compressible flow had been created.

The above corrections apply to subsonic and supersonic flow. They allow statements concerning wing characteristics at high speeds based on results obtained at low speeds – for instance in subsonic wind tunnel tests. One speaks therefore also of "similarity rules". For $M_\infty \to 1$, there exists, however, no meaningful result (singular behavior of the relations) for lift, pitching moment and, at supersonic speeds ("Ackeret-rule") for wave drag. One can easily show [2] that these relations, which were obtained with the aid of linearized theory ("small-disturbance theory"), are only valid within certain Mach number ranges, the Prandtl-Glauert rule for 0 M 0.8 and the Ackeret rule for 1.2 M 5. The Mach number range of 0.8 M 1.2 is also designated as "Transonic Range", which shall subsequently be considered somewhat more closely.

In today's view, there exist for a slender body, dependent on the Mach number, three flow scenarios as schematically depicted in the accompanying figure. The "Prandtl-Glauert rule" is valid in the compressible subsonic range. The transonic flow regime starts at the lower critical Mach number, hereafter called critical Mach number, where, for the first time, with increasing Mach number an embedded supersonic region occurs, which is terminated by a shock wave. Totally supersonic flow exists, except for a small area around the blunt nose, once the upper critical Mach number has been exceeded. Here, the "Ackeret rule" is valid. The (lower) critical Mach number (M_{crit}) is very important since a very strong increase in the drag of a body occurs shortly after this Mach number has been exceeded (for $M < M_{crit}$ drag is almost constant, i.e., independent of the Mach number). This "drag divergence (DD)" at $M_{DD} > M_{crit}$ is due to the fact that at $M > M_{crit}$ wave drag is added to the body drag and, in addition, the friction-induced pressure drag is increased. The latter occurs for a wing with an embedded supersonic area, whose terminating shock wave thickens or even separates the boundary layer on the body due to the pressure rise associated with the shock. This shock/boundary layer interaction can be highly unsteady (shock oscillations, "buffeting"); moreover, the pitching moment is changed since the center of pressure (location where the integrated forces act on the body) moves downstream. However, at special conditions, e.g., at a large scale separation at transonic flight conditions, also a rearing-up of the aircraft, called "transonic pitch up", may be

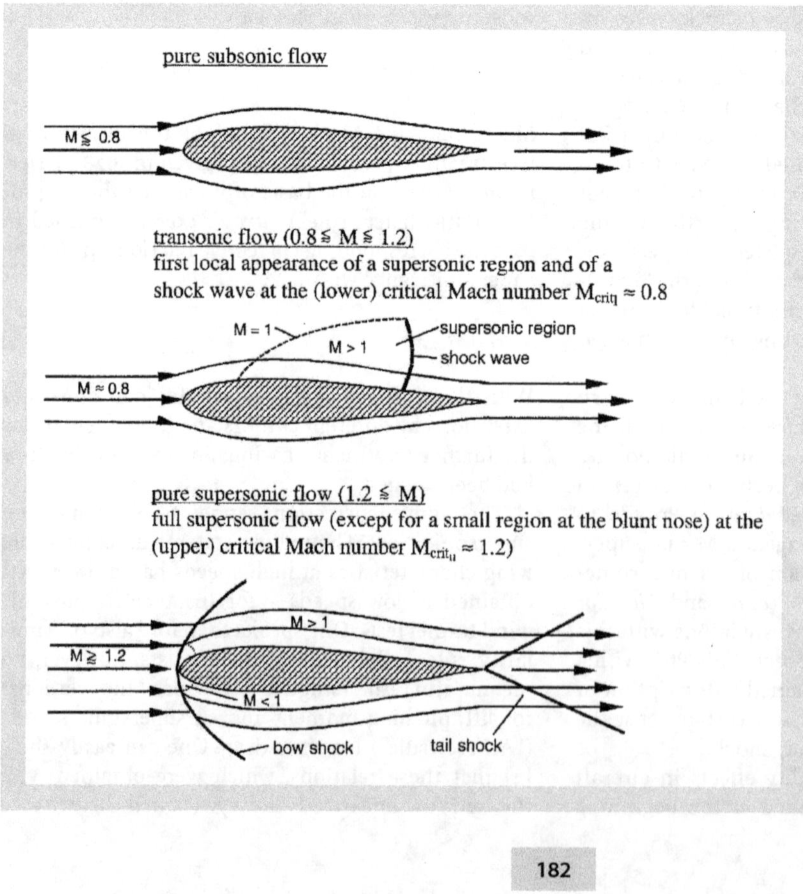

Classification of airfoil flow (inviscid, schematic).

Adolf Busemann (1901–1986).

possible due to a forward movement of the center of pressure.

If one wants to enter into the supersonic range, one must first overcome the drag rise associated with the transonic range. If one wants to cruise at high subsonic speeds, as is the case for today's commercial aircraft, M_{crit} respectively M_{DD} should, on the other hand, be as high as possible. The critical Mach number is dependent on the excess speed (relative to the freestream velocity), which is roughly dependent on the ratio of the body thickness to the body length

Adolf Busemann, born in Lübeck.

1924 PhD with *E. Föppl* in Munich with a thesis on torsional oscillations of bars.

1925 he joins the "Kaiser-Wilhelm-Institute of Fluid Research" ("Kaiser-Wilhelm-Institut für Strömungsforschung") in Göttingen.

1930 habilitation at the University Göttingen.

1931 lecturer in fluid dynamics at the Institute of Technology (TH) Dresden.

1935 he presents at the Volta-Congress in Rome his paper on swept-wing effects.

1936 head of the institute of gasdynamics of the DFL; here, the essential measurements on swept-wing aircraft, including the Ju 287, were performed.

1946 he joins the Royal Aircraft Establishment in Farnborough, England.

1947 he transfers to the USA to join the Langley Research Center of the NAETWA.

1963 he receives a professorship at the University of Colorado at Boulder.

1966 honorary doctor of the TH Aachen; recipient of the Ludwig-Prandtl-Ring.

1968 he becomes a fellow of the American Astronautical Society.

1970 he becomes a member of the National Academy of Engineering.

1986 he dies in Boulder, Colorado on November 3.

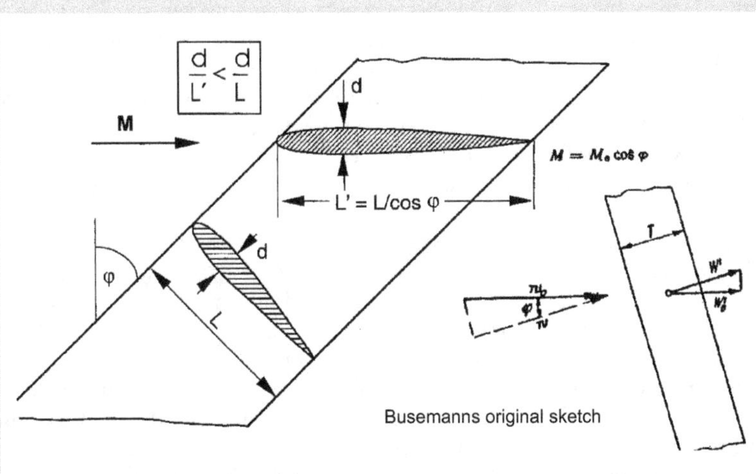

Principle of a swept wing: the flow "sees" a wing more slender than in the unswept case or: the "effective" Mach number M is lower than the flight Mach number M_0 (right: *Busemann's* original sketch).

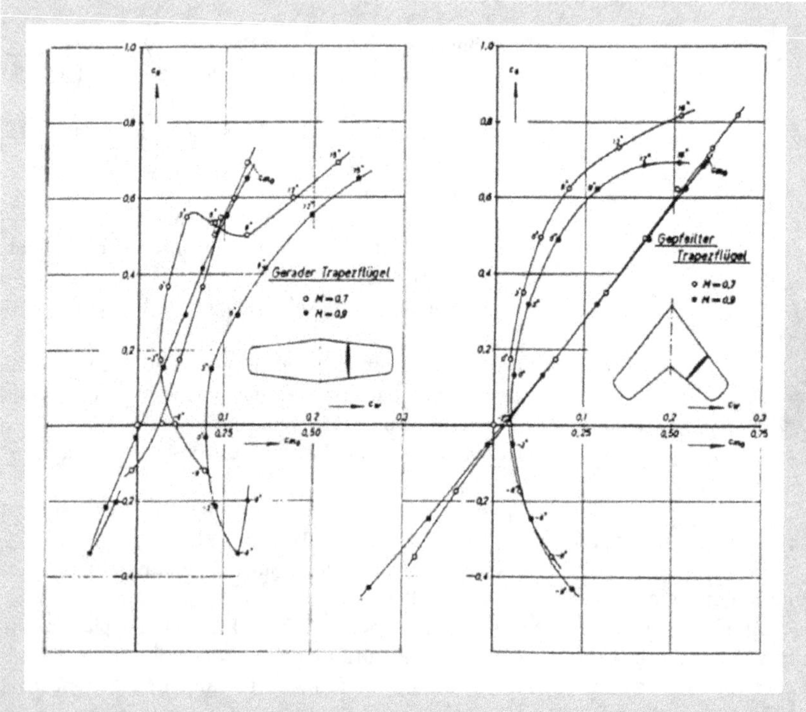

The results of top-secret wind tunnel measurements on swept wings at the AVA were reported for the first time by *Hubert Ludwieg* in 1940 at a meeting of the Lilienthal-Society in Göttingen.
(Left graph: straight wing; right graph: swept wing, in both graphs lift (C_a) as function of drag (C_w) and pitching moment (c_{m_0})).

(d/l). For a cylinder in cross flow (d/l = 1) the critical Mach number M_{crit} is already reached at a freestream Mach number of $M_\infty \approx 0.5$. The more slender a body is, the higher is M_{crit}. Hence one is able to shift by means of a very thin wing M_{crit}, and thus the drag divergence Mach number M_{DD}, to higher flight Mach numbers. Since a wing needs a certain thickness for strength reasons (and frequently to provide the required volume to accommodate fuel), this approach can only be pursued in exceptional cases (see F-104 "Starfighter"). It was *Adolf Busemann*, who, in 1935, showed a simple solution to this problem, namely, to "sweep" the wing. *Busemann's* original sketch is depicted in the figure on the preceding page. By sweeping the wing, the flow "sees" a wing with a thickness ratio reduced by $\cos \varphi$ (φ is the sweep angle), that means a "thinner" wing, so to speak, and M_{crit}, hence the drag divergence, occurs at a higher flight Mach number than in the case of the unswept wing. *Busemann* presented his results (originally for the low supersonic range) in 1936 at the 5th Volta-Congress in Rome [42]. Strangely enough, they were not perceived abroad, especially not in the USA (work on this topic by *R.T. Jones* at NACA [43] originated only 1944), so that there was a great surprise when these investigations became known in 1945 [5]. The first jet in quantity production, the Me 262, had a – slightly – swept wing still exclusively for center-of-gravity reasons. Military aircraft projects incorporating swept wings were, however, since 1943 on the drawing boards of all German aircraft manufacturers (see also [43a]).

Besides wing sweep, there exist two further possibilities to meet the drag rise once the critical Mach number has been exceeded, i.e., to shift the drag rise to transonic Mach numbers as high as possible. One relies on appropriately shaping the airfoil (independent of whether the wing is unswept or swept) today, however, mostly in combination with sweep, if a certain minimum thickness must be maintained. The other is the appropriate shaping of the overall configuration of the aircraft.

The availability of high-thrust jet engines at the beginning of the forties and the demand for high-speed aircraft triggered a multitude of research projects, which led to the "high-speed airfoil", today the "supercritical" or "transonic airfoil", and to the "area-rule" for the shaping of the overall aircraft configuration.

K.A. Kawalki at DVL Berlin-Adlershof suggested in 1940 "high-speed airfoils" ("Schnellflugprofile"), which were characterized by elliptical leading edges (noses), a thickness maximum located further downstream (up to 50 per cent chord) compared to common airfoils, and "flat" upper und lower surfaces (in the case of symmetrical airfoils) [44]. The excess velocities were strongly reduced due to this shape. Though *Kawalki* considered only high subsonic flows, his airfoil shapes were identical to *R.T. Whitcomb's* "supercritical airfoils" proposed in the USA at the beginning of the fifties. (Up to now it was not possible to re-trace the transition process from the "high-speed airfoils" at high subsonic flow to the "supercritical" ones at transonic flow.)

B. Göthert and *K.A. Kawalki* reported in 1944 airfoil measurements at transonic speeds [45]. From a report of the allied "Combined Intelligence Objectives Subcommittee, CIOS" concerning Focke-Wulf, one may conclude that in 1944/45 industrial design groups in Germany knew all aspects of flight at supercritical (transonic) Mach numbers including the problems related to wind tunnel simulation [46]. At that time one had already fighter aircraft – also ones driven by propellers – entering this speed range in flight testing. *Klaus Oswatitsch* cites in his book "Gas Dynamics" [47] many theoretical investigations concerning the overall problem. The search for supercritical airfoils with an isentropic recompression, i.e., the shock-free return to subsonic flow, continued into the seventies, but has, however, today for practical reasons been abandoned [48] (The shock-free recompression has been replaced by an isentropic compression followed by a weak low-wave-drag shock.) The first transport aircraft utilizing supercritical (transonic) airfoils was at the beginning of the eighties the Airbus A 310.

The "area-rule" was discovered experimentally [48a] by *Otto Frenzl* at Junkers in Dessau employing a transonic wind tunnel ($M \approx 1$) and patented together with *Heinrich Hertel* and *Werner Hempel*. The area-rule states that the distribution of the cross sections of the overall aircraft (that is fuselage, wing, nacelle and empennage) from the tip of the fuselage to the tail should steadily increase, then steadily decrease to avoid unnecessary wave-drag increments. This was realized on supersonic aircraft after 1945 by corresponding local reductions in the fuselage cross-section. Of course, the area-rule is still today, as is the swept wing and the supercritical (transonic) airfoil, an essential aerodynamic design element of, say, Airbus-type aircraft, although fuselage cross-section reductions are no longer being used. The effect is achieved by local shaping measures, for instance in the area of the wing-fuselage junction or in the tail area, if not already achieved by the swept-wing geometry and the engine nacelles. The theoretical proof of the area-rule was given by *Friedrich Keune* and *Klaus Oswatitsch* at the beginning of the fifties with the "Equivalence Theorem" ("Äquivalenzsatz") [49], which also allows to obtain quantitative results by considering the "equivalent body of revolution". At about the same time, *R.T. Whitcomb* again confirmed the area-rule experimentally [50].

The swept wing, high-speed airfoils and the area-rule are the most important discoveries prior to 1945, which still today determine the aerodynamic shape of aircraft. Germany was in the thirties and forties up to 1945 also generally and very intensively concerned with high-speed aerodynamics. *A. Busemann* and *O. Walchner* published first results of airfoil measurements at supersonic speeds in 1933. The "Committee of General Fluid Research" ("Ausschuß für allgemeine Strömungsforschung") of the „Lilienthal-Society of Aeronautical Research" organized in 1940 in Göttingen and Brunswick a meeting on the subject "High Speeds". Main topics were experimental methods, wind tunnel corrections and general compressibility effects. The high-speed measurements on swept wings by *Hubert Ludwieg* [51] should be highlighted. He investigated in 1939/40 several swept-wing geometries, which had been derived from an unswept wing with a thickness ratio of 12 per cent, and which were swept back, but partly also swept forward, and were tested with and without nacelles at Mach numbers of $M = 0.5, 0.7, 0.8, 0.9$ and 1.21. Also to be emphasized is the presentation by *Bernhard Göthert* concerning the computation of the velocity field on swept wings at high subsonic speeds, introducing a further correction rule, in addition to the "Prandtl-Glauert rule", today known as "Göthert rule". Furthermore to be mentioned is a contribution by *K.A. Kawalki* on the design of high-speed airfoils. This kowledge has been used by designing the super-critical wing of the A 310.

Wind Tunnel Technology

Hans Ulrich Meier

The Development of the Wind Tunnel at the Beginnings of Flight Research

Flight research is unthinkable without wind tunnels [52–55]. The first attempts to fly with muscle power like a bird had to fail since, at that time, the understanding of the flow past a wing or a body and the forces generated, such as lift and drag, was lacking. *Leonardo da Vinci* and *Isaac Newton* already provided the fundamental ideas for the systematic investigation of the flow past bodies. Physically, it is the same whether a body moves through air at rest or whether a body at rest is embedded within the moving air. In spite of this knowledge, aerodynamic tests were initially carried out with the aid of rotating rigs since these were uncomplicated and therefore inexpensive. It was *Benjamin Robins* (1702–1751), an English mathematician, who was the first to notice that differently shaped bodies of equal cross sections "A" attached to a rotating arm of 1.26 m (4 ft) led to different circumferential velocities even if the weight "G" driving the cylinder via the cord was the same. *Robins* had this idea when contemplating the experience that the flight of projectiles depends on the drag of the projectiles. We know today that one reason for this behavior relates to the fact that the model, after each revolution, enters the disturbed flow field of the preceding revolution. Due to the non-uniformity of the freestream thus generated (turbulence) it is possible to appreciably change drag, hence the circumferential velocity. *Robin's* observation (or conclusion) that drag must be dependent on the form of the body and not only on the cross sectional area was of great importance. Similar considerations hold for the first investigation of different wing models attached to a rotating arm conducted by *Sir George Cayley*. Based on his experience, *Cayley* was able to build a hang glider with a wing area of more than 20 m² (215 ft²) and, for the first time, to successfully fly it.

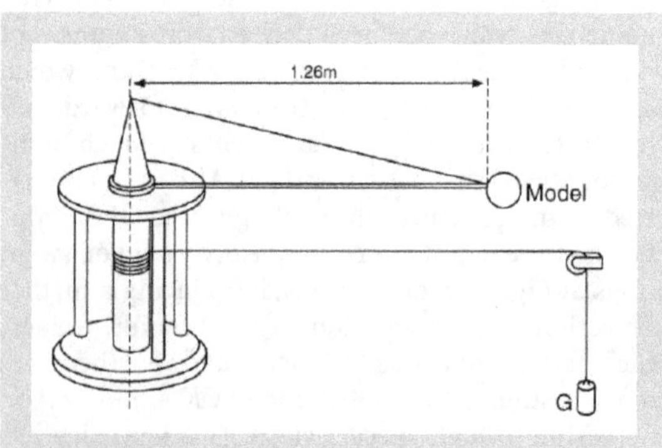

Rotating arm apparatus for the investigation of drag according to *Benjamin Robins* (1746).

Otto Lilienthal also used a rotating rig for his experimental investigation and performed with it first systematic lift and drag measurements on wings. With rotating arms of 2–7 m (6.6–23 ft) length he reached circumferential velocities of 10 m/s (33 ft/s) and was able to demonstrate for the first time the advantage of cambered wings. Although *Lilienthal's* experiments still were tainted with considerable errors, his aerodynamic tests – and based thereupon his more than 2,000 hang-glider flights – won him the acceptance and admiration as the first aviator.

In 1871, two English engineers, *F.H. Wenham* and *J. Browning*, constructed an apparatus consisting of a steam-powered fan, which blew air across a model mounted in a tube. The Wenham-Browning wind tunnel concept (however, with an electric motor) was in 1901 also employed by the brothers *Orville* and *Wilbur Wright* in the design of their wind tunnel, after having previously performed tests on airfoils utilizing a bicycle and the natural wind generated by going downhill at 25 km/h (15.5 mph) in Dayton/Ohio. The reason for the wind tunnel was the recognition that their glider, built according to data gathered by *Lilienthal*, did not generate the expected lift. The wind tunnel, designed and constructed by the *Wright Brothers*, had a test section area of 0.4 m × 0.4 m (1.3 ft × 1.3 ft) and reached a maximum wind speed of 60 km/h (37 mph). With the aid of this facility, they generated their personal handbook of aerodynamic data for various airfoil shapes. With their experience they consequently – as did *Otto Lilienthal* – realized an aircraft, the famous Wright Flyer, completing on December 17, 1903 their first successful powered flight. It is interesting to note that the industrial commercialization of such flight vehicles was, at that time, not possible in the USA. The *Wright Brothers* went to Europe and demonstrated in August of 1908 in France successfully the functionality and usability of their flight vehicle for military purposes. Due to the interest of the military in all industrialized European countries, substantial research and development funds were quickly made available by the respective governments for the foundation and built-up of aerodynamic research and test organizations.

In Germany, during the planning of a "Powered Airship Study Society" ("Motorluftschiff-Studiengesellschaft, MStG") in Göttingen, *Ludwig Prandtl* was in charge in 1906 to work out proposals for a wind tunnel to measure drag and the pressure and velocity distributions on airship hulls [55]. *Prandtl* presented two solutions: an Eiffel-tunnel with an open air passage and, alternatively, a novelty, a wind tunnel with a closed circuit return, where always the same air circulates. *Prandtl's* design of a closed-loop wind tunnel was based on the consideration that, in case of an Eiffel-tunnel, the total volume drawn in had continuously and over and over again to be accelerated, while for a closed circuit tunnel only the losses in the test section, the diffuser, the turning vanes, the flow-straightener and the screens had to be compensated. The energy consumption of a closed-loop wind tunnel is, therefore, considerably less. The wind tunnel with closed air ducting finally built in Göttingen in 1908 was the first of this type and marks a milestone in the history of wind tunnel technology. The tunnel consisted of a closed-loop air duct with a quadratic cross section of 2 m × 2 m (6.6 ft × 6.6 ft) throughout and was housed in a wooden half-timbered building with a floor area of 10 m × 10 m (32.8 ft × 32.8 ft). To improve the uniformity and to steady the flow, a large guide system was installed immediately downstream of the fan for the transition from the circular area of the 2-meter diameter to the quadratic tunnel cross section. In addition, there were a coarse and a fine honeycomb-structured flow straightener and a screen with a 2 mm mesh size inserted into the air stream. Guide vanes accomplished the turning of the flow within the four corners. The technical build-up

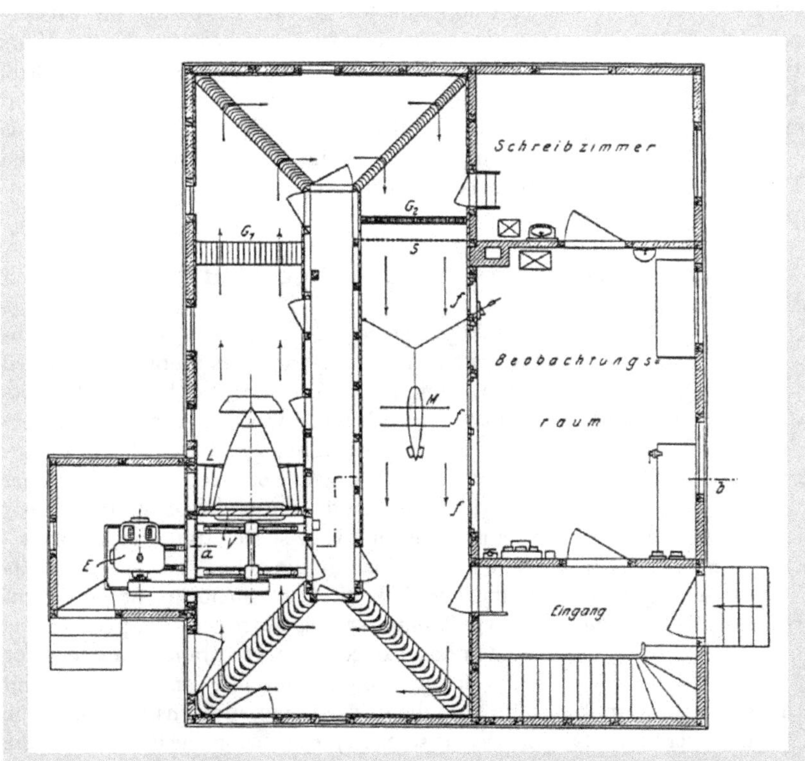

Plan view of *Prandtl's* model test institution in 1907. The electric motor "E", housed in an annex, drove the fan "V" via a belt. L = the large guide system, S = screen. In the four corners, the guide vanes turning the flow are visible. (Eingang = entrance, Beobachtungsraum = control room, Schreibzimmer = data room).

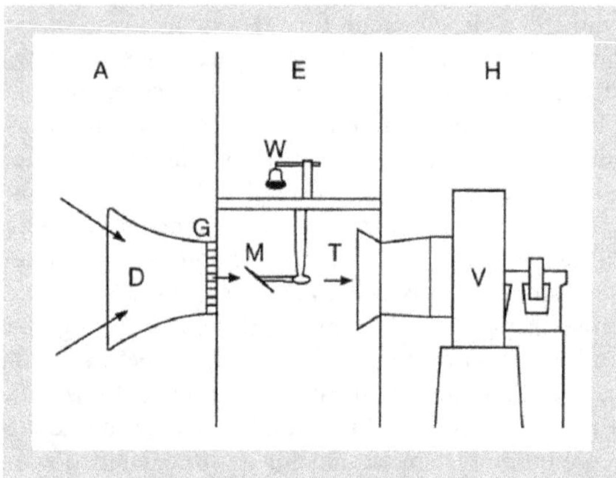

Schematic depiction of the test installation of *G. Eiffel*:
A: Suction chamber with nozzle **D** and flow straightener **G**.
E: Test section with model **M**, balance **W** and air scoop **T**.
H: Chamber housing the fan **V**.
In case of the Eiffel wind tunnel air is sucked into the evacuated room A, which must be carefully sealed to the outside. The flow into the suction chamber may, depending on the design, be subject to large disturbances.

of this novel wind tunnel and the development of a balance for the measurement of lift, drag and pitching moment as well as other instruments is very much to the credit of *Georg Fuhrmann*.

Up to the beginning of World War One, many fundamental research and development projects were performed in this first wind tunnel with closed (circuit) air ducting. Concerning the objectives of the envisaged wind tunnel activities *Prandtl* commented as follows [55]:

"We want with our tests
1. *directly serve practical applications by delivering data which are needed;*
2. *through systematic test series explore the laws of drag;*
3. *to check the results of theoretical investigations;*
4. *to put our own results on a secure basis by carefully studying the measurement of velocity and of pressure and of errors due to the wind tunnel walls, and by studying model similarity laws."*

These objectives still hold, in a figurative way, until today as permanent guidelines for the operation of wind tunnels. *Prandtl* consciously emphasized the importance of the wind tunnel for industrial aircraft development, but also clearly expressed its importance as experimental tool for research in fluid mechanics. As outstanding scientific activities within the "Model Test Establishment Göttingen" up to 1924, the following must be named:

- Theoretical and experimental investigations on balloons.
- Measurement of forces on flat and cambered plates.
- Measurement of forces on wings in the proximity of a horizontal plane (ground effect).
- Measurement of forces on a biplane.
- Tests with propeller models.
- Systematic experimental investigations concerning the laws of drag.

In order to explore the laws of drag, a nozzle with a diameter of 1.015 m (3.3 ft) was introduced into the wind tunnel, similar to the layout of the large Eiffel-wind tunnel in Paris [56], which generated a free jet within the test section increasing the existing maximum velocity from just below 10 m/s to 23 m/s (33 ft/s–82 ft/s). The nozzle also surprisingly resulted, due to the (jet) contraction, in a noticeably improved flow uniformity and a reduction in vorticity thus leading to a decisive improvement of *Prandtl's* closed air-duct wind tunnel concept.

The Göttingen-Type Wind Tunnel

Prandtl presented the design of a considerably improved facility already in 1911. Funding was provided at the beginning of World War One, but the construction of the wind tunnel was only completed in 1917 due to the chaos generated by the war. This wind tunnel concept – also called "Göttingen-type" ("Göttinger Bauart") – is until today the model for all continuously operating subsonic (low-speed), transonic and supersonic wind tunnels. In the publication series "Results of the Aerodynamic Test Establishment Göttingen" *Ludwig Prandtl* [57] has described all aspects of this wind tunnel concept. These comprise: reasons for the selection of the wind tunnel concept, design and control of the drive, design and construction of the nozzle, the screens, the air scoop, the diffuser and the turning vanes. *Prandtl* gave as reason for the selection

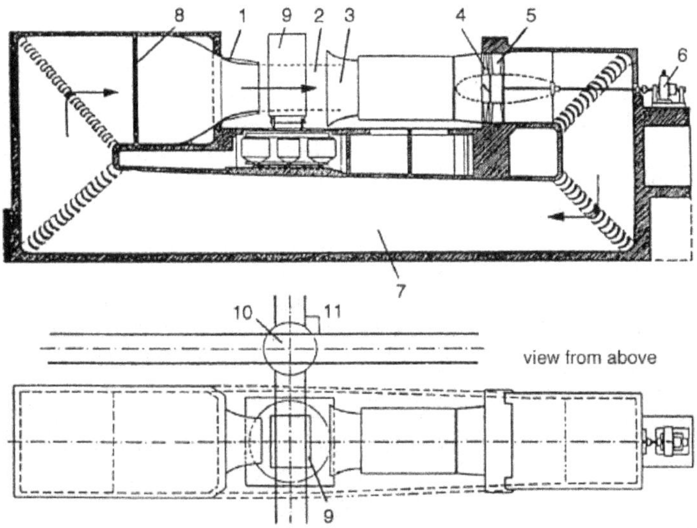

Prandtl's wind tunnel, 2nd generation (1917).
1 Nozzle, *2* Open test section, *3* Air scoop, *4* Fan, *5* Guide system, *6* Motor, *7* Return leg (in the basement), *8* Flow straightener, *9* Traversing rack with instrumentation, *10* Turntable, *11* Tracks.

of the free-jet (open) test section – in comparison to the closed 2 m × 2 m (6.6 ft × 6.6 ft) test section of the first "Model Test Establishment" – the easy access and the superiority of the free jet with regard to the size of the models, which "*may be utilized without excessive errors due to the finite air-stream (jet) dimensions*".

Through the application of wall correction methods it is possible today to correct reliably for wall effects, so that for force and moment measurements the closed test section – at least in the case of large facilities – prevailes.

The large wind tunnel of the "Model Test Establishment Göttingen" ("Modellversuchsanstalt Göttingen, MVA"), conceived by *Prandtl* was finally realized with a 15-cornered nozzle with a 2.25 m (7.4 ft) diameter and a maximum drive performance of 355 kW (475 hp) attaining a maximum velocity of 55 m/s or 200 km/h (180 ft/s or 124 mph).

In this wind tunnel, equipped with a six-component balance, many fundamental model tests, mostly under contract of aircraft companies, were carried out at the MVA until the end of World War One. Tested were airfoils ("Göttingen Airfoils"), wings of different geometries (among other activities the verification of *Prandtl's* lifting line theory), fuselages and coolers. In 1920, the MVA was, as "Aerodynamic Test Establishment Göttingen" ("Aerodynamische Versuchsanstalt Göttingen, AVA") incorporated into the "Emperor-Wilhelm-Society" ("Kaiser-Wilhelm-Gesellschaft, KWG") and the large wind tunnel was now given the name "Tunnel I" ("Kanal I") and the older, smaller wind tunnel was named "Tunnel II". In both wind tunnels, since 1925 in unison with the newly founded "KWG-Institute of Fluid Research", further scientifically important fluid mechanical investigations, also of non-aeronautical character, were conducted before both historic wind tunnels were closed in 1948 by order of the British occupation force.

Fluid Mechanical Similarity Laws Related to Wind Tunnel Measurements

Results of model investigations in a wind tunnel may only be transferred to the full-scale configuration, for instance an aircraft, if, besides the model similarity, also a fluid mechanical similarity exists. Here, fluid mechanical similarity laws or rules are applicable:

Reynolds' similarity law. According to this law, physical similarity of the flow about the full-scale configuration and about the wind tunnel model exists, if the ratio of the fluid mechanical inertia forces and the friction forces is the same (*Osborne Reynolds*, 1833). The similarity is expressed by the so-called Reynolds number:

$$Re = \rho Ul/\mu .$$

Here, l is a characteristic length (e.g., a wing chord), U is the freestream velocity, ρ the density and μ the dynamic viscosity of the fluid. Hence fluid mechanical similarity is given if $Re_{model} = Re_{full\text{-}scale}$. Reynolds numbers of large aircraft are in the order of 10^7–10^8 and cannot be realized even in large wind tunnels without special measures.

Possible measures can directly be deduced from the definition of the Reynolds number, viz., by changing the quantities of state ρ and μ. For the latter, the following thermodynamic relations exist:

$$\rho = p/RT \text{ and } \mu = f(T),$$

this means that a variation in the wind tunnel Reynolds number is, besides by U and l, also possible by changing the pressure p and/or the temperature T. (Note that an increase in Re is achieved by increasing the pressure p or by decreasing the temperature T.) "R" is the gas constant of the test gas.

We will later return to the subject in the chapter "Advanced Wind Tunnel Technology".

Mach's similarity law. The law is reflected in the so-called Mach number (*Ernst Mach*, 1838–1916):

$$M = U/a.$$

The Mach number M describes the ratio of the local flow velocity U to the speed of sound a, which decreases with decreasing temperature as is the case at high altitudes. The Mach number is also directly related to the compressibility of the fluid. For air, compressibility effects must be accounted for at Mach numbers above $M = 0.3$; below, the Mach number is, as far as the flow similarity is concerned, for attached flow hardly of importance. However, in the range of applicability *Mach's* similarity law must be satisfied exactly, i.e., wind tunnel tests must be carried out at the original flight Mach number. Especially in transonic flow is the Mach number, together with the Reynolds number, a fundamental characteristic quantity. The adherence to both similarity requirements leads to elaborate and correspondingly expensive test facilities, which have only been realized during the last two decades.

Model similarity in wind tunnel investigations related to aeroelastics. In the case of dynamic aero-elastic (flutter)-model investigations in a wind tunnel, one has to meet, besides the Mach and Reynolds number similarities, also model-related flexibility requirements. These are expressed by the so-called Cauchy number (characteristic stiffness number), the Froude number (gravitational parameter) and the density ratio (mass density to air density). These non-dimensional similarity parameters must, for the selected mass, length and velocity scales (hence a fixed time scale), be satisfied for the dynamic aero-elastic process on the model to be investigated to be the same as on the full-scale configuration. The agreement in the reduced frequency, a further similarity parameter, which is related to the unsteady flow occurring on an oscillating model and which is important for the simulation of the flutter process, is implicitly accounted for by the Cauchy number.

The simultaneous attainment of the Reynolds number, the Froude number and the Mach number for a given aero-elastic model in the wind tunnel is, however, not possible. The Froude number is, on the other hand, only needed for the investigation of dynamic flight stability (simulation of flight maneuvers). It is, in addition, generally also not possible to duplicate the Reynolds number, except in cryogenic wind tunnels, without violating other similarity parameters. The Reynolds number is, however, for the flutter process in shock-free, attached flow of minor importance but it is significant for the transonic flutter process. The duplication of the Mach number is in any case a must, as is, for complete models, the realization of a free-floating model condition, which can be achieved by a correspondingly flaccid suspension mechanism.

It should finally be mentioned that, of course, the full-scale geometry of the aircraft (respectively of its components) as well as the mass and stiffness distributions must be realized on the aero-elastically similar model as closely as possible. Altogether, one realizes that the design and construction of models for aero-elastic investigations in the wind tunnel, and the wind tunnel tests themselves, require much more effort than is the case for steady aerodynamic investigations with rigid models.

Carl Wieselsberger (1887–1941).

Carl Wieselsberger, born in Ebersthal (Lower Bavaria) on November 3, 1887.

1906 studies at the Institute of Technology (TH) Munich mechanical engineering.

1910 graduates with the Masters Degree (Diplom-Ingenieur).

1912 he receives his PhD from the TH Munich with a thesis related to the longitudinal stability of kite aircraft and then turns to aerodynamics. He transfers as assistant to *Prandtl* to the "Model Test Establishment Göttingen", where he conducts the decisive tests on laminar-turbulent boundary layer transition on spheres (trip-wire tests), which allowed interpreting the different test results of *Eiffel* and *Prandtl* concerning the drag of spheres.

1914 he joins the war as volunteer, is trained as a military pilot and is detailed in the spring of

1915 to his old location of occupation in Göttingen, where he continued to work as department head. Due to his skills in measurement technology and his quiet enthusiasm for his work, he was able to make contributions concerning the effect of form, surface design and suspension of engines on drag.

1922 he follows an invitation of the Japanese government and directs in Tokyo and at several other locations the set-up of wind tunnel facilities. He, furthermore, builds up the aeronautical research institute at the University of Tokyo.

1929 he publishes a paper concerning the determination of the change in aerodynamic forces on a wing due to engine installations.

1930 he receives a call to the Institute of Technology (TH) Aachen and takes over as head of the institute of aerodynamics which had been established by *von Kármán*. The first supersonic wind tunnel with a variable Mach number is built under his supervision.

1940 he becomes, in addition, head of the "Aeronautical Research Establishment Munich" ("Luftfahrt-Forschungsanstalt München, LFM") planning the then largest wind tunnel in the world to be constructed in the Ötz valley (Ötztal) in Austria.

1941 he dies on April 26, much too early.

The Effect of Wind Tunnel Turbulence on the Transfer of the Results to Full-Scale

An aircraft flies to first order through quiescent air, i.e., disturbances in the freestream are essentially zero. Within a modern wind tunnel, turbulent fluctuations are today, dependent on the design, minimized, as already shown and realized by *Prandtl* for his second wind tunnel. The considerable importance of wind tunnel turbulence for boundary layer separation was discovered by *Prandtl* due to contradictory drag measurements on spheres [55]. A comparison of drag coefficients obtained at the laboratory of *Eiffel* and gained at the "Model Test Establishment Göttingen" showed that, at certain Reynolds numbers, some of he latter drag values were two and a half times the ones measured at *Eiffel's* laboratory. In order to rule out measurement errors and to clarify the contradictions, *Eiffel* conducted further tests with spheres of different diameters; in the process, it was discovered that, when increasing the freestream velocity to a certain critical value, drag suddenly appreciably decreased. For the sudden drop in drag, *Prandtl* soon found an explanation based on his boundary layer theory. It was known that in the case of pipe flow a sudden change from laminar to turbulent flow occurs at a certain Reynolds number. *Prandtl* surmised that for a laminar boundary layer a separation occurs at a certain rearward position due to the shape of the pressure distribution. If, however, at higher freestream velocities laminar-turbulent boundary layer transition takes place, the boundary layer stays attached longer and the dead-air region downstream of the crest of the sphere becomes smaller which, in turn, leads to lower drag. In order to confirm this idea, *Prandtl* had his collaborator *Carl Wieselberger* conduct experimental investigations with spheres of different diameters.

The results were consistent with *Reynolds'* similarity law and showed in a Reynolds number range between 2×10^6 and 2.7×10^6 the sudden drag decrease due to the laminar-turbulent boundary layer transition. Since *Prandtl* still had to argue with skeptics, he forced boundary layer transition with the aid of a trip wire upstream of the crest of the sphere. This caused also at lower Reynolds numbers the sudden drop in drag [58]. In order to find an explanation for the fact that in *Eiffel's* sphere investigations the Reynolds number at the drop in drag was noticeable lower than in *Prandtl's* experiments, *Prandtl* artificially increased the flow turbulence by introducing a screen at the exit of the nozzle. The results confirmed the pronounced effect of the wind tunnel turbulence on laminar-turbulent boundary layer transition, hence on the drag behavior of bodies immersed in a flow field. For a long time, investigations with spheres were, therefore, utilized to determine the flow quality (e.g., turbulence level) of subsonic wind tunnels.

Research and Test Establishments with Large-Scale Wind Tunnels

While the development and the construction of the first wind tunnels were still in the responsibility of individual researchers, universities, and industry, was the construction of large-scale facilities from the very beginning carried out by newly founded research institutions supported by government and private industry. The knowledge gained in these research institutions was available to the entire aeronautical industry. This trend can clearly be observed in all industrialized nations up to World War Two. Due to this historical development, the demands of science, industry and politics were considered in the construction especially of large-scale facilities. Market and economic considerations related to the later operation of the facilities were judged secondary. Based on the experiences gained during World War One, played the development of military aeronautics after the power seizure through the National Socialists and the forced military armament after 1933 the main role.

As *Adolf Baeumker* [59] cites in his report on the occasion of *Prandtl's* 70th birthday in 1945 "Concerning the history of German aeronautical research", increased the number of employees between 1933 and 1945 within all research organizations from about 550 to 6,500. A diagram in the appendix shows the development of the personnel capacity of the research institutions AVA, DFL and DVL. The technical build-up of the existing and the new research facilities, and hence the completion and construction of new large-scale research facilities, were, at the same time, strongly accelerated (see the table on pages 194/195). *E. Meyer* [52] has established a complete catalogue of all wind tunnels since *Prandtl* until today. In what follows, wind tunnels, which were in operation until 1945 and which provided essential research results, shall be described.

However, first German test establishments and research institutions are listed, some of which were and still are substantially responsible for the development, construction and operation of wind tunnels, see page 193. Included are also the rocket test facilities in Peenemünde and Kochel since here supersonic and hypersonic wind tunnels were developed and operated; these were for the later development of military and civil supersonic aircraft as well as for hypersonic technology projects, such as "Sänger", of decisive importance. It should be mentioned here that before, during and after World War Two, aeronautical industry had, and still has, for its project pre-developments smaller in-house and external facilities at its disposal. The companies Heinkel, Junkers, Messerschmitt, Focke-Wulf and Dornier thus operated already their own small wind tunnels. In the Department of Flight Physics the DaimlerChrysler Aerospace Airbus Company (DA) in Bremen has in the LST (see No. 17 of the wind tunnel table) a low-speed facility at its disposal, which allows a quick access, for instance, for half-model investigations within the framework of pre-development work for new projects.

Foundation Dates of German Aeronautical Research Organizations

July 31, 1906
"Powered-airship Study Association" ("Motorluftschiff-Studiengesellschaft") in Göttingen, which assumes in March of 1921 the name "Aerodynamic Test Establishment" ("Aerodynamische Versuchsanstalt, AVA") within the "Emperor-Wilhelm-Society" ("Kaiser-Wilhelm-Gesellschaft, KWG").
First head: *Ludwig Prandtl*

1908
"Radiotelegraphic and Air-electricity Test-station Gräfelfing" ("Drahtlostelegraphische und Luftelektrische Versuchsstation Gräfelfing, DVG") founded near Munich.
First head: *Max Diekmann*

June 3, 1912
"German Test Establishment of Aeronautics" ("Deutsche Versuchsanstalt für Luftfahrt, DVL") founded in Berlin-Adlershof.
First head: *Friedrich Bendemann*

1924
"Research Institute of the Rhön-Rossitten-Society" ("Forschungsinstitut der Rhön-Rossitten-Gesellschaft, RRG"), which is in 1937 transferred into the "German Research Institute of Gliding" ("Deutsches Forschungsinstitut für Segelflug, DFS").
First head: *Walter Georgii*

1929
Foundation of the "Flight-technical Institute Stuttgart" ("Flugtechnisches Institut Stuttgart, FIST") and in 1940 of the "Research Organization Count Zeppelin" ("Forschungsanstalt Graf Zeppelin, FGZ").
First head: *Georg Madelung*

February 10, 1936
"German Research Establishment of Aeronautics" ("Deutsche Forschungsanstalt für Luftfahrt, DFL") founded in Brunswick.
First head: *Hermann Blenk*

1936
"Army Test Establishment Peenemünde" ("Heeresversuchsanstalt Peenemünde, HVA").
First heads: *Walter Dornberger/Wernher von Braun*
"Institute of Aerodynamics Peenemünde".
First head: *Rudolph Hermann*

July 1936
"Institute of Radio Technology and Air-radio Transmission" ("Institut für Radiotechnik und Flugfunkwesen, IRA") founded at the Institute of Technology (TH) Munich
First head: *Max Dieckmann*

June 22, 1938
"Air Radio Research Institute Oberpfaffenhofen" ("Flugfunkforschungssinstitut Oberpfaffenhofen, FFO") founded near Munich.
First head: *Max Dieckmann*

1940
"Aeronautical Research Establishment Munich" ("Luftfahrtforschungsanstalt München, LFM").
First head: *Carl Wieselsberger*

Large Subsonic Wind Tunnels during the Time Period 1935 to 1945

In the process of the tremendous air armament of the Third Reich since 1935 and the rapid increase in flight speeds up to the speed of sound, arose for the German aircraft development the urgent need for larger wind tunnels with essentially larger air velocities and cross-section areas of the test-sections. This led to the construction of several large-scale subsonic wind tunnels, which were all of "Göttingen-type" design and established at the aeronautical research institutions. Already in 1932, the planning of a large subsonic wind tunnel commenced at DVL-Adlershof, which became operational in 1934. The tunnel had an elliptical nozzle with a cross section of 5 m × 7 m (16 ft × 23 ft), an open test section, and reached a flow velocity of 65 m/s (213 ft/s) at a drive performance of 2,000 kW (2,680 hp). Later, this wind tunnel was provided with an even larger nozzle of 6 m × 8 m (19.7 ft × 26 ft) reaching a velocity of 50 m/s (164 ft/s).

No.	Internal designation	First location of operation	Start of operation	Operator	Type	Test section	Width b [m]	Height h [m]	Contraction ratio	Speed [m/s]	Max. Mach number
1	Wind tunnel II	Göttingen	1908	MVA	Göttingen-design	open (free jet 8-cornered)	d = 2 m		1,0	10	0,03
2	Wind tunnel (large nozzle)	Göttingen	1917	MVA	Göttingen-design	open (free jet 15-cornered)	d = 2,25 m		4,0	58	0,17
3	Large wind tunnel	Berlin-Adlershof	1934	DVL	Göttingen-design	open	7,0 8,0	5,0 6,0	4,0 2,91	65 50	0,20 0,15
4	Supersonic facility 2	Aachen	1936	TH Aachen	Storage principle	closed	0,2	0,2	–	–	3,10
5	Wind tunnel VI	Göttingen	1936	AVA	Göttingen-design	open (free jet elliptical)	7,0 5,4	4,7 4,0	3,0 4,5	38…96 48…117	0,29 0,35
6	Large Junkers tunnel (Nozzle III)	Dessau	1937	Junkers	Göttingen-design	open	2,5	2,5	5,6	51	0,15
7	High-speeed wind tunnel (HWK)	Berlin-Adlershof	1938	DVL	Göttingen-design	closed circular	2,7	2,7	7,2	230	0,95
8	Supersonic-tunnel	Göttingen	1938	AVA	Storage principle	closed rectangular	0,13	0,11	–	–	3,2
9	High-speed wind tunnel	Peenemünde	1939	Army Research Organiz.	Storage principle	closed	0,4	0,4	–	–	4,4
10	Hermann-Göring-tunnel A3	Brunswick	1940	LFA	Göttingen-design	open (free jet circular)	d = 8 m		4,0	95	0,28
11	High-speed wind tunnel WK A2	Brunswick	1939	LFA	Göttingen-design/Eiffel (variable)	closed circular	2,8	2,8	6,8	230	0,69
12	Ötz valley wind tunnel	Modane	1951	ONERA	Göttingen-design	circular	d = 8 m		9,0	330	1,03
13	Junkers high-speed wind tunnel	Dessau	1941	Junkers	Ejector principle	open or closed	0,3	0,3	–	275	0,83
14	3-m-tunnel NWB	Brunswick	1960	DLR/NLR	Göttingen-design	open or closed (8-cornered)	3,6	2,8	5,6	60	0,18
15	3-m-tunnel 2.4 KKK	Cologne	1962 (1982)	DLR/NLR	Göttingen-design	closed	2,4	2,4	10,4	115	0,35
16	1-m transonic wind tunnel TWG	Göttingen	1963	DLR/NLR	Göttingen-design	closed	1,0	1,0	16,0	–	0,3…2,25
17	DASA low-speed tunnel (LST)	Bremen	1970	DASA	Type Eiffel	closed squared	2,1	2,1	4,82	75	0,22
18	DNW-LLT (large low-speed tunnel)	Nordostpolder, NL	1980	DLR/NLR	Göttingen-design	rectangular 1 closed 2 closed 3 open 4 closed	 6,0 8,0 8,0 9,5	 6,0 6,0 6,0 9,5	 12,0 9,0 9,0 4,8	 150 120 80 60	 0,45 0,36 0,24 0,18
19	Cryogenic Ludwig-tube KRG	Göttingen	1991	DLR/NLR	Storage principle	closed rectangular	0,4	0,35	–	–	0,3…0,95
20	ETW	Cologne	1993	ETW GmbH	Göttingen-design	closed	2,4	2,0	12,0	–	0,15…1,3

Total press. [bar]	Total temp. [K]	Drive perform. [kW]	Operation	Main tasks	Special instrumentation	Remarks
ambient	ambient	25,5	continuously	airship and balloon measurements	simple balances	first German wind tunnel with closed-circuit return
ambient	ambient	400	continuously	force measurements, Göttingen airfoils	3-component balance, 6-comp. balance since 1928, pressure control	first German wind tunnel of Göttingen design
ambient	ambient	2.000	continuously	measurements on aircraft and aircraft-airfoils	3-component balance	first German large-scale wind tunnel of Göttingen design
ambient	ambient	–	6 sec	basic research, missile measurements	inductive 3-component balance, Schlieren optic	discovery of the condensation shock
0,25… 0,3	ambient	1.800	continuously	wing measurements, propeller investigations	6-component balance	Reynolds number variation at constant Mach number
ambient	ambient	220	continuously	model tests for aircraft development	standing balance	industrial tunnel with exchangeable nozzles 3.2 $3.2\,m^2/4$ $4\,m^2$
ambient	ambient	13.000	continuously	boundary layer observation, aircraft measurement, propeller examinations	3-comp. balance, multi-tube manometer, momentum rake, Schlieren optic	velocity dependent on total temperature
–	–	–	10…20 sec	basic research	3-component balance	first swept-wing measurements (1939)
ambient	ambient	–	intermittent	rocket research	electronic balance, interferometer, Schlieren optic, high-speed camera	air dryer
ambient	ambient	8.500	continuously	aircraft measurements on original and full-scale models	standing balance, 6-component balance	–
ambient	ambient	12.000	continuously	aircraft measurements	3-component balance, multi-tube manometer, Schlieren optic, interference measurement	air cooling, dust-layer observation
ambient	ambient	88.000	continuously	aircraft models, engines, rotors, missiles, space vehicles	internal 6-component balances, TPS (Turbine Powered Simulator), icing tests, Laser optics	water-turbine propulsion, 1951 moved from the Ötz valley to Modane (LFM-tunnel)
–	–	–	approx. 3 min	high-speed aerodynamics for aircraft projects	3-component balance	development of the area-rule (1943)
ambient	ambient	1.200	continuously	aerodynamic fundamentals/ VSTOL/ high lift, air-intake measurements	6-component balance, strain gage balance (DMS), TPS, multi-tube manometer, rolling floor	also used for automobile and railroad investigations
ambient	100… 300	1.000	continuously	half- and full-model measurements, airfoil investigations	6-component balance, half-model balance, strain gages	modification in 1987, independent control of Re and Mach number possible
0,3…1,5	ambient	12.000	continuously	aircraft and missile measurements for industry and research	DMS and task balances, PSI pressure measurements, Schlieren optic	adaptive-wall test section for $0.3 < M < 0.9$
ambient	ambient	500	continuously	half-model measurements for Airbus development	internal 6-component balance, half-model balance, TPS, DMS	industrial wind tunnel
ambient	ambient	12.700	continuously	1. high-speed range, 2. aircraft-, helicopter-, automobile-, engine-investigations, 3. helicopter/aircraft performance, acoustics, 4. truck 3-component measurements	internal and external balances, TPS, DMS, LDV, PIV for flow field investigations	1. utilized for missile investigations, 2. optimized for Airbus take-off and landing phase, 3. hot gas simulation possible, microphone array (acoustical mirror), 4. designed for investigations on vertical take-off configurations
2…10	100… 300	–	0,6… 1,1 sec	airfoil design	force and pressure measurement systems, adaptive-wall test section	facility for large Reynolds numbers
1,25… 4,5	90…313	50.000	continuously	half- and full-model measurements on aircraft, missiles and space vehicles	6-component balances, half-model balance, DMS, infrared detection of transition	most modern industrial facility for transonic speeds at high Reynolds numbers

In 1938, the high-speed wind tunnel ("HWK") with a closed circular test section of 2.7-meter (8.9 ft) diameter became operational at DVL, reaching, at a drive power of 13,000 kW (35,100 hp), an air velocity of 280 m/s (919 ft/s) equivalent to a Mach number of M = 0.85.

At the AVA in Göttingen the new large-scale wind tunnel "VI" was put into operation in 1936. The tunnel had an elliptical nozzle cross section of 4.7 m × 7 m (15.4 ft × 23 ft), an open test section and a maximum air velocity of 96 m/s (315 ft/s). With an alternate nozzle of 4 m × 5.4 m (13 ft × 17.7 ft) cross section velocities of up to 117 m/s (384 ft/s) were reached at a drive performance of 1,800 kW (2,414 hp). To vary the Reynolds number, total pressure could be changed between 0.25 and 3 bar.

After the foundation of the "German Research Establishment of Aeronautics" ("Deutsche Forschungsanstalt für Luftfahrt, DFL") in 1936 in Brunswick, the Hermann-Göring-Tunnel "A 3" was put into operation there in 1940. This low-speed wind tunnel with a circular open test section with a diameter of 8 m (26 ft) attained, at a drive power of 8,500 kW (11,400 hp), an air velocity of 95 m/s (312 ft/s). Also into operation went already in 1939 a high-speed wind tunnel with a closed circular cross section of 2.8 m (9.2 ft) diameter, which allowed, at a drive power of 12,000 kW (16,100 hp), velocities of 230 m/s (755 ft/s) equivalent to a Mach number of 0.69. A gigantic high-speed wind tunnel project was, still in 1942, started by the "Aeronautical Research Establishment Munich" ("Luftfahrtforschungsanstalt München, LFM") in the Ötz valley in Austria. Counting on the water resources available at this location for energy, the construction of the up to that time largest wind tunnel was begun, however, not completed before the end of the war. Within a circular jet from a nozzle with a diameter of 8 m (26 ft), it was anticipated to reach at a drive power of 88,000 kW (118,000 hp) – corresponding to the power output of a medium power station – an air velocity of 330 m/s (1,080 ft/s) corresponding to a Mach number close to one. The drive power for the facility (compressor) was to be provided directly by a water turbine. The movable parts of this wind tunnel were in 1946 transported to Modane in the French Alps and utilized there to construct, based on the same principle, the ONERA S1-MA wind tunnel. This wind tunnel, since 1952 in operation, was (and is) several times successfully used in the Airbus development.

All these large wind tunnels were, at the time, unique test facilities and had in many aspects model character. They were primarily utilized for large project investigations, which provided, however, also many new scientifically important results. Among others, pressure measurements were carried out on large wind tunnel models with up to 8 m (26 ft) span, hence with very large Reynolds number, and aerodynamic investigations were even performed on full-scale aircraft. These large subsonic wind tunnels were after World War Two by order of the Allies dismantled and in parts reused abroad.

As already mentioned, also the aircraft companies had up to the end of the war smaller subsonic wind tunnels. Especially to be named is the high-speed wind tunnel of the Junkers Company in Dessau. The tunnel operated intermittently, based on the ejector principle, and reached in its 0.3-meter (0.89 ft) diameter open or closed test section, respectively, at a maximum run time of three minutes, velocities of 275 m/s (902 ft/s), i.e., a Mach number of 0.83. Here, *Otto Frenzl* discovered in 1943 the "area rule". Also several university institutes operated small subsonic wind tunnels for educational purposes.

First Supersonic Wind Tunnels

With the flight speeds approaching the speed of sound but also due to the development of supersonic missiles (rockets), also the construction of supersonic wind tunnels became urgent. However, the realization of such wind tunnels was coupled with a multitude of problems. These tunnels essentially needed closed test sections and the concept of a closed-loop air return according to the "Göttingen-design" could not directly be adopted. In the case of a subsonic wind tunnel, the closed test section has the smallest cross section area of the tunnel circuit; here, the maximum attainable velocity would be the speed of sound, which means that the tunnel "blocks" so that supersonic flow cannot be established. In a supersonic wind tunnel one must place, therefore, upstream of the test section proper in order to generate a supersonic flow, a Laval-nozzle, where the cross section area first decreases and then increases again. The potential (pressure) energy present at the nozzle entry is converted to kinetic energy (velocity), with the velocity reaching the speed of sound at the throat of the nozzle and the desired supersonic velocity at the nozzle exit (entrance of the test section). The supersonic Mach number can be controlled by a change in the nozzle contour, more accurately by a change in the ratio of the nozzle throat area to the nozzle exit area (variable nozzle). The attainable Mach number is ultimately dependent on the pressure ratio as long as a certain critical pressure ratio has not yet been exceeded. While for a low-turbulence closed-circuit subsonic wind tunnel the fan has to generate a 10 per cent increase in pressure to overcome the losses in the flow, is for a supersonic wind tunnel at a Mach number of $M=2$ a pressure ratio of 2 and at $M=5$ a pressure ratio of 20 required to generate supersonic flow. This is only possible with the aid of multi-stage compressors with tremendous drive performances. In addition energy (total pressure) losses due to shock waves present downstream of the test section must be taken into account.

An essential difference compared to a subsonic wind tunnel is also due to the fact that the air must have - dependent on the magnitude of the Mach number - a very low dew point in order to avoid water condensation. This requires air dryers, which constitute, should they be installed inside the tunnel circuit, a large resistance causing additional losses. The technical prerequisites to realize such large supersonic wind tunnels, with a closed air return and continuous operation, and especially the availability of corresponding axial compressors to generate the required pressure ratios, were not given up to the end of World War Two. One had, therefore, to fall back on other possibilities, viz., on the utilization of the storage principle and an intermittent operation. Since the required energy per unit time must only be available briefly, are the large drive powers no longer needed. In applying the storage principle, there basically exist two modes of operation:

- Utilizing a pressure storage tank and blowing the air through the test section into the atmosphere after opening a quick-opening valve.
- Utilizing a vacuum tank and sucking air from the atmosphere through the test section after opening a quick-opening valve

Also a combination of the two approaches is possible. The first suggestion to build a supersonic wind tunnel according to the vacuum-storage principle was already made by *Ludwig Prandtl* in 1917. It was envisaged to have a test section with a cross-section of 100 cm^2 (15.5 in^2) for projectile investigations; however, the tunnel could not be realized until the end of World War One. During the same year, the National Physical Laboratory (NPL) in Teddington/England started operating, for the first time ever, a supersonic wind tunnel based on the pressure-storage principle, which allowed, however, only to investigate projectiles with a diameter of a few millimeters.

The first supersonic wind tunnel based on the vacuum-storage principle in Germany was put into operation in 1936 at the TH Aachen by the Institute of Aerodynamics headed by *Wieselsberger*. It was utilized by *Rudolf Hermann* [60] for the first aerodynamic investigations on models of the Peenemünde Rocket A3, which also led to the discovery of condensation shocks in supersonic wind tunnel nozzles. *Ludwig Prandtl* and *Carl Wieselsberger* [61] already pointed out their possible existence in 1935. This also led to the recognition that in supersonic wind tunnels air-drying is a necessity.

In 1938, the AVA in Göttingen started operating the "High-Speed Wind tunnel", which was based on the vacuum storage principle, had an open-jet cross section of 0.11 m × 0.13 m (0.36 ft × 0.43 ft) and reached a Mach number of 3.2. The run time varied, dependent on Mach number, between 10 and 20 seconds. *Hubert Ludwieg* [51] conducted in this wind tunnel in 1939, following a suggestion of *Albert Betz*, tests on swept wings to confirm the $\cos\varphi$-law, which states that on a wing with a sweep angle of φ the critical Mach number is increased by $1{:}1/\cos\varphi$. This knowledge, already presented in 1935 by *Adolf Busemann* at the Volta-congress, has revolutionized high-speed aircraft design. This supersonic wind tunnel received in 1943, with the installation of a bigger vacuum vessel of 400 m^3 (14,100 ft^3), a test section with a cross section of 0.25 m × 0.25 m (0.82 ft × 0.82 ft).

After the foundation of the "Army Test Establishment Peenemünde" ("Heeresversuchsanstalt Peenemünde"), the rocket development and test center of the army and the air force, a supersonic wind tunnel was, under the supervision of *Rudolph Hermann*, put into operation. This tunnel was, according to today's knowledge, the most advanced supersonic wind tunnel up to the end of the war. It operated according to the vacuum-storage principle, had a vacuum vessel of 1000 m^3 (35,220 ft^3), a test section with a cross section of 0.4 m × 0.4 m (1.31 ft × 1.31 ft), and reached at

Rudolph Hermann (1904–1991).

Rudolph Hermann, born on December 15, 1904, received his PhD in Leipzig and designed at the Institute of Technology (TH) Aachen under the supervision of *Carl Wieselsberger* the first supersonic wind tunnel with a variable Mach number, having a test section with a cross section of 10 cm × 10 cm (3.9 in × 3.9 in).

1935 he has contacts with *Wernher von Braun* concerning the design of the A3 and A4 rockets; first measurements concerning the stabilization of the rockets at supersonic speeds were performed in the Aachen tunnel.

1936 *Hermann* suggests the construction of the Peenemünde supersonic wind tunnel, which is initiated by *Dornberger* on November 30.

1937 in April, *Herrmann* transfers to Peenemünde and builds up the Aerodynamic Institute, whose core is the supersonic wind tunnel with a cross section of the test section of 0.4 m × 0.4 m (1.31 ft × 1.31 ft) and Mach numbers up to 4.4. *Hermann* calls *Hermann Kurzweg* and his chief assistant to Peenemünde. The compact design of the rear stabilizers of the A4 rocket is a result of theoretical considerations and first measurement in this wind tunnel. The completion of the Peenemünde tunnel trails the development demands.

1938 it is decided to design and construct the A5 rocket.

1939 the tunnel goes into operation; the highest velocity during the development of the A4/A5 rockets corresponded to M = 2.5.

1941 the M = 3.1 nozzle goes into operation.

1942/43 the nozzle for the highest velocity functions. The number of employees of the institute increases in the meantime to 200.

1943 after the bombing of Peenemünde in August, the two wind tunnels are transported to Kochel near the "Lake Walchen Power Station" ("Walchensee-Kraftwerk") and for disguise reasons subordinated to the private enterprise "Hydraulic Engineering Test Organization Kochelsee, Ltd" ("Wasserbau-Versuchs-Anstalt (WVA) Kochelsee GmbH").

1944 in November, the tunnels are again ready for measurements.

1945 *Hermann* joins the Wright-Patterson Air Force Flight Dynamics Laboratory in Ohio, USA. The supersonic wind tunnel of the WVA is taken over by the Naval Ordnance Laboratory (NOL) in White Oak near Washington.

1947 tests are again resumed. *Hermann* transfers to the University of Minnesota; during this time, pioneering work related to the heating of the air in supersonic wind tunnels was performed.

1956 he becomes honorary member of the DGLR.

1962 he transfers to the University of Alabama in Huntsville, USA and becomes head of the research institute.

1991 he dies in Huntsville on May 17.

run times between 10 and 14 seconds Mach numbers between 1.2 and 4.4. Preceding the air intake of the tunnel was a Silica Gel air dryer, which could be continuously regenerated by means of a heater. This wind tunnel, supervised by *Hermann Kurzweg*, was, most of all, characterized by its modern instrumentation. It had, among others, a powerful Schlieren system and an interferometer built by the Carl Zeiss Company in Jena. Lift and drag were measured with a modern electro-mechanical balance developed by *Carl Wieselsberger.*

The balance contained an inductive signal converter, which allowed short-time force measurements in a "storage" wind tunnel with high accuracy. This balance principle was also used in other (smaller) Peenemünde wind tunnels. For longitudinal-stability measurements on rocket models, high-speed cameras were employed.

After the heavy bombing by the Allies on August 17/18, 1943, all undamaged test facilities were evacuated to Kochel/Bavaria, where new wind tunnels were to be installed. *Siegfried Erdmann* was able to con-

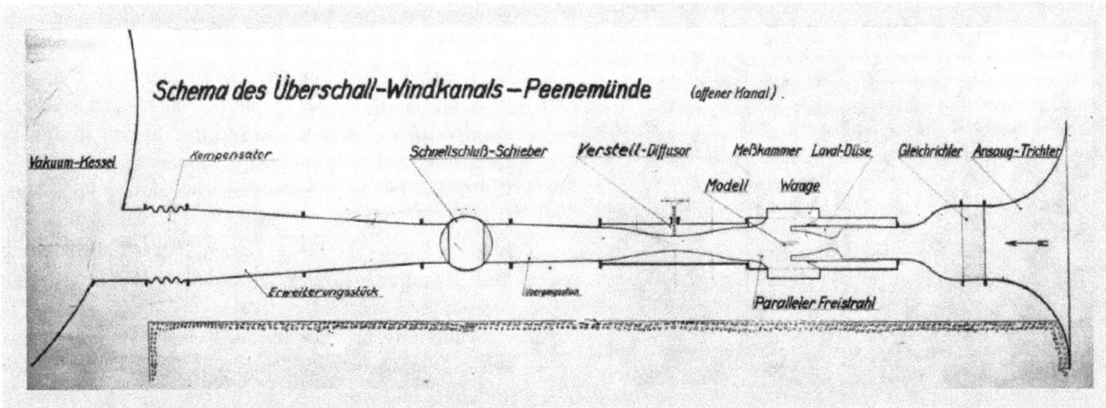

Historical sketch of the supersonic wind tunnel Peenemünde, which allowed short-time measurements up to a Mach number of 4.4.
Right to left, top to bottom: scheme of the supersonic wind tunnel Peenemünde (open tunnel), bell mouth, flow straightener, Laval nozzle, test section, variable diffuser, quick-opening valve, compensator, vacuum vessel, balance, model, parallel free jet, transition section, diverging section.

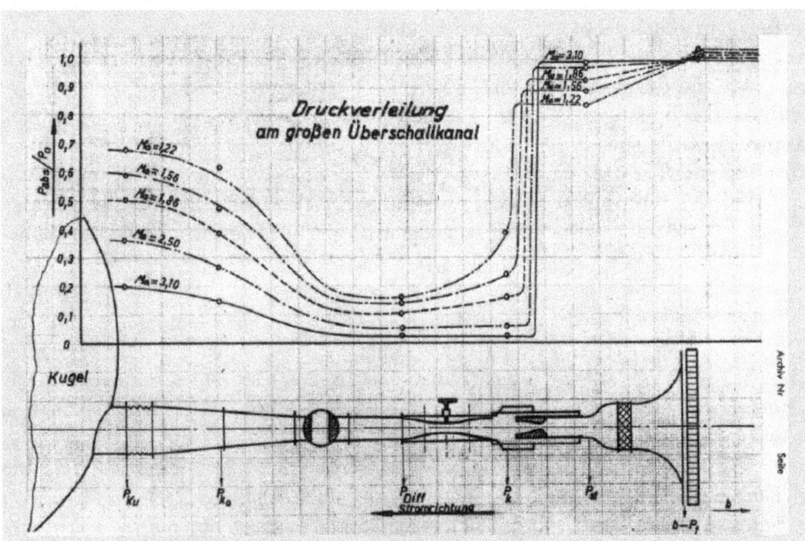

Original drawing of the supersonic wind tunnel Peenemünde with pressure distributions for various Mach numbers.
Top to bottom: pressure distribution for the large supersonic tunnel, sphere, flow direction.

vince, among others, *Wernher von Braun* of the necessity to realize, before the transfer of the said tunnel, a modification to accommodate a M = 9 hypersonic flow [62]. He succeeded to take the first Schlieren pictures of a hypersonic flow at M = 9.5. The experiences gained in the process were to be applied to a new large supersonic wind tunnel in Kochel, which could, however, not be realized before the end of the war. The 0.4 m × 0.4 m (1.31 ft × 1.31 ft) test section, including the optical measurement systems, dismantled in Kochel, was later again used with a supersonic tunnel of the Naval Ordnance Laboratories (NOL), Silverspring/Maryland.

In summarizing, it may be stated that in Germany, concerning the development of wind tunnel technology until 1945, trend-setting and pioneering work was

Numerous mercury manometers were used for pressure measurements in the Peenemünde wind tunnel.

performed. Although the theoretical fundamentals of aerodynamics were largely worked out at that time, numerical computations could, with the mechanical desk calculators available, only be performed in a very limited way. All the more became the wind tunnel the standard tool of aerodynamic research and development to cope with the large aerodynamic challenges present when the flight speed approaches the speed of sound. In that process, many fundamentals of gasdynamics were worked out in supersonic wind tunnel tests.

References

1 Th. von Kármán (with Lee Edson), The Wind and Beyond: Theodore von Kármán, Pioneer in Aviation and Pathfinder in Space, Little, Brown and Company, Boston, Mass., USA, 1967

2 J.D. Anderson, A History of Aerodynamics and its Impact on Flying Machines, Cambridge University Press, New York, 1982

3 G.Ch. Lichtenberg, Schriften und Briefe, Carl Hanser, München, 1992

4 L. Prandtl, K. Oswatitsch, K. Wieghardt, Führer durch die Strömungslehre, Vieweg, Braunschweig, 1984 (8. Aufl.)

5 Th. von Kármán, Aerodynamics: Selected Topics in the Light of their Historical Development, Cornell University Press, Ithaka, N.Y., 1954

6 L. Schiller (Herausgeber), Hydro- und Aerodynamik, in: Handbuch der Experimentalphysik (W. Wien, F. Harm, Herausgeber), Band IV, Teil 1–4, Akademische Verlagsgesellschaft, 1932

7 C. Truesdell (Mitherausgeber), Strömungsmechanik 1 bis 3, in: Handbuch der Physik (S. Flügge, Herausgeber), Band VIII/1 und 2, Band IX, Springer, Berlin/Göttingen/Heidelberg, 1959

8 L. Bölkow (Herausgeber), Ein Jahrhundert Flugzeuge, VDI-Verlag, Düsseldorf, 1990

9 F. Dubs, Aerodynamik der reinen Unterschallströmung und Hochgeschwindigkeits-Aerodynamik (2 Bände), Birkhäuser, Basel/Stuttgart, 1954

10 H. Schlichting, E. Truckenbrodt, Aerodynamik des Flugzeuges, Band 1 und 2, Springer, Berlin/Heidelberg/New York, 1958 and 1960

11 G. Cayley, On Aerial Navigation, Nicholson's Journal, 24, 1809, pp. 164–174, 25, 1810, pp. 81–87, 161–173.

12 A. Pénaud, Aéroplane Automoteur, L'Aéronaute, 5, 1872, pp. 2–9

13 I. Newton, Philosophiae Naturalis Principia Mathematica, London, 1726

14 J.-P. d'Alembert, Essai d'une nouvelle théorie de la résistance des fluides, Paris, 1752

15 F. von Loessl, Die Luftwiderstandsgesetze, der Fall durch die Luft und der Vogelflug, Wien, 1896

16 Lord Rayleigh, On the Irregular Flight of a Tennis-Ball, Messenger of Mathematics, 7, 1878, pp. 14–16

17 H. von Helmholtz, Über Integrale der hydrodynamischen Gleichungen, welche den Wirbelbewegungen entsprechen, Journal für die reine und angewandte Mathematik, 5, 1858, pp.25–55

18 L. Prandtl, Tragflügeltheorie, 1. Mitteilung. Nachr. d. Ges. d. Wiss. zu Göttingen, Math.-phys. Klasse, 1918, pp. 451–477, PGA p. 321

19 L. Prandtl, A. Betz (Editor), Ergebnisse der Aerodynamischen Versuchsanstalt Göttingen, 1. Lieferung 1925, 2. Lieferung 1927

20 M.M. Munk, Isoperimetrische Probleme aus der Theorie des Flügels, Dissertation, Göttingen, 1918

21 H. Multhopp, Die Berechnung der Auftriebsverteilung von Tragflügeln, Luftfahrtforschung, 15, 1938, pp. 153–169

22 H. Blenk, Der Eindecker als tragende Wirbelfläche, Dissertation, Göttingen, 1923

23 H. Glauert, The Elements of Aerofoil and Airscrew Theory, 1926

24 R.T. Jones, Properties of Low-Aspect-Ratio Pointed Wings at Speeds Below and Above the Speed of Sound, NACA R 835, 1946

25 M.M. Munk, The Aerodynamic Forces on Airship Hulls, NACA R 184, 1923

26 L. Prandtl, Über Flüssigkeitsbewegung bei sehr kleiner Reibung, Verhandlungen des III. Internationalen Mathematischen Kongresses Heidelberg, 1904, pp. 484–491
27 S. Goldstein, Fluid Mechanics in the First Half of the Century, Annual Review of Fluid Mechanics, Palo Alto, 1, 1969, pp. 1–28
28 H. Blasius, Grenzschichten für Flüssigkeiten kleiner Reibung, Dissertation, Göttingen, 1907
29 H. Schlichting, Boundary-Layer Theory, 7th edition, McGraw-Hill, New York, 1979
30 O. Reynolds, An Experimental Investigation of the Circumstances, which Determine whether the Motion of Water shall be Direct or Sinous and of the Law of Resistance in Parallel Channels, Philosophical Transactions of the Royal Society of London, Series A, 174, 1883, pp. 935–982
31 L. Prandtl, Über die ausgebildete Turbulenz, Zeitschrift für Angewandte Mathematik und Mechanik, 5, 1925, pp. 136–139
32 Th. von Kármán, Mechanische Ähnlichkeit und Turbulenz, Göttinger Nachrichten, mathematisch-physikalische Klasse, 1930, pp. 58–76
33 G.H.L. Hagen, Über den Einfluß der Temperatur auf die Bewegung des Wassers in Röhren, Abhandlung der Königlichen Akademie der Wissenschaften zu Berlin, mathematische Klasse, 1854, pp. 17–98
34 L. Prandtl, Bemerkungen über die Entstehung der Turbulenz, Zeitschrift für Angewandte Mathematik und Mechanik, 1, 1921, pp. 431–441
35 O. Tietjens, Beiträge zur Entstehung der Turbulenz, Dissertation, Göttingen,1922
36 W. Tollmien, Über die Entstehung der Turbulenz, 1. Mitteilung Göttinger Nachrichten, mathematisch-physikalische Klasse, 1929, pp. 21 –44
37 G.B. Schubauer, H.K. Skramstad, Laminar Boundary Layer Oscillations and Stability of Laminar Flow, NACA War-Time Report W-8, 1943
38 E. Mach, P. Salcher, Photographische Fixierung der durch Projektile in der Luft eingeleiteten Vorgänge, Sitzungsberichte der Wiener Akademie der Wissenschaften, Abt. II, 95, 1887, pp. 764–780
39 Th. Meyer, Über zweidimensionale Bewegungsvorgänge in einem Gas, das mit Überschallgeschwindigkeit strömt, Dissertation, Göttingen, 1908
40 H. Glauert, The Effect of Compressibility on the Lift of an Airfoil, Proc. Royal Society, A, 118, 1928, pp. 113–118
41 J. Ackeret, Luftkräfte an Flügeln, die mit größerer als Schallgeschwindigkeit bewegt werden, Zeitschrift für Flugtechnik und Motorluftschiffahrt, 16, 1925, pp. 72–74
42 A. Busemann, Aerodynamischer Auftrieb bei Überschallgeschwindigkeit, Volta Kongress, Rom, 1935, pp. 328–360
43 N.N., Collected Works of Robert T. Jones, NASA TM X-3334, 1976
43a P. Hamel, The Birth of Sweepback – Related Research at LFA-Germany, AIAA Atmospheric Flight Mechanics Conference, Austin, Texas, August 12, 2003
44 K.A. Kawalki, Theoretische Untersuchungen von Schnellflugprofilen, die aus Ellipsenprofilen entwickelt sind, 1. Mitteilung: Symmetrische Profile bei Nullanstellung, 2. Mitteilung: Symmetrische Profile bei Anstellung und Profile mit überlagerter Wölbung, Forschungsberichte (FB)1224 und 1224/2, 1940
45 B. Göthert, K.A. Kawalki, Die kompressible Strömung um verschiedene dünne Profile in der Nähe der Schallgeschwindigkeit, UM 1471, 1944
46 V.W. Peterson, G.B. Stephens, P. Cherney, Focke Wulf Designing Offices and General Management Bad Eilsen; Appendix XVIII: Notes on Discussion with the Staff of Focke-Wulf Ltd by R.Smelt, CIOS Item No. 25, File No. XXVI-6, May 1945
47 K. Oswatitsch, Gas Dynamics, Academic Press, New York, 1956
48 G.Y. Nieuwland, B.M. Spee, Transonic Airfoils: Recent Developments in Theory, Experiment, and Design, Annual Reviews of Fluid Mechanics, Palo Alto, 5, 1973, pp. 119–150
48a O. Frenzl, Motorgondel-Flügel-Interferenz bei hohen Geschwindigkeiten (die Flächenregel), Zeitschrift für Flugwissenschaften 5, 1957, pp. 181–183
49 F. Keune, K. Oswatitsch, Nicht angestellte Körper kleiner Spannweite in Unter- und Überschallströmung, Zeitschrift für Flugwissenschaften, 1, 1953, pp. 137–145
50 R.T. Whitcomb, A Study of Zero-Lift Drag Rise Characteristics of Wing-Body Combinations Near the Speed of Sound, NACA Research Memorandum L52 H108, 1952
51 H. Ludwieg, Pfeilflügel bei hohen Geschwindigkeiten, in H. Blenk (Herausgeber), Bericht über die Sitzung Hochgeschwindigkeit, Lilienthalgesellschaft für Luftfahrtforschung, Berlin, Bericht 127, 1940
52 E. Meyer, Windkanäle in Deutschland von Ludwig Prandtl bis heute, Studienarbeit, TU Clausthal,1998
53 W. Heinzerling, Windkanäle, in: L. Bölkow (Hrsg.), Ein Jahrhundert Flugzeuge, VDI-Verlag GmbH, Düsseldorf 1990
54 D.D. Baals, W.R. Corliss, Wind Tunnels of NASA, NASA SP 440, Washington, 1981
55 J.C. Rotta, Die Aerodynamische Versuchsanstalt in Göttingen, ein Werk Ludwig Prandtls, Vandenhoeck & Ruprecht, Göttingen, 1990

56 G. Eiffel, La resistance de l'air et l 'aviation, Dunod et Pinat, Eds., Paris,1910
57 L. Prandtl, Ergebnisse der Aerodynamischen Versuchsanstalt zu Göttingen, Erste Lieferung, Verlag R. Oldenbourg
58 L. Prandtl, Der Luftwiderstand von Kugeln, Nachrichten der Gesellschaft der Wissenschaften zu Göttingen, Mathematisch-physikalische Klasse 1914, pp. 177–190
59 A. Baeumker, Zur Geschichte der deutschen Luftfahrtforschung, Ludwig Prandtl zum 70. Geburtstag gewidmet, München 1944, DLR-Interner Bericht UB 183, 1985
60 R. Hermann, Der Kondensationsstoß in Überschall-Windkanal-Düsen, Luftfahrtforschung 19, 1942
61 C. Wieselsberger, Diskussionsbemerkung in: Convegno Di Scienze Fisiche. Mat. e. Nat., Roma,1935
62 P.P. Wegener, The Peenemünde Windtunnels – A Memoir, Yale University Press, New Haven and London, 1996

Aeroengines – Performance and Speed Increase as Impetus for Development*

Piston Aeroengines – Technological Milestones of Development

Kyrill von Gersdorff

Apart from the fundamental research activities in aerodynamics, engine development was a decisive area of work. The science of heat engines was at the beginning of the aircraft engineering no longer young, but the requirements regarding a very much improved weight/power ratio or the improvement of the altitude capability of fast rotating Otto and Diesel engines represented a considerable challenge. As regards the test facilities, new ground had to be broken.

At the beginning of World War One 1914 the standard aeroengine of the German military aircraft was a water-cooled in-line four cylinder engine of 100–120 hp (74–88 kW) take-off power, which was produced in this construction by several companies. These engines were relatively heavy, their weight/power ratio amounted to approx. 1.6 kg/PS (3.5 lb/hp) and their power/cylinder capacity ratio slightly more than 11 hp/l. One used to fly at low altitudes with speeds from 100–120 km/h (62–75 mph).

The Kaiser Award Competitions

Notable development impetuses were given by both Kaiser award competitions which were announced and carried out by the newly founded "Deutsche Versuchsanstalt für Luftfahrt, DVL". In these competitions the so-called construction weight was assessed for an endurance run of seven hours, that is the weight of the engine with full water cooler, with oil cooler, lubricant and fuel. The DVL formulated the test instructions, erected the test beds and carried out the competition. At the first Kaiser award competition, which started on October 1, 1912, initially 68 engines were registered, of those 43 were supplied, but only 17 fulfilled the test requirements. The Benz FX engine with 9.6 l (2.1 Imp. gal.) cylinder capacity and

The air-cooled rotary engine Siemens Sh III, 118 kW (160 hp) take-off power, which was produced in series in the war years 1917/18, was the most powerful German aeroengine of this type.

* see Volume 2 "Flugmotoren und Strahltriebwerke" (Die deutsche Luftfahrt)

A succession of research and testing activities resulted at the end of World War One in the altitude engine BMW IIIa with 165 kW (225 hp) take-off power, the top product of the German aeroengine development at that time.

105 hp (77 kW) take-off power constructed by *Arthur Berger* gained the first prize. All air-cooled rotary radial engines failed to comply with the requirements because of the high fuel consumption.

The 2nd emperor award competition, which was envisaged for September 1914 and for which the registration deadline was May 1, 1914, could not be carried out because of the outbreak of war [1]. Registered were 35 engines from 15 engine works, mostly six cylinder engines of in-line construction with take-off power from 120–150 hp (88–110 kW).

The Rotary Engines Reach their Performance Limits

During World War One the rotary engine with revolving radial cylinder configuration and stationary crankshaft was held for a long time as the best engine for fighter aircraft. Its weight/power ratio amounted only to approximately 1 kg/PS (2.2 lb/hp). Because of the unfavourable mixture processing, the specific fuel consumption was relatively high. Starting with the manufacture of the Gnôme under licence, derived from this was the Oberursel U 1 with 81 kW (110 hp). It was followed by the UR II having 103 kW (140 hp), which was further developed to 132 kW (180 hp). Siemens manufactured 1917/18 the Sh III with 118 kW (160 hp). Rotary twin-row radial engines of Oberursel and Siemens with a somewhat higher power had cooling difficulties and did not obtain the type certification despite of several attempts. One had come to the limits of this design.

Beginnings of the Altitude Engine Development

The air combat moved in the course of the war continually to higher altitudes. Often on the Western Front fighter formations flew briefly at approx. 6,000 m (19,700 ft) above the front, that is at an altitude where it was still possible to fly without breathing apparatus which was not yet in common use. Higher rates of climb, that is faster climb to the operational altitude was in demand and with it also sufficient excess power. This led to engines with increased compression ratio (super-compression) and to enlarged cubic capacity (oversize). These engines had to be throttled up to altitudes of 2,000–2,500 m (6,560–8,200 ft) in order not to overload or overheat them. The first altitude engine of this kind was the Maybach Mb IVa, an in-line six cylinder engine with 205 kW (280 hp). Soon followed altitude engines of Benz, BMW and Daimler. The most reliable and technically most matured German altitude engine was the BMW IIIa [2] which received the type certification with 166 kW (225 hp) in April 1918. More than 700 of these engines were delivered up to the end of the war. Especially powerful was with this the single-seat fighter Fokker D VII F, reaching a maximum speed of 200 km/h (124 mph) and climbing in 1.75 min to 1,000 m (3,281 ft). Tests with centrifugal superchargers for supercharging of altitude engines took place 1917/18 only at Siemens, AEG and BBC.

Restart under the "Definitions"

The Treaty of Versailles prohibited the production and import of aircraft of all kinds in and into Germany. Only with implementing the "definitions" of May 5, 1922 it was again possible to develop and build sporting aircraft and light commercial aircraft, however limited to a maximum speed of 170 km/h (106 mph) and a payload of 600 kg (1,323 lb). For single-seat aircraft only engines with a maximum of 45 kW (60 hp) were allowed to be used. Engines for commercial aircraft were permitted to be manufactured only up to 147 kW (200 hp), after loosening of the definitions in the year 1923 up to 220 kW (300 hp). Most restrictions were abolished not until May 1926 at the conference of ambassadors in Paris. The manufacture of military aircraft, however, remained prohibited as before.

The Siemens & Halske works had already – that is still within the period of production prohibition – started the development of light sporting aircraft engines using the so-called modular construction system. The three types of engine Sh 4, Sh 5 and Sh 6 had the same cylinders but different in number, the same connecting rods and pistons as well as the same parts of the control. For the first time a double ignition system was used, that is two spark plugs per cylinder and two magnetos. Two years later followed the engines of the second development stage Sh 10, Sh 11 and Sh 12. In order to improve the heat removal in the region of the cylinder head, a finned steel liner was screwed to the cylinder head of light metal casting, the semi-spherical combustion chamber of the cylinder head contributing to a favourable mixture processing. With enlargement of the cylinder bore, connecting rod strengthening and high pressure lubrication the next step followed to the Sh 13 and Sh 14 and then with the crankcase of elektron, higher compression and compressed air starter the breakthrough to success with the Siemens Sh 14A [2] of 110 kW (150 hp) and 117 kW (160 hp) respectively, of which over 3000 units were produced. The here described development of radial engines for sporting aircraft at Siemens – from Sh 4 (1922) to Sh 14A (1931) – is exemplary and shows with what rigour the German engine companies succeeded, not in the least by determined research work, to catch up with the international standard of light engines. The air competitions, financially supported by the Ministry of Transport, like the "Deutsche Rundflug" 1925 (German Air Rally 1925) with the "BZ-Preis der Lüfte" (BZ-Prize of the Air) yielded clearly discernible impetuses for the German sporting aircraft industry and with it for the German aeroengine industry and that against strong competition from France and England. A remarkable example of this period of time concerning the lowest power category was the air-cooled two cylinder opposed – cylinder engine of Daimler, the F 7502 with 15 kW (20 hp) developed by *Ferdinand Porsche*, which was mainly installed in the sporting and basic training aircraft Klemm L 20.

The "definitions" in spring 1922, which followed the restrictions of the Treaty of Versailles, gave rise in Germany to the development of several from the technical viewpoint remarkable light aeroengines for sporting and commercial aircraft of which the radial engine Siemens Sh 14A with 110 kW (150 hp) was a progressive example. The Siemens Sh 14A engines are today still in use for old-timer aircraft.

The company Argus developed in 1928 with support from the Ministry of Transport the first air-cooled aeroengine with inverted cylinders, the As 8 having a power of 70 kW (95 hp). The As 8 was followed in 1930 by an eight cylinder inverted – Vee engine. This engine had a dry sump circulation pressure lubrication as well as a ram cooling with a central cooling duct and air guide shrouds on each cylinder. Responsible for these developments, which required a lot of research and test work, was *Manfred Christian*. In an improved version with more cooling fins, higher compression and a take-off power of 175 kW (240 hp) the As 10C went into quantity production in 1934 [2]. Of this engine about 15000 units were manufactured. It was held as the most successful and reliable engine in its class.

Being the third German company for light engines, the Hirth-Motoren GmbH in Stuttgart-Zuffenhausen, which was managed by *Hellmuth Hirth*, produced in 1930 the HM 60 with a power of 52 kW (70 hp) [2], an air-cooled inverted-Vee four cylinder engine which proved itself on the "Deutschlandflug" 1931 (German Air Rally 1931). The technological special feature of the Hirth aeroengines was the roller bearing of the complete connecting rod assembly and the crankshaft screwed to the "Hirth" joint. In the years 1934 and 1935 the Hirth engine series HM 504, 506, 508 and later 512 arose with the same connecting rod assembly as four, six, eight and twelve cylinder engines.

The Argus As 10C with 175 kW (240 hp) was the first German air-cooled aeroengine with inverted cylinders in Vee arrangement, which was installed in many basic and advanced training aircraft as well as commercial aircraft.

The inverted cylinder arrangement, preferred by the German aeroengine manufacturers Argus and Hirth, came to use in several development series. Top products of this type, were the 12 cylinder Vee engines Argus As 410 and 411 which were installed in training and commuter aircraft. Numerous research and test activities were especially concerned with the cylinder cooling.

At Argus, under the direction of *Manfred Christian* and with the collaboration of *Fritz Goßlau*, in 1936 a new generation of fast rotating, technically demanding, supercharged air-cooled twelve cylinder inverted – Vee engines with reduction gear were developed, these being the As 410 with a power of 340 kW (465 hp) and the As 411 with 445 kW (600 hp).

Beginning of the Secret Armament

The first plans for establishing again a new German Air Force were already made in 1927 at the "Reichswehrministerium" and at the "Heerswaffenamt" in Berlin. One expected in those days (1927) flight speeds of 220 km/h (137 mph) for bombers, 250 km/h (155 mph) for reconnaissance aircraft and 350 km/h (217 mph) for fighters. For the aeroengines, which could be water-cooled or air-cooled, a low weight/power ratio of less than 0,5 kg/PS (1.1 lb/hp) was specified, they were to have a supercharger and had to be equipped with a variable-pitch propeller. A power of up to 2,000 hp was demanded and the capability of altitude flights up to 10,000 m (32,810 ft). In the year 1929 a sort of invitation of tenders for aeroengines with inverted cylinder arrangement and a size of the cylinder capacity of 20–30 l (4.4–6.6 Imp. gal.) was announced. This, for the integration of powerplant and airframe favourable constructional feature, had been proposed by *Georg Madelung* already in 1918 because of the better forward view for the pilot and was later supported especially by *Wolfram Eisenlohr*. In this programme in the 20 litre class (4.4 Imp. gal.) the Junkers Jumo 210 [3] and the 30 litre class (6.6 Imp. gal.) the Jumo 211 as well as the Daimler-Benz DB 600 and 601 respectively were developed and brought to readiness for production. To achieve this, a lot of research and testing work has been performed by the engine industry as well as by the DVL.

For the air armament programme of 1933 available were the liquid-cooled in-line-Vee engines BMW VI, Jumo 210 and DB 600, the air-cooled radial engines Siemens SAM 22 and BMW 132 as well as for the basic and advanced training aircraft the types Argus As 8, Hirth 60 and Siemens Sh 14, later also the Diesel aeroengine Junkers Jumo 205 and the Argus As 10.

In the so-called "Battle of Britain" (June–September 1940) clear performance deficiencies in terms of speed, altitude and range of the German front-line aircraft showed up. The "development cancellation" ordered in spring 1940 by *Hermann Göring* – among other things because of a misjudged sense of superiority – began to show effects. There followed numerous often fairly quickly again superseded development and production programmes which suffered increasingly from raw material shortage (copper, tin, molybdenum, chromium and temporarily aluminium). Nevertheless it is very remarkable what has been developed in that period of time, tested, and produced by the German industry supported by the institute research.

The essential development progress of that time – apart from the revolutionary general introduction of the direct injection for all larger aeroengines – were the rotational speed and boost pressure increase, the improvement of the pressurized cooling as well as the two methods of power increase, that is the methanol-water injection (MW50) [2] and the oxygen enrichment (GM1). For improvement of the altitude performance use was made of exhaust gas thrust nozzles, two-stage superchargers with two-speed or three-speed gears as well as combinations of exhaust-driven turbo-superchargers and gear superchargers. To cover the demand for higher power, integrated double engines were used. Under the term operational improvement one can categorize the development of the fully automatic one-lever power control with a kind of analogue-mechanical "computer", the quick replacement and standardized engines respectively and the cold starting procedure.

The German Aeroengine Industry Seeks to Catch Up with "The State-of-the-Art Technology"

The period of production prohibition (1919–1925) had an especially strong influence on the German aeroengine development. Deficiencies were clearly recognizable in the power class higher than 220 kW (300 hp) until the thirties [4]. Whilst the companies Siemens and BMW made the resumption easier for themselves by manufacturing under licence (Bristol/Gnôme et Rhône "Jupiter" and Pratt & Whitney "Wasp" and "Hornet" respectively), at Junkers and Daimler-Benz new developments were started with considerable effort. The types of engines dealt with exemplarily in the following illustrate the arduous path to best performance. They however show at the same time also the attainment of the limits of the piston aeroengine of the upper power class. It followed the logical replacement of this method of propulsion by the turbo-jet engines.

The in 1923 reactivated Junkers engine production began with the development of the six cylinder in-line engine L 2 with 19 l (4.2 Imp. gal.) cylinder capacity and a power of 147 kW (200 hp). After loosening of the definitions followed the L 5 with 23 l (5.1 Imp. gal.)

As especially reliable and therefore well suited for air traffic with the newly developed Junkers commercial aircraft was considered the since 1925 produced Junkers aeroengine L 5 with 228 kW (310 hp) take-off power. As engine of the Atlantic aircraft Junkers W 33 "Bremen" it proved its reliability.

and 228 kW (310 hp) take-off power. It was installed in 1925 in the Junkers aircraft F 13, W 33, G 24, G 31, A 20 and A 35 and was considered to be especially reliable. The L 52 was in 1934 still in production for the German air armament as L 5G with 275 kW (375 hp) for the training and commuter aircraft W 33. The L 5 engine also became to be known for the Atlantic crossing of the W 33 "Bremen" in 1928. As result of the development of an oil-hydraulic vibration damper, the operational rotational speed could be increased and the take-off power raised up to 312 kW (425 hp). By doubling the L 55 arose with power up to 515 kW (700 hp) and then via the L 8 the L 88 with 588 kW (800 hp) and reduction gear. By order of the DVL, Junkers developed and manufactured the altitude research aircraft Ju 49 with pressurized cabin, which reached with the L 88 supercharged engine altitudes over 10,000 m (32,800 ft) in 1931. Without supercharger but with long transmission shaft the L 88 engines were installed as inboard engines in the four-engined large aircraft G 38.

The "Bayerische Motorenwerke, BMW" resumed in 1922 the aeroengine production, initially with the BMW IVa with a power of 220 kW (300 hp). By doubling of the BMW IV, the twelve cylinder BMW VI arose. The final production series had a take-off power of 550 kW (750 hp). One special feature of this engine was its so-called hot cooling with ethylene glycol instead of water. The allowable operating temperature was thereby 140°C and permitted smaller and retractable radiators, as for example for the fast commercial aircraft Heinkel He 70 "Blitz". The BMW VI aeroengine was installed in numerous commercial and military aircraft, with them world records were flown and the airmail service across the South Atlantic was commenced.

The nine cylinder radial engine BMW 132 was the standard engine of the three-engined commercial aircraft Junkers Ju 52. It arose from a follow-on development of the Pratt & Whitney "Hornet", for which BMW received the licence for manufacture in 1933. The BMW 132 was produced in various versions in large numbers of units until 1944 and was a success-

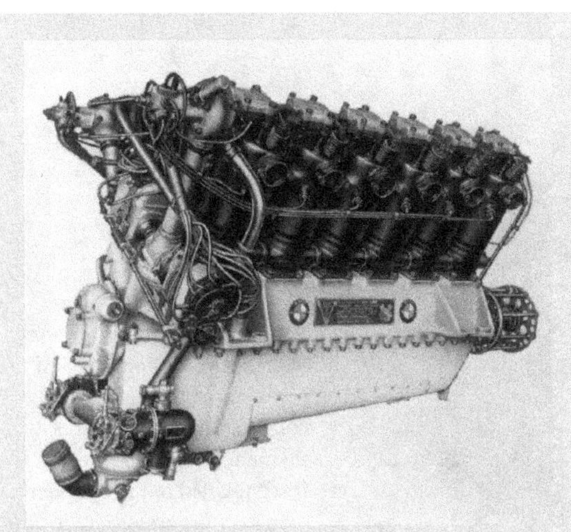

For the 12 cylinder aeroengine BMW VI with 550 kW (750 hp) power for the first time the so-called hot cooling came into use, which led to smaller radiator dimensions and thus lower drag.

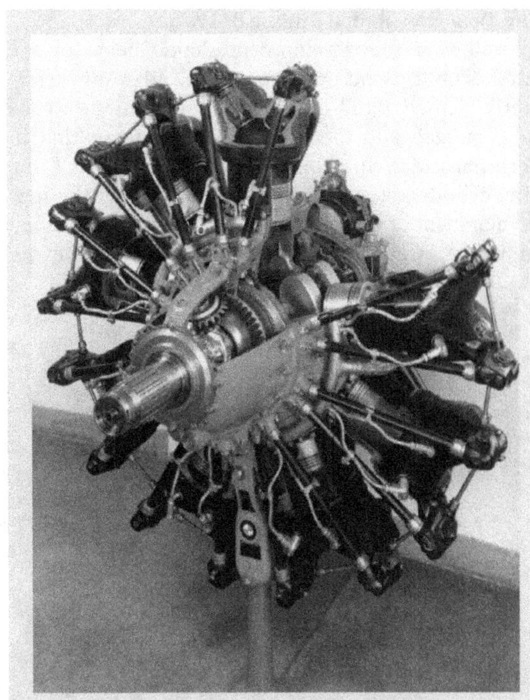

An often used reliable radial engine was the BMW 132 which, as follow-on development of the Pratt & Whitney licence engine "Hornet", was improved from initially 485–880 kW (660–1,200 hp). In order to optimize the engine cooling, the BMW 32 was installed together with Townend ring or NACA cowling.

ful engine of high reliability. Its take-off power of initially 485 kW (660 hp) was raised by improvements of various kinds to 880 kW (1,200 hp). Especially known became this engine also for a number of long distance flights, among others for the flight of the Focke-Wulf Fw 200 "Condor" from Berlin to New York and back. With the BMW 132 the "Townend ring cowling" and the "NACA cowling" came in use on many German aircraft. These "aerodynamic" engine cowlings reduced cooling drag compared with the usual unshrouded installation of radial engines and produced a speed gain. In combination with strong finning of the cylinders and adjusted air guide shrouds, optimum cowlings were developed. The research and the testing took place in this connection in close cooperation of aircraft manufacturer, engine developer and research institutes.

The Twin-Row Radial Engine BMW 801

In mid 1939 the development departments of the "Brandenburgische Motorenwerke, BRAMO" (formerly Siemens) and of the BMW aeroengine works were merged and the programs split up so that in Berlin-Spandau the follow-on development of the nine cylinder radial engines and in Munich the new development of the fourteen cylinder twin-row radial engine BMW 801 took place. Making use of the experience with the predecessor test engines BMW 139 and Bramo 329, in autumn 1938 the development of the twin-row radial engine commenced under the direction of *Martin Duckstein*. The RLM ordered the complete powerplant, that is with entire cowling, with cooling fan, air guide shrouds as well as all accessories up to the interface at the fire bulkhead. Added to the powerplant was also the exhaust gas assembly and the mounting structure. It concerned the so-called standard engines with standardized interfaces, which were thus installable on various types of aircraft. The fighter aircraft Focke-Wulf Fw 190 initially received the BMW 801C and later the follow-on development 801D-2 with a power of 1,270 kW (1,730 hp) and then the 801E with 1,470 kW (2,000 hp) with which a maximum speed of 630 km/h (391 mph) at an altitude of 5,500 m (18,040 ft) could be attained. One special feature of the BMW 801 was the so-called command unit, an operating and control device for one-lever power control. The basic idea for this with considerable mechanical effort associated equipment was the relief of the fighter pilot from many separate manual operations as, for example, boost pressure, rotational speed, mixture, moment of ignition and throttle valve setting. The injection pump had been developed by the company Deckel.

Extensive research work, connected with component and complete engine testing in cooperation with the aircraft manufacturers and research institutes, led to the only German very progressive twin-row radial engine BMW 801 with forced cooling by front cooling fan, command unit for engine control and thus to a quickly exchangeable standard engine.

The Engine Series Jumo 210 to 213

The development of the 20 l (4.4 Imp. gal.) aeroengine Jumo 210 began at the Junkers company in the year 1931 under the direction of *Otto Mader* and did comply with the later specified power category and version. The RLM specified twelve cylinders in inverted construction, with spur wheel reduction gear, provision for the installation of an engine cannon, with two-gear radial supercharger, petrol injection and automatic mixture control. The first test engine – still with carburetor – did run on the test bed in October 1932. The construction series 210A and 210D, the latter with 500 kW (680 hp) take-off power, were installed in the Messerschmitt fighter aircraft Bf 109. Under the direction of *August Lichte* the petrol injection equipment and the fully automatic mixture control was developed for the 210G. The Jumo 210 was thus the first in high quantities produced injection aeroengine. With the requirement for higher power, it then came to a kind pantographic enlargement yielding the Jumo 211 with a 35 l (7.7 Imp. gal.) cylinder capacity. The design leadership for this engine had *Franz Neugebauer*. Because the RLM wanted to limit the weight of the engine to 500 kg (1,103 lb), a costly weight reduction action had still to be performed. The first Jumo 211A with 735 kW (1,000 hp) was produced in late summer 1937, it soon followed the 211B/D with 882 kW (1,200 hp) and 211F with 985 kW (1,340 hp) and finally the construction series P with 1,103 kW (1,500 hp). Of the Jumo 211 a total of over 68,000 units have been produced. It was the standard engine of several construction series of the He 111, the Ju 87 and the Ju 88. For the top product of the German aeroengine development, the Jumo 213, *August Lichte* was responsible from the start. More than 9,000 units have been produced from 1942 to the end of the war. They were installed in the aircraft types Fw 190D, Ta 152, Ta 154, Ju 188 and He 219. The Jumo 213 was delivered as powerplant, that is either as "fighter engine" or as "engine for twin-engined aircraft", exchangeable with the standard engine Daimler-Benz DB 603. The Jumo 213 had the same basic construction as the 211, however, being more progressive, that is developed further for highest thermal and mechanical loading. It had as 213A [2] a take-off power of 1,285 kW (1,750 hp), as 213B with C3 fuel (95 octane number) it produced 1,470 kW (2,000 hp) at a crankshaft rotational speed of 3,250 rpm and as 213E altitude engine with two-stage three-gear supercharger and supercharger air intercooler the full pressure altitude was 9,800 m (32,144 ft). Short period power increases with water-methanol injection or oxygen enrichment by GM1 injection were possible. A four-valve version, which because of the end of the war could no longer be realized, was planned with a take-off power of 1,645 kW (2,240 hp).

The Jumo 213 engines possessed a very low specific fuel consumption between 0.265 kg/kWh and 0.272 kg/kWh (0.584 lb/kWh and 0.599 lb/kWh) that is 0.195 kg/hph and 0.200 kg/hph (0.430 lb/hph and 0.441 lb/hph) at cruise condition. The Jumo 213 is to

As top product of the German aeroengine development of World War Two is considered the 12 cylinder aeroengine Jumo 213 which was derived from the engine series Jumo 210 and 211. The last version had a take-off power of 1,470 kW (2,000 hp) and a full pressure altitude of 9,800 m (32,144 ft). For this engine too the GM 1 injection, that is the oxygen enrichment to improve the altitude power, was used.

be considered as final stage of the German piston aeroengine development in World War Two.

The In-Line Radial Engine Jumo 222

The development of the twenty-four cylinder in-line radial engine Jumo 222 was connected with the planning of the "Bomber B" and stood in a competitive-like relation to the Daimler-Benz integrated twin-engine DB 606. In April 1937 *Ferdinand Brandner* at Junkers received the order for the development of a 1,470 kW (2,000 hp) engine, water-cooled with twenty-four cylinders in six banks at four cylinders each and radially arranged. When in April 1941 the first 100-hour run was performed with this power and also the flight testing in a Ju 52 modified as a flying test bed proceeded satisfactorily, one could assume that this engine would soon be ready for high quantity production and commenced with the planning of the aeroengine works Ostmark. However, then in the testing of the Jumo 222 numerous difficulties like vibration fractures, bearing damage and casing corrosion occurred, which led to more than one year of development delay. Probably the state of technology covered by research had been overdrawn. In addition, the bomber Ju 288, for which the Jumo 222 had been the first application, had been changed to a higher take-off weight and now a take-off power of 1838 kW (2,500 hp) was demanded. Test bed runs with increased power brought again new problems due to bearing damage and piston seizure. Extensive constructive changes became necessary, especially an enlargement of the bore, initially from 46.6 l (10.3 Imp. gal.) cylinder capacity to 49.8 l (11 Imp. gal.) and later – when 2,205 kW (3,000 hp) were requested, again to 55.5 l (12.2 Imp. gal.). The Jumo 222 was temporarily reduced in priority, then, however, when difficulties arose with the DB 606 and DB 610 integrated double engine installed in the bomber Heinkel He 177, again as 2,205 kW (3,000 hp) engine it was developed further to the 222E altitude engine with two-stage supercharger. The Jumo 222 is an example for a partially war related uncoordinated planning on the one hand and technical excessive demand on the other hand.

The Diesel Aeroengines of Junkers

A purposeful development of Diesel aeroengines began at Junkers in the year 1922, making use of experience gained with the already in 1918 manufactured opposed cylinder test engine FO 2 with a power of 365 kW (500 hp). This engine had thus early all essential characteristics of the later Diesel aeroengines of Junkers like double-crankshaft construction, opposed

The 24 cylinder in-line radial engine Jumo 222 arose in connection with the "Bomber B" programme in a very progressive design. From the technical viewpoint is was a top-class performance. The development was, however, war related delayed. Initially a take-off power of 1,470 kW (2,000 hp) was intended, this was then to be 1,838 kW (2,500 hp) and finally 2,205 kW (3,000 hp) was demanded because the take-off weight of the planned bomber Ju 288 had been increased repeatedly.

Result of a very long development time was the Junkers Diesel aeroengine Jumo 205 with 440 kW (550 hp) take-off power, the worldwide first operational Diesel aeroengine which distinguished itself by an especially low fuel consumption.

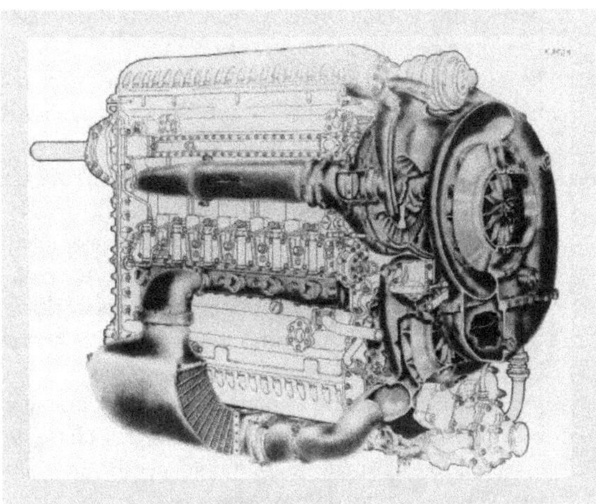

The Diesel altitude engine Jumo 207 with a two-stage supercharger, a follow-on development of the Jumo 205, allowed the operation of the Junkers reconnaissance aircraft Ju 86 P/R at altitudes up to 14,000 m (45,920 ft).

pistons, airless injection and spiral scavenging process. After a series of tests with single cylinder engines, in 1926 the five cylinder test engine FO 3 originated, which reached in 1928 a power of 610 kW (830 hp) on the test bed. For the development *Josef Gasterstädt* was responsible. The FO 3 was followed by the FO 4 with six cylinders and 28.6 l (6.3 Imp. gal.) cylinder capacity from which later the Jumo 204 with 552 kW (750 hp) was derived. Approximately seven years of intensive testing passed, connected with a lot of research effort, until a Jumo 204 engine installed in a single-engine Junkers F 24 flew for the first time on February 4, 1929. The RLM and the Deutsche Lufthansa were very interested in the economical Junkers Diesel aeroengine which consumed at cruise power only 0.218–0.231 kg/kWh (0.481–0.509 lb/kWh) or 0.160–0.170 kg/hph (0.353–0.375 lb/hph) of Diesel fuel (heavy oil) respectively. Because the constructional vertical dimension of the Jumo 204 was too large for multi-engined aircraft and the power of 552 kW (750 hp) was not required, the Jumo 5 – later 205 – arose with 16.6 l (3.7 Imp. gal.) cylinder capacity and 440 kW (550 hp) as well as smaller dimensions and higher rotational speed [2]. The operational testing in the Dornier flying boats Do 18 and Do 26 on the mail routes over the South and North Atlantic; in a Focke-Wulf "Möve", since 1934 in several Ju 52 and since 1935 in the Ju 86 of the Lufthansa were successful. The DLH had shown great interest in the operation of the Junkers Diesel aeroengines. In the year 1937 there were already 21 aircraft of the DHL flying with Jumo 205 Diesel engines which had in the meantime been improved and as 205D raised to a take-off power of 647 kW (800hp). In the Air Force the Jumo 205 came into service in the aircraft types Ju 86, Do 18, Do 26 and BV 138. It was succeeded by the follow-on developments Jumo 206, 207 and 208, all based on the same design principle. With the exception of the altitude aeroengine Jumo 207, it came, however, to no further high quantity production. In autumn 1939 there arose an urgent demand for the altitude engine for reconnaissance aircraft which were to come into service above the ceiling of enemy flighter aircraft. For this, Ju 86 bombers were modified and equipped with a pressurized cabin. With highest priority the Jumo 207A with 647 kW (880 hp) take-off power was tested, which had with exhaust-driven turbo-supercharger a full pressure altitude of 10,000 m (32,800 ft), followed by 207B and 207C with 735 kW (1,000 hp). With the Ju 86P operational altitudes of 14,000 m (45,920 ft) were reached.

The Junkers development of Diesel aeroengines was unique and even today there exists no successful successor despite of demand from the general aviation.

Otto Mader (1880–1944).

Otto Mader, born in Nürnberg on September 17, 1880

1903 masters examination at the Institute of Technology (TH) Munich.

1904 employment by a mechanical engineering company in Munich.

1905 change to Maschinenfabrik Augsburg – Nürnberg in Augsburg, where he is working in the field of Diesel engines.

1906 he becomes assistant at the TH Munich and receives

1909 his doctorate for a thesis "Resonance undograph, a means for measuring the degree of distortion". He changes to *Junkers* at the TH Aachen. There he heads the Mechanical Engineering Laboratory at *Junkers'* Chair of Theory of Heat

1912 he qualifies as a university lecturer in the field of measurement technology and assumes the management for the "Research Establishment Prof. Junkers", which is soon transferred to Dessau to become the basis for the development activities of Junkers.

1923 he becomes the first managing director of the "Junkers Motorenbau GmbH" and devotes himself also to the development of the Diesel aeroengine which promises benefits especially for long range flight

1927 he is appointed to Full Professor of Technical Mechanics and Aircraft Structures at the TH Munich, an appointment which he relinquishes already

1928 for health reasons in order to manage again his principal field of work, the engine development. In an advisory capacity he is active in the aircraft development. He is appointed to the board of directors of the company Junkers. Temporary culmination of his work are the high performance engines Jumo 210 and Jumo 211 for the German Air Force.

1933 *Junkers* must leave his own company under pressure from the government. *Mader* plays in this phase an essential role as amalgamation manager for the "Junkers Flugzeug- und Motorenwerke AG".

1940 there appear signs of a serious illness more and more distinctly. He shows commitment despite his illness especially for the further development of the aeroengines Jumo 211, 213 and 222, of the variable – pitch air screws as well as the development and maturing of the revolutionary jet powerplant Jumo109-004 under *Anselm Franz.*

1944 he dies in a sanatorium in Landeck/Tirol on September 9.

The Daimler-Benz
Record Aeroengines DB 601 Re

With the second invitation to tender for fighter aircraft by the RLM of December 1933 a kind of competition arose between the aircraft companiess Heinkel and Messerschmitt. The young German Air Force had in Zürich-Dübendorf in summer 1937 the opportunity to demonstrate its new military aircraft; *Ernst Heinkel* [4] and *Willy Messerschmitt* urged the RLM and the engine company Daimler-Benz to power increases of the DB 601 engine, especially for record purposes. One wanted to gain the speed world record for Germany and demonstrate the technological progress of the German aviation industry. The RLM supported these plans generously.

The Heinkel He 112 V4, flown by *Ernst Udet* in Zürich, had a DB 601 aeroengine of increased performance with 1,005 kW (1,366 hp) take-off power using alcoholic special fuel. As further step arose the DB 601 Re III with 1,221 kW (1,660 hp), which was made available to both aircraft companies in spring 1938. At Heinkel in Warnemünde the He 100 V2 was prepared with this engine. On the airframe several modifications were undertaken. Surface cooling replaced the radiator protruding from underneath the engine, the cooling water temperature of the pressurized system could increase up to 120 °C. The steam separator separated the water from the steam, with the steam after cooling and condensation being fed back to the cooling system. The Re III had an increased compression (8.9 instead of 6.9), a raised rotational speed (2,650 rpm instead of 2,400 rpm) and the boost

The 12 cylinder Daimler-Benz aeroengine DB 601 with initially 805 kW (1,100 hp) take-off power, the standard engine of many German fighter and bomber aircraft, was boosted for record purposes up to a short duration power of 2,035 kW (2,770 hp). With this engine a Messerschmitt Me 209 established in spring 1939 the speed world record with 755 km/h (469 mph), which was only broken many years later.

pressure was set to a maximum of 1.59 bar (instead of 1.37). With this engine, installed in the He 100 V2, *Ernst Udet* flew on Whitsun 1938 a new world record over the 100 km (54 nm) distance at optimum altitude with 634,73 km/h (394.42 mph) and exceeded the existing record by 80 km/h (49.7 mph). *Hermann Wurster* did fly a little later with the same engine, installed in the Messerschmitt Bf 109 V13, over the 3 km (1.6 nm) distance with 611 km/h (379.68 mph) near ground level the FAI world record for landplanes. This again did not allow *Ernst Heinkel* to rest, for Daimler-Benz had in the meantime boosted the record engine for a short time performance to 2,035 kW (2,770 hp). In addition to the special fuel, use was made of an alcohol injection upstream of the supercharger as internal cooling, corresponding to the MW50 process introduced later. This racing engine of 1939 with the designation DB 601 Re V was increased to a rotational speed of 3,100 rpm and a boost pressure of 2.4 bar, limited to five minutes. With this engine *Hans Dieterle* established for the 3 km measuring distance near Oranienburg on March 30, 1939 a new FAI world record of 746,60 km/h (463.94 mph). The same engine was now installed in the newly developed record aircraft of Messerschmitt, the Me 209. Because difficulties were experienced with the cooling system, one did without the condensation and the return of the coolant and accepted instead a water tank of 450 l (99 Imp. gal.). On April 26, 1939 *Fritz Wendel* did fly near Augsburg a new world record with 755,138 km/h (469,2 mph) which lasted more than 30 years. As Heinkel now – with reference to the 450 m (1,476 ft) higher situated measuring course near Augsburg – wanted to undertake a new record attempt, the RLM stopped all further record flights. The Daimler-Benz DB 601 R V world record engine [5] was only built in three units. It had in the view of the development engineers a life of 30 min to a maximum of 60 min, was thus a souped up "racing engine" not suited for a normal flight operation. Remarkable are, however, the specific parameters which allow an assessment regarding the technological state achieved. With 2,035 kW (2,770 hp) a power/cylinder capacity ratio of 60.9 kW/l and 81.7 hp/l (372 hp/gal) respectively and a weight/power ratio of 0.30 kg/kW (0.66 lb/kW) and 0.22 kg/hp (0.49 lb/hp) respectively was attained. The experience from this record flight action benefited later Daimler-Benz aeroengines in diverse ways.

Otto Lutz (1906–1974).

Otto Lutz, born in Stuttgart-Bad Cannstatt on April 8, 1906

1931 doctorate at the Institute of Technology (TH) Stuttgart for a thesis "Investigations into the scavenging of two-stroke engines" for the clarification of the flow phenomena, resonance vibrations, processes regarding the compression and expansion in piston engines.

1934 postdoctoral lecturing qualification at the Institute of Heat Engines of the TH Stuttgart and concentration on the unsteady gas dynamics and thermodynamics

1936 with the founding of the DFL in Brunswick he commences his work at the Institute of Engine Research. Still today in use are his enthalpy charts for air and combustion gases. Publications demonstrate the wide spectrum of his work from the screw piston principle and swing piston engine via the jet engines to the rocket powerplants. This period of time also covers the investigations into 1100 self-igniting fuel combinations, hypergolic fuels (with *Hansemann* and *Noeggenrath*). Here too was developed the GM1 method for performance increase of piston engines through addition of oxygen carriers.

1944 he receives the Lilienthal prize.

1945 summary of the results in the Völkenroder monographs.

1947 prohibition of aeronautic research activity bridged with the mass production of own developments of two-stroke auxiliary engines and scooters.

1953 he founds the Institute of Jet Propulsion within the DFL. He originates the reaction thermodynamics which allows the novel calculation of the changes of state in reacting gas mixtures. The fields of gas dynamics and thermodynamics are united.

1954 in addition he takes over the Chair of Engineering Elements, Conveyor Technology and Jet Engines at the TH Brunswick.

1956 establishment of the Space Research and Testing Ground Trauen together with Entwicklungsring Nord (ERNO). He is appointed "Fellow" of the Royal Aeronautical Society in England and Member of the AGARD Panel "Combustion and Propulsion" which he chairs from 1963–1965.

1967 Great Merit Cross of the Order of Merit of the Federal Republic of Germany.

1974 he dies in Brunswick on May 2.

Daimler-Benz DB 601 to DB 628 – Engines of the Fighter Aircraft Bf/Me 109

From the production series E onwards, all Messerschmitt Bf/Me 109 used Daimler-Benz aeroengines, initially DB 601, later DB 605 and finally the DB 628 [2] as altitude engine. This one, however, only for testing, because it is not yet for operation. In those seven years – war influenced in "top gear" – an exemplary engine development took place. The maximum speed rose from 570 km/h (354 mph) for E-3 to nearly 700 km/h (473 mph) for K-4, the take-off power of the installed DB engines from 805 kW (1,100 hp) (DB 601A) to 1,324 kW (1,800 hp) (DB 605D). The full pressure altitude of the DB 601A was 4,000 m (13,120 ft), that of the DB 605D 6,500 m (21,320 ft) and that of the DB 628 test engine with two-stage supercharger over 11,000 m (36,080 ft). Of the DB 601 altogether more than 19,000 and of the DB 605 over 42,000 units were produced by the end of war.

From the production series B onwards, for the DB 601 a novel hydraulic supercharger clutch was introduced, which worked in a way similar to the two-speed gear. Starting with the production series E the variable-pitch gear of the VDM air screw was integrated in the reduction gear. Initially as kit, but soon afterwards planned for the series, the GM 1 injection, that is addition of nitrous oxide as oxygen carrier, was used for the performance increase above the full pressure altitude. This method, applied to German single-engined fighter aircraft and twin-engined heavy fighter aircraft since 1941 for Daimler-Benz and Junkers engines, gained for the front-line aircraft of the Air Force for an extended period of time a marked superiority at higher altitudes.

In the quantity production the DB 601 was followed since 1942 by the DB 605 [2] with the same construction but with an enlarged cylinder capacity from 33.9–35.7 l (7.5–7.9 Imp. gal.). For these engines the methanol-water injection (MW50) was intended, which increased the normal take-off power of the production series A–C from 1,085–1,320 kW (1,475–1,800 hp). The production series D received the larger supercharger of the DB 603 and developed with 100 octane fuel 1,140 kW (1,550 hp). The full pressure altitude was 6,500 m (21,320 ft). The last production series L, of which only two were installed for testing, possessed a two-stage supercharger with 9,600 m (31,488 ft) full pressure altitude. With this, the Me 109 K14 reached nearly 700 km/h (435 mph). For the series, which could not be started because of the end of war, the altitude engine DB 628 with two-stage supercharger (full pressure altitude approx. 11,000 m (36,080 ft)) and a power of 1,175 kW (1,600 hp) was intended.

The interest of the Allies, especially specialist engineers from England, France and the USA, for the German high-performance injection engines was very high. Undamaged engines from emergency landings or shot down German aircraft were thoroughly examined, put on the test bed and occasionally also installed in own fighter aircraft for test purposes.

Large Aeroengines of more than 1,838 kW (2,500 hp) Power

When in 1936 *Heinrich Hertel* at the company Heinkel realized in a design the military requirement of the RLM for the bomber Heinkel He 177, it was based on a 500 kg (1,103 lb) bomb load, 5,000 km (2,700 nm) total flight range and 500 km/h (311 mph) cruise speed at the nominal altitude of the engines. Because at that time no suitable engine for long range bombers was available in Germany, one fell back upon a proposal of *Siegfried Günter*, who had the idea to make two existing engines to work on one air screw, that is to create an integrated double engine by coupling of two single engines via the in any case required reduction gear. The thus resulting Daimler-Benz DB 606 (1,985 kW/2,700hp) [2], which had proved itself in the long range reconnaissance aircraft He 119, consisted of two DB 601E. *Gerhard Nitschke* flew with the He 119 V4 on November 22, 1937 a world record with 504.98 km/h (313.79 mph) over the 1,000 km (540 nm) distance. The maximum speed of the V4 was 620 km/h (385.27 mph) and was thus higher than that of the fighter aircraft used in those days.

The long range bomber Heinkel He 177 received for the mass production two DB 606 with each working on one four-blade air screw. The production series A-3 and A-5 used the more powerful DB 610 consisting of two DB 605 (2,165 kW/2,950 hp) each. The aerodynamic benefit of the integrated double engine was unfortunately associated with a persistent susceptibility to engine fire during flight operation. The DB 613 (2,750 kW/3,500 hp), consisting of two D 603 engines, just reached testing.

Concerning the invitation to tenders for the "Bomber B", which was issued by the RLM in the first version in mid 1939, the company Junkers had a timewise advantage due to previous work on the bomber Ju 288 and the in-line radial engine Jumo 222.

Up to the end of the World War Two the Daimler-Benz integrated double engines DB 606 and 610 with 1,985 kW (2,700 hp) and 2,165 kW (2,950 hp) respectively, consisting of two coupled DB 601 and DB 605 respectively, were the only German large aeroengines which were installed in the bomber Heinkel He 177.

When in the development and testing of the Jumo 222 (intended were 1,840 kW/2,500 hp) considerable de-lays occurred and the hope of an early start of the mass production was timewise constantly postponed, both the other large engine companies, the "Bayerische Motorenwerke, BMW" and "Daimler-Benz, DB", increased their development efforts for large engines with the support of the RLM. At BMW in Munich already in 1939 a project of an eighteen cylinder double-row radial engine had been launched, which received the designation BMW 802 and possessed with a cylinder capacity of 60 l (13.2 Imp. gal.) the same stroke and the same bore as the BMW 801. At the time of the development stop in 1942 only some test engines with a power of 1,800 kW (2,450 hp) had been run on the test bed. Likewise in 1939 development work began on a 2,940 kW engine (4,000 hp), the water-cooled BMW 803 [6] with 28 cylinders as integrated double engine consisting of two fourteen cylinder double-row radial engines situated one behind the other. In the fifties this type of construction for commercial aircraft was widely used in the USA before the advent of the "jet age". The power of the component engines, which could alternatively be switched-off, was transmitted to a counter-rotating double three-blade propeller. Of this 84 l (18.5 Imp. gal.) engine only a few test objects were manufactured and tried on the test bed.

The peak of the German large aeroengine development was the 24 cylinder opposed piston Diesel aeroengine Jumo 223 with 1,835 kW (2,500 hp) which, however, could not any more be flight tested before the end of the war.

A parallel development to the Jumo 222 was represented by the Daimler-Benz DB 604 with 24 cylinders in X arrangement, which reached with a cylinder capacity of 46.5 l (10.2 Imp. gal.) 1,835 kW (2,500 hp). Five test engines were manufactured, which were tried on the test bed and a Ju 52 flight test vehicle. On instruction from the RLM, the development was discontinued in 1942.

A further new development of Daimler-Benz was the DB 609 with 16 cylinders on the basis of the DB 603, which was for the first time on the test bed in 1942. The DB 609 had a take-off power of 1,955 kW (2,660 hp). It was taken off the development programme in 1943.

Apart from these test engines, there was still a number of designs, pre-projects and test engines of the higher power class with quite different concepts. For the development work on these engines since 1942 and partially only since 1943 no priority was assigned. The companies have in many cases continued the work with their own resources.

To these projects also belonged the Argus As 413, a water-cooled 24 cylinder H engine which was developed under the direction of *Fritz Goßlau*. For this 2,940 kW (4,000 hp) engine the cylinders and the connecting rod assembly of the Jumo 213E were used. The power was transmitted to two counter-rotating air screws. The As 413 stood shortly before the commencement of test bed runs at the end of the war. At Daimler-Benz work was progressing on a 2,940 kW (4,000 hp) engine with 36 cylinders in W arrangement.

At the "Forschungsinstitut für Kraftfahrwesen und Fahrzeugmotoren" ("Research Institute of Motor Vehicle Technology and Motor Vehicle Engines"), which was managed by *Wunibald Kamm*, the development of group engines [2] consisting of four 12 cylinder Vee engines was undertaken. The Type A, for which the main constructional components of the Hirth HM 512 were used, achieved 1942 on the test bed a power of 1,400 kW (1,900 hp). The type C, with single cylinders of 1.1 l (0.2 Imp. gal.) cylinder capacity, was expected to produce 2,650 kW (3,600 hp).

At Junkers a two-stroke opposed piston Diesel aeroengine was developed, the Jumo 223 with 24 cylinders and four crankshafts and with a power of initially 1,470 kW (2,000 hp), which as so-called fast runner with a crankshaft rotational speed of 4,400 rpm finally produced 1,835 kW (2,500 hp).

Counter-rotating propellers for large aeroengines were developed at VDM and at Junkers and were tried on the test bed on powerful development engines, but they could not any more be flight tested.

In summary it can be established that with the German large engine development in the period of time of World War Two the limits of the piston aeroengine were reached. They showed up regarding the compression, the rotational speed and piston speed, the boost pressure and the size of the single cylinder, but also regarding the number of cylinders and the complexity. With a few exceptions none of these large engines reached the readiness to go into production, however their development brought valuable experience. For these engines a lot of research and development work was expended and many remarkable engineering feats were achieved.

Aeroengines – Altitude Test Beds

KYRILL VON GERSDORFF

The first altitude test beds for aeroengines were established in 1916/17 by the engine companies Maybach at the Wendelstein (Oberbayern) at an altitude of 1,800 m (5,904 ft) and by Benz – as mobile facility – at an altitude of 1,600 m (5,248 ft) at the Schneeberg (Niederösterreich). They served as test beds for altitude engines with increased compression ratio (supercompression) and enlarged cylinder capacity (oversize), which up to an altitude of 2,000 m (6,560 ft) had to be throttled in flight. These mountain test beds, however, were not satisfactory on the one hand because of the low altitude and on the other hand because of the frequent interruptions due to weather effects and difficult accessibility [2].

In the twenties most German engine companies started with the development of supercharged engines and from 1932/33 with altitude engines equipped with multi-gear superchargers. Similar developments also took place in foreign countries. In several countries an acute need for altitude test beds arose. Thus, also at Fiat in Guidonia (Northern Italy) a test bed facility for a complete simulation of the altitude operation was built, which in its performance limit for air-cooled engine was designed for 368 kW (500 hp) up to 4,900 m (16,072 ft) and which was regarded for a long time as model facility [7].

An altitude testing of an aeroengine was in 1927 performed with an altitude research free balloon which was designed by *Hermann Naatz* and built by the "Luftfahrzeug-Gesellschaft" ("Air Vehicle Society"). The idea for this originated at the "Deutsche Versuchsanstalt für Luftfahrt, DVL" in Berlin-Adlershof. Used was a single cylinder test engine with 7.35 kW (10 hp), which had a rotatable bearing. The performance was measured at various altitudes for variable compression conditions taking into account the temperature, moisture and air pressure.

The DVL Facility in Adlershof

After in-depth investigations the board of the "Deutsche Versuchsanstalt für Luftfahrt, DVL" decided in August 1933 to realize an engine altitude test bed in Berlin-Adlershof with which a partial simulation of the altitude conditions took place up to an altitude of 24,000 m (78,720 ft) at the most. The test facility was planned for an air mass flow rate of 1.2 kg/s (2.6 lb/s) with a possible increase to 3.0 kg/s (6.6 lb/s) later [8].

With putting this altitude test bed into operation also a transition from experimental test and trial runs to research investigations took place. The tasks and the layout were described in the DVL Yearbook 1938 as follows:

"The altitude test bed for aeroengines serves the implementation of investigations and test runs especially endurance runs of altitude aeroengines and superchargers for an extensive adaptation of the working conditions to the real operating conditions of altitude flight. The candidate stands in this facility on a normal test bed for liquid-cooled engines or on a blower test stand for air-cooled engines. The braking is achieved by swinging generators. The combustion air is supplied under altitude conditions via a twin duct. The cooling and control system is situated in front. On the exhaust gas side the suction takes place via a cooling tower and via three stages of an exhaust gas compressor with interspaced coolers.

The air for the actual altitude facility enters via a filter and a compressor into the three-stage cooling system with sprinkler cooling (1st stage water, 2nd and 3rd stage cooling brine) and is discharged very dry with a temperature of −30 °C. A further temperature reduction to −60 °C takes place in a cooling turbine in which the expansion causes a further cooling. Via bypass ducts an addition of warmer air is possi-

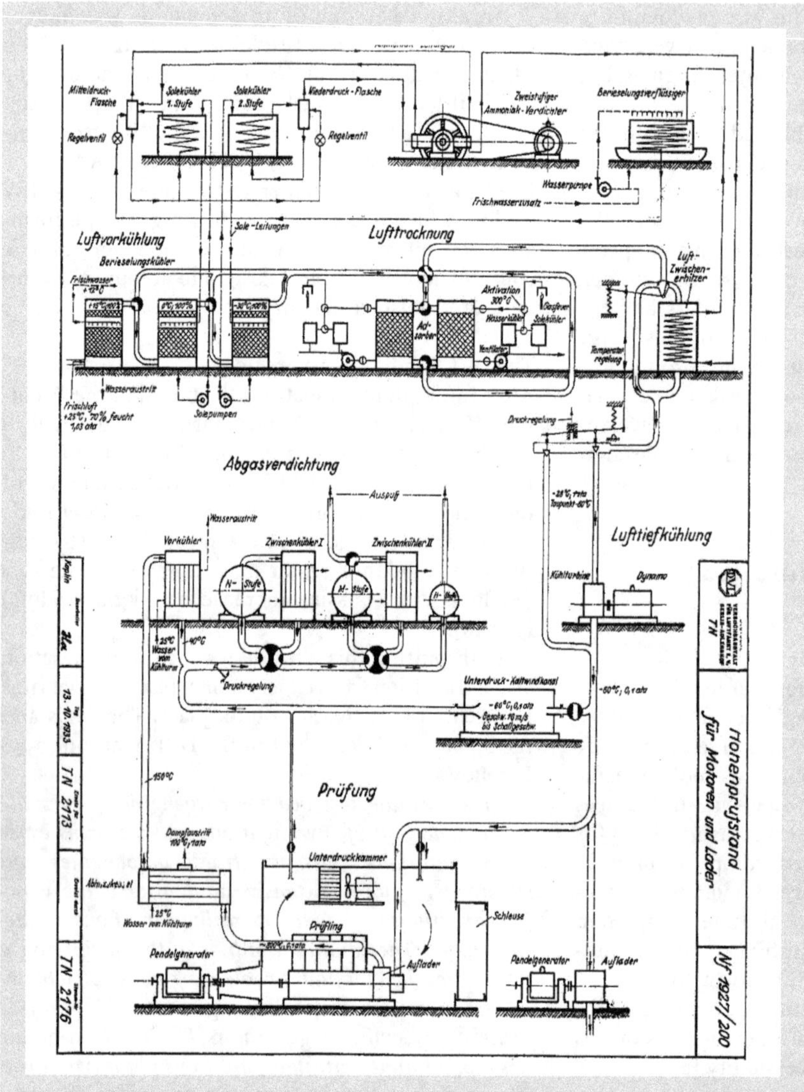

Schematic of the operation principle of the DVL altitude test bed in Adlershof for engines and superchargers (Drawing dated 13.10.1933).

ble in order to maintain in this way the required altitude temperature".

The performance limits of the DVL altitude test bed in Adlershof were 1.2 kg/s (2.6 lb/s) air mass flow rate and 19,000 m (62,320 ft) altitude or 0.35 kg/s (0.77 lb/s) and 24,000 m (78,720 ft) altitude. For normal excess air values the measureable performance limit was 882 kW (1,200 hp) at 19,000 m (62,320 ft) altitude. The safety equipment for explosion pressures consisted among other things of spring-loaded safety valves. For an accurate and quick attainment of the pressure and temperature parameters of the air supplied for combustion to the engine and for the exhaust gas pressure a special altitude control was necessary, which allowed to set the temperature between +55 °C and –30 °C. In this way the effects of the air pressure and the air temperature in front of the engine and the exhaust pressure behind the engine could be investigated separately by adjustment of the automatic control system. An extensive measure-

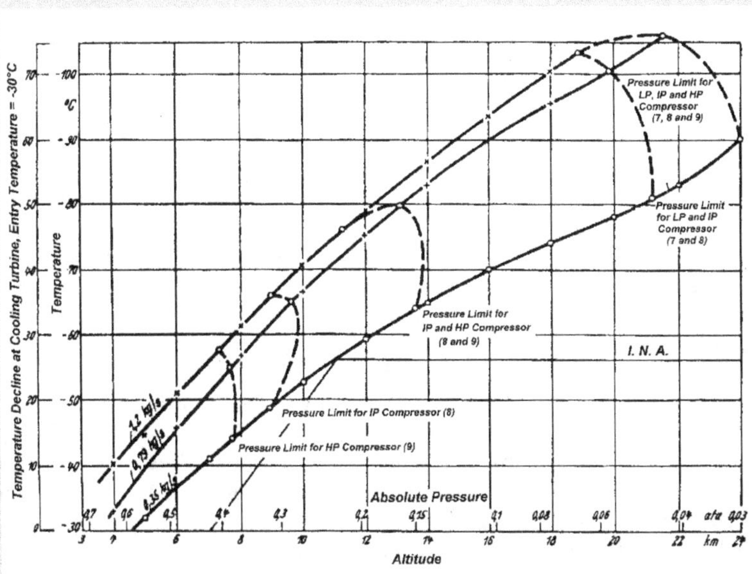

Performance limit of the DVL altitude test bed versus the simulated altitude.

ment system allowed the determination of all important operating parameters. The idea for the use of a cooling turbine for the reduction of the temperature of the combustion air originated from *Asmus Hansen* (DVL) who also participated decisively in the development of further altitude test beds. Main contractor was the company Brown, Boveri & Cie (BBC) which also possessed the right to offer this DVL standard facility to the testing sites (for example Rechlin) and the engine companies. Thus, also such a facility with 1.8 kg/s (4.0 lb/s) air mass flow rate was built at BMW in Munich-Milbertshofen and put into operation there at the end of the thirties. The DVL altitude test bed facility remained in operation for nearly eight years and proved itself very worthwhile for the research of the engine behaviour under altitude conditions. Already two years after being put into operation 1000 operating hours were recorded. The facility was available to four institutes of the DVL Engine Directorate (state 1937) for research work: the Institutes of Engine Layout, Engine Working Processes and Thermodynamics, Engine Mechanics and Fuel and Lubricant Research. With the DVL engine test beds and especially with the altitude test bed facility a number of personalities are connected, among others: *Karl Leist, Werner von der Nüll, Eugen Knörnschild, Fritz A.F. Schmidt, Karl Lürenbaum, Oskar Kurtz, Hermann Oestrich, Kurt Löhner, Helmut Caroselli and Alexander von Philippovich* [9].

In practice the test bed results were of considerable importance for the behaviour of aeroengines under altitude conditions. They led within the framework of research among others to calculation methods with whom from performance measurements at ground level conditions the altitude performance of aeroengines could be determined. Thus, it was possible to calculate with satisfactory accuracy the altitude performance and consumption curves without

Building of the former DVL altitude test bed in Adlershof. The photograph was made in 1994.

taking measurements of the engine on the altitude test bed.

The BBC Facility at the Testing Site Rechlin [10]
At the testing site of the Air Force in Rechlin (Müritz) a first aeroengine test bed was put into operation in the year 1934, which allowed the testing of aeroengines under ground level conditions up to a power of 735 kW (1,000 hp). Soon, however, the wish and the requirements arose to test military equipment of aviation technology under winter climate and altitude conditions.

About the same time, when at the DVL in Adlershof the aeroengine altitude test bed for research purposes was planned and built, in Rechlin a similar facility was established, in which, however, a different – in those days novel – system of refrigeration generation was used. The initiative for this came from *Helmut Sachse*, in those days head of the engine development in the Technical Department of the RLM. Planned was the erection of several facilities for the German aeroengine industry in three sizes characterized by an air mass flow rate of 0.6, 1.2 and 2.4 kg/s (1.3, 2.6 and 5.3 lb/s) referred to the suction air quantity for the engine to be tested. The development and supply of these facilities was assigned to the company BBC in Mannheim by the RLM. The facility intended for Rechlin was designed for an air mass flow rate of 1.2 kg/s (2.6 lb/s) for normal operation and up to 1.8 kg/s (4.0 lb/s) for the overload point as well as for the operation up to an altitude of 18,000 m (59,040 ft). The temperature of the sucked in combustion air could be controlled between +15 °C and −60 °C and the pressure between 1.1 and 0.1 bar (16.5 and 1.5 lb/in^2). On the exhaust gas side the pressure was required to correspond to that of the altitude. Different from the facility in Adlershof, the pre-compressed air was expanded in a cooling turbine and then sucked in from the engine under test. The temperature then adjusted itself – thermodynamically related – independently. Since for the strong cooling the precipitation of snow was to be expected, initially a large vessel as snow precipitator was installed. However, only a silica gel air drier could solve the problem. Considerable difficulties arose several times with the control system for whose electrical part the company Siemens & Halske was responsible. Only a new pressure controller led to a stable operation. The BBC facility was installed since 1936 in Rechlin in Building 405 and used till 1945.

Apart from the large altitude test bed for aeroengines, there existed in Rechlin since 1939 also a large refrigeration chamber of 6 m (19,7 ft) length, 4 m (13.1 ft) width and 4 m (13.1 ft) height into which a fighter aircraft could be moved and the engine put into operation. In this refrigeration chamber the cold starting method by petrol addition to the lubricating oil – in the Air Force known as "Rechlin cold starting method" – was tested experimentally [10].

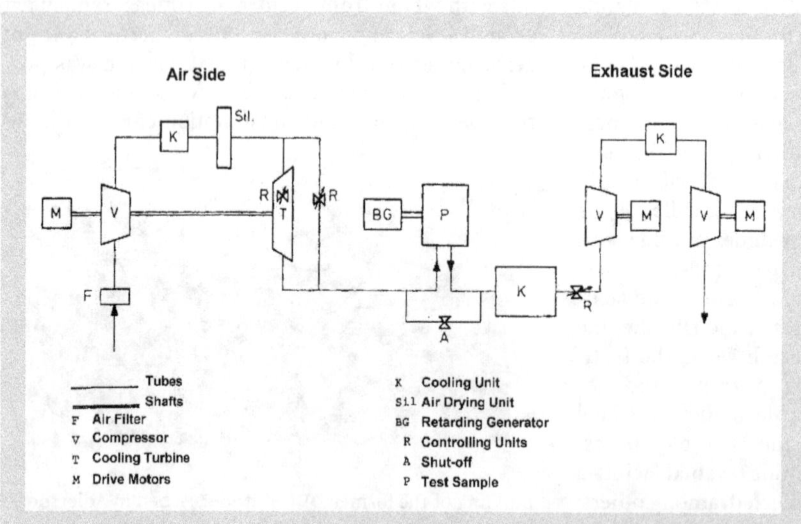

Schematic of the BBC aeroengine test bed which was put into operation in Rechlin in 1938.

The FKFS Altitude Test Bed

At the "Forschungsinstitut für Kraftfahrwesen und Fahrzeugmotoren, FKFS" in Stuttgart-Untertürkheim, called in brief "Kamm-Institut" after the long standing head *Wunibald Kamm*, an altitude test bed for aeroengines was built at the end of the thirties, which allowed the investigation into engines up to an air mass flow rate of 1.8 kg/s (4.0 lb/s) and up to altitudes of 12,000 m (39,360 ft) [11]. The power was absorbed by a swinging generator of 1,100 kW (1,497 hp), which was later supplemented by a water brake of 1,000 kW (1,361 hp). A fan of 600 kW (616 hp) was available for the operation of air-cooled aeroengines. This facility was part of an extension of the FKFS institute and was related to the development of multi cylinder air-cooled large aeroengines of the 2,940 kW (4,000 hp) class.

The Herbitus Facility in Munich

In coordination with the engine experts of the Testsite Rechlin and the DVL Adlershof, already 1941 the awareness grew that for the new altitude engines one could not do without an altitude chamber and thus a complete simulation of the ambient altitude conditions. Consequently the "Reichsluftfahrtministerium, RLM" placed an order with the "Bayerische Motorenwerke, BMW" for the planning and establishing of a pilot facility for 20 kg/s (44 lb/s) combustion air mass flow rate with altitude chamber. With the supply of the engineering facility the company BBC was engaged.

Cause for the new altitude test bed with complete altitude simulation was the BMW 801 J, an air-cooled twin-row radial engine of the 1,470 kW (2,000 hp) class with exhaust-driven turbo-supercharger – as powerplant – and with a full pressure altitude of 1,2000 m (39,360 ft). The development of this altitude engine – initially meant as a study – had started in 1937 under the leadership of *Kurt Löhner* and *Alfred Müller-Berner* and had advanced in 1943 to the point where a quantity production was possible. The engine was, however, not included in the production programme. To a considerable extent difficulties had arisen in flight at high altitudes with the flight test vehicle Junkers Ju 88 D, particularly regarding the cooling of the cylinders and the ventilation of the space behind the engine. The BMW 801 TJ only went towards the end of the war (1944) in a small quantity into production for the Junkers Ju 388.

As soon as the development of the German turbojet engines – BMW 003 and Jumo 004 – had advanced far enough so that an early installation in fighter and reconnaissance aircraft could be expected, the RLM ordered in 1943 to modify the facility so that also the BMW, Jumo and Heinkel-Hirth jet engines could be tested in it. Extensive modifications were, however, not necessary, since the design air mass flow rate of 20 kg/s (44 lb/s) was sufficient. Because of the tight deadlines specified, the proven operating principle of the DVL altitude test bed facility was adopted with a corresponding enlargement of constructional elements [12].

At the BMW facilities in Munich-Milbertshofen, on whose premises the altitude test bed facility – with the cover name "Herbitus" – was built, there still were, apart from the design, ordering of machines and building site supervision, a number of development activities to be performed for which partly little experience was available, among other things for the pressure and temperature control system, the thrust measuring system and the excess pressure safety devices.

The mode of operation of the Herbitus facility can be explained with reference to the schematic: The sucked in air is compressed in a centrifugal compressor up to a pressure of 2.5 bar (37.5 lb/in^2) and is cooled to –24 °C via two sprinkler coolers – the first one uses water and the second one a cooling mixture (cooling brine). The further cooling takes place in a cooling turbine down to –70 °C. Via a different duct pressurized air at +55 °C is routed to the regulating valves. By changing of the mixture ratio the temperature in the altitude chamber can be controlled in the range of +55 °C to –70 °C. With valve arrays the kinetic pressure, that is the speed of the entering air, can be regulated. Two centrifugal compressors, acting as exhaust gas fans and having different suction volumes but the same pressure ratio, produce, depending on being switched in parallel or in series, the required pressure in the altitude chamber with a controlled throttle valve at the fan inlet keeping the pressure constant.

For the preparation of equipping the altitude chamber and the installation of the measuring equipment, a 1 : 1 scale dummy of the altitude chamber was built from wood. The altitude chamber had the form of a horizontal cylinder of 3.8 m (12.5 ft) diameter and 8 m (26.2 ft) length. The thrust measuring trolley, on which the candidate was mounted outside of the alti-

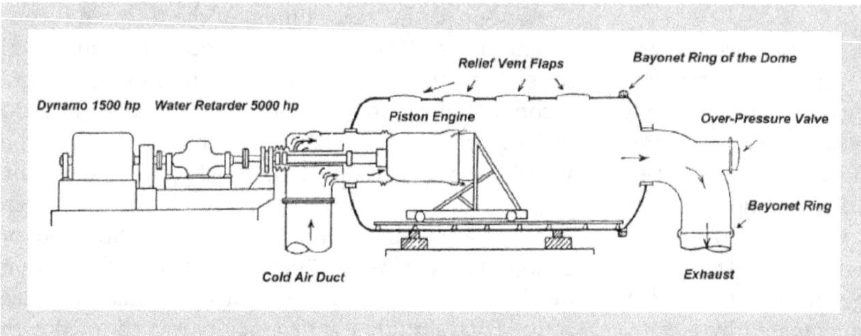

The Herbitus facility in Munich with the layout of the altitude chamber with brake system for the investigation of piston aeroengines (1944).

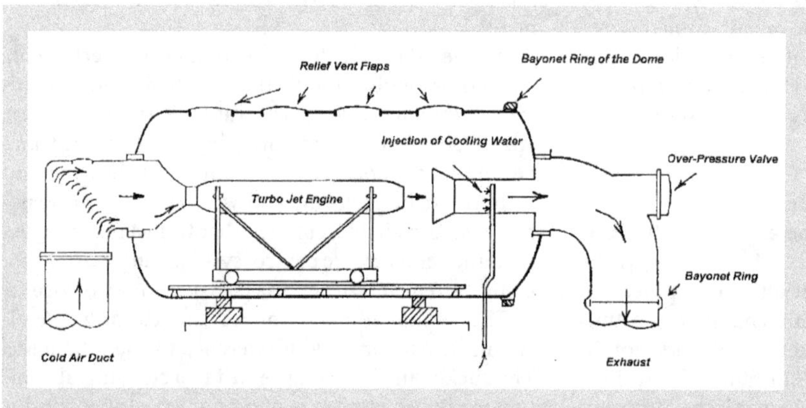

The Herbitus facility with the layout of the altitude chamber for turbo-jet engines.
The thrust measuring trolley, on which the engine was mounted, could be wheeled out after opening the dome-shaped access (right hand end of chamber).

tude chamber, could be wheeled into the chamber after opening the hinged dome-shaped access with integrated exhaust gas duct. The dome-shaped access could be attached to the altitude chamber with a sealed bayonet socket. Since no experience existed with explosions and detonations for such a large chamber, the chamber was designed for 10 bar (150 lb/in^2) maximum pressure and a number of large openings were closed with light metal foils.

The Herbitus facility, originally designed for the testing of piston aeroengines, could in the middle of the year 1944 commence with the test operation of a BMW 003 jet engine. Because of the very high energy requirement of a maximum of 30,000 kW (40,816 hp), the operation had to be moved to the night hours. After elimination of some teething troubles, then night after night measurements on BMW 003 and Jumo 004 jet engines were performed up to speeds of 900 km/h (539 mph) and altitudes of 13,000 m (42,640 ft).

Mainly measured were the engine thrust and the fuel consumption in dependence on the altitude and the flight speed. Thus, complete thrust-consumption charts were obtained, which represented a valuable and exact basis for the performance calculations of jet propelled aircraft. In addition the flame-out limit of the engines could be determined, thus enabling the engine development to undertake improvements on the combustor and the injector. On the other hand the pilots received exact instructions for the rotational speed and the flight speed respectively which were not to be allowed to fall below a certain lower limit. The in-flight re-lighting behaviour, that is the engine re-starting in flight, could also be investigated and explored. This knowledge was for the Me 262 aircrews of the fighter units of vital importance during the last months of the war, for the losses after a flame-out of the combustion chamber, that is without enemy action, were temporarily very high.

The operating limits of the Herbitus facility were determined by the exhaust gas compressors, however they were satisfactory for the tests performed. An exhaust gas mass flow rate of 24 kg/s (53 lb/s) could be sucked off up to an altitude of 8,000 m (26,240 ft)

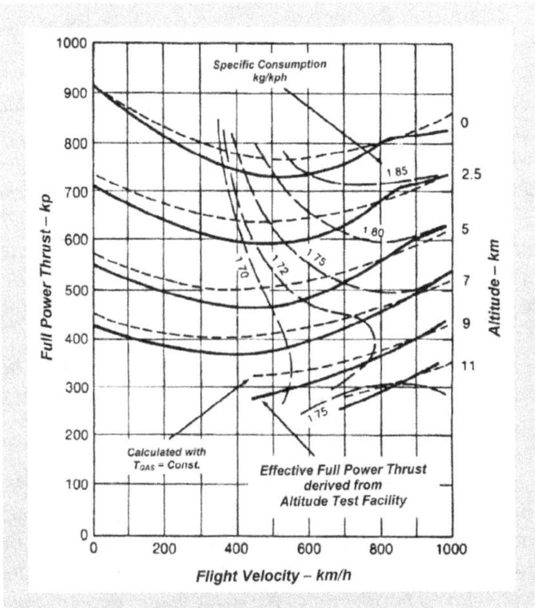

Measurement chart of the turbo-jet engine Jumo 004 for full throttle thrust and specific fuel consumption in dependence on flight speed and altitude. Shown are also the beforehand predicted thrust values.

and 12 kg/s (26.5 lb/s) still up to 11,000 m (36,080 ft). It was possible to simulate also high flight speeds, the measurable maximum thrust was 9.8 kN (2,205 lbf).

The Americans Take Over the Herbitus Facility

In May 1945 American troops occupied the BMW works in Munich and took over the Herbitus facility. Two weeks later several groups of surprised experts of the US Air Force and the US engine industry arrived for inspection, questioning and stocktaking. In the course of this it was concluded that the Herbitus facility was the first altitude test facility for jet engines and that the Allies had beforehand no knowledge of the facility. The interest of the American side was very high. It was decided to dismantle the machine facility and to reconstruct it in a development centre in the USA. From the British side the condition about that was that, before the dismantling, English turbo-jet engines were to be tested in order to resolve critical problems that had cropped up in flight, especially during re-starting. A part of the BMW Herbitus team was gathered and conducted in autumn 1945 together with English experts measurements on a Rolls-Royce "Derwent" under altitude conditions.

The manager of the Herbitus facility *Christoph K. Soestmeyer* was requested by the US Air Force to

The Herbitus facility was shipped to the USA in 1946 and is still today in operation at the Air Force Arnold Engineering Development Centre in Tullahoma/Tennessee for testing American military engines.

investigate the reconstruction of the facility including at the same time an increase in the capacity and to make relevant proposals. Mid 1946 the dismantling, packing and shipment to the USA started. The enlarged facility was erected in Tullahoma/Tennessee. There *Christoph Soestmeyer* worked for several years as "Manager of Operations" of the ETF (Engine Test Facility) at the Arnold Engineering Development Centre. It came later to a considerably larger altitude test bed facility – based on the same operating principle – which was put into operation in 1985 [13].

The Beginnings of Jet Engines

Horst Prem

In the logical development of the test beds there was a smooth transition from testing the piston engine to testing the jet engine. This transition, however, did not take place so smoothly and must therefore be examined more closely.

The piston aeroengines with direct injection, turbo-supercharger and double spark plugs had as early as the mid thirties reached limits, which was especially apparent in the fuel consumption. The BMW IVa engine at the beginning of the twenties already had a minimum fuel consumption which was only insignificantly lowered by the best engines in the mid forties. Only the Junkers Diesel engines did exhibit compared with the Otto engines further consumption reductions of 15 per cent, which, however, had remained since the beginning of the thirties at a specific consumption of 0.160 kg/hph (0.353 lb/hph).

These relations are presented in the chart below. In order to keep this easily comprehensible, only engines are included which characterize the upper limit of the feasibility at that time.

If higher flight speeds were to be attained, even with the unpleasant side-effect of higher consumptions, then in the engine field something new had to take place in the mid thirties. This trend is deducible from the typical S-shaped performance curve for piston aeroengines whose point of inflexion lies at about 1935.

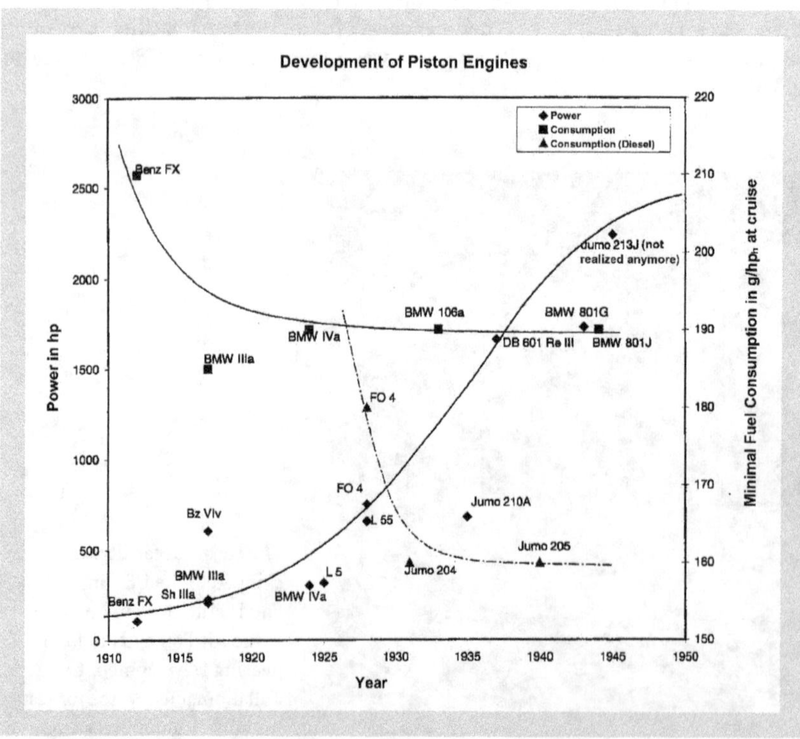

Piston engine development, performance and minimum specific fuel consumption versus time. Clearly discernible are the minimum specific fuel consumptions of the Junkers Diesel engines.

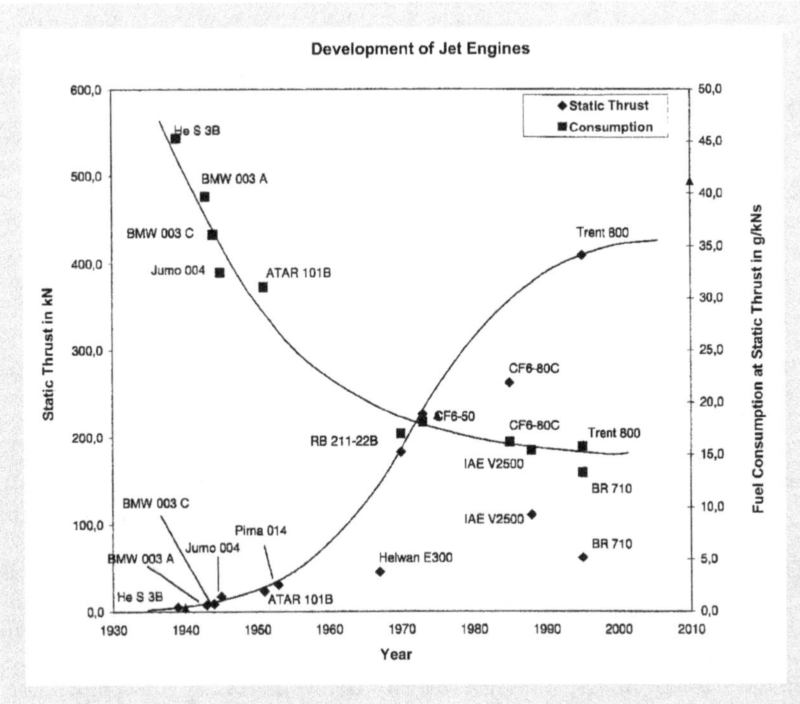

Jet engines, take-off thrust and specific fuel consumptions at take-off thrust. Since the beginning of the seventies with the introduction of large fan engines only small improvements of the specific fuel consumption are seen, since the mid eighties only insignificant improvements.

Nearly at the same time of the point of inflexion for piston engine power the investigations of Herbert Wagner and Hans-Joachim Pabst von Ohain commenced into the jet engines for which the intermittent combustion of the piston engines was replaced by a continuous one, admittedly with a higher consumption. For the jet engines it is also possible to draw the S-shaped performance curve up to the present time. Here too it is discernible that since the mid eighties the specific consumptions do only insignificantly improve even for engines growing in size. With further increasing bypass ratio, by which we approach the conditions of propeller engines, it is perhaps possible to reduce the consumption of the engine alone. The growing diameters of the engine nacelles, however, increase the drag because of the larger air-washed surface and interference problems so the complete system wing/engine leads to higher consumption due to the larger total drag.

Must therefore in the wing/engine field a new innovation step take place because the optimization of the engine as aircraft sub-component no longer suffices for the total optimum of the complete aircraft?

Axial Flow Compressors

Heinz Hoheisel

The Compressor Problem and the First Axial Flow Compressors

New powerplants, as was already outlined by *Junkers* in his patent for an ejector in 1912, were intended to basically revolutionise the propulsion scenario. A continuously sucked in and accelerated airflow was to supersede the piston engines.

At the beginning of the jet propulsion development in Germany a fundamental decision had to be made regarding the type of machinery to be selected with which the thermodynamic cycle of a gas turbine, namely compression, heat addition and expansion, could be best achieved. Although long before the appearance of the piston engine the Englishman *John Barber* took out a patent for a gas turbine (1791), the piston engine was applied earlier in praxis. The success of the gas turbine was modest [14]. It was only with the knowledge of aerodynamics and the materials research that, beginning in the first decades of the 20th century, useable results for the realisation of the gas turbine appeared. The really great success of the gas turbine came with proposals for their application as powerplant for aircraft. Related first patents show different concepts regarding the type of machinery in which the compression of the air was to take place. One of the first patents on an axial flow compressor was obtained by *Sir Charles Parsons* in 1901. The Frenchman *Maxime Guillaume* showed in his patent of 1921 a jet engine with a multi-stage axial flow compressor. The Englishman *Frank Whittle* was granted in 1930 a patent for a turbo-jet engine with axial front stages and radial rear stage of the compressor. In Germany *Hermann Oestrich* described in a DVL-report 1931 the prospects of the turbo jet engine for aircraft [15]. *Herbert Wagner*, Professor at the Institute of Technology in Berlin, was in 1934 the first in Germany to develop a concept for a purely axial flow jet engine as powerplant for a propeller. On August 14, 1938 on his behalf a German patent was granted for a propulsive unit for air vehicles, which describes today's configuration of jet engines. Independently, during the same period of time in Göttingen the physicist *Hans-Joachim Pabst von Ohain* started to realize a gas turbine as aeroengine. With the concept of the application of a radial flow compressor and a radial flow turbine the worldwide first flight with a gas turbine (He S 3B) succeeded on August 27, 1939 with the aircraft He 178 built by the company Heinkel.

The decision regarding the most effective type of machinery for an aircraft gas turbine was influenced by the Flugbaumeister *Helmut Schelp* working at the "Reichsluftfahrtministerium, RLM" ("German Air Ministry"). From there, namely, in 1939 the request was made to industry to develop jet engines with a thrust of 2,000 kp/m^2 (19.6 kN/m^2, 410 lbf/sqft) frontal area [16]. This requirement was necessary in order to keep the drag of the engine nacelle as low as possible. Such value as guideline could hardly be achieved using radial flow compressors. This is comprehensible from the mode of operation of turbomachinery (compressors and turbines) which, based on the momentum theorem of physics, work according to the principle of velocity changes. For the flow over the principal components of such machinery, i. e., the blades being mounted on a rotor (because of their cascade-like arrangement also known as blade cascades), what matters, is in which main flow direction – axial or radial – the necessary velocity changes are achieved.

For all turbomachinery one differentiates between relative velocities (w), circumferential velocities (u), and absolute velocities (c). The relative velocity is in this context the velocity relative to the moving coordinate system, that is the rotor. The absolute velocity is referred to a fixed coordinate system. In a radial compressor stage, for the flow path from entry to exit, all three components u, w and c are involved in producing the velocity change necessary for compression. For the axial flow, the circumferential velocity is the same in the streamtube in front and behind the rotor, yielding here a pressure increase which is lower than for a radial rotor. If with an axial rotor a high pressure increase is to be achieved, a multitude of compressor stages (consisting of rotor and stator) must be arranged one behind the other, leading to a multi-stage axial compressor. Compared with the axial flow compressor, the radial flow compressor has the disadvantage to possess a smaller air intake area for the same frontal area. Because of this, the achievable

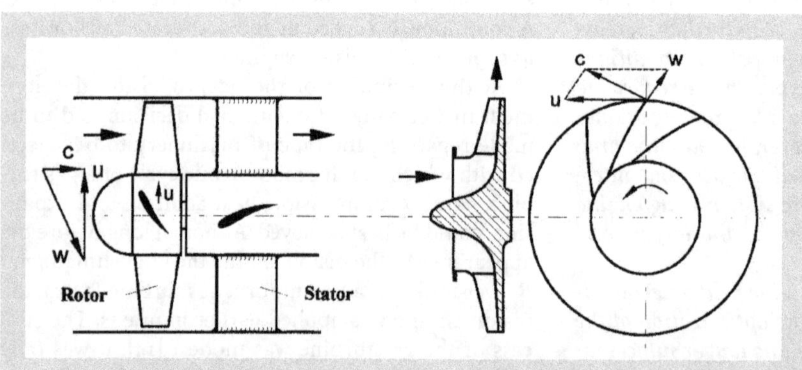

Flow through an axial flow compressor stage (left hand) and through a radial flow compressor stage (right hand) including the velocity triangles with the components. *u* circumferential velocity, *c* absolute velocity, *w* relative velocity, → path of a fluid particle.

sucked-in air mass is for the radial flow compressor smaller than for an axial flow compressor. At the time of the start of the development of jet engines, the fact that the aerodynamics of the radial flow compressor was easier to control than the still largely unexplored aerodynamics of the axial flow compressor was in favour of the application of the radial flow compressor [17]. Although the knowledge regarding the aerodynamics of the axial flow compressor – in particular of the multi-stage one – was still very modest, nevertheless the correct decision was made and engine developments with axial flow compressors were initiated [18]. Considering the projects originated in their chronological sequence, each engine is found to be marked with special features:

- For the design of the 14-stage axial flow compressor of the *Wagner/Müller* engine, the rotor and stator blades were equally employed for producing the pressure rise, a principle still to be addressed.
- The engine with counter-rotating compressor designed by *Hellmut Weinreich* and tested on a test bed, an engine design consisting only of rotors, led to a small length of the compressor. This was a significant aspect. Because of the approach of the speed of sound, one was restricted regarding the circumferential velocity u. The concept of counter-rotation was not successful with reference to engine compressors, particularly because of considerable mechanical and constructional problems. For single stage fans with a low pressure rise counter-rotation may be of advantage.
- The development of the axial flow compressors for the BMW 003 and Junkers 004 engines was started already according to the requirements of the RLM. In this context a statement of *Helmut Schelp* is of significance, saying:
  "Because in 1939 the development difficulties could not be foreseen, it was decided to assign as task a parallel development. Junkers was to work on a more conventional basis and BMW on a more advanced one with better technology [18]".
  For BMW this meant a compressor with seven stages, for Junkers one with eight stages for the same pressure rise. Both these compressors will be addressed later on.
- Since the RLM had initiated promising development projects for turbo-jets and with a thrust of 8–9 kN (1,800–2,025 lbf), Daimler-Benz was given the task to develop an engine with higher thrust and a lower fuel consumption. The engine proposed under the direction of *Karl Leist* and tested 1943 on a test bed was one of the first bypass engines of the world [18]. For the required pressure ratio of eight, based on proposals made by *Walter Encke,* a compressor was designed consisting of two counter-rotating drums. A parallel development of the compressor was also undertaken and the compressor constructed by the company Voith in Heidenheim.
- The development of the engine He S 30 under the direction of *Max Adolf Müller* was the first engine of the company Heinkel with purely axial flow components: compressor, combustion chamber and turbine. The compressor, based on proposals by *Rudolph Friedrich* [18], reached a circumferential velocity of 313 m/s (1,027 ft/s) and a notable high efficiency of 87 per cent. The design of the compressor was based on the same considerations as for the *Wagner* engine, namely pressure transformation in the rotors and stators (degree of reaction 50 per cent). The He S 30 engine was thus superior at that time to all other designs regarding the important parameters for an aircraft like frontal area, weight and specific fuel consumption. By order of the RLM this engine development was however stopped at a time when a Me 262 with Jumo 004 engines took-off on its first flight. Similar decisions of the RLM did also affect other developments.

The engines with axial flow compressors addressed here were followed by further ones with higher thrust. Apart from turbo-jet engines (TLs), developments were under way of turbo-propeller engines (PTLs) with axial flow compressors (BMW 028, Jumo 022). Regarding the axial flow compressor, similar peculiarities and problems were experienced as with the turbo-jet engines.

Details of the Compressor Development

The final decision to select an axial flow compressor for the engine concepts BMW 003 and Junkers 004 was strongly supported when towards the end of 1938 the "Aerodynamische Versuchsanstalt Göttingen, AVA" presented research results on multi-stage axial flow compressors. In the case of Junkers there existed an extensive experience in the cooperation with the AVA regarding the development of four-stage axial

Chronological order of development of turbo-jet engines with axial flow compressors.

Item No.	Description Development head Company/Place	Year of design	Thrust [kN/lbf] Frontal area thrust [kN/m²] (lbf/sqft)	Remarks
1.	RTO *H. Wagner/M.A. Müller* Werkzeugmaschinenfabrik Magdeburg (later Junkers)	1936		Engine with 14-stage axial flow compressor based on proposals by *Herbert Wagner* (TH-Berlin). 1939 test-bed testing, after that development stop. *M.A. Müller* moves with some co-workers to Heinkel.
2.	P 3304 *H. Weinrich* Chemnitz	1938	6.0/1,350	Engine with counter-rotating axial flow compressor. 5 stages on external drum, 4 stages on internal shaft. Test-bed testing, 1942 development stop. Experience used by BMW.
3.	BMW F9225 *K. Löhner/Müller-Berner* BMW Munich	1938	6.0/1,350	Engine project with 7-stage axial flow compressor. Concentration of all development activities under *Oestrich* in Berlin-Spandau.
4.	BMW 003 (P3302) *H. Oestrich* Bramo/BMW Berlin-Spandau	1939	7.8/1,755 [21.8] (454)	Production standard engine with 7-stage axial flow compressor. Blades based on AVA profile data. Production of 450 units in Germany, about 1,200 units in the USSR. Licence 1944 to Japan.
5.	Jumo 004 *A. Franz* Junkers, Dessau	1939	9.0/2,025 [19.9] (415)	Production standard engine with 8-stage axial flow compressor. Blades based on AVA profile data. Production of 6,000 units in Germany, about 800 units in the USSR.
6.	DB 007 *K. Leist* Daimler-Benz Stuttgart	1941	11.3/2,543 [25.7] (535)	First bypass engine with counter-rotating compressor, based on proposals by AVA, consisting of two drums with 9 external and 8 internal stages. Counter-rotating planetary gear. Bypass ratio 2.42. Test-bed testing 1943, after that development stop.
7.	He S 30 *M.A. Müller* Heinkel/Hirth Stuttgart	1941	8.5/1,913 [30.8] (642)	First completely axial Heinkel engine with 5-stage compressor. Pressure ratio 3. Test-bed testing. High technology development stage.
8.	Jumo 012 *A. Franz* Junkers, Dessau	1944	30/6,750 [33.4] (696)	Project with 11-stage compressor. Test-bed testing with 12 stages in the USSR with several endurance runs. 8 development engines.
9.	BMW 018 *H. Oestrich* BMW Berlin-Spandau	1944	33/7,425 [29.1] (606)	Project with 12-stage compressor derived from PTL 028 omitting the propeller drive. Test-bed testing in the USSR.

flow compressors for superchargers of piston engines. The knowledge of the AVA could be traced back already to the twenties when fans for wind tunnels were required [19]. But also in the ventilation technology AVA fans already found application such as the so-called Siemens-Betz axial fan which was produced in large numbers. Regarding the engine concept decisions therefore the following remarks by *Albert Betz* at a scientific meeting of the "Akademie der Luftfahrtforschung" in January 1938 were of significance. He stated: "*We have gained the impression from the supercharger developments that one can obtain quite good efficiencies even if one would use the wheels for large pressures. One should only arrange more wheels after another and increase the speed of rotation as much as the speed of sound allows* [20]."

This confidence arose from research results of the AVA which had been obtained by *Walter Encke* and his co-workers [21].

From systematic investigations of various single-stage model wheels measured performance lines provided data, allowing the design of multi-stage axial

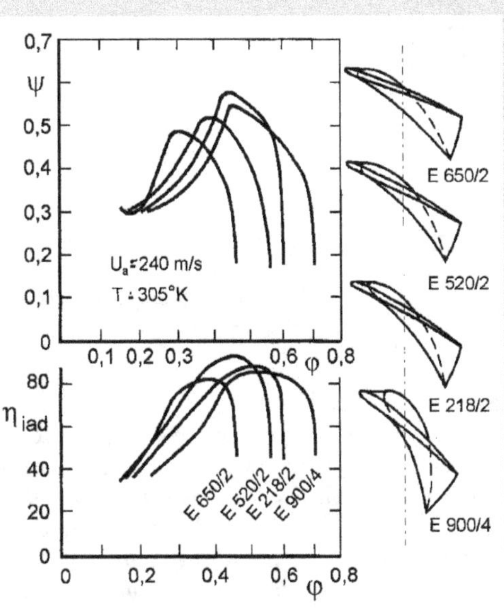

Performance lines of various axial blade wheels investigated at AVA. Stage loading ψ and efficiency η versus the flow coefficient φ (mass flow rate).

flow compressors taking the similarity laws [22] into account. Each stage of such a compressor was to work in its normal operating range in the vicinity of the model wheel design point. Care was also taken that the normal operating point of the compressor did not come to lie too close to the stall limit (analogue to the aerofoil problem). This is defined in the normal case by the individual performance line at maximum pressure rise (stage loading ψ_{max}). In the design of a compressor the direction of the approaching flow to the first rotor plays an important role. By placing a stator in front of the rotor (so-called inlet guide vanes), the air can be influenced in the sense of the direction of rotation of the rotor (co-swirl) and can thus allow quite high circumferential velocities. The other possibility, to make the flow leave the inlet guide vanes in the opposite direction to the direction of rotation (contra-swirl), leads to a strong increase of the approaching flow velocity w_1 to the rotor. With this the danger arises to reach the speed of sound very early at still very low circumferential velocities, which was at that time not desirable owing to a lack of knowledge regarding transonic aerodynamic phenomena. In Göttingen in those days most of the designs were investigated with co-swirl. For the determination of the design concept of an axial compressor stage, a further point of view was applied, namely that the pressure transformation is in general more favourable in a rotor (influence of the centrifugal forces on the blade boundary layer) than in a stator. These points of view led in Göttingen for the design of an individual stage to a degree of reaction $r = 1.0$ with a constant axial velocity c_m over the radius as well as a constant total pressure rise. Use was made of a swirl distribution across the radius corresponding to the so-called potential flow vortex law $c \cdot r$ = constant (free vortex design). This law is known in fluid dynamics as hollow vortex and can be observed in nature as the so-called bathtub vortex. It shows in the centre a velocity distribution differing from that of the specified swirl law. For turbomachinery this behaviour is of significance under the keyword "hub vortex flow separation (hub dead water) [23–25]".

The results of the Göttingen axial wheels were obtained with a compressor test bed erected already in 1927. On this test bed, model wheels with an external diameter of 150 or 200 mm (5.9 or 7.9 in) could be tested as single-stage or two-stage versions at a speed of rotation up to 33,000 rpm and variable density.

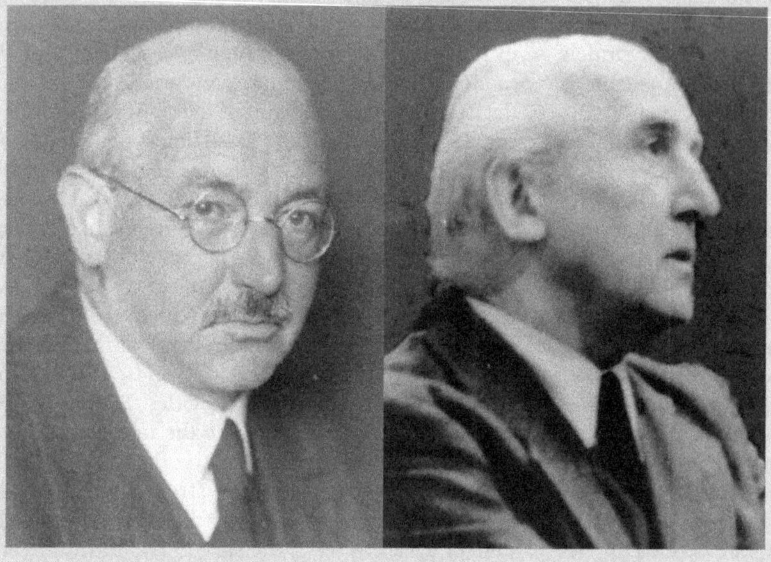

The pioneers of the axial flow compressor development: *Albert Betz* 1885–1968 (left), *Walter Encke* 1888–1982.

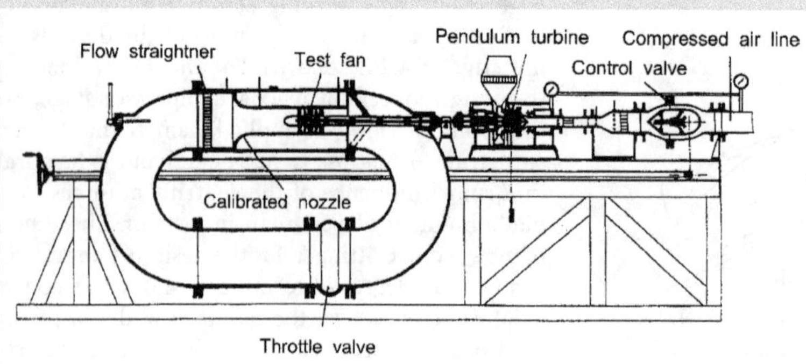

Göttingen axial flow compressor test bed with 70 kW (95.2 hp) transmission power, erected 1927.

The power for driving the compressor wheels was obtained from a pendulum supported two-stage compressed air turbine with a power of 70 kW (95.2 hp). By systematic research it was possible to develop such blade geometries for axial wheels that the efficiency as well as the pressure rise could be steadily improved. For a wheel efficiencies up to 90 per cent and pressure rises up to a factor 1.25 were achieved. For the design of axial flow compressors, which was started at Heinkel, BMW, Junkers and Daimler since about 1938, *Walter Encke* was involved as consultant up to 1945.

The Jumo 004 Compressor

At the time of the start of the Jumo 004 development there existed no possibility to test a multi-stage compressor on a suitable test bed. *Anselm Franz*, responsible for the complete Jumo 004 development, reported: "*The first decision to build an experimental compressor in half size led to failure. The compressor was completely destroyed owing to blade vibration failure.*"

Further testing of the Jumo compressor had to be undertaken on a running engine because of lack of time. The possibility to test an engine compressor on a test bed first arose in 1944 at the Gutehoffungshütte

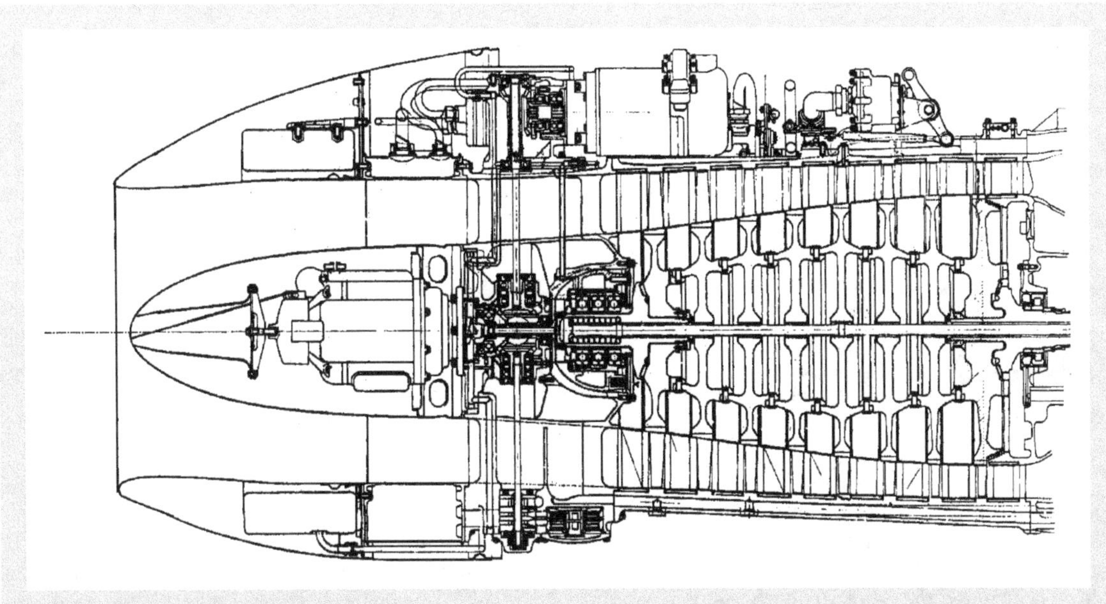

Reduced scale original drawing of the Jumo 004 compressor. Tip diameter of the first stage 544 mm (21.4 in), of the last stage 582 mm (22.9 in). Realized with inlet guide vanes.

(GHH) in Oberhausen and in Dresden at the company Brückner & Kains, here with a shaft horse power of 15,000 PS (15,000 hp) [18]. The high performance of the Dresden test bed powered by a steam turbine allowed tests with free entry to the compressor and throttling at its exit, that is for flow conditions existing in the engine.

Based on designs by *Walter Encke*, the compressor built for the engine Jumo 004 was specified for the following data:

mass flow rate $\dot{m}$	= 21.2 kg/s (46,7 lb/s)
pressure ratio π	= 3.14
rotational speed n	= 8700 rpm
number of stages z	= 8
circumferential velocity u	= 248 m/s (813 ft/s)
efficiency η	= 78 per cent
degree of reaction r	= 1.0.

The external diameter of the compressor stages increases from front to rear and in this way the circumferential velocity grows also. Because of the rise of the speed of sound with $\sqrt{T}$, where T is the static temperature in K, one obtains thereby a rise of the stage work from front to rear at a nearly constant Mach number level.

The design and manufacture of the blades for an axial flow compressor require a high degree of experience and certain machine tools which were not yet available in the companies. Therefore, the blades for the first jet engines were produced in the model workshop of the AVA. The production of the blades was performed by copy milling of very high quality. The manufacturing accuracy of the blades is also responsible for the scatter in the performance data of the manufactured compressors. Apart from the accuracy of manufacture, care must also be taken in the compressor design that no vibrations are induced by support ribs which could lead to fatigue failures. Fatigue failures have often caused large delays in engine development.

The BMW 003 Compressor

The planning of the first experimental engines of the BMW 003 (V1–V10) had originated from research results that were obtained at the AVA. The transfer of such results to the conditions of a jet engine, particularly also the constructional details still gave rise to

Anselm Franz (1900–1994).

special requirements. In this context, the head responsible for the development of the engine 003 *Hermann Oestrich* reported: "*The construction details of the design confronted the designers with many new problems. Ideas stemming from stationary turbine manufacture were few, because there mostly higher weights and volumes were allowed for the same performance. Furthermore, from the start the point of view of an economic mass production had to be taken into account* [18]."

The blades designed jointly with the AVA for the first 003 compressors were to deliver a pressure ratio of 2.9 with six stages. The pressure rise here too took place, like for the Jumo compressor, only in the rotor with r = 1.0 and appropriately designed blades. The stators only changed the flow direction (no velocity change) and were made of sheet metal in the form of circular arc profiles rounded at the leading edge (low manufacturing costs). With this compressor version, the specified pressure ratio could not be realized. Only after adding a further stage, the design pressure ratio could be obtained. However, the engine did not yet produce the desired thrust.

Based on the experience gained, a new compressor design with 30 per cent higher mass flow and seven stages was undertaken. The design calculations were now performed by BMW itself where in the meantime an own computation group under the responsibility of *Hermann Hagen* had been established. For this new design, the AVA design concept was given up and the new compressor designed with a degree of reaction of 70 per cent, that is the stators also contributed to the

Anselm Franz, born in Schladming, Austria on January 21, 1900.

1924 completion of studies of mechanical engineering at the Institute of Technology (TH) Graz, there then assistant at the chair of turbomachinery.

1928 joining the company Schwarzkopf, Berlin, department of hydraulic gears.

1936 start of work for the Junkers engine development, initially on aeroengine superchargers.

1939 taking over the development leadership for a turbojet engine. The Jumo 004 jet engine is created. About 6,000 units are produced. Furthermore, the development engines TL 012 and PTL 022 are designed.

1940 PhD with a thesis on exhaust jet development at the TH Berlin.

1945 change to the USA, initially to the Wright-Patterson Air Force Base in Dayton/Ohio.

1951 head of the gas turbine development at Lycoming in Stratford/Connecticut. In the end Vicepresident of AVCO-Lycoming Division. Under his leadership the concept is created for the successful Lycoming engines T53, T55 on the basis of axial flow compressors with radial final stage and return flow combustion chamber. This configuration characterizes the Lycoming engines as well as the worldwide first high bypass engine PFL 1A. Also originated is the first tank gas turbine built in series for the M-1.

1968 retirement, furthermore active as consultant.

1969 honorary doctor of the Technical University Graz.

1994 he dies in Stratford/Connecticut on November 18.

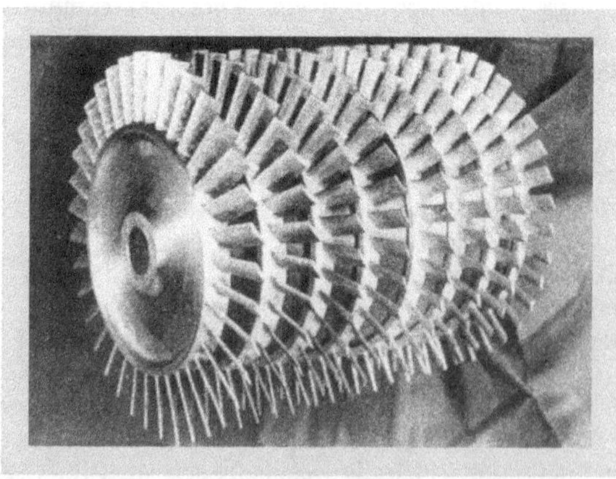

Rotors of the eight-stage axial flow compressor for the engine Jumo 004, produced at the AVA.

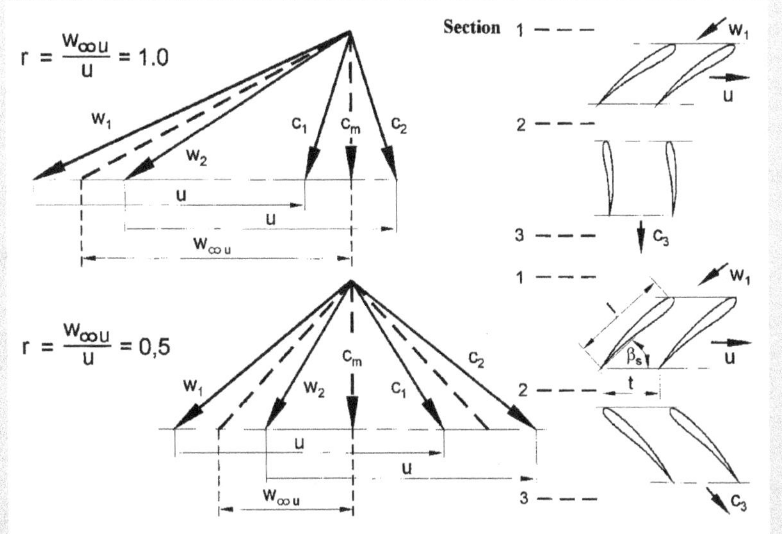

Velocity triangles and blade settings for a compressor stage with 100 per cent reaction (top) and with 50 per cent reaction (bottom).

pressure rise. This method was necessary, because for a jet engine compressor with the lowest possible weight the highest possible stage pressure rise is to be aimed at. A change of the degree of reaction to r < 1.0 reduces the axial thrust of the rotor because part of the blade forces are transferred to the casing via the stators. What, however, becomes more important is the fact that out of consideration for the Mach number influence, for the same entry velocity w_1 a larger circumferential velocity u in the rotor can be allowed, which leads, according to the Euler equation of motion, to a higher pressure rise per stage. This fact can be demonstrated using the velocity triangles. The reaction degree reduction for the 003 compressor from V11 onwards and for all those produced thereafter was the fundamental measure to achieve with only seven stages instead of eight, like the Jumo compressor, a pressure ratio of 3.1. In this context it should be mentioned that already for the Wagner/Müller engine and the Heinkel He S 30 the compressor was designed with 50 per cent degree of reaction and significantly high efficiencies were reached. *Rudolf Friedrich* showed later that with 50 per cent degree of reaction the highest efficiency could be achieved in an axial flow compressor stage [26]. This statement is also valid for turbine stages [27]. A compressor design ordered by BMW at BBC Mannheim was also produced with 50 per cent reaction. The Göttingen design concept of 100 per cent reaction was strongly influenced by aerofoil aerodynamics. To what extent the aerofoil point of view is sensible for compressor blades will be explained later. The new compressor design by BMW with seven stages yielded the following data:

mass flow rate $\dot{m}$	= 18 kg/s (39,7 lb/s)
pressure ratio π	= 3.14
rotational speed n	= 9,500 rpm
efficiency η	= 78.5 per cent max. 80 per cent
circumferential velocity u	= 265 m/s (869 ft/s)
degree of reaction r	= 0.7.

The result is to be regarded as being very favourable if it is considered that the design of the compressor took place without pretests and with a hardly validated blade cascade theory. Subsequently also at BMW systematic research work in the area of axial flow compressors was undertaken. Thus, testing of a compressor on the test bed of GHH showed that the efficiency could still be improved by a small change of the blade angle. In this context it is appropriate to point out the high share of contribution the component development has made to the progress in engine development. For each engine component there was

Hermann Oestrich (1903–1973).

at least one test bed available [28]. With reference to the blades, it remains to be mentioned that the manufacture of blades from magnesium alloy was accomplished by extrusion of each single blade including the root. The manufacturing accuracy was initially not very good and only gradually the design and manufacturing departments could be convinced of the need of a careful production. Blade vibration failures in the first compressor stage, caused by support ribs in the front section of the compressor, have negatively influenced the development time.

Hermann Oestrich, born in Duisburg on December 30, 1903. Studies at the Institute of Technology (TH) Hannover and (TH) Berlin.

1928 joining the DVL engine department.

1929 work on "The prospects of jet propulsion for aircraft taking into consideration the exhaust gas jet propulsion".

1935 joining the Siemens aeroengine works in Berlin-Spandau, which are taken over by BMW in 1941.

1937 PhD with a thesis "Gas exchange processes in altitude engines" at the TH Berlin-Charlottenburg.

1939 head of aero-gas turbine development in Spandau. The experimental engines BMW 018 and BMW 028 are created as well as the production standard engine BMW 003. Of this 450 units are built up to 1945.

1944 the Berlin development works are transferred to Staßfurt. In April 1945 *Oestrich* and parts of his team are moved to Munich by English and American troops.

1946 *Oestrich* goes with about 120 engine specialists and about a dozen aircraft designers to France. The "Atelier Aeronautique de Rickenbach, ATAR" engine is created, of which over 5,000 units are produced up to the end of the sixties.

1948 *Oestrich* becomes a French citizen.

1950 he becomes Technical Director of SNECMA and hands over 1960 this task to his deputy.

1973 he dies in Paris on April 2.

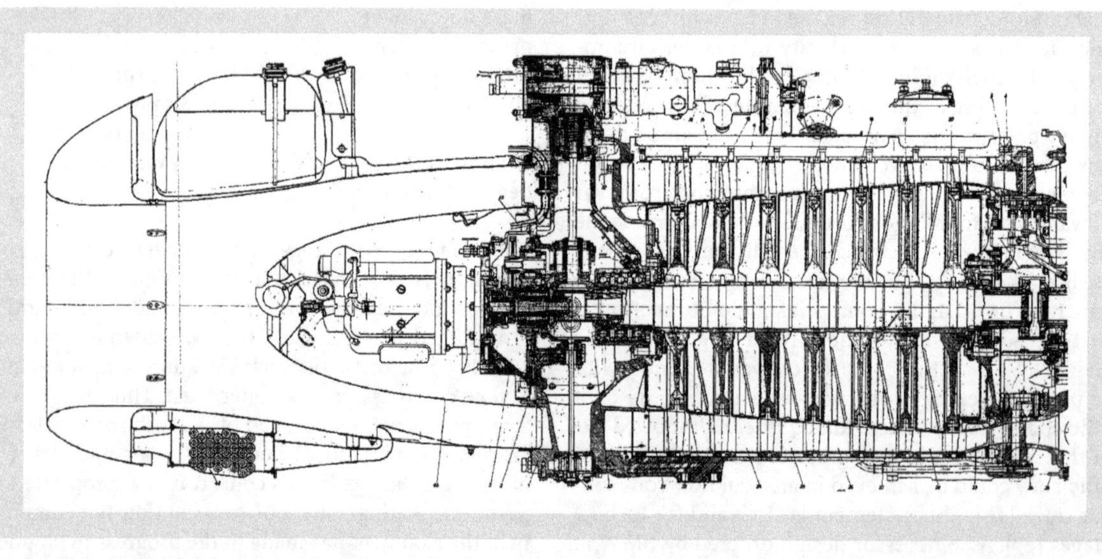

Reduced scale original drawing of the BMW 003 compressor. Realized with constant tip diameter and inlet guide vanes.

The RLM specified requirement for a progressive design of the BMW engine was implemented in many aspects. If one compares the design details of the BMW 003 compressor with those of the Jumo 004 compressor, one notices the massive disc construction of the rotors of the Jumo compressor. The BMW compressor already had a considerably slimmer implementation, which certainly had contributed to the fact that the BMW engine possessed a remarkably low mass (570 kg, 1,257 lb) compared with the Jumo engine (745 kg, 1,643 lb). On the other hand there exists the statement by *Anselm Franz* regarding the Junkers engine:

"*In consideration of the novelty of the object, in order to guarantee as much as possible the greatest chance of success, it was deliberately avoided to aim at the maximum possible*".

The over 6,000 manufactured engines Jumo 004 confirm the farsighted approach to the solution of the task [29]. Further aspects of the axial flow compressor topic that were already of significance for the first jet engines are addressed in the chapter "The Evolution to Modern Jet Engines".

If, however, one compares the technological status at the cessation of the work in Spring 1945 with the status in foreign countries, there existed a considerable lead in this field.

Combustion Chambers

Dieter Rist

From the Hydrogen Test Combustion Chamber to the Diesel Annular Combustion Chamber

The combustion chamber of a jet engine is the component in which the thermal power is supplied in the form of fuel and thrust is gained for the propulsion of the aircraft by combustion. The usable power corresponding to the specified design points of the engines, the turbine entry temperature and the compression pressure ratio, can only be a fraction of the thermal power, which was at the beginning of the jet engine development at relatively small thrusts already quite high. Thus, the combustion power of the first flown jet engine of the world, the Heinkel He S 3B, was at a take-off thrust of 4.4 kN (990 lbf) about 11.9 MW (16,190 hp). For the in quantity produced BMW jet engines 003A and 003C there were 13.2 MW (17,959 hp) and 14.0 MW (19,048 hp) respectively converted into 7.85 kN (1,766 lbf) and 8.83 kN (1,987 lbf) respectively take-off thrust. The Junkers jet engine Jumo 004B had 8.93 kN (2,009 lbf) take-off thrust at about 15.6 MW (21,224 hp) combustion power. By the end of World War Two the completed but not yet operational more powerful engine BMW 018 with 33.35 kN (7,504 lbf) take-off thrust required an even considerably higher combustion power of about 45.5 MW (61,905 hp).

For the engineers and designers thus the task was first of all to extract this large thermal power from the fuel by continuous and as complete as possible combustion in as small and short as possible combustion chambers. In the year 1938 there existed already a stationary gas turbine of the Swiss company BBC (Brown, Boveri and Cie, today in the ABB group) with a core part whose concept with the one behind the other layout for continuous flow through the components: compressor, combustion chamber and turbine, corresponded to the gas generator of the turbojet engines, but the specific combustion space loading characterizing the combustion chamber size was much too small (combustion space loading is the heat quantity per second produced referenced to the flame tube volume and the mean combustion chamber pressure). Even the best thermal power stations at that time had combustion chambers with specific combustion space loadings which were only about 1/30 to 1/25 as high as those that were required for the aeroengines [30]. One of the essential aims of the relevant applied research was therefore at that time to develop compact combustion chambers for high thermal power, whose size and weight represented a considerable technological step. The jet engines of the first generation aimed at maximum values of the specific combustion space loading of about 30–40 MW/($m^3 \cdot$ bar) = 300–400 s^{-1}. Moreover, regarding the operational behaviour further conditions and criteria had to be fulfilled, which represented additional great challenges.

What difficulties the combustion in a relatively small combustion chambers could bring about, *Hans-Joachim Pabst von Ohain* experienced already in the testing of his first engine model in the year 1935 [18]. The combustion did not take place inside the combustion chamber, but inside the centrifugal turbine and behind it. A long yellow flame shot out of the engine exhaust. Based on this experience as well as on results of other researchers and industry engineers, it was recognized that the combustion chamber is a very critical element of a turbo-jet engine. Because of time pressure, one started to proceed with the industrial combustion chamber development parallel to the development of the other turbo-jet engine components.

Combustion Chamber Developments at Heinkel and BMW

In order to quickly build an appropriately functioning test engine as precursor of an airworthy powerplant, *von Ohain* decided in February 1936 to use gaseous hydrogen as energy source instead of liquid fuel. The former has various advantages: The timewise and energetic expenditure of the liquid vaporization does not apply, the calorific value per unit mass is about 2.7 times as high as that of initially normally used aviation petrol, which is why less hydrogen must be injected into the air stream; the mixing of the hydrogen gas with the air is improved because of the considerable smaller density and the flame speed is larger by an order of magnitude than that of the aviation petrol. In addition also a better operational behaviour at engine partial load could be expected. With the knowledge of all these advantages, the test engine He S 2, consisting essentially of the centrifugal turbomachinery and the short hydrogen combustion chamber, was designed, sized, developed, and built by *von Ohain* with important contributions by *Max Hahn* and *Wilhelm Gundermann* at the company Heinkel in Rostock-Marienehe and in March 1937 put into operation

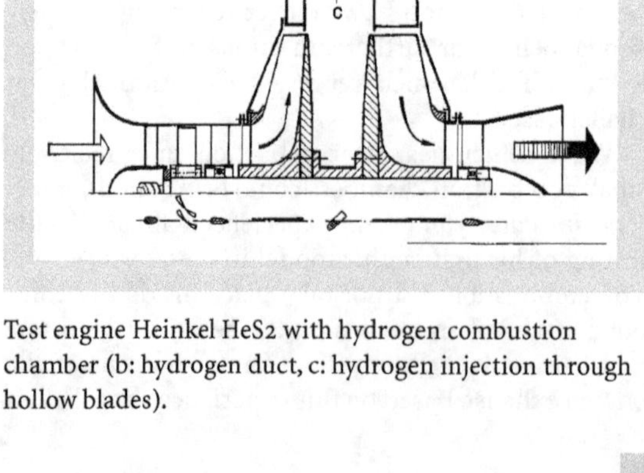

Test engine Heinkel HeS2 with hydrogen combustion chamber (b: hydrogen duct, c: hydrogen injection through hollow blades).

on a test bed. This new combustion chamber consisted of an annular channel with about 60 slender hollow blades as injection elements. The hollow blades had a blunt trailing edge with evenly distributed small orifices through which the compressed hydrogen gas was injected into the turbulent wake of the blades. The test engine functioned from the start and had a very good operating behaviour. This was the result of the excellent hydrogen combustion covering a wide operational range with a very low total pressure loss in the combustion chamber because of the relatively low through flow, Mach number level, and the approx. 400 favourably located hydrogen injection nozzles. Mainly because of the drawbacks of the relatively large space requirement, the required insulation of the tank for the deeply cooled hydrogen and the danger of the handling in the "rough" practice of flight operation, it was obvious that for an airworthy turbo-jet engine initially only liquid fuel would be suitable as possible energy source. At Heinkel the development of an aeroengine for such a fuel commenced in May 1937. At the same time several tasks were tackled, among these also a combustion chamber for liquid fuel. Initially the fuel was brought into the combustion chamber in an already gaseous state and then as free mist. In the course of this work the problems of fuel vaporization were explored intensively. This is part of the fuel preparation consisting of injection and spraying, useful distribution and vaporization, and together with the air mixing, ignition as well as primary reaction an important partial process of the stoichiometric combustion in the primary zone of the combustion chamber. Because the temperature in the primary zone is much too high for the turbine, to the hot gas in the adjacent mixing zone of the combustion chamber more cold air for cooling must be added. All these parallel and successive processes taking place require a certain duration of stay of the flowing gases in the combustion chamber and they determine its length.

In order to fulfil these requirements in the corresponding space between compressor and turbine rotors and yet to keep their spacing short, based on a proposal of *Max Hahn*, a (later so-called) reverse-flow annular combustion chamber was installed. The longitudinal section of the engine He S 3B, which was used on August 27, 1939 for the first jet engine flight, is shown on page 238 with the combustion chamber which determines the diameter. Discernible is

Hans-Joachim Pabst von Ohain (1911–1998).

Hans-Joachim Pabst von Ohain, born in Dessau, 1911.
1934 first theoretical considerations concerning turbine jet propulsion for aircraft.
1935 in a motor vehicle workshop in Göttingen a test gas turbine is built at own expense. In the same year doctorate at the Physikalischen Institut.
1936 change to Ernst Heinkel in Rostock as development head for jet engines.
1939 first flight of the He 178 on August 27 with the turbo-jet engine He S 3B. Under his leadership further turbo-jet powerplants are developed at the Heinkel-Hirth GmbH in Stuttgart, in the end the turbo-jet engine He S 011.
1947 change to the Wright Patterson Air Force Laboratory in Dayton/Ohio.
1963 Chief Scientist at the Aerospace Research Laboratory.
1966 Goddard Prize of the American Institute of Aeronautics and Astronautics (AIAA).
1970 Fellow of AIAA.
1979 chair at the University of Dayton and consultant of many institutions. Honorary doctor of the West Virginia Morgantown University.
1985 Aachen and Munich Prize for Technology and Applied Natural Sciences. Corresponding member of the DGLR.
1992 receives the Ludwig-Prandtl-Ring.
1998 he dies in Mebourne, Florida on March 13.

the external partial branching of the primary air and the external secondary air behind the diffuser of the centrifugal compressor. The guidance of this air by reverse baffle plates up to the fuel injection nozzles is important. To the end of the curvature, the fuel supply with the vaporization system (later replaced by a "cold-misty" fuel system), the inner pressure loaded metal sheet cowl on the external and internal side of the flame and mixing region for cooling the pressure loaded casing secondary air from the centrifugal compressor exit is seen. The fuel vaporizers of the 16 burners, whose oxygen was only supplied from the middle portion of the compressor air, worked according to the principle of oblique tube boilers and were directly heated by the flames. All combustion chamber metal plates and the vaporization tubes, drilled from solid material, consisted of heat-resistant steel with 38 per cent nickel. To guarantee a uniform fuel distribution around the circumference, the fuel lines were designed in such a way that all burners were at the same distance from the supply pumps [31]. The target value of the average combustion chamber exit temperature of this engine was 970 K (697 °C).

The Axial Annular Combustion Chamber

At about the same time as *Hans-Joachim Pabst von Ohain* in Rostock, *Herbert Wagner* at the Magdeburger Werkzeugmaschinenfabrik, at that time a branch of the Junkers-Motorenwerke, Dessau, started to realize his concept of an axial aircraft gas turbine. On April 1, 1936 he explained to three young engineers, among them *Rudolf Friedrich* [18], his ideas concerning the development of a novel aircraft powerplant based on the British patent No. 495469. The most important characteristics of this powerplant were the axial construction of the compressor and of the turbine as well as the axial annular combustion chamber located in between. Apart from the multi-stage axial compressor, especially the annular combustion chamber with the necessary unusually high combustion space loading and the requirement for a relatively short construction length, an as complete as possible combustion and low fluid mechanic pressure losses were new ground. In spring 1939 the developed test object RTO was installed on an engine test bed in Magdeburg and prepared for the combustion of propane [32]. After the shortly following order to stop the test work, the engine group handed in its notice to Junkers

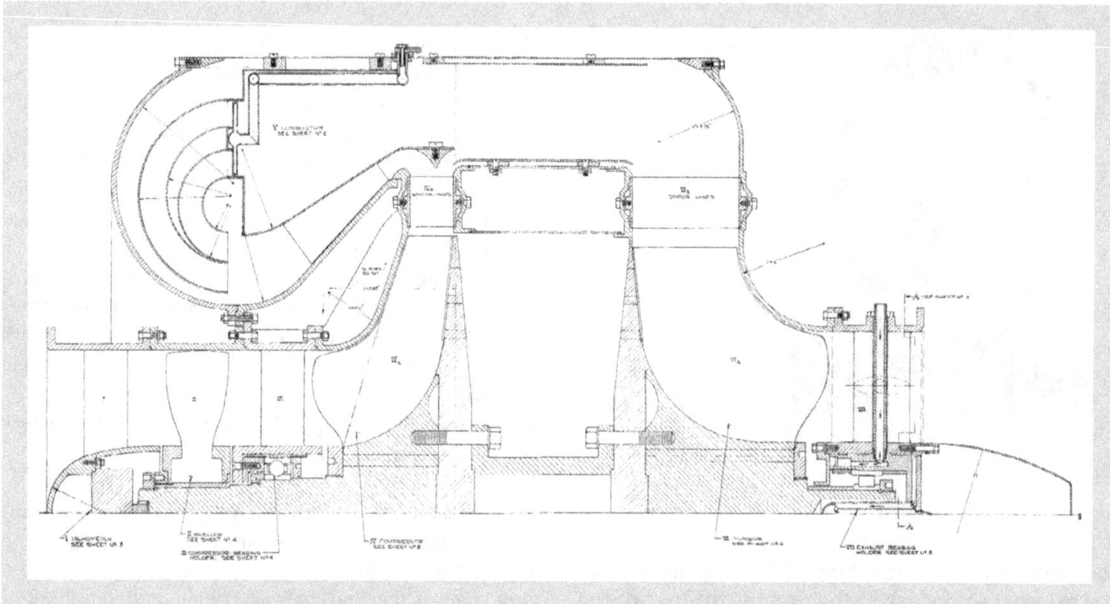

First airworthy turbo-jet engine Heinkel He S 3B with reverse-flow annular combustion chamber.

and moved as second jet engine development team to Heinkel in Rostock. From the RTO engine the axial engine He S 30 was developed.

An essentially different form as for the He S 3B received the combustion chamber of the Heinkel-Hirth engine He S 011, belonging to a new, modified design and more powerful jet engine class, whose development was started in autumn 1942. On request by the "Reichsluftfahrtministerium, RLM", this engine was designed with a short annular combustion chamber enclosing the compressor exit guide vane cascade and the air-cooled turbine entry guide vane cascade [18]. Generally valid methods to achieve a combustion chamber optimum based on primary combustion zone, mixing space size, combustion chamber length, degree of complete combustion and pressure loss did not yet exist at that time. For the application of test results, which had been obtained with small sector combustion chambers on research test beds, discrepancies existed due to the effects of the side walls which were missing in the complete engine and the not simulated flow profile at the compressor exit in the test bed investigations. Initially the combustion chamber concept of a different axial engine, the He S 8A, was used, in which fuel was sprayed at low pressure through 128 small tubes onto impingement plates and atomized by the air stream. Large problems arose from the local extinction of the combustion, the low degree of complete combustion and unfavourable temperature distributions. This meant that the above principal requirements could not be fulfilled. Added were additional ones like a large control range, a good starting and altitude behaviour, a radial temperature profile compatible with the thermal-mechanical loading of the first turbine rotor blade cascade, and an as far as possible uniform temperature distribution in the circumferential direction as well as a sufficient durability of the flame tube. After eight basically modified versions at last one configuration was found in which the compressor exit flow was led through a triple-guide cascade in the axial direction and retarded, in the subsequent internal and external diffusers further retarded and distributed among the combustion chamber zones. The fuel was centrally, conically injected through 16 two-mass flow nozzles and mechanically atomized. Vortex flow fields behind the front small metal plate fingers allowed a stable combustion. The large metal plate fingers in the mixing zone provided the rest of the combustion and the required temperature distribution by the metered and guided air addition. For a full load run on a ground test bed in December 1944 the combustion chamber achieved a specific thermal power referenced to the characteristic combustion chamber volume of 90.7 MW/m^3 (2.55 MW/ft^3) and a specific combustion space loading of 21.6 $MW/(m^3 \cdot bar) = 216\ s^{-1}$ at a pressure of 4.2 bar (63 lb/in^2) and an average duration of stay of the gas flow of 6.7 msec. With a fuel-air mass ratio of about 1.5 per cent, that is the equivalence ratio of 21.3 per cent at the exit, the mean temperature of about 470 K (197 °C) at the entry was increased to 1,048 K (775 °C) at the exit. The highest degree of complete combustion of 95 per cent was reduced at lower and higher power to approx. 90 per cent. The pressure loss of the actual combustion chamber was 6 per cent at the maximum and that of the combustion chamber including the three compressor exit guide vane cascades – between measuring stations 11 and 20 – about 5 per cent higher, that is 11 per cent. At full load static operation a thrust of 13.1 kN (2,948 lbf) was produced with a thermal power of approx. 21 MW. The degree of complete combustion as well as the lifetime were considered unsatisfactory. One of the reasons for the too low values were the inadequate test facilities.

BMW began at the end of the thirties with the development of axial jet engines under the leadership of *Hermann Oestrich* [28]. The project, which led to readiness for quantity production, was the turbo-jet engine BMW 003 A. In contrast to the compressor and the turbine for which some experience was available, the combustion chamber had to be developed completely from scratch [33, 34]. With the requirement for an as small as possible engine diameter, it was from the beginning conceptionally specified to develop the combustion chamber as annular combustion chamber. In contrast to the annular arrangement of several single-tube combustion chambers, this (modern) constructional form had among others the advantages of a better use of space, of a simpler structure and a lower weight. By basic tests on single-tube combustion chambers it was found that for through flow speeds of about 70 m/s (230 ft/s) a quasi-steady combustion could only be maintained if a vortex field as reverse flow zone of the fuel-air mixture is present in the primary zone of the combustion chamber. Initially, therefore, a circular impingement plate with numerous single bore holes was used as simple vortex generator. Behind each of the 16 fuel nozzles in the annular combustion chamber one of these ceramic plates, heated by the combustion, was fitted. The fuel impinging on the hot impingement plates atomized further, evaporated and did flow partly through the small orifices in the plates and partly around the edges of the plates into the wake. Owing to the separation of the air flow at the plate edges, on the back side a vortex region was established where the fuel vapour could mix itself intensely with the air and could, because of the considerable lower velocity level, stably burn. Farther behind the impingement plates, the external air stream, which mixed itself with the hot flames, was responsible for

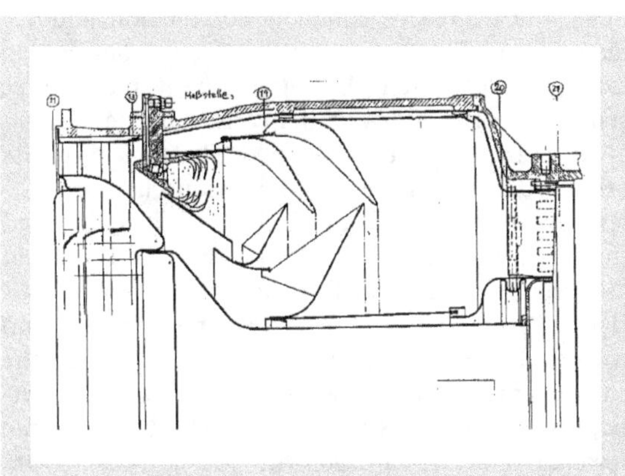

Annular combustion chamber of the Heinkel-Hirth engine He S 011 with compressor exit cascade and cooled turbine inlet guide cascade.

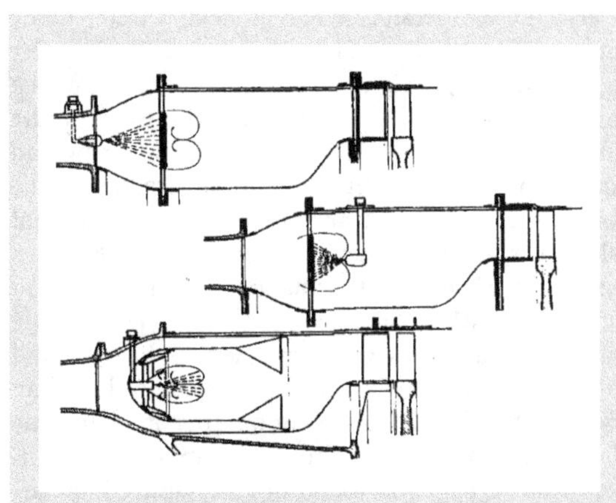

Schematic of the combustion chamber development for the jet engine BMW 003.

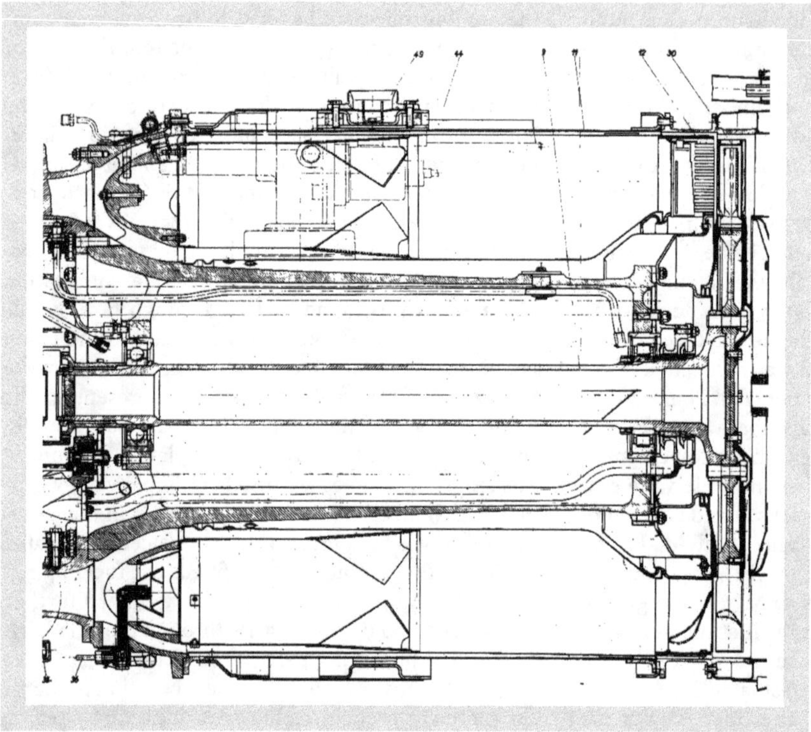

Longitudinal section installation drawing of the BMW 003 jet engine.

the establishment of a mean temperature level for the turbine guide vanes. Although the flames did not come directly in contact with the combustion chamber walls, these were nevertheless subjected to relatively high temperatures. The inner and outer flame tube wall was therefore only fixed at one side in order to allow free expansion towards the other side. Whilst the outer wall could slide into the turbine casing, the inner wall was attached to the inner turbine ring and could expand towards the front in the direction of the compressor. The force transmission, therefore did not take place via the flame tube walls but via the cooler, due to the in between flowing air portion, inner and outer combustion chamber casing walls. The turbine nozzle mounting ring was supported on the one hand by the inner combustion chamber casing wall and on the other hand by the bearing mounting of the turbine shaft. During the testing of the first combustion chamber of this type it was found that the desired favourable temperature distribution could not be obtained, the flames extinguished at small air mass flow rates and the ceramic plates broke easily.

In the next development stage, therefore, an annular steel impingement plate with numerous orifices and a fuel contra-flow injection was installed. In this way the fuel preparation and mixing, the combustion stability, the degree of complete combustion and the temperature distribution were improved, but not yet to a sufficient extent. Essential improvements were achieved in the end with the conical burners developed on the single-tube test bed. The conical inserts produced in the annular combustion chamber vortex regions into which the fuel was directly injected (figure on page 239 righ). In the sketch the trace of the inner flame tube wall highlights that the air discharged from the compressor was now divided into two streams. The primary air did flow through the conical burner and served for a nearly stoichiometric, very hot combustion of the fuel. Thereafter the secondary air was mixed with the hot gas stream using a number of mixing fins in order to yield for the following turbine an acceptable temperature level. In the longitudinal section drawing the surroundings are shown into which the combustion chamber

Outer and inner flame tube wall of the BMW 003 A combustion chamber.

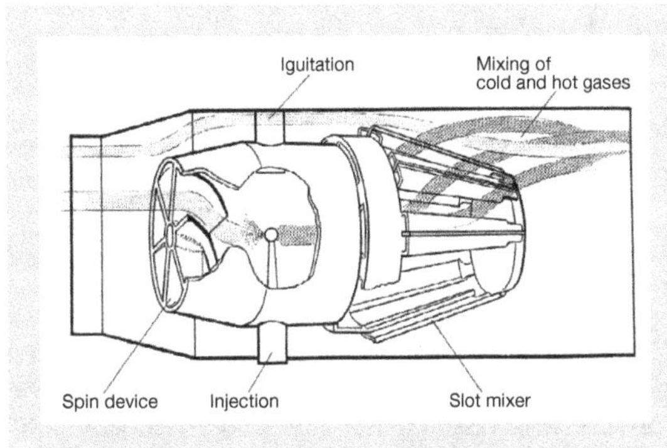

Flow, fuel injection and construction details of the Jumo 004 combustion chamber.

had been installed in the jet engine. This version was as yet increased in length by 150 mm (5.9 in) in order to improve the degree of complete combustion further. The photography above shows the outer and inner flame tube wall of the BMW 003 A combustion chamber [35]. The BMW combustion chamber configuration was after World War Two used and further developed for decades in the French engine company SNECMA for the ATAR engine family.

Combustion Chamber and Afterburner Developments at Junkers

For the same requirements – combustion stability in the whole operating range, a high degree of complete combustion, low pressure losses, suitable temperature distribution in the combustion space and in front of the first turbine stage – the technical solution for the combustion chamber of the first large scale quantity production jet engine Jumo 004 of the Junkers Flugzeug und Motorenwerke, Dessau, differed essentially in some details from that described above. Although the annular combustion-chamber configuration of all today's aeroengines appeared at that time to be superior for the engine technology, it was decided for this engine the installation of six single-tube combustion chambers interconnected by channels because the contribution to the air mass flow rate of a single combustion chamber could be realized with available test facilities. Furthermore, based on the experience with other tube combustion chambers at other places and for a better application of the test results, fewer development problems were expected.

The connecting channels between the six tube combustion chambers served primarily for ignition transmission when starting the engine. The figure above shows in the upper part a simplified representation of the flow and of the fuel contra-flow injection in the combustion chamber [27] and details of the construction are seen. The combustion took place in the centrally arranged combustion muzzle and with a nearly stoichiometric mixing ratio. Into this primary zone of the combustion chamber entered a portion of the air with swirl and the fuel was injected via a swirl nozzle against the air flow in order to achieve a good mixing and a longer duration of stay of the fuel droplets in the muzzle. The combustion chamber of the Jumo 004 as well as the fuel supply system, that is the ducting, pumps and controller, were designed for Diesel oil (later designation: J 2). When later this fuel was also laid down for the jet engines of BMW and Heinkel and their pressure difference control systems did not work accurately enough, the Jumo control systems were also used for the BMW 003 and Heinkel S 011 engines by order of the RLM. The change from the expensive fuel B 4 (normal aviation petrol) to cheaper J 2 (unrefined Diesel oil) took place for the jet engines not coming from Junkers in 1944. It also entailed a change of the starting ignition system [36] apart from the change of the fuel system. The first 004B series engine was delivered in June 1943 to the Air Force for the fighter aircraft and fast bomber Messerschmitt Me 262. At full rotational speed of 8,700 rpm, this engine had on a standard day on the ground test bed the following performance data:

- Air mass flow rate: 21.2 kg/s (46.7 lb/s)
- Compressor pressure ratio: 3.14
- Combustion chamber exit temperature: 775 °C (1,048 K)
- Combustion chamber efficiency (degree of complete combustion): 95 per cent
- Specific fuel consumption: 1.4 kg/kph (143 kg/kNh)
- Thrust: 8.9 kN (2,003 lbf)

The Junkers combustion chambers distinguished themselves by a remarkably low pressure loss. The pressure loss coefficient referenced to the dynamic pressure of the cold combustion chamber inlet flow was only 13.8 and was thus considerably lower than that of comparable British single-tube combustion chambers. Initially, this was however obtained at the cost of a relatively poor combustion chamber efficiency (degree of complete combustion) of 90 – 92 per cent [37].

A later type of this engine, the Jumo 004E, was already equipped with an afterburner (reheat pipe) behind the turbine. It attained at a combustion chamber outlet temperature of 870 °C (1,143 K) a ground static thrust of 9.8 kN (2,205 lbf) without and about 11.8 kN (2,655 lfb) with afterburner. In 1945 several such engines had completed 100 hour runs. However, an introduction into the production could not be accomplished any more.

Combustion Chamber Development for a Bypass Engine at Daimler-Benz

Daimler-Benz started after initial hesitation somewhat later with the jet engine development because of the large programme variety for piston aeroengines. From *Helmut Mauch* of the RLM the company obtained the difficult task to develop an engine with larger thrust and considerably lower fuel consumption. Under the leadership of *Karl Leist*, who had joined Daimler-Benz from the "Deutsche Versuchsanstalt für Luftfahrt, DVL", in September 1939 studies were commenced for the very advanced and very complex bypass engine DB 109-007 (later RLM designation) proposed by him, one of the first bypass engines of the world. After the start of the actual development work in the year 1941 the first of three test engines built was put on April 1, 1943 on a test bed in Stuttgart-Untertürkheim and reached in autumn of that year the full load rotational speed of 12,600 rpm. The combustion chamber module consisted initially of four single combustion chambers, later of five. In ground static operation a total of 50.6 kg/s (111.6 lb/s) of air flowed through the engine, of which 14.8 kg/s (32.6 lb/s) went through the gas generator with the combustion chambers. The design target for the combustion chamber outlet temperature had a very high value of 1,100 °C (1,373 K), requiring for the turbine an especially sophisticated cooling. At the end of 1943 the development was discontinued on order by the RLM [38].

Ram Jet Engines

From the viewpoint of construction volume, a proportionately considerably larger role than in aircraft gas turbines plays the combustion chamber in a different kind of aircraft powerplants, the ram jet engines. These can operate only effectively at higher flight speeds because then the in-flight stagnation pressure – without turbo-compressor – suffices for a higher pressure level in the engine. In 1936 *Hellmuth Walter* in Kiel took up the idea of *René Lorin* published in 1913 in France and constructed a test engine whose combustion chamber was fed with natural gas. Theoretical studies were carried out at BMW and Junkers as well as the "Deutsche Forschungsanstalt für Luftfahrt, DFL". For a new interceptor *Eugen Sänger* of the DFL proposed in 1941 a ram jet engine with a very high combustion chamber temperature. Because of problems in the application of model test results to the original configuration he constructed very large ram jet pipes with 500 mm (19.7 in) and 1,000 mm (39.4 in) diameter as well as 5.50 m (18 ft) and 8.60 m (28.2 ft) length. With these, combustion tests were carried out on a lorry with 25 m/s (82 ft/s) driving speed and on a strut rig on the aircraft Dornier Do 17Z with flight speeds up to 200 m/s (656 ft/s) (in a shallow dive). These engines were operated with petrol, later, because of petrol getting scarcer in Germany during World War Two, combustion tests with foam coal (blocks of pressed coal dust) were carried out, which, however, were not successful on account of the slow combustion. For the application of *Sänger's* ram jet engines on military subsonic aircraft there existed several proposals as well as for the delta-wing supersonic aircraft project of *Alexander Lippisch* in Wien. Like the projects of ram jet engine equipped fighter aircraft at the companies Dornier, Heinkel and Focke-Wulf, none of these projects were realized for operation during the war. Only afterwards such engines were built by foreign companies for missiles, aircraft and helicopters as test engines and also in quantity production. *Eugen Sänger* was working from 1946 onwards for some years in France and was among other things involved in the test aircraft "Griffon II" of the company Nord Aviation, whose propulsion system consisted of an axial turbo-jet engine enclosed by a ram jet engine.

About 1942 in the gas dynamics department of the company Focke-Wulf under the leadership of *Otto Pabst* the development of a subsonic ram jet engine was started, which was intended for the application on the so-called "driven-wing fighter". The rotor of this aircraft, the aircraft being very similar to a larger helicopter of today, was to be driven only by the ram jet engines mounted at the blade tips and it thus exerted no torque on the aircraft fuselage. The engine was designed for the freestream Mach number of 0.85 and a thrust of 8.2 kN (1,845 lbf). Its diameter was to be 0.7 m (2.3 ft) and its length 1.4 m (4.6 ft). The necessary circumferential velocity for an effective operation of the ram jets at the blade tips was achieved with the aid of installed small hydrogen peroxide rocket engines. In the figure below details of the corresponding model engine having smaller dimensions are shown. This had a diameter of 136 mm (5.4 in) and a length of 480 mm (18.9 in) and was, therefore, not geometrically similar. After functional tests with coal gas and petrol, a very suitable vaporization burner was developed for this model engine, with which the combustion length could be kept relatively short. By the numerous burner tubes in the engine pipe and the in this way produced strong turbulence the desired distribution for the heat release and temperature were achieved [39]. During tests in the high speed wind tunnel of the "Luftfahrtforschungsanstalt Brunswick, DFL" in the year 1944 the freestream Mach number of 0.95 was achieved. In the operating range with combustion space temperatures between 220 °C and 820 °C (493 and 1,093 K) a stable combustion could be realized [2]. Ram jet engines for supersonic and hypersonic aircraft – especially those with supersonic combustion – are today subject of research in several countries. To the same category as the ram jet engines belong the jet engines with pulsating combustion because these too contain no rotating compo-

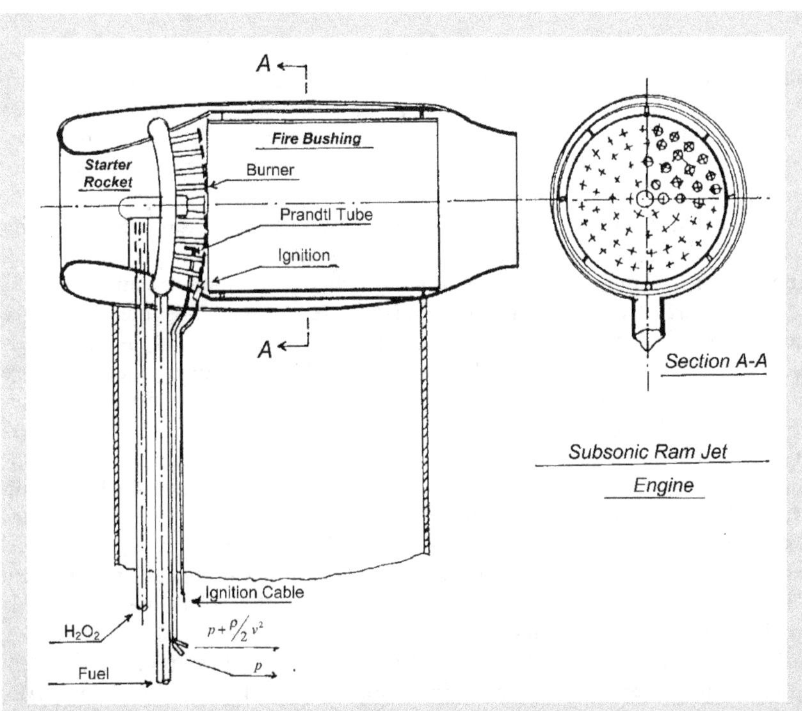

Subsonic model ram jet engine of the company Focke-Wulf for measurements in the high speed wind tunnel of the DFL, Brunswick.

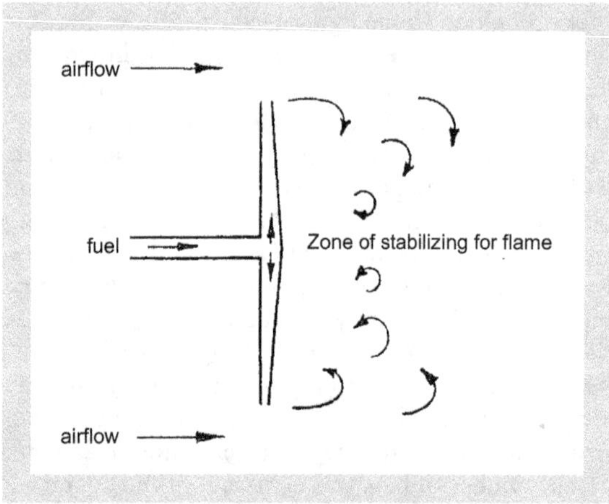

Schematic of the disc-like Pabst combustion nozzle.

nents. The principle of the intermittent combustion originates with a proposal of *Marconnets* in France in the year 1908. A technical implementation, which reached in World War Two readiness for mass production and application with the unmanned bomber V1 (flying bomb), is the Schmidt pipe (Argus-Schmidt pipe 014).

Further Contributions to Combustion Chamber Technology Prior to 1945

Up to the end of World War Two the German aero-engine manufacturers – Heinkel, BMW, Junkers and Daimler-Benz – worked independently of each other and practically without bringing in German research institutes concerning the combustion chamber development. In this respect the action in this field differed fundamentally from that of the Anglo-American side. The contributions of the German research institutes to the development of combustion systems had only slight relation to the actual engine projects. Experimental work at the DVL and DFL concerned mainly the fuel preparation in combustion chambers. At the DVL in Garmisch a photographic method was developed for the investigation into fuel spray cones with the aid of a high tension ignition spark with repetition time of 30 seconds for short duration exposure. In this way the vaporization rates and the droplet speeds – however not the droplet size distributions – could be approximately determined. The DFL in Brunswick-Völkenrode developed a method for improvement of the fuel atomizing by means of reduction of the droplet size by spraying on sooty surfaces and for obtaining a suitable spatial fuel mist distribution.

In the aircraft company Focke-Wulf in Bremen, which produced no engines or engine components, also some research and development work on combustion was carried out. Under the leadership of *Otto Pabst* a very efficient system for a stable combustion was developed, whose schematic is represented in the left figure. The liquid fuel vaporized in the fuel supply system prior to entry into the combustion chamber and was then sprayed as vapour through the annular burner nozzle exit transversely to the air flow into the combustion zone. In the wake behind the disc-like burner nozzle the fuel vapour could thoroughly mix itself with the air and this mixture could burn with a stable flame. In this way a short combustion chamber length was possible.

Turbine – The Hollow Metal Blade as Solution for Material Shortage

Helmut Schubert

The Preliminary Work by the DVL

The development work and the associated necessary basic investigations like cascade investigations for the individual components of the first German turbo-jet engines represent a good example for the successful cooperation of aviation companies including the suppliers and of the institute and university research in Germany in the thirties. Not only the frequently practised change of engineers and researchers from industry to research establishments and vice versa, but especially the broad use of all research facilities like test beds and measurement equipment are an example how at that time in a short period successful research work was jointly performed. Important was also the coordinated planning and technology support of the RLM in Berlin under the leadership of *Wolfram Eisenlohr* with his competent colleagues *Hans Mauch* and *Helmut Schelp*.

The theoretical and experimental preliminary work for the realization of the turbine blade cooling was carried out at the DVL in Berlin-Adlershof, the practical implementation took place at the German engine companies. At the DVL in Berlin *Christian Lorenzen* performed already in 1929 tests with

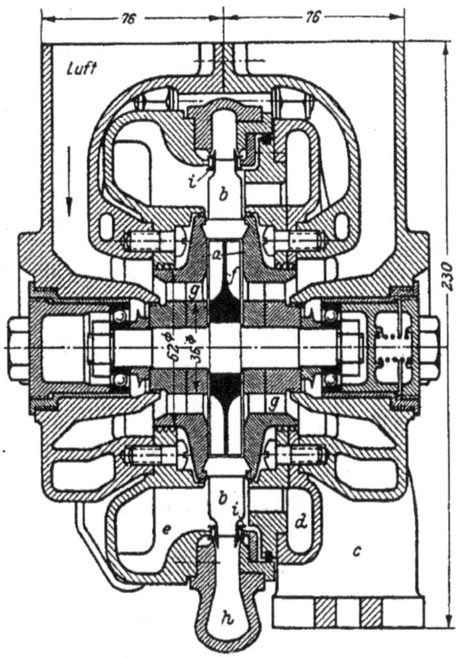

a Turbinenlaufrad mit zwei Stahlscheiben,
b hohle Turbinenschaufeln,
c Eintrittsstutzen für die Abgase,
d Gasraum vor dem Turbinenlaufrad,
e Auspuffkammer,
f Laufrad des Gebläses,
g Leitkanäle für Luft,
h Diffusor,
i Labyrinthringe.

DVL exhaust-driven turbo-supercharger developed by *Christian Lorenzen.*

cooled turbine blades of exhaust-driven turbo-superchargers which were at that time manufactured by BBC in Mannheim [40, 41]. With this experience it was possible to allow for these axial flow turbines mean circumferential velocities in the middle of the blade of u_m = 250 m/s (820 ft/s) and blade temperatures of 800 °C (1,073 K). With highly improved cooling it appeared that temperatures of 900 °C (1,173 K) were allowable.

The research work concerning the cooling of exhaust gas turbines at the DVL in Berlin-Adlershof were represented in detail by *A. Fredrichs* and *H. Quentin* [42]. The results of these theoretical and experimental investigations [43–45] were used by Daimler-Benz in Stuttgart as well as by BMW in Munich and BRAMO in Berlin-Spandau for the design of the turbine cooling of exhaust-driven turbo-superchargers. The figure above shows the construction layout of the turbine blade cooling by BMW for their exhaust-driven turbo-superchargers.

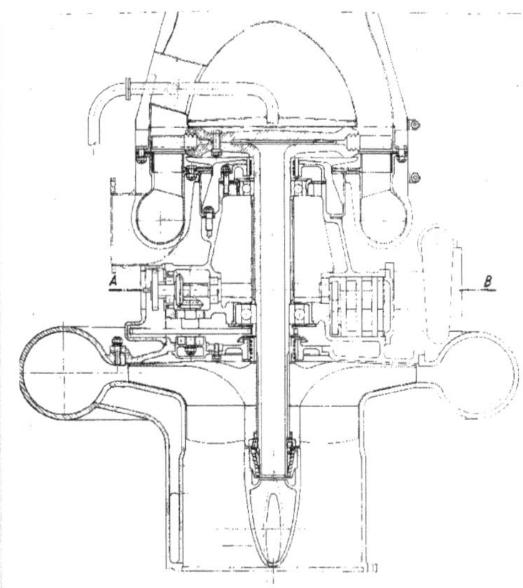

BMW exhaust-driven turbo-supercharger of 1942, applied to the twin-row radial altitude aeroengine BMW 801 TJ.

Turbine Blade Cooling of Heinkel Turbo-Jet Engines

The turbo-aeroengine successfully developed by *Hans von Ohain* and *Wilhelm Gundermann* since 1936 at the company Ernst Heinkel in Rostock-Marienehe possessed an uncooled single-stage radial turbine [31]. The turbine guide vanes and the rotor were of highly heat-resistant materials from Krupp (Cromadur). Under the leadership of *Max Adolf Müller* still in 1939 at Heinkel the development of the He S 011A jet engine commenced, apart from the team that was furtheron concerned with radial compressor engine developments.

This engine had a cooled two-stage axial turbine. The hollow turbine blades, shown next page, were manufactured as so-called pot blades in several forming stages. By using an insert, the compressor air quantity required for the cooling could be kept small. The installation of the cooled turbine guide and rotor blades in the He S 011A is shown on the next page, too.

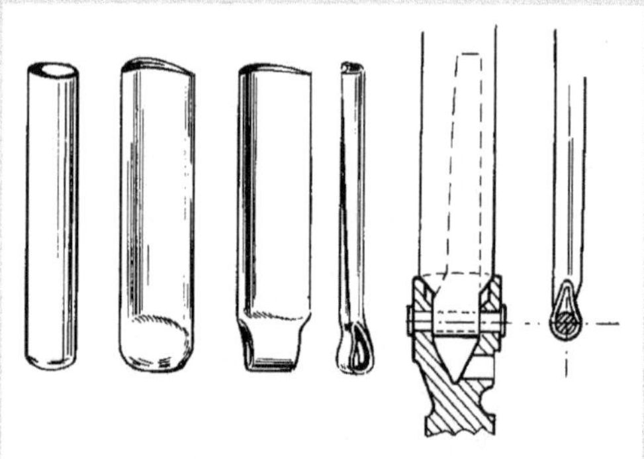

The hollow turbine rotor blades of the He S 011A engine as pot blade. By an insert the cooling air quantity could be kept small.

Two-Stage turbine of the He S 011A engine with cooled turbine guide vanes and rotor blades.

The thrust was 12.75 kN (2,866 lbf). The first run was in July 1944. The development and the preparation for quantity production could not be concluded by the end of the war.

Turbine Blade Cooling for the BMW 109-003

Preliminary investigations in the year 1938 at BRAMO in Berlin-Spandau concerning the new turbojet project P 3302 had shown that a compressor pressure ratio of about 3 : 1 would just suffice to obtain a useful specific thrust and an acceptable specific fuel consumption. The turbine design estimate furthermore yielded the possibility, that, for a not too large thermodynamic drop and for allowable circumferential velocity, a one-stage turbine is sufficient. The requirements were:

- Entry temperature maximum of 800 °C (1,073 K) corresponding to the development state at BMW for exhaust-driven turbines. *Hermann Oestrich* [46] considered at BRAMO in Spandau a gas temperature of up to 1,000 °C (1,273 K) for exhaust-driven turbines.
- High air mass flow for an as small as possible diameter.
- Stage number as small as possible, preferably only one turbine stage.
- Turbine efficiency as high as possible.

Turbine Blade Cooling at Junkers

The development work concerning turbo-jet engines started already in 1936 at the company Junkers under the leadership of *Herbert Wagner* in Magdeburg. The result was a small axial turbo-jet engine prepared in summer 1939 for test bed runs. The engine could, however, not be made to run independently because of difficulties of principle nature [47]. Under the leadership of *Anselm Franz* in 1939 a new development team was formed in Dessau and in rivalry to BMW the development of an axial engine for 5.8 kN (1,304 lbf) take-off thrust was launched. The designation for this was Jumo 109-004 [2, 48].

The single-stage impulse turbine based on a design of the AEG Turbinenfabrik was fitted with uncooled turbine guide and rotor blades. Sixty-one rotor blades were mounted with the rectangular root perpendicular to the turbine disc plane using two pins. The turbine wheel was floatingly supported by two roller bearings. In accordance with the Junkers' design aim of the maximum safety possible, in the design of the Jumo 109-004A all thermally highly stressed components were manufactured from the best materials available at that time. The turbine blades were of Tinidur, a austenitic Krupp steel with 30 per cent nickel, 15 per cent chromium and 2 per cent titanium. Tinidur had at 600 °C (873 K) the remarkable permanent mechanical strength of 32–40 kp/mm² (45,481–56,851 lb/in²). The turbine guide system, as

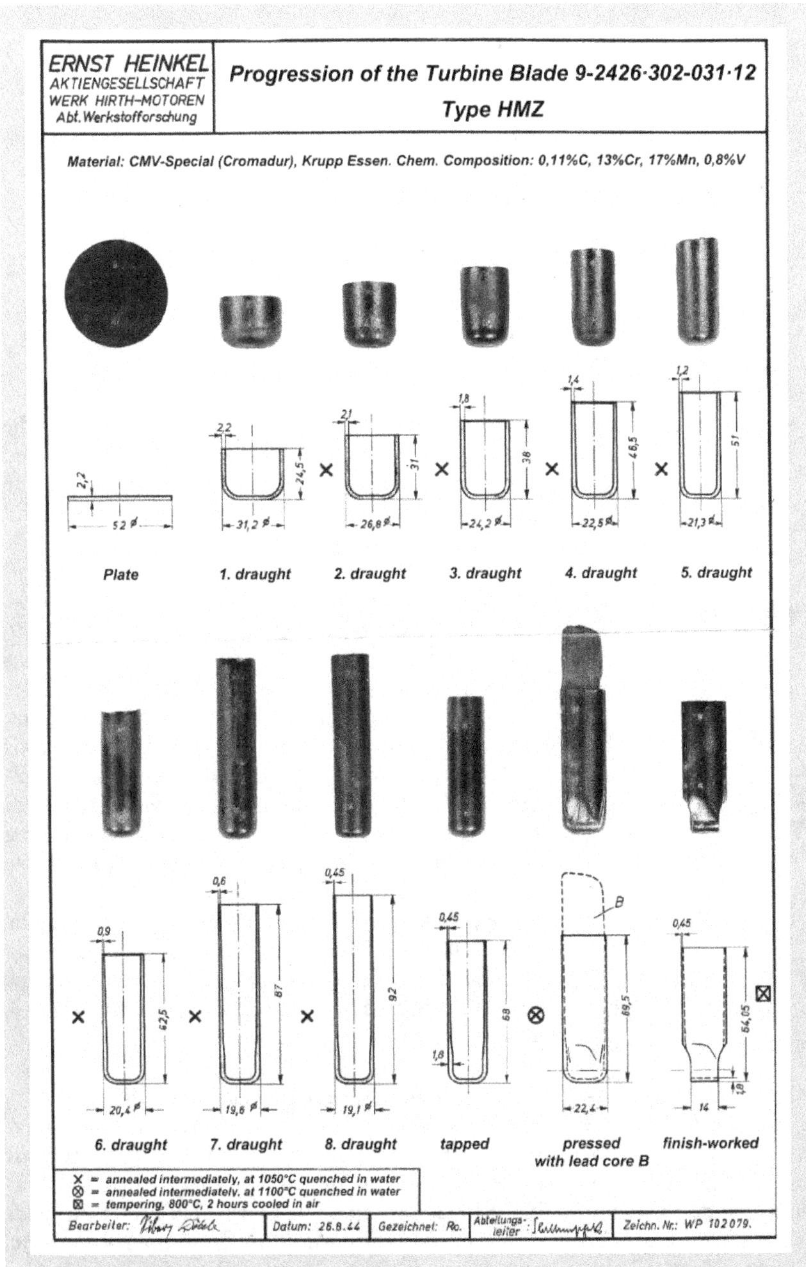

Development of the hollow turbine blade of the Heinkel He S 011 engine with eight steps of draughts (deep drawing) and subsequent final finish processing.

well as other metal parts, consisted of C9M18. After some minor initial difficulties, the engine achieved on August 6, 1941 the required take-off thrust of 5.8 kN (1,304 lbf). The non cooled turbine wheel hardly made any problems. Only the casing ring above the turbine wheel was deformed despite a very stiff construction, possibly due to an uneven thermal expansion. The radial gap width was therefore increased from

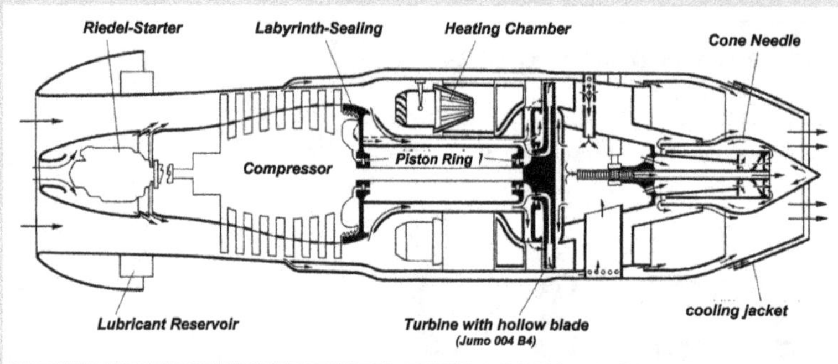

Air cooling system exemplified for the turbo-jet engine Jumo 004 B4.

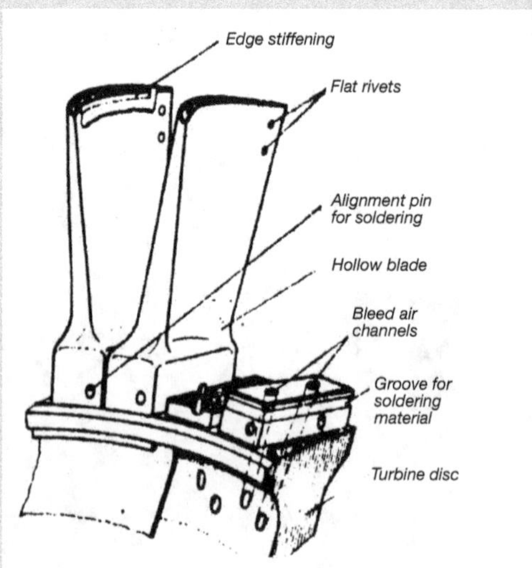

Mounting of the hollow turbine blade for the Jumo 004 B engine.

2–6.5 mm (0.079 0.256 in). Furthermore, the cross-section of the turbine guide system had to be increased from 480–550 cm^2 (74.4–85.3 in^2) since the compressor mass flow rate was higher than predicted. Because the A-version of the Jumo 109-004 was too expensive in the quantity production, from summer 1941 a more cost-effective version Jumo 109-004B was constructed using economic materials. In the turbine region therefore the following changes took place: For the solid blade turbine a modified root fixing by means of turned grooves in the circumferential direction of the wheel, correspondingly turned blade roots and shear pins were introduced (see figure). From the start also at Junkers the introduction of air-cooled hollow blades were planned for the turbine wheel, the quantity production, however, commenced with solid blades, while two different hollow blades were developed: One blade from the company Prym produced by deep-drawing Tinidur, and a blade free from nickel, pressed from taper-rolled Chromadur steel sheet, folded and welded at the trailing edge, which was manufactured by the company Wellner. The blades were fixed to the turbine disc by means of a box root and brazing using silver-manganese solder, as shown. Sufficient cooling was achieved without a cooling insert in that the cooling air tapped at the compressor exit was blown into the leading and trailing edge of the blade through two channels. Both hollow blade types have been introduced into the Jumo 109-004 B4 in the quantity production in the last months of 1944. It is of interest that at Junkers already in 1944/45 under the leadership of *Anselm Franz* and *Heinrich Adenstätt* turbine blades made from ceramic material were used [42].

Turbine Cooling for the Daimler-Benz DB 109-007

The development of turbo-jet engines at Daimler-Benz in Stuttgart-Untertürkheim began on request by the RLM in 1940/41. Preliminary work concerning the blade cooling was carried out under the leadership of *Karl Leist*, who came from the DVL in 1939, on exhaust-driven turbo-superchargers for the Daimler-Benz altitude engines. Because the bypass engine DB 109-007 possessed a turbine entry temperature of 1,100 °C (1,373 K), the single-stage axial turbine had to be cooled. A partial bleed of cooling air from the bypass flow duct was chosen, which was supplied via slots from the secondary air duct. Preliminary tests for these methods were performed at Daimler-Benz on exhaust gas turbine wheels also already at high gas temperatures. For supplying the cooling air a radial blade ring on the turbine rotor was responsible, a kind of radial fan.

Design and Testing of the Cooled Turbine Rotor Blades

For reasons of mechanical strength and diameter size, at BMW for the turbine a mean circumferential velocity of 250 m/s (830 ft/s) was chosen. The rotational speed at full load was determined by the axial flow compressor (9,000 rpm, later 9,500 rpm). From the start for the turbine rotor the application of hollow blades was intended like those that were successfully used for the BRAMO/BMW exhaust gas turbines [28]. They relieved the turbine rotor and permitted a lighter design. For the aerodynamic design of the turbines, experience was available at BMW as well as Heinkel and Junkers from companies of the German steam turbine industry, as for example AEG. Besides, an accelerated flow as for turbines is basically easier to have control of. As shape of the aerodynamic section for the turbine rotor blades constant pressure profiles in classical form were intended. The individual turbine blade halves were welded together from two parts and resulted in the hollow blades already mentioned, which were completed to a finished product after the welding. For the first test wheels, the blades were welded to the wheelbody as was done for the exhaust gas turbines. Because initially it was not possible to have control of this welding, the change was made to the root fixing shown. The rotor blade was secured by six cylindrical pins of which three pieces were knocked in on each side of the turbine rotor. The two-piece blade construction was given up because not having control of the welding at that time, which again and again led to failures. Forged rotor blades using an awl were actually better but more expensive. This version allowed already a life time of at least 100 operating hours. A more favourable shape of the aerodynamic section, as shown on page 250, yielded for the engines BMW P 3302 V11-V14 higher degrees of turbine efficiency and thus a larger engine thrust. With some selected, faultless forged blades one obtained turbine rotors with a reliable operational life of up to 200 hs. In this way the design target of the BMW P 3302 with a thrust of 5.5 kN or 1,238 lbf (target value 6.0 kN or 1,350 lbf) was nearly reached. For a reduction of the cooling air requirements and an improvement of the cooling a displacement insert, based on a proposal by *Sonntag*, was welded to the forged turbine blade in the process of manufacture and for the avoidance of vibration cracks so-called small bowls, as proposed by *Hermann Oestrich* [18]. The continuation on the development path described above would have been more reliable than that which had to be followed due to shortage of materials. To decrease the portion of alloy metals, the development of turbine rotor blades from metal sheet was started. The newly equipped BMW production test department in the factory Berlin-Spandau, which was established in order to fulfil the numerous technological production requirements of the turbo-jet and rocket engine developments, produced from metal sheet a blade which contained a cooling insert. For improving the mechanical strength conditions, based on a proposal of *Hermann Hagen* and then by *Paul Leistritz* in Nürnberg taper-rolled steel sheets were produced from which in several steps blades were manufactured [49]. The figure on page 250 shows the mounting of the metal sheet blades built by *Leistritz*.

For the quantity production blades of the BMW 109-003-A2 later the metal sheets at the root were strengthened and in this way a longer life achieved.

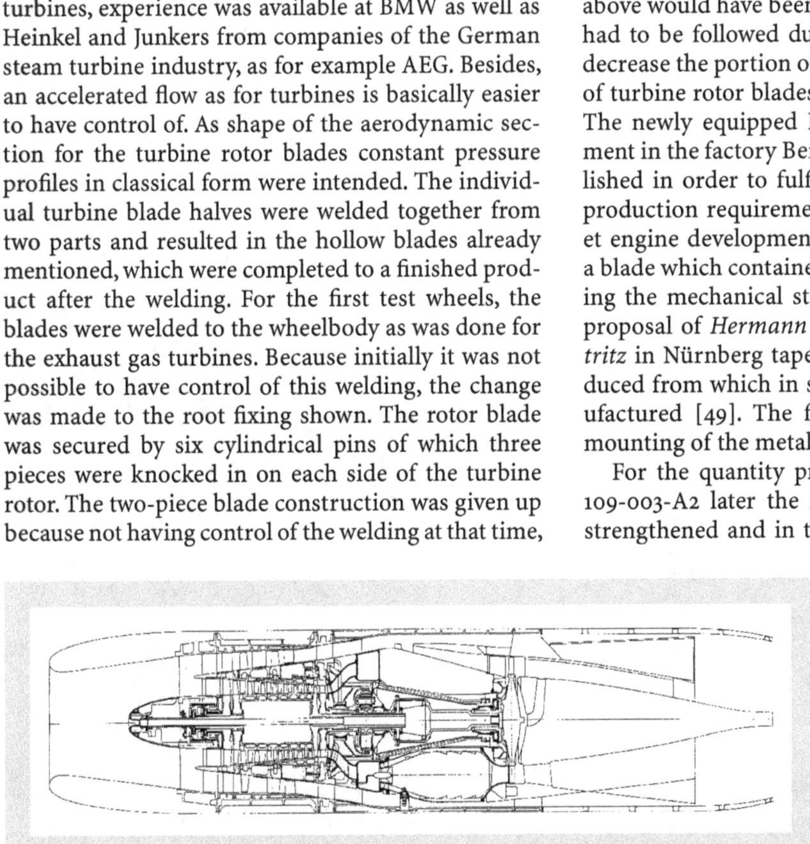

Bypass turbo-jet engine Daimler-Benz 007 with single-stage cooled turbine by partial bleed.

The small bowls at the tip could be dispensed with, as experience had shown. The cooling air requirement was with 2 per cent of the engine total mass flow very low. The required cooling air was tapped behind the fourth stage of the compressor. Up to the end of the war in 1945 a hollow blade developed by the Württembergische Metallwaren Fabrik in Geislingen (WMF) from deep drawn metal sheet spouts, for which the welded joint at the blade trailing edge could be avoided and which promised to become more reliable and cheaper than the welded metal sheet blade [50], did not any more come into operation at BMW. As life time for this blade version for the time being 50 hs were stated. After a high or an unintentional too high temperature loading of the turbine, a crack inspection was stipulated. The blade exchange was without problems. Because the turbine was sub-critically mounted, a blade damage was often only discovered after the engine had been switched off.

The improvement of the efficiency of the turbine was undertaken systematically. From cascade investigations in the BMW-owned cascade flow tunnel in Spandau, a favourable cascade and turbine blade section respectively was determined. In the high speed cascade tunnel of the DFL Brunswick in addition the model cascades were once again re-tested. A further control took place for 1 : 1 scale models of the turbine wheels on the turbine test bed in Berlin-Spandau and on the steam turbine test bed at the Institute of Technology (TH) in Dresden. With this procedure it was possible to achieve a reduction of the number

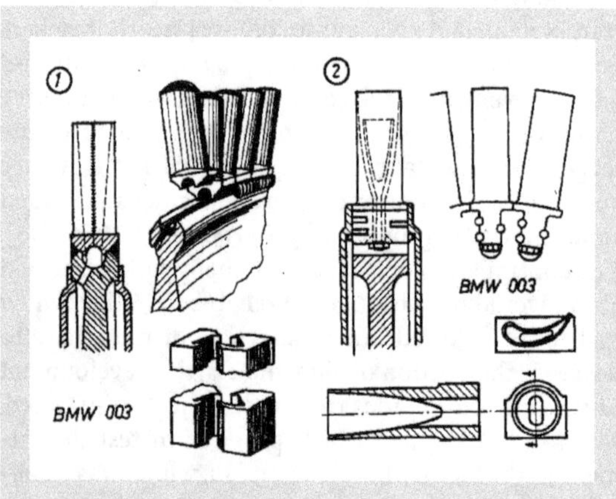

Concerning the development of the cooled hollow turbine blade of the BMW 003 turbo-jet engine: left (1) the first version, right (2) the improved version.

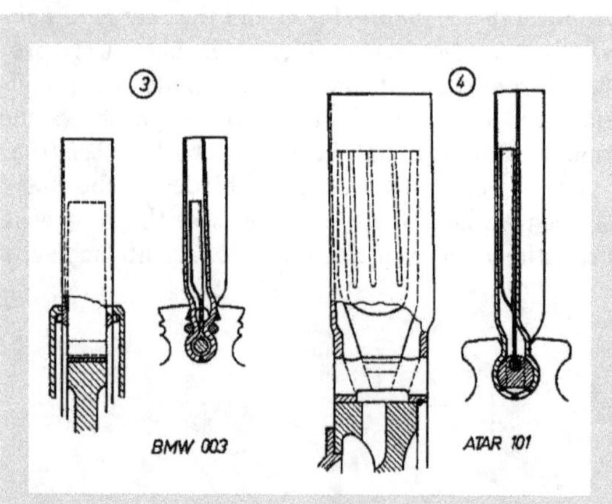

Last production version of the Leistritz hollow turbine blade of the BMW 003 engine (3) in comparison with the hollow turbine blade of the ATAR 101 engine (4).

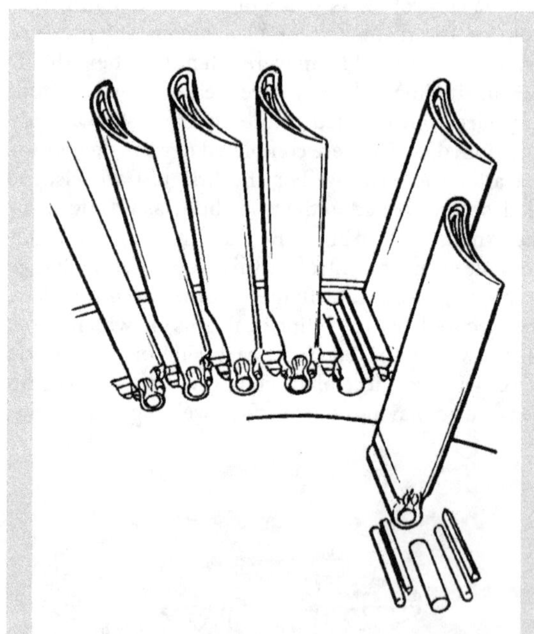

Mounting of the turbine metal sheet hollow blade produced by the company Leistritz for the BMW 003 engine.

of blades on the turbine wheel for a constant wheel width and a strengthening of the root mounting.

As for the turbine rotor, for the cooling of the turbine guide vanes 1 per cent of the total engine mass flow was tapped and this from the combustion chamber secondary air guide duct. The cooling was thus not very effective, it was, however, helpful to just relieve the particularly endangered blade zones.

Because of the cooling, already a mean turbine entry temperature of 1,050 K (777 °C) could be employed, although the temperature distribution of the combustion chamber was still quite unfavourable and in addition the highest rotor blade temperature of 970 K (697 °C) was not permitted to be exceeded. The temperature of the rotor was here measured by means of a pyrometer aimed through the propulsion nozzle by hand.

Turbine Blade Cascade Test Bed, DFL Brunswick

The RLM acquainted BMW with the possibility to visualize the turbine blade cascade flow. *Helmut Schelp* of the RLM provided the contact with *Ernst Schmidt.* At the DFL Brunswick under the leadership of *Schmidt* an air supply facility was available to which could be connected a modern cascade test bed. Through a schlieren optics the flow at high Mach number could be made visible. Of the turbine cascade, because of the untwisted turbine blades, flow photographs could be taken corresponding to the blade pitch on various diameters. Flow separation phenomena due to high excess velocities could be visualized. By aerodynamic section corrections or by changes of pitch, losses could be avoided. The experimental validation of the flow conditions was the only possibility for the flow optimization. Thus, also the engine companies themselves were very well equipped with test beds.

One of the available pre-development test beds at BMW in Berlin-Spandau for single combustion chamber testing was modified into a turbine blade cooling test stand. A stator cascade system or a rotor cascade segment was put in the hot gas flow produced by a single combustion chamber and subjected to the cooling air corresponding to the cooling air inlet flow into the stator or rotor blades and measurements performed. The hot gas mass flow rate and its temperature were adjustable. Controllable was also the cooling air quantity, however not its temperature. This was determined by the flow state associated with the supercharger (air supplier).

Turbine Blade Developments at MAN

At MAN in the factory Augsburg from 1940 till 1945 research work was carried out under the leadership

At MAN in Augsburg research work was carried out until 1944/45 on turbine blades from ceramic material.

of *Walter Handschuber* concerning gas turbine blades from ceramic materials. Already also cooled turbine guide vanes and rotor blades were produced and tested. The results were presented at the RLM ceramics conference at the University of Jena on November 29/30, 1944 [51].

The Use of Highly Heat-Resistant Materials

Already at the end of 1935 on the initiative of *Kötzschke* of the RLM in the German steel industry tests were started to develop improved highly heat-resistant steel for steam and gas turbines preferably without using materials subject to economy measures such as tungsten, nickel, cobalt, chromium and so on, which were not available in the Germany of those days. Sufficiently available at that time was molybdenum, manganese and silicon. The participating companies in particular were Krupp, DEW, Hakort-Eicken, Böhler and Heraeus. Largest progress was made by Krupp in Essen. On the basis of their NCT3 the highly heat-resistant material Tinidur was developed. It could withstand a temperature up to about 720 °C (993 K). However, the material was very brittle and the mechanical processing difficult.

Fuel and Lubricant Research

EILHARD JANTZEN

In Germany since World War One research was performed to a considerable extent in the field of fuels and lubricants for aviation. Principal reasons for this were on the one hand problems and increasing requirements of the engine manufacturers as well as users and on the other hand the nearly complete dependence on foreign countries due to insufficient own sources. The latter led to specific research activities in Germany in the field of synthetic aviation fuels and oils. The large scale technological manufacture of these products resulting from this was the basis for making the extensive and over several years lasting operation of the Air Force possible.

From the point of view of time, in Germany since nearly eight decades research on fuels and lubricants for the use in aviation has been undertaken. In this connection there are three periods which were determined by the technological requirements and especially the prevailing politics. The first period covered the space of time from 1912 until the end of World War One, the second one the period of time from 1925–1945 and the third one lasted from 1955 to the present. In all periods of time basic research as well as research close to application was conducted but to a different extent. However, both these fields did lie in part so close together that they cannot be considered separately. The transition between both fields is also often fluid.

Predominantly the research in this field was carried out by the "Deutsche Versuchsanstalt für Luftfahrt, DVL" until 1944 in Berlin-Adlershof and after the war in Munich. In addition until 1944 extensive research was conducted on the synthesis in the laboratories of the IG-Farben company with centres in Ludwigshafen as well as Leuna. At the technical universities there existed no subject-specific chairs or institutes, however, especially in the Third Reich, research contracts were awarded to them to a considerable extent by the RLM.

In the following a survey of this field is presented. For a detailed study, reference should be made to the relevant specialist literature. Concerning this, up to the present more than 200 references are known.

From the Beginnings to the End of World War One

In the initial years of aviation the fuels and engine oils available for the time being were only those products used in the internal combustion engines of motor vehicles. These were predominantly obtained from foreign countries. Already before the begin of World War One a first differentiation in supply as well as demand took place with reference to aviation-specific products. A first own research then began in World War One, this especially to counteract the shortage of home products and raw materials.

Fuels. In the early days of motorization petrols were exclusively traded and classified according to the specific weight, the value of the products used in aviation being required to be particularly low. As initial stage to a first quality research must be regarded the work of *Karl Dietrich* around 1913. Apart from diverse investigations into the products available at that time, he held the view that, instead of the not very meaningful specific weight, the boiling point should be taken as measure [52]. He also advocated already benzol as fuel, which was to be used later in the aviation to a large extent. These ideas were taken up by the High

Command and *Alois Riedler* of the Institute of Technology in Berlin received probably the first research contract in this field on the condition to investigate fuels and especially benzol concerning their suitability in aeroengines. Since 1917 this work was then continued in the "Flugzeugmeisterei Adlershof" and extended to further substances, this especially in order to counteract the shortage arising regarding aviation petrol.

High expectations with respect to a fuel supply independent of foreign countries were placed by the High Command in the research work of *Friedrich Bergius* in Hannover, who was concerned since 1912 with the conversion of coal into liquid products [53]. This work he then could continue on a broader basis at the company Goldschmidt in Essen thanks to a considerable governmental sponsoring. Based on the results obtained there, in 1917 for the first time in a technological facility in Mannheim-Rheinau petrol was obtained from coal. The in this way gained small quantities of aviation petrol, however, brought no relief for the Flying Corps. *Bergius* received in 1939 for his pioneering research work together with *Carl Bosch* the Nobel prize for chemistry.

Engine oils. At the beginning of 1917 emerged a growing shortage also for the engine oils required by the Flying Corps. Therefore, from the High Command research work was initiated with the aim of finding alternatives to the products of foreign countries used up to now. This concerned the oils for the "fixed engines" employed to a large extent as well as those for the "rotary engines" which were predominantly used for fighter aircraft. For the latter the problem was especially difficult because for them only the fuel insoluble castor oil could be employed in the past and which could no longer be obtained. An acceptable solution could finally be found in the form of the voltoles. Furthermore, because of the difficult situation it had become necessary to conduct extensive tests in order to specify new minimum requirements concerning oils for aeroengines. The work addressed was predominantly carried out by the fuels and lubricants department of the "Flugzeugmeisterei".

Friedrich Bergius (1894–1949).

The Period from 1925 until 1945

The fuel and lubricant research had come to a complete standstill after 1918. Since 1925 a new start commenced with a slowly increasing trend. From 1935 then a steep rise of the research activities followed. These involved basic research as well as to a growing extent application-specific work. From the engine manufacturers and users at that time the following main topics among others were posed regarding the fuel and lubricant research:

1. Safe functioning and longer operating life of the engines.
2. Higher performance of the engines and thereby of the air vehicle.
3. Production of large quantities of optimal suitable products from home raw materials.

In all three fields in most cases at first fundamental work had to be performed and methods of investigation and equipment acquired in order to be able to solve the problems to be dealt with. In 1925 at the "Deutsche Versuchsanstalt für Luftfahrt, DVL" in the engine department a chemical laboratory was established in order to be able to tackle the first of the above topics. Since the thirties this field was increasingly extended under the leadership of *Alexander von Philippovich* and in 1936 changed into an Institute of Fuel and Lubricant Research [54]. Coincident with this time was the start of work concerning item 2., into which with sub-topics increasingly also institutes of the technical universities were incorporated, like especially in Berlin (*Richard Heintze/Maximilian*

Alexander v. Philippovich (1895–1968).

Hermann Zorn (1896–1983).

Matthias Pier (1882–1965).

Marder), in Munich (*Albert Wolfgang Schmidt/Fritz Kneule*), and in Stuttgart (*Wunibald Kamm/Otto Widmaier*).

For the third of the above items, since 1927 in the research laboratories of BASF in Ludwigshafen-Oppau a first centre was established. There under the leadership of *Matthias Pier* the previously addressed work of *Bergius* concerning the conversion of coal to fuels was continued on a broad basis, and on the other hand *Hermann Zorn* commenced with the research on structure and properties of oils. Both research fields were then, after technological implementation on a large scale, to gain the highest significance for the Air Force in World War Two.

Aviation Fuels. The extraordinary increase of aeroengines and their further development as well as the diverse fuels in World War One had made problems apparent, which were not known at the start. Most urgent was here the pinking in engines taking increasingly place in dependence on the fuel as result of the growing compression. In the USA and England this was already known since 1916 and extensively explored. In Germany one started since 1927 to investigate the pinking. With reference to the significance for the aviation, at the DVL work commenced lasting over several years among other things for exploring the phenomena in the engine, the composition of the fuels and the engine operating conditions regarding the pinking [55, 56]. In addition engine test methods for research and quality control were worked out [57]. In this connection the influence of a multitude of basic substances especially on synthetic basis were investigated. Questions concerning also possible side effects associated with relevant substances were examined like overheating, deposits and corrosion [58]. Since 1938 at DVL fuels for highly compressed and supercharged aeroengines were inves-

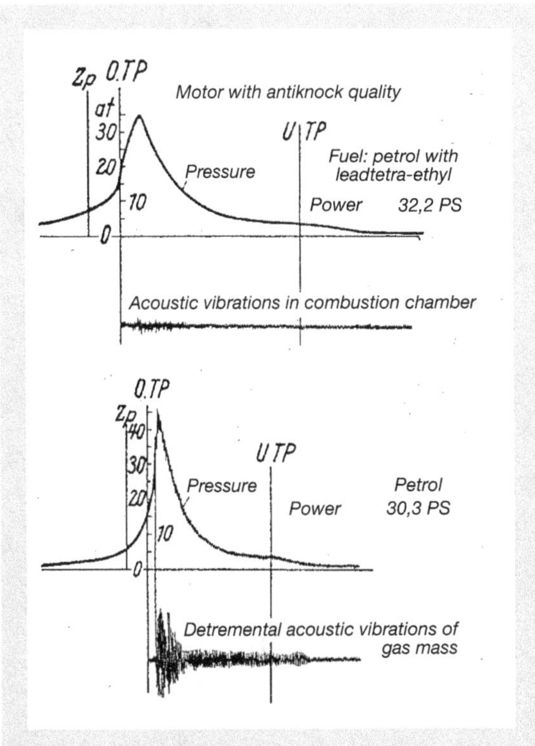

Pressure distribution of a non-pinking (top) and a pinking combustion (bottom) process.

tigated and in this connection a relevant engine test method (DVL supercharging method [59]) was developed. Because of the great significance of the leadtetraethyl for the anti-knock quality especially of the aviation fuels, until into the forties a number of laboratories investigated this topic [60]. This had become necessary regarding the pinking behaviour as consequence of different fuel components which had to be processed to obtain aviation fuels.

At the same time one had started again in the research laboratories of the BASF (later IG-Farben) to deal with the Bergius method of coal hydrogenation and thereby among other things found ways to obtain high-octane fuels (special catalysts and conditions). It was also possible by further research work to hydrogenate apart from brown coal hard coal as well to fuels which were, due to their higher octane rating, to a large degree better suited for aviation fuels.

Aeroengine Oils. For the research in the field of the aeroengine oils one of the first topics was the formation of deposits in aeroengines and their causes [61]. However, one required still years for the solution of the problem of the piston ring sticking [62]. As similarly difficult and time consuming proved the question concerning the causes for the oil ageing under engine operating conditions and its suppression by relevant inhibitors. Further important topics of the thirties were the viscosity-temperature properties and the cold behaviour of oils [63]. In the forties then also the lubricating behaviour in bearings and gears

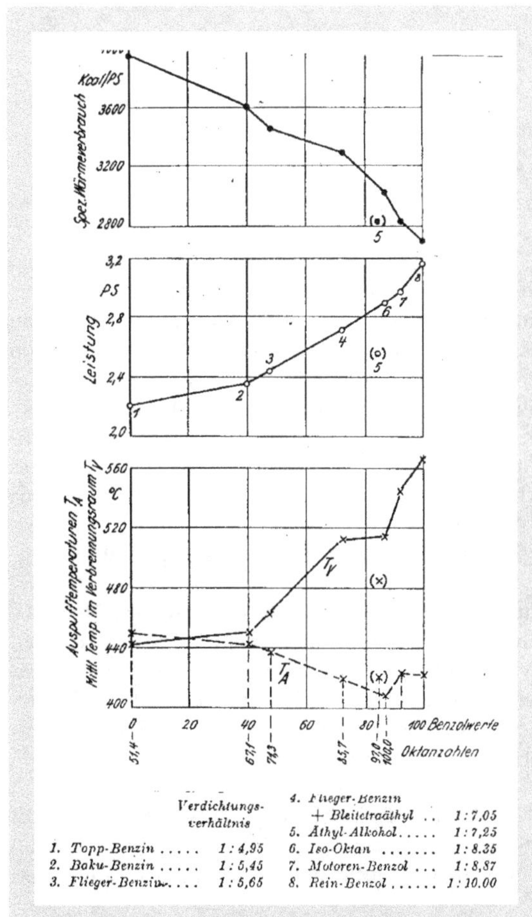

Heat consumption, power, and temperature in dependence of the octane number.

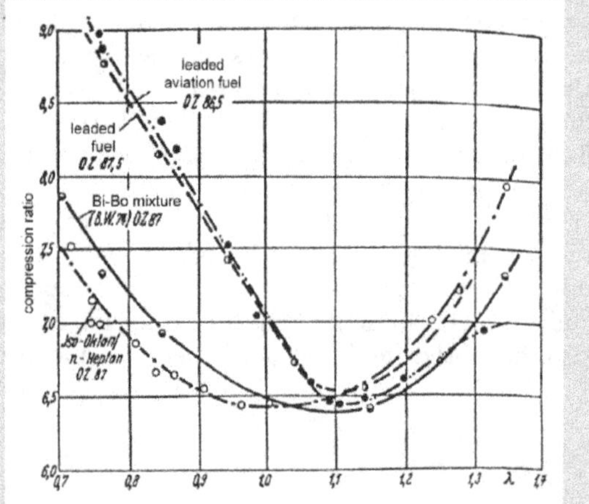

Determination of the anti-knock quality for various excess air values.

at higher temperatures and mechanical loadings was an important task [64].

Already in 1927 *Hermann Zorn* at BASF in Ludwigshafen-Oppau started to address questions concerning the suitablility of synthetic oils in engines. In an extensive fundamental work he synthesized about 300 hydrocarbons in order to explore for these the relation between chemical structure and properties [65]. In addition he dealt with questions regarding a definite improvement of certain properties by additives which he synthesized and tested [66]. Based on this work, Germany could as first country produce and use synthetic oils for aeroengines on a large scale.

Fuels and Lubricants and the Cold Starting Method
The cause for the cold starting problems experienced by the Air Force in North and East Europe in the winter 1941 was the insufficient low temperature behaviour of the engine oils used. For the solution of this problem work was started on two topics:

1. Lowering the solidifying point of the oils by addition of diluents.
2. Creation of special oils for low temperatures.

As first measure, for which in a short time usable results could be expected, diluting was dealt with. Here the focus was, also for logical reasons, on avi-

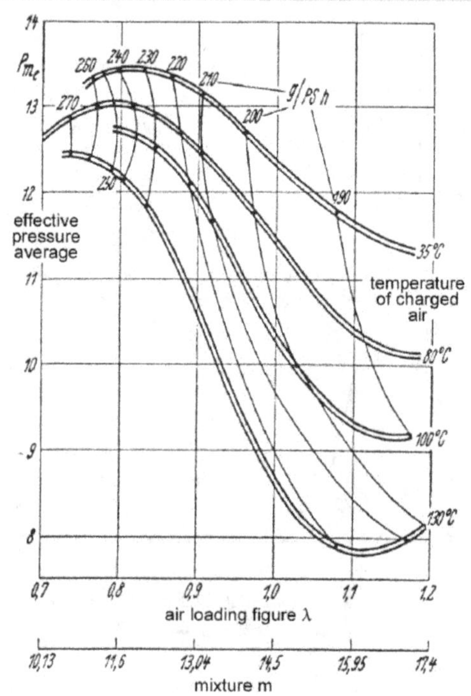

Testing of an octane number 87 fuel using the DVL supercharging method.

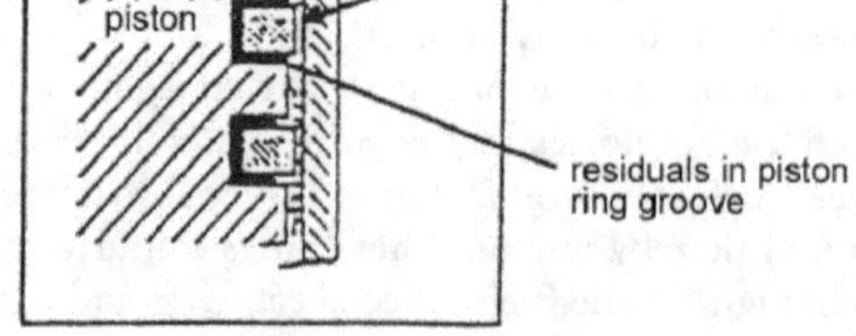

Piston ring region with deposits as result of the piston ring sticking.

ation fuels which were added in different quantities depending on the ambient temperature. Furthermore, it had to be clarified that this addition would not become a danger to engine and personnel. In the forces the acquired method was introduced as "Rechlin method".

The research carried out in Leuna aiming at a good temperature behaviour showed that paraffin chains with middle-class branching were especially suitable. These oils were synthesized from ethylene with the aid of catalysts and led then to the first synthetic aeroengine oils with a solidifying point of −30 °C (243 K). Subsequent research undertaken on a group of esters resulted in oils with in part much lower solidifying points. On the basis of the available raw materials then a second aeroengine oil with a solidifying point of −45 °C (228 K) could be technologically synthesized. This, however, came first into use in 1944. Both were the work of *Hermann Zorn*.

References

1 J. A. Gilles, Flugmotoren 1910 bis 1928, Verlag Mittler & Sohn 1971, 34/35

2 K. von Gersdorff, K. Grasmann, H. Schubert, Flugmotoren und Strahltriebwerke, Die deutsche Luftfahrt, Vol. 2, Bernard & Graefe Verlag, Bonn, 3rd edition 1995.

3 K. Grasmann, Aufschwung und Wende – Vom Kolben-Großflugmotor zum Strahltriebwerk, Interview von Wolfram Eisenlohr dated 10. April 1987, DGLR-Report 88-01, pp. 20–23

4 H. D. Köhler, Ernst Heinkel – Pionier der Schnellflugzeuge, Heinkel-Memorandum von 1936, 148–150, DdL Vol. 5, Bernard & Graefe Verlag, Bonn, 1983

5 Z./Luftfahrt international Nr. 7, Jan./Feb. 1975, Wettbewerbserfolge und Rekorde des DB 601 und DB 606, 1041/1042

6 Z/Luft- und Raumfahrt 5/92: BMW 803 A – Endprodukt der Kolbenmotor-Ära, 37–39

7 C. Himmler, Prüfstände für Höhenuntersuchungen, DVL-Jahrbuch 1938, pp. 349–352 u. Luftwissen 5/1938 pp. 432–436

8 O. Kurz, neuzeitliche Einrichtungen und Hilfsmittel der Triebwerksforschung, Luftwissen 9/1936, pp. 247–255, und R. Lucht, 30 Jahre Deutsche Versuchsanstalt für Luftfahrt, Luftwissen 4/1942, pp. 107–109, und
K. Graichen, Technische Denkmale der Luftfahrtforschung in Berlin-Adlershof – Höhenprüfstand für Motoren, Schriften zur Luftfahrtgeschichte, GBSL, Heft 3 (1994) pp. 27/28, und
H. Barth, Zur Entwicklungsgeschichte der deutschen Flugtriebwerks-Höhenprüfstände, DLR-Paper for the 75th anniversary of Adolf Baeumker

9 J. Zeyns, H. Caroselli, Bestimmung der Höhenleistung von Flugmotoren auf Grund von Leistungsmessungen bei Bodenbedingungen, DVL-Jahrbuch 1938, pp. 353 ff.

10 K. Pflügel, Klimaanlagen zur Prüfung von Flugtriebwerken in der Gruppe Süd der E-Stelle Rechlin, Rechliner Briefe, Erste Folge (1976), pp. 17/1–13

11 C. Schmid, R. Gauger, 50 Jahre Forschungsinstitut für Kraftfahrwesen und Fahrzeugmotoren Stuttgart (FKFS), ATZ-Automobiltechnische Zeitschrift 7/8 1980, pp. 359–366

12 Erinnerungen 1934–1984 – Flugtriebwerksbau in München: Der erste Höhenprüfstand der Welt: Herbitus, MTU 5. 1994, pp. 61/62

13 Ch. Soestmeyer, Betrachtungen zu Entwurf und Betrieb des BMW Höhenprüfstandes "Herbitus", DGLR-Vortragsveranstaltung dated 19. 3. 1991 in Munich, Schriften zur Geschichte der Deutschen Luft- und Raumfahrt Heft V (1992), pp. 141–164

14 D. Eckardt, P. Rufli, Advanced Gas Turbine Technology – ABB/BBC Historical Firsts. Contribution to the ASME Turbo Expo 2001, Session TE 01 EP 06-01, New Orleans, Louisiana, USA, June, 2001

15 H. Oestrich, Die Aussichten des Strahlantriebes für Flugzeuge unter besonderer Berücksichtigung des Abgasstrahlantriebes. DVL-Report 239 – DVL-Jahrbuch 1931

16 H. Schelp, Planung und Steuerung der Strahltriebwerksentwicklung. In: 50 Jahre Turbostrahlflug. Vol. II, DGLR-Report 92-05, 1992

17 A. Franz, K. Leist, K. Löhner, M.A. Müller, H. Oestrich, H. v. Ohain, H. Schelp, Strahltriebwerke, Vorträge der Arbeitssitzung der Akademie der Luftfahrtforschung dated 31.1.1941

18 Different Authors, DGLR-Symposium „50 Jahre Turbostrahlflug", Munich, 26./27 October, 1989, Vol. I DGLR-Report 89-05 (1989) and Vol. II DGLR-Report 92-05 (1992)

19 A. Betz, Patentschrift DRP 319413 vom 31.1.1918

20 A. Betz, Axiallader, Jahrbuch der deutschen Luftfahrtforschung, Teil 2, 1938, p 183

21 W. Encke, Tätigkeitsbericht des Institutes für Strömungsmaschinen der AVA Göttingen, AVA Report 45/S/2, June 1945

22 H. Kühl, Ähnlichkeitsbetrachtungen an Kreiselverdichtern, Forschung Ingenieurwesen, Vol. 13, 1942

23 E. Meldau, Drehströmungen im Drehhohlraum, Dissertation, TH Hannover, 1935
24 K. Bammert, H. Kläukens, Nabentotwasser hinter Leiträdern von axialen Strömungsmaschinen. Ing.-Arch. 17, No. 5, 1949
25 M. Strscheletzky, Gleichgewichtsformen der rotationssymetrischen Strömungen mit konstanten Drall in geraden, zylindrischen Rotationshohlräumen, Voith Forschung und Konstruktion, Issue 5, October 1959
26 R. Friedrich, Vergleich verschiedener Bauarten von Axialverdichtern. Dissertation TH Hannover 1949
27 G. Cordes, Strömungstechnik der gasbeaufschlagten Axialturbine, Springer 1963
28 H. Hagen, Zur Geschichte des Strahltriebwerkes – die Entwicklung bei BMW in den Jahren 1938–1945. Jahrestagung der DGLR, DGLR-Vortrag No. 81-075, Aachen, 11–14 May, 1981
29 A. Franz, Zur Geschichte des Strahltriebwerkes – die Entwicklung bei Junkers und seither erzielter Fortschritt. DGLR-Jahrbuch 1981, Report 76
30 H.-G. Münzberg, 50 Jahre Forschung und Entwicklung auf dem Gebiet der Turbostrahlantriebe (1938/39–1988/89). Sudetendeutsche Akademie der Wissenschaften und Künste, Naturwissenschaftliche Klasse, Sitzungsberichte, Jahrgang 1989, Heft 7
31 W. Gundermann, Zur Geschichte des Strahltriebwerkes – die Entwicklung bei Heinkel von 1936–1939. Jahrestagung der DGLR, DGLR-Vortrag No. 81-074, Aachen, 11.–14. May, 1981
32 K. v. Gersdorff, Die Anfänge der Turbostrahltriebwerke. In: „100 Jahre Menschenflug – Ein Jahrhundert Flugzeuge. Technik und Geschichte des Fliegens". Edited by L. Bölkow, VDI-Verlag, Düsseldorf, 1990
33 H. Oestrich, Die Entwicklung der L-Triebwerke bei BMW-Spandau. Deutsche Akademie der Luftfahrtforschung, Arbeitstagung „Strahltriebwerke", 31. January, 1941, pp. 135–144
34 H. Oestrich, Entwicklung der Strahltriebwerke. Interner Vortrag bei der SNECMA, France, 1960
35 R. Gunston, Lexikon der Flugtriebwerke, Motorbuch Verlag, Stuttgart, 1991
36 K.W. Streit, J.W.R. Taylor, Geschichte der Luftfahrt. Sigloch Service Edition, Künzelsau, 1975
37 J.W. Adderley, German Gas Turbine Developments during the Period 1939–1945. B.I.O.S. Overall Report No. 12, London, 1949
38 Erinnerungen: 1934–1984 Flugtriebwerkbau in München. Herausgegeben von der MTU Motoren- und Turbinen-Union München GmbH, Munich, May 1984
39 O.E. Pabst, Entwicklungsarbeiten an einem Lorin-Triebwerk. ZfW, Zeitschrift für Flugwissenschaften, 2. Jg., Heft 2, Februar 1954, pp. 29–40
40 A. Keller, Die Abgasturbine von C. Lorenzen, VDI-Z 72 (1928) p 1869
41 C. Lorenzen, The Lorenzen Gas Turbine and Superchargeres for Gasoline and Diesel Engines, Mechanical Engineering, 52 (1930), July; 665–672
42 A. Betz, (Hrsg), Monographien über Fortschritte der deutschen Luftfahrtforschung (seit 1939). Teil O, Strömungsmaschinen Aerodynamische Versuchsanstalt, Göttingen, 1946
43 K. Leist, Probleme des Abgasturbinenbaues, DVL-Jahrbuch 1938, pp. 399–412
44 K. Leist, Der Laderantrieb durch Abgasturbinen, DVL-Jahrbuch 1937, pp. 385–390
45 E. Knörnschild, K. Leist, Untersuchungen an Turbinenschaufelgittern, DVL-Jahrbuch 1939, pp. 424–432
46 H. Oestrich, Entwurf und Ausreifung des Triebwerkes BMW 003, DGLR-Veröffentlichung, 89–223, 1989
47 H. Wagner, Dokumentation zu Leben und Werk, DGLR-Gedenkschrift 1988, mit Beitrag v. R. Friedrich, pp. 81–94
48 C.B. Meher-Homji, The Development of the Junkers Jumo 004 B, ASME-Paper 96-GT-457, 1996
49 K. Gradl, M. Baumgärtner, Triebwerkschaufeln im Wandel der Zeit, Bauart, Werkstoffe, Fertigungsmethoden, DGLR-Vortrag 94-G1-109, 1994
50 A. Burghardt, H. Beisswänger, Über die Herstellung von Hohlschaufeln für Abgas- und Gasturbinen, Teil 1 – Werkstattstechnik 51 (1961), 8, pp. 413–418, Teil 2 – Werkstattstechnik 51 (1961), 11, pp. 58–590
51 A. Garve, Geschichte der Augsburger MAN – Gasturbinen-Entwicklung 1940–1964, Interner Bericht MAN AG, Werk Augsburg, Mai 1975
52 K. Dietrich, Die Analyse und Wertbestimmung der Motorenbenzine und -Benzole des Handels. Berlin: Verlag des Mitteleuropäischen Motorwagen-Vereins, 1915
53 F. Martin, Friedrich Bergius zum Gedächtnis, Chem. Ing. Technik 21 (1949), pp. 329–331
54 E. Jantzen, Das Institut für Betriebsstofforschung der Deutschen Versuchsanstalt für Luftfahrt, Berlin-Adlershof, Blätter zur Geschichte der Deutschen Luft- und Raumfahrt (1998)
55 A. Philippovich, Der Verbrennungsvorgang im Explosionsmotor. Luftfahrtforschung 13, (1936), pp. 199–205
56 F. Seeber, Über die Beziehungen zwischen dem motorischen Verhalten von Kraftstoffen und den Betriebsbedingungen. Luftwissen Vol. 3, (1936), pp. 326–328

57 A. Philippovich, Über die motorische Betriebsstoffprüfung. 01 und Kohle, (1939), pp. 551–557
58 O. Widmaier, Siedeverhalten und Rückstandsbildung von Flugmotoren-Kraftstoffen. Jahrbuch der Deutschen Luftfahrtforschung 11, (1939), pp. 356–360
59 F. Seeber, Neuere Verfahren zur Kraftstoffprüfung. Luftfahrtforschung Bd 16, (1939), pp. 431–437
60 O. Widmaier, L. Neuninger, Entwicklung und Bedeutung des Bleitetraäthyls als Klopfgegenmittel. Technische Berichte 9, (1942), pp. 109–121
61 A. Philippovich, Neuere Untersuchungsverfahren zur Bestimmung der Rückstandsbildung in Motorenölen. DVL-Jahrbuch 1933, pp. 36–42
62 W. Glaser, Der Einfluß der Betriebsbedingungen auf das Kolbenringstecken. Luftfahrtforschung 16, (1939), pp. 438–446
63 K. Siebald, Das Verhalten von Schmierölen in der Kälte. Luftwissen 8, (1941), pp. 221–229
64 C. Krienke, Die Prüfung von Schmierölen durch Reibungs- und Verschleißversuche. Jahrbuch der deutschen Luftfahrtforschung (1942), pp. 156–161
65 E. Jantzen, The Origins of Synthetic Lubricants: The Work of Hermann Zorn in Germany - Part I Basic Studies of Lubricants and the Polymerisation of Olefins. J. Synth. Lubr. 12, (1996), pp. 238–301
66 E. Jantzen, The Origins of Synthetic Lubricants: The Work of Hermann Zorn in Germany - Part II Esters and Additives for Synthetic Lubricants, J. Synth. Lubr. 13, (1996), pp. 113–128

Flight Mechanical Stability and Control

Inherent Stability and Controllability

Andreas Hafer, Gottfried Sachs*

At the beginning of the development of flight devices heavier than air, the question of how to create lift to overcome gravity naturally stood in the foreground. Another important role in the 19th century played the problem of gaining inherent stability, mainly longitudinal stability [1]. *It was no coincidence that the first thorough theoretical investigations of the characteristics of aircraft during flight dealt indirectly (F. W. Lanchester* [2]*) or directly (Bryan and Williams* [3]*) with longitudinal stability.*

The Development of Efficient Control Systems

There is probably a connection between the big importance of longitudinal stability for flight technical research and the fact that the flight experience from experimental research was exclusively based on flight models. In this sense it is remarkable that subsequently neither *Lilienthal* nor *Weißkopf* or the *Wright Brothers* paid much attention to the problem of longitudinal stability. *Orville* and *Wilbur Wright* directed their main attention to the controllability of their aircraft. Their aircraft proved, despite the inherent longitudinal instability, under the chosen test conditions, securely flyable, and the entire "man-aircraft system" proved to be stable [4].

Here, the prerequisite was the development of efficient control systems [5, 6]. The early Wright aircraft were designed as canard configurations; a shaft supported along the leading edge of the wing moved the elevator. Roll control was realized by wing twist combined with a deflection of the rudder. *P. Mouillard* already patented the principle of wing twist in 1897 [7], but it is questionable whether the *Wright Brothers* knew about this. The activation of this combined wing-twist-rudder-control was realized for the early Wright-aircraft by the "hip cradle", a device, which could freely be shifted laterally on the lower wing by hip movements of the pilot resulting in a left or right roll of the air vehicle.

The short description of the control of the Wright aircraft shows that today's common steering elements, like wheel steering, control stick steering, pedal steering, as well as the utilization of fixed rudders, altogether did not appear on these early aircraft. But they were developed only a short time later. The control stick, which could simultaneously operate elevator and aileron was introduced in 1904/05 (*Esnault-Pelterie* [8]). *Santos-Dumont* used in 1906 for the first

---

* Text corresponds to a slightly shortened version of a contribution to the DGLR Yearbook 1988, I, pp. 659–669

time the aileron. Approximately at the same time, the pedal steering was introduced for the operation of the rudder. It is remarkable that these modern control elements were developed in France independently of the Wright-aircraft.

In France, *Ferdinand Ferber* began in 1899 – inspired by the publications of *Otto Lilienthal* – to consider flight technical problems and to conduct flight tests. After 1904 he developed a new self-stabilizing aircraft with a long tail section with a horizontal control surface and an attached rudder. With this aircraft, which the pilot controlled sitting upright, *Ferber* conducted his first successful flight on May 27, 1905 in Chalais Meudon. He was the first to transfer, theoretically, *Bryan's* approach to longitudinal stability to the lateral motion [9]. In his paper [10], published in 1907, he writes in the introduction: "*...it is the first time that we surpass in our calculations the honorable rules of the three V*". These rules concern the instruction of *Penaud* to apply dihedral to the wing and the horizontal stabilizer with respect to each other, the latter having been called by *Zahm* "longitudinal dihedral". The third "V" refers to the sweep of the leading edges, whose flight mechanical importance could possibly be related to the stall characteristics. *Ferber* took the existence of static longitudinal stability as well as the application of positive wing dihedral for granted.

First Attempts to Define the Desired Flight Characteristics

First attempts to bridge the gap between science and practical application with regard to flight characteristics can be observed after 1913. Here, the foundation of corresponding research institutions may have been very supportive: the RAE in Farnborough (1910), the DVL in Adlershof (1912) and NACA in Langley Field (1915). There was, in addition, a further development. More and more pilots were employed who did not understand too much about the physical aspects of flight. This was especially true in the case of the emerging military corps, but also for the aircraft companies, which employed an increasing number of test pilots.

Bendemann (himself a member of the of the DVL in Adlershof) considers in 1913 in a presentation "Today's status in air vehicle engineering". Here, he used the expression "flight characteristics" but meant "stability and dynamic economics", i.e., also the optimization of engine performance.

On the other hand his statement "*concerning stability and controllability properties, it is the overwhelming opinion of our pilots that the "pigeons" ("Tauben") are the most comfortable to fly*" aims at the problem of the flight characteristics of aircraft and the inherent factors of stability, controllability and the pilot's susceptibility [11]. During the same year, *L. Bairstow, B.M Jones* and *A.W Thompson* wrote in the introduction to an article: "*It has to be the absolute objective to design an aircraft as airworthy as possible reacting satisfactorily to any possible practical circumstances*" [12]. In his contribution to a discussion of a presentation of *von Kármán* concerning the topic of longitudinal stability [13], *Ludwig Prandtl* demands in 1914 the continuation of the treatment of the questions considered by *von Kármán* in two directions: on the one hand, the refinement of the stability models should be advanced, and on the other hand:

"*What stability conditions should be strived for, high or low stability, moderate or strong damping. The answer will, on the one hand, possibly be gained by investigating realized aircraft with respect to their stability properties, on the other hand, by collecting the pilots' judgments concerning minor or major comfort during controlling the aircraft in calm or choppy air*".

Here, he anticipated efforts, in questions and approach, which were carried out only twenty years later. Another aspect seems to be remarkable: with *Prandtl* assigning the engineers to interview the pilots, he is underlining the importance of science for the further development of aeronautics.

Already prior the outbreak of World War One in August 1914, the military turned out to be the main party of interest in the aeronautical technology. The military use of aeronautics, not only in Germany, was further advance by the outbreak of the war. The occupation of the installations of the DVL in Adlershof by the "Inspection of the Flying Corps" ("Inspektion der Fliegertruppe, IdFlieg") was evidence for this development. The latter was responsible for testing and selecting front-line aircraft during the war.

Still during the war, the flight tests concentrated mainly on the determination of flight performances: these were decisive for selecting an aircraft. Controllability and stability characteristics were thought to be secondary in judging the quality of an aircraft for front-line duty. This was probably not only due to the attitude of the inspector, but also reflects the fact that flight performance was the only measurable, hence unbiased aircraft property.

The selection criteria led during the war, where aircraft were utilized in increasing numbers, to complaints of the front-line pilots concerning the nearly useless aircraft [14], with regard to handling qualities under combat conditions.

To solve the problem, the IdFlieg decided to perform flight comparisons, where aircraft offered by industry could be selected during a direct comparison. Also combat pilots participated in these comparative flights.

The comparative flights were performed in January, June and October 1918 and each time two fighter aircraft of different performance were to be selected. For this purpose, the same pilots flew the aircraft presented by industry, and mock air battles with changing crews were carried out. The judgment of the pilots together with the evaluation of the performance then led to the selection of the aircraft.

This method is remarkable: lacking reliable objective statements about the "flight characteristics", a very empirical procedure was chosen for their evaluation. This procedure had, of course, predecessors especially in general and military flight contests. "Flight characteristics" in today's sense were, of course, not assessed in an isolated way.

In addition to the comparative flights organized by the IdFlieg, a "meeting of aircraft designers" took place in Adlershof in August 1918 concerning the topic "flight characteristics". Participants were representatives of the military and of aircraft companies. It is remarkable that the leading personalities of the flight sciences were not yet present [15].

Examination of Flying Qualities during the Twenties

Examinations of flying qualities were for the first time performed by the DVL in 1926 within the framework of type-rating [16]. However, guide lines, let alone requirements or regulations, were entirely lacking. Accordingly was the practical examination of the following year accompanied by attempts to gain clarity about what and how one should examine.

In 1928, *Walter Hübner*, responsible for the determination of flying qualities, reports for the first time on experiences and work related to the test procedure [17]. Without considering an individual type of aircraft, he states that with regard to stability and controllability about the vertical and lateral axes, a satisfactory status had been achieved, on average, and that the differences between the individual types were not large.

Difficulties were still related to the controllability of stalled flight, something that was still considered as an unsolved task.

Hübner, when talking about flight characteristics, lists the following properties that determine the flying qualities: load distribution, stability, controllability, maneuverability, and spin characteristics.

An inspection plan was set up, which also included those characteristics of the aircraft that influence the operation by the pilot such as sight, control of the engines, instrumentation layout, et cetera. The examination of the flight characteristics during type testing was initially of only little influence on the development since research results were lacking, which would provide precise design rules, as well as the corresponding design and construction regulations.

There was thus no lever to demand certain flight properties from the manufacturer. Starting in 1927/28, there were requirements concerning the flight characteristics, which were put down in the design and construction regulations [18], but they remained permanently in a state of development to allow, with increasing experience from practical tests, further additions. They are very general and contain nearly exclusively qualitative statements. In how far individual quantities are based on more or less systematic tests is hard to decide. Only occasionally, after 1928 more systematically, began the Flight Department of the DVL to develop in-flight measuring procedures and to conduct flight measurements. *Hübner* pointed out that, due to the quantitative statements to be expected from measurements, "*... the designer and the theoretician would undoubtedly be provided with important documentation... The measurements will probably be able to supply precise parts out of the overall flying qualities of an aircraft, whose entirety, can only be provided by the assessment of the pilot*".

In this sense, the measurements served "*only as a supplement to and evidence of the observations of the pilots*".

The reason for *Hübner's* efforts was to enforce basic safety standards for German aircraft: for that purpose all flying qualities guidelines respectively regulations, should contain useful and easily verifiable criteria. Quantitative information played only a mi-

Specification of flight characteristics within the German design and construction regulations of 1928

4515 Stability: with the center of gravity being in the most rearward position and with constant stabilizer position, it is required with a
a) free elevator and full engine power: continuous climb,
normal engine power: straight and level flight, throttled engine: continuous glide (velocity no more than 110 per cent of horizontal flight),
b) free rudder: directional stability and straight flight, also for multi-engine aircraft with one engine failing, with the aid of rudder deflections,
c) free aileron: the banking aircraft has to return into a normal position by itself and remain there.

4516 Friction in control and rudder bearings such that released control devices return automatically into their central position.

4517 Maneuverability such that the aircraft can, from any attitude, be quickly and safely returned into the normal position, especially out of a spin.

4518 Elevator steering forces at the control stick at full power and strait and level flight with the elevator in cruise position no more than 6 kg (13.2 lb).

4519 Equal steering forces at equal deflections of the aileron, rudder and elevator.

Additional demands for aerobatics

4541 Height loss: In case a spin motion is not prevented by aerodynamic or constructive means, the height loss from the beginning of the control movement to pull out of the spin to normal horizontal flight must not exceed 70 m (230 ft). The engine must here be in idle.

4542 Pulling out of a spin motion (in case it is not excluded) after at least five steady consecutive turns with throttled engines must faultlessly be possible. The slope of the longitudinal axis of the aircraft must during the spin motion never be smaller than 45°.

4543 The aileron effectiveness must be sufficient to perform in a non-stalled flight a steady turn about the longitudinal axis of 360° at a height loss of no more than 20 m (66 ft).

nor role. He correspondingly restricted himself in his instructions for flying qualities examinations, which were meant for the manufacturer, to qualitative examinations [19]. *Hübner* stated in 1930 [20]: "*... only the aircraft is safe, which can easily and under all circumstances be controlled by the pilot, i.e., an aircraft with good flight characteristics. The question of flight characteristics is a question of safety. That is why for civil aviation the provision of perfect flight characteristics is a question of survival ...*" [21].

The research activities of the Flight Department regarding the flight characteristics was stopped in 1933. The flying qualities examinations were transferred as part of the general type rating to the "Flight-Test Center of the Air Force" ("Erprobungsstelle der Luftwaffe") in Rechlin.

The transfer of the flying qualities examinations also for civil aircraft to a military installation had not only many advantages, it also documents the priorities for future aircraft development. Also other countries proceeded in that way.

The Development of Flying Qualities Regulations during the Thirties and Forties

In 1935, *August Kupper* (he had received a PhD with a thesis related to mathematical physics) was transferred as an employee of the DVL to Rechlin, obviously with the objective to systematize the knowledge concerning flying qualities criteria and to further develop them, in the sense of a theoretical reconditioning, especially in the direction of quantification.

The first result of this work was a report "Guidelines for the assessment of flying qualities" [22], which represented an inventory of the criteria developed up to that time. *Kupper* explained the objective of this work in the introduction:

"*It is being tried to express, as far as possible, flying qualities in hard numbers and to replace the more or less biased and intuitive judgment of flying qualities with these numbers*".

The aim of this work was obviously another than the preceding work of *Hübner*: it was not the "breeding" of good and hence safe aircraft, but objectivity. Only a sufficient quantification could, according to *Kupper*, contribute to this.

Kupper's demand could, however, not conceal that "*reliable numbers were lacking at every turn*", as *K.H. Doetsch* et al. concluded [23] in 1943. It is remarkable in this context that in *Kupper's* report there is no mentioning of criteria for dynamic stability. Did he consider them "not sufficiently measurable"?

Kupper died 1937 in a flight accident. His work was continued only slowly, which was mainly due to the fact that his approaches were already ahead of his time. It was not only that measured values were, as mentioned, lacking; it was more because of the lack of practical measuring devices.

Thoenes (who worked together with *Kupper* in 1936) thus comes in his contribution to the "Ringbook of Aeronautics" ("Ringbuch der Luftfahrttechnik") still in 1940 to the conclusion that the possibilities for deriving quantitative data for the establishment of flying qualities regulations were still very limited despite some progress since 1936.

Also in England work was carried out, which intensified the effort towards a mainly quantitative assessment of the flying qualities. In the Netherlands, *Baumhauer* worked out extensive catalogues of flying qualities criteria [24], and also in the USA more effort into a research related approach was invested.

This simultaneousness, without communication across borders and particularly since neither German nor British or French research results were publicly accessible, shows the growing appreciation of a solution to this problem by all researchers and engineers. Of greater importance were probably changes in the structure of flight technology and the utilization of aircraft. *Melville Jones* saw already in 1939 the growing interest in flying qualities in conjunction with the "*mass production of aircraft and pilots*" [25]. On the one hand became the production of aircraft more expensive, economic reasons made it more important to be able to plan the flying qualities of an aircraft well in advance, especially to shorten the test phase, on the other hand the education of the pilots had to be standardized requiring trainer aircraft, which, in their flying qualities, were not to deviate too much from the future operational aircraft. Generally, a better compatibility of the individual aircraft types had to be achieved to ease the transfer from one aircraft to another.

Concerning the approach to the determination of comprehensive flying qualities regulations, which are essentially based on quantifiable criteria, one can observe large differences between different countries; these differences shall be demonstrated by looking at the developments in Germany and in the USA [26–28].

NACA started in 1936 in the USA with a program to determine flying qualities criteria. The starting point was the cooperation between NACA, leading airlines and Douglas regarding the determination of desired flying qualities for the DC-4. *E.P. Warner* established a catalogue of qualities criteria for a larger transport aircraft, which showed where quantified data were available and where they were completely lacking. This work (that *C. Milliken* already discussed in detail at a meeting of the *Lilienthal* Society in 1936 [29]) was essentially based on conversations with airline pilots and engineers of the aircraft manufacturers. It had on purpose a preliminary character and was considered as starting point for further research. The deficits outlined in this work resulted in further research steps: the development of instrumentation and test procedures, flight tests with numerous aircraft (all in all 24 different types) to determine flying qualities and systematic comparisons of the quantitative test results with an assessment by pilots. Here, rating scales were used similar to the ones *Soulé* had already worked out in 1936 for the comparative assessment of quantitative test results and pilot perceptions concerning dynamic longitudinal stability [30].

The flight tests included all types of aircraft, military as well as civil. The results achieved were presented in the 1940 report of *H.A. Soulé* "Preliminary Investigation of the Flying Qualities of Airplanes" [31] and in the report of *R.R. Gilruth* "Requirements for Satisfactory Flying Qualities of an Airplane" [32], written in 1941 but only published in 1943.

The fundamentals of the assessment of flying qualities, which were established by NACA in cooperation with military agencies and aircraft companies, were (updated because of the USA's entrance into the war in 1941) relatively quickly revised for military purposes and published as NAVAER SR 119 for the Navy in 1942 and as AAF-C-1815 for the Air Force in 1943.

These led after the war to a version, which was published as Mil Spec 1815-B in 1949, and which became the basis for the development of the flying qualities regulations in the time thereafter.

In Germany, the assessment of flying qualities remained essentially a matter of the Test Centers and their employed test pilots and engineers. In this sense it is characteristic that the first (and for a long time the only) attempt of a systematization of the examination of flying qualities in 1937, half-published, was carried out by employees of the Test Center for Seaplanes in Travemünde [33]. The result is contained in a "Ring-book", which was envisaged by the Test Center (E-center) as communication basis especially with other E-centers. The chapter on flying qualities regulations is, accordingly, laid out as a flight-test guide. For every single checkpoint – balance, stability, heaviness (distinguishing between axes), rudder (differentiating according to forces and effectiveness), and occurrence of flow separation – it is distinguished between test type and test quantity and again between qualitative and quantitative results. Slightly more general are the test regulations for "aerobatics", "take-off and landing", "maneuvers on the ground", "spin", "dive" and "vibration phenomena". Within the "Travemünde Ring-book" [34], it is, in considering the weighting of the criterion "stability", very accurately distinguished between the aircraft type and the "type of stability".

The Travemünde Ring-book could only partly fulfill its intended task, to serve as guide for test pilots, since it was obviously (probably due to secrecy requirements?) not widely distributed. Otherwise was the work on flying qualities regulations continued at the DVL under the supervision of *K.H. Doetsch*, who was until *Kupper's* death his collaborator.

The individual steps of these activities were in principal similar to the American ones: they encompassed flight tests, improvement of the instrumentation, and comparisons with the pilot's opinions. The extent of these activities was, of course, somewhat smaller. Especially the analysis of the pilots' assessments was not based on rating scales but on conversations with established test pilots mainly of the Test Center in Rechlin. Also the number of aircraft employed was less. After all, the outside did not especially support the projects. This holds especially for the "Reichs Ministry of Aviation" ("Reichsluftfahrtministerium, RLM"), which was not very much interested in basic research [35]. Also the employees of the Test Centers were skeptical about the establishment of regulations since they feared a transfer of competence and more bureaucracy. The German Air Force obviously did not highly appreciate the value of such regulations either.

In 1943, DVL issued the "New Draft of Flying Qualities Regulations". The proceedings of the conference, were published as Lilienthal Report No. 163 [36]. This new draft constitutes in so far a new system of flying qualities examinations as a new grading according to the utilization was introduced, replacing the old one according to take-off weight, which means that the qualities requirements were now directly linked to the field of application.

On the other hand "*seems the controllability to be the dominating demand*". This corresponds to the conceptual development already outlined in the introduction. While *Hübner* mainly derived his flying qualities criteria from stability properties and steering forces – also in the Travemünde Ring-book stability conditions are given the widest space – means the altered focus in the new draft of the regulations a significant shift in the basic attitudes and thus a further development in the field of flying qualities. In this sense it is characteristic that in the new version the new expression "Response" was introduced into the technical terminology.

The investigation now is divided into a predominantly qualitative basic examination where statements are made based on flight perceptions or according to the indications of the common onboard instruments. Here, the result is simply "passed" or "failed". To pass this basic examination was, among other things, to guarantee that the aircraft types tested were able to fly with the necessary safety under all possible flight conditions.

Only if this requirement was met, the test aircraft was admitted to the examination by measurement. This was meant to provide quantitative results in all instances where a comparison with reference values was required and possible. Here, self-recording instrumentation was employed whose recordings were analyzed after the flight. From the recordings, the test data were extracted, following given processing instructions. Data had to be compared to the reference values for an assessment. For the most important characteristics, quantitative reference values were given, under consideration of the grading steps already mentioned, which allowed an assessment of the flying qualities data derived from the flight measurements.

Since the examination was to be carried out by self-recording instrumentation, the role of the pilot with regard to the flying qualities examinations changed. The evaluation was performed afterwards, namely by the systems engineer. Considering the history of the development of the regulations in the USA and in Germany, it is astonishing that the two reports published in 1943 by *Doetsch* and his co-workers and by *Gilruth*, respectively, essentially came to the same results despite the lack of scientific communication due to the war.

Control Configured Stability

Horst Prem

In the context of the trend setting investigations related to the determination of flying qualities one has to mention the 7th Scientific Meeting on September 20, 1940 [37, 38], which dealt, under the chairmanship of *Ernst Heinkel*, with questions of static longitudinal stability at high-speed flight and with control configured stability. *Heinkel* initiated the meeting with the following words:

"The design and construction regulations for aircraft require that an aircraft trimmed for a normal flight attitude and left to itself will not assume any dangerous attitudes and will automatically return to its normal attitude. 'Automatically' means without action of the pilot or by an automatic control system/ autopilot. Such an aircraft possesses an inherent stability.

At the beginning of aircraft engineering the requirement of inherent longitudinal stability was rarely called for and only seldom realized. The French school actually taught that inherent stability – as it was called at the time – was detrimental to the controllability and maneuverability of the vehicle, and so there existed at the beginning of World War One some successful aircraft types, which were, with the knowledge and the intention of their builders, unstable. To fly these machines might not have always been easy. In any case, training in these machines was difficult. The necessity to school a large number of pilots quickly and safely, soon led during World War One to a general acceptation of the demand for an, if only small, longitudinal stability. When after the war passenger transport aviation arose and an increasingly greater regularity and safety of long-range air traffic was demanded, the requirement for inherent longitudinal stability became law. IFR (Instrument Flight Rules) schooling and IFR equipment, initially without automatic control, soon were taken for granted. It was rightfully demanded that, when IFR instrumentation failed, e. g., due to icing, the aircraft finally would automatically assume a normal and safe attitude. In the first years after the war, it was obviously not very difficult to meet the stability requirements. Difficulties only arose when the performance of the engines increased manifold in comparison to the performance at the end of the war, since the propeller slipstream had the property of reducing the longitudinal stability. Another reason for a reduced longitudinal stability could be found in the design of faster aircraft where the compressibility of air could no longer be neglected. Of importance was here an increase in downwash and in the unstable fuselage moment with the increasing influence of compressibility. This is something Mr. Helmbold shall talk about. Further stability problems occurred when it was attempted to reduce the friction drag of the airfoils to a fraction of its turbulent value by laminarization of the boundary layer. The laboratory tests showed for especially low-drag airfoils a stability-wise extremely unfavorable pitching moment behavior. But there is still hope that this unfavourable trend was created by the turbulence remaining within the artificial airflow of the wind tunnel and that this trend does not occur in the practically turbulence-free atmosphere.

The further advancements in aircraft engineering, characterized by an increase in engine performance and speed, necessarily lead to a rapidly diminishing inherent longitudinal stability. The transition to control configured stability, which will be created by an automatic control system, will, therefore, become an unavoidable necessity in near future. It is hence valuable to know that the generation of control configured (artificial) longitudinal stability is principally and generally possible. Mr. Fischel will address the possibility of creating control configured longitudinal stability, the necessary properties of control systems and the control methods, which are being applied or are still applicable."

The fundamental stability problems related to the rigid aircraft had been correctly recognized at the beginning of the forties, and also the means to avoid these problems are already described in contributions of *Helmbold* [37] and *Fischel* [38]. The practical realization occurred only 35 years later through *Wolfgang Kubbat* and *Gerhard Löbert* with the CCV-F-104, the first aircraft with control configured instability about the pitch axis, which performed its first flight in 1977 with a fully digitalized and quadruplex redundant flight control system. Only 50 years later one was able to demonstrate with the X-31 that also an aircraft, which has surpassed its longitudinal stability limits and has left the domain of attached flow, is controllable. The unstable lateral movement is today investigated within the VECTOR-program (Vectoring Extremely Short Take-off and Landing, Control and Tailless Operations Research).

References

1 A. Hafer, X. Hafer, G. Sachs, Die Anfänge der Stabilitätsbetrachtung in der Flugmechanik, ZFW 12 (1988), pp. 27–44
2 F.W. Lanchester, Aerodynetics, London, 1909 (In German by C.and A. Runge, Leipzig 1911)
3 G.H. Bryan, W. E. William, The longitudinal stability of serial gliders, Proc. Roy. Soc. London 73 (1904), pp. 100–116
4 C.S. Draper, Flight Control, J. Roy. Aer. Soc. 59 (1955), pp. 451–477
5 H.R. Jex, F.E.C. Culick, Flight Control Dynamics of the 1903 Wright Flyer, AIAA-Paper 85-1804, 1985, p. 541
6 R. McLaren, The Wright Flyer, Aero Digest, Vol. 67, July 1953, pp. 28–32
7 P. Mouillard, US-Patent Nr. 582737 dated 18 May, 1897, The warpening of the rear of the wing
8 R. Esnault-Pelterie, Experiences d'aviation exécutées en 1904, en verification de celles des frères Wright, L'aérophile, June 1905, pp. 132–138
9 F. Ferber, Les Progrès de l'Aviation par le Vol plané, Revue d'artillerie, Novembre 1905, pp. 81–105
10 F. Ferber, Les Progrès de l'Aviation par le Vol plané, Revue d'artillerie, Octobre 1905, pp. 44–74
11 F. Bendemann, Der heutige Stand der Flugmaschinenkonstruktion, Jb. der WGF 2 (1913/14), pp. 118–142
12 L. Bairstow, B.M. Jones, W. Thompson, Investigation into Stability of an Airplane, R&M, 77.3 (1913), pp. 135–171
13 Th. Von Kármán, E. Trefftz, Über Längsstabilität und Längsschwingungen von Flugzeugen, Jb. WGF 3 (1914/15), pp. 116–137
14 A.R. Weyl, Fokker, The Creative Years, London (Putnam), 1965, 150
15 Besprechung der Flugzeugkonstrukteure in Adlershof am 23. August 1918, Techn.Berichte III (1918), pp. 185–186
16 Jb 1927 der DVL, S. XI
17 W. Hübner, Erfahrungen bei Flugeigenschaftsprüfungen im Jahre 1927/28, ZFM 20 (1929), pp. 189–195
18 Bauvorschriften für Flugzeuge, 1928
19 W. Hübner, Prüfung der Stabilitätseigenschaften von Flugzeugen, Annual publication of DVL 1929, p. 23
20 W. Hübner, Anweisung für die Prüfung der Eigenschaften von Flugzeugen, ZFM 21 (1930), pp. 529–533
21 A. Thoenes, Über Wesen und Bedeutung der Flugeigenschaften und ihre Prüfung, Ringbuch der Luftfahrttechnik, 1940, Vol. 11
22 E-Stelle Rechlin (A. Kupper), Richtlinien für die Beurteilung von Flugeigenschaften, 1936 (unpublished)
23 K.H. Doetsch, G. Friedrichs, P. Höhler, W. Liebe, W. Lungner, Neuentwurf von Flugeigenschaftsrichtlinien, DVL, Institut für Flugmechanik, 1943
24 A.G. von Baumhauer, Testing Stability and Control of Aeroplanes, J. Roy. Aer. Soc. 43 (1939), pp. 334–369
25 G.T.R. Hill, The Flying Controls of the Aeroplanes, J. Roy. Aer. Soc. 43 (1939), pp. 411–432
26 H. Herb, Zur Entwicklungsgeschichte der Flugmechanik, DLR FB 67-44, pp. 95–107
27 D.J. Moorhouse, The History and Future of US Military Flying Qualities Specifications, AIAA-Paper 79-0402, 1979
28 R.P. Harper, G.E. Cooper, Handling Qualities and Pilot Evaluation, J. Guidance Control and Dynamics, 9 (1986), pp. 515–529
29 C.B. Milliken, On the Results of Aerodynamic Research and their Application to Aircraft Construction (Vortrag auf der Jahresversammlung der Lilienthalgesellschaft, 1936), J. of Aer. Sciences 4 (1936), pp. 43-53, (deutsch: Luftwissen 3 (1936), pp. 277–287)
30 H.A. Soulé, Flight Measurement of the Dynamic Longitudinal Stability of several Airplanes and a Correlation of the Measurement with Pilot's Observation of Handling Characteristics, NACA Rep. 578 (1936)
31 H.A. Soulé, Preliminary Investigations f the Flying Qualities of Airplanes, NACA Rep. 700 (1940)
32 R.R. Gilruth, Requirements for Satisfactory Flying Qualities of an Airplane, NACA Rep. 755 (1943)
33 E-Stelle Travemünde, Ringbuch für die Entwicklung und Erprobung von Flugzeugen im Bereich der Erprobungsstellen 1937 (unpublished)
34 H. Beauvais, K. Köhler, M. Mayer, C. Regel, Flugversuchsstellen bis 1945, Johannisthal, Lipetzk, Rechlin, Travemünde, Tarnewitz, Peenemünde-West, Bernard & Graefe Verlag, 1998
35 H. Boog, Die deutsche Luftwaffenführung 1935–1945, Führungsprobleme, Spitzengliederung, Generalstabsausbildung, Stuttgart, DVA, 1982
36 Lilienthalgesellschaft für Luftfahrtforschung, Bericht 163, Bericht über die Sitzung Flugeigenschaften am 8. und 9. April 1943, Berlin
37 H. Helmbold, Statische Längsstabilität bei hohen Fluggeschwindigkeiten, 7. Wissenschaftssitzung 20. September 1940, Sitzungsperiode 1940/41
38 E. Fischel, Künstliche Längsstabilität, 7. Wissenschaftssitzung 20. September 1940, Sitzungsperiode 1940/41

Light-weight Construction – a Prerequisite for Flight

All flight pioneers at the end of the 19th century had the common desire to imitate the gliding of the large long-distance flyers as storks or swans. So what was initially closer than to look for constructions that would spread-out the wing area and not consider wing profiles? Lilienthal's wing construction of the fiber material wood with a fabric covering and wire bracing was the attempt to realize in light-weight construction the necessary wing area and to give the wing the shape of a cambered plate with a rounded nose.

First Scientifically Sound Formulation of the Principles of Light-Weight Construction

Heiner Dörner, Horst Prem

At the beginnings of aeronautics there were supporters and patrons, probably called sponsors today, who financially supported the idea of flying. In 1911, aeronautical enthusiasts collected from industry in Stuttgart the amount of 59,000 Gold Mark to establish a chair for Aeronautics, Flight-technology and Motor Vehicles at the Royal Institute of Technology ("Technische Hochschule, TH") Stuttgart. *Alexander Baumann* became the first full professor at this chair. *Alexander Baumann*, born on May 15, 1875 in Heilbronn at the Neckar river, taught already since April 1, 1910 as a lecturer at the TH. *Count Zeppelin* had founded, just for the collection of the money, the "Society for the Support of Airship Aviation in Stuttgart AG" ("Gesellschaft zur Förderung der Luftschiffahrt"). The money collected was given to the "Royal Württemberg Ministry of Church and Educational Systems" ("Ministerium für Kirchen- und Schulwesen"). Today, this procedure is known as "Establishment of a foundation chair".

The introduction of technical aeronautical research and education at the TH Stuttgart in 1910 was probably the first action of this kind worldwide. Since this time, it was possible to attend in Stuttgart aeronautics-related lectures in continuity, the exception being the time during the war.

Little known is *Alexander Baumann's* textbook which summarized the state of the art in research at that time and which was published in two volumes in 1913 under the title "Mechanical Basis of Aircraft Engineering [1]" by the publishing house of *R. Oldenbourg*. These volumes were up into the twenties a standard reference of aeronautical technology. The following topics were treated: "drag", "work effort for gliding", "the completed vehicle", "materials for construction", "the interrelation of propeller and engine and its influence on the overall aircraft". It was the desire of the publisher that within the book only elementary mathematical tools should be used. Nevertheless the first volume treats problems related to the stability about all three axes as well as ascending and descending trajectories with and without engine, respectively. Related to the materials for construction, wood, bamboo, steel and aluminum are addressed, even the effect of deformation energy is described.

In Volume 2, *Alexander Baumann* explains first "starting (take-off) and landing" and treats subsequently within a larger chapter "control" (elevator, control in yaw, wing twist, curved flight). After the chapter on "engine torque", practical aspects of flight are considered. The author also explains individual parts of the aircraft as well as their design elements. At the end of Volume 2, several tables exhibit drag coefficients as determined in tests by *Frank, Eiffel* and

Lilienthal's glide from the "Flight Hill" ("Fliegeberg") with his normal gliding apparatus in the summer of 1895.

Föppl, as well as results of wing tests of *Bleriot, Voisin, Wright* and *Farman* gathered by *Eiffel*.

Already during the first decade of the 20th century, *Alexander Baumann* put down in his book initial formulations of the principles of aircraft control:

"The final objective in this direction would then be a vehicle, which would remain in straight and level flight without changing its attitude, independent of outer influences and without the help of the pilot, or would return to this attitude when forced from it by outside influences.

Now an aircraft can pitch forward or back, that means turn an axis, which is horizontal and normal to the direction of the movement. It may also roll about a horizontal, in flight direction lying axis; or, finally, it may turn about a vertical axis. So there are three different kinds of possible rotational movements, which require correspondingly three different types of control: control in pitch, roll and yaw. With a suitable size and design of the control surfaces and corresponding practice of the pilot, it should be possible to always balance the aircraft".

In the Chapter "Materials for Construction – Effect of the Material on Weight and Drag" – one may read:

"... that it is at least concerning weight of importance to use a material as light as possible but strong. However, it is not strength by itself, but also, as shall be described, tenacity and elasticity of the material and, above all, the uniformity of the material".

Concerning the utilization of metal tubes and the construction of layered materials, even composites, he explains:

"... if it is bamboo, one has to calculate with the inherent wall thickness; if it is drilled wooden tube, one will not be able to produce very thin walls, whereas this is easily possible with metal tubes. If wooden tubes are produced by gluing and rolling up of veneer, possibly with fabric layers in between, one will still not be able to go below a certain wall thickness".

Alexander Baumann presents test results of the material testing laboratory of the TH Stuttgart for different kinds of wood. Interestingly enough, this laboratory was headed by his brother *Richard Baumann*.

Principle statements concerning compromises between aerodynamics and light-weight construction may also be found in his textbook: *"Elliptical cross-sections always result in higher weight. If they are applied anyway on wooden rods, it is due to the lower drag of elliptical cross-sections compared to circular ones.*

In the case of a steel tube there is the possibility to make a fairing for it with a suitably thin aluminum sheet or wooden veneer, etc., which would further reduce drag at a weight gain, dependent on wall thickness, of only 5–10 per cent; however, drag could be reduced, depending on the shape of the fairing, of up to 50 per cent, thus an additional superiority of the steel tube becomes obvious."

Alexander Baumann realized (already before 1913) that the strength of a material does not necessarily play a role in selecting it: *"As you can see, wood is able to take up more energy than metal. All materials are inferior to the tough ash-tree."*

However, also safety related questions in conjunction with accidents – today we would call it "crash worthiness" – were important to him:

"At the very moment when the wooden construction breaks, its capability to take up energy is exhausted... the still existing vital force can now only be taken up by the deformation energy potential of the body of the pilot. A steel vehicle can, under the same circumstances, take up ten times the deformation energy. Under these conditions, a steel construction would guarantee a greater safety".

Alexander Baumann did not only set first milestones in teaching in Germany, he was also successful in realizing his research results. His giant aircrafts proved this during operational service between 1915 and 1918.

How much *Alexander Baumann* was associated with the light-weight construction principles shows a small but very consequential detail. During the design of his giant aircraft, he decided to position the engines between the two wing decks in their own nacelles. He argued that the wing spars could be built lighter due to this distribution of the main loads resulting in a weight saving of 1–2 per cent. *Baumann's* hunt for percentages in favour of an optimized light-weight construction through an ingenious arrangement of construction modules and subsystems relative to each other already lets one foresee today's approach in applying the modern design philosophy of "system light-weight construction".

When the allies captured a German giant aircraft in 1918, a French newspaper wrote: *"The Germans have a sense for colossal ... This apparatus means a new stage in the development of flight. The interest it must receive is not solely military. One has to think of the role the aircraft may play during the ensuing peace. It*

Alexander Baumann (1875–1928).

Alexander Baumann, born in Heilbronn on May 15, 1875.

1899 graduation as "State Construction Supervisor" ("Regierungsbauführer") after his studies of Mechanical Engineering at the TH Stuttgart.

1899 employment as engineer at the "Saxony Machine Company" ("Sächsische Machinenfabrik") in Chemnitz and at the "Dresden Machine Company and Shipyard" ("Dresdner Maschinenfabrik und Schiffswerft") in Dresden-Neustadt.

1902 teacher at the engineering school Zwickau.

1907 receives various patents for flight vehicles, flight control, slotted wings.

1908 employment at the "Physical-technical Reichs Establishment" in Berlin-Charlottenburg.

1910 lecturer at the TH Stuttgart.

1911 full professor of airship aviation, flight technology and motor vehicles at the TH Stuttgart. Lectures: Airships and flight vehicles; Motor vehicles; Flight vehicles and design details; Free-balloon and airship engineering; Engines for land, water and air vehicles; Mechanical engineering design; Detailed treatment of important questions of motor-vehicle engineering.

1912 founding member of the "Scientific Aeronautical Society" ("Wissenschaftliche Gesellschaft für Luftfahrt, WGL"), today's DGLR. Presentations on the development of world air transportation with visionary statements. Development, construction and testing of a biplane together with the manufacturer *Freytag.*

1913 two-volume textbook: "Mechanical Basis of Aircraft Engineering", published by R. Oldenbourg, Munich/Berlin.

1914 development, design, manufacturing and testing of the Zeppelin-Staaken giant aircraft. 48 series aircraft in different versions (take-off weight up to 14.65 tons, payload up to 6 tons, up to 10 crew members, up to 10 hrs flight duration, cruise speed up to 140 km/h (76 kn), engine performance up to 1,500 hp from up to 5 coupled engines).

1917 award of the "Royal-Württemberg Wilhelm Cross without Swords" ("königlich-württembergisches Wilhelmkreuz ohne Schwerter") for his achievements in aviation.

1924 chief designer at Mitsubishi Jukogyo Kabushiki in Nagoya, Japan. Development of three small military aircraft, e. g., the reconnaissance aircraft Mitsubishi 2MR1 "Tobi" and fighter Mitsubishi 1MF2 "Hayabusy" ("Falcon").

1928 he dies in Stuttgart on March 23.

is up to industry to design aircraft that would satisfy the arising intercontinental transport."

E. Offermann writes in his book "Giant Aircraft [2]" about flights against the enemy, carried out in September 1917 to London in England, where a distance of 800 km (432 nm) was covered with bomb loads of up to 2,000 kg (4,500 lbs) in seven hours flight time: *"The glory of having made this performance possible is almost exclusively due the aircraft company Staaken and the lucky hand of its designer Professor Alexander Baumann … Today it is clear that in Germany the technical and flight mechanical basis of the G-aircraft (G stands for giant) for air traffic, also for intercontinental traffic, was established during these years of war, one may think of its value as a military tool as one likes …"*

The giant aircraft of *Alexander Baumann* of the years 1915–1918 may, looking back, be considered as masterpieces of aircraft engineering of that time. *Adolf Rohrbach*, a pupil of *Baumann*, wrote in the obituary after the death of his teacher on March 23, 1928:

"The giant aircraft were the first of their time that were capable, due to their overall construction and instrumentation, of performing long and difficult flights with heavy loads also during the night. Consid-

ering this, Baumann helped in many ways to establish the basis for today's transport aircraft. With his death, aeronautics is losing one of those, who worked from the very beginning for the application of the state of the art of engineering in aircraft construction."

This conclusion is still valid today. Furthermore, this pioneer of aeronautics wrote down already in 1914 fascinating and progressive thoughts concerning transatlantic air traffic and imaginations and visions about civil aviation, which have, figuratively, become reality today at the age of wide-body aircraft and the envisaged "new generation" aircraft with more than 650 passengers.

From the Fiber Material Wood to Metal Constructions

Heiner Dörner, Horst Prem

Many flight pioneers had in common that they came from engine, heating or ventilation-system engineering before they concerned themselves with wing design and construction. *Hugo Junkers* was probably the first, who recognized the cantilever, thick wing and the all-metal design as the key elements of future aircraft. However, initially he could only realize the cantilever wing as steel construction and had to accept the weight disadvantages. But this forced him to search for new metallic materials.

Parallel to and independently of the development of aircraft, *Alfred Wilm* applied in 1903 (DRP 170 085), 1907 (DRP 204 543), and 1909 (DRP 244 554) for patents concerning the refinement of aluminum with copper and magnesium; for the latter two alloys, he invented the name "duralumin [3]" (1906). The "Dürener Metal Works AG" bought the last patent and had the name "Dural" registered as trademark in October 1909.

Starting in the spring of 1914, duralumin became the only material used for the construction of the airships of *Count Zeppelin*, and it was only consequent to reduce the performance problems of the Junkers steel aircraft by using the new, much lighter material beginning in 1917 with the project J 4. In the process, it was possible to increase the payload, in percentage of the gross mass, from 17 per cent for the earlier design to more than 50 per cent for the F13. To avoid the buckling of the skin areas of his aircraft made of duralumin, *Junkers* stiffened these areas by employing corrugated sheet metal, which became a characteristic of his aircraft.

In an "Aircraft Handbook" ("Fliegerhandbuch [4]") of 1918, a comparison of the strengths of steel, duralumin and the fiber material spruce, which were the basic materials of aircraft construction at that time, can be found.

Since the "web theory" ("Schubfeldtheorie") for the calculation of thin, stiffened plates and monocoque constructions had not yet been invented, the aircraft designers were restricted to lattice constructions stiffened with diagonal rods or had to evade to *Junkers* corrugated sheet metal construction. The fuselage was made alternatively of a kind of plywood monocoque. *Georg Madelung* designed in 1921 for the first time the load-bearing structure of the wing of the glider "Vampyr" on the basis of *Bredt's* torsion tube theory, as *Hans Zacher* describes in detail in the chapter "Highlights in the Evolution of Sailplanes until 1945".

The next innovational step in airframe construction was made with the web theory of *Herbert Wagner*, which allowed a consequent utilization of duralumin in very thin sheets up to the buckling limit. Looking back at his life, *Wagner* writes in 1980:

"Apart from the corrugated sheet metal construction of Junkers, there existed at the early twenties metal aircraft "covered" with plain sheet metal. For smaller parts and most of all for the floats, the outer sheet metal cover was load carrying, but design objectives based on precise theoretical aspects were apparently lacking. Rohrbach covered all outer-wall frame structures, which were subjected to higher loads, with thin sheet metal, similar to the fabric-covered frame structures. I suggested to omit the diagonal rods within the frame and to mount the vertical rods closer together, so that under load the obliquely folding sheet metal skin takes up the lateral (vertical) forces. Comparative tests showed, for the same weight of the wall, a nearly doubling of the allowable lateral force; the workshops welcomed the simplification. The German Patent No. 547624, which was very lucrative for the company heads, and most of all the publication "Flat sheet metal beam with very thin web sheet metal [5]" in the ZFM, 1929 described the design, known in the English-speaking world as "Wagner web".

During my time at Rohrbach also occurred the transition from square to oval fuselage cross-sections and the creation of non-braced inner fuselages as passenger cabins. This led to studies of monocoque structures. Very useful seemed to me the thesis of Dschou "The compressive strength of stiffened cylinder shells", ZAMM, 1936 and Wagner/Simon "About the introduction of forces into thin-walled cylinder shells", Lufo 1936. I am a bit proud of my work "Torsion and buckling of open profiles" in the commemorative publication [6] "25 Years of the Danzig (TH) Institute of Technology", in 1929."

Wagner's pioneering idea of the "flat sheet metal beam" is based on the knowledge that in case of a bending beam under a vertical load, the bending moment is essentially taken up by the longeron flanges while the vertical force is taken up by the longeron vertical plate. If the upper and lower longeron flanges support each other by vertical rods, as proposed by *Wagner* for his web structure, and if the web sheet metal is very thin so that it buckles under shear, folding diagonally, then a "diagonal tension field" is being created. The "monocoque construction" based on this principle, with the integration of the sheet metal planking into the load bearing structure, was a breakthrough towards an extreme light-weight construction, which revolutionized aircraft engineering. Since the mid thirties, it was now possible to design weight-optimized aircraft with flat smooth surfaces as aluminum constructions. It should be mentioned that already in 1925 was among the "with flat sheet metal covered metal aircraft", mentioned by *Wagner*, a small transportation aircraft (M 18) of duralumin flat-sheet-metal construction, designed and built by the 27 years old *Willy Messerschmitt* – lacking a theory, but based on his own systematic tests – and turned successfully into a small series.

A further pioneering work of *Herbert Wagner* was the theory of the "shear center" and its application to the calculation of the bending-torsion behaviour of open profiles (e. g., u-profiles), which was described

Old-fashioned engine installation within a lattice-type frame fuselage without load-bearing skin.

in 1929 in the Danzig commemorative publication [6]. These fundamental basics of modern light-weight construction were summarized in 1940 in the book "Construction Elements of Aircraft Engineering [8]". It is based on *Wagner's* lecture notes and belongs to the first standard textbooks on modern light-weight construction.

However, the static calculation of such light-weight constructions in aircraft engineering encountered initially large difficulties. Although the theory of rod and frame trusses was already developed, and at the end of the thirties also the fundamentals of matrix theory of structural mechanics were established, the application to static and dynamic calculations of such complex aircraft elements as wings and fuselages in monocoque construction was hardly possible with the calculation tools of that time. The solution of systems of linear equations with more then ten unknowns, i. e., the mathematical calculus of the corresponding determinants and matrices, was hardly manageable. The modeling of simplified structural-mechanical substitute systems required great engineering skill and physical understanding. The calculation of elastic

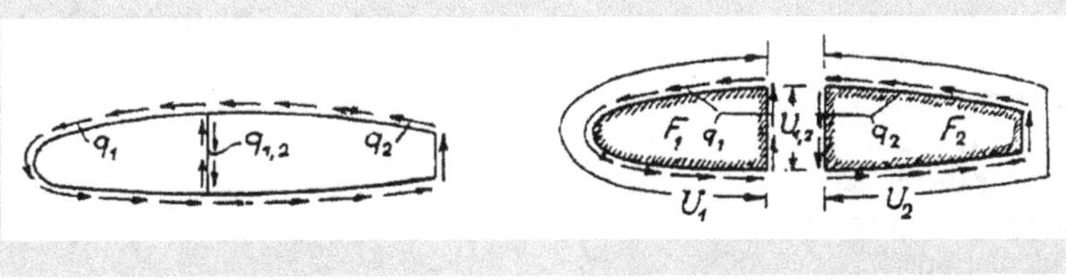

Prismatic hollow body with center beam – *Bredt's* torsion tube.

Herbert A. Wagner (1900–1982).

Herbert A. Wagner, born in Graz, Austria on May 22, 1900.
1914 cadet in the Austrian navy.
1919 studies of shipbuilding at the TH Berlin. PhD Thesis "On the origin of the dynamic lift of wings", which became the basis for the later flutter investigations.
1924 development engineer at Rohrbach in Berlin. During this time a great number of papers originate, which becomes the foundation of the shear panel (Wagner web) theory for flat sheet metal beams with very thin vertical plates. Here also originates the paper on the hydrodynamics of landing and starting seaplanes, which *Prandtl* judged the best hydrodynamics paper since *Helmholtz.*
1927 professor of Aircraft Engineering at the TH Danzig.
1930 professor of Aircraft Engineering at the TH Berlin. Numerous papers concerning design and construction elements of aircraft engineering evolve, which anticipated the basic idea of the finite-element-theory.
1935 head of the aircraft development at Junkers in Dessau. Under his supervision originates the high-altitude aircraft EF 61 with pressure cabin, as well as since 1936 the RTO experimental jet engine with axial compressor [7], which was tested in 1939 at the Junkers branch in Magdeburg.
1940 development of guided missiles at Henschel in Berlin-Schoenefeld, starting with the air-ground-gliding bomb HS 293 up to HS 298, a ground-air and an air-air missile with optical control, which were only completed after the end of the war in the USA.
1945 studies and theories concerning the control and maneuverability of missiles. Development of a radar-based control system for bomber aircraft, which was utilized during the Korean War.
1952 *Wagner* founds his own company in Van Nuys, California, and develops for the US-Navy optical control and target-tracking methods for ground-ground-missiles. Together with Aero Physics in Santa Barbara, he develops control methods for antitank missiles for the US-Army.
1957 *Wagner* takes over the professorship of Technical Mechanics and Space Technology at the TH Aachen. In 1960 he receives a honorary PhD from the Technical University Berlin.
1966 consultant for many American and German companies.
1982 he dies in Vienna.

influence, respectively stiffness matrices, was, anyway, connected with large computational effort and the statics departments of aircraft companies employed huge calculation offices, equipped with mechanical desk calculators. Final certainty with regard to the required static strength was provided – just as today – by fracture tests.

On the other hand the integral monocoque construction had the advantage to model rather easily the respective elements, such as longerons, girders, ribs and planking by rod, beam, plate and diaphragm elements. This model-like dissection of the monocoque structure and the subsequent description by the matrix methods of elasto-mechanics became finally the basis for the "finite elements method, FEM", which was vigorously developed since the early fifties and which gained its practical break-though by the occurrence of the electronic calculator. FEM became today's standard method in the field of structural analysis and the impulses for this development came primarily from aircraft engineering. *Konrad Zuse* received for the development of his first electronic calculators Z3 and Z4 considerable financial support from the DVL. Bombs destroyed the Z3 in 1944, but the Z4 was, after its transfer from Berlin to Göttingen in March 1945, utilized, under the direction of *Prandtl* and *Küssner*, for the first program-controlled aerodynamic and aeroelastic calculations.

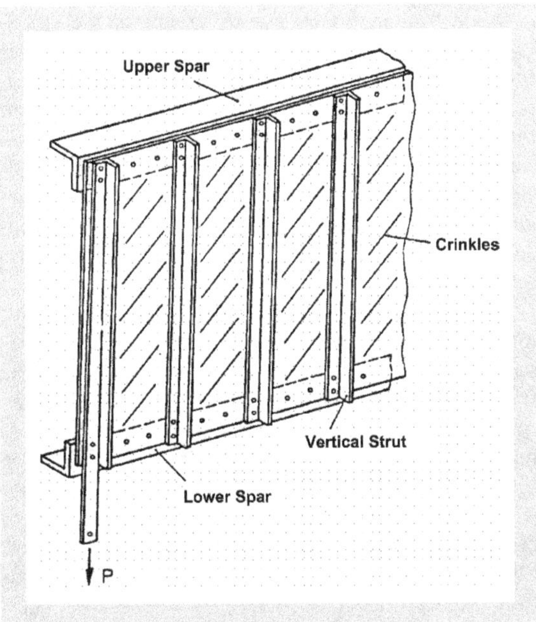

Herbert Wagner's diagonal tension field.

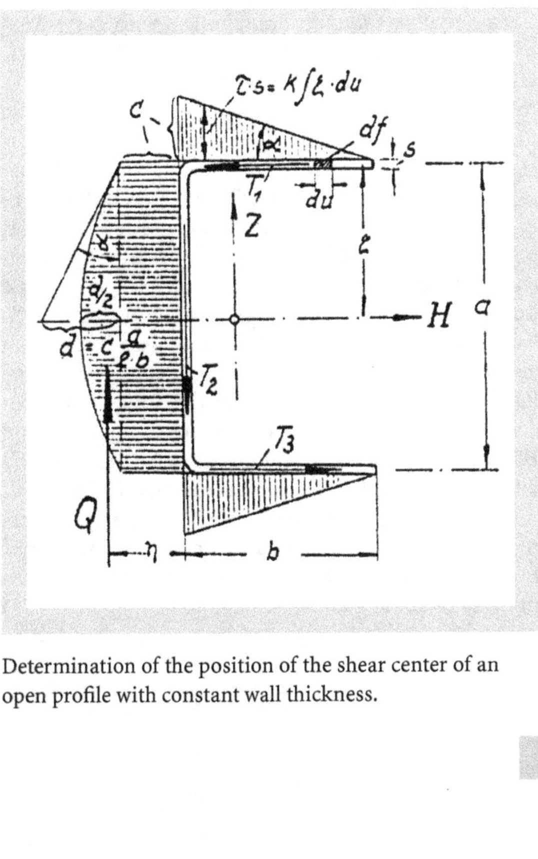

Determination of the position of the shear center of an open profile with constant wall thickness.

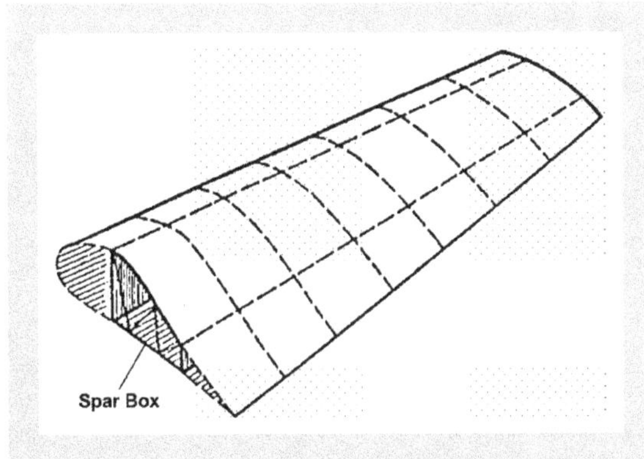

Airfoil in monocoque construction (schematic).

Fatigue and Structural Endurance

HANS FÖRSCHING

Aircraft are much more than automobiles or rail vehicles subject to numerous dynamic loadings, for instance, during flight by atmospheric turbulence and shear winds, during taxiing on runways, by the landing shock or acoustically by propeller and engine noise. These dynamic loads have far reaching consequences regarding operational safety and reliability. The establishment of corresponding strength concepts, damage hypotheses and design and construction regulations based on reliable load assumptions became a stringent necessity for the structural design of operationally safe aircraft, especially for civil aviation. This required intensive research to which Germany contributed considerably in the years between 1929 and 1945. The milestones of these research activities are here only briefly summarized since there exists extensive literature as Walter Schütz has shown in his contribution "On the history of dynamic strength" [9].

First Contributions to Fatigue Strength

The expression "fatigue" first appeared in 1850 in English publications, where many fatigue cracks due to dynamic loading on levers, water pumps, boilers, crankshafts, railway axels, cranes, etc., are being described, which partly led to severe accidents. In Germany at that time it was *August Wöhler*, the "Royal Chief Machine Engineer" ("Königlicher Obermaschinenmeister") of the "Niederschle-

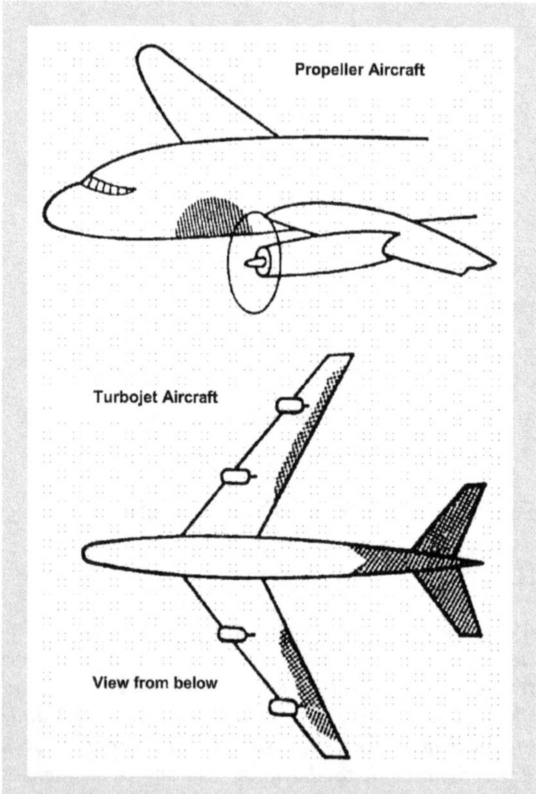

Typical zones of acoustical fatigue on aircraft.

sisch–märkische Railroad" in Frankfurt/Oder, who conducted the first systematic fatigue-strength investigations and measurements in conjunction with the dynamic loading on railroad-car axels. As a result of his findings, *Wöhler* suggested already in 1860 the dimensioning of the axels according to the time-wise, respectively operational strength requirements [10]. By axial vibration tests, bending vibration tests and torsion vibration tests on notched and non-notched smooth rods, utilizing test benches of his own design, he gained knowledge, which became known as "*Wöhler's* law" and is reflected in the "Wöhler-curve". The latter describes for a given average material stress σ_m the maximum allowable stress amplitude $2\sigma_a$ as function of the number of load cycles N until fracture, and as asymptotic limit the number of load cycles N_D where fatigue strength exists. He stated:

"The fracture of the material can also be provoked by numerous load cycles, neither of which reaching the absolute fracture limit. The difference in the stresses, which limit the amplitude, is here responsible for the destruction of the cohesion. The absolute magnitude of the stress limit is only of influence since, with increasing stress level, the differences that lead to fracture decrease". With regard to the time-related dimensioning he states: *"It must also be considered whether an unlimited life time of the construction is demanded or whether it shall be utilized only a certain period of time."*

Wöhler also already recognized the important influence of notches and (for later aircraft constructions important) riveted joints on the fatigue strength and proposed corresponding safety factors. Absolutely correctly he is considered the founder of modern fatigue research; he also initiated the set up of the "Royal Prussian Test Establishments", from which evolved, after numerous name changes, today's "Federal Institute of Material Research and Testing" ("Bundesanstalt für Materialforschung und -prüfung") in Berlin.

Up to the beginnings of flight technology, *Wöhler's* pioneering work in the area of structural fatigue and operational strength were the basis for further important research mainly in England and in the USA. In Germany it was *Johann Bauschinger*, who discovered in 1886 the reduction in the elasticity limit of materials by often-repeated load cycles (*Bauschinger* effect). In 1910, *O.H. Basquin* introduced in the USA with the double-logarithmic representation of the Wöhler-curve, log σ_a on the ordinate and log N on the abscissa, the so-called "Wöhler-line" and showed that there exists a logarithmic law. First fatigue tests on aircraft parts during World War One were described in English literature in 1918. From 1920–1945 nearly all basic knowledge concerning fatigue strength was established. In the process, essential impulses came from aircraft technology due to the necessity of extreme light-weight construction and the high standards concerning operational safety of passenger aircraft. Here, investigations concentrated most of all on the air-vehicle material aluminum. During this time period, the field of fatigue strength was a domain of German research, which is documented by the number of publications on fatigue strength – not only in the area of flight technology.

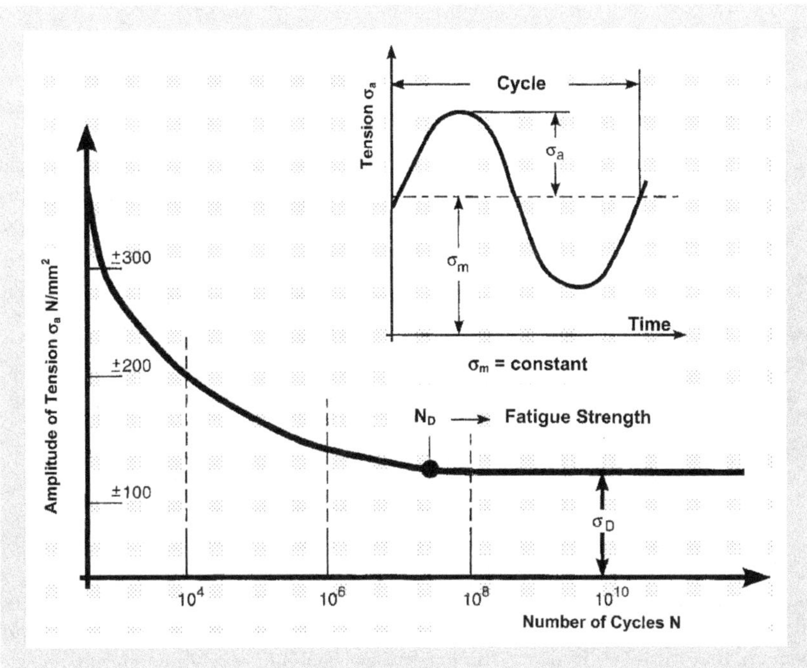

Wöhler-curve (schematically), number of load cycles N on a logarithmic scale.

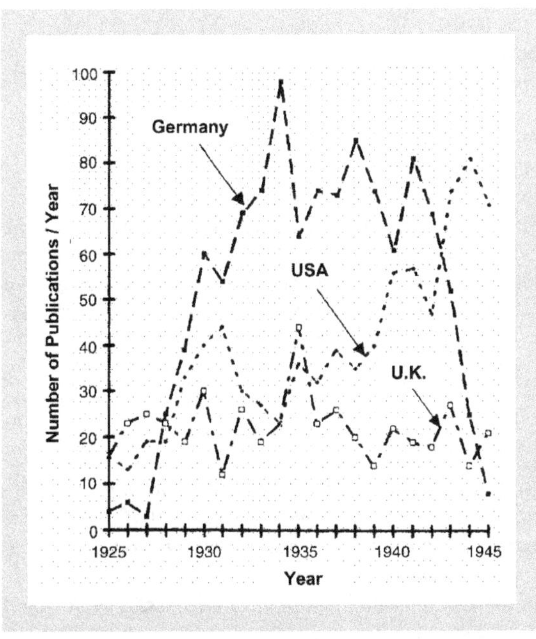

Number of publications on fatigue strength in Germany, England and the USA from 1925–1945.

Flight-Technology Relevant Research Between 1920 and 1945

One of the main reasons for the intensive research on fatigue strength and fatigue characteristics of aircraft structures was in the mid-twenties the shift to all-metal constructions, and the crash of a Dornier-"Merkur" aircraft on September 23, 1927 due to fatigue of a longeron support. This crash was also the reason for the first fatigue-strength measurements of the DVL on wing longerons of steel, aluminum and wood, respectively. Since a sufficient dimensioning of structural elements with respect to fatigue strength and reliability is only possible by knowing the real operational loads – as already pointed out by *Wöhler* – DVL started in the thirties with first stress measurements on aircraft structural elements during flight operations. Stress oscillations during flight do not occur as regularly as during laboratory tests (with a periodic time dependence), but as irregular flight load collectives, as was pointed out by *H.G. Küssner* in 1931 in connection with aeroelastic gust problems. *H.W. Kaul* [11] and *H. Freise* [12] published such measured flight load collectives in 1938. *Freise* used the inductively controlled DVL "Glass-scratch Strain Recorder", described by

Ernst Gaßner (1908–1988).

Ernst Gaßner, born in Bingen/Rhine on May 22, 1908.
1928 studies of engineering at the TH Darmstadt.
1934 he joins the Department of Statics of the DVL in Berlin-Adlershof.
1936 under his supervision as associate scientist, basic work on the operational strength (fatigue strength) of aircraft is being performed; he is the founder of the fatigue strength test.
1941 PhD at the TH Darmstadt with the thesis: "Effects of operation-like load sequences on the strength of aircraft construction elements".
1946 foundation of the "Physical-Technical Laboratory" ("Physikalisch-Technisches Laboratorium") in Kempten.
1948 he moves to Darmstadt.
1950 merger with the "Bautz-Bergmann Material Consulting Company, Ltd." ("Bautz-Bergmann-Werkstoffberatungs GmbH") to the "Laboratory for Operational Strength" ("Laboratorium für Betriebsfestigkeit, LBF").
1962 integration of the LBF into the "Fraunhofer-Society, FhG" as "Institute of Operational Strength" ("Institut für Betriebsfestigkeit").
1963 managing director of the FhG; he gains high merits in setting up the FhG as holding society for applied research.
1964 honorary professor at the TH Munich.
1966 honorary member of the board of directors of the FhG.
1971 member of the senate of the FhG.
1973 retirement.
1978 he receives the Fraunhofer commemorative coin.
1979 he receives a honorary PhD (honoris causa) from the TU Munich.
1985 he receives the Erich-Siebel commemorative coin of the "German Association of Material Testing" ("Deutscher Verband für Materialprüfung").
1988 he dies in Darmstadt on November 25.

F. Seewald [13], to measure the elongations of wing longerons on two aircraft during Lufthansa flight operations over roughly 600 flight hours. The electrical resistance strain gage was at that time still unknown. It was only invented in the USA at the end of the thirties and extensively used during World War Two. In the case of *Kaul's* measurements, acceleration load collectives were determined on six other civil aircraft over a total of about 700 flight hours. Starting in 1941, front-line measurements were conducted on military aircraft, e. g., on the Ju 87 and Ju 88. Further load collectives of such front-line measurements covering approximately 1,000 flight hours were published early 1945. They were used, for lack of any own data, during the forties in the USA for design purposes. On the basis of such load collectives and Woehler-lines during the crack propagation phase, *B.F. Langer* [14] published in 1937 a damage accumulation hypothesis, which was further developed in the USA by *M.A. Miner* especially for aluminum.

Measured load collectives were then the basis for the "operational strength test" (fatigue strength test) worked out by *Ernst Gaßner* at the DVL Institute of Strength, which made him the founder of the theory of fatigue strength. In his PhD-thesis [15[, he describes the fatigue strength test as follow:

"It is mainly about applying during a test a certain number of load steps and imitate the alternation of higher and lower loads as it occurs during the actual operation. More detailed, this concept of fatigue strength is generally based on dimensioning the structure according to a limited but sufficient lifespan under variable loads gained in the following way:

- *measurement of the operational stresses in the form of a stress collective,*
- *this is used to gain generally valid load assumptions for similar designs; standardization of specific col-*

lective forms for aircraft engineering, e. g., for civil and military aircraft,
- *simulation of the load collectives by multiple-stage tests with load sequences that also contain rare high load peaks,*
- *presentation of the results as Gaßner-lines, i. e., double-logarithmic presentation of the maximum stress associated with the load collective as function of the total number of load cycles,*
- *accounting for fatigue-life scatter by a sufficiently high safety factor on a statistical basis to gain the necessary failure probability dependent on the structural element."*

The result of investigations is the lifespan in flight hours, number of flights, cruise-range in kilometers or similar. To conduct the Gaßner block test program, new test benches were necessary, which were built by the company Schenck in Darmstadt, according to proposals of *Gaßner*, and patented in 1940. Before, unbalance generators had to be used with limitations in the frequency height. Even today, the Schenck Company is recognized worldwide for its vibration test facilities.

The Gaßner test concept represented the state of the art concerning fatigue strength until the beginning of the sixties, before it was, by more systematic research and with the aid of improved experimental and computational means, further developed into today's stochastically excited fatigue strength test. German aircraft industry accepted the concept enthusiastically and applied it a thousand times during World War Two for the determination of the operational and fatigue strength of aircraft constructional elements of any kind. After 1945, entire airframes were tested in England and the USA according to this concept. Here, it is interesting to note that alone twenty English Vickers "Wellington-bombers" crashed due to fatigue fractures. But also ten Me 110-destroyers were lost due to fatigue damage on a horizontal-stabilizer rib and the resulting horizontal stabilizer flutter.

Within the context of the work on fatigue strength of aircraft, performed at the DVL until the end of World War Two, also the basic work performed until 1945 by *August Thum* at the TH Darmstadt, *Otto Föppl* at the TH Brunswick and *Otto Graf* at the TH Stuttgart has to be mentioned. *Thum* was in Darmstadt since 1927 the head of the first chair for material research in Germany, and he established the theory of form strength. The latter is described by a strength value, which depends on the magnitude and kind of stress as well as on the material and the structural form. Here, *Thum* introduced the notch influence factor β_k, which allows the transfer of a material test result to the completed construction element. *Thum* and his students published 574 scientific papers between 1922 and 1956; they are the best evidence for his superior work in the field of fatigue strength in nearly all technical areas. *Föppl* and his students mainly dedicated themselves in their research work in Brunswick to mechanical procedures to improve vibration strength, which means rolling, shot preening, and pressing, while *Graf* mainly investigated the vibration strength of riveted and welded joints utilized in structural engineering (high-rise structures). The results gained here are also of great interest to aircraft engineering. Finally, the book "Notch-stress Technology" ("Kerbspannungslehre [16]"), published by *H. Neuber* in 1937, has to be mentioned, where the theoretical fundamentals for the calculation of "form factors" ("Formzahlen") and "notch influence coefficients" ("Kerbwirkungszahlen") are presented. The book was already translated into English in 1946 and was the starting point for a large test program of the NACA.

References

1 A. Baumann, Mechanische Grundlagen des Flugzeugbaus, R. Oldenbourg, Berlin und München, 1913, Part I und II

2 E. Offermann, Riesenflugzeuge, R. C. Schmidt & Co, Berlin, 1927

3 H. Joliet, Silber aus Lehm, Kultur & Technik, 1/1990, Deutsches Museum, München

4 R. Eyb, Fliegerhandbuch, ein Leitfaden der gesamten Flugtechnik, 3. Ed., Berlin W. 62, Richard Carl Schmidt & Co, 1918

5 H. Wagner, Ebene Blechwandträger mit sehr dünnen Stegblechen, Zeitschrift für Flugtechnik und Motorluftschiffahrt 20 (1929), Heft 8–12, pp.200–207, 227–233, 256–262, 279–284, 306–314

6 H. Wagner, Verdrehung und Verknickung von offenen Profilen, In: Fünfundzwanzig Jahre Technische Hochschule Danzig, pp. 329–344, A.W. Kafermann GmbH, Danzig, 1929

7 K. von Gersdorff, K. Grasmann, H. Schubert, Flugmotoren und Strahltriebwerke, Vol. 2, Bernard & Graefe Verlag, Bonn, 3. Ed., 1995
8 G. Kimm, Bauelemente des Flugzeugbaues, R. Oldenbourg, München und Berlin, 1940
9 W. Schütz, Zur Geschichte der Schwingfestigkeit, in: Materialwissenschaft und Werkstofftechnik, 24 (1993), Darmstadt, pp. 203–232
10 A. Wöhler, Versuche zur Ermittlung der auf die Eisenbahnwagenachsen einwirkenden Kräfte und die Widerstandsfähigkeit der Wagenachsen, Zeitschrift für Bauwesen, Vol. X (1860), pp. 583–616
11 H.W. Kaul, Statistische Erhebungen über Betriebsbeanspruchungen von Flugzeugtragflügeln, Jhb. 1938 der deutschen Luftfahrtforschung, Ergänzungsband, pp. 307–313
12 H. Freise, Spitzenwerte und Häufigkeit von Böenbelastungen an Verkehrsflugzeugen, Jhb.1938 der deutschen Luftfahrtforschung, pp. 289–302
13 F. Seewald, Messungen mit Glasritzdehnungsschreiber der DVL, Maschinenbau, Vol. 10 (1932), No. 23, pp. 725–727
14 B.F. Langer, Fatigue Failure from Stress Cycles of Varying Amplitudes, Trans. ASME Journal of Applied Mechanics, Vol. 59 (1937), pp. A160–A162
15 E. Gaßner, Auswirkungen betriebsähnlicher Belastungsfolgen auf die Festigkeit von Flugzeugbauteilen, Jhb. 1941 der deutschen Luftfahrtforschung, pp. 1472–1483
16 H. Neuber, Kerbspannungslehre, Springer-Verlag, Berlin, 1937

Aeroelasticity – The Interaction of the Elastic Aircraft Structure with Aerodynamic Forces

Hans Försching

Classification of Aeroelastic Problems

The science of aeroelasticity encompasses those physical processes and problems that result from the interaction between elasto-mechanical systems and the surrounding airflow. Aeroelastic processes, like the swaying movement of trees, plants and oat fields in the wind, daily occur in nature. But also in many technical fields significant aeroelastic phenomena happen in numerous forms such as the wind-induced flutter of towers, masts and bridges, the flutter of tube clusters in heat exchangers, blade flutter in turbo machinery or the oscillation of ocean platforms as well as shields and dams in rivers. Generally, one speaks here of fluid-elastic processes.

Aeroelastic problems in aircraft engineering became increasingly important during developments. Aeroelastic aspects became limiting factors and the newly created corresponding expression "aeroelasticity" appeared for the first time in the mid-thirties in English publications. The expression was, however, initially not adopted by the German language. Aircraft are forced to be extremely light and hence relatively flexible; the elastic yield of the aircraft structure increases progressively with increasing size. The static and dynamic deformations (flutter) and the induced steady and unsteady aerodynamic forces may in their interaction lead to problems with far-reaching flight mechanical consequences; here, the question concerning the aeroelastic stability is of special importance.

Aeroelastic problems already occurred at the beginnings of flight technology. Already the *Wright Brothers* and other aeronautical pioneers were confronted with these problems. They found solutions intuitively rather then by recognizing the actual physical relations. Most of all did aeroelastic problems increasingly occur with the development of the monoplanes towards the end of World War One. At this time also specific aeroelastic investigations commenced. With the rapid technological development of aircraft engineering towards ultra light-weight construction and transonic and supersonic flight, the field of aeroelasticity became more and more important. New aeroelastic phenomena and problems arose and new technical solutions had to be found. This development is today not at all complete, and by far not all aeroelastic problems of aircraft engineering have been solved satisfactorily, although the development of the computer technology in newer times has resulted in huge progress.

Generally, three different forces interact in aeroelastic processes:

E elastic forces,
A aerodynamic forces,
I inertial forces (mass forces).

These forces, which are continuously distributed along the aircraft, are in a state of equilibrium. So the following aeroelastic equilibrium equation holds:

$$E(q) + A(q) + I(q) = Q,$$

where Q denotes explicitly known and system-independent external disturbance forces, e.g., gust loads or landing bumps. E, A and I are each a function of the elastic deflections q. If Q = 0, this equation expresses the general (dynamical) aeroelastic stability equation. The problem in the analytical treatment of aeroelastic events hence consists in the determination of the system forces as function of the local elastic displacements for the whole aircraft or its components, and in the solution of the aeroelastic equilibrium equation with respect to q. The here existing extraordinarily complex analytical relations may be algebraic equations, differential equations, integral equations or integral-differential equations.

The variety of aeroelastic phenomena and problems in aircraft engineering, resulting from the various possible interactions between these system forces, can be demonstrated with the aid of the triangle of aeroelastic forces. It should be mentioned here that the aeroelastic problems can be divided into two groups, static and dynamic aeroelastic problems. In the case of static problems is I = 0, and the occurring aerodynamic forces A are steady and can be determined with the traditional methods and procedures of steady aerodynamics. The static aeroelastic problems are located outside the triangle at the upper left side, while the dynamic aeroelastic problems lie inside the triangle since all three system forces are involved. Overall, these problems can be classified as follows:

Divergence **D**: Static aeroelastic stability problem where a lifting system is, after reaching a critical velocity, abruptly deformed up to fracture.

Rudder effectiveness **R**: Effect of an elastic deformation on the effectiveness of control surfaces.

Distribution of lift **L**: Effect of an elastic deformation on the aerodynamic pressure distribution.

Static flight stability **SS**: Effect of an elastic deformation on static flight stability and control behaviour.

Flutter **F**: Dynamic aeroelastic instability where, after reaching a critical velocity, abruptly self-excited flutter of a lifting component or the whole aircraft occurs with rapidly growing amplitudes and generally catastrophic consequences.

Dynamic Response **DA**: Externally excited flutter of individual components by outer dynamic loads, e.g., by gusts, landing bumps, abruptly activated control surfaces or similar.

Dynamic flight stability **DS**: Effect of structural deformations on dynamic flight stability.

Buffeting **B**: Flutter of the aircraft or individual components due to flow separation at transonic speeds or due to aerodynamic interferences, such as the empennage being affected by the wing wake.

Dotted lines outside of the triangle indicate those fields of knowledge that are closely related to aeroelastics. The knowledge of the natural flutter (V) of the aircraft or individual components (wing, empennage) at rest (A = 0) is the basis for the analytical treatment of all dynamic aeroelastic problems. On the other hand leads the consideration of aeroelastic effects, necessary for large-scale aircraft, in the analytical investigation of the flight mechanical behav-

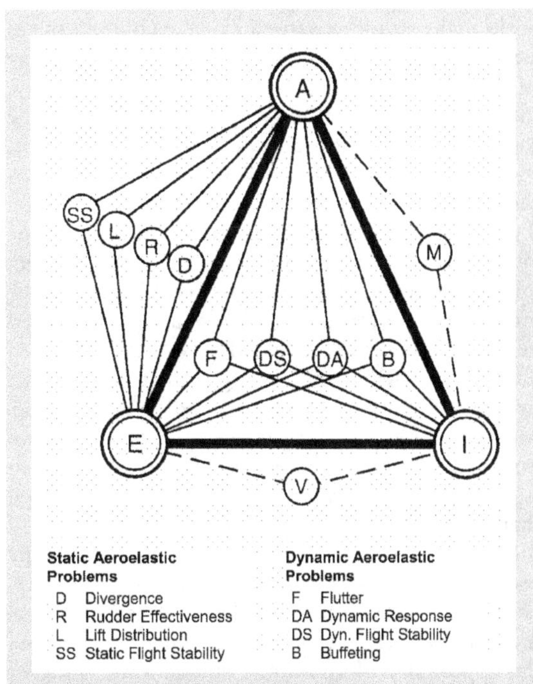

Triangle of aeroelastic forces.

iour, as well as for the application of modern active flight control and stability systems, to the interdisciplinary coupling of classical flight mechanics (M) and aeroelastics (aero-servo elastics).

Aeroelastic Research until 1935

Experiences and Findings until the End of World War One

The first known flight accident caused by aeroelastic circumstances already occurred nine days prior the first flight of the *Wright Brothers* on December 9, 1903, when on the monoplane of *Samuel P. Langley* the wings dismounted after a catapult take-off from a houseboat on the Potomac river near Washington. A later investigation showed that the accident was caused by a static torsion divergence of the wings. After their stiffening, the apparatus was flown successfully. In general the static torsion divergence was at the beginning of flight a main problem since the torsional stiffness of the mostly braced wings was too small. However, the problem was phenomenologicaly correctly recognized, and it was recommended to position the elastic axis (torsional axis) of the wing as far as possible in the direction of the wing leading edge.

During World War One, flight corps of either side were confronted with massive aeroelastic problems with the flutter problem becoming of prime concern as the flight speed increased. Many pilots, among them *E. Udet*, reported intense wing and empennage vibrations, sometimes also control surface vibrations, for instance, on the biplane Albatros D-III and on the monoplane Fokker D-VIII. Many accidents caused hereby ended fatal. In England, *F. W. Lanchester* conducted first theoretical investigation in 1916 on the two-engine bomber aircraft Handley-Page 0/400, where elevator flutter in conjunction with intense torsional vibrations of the fuselage occurred. From a simple energy consideration *Lanchester* concluded that the observed oscillation amplitudes could neither have been excited by engine unbalances nor by propeller slipstreams and that apparently other excitation mechanisms must have been the cause. In Germany, *H. Blasius* considered for the first time in 1918 the appearance of vibrations on the lower wing of the Albatros D-III, but without coming to a conclusive result.

Summarizing it can be concluded that towards the end of World War One a serious problem was presented by fractured wings due to static torsion divergences and, increasingly, by intense vibrations of the lifting surfaces (flutter). It was not yet recognized, that in both cases aeroelastic instabilities were the cause. The excitation of the oscillations occurs in flight. The difficulties were still rather considered as related to strength problems. In the meanwhile, the expression "flutter" had been established within the flying communities for the destructive flutter. It was also believed that cantilever-wing monoplanes were especially endangered, an assumption that was refuted by later experiences. It also became obvious, that an analytical solution of the flutter problem could only be reached with enormous effort, and would require, besides a sufficient knowledge of the natural oscillation characteristics of the wings, also information concerning the induced unsteady aerodynamic forces resulting from the wing flutter; this required fundamental research. Especially in the area of unsteady aerodynamics, pioneering work was then performed in Germany until the end of World War One.

Prandtl's Fundamental Considerations Concerning Unsteady Aerodynamics

The physical foundation concerning the computation of unsteady aerodynamic forces on a (harmonically) oscillating airfoil was presented by *Prandtl* [1] during a presentation at a scientific conference in Innsbruck in September 1922. He considered an airfoil, which during flight *"carried out a periodic up and down movement, where, simultaneously, the angle of attack may be changed by a rotation of the airfoil"*.

On the basis of his theory of the lifting vortex, he continued:

"If the lift is changing, vortices of a strength corresponding to the change in circulation have to come off (Kelvin's vortex conservation theorem, author's remark)... This results in a wave-like dividing sheet of an alternating positive and negative rotational sense."

Prandtl replaced the airfoil by a flat plate with a distribution of bound oscillating vortices of the density γ (x, t) along the mean camber line, and an arrangement of free vortices of density ε (x, t) on the dividing sheet (vortex sheet) in the wake, which separate tangentially from the trailing edge with the flight velocity V. From the requirement of temporal constancy of the sum of bound and free vorticies, *Prandtl* arrived at a

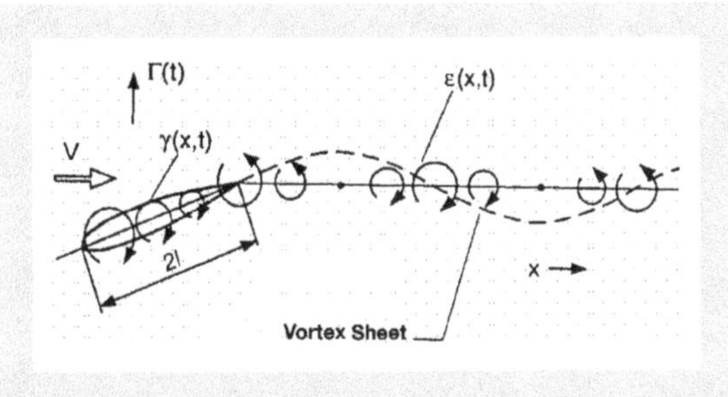

Bound vortices γ (x, t) and free vortices ε (x, t) on an oscillating wing with a vortex sheet (dividing sheet).

formulation for the time-dependent circulation ε (t), and hence for the unsteady lift of the oscillating airfoil. In his well-noticed presentation, he also remarks that "*apart from the method of bound and free vortex distributions (singularity method, author's remark), the task in question may also have been solved, according to a proposal of A. Betz, by the method of conformal mapping*".

Indeed, both methods for the computation of the motion-induced unsteady aerodynamic reactions on oscillating wings were soon applied.

PhD-Thesis of *Walter Birnbaum* on the Motion-Induced Lift of Airfoils

On the basis of the fundamentals described by *Prandtl* obtained his pupil *Birnbaum* [2] at the University of Göttingen in 1922 his PhD with the thesis: "*The two-dimensional problem of the flapping wing*". Here, he investigated, as described in the introduction, besides the problem of the wing and its elementary movements of "flapping" and "rotating" also "*an interesting case of the dynamic instability of a spring suspended wing, a phenomenon that was confirmed by experiments at the "Aerodynamic Test Establishment Göttingen, AVA*". Here, wing flutter is being addressed for the first time as an aeroelastic stability problem. In the process, *Birnbaum* introduced for harmonic flutter with the angular frequency ω, the reduced frequency $\omega^* = \omega\, l/V$ as a characteristic quantity into unsteady aerodynamics. It is a measure for the extent of "unsteadiness" of aerodynamic processes and also an important aeroelastic similarity parameter.

With *Prandtl's* congenial approach of replacing the lifting surface with a distribution of bound vortices, and on the basis of *Prandtl's* formulation of the preservation of vorticity in the case of an oscillating wing, *Birnbaum* developed an integral equation for the unsteady aerodynamic pressure distribution on a two dimensional flat plate oscillating harmonically with a small amplitude (linearization) in inviscid incompressible flow. For its solution, he used vortex density distributions, which he took from a document of *W. Ackermann*, which remained at the AVA after the end of the War; the distribution became later known as *Birnbaum-Ackermann* normal distribution. Despite great mathematical difficulties *Birnbaum* was able to solve this integral equation for reduced frequencies up to $\omega^* = 0.12$, which were not so relevant for practical purposes but led to a number of fundamental findings. He could show that, due to the effect of the free vortices leaving the trailing edge of the oscillating plate, lift and moment were phase-shifted with respect to the motion of the plate and that they also depended strongly on the reduced frequency. This was the key to the physical understanding of the flutter problem and its analytical treatment as a dynamic aeroelastic stability problem. A closer theoretical investigation led to further fundamental results, which became, as design rules for the prevention of the (classical) wing flutter, universally valid:

- Arrangement of the center of gravity of the wing cross-section and the elastic axis (axis of torsion) as close to the leading edge as possible,
- Avoidance of neighboring frequencies of wing bending and wing torsion.

With this pioneering work, it was possible to give a complete physical description of the flutter phenomenon about 20 years after the beginning of powered flight. In his work, *Birnbaum* also recognized the static torsion divergence problem as an aeroelastic stability problem.

PhD-Thesis of *Herbert Wagner* on the Origin of Dynamic Lift

In 1924, *Herbert Wagner* graduated from the TH Berlin with another pioneering work in the field of unsteady aerodynamics, his PhD-thesis: "*Origin of the dynamic lift of wings* [3]". In his work, known as the "Wagner problem", he investigated the origin of lift on a wing in an inviscid incompressible fluid, if the wing, at an angle of attack α, is abruptly accelerated from rest to a streamwise velocity V = constant. *Wagner* solved this problem for the two-dimensional flat plate by *Betz'* method of conformal mapping, where the increase in lift A(s) is a function of the non-dimensional time s = Vt/l characterized by the so-called *Wagner*-function W(s). The real practical importance of this solution is, that the unsteady aerodynamic reactions (lift and moment) can be computed for arbitrary airfoil movements by super position, for instance, in the case of non-harmonic flutter or when entering in a sharply-defined gust. *Wagner's* solution also contributed substantially to the further understanding of the basic physical relations concerning the origin of dynamic lift. However, the method of conformal mapping, elegantly applied by *Wagner* for the solution of his problem, remained subsequently restricted to the incompressible velocity range, since it resulted at higher velocities in insurmountable mathematical difficulties. Further theoretical progress in the field of steady as well as unsteady aerodynamics remained, therefore, the domain of the singularity method based on *Prandtl's* theory of the lifting vortex surface.

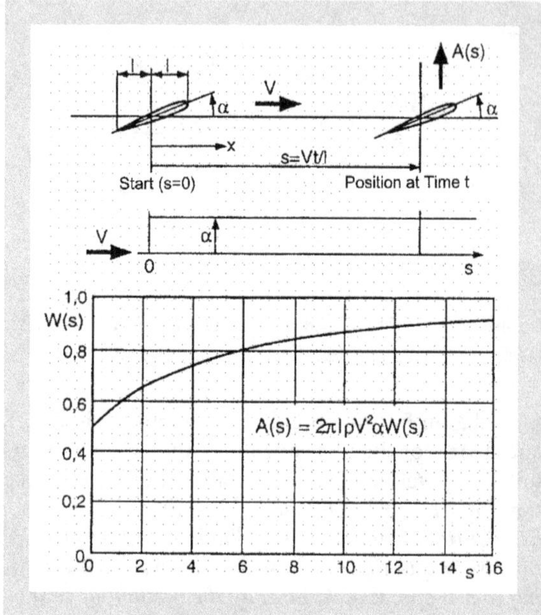

Schematic depiction of the "Wagner-problem" and the Wagner-function W(s).

Küssner's Theory of the Harmonically Oscillating Airfoil with Control Surfaces

In 1929, *Hans-Georg Küssner* published [4], after receiving his PhD from the TH Danzig and working since 1928 at the DVL in Berlin-Adlershof, his first pioneering work in the field of aeroelastics, which was followed later by a lot more. Starting point of his work with the title "Flutter of aircraft wings" was *Birnbaum's* work, whose model of the elastically supported two-dimensional flat plate with the degrees of freedom flapping and rotation he adopted, but who also added the degree of freedom of a co-oscillating control surface (control-surface rotation). In a consequent application of *Prandtl's* interpretation of the vortex conservation theorem, he succeeded, with a Fourier approach to the distribution of the bound vortex density, to find the general solution for the integral-equation problem, and to compute the motion-induced aerodynamic forces in the form of the unsteady aerodynamic coefficients for harmonic flutter up to $\omega^* = 1.5$. Since the harmonic motion represents the limiting case of dynamic stability, the basis for specific practical flutter calculations was thus given, at least for the case of two-dimensional wings with control surfaces in incompressible flow.

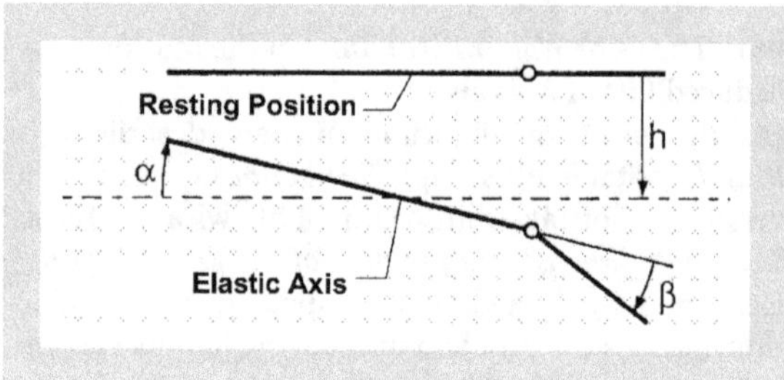

Küssner's oscillating two-dimensional (rectangular) wing with the degrees of freedom flapping h, rotation α and control surface deflection β.

Research in the Area of Wing Flutter since 1925

After *W. Birnbaum* and *H. Wagner* had achieved, with the above mentioned pioneering work, globally recognized breakthroughs in the field of unsteady aerodynamics in Germany after World War One, despite very considerable restrictions in aeronautics, there followed, after 1925, a number of important theoretical and experimental investigations concerning aircraft flutter, which occurred more frequently with increas-

H.G. Küssner (1900–1984).

Hans Georg Küssner, born in Bartenstein/East Prussia on September 14, 1900.

1919 studies at the TH Danzig; in 1923 graduation as (Diplom) Engineer, thereafter assistant at the Institute of Machine Elements and Water Turbines; in 1928 PhD with the thesis "The Economic Ocean Aircraft".

1928 assistant at the DVL Berlin-Adlershof at the Department of Statics. During this time, he produced several basic papers on the aeroelastics of aircraft, especially an unsteady aerodynamic theory of the oscillating two-dimensional wing with control surface, and trend setting contributions to the aeroelastic flutter and gust problem.

1934 transfer to the "Aerodynamic Test Establishment Göttingen, AVA". There he conducted further basic investigations on the unsteady aerodynamics of oscillating lift systems, which culminated in 1940 in his "General Wing Theory".

1939 head of the AVA "Institute of Unsteady Processes" ("Institut für Instationäre Vorgänge"). 1940 associate professor in the services of the State (Reich).

1947 employee of the MPI (Max-Plank Institute) of Fluid Research. Publisher of the book "Principia Physica".

1955 head of the AVA Department "Aeroelasticity" in Göttingen. In this position, he made further extraordinary contributions to the unsteady aerodynamic airfoil theory, to the analytical solution of the flutter problem (derivation of the generalized aeroelastic equations of motion), as well as to static vibration tests on aircrafts.

1960 associate professor of Aeroelasticity at the TH Brunswick.

1968 retirement.

1976 publisher of the book "Basics of a uniform theory of physical particles and fields".

1981 recipient of the Ludwig Prandtl-Ring of the DGLR together with *Herbert Wagner*.

1984 he dies in Göttingen on March 21.

ing flight speeds. Already in 1923, *A.G. von Baumhauer* and *C. Koning* [5] could prove in wind tunnel tests in Amsterdam that a static rudder (control surface) mass balance with respect to the axis of rotation avoided rudder flutter. *H. Blenk* and *F. Liebers* [6, 7] have then conducted between 1925 and 1929 at the DVL extensive experimental and theoretical flutter investigations on three-dimensional wing models. Since in their opinion, the *Birnbaum* unsteady aerodynamic-force theory "*would face great mathematical difficulties and would lead, via a non-elementary integration, to unwieldy equations*", they tried to come to a more manageable formulation by applying a quasi-stationary aerodynamic-force theory. In the process they neglected the important influence of the free vortex wake and used in their calculations steady force coefficients and the dynamic angle of attack, i.e., the angle between the chord line and the momentary direction of motion of the oscillating wing. The unsteady forces on the three-dimensional wing were calculated by applying strip theory, the oscillation forms of bending and torsion determined experimentally on the model or assumed, and introduced into the corresponding differential equations of motion. The knowledge resulting from these investigations was, as experience showed, sufficient to explain the flutter process, and it was also possible to verify *Birnbaum's* flutter suppression means and the positive effect of a static rudder mass balance, but the accuracy of those flutter calculations was insufficient and generally missed the target. Others, also outside Germany, came to the same conclusion when applying quasi-steady aerodynamic-force theory for their flutter investigations. Nevertheless, this simplified aerodynamic-force theory was still applied for some time.

The successful but by far more difficult and elaborate approach to a theoretical prediction of the flut-

Statistic of buffeting accidents recorded in Germany up to about 1934 (*H.G. Küssner*) [12].

	l_m [m]	Frequency [1/min]	V_k [km/h]	ω^*	Oscillation form	Stiffness of control cables	Kind	Wing material
DP 9	1,5	520	180	0,82	symmetric	small	bad	Wood
He 8	3,0	540	350	0,87	symmetric	small	bad	Wood
L 78	1,36	860	210	1,05	anti-symmetric	large	good	Wood
He 60	12,2	780	350	0,93	anti-symmetric	large	bad	Metal
He 46c	1,4	845	268	0,87	symmetric	small	good	Wood
KL 1A	1,3	675	145	1,14	symmetric	small	bad	Wood
Do 10	1,6	1450	450	0,97	anti-symmetric	large	bad	Metal
L 102	1,56	835	290	0,85	anti-symmetric	small	bad	Metal
AC 12E	1,4	815	220	0,98	symmetric	small	good	Wood
Do 12	1,3	580	180	0,79	anti-symmetric	small	bad	Metal
Me 28	1,05	770	220	0,69	anti-symmetric	large	good	Metal
S 24	1,18	1215	280	0,97	anti-symmetric	small	bad	Wood
Ar 66C	1,65	790	340	0,72	symmetric	small	bad	Wood

ter behaviour of wings with control surfaces was then given by *H.G. Küssner* in 1929 in his already quoted pioneering work [4]. For the idealized three-degrees-of-freedom system, considered there, he established, on the basis of his unsteady aerodynamic-force theory, the aeroelastic equations of motion also considering structural damping; he not only solved the complex linear algebraic flutter equations for harmonic flutter as was the custom (especially in England) according to the awkward stability criterion of *Routh*, but via the condition of a vanishing denominator determinant. This provides two real equations, whose solutions lets one determine, in a relatively simple way, the critical velocity, the flutter frequency and also the amplitudes of the oscillation forms. This also physically very clear method of solution became in the years that followed the standard method. In his work, *Küssner* also performed a number of parametric investigations to determine the effect of mass distribution, natural frequencies of the degrees of freedom, position of the elastic axis and of structural damping. Finally, he investigated, with measured natural oscillation forms and the application of the strip theory, in a computational analysis two flutter cases, which had occurred on the wings of a Junkers A 20 and a Messerschmitt Me 20, respectively; he obtained a rather satisfying agreement. Further checks on *Küssner's* calculation method, later also with results of wind-tunnel tests at the DVL, led to similarly good results.

First Basic Investigations Concerning the Aeroelastic Buffeting Problem

A mysterious crash of a Junkers F 13 postal-service aircraft on September 29, 1930 in England initiated in Germany at the DVL, and also in England, great efforts to clarify the "buffeting" process (in German "Schütteln"), which may be caused by a periodic vortex separation at high angles of attack, by gusts and on empennages due to the interaction with the wing wake. None of the three possibilities could be ruled out. Although the real causes for the accident could never be clarified, the corresponding research led to a number of further fundamental results concerning the dynamic response of aircraft. *H.G. Küssner* [8] dealt for the first time in a theoretical investigations with the load on a wing induced by gusts and calculated the load factor of the wing bending at the entry into a sharply defined vertical gust; here, he also fell back on *H. Wagner's* work on the origin of dynamic lift. *E. Peterson* [9] and *M. Kramer* [10] investigated experimentally in the wind tunnel the lift behaviour of wings, as well as the pressure fluctuations within the wake by means of a piezoelectric quartz (!), at a pullout process beyond the stall angle of attack. They found that it increases and that the maximum lift coefficient c_l enlarges dynamically. This so-called "dynamic-stall effect", also sometimes named "Kramer effect" in the English literature, later reached considerable importance in modern fighter aircraft engineering. Finally, *H. Blenk et al.* [11] considered the wake-vortex resonance problem on empennages. They conducted wind-tunnel experiments at the DVL on wing models at angles of attack up to 90 degrees and found in the turbulent wake dominant vortex frequencies of certain Strouhal numbers; if these Strouhal numbers coincided with the natural frequency of a tail plane embedded in the wing wake, aeroelastic resonance occurred leading to considerable oscillation amplitudes. These very important and basic results on aeroelasticity-induced wing flutter were the reason for the demand of corresponding design and construction regulations.

First Design and Construction Regulations to Prevent Wing Flutter

H.G. Küssner summarized the mid-thirties state of the art with respect to wing and empennage flutter of aircraft in another well-recognized paper [12] in 1935, also giving recommendations for corresponding design, construction and strength regulations. It should be mentioned that this was his last publication at the DVL in Berlin-Adlershof before he transferred to the "Aerodynamic Test Establishment, AVA" in Göttingen. His regulations and recommendations concerning the prevention of flutter are based on a statistic of the flutter occurrences recorded in Germany and England. Starting from the observed flutter velocities, the wing dimensions and the natural frequencies of the oscillation forms involved in the flutter process, he derived at a critical reduced frequency $\omega^* = 0.9 \pm 0.12$ where flutter was observed on the aircraft types of that time.

Here, mainly flutter of the ailerons were the cause, with the critical reduced frequency then being significantly below the average. To avoid aeroelastically-induced wing flutter, he recommends to achieve

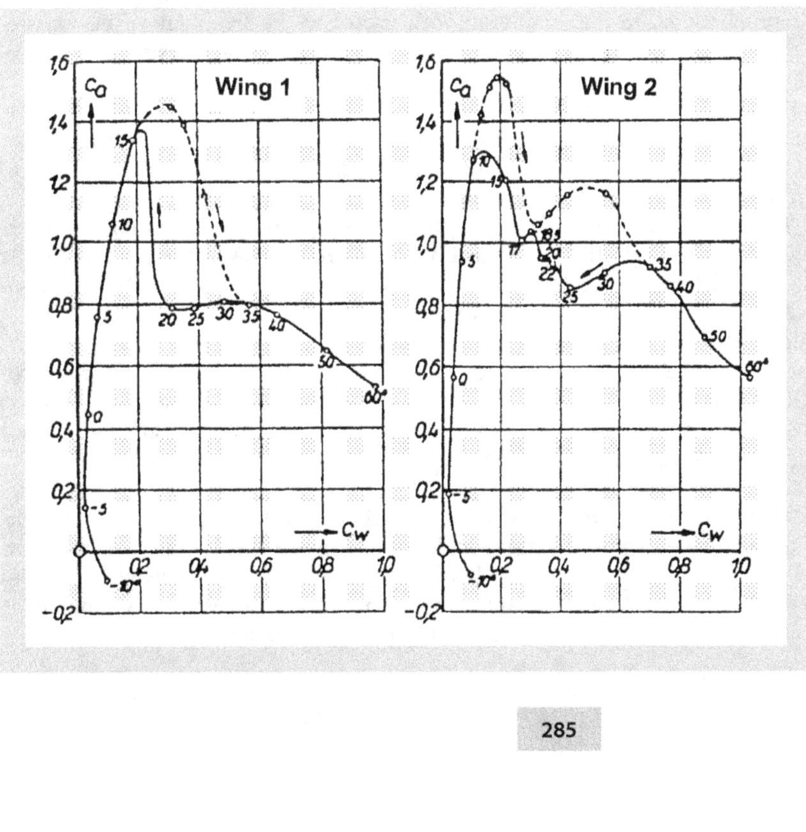

Static and dynamic polars (dynamic-stall effect) according to *E. Peterson*.

Shake tests (1930) at the DVL in Berlin-Adlershof on an aircraft elastically suspended by rubber ropes.

the critical velocity demanded in the design and construction regulations at high rotational frequencies for the control surfaces of $\omega^* > 0.8$, and with a careful mass balance at $\omega^* > 0.27$. For empennage-surface flutter caused by vortex separations on the wing, the fuselage, or on propeller blades, he points out that the dominating frequency of the vortex streets behind these respective parts, which is proportional to the flight speed, should not coincide with any empennage bending frequencies or with the fuselage rotational frequencies (resonance). Since the theoretical determination of these natural frequencies with the computational tools available at that time was hardly possible he recommends, to obtain the necessary natural oscillation parameters by a "shake test". *Heinrich Hertel* [13] conducted such measurements on elastically suspended aircraft already since the end of the twenties at the DVL in Berlin-Adlershof; such tests became later known as so-called "static vibration tests" ("Standschwingungsversuche"), a standard procedure, mainly in Germany, within the framework of type certification. Concerning the enlargement of the stiffness, *Küssner* added in conclusion: *"At a further increase in flight speed, one reaches the point where a simple increase in stiffness is for weight-reasons no longer viable, and one is forced to combine all known means to prevent flutter."*

To reach this point, decades of still more basic research were necessary.

Aeroelastics Research Between 1935 and the End of World War Two

Since 1935, aeronautical research in Germany experienced a considerable boost in terms of material and personnel; research in the field of aeroelastics encountered great challenges due to the rapid progress in the development of a new generation of high-performance aircraft. From this time on, it is no longer possible to describe in detail the further development of aeroelastic research in Germany via individual publications. New centers of focused research formed, partly within newly founded institutes, and also aircraft industry and universities provided increasingly important contributions. With the move of *H.G. Küssner* to the AVA in Göttingen in 1935, a new center focused on aeroelastic research was established. At the DVL in Berlin-Adlershof aeroelastic research was intensively pursued, too. However, the English expression "aeroelasticity" still did not enter the German language. The institute, founded on January 1, 1939 at the AVA and headed by *H.G. Küssner*, was named "Institute of Unsteady Processes" (" Institut für instationäre Vorgänge").

At the latest in 1935, it also became obvious that for the emerging new generation of high-performance aircraft with regard to the treatment of new aeroelastic problems three requirements existed to concentrate aeroelastics research to:

- An extended three-dimensional unsteady aerodynamic wing theory for compressible flow up to near-sonic speeds.
- Improved methods and procedures for the determination of the natural oscillation characteristic parameter of aircraft and their structural components in static tests.
- New analytical methods for the solution of the aeroelastic equations under consideration of many elasto-mechanical degrees of freedom (natural oscillation forms) within the aeroelastic equations.

The mastering of these tasks concerned with the reliable prediction of the aeroelastic behavior of aircraft already during the development stage was of fundamental importance to the successful development of modern high-performance aircraft.

Continued Development of the Unsteady Aerodynamic Airfoil Theory

H.G. Küssner summarized the 1935 state of the art in unsteady aerodynamics of oscillating wings in a further basic paper [14]. He includes in his considerations all known foreign investigations, and in the main chapter he again treats the two-dimensional problem of the oscillating wing with the degrees of freedom being flapping, rotation and control surface rotation. In his closed analytical solution of the problem, the circulatory contribution of the aerodynamic forces, resulting from the free vortices, is expressed by the so-called T-function, which became later known as *Küssner's* "Wake Vortex Function". *Küssner* also shows the identity of his solution with the also exact solution of *Th. Theodorsen*, which was simultaneously developed in the USA with the method of conformal mapping. *Küssner* also determined elegantly the Wagner-function W(s) with the aid of the Fourier-integral of his harmonic solution and solved this way the problem of the unsteady lift of a wing entering a sharply defined vertical gust. The circulation function K(s) appearing here, was later named *Küssner*-function.

But one was still far away from the ability to determine the aeroelastic processes during the development of a new type of aircraft. Flutter occurred again and again as the statistic of *K. Schwarzmann* [15] shows for 95 cases up to 1940. During the further development of the aircraft, the control surfaces received an aerodynamic balance (auxiliary control flap) to reduce the manual actuation forces; with a smaller aspect ratio and more complicated oscillation forms, the approximation given by the strip theory (two-dimensional flow within chord-wise sectors) became increasingly more inaccurate and a spatial theory for the oscillating wing was required, which had also to be extended to compressible flow, as the flight speed approached the speed of sound. Since 1937, *H.G. Küssner* and his colleagues (*L. Schwarz, P. Jordan* et al.) attacked all these problems at the AVA. At the DVL, it was mainly *H. Söhngen, F. Dietze* and others, at the Flight Technological Institute of the TH Berlin it was *S. von Borbely, R. Kassner* and *H. Fingado* who worked on the same problem.

The severe problem of the harmonically oscillating control surface with auxiliary flap was approached simultaneously, and with different methods, by *L. Schwarz* [16], *H. Söhngen* [17] and *F. Dietze* [18], and

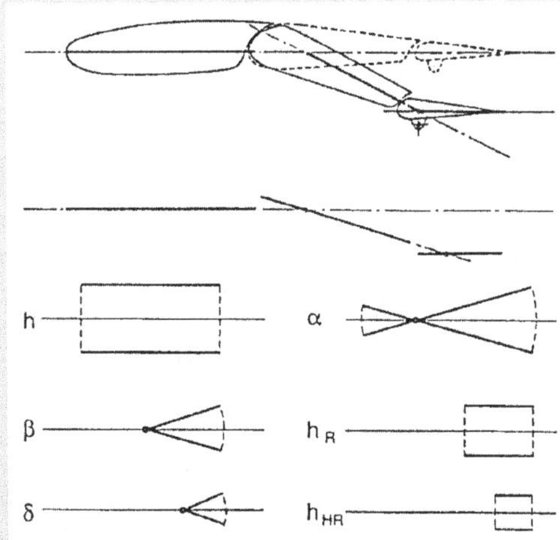

Modeling of the oscillating wing with aerodynamically balanced main and auxiliary control surfaces.

partially solved. *Küssner and Schwarz* [19] then presented in 1940 the solution for a total of six degrees of freedom, for fin, control surface and auxiliary flap, in numbers and equations; with their superposition one can calculate practically all interesting cases including the inner and outer aerodynamic control surface balance. The fundamental work of the "Göttingen School" in the field of unsteady aerodynamics until the end of World War Two climaxed at the end of 1940 when *H.G. Küssner* published his "General Airfoil Theory" [20], a first class pioneering paper. It holds for three-dimensional wings of arbitrary contour in compressible subsonic flow; it is based on *Prandtl's* concept of the acceleration potential [21] and the classical acoustic wave equation with its solution for the sound source at rest. The congenial idea of *Küssner* was to transfer this solution via the Lorentz-transformation to moving, pulsating sources as a basic solution for the compressible unsteady flow. The determination of the unknown strength of the source distribution on the upper and lower wing surfaces, while meeting the kinematical flow condition, which results from the oscillatory motion leads to an integral equation by applying the singularity method. *Küssner's* integral equation contains as integrand the requested pressure jump on the wing as product with a known Kernel function, which is singular at the control point. *Küss-*

ner showed that the integral equation of his general airfoil theory encompassed all special cases known at that time. With the exception of a few cases, this equation could numerically only be solved approximately; an exact solution became possible only two decades later with the occurrence of high powered electronic computers (Kernel function method). *Küssner* discovered later (1944) that with the addition of advancing potentials corresponding solutions for the supersonic range were possible. Based on this discovery, *I.E. Garrick* developed in 1947 in the USA an unsteady three-dimensional wing theory for supersonic flow.

In Germany further research in the field of unsteady aerodynamics focused until 1945 on solutions of the two-dimensional problem of the oscillating wing in compressible subsonic and supersonic flow on the basis of an integral equation developed in Italy in 1938 by *C. Possio*, which is, as a special case, also con-tained in *Küssner's* integral equation. Further research was focused on the oscillating wing in supersonic flow with a first publication by *S. von Borbely* in 1942 [22]. Here, valuable contributions were made by *Th. Schade* (1940), *L. Schwarz* (1943) and *P. Jordan* (1944/45) from Göttingen, and by *F. Dietze* (1943) of the DVL. Since there were only mechanical calculators available for the practical determination of the unsteady aerodynamic forces on oscillating three-dimensional wings and empennages the little satisfying, quasi-two-dimensional aerodynamic strip theory remained the only practical solution until the end of World War Two.

Flutter Computations with many Simultaneous Elasto-Mechanical Degrees of Freedom

The rapid progress in aircraft engineering since 1935 led not only to an increase of flight speed up into the near-sonic speed range, but also to much more complicated designs. Divided tail planes, more slender airfoils, a larger number of cut-outs on the wings, heavier engines and the development towards an extreme light-weight construction, were all factors that led to more flexible aircraft structures; this reduced the margin between the critical flutter speed and the demanded maximum speed and thus increased the danger of the occurrence of flutter. This required an increase in the number of elasto-mechanical degrees of freedom in the theoretical flutter investigations and, by necessity, an improvement of the elasto-mechanical fundamentals for the computation of the static natural oscillation parameters, and, furthermore, the development of mathematical methods for the solution of the complex eigenvalue problem.

The elasto-mechanical substitute for wing, fuselage and empennage was the elastic beam with the degrees of freedom "bending" and "rotation", sometimes also rigid, elastically connected construction elements. But already the correct dimensioning of the structure of these models was a difficult engineering problem, while the computation of the requested real eigenvalues λ was, at that time, an extraordinarily difficult mathematical problem since these can, in the case of more than two degrees of freedom, only be determined by approximative methods. Depending on the respective solution procedure, the eigenvalue problem led either to the determination of the zero points of a polynomial of the characteristic secular equation (denominator determinant) or, if applying the *Ritz-Galerkin* method with selected functions and free coefficients, to the iterative solution of a characteristic system of equations in matrix form. With the computational tools at that time, a huge time-wise effort was necessary, which led in research institutions and industry to the occupation of entire "calculation offices". For more than four degrees of freedom, only an iterative solution was possible. Especially at the DVL, these problems and the mathematical solution of the flutter equations were subject of intensive work (*A. Teichmann* [23], *R. Kassner* and *H. Fingado* [24]).

If the eigenfunctions without aerodynamic forces are known, they can be introduced into the flutter equations together with the unsteady, on the reduced frequency $\omega^* = \omega\, l/V$ time dependent aerodynamic forces. In the case of harmonic flutter, as limit of the dynamic stability, one obtains a complex linear algebraic eigenvalue problem, whose solution became even more difficult since the necessary mathematical basics were not yet available. Requested were the only physically meaningful pure real eigenvalues (flutter frequencies), if they were at all present. With these and the given respective reduced frequency, the corresponding critical flutter speed V_k was then known; it appeared, however, not explicitly in the flutter equations. Dependent on the way that the algebraic conversion was performed, different flutter criteria were applied in order to simplify the laborious iteration process. Concerning the aeroelastic coupling mechanism, these criteria led, however, to more confusion

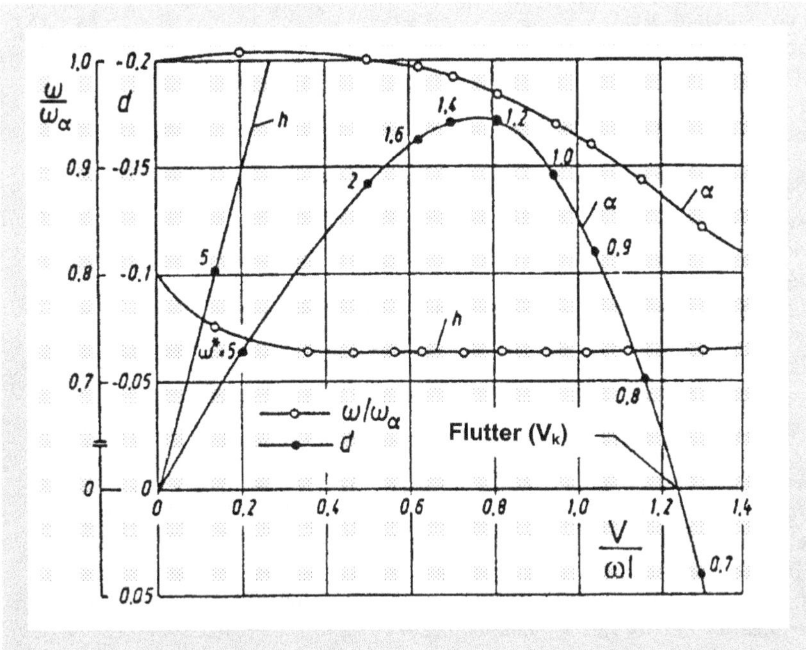

Flutter stability characteristic of a two-dimensional wing with the degrees of freedom flapping h and rotation α.

rather than contributing to an understanding of the physical background and, furthermore, to many controversial discussions among the "flutter experts".

Fundamental contributions to the solution of the complex flutter equations with many degrees of freedom were then provided at the beginning of the forties at the AVA by *P. Jordan* [25] and *H. Wielandt* [26]. They established the basis for the solution of the mathematically not self-adjunct complex eigenvalue problem, especially the mathematical convergence proof of the iteration procedure (*Wielandt*) and the physical interpretation of the complex eigenvalues in relation to the coupling mechanism of the eigenforms of the respective flutter form (*Jordan*). The complex eigenvalues describe physically a real oscillation frequency ω (real part) and a damping d (imaginary part) and the development of both of the corresponding branches of the solution of the stability characteristic, with the reduced frequencies as parameter. A clear physical insight into the aeroelastic coupling process is possible. Flutter occurs if for one of the damping branches d = 0 is given.

The fundamentals underlying the solution of the dynamic aeroelastic flutter-stability problem as a complex eigenvalue problem, which were established shortly before the end of World War Two at *Küssner's* "Institute of Unsteady Processes" in Göttingen, led to another breakthrough in aeroelastics. The solution of the complex eigenvalue problems, which occurred here and in many other areas of modern physics, became a routine as more efficient computers arose. It is interesting to mention that *Konrad Zuse* received valuable advice and also material support from the aeroelastics researchers of the DVL (*A. Teichmann*) during the development of his first electronic computer early 1940 in Berlin.

Experimental Research

Static Vibration Tests (Shake Tests). As already mentioned, the DVL in Berlin-Adlershof conducted already since 1929 the so-called "shake tests" to determine experimentally the natural oscillation behaviour of stationary aircraft. For that purpose, the aircraft was elastically suspended on rubber ropes to simulate a free-floating flight condition. The excitation of the flutter was accomplished via a rotating unbalanced disk attached to an electric motor with a continuous rpm-control; the oscillation amplitudes were determined by so-called "measurement wedges".

This measurement procedure was applied at DVL, with certain refinements, on numerous aircraft up to the end of the war. The measurement results were mainly used to validate natural-oscillation computations, but were also directly introduced into flutter calculations.

Also the aircraft manufacturers carried out such static vibration tests, as they were later called. At Dornier, *G. de Vries* [27] considerably improved the static-vibration-test instrumentation in many details during the early forties. He developed a multiply-unbalanced excitation device, which was able to systematically excite symmetric and asymmetric flutter, and an electrical procedure to check the resonance criterion as well as the purity of the phases of the excited oscillation forms. After 1945, *de Vries* established at the ONERA in France with the development of electro-dynamic exciters and oscillation sensors, and further electronic devices for checking the phase-resonance criterion, the basis for the modern static vibration test. This test has become an indispensable tool for the flutter verification for aircraft within the type-certification process; it was relatively late also applied in England and the USA.

In-Flight Vibration Test. Despite all the imaginable effort to determine the flutter characteristics prior to the first flight, it was not possible to exclude flutter during flight with last certainty due to the many remaining unknowns. The final answer had to be provided by the controlled test flight; for the latter, a flight vibration test was being developed since 1936 at Junkers by *B. von Schlippe* [28]. His concept was to closely approach, during a flight test with artificially excited flutter and by monitoring the amplitude, the flutter boundary – meaning the critical speed – and to determine the latter by an asymptotic extrapolation. The DVL was charged with the development of a corresponding measurement technique. For the excitation, two "oscillating clubs" were used arranged internally at the respective wing end; with these and the aid of electric motors, symmetric and asymmetric flutter could be excited. As test vehicle for these experiments, a Ju 52 was used. With this installation, the "Schlippe method" was employed to in flight "flutter-test" successfully for several prototypes, before in 1938, during a flight vibration test with the prototype of the Ju 90, suddenly flutter occurred on the empennage and the aircraft crashed killing the entire crew. The knowledge gained here, namely, that during vicious flutter close to the critical speed the slope of the amplitude-versus-speed changes rapidly and that therefore an asymptotic approximation to the critical speed requires extreme caution, allowed many further prototypes to be successfully subjected to flight oscillation tests up to the end of the war. In order to prove the demanded flutter-free behaviour up to a speed corresponding to 1.35 times the maximum horizontal speed, this flight vibration procedure also became a standard procedure.

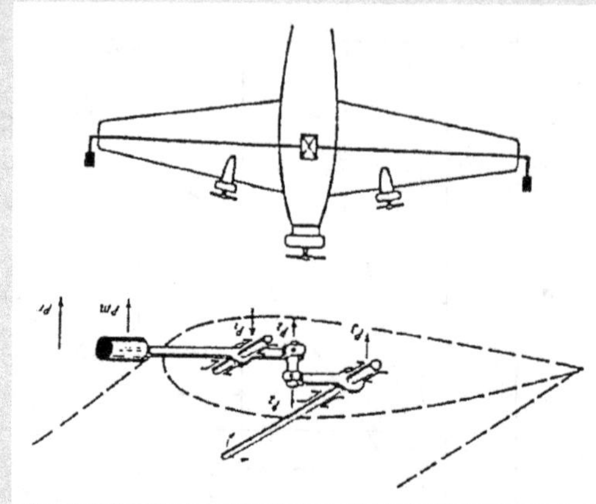

Unbalance weights (clubs) mounted at the wing ends for the excitation of in-flight oscillations.

Wind Tunnel Model Tests. The performance of flutter tests in the wind tunnel beginning in the early twenties with *W. Birnbaum* and continuing with *H. Blenk* and *F. Liebers*, had up to 1935 already a certain tradition. Up to then, these investigations were purely phenomenon-oriented, i.e., they were directed towards the investigation of fundamental questions and were not project-related. Here, DVL in Berlin-Adlershof had established a focused research area. Since approximately 1940, the aircraft industry also conducted more and more project-related aeroelastic wind tunnel model tests, among others, on the He 177 (wing), the He 219 (complete model), the Ju 288 (complete model with 7-meter (23 ft) span), the Me 264 (wing), and the Ar 232 (empennage). For these tests, the model has to be dynamically similar to the full-scale configuration, it must satisfy certain similarity laws related to geometry, stiffness and mass distribution, and it must be, with regard to its inner structure, as far as possible similarly built. Of course, the Mach number has to be identical. Here, plastics (Vinidur) were for the first time applied. The excitation of flutter was realized by pulling by hand on strings attached to the wing and empennage surfaces; flutter could be stopped by rapidly tightening thin wires.

Such wind tunnel model tests proved in many regards very valuable, especially in cases, where the conclusiveness of the flutter computations was not ascertained. *P. Jordan* conducted with the complete model of the twin-engine He 219 of the Heinkel company, mentioned above, in 1944 in the large wind tunnel of the AVA in Göttingen flutter tests with systematic parameter variations, and summarized their investigation in an educational movie. This movie shows control surface and empennage flutter in many variations accompanied by scientific explanations; today, it has not lost its actuality and caused a sensation when discovered by the Allies in 1945.

Aeroelastics of the Sweptback Wing and Skin Flutter

The "Institute of Unsteady Processes" of the AVA received a confidential letter, dated September 12, 1944, from the "Elektromechanische Werke GmbH", Karlshagen (Pommern) with an order to conduct the following aeroelastic investigations,

- flutter of sweptback wings,
- flutter of thin, flat metal panels,

related to the fin of the A4-missile and the wing of the A9 missile, respectively. Frankly speaking, it meant to conduct several flutter investigations for the V2 and its derivative with sweptback wings for the Peenemünde test center. As one knows from a later report of *Wernher von Braun*, a "bursting" of the skin on the V2 occurred several times during the experimental phase, presumably due to skin flutter. For that time, both of these tasks represented an extraordinary challenge since they boiled down to the problems of flutter of a 45-degree sweptback wing at Mach numbers up to 5 and of skin flutter, respectively; there was up to that time no knowledge or experience related to these problems at all.

Already on November 7, 1944, *P. Jordan* told the customer in Peenemünde with respect to the skin flutter that this problem had, in general, theoretically and through wind tunnel tests been solved. The danger of flutter was given if the approach velocities were higher than the wave propagation velocity within the sheet-metal skin, which is mainly to be expected at supersonic speeds. *P. Jordan* came to that conclusion theoretically, and correctly, as was shown later, in following his Göttingen teachers *L. Prandtl* and *A. Betz* in a congenial physical modeling of the problem, which led him to a partial differential equation. Into this equation, he introduced as the unsteady aerodynamic force generating mechanism the linearized, quasi-stationary supersonic flow according to *J. Ackeret*, a concept that became ten years later the basis for the supersonic "Piston Theory" developed in the USA. With the solution of this differential equation, *Jordan* gained far reaching basic knowledge, which he presented, shortly before the end of the war, in a secrete AVA-report [29].

In a further interim report, dated January 8, 1945, he reported to Peenemünde the results of his investigation related to the aeroelastic behavior of sweptback wings gained up to then, which were published only after the war (1946) in the so-called "Göttingen FIAT Monographs" [30]. To keep the contact to the straight wing, he considered the sweptback wing as a straight wing rotated backwards at the fuselage junction, and he recognized in the process that, when applying the aerodynamic strip theory, the freestream velocity (Mach number) had to be reduced by the cosine of the sweep angle, and that a bending of the sweptback wing caused kinematically-coupled a reduction of the flight angle of attack. Consequently he drew the conclusion that the sweptback wing under certain circumstances, also at a large torsional stiffness, may flutter with only one degree of freedom "bending", and that the danger of this happening was greatest in the transonic speed range, a fact that is reflected in the so-called "transonic dip" in the flutter boundary. *P. Jordan* was very likely the first to correctly recognize the principal aeroelastic characteristics of the sweptback wing. A further static aeroelastic cardinal problem related to the sweptback wing with control surfaces, which was not of his concern, but was repeatedly reported by many pilots when approaching the speed of sound, must be added, the loss of control-surface effectiveness up to a control reversal. This problem was then in the fifties a matter of intensive aeroelastic research in England and in the USA.

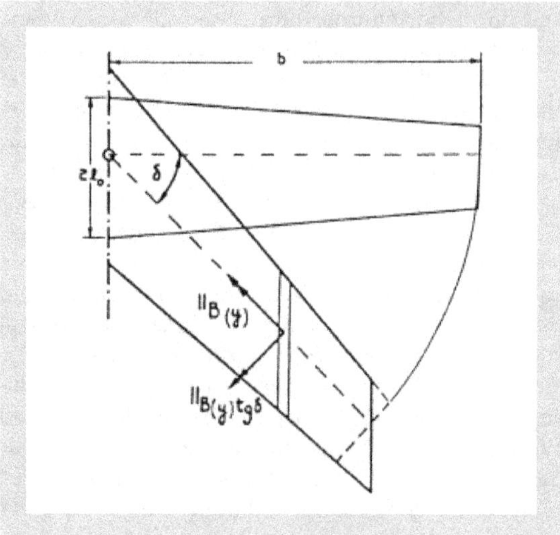

P. Jordan's explanatory sketch of the kinematical coupling of rotation and bending on a sweptback wing.

Aeroelastic Research in Germany until 1945 – an International Comparison

Although such a comparison always seems to be subject to reproach and the danger of national subjectivity, one may summarize, based on the stated scientific work and the opinion of well-recognized foreign experts, that Germany has at the end of World War Two certainly played a leading role in the field of aeroelastics, especially in the areas of compressible unsteady aerodynamics, static and in-flight vibration tests, wind tunnel tests with dynamically similar (plastic) flutter models, and the aeroelastics of sweptback wings. But also earlier German work in this field, say, since the twenties, has found great international recognition, hence documented this leading role. If one takes the number of quotations in the respective international literature as basis, the papers of *L. Prandtl* [1, 21], *W. Birnbaum* [2], *H. Wagner* [3], *H.G. Küssner* [14, 19, 20], *H. Blenk et al.* [11], and *P. Jordan* [29, 30] are certainly milestones in the development of aeroelastics of aircraft.

Looking back, it can be concluded, that at the end of World War Two all essential aerodynamic and elasto-mechanical fundamentals and known aeroelastic problems in aircraft engineering, especially the problem of flutter, had scientifically been worked out. But it took decades before large-scale electronic computing systems made it possible to take full advantage of the potential of this knowledge in modern aircraft engineering. However, with the progress of flight technology, many new and more complex aeroelastic problems arose, which again added new challenges to research.

References

1 L. Prandtl, Über die Entstehung von Wirbeln in der idealen Flüssigkeit, mit Anwendung auf die Tragflügeltheorie und andere Aufgaben. Berlin, J. Springer, 1924, pp. 18–33
2 W. Birnbaum, Das ebene Problem des schlagenden Flügels. Zeitschrift für angewandte Mathematik und Mechanik (ZAMM) 4 (1924), pp. 277–292
3 H. Wagner, Über die Entstehung des dynamischen Auftriebs von Tragflügeln. Zeitschrift für angewandte Mathematik und Mechanik (ZAMM) 5 (1925), pp. 17–35
4 H.G. Küssner, Schwingungen von Flugzeugtragflügeln. Luftfahrt-Forschung 4 (1929), pp. 41–62
5 A.G. von Baumhauer, C. Koning. On the Stability of Flutter of an Airplane Wing. Ryks-Studiedienst voor de Luchtvaart, Amsterdam, Rapport A48 (1923)
6 H. Blenk, F. Liebers. Flügelschwingungen von freitragenden Eindeckern, Luftfahrt-Forschung 1 (1928), pp. 1–17
7 H. Blenk, F. Liebers, Gekoppelte Biegungs-, Torsions- und Querruderschwingungen von freitragenden und halbfreitragenden Flügeln. Luftfahrt-Forschung 4 (1929), pp. 257–281
8 H.G. Küssner, Beanspruchung von Flugzeugflügeln durch Böen. Zeitschrift für Flugtechnik und Motorluftschiffahrt (ZFM) 22 (1931), pp. 605–615
9 E. Petersohn, Abwindmessungen hinter Tragflügeln mit abgerissener Strömung. Zeitschrift für Flugtechnik und Motorluftschiffahrt (ZFM) 23 (1932), pp. 289–300
10 M. Kramer, Die Zunahme des Maximalauftriebes von Tragflügeln bei plötzlicher Anstellwinkelvergrößerung (Böeneffekt), Zeitschrift für Flugtechnik und Motorluftschiffahrt (ZFM) 23 (1932), pp. 185–189
11 H. Blenk, D. Fuchs, F. Liebers, Über Messung von Wirbelfrequenzen, Luftfahrtforschung 10 (1932), pp. 38–41
12 H.G. Küssner, Augenblicklicher Entwicklungsstand der Frage des Flügelflatterns, Luftfahrt-Forschung 12 (1935), pp. 193–209

13 H. Hertel, Die dynamische Festigkeit der Flugzeuge, DVL-Jahresbericht 1931
14 H.G. Küssner, Zusammenfassender Bericht über den instationären Auftrieb von Flügeln, Luftfahrt-Forschung 13 (1936), pp. 410–424
15 K. Schwarzmann, Flatterfälle und Flatterkonstruktionsvorschriften, Jahrbuch 1942 der deutschen Luftfahrtforschung, pp. 1541–1548
16 L. Schwarz, Berechnung der Druckverteilung des schwingenden Flügels mit Ausgleichsruder, LGL-Tagungsbericht 096/012 (1938), pp. 50–54
17 H. Söhngen, Auftrieb und Moment der geknickten Platte mit Spalt, Luftfahrtforschung 17 (1940), pp. 17–22
18 F. Dietze, Zum Luftkraftgesetz der harmonisch schwingenden, knickbaren Platte (Flügel mit Ruder und Hilfsruder), Luftfahrtforschung 18 (1941), pp. 135–141
19 H.G. Küssner, L. Schwarz, Der schwingende Flügel mit aerodynamisch ausgeglichenem Ruder, Luftfahrtforschung 17 (1940), pp. 337–354
20 H.G. Küssner, Allgemeine Tragflächentheorie, Luftfahrtforschung 17 (1940), pp. 370–378.
21 L. Prandtl, Theorie des Flugzeugtragflügels im zusammendrückbaren Medium, Luftfahrt-Forschung 13 (1936), pp. 313–319
22 S. von Borbely, Über die Luftkräfte, die auf einen harmonisch schwingenden zweidimensionalen Flügel bei Überschallgeschwindigkeit wirken, Zeitschrift für Angewandte Mathematik und Mechanik (ZAMM) 22 (1942), pp. 190–205
23 A. Teichmann, Das Flattern von Trag- und Leitwerken, Schriften der deutschen Akademie der Luftfahrtforschung (1942), Issue 49, pp. 1–49
24 R. Kassner, H. Fingado, Flügelschwingungen. Jahrbuch 1935 der Vereinigung für Luftfahrtforschung (VFL), Berlin 1935, pp. 54–66
25 P. Jordan, Das Flügelflattern als komplexes Eigenwertproblem, 1. Vergleichsrechnungen an einem 2-kieligen Leitwerk. AVA-Bericht 42/J/8 (1942)
26 H. Wielandt, Beiträge zur mathematischen Behandlung komplexer Eigenwertprobleme, AVA-Berichte 43/J/9 (1943), 43/J/21 (1943), 44/J/21 (1944) und 44/J/37 (1944)
27 G. de Vries, Zur Technik des Standschwingungsversuchs, Jahrbuch 1942 der deutschen Luftfahrtforschung, pp. 1590–1595
28 B. von Schlippe, Ermittlung der kritischen Geschwindigkeit durch Flugschwingversuche. VFL-Tagungsbericht 035/001 (1935), pp. 15–31
29 P. Jordan, Über das Flattern von Beplankungen, AVA-Bericht 45/J/3 (1945)
30 H.G. Küssner, Flügelflattern, FIAT Review of German Science (1939–1946), Applied Mathematics Part 11, Mechanics of Rigid and Elastic Bodies (1946), SA 27–180

First Helicopters and Rotor Systems*

Kurt Knobling

Until 1936 when the first helicopter Fw 61 of Focke Wulf came to fly, many inventers and designers tried to develop a practicable helicopter. But the designers failed due to the lack of scientific and technological knowledge so that their devices showed a very unstable flight behaviour and could hardly leave the vicinity of the airport. Henrich Focke and his colleagues constructed with the Fw 61 the first helicopter worldwide that was really operational. Further designers were in Germany Anton Flettner and Friedrich von Doblhoff, who developed pioneering propulsion and design principles.

The developments were related to these three pioneers in person. They were the carriers of the ideas. Scientific research is connected with their names. Together with a small staff they were able to develop helicopters, whose technology was taken up by foreign countries after the war.

Visions and Technical Requirements

Controllable rotor systems, which allow vertical takeoff as well as horizontal flight, became possible only in the thirties when the flight mechanical relations were understood and the needed materials were available. Rotor systems were therefore also always the pacemakers for high performance structures including aeroelastic design methods for wings.

The idea and fascination of performing vertical climbs and hovering flights with rotating wings is very old. A flight model is known from China dat-

* Also see Volume 3 of this series "German Aeronautics – Helicopters and Autogyros" ("Die deutsche Luftfahrt – Hubschrauber und Tragschrauber"), (3. Ed. 1999), Bernhard & Graefe Verlag, Bonn)

ing back to around 400 BC. A rod with a propeller made of bird's feathers pinned together at one end is set into rotation by the palms and thus able to lift up vertically. *Leonardo da Vinci* sketched a helicopter (late 15th century) thus providing stimulation for helicopter pioneers. During the 18th and 19th centuries, models were presented in various countries and there mainly at scientific academies: *Michail V. Lomonosow* in Russia (1754), *Launoy* and *Bienvenue* in France (1784), *George Cayley* in England (1790); but these had only little effect on the helicopter development. During the second half of the 19th century, many inventors concerned themselves with the helicopter. There was progress but no viable air vehicle evolved. In the beginning the problem was the lack of a light and efficient engine. The second problem was the development of designs for the rotor, the rotor head, and the rotor blades, which would meet the strength and oscillation requirements; furthermore of concern was the aerodynamic quality of the blades. Finally, there was a lack of knowledge concerning the control and dynamics of the helicopter rotor including torque compensation.

Propulsion by steam engine was attempted, but the ratio of mass to performance was too large. At about 1900, combustion engines were available that permitted powered flight and also offered possibilities for helicopters. The first manned hovering flights with helicopters were accomplished at the beginning of the 20th century: *Louis* and *Jaques Breguet, Charles Richet, Paul Cornu* in France in 1907. World War One accelerated the development of efficient light engines so that during and after the war some designs were made to hover and to fly horizontally over short distances: *Stephan von Petroczy, Theodore von Kármán, Wilhelm Zurovec* in Austria (1916), *George de Bothezat* in the USA (1922), *Etienne Oehmichen* in France (1924), *Raul Pateras Pescara* in Spain (1924), *Corradino d'Ascanio* in Italy (1930), *Boris N. Jurjey* at ZAGI, Soviet Union (1931).

For all these designs, stability and controllability were limited, and the helicopters could hardly leave the airfield. For the development of an airworthy device, it was necessary to establish the scientific fundamentals concerning rotor aerodynamics, rotor dynamics and oscillation phenomena; the established theories had to be validated by systematic experiments. Here, pioneering work was conducted in Germany, which provided a worldwide top position, and the capability to build operational helicopters before and during World War Two. Three German pioneers shall be given prominence here and their research and development work shall be described in detail: *Henrich Focke, Anton Flettner* and *Friedrich von Doblhoff.*

Henrich Focke's Pioneering Work

Research and Development Objectives

Henrich Focke [1, 2] first considered the development and construction of fixed-wing aircraft. A number of successful aircraft was built at the "Focke-Wulf Aircraft Engineering, Ltd." ("Focke-Wulf-Flugzeugbau GmbH"), founded in Bremen in 1924 by *Focke* and *Georg Wulf.* Due to his involvement with the autogyro of the Spaniard *Juan de la Cierva,* also called rotorplane, built under license at Focke-Wulf, *Henrich Focke* got acquainted with rotorcraft. An autogiro, whose rotor is driven by the air stream due to a forward movement, cannot hover on the spot and cannot take-off vertically. *Henrich Focke,* however, considered these capabilities as especially essential. Since about 1930, he occupied himself with the idea of the helicopters. The impulse given by *Juan de la Cierva* was reinforced by a statement of *Friedrich Seewald,* at the time head of the Department of Propellers at the "German Aeronautical Test Establishment, DVL", who predicted from the safe functioning of a propeller as a lifting device a decisive influence on the development of a real helicopter. There were hardly any test reports from the earlier experiments with helicopters, let alone reports of scientific quality. He and his colleagues, who were thus far concerned with fixed-wing aircraft, were lacking fundamental knowledge concerning helicopters. The theoretical and technical prerequisites had first to be accurately and reliably established. *Henrich Focke* [1, 3] put forward six theses as research and development goals:

1. A basic prerequisite of a useful design is the capability to land safely after a failure of the engine or the power transmission, i.e., the capability of auto rotation.
2. Assurance of controllability and stability without higher demands on the pilot.
3. The general operational safety, which could not be realized with past helicopters.

4. Simplicity of control by the pilot. Steering forces of the helicopter should have a similar characteristic as the ones of fixed-wing aircraft.
5. Acceptable performance, approaching the one of fixed-wing aircraft (kite aircraft) especially in horizontal flight, however, with acceptable system inherent deficits concerning flight speed.
6. Maintenance should not be more difficult than for normal aircraft.

These demands show how diligently *Henrich Focke* [4, 5] and his staff approached the development of the helicopter. Broad and extensive scientific investigations had to be conducted. Wind tunnel tests in the in-house facility and at *Ludwig Prandtl's* "Aerodynamic Test Establishment Göttingen, AVA" as well as free-flight model tests were conducted prior to the realization of the actual "machine". The free-flight model was not remotely controlled and had frequently to be repaired after hard landings. Parts of the helicopter were tested in ground-test facilities, where data concerning performance requirements (lift per unit performance) and blade forces were, among others obtained. Together with the scientific studies and computations, practical trials had to be performed. The problem of torque compensation – the torque acting on the rotor is supported by the fuselage via the propulsion system and would force the fuselage to rotate in the opposite direction – vanishes if two counter-rotating rotors are installed. *Focke* chose a design where the rotors are arranged on lateral outriggers, which constituted a progress in the development of helicopters. After the tests with the free-flight model were successful – in November 1934 a flight altitude of 18 m (59 ft) was reached – a prototype was built after intensive testing. The helicopter obtained the type designation Fw 61. *Ewald Rohlfs* undertook on June 26, 1936 the first free flight. On June 25 and 26, 1937, *Ewald Rohlfs* was able to bring all existing helicopter records to Germany (flight duration: 1 h 20 min 49 s, altitude: 2,439 m (8,000 ft), velocity: 112.5 km/h (61 kn)). On May 10, 1937, *Ewald Rohlfs* conducted the first auto-rotation landing from a height of 400 m (1,312 ft). None of the experimental helicopters flown up to that time had ever proven to be able to auto-rotate. The extraordinary and impressive maneuverability of the helicopter Fw 61 was demonstrated publicly at flight shows in February and March 1938 in the "German Exhibition Hall" ("Deutschlandhalle") in Berlin. This was the result of consequent research work. With these flights the *Focke* helicopter proved to be the first flyable and controllable helicopter worldwide.

The Focke Rotor System

Henrich Focke chose a two-rotor lifting system with the rotors arranged laterally next to each other [6]. Counter-rotating rotors provided torque compensation. The development of the transmission was scientifically and experimentally supported, and built by Siemens, respectively later BMW in Spandau. At the rotor heads, developed by *Henrich Focke*, the rotor blades were cyclically controlled, which was accomplished by a swash plate, which could be tilted in any direction by the control stick. The angle of attack of the blades could be changed in unison by lifting or lowering the swash-plate so that the rotor thrust could be increased or decreased. The prerequisites for a controllable helicopter were thus realized in case of the Fw 61 for the first time.

The blades were connected to the rotor head by flapping and drag hinges. According to the design thesis No. 1 of *Henrich Focke* an operational helicopter must be able to switch over to auto rotation in the case of an engine failure or shaft breakage. It is, therefore, necessary to reduce the blade angle of attack from approx. 15°–8° in a very short time to prevent the rotor from stopping. The automatic switchover was realized by a centrifugal controller, which severed the connection between shaft and rotor if the number of revolutions fell short of a certain value by disengaging a clutch, and which simultaneously adjusted the angle of attack of the helicopter blades to the autogyro position.

Also at a certain difference in the number of revolutions of the (left/right) rotor heads, the switchover automatic engages and sets the rotor blades to the autogyro position. The switchover is carried out hydraulically. The switchover automatic of the helicopter Fw 61 was a first; it was later also employed on the cargo helicopter Fa 223.

A specialty on the rotors of the Fw 61, and also on later *Focke* helicopters, were balance cables, called sway-cables ("Eierseile"), which establish a loose connection between the three blades of the rotor. In the center of the individual connecting cables, centrifugal weights were mounted. The weights, subject to centrifugal forces, ensure that the main rotor blades come back to their symmetric 120°-position after a distur-

Henrich Focke (1890–1979).

Henrich Focke, born in Bremen on October 8, 1890.

1911 he became acquainted with *Georg Wulf*, born 1895, during his studies in Hanover.

1924 *Focke* and *Wulf* reported to the (army) pioneers. *Focke* was deferred because of a heart ailment.

1914 *Focke* was, nevertheless, drafted into an infantry regiment; after serving in the East, he was transferred to the flight corps.

1920 *Focke* graduated (as Diplom-Engineer, MS) with distinction. He obtained his first position with the "Francke Company" of Bremen as a designer of water-gas systems.

1924 the "Focke-Wulf Aircraft Engineering, Ltd." ("Focke-Wulf-Flugzeugbau GmbH") takes up its operations. In 1930, *Focke* is offered a chair at the Institute of Technology (TH) Danzig but he declined.

1930 merger of Focke-Wulf with the Albatross-Aircraft Company.

1931 the Senate of the city of Bremen bestows him with the title Professor.

1930–1937 Focke-Wulf builds the autogyro of *Juan de la Cierva* in license. *Focke* considers the conversion of the autogyro into the helicopter. On June 26, 1936, the first flight of the world's first flyable and controllable helicopter, the Focke-Wulf Fw 61, took place.

1937 foundation of the Focke-Achgelis & Company, Ltd. on April 27.

1945 on September 1, he signs a contract with the SNCASE in France.

1950 on March 10, he starts work at the "North German Automobile Company" ("Norddeutsche Fahrzeugwerke") in Wilhelmshaven.

1951 since April 1, *Focke* again develops helicopters, this time in Brazil.

1954 he is offered a full professorship at the TH Stuttgart, but he declines.

1956 *Focke* starts with the development of a three-seat helicopter at the "Automobile Company Borgward" in Bremen.

1961 Borgward collapses; *Focke* becomes consulting engineer.

1979 he dies in Bremen on February 25.

bance within the plane of rotation (in-plane motion), which resulted in a decay of the unbalances. The "sway-cables" were later also utilized on the French helicopter Alouette II, which was built in large numbers. *Henrich Focke* founded on April 27, 1937 together with *Gerd Achgelis* the company Focke-Achgelis & Co., Ltd., which erected in Hoyenkamp near Delmenhorst a flight field with a hanger and a wind tunnel. In the summer of 1938, *Focke-Achgelis* received a contract to build for Lufthansa a helicopter with a payload of 700 kg (1,570 lbs), with a separate compartment for the pilot and the flight attendant and a cabin for four passengers. In comparison to the single-seat helicopter Fw 61 with a gross mass of 950 kg (2,136 lbs), meant this increase up to a flight mass of 4,300 kg (9,700 lbs) in case of the helicopter called Fa 223 a giant leap. With the usual scientific preparatory work at Focke, the Fw 61 was modified to an in-flight laboratory, where essential and basic parameters were measured. The development of the instrumentation of the test vehicle and the acquisition and processing/analysis of the scientific test results is essentially the merit of *Carl Bode*. The employees charged by *Henrich Focke* with the theoretical development extended basic work on rotor aerodynamics, on the theory of statics and dynamics of the rotor, and on airframe and blade oscillations. On March 8, 1940, the first free hovering flight of the Fa 223 – the civil version of the Fa 266 – took place; with this helicopter, basic and trend-setting operational tasks, such as the transportation of loads in high-mountain regions, rescue operations, submarine surveillance, flight reconnaissance and training, were later to be

The Focke-Wulf-helicopter Fw 61 lifted off for the first time on June 26, 1936 at the company airport in Bremen; already on its fourth flight, it remained airborne for 16 min.

tested and conducted. The system of the rotors, mounted side by side, was the same as for the Fw 61. Also the propulsion system was adopted and adapted to the larger helicopter.

The transition to a helicopter with a four times larger flight mass created a number of further problems, whose mastering required basic investigations. For instance, the blades moved upwards in an uncontrolled fashion when starting and accelerating. It turned out that the effective angle of attack increased during rising and lagging of the blades, so that the moment of the lift forces became larger than the moment of the blade-stretching centrifugal forces. So the blade angle of attack had to be reduced while the blade was rising; this was accomplished by a new design of the, at first rectangular, flapping-hinge. With a 16-mm camera, mounted on the rotor head, it was possible to film, with respect to one blade, the cone angle, the blade angle, the lag angle, and deflection and torsion at all flight conditions; during the process, one became aware of the effect of the Coriolis force on the flapping blade. The new procedure of filming the rotor in operation decisively supported the research work.

Henrich Focke also tried new methods during the development of non-powered towed autogyros. In 1942, the freight sailplane DFS 230 of the "German Research Establishment of Gliding" ("Deutsche Forschungsanstalt für Segelflug") in Darmstadt without wings was equipped with rotor head and rotor of the Fa 223. The single-rotor autogyro, designated Fa 225, with a gross mass of 2,000 kg (4,400 lbs), was able to carry out several successful towed flights. The small single-rotor towed autogyro Fa 330 "Bachstelze", whose development started in 1941, was envisaged to operate off submarines. At a towing height of 200 m (656 ft), the range of visibility was about 50 km (31 mi). Trials with the Fa 330 "Bachstelze" were performed in the spring of 1942 in the large wind tunnel in Chalais-Meudon near Paris. Further trials took place on ships. Of the approximately 120 autogyros

The cargo-helicopter Focke-Achgelis Fa 223 – first flight on March 8, 1940 could transport external freights up to 1,000 kg.

built at the factory in Hoyenkamp under the supervision of the chief designer *Paul Klages*, only relatively few entered operations.

Focke was able to gather an excellent team of engineers, engineering pilots and scientists, which build up the success of the Focke helicopters, and which greatly contributed, after the end of the war, to the further development of helicopters abroad. Basic knowledge in a new field of flight technology was produced in research and development.

Co-workers and employees of *Henrich Focke* in development, manufacturing and testing of helicopters were:

- *Carl Bode* (Test pilot)
- *Harry Duda* (Mechanical design, manufacturing)
- *Hans Gerd Eyting* (Mechanical design, manufacturing)
- *Reinhold Gensel* (Design)
- *Hans H. Gerstenhauer* (Test flights)
- *Helme* (Mechanical design, control)
- *Walter Just* (Scientist associate)
- *K. Jaeckel* (Scientist associate)
- *Paul Klages* (Mechanical design, overall development)
- *Heinrich Papenhausen* (Statics, material tests, test department)
- *Ewald Rohlfs* (Test pilot)
- *Friedrich Schaper* (Statics)
- *Erich Schweym* (Aerodynamics)
- *Herbert Spranger* (Development, aerodynamics)
- *Enno Springmann* (Mechanical design).

Schweym, Just and *Jaeckel* gained extraordinary merits in developing the aerodynamic theory and the statics and dynamics of rotor blades, as well as in investigating oscillations of airframe and rotor blades. In the fifties, *Walter Just* took over the scientific lead of the "German Study Association of Helicopters" ("Deutsche Studiengemeinschaft Hubschrauber e.V.") in Stuttgart; he held a lectureship at the local Institute (TH) of Technology. His work in Stuttgart constituted the nucleus of the again developing helicopter industry in Germany.

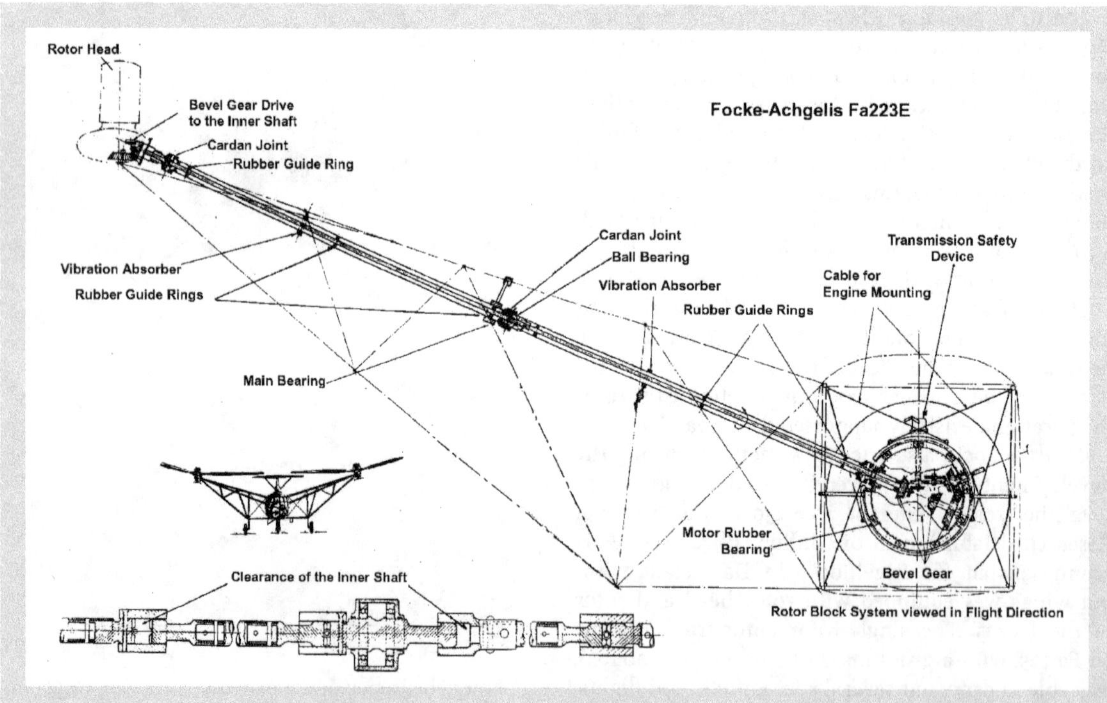

Arrangement of the left outrigger of the FA 223-helicopter with transmissions, shafts and bearings.

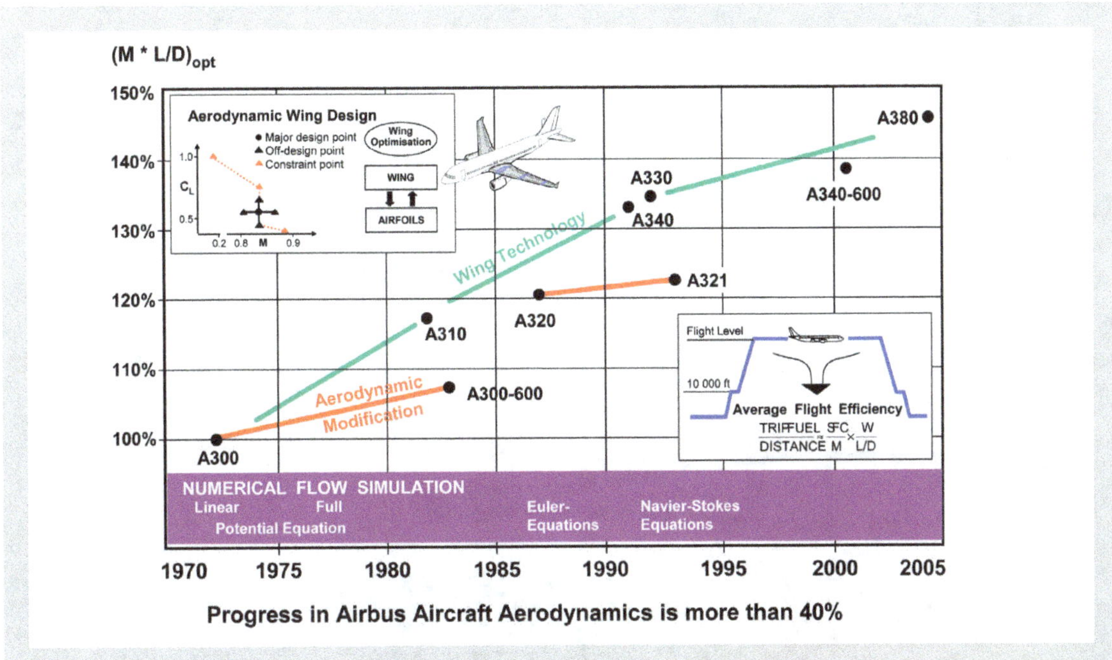

Progress through aerodynamic developments: Cruise-flight aerodynamic quality.

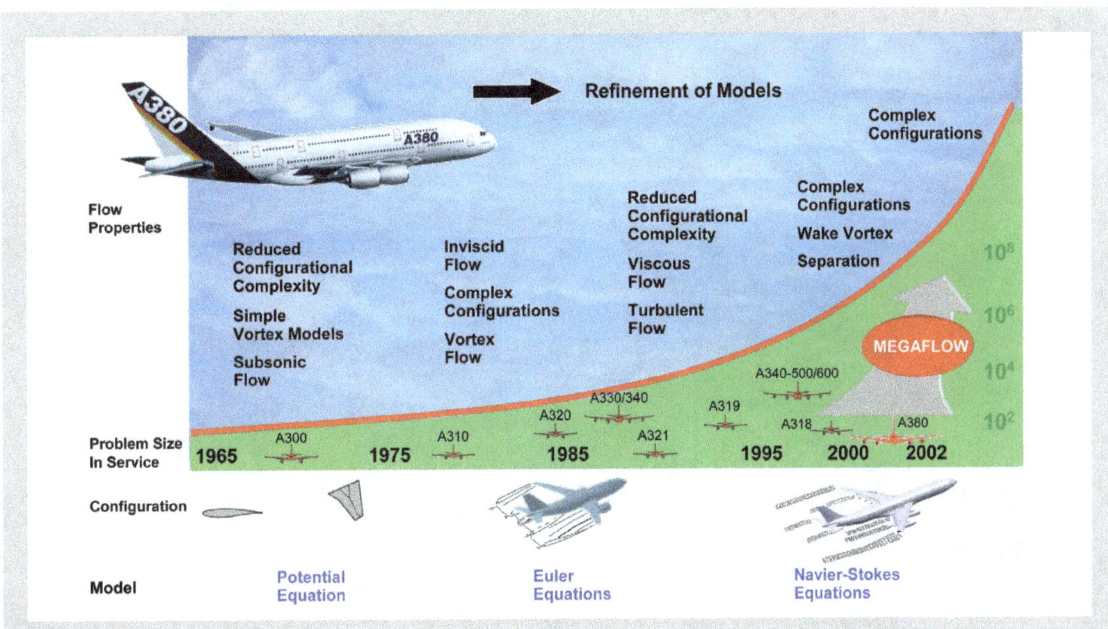

Numerical aerodynamics related to Airbus: development and requirements.

Numerical aerodynamics allows today the description of the full configurational and fluid mechanical complexity. The example depicts the pressure distribution on a "Eurofighter" model.

Computation of transonic (supercritical) airfoil flow at M = 0.73 (top: discretization of the computational space; box: details of the flow at the trailing edge; bottom: Iso-Mach-lines, the shock Wave is located at about 55 per cent of the chord).

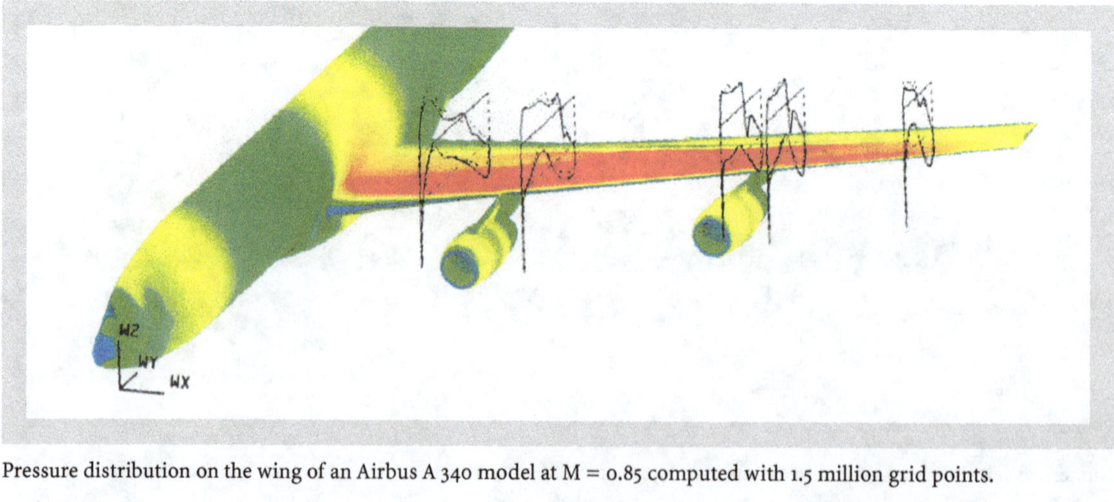

Pressure distribution on the wing of an Airbus A 340 model at M = 0.85 computed with 1.5 million grid points.

DLR-ALVAST model with engine simulators in the test section of the German-Dutch Wind Tunnel (DNW).
Top: CRUF simulators on the wing in landing configuration.
Bottom: Turbofan simulators on the wing in cruise configuration.

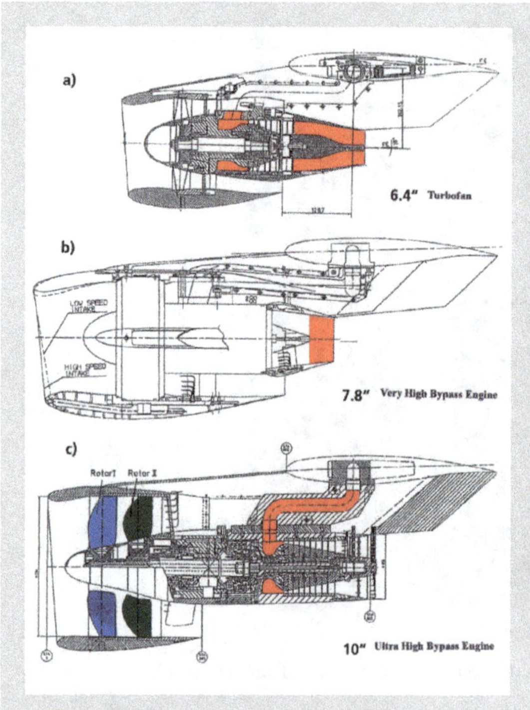

Engine simulators: a) 6.4″ Turbofan, n = 50,000 rpm, b) 7.8″ Very high bypass turbofan, n = 38,000 rpm, c) UHB-CRUF simulator, n = 16,350 rpm. The numbers in inches represent the fan diameters.

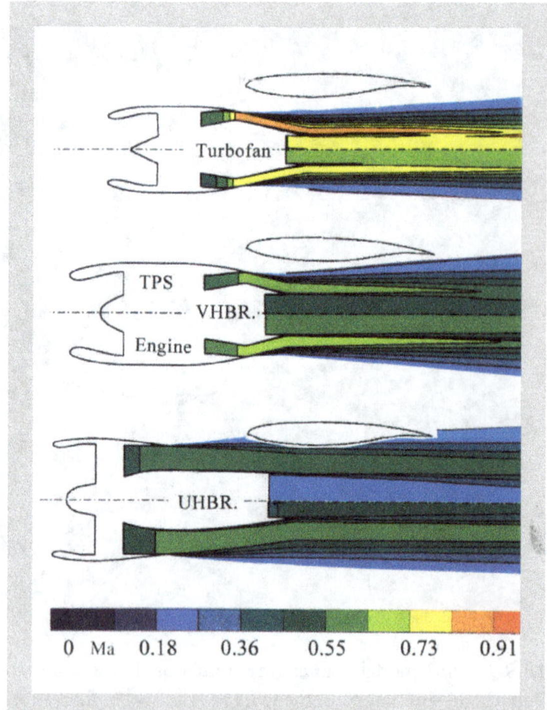

Viscous computation of the development of the engine jet in the area of the wing for turbofan, VHB and UHB engines. Computation without wing is in the upper half for the simulator (TPS), in the lower half for the actual engine.

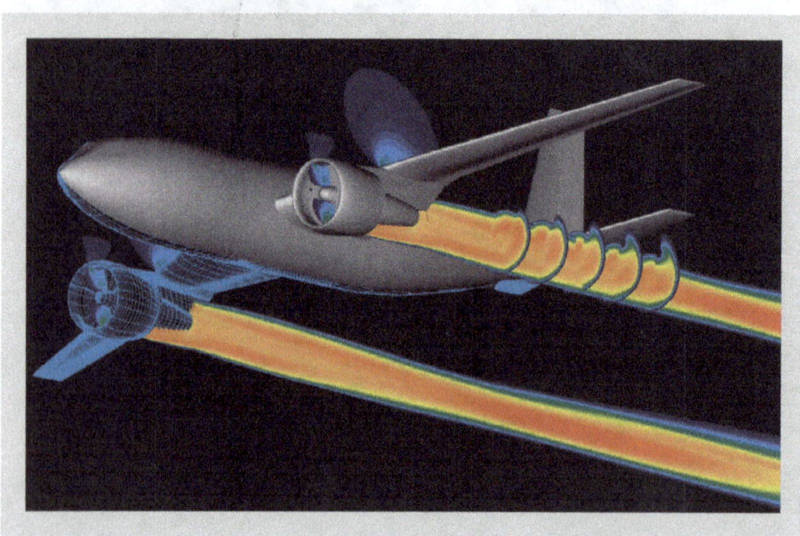

Total engine/wing simulation. Computed with the DLR code MEGAFLOW.

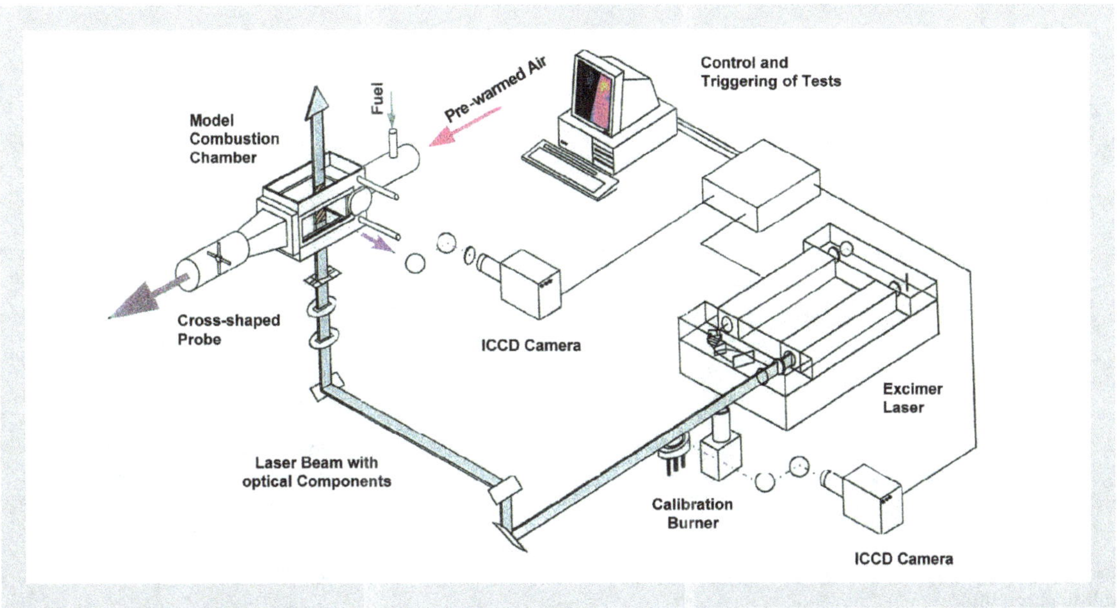

Test setup for two-dimensional measurements of the laser-induced fluorescence (LIF) of intermediate products (OH, NO) of a combustion process within a model combustion chamber.

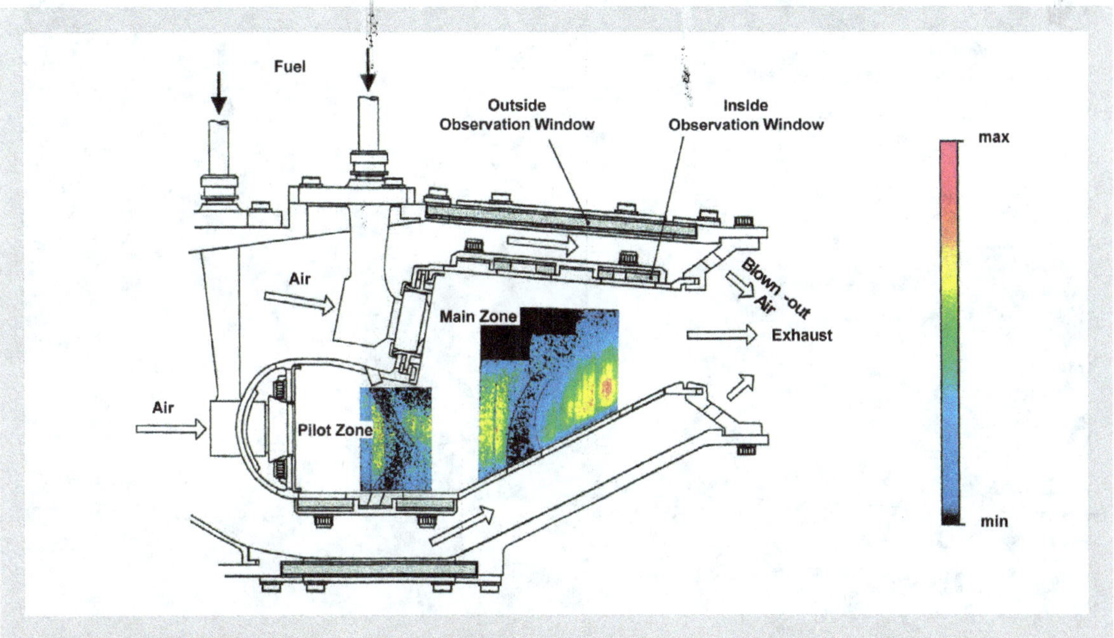

Determination of the effect of dilution-air jets on the combustion process by laser-spectroscopic measurements (LIPF) of the OH-radicals distribution within an optically accessible step-like experimental combustion chamber.

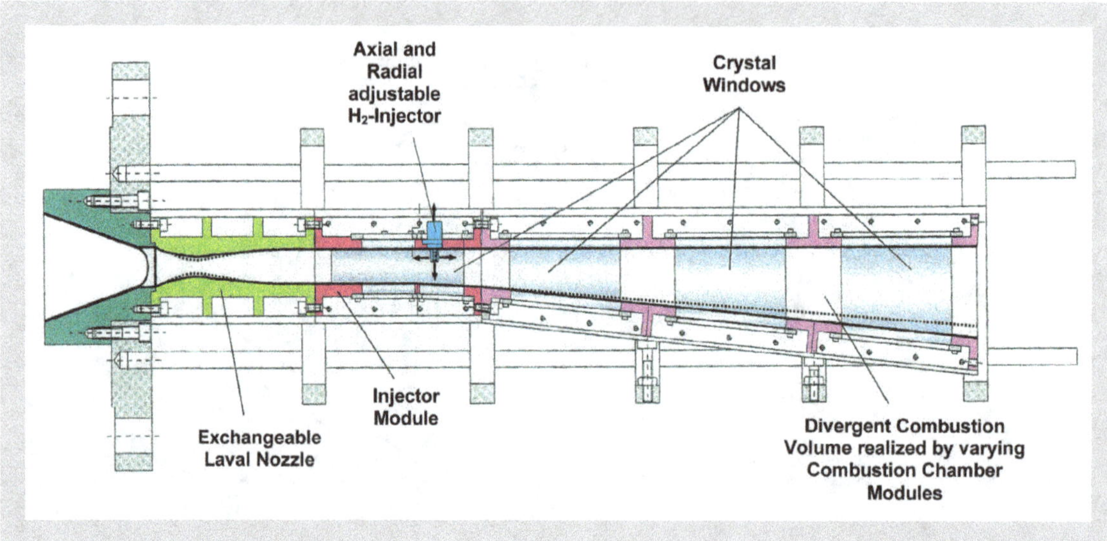

Basic concept of a supersonic-combustion model chamber for fundamental experimental investigations at partial- and full-load conditions.

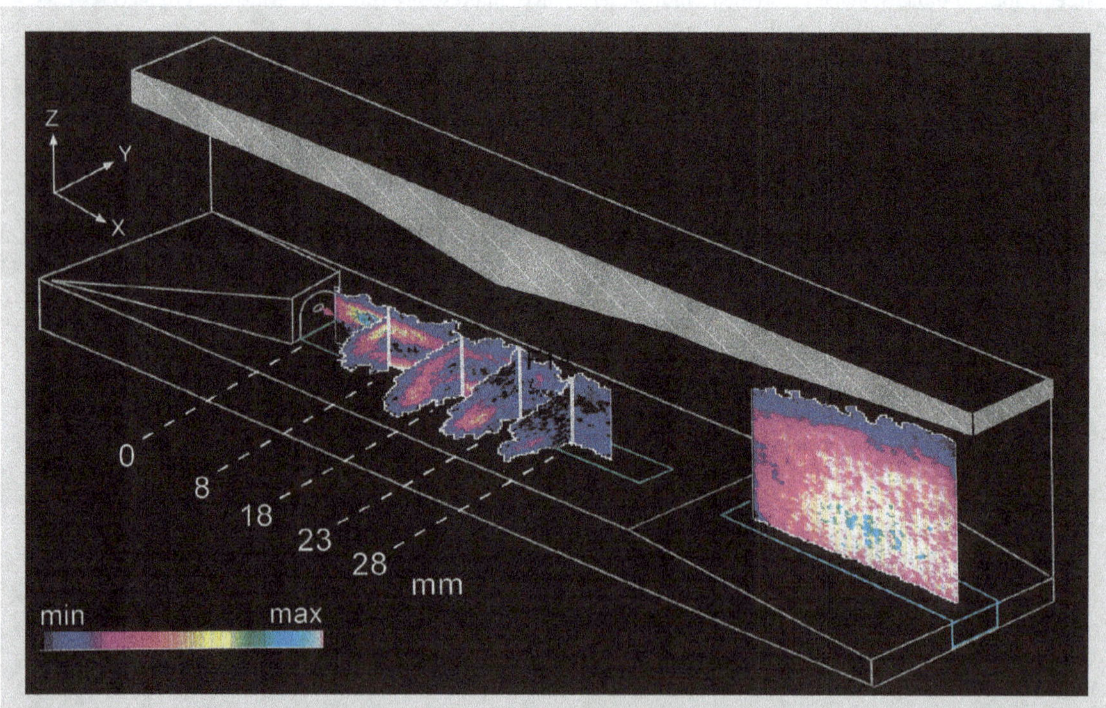

Example of the spatial flame structure in a supersonic-combustion model chamber with hydrogen ramp injector, determined by the laser-spectroscopic measurement of the hydroxide-radicals distribution.

The F-104 G of the German Air Force modified by the addition of a canard (CCV-F-104 G).

The first flight of the "Experimental-Aircraft-Program, EAP" demonstrator took place on August 8, 1986 in Warton.

The prototype JP 001 of the "Eurofighter" 2000, whose final assemply was carried out by Dasa in Manching, was also flight-tested by the German Air Force.

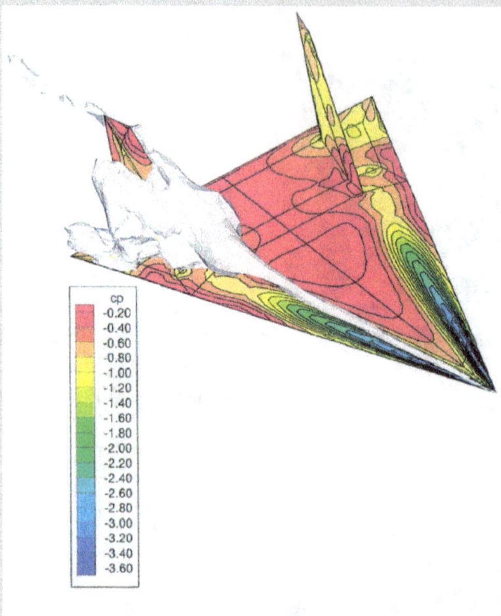

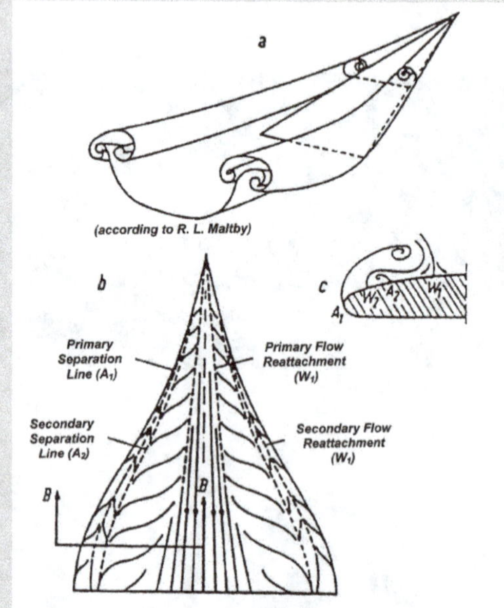

Flow past a delta wing in the presence of separation at high angles of attack and low speeds, left: Numerical representation, right: Principal surface-flow sketch (after Küchemann and Maltby)

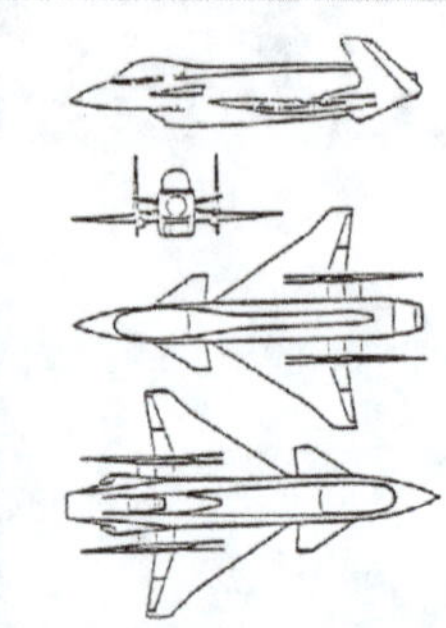

Left: Free-flight model of SNAKE configuration in the 60 ft x 30 ft Wind tunnel of the NASA Langley Research Center (Photograph of NASA).
Right: Three-view drawing of the SNAKE

German-American flight demonstrator X-31 with paddle vector nozzle during test flights in the USA.

Grob "Strato 2C" wing as CFRP structure with a span of 56.5 m (185 ft).

Nature as model: Highest possible strength at the lowest possible weight can be found in structures made of fiber-composite materials, especially in the case of carbon-fiber reinforced plastics (CFRP).

Grob "Strato 2C" high-altitude research aircraft in carbon fiber reinforced plastics (CFRP) structure.

Manufacture of a CFRP rotor blade for the Eurocopter Company helicopter EC 135 in a heated metal mould.

Top left: The CFRP horizontal stabilizers of all Airbus models are being manufactured since 1985 at the EADS-Airbus plant in Stade. The photograph shows the outer skin during the integration of the (stiffening) ribs.

Left: Assembly rig for the center box of the Airbus CFRP vertical stabilizer.

Fracture test (ultimate-load test) of a CFRP wing after 90,000 flights (ultimate load: 2.5 m (8.2 ft) deflection at the wing tip).

Eurocopter EC 135 helicopter with a hingeless and bearingless rotor system.

German-Indian light helicopter with an integrated dynamic rotor system. The first flight occurred in 1991 in Bangalore, India.

Integrated dynamic system of the ALH helicopter with fiber composite rotor head and rotor blades.

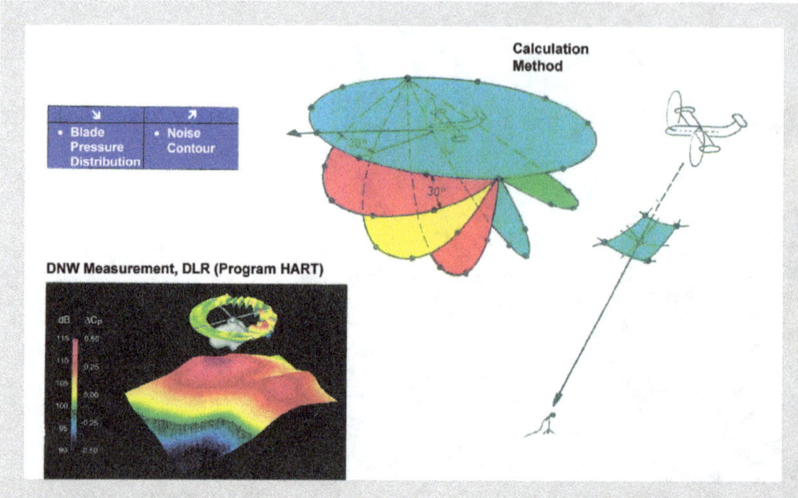

Modular wind-tunnel model (MWM) during noise measurements in the DNW tunnel.

Left and above left: DNW measurements of the DLR (Program HART) and external noise emission of a tilt-rotor aircraft.

Bottom: Individual blade control by "trailing edge flaps" activated by piezoelectric-stack actuators.

BO 105 with individual blade control.

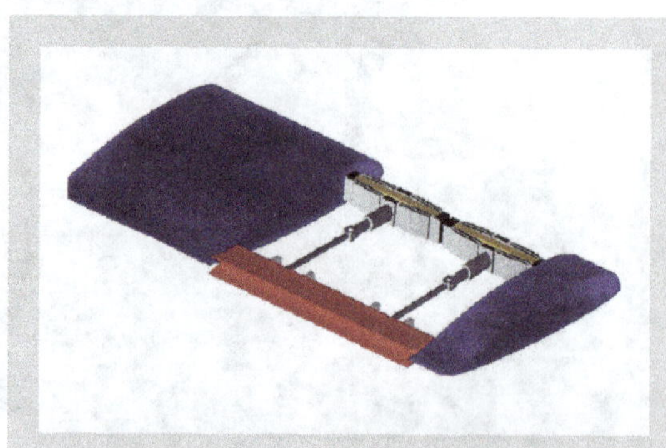

Structural tests on fuselage center parts and wings of all Airbus models, starting in 1973 with the A 300 and proceeding up to the A 340 in 1996, were carried out at the IABG in Ottobrunn. The photograph shows in a double exposure the deflection of an A 340 wing with a difference of 2.50 m (8.2 ft) at the wing tip.

Static structural tests on the "plastic" frame of the helicopter BK 117 were carried out in the spring of 1988 at the IABG in Ottobrunn.

The entire frame of the "Eurofighter" was subjected at the IABG to the prescribed fatigue tests. Here was, for the first time, a digitally controlled servo-hydraulic system with digital closed-loop control employed.

The Airbus A 330/A 340 carbon-fiber vertical tail was subjected at EADS-Airbus in Hamburg to the prescribed static and dynamic tests prior to flight-testing.

The IABG owns a "barrel test rig" for dynamic tests on aircraft landing gears, which can also be utilized for special roll tests on wheels and tires as well as for wheel-brake tests up to speeds of 400 km/h (216 kn).

At the Liebherr-Aerospace Company in Lindenberg/Allgäu exists a special Airbus wing-flap test rig, where the entire flap drive system, in an aircraft-conform arrangement, can be subjected to endurance tests.

In-flight simulator HFB 320 "Hansa Jet" of the DFVLR. The first flight tests on "fly-by-wire" control took place in 1973, as did the first flights with a side-grip control stick.

The test helicopter BO 105 "ATTHeS" of the DLR in fully automatic display-supported hovering flight. (Photograph: Braunschweiger Zeitung)

Left: The DLR test aircraft Dornier Do 27 served since 1965 to investigate digital control and sidegrip control arrangements.

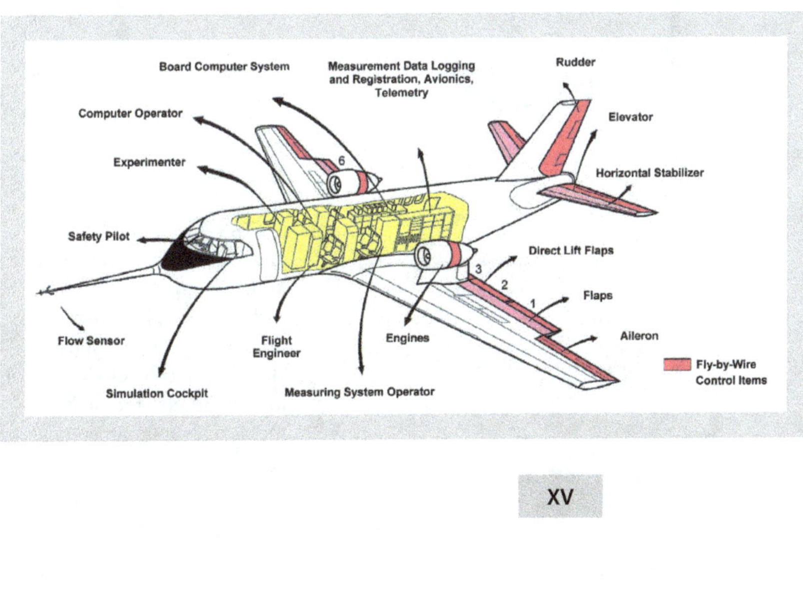

In-flight simulator and technology carrier "ATTAS" of the DLR. Modifications and equipment and control surfaces (red) addressable by "fly-by-wire".

The experimental variable stability helicopter EC 135 ACT-FHS in flight.

DFVLR obtained in 1981 a VFW 614 that was converted into a flying simulator and technology carrier ("ATTAS"). The aircraft was since 1986 available for research work and for the first time also equipped with electro-optical control ("fly-by-light").

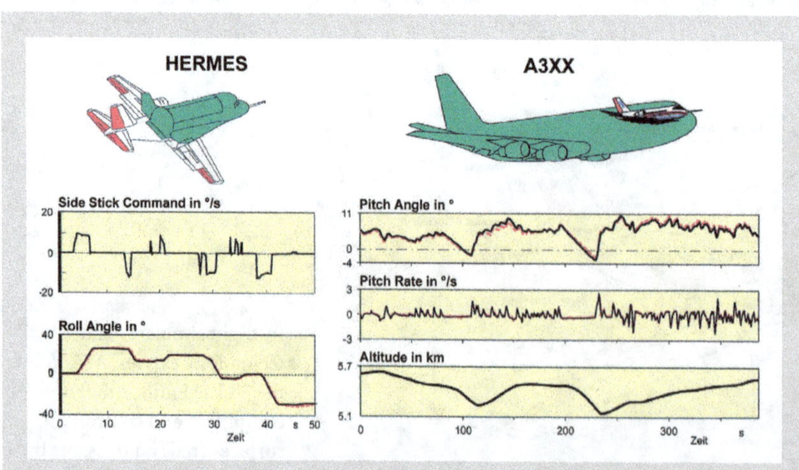

"ATTAS" in-flight simulation of different flight vehicles, such as the "Hermes" space plane and Airbus A3XX-like configurations, by model-following closed-loop control. Compared here are model and flight-measurement parameters.

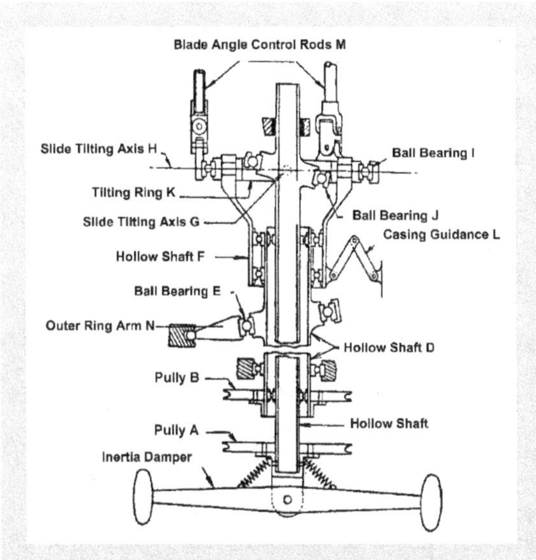

Schematic of the rotor blade pitch (angle of attack) control of the Fa 223 helicopter.

Detailed reports on the development of helicopter technology in Germany during the war were established by the investigative committee of the Allies charged with the sifting of the German scientific and technical status at the end of the war. Especially detailed is the report: "Rotating Wing Activities in Germany during the Period 1939–1945 [7]".

On May 11, 1945, *Focke* and his colleagues fell into French captivity. This ended the work of one of the most outstanding helicopter pioneers for the time being.

After the end of the war, *Focke* and his colleagues again developed helicopters, at first in France (SNCASE), then in the Netherlands (Textile merchant *Keyser* 1950), from 1952–1960 in Brazil (Centro Technico da Aeronàutico), and since 1956 in Germany (Carl F.W. Borgward, Ltd., Bremen).

Anton Flettner and his Rotor System

Anton Flettner was an ingenious inventor, who was already in high school concerned with the development of a remote radio control for riverboats and its demonstration. He became known for the invention of the Flettner rudder and a servo control for large air vehicles and later for watercraft. He aroused worldwide attention with his proposal to replace the sails of ships with upright standing perpendicular rotating cylinders, which produce, in the presence of an airflow (wind), thrust by utilizing the *Magnus* effect (discovered by *Gustav Magnus*, 1852). The *Flettner* experimental boats did, however, not achieve the success hoped for. In 1925, he founded his own company aiming at the application of the *Flettner*-rotor as wing replacement on large wind turbines. *Albert Einstein* appreciated *Anton Flettner's* sense for utilizing physical knowledge in practical applications in his publication "My View of Life" ("Mein Weltbild"), of 1934: "*Flettner saw Ludwig Prandtl's experiments (remark: experiments on rotating cylinders in an incoming air flow) and thought that one could use this installation to replace sails. Who knows, whether anybody else would have thought of that?*"

During the thirties, *Anton Flettner* turned his ingenuity towards the design of rotary wing aircraft. In 1927, he developed and built a rotary aircraft with propellers, driven by small engines, on two of the four-blade rotor of 30 m (98 ft) diameter. Because of the reaction-moment-free propulsion, the installation of a moment-compensation system could be left out. This helicopter got airborne in 1934.

In 1935, *Anton Flettner* founded his own company, "Anton Flettner Aircraft Engineering, Ltd." ("Anton Flettner-Flugzeugbau GmbH"), and built a factory in Berlin-Johannsithal. He engaged two helicopter-orientated scientists of the "Aerodynamic Test Establishment Göttingen, AVA", *Gerhardt Sissingh* [8, 9] and *Kurt Hohenemser* [10], who greatly contributed to the development in the years to come. Initially, *Anton Flettner* developed in 1934 an autogyro with the type designation Fl 184. The design was closely related to the autogiro C 30 of *Juan de la Cierva*, however, had a cyclic blade-angle control instead of the rotor-head pitch control of the Cierva autogyro. Between 1936 and 1938, work concentrated at Flettner's on the single-rotor helicopter Fl 185. The engine power was distributed via a gear box to the three-blade rotor and to propellers mounted on cantilevers on either side of the fuselage with axes parallel to the fuselage axis. The lateral propellers served as reaction-moment compensators, but also produced thrust in the direction of flight. This principle was later also copied by other designers. The essential characteristic of the helicopter was a yaw-damping gyro, which assumed the rotor-moment compensation by the pitch setting

Anton Flettner (1885–1961).

Anton Flettner, born in Eddersheim at the Main on November 1, 1885.

1905 lectureship of mathematics and physics at technical schools in Frankfurt; development of remote controls for airships in cooperation with *Count Zeppelin*.

1914-1918 employment at the German Ministry for War; development of remote controls for air, water, and land vehicles.

1915 presentation of a remotely controlled armored vehicle; invention of a servo control for large-scale aircrafts. After the war, he applies servo control surfaces in shipbuilding ("Flettner-Rudder").

1921–1926 "Flettner-Ship-Rudder-Society" ("Flettner- Schiffsruder-Gesellschaft", Berlin and Rotterdam.

1918–1931 director of the Institute of Aero- and Hydrodynamics, Amsterdam.

1922–1926 development of the "Flettner-Rotor Ships".

1926–1945 "Anton-Flettner-Aircraft Engineering" in Berlin-Johannisthal; beginning in

1930 development and production of rotary-wing aircraft.

1931–1932 helicopter giant with blade propeller propulsion.

1934 autogyro Fl 184 with flight-tests in 1937.

1936 helicopter Fl 185 with first flight in 1938.

1937 helicopter Fl 265 with first flight in May 1939.

1938 helicopter Fl 282 with first flight on October 3, 1941.

1943 transfer of his factory to Schweidnitz in Silesia.

1945 transfer of his factory back to Berlin in January with a renewed transfer in March to Bad Tölz.

1947 emigration to the USA. *Flettner* becomes an American citizen.

1947–1948 consultant to the US Navy.

1958 foundation of the Flettner Aircraft Corporation in New York.

1961 he dies in New York on December 29.

of the lateral propellers. The yaw-damping gyro has since found a widely distributed application. *Anton Flettner's* greatest contribution to helicopter technology was, however, the development of the helicopter Fl 282 "Kolibri" ("Hummingbird"), whose design commenced during 1938.

The lifting system consisted of two counter-rotating rotors with two or three blades. The axes of the two rotors protruded, in close vicinity, from the upper gearbox and were inclined outward under 12° with respect to the vertical. The Flettner-rotor principle is a compromise between the Breguet helicopter, with two rotors in a coaxial arrangement, and the Focke-concept with two rotors arranged laterally next to each other. Breguet encountered problems due to the blade-tip distance, Focke due to controllability, maneuverability and performance. The Flettner arrangement had the advantage of symmetry, just as the other systems, but without their disadvantages. The blades are mounted via flapping and drag hinges on the respective rotor head. The blade angle setting is accomplished via a rotating hinge and a control rod, the latter being connected via a further hinge to the swash-plate. The blades intermesh during rotation. A gear takes care that the rotors turn in phase and that there will be no touching of the blades. The in-plane motion, enabled by the lag (or drag) hinges, is dampened by adjustable frictional lag dampers. The motion about the drag and flapping hinges is limited by stoppers. The control about all three axes is achieved by blade adjustments by means of the control stick, blade adjustment lever and rudder pedal, whose motions are transferred via control rods to the swash-plates mounted underneath the rotor heads. The plates are connected via bridges in a way that they perform co-directional motions. Vertical control is achieved by raising or lowering the swash-plates. Lateral control by means of the foot pedal results, via an eccentric, in an opposing up and down movement of the swash-plates and hence in a change of the average blade setting of one rotor with respect to the other. The resulting difference in torque rotates the "machine" about its vertical axis ("torque control"). The rudder pedal is also connected to the rudder at the end of the vertical stabilizer.

The Flettner helicopter Fl 282 "Kolibri" ("Hummingbird") – characterized by its intermeshing twin rotors, took off on October 30, 1941 for its first flight. Also a two-seat version with an observation seat in the fuselage existed.

Prior to flight tests, extensive wind tunnel measurements were conducted in August 1940 in the French wind tunnel in Chalais Meudon, which was at the time the continent's largest wind tunnel; test supervisor was *Gerhardt Sissingh*.

The Flettner helicopter showed a high maneuverability and good control about all axes. In the case of an engine failure, a rotor-driven regulator and an adjustment gear automatically switched from the helicopter setting to the autogyro setting. The resetting from autogyro to helicopter flight was solved constructionwise in a way that enabled the transition from autogyro to hovering flight by "stepping on the gas", a huge success for the Flettner designers.

The Fl 265 reached 126 flight hours and performed 1,180 take-offs and landings. No other helicopter worldwide had reached this operating time before.

With the Fl 265, which served as a test-flight vehicle, many detailed experimental flights were performed looking at ground clearance of the rotor blades, blade angles at different flight conditions, such as autogyro flight and helicopter flight at different flight altitudes. With the results of the flight tests and tests on the endurance test rig, design and construction regulations were developed and insight was gained into the complicated stress situation on helicopter parts, e. g., with regard to the bending fatigue strength of the helicopter blades. With the internal stabilization of the blade spars by shot peening, Flettner gained a significant increase in the bending fatigue strength.

In 1941, perfect flights were conducted with the experimental types V-4 and V-5, which contained the switch automatic. These investigations were accompanied by measurements of the height-control (elevator) moments, of the flapping angles in the helicopter and autogyro modes and by filming. The Type Fl 282 was equipped with a nose wheel, which was a novelty on helicopters. Investigations and constructive means to prevent nose-wheel flutter (shimmy) were successfully performed in cooperation with *Wunibald Kamm* at the "Institute of Motor Vehicle Technology" ("Kraftfahrttechnisches Institut") in Stuttgart. Through tests with the V1, detailed knowledge was gained concerning resonance oscillations of the propulsion – rotor system. The helicopter type Fl 282 B was built in two versions, viz., with two- and three-blade rotors; especially the latter ran very smoothly.

A comparison between the Flettner-helicopter and the Focke-helicopter, both being equipped with the same engine and having the same rotor disk area, showed the better performance of the Flettner-type, which was due to the intermeshing rotors.

After flight-testing the Fl 282 at the test center in Travemünde, the helicopter was assigned to operational trials at the Mediterranean. Operations related to submarine detection and convoy protection under real conditions were tested. The technique of landing on-deck was developed during cruises of the cruiser "Cologne" in the Baltic. The army showed interest in the Flettner helicopter as observer for the artillery and for fighting tanks.

In the spring of 1944, a remarkable test was performed. Two fighter aircraft (Me 109 and Fw 190) were ordered to "shoot down" a Flettner Fl 282 with a camera within 20 min. Not a single hit was landed. During low-level flight, the helicopter could not be detected. Due to its maneuverability, the Fl 282 was already at that time able to perform contour flights.

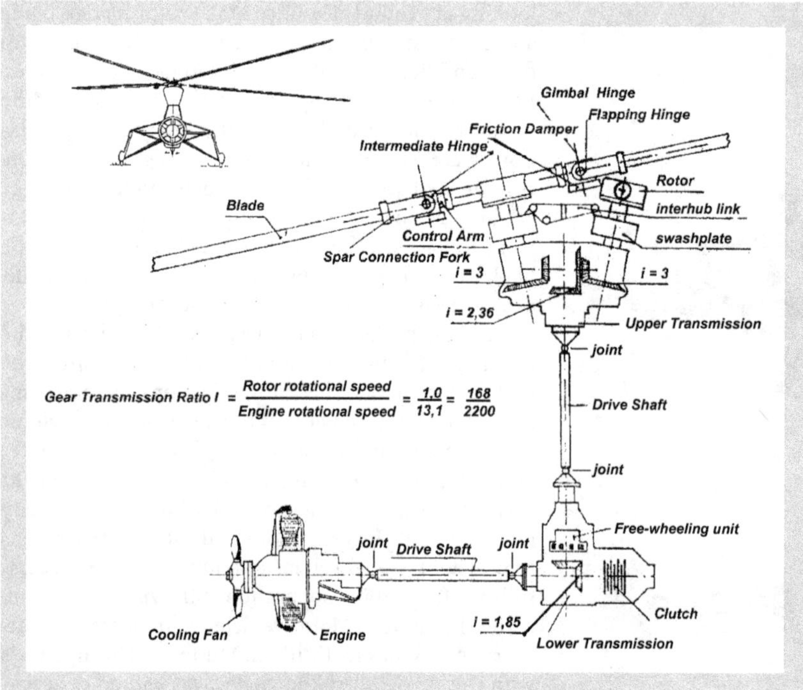

Flettner intermeshing synchronised rotors. Arrangement of rotors, hub, gear, transmission and motor.

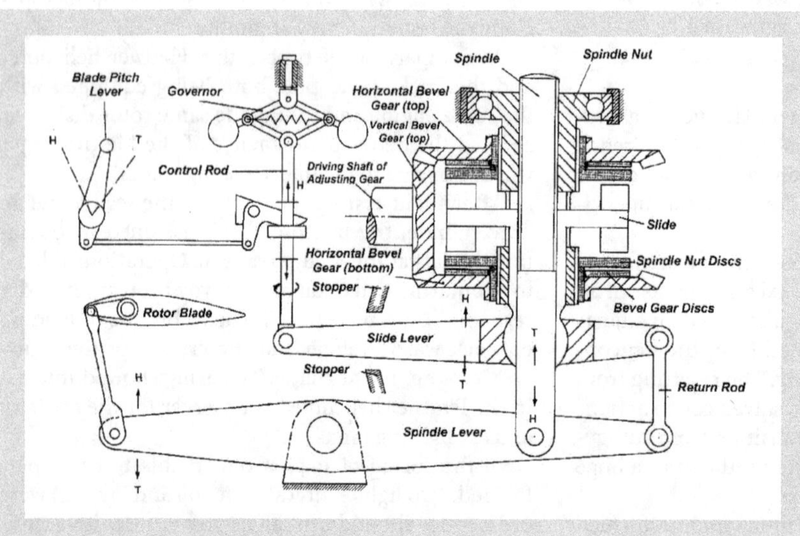

Schematic diagram of governor and blade pitch gear of Fl 282.

The good stability of the helicopter, resulting from the rotor system, was demonstrated by the following test: The helicopter performed well with respect to longitudinal and directional stability even without tail girder and empennage. Since the wheel-type landing gear did not prove to be an optimum in all instances, also landing-gear variations, e. g., with skids and floats were investigated.

The operational skill and operational possibilities of the Flettner helicopter convinced the army, and especially the navy, that an order for a large series of 1,000 Fl 282 was placed. Since the capacity of the factory in Schweidniz was limited, it was envisaged to carry out the manufacturing in Eisenach/Thüringen at a manufacturing-branch of the "Bavarian Motor Company" ("Bayerische Motorenwerke, BMW") Munich. The war delayed the start of production so that until the end of the war only 30 copies of the type Fl 282 were completed and only 22 of them were finally accepted as airworthy.

In 1944, *Flettner* started the development of a multi-purpose helicopter with the type designation Fl 339. The payload was to be increased compared to the Fl 282. The gross mass was envisaged to be 1,300 kg (2,900 lbs) (Fl 282: 1,000 kg (2,250 lbs)). The six-cylinder in-line engine Argus As 411, with a performance of 397 kW (540 hp) at 2,000 rpm, was to be assembled in one unit with the rotor drive, gear and cooler drive above the fuselage. The propulsion of the rotors was accomplished by one worm gear each. Due to this arrangement, the fuselage remained free of all propulsion elements. It could thus also be isolated from any oscillations/vibrations possibly resulting from the propulsion unit. The following operational tasks were envisaged:

- On-board reconnaissance aircraft, pilot reclining.
- Reconnaissance and observation aircraft, twin seats.
- Courier aircraft, three seats.
- Ambulance aircraft with two stretchers.
- Transport aircraft for 300 kg (670 lbs) payload.
- Crane helicopter with winch.

The scientists *Kurt Hohenemser* and *Gerhardt Sissingh* [10–16] had a decisive influence on the development of helicopters. *Kurt Hohenemser* published basic treatises on strength, on the computation of flight performances of rotary aircraft, on dynamic stability, and, generally, on dynamic problems, on rotor-blade flutter, on oscillations and other helicopter-related problems. He was, after the war, a leading capacity in the helicopter development in the USA, and held here, at the end, a professorship.

Gerhardt Sissingh wrote in numerous reports about his research in aerodynamics and about his computations concerning the performance and the dynamic stability of helicopters with hinged blade attachments. In 1943, he was transferred by the Flettner company to *Hans Georg Küssner* in Göttingen.

Gerhardt Sissingh went after the war to the USA, initially to the Kellet Aircraft Corporation in Camden (New York), where helicopters were built according to the Flettner principle. Later he changed to the United Helicopter, Hiller Corporation in Palo Alto.

Similar to *Henrich Focke*, was *Anton Flettner* able to engage a team of excellent engineers and scientists for his company, whose accomplishments and findings are still valid today in rotary aircraft technology. Co-workers of *Anton Flettner* were:

- *Emil Arnold* (Chief designer),
- *Willi Deilitz* (Test facilities),
- *Hans Ehrenfried Fuisting* (Test pilot),
- *Heinz Gundlach*, (Designer),
- *Kurt Hohenemser* (Dynamics),
- *Ludwig Hofmann* (Test pilot),
- *Kuchler* (Transmissions),
- *Oskar Nagel* (Project designer),
- *Richard Perlia* (Test pilot),
- *Otto Reder* (Technical Liaison Officer),
- *Xaver Schleicher* (Flight mechanics, Pilot),
- *Joseph Schmidt* (Flight-test Engineer),
- *Gerhard Siegel* (Statics),
- *Gerhardt Sissingh* (Aerodynamics).

Because of the stringent regulations concerning secrecy, the public knew hardly anything about the Flettner helicopter during the war. The "victorious powers" investigated the captured helicopters in detail and put down their investigative results in detailed descriptions and illustrations in respective reports, e. g., by *Captain R.N. Liptrot* [7], *J.A.J. Bennett* [17], *Bartram Kelley* and *Hugh J. Mulvey* [18].

According to these reports was the Fl 282 the most advanced German helicopter that had more flight hours than any other (*Liptrot*). It is highly maneuverable and was flown during bad weather, rain, gust conditions, and under instrument-flight conditions. Numerous tests were conducted with respect to auto rotation (autogyro), high-altitude flight, velocity, stability, oscillations and oscillatory stress; hundreds of autogyro landings were carried out (*Bartram Kelley* and *Hugh J. Mulvey*). The Fl 282 had a very good control response, a compact design, and was relatively stable even during hovering flight. During cruise, it could be flown with control stick released (*R.H. Prewitt*). *Dave Driskill*, an experienced helicopter pilot, said after test flights with the US Air Force that the Flettner helicopter was the best he had ever flown. *Charles H. Kaman*, president of the Kaman Aircraft Corporation, wrote in a contribution for the American Helicopter Society in February 1953: "*The Flettner-helicopter was an extremely interesting machine. It had excellent controllability and stability characteristics.*"

The design of *Flettner's* intermeshing rotors had an effect on *Charles H. Kaman*, and the Kaman Corporation, which was founded 1945, developed very successful helicopters utilizing this rotor concept, e. g., the rescue helicopter HH-43 "Huskie" (1958) and the cargo helicopter "K-Max" (1991).

After the end of the war, *Anton Flettner* and his staff remained in Bad Tölz until 1947, still being concerned with his helicopter system until he emigrated to the USA and became an American citizen. In 1949, he founded the Flettner Aircraft Corporation in New York, whose president he was. As chief engineer and vice-president for his company, he won the helicopter engineer *E.K. Liberatore*. *Anton Flettner* considered the association of his Flettner Aircraft Corporation with the Kaman Corporation. At that time, the prototype of a twin-rotor helicopter, with the Flettner-arrangement of the rotors, was in operation. However, he did not associate since he was, at that time, concerned with the jet propulsion for a crane helicopter.

Anton Flettner died in a New York hospital on December 29, 1961. He and his co-workers had developed up the end of the war the most advanced helicopter, the Fl 282. *Flettner* pointed out a wide field of potential applications for helicopters; the principles he developed at the time are still effective today.

Von Doblhoff and his Rotor System

Friedrich von Doblhoff, born in 1916 in Budapest, deceased in the year 2000, started in 1942 with the development of a jet-propelled helicopter with only a few co-workers at the "Wiener Neustädter Flugzeugwerke, WNF". His plans gained the interest of the "State Ministry of Aeronautics" ("Reichsluftfahrtministerium"), which provided him with the money for the development, about half a million Mark, and allowed the installations of the aircraft company to be used. *Friedrich von Doblhoff* envisaged that his helicopter could be operated off submarines to enlarge their range of sight.

The advantages of jet propulsion [19] consist of a substantial simplification of the propulsion system and the elimination of the torque-compensation equipment. The weight savings are approx. 20 per cent of the gross weight. The larger number of moving parts in the case of a helicopter with mechanical propulsion compared to a jet-propelled helicopter means for the latter reduced manufacturing and maintenance costs; in addition, there is a 20 per cent increase in payload based on gross weight. However, jet propulsion has a system-inherent lower efficiency.

All helicopter designs known at that time had great difficulties to overcome with respect to eliminating the reaction moment and sustaining stability. A second, counter rotating rotor, or a rear rotor can be eliminated in the case of jet propulsion. Fore these reasons and reasons of constructional simplicity, *Friedrich von Doblhoff* started in 1942 with the development of a helicopter, whose rotor torque was provided by the reaction force of gas jets emerging from the blade tips (tip-driven rotor). The propulsion element was a "jet producer" at the blade tips that was fed with a compressed-air-fuel-mixture through the hollow blades.

The combustion chambers were made of a chromium-nickel-steel alloy. The combustion takes place at a stoichiometric fuel/air ratio of 1 : 20. The combustion temperature was slightly above 2,000 °C. The large Coriolis accelerations imposed on the flow inside the rotor blades caused combustion problems.

Four helicopters of the model WNF 342 were built. In the process, knowledge gained with the predecessor model was consequently used in the design of the follow-on models.

- **WNF 342 V-1** Flight mass 240 kg (540 lbs), 44 kW (60 hp) Walter-Mikron engine for driving the turbocharger. The envisaged head-pitch control did not meet the requirements; therefore, a cyclic blade control was employed. Hovering-flight trials were performed with good stability.
- **WNF 342 V-2** (1944) Generally the same design as the V-1, but with a 66 kW (90 hp) Walter-Minor-

Friedrich von Doblhoff (1916–2000).

engine and a net-mass of 340 kg (760 lbs). The models V-1 and V-2 were extremely small so that they could operate from submarines and small ships.

- **WNF 342 V-3** (1944): Single-seat model with cabin and an additional pusher propeller for propulsion, which could be engaged during cruise. A stronger engine with 100 kW (130 hp) was installed. The flight mass was increased to 540 kg (1,200 lbs). Due to oscillatory resonance, the vehicle fractured.

Kurt Hohenemser, at that time at Flettner responsible for aerodynamics and dynamics, was consulted on the investigation of the oscillation problems during hovering flight of the heavy V-3. He diagnosed that the difference between the natural frequency of the rotating blades and the rotational frequency of the rotor had become so small that it coincided with the pendulum frequency of the fuselage below the rotor, which causes a dangerous instability of helicopters. By decreasing the oscillatory frequency of the blades in the plane of rotation of the rotor this resonance could be avoided.

- **WNF 342 V-4** (1945) Model V-3 as two-seat version rebuilt with the same engine. The V-4 conducted hovering flights with a total flying time of 25 hours. Flights with a lower forward speed were performed to a lesser degree.

Friedrich von Doblhoff originally wanted a drive with Lorin-nozzles. However, before the model V-1 was built, he chose compressed air as propulsion, which was produced by a piston-engine driven charger. The compressed air was mixed with fuel and the combustible mixture was led through hub and blade spar to the combustion chambers at the blade tips of the three-bladed rotor. The control of the helicopter about its vertical axis was accomplished by a blower generated jet impinging on the rudder.

Due to the high fuel consumption of the jet propulsion, it was anticipated to use it only for a few minutes during hovering flight at the start and the end of the horizontal flight. For cruise, the *Doblhoff*-helicopter should fly as an autogiro. A propeller, which could be connected to the engine, provided the propulsion during horizontal flight.

The Doblhoff-system could also be applied to a gyrodyne (e.g., a winged helicopter), if the power is distributed to the lifting system as well as to the thrust system. A variable transition from a pure helicopter to a gyrodyne and finally to an autogiro would be possible by switching the power gradually from the rotor to the pusher propeller.

Friedrich von Doblhoff also suggested as operational task for his jet-propelled helicopter, the transportation of cargo of up to six tons over a medium distances at low speeds.

Research topics related to the rotor system were:

- Compressors: *Friedrich von Doblhoff* developed a calculation method for the determination of design data for a compressor. For lack of time, he chose the charger Argus As 411 available on the market.
- Gas generator: The combustion chambers or nozzles at the blade tips are for low-drag reasons necessarily very small. The result is a relatively small mass flow through the blade. This, in turn, results in difficult combustion problems since the combustion, for a given fuel consumption and thrust, has to be extraordinarily fast. The fuel-air ratio has to be kept constant and mixing complete. At first, the direct injection of the liquid fuel was unsuccessfully tried. Only 40–50 per cent of the caloric contents could be transferred to the air. A suitable procedure was to evaporate the fuel at 160 °C and to overheat it, and to lead the fuel vapor into the system. The heat addition took place in a heat exchanger, which was heated by the hot motor exhaust gases. The fuel vapor could be introduced

The first experimental model of the Doblhoff-helicopter WNF 342 V-1 at free flight in October 1943.

into the air stream either upstream or downstream of the turbocharger. The introduction on the high-pressure side proved to be more secure for the operation and control of the rotor.

- Rotor blade: To keep the losses inside the rotor blade as small as possible, a thick hollow metal blade is to be used. An aluminum tube is pulled through a metal female die and by attaching an oak trailing edge, the airfoil NACA 23018 is realized. The end of the blades leads to five tubes through and around which the fuel mixture flows. They act as flame arresters to prevent backfiring into the blade. At the end of each tube, a small wedge is mounted as a turbulence generator to prevent a blowout of the flame. The initial ignition inside the combustion chambers is accomplished via an ignition cable that is led through the blade trailing edge to the combustion chambers and the spark plugs. The combustion temperature is approx. 2,000 °C.
- Rotor hub and blade suspension: The transfer of the fuel mixture from the fixed hub to the rotor blades proceeded through a flexible tube. The blade attachment is realized via a four-layer leaf-spring block screwed to the upper part of the rotating and swiveling gas-collector reservoir. The latter is mounted, spherically sealed, inside the lower, fixed part of the reservoir. In case of a cyclic blade control, the upper reservoir tilts inside the lower one. It was planned to perform the collective blade adjustment via the pressure of the turbocharger. The helicopter is flown via the gas lever. An increase in the rpm of the drive results in an increase in the charge pressure and thus, via the collective-pitch-controller, in an increase in the set angle of the rotor blades. This keeps the number of revolutions of the rotor constant and eases the control substantially. However, the planned automatic adjustment of the collective set of angle of attack was never put into operation. To avoid frictional forces within this control system, time-consuming developments would still have been necessary. For that reason, the rotor thrust was adjusted at fixed blade angles by rotor rpm variations. The blade attachment springs are so flexible that the rotor blades can be folded back.
- Control about the vertical axis: To remove the problems associated with the directional control, a small pusher propeller was mounted on the V-3, whose jet was directed towards a rudder located downstream.

Friedrich von Doblhoff and his approx. 20 colleagues at the Wiener-Neustädter Flugzeugwerke performed pioneering work in the development of jet-propelled helicopters. His closest co-workers were:

- *Theodor Laufer* (Research associate, dynamic and aerodynamic computations),
- *August Stepan* (Designer and test pilot),
- *Vordren,*
- *Kurt Löffler,*
- *Alexander Czernin.*

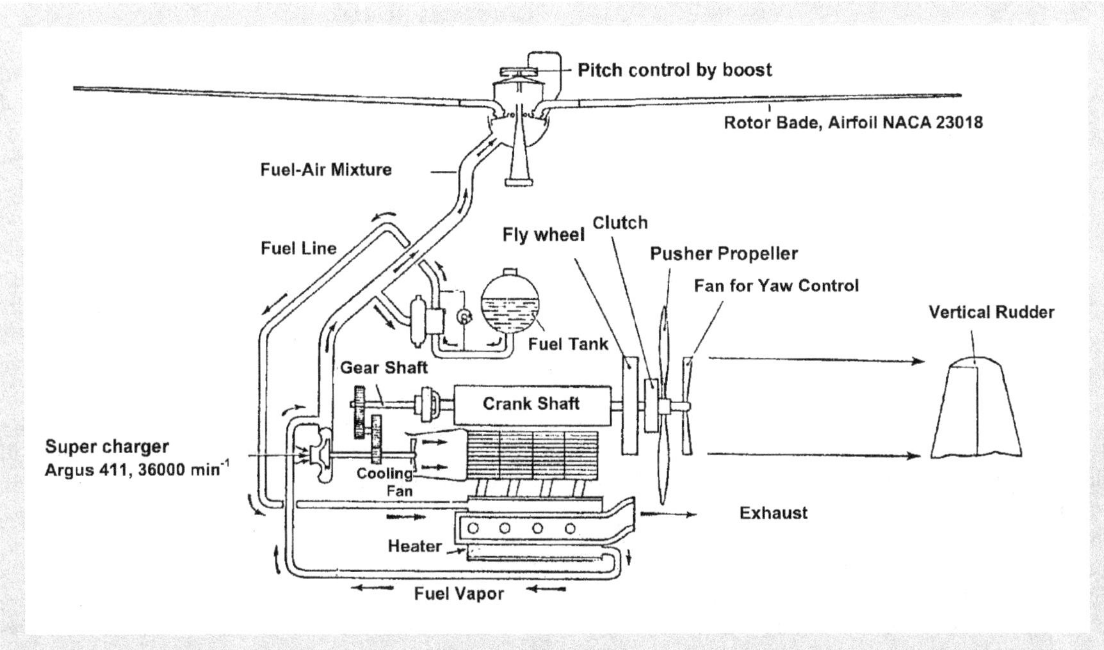

Schematic of the preparation of the fuel-air mixture of the *Doblhoff* helicopters. The fuel evaporated in a heat exchanger heated by the motor exhaust gases and mixed with the compressed air behind the charger. The fuel-air mixture was led through the rotor head and the hollow rotor blades to the combustion chambers at the blade tips. Due to the high fuel consumption, the blade-tip drive was to be switched off during cruise and the rotor set to autogiro conditions; thrust was to be provided by a pusher propeller. Directional control was to be achieved by the air jet of the fan for yaw control.

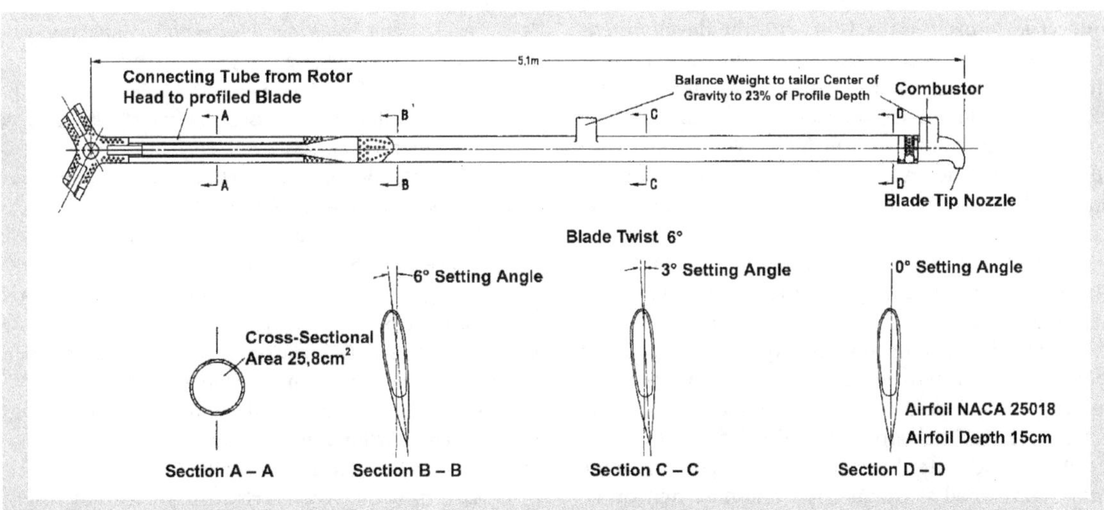

Hollow rotor blade of the WNF-342 helicopter with combustion chambers at the blade tips.

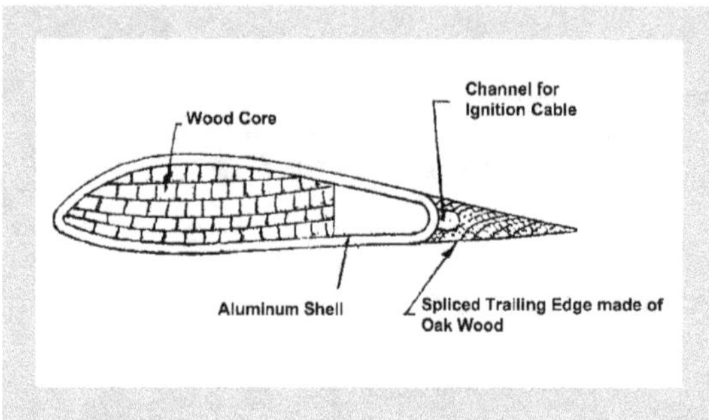

Cross-section of the rotor blade of the Doblhoff-helicopter WNF 342.

Before the end of the war, *Friedrich von Doblhoff* planned a jet-propelled helicopter without a "tip combustor", whose propulsion was provided by a gas turbine with the compressor not only supplying the air for the turbine but also, by tapping, for the jet propulsion at the blade tips. The exhaust heat is transferred to the tapped air by a heat exchanger to increase efficiency. For directional control, the exhaust gases of the turbine were to be directed towards a rudder.

The rights for this design went after the war by contract to the Société Nationale de Construction Aéronautiques du Sud Ouest (SNCASO), which built a two-seat helicopter, the "Djinn", according to this principle under omission of the heat exchanger in series.

After the end of the war, *Friedrich von Doblhoff* concerned himself in the USA with the development of jet-propelled helicopters, first at Kellet and later at McDonnell. *August Stepan* and *Alexander Czernin* received employment contracts from Fairey Aviation in England to work in the helicopter department; here, they considered the development and construction of a large compound helicopter and of a light helicopter with blade tip propulsion.

August Stepan returned to Germany in 1963, where he took over the supervision of the overall test activities of helicopters at Bölkow KG, later MBB.

Theodor Laufer moved with two colleagues in May 1946 to France, where he was again able to work in his field; he participated in the development of the jet-powered "Ariel" helicopter and the "Djinn" helicopter of SNCASO. After his return to Germany, *Theodor Laufer* developed at Dornier the one-man helicopter DO 32 E (first flight on June 29, 1962), in design similar to the "Djinn" helicopter.

Concluding Remarks

Summarizing it may be stated that the helicopter technology in Germany at the end of the war had reached a high standard not existing in countries of the former enemies. Basic scientific knowledge concerning aerodynamics and dynamics of helicopters was obtained in the development and design offices of *Henrich Focke, Anton Flettner* and *Friedrich von Doblhoff*. The rotor systems were developed with scientific precision, tested and flown. The literature seized, the design and construction drawings and the equipment found, including its testing, as well as the engagement of German scientists and engineers have saved the developers in the USA, England, France, and in the Soviet Union a large amount of work. The practical and scientific knowledge gained at the time still has an effect today.

References

1 H. Focke, Mein Lebensweg, Kurze-Schönholz und Ziesemer Verlagsgesellschaft, 1996
2 E. Springmann, Focke, AVIATIK Verlag, Oberhaching, 1997
3 H. Focke, Das Trag- und Hubschrauberproblem, Vortrag gehalten am 26.11.1937, Schriften der Deutschen Akademie der Luftfahrtforschung, 1937/1938
4 H. Focke, Der Focke-Hubschrauber, Luftwissen Vol. 5, 1938, pp.33–39
5 H. Focke, Neue Wege der Flugtechnik, Abhandlungen und Berichte, 10. Jhg., Heft 6, 1938, VDI-Verlag, Berlin
6 K. von Gersdorff, K. Knobling, Hubschrauber und Tragschrauber, Vol. 3 „Die deutsche Luftfahrt", Bernard & Graefe Verlag, Bonn, 1985
7 R.N. Liptrot, Rotating Wing Activities in Germany during the period 1939–1945, B.I.O.S. Overall Report No. 8
8 G. Sissingh, Beitrag zur Aerodynamik der Drehflügelflugzeuge, Luftfahrtforschung, 1937, pp. 290–302
9 G. Sissingh, Beitrag zur Aerodynamik der Drehflügelflugzeuge, Luftfahrtforschung 1938/6
10 K. Hohenemser, Gültigkeitsgrenzen der Glauert'schen Drehflügeltheorie, Techn. Berichte 1942
11 G. Sissingh, Beitrag zur Aerodynamik der Drehflügelflugzeuge, Luftfahrtforschung, 1940/7
12 G. Sissingh, Die Berechnung der Rotoren von Drehflügelflugzeugen bei höheren Fortschrittsgraden mit besonderer Berücksichtigung des Schnellfluges, Jhb. der Luftfahrtforschung 1941, I

13 G. Sissingh, Ein Verfahren zur direkten Bestimmung der Schlagbewegung von Flugzuständen mit hohen Blattbelastungen, Jhb. der Luftfahrtforschung 1941, I
14 G. Sissingh, Luft-und Massenkräfte bei Steilschrauben mit Schräggelenk, Ergänzungsbericht UM 3066, AVA-Bericht 344/J/2, 1944
15 G. Sissingh, Untersuchungen über die Stabilität der Schlagbewegung von Steilschrauben, TB 44, Heft 10
16 A. Betz, Monographien über Fortschritte der Deutschen Luftfahrtforschung, Überblick über die deutsche Entwicklung und Forschung auf dem Gebiete der Drehflügelflugzeuge, redigiert von G. Sissingh, AVA Göttingen, 1946
17 J.A.J. Bennett, Notes on a B.I.O.S. Mission to Germany and Austria to Investigate German Flettner Helicopter Fl 282 'Kolibri' Developments, Januar 1945
18 B. Kelley und H.J. Mulvey, Combined Intelligence Objectives Sub-Committee (C.I.O.S.) Target No. 25/488, August 17, 1947
19 F. von Doblhoff, The Helicopter Pressure Jet, The American Society of Mechanical Engineers, Semi-Annual Meeting, St. Louis, 19–23. June 1950

Rescue and Safety

GERHARD SEDLMAYR, HOLGER HANSEN

Ribbon Parachutes as Aircraft Brake and Rescue System

The Parachute and its Realisation

Parachutes serve to slow down objects (for instance human beings, cargo, aircraft and capsules) in the air, whereby equilibrium is established between the load attached to the parachute and the drag of the parachute. Depending on the design of the parachute, a steady sinking or flight-path speed, respectively, with an unsteady landing/roll air bubble (canopy) results, which guarantees a safe landing of man and material. According to the application, one distinguishes between aircraft brake parachutes, stabilising parachutes, cargo parachutes, and man-rescue parachutes.

Since the beginning of the development, textile parachute canopies have the shape of a flattened semisphere and the plan-view of a circle. The shape of high performance parachutes can, however, not be traced to simple geometric bodies. Most parachutes consist of a simple spherical canopy, which has its highest drag when the flow is directed towards its opening. The pressure conditions associated with the flow cause the canopy to assume and keep its form, a process also supported by the rigging lines. This is why the smallest projected area will be sufficient and stiffening is not needed. This means that the canopy can be packed and stowed away in a handy format.

To bring the parachute into the air stream, two different procedures were already utilised very early.

Procedure No. 1: Since the beginning of the development of parachutes, the so-called static opening was applied; here the parachute-pack cover is immediately removed when exposed to the air during a bailout and the parachute opens.

Procedure No. 2: The second procedure – a stepwise deceleration – was developed by *Waldemar Müller* in 1924; here, a small auxiliary parachute is applied (free bag deployment), which, after being exposed to the flow, deploys and pulls the main parachute from the packing case and stretches it. The auxiliary parachute is connected to the apex of the main parachute by a short interconnecting cord. To improve the automatic procedures, the auxiliary parachute is, for instance, equipped with a steel spring so that it opens immediately after being released from the packing case.

Auxiliary parachutes are not absolutely necessary, but they considerably reduce the time to fill the main parachute canopy and enable a more secure deployment of the parachute during free fall. Most of the parachutes consisted, since the beginning of parachute technology, of simple spherical canopies, whose fabrics exhibit an air permeability of a small to medium porosity. Already in the early days of balloon flight, nearly all balloonists carried along parachutes for safety reasons. However, during the first jumps an especially unpleasant characteristic of the parachutes was experienced, namely a considerable tendency to pendulum-like oscillations (hunting).

Spherical canopies tend to oscillations if they are excited by small disturbances. At especially unfavourable circumstances, spherical canopies may even amplify these movements. To let air escape from the parachute canopy, an opening was placed at the apex of the parachute canopy, which resulted in a reduction of the pendulum-like oscillations.

First Scientific Investigations

The beginning of the parachute technology goes back to the performances of reckless showmen, who jumped at the end of the 18th century from buildings and balloons with textile braking devices for the entertainment of others. They gained their knowledge through pure experimentation and testing. Up to the first decade of the 20th century, there were hardly any engineering-science investigations; solely the expertise of tailoring was asked for. With the start of military aviation, also the importance of the parachute technology increased since parachutes were applied to rescue pilots out of aircraft distress, but also to drop – the so-called paradrop – of cargo such as medication, goods and military products.

Real engineering/scientific investigations can be traced back to the time of 1933–1945, when in Germany scientific development of parachutes started [1]. *Georg Madelung, Theodor W. Knacke* and *Helmut Heinrich* conducted parachute research and experimental developments at the "Flight Technical Institute of the Institute of Technology in Stuttgart" ("Flugtechnisches Institut der Technischen Hochschule Stuttgart, FIST). The clear objective was to decelerate (slow down) bombs, mines, torpedoes and projectiles, to stabilise them and hence to bring them closer to the target; it was not the rescue of humans. The first stabilising parachutes were development with the idea of replacing the relatively large and bulky empennage of conventional bombs by a parachute, having the same function but requiring less space, to keep the size of the hatches of the aircraft small. This would also reduce the space requirements within the aircraft. These theoretical and practical research activities [2] led to various parachute types and sizes. Parallel, also brake parachutes for the already developed cargo-glider technology were investigated and systematically developed. Two essential brake parachutes originated which are still widely applied today. In 1933, *Theodor W. Knacke* started his theoretical and practical investigations to achieve the most effective deceleration of an aircraft in the air and during landing [3]. In 1938, the development of the FIST-ribbon parachute was carried out by *Georg Madelung* and *Theodor W. Knacke* (first landing in 1939), and in 1941, the development of the guide-surface parachute by *Helmut Heinrich*. For the first time, parachutes became technically very efficient due to extensive theoretical research [3] and numerous systematic experimental developments.

Georg Madelung (1889–1972).

Experimental Development of the Parachute Technology

During 1933 and 1945, practical investigations of about 10,000 parachute drops were performed at the test center Rechlin [1] from aircraft. More than half of these drops were filmed by high-speed cameras [4, 5] and documented. Due to the large number of drops, a scientific analysis could only be performed by setting priorities. Unfortunately, these unique documents were lost in 1945. A complete analysis of these measurements after 1945 would certainly have constituted a comprehensive scientific treatment of parachute technology.

The capture and investigation of the deployment and descent characteristics of a parachute are impossible by observations with the bare eyes. Therefore, high-speed cameras recorded deployment and filling of a descending parachute. The determination of the flight path of a parachute in space, after its drop, was carried out by means of a cine-theodolite, a device, which allows the simultaneous measurement of the vertical and horizontal angle of a point between two planes positioned normal to each other. In case of a cine-theodolite, a camera records simultaneously both angles and the object to be tracked.

FIST-Ribbon parachute manufactured by the Kostelezky Company.

Effect of Parachute Fabrics on Sinking Characteristics

To reduce the parachute oscillations, numerous changes were made to the parachute canopies. This led to research on parachute canopies [6] made of fabrics of increased porosity (air permeability) or parachute canopies with holes or slots, respectively, which were arranged equidistantly and provided ring-shaped wider openings in the canopies to increase air permeability. That way it could be demonstrated that the pendulum-like oscillations could be reduced by ±15 per cent by increasing the porosity. But this did not yet correspond to the ambitious objective of oscillation-free behaviour, i.e., oscillations of less than ±5 per cent.

The greatest difficulty occurred when mass-producing the fabrics started. It was tried to adhere to the test determined value of porosity and to validate this porosity. As soon as the permissible tolerances were exceeded, the parachutes encountered either premature stress fractures or, if the porosity was too large, difficulties in filling up the canopy – in extreme cases, the canopy did not fill up at all.

Nevertheless, the tests did confirm [6] that an increase in porosity was the right direction to restrict or even suppress the pendulum movement of a parachute.

To prevent fractures, the experimental development was extended to the application of additional ribbons [3, 6] to reinforce the porous canopies. In about 1937, the employees of the FIST (*Georg Madelung, Theodor W. Knacke, Rudolf Isermann* and *Albert Keller*) had the idea to build a parachute whose canopy consisted only of ribbons. This way, the first ribbon parachute was built. This type of parachute only had a maximum pendulum-like oscillation of ±3 per cent! It still bears the name of the institute where it was developed, viz., "FIST-ribbon parachute"; it is still being utilised today as recovery parachute, e.g., for reconnaissance drones, and as brake parachute for aircraft, landing space capsules and the Space Shuttle.

After development work on a parachute canopy that consisted only of ribbons, the first trials started in 1938. Initial difficulties with the various ribbons were overcome by the 50-mm wide silk ribbons of the Kostelezky Company in Stuttgart, which were originally used as hat ribbons. These ribbons were characterised by reinforced edges, which gave them extra strength. In 1939, the Kostelezky Company also took over the manufacture of the first FIST-ribbon parachutes.

The result of the systematic development of the FIST-ribbon parachute was a parachute that showed, if displaced from the normal, a large restoring moment and provided that way the desired very small pendulum effect. The design was also characterised by a considerably reduced opening shock, which, in turn, meant a much smaller load on the flight vehicle. A further advantage was the reliable controlled deployment and opening of the parachute. The large brake effect was reached by the strong turbulence of the air at the edges of the ribbons. Overall, the ribbon parachute constituted a simple design with good maintainability, especially since it could be repaired with little effort. Besides the application as aircraft brake parachute, this type of parachute was also applied as stabilising and cargo parachute due to its small pendulum effect. Some modern ribbon parachute designs reached only ±2 per cent and were thus far superior compared to normal designs.

The disadvantage compared to the standard spherical canopies was the approx. 30 per cent lower efficiency, this corresponds in parachute technology to a parachute-characteristic technical parameter, the so-called drag coefficient C_{do}, comparable to the C_d-value, which was 30 per cent lower.

The problem to accurately achieve the prescribed porosity during manufacturing was now solved: the ribbons of the ribbon-parachute design are nearly impermeable. The air permeability of the canopy is nearly entirely achieved by the respective arrangement of the ribbons and their distance to each other, i.e., by the so-called geometric porosity.

The material of the canopy consists today of chemical-fiber materials or cotton (earlier also of silk). The canopies of personnel parachutes consist of fabrics, the ones of brake and stabilising parachutes mainly of ribbons. The investigations conducted at FIST, which merged in 1939 with the "Research Establishment Count Zeppelin" ("Forschungsanstalt Graf Zeppelin, FGZ"), proved that the elasticity and the porosity of the material as well as also the so-called geometric porosity have a decisive influence on the overall behaviour of a parachute.

The "utilisation-speed" – that means the speed when leaving the flight system – representative of the load limit for standard fabric-type parachute canopies [3], is at about 400 km/h (216 kn), the one for ribbon-parachutes is approximately two to three times the speed of sound. These high velocities generate high skin friction on the ribbons, especially at the edges of the ribbons, due to the high air turbulence resulting in high temperatures. Polyamide ribbons only have very limited temperature stability and thus fail quickly. Modern materials, as, e.g., Aramid fibers and carbon fibers – used for ribbons as well as for rigging lines – are increasingly employed in modern designs for these specific applications.

To clarify basic questions, for instance regarding the stability characteristics of parachutes, and to determine drag, deployment shock, and friction forces, further research was conducted at the aerodynamic institutes of the DVL, the AVA and others in conjunction with wind tunnel tests.

To produce the same flow conditions on the models, it was recognised that, as characteristic parameter, mainly the Reynolds number had to be kept constant, i.e., the velocity of the flow in the wind tunnel had to be increased corresponding to the reduced scale of the model. Another way to investigate a model is the test in a water tunnel. Here too, the model scale can be sustained via the Reynolds number, despite the lower flow velocity, due to the much higher density of the water. This was only applicable for velocities of up to 400 km/h (216 kn), the application velocity of standard parachutes. At velocities above 540 km/h (290 kn), another characteristic similarity parameter becomes important, viz., the Mach number as a measure for the compressibility of air.

It became quickly clear that it is, in the case of most parachute models, impossible to scale down the fabric, i.e., the fabric porosity, in accordance with the model so that results are strongly distorted. Most wind tunnel tests on fabric parachutes are, therefore, limited to general flow visualisations and comparative testing. However, concerning ribbon parachutes, with their geometric porosity, the favourable characteristics of the FIST-ribbon parachutes could in wind-tunnel tests at the FGZ in Stuttgart be confirmed as being very good.

Ribbon-Parachutes as Aircraft Brake Parachutes

In 1940, ribbon parachutes were first employed as aircraft brake parachutes [7, 8], within an experimental development program during the penetration of the Fort Eben-Emael in Belgium, to stabilise uncontrolled flight conditions, to decelerate during flight in preparation for landing, and as brake parachutes proper after the landing of cargo gliders. Here, these mature textile devices proved their high reliability and efficiency especially in the case of aircraft and glider landing on extremely short runways.

In 1943, this brake parachute type [6] was used for dropping the 500 kg (1,124 lbs) bombs "Liesele" at a speed of approx. 750 km/h (405 kn). In 1944, also the rocket/missiles V1 and V2 and their constructive elements were recovered by FIST-ribbon parachutes during flight tests.

For the Do 335, which had entered tests in 1943, an aircraft brake parachute was requested which was

Ribbon parachutes were used for the first time when the DFS 230 cargo glider was employed during the conquest of the fort Eben-Emael in Belgium.

The DFL conducted in 1962 a comparative study with aircraft brake parachutes of different makes, among them also parachutes of World War Two of the companies Autoflug, Kohnke and Kostelezky.

supposed to decelerate the mass of the aircraft of 6 t (13,500 lbs) within approx. 10 sec from 1,050–500 km/h (567–270 kn) so that the crew could, in case of an emergency, abandon the aircraft with less risk (with ejector seats), or that, alternatively, the aircraft could possibly be brought back under regular control.

For applied research [9], five ribbon parachutes were available between March and May 1944. Six test drops from a Me 323 with a cargo of 6 t (13,500 lbs) at a release speed of 180 km/h (97 kn) and altitudes varying between 4,000 and 6,000 m (13,150–19,700ft) were performed. The task, unusual for that time, required the utilisation of drop-bodies, which deployed the parachutes only after a corresponding free-fall and the attainment of the desired velocity.

Under the supervision of *Theodor W. Knacke* [3], the FIST-ribbon parachutes were further developed after 1946 in the USA, and various versions of the ribbon parachute resulted and were employed as brake and recovery parachutes. Probably the most attention-getting application of the further advanced ribbon parachute was the combination of several such chutes to a landing system [6] for manned space flight up to the Apollo-program of NASA.

The Guide Surface Parachute

A further parachute design with extremely small pendulum effects was the guide surface chute developed by *Helmut Heinrich* at the FGZ. In contrast to *Theodor W. Knacke*, who increased the permeability of the parachute within controlled limits, *Helmut Henrich* approached the task of decreasing the pendulum effect strictly aerodynamically. In his design of the guide surface parachute, the fabric canopy consisted of a spherical cup with a downward pointing frustum at its base acting as guide surface.

Also this parachute design had, when deflected from the normal, a restoring moment and hence little tendency to oscillate. Due to the extraordinary stability of this type of parachute, this brake parachute was used in 1943 for the first time for the stabilising control of bombs and torpedoes. The guide surface parachute of *Helmut Heinrich* was utilised in the unmanned spaceflight programs [6] in conjunction with the Mars-probe Viking and the Venus probe.

Great-Circle Ribbon Parachute (Wako Canopy Parachute)

Based on the experiences with ribbon parachutes, the concept was again revised at FIST, and in 1940, *Georg*

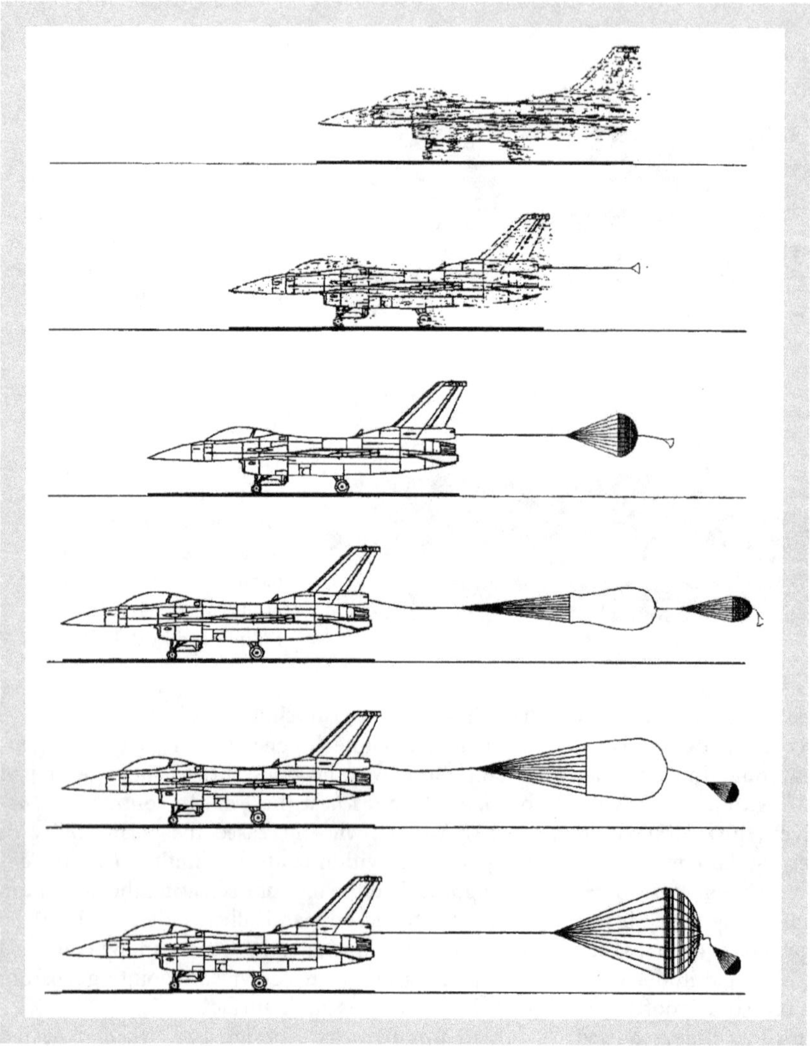

Deployment sequence of a modern ribbon parachute braking the "Eurofighter" during landing.

Ribbon aircraft brake parachute during landing roll of the "Eurofighter".

Pirzer and *Friedrich Weinig* started with the development of an entirely new ribbon parachute, the so-called "Great-circle Ribbon Parachute (Wako Canopy Parachute)". This design exhibits an up to now unprecedented performance. It allows an optimal introduction and distribution of forces. The maximal application velocity of the great-circle ribbon parachute was 700 km/h (378 kn) roughly twice as high as the one of known designs; this velocity could not be reached with the available test aircraft. Therefore, bomb-like loads were dropped which only deployed the parachute when reaching the desired velocity after a corresponding free-fall. The great-circle ribbon parachute was approved as a personnel-rescue parachute in 1944 and introduced into quantity production for jet/high-speed aircraft, such as the Me 262, where it proved extremely successful.

As already for the first FIST ribbon parachutes, the large porosity allowed a nearly oscillation-free descent. For this derivative of the ribbon parachute, the forces of the deployment shock were only approx. 30 per cent of the one of the conventional parachutes; however, the packing volume was noticeably enlarged.

Specific Forms of Ribbon Parachute

Modern ribbon-parachute designs, with special applications at speeds above sound, exhibited phenomena that required special investigations. Specific demands on the fabric material created new problems – among others due to ribbon flutter and temperature effects – because of the strong vortices of the air at the ribbon edges.

In 1987, the Autoflug Company developed and manufactured the so-called "Hemisflo-ribbon parachute" for stabilising and braking scientific descending capsules. Here, the ribbon parachute safely recovers a 6-m (19.7 ft)-long falling body of up to 500 kg (1,120 lbs) weight at a speed corresponding to a Mach number of 2.4 after a 20-km (14.9 mi)-free fall. This ribbon parachute has under flow conditions a spherical canopy, is extremely robust and built up of carbon fiber at its base and apex opening. The parachute exhibits only small oscillations of less than 2 per cent and is extremely well suited for this kind of applications. A further version of the ribbon parachute, the conical ribbon parachute, has an approx. 20 per cent higher drag coefficient than the plane ribbon parachute and allows a maximum application speed corresponding to a Mach number of 1.5.

Recent Research and Developments

Besides the controlled oscillation behaviour, a safe deployment of the parachute is demanded; here, the deployment sequence is of main importance. Especially for the application at high speeds, the standard deployment sequence was changed in that now the rigging lines are first being pulled out of the parachute pack assembly (lines first) and not – as usual – the apex. During newer experimental investigations of the Autoflug Company, new packing procedures are being explored, which are to allow a sequential deployment of the parachute. This requires that the parachute be packed according to its progressive opening behaviour. This should guarantee that,

Space Shuttle "Endeavour" with a brake parachute (drag chute) analogous to the system "FIST-ribbon parachute" at landing.

depending on speed, only those parts of the parachute are deployed, which are, due to their partial design, accordingly dimensioned. Since this deployment has to occur inversely proportional to the speed – high speeds demand a small parachute diameter first, respectively low speeds a large one – the deployment is controlled by a corresponding control device. The control can be mechanically parameterised in advance, with the deployment conditions such as height and speed as input, and the actual flow speed can in real time be electronically converted to control signals for the sequential parachute deployment; all other data can be preset or transferred in real time to the control device via a data bus. These parachute systems are suitable for the deployment at the lowest levels as well as at highest speeds.

Ejection Seats for a Safe Bailout

The Necessity of Bailout Support

Since the beginning of manned flight, there exists the necessity to be able to bailout from the aircraft in time in the case of an emergency. Prerequisite for a full parachute deployment before touching the ground is, however, a sufficient flight altitude.

It is unquestioned that with the development of the ejector seat during the years 1938–1945, the largest contribution to safety in aviation was made. All basics, still valid today – including the many detailed solutions –, were already established during that time. The prognoses of that time, that for the future equipment of all high-speed military aircraft the ejector seat will be indispensable, is still entirely true today. This prediction was not made by chance; it was, indirectly, already made in the years between 1914 and 1918 when, in contrast to other nations, exclusively German balloon and aircraft pilots were for safety reasons equipped with a parachute. It is consistent with this tradition that since 1939 ejector seats were initially only tested on German aircraft. The equipment with ejector seats was such a great success that since 1944 regulations demanded that all quantity-produced fighters had to be equipped with ejector seats.

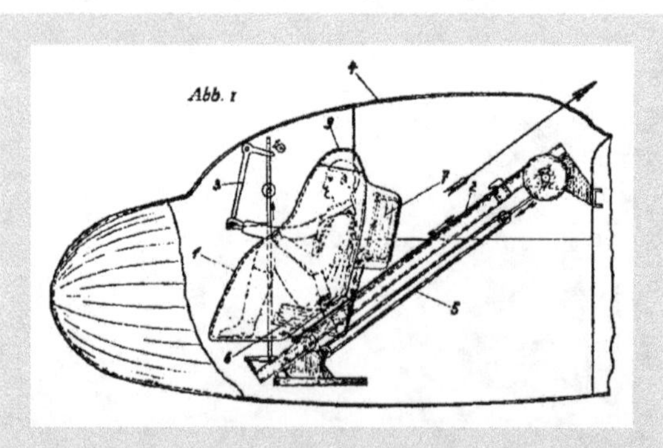

The Junkers Patent No. 722 045 of December 23, 1938 is related to a seat, equipped with a parachute, which can be removed by an energy storing mechanism (power storage), and to a detachable parachute connected to the passenger.

Even though the idea of a bailout support, like the ejector seat, was, so to speak, "in the wind" at all nations concerned with aviation, German scientists better anticipated the problem due to their long experience in the use of rescue parachutes; consequently they started in 1939 with the exploration of the fundamentals and the successful realisation of a bailout support – the seat-catapult, as the ejector seat was called at that time.

Research on the Interactive System: the Ejection-Seat Aircraft

During the years prior to 1939, most emergency exits (bailouts) occurred at velocities where the pilot's muscular strength was sufficient to separate from the aircraft. During the years 1939–1944, an increasing number of more severe and fatal accidents occurred during bailouts [1]; the cause of the injuries was soon recognised, frequently being a collision of the pilot with the empennage. The analyses emphasised the alarming increase in accidents, consisting of approximately 45.5 per cent accidents with severe injuries and of about 12.5 per cent fatal incidences. This trend increased with increasing flight speed, or expressed differently, the probability to survive a bailout without assist decreases with increasing flight speed, with a high sinking speed further increasing the risk of injuries. Movie pictures of a test person, who tried to leave the cockpit (for security reasons, a net was attached to the empennage), demonstrated, in ground tests in the wind tunnel, the relation between the velocity and the probability of survival. Already from approximately 280 km/h (151 kn) on, the probability of survival stayed < 50 per cent; at approx. 320 km/h (173 kn), it dropped to 25 per cent. A probability of 0 per cent was reached at approx. 430 km/h (232 kn).

An analysis of more than 2,500 bailouts [1] as well as experience gained in flight tests with new aircraft led to the idea to develop seat catapults (seat ejection). The proper contact person was *Heinkel* who already had extensive experience with large-scale catapults. The Institute of Flight Medicine in Berlin-Adlershof already dealt with parachute-related deployment shocks and with seatbelts in aircraft. The analysis of accidents resulting from parachute jumps, carried out at the Institute since 1935, led to research on the effect of shock-like accelerations/decelerations on the human body [10].

The Seat Catapult

The danger associated with flight-testing the *Junkers* Ju 88 led in 1938 to the idea to install a seat catapult [11] within the aircraft for an improved safety of the test pilot. Originally, the Ju 88 was designed as a "horizontal" bomber, but since 1938 it was developed as a dive-bomber. The high speeds during a dive, equivalent to a high sink rate, and the associated aileron flutter did no longer allow the control of the aircraft under all conditions. It is therefore not surprising that the test experts were not "really enthusiastic" to conduct flight tests in an emergency case "only" with a parachute for safety [12]. The idea *"one should be able to 'shoot out' the pilot!"* was born; this raised the question whether humans were able to bear the conditions at ejection with about 12 g for a duration of 0.1 sec. Immediately first catapult tests were performed with elastic ropes like they were used for the launch of gliders. The tests with sand bags as replacement for the human body confirmed the concept.

In 1938, employees of *Junkers* were assigned the patent [13,14] for a seat catapult equipped with a parachute. In 1943, a further patent followed, which essentially describes the seat catapult of the He 280. Parallel to the basic technical solution of the seat catapult, which has not changed until today, consequent research to determine the compatibility of humans with short time accelerations was pushed ahead. The results available up to then were related to the effect of curved flights of longer duration, and to crash landings and strong deployment shocks on parachutes (up to about 20 g at a duration of < 1 sec).

For a more detailed investigation of the compatibility during short time accelerations, planning of a catapult test track [15] was initiated in 1939; here, a cabin on support tracks with a variable braking distance was to be catapulted. It was built in 1943 but was not put into operation due to damages received during an air raid. (The USA took over the catapult track and gained sensational results with it.) In 1940, *Heinkel* already owned a catapult-launch tower, which was operated by a compressed-air gun [16] for the stepwise variation of the acceleration at launch. Parallel, investigations were conducted at the Institute of Flight Medicine on a horizontal track – also operated by a variable compressed-air gun – concerning the changing acceleration directions during entrance into an air-stream.

The focus was on determining the dynamic-compatibility limits of humans during shock-like accelerations/decelerations. Up to that time, only results on the effect of long-term accelerations on the blood circulation were available. First comparative tests were performed with inorganic materials. In 1939, pendulum tests (shock tests) started, which confirmed a high compatibility of humans with regard to shocks. Up to that time, only results concerning the static behaviour of organs in comparison to mechanical materials were available.

Since 1940, during the continuation of the investigations, it became obvious that the effects and interrelations of acceleration changes, impulses and dy-

The Heinkel seat catapult, developed in 1940, was produced during World War Two in larger numbers and as standard installed in the fighters Heinkel He 219, Heinkel He 162, and Dornier Do 335.

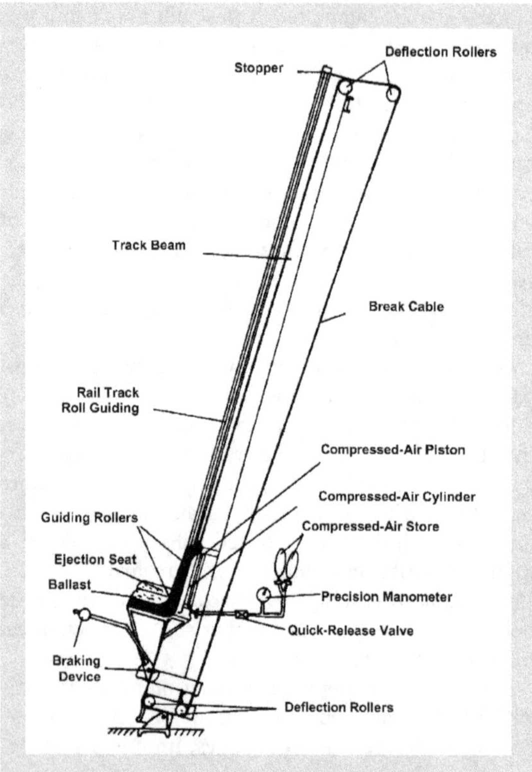

At the Heinkel Company, a compressed air driven catapult system was created in 1940, which was utilized in many tests.

namic effects on the human body represent stresses, which are not directly comparable to the ones on test dummies (much higher shock forces occurred on the dummies). Therefore, comparisons with tests on deployment shocks of parachutes were utilised since it was known that during bailouts from fast flying aircraft fatal injuries already occurred in the air due to the deployment shock of the parachute.

Tests at the drop tower with high shock forces during deployment, stretching and filling of the parachute followed. Good results were obtained at about 21 g acting over 0.1–0.2 sec. These values significantly exceed the maximum load on standard rescue parachute canopies. Tests with humans showed loads with peak values of 1,290 kp (2,846 lbf), the ones determined with dummies about 1,900 kp (4,190 lbf). Further differences in the case of humans could be traced to tightened muscles with 1,330 kp (2,930 lbf) having been registered when the shock was expected and only 830 kp (1,830 lbf) under unprepared conditions (soft muscles result in a good attenuation during the introduction of forces into the body). It could now be concluded that there is no comparability with standard dummies. One could, therefore, assume that humans are much more resistant.

Experimental Developments Related to the Ejector Seat

In 1939, first considerations were made with regard to the planned He 280 to be equipped with a seat catapult [16]. Between June 15 and November 7, 1940, experimental developments regarding the catapult cylinder and the quick-release valve were conducted with a newly developed seat catapult. In addition, seat catapult launches, with a sand bag as substitute for a human, were performed from a mock-up of the He 280. Of these seat catapult launches, slow-motion pictures were taken while still on the vertical track, and various acceleration measurements [17] were preformed. The catapult cylinder unlocks the locking key at 3.5 bar and subsequently the rollers of the seat. The quick-release valve withstood a pressure of 185 bar. The catapult cylinder (piston) was able to lift a mass of 300 kg (674 lbs) > 168 mm (6.6 in) within about 1 sec (the total stroke was 178 mm (7 in)), and, from that point on, a mass of 125 kg (280 lbs) > 1,000 mm (39 in) within 0.9 sec (the total stroke was 1,068 mm (42 in)). The compressed air tank, equipped with a valve, lost during an 8-day endurance test only 1 bar of its 125 bar charge pressure. A second quick-release valve, which was activated 150 times during tests with manned seat catapults [18], showed an impeccable sealing.

A complete seat catapult system was installed in a mock-up of the He 280 [19] in 1940. It was investigated at what pressures the seat, with a sand bag as substitute mass (total ejection mass 120 kg (270 lbs)), leaves the mock-up properly and whether the arrangement of tracks and roller guides on the seat catapult meets the requirements in the presence of wind loads. For that purpose, the catapult compressed-air bottle (volume 2 l), installed inside the aircraft, was filled with compressed air of various pressures from a larger bottle of compressed air of 130 bar (volume 40.2 l). The pre-tests started with a 30° (with respect to the horizontal) inclined ejection path and ended, after a gradual increase of the mass as well as the ejection pressure, at a mass of 120 kg (270 lbs) and 125 bar ejection pressure at a standard inclination of 63° (with respect to the horizontal). During the last test, at a mass of 120 kg (270 lbs) and 125 bar ejection pressure, also wind-force moments, corresponding to a speed of 900 km/h (486 kn), were applied to the seat catapult.

After completing this test series, further tests were performed with test persons. During these catapult tests, the compressed air pressure was increased corresponding to accelerations of 2 g up to the "subjectively tolerable"; here, 19.5 g proved to be the limit. High flight speeds require high energies during the ejection due to the limited acceleration path of only about 1.5 m (4.9 ft), also requiring a post-exit trajectory that carries the seat catapult also at maximum flight speed safely over the top of the empennage. Since 1941, the necessary accelerations for various trajectories were determined for aircraft (known at that time) during flight measurements.

Research on Ergonomics (Human Factors Engineering)

In 1941 employees at Heinkel tinkered without permission with the seat catapult and conducted a self-test [20], without medical supervision, manipulating the release valve and increasing the ejection pressure equivalent to an acceleration of 26 g. This was, without doubt, considerably exceeding the "objective compatibility". X-ray photographs showed fractures in the lumbar vertebra and the dorsal vertebra areas, respectively. This "self-test" was repeated with barite (Baryt), which exhibits nearly the same attenuation as the human body. The spinal column was intensively investigated for an ideal and a wedged position, respectively.

It became clear that the measurement of short-time accelerations was the most difficult measurement task in the first place. For first investigations, deployment-shock recorders were attached to the parachute restraining strap, which are, however, only suitable for tensile loads, to record rapidly changing lengths in the order of tenths of millimeters. During the acceleration, considerable peaks occurred, which was traced to the different elasto-mechanical behaviour of the compressed-air gun and the seat bucket. Finally, measurements were conducted with a cathode-ray oscilloscope, which allowed determining permissible accelerations before wedge-fractures of the spinal column occurred. For the first time, an "objective compatibility" of 23 g was proven as acceleration limit.

The next step was to determine an ejection acceleration [21, 22], which was tolerable by humans but also provided the required launch (bailout) velocity. To prevent peak loads (strain) on the spinal column, the suppression of short-time impulses during the acceleration phase was tested. As a first means, a reduction in the piston area was realised by a "restrictor mandrel". Also the installation of an additional plenum-chamber (wind vessel) was tested, which, however, meant a large loss in performance at a large weight penalty. The most appropriate method finally proved to be the installation of a time-mass-control device, which allowed reaching the required launch velocity at the lowest possible acceleration values or peaks, respectively.

It must be stressed at this point that the compressed-air catapult was mainly used because of the possibility to vary the launch pressure. Simultaneously, investigations were performed with a cartridge catapult, which yielded, however, a large scatter of the results with a low repeatability. The cartridges still required a large amount of development work. The He 162 was already equipped with a cartridge seat-catapult. This resulted in a weight reduction, less maintenance effort, and a higher safety. (Compressed air catapults suffered from a reservoir pressure loss over a longer period of time, and quick-release valves experienced icing when operating at high altitudes.)

Tests were performed at Heinkel and Focke-Wulf on catapult tracks and the handling of the cartridges [23] essentially improved. Unfavourable peaks during accelerations could be excluded and an excellent launch characteristic was achieved. There existed already 266 results of unmanned ground launches. For the *Focke-Wulf* Ta 154, being under development, the cartridge seat catapult was envisaged. The increasing flight speeds (especially of fighters) soon made one realise that a successful bailout by means of a seat catapult required ejection values beyond the limits of compatibility of the spinal column. Accelerations up to 23 g, which were considered tolerable, were here not sufficient.

Already in 1942, the Do 335 (first flights in 1943) was designed for 800 km/h (432 kn) [24] thus coming close to the speed of the jet aircraft of that time. The Do 335 was equipped with one, respectively two ejection seats; the rear propeller could be jettisoned. The required acceleration already exceeded, with a maximum of 26 g, the limit of the "objective compatibility" determined to be 23 g. Since 1942, it was, therefore, investigated whether humans would be able to tolerate peaks of up to 28 g.

Of concern was here mainly the determination of the bone and spinal column strengths [20]. Again comparisons with the parachute deployment shock were utilised. The bone strengths of the spinal column was known but could only be used in a limited way for the identification of the dynamic characteristics. The results firstly proved the already known 23 g as maximum compatibility. This confirmed what tests on living organisms had shown, viz., a subjective compatibility of 19.5 g. In earlier measurements, with inadequate equipment, a value of 17 g was established as limit concerning spinal column fractures. Only the exact measurement with a cathode-ray oscilloscope resulted then in the 26 g as maximum acceleration. Tests were conducted at Heinkel with five test persons and increasing loads. The results showed no disturbances or hardships on the human organism and could, indeed, be compared to findings in conjunction with falls off horses or similar.

Further tests confirmed the quantitative benefits of armrests in reducing the loads on the spinal column [25] enabling an increase from the 23 g (without armrest) to 28 g (with armrest). Simple tests proved that the spinal column is relieved by about 5 g (0.011 lbs). These investigations were related to the contribution of the body weight to the respective vertebra under strain. Here too, 23 g were established.

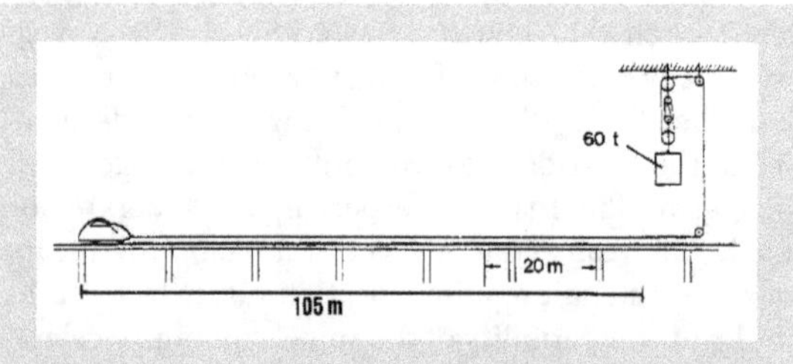

At the DVL Institute of Flight Medicine, a catapult track of 105 m (344 ft) length was already planned in 1939 and built in 1943; a small cabin could be accelerated with 40 m/s^2 (131 ft/s^2) and stopped within 20 m (66 ft). The installation was brought to the USA after the war and operated there in acceleration tests.

A uniform toleration over the entire spinal-column range was shown to exist, which allowed allocating to each lumbar vertebra a force distribution.

The 28 g, with armrests and without injuries, were also confirmed in 27 test launches on the Heinkel catapult track. With such a load on the spinal column, the limit in human compatibility should have been reached [20]. This sufficiently determined the effect of shocks and of an increase in the acceleration on injuries to the spinal column. In case of the expected higher speeds of jet aircraft in conjunction with a high sink rate, a longer engagement of the launch power was recommended. This was to be accomplished by a rocket motor, mounted underneath the seat bucket, in addition to the seat catapult. However, safety considerations put forward prevented the introduction of such a device.

But more important than the extension of the power-engagement time was the influence of the dynamic pressure [26] during the ejection. The effect of the dynamic pressure becomes essential as soon as the protective cover of the cockpit is being left during the ejection. Here, the exiting person experiences a considerable slow-down due to the deceleration at the given speed. The acceleration of the human body is reduced to tolerable levels after approx. 2 sec due to the effect of the dynamic pressure, and it is further reduced so that after 6–8 sec the sinking speed is reached.

To determine the load on the occupant, investigations were conducted after the actual ejection [27, 28]. It turned out that, as a further factor, the speed of the aircraft at the moment of the ejection of the seat catapult has to be considered. In addition, acceleration forces change in strength and direction. Becoming subject to the dynamic pressure causes drag and negative lift. Both are largest at the moment the dynamic pressure takes effect. The ejector seat experiences an acceleration in the direction of the empennage due to the drag, a process dependent on flight speed and on drag. It was shown that a human, exiting from a regular aircraft flying at a speed of 850 km/h (459 kn), has a relative velocity with respect to the empennage, which he passes, of 145 km/h (78 kn). In the (human) breast and back areas, accelerations of 10–15 g (at flight speeds of 500–600 km/h (270–324 kn)) over 2–3 sec were shown to exist. It was also shown that gravity, acting simultaneously, causes for 1–2 sec a downward force, which results in a negative acceleration of about 6–10 g and a flattening of the ejection trajectory.

The investigations related to the stagnation pressure [29] were initially conducted with a wooden head equipped with woolen threads. Results of these tests led to the introduction of a headrest to reduce the effect of the stagnating air in front of the face. Subsequently, tests with persons with unprotected heads followed on a centrifuge in a wind tunnel at speeds of up to 800 km/h (432 kn). Later, flight tests were performed with a Ju 87 at speeds of 430–480 km/h (232–259 kn). The hood of the Ju 87 was removed, the test person got up when reaching the test speed, and a movie camera recorded this process. The eyes of the test person were unprotected for 30–60 sec; the test person had no headrest. Good compatibility (tolerability) was given up to 252 km/h (137 kn), confirming the experiences of parachutists in free fall. The tests were stopped at 385 km/h (208 kn) after 6.3 sec since a considerable effect on the respiration and circulation systems was observed. These tests were repeated with the test person wearing pilot's goggles. No disturbances were observed up to 485 km/h (262 kn). But problems started at 250 km/h (135 kn) in the presence of side wind (yaw).

Since these speeds were not sufficient for future aircraft, tests were performed in 1942 in the DVL high-speed wind tunnel [30] at maximum speeds of approx. 720 km/h (389 kn). Tubes connected to the wooden-head model transferred the respective pressures. The investigations were conducted within the speed range of 100–720 km/h (54–389 kn). It was observed that the head lifted off of the concave head cushion. The highest pressure occurred at the chin with differences in velocity over the face amounting to up to 100 km/h (54 kn). In addition, unpleasant phenomena occurred when yawing, without severe consequences, in the form of oscillations of the head; but also annoying oscillations in the areas of breast and abdomen were determined. It was suspected that possibly natural oscillations of the wind tunnel flow had been transferred and excited the seat and the test person. The results showed that the unprotected body was able to withstand still higher dynamic pressures. But more important were the effects of a very abrupt exposure to dynamic pressures in the speed range of 400–1,050 km/h (216–567 kn). Tests with persons were performed at speeds of up to 850 km/h (459 kn), at the time the largest envisioned flight speed. The tests passed without detrimental effects on the unprotected head, the inner organs and especially on the respiratory system (lung). But it also became clear that, above a speed of 500 km/h (270 kn), the human head should not be exposed to the dynamic pressure suddenly and unprepared – especially not with an open mouth.

One of the first test ejections of a catapult seat from a Ju 87 at the DVL Institute of Flight Medicine (1941).

Parachute-deployment sequence of a modern ejection seat. The sequence starts on the lower left and ends on the lower right shortly before the landing of the pilot suspended from the parachute.

Applied Research Related to the Ejection Seat

These tests were followed by flight tests [31] with the seat catapult of the Heinkel He 280 to check the computed and in laboratory tests determined results, and to check the required compressed-air pressure. All previously measured values were confirmed as expected. However, it turned out that for a dive, the ejection requirements determined in level flight were insufficient. In curved flight (banked turn), centrifugal forces resulted in an increase in the ejection weight from 120–360 kg (270–809 lbs). At a standard setting of the compressed-air pressure, the seat collided with the empennage. Therefore, a higher energy or a prolonged engagement, respectively, of the ejection power by means of an additional (to the seat catapult) rocket motor was recommended. Movies taken from escort aircraft and from the ground also showed a turnover of the seat during the ejection due to the sudden exposure to the dynamic pressure. Here, the seat together with the occupant experiences drag and negative lift, which results in a backwards pointing (clockwise) moment (pitch-up) about the lateral axis of the seat.

To prevent this overturn, a small stabilising parachute was attached to the seat, which deployed at the instant the seat became subjected to the dynam-

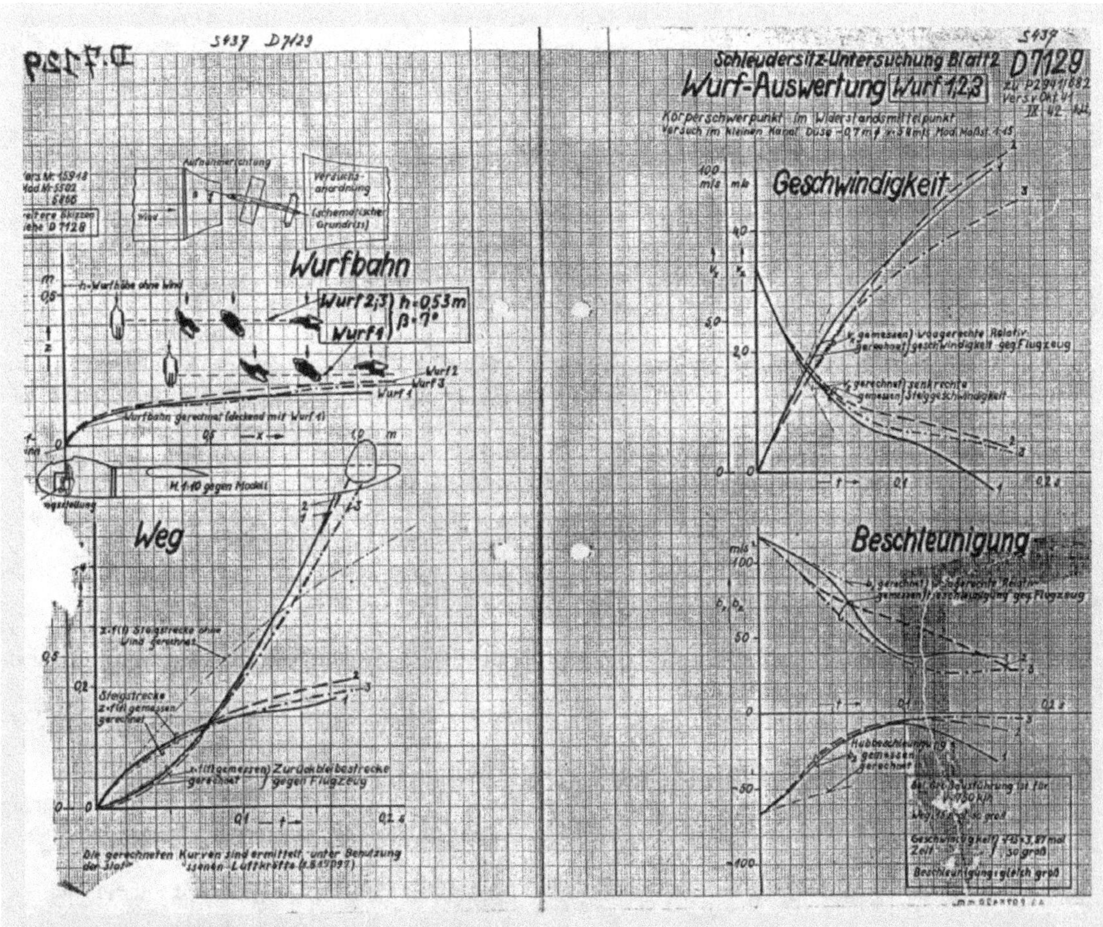

Analysis of wind tunnel measurements conducted in the small wind tunnel of the DVL at Adlershof during October 1941 with ejection-seat models at a speed of 54 km/h (29 kn); the document was saved over the war.

ic pressure. Initially, tests with a small stabilising parachute still showed the overturn; this resulted in the introduction of a larger stabilising parachute (ribbon parachute). This parachute exhibited a larger stability than standard parachutes and met, also at higher speeds, all expectations, especially with regard to the control surface, having at the same time lower deployment shocks. This stabilising parachute effectively prevented overturns of the seat including occupants. Spin could not be observed in any of these cases.

For the Ju 290, an extension to the seat above the head was attached, which enabled an ejection through the roof of the cockpit [25]. The controls on the armrests were extended by a two-part triggering device on one of the armrests with the first one releasing the lock of the cockpit canopy and the second one triggering the then following ejection of the catapult seat. The development of rescue parachutes, which deployed automatically after a set time followed later; here, the occupant was separated from the seat after having been catapulted.

Successful Rescue Ejections – Successful Ejection-Seat Research in Germany

The first real test ejection of a seat catapult from an aircraft was performed in 1941. The first manned seat-catapult ejection was conducted in 1942 by the test parachutist *Wilhelm Buss* (who, according to his own account, performed 248 test jumps with parachutes). The aircraft pilot *Rudolf Schenk* carried out the first rescue ejection with a seat catapult on 13 January, 1943 from a He 280 [32]. In 1944, *Wilhelm Buss* performed three more successful test-ejections with a seat catapult.

References

1 M. Ruck, Erfahrungen aus etwa 2500 Meldungen über Fallschirmabsprünge 1939 bis 1944. Chef der technischen Luftrüstung FI/ES Nr. 27367/44 geh
2 Flugunfallbericht, mit einer Albatross L76, 5.12.1933
3 T.W. Knacke, Parachute Recovery Systems, Design Manual, 1992 Para Publishing
4 Flugunfallbericht mit einer He 111, 4.5.1940
5 Zeitdehnerfilmaufnahmen, Fallschirmerprobung Kugel-Großkreis-Bänderschirm, Abwurf-Nr. P4327
6 AGARD-AG-319, Design and Testing of High Performance Parachutes, 1991
7 Flugunfallbericht, mit einer Ju 52, 5.7.1942
8 Flugunfallbericht, mit einer Do 215, 14.5.1943
9 Flugunfallbericht, mit einer Me 323, 30.9.1944
10 S. Ruff, Über die Beschleunigungsfestigkeit des Menschen gegenüber einigen in der Fliegerei vorkommenden stoßförmigen Beschleunigungen. Schriften der Deutschen Akademie der Luftfahrtforschung, 1941, Issue 47
11 Heinkel: Mitteilung Nr. TD 353/39 vom 20.6.1939. Betr. He 180
12 Heinkel: Mitteilung Nr. 436/39 vom 25.6.1939. Betr. He 180
13 Patentschrift Nr. 711 045 vom 25.9.1941: Aus dem Innenraum eines Luftfahrzeuges durch einen Kraftspeicher entfernbarer, mit einem Fallschirm ausgerüsteter Sitz
14 Patentschrift Nr. 918 006: Aus dem Innenraum eines Flugzeuges durch einen Kraftspeicher herausschleuderbarer Sitz
15 Schriftwechsel vom 27.6.1939 zwischen DVL und Fa. Dingler, Zweibrücken über die Lieferung einer Katapultbahn
16 Aktenvermerk KOM/79/40 vom 14.5.1949. Betr. Funktionsbeschluß der Konstruktionsattrappe He 280
17 Sitzkatapult He 280, Versuchsbericht Nr. 3009, 21.10.1940
18 Versuchsbericht 3009 Nachtrag 1, 29.10.1940
19 Versuchsbericht Nachtrag 2, 8.11.1940
20 S. Ruff, Flugmedizinische Grundlagen zum Bau von Katapultsitzen. Schriften der Deutschen Akademie der Luftfahrtforschung, 1944
21 R.K. Pawlas, Erprobung des Sitzkatapultes He 280, vom 21.10.1940 Publiz. Arch. Abt. Luftfahrt H.7
22 S. Ruff, H. Strughold, Grundriß der Luftfahrtmedizin, 3. Auflage. München: Johann Ambrosius Barth, 1957
23 Reinelt, NN.: Katapultsitzanlage für das Muster Focke-Wulf Ta 154. Bericht Nr. 20.009, 1944
24 Dornier: Bericht Nr. 3240, 1943 und 1944. Betr. Katapultsitzanlage Do 335
25 H. Wiesehöfer, Über den Bau von Schleudersitzen. Deutsche Luftfahrt-Forschung, Untersuchungen und Mitteilungen, DVL-Bericht Nr. 1175, 1944
26 W. Knackstedt, Untersuchungen am menschlichen Körper bei hohen Staudrücken, LFA-Bericht Br.B.No G 14/43 vom 20.1.1943
27 A. Geertz, Grenzen und Sonderprobleme bei Anwendung von Sitzkatapulten. Dissertation, TH Stuttgart, November 1944
28 Funktionsprüfung und Flugerprobung der Preßluft-Katapultanlage der He 219, vom 3.8.1944.
29 Schütze, NN.: Untersuchungen über den Einfluß hoher Windgeschwindigkeiten auf den Kopf, DVL-Bericht, Rf 305/2 vom 10.11.1940
30 E. Thaler, Untersuchungen über den Einfluß plötzlich auftreffender hoher Windgeschwindigkeiten auf den ungeschützten Kopf, DVL-Bericht, Rf 305/11 vom 10.9.1943
31 J. Eisermann, Flugversuche mit dem Schleudersitz in Rechlin. VFW-Fokker-Journal 11, 1978
32 Bericht über den 1. Rettungsabschuß, Flugkapitän Schenk, 13.1.1943

Transfer of German Aeronautical Knowledge After 1945

Kyrill von Gersdorff

Within the subsequent chapter, many statements of contemporary witnesses are being used. Thus, it represents a summary of the situation during the years after World War Two, when German aeronautical research was suspended, and many scientists and engineers went abroad. Many of them returned after 1955 to the German Federal Republic or the German Democratic Republic, respectively, and were able to put, in continuation of their activities, their accumulated knowledge and experience to use. Thus, a renewed transfer of knowledge occurred.

German aircraft industry and aeronautical research still had during the final years of World War Two – despite the effects of war – a high technical and scientific potential. The interest of the Allies, i.e., primarily France, Great Britain, the Soviet Union and the United States of America, but also of other aeronautically-active nations, concerning this potential was therefore considerable. The USA was, after the surrender of the German Armed Forces, i.e., after May 8, 1945, still at war with Japan and, therefore, initially very much interested in weapon systems which could be employed on a short-term basis. France looked for possibilities to rebuild her aeronautical industry and research, which had during the German occupation in essence – but not on the highest technical level – been incorporated into the German armament efforts, and to overcome the technical "backwardness" as quickly as possible. Great Britain was mainly interested in special research areas where she thought to be behind or lacking knowledge. The Soviet Union, on the other hand, was interested to learn from German armament projects and to transfer as many facilities as possible to Russia.

The allied troops and the trailing information gathering organizations (CIOS, FIAT) found several large centers of the aircraft industry and of aeronautical research (Berlin-Adlershof, Brunswick, Göttingen) with facilities, scientists, engineers, operational staff and a large amount of knowledge, in form of scientific and technical documents, and hardware. The subsequent realization of this knowledge, beginning in 1945, constitutes a continuation of German aeronautical developments, based on German research results, for the benefit of foreign nations and with applications abroad. This continuation, which is sketched in the following, proceeded in very different ways.

Great Britain

After the end of World War Two only a relatively small number of German scientists worked for a longer time period – that is for more than one or two years – in England. There was a thorough interrogation of many German specialists who were brought over to England for a few weeks. The publications and the extensive number of documents found in Germany were carefully registered, assessed, and brought to the knowledge of interested parties. Numerous captured German aircraft were, partly already during the war but mainly thereafter, flown at the Royal Aircraft Establishment (RAE) in Farnborough, and the results evaluated [1].

Now, what of the collected findings was realized, exploited, applied or directly used? Most of the in Britain remaining German scientists were advisors, only a few advanced to a leading position [2]. As far as can be established today, the realization occurred mainly in the following areas:

Group photograph of the German scientists, who were accommodated in 1947 at the "Farnborough Court"; seated from left to right in the lower row: *Gerald Klein* (2.), *Adolf Busemann* (3.), *Wilhelm Tolmien* (4.), *Ernst Schmidt* (6.), *Herman Schlichting* (8.). in the center row: *Karl Doetsch* (2.), *Hermann Neubert* (3.), in the upper row: *Herman Jordan* (5.), *Gerhard Sissingh* (6.) *Dietrich Küchemann* (7.).

- Flight control, e. g., by Siemens: Range-rate-principle for autopilots and yaw-damping (RAE Autostabilizer).
- Spoiler as an auxiliary rudder (tested on "Meteor" and F-86).
- Electric aircraft control (fly-by-wire), on Avro 707 for the first time.
- Free-flight measurement techniques in the sonic and subsonic speed ranges.
- Boundary layer control in various speed ranges.

Especially the following German scientists were involved in these activities: *Karl Doetsch, Dietrich Küchemann, Hans Multhopp* and *Heinrich Voepel.* Their work left its marks, for instance in the fly-by-wire control of the supersonic transport "Concorde" (first flight in 1969). The "tilt-nose" ("Kippnase"), replacing the slat, and invented by *Ludwig Bölkow* at Messerschmitt [3] in 1939, was first applied on the

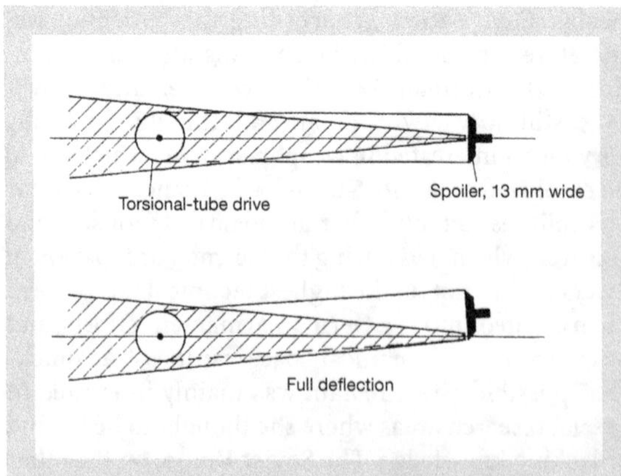

In order to dampen yaw oscillations, small spoilers were, as proposed by *Karl Doetsch*, successfully attached to the rudder of the British twin-jet fighter Gloster "Meteor". The empennage was later modified.

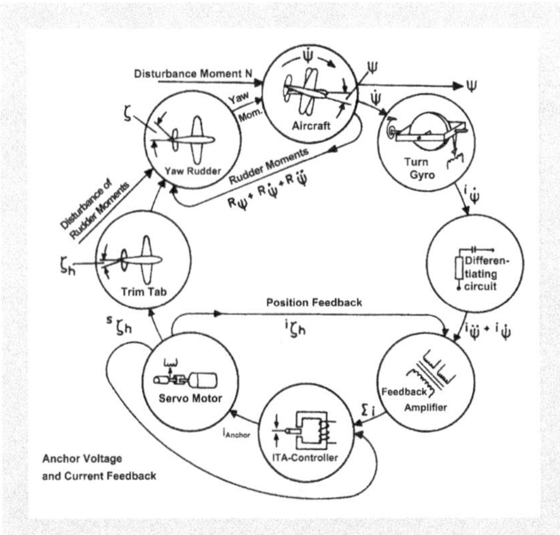

The RAE yaw-stabilizer, developed in Farnborough under the supervision of *Karl Doetsch*, was, after extensive tests, installed in all English fighter and bomber aircraft of that time.

Dietrich Küchemann had considerable influence on the design and form of the delta wing – the "slender wing" – of the English-French supersonic transport "Concorde".

English bomber Handley-Page "Victor" (first flight in 1952). *Dietrich Küchemann* participated decisively in the aerodynamic design of the delta wing of the "Concorde". The slender wing for economic supersonic flight originated with *Dietrich Küchemann* and was developed by him up to design maturity. He also

The delta-wing test-aircraft "Avro 707" was for the first time, for trial reasons, equipped with an electric flight control-system, called "Fly-by-Wire". Later, the RAE autostabilizer was used in the English VTOL test-vehicles and the supersonic transport "Concorde".

Several members of the Austrian team of "*Friedrich von Doblhoff*", among them *August Stepan*, went in 1946 to Fairey and contributed to the development of the compound helicopter "Rotodyne".

The "tilt-nose" ("Kippnase"), invented and patented by *Ludwig Bölkow* at Messerschmitt, was used on the outer wing leading edge of the English bomber Handley-Page "Victor", later also on the supersonic fighter Lockheed F-104 "Starfighter".

introduced the utilization of the leading-edge vortex of such a wing to considerably increase lift during low-speed flight and landing [4]. Not realized was the 1947 design of a supersonic fighter by *Hans Multhopp* and *Martin Weber*. It constituted a type of blend between the Focke-Wulf fighter "Ta 183" and the transport aircraft "DSF 346" (with the pilot in prone position) and was envisaged to reach a Mach number of 1.24 at an altitude of 11,000 m (36,090 ft). The "Lightning" fighter [5] was essentially based on this design. *August Stepan* of the Austrian helicopter team "*Friedrich von Doblhoff*" went in 1946 to Fairey and contributed decisively to the development of the compound helicopter "Rotodyne" which went into flight-testing in 1957 [6].

The first test-rig run of the SNECMA "ATAR 101" turbojet engine, realized by the Group O under the supervision of *Hermann Oestrich*, occurred in March 1948 in Villaroche.

France

German engineers and scientists worked in the years after the war closely and successfully together with their French colleagues. With time, the French engineers, who partly finished their studies only after the war, increasingly took over responsibility. An exemplary implementation of the German industrial and scientific knowledge occurred during the realization of the military turbojet engine "ATAR 101" of SNECMA. *Hermann Oestrich*, up to the end of the war head of the BMW turbojet-engine development, had succeeded in gathering, within a few months, in the French occupational zone in Lindau-Rickenbach a development team for turbo engines. The first batch, Group O, of that team transferred in July 1946 to France. The turbojet engine ordered by the French Ministry of Aviation was supposed to have an initial thrust of 16.6 kN (3,732 lbs), be as light as possible and small in diameter. The basis for the design, established in Rickenbach, was a larger version of the BMW 003 engine [7].

The first ATAR test engine already operated on the test rig in Villaroche (Seine&Marne) in March 1948. With the engine V 2, a thrust of 21.6 kN (4,855 lbs) was reached in May of the same year [8]. After enlarging the turbine diameter and switching to an eight-stage compressor with a larger mass flow and a higher pressure ratio, as well as turbine blades of Nimonic-75 alloy, the engines SNECMA "ATAR 101 E" and "ATAR 101 G" with afterburner, built in quantity, followed. These engines were installed in many French military aircraft – also those marked for export. An "ATAR 101 E-3" was utilized in 1958 for the vertical-take-off test aircraft [9] "Coleopter" (with a circular wing). The development of the "ATAR 101" was continued over many years up to Version 9, which delivered with afterburner a thrust of 70.6 kN (15,870 lbs)

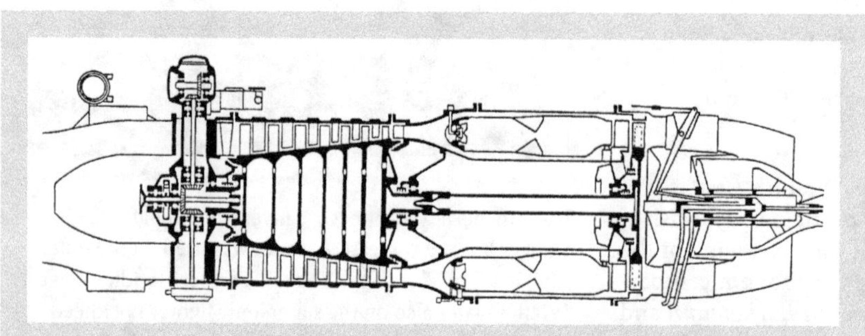

The ATAR-engine, manufactured in France in large numbers, was an advanced development of the BMW 003 engine; it was based on a design drawn up in 1946 in Rickenbach (near Lindau).

in the end. The "ATAR" engine belongs with a production runtime until 1994 and 5,000 copies delivered, to the most successful turbojet engines in the world.

Especially extensive research and test effort was required by the gradually implemented thrust increase, the reduction in specific fuel consumption and the adjustment to high flight speeds, here up to a Mach number of 2.3.

The Oestrich-group also participated in the development of the engine for the supersonic transport "Concorde". Here, part of the development and manufacturing program of SNECMA were the afterburner unit, the adjustable-thrust nozzle and the thrust reversal unit of the "Olympus 593" engine.

A close cooperation existed between the Bureau Technique Zborowski (BTZ) and SNECMA in case of the experimental vertical-take-off aircraft "Coleopter" and the preliminary test vehicle "ATAR-Volant". Here, largely new ground was broken with research in the area of vertical-take-off technology. The exchange of experience – which may also be called knowledge transfer – back from the SNECMA and BTZ employees was particularly extensive. Almost a dozen of these employees later took over chairs and professorships at German institutes of technology and universities. Many engineers of these groups transferred between 1955 and 1964 to the slowly recovering German aircraft companies, others went again to a foreign country where they developed engines for supersonic aircraft.

In cooperation with the Bureau Technique Zborowski resulted at SNECMA – via the preliminary "ATAR-Volant" – the vertical-take-off experimental aircraft "Coleopter".

An engineering group, established in Bregenz by *Fritz Nallinger*, the former head of the Daimler-Benz engine development, worked initially on the project of a long-range transport aircraft with six clustered turbojet engines of the 7,000 kp (16,000 lbs) category, whose design, UR 1, had been brought along by Focke-Wulf engineers headed by *Ludwig Mittelhuber* [10]. Soon after the transfer of the Bregenz-group to the company Turboméca in the south-west of France, *Fritz Nallinger* went back to Germany and *Erich Übelacker* took over as technical head of the group. The work on the transport aircraft and on the large jet engine was suspended after two-and-a-half years. As head of the small-gas-turbine group followed, after the retirement of *Erich Übelacker*, *Georg Oberländer* (until 1951) and *Willy Syring* (until 1971). The compressor group under the supervision of *Bruno Eckert* was, among others, responsible for the compressor design of the small bypass engine "Aspin". After the break-off of the pre-development of the large jet engine mentioned above, a reduced engineering team under the supervision of *Erich Übelacker* concentrated on small gas turbines for various applications, especially for trainer aircraft and helicopters. It started with the project 60 CV, a turboshaft engine with radial compressor and axial turbine, for which a rotating injection system and a special form of an annular combustion chamber, with the combustion air being supplied through the hollow turbine guide vanes, was developed. The first quantity-produced Turboméca turboshaft engine was the "Oredon" with 118 kW (160 hp), which was, at the same time, the start of the "Artouste"-series. A smaller jet engine of 1.08 kN (343 lbs), the "Pimené", followed and was further development into the "Palas" and "Marboré". The latter had a thrust of 3.9 kN (877 lbs) and was man-

A larger German engineering group, which was initially headed by *Fritz Nallinger*, developed at the French engine company Turboméca a number of small gas turbines, here the "Marboré".

ufactured in large numbers, among others, for the jet trainer Fouga "Magister", as was the turboshaft engine "Artouste" for the helicopter "Alouette". Also today's Turboméca gas turbines for helicopters contain design elements from the immediate postwar period when former Daimler-Benz engineers were working in France [7].

The Turboméca small gas turbines of the fifties were the basis for a very successful worldwide utilization and also for the repeated cooperation in developmental tasks with European engine manufacturers such as Rolls-Royce, SNECMA, KHD and MTU.

During the German occupation, France did not have the opportunity to develop modern engines. This also held for rocket propulsion, where one had to start from zero. The SEPR (Société d'Etude de la Propulsion par Réaction), already founded in 1944 in Paris, obtained after the war several Walter and BMW rocket motors and individual parts, especially pumps, and used them for bench tests. A systematic and purposeful research and development program was installed at SEPR when *Hans Schneider* (formerly BMW) joined the company in 1946, first as adviser, then as test supervisor and, since 1955, as chief engineer. At SEPR, a number of high-performance liquid-fuel rocket motors, with nitric acid as oxygen carrier and hypergolic-ignition fuels, were developed – a continuation of the BMW rocket-engine development.

In case of the non-autonomous engines of the SEPR – as for the BMW 003 R – were the sup-

The supersonic interceptor SO 9050 "Trident" attained in 1958 with a SEPR rocket motor almost twice the speed of sound (M = 1.95).

Under the technical supervision of *Hans Schneider*, the newly founded company SEPR started the development of liquid-fuel rocket motors – on the basis of BMW-technology – and applied it several times. Photograph: Dassault "Mirage III C".

ply pumps of the turbojet engines driven by interconnecting shafts. The flight tests commenced in 1952 with the experimental aircraft SO 6025. A larger number of the French fighters "Mirage III A and III C" were equipped with booster engines SEPR 841 or 844 together with the SNECMA "ATAR 9" turbojet engine. For the supersonic interceptor SO 9050 "Trident", SEPR rocket motors were developed, whose pumps were driven by gas generators [7].

At the Arsenal de l'Aéronautique, the large aeronautical research center was revived after 1945 and equipped with in-house test facilities, workshops and possibilities to manufacture prototypes. A number of German scientists were involved in activities in various areas. Among them were *Walter Georgii, Otto*

Eugen Sänger contributed as adviser to the design and layout of the French test aircraft Nord 1500 "Griffon" with a combined turbo/ramjet-engine.

Hohlfelder, Eugen Sänger and *Irene Bredt.* Their status was of an advisory nature with orders to establish research reports and work out suggestions for new applications [11]. *Eugen Sänger,* for instance, contributed, among others, to a number of Arsenal-projects involving rocket motors and ramjet and turbojet engines applied for instance for the test aircraft Nord 1500 "Griffon" [12]. Advisory activity was also related to the ramjet test aircraft "Leduc 021" und the supersonic test aircraft "Trident".

From a rebuild of the missile Fieseler Fi 103 (V1), the homing-missiles Nord CT 10 and CT 20 with pulsejet, respectively turbojet engines were realized under participation of German engineeres [13]. The advice of *Otto Hohlfelder* was especially related to the Arsenal flight engines (24 H), which were developed and built on the basis of the JUMO 213 engines.

Heinrich Hertel and a small group of German aircraft engineers worked during the years 1946–1950 at the SNCASE in Marignane (near Marseille) on transport-aircraft designs and air-traffic systems. The project studies concerned a large "flying wing" transport aircraft with a wing area of 1,000 m^2 (10,700 ft^2) and 8 or 16 jet engines; for this aircraft also wind tunnel measurements were performed.

Of the many research projects, which were treated at the ONERA (Office National d'Etudes et de Recherches Aéronautiques) under the participation of German scientists since 1947, especially the activities related to static vibration tests must be mentioned, which were conducted by *Gerhard de Vries* and his co-workers. In order to avoid larger changes on the completed aircraft, it was necessary to perform reliable calculations. *Wilhelm Flügge* and *Kurt Maguerre* dealt with computational methods for swept wings, which were utilized on some of the French military aircraft developed at the time. Also the computational methods of the physicist *Johannes Dörr,* related to the flutter characteristics of aircraft, were immediately applied [14]. Of the German scientists working at ONERA, many went to German universities and research institutions; some went to the USA, to Canada or to Switzerland.

At today's "German-French Research Institute Saint-Louis" ("Deutsch-Französisches Forschungsinstitut Saint-Louis, ISL") – at the end of the war a branch of the "Laboratoire Central de l'Armement" in Saint-Louis (Alsace) – initially 35, later 55 German scientists under the supervision of *Hubert Schardin* were able to continue in August of 1945 their scientific work which they had started at the Air Force Academy in Gatow. The Institute was later renamed into "Laboratoire de Recherches Balistiques et Aérodynamiques, LRBA". A German-French treaty concerning a joint institute was signed in 1958 and became effective in June of 1959. Since then, all leading positions are proportionally filled.

Of special interest to the French scientists was the "Cranz-Schardin-Camera", which had been developed in Gatow and was moved to Biberach in early 1945, together with the research group. With this "Spark Flash Camera" it was possible by means of cinematographic pictures to visualize rapidly passing events.

Eugen Sänger also advised in case of the experimental ramjet aircraft "Leduc 021", which was brought up piggyback to its launch altitude by a four-engine carrier aircraft.

Tasks and objectives of the institute were, and still are today, fundamental research, scientific investigations and basic predevelopments related to defense engineering especially in the following areas: aerodynamics, ballistics, blast physics, chemistry of explosives, laser applications and instrumentation. An important project of the ISL was the anti-tank missile "ENTAC" [15], which was supervised by *Hans Bender*. The German-French Research Institute Saint-Louis developed over the years into an important European research institution [16].

In cooperation of French engineers and German engineers of the "Henrich-Focke Group", the light helicopter SE 3101, with a "Mathis" engine of only 73 kW (100 hp), and thereafter the helicopter SE 3120 with a more powerful "Salmson" engine were developed at the SNCASE (Société National de Construction Aéronautique de Sud-Est). Both test helicopters had twin tail rotors, as they were later also used by *Focke* on other rotorcraft. The SE 3120 was later designated "Alouette I". With the switch-over to the turboshaft engine "Artouste" of Turboméca and the use of a normal tail rotor, the first and most successful turbine-powered helicopter of the time, the SE 3130 "Alouette II", was developed. The "Alouette II" was built in quantity production for almost 20 years nearly unaltered.

Also at the SNCASO (Société National de Construction Aéronautique de Sud-Ouest), which was joined in 1946 by *Theodor Laufer*, helicopter development commenced soon after the war. During the war, *Laufer* was a co-worker of *Friedrich von Doblhoff*, who had developed at the "Wiener-Neustädter-Aircraft Company" a reaction-propulsion helicopter. Under the supervision of *Theodor Laufer*, the experimental helicopters "Ariel" I, II and III, with blade-tip propulsion, were developed at SNCASO. When the light gas turbine "Palouste" of Turboméca became available as air supplier, the light helicopter SO 1221 "Djinn" with cold-jet propulsion was developed incorporating the "Palouste". This small helicopter was built in quantity production and was operated in more than 20 countries [6].

Under participation of a German group of engineers, headed by *Henrich Focke*, a helicopter development program was started at the SNCASE with the test vehicle SE 3101, which led to the successful French helicopter "Alouette II".

In continuation of the "Doblhoff-concept" of blade-tip propulsion, the light helicopter SO 1221 "Djinn" was developed at SNCASO under the technical supervision of *Theodor Laufer*.

Of the about 500 German and Austrian scientists, engineers, foremen, female calculators and secretaries working after the war in France, most went gradually back to the German Federal Republic and were able to essentially continue their professional activity utilizing their acquired knowledge and experience. They worked in industry, at universities and research institutions, became self-employed or public servants in ministries or at the patent office, or returned again to a foreign country. Some stayed, as advisors, in touch with their former French company or institution over many years.

USSR (Soviet Union)

During the cover-of-night action on October 22, 1946, 250 specialists of BMW in Staßfurt and 350 of Junkers in Dessau were brought into the then Soviet Union. The BMW-group had, at first, the task to support the remanufacturing of the BMW 003, which had, in the meantime, in Stassfurt been brought up to a static thrust of 9.8 kN (2,200 lbs). There were appreciable difficulties regarding the turbine blades due to a deviating material quality. Only a small number was built. The Junkers-collective under the supervision of *Ferdinand Brandner* was charged with the development of a 3,000 kp (6,700 lb) turbojet engine. Here, the concept of the Jumo 012 was essentially adopted. The combustion chamber was newly developed as a combination of an annular combustion chamber (BMW) and a single chamber (Jumo). Shortly before the end of the so-called state-examination, turbine-blade breaks occurred. The development project 012 was, as a consequence, terminated in 1948. The new task for the combined BMW/Junkers specialist team concerned the 6,000 hp turboprop engine 022. The type supervision was in the hands of *Alfred Scheibe*; *Josef Vogts* supervised the predevelopment, *Ferdinand Brandner* the construction and *Karl Prestel* the test-stand trials. New test rigs had to be erected, and their rapid construction was greatly supported by the chief designer *Nikolai Kusnezow*. Since no suitable starter was available, a new turbine starter had to be developed in the shortest of times. After the successful state-examination in October of 1950, with 100 hours of operation, the 6,000 hp engine – now designated TW-2 – was transferred to another factory to be readied for quantity production and for the quantity production itself – this without participation of the German engineers. The engine, now designated NK-4, was thereafter, with a simple four-blade propeller instead of the counter-rotating double-propeller and a take-off performance of 5,000 hp, produced in large numbers and installed in Russian transport aircraft [17].

After an interim development consisting of the coupling of two 6,000 hp engines (TW-2F) to a common propeller gear, the BMW-Jumo-collective was charged with the development of a large 12,000 hp (8,820 kW) engine. The result was the engine NK-12, respectively TW-12, which was tested early in 1953 and soon reached its demanded performance [18]. Flight testing commenced in the summer of 1954 on a four-engine "Tupolew 95" bomber. Soon after, the quantity production started at a factory off-bounds to the German engineers. A noteworthy world record was set in 1957 by a four-engine Tu 114, namely, a flight with a payload of 25 tons over a distance of 1,000 km (612 mi) at a very high average speed of 871 km/h (541 mph). The performance of the NK-12 was increased to an equivalent shaft performance of 15,000 hp (11,000 kW) and is today still installed in the long-range reconnaissance aircraft Tupolew Tu 95 "Bear" [19].

The BMW-Junkers-collective was in 1953 combined with the aircraft specialists still remaining in Russia and charged with the predevelopment of a civil turbojet engine for a four-engine transport aircraft. These evolved in the GDR as the engine Pirna 014 and the transport aircraft 152, as is described in the section "The Aircraft Development in East Germany" in this book.

Of the Junkers' engineers and scientists working in the Soviet occupational zone at the OKB 1 (Opytno-

The end product of a remarkable engine-development effort of a larger German engineering team, headed by *Ferdinand Brandner*, in the Soviet Union was the 12,000 hp turboprop engine NK-12, which provided in its final version an equivalent shaft performance of 15,000 hp. The engine is still in operation in long-range reconnaissance and transport aircraft in Russia.

The transport aircraft Tupolew Tu 114 was powered by four NK-12 engines with counter-rotating propellers.

je Konstruktorskoje Bjuro) (Dessau), two groups had already prior to the so-called "transfer day" (October 22, 1946) been brought to Russia. A first small group went in conjunction with the transfer of a flying test bed, Ju 88, equipped for flights with an attached pulsejet engine for the transport aircraft Junkers EF 126. A second, somewhat larger group followed a railroad transport of an EF 131 prototype in September 1946. The transport was for deception purposes first directed towards Rechlin, but was actually bound for Moscow. The Junkers engineers and technicians – their number had grown to >500 – were accommodated in Podberesje. The OKB 1 was headed by *Brunolf Baade*; the accompanying *Hans Wocke* and *Fritz Freytag* were responsible for the Ju 287. In the Junkers wind tunnels in Dessau, extensive investigations were carried out over a longer period of time, under the supervision of *Hans Wocke*, on models with different engine arrangements and swept-wing variations [20]. Contrary to the first prototype of the Ju 287 with four Jumo 004 jet engines – two on the wings left and right and two at the front fuselage – had the advanced EF 131 six Jumo 004 engines mounted in two triple-pods under the wings. Characteristical for the Ju 287 and the EF 131, respectively, was the forward-swept wing (23°) designed for a speed of over 800 km/h (497 mph). The flight tests in Russia seemed to have met the expectation, but the OKB 1 soon received new tasks [21]. Considered was first the project EF 132, a bomber with six Jumo 012 engines mounted inside the swept-back wings, still designed in Dessau in 1946. The bomber 140, with two Mikulin engines and swept-forward wings, whose prototype was realized by modifying an EF 131 frame, followed. First flight tests were performed in 1948 still with a German crew, later flights only with Russian pilots. Under Russian direction followed in 1949 the development of the bomber 150 with swept-back wings and two Russian jet engines. Flight-testing commenced in 1952 and ended in 1953 with a crash landing. A quantity production was never realized, but the 150 was the starting point for the transport aircraft 152, which was later in the GDR derived from the 150 [22]. Russian interest existed in the Junkers experimental aircraft EF 126 with the Argus-Schmidt pulsejet engine As 044 and Walter take-off assisting rockets. After some test flights with the five prototypes, the project was dropped [23].

The OKB 2 in Halle worked at the Siebel Company, under the supervision of *Hans Roessing*, on supersonic aircraft with rocket propulsion. Following the "displacement" to Russia in October of 1946, one concentrated on the supersonic rocket aircraft DFS 346, with the pilot in a prone position, a Walter 509 A2 rocket engine and 45-degree swept wings, which was designed for speeds corresponding to Mach numbers between 1.5 and 2.0. A full-scale dummy was investigated in the wind tunnel of ZAGI. The launch, respectively the drop of the attached aircraft, was realized from an American B-29 bomber, which had, after an emergency landing, not been returned to the USA. Later, a Tu-4 was utilized. Instead of a regular landing gear, a landing skid was attached to the aircraft. After a number of successful test flights and a hard landing with the 346-1, the pilot *Wolfgang Ziese* had to engage the rescue system, i.e., eject the cabin and land by parachute. During the test flights, speeds of up to 1,100 km/h (684 mph) were reached accompanied by flow separation and buffeting. With the test aircraft 346, Russian scientists gained knowledge and experience related to high-speed flight and rocket propulsion. A further reaching cooperation or an exchange of experience, respectively, did not materialize since the Germans were not admitted to Russian research institutions.

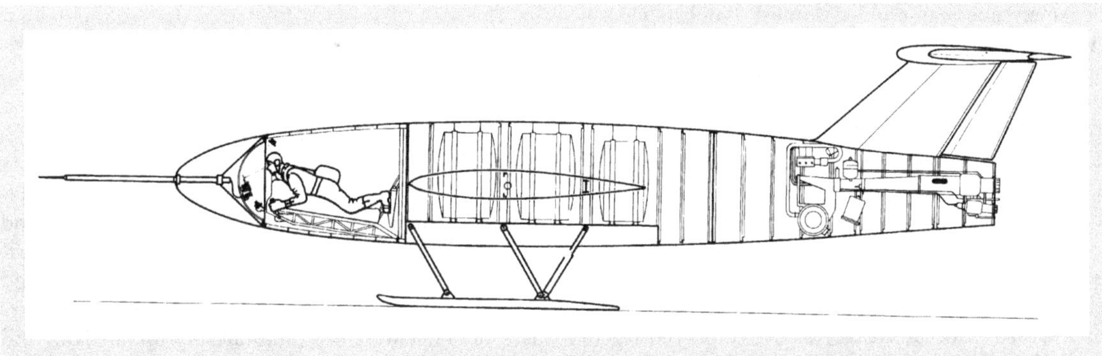

The supersonic rocket aircraft DFS 346, built at the Siebel-Aircraft Company in Halle in a test version, was, after World War Two, extensively tested in the Soviet Union.

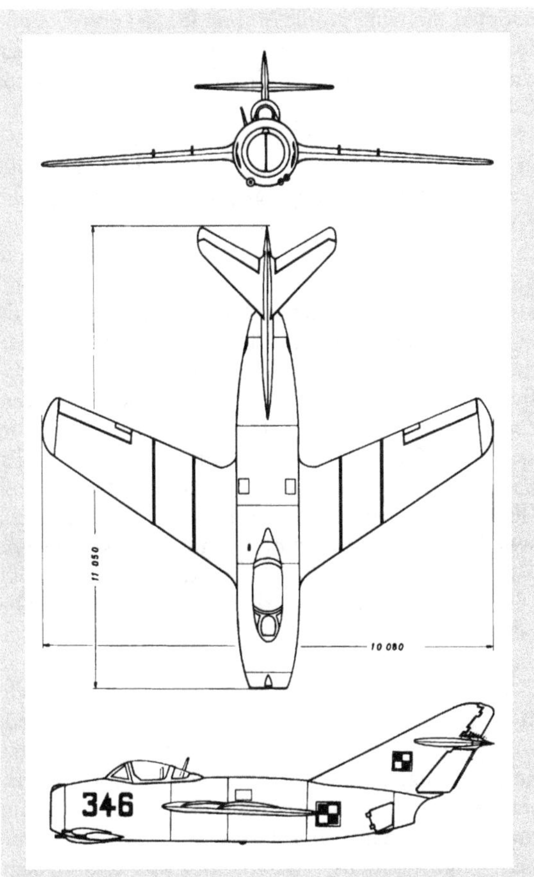

Siegfried Günter advised during the development of the first Soviet swept-wing fighter aircraft Mikojan MiG 15.

Towards the end of World War Two, the advancing Soviet troupes had captured several German aircraft of interest to aircraft designers, which were hurriedly brought to Moscow. Among them were a Me 262 A-1, a Me 163 B, and later a He 162 A-2 captured in Austria. These aircraft were flown and investigated in every detail. Highly interesting documents were found by the Russian specialists in Berlin-Adlershof (DVL), in Halle (Siebel) and in Dessau (Junkers). The strong interest of the Soviets in *Siegfried Günter*, the long-time chief project engineer of Heinkel (from the He 64 to the He 162), can also be explained in that regard. He remained in Podberesje until 1954 and participated there in various endeavors and aircraft projects or in an advisory capacity. During the design and development of the first Russian swept-wing fighter, the MiG 15, he has certainly been asked for his expert opinion. The MiG 15 was developed using the captured documents, especially the ones related to the swept-wing measurements, which were also discovered by the Americans and the English. Regarding the outer form, the Soviet MiG 15 (first flight on December 30, 1947) and the American F-86 (first flight on October 1, 1947) look quite similar; they engaged for the first time during the Korean War.

United States of America (USA)

Already in 1944, a first realization of German research and development results occurred in the USA. From debris of duds, found in 1944 in southern England, the Americans reassembled some V1 rockets (Fieseler Fi 103) and Argus 014 pulsejet engines. Following wind-tunnel measurements at velocities of up to 640 km/h (398 mph), a quantity production rapidly commenced. Envisaged were initially 1,200 copies of the "Jet-bomb 2" ("Loon"), but after the surrender of the Japanese, the fabrication was terminated. Only about 300 copies were delivered.

One of the most spectacular examples of the continuation, realization and application of German research is the swept wing. When *Theodore von Kármán* (then a Professor at Caltec) and *Georg Schairer* (a Boeing development engineer) discovered early Mai 1945 in Brunswick-Völkenrode models of swept-wing aircraft and documents of wind-tunnel measurements on swept wings, they were both surprised and reacted promptly. *Georg Schairer* wrote – dated May 10, 1945 – a letter to his company, the Boeing Aircraft Company in Seattle, and caused a project to be stopped and the design of the bomber B-47 to be changed to incorporate a swept wing [24]. The documents found in Brunswick, and others discovered in Göttingen and at Messerschmitt in Oberammergau, as well as supplementary interrogation reports, were brought to America and were soon made available to aircraft companies involved in the development of military aircraft. The Boeing B-47 had its first flight in 1947; more than 2,000 aircraft were built.

The American four-jet bomber Boeing B-47 was, in the spring of 1945, at an advanced stage in its development. Its swept-wing form finally realized was directly influenced by the results of wind-tunnel measurements on swept wings discovered in Brunswick and Göttingen.

The RLM issued in July 1944 the second invitation to bid for a jet fighter. For the fighter, the Heinkel turbojet engine He S 011 with 12.7 kN (2,855 lbs) was envisaged as propulsive unit and the aircraft was supposed to reach a speed of 1,000 km/h (622 mph) at an altitude of 7,000 m (22,960 ft). In response, six German aircraft companies set up projects involving swept wings with sweep angles between 35° and 45°. Many documents concerning these air-superiority fighters were found by the interrogation teams and sent to the aircraft industry of their respective county. Only later were the summarizing BIOS/CIOS- and FIAT-reports generated. Especially fast was the reaction of the US company North-American, whose new fighter F-86 "Sabre", with a wing-sweep of 40°, took off for its first flight on October 1, 1947. The F-86 had adopted the wing profile of the Me 262 and was built in large numbers over many years. Also the fighter units of the new German Air Force started with some Canadian-built F-86.

The first design of the Messerschmitt jet fighter P 1101, in response to the RLM invitation to bid, was presented on July 24, 1944 by *Hans Hornung* and revised until the end of August. In order to save time, the prototypes were envisaged to be equipped with quantity-produced Me 262 wings, which were to be attached to the new airframe under a sweep angle of 40°. At the same time, an adjustment mechanism for the wing sweep (on the ground) to 35° and 45°, respectively, was provided. An in-flight adjustment had been contemplated at Messerschmitt already earlier. The assembly of the first P 1101 started in January 1945 in a hangar in Oberammergau. Since the intended Heinkel He S 011 engine was not yet available, the engineers switched to the Junkers 004 engine, whose delivery was, however, delayed. When the Americans occupied Oberammergau in late April, they discovered the 80-per-cent-complete aircraft in an underground hangar – ready to be blown up.

On May 17, 1945, the first experts of the "USAF-Air Technical Intelligence Team" arrived and inspected, with great interest, the Messerschmitt P 1101, the first nearly realized German swept-wing aircraft, which was scheduled to take-off a few weeks later. Among the American aeronautical experts was the Bell-company engineer *Robert Woods*, who was also present when *Ludwig Bölkow* and *Richard Voigt* were interrogated. Before, a group of French engineers had taken along some P 1101 documents to Paris, but later had to

Also in case of the American fighter North American F-86 "Sabre" were German research results applied for the airfoil as well as for the swept wing design.

The jet fighter P 1101 was at its discovery at Messerschmitt in Oberammergau about 80 per cent complete. It led to the construction of two models of the American test aircraft Bell X-5 with variable-sweep wings.

return them to the Americans. Since a completion of the discovered P 1101 was not possible on location, the aircraft was shipped for investigation to Dayton [25]. The Messerschmitt P 1101 was trend setting for a first generation of swept-wing jet fighters, especially for the types F-86, MiG 15, and Saab J 29. *Robert Woods* convinced the Bell Company to propose to the US Air Force the construction of 24 fighters, with in-flight adjustable wing sweep (variable-sweep), on the basis of the P 1101. It followed an order for two test aircraft of this type, the Bell X-5, with the first take-off in June of 1951. The Bell X-5 was, with its variable-sweep wing, the predecessor of the navy aircraft Grumman XF-10 (1952), the fighter-bomber General Dynamics F-111 (1964), and, in a wider sense, of the European multi-purpose fighter aircraft Panavia "Tornado" (1974).

German research results and experience, related to the aircraft development at the Arado Company, were brought to Northrop by *Rüdiger Kosin*, who joined this company in August 1955. He worked during 1957/58 at the design office on the two-engine jet trainer T-38 and the derivative thereof, the fighter F-5, which was produced in the following years in greater numbers. With his Arado-experience in boundary layer control (suction), he also contributed to the experimental "laminar" aircraft Northrop X-21 [26].

In conjunction with the "Operation Paperclip" came *Friedrich von Doblhoff*, during the war head of the helicopter development at the "Wiener-Neustädter Aircraft Company", in 1946, after an interim period at Wright Field, to McDonnell in St. Louis. Here, he developed together with *Kurt Hohenemser* the compound helicopter XV-1 with blade-tip propulsion and a so-called "idling" rotor as a test vehicle. The XV-1 reached a flight speed of 320 km/h (199 mph); the project was, however, not pursued.

Rüdiger Kosin contributed to a considerable degree to the development of the Northrop jet aircraft, the trainer T-38 (*top*) and the light fighter F-5 (*bottom*).

The helicopter development at McDonnell was considerably influenced by *Friedrich von Doblhoff*, who was able to apply his reaction rotor with blade-tip propulsion, among other things, to the light helicopter "Model 120".

A small test crane-helicopter with three AiResearch air suppliers and a three-blade reaction rotor of 10-m diameter (3.21 ft) carried, as "Model 120", in 1957 at flight tests a payload far greater than its net-mass. A derivative project with a 23-m diameter (75 ft) reaction rotor was not realized. In the case of both vehicles, also research results and experience gained with the helicopter WNF 342 were employed [6].

Under the supervision of *Anselm Franz*, during the war head of the turbo-engine development at Junkers (Jumo 004), the new division "Gas Turbines" was set up, starting in 1951, at the flight-engine manufacturer Lycoming in Stratford/Connecticut and the factory accordingly enlarged. *Anselm Franz*, who had come to the USA in conjunction with the Operation Paperclip, obliged several former Junkers employees to join him in the USA. With a development team of finally about 150 engineers, the well-known and successful turboshaft engines T-53 and T-55 were developed, tested and manufactured in large numbers over many years. Worldwide, thousands of helicopters are powered by these engines. A special design principle, which was already applied in the case of the Jumo 004, is the so-called modular design, which was consequently applied to the Lycoming engines and is, in the mean time, state-of-the-art. Also the principle "overhaul on condition", i.e., not dependent on operating hours, was, for the first time, applied to the T-53 and T-55 engines [7].

German scientists and engineers have decisively contributed to the extensive development and manufacturing program of the Los Angeles-based "Garett-AiResearch Company [7]". Among them was first *Werner von der Null* (formerly DVL-Adlershof) and since 1951 *Helmut Schelp* (formerly RLM Special Engines).

It would be going too far to list, in detail, the activities of German scientists and engineers in American industry and also at research institutions. There were many who left their traces in aeronautics, either in scientific endeavors or in products. Besides the experts already mentioned, the following should be named: *Bruno Bruckmann* (General Electric), *Willy Fiedler* (Lockheed), *Eduard Fischel* (Navigation-Sys-

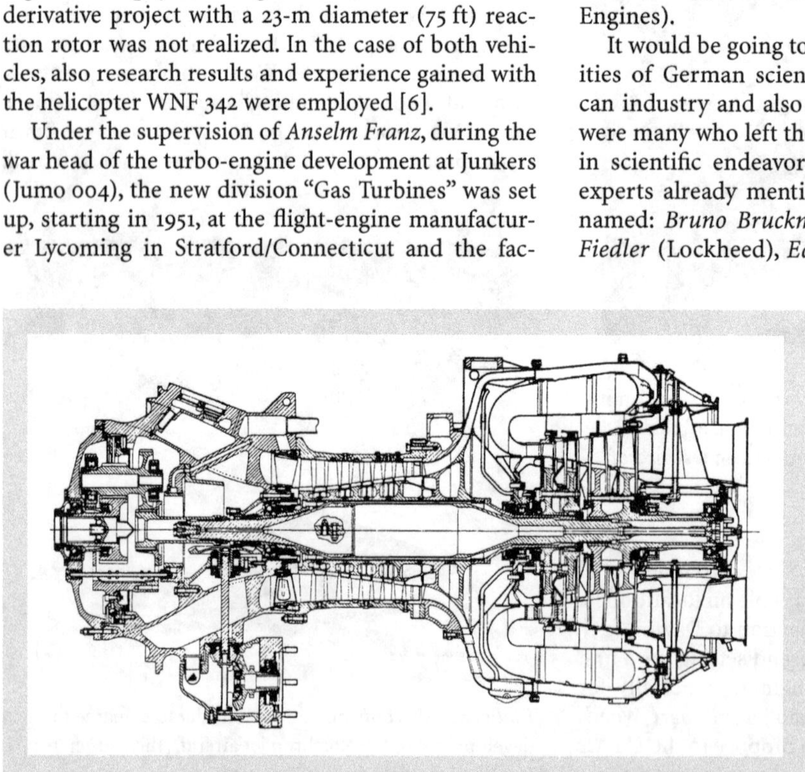

A group of German engineers, headed by *Anselm Franz*, developed at the Gas Turbine Division of Lycoming, among others, the successful turboshaft engines T-53 (figure) and T-55, which were built in quantity production over many years.

Anselm Franz (right) with his collegues *Heinz Moellmann*, *Siegfried Decher* and *Wolfgang Stein* in front of a T-53 engine.

tems), *Anton Flettner* (Flettner Aircraft Corp.), *Helmut Heinrich* (Parachutes), *Peter Kappus* (General Electric), *Theodor Knacke* (Parachutes), *Siegfried Knemeyer* (Wright Field), *Alexander Lippisch* (Collins Radio Corp.), *Hans Multhopp* (Martin), *Hans von Ohain* (Wright Field), *Gerhard Sissingh* (Lockheed), *Christoph Soestmeyer* (Arnold Engineering Development Center), *Waldemar Vogt* (Martin), *Richard Vogt* (Aerophysics Development Corp.), *Herbert Wagner* (Naval Air Missile Test Center). Of the German scientists and engineers, only a few have later returned to Germany.

Egypt

Egypt purchased, at the instigation of President *Gamal Abdel Nasser* and arranged by *Willy Messerschmitt*, in 1959 from the Spanish Hispano Aviation S.A. (HASA) the production license for the HA 200, and the development documents and rights concerning the light supersonic fighter HA 300 [27]. Although the contractual arrangement for the development occurred only in the spring of 1964, this development may be considered a continuation of German research and development activities. Delivered were ten HA 200 aircraft as well as aircraft elements and components for three HA 300 prototypes.

In 1953, *Ernst Heinkel* signed a development contract with the Egyptian government concerning a fighter and a corresponding turbojet engine of the 9,000 kp (20,000 lb) class. Development work for the Heinkel He S 053 started in 1953 in Stuttgart-Zuffenhausen. After trial runs, Heinkel delivered three development-stage engines to Egypt. The contract was cancelled in 1956 during the second Egypt-Israeli war. The German team, partly already settled in Egypt, returned to Germany [7].

A Bristol "Orpheus 12 SR" engine was initially considered for the Messerschmitt fighter HA 300, which

Based on the design of a light supersonic fighter, revised in Spain by *Willy Messerschmitt*, an independent development started in Egypt in cooperation with German engineers, which led to three prototypes of the jet fighter aircraft HA 300 and to the Brandner jet engine E 300.

did, however, not go into quantity production. *Ferdinand Brandner*, rather recently returned from the Soviet Union, agreed to take over the engine development together with a hired German-Austrian team, which was joined by some Swiss people. At first, a small turbojet engine, the E 200 (similar to the "Marboré"), was realized, subsequently a larger one, the E 300, for the fighter HA 300. The newly developed E 300 engine flew for the first time as left engine on an Antonow An 12 in January 1966. Further flight tests were conducted in conjunction with an Indian aircraft HF 24, a development of *Kurt Tank*. Due to the deteriorated economic situation in Egypt, the entire development was terminated in June 1969; thereafter, all engineers returned back to Germany or to their respective home countries [7].

Argentine

Argentine, who had already an aircraft company since 1927 for license productions and an independent development, was, after World War Two, strongly interested in modern aircraft. At first, the French aircraft designer *Dewoitine* was charged with the development of a jet fighter. The aircraft "Pulqui (Arrow)" flew in 1947 but was unsatisfactory with respect to its performance and flight characteristics. By intervention of President *Juan Peron*, *Kurt Tank* was then charged in late 1947 with the development of an efficient fighter aircraft. With a number of former Focke-Wulf employees, who came to Argentine, *Tank* had soon established an expert team. Thus the fighter "Pulqui II", with a Rolls Royce "Nene II" engine, was realized in Cordoba at the "Instituto Aerotécnico de Cordoba". The first flight was performed in June of 1950. *Kurt Tank* had strongly followed the design of the Ta 183 of *Hans Multhopp*; however, the fuselage was stretched and became more slender, the 40-degree swept wing was realized as high-wing and a T-tail was installed. It was the T-tail that brought *Kurt Tank* during flight tests into a critical situation – a condition today called super stall. A report concerning flow-field measurements at the "Kaiser-Wilhelm-Institute of Fluid Research" during the thirties rendered an explanation: the slipstream (wake) off the wing had impinged on the T-tail. Changes to the wing nose near the wing-fuselage junction, causing a controlled flow separation, corrected the situation. German research results – although of an earlier research era – had helped to find the correct explanation and hence a remedy for the problem. Six "Pulqui II" aircraft were built; however, with the overthrow of *Peron* (1954) this development ended. *Kurt Tank* and several of his co-workers went to India [28].

Reimar Horten, had, together with his brother *Walter*, since the thirties developed and built all-wing (flying-wing) sailplanes and powered aircraft, and had, in the process, performed considerable research work. He went after the end of World War Two to Argentina. There, he generated a number of designs for flying-wing aircraft, from hang-gliders to four-engine transport aircraft. Under his supervision, some sailplanes were built, which remained, however, unique specimen.

A small number of German engineers, among them *Carl Diederich, Helmut Kloos, Julius Henrici* and *Götz Mandel*, worked for some years in the vicinity of Buenos Aires on an Argentine version of the Hs 293 missile and on other devices with rocket propulsion.

In Argentine, a German engineering team under the supervision of *Kurt Tank* developed the jet fighter "Pulqui II", with swept wings, based on the design of the Ta 183 of *Hans Multhopp*.

Brazil

The Brazilian embassy in Amsterdam approached in 1952 *Henrich Focke* and offered him, and some of his co-workers, contracts for an occupation in Brazil. Initially, their task was to continue the previously interrupted development of some sort of a vertical take-off aircraft, the "Helconair", at the "Centro Technico de Aeronautico" in Sao Paulo. However, the venture was soon abandoned. The Focke-group was then charged with the development of a two-seater light helicopter with a 165 kW (225 hp) engine. The prototype of the "Beijaflor" made its first flight in 1952. The three-blade rotor had only flapping hinges; the lag hinges were still missing. Due to a special control mechanism, the helicopter was in flight self-stabilizing. Only a few vehicles were built.

Henrich Focke and some of his co-workers realized in Brazil the light helicopter "Beijaflor".

India

Kurt Tank was already in early 1955, while still in Argentine, approached by the Indian Government. He went in 1956, with some co-workers, to Bangalore and developed at the "Hindustan Aeronautics Ltd., HAL" the interceptor HF 24 "Marut", whose prototype took-off in June 1962 for its first flight. The "Marut" was manufactured in quantity production until 1977. The Indian Government gave one copy to the "Deutsches Museum" in Schleissheim where it is on exhibit [28].

In India, *Kurt Tank* developed with a small group of German engineers the interceptor HF 24 "Marut", which was manufactured in quantity production by the "Hindustan Aeronautics Ltd.".

Shortly after the end of World War Two, Sweden developed the swept-wing fighter Saab 29, utilizing German wind-tunnel measurements. The Saab 29 flew already in 1948 as prototype.

Sweden

Between companies in neutral Sweden and German industry and research existed also during World War Two – although monitored – some contacts. Probably due to these long-standing relations was Sweden able to obtain, shortly after the war, German reports and results of wind-tunnel measurements on swept wings via Switzerland [25]. These documents could then be used for the development of the jet fighter aircraft Saab 29, whose prototype flew in September 1948 for the first time. The Saab became thus the first Western European fighter aircraft with swept wings. The aircraft was built in quantity production until 1956.

Spain

Claude Dornier founded in January of 1951 in Madrid a technical office, hired some German engineers, mostly former employees, and bid for development contracts from the Spanish "Ministry of Aviation". He entrusted his son *Claudius* with the management of the office. In 1951, the Dornier office received a development contract for a commuter aircraft with short-take-off characteristics – comparable to the Fieseler "Stork". The Spanish 110 kW (150 hp) "Tigre" was intended as engine but soon turned out to be too weak. The first flight of the Do 25 occurred in July 1954. The second prototype obtained a Continental O-470 engine of 165 kW (225 hp) with a variable-pitch propeller; the aircraft was utilized for demonstration flights for the later Do 27 [9].

In Spain, the short-take-of aircraft Do 25, the immediate predecessor of the German Do 27, was developed under the supervision of *Claude Dornier.*

The development of the Spanish jet trainer HA 200 was carried out under the supervision of *Willy Messerschmitt* by the Messerschmitt office in Seville. The construction of the prototypes and the subsequent quantity production was carried out by the Spanish company HASA, later designated CASA, now a part of EADS.

The contact between *Willy Messerschmitt* and the Spanish Government, which had its beginnings during the production of the Me 109 under license, was reestablished in 1951. The occasion was initially only consulting with regard to the installation of the Rolls-Royce "Merlin" engine instead of the Daimler Benz DB 605. Since this was a structural problem, *Julius Kraus*, at the time professor at the TH Munich, came to Madrid. A little later, a Messerschmitt office was established at HASA – later designated CASA – in Seville, which was gradually joined by a number of former Messerschmitt employees. The first contract involved the development of the reciprocating-engine aircraft Ha 100 for schooling and training; the first flight was in December 1959. It followed the development of the jet trainer HA 20 with two Turboméca "Maboré" engines; the first flight was performed in August 1955. With a reinforced crew, the design of the supersonic fighter HA 300 was approached starting in the mid-1954. The development was delayed due to the deteriorating economic conditions in Spain and difficulties in procuring the engine. The Egyptians were very much interested in the advanced light supersonic fighter aircraft and signed with the Spanish Government and CASA a takeover and licensing contract. All documents and the parts of the HA 300, manufactured up to that time, were shipped to Cairo, respectively to the factory at Heluan [27]. A prototype, resulting from the continuation of this development and the corresponding trials in Egypt, is exhibited at the "Deutsches Museum" in Oberschleissheim near Munich as a further witness to the continuation of the German research and development originating during World War Two.

References

1 E. Brown, Wings of the Luftwaffe, Flying German Aircraft of the World War Two, Jane's Publishing Company, London/Sydney, 1977

2 K.H. Doetsch, Deutsche Luftfahrtforscher nach 1945 in England, DGLR-Blätter zur Geschichte der Luft- und Raumfahrt, V (1992), pp. 165–188

3 L. Bölkow, Erinnerungen, Herbig Verlag, 1994, pp. 56-60

4 A.C. Kermode, "Aerodynamik und Flugverhalten", Motorbuch Publisher, 1992, pp. 374–377

5 F. Käsmann, Die schnellsten Jets der Welt, Aviatic Publisher, Oberhaching, 1994, pp. 42

6 K. von Gersdorff, K. Knobling, "Hubschrauber und Tragschrauber", Vol. 3, Die deutsche Luftfahrt, Bernard & Graefe Publisher, Koblenz, 1985, pp. 101–107

7 K. von Gersdorff, K. Grasmann, H. Schubert, "Flugmotoren und Strahltriebwerke", Vol. 2, Die deutsche Luftfahrt, Bernard & Graefe Verlag, Bonn, 1995 (English ed., K. von Gersdorff, K. Grasmann, H. Schubert, German Aviation Powerplants (The Beginnings – Aircraft Engines of WW II – Postwar Developments), SAE Aerospace International, Warrendale, PA, USA, 2003/2004)

8 H. Oestrich, Aus der Entwicklung des Strahltriebwerkes ATAR 101, WGL-Jahrbuch 1954, pp. 135–139

9 O. Pabst, "Kurzstarter und Senkrechtstarter", Vol. 6, Die deutsche Luftfahrt, Bernard & Graefe Verlag, Bonn, 1984, pp. 179–188

10 G. Oberländer, Die ersten Kleingasturbinen für die Luftfahrt, DGLR – Blätter zur Geschichte der Luft- und Raumfahrt V (1992) pp. 4–26

11 U. Albrecht, Rüstungsfragen im deutsch-französischen Verhältnis (1945–1960), ZFSL-Beiheft 23 (1997), Franz Steiner Verlag Stuttgart, pp. 120–123

12 H.E. Sänger, Eugen Sänger, wie ihn Mitarbeiter, Freunde und Zeitgenossen kennengelernt haben, DGLR-Jahrbuch 1993 II, pp. 717–727

13 H.E. Sänger, Prof. Eugen Sänger: Seine Tätigkeit auf dem Antriebsgebiet in Frankreich, DGLR-Blätter zur Geschichte der Luft- und Raumfahrt, V (1992), pp. 189–196

14 F. Sorg, "Nachkriegsaktivitäten deutscher Wissenschaftler und Ingenieure beim Office National d'Etudes et de Recherches Aéronautiques (ONERA) in Paris", DGLR-Blätter zur Geschichte der Luft- und Raumfahrt V (1992) pp. 104–111

15 ENTAC Engin Teleguidé Anti-Char, drahtgelenkter Panzerabwehr-Flugkörper mit Hohlladungs-Gefechtskopf

16 G. Krauth, Das deutsch-französische Forschungsinstitut Saint-Louis (ISL), Wehrtechnik 10/95, pp. 76–78 and Festschriften zum 25., 30. und 50. Jahrestag der Gründung des Deutsch-französischen Forschungsinstitutes Saint-Louis (1984, 1989, 1992, 1995) und U. Albrecht, "Rüstungsfragen im deutsch-französischen Verhältnis (1945-1960)", ZFSL-Beiheft 23 (1997), Franz Steiner Verlag Stuttgart, pp. 116/117

17 J. Michels, S. Kuwschinow, W. Srelow, J. Woronkow, "Deutsche Flugzeugspezialisten im sowjetischen Russland – Leben und Arbeit 1945–1954", Vol. 1 and 2, Verlag Russland-Deutschland 1996

18 N.D. Kusnezow, "Das Konstruktionsbüro Trud und seine Triebwerke", DGLR -Jahrbuch 1992 I, pp. 13–25

19 J. Michels, J. Werner, "Luftfahrt Ost", Vol. 22, pp. 65–71 und U. Unger, Deutscher Start für sowjetische Triebwerksepoche, Edition 06, Flieger Revue 1996

20 W. Wagner, "Hugo Junkers – Pionier der Luftfahrt – seine Flugzeuge", Vol. 24, Die deutsche Luftfahrt, Bernard & Graefe Verlag, Bonn, 1996, pp. 521–527/542–544

21 W. Wagner, "Die ersten Strahlflugzeuge der Welt", Vol. 14, Die deutsche Luftfahrt, Bernard & Graefe Verlag, Bonn, 1989 (English Ed.: W. Wagner, The First Jet Aircraft, Schiffer Publishing Ltd., Atglen, PA, USA, 1998)

22 K.H. Mewes, Pirna 014 - Flugtriebwerk der DDR, Aviatic Verlag, 1997, and J. Michels, J. Werner, Luftfahrt Ost, Vol. 22, pp. 46–53
23 Th. Benecke, K.H. Hedwig, J. Hermann, "Flugkörper und Lenkraketen" Vol. 10, Die deutsche Luftfahrt, Bernard & Graefe Verlag, Bonn, 1986, S. 76
24 L. Bölkow, Mit dem Pfeilflügel zum Hochgeschwindigkeitsflug, DGLR-Bericht 89-05, 50 Jahre Turbostrahlflug, pp. 247–287, and G. Schairer, On the design of early large swept wing aircraft, DGLR-Bericht 89-05, 50 Jahre Turbostrahlflug, pp. 289–321 and Th. von Kármán (with Lee Edson), The Wind and Beyond: Theodore von Kármán, Pioneer in Aviation and Pathfinder in Space, Little, Brown and Company, Boston, Mass., USA, 1967, pp. 326–332
25 W. Radinger, W. Schick, Messerschmitt - Geheimprojekte, Aviatic Verlag, 1991, pp. 120–133, and H.J. Ebert, J. Kaiser, K. Peters, "Willy Messerschmitt - Pionier der Luftfahrt und des Leichtbaues", Vol. 17, Die deutsche Luftfahrt, Bernard & Graefe Verlag, Bonn, 1992, pp. 281/282
26 R. Kosin, "Beiträge zur Luftfahrtentwicklung und deren Anwendung im Ausland", DGLR-Blätter zur Geschichte der Luft- und Raumfahrt, V (1992) pp. 112–140
27 H.J. Ebert, J. Kaiser, K. Peters, "Willy Messerschmitt - Pionier der Luftfahrt und des Leichtbaues", Vol. 17, Die deutsche Luftfahrt, Bernard & Graefe Verlag, Bonn, 1992 (English ed., H.J. Ebert, J. Kaiser, K. Peters, "Willy Messerschmitt - Pioneer of Aviation Design, Schiffer Publishing Ltd., Atglen, PA, USA, 2000)
28 W. Wagner, "Kurt Tank - Konstrukteur und Testpilot bei Focke-Wulf", Vol. 1, Munich, 1980, and O. Pabst, "Kurzstarter und Senkrechtstarter", Vol. 6, Die deutsche Luftfahrt, Bernard & Graefe Verlag, Koblenz, 1984 (English ed.: W. Wagner, Kurt Tank - Focke-Wulf's Designer and Test Pilot, Schiffer Publishing Ltd., Atglen, PA, USA, 1999)

Research Results of the Period After 1945

A NEW BEGINNING

Licensing and Cooperation – Bridges to a New Industrial Structure

Horst Prem

During the ten years of forced abstention in Germany after the war, numerous important developments occurred without the participation of German aeronautical industry and research. Although the fundamentals of this development had, to a large degree, been established before 1945 by German research institutions and industry, they had no more been applied in Germany. These fundamentals included:

- *the aerodynamically clean design of high-speed aircraft,*
- *the swept-forward and swept-back wing for the high subsonic and supersonic speed regimes,*
- *the supercritical wing profile,*
- *the Delta-wing with control flaps at the trailing edge,*
- *the aircraft with variable-sweep wings,*
- *the jet engine,*
- *the arrangement of jet engines in pods below the wing,*
- *the controlled helicopter flight.*

Developments based on German research results were found in numerous products of the victors of World War Two after 1945.

Timid aeronautical research activities already started again in 1951 with respect to sailplane design and construction. Here, the initiatives were with the individual German states as is documented by the revival of the research institutions. While the assets of the AVA were confiscated, the DVL and DFL, as registered associations, never stopped to exist and had never been dissolved by the Allies. There existed an uncertainty of law, with the authorities responsible, as to the status of these organizations, which were subject to private law. The problems and new beginnings are sketched in Part 2. They are exemplarily clarified by the curriculum vitae of *Heinrich Koppe*.

With the full sovereignty of the Federal Republic of Germany in 1955, specifically aeronautics-related developments and productions became again possible. In the civil domain, responsibility for the national aeronautical development was given to the Ministries of Economics and Transportation, respectively, in a way that made it initially almost impossible for industry to find contacts with the authority to award development and construction contracts [1]. Priority with the ministries had initially the establishment of air traffic and the corresponding infrastructure, like the rebuilding of German Airlines Lufthansa, the extension of airports, flight control, and regulations concerning sovereignty and safety related aspects.

The aeronautical industry did not yet have a national contact person, and it had to reestablish itself. The known aircraft companies were involved in automobile engineering or had transferred their development and manufacturing capacities to foreign countries.

The situation was much better in the military sector. There was initially the "Office Blank" ("Dienststelle Blank"), which was, however, quickly transformed into the "Ministry of Defense". Now, industry had a national contact and a "client".

Here, not all facts that contributed to the rapid development of aeronautics in Germany since

Heinrich Koppe (1891–1963).

Heinrich Koppe, born in Nordhausen, Harz mountains on March 26, 1891.

1911 starts his studies of natural sciences and mathematics in Jena and Halle. In Halle, the aerologist *Albert Wigand* introduces him to piloting airships and airplanes and *Emil Everling* to air-electrical discharges.

1914 he is sent to the western front as airship aviator and meteorologist.

1915 he is charged to organize the Ottoman weather service. He experiences the end of the war in Syria.

1920 receives his PhD with *Wigand* in Halle. Subsequently he takes over the Department of Physics from *Everling* at DVL with the subject areas aeronautical instrumentation, flight meteorology and navigation.

1925 lecturing qualification (habilitation) at the TH Berlin in the areas navigation and meteorology for aviation and shipping. His objective was from the beginning to render air traffic suitable for night and blind flight navigation.

1930 he receives at the TH Brunswick a lectureship in flight navigation and instrumentation. Parallel to these activities, he sets up the "Brunswick Institute of Aeronautical Measurement Technology and Flight Meteorology", which, in

1934 is taken over by the TH Brunswick. *Koppe* is appointed full professor for these subject areas.

1935 he becomes dean of the independent Department of Aeronautics at the TH Brunswick. In the following ten years, Brunswick becomes the aeronautical instruction center of Germany. *Koppe* suffers, as many others, under the ruthless political and military constraints of the Third Reich regime.

1945 towards the end of the war, he rejects the advice of high military officials to blow up and burn down the main building of the institute located near the airport Brunswick-Waggum.

1953 after the relaxation of the ban on aeronautical research, six heads of the original aeronautical institutes resume their work within the DFL. *Koppe* resurrects his "Institute of Air Vehicle Control" ("Luftfahrzeugführung").

1963 he dies in Brunswick on November 9.

1955 can be addressed. In numerous books [1, 2] and reports [3, 4], the propelling forces are extensively described, among them being:

- national defense and security policy,
- set up and preservation of a national aeronautical industry,
- aviation as motor of the economy and basis for an industrial globalization,
- establishment and support of a civil aeronautical industry as export motor and provider of secure high-quality jobs.

Here, we only describe in a short survey the market and the socio-political environment and the resulting product ideas derived thereof since they were the propelling forces for the financial support and the continued development of research.

Continuation and new foundations of German aeronautical research organisations after 1945

January 23, 1952
On an initiative of the state of North-Rhine Westphalia, the reconstruction of DVL in Aachen and Essen-Mühlheim and later in Porz-Wahn near Cologne commences in 1952.
First head: *August Wilhelm Quick*

1953
The DFL in Brunswick resumes, on the initiative of six institute directors, its activities.
First head: *Hermann Blenk*
At the same time activities at the AVA in Göttingen were again taken up.
First head: *Albert Betz*

1956
The "German Study Society of Helicopters" ("Deutsche Studiengesellschaft für Hubschrauber, DHS") and the "Research Institute of Jet-propulsion Physics" ("Forschungsinstitut für Physik der Strahlantriebe, FPS") are founded in Stuttgart and Lampoldshausen; they are later integrated into DFL and DVL, respectively.
First heads: *Walter Just/Eugen Sänger*

April 29, 1969
Merging of all research institutions into the "German Aerospace Research and Test Establishment" ("Deutsche Forschungs- und Versuchsanstalt für Luft- und Raumfahrt, DFVLR") with headquarters in Porz-Wahn near Cologne.
First chairman of the Board: *Volker Aschoff*

Aircraft Development in West Germany

Dieter Schmitt

Military Developments

First equipment for the German Armed Services, as far as air vehicles are concerned, was mainly obtained from NATO partner countries. Since in Germany neither development nor manufacturing capacities with the necessary skilled workers were available, the following aircraft were – with political considerations in the background – initially selected and build in Germany under license:

- Fouga "Magister" CM 170R as a light jet trainer,
- FIAT G.91R3 as ground-support fighter,
- Lockheed F-104 G as interceptor.

The contract for the manufacture (under license) of the Fouga "Magister" went to the Ernst Heinkel Fahrzeugbau GmbH and to the Messerschmitt AG; a total of 194 aircraft were built between 1957 and 1961.

In the autumn 1958, the first discussions concerning the license-production of the ground-support aircraft Fiat G.91R3 started under the chairmanship of the "Federal Office of Defense Technology and Procurement" ("Bundesamt für Wehrtechnik und Beschaffung, BWB"), with the companies Dornier, Heinkel and Messerschmitt. The three companies formed a working pool "G.91" and manufactured 294 aircraft. During the same year, also preliminary discussions concerning the license-production of an interceptor commenced. The Lockheed F-104A "Starfighter" was selected and numerous other countries followed the German decision. Within Western Europe, four working pools existed, which participated in the "Starfighter" manufacturing. In Germany a "North Group" participated, consisting of the German companies VFW and HFB and the Dutch company Fokker. In a "South Group" were the companies Messerschmitt, Dornier and SIAT Cooperation. The northern and southern consortia started to become apparent.

Parallel to the build-up of the manufacturing capacities the reassembly of the development teams occurred. The Ministry of Defense called already in 1956 for a project competition concerning a future interceptor. Vertical take-off capability was initially not envisaged. But new considerations in view of the vulnerability of German runways led finally to the additional requirement of V/STOL capabilities. The development offices of the companies Bölkow-Developments KG, Ernst Heinkel Aircraft Engineering, Ltd., and Messerschmitt AG started then with development studies and conceptual considerations. The necessary steps in the development, and problems related to the development of the vertical-take-off concept shall be comprehensively described starting on page 372.

In the early sixties, a change in the defense doctrine evolved. While the equipment of the NATO air forces in the fifties, up to and including the introduction of the F-104 G, corresponded to a strategy of the so-called "massive – possibly nuclear – retaliation" against a potential aggressor, led the nuclear stalemate with the Soviet Union to a change in this strategy. Considering the mutual nuclear destruction capabilities, it was essential to develop a deterrent against the superiority of the Warsaw Pact in the area of conventional weapons. Thus, the strategy of the "Flexible Response" was born. It envisaged an initial direct defense against an enemy attacking with conventional weapons, but it made, by a willingness and ability to deploy nuclear defense, any aggression to an incalculable risk for an aggressor. The new doctrine of "Flexible Response" required an increase in military payload that was no longer negotiable [1] for a vertical take-off aircraft. The capability of short-take-off remained, however, as essential requirement. In addition, versatility (multi role combat capability) should be guaranteed. The application of the variable-sweep wing – the so-called variable geometry (VG) of the wing – in conjunction with a thrust-vectoring nozzle, developed in conjunction with vertical take-off, was promising in that regard. First experience was gained by German industry during the years 1964–1967 in a joined study with Boeing under the project name AVS (Advanced Vertical Take-off and Landing System). The technical and organizational know-how thus gained, was utilized in a new national program, which was first designated NKF (new fighter aircraft), then turned into the tri-national program of the "Panavia" as MRCA (Multi Role Combat Aircraft); the preliminary "final" product was the "Tornado"-prototype, which demonstrated on its first flight in August 1974 in Manching its new and superb flight performance as well as its operational versatility.

Within the industrial environment, several mergers had taken place. Many of the old aircraft companies, with well-known names like Blohm, Focke-Wulf, Heinkel, and Messerschmitt, had merged into new company structures: Messerschmitt-Bölkow-Blohm (MBB) and Vereinigte Flugtechnische Werke (VFW). Only Dornier remained an independent individual company, because the company founder, *Claude Dornier*, wanted to keep system capability.

At the beginning of the development of the MRCA, the meanwhile increased technical complexity of the development task became apparent[3, 4]. This shows the need to develop new technologies with an effective integration into the system "aircraft".

In aerodynamics the contradictory demands on MRCA, as supersonic capability, flight at high subsonic speeds close to the ground, and short take-off and landing distances were to be combined. For this reason, the concept of variable-sweep wing was used, which allows changing sweep during flight between 25° and 67° depending on the flight mission. The change in sweep angle also changes the aspect ratio (ratio of wingspan to wing chord) and hence aerodynamic forces and characteristics decisively. New aerodynamic challenges are the various outer loads attached to the wing, the separation of these loads in critical flight regimes, the reliable operation of the engines and a sufficient air supply during critical flight phases.

The structure of the aircraft must meet the classical as well as the additional demands. The demand for weight reductions, higher g-loads due to an increased maneuverability, increased durability and hence less maintenance effort are among the classical requirements. They increase and become more demanding with every new product. The new attributes "invisible" and "invulnerable" are added as further design parameters to the classical and well-known general requests "faster", "higher", "more agile".

The new desire to "remain undetected" (stealth capability) requires an unconventional shape, which exhibits a low radar cross-section and radar-absorbing materials and paints. Besides the radar signature the infared signature must be minimized.

But also the pilot's tasks have become considerably more complex. Automatic systems are supporting the pilot by taking over many routine activities so he can concentrate on his monitoring function. His various tasks, such as flight control, navigation and target recognition, armament control and friend/foe detection, frequently lead to new developments. They became realisable due to the rapidly developing computer and information-processing capacities.

In supporting the pilot, the flight control system (FCS) has to reliably stabilize the aircraft during all phases of flight and must, at the same time, implement control inputs of the pilot without bringing the aircraft into dangerous flight attitudes. However, it must still permit to fully utilize the aerodynamic and structural reserves available for the required flight maneuvers. To support aircraft control (e.g. for low-level flight), additional sensors and electronic systems (FLIR, ODIN, etc.) must be integrated into the autopilot, respectively must be able to communicate with it.

Additional sensors, which utilize various wavelengths, such as radar, infrared and ultrasound, are being further developed to recognize, on the one hand, incoming weapons and air vehicles, and, on the other hand, avoid being detected. Here, new materials, more computer capacity and new algorithms are demanded. All information concerning flight condition, flight trajectory, target recognition, et cetera, must be displayed to the pilot in a clear and unambiguous way.

The development of the EF 2000 is for Germany, at the time, the ultimate in fighter-aircraft development, with all the complexity demanded by an unstable canard-type fighter operating at high angles of attack and equipped with highly integrated weapons-guidance systems. But also the public discussion of financing and economic sense of ever more complex fighter aircraft, as well as the search for alternatives (observation from outer space, unmanned vehicles, etc.) are of importance and are also a motivation for further research activities.

The weakest and most susceptible link in the steadily growing demand for a higher load-carrying capacity is man. If the rapid development in computer technology and storage capacity continues, one may ask whether fighter aircraft have reached a critical phase due to man. Could the tasks of fighters in the future not be taken over by unmanned flight vehicles, with artificial intelligence, possibly at lower costs?

Here a new catalogue of demands emerges, which will simultaneously lead to new questions and further research activities.

Parallel to the need for fighter aircraft with different operational scenarios, there is a need for trainer aircraft.

In May 1969, government officials in France and the Federal Republic announced an open competition for a subsonic aircraft with jet propulsion, which could also be employed as ground-support aircraft. In 1970, a consortium of Dornier and Breguet was awarded the contract. On October 6, 1973, the prototype 01 of the "Alpha Jet" made its first flight in Marseille, the prototype 02, with its final assembly carried out at Dornier in Oberpfaffenhofen, followed with its maiden flight on January 9, 1974. From 1975–1983, 500 copies of the "Alpha Jet" were produced and delivered to ten countries. Thus, the Alpha Jet advanced to a very successful aircraft program for Dornier, even if today the aircraft is no longer operated by the German Air Force.

In addition to the demand for high-performance fighter aircraft, there is within the military units a large requirement for transport and surveillance aircraft. The tasks of military transport aircraft are manifold, however, they essentially concern the transportation of troupe units, military equipment and many kinds of vehicles, and the drop of parachutists and loads under different external conditions. An additional requirement is that they must be able to unload without ground-based equipment and, furthermore, take-off and land on short, unprepared run ways. The historic development of military transports is traced in the paper by *J.R. Roeder* "Civil Freighters and Military Transports [5]". After World War Two, mainly the configuration with tail gate was successful and prevailed. Characteristical for it is a wing in high-wing arrangement. That way, the fuselage is closer to the ground, the landing gear is stowed laterally in the fuselage, and for a ground-independent loading a rather large tail gate can be utilised. As propulsion device, frequently a turboprop engine is chosen, which is arranged on the wing in a way that the propeller stream passes the wing with extended high-lift devices further increasing lift, thus achieving shorter take-off and landing distances. In Germany, the first military transport acquired was, in 1955, the NORD-2501 (better known as "Noratlas") developed in France. A total number of 186 aircraft were manufactured under license by the companies Siebel, Weser Flugzeugbau and Hamburger Flugzeugbau, combined in the consortium Flugzeugbau Nord GmbH.

This helped to establish know-how in manufacturing. Subsequently, the new military transport "Transall" was developed in a joint German-French venture. Here, young German engineers were able to gather experience in two ways, which later benefited the civil aircraft development: Firstly, the development of a military transport is very similar to the development of a civil one; the haul of a large payload must always be effective and economic, only the operating conditions are different. Secondly, the development of the "Transall" represented the first successful cooperation between Germany and France within the aeronautical industry after World War Two. This has certainly been an important and preparatory contribution to the later successful German-French cooperation within the Airbus programmes.

Civil Aircraft Developments

The development of jet engines has not only decisively affected the military development of fighter aircraft; also civil aeronautics has experienced essential changes due to the introduction of jet engines. One of the early visionaries was *Wilhelm Bansemir* who wrote already in 1954 [6]: "*At the moment when turbo engines with proven operational safety became available, one could no longer ignore the advantages these engines also promised for the civil domain. …It also seems to be proven that transport aircraft with pure jet propulsion are economical on long flight distances.*"

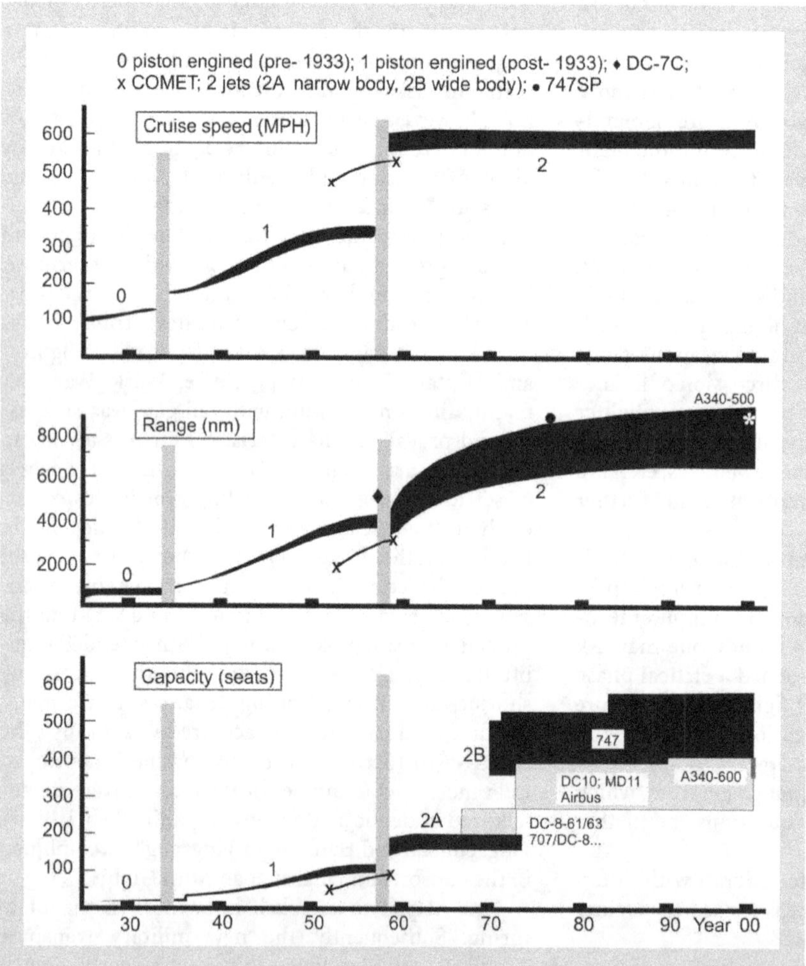

History of design parameters.

Today, this has been confirmed, it could, however, in 1954 not yet be taken for granted.

If one considers the development of characteristic aircraft parameters such as speed, range and capacity, one observes that with the introduction of the jet engine speed and flight altitude dramatically increased (see figure p. 350). The cruise speed increased from the common 500 km/h (270 kn) to 900–950 km/h (486–513 kn) in case of the B-707 and DC-8, which went into operation in 1958; instantly a new transportation quality was offered.

The higher speed provided by the introduction of the jet engine brought for the passengers the advantage that now larger distances, such as an Atlantic crossing, could be accomplished easily in one day. The better economic efficiency of jet engines at high altitudes allowed flights at altitudes of about 10,000 m (32,800 ft), i. e., a range with less turbulences and thunderclouds. The higher altitude allowed better comfort; only the cabin had to be designed for a larger pressure difference. Despite the fact that already in 1939 the problems of pressurized cabins had been declared solved, showed the introduction of the first jet transport aircraft larger difficulties than initially expected. The mysterious crashes of the British "Comet" were followed by intensive and systematic endurance tests with original airframes. From the results it could be concluded that the crashes had been caused by fatigue cracks in the fuselage. These findings, which were openly presented, then led to new design philosophies and certification rules and to operational-strength proofing of complete airframes. The corresponding capacities for the life-tests were in Germany established at the "Industrieanlagen Betriebsgesellschaft, IABG". These test facilities will be considered in a separate section.

An important element for the acceptance of air traffic by the public is, naturally, the question of safety. We enter road traffic knowing well, or suppressing, that today there are almost 8,000 (road) traffic deaths per year just in Germany. In air traffic this result would not be tolerable. The average in air traffic deaths worldwide over the last 10 years is about 1,000, i. e., about 10 per cent of the yearly road-traffic casualties in Germany. If one refers the number of deaths to transport performance, normally defined as passenger-kilometers (or nautical miles) per unit of time, one comes to the conclusion that the aircraft is a very reliable means of transportation. But here too, one must strive to maintain the high level of safety and possibly further improve it, especially in view of the twofold increase in traffic over the next 10–15 years and the fact that a doubling of the casualties in air traffic is not acceptable. An analysis of the accidents shows that man, as the weakest link, is the cause of 75 per cent of all accidents. Preceding the accident

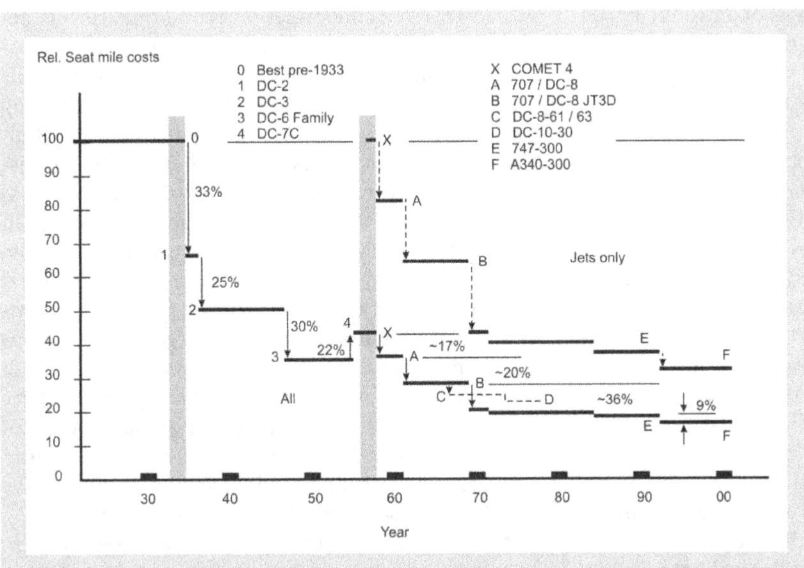

Development of seat mile costs versus time.

is generally a technical fault, whose consequences are misjudged by the pilot, who then brings the aircraft, due to an improper reaction, into a critical situation. There exist, in the meantime, numerous additional indicators in the cockpit, such as an indication of cross winds, warnings when approaching the ground, warnings in case of an eminent collision with other air vehicles, indication and warnings in the case of system or subsystem loss.

With this, new sources of pilot stress have been established within the cockpit by systems, which are more complex and multifarious, thus a simpler and explicit presentation of the information is required. In addition, almost every aircraft offers a different display of the information within the cockpit. The concept of "Cross Crew Qualification" at Airbus, which offers for all new aircraft-types a nearly identical cockpit lay-out and information display, is certainly an important step towards the standardization of the pilot's "work bench" and also an important contribution to flight safety.

The higher cruise speed and the smoother flight conditions were associated with a better planning ability and reliability of the flights. Although the jet engines generated much more noise during take-off and landing, this noise was being accepted during the post-war period since it also represented the "roaring of the economy" and the unbroken confidence in the technical possibilities signaling the beginnings of an enormous economic upturn.

The second decisive advantage of introducing jet engines resulted from the higher thrust at cruise, which, in turn, resulted in higher payloads, represented by the number of passengers. The Boeing B-707 and the DC-8 were designed for about 175 passengers and their payload capacity was 75 per cent higher than the one of preceding models like the DC-7 and the Lockheed L-1049, better known as "Super Constellation". The jet engine, which efficiently develops higher thrust at high altitudes and high subsonic speeds, made this progress possible. With the larger cabin also the operating costs per passenger could be reduced, for instance, by about 46 per cent with respect to the forerunner models such as the DC-7 (see picture on p. 351). Now, the triumphant advance of the jet-engine-propelled passenger aircraft could no longer be stopped [6].

The American airlines carried already in 1956 more passengers than the railroad; in 1957 more passengers traveled across the North-Atlantic by air than by boat. Ten years later, the passenger service by boat between Europe and North America practically ceased [2]. Air traffic exhibited in the sixties and seventies growth rates of 6–10 per cent per year, doubled its transport capacity every ten years, preferred and was interested in more powerful and more economic aircraft. Economic efficiency is certainly the most important criterion in air traffic and the incentive for further research activities in all aeronautical technology disciplines.

The further steps in the development were inevitable. The larger cabin required larger wing areas, the larger wing areas, in turn, provided a progressively larger fuel capacity and hence larger cruise ranges. The continued development of the engines to higher thrust and higher bypass ratios resulted in an improvement in the specific fuel consumption and allowed to build still larger aircraft. The Boeing B-747, originally subject of a conceptual proposal for a military transport, marks an important milestone in civil aeronautics since it represents a doubling of the transport capacity. New airport terminals were necessary, new flight control and operational procedures had to be considered and adapted. New direct links between the continents were now possible due to the range increases.

The doubling of the passenger capacity allowed to reduce airfares, which, in turn, made air travel affordable for a wider community. The forces of the market, based on supply and demand, and the perpetual demand of the airlines for more economical flight vehicles thrived and rendered civil aeronautics attractive. Contrary to other transport means, as automobile and railroad, the passenger pays all costs for air traffic control, airport and flight transport, including on-board service, with the price for his ticket. Railroad travel, which presents itself in public discussions as the more environmentally friendly alternative, and which is, as favorite of politics, supported by subsidies, is on longer distances very much slower, uses more energy and is, in addition, more expensive. The privately organized air traffic is more customer friendly and attractive than the government-supported national railroad traffic. Today's aircraft, with cruise ranges of more than 14,000 km (7,560 nm) are able to connect two cities, located on opposite sides of the globe, in a direct flight. Here, a natural limit in range for an atmospheric flight vehicle seems to have been reached solely by the duration of the flight.

Now back to the development of the German aeronautical industry and its restart after the war. There existed various ideas to introduce innovative products into the civil market and be successful. The HFB 320, a business aircraft for 8–12 passengers, employed a forward-swept wing. The forward-swept wing allowed, for the small fuselage cross section, to place the fuselage-wing junction downstream of the cabin and still realize a reasonable center-of-gravity range. Concerning flight mechanics and aerodynamics, some difficulties arose, which could, however, all be solved technically. The technical concept was interesting and successful, but the economic success did not materialize.

Something similar happened to the VFW 614 programme. Here, the task was to develop a 40-passenger transport aircraft with jet engines. To optimize the configuration by means of a short landing gear, a better circulation about the wing and low noise, an engine position above the wing was chosen. Unfortunately, a mature engine of the required thrust class was not available and the necessary funding and market experience to help this technically rather good aircraft to sufficient acceptance was lacking. Though the aircraft was in its flight performance convincing and found some customers, the market wanted larger aircraft. Technology by itself could not be sold. Today, one needs, besides the technically innovative product idea and the industrial product concept, sufficient competence in the other three critical areas: market knowledge, market admission and financial power. Global marketing research, technical know-how and financial endurance must come together. A civil program is, at the earliest, after 15 years profitable, provided the number of aircraft produced is sufficiently high.

Together with the French company Breguet and the Dutch company Fokker, Dornier formed already in 1955 a working group "Sea-Flight" ("Seeflug") and bid for a sea-reconnaissance aircraft in response to a NATO request for proposals. The project Breguet "Atlantic" became a successful European aircraft program. Based on the good cooperation with France, the project "Alpha-Jet", a trainer and light combat aircraft, followed.

Dornier pursued its ideas of the twenties concerning short and steep take-off aircraft, today termed Short Take-off and Landing (STOL) aircraft. A STOL design requires that one pays special attention to the high-lift system and the propeller/wing interaction. The Do 27, a short take-off, single-engine multiple-purpose aircraft, was introduced on October 17, 1956. It attracted international interest due to the short take-off distance of only 260 m (853 ft). The aircraft served for a long time as commuter aircraft for the air force. The concept of the Do 27 was then employed for the development of a two-engine version with the designation Do 28. The version Do 28D "Skyservant" originated in the mid-sixties; it had a cruise speed of 300 km/h (162 kn) and a cruise range of almost 3,000 km (1,619 nm). The quantity production of the Skyservant commenced in 1967. The aircraft turned out to be a successful product and a trademark of Dornier. The figures below depict the various Dornier projects related to V/STOL. The culmination of the VTOL activities of Dornier is represented by the

Multi-purpose aircraft Dornier Do 27 (1956).

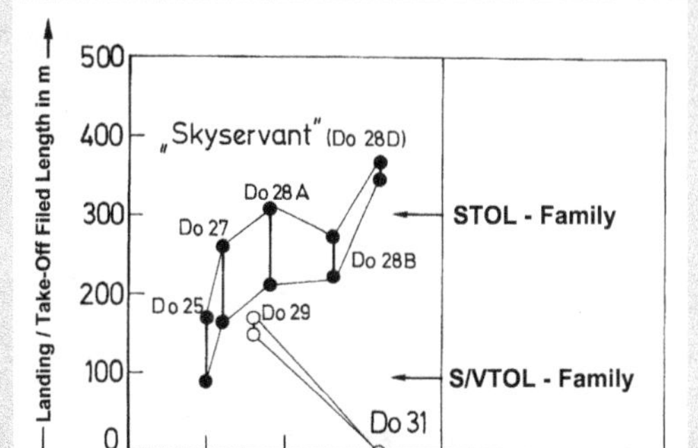

STOL and VTOL aircraft of the second Dornier aeronautical-technology-related work period.

Do 31. It was conceived as a test vehicle for a vertical-take-off transport jet aircraft. Three prototypes were built and successfully tested. The Do 31 is the only vertical-take-off transport jet in the world realized. One of the prototypes can today be admired at the flight yard in Oberschleissheim, a branch of the "Deutsches Museum" near Munich.

After many failures of the German and European aeronautical industry, which had developed technically excellent aircraft but had, in their respective home markets, only insufficient sales possibilities, and had, different from America, also not developed global market and sales strategies, European industry tried to establish cooperations. In 1969, the Airbus consortium was founded, which, after initial problems with being accepted by the market, was able to start a very successful development and is today a model for a successful and innovative cooperation in Europe. The cooperation has proven that also in a very difficult market, as the civil aircraft market certainly is, modern and very complex airplanes can be developed and sold globally despite the many very different partners and different mentalities. The different new technologies, ready to go

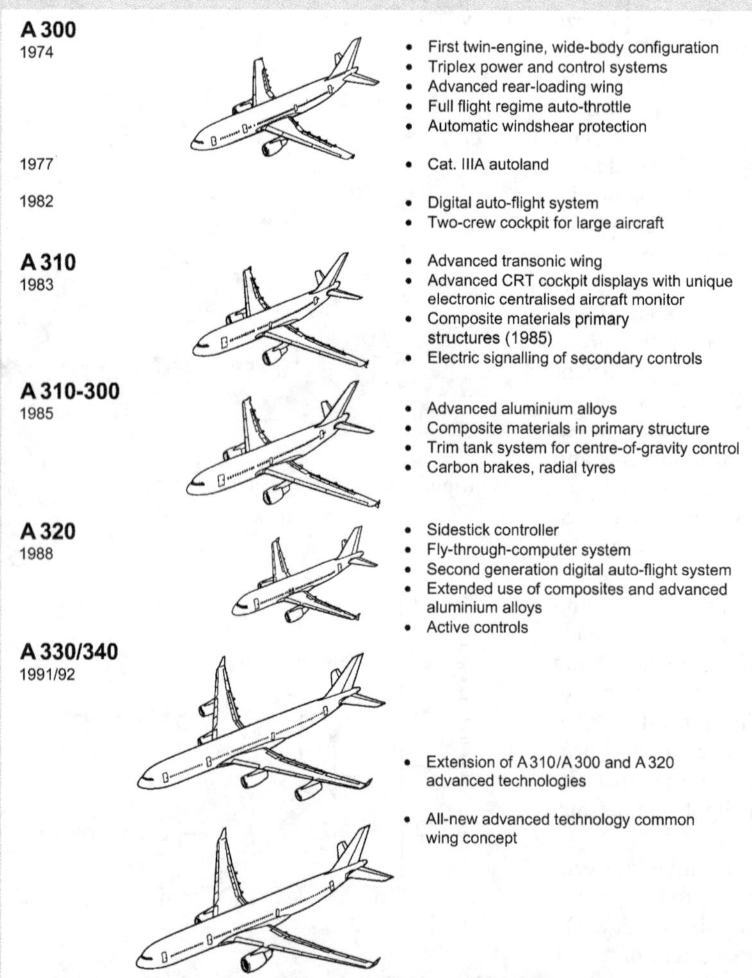

Technical "Firsts" within the Airbus program.

into quantity production, which were first utilised on an Airbus product, are summarised in the figure on the left page. All these new elements are a result of European research efforts. Here, a treatment of the technology programs in competition with others has led to competitive solutions for the final product. But one needs successful aircraft programs and consequent program strategies to apply these new technologies, which are certainly of benefit to the customer. This stimulates research and development. Here, Airbus is for the last 30 years an important motor.

On the German side, a working consortium, Deutsche Airbus GmbH, was established in 1969, where the various companies, such as VFW, HFB, Dornier and Messerschmitt, were represented. But only a few believed at that time in a sustainable success of the Airbus programmes. The basic configuration of the Airbus aircraft was definitely demanding and new; for the first time it was decided to employ for a large-capacity wide-body aircraft a configuration with only two engines. A first small success was an order by the American carrier Eastern Airlines, which became, after an extensive test phase of one year, the first customer of Airbus in North America. But the question concerning a further family member arose very soon. Airbus had, for that purpose, investigated various derivatives which were designated B1–B12. A shorter version, B10, was investigated with high priority. The question arose whether the wing should be kept and only the fuselage shortened or whether also a new wing should be developed. The latter solution was supported by the progress in transonic aerodynamics, which allowed either thicker wings at constant sweep or less sweep at constant thickness for the flight at high subsonic speeds. Airbus finally decided to go for the project A 310 with a new wing and a shortened fuselage. The customers – among

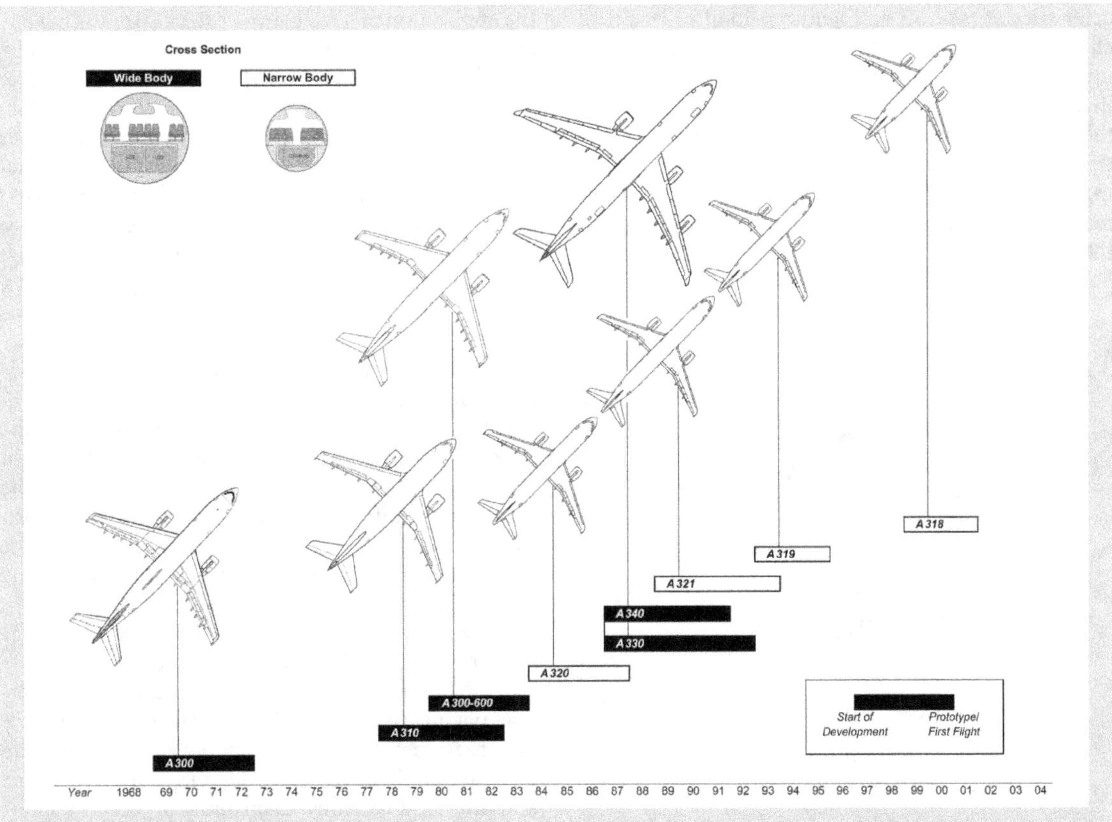

The concept of the Airbus-family.

them Lufthansa and Swissair – wanted an optimized wing, however, without stretch potential.

Among the Airbus partners, a thrilling competition arose concerning the best and most efficient wing design. The German side participated very successfully in preparing project proposals for the A 310 wing. In a direct comparison with the French wing proposal, the German design showed clear advantages at cruise conditions with equally good low-speed characteristics. Even in a direct comparison with the design of the traditional European supplier of wings, Hawker-Siddley in Hatfield, Great Britain, the German wing turned out to be on par. The German contribution within the international wing team, which initially met in Bremen and then in Hatfield, showed that also in Germany in the area of wing design, with all its difficult boundary conditions, such as a good buffet boundary at sufficient wing thickness, a high maximum lift (C_{Lmax}), and good-natured stall characteristics at take-off and landing, good roll-control characteristics and low drag at cruise, one had caught up with the world's vanguard. Finally realized was the English wing proposal; this was not so much due to the better design but a result of the greater industrial experience in production and due to the existing infrastructure for manufacturing the wing elements. During the further development of the A 310 into a long-range version, A 310-300, essential contributions of the German partner were, however, still realized such as, for instance, the vertical tail (rudder) in a fiber-reinforced design, the integration of the trim tank into the horizontal tail (elevator), and a modified wing flap, which led to a larger camber of the wing, hence to a higher buffet boundary. All three measures, developed within three technology programs, led to a considerable improvement in fuel consumption and, together with other modifications, to an almost doubling of the range. They were introduced into the aircraft series under the supervision of *Heribert Flosdorff* together with his staff *Dieter Volk, Günter Krenz* and *Reinhard Hilbig*. The wing designs for the Airbus show large aspect ratios (figure below), an indication of the aerodynamic quality. This design philosophy started in Bremen with the Fw 200 and could, via the Transall, also consequently be introduced into the Airbus family.

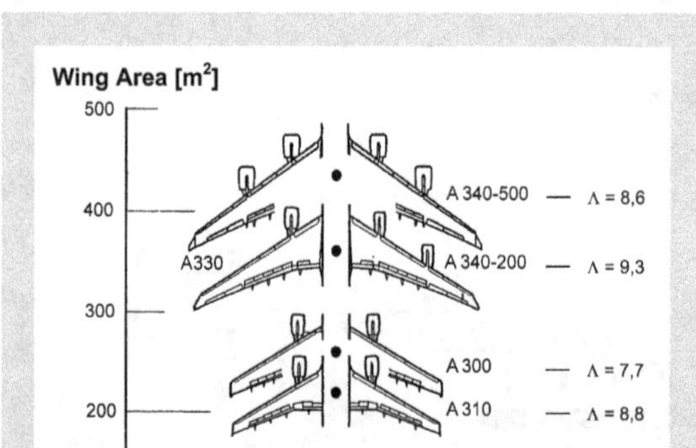

Wing areas and aspect ratios of the Airbus family.

The first flight of the carbon-fiber vertical stabilizer on the A 310 on July 8, 1985 represents a milestone in the application of such primary structures in civil aircraft engineering. The element was developed, built, tested and certified under the supervision of *Dieter Schulz* and has been installed, in quantity production, on the aircraft in airline service since December 1985. Following this success, a license for the manufacturing process was given to the Spanish Airbus partner CASA for the manufacture of the horizontal stabilizer (elevator) so that today all Airbus empennages could be made of carbon-fiber-reinforced plastics. Also the trim tank is based on suggestions and concepts of the German partner, even if its construction was not in its responsibility. The trim tank has a twofold effect. It allows using the inner volume of the horizontal tail box as tank – i. e., it allows the installation of an additional tank with little weight penalty – and, at the same time, one can keep the center of gravity at the most aft permissible position by pumping fuel between the wing and the tail tank thus decisively reducing trim drag during cruise. This technology, first used in the Junkers F 13, is in the meantime utilized with almost all long-range aircraft including the Boeing B-747-400.

A further important step was achieved by the German partner in establishing the final assembly line for the program A 321/A 319 in Hamburg. Decisive was an intensive study of the assembly of the aircraft frame and the simultaneous furnishing of the cabin for the respective costumer, which showed a considerable potential for rationalization. *Hartmut Mehdorn*

Feeder aircraft Do 328 and Do 328 Jet with an extremely good climb performance and a high percentage of the structural weight being fibre-reinforced plastics.

has essentially contributed to the success of celebrating in 1991, for the first time, "the roll-out of a European transport aircraft" from a German aircraft factory, located in Hamburg-Finkenwerder. At the same time, new qualifications for technicians and engineers were required to cope with the new tasks, but also the "know-how" was considerably enlarged, and the testing of all aircraft functions provided a better understanding of the structural and system responsibilities of the other Airbus partners.

However, the Airbus system required that each partner put in further effort to improve the aircraft component in his responsibility. The direct potential for an improvement of the economic efficiency is, at the time, strongly related to the production costs. New research foci can, however, also be found in areas, which seemed not so attractive up to today. Thus, the aircraft cabin represents a new challenge for engineers. Each seat of the aircraft is supposed to provide, for entertainment on long-range flights, several video programs in various languages as well as radio, respectively music programs. Each passenger should possibly be able to individually select her or his movie, entertainment or music program, or communicate via the Internet with the world. For an aircraft with 400 seats in the A 330/A 340 long-haul class, this results in a gigantic data stream, the cabin thus becoming the high-tech area of the aircraft. Also modern vacuum toilet systems, with the demand for reliability, low weight and low maintenance costs, have advanced to technical masterpieces. Within new national aeronautical research programs, the cabin system, with all its challenges, assumes an important position.

Dornier had left the Airbus program, driven by the intention to maintain system capabilities and not to bear the Airbus risk, but remained a component supplier. Simultaneously, the turboprop passenger aircraft Do 228, without a pressurized cabin, was developed based on the Do 28 "Skyservant"; it was later also built in India under license. Then in 1991, production of the turboprop aircraft Do 328 commenced, which was technologically well designed and extremely economic, but still had to cope with the bad image of propeller aircraft in passenger air traffic.

After the merger of the aircraft-engineering activities of Dornier and Fairchield, the Do 328 jet evolved, which may be the starting point for a successful jet-aircraft family in the below-50-seat range.

The Aircraft Development in East Germany

Horst Prem

The timetable of selected key milestones of the German aviation history shows that the aircraft development in East Germany restarted much earlier than in the West. The 152 performed its first flight on December 4, 1958, and reached a further decisive milestone when being equipped on August 26, 1960 with four Pirna 014 engines developed in the GDR. The first post-war product of West Germany, the VJ 101C-X1, flew on April 10, 1963 for the first time. The aircraft was, however, from the start planned as joined European product.

It is not surprising that the reconstruction happened faster in the East since 60 per cent of the aircraft engineering capacity up to 1945 was located within the territory of the then called "Soviet Occupation Zone", and that, since October 22, 1946, a large part of the important employees were brought into the Soviet Union and released back to the GDR since the early fifties. These were the aircraft engineers of Junkers, Heinkel, Siebel and Arado as well as the engine designers of Junkers and BMW. These events are described in detail in several books published in the meantime [7–9]. Nevertheless, we will here briefly address the development history of the 152 since this medium-range aircraft goes back to a Junkers line of aircraft pursued since 1940. It also becomes clear that the experimental approach dominates the configuration-finding process, which only afterwards provides reliable interpretations.

The pictures of the Ju 287 with open bomb bay in the Brunswick wind tunnel show that this aircraft was initially configured as a bomber. The design had, nevertheless, with forward-swept wing and engine pods under the wing, the potential for a long-range turbojet aircraft. However, for transportation of passengers a pressurized fuselage was still lacking since the unaltered fuselage of the He 177 had been adopted. However, the responsible people within the Ministry of Aviation in Moscow agreed with this concept. The experimental configurations Ju 287 V2 and Ju 287 V3 were, therefore, modified and flight tested in Dessau under Soviet supervision before they were transferred to the USSR. The experimental configurations had three engines, type Jumo 004D, mounted under each wing so that basically the design of the Junkers EF 131, respectively EF 140 had been anticipated. The latter were still being developed in Dessau under the supervision of *Brunolf Baade*, respectively in the Soviet Union under *Hans Wocke*.

The first flight of the EF 140 occurred on October 12, 1949 near Sawjelowo. Characteristical for these aircraft were the forward-swept wings, which are statically and dynamically, due to an increase in angle of attack under load, harder to control. They are aerodynamically more favourable since the cross flow along the wing span turns towards the fuselage resulting in a more stable flight behaviour than for the swept-back wing. Nevertheless, also the EF 132 with swept-back wing was conceived in Dessau, a six-jet high-wing bomber aircraft with three engines mounted in each wing root. The continuation of the development was also transferred to the Soviet Union, however, the development was terminated in June 1948. The configuration selection process was carried out without being accompanied by systematic fundamental research.

The swept-back wing with boundary layer fences, which hinders the flow (essentially the boundary layer) to turn towards the wing tip, thus avoiding tip stall, became a design characteristic of the Soviet long-range bomber 150 (note that the Junkers terminology was continued). The engines were again, as in the case of the Ju 287, suspended under the wing, but now a single large engine under each wing-half. This was the last design of the Junkers team in the Soviet Union. Part of the Germans was already allowed to return home beginning in 1950. The specialists had to stay longer. It was the intention that they lose all contact to international flight sciences. They were in 1953 resettled in Sawjelowo. Here, the design of the 152 was worked out since 1953 under the supervision of *Baade*. This design was to become, after the return of all specialists to the GDR, the nucleus of the aircraft industry in Dresden.

Returning to the GDR, the Germans were not allowed to take along the design documents of the 152. The similarity between the 152 and the long-range bomber 150 was obviously too large, as indicated in the figure next page. The 152 had, therefore, to be redesigned in Dresden, starting in 1955. The Russians released the documents only after the development had reached the Soviet status.

The first flight on December 4, 1958 was followed by a grave accident on March 4, 1959 on the approach to the Dresden airport. Obviously the pilots were lacking flight technical knowledge concerning the delayed response of turbojet engines so they encountered, at too low speed, a stalled aircraft condition, which could not be recovered and led to a crash. The crash is sometimes named as cause for the end of the project 152.

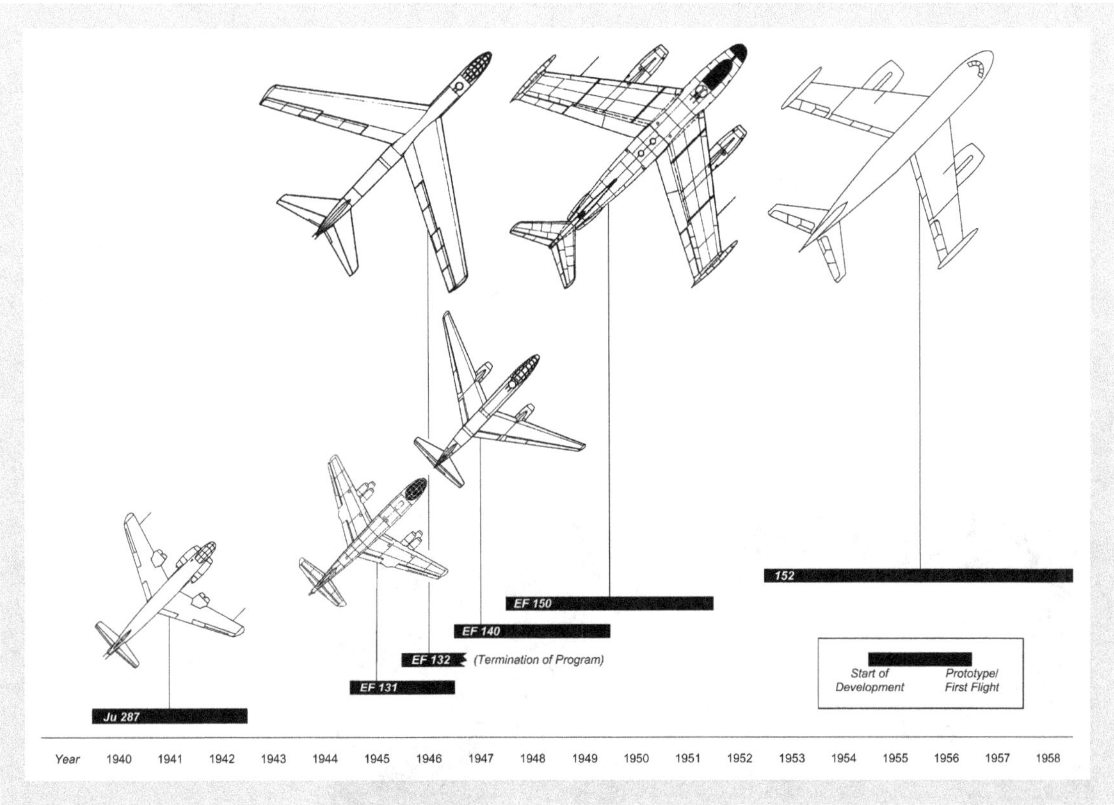

From the Ju 287 to the 152.

The medium-range transport aircraft 152 was projected in 1953 by *Brunolf Baade* and his engineering team in Sawjelowo. The development commenced late 1954 in the GDR and led to two prototype aircraft, which performed their first flights in December 1958 and in August 1960, respectively.

At about the time of the first flight of the 152, the swept-wing jet aircraft developed within the western hemisphere, the Comet, the Boeing 707 and the Caravelle, had already entered airline service. The much larger capacity, for instance, of the Boeing 707 in comparison to all other jet aircraft, and the modern design as a low-wing aircraft with four separate engines attached to the wing, has led to the triumphant advance of this type of aircraft since the direct operational costs were not to be undercut. Contrary to all other nations, US companies had conducted systematic market analyses and designed their aircraft accordingly. Also the Soviet Union wanted to participate with its own products in the international com-

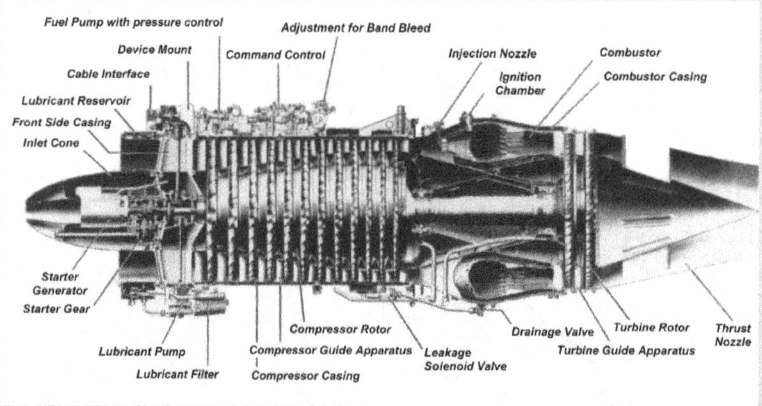

The development of a jet engine for the transport aircraft 152 started in the GDR in 1954 leading to the engine Pirna 014 with a 12-stage axial compressor, a tubular combustion chamber and a two-stage turbine. The first trial took place in October 1956.

In front of the 152 II V4: *Brunolf Baade* (left) and the chief designer of the 152, *Fritz Freitag* (right).

petition and considered as disturbing the build-up of a competing GDR aircraft industry. The promise concerning the purchase of 152 aircraft was not honored by the USSR so that the acceptance of this aircraft by the market stagnated. Although the first flight of the 152 II in August 1960, with especially designed Pirna 014 engines, could be performed successfully, this success was not able to prevent the de facto termination of the essentially self-supporting aircraft engineering in the GDR in late autumn of 1960.

On the instruction of the Soviet military authorities, the official resolution of the SED (Socialist United Party of Germany) to dissolve the independent GDR aeronautical industry was issued on February 28, 1961. This example more than clearly shows the dominance of international integration and market research. Already in the mid-fifties technical excellence by itself was not sufficient to create an aircraft that could be sold within a global market.

Rollout of the 152 II V4 on August 26, 1960 before its first flight.

The selected key milestones of the German aviation history show that the West-European countries still needed more than a decade to introduce into market an internationally competitive product in the category "wide-body short- to medium-range aircraft" with the first flight of the A 300 in October 1972. Even after this success, a long lean period had to be overcome before the Airbus could be turned into a commercial success. This (also national) success could only be reached via the European cooperation of the German aeronautical industry.

So not technology but the island-like situation of an aircraft industry in the GDR and the lack of a systematic market analysis during the product definition phase were the real reason for the failure of establishing an independent, and with the USSR competing, aircraft industry.

Thereafter, aircraft engineering in the GDR had to concentrate on maintenance and repair work primarily for Russian products. There remained no leeway for aeronautical research in the GDR.

The Development of Sailplanes and its Highlights after 1951

Hans Zacher

After the end of World War Two, gliding was again permitted by the Allied Control Council in 1951. Many gliding associations, and most of the "Akafliegs" and "Idafliegs", were newly founded during that year. The DVL was able to build up an "Inspection Office for Aeronautical Equipment" ("Prüfstelle für Luftfahrtgeräte, PfL") in Munich, however, initially only for non-powered aircraft. Similarly, offices were established at the DFL in Brunswick and at the TH Darmstadt, however, either on a temporary basis or only responsible for particular areas. The offices had to examine sailplanes, including equipment, according to the old construction regulation of 1939; gliders appeared everywhere, were either bought in foreign countries or had previously been built in Germany and hidden; even new designs emerged.*

The "Organisation Scientifique et Technique du Vol à Voile, OSTIV", the successor of ISTUS, with its international relations and the "friendship among glid-

* Further detailed remarks on sailplanes and sports and cruise aircraft can be found in Vol. 19 and 23 of the book series "The German Aviation".

er pilots", was a great help in rebuilding the German sailplane activities. It was especially the American *August Raspet* [10], who encouraged the designers by his research and his reports and lectures, to think about various improvements on sailplanes. He himself had improved the glide ratio of the RJ-5 of *Harland Ross* and *Richard Johnson* from 30–40 by removing waviness, roughness, edges, gaps and unshapely cockpit covers, etc. *"Small things create perfection, but perfection is not a small thing."*

The HKS [11] of *Ernst-Günter Haase, Heinz Kensche* and *Ferdinand Schmetz*, with its laminar airfoil, achieved a glide ratio of almost 40 due to an intensive surface treatment, elastic wing controls without edges and gaps, and a streamlined cockpit following the ideas of *August Raspet*. The Ka 6, a sailplane of the 15-meter standard class, built in quantity production, achieved without such a refined (wooden) construction already glide ratios of 30. Now, former sailplane manufacturers started anew, new companies were established, and also the DFS was founded anew in 1954 in Munich-Riem, although with a reduced program [12, 13]. Soon there was success in the production and in competitions; in 1958, the first OSTIV-prize for the best standard-class sailplane (15 meter span wing without flaps) was awarded to the Ka 6 of *Rudolf Kaiser*. This gave additional impetus to the manufacturers.

Already in the mid-fifties, the fundamentals were worked out for exceptional performance increases of sailplanes made of "plastics", which are closely related to the names of *Wolfgang* and *Ulrich Hütter, Eugen Hänle, Richard Eppler, Hermann Nägele* and the Akafliegs (Academic Flying Groups) of Stuttgart, Brunswick and Darmstadt. The history of this development and of the designs and the laminar airfoils shall here briefly be considered.

Important preparatory work had been performed by *Ulrich Hütter* [14, 15], with his research results and experience from wind-energy engineering, where he used glass fiber roves for the rotor blades, and solved in the process also attachment and force-transmission problems, and by *Eugen Hänle* with his talent to correctly apply essential results to aircraft construction. *Richard Eppler* [16] and *Hermann Nägele*, two experienced model-aircraft builders from the Akaflieg Stuttgart, designed a sailplane made of glass-fiber-reinforced polyester resin (GFRP), which they named FS 24 "Phönix". The GFRP was used as balsa-wood supported outer skin for the inner sandwich

The first composite sailplane, "Phönix" (1957).

construction; the outer skin could be brought into any desired shape.

For the construction of the wing, the revolutionary method of the "negative mould" was developed. Contrary to the construction procedures of aircraft employed up to then, where first the inner structures, consisting of frames, ribs, stringers and spars, are assembled and then the skin is attached, starts the construction process in case of the new method on the very outside (today with a coat of varnish) with the outer skin, followed by the supporting material and the inner part of the sandwich structure and the spar. Finally ribs, frames and the control rods are placed into the completed outer half shells. First, the glass-fiber mats (generally Velcro-reinforced) are placed into the moulds, coated with resin and covered with balsa wood, and, finally, again covered with glass-fiber mats. These sandwich shells are placed in positions favourable with regard to stresses and their direction, which results in considerable weight reductions. This particular monocoque wing had three spars and only 13 ribs plus one stronger rib at the wing root. A large part of the glass-fiber layers merged into a wing stub, which led into a fuselage-fixed box-like main structure. Also the fuselage, with its spherically curved surface, was manufactured in the sandwich-monocoque fashion utilizing a negative mould.

For the construction of the sailplane (Phoenix) and its further development, the support of the State of Baden-Württemberg could be secured in 1956 so that the necessary weight and strength investigations

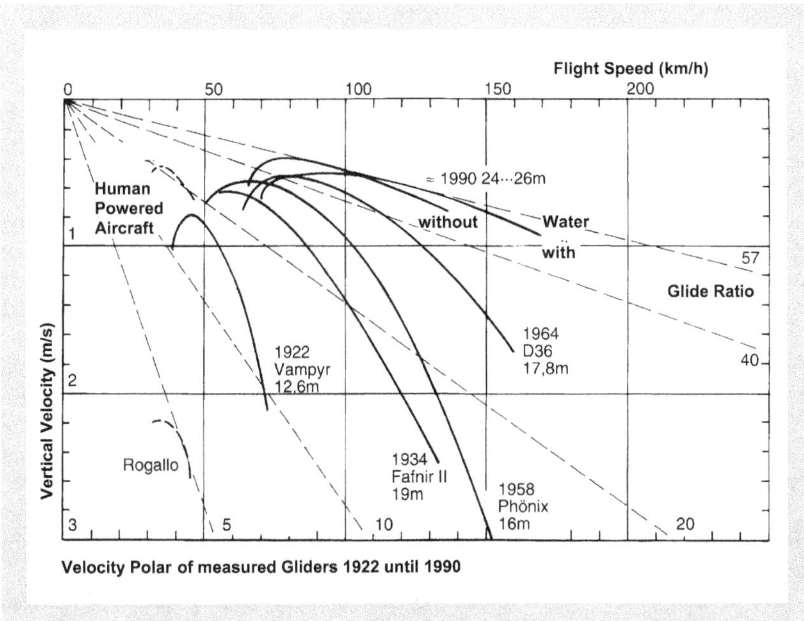

Historic and modern polars.

could be undertaken, and auxiliary equipment for the construction could be procured. Difficulties were still attached to gaining the goodwill and finally the permission of the Inspection Office for the test flights. After all, this plane was something completely new, "an entire airplane made of glass". But permission was granted due to favorable circumstances but, most of all, due to the intervention of open-minded responsible engineers.

At the same time, new laminar airfoils were developed which were adapted to the wide C_L-range where sailplanes generally fly. That these airfoils (wings) could be realized true-to-form and with a smooth surface in "plastics" was a happy coincidence, which considerably boosted performance. Already during 1935 and 1938, the D-30 was provided with a "laminar" airfoil composed of the Gö-600 droplet and the skeleton line of the NACA24xx. But the construction of the wing of wood and duralumin did not provide a true-to-form and smooth surface as could later be realized using plastics. During and after the war, NACA laminar airfoils became well known, and they were already employed on the HKS and the Ka 6. For the "Phoenix", *Richard Eppler* [17] had developed a special airfoil, EC 86(-3)-914, based on NACA airfoils, but utilizing his own computational method, with a large laminar bucket (region within a polar where laminar flow prevails). Subsequent wind-tunnel measurements in Göttingen confirmed the theoretical results and showed, furthermore, a maximum lift coefficient of $C_{Lmax} = 1.75$ (without flaps extended).

Franz Xaver Wortmann [18, 19] had, in 1955, published "a contribution concerning the design of laminar airfoils for sailplanes and helicopters". Together with *Dieter Althaus*, he erected the laminar wind tunnel in Stuttgart [20], where he investigated since 1962 his analytically designed FX airfoils for wings and tail planes with and without flaps. The results were published in a first "Stuttgart Airfoil Catalogue". Following *Richard Eppler* and *Franz Xaver Wortmann*, essential contributions to the development of airfoils and to fuselage designs were provided by *Karl-Heinz Horstmann* and *Armin Quast* (HQ) [21] at DLR as well as *Loek M.M. Boermans* at Delft University.

The first test flight of the "Phönix" was carried out on November 27, 1957 and the tests met the expectations. Already in mid-1958, a license was granted and the construction of eight planes on a experimental manufacturing basis permitted in order to gain quick and reliable experience with the new method of construction. But it still took seven years before the first larger series could be launched and further copies of the "Phönix", and also of the "Phöbus", were manufactured at the Bölkow plant in Laupheim, since the established aircraft manufacturers had met the "plastic" airplanes with skepticism.

But the Akafliegs Stuttgart, Brunswick and Darmstadt initiated further improvements by elaborate tests and intensive research. In Brunswick, *Björn Stender* designed in 1961 the SB-6, which exhibited as first sailplane the fork-tongue (yoke) connection of the two wing halves. After his studies he became independent and developed in 1962 the BS 1, the most efficient sailplane at that time. At the Akaflieg Darmstadt, the D 34 received, after some tests, in 1961 a new wing whose plastic shells were honeycomb supported. Thus commenced under the supervision of *Alfred Puck* [22] a series of studies and thesis-related investigations at the TH Darmstadt, partly in conjunction with the "German Institute of Plastics" ("Deutsches Kunststoffinstitut"), which researched many problems related to the new construction method, such as, for instance, the introduction and transmission of forces, the influence of temperature, humidity and aging, and the resistance to pressure, buckling and peeling. Herewith, a better basis for the design of the D 36 had been created, which then led, together with the BS 1, to the second generation of GFK sailplanes (GFK: Glasfaserverstärkter Kunststoff = Glass fiber reinforced plastics, GFRP). The D 36 "Circe" had a slender waisted fuselage with a long streamlined cockpit cover and a Wortmann airfoil FX 62-K1 31 with a flap, especially designed for the D 36, but later also used on other glider types. The flap provided variable camber for the laminar airfoil and hence an extend C_L-range (lift-range) for the laminar bucket. The sailplane manufacturers were now, seven years after the first flight of the "Phönix", so convinced of the quality of GFRP airplanes that they hired the designers of the D 36 after their examinations. They initiated new series of better and more successful gliders. The young designers in question were *Gerhard Waibel* at the Schleicher company, *Wolf Lemke* at Schneider, and *Klaus Hollinghaus* at Schempp-Hirth. The fourth one, *Heiko Friess*, joined the "Federal Aviation Office" ("Luftfahrtbundesamt"), where he was able, with his high technical expertise, to further the evolution of the GRFP sailplanes also with respect to air-law-conformity and technical licensing.

Wilhelm Dirks was later employed by the Glaser Company, where he introduced the D 38 of the Akaflieg Darmstadt as experimental model for the later DG 100. In 1964 the "Glasflügel (glass wing) Company" of *Eugen Hänle* had brought out the H 301 "Libelle" ("Dragon Fly"), designed by *Wolfgang Hütter*, as the second GRFP quantity production sailplane ever.

There were now sufficient manufacturers and development-enthusiastic designers, who applied the numerous research results in the areas of laminar airfoils, plastics and construction in aircraft engineering. Competing with each other, they introduced outstanding sailplanes into the market, which also stimulated exports since also foreigners had noticed, in the meantime, that the new German "Glass Sailplanes [23]" were safe and efficient. At the DLR in Brunswick and Stuttgart, at the manufacturers' and, of course, also at the Akafliegs, research was continued with the investigation of other fibers, roves, fabrics, resins, support material, etc. Within the committee "New Fiber Materials", headed by *Christoph Kensche* (DLR), results are being discussed and presented at the yearly "Symposium on the Development of Sailplanes", which was called into being by *Fred Thomas* [24] (DLR) in 1977.

That the slender wings and the airfoils, steadily getting thinner, may lead to flutter problems was early on recognized by *Walter Stender* [25, 26], for instance, who pointed out the difficulties possibly arising at high speeds. At the DLR in Göttingen, theoretical and practical investigations were carried out on numerous glider designs. *Helmut Treiber* finally demonstrated in test flights with the sailplane SB 9 of the Akaflieg Brunswick, after careful preparations, in an impressive way the excitation of various forms of wing flutter. This endeavour has been preserved in the famous flutter movie of *Horst Prem*.

In July 1972, the Brunswick SB 10 was completed, a two-seater with a span of 29 m (95 ft), for which, for the first time, carbon fibers were used. These were initially very expensive but necessary, for strength and stiffness reasons, for the center part of the extremely slender wings. The experience gathered by the members of the Akaflieg with respect to the new material and its processing became soon of benefit to industrial sailplane manufacturing, with its high aspect-ratio wings, and was later also applied in structural elements of large transport aircraft.

The two highlights in the evolution of sailplanes, namely, since 1921 the cantilever single-spar wing with torsion-nose and, since 1957, the smooth true-to-form GFRP sailplanes with laminar airfoils, were

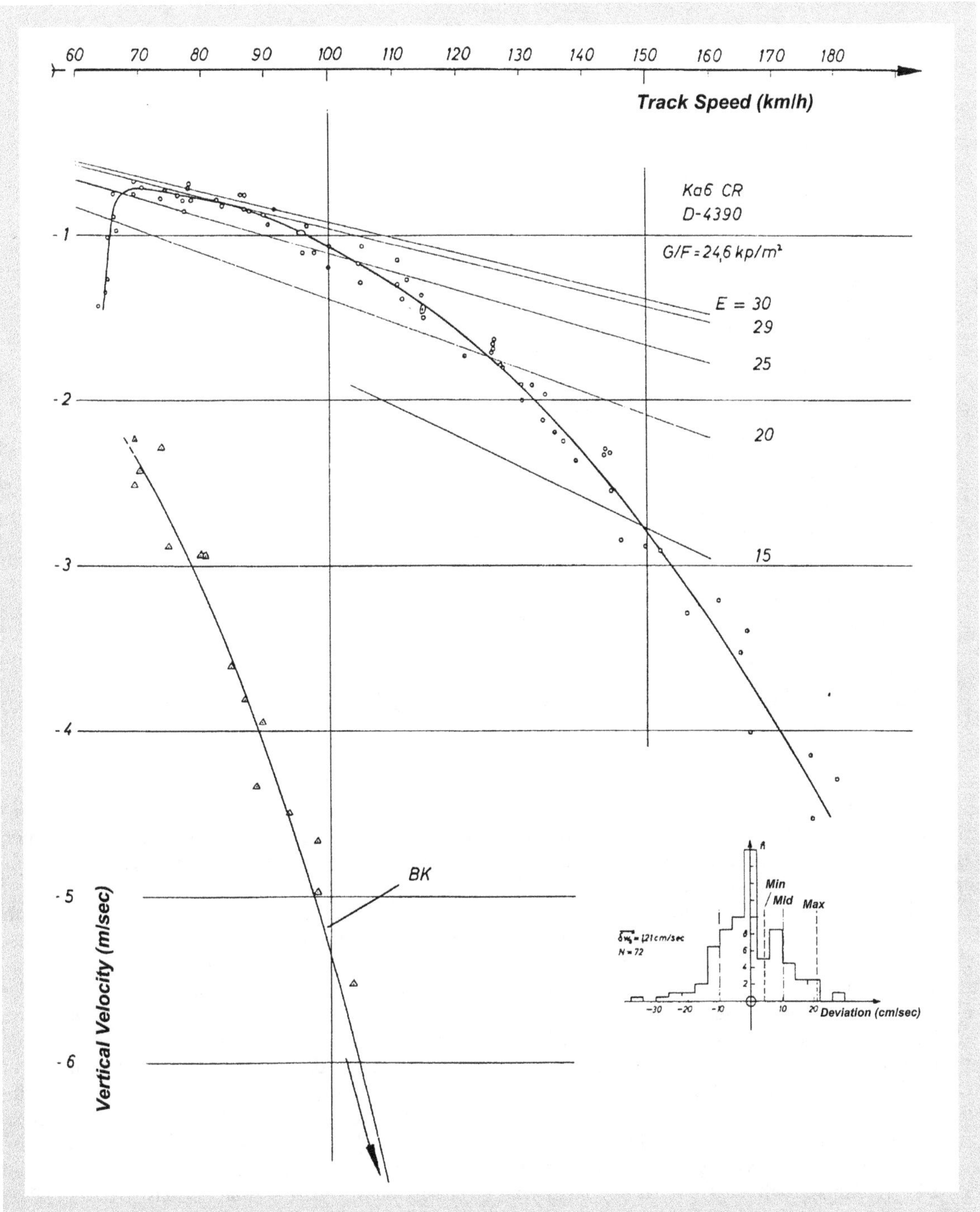

Polar measurements for the Ka 6 CR (1960); G/F: wing loading, E: glide ratio, BK: polar with air brake.

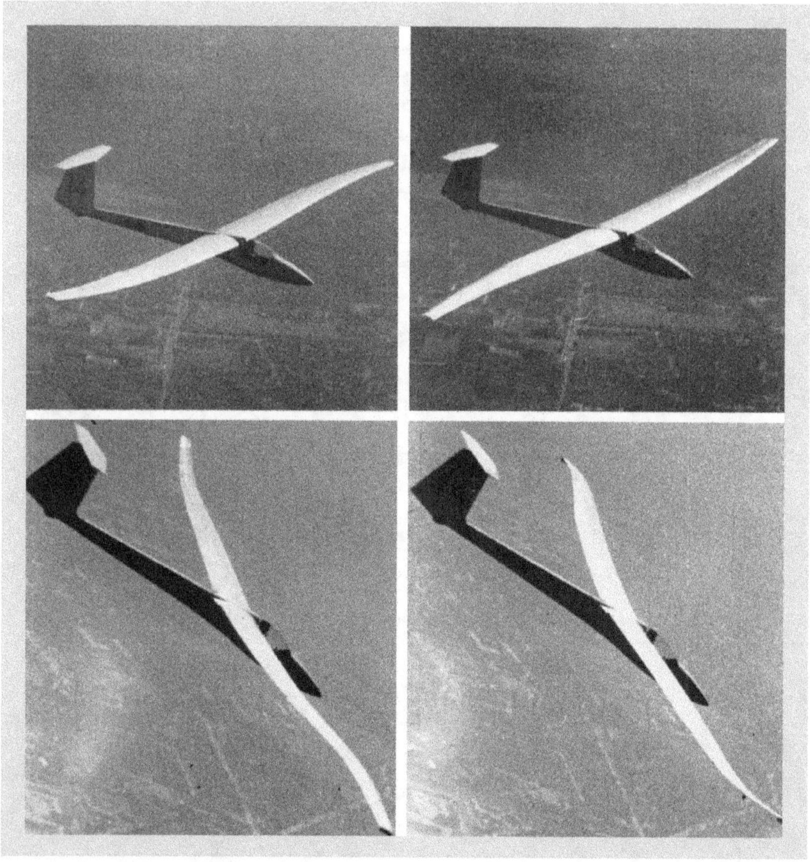

SB 9 wing flutter (1970), fundamental and first-mode oscillations.

quite spectacular improvements in aerodynamics and structural design. Further highlights are waisted fuselages, optimized wing-fuselage junctions, improved air brakes and cambered flaps, T-tails, automatically coupled control rods, and drainable water ballast. Finally to be mentioned are the detailed and precise endeavours, which made flying safer, simpler, more convenient and efficient: ergonomic pilot's seats, pilot rescue systems, variable-geometry wings, retractable cushioned landing gears and winglets. Here, considerable research and development work has been invested, which also contributed to push the glide ratio (until now) up to 60.

"Pure" gliding depends on take-off aids and retriever teams. Therefore, the trend is towards motor-gliders, i. e., a sailplane with a (retractable) auxiliary engine which allows to take off, overcome upwind calms or return to the take-off location. But it is necessary to perform more research and development work to render these engines less expensive, lighter, smaller, less noisy, safer, more reliable, and more reliably to restart even in flight.

After up to now only discussing airplanes with respect to their (sporting) performance, one must also mention that part of the success has to be ascribed to the equipment. Especially the use of the climb indicator (variometer) since 1927 has helped to detect and utilize thermal lift. All kinds of total-energy climb indicators and set-speed transducers have relieved the pilot or eased his decisions. Also the onboard computer and the Global Positioning System (GPS) are helpful and reduce the risks. The now possible digital recording of the flight path supports the conduction of sporting competitions. Meteorological research and consultation has greatly contributed to the successes.

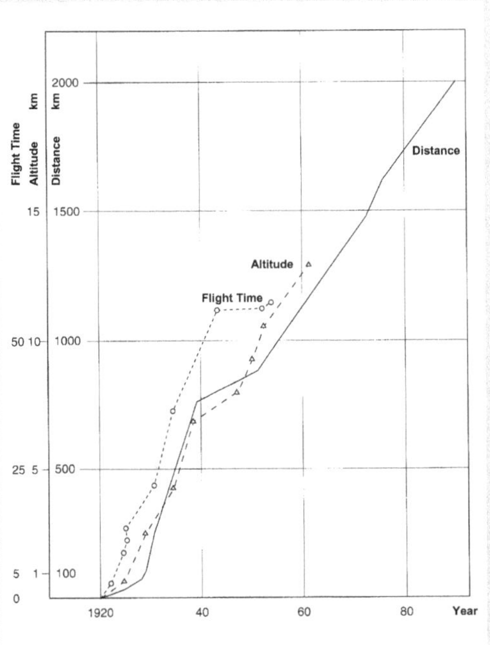

Soaring performances in duration, altitude and distance.

The sailplane "Janus" of the DFVLR (now DLR) equipped for airfoil measurements.

Since more than 60 years, the Idaflieg comparative flights [13, 27, 28] are held during the semester breaks, supported earlier by the DVL and the DFS, today by the DLR. Here, in-house designs of the Akafliegs as well as industrial products are investigated in flight. The velocity polars are determined by means of the descending altitude-step method or by a comparison with a calibrated aircraft; the flight characteristics are examined according to a given program with fixed methods and simple instruments [28] since inside gliders space, allowable weight and energy are sparse. Essential are, most of all, low-speed behavior, wing-tip stall, controllability, maneuverability and flap effectiveness. In addition, the following features and characteristics are investigated:

- wing-fuselage junction,
- stall characteristics,
- extent of laminar regions on wing and fuselage,
- location of separation bubbles,
- effect of suction and blowing,
- effect of burl and saw-tooth ribbons,
- effect of riblets (e. g., shark-skin),
- wing-tip shape, winglets,
- dust and mosquito contamination of the wing surface.

Also new gliders with variable geometry, for instance, variable wing chord or span (SB 11, Mü-27, D 40, fs-29), as well as flying wings (SB 13) were tested and examined. It is thus continuously attempted to find, sometimes in very small steps, the optimum and to contribute to the increase in performance and efficiency. By trial and error, the young engineers thus gain a treasure of experience.

The Idaflieg comparison flights are frequently described as (inexpensive) trial centers and the Akafliegs as development organizations for the sailplane industry. This research base has certainly substantially contributed to the fact that German manufacturers are the undisputed market leaders. At the same time they (the Akafliegs) are educational institutions, which well prepare the new industrial generation of engineers already during studies for their future tasks by combining theory and practical application. The research associated with gliding is also today, as it was since the time of *Otto Lilienthal*, of considerable benefit to the entire aeronautical research.

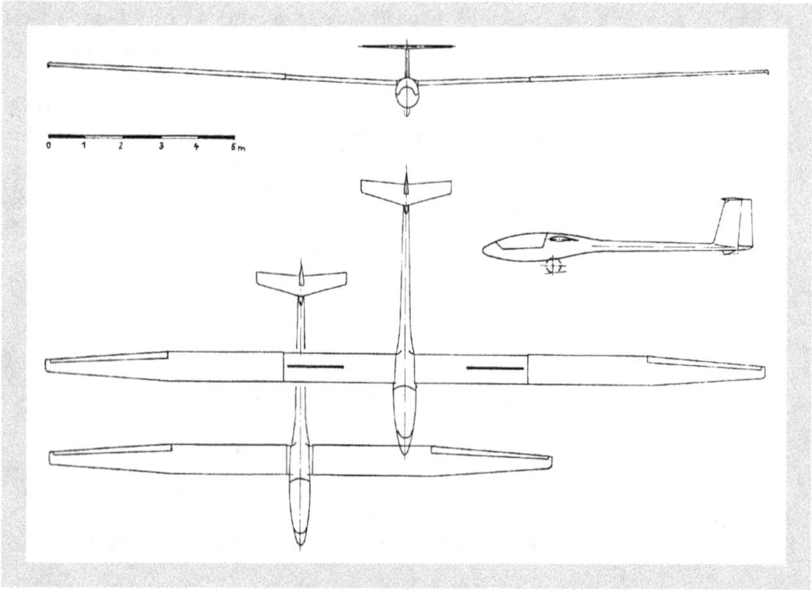

High-performance sailplane fs-29 (1975) of the Akaflieg Stuttgart with in-flight variable span (13.3–19 m; 44–62 ft).

Flying-wing glider SB 13 (1988), a result of numerous study papers of the Akaflieg Brunswick.

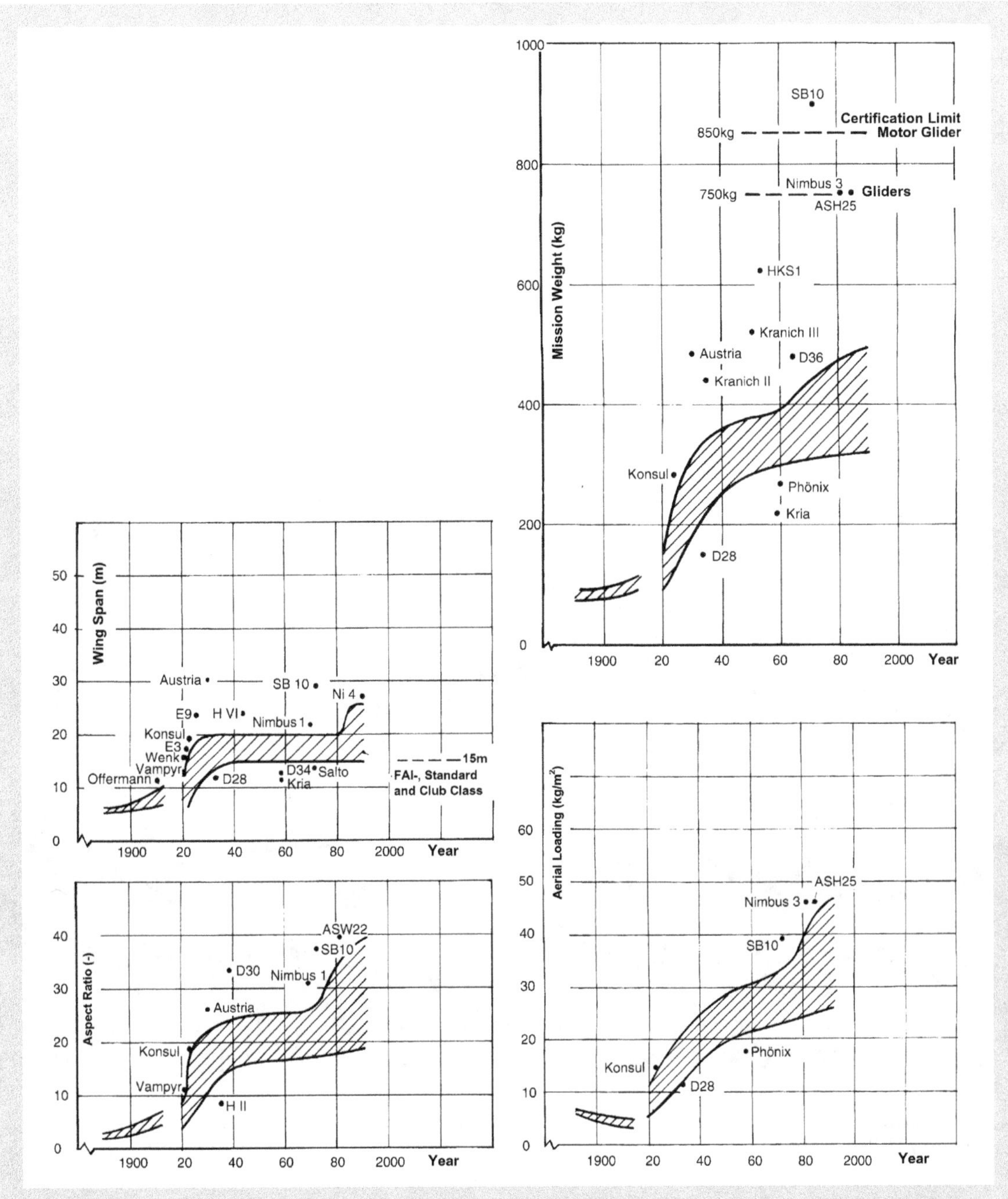

Development with time of wingspan, aspect ratio, mission weight and wing loading for single-seat high-performance sailplanes. The span of 15 m (49 ft) is the maximum for the Olympic Sailplane of 1938, the standard sailplane of 1958 and also for the FAI racing-class planes.

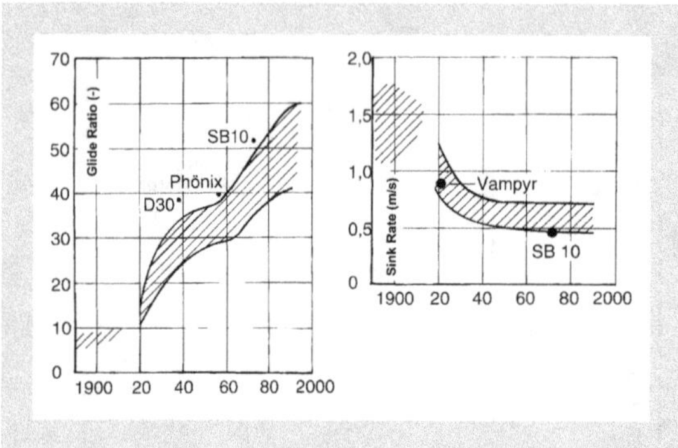

Development of the glide ratio and the sinking speed versus time. Between 1920 and 1923, span, aspect ratio, mass and glide ratio increased rapidly while, at the same time, the sinking speed decreased. After 1960, there was again an important increase, especially in the glide ratio, due to the utilization of laminar airfoils, new materials (GFRP) and designs, and fabrication methods, which allowed a better shaping. With the introduction of carbon-fiber composite structures (CFRP), span and aspect ratio could be increased again.

In the mid-sixties, several German manufacturers began the quantity production of sailplanes utilizing glass-fiber reinforced plastics (GFRP). The photograph shows the production of the "Cirrus" at Schempp-Hirth in Kirchheim/Teck.

Essential data of trend-setting and industrially produced German sailplanes during the time period 1955–1989.

Sailplane type	Designer/ manufacturer	Year	Span [m]	Wing area [m²]	Aspect ratio	Net-mass [kg]	Gross-mass [kg]	Wing loading [kg/m²]	Sinking speed [m/s]	Glide ratio	Remarks
HKS 3	*Haase, Kensche, Schmetz*	1955	17,2	14,8	20,2	242	380	25,7	0,55	37	Camber flaps, wooden construction, laminar airfoil
Ka 6A	*Kaiser, Schleicher*	1956	14,0	12,0	16,3	170	320	25,0		29	Wooden construction, laminar airfoil, quantity production
fs 24 "Phönix"	Akaflieg Stuttgart	1957	16,0	14,4	17,8	164	265	18,5	0,51	37	GFRP construction, laminar airfoil
BS 1	*Björn Stender*	1962	18,0	14,1	23,0	320	450	28,4–31,9	0,56	44	GFRP construction
D 36 "Circe"	Akaflieg Darmstadt	1964	17,8	12,8	24,8	285	410	32,0	0,53	44	GFRP construction
H 301 "Libelle"	*Hütter, Hänle*	1964	15,0	9,5	23,6	185	300	31,6	0,58	40	
ASW 12	*Waibel, Schleicher*	1966	18,3	13,0	25,8	324	430	33,1		46	
SF 27 M	*Scheibe*	1969	15,0	12,0	18,6	260	370	32,0	0,7	32	Motor glider, retractable engine
"Nimbus 1"	*Holighaus, Schempp-Hirth*	1969	22,0	15,8	30,6	370	500	28		51	
"Nimbus 4"	*Holighaus, Schempp-Hirth*	1990	26,4	17,8	39,1	480	750	31,0		60	
LS 1e	*Lemke, Schneider*	1972	15,0	9,8	23,1	235	341	32		38	
SB 10	Akaflieg Brunswick	1972	29/26	23/21,8	36,6/31,0	638/630	841/888	32–37	0,41/0,43	53/51	Two-seater, partly CFRP construction elements
fs 29	Akaflieg Stuttgart	1975	19,0/13,3	12,7/8,6	28,5/20,7	365	450	35,6/52,6	0,56	44	In-flight variable span
"Discus a"	*Holighaus, Schempp-Hirth*	1984	15,0	10,6	21,3	230	525	30,0		43	
ASH 25	*Heide, Schleicher*	1986	25,0	16,3	38,3	480	750	33,7		58	Two-seater
SB 13	Akaflieg Brunswick	1988	15,0	11,6	19,4	267	357	30,7			Flying-wing with winglets
DG 600 M	*Dirks, Glaser*	1989	15,0/17,0	11,0/11,6	20,4/24,9	305/312	525/480	34,0	0,6/0,51	45/49	Motor glider

For further data see: G. Brinkmann, H. Zacher, Evolution der Segelflugzeuge, Bernard & Graefe Publisher, Bonn, 1992 and 1999 (2. Ed.)

Powered-Lift Technology

Xaver Hafer

Definition

Powered-lift technology enables an aircraft to operate from very short run-ways, to climb and descend steeply for noise abatement during take-off and landing and further to operate in manoeuvre and cruise flight-missions as powerful as conventional aircraft.

Tailsitter

After the successful vertical take-off of the A4-missile in Germany in 1942 one believed, that it could be suitable to use this method of take-off also for aicraft to enable them to support the air defence against the allied bomb attacks.

Rocket-powered aircraft Bachem Ba 349 "Natter" ("Viper"), 1944/45. Propulsion: Walter liquid-propellant rocket motor HWK 109-509.

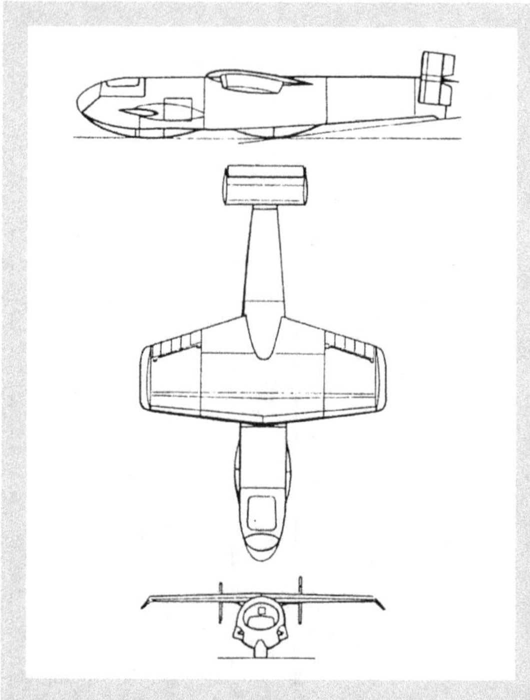

Vertical take-off project "Julia" of the Heinkel Company, 1944/45. Propulsion: Walter liquid-propellant rocket motor HWK 109-509.

The first project of this category was the rocket-aircraft "Natter"designed by *Erich Bachem* [29]. After the succesful mission the pilot should be shoot-off together with his cockpit and saved by parachute. Some take-off tests could be performed but the project did not lead to any military applications.

In the Heinkel design-office at Vienna some projects with vertical-take-off capability were created. The P-1077 "Julia" [30], powered by a Walter-rocket with liquid-propellant, was equipped with skids for horizontal landing.

Two other projects using two anticyclically running propellers in a circular wing were planed by *Karl Reiniger* [31] and *Gerhard Schulz*: The "Wespe", which should be powered by a Heinkel-Hirth turboprop engine He 21, and versions of the "Lerche", which should be powered by three different Daimler-Benz engines. These projects were also designed for vertical landings. Finally in the Focke-Wulf company *Heinz von Halem* [32] proposed a propelling wing aircraft with

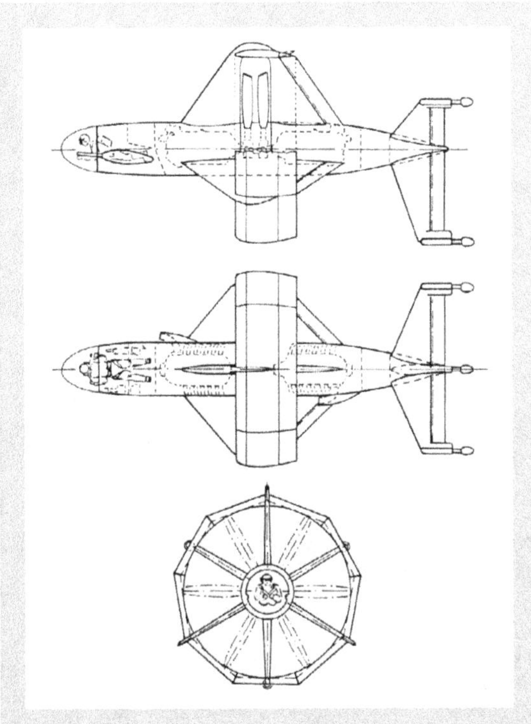

Ducted propeller (circular-wing) project "Lerche" ("Lark") of the Heinkel Company, 1944/45. Propulsion: Two flight engines DB 605 D.

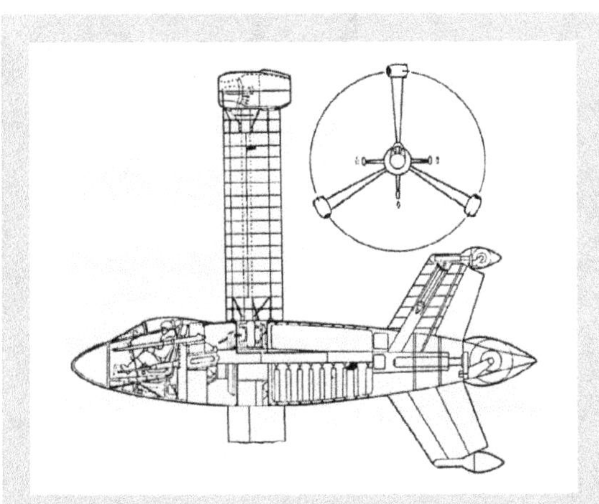

Rotary-wing propelled aircraft of the Focke-Wulf Company, 1944/45. Propulsion: Three wing-tip mounted Focke-Wulf Lorin engines.

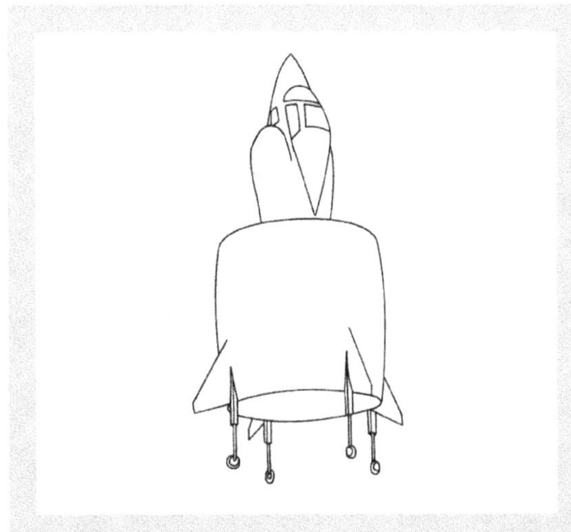

Annular-wing aircraft "Coléopter" of the SNECMA Company, 1959. Propulsion: Turbojet engine "ATAR" 101 E-5V.

vertical take-off capability. Three wings which should rotate around the fuselage were powered by ramjets on the tips of the three wings.

After World War Two in the early 1950's the idea of vertical take-off technology was revived in the USA and a number of projects with propeller and jet engine propulsion were tested as experimental aircraft in flight successfully. Here the notion „VTOL" for "Vertical Take-off and Landing" was first used for this new technology.

The last experimental powered-lift aircaft with take-off in vertical position was the "Coleopter", equipped with a circular wing and powered by an "ATAR" turbojet-engine. It was developed in France by *H. von Zborowski* [33] and his team of mostly German engineers. It performed several successful flight-tests but finally a vertical descent ended with a crash. Fortunately the pilot was saved by the ejection-seat. Later on the program then was cancelled.

The result of numerous flight-tests with tailsitter powered-lift aircraft was, that the performing of vertical descents made a lot of problems. The pilots often failed in finding out the correct coordination between altitude and sinking rate in the landing-phase, while the transition from vertitical take-off to horizontal flight was relatively easy to perform. This is the decisive reason, that later on generally powered-lift aircrafts were designed to be able to take-off and

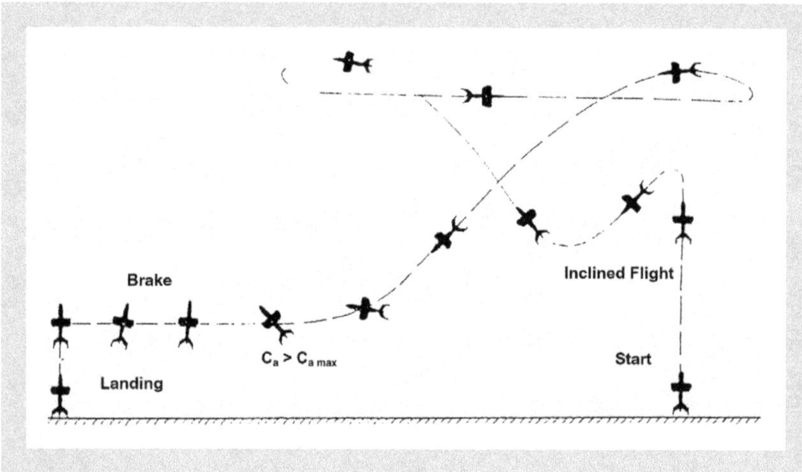

Take-off, flight and landing of a vertical take-off "tail-sitter".

descend in horizontal attitude like conventional aircraft.

The main difference of the described two systems of powered-lift aircraft can be demonstrated very impressively by considering their different take-off and landing transition. In the case of the tail-sitter system, the longitudinal axis of the airplane and with it the attitude of the pilot has to be changing during the take-off and the landing-transition about 90°, while in the other system only the direction of the thrust vector has to be changed and the pilot can remain his usual position.

The German VTOL-Program

Powered-lift aircraft with horizontal-attitude take-off ability differ in the way how the position of the thrust vector will be changed during the transition phase. Good examples for this are the three governement

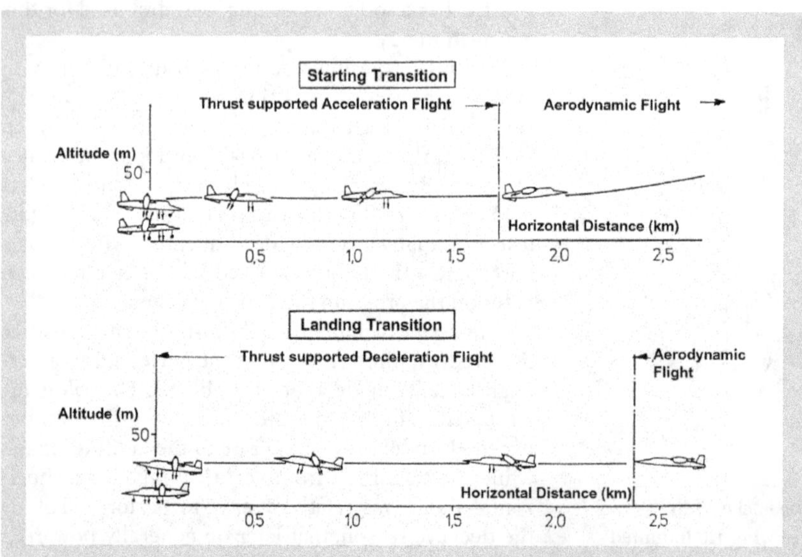

Take-off and landing transition of a horizontally starting and landing VTOL aircraft.

Vertical take-off and landing aircraft VJ 101 C of the "Development Group South"; first flight in 1963. Propulsion: X1: six jet engines RB 145, X2: two RB 145 inside the fuselage and four RB 145 with afterburners inside rotatable nacelles.

sponsored German programs of powered-lift aircraft, which will be described shortly in the following. Detailed presentation of the complete programmes are published by *Otto E. Pabst* [34], *Karl Schwärzler* [35], *Gero Madelung* [36], *Hans Pasche* [37], *Rolf Riccius* [38], *Fritz Hoffert* [39] and *Johann Schäffler* [40].

Activities in the field of powered-lift technology in Germany were started from the German Ministry of Defence (MOD) in 1957. The first project in question should be a supersonic fighter with VTOL-capability as successor of the F-104G.

By order of the MOD a cooperation of the companies Bölkow, Heinkel and Messerschmitt, the "Entwicklungsring-Süd", was established and began its activities on April 1959 at Munich.

Two versions of the new project, the VJ 101, were in discussion until it was decided to continue further with the C-Version, which should be powered by twice two jet engines with after-burner in tiltable nacelles at the wing-tips and two lift-engines in the fuselage behind the pilot-seat.

Besides the two experimental aircraft, powered with available Rolls-Royce RB 145 engines, five prototypes of the final version should be equipped with four jet-engines RB 153 for the tiltable nacelles and two lift-engines RB 162 with a gross weight of 13.4 t to be able to accomplish the demanded military missions. In 1962 the program was reduced: now only the less powered experimental aircraft VJ-101 C X1 without and the X2 with afterburner nacelle-engines could be built and tested in flight.

After a lot of detailed tests on the simulator and in flights with simple test-rigs, performed successfully by the test-pilot *George Bright*, he demonstrated with the VJ-101 C X1 on April 10, 1963 the first hover-flight. In September 1963 the complete transition including vertical take-off and landing was flown. It was also demonstrated, that each phase of the transition could be performed in a stationary flight.

Finally in July 1964 for the first time the sonic Mach-number could be exceeded by the X1 and in October 1964 the complete transition of the X2 with afterburner nacelle-engines was performed. Until 1971 the X2 was employed as powered-lift test airplane in several international flight-test programs mostly in the VTOL-phase.

The second powered-lift project, the VTOL transport aircraft Do 31, designed by the Dornier company, started in 1961. It was powered by two Rolls-Royce "Pegasus" lift-cruise engines and four lift-engines in two nacelles on the the outer wing. Two test rigs were built, from which the second was powered with the original set of engines and which started its first hover-flight with *Drury Wood* as pilot in January 1967. From the two built airplanes the first should be used for aerodynamic tests only, for which neither lift-engines nor VTOL-systems were necessary. It made its first flight in February 1967.

In July 1967 the first hover-flights began with the final VTOL version of the Do 31. The complete transition flight succeded in February 28, 1968. In October 1969 the program was canceled. At the Hannover air show

Vertical take-off aircraft Do 31 of the Dornier Company; first flight 1967. Propulsion: Two Bristol-Disseley "Pegasus" lift/cruise engines and twice (i.e., at each wing tip) four Rolls-Royce RB 162-4D lift engines.

1970 the Do 31, as well as the VJ 101 six years before, could demonstrate its excellent flight-behaviour in numerous conventional and VTOL-flight-presentations.

The "NATO Basic Military Requirements (NBMR 3)" for a low-level reconaissance and strike aircraft was the base for the German MOD to replace the low-level-strike aircraft G 91 by a new project with VTOL-capability.

Focke-Wulf at Bremen started with a new approach for this competition, the FW 1262, which was powered by a lift-cruise engine RB 193, a lighter version of the "Pegasus"-engine, and two lift engines RB 162 in front of and behind the main engine.

The project should be managed as a German-Italien cooperation. In May 1963 the requirements as a base for a competition were available. The project winner was the FW 1262, later on called VAK 191 B. Responsable for the project-management was VFW-Bremen, an association of the companies Weserflug and and Focke-Wulf, founded in the end of 1963. The number of aircraft was reduced to three from originally planned six. Also here test rigs were built to check the VTOL-systems in flight. On September 10, 1971, the first hover-flight was performed by *Ludwig Obermeier*.

For the further flight-testing the three VAK 191 B were transported to the German air force test-center E 61 at Manching, near Munich, were in October 1972 the complete transition-flight succeeded. The further flight-testing, partly in cooperation with the

Vertical take-off aircraft VAK 191 B of the "United Flight-technological Company" ("Vereinigte Flugtechnische Werke, VFW"); first flight 1971. Propulsion: one Rolls-Royce-MTU thrust/lift engine RB 193-12 and two Rolls-Royce lift engines RB 162-81.

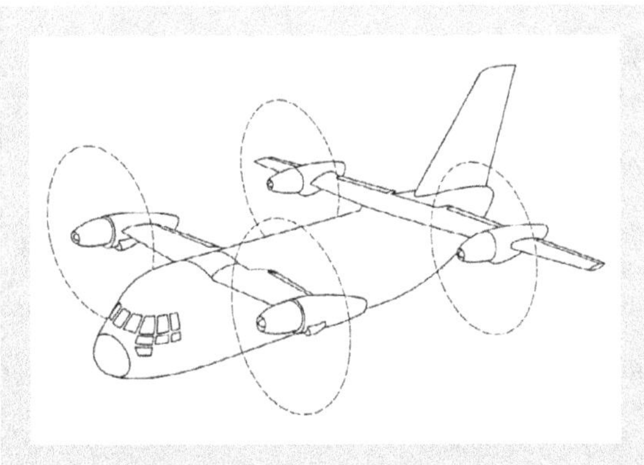

Vertical take-off project VC 400 of the "United Flight-technological Company" ("Vereinigte Flugtechnische Werke, VFW"). Propulsion: four General-Electric turboprop engines T-64-GE16 with four propellers connected by a shaft system.

US-navy, ended in October 1975 as the program was cancelled.

Beginning in 1963 *Siegfried Günter* created the idea of a powered-lift aircraft with tandem wings and four rotors at the wing tips. They could be tilted about 90° together with the wings to reach the vertical flight position.

In 1964, the Heinkel-team changed from EWR to VFW, working now on this project, the VC-400, in a separate group at Munich very successfully. Also sponsored by the MOD, in the years until 1969 intensive activities in research and development could be performed including numerous windtunnel tests, the definition of the different systems and the construction of the structure. Rotors, engines and gears were ordered from Hamilton Standard, USA. In a further phase it was planned to start with ground-tests of a test-rig, which later on should be able to execute hover flights, but in spring 1971 the total program was cancelled.

Research Subjects

While developing the German VTOL projects, a great number of new problems had to be solved, which were unknown from earlier conventional aircraft developments. Besides the particular research efforts of the scientists and engineers in the aircraft industry valuable contributions by the German research establishment followed later.

Basic information on general VTOL research problems are given in a text book of *X. Hafer* and *G. Sachs* [41]. In the following a summary of the most important reseach subjects and the correspondent published contributions will be described.

Ground Effects

Ground effects of a VTOL airplane involve several influences. These include the base pressure on the lower portion of the surface of wing and fuselage and circulation beneath the wing. This can reduce the available lift and thrust considerably in the vertical take-off phase. When approaching the ground this effect will be increased. Therefore it is called jet-induced or ground-effect.

A further influence on the effective lift-thrust can appear due to reingestion of hot gases into the air-intakes, when approaching to the ground, called "thermal ground-effect".

To the problem "ground-effect" a lot of investigations were performed. In 1968 *Eckart Schwantes* [42] gives a survey of 157 papers, which were published since 1960. *Heinrich Hertel* [43] reports on this effect in

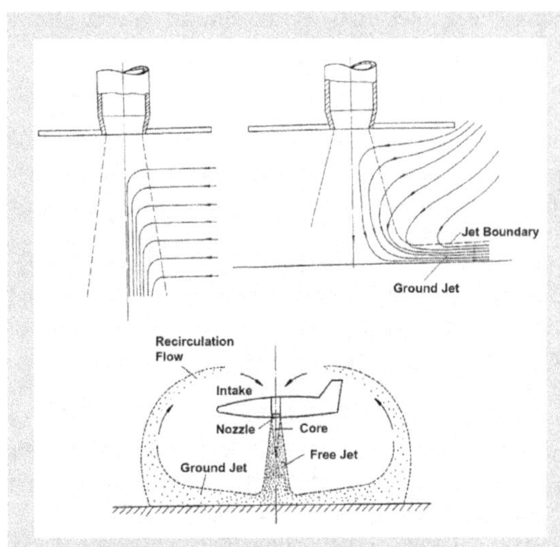

Jet effects. *Upper left:* Air entrainment in case of the free jet, *upper right:* Air entrainment in the case of ground proximity, *below:* Re-circulating flow in the case of ground proximity.

Heinrich Hertel (1901–1982).

connection with basic approaches to vertical take-off and landing and *Wolfgang Herbst* [44] from the same institute carried out experimental investigations.

Wilhelm Seibold [45] presented fundamental experimental results how the ground effect will be influenced by different arrangements of the jet engines and gives first ideas for its estimation. *Helmut Langfelder* [46] published relevant results of the EWR and *Rudolf Leistner* [47] is the author of an EWR-Report on wind tunnel results of a project with engines arranged along the center-line of the aircraft which shows extreme high lift losses by induced ground effects.

Dieter Welte [48] reported on the experience with ground effects when developing the Do 31 and demonstrated its importance for future applications, while *Peter Dick* and *Peter Kühl* [49] gave a detailed survey on the problems of hot-gas recirculation for the same aircraft.

In the DVL-Porz-Wahn *Gerhard Schulz* and *Günter Vieweger* [50] developed ideas to estimate the ground effect using methods of potential theory. *Eckart Schwantes* [51] started in the laboratories of the DFL-Brunswick an extensive program to investigate effects due to hot gas recirculation using a lift-engine as hot air generator. He also indicated methods for their estimation. The same problem was the matter of an experimental program perfomed by *Manfred Strauber* [52] at the TU Darmstadt, sponsered by the DFG. Here especially the influence of the temperature on the induced ground effect should be investigated.

Finally, *Jürgen Barche* [53] developed a procedure to estimate the induced lift losses of jets, arranged in line, using the method of singularities.

Heinrich Hertel, born in Düsseldorf on November 13, 1901.
1926: graduation (Diploma) from the Institute of Technology (TH) Munich in Aeronautical Engineering.
1931: PhD from the TU Berlin.
1932: start of scientific work at the "German Aeronautical Test Establishment" ("Deutsche Versuchsanstal für Luftfahrt, DVL") in Berlin-Adlershof.
1933: Technical Chief Assistant at the Heinkel Aircraft Company in Rostock.
1934: Technical Director at Heinkel. Under *Hertel's* supervision the aircraft H 111, He 100, He 76, He 177 and He 178 originate.
1939: appointment to the Board of Directors of the "Junkers Aircraft and Engine Company" ("Junkers Flugzeug- und Motorenwerke AG") in Dessau.
1940: Technical Director and Chief of Development at Junkers. Development and construction of the Ju 287 and other jet-propelled bomber aircraft.
1955: appointment as full professor at the Technical University Berlin; up to his retirement he is head of the "Institute of Aircraft Engineering" ("Institut für Luftfahrzeugbau") and technical adviser to the "United Flight-technological Company" ("Vereinigte Flugtechnische Werke, VFW") in Bremen. Honorary PhD from the TH Aachen and Honorary Citizen of Huntsville/USA (Space Center); author of numerous scientific publications concerning design and construction in modern aircraft engineering with the focus on biomechanics. His wealth of ideas made him to one of the leading personalities of German aircraft engineering.
1982: he dies in Berlin.

Ground Erosion

Hot gases ejected from the turbojet engines vertical to the ground can be the reason for ground erosion, which depends on the ground nature, the time of influence and the temperature of the gas.

Early investigations to this problem are published by *J.P. Versigot* and *E. Gire* [54] in 1960. *Fritz Hoffert* [55] and co-authors decribe a ground testing rig with which they investigated the influence of different parameters on the ground erosion.

Hans Brinkmann [56] reports on activities to protect the runway by premanufactured plates of plastics, which were tested on the airfield of Manching using the afterburner jets of the VJ 101 C X2 standing on the runway with tilted nacelles.

Rotor-Wing Interference

For powered-lift aircraft with rotors or propellers the deflection of the thrust vector can be achieved by tilting the propeller at fixed wing or together with the wing, as it was proposed for the VC 400. In both cases a complex air flow will be generated for the wing propeller system.

Detailled investigation were performed by *Hans Martin Dathe* [57] from Dornier, considering mainly propellers at high angle of attack.

Gerhard Löbert [58, 59] investigated flow separating effects of tilting wing aircraft during the landing transition and developed methods to calculate the air flow of a wing in the slipstream of a propeller.

At the TH Darmstadt *Bernd Sträter* [60] solved the problem of the interaction between wing and propeller at arbitrary angles of attack with unequal flow distribution in the propeller plane and considered also non-linearities of the wing flow. *Friedrich Kowalke* [61] reports on flight-physical phenomena of tilt-wing tandem configurations.

Jet-Engine Air Intakes

To guarantee an optimal thrust-generation of the jet-engines flow-separation in the air intakes is to be avoided reliably in all phases of the flight. *Gerhard Richter* [62] gave a general survey on the problems of air intakes of vertical take-off airplanes. Lift-engines, normally installed in vertical position, need for a good air-supply deflectors in front of the intakes. This may be cascades as applied to the Do 31, or a system of intake flaps, which are closed during the conventional flight as it was applied for the VJ 101 C aircraft. This system is described in detail by *Helmut Langfelder* [63].

For tilting engines the design of the inlet requires special carefulness, particulary if supersonic intakes are in question as with the VJ 101 C. In the begin of the design work, several specialists did not believe, that this problem could be solved. *Siegfried Günter* [64] proposed a solution where the supersonic front part of the intake should be shifted forward and by this a well-rounded secundary intake was activated. With wind tunnel tests it could be demonstrated, that this system did not show any flow-separation in the complete transition phase. This also later on could be proved in flight. *Xaver Hafer* [65] presented these results first in 1960 on an international meeting at Bonn. Further details of this intake-design are described by *Karl Knauer* [66], *Alexander Erismann* [67], *Felix Aulela* and *Kurt Lotter* [68].

J.S. Hart [69] from Rolls-Royce confirmed on a meeting at Munich the excellent behavior of this intake design in reporting about full-scale wind tunnel tests with the original engines and intakes at Hucknell-UK.

Control and Stabilisation

In hover-flight and parts of the transition the conventional control surfaces have no or no sufficient effectiveness, respectively. Reaction controls are necessary for which the control nozzles are working at the wing tips for roll and yaw control and at the rear fuselage for pitch control and for which a supply of bleed air from the engines is necessary. Normally the control-nozzles are joined with the lever of the conventional controls.

VTOL-aircraft with excentrically arranged engines offer as advantageous solution to use the engines as controls, where an anticyclic thrust change generates control moments. This method of control, also proposed by *Siegfried Günter* [70], called "thrust-modulation", could be proved by means of a test-rig, called "see-saw", which was investigated successfully with an original lift engine at Hucknell-UK.

Reaction controls generate control accelerations, which lead at certain control inputs to increasing attitudes. To obtain a certain attitude it is necessary to let follow a reverse control input of the same magnitude.

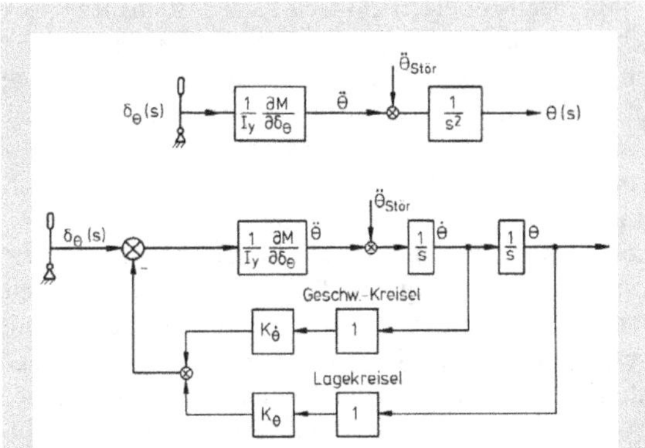

Control schemes applied during hovering flight. *Above:* acceleration control, *below:* closed-loop attitude and speed control.

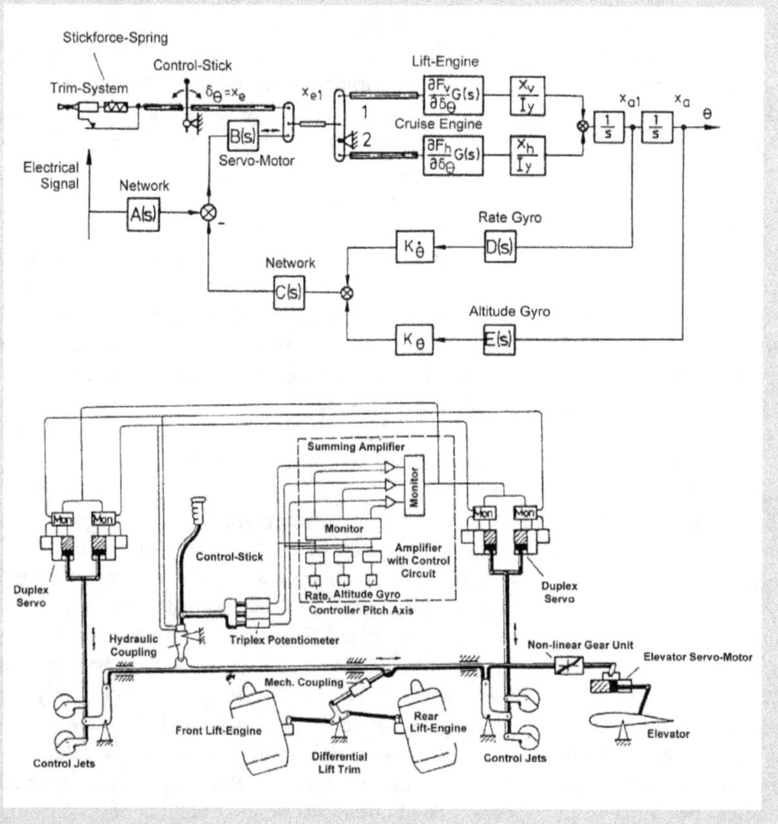

Schematic build up of the pitch-control system of the VJ 101 C (*above*) and the VAK 191 B (*below*).

By means of electrical feed-back systems it is possible to reduce the control efforts for the pilot, to the effect, that a control input will cause either a change of the angular rate or of the attitude. These kind of control systems are called "rate-control" or "attitude-control". The last one was preferred for the pitch and the roll axis. For the VJ 101 and the Do 31 pilot's control had a predominance, if a failure in the electrical system would appear. The VAK 191 was equipped with a fully electrical control system. If a failure in the system occurs, the pilot could engage a stand-by mechanical control system, which normally was inactive.

A lot of simulator and flight-test research had to be done, to get an idea of the optimal control power to be installed in the three axes at hover flight. One of the results were the "Recommandations for V/STOL Handling Qualities [71]" which were published by AGARD in 1962 and in a revised form in 1970.

A review of the control by thrust modulation was given by *Xaver Hafer* and *Hans Martin Franke* [72]. As result of evaluating a great number of flight tests with the VJ 101 C and the Do 31 respectively a survey of the statistically required control-power is given by *Johann Schäffler* [73] with collegues and by *Gerhard Schweizer* and *Hans Seelmann* [74], respectively.

In hover flight and parts of the transition powered-lift aircraft do not have inherent stability. Therefore systems are needed to stabilize the aircraft artifically. To clear up these problems a lot of research work was done of the system engineers charged with these problems in the aircraft industry in cooperation with national and international specialists.

First publications on these problems are given by *Waldemar Möller* [75] from the equipment industry and *Günter Ernst* [76,77] reporting from the flight experience with the coleopter. *Rolf Staufenbiel* [78] from

VFW reports on lay-out problems of artifical stabilisation systems, specially considering failures. Other noteworthy contributions are given by *Wolfgang Burghardt* [79], *Gerhard Kissel* [80], *Hans Martin Franke* and *Georg Döppner* [81] from the EWR and *Gerhard Schweizer* [82] from Dornier. *Wolfgang Burghardt* and *Gerhard Kissel* [83] developed in cooperation with the "Bodensee Gerätewerke" an altitude control system for VTOL applications.

Later on contributions followed from the DFL Brunswick by *Manfred Rodewald* [84] as well as a comparison study from stability augmentation systems of different VTOL aircraft by *Wolfgang Kubbat* [85]. In a final Dornier-Report *Günter Schmidt*, *Heribert Friedel* and *Gerhard Schweizer* [86] gave a survey on problems of stabilization and guidance concerning future VTOL-landings. The investigations and research results are mainly considering aspects of the cooperation between pilot and control system of the aircraft and contain also experiences with certain types of automatic landing systems.

Hans Georg Schumann [87] and *Hubertus Schmidtlein* [88] gave a report on the importance of the flight simulator in cooperation with the pilot with respect to the layout of the VTOL systems in accordance to the dynamics of the aircraft and in preparing the flight-tests.

Aircraft Noise

The noise level of VTOL-aircraft at take-off and landing is much reduced in comparison to convetional aircraft, in spite of their higher level of engine-power, due to the capability to climb and descend steeply. This results in greatly reduced ground-areas of high noise-level in the surroundings of the air-field [89].

Peter Hamel [90] of the DLR-Brunswick published a comparison of the reduced ground areas of high noise level of the landing of VTOL and conventional aircraft. At the TH-Darmstadt *V. Nitsche* [91] investigated – sponsered by the DFG – the influences of the most important parameters and the kind of power-system of VTOL aircraft at take-off and landing on the size of the ground area of high noise level.

Peter Bartels [92] presents test results of the noise field around the Do 31 and *Friedrich H. Schmitz* [93] discusses the coordination between noise measurements below the flight path in comparison with the total noise level in the surroundings of the airfield. Finally, *Alfred Schneider* [94] compares the problem of noise reduction with future aircraft developments.

VTOL Air Traffic Realization

The German programs of different military aircraft with VTOL capability initiated a lot of activities in research and development of the VTOL technology and proved its feasibility and effectiveness.

At this time also the question of an application of the VTOL technology to civil air transport was often discussed. *Rolf Stüssel* [95] reports on systematic investigations and project studies of feasible engine aircraft combinations at Focke-Wulf and at the TU-Berlin. *Dieter Stüssel* [96] describes the flight coordination of future VTOL-aircraft with the environment of the airport.

Also the project VC-400, proposed by *Siegfried Günter*, was mainly designed for future civil applications although it was sponsored by the MOD. In the scope of this project *Alfred Schneider* [97] investigated the special problems of a VTOL air traffic, considering also the project FW 300 of Focke-Wulf.

A hopeful input for a realization of the civil application of VTOL transport aircraft came from an interministry aerospace board at Bonn on January 13, 1969 [98]. *Karl Thalau*, at this time the consultant of the government for aerospace questions, was invited to convene a commission of experts as advisory board for the discussion of a future civil and military application of the powered-lift technology.

Some weeks later, the representatives of the ministry, the air-force and the Lufthansa [99] discussed plans for the development of a short-haul transport aircraft with V/STOL capability, for which an introduction was expected at the end of the 1970's.

The specification for the civil version was to develop a transport-aircraft for 80–100 passengers to be transported under VTOL conditions from an airport at 600 m elevation over a distance of 800 km with a cruise speed of 740 km/h.

This specification was sent on May 30, 1969 to the companies of the German aircraft industry with the demand, to offer project studies until October 1, 1969 [100].

Five projects with different powerplants and thrust vectoring systems were evaluated by the commission. The result was summarized in a document [98] of June 26, 1970:

A) Bo 140 from MBB, Ottobrunn

A wing of large aspect ratio with four propellers, interconnected by shafts and gears, powered by prop-

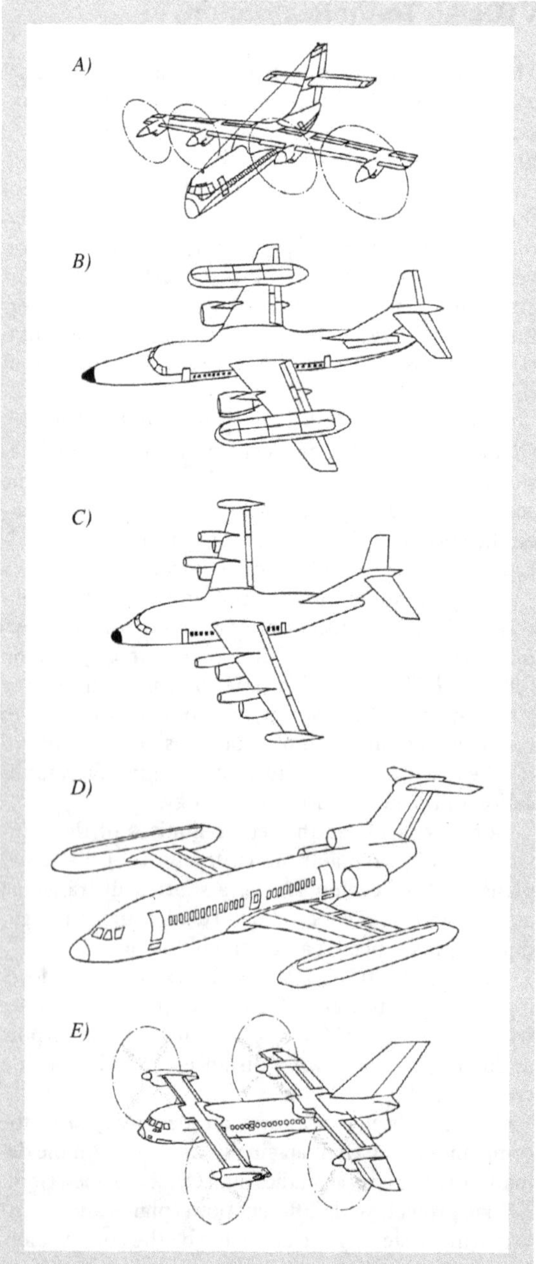

◀ VTOL-projects for air transportation in 1970.
A: VTOL-project Bo 140 of MBB powered by 4 propjet-engines GE 1 with shaft coupling.
B: VTOL-project Do 231 of Dornier powered by two lift/cruise RB 220 engines and twelve lift engines RB 202.
C: VTOL-project HFB 600 of Hamburger Flugzeugbau powered by light gas generators GE 1/10 coupled to lift/cruise fans.
D: VTOL-project VC 180 of VFW powered by three cruise engines RR TwC and two-times five lift engines RB 202.
E: VTOL-project VC 500 of VFW powered by four double propjet-engines GE T 64 uncoupled with tandem-wing configuration.

turbines GE 1, can be tilted to 90° for vertical take-off. In hover the longitudinal control is performed by a cyclic-pitch system which eleminates the need of a tailrotor. For the roll control the method of thrust modulation is applied and yawing is generated by anticyclic deflection of wing flaps behind the propellers. The gross weight amounts to 48.7 t for the civil and 62.4 t for the military version.

B) Do 231 from Dornier, Friedrichshafen
This project is a similar but enlarged version of the Do 31. It is powered by 12 turbofan lift engines RB 202 of which eight are generating lift-thrust in two nacelles at the outer wing and two are arranged in the front and in the rear of the fuselage. The two cruise engines are fastened with pilons at the inner wing and can participate in the hover-lift by activating a system of externally blown-flaps. The symmetric position of the lift engines simplifies the control-system in hover-flight. For pitch and roll control a separate group of lift-engines can be activated working with thrust modulation. For yaw control the lift engines in the nacelles will be swivelt in opposite direction. The gross weight amounts to 59 t for both versions.

C) HFB 600 from Hamburger Flugzeugbau (MBB)
This is the only offered project with lift-fan technology powered by eight gas generators GE 1/10 with pneumatic connections to four lift and cruise fans respectively. The four cruise fans are installed in nacelles under the wing. By means of cascades a thrust deflection up to 110° is possible.The further gas generators are positioned at the bottom of the fuselage. For hover control four control fans are available from which two are positioned in the rear fuselage and the

others in nacelles at the wing tips. For the pneumatic connections isolated pipelines are used. The rear fans generate pitch or yaw control accelerations depending of which control nozzle is activated. The gross weight amonts to 50,5 t and 56.5 t, respectively.

D) VC 180 from VFW Bremen
For this project separate thrust generation for hover and cruise flight is proposed. Lift-thrust will be generated by ten turbojet engines installed inclined at 15° in two nacelles at the wing tips. In the rear fuselage three cruise engines RR-Tw "C" are available. For the transition phase the acceleration is increased by the thrust component of the lift engines. Roll control in hover will be activated by thrust modulation of the lift-engine groups while for the yaw control the vanes under the nacelles will be actuated anticyclicly. The cold jet of the outer rear cruise engines can be activated up or downwards to generate pitch-control. The gross weight amounts to 55,3 t for both versions.

E) VC 500 from VFW, Munich-Department
This is a tandem tilt wing aircraft, similar to the smaller VC 400 described earlier. The main difference – apart from the size – is the fact, that each of the four propellers is powered with two engines GE T 64/S5C and so no connection by shafts between the propellers was applied. In the transition from hover to conventional flight the wings will be tilted together with the propellers to conventional position. The controls for pitch and roll are performed by thrust modulation. For the yaw control flaps behind the propellers will be actuated anticyclicly. Of all the projects, described above, the VC 500 has with 41,3 t and 46,8 t, respectively the lowest gross weight.

Final Remarks and Prospects

When examinating the five projects the commission followed the procedure, which is usually applied by the Lufthansa, when evaluating new aircraft. It contains all aspects of technics, operating and economy including the fulfillment of the dedemanded flight performance of both, the civil and the military version.

The result of the evaluation [101] was that the Do 231 was found to be the best proposal considering the technical and economical aspects, followed by the VC 180 and the VC 500. The recommendation of the commitee was to develop an VTOL aircraft on the base of the Do 231 technology.

In spite of these positive statements of the committees members about a future application of the VTOL technology for the civil and military transport the MOD failed in finding a positive decision.

As answer to a letter of *Karl Thalau* of July 12, 1971 [102] the airforce command indeed documentated its further interest for future application of VTOL technology but regrets that the budget does not allow procurements of the size to be expected for the next ten years. With this decision of the military part also the Lufthansa did not see any possibility for a further support of the VTOL idea.

The commercial air traffic is increasing with constant rates every year. The large airports are overcharged and the political problems for building new runways are well known.

In the long run there should be possibilities for an approvement by introducing an integrated air-traffic system including aircraft with VTOL technology at least for the European air traffic.

Besides it is to consider, that VTOL aircraft with todays state of the technology would be essentially more effective, reliable and noiseless than could be expected from those projects designed more than 30 years ago.

Nevertheless the efforts in the field of VTOL technology in the 1960's and the 1970's in Germany had the positive effect [103] to create teams of engineers and scientists with high level of skill to work successfully in modern aircraft developments.

Most of the systems concerning originally the advanced VTOL technology as stability augmentation systems, fly-by-wire controls or thrust vectoring for example, in the meantime found applications in conventional modern aircraft developments.

References

1 H. Ebert, J. Kaiser, K. Peters, Willy Messerschmitt – Pionier der Luftfahrt und des Leichtbaus, Die deutsche Luftfahrt, Band 17, Bernard & Graefe Verlag, Bonn 1992
2 L. Bölkow (Hrsg.), Ein Jahrhundert Flugzeuge, VDI Verlag, Düsseldorf 1990
3 H. John, The ARA Contribution to the Aerodynamic Design & Development of Military Aircraft at MBB, ARA

40th Anniversary Symposium, Held at Woburn Abbey, Bedford on May 18th 1992
4 O. Friedrich, Warum Flügel mit variabler Pfeilung?, EWR Nr. 9268, Vortrag gehalten anläßlich der Luftfahrtschau Hannover 1968
5 J. Roeder, Civil Freighters and Military Transports, Their Origins and Development, International Symposium on the Future of Cargo Transport Aviation, Straßburg, March 1993
6 W. Bansemir, Ein Beitrag zur Frage der Wirtschaftlichkeit von Verkehrsflugzeugen mit reinem Strahlantrieb auf Kurz- und Mittelstrecken, Cordoba, Argentinien, 1954
7 J. Michels, J. Werner, Luftfahrt Ost 1945–1990, Bernard & Graefe Verlag, Bonn, 1994
8 Hrsg. Lothar Brehmer, Luftfahrt in Sachsen, Ein historischer Abriß, UniMedia GmbH, Leipzig, 1998
9 K.-H. Mewes, Pirna 014, Flugtriebwerke der DDR, Aviatik Verlag, Oberhaching, 1997
10 A. Raspet, Leistungssteigerung von Segelflugzeugen durch Berücksichtigung der Grenzschichtforschung, in: W. Hirth, Handbuch des Segelfliegens, Franckh'sche Verlagsbuchhandlung, Stuttgart, 6. Aufl., 1952
11 H. Kensche, Die Entwicklung des Segelflugzeuges HKS 1, ZFW 1954
12 H. J. Merklein, Bestimmung aerodynamischer Beiwerte durch Flugmessungen an 12 Segelflugzeugen, FFM-Bericht Nr. 63, 1963
13 H. Zacher, Flugmessungen an 12 Segelflugzeugen, Aerokurier 12/1963 und 1/1964
14 U. Hütter, Neue Wege im Segelflugzeugbau, ZFW 1954
15 U. Hütter, Tragende Flugzeugteile aus glasfaserverstärkten Kunststoffen, VDI Kunststofftechnik 1960
16 R. Eppler, Kunststoff-Flugzeugbau, Z. Kunststoffe 1961
17 R. Eppler, Laminarprofile für Segelflugzeuge, OSTIV Publication III, 1954
18 F.X. Wortmann, Ein Beitrag zum Entwurf von Laminarprofilen für Segelflugzeuge und Hubschrauber, ZFW 1955
19 F.X. Wortmann, Experimentelle Untersuchungen an neuen Laminarprofilen für Segelflugzeuge und Hubschrauber, ZFW 1957
20 D. Althaus, Stuttgarter Profilkatalog I, Techn. Hochschule Stuttgart 1972
21 K. H. Horstmann, A. Quast, Tragflügelprofilentwurf für das Weltmeisterschaftssegelflugzeug SB 11, DFVLR-Nachrichten 1979
22 A. Puck, Untersuchungen über die Verwendungsmöglichkeit von Kunststoffen beim Bau von Tragflächen für Segelflugzeuge, Schweizer Aero Revue 1960
23 D. Geistmann, Die Entwicklung der Kunststoff-Segelflugzeuge, Motorbuch Verlag, Stuttgart, 1976
24 F. Thomas, Grundlagen für den Entwurf von Segelflugzeugen, Motorbuch-Verlag Stuttgart, 1979
25 W. Stender, Entwurfsgrundlagen für Segelflugzeuge, Weltluftfahrt 1952/53
26 W. Stender, Praxisnahe Abschätzungs- und Vorbeugungsmöglichkeiten gegen die Flattergefährdung von Segelflugzeugen und kleinen Motorflugzeugen, DFVLR-Berichte I, II, II, 1974
27 H. Laurson, H. Zacher, Flugmessungen an 25 Segelflugzeugen und Motorseglern, OSTIV-Publication XIII, 1974
28 Idaflieg, 50 Jahre Vergleichsfliegen, Idaflieg Aachen, 1987
29 E. Bachem, Einige Probleme des Starts bemannter Raketen, Weltraumfahrt 3/1953
30 Heinkel-Projekt "Julia", Projekt 1077, Luftfahrt International, 1977/20
31 K. Reiniger, Vorschläge zur Geschwindigkeitssteigerung durch Verwendung eines Luftschrauben-Mantelringes als Tragflügel unter Ausnutzung des Senkrechtstarts, Heinkel-Bericht, Wien, 8. März 1945
32 H. von Halem, Entstehung des Focke-Wulf-Triebflügelflugzeugs mit Lorin-Düsenantrieb, Flugwelt 3/1950
33 H. von Zborowski, Über die Konzeption von Coleopteren, Jb.WGL 1954, p. 109
34 O. E. Pabst, Kurzstarter und Senkrechtstarter, Bernard & Graefe Verlag, Koblenz, 1984
35 K. Schwärzler, Die Entwicklung von Senkrechtstartflugzeugen mit Turbinenstrahltriebwerken in Deutschland, Jb.WGLR, 1963, p. 25
36 G. Madelung, Zur Reorientierung der VJ 101C-Entwicklung in die Version "D", Jb. 1996, p. 1351
37 H. Pasche, Beiträge zur Geschichte von VFW-Focker, Denkschrift vom 31. Dezember 1974
38 R. Riccius, W. Sobotta, VAK 191B Experimental Program for a V/STOL Strike Recce Aircraft, AGARD-CP-126, pp. 6–1, 1973
39 F. Hoffert, Senkrecht startende Strahlflugzeuge (Das Do-31 Programm und andere dt. Entwicklungen) Jb.WGLR, 1968, p. 115
40 J. Schäffler, VC400 – Grundentwurf eines vielseitig einsetzbaren V/STOL-Transportflugzeugs, Luftfahrttechnik/Raumfahrttechnik, Bd. 14, 1968, p. 95
41 X. Hafer, G. Sachs, Senkrechtstarttechnik, Springer-Verlag, Berlin-Heidelberg-New York, 1982
42 E. Schwantes, Übersicht über den Bodeneffekt bei strahlgestützten V/STOL-Flugzeugen, Literaturbericht, DLR-Mittg. 68/28, 1968

43 H. Hertel, Vertikal- oder Kurzstart und Kurzlandung, Einige grundsätzliche Betrachtungen, Jb.WGLR 1960, p. 23

44 W. Herbst, Experimentelle Untersuchungen über den Austritt eines Luftstrahls hoher Unterschallgeschwindigkeit aus kreisförmigen und tragflügelähnlichen Flächen senkrecht gegen eine Wand. Institut für Luftfahrzeugbau, TU Berlin, 1962

45 W. Seibold, Untersuchungen über den Einfluß von Hubstrahlen von Senkrechtstartflugzeugen, Jb.WGLR 1962, p. 257

46 H. Langfelder, Bodeneffekte bei Senkrechtstartflugzeugen, 5. Europ. Luftf. Kongress, 1962

47 R. Leistner, Strahleinfluß- und Bodensogversuche am 1:10-Modell des Typs 320 im kleinen Windkanal Emmen, EWR-Ber. 109, 63, 1963

48 D. Welte, Die Bedeutung der aerodynamischen Strahlinterferenz bei der Entwicklung und Erprobung des V/STOL-Transportflugzeugs Do 31, Vortrag 72-06, DGLR Jahrestagg. 1972

49 P. Dick und P. Kühl,Heißgasrezirkulation beim Strahltransporter Do 31 und ihre Bedeutung für zukünftige V/STOL Entwicklungen, BMVg-FBWT 72-29, 1972

50 G. Schulz, G. Viehweger, Sekundärkräfte an einem Senkrechtstarter beim Schwebe- und Transitionsflug mit Berücksichtigung des Bodeneffekts, Jb. WGLR, 1967, p. 152

51 E. Schwantes, Berechnung der Rezirkulationsströmungen bei VTOL-Hubtriebwerken, Jb. DGLR, 1972, p. 306

52 M. Strauber, Untersuchungen zum Temperatureinfluß auf induzierte Strahleffekte, Bericht z. DFG-Vorhaben Ha 514/3, 19, 26, 1972

53 J. Barche, Developing a Prediction Method for Jet-Induced Effects of Arbitrary Jet-Lift V/STOL-Configuration, VFW-Rep,Topic 15, Contract Nr. NO 0140-74C.0113, 1975

54 J. P. Versigot, E. Gire, Terrain de Decollage ou Atterissage pour Avion V/STOL, AGARDograph 46, p. 335, 1960

55 F. Hoffert, N.N. Riemann, Entwicklung des Transportflugzeugs Do 31 und die Herstellung und Erprobung von zwei Experimentalflugzeugen, Dornier-Bericht GE 51-481/69, 1969

56 H. Brinkmann, Herstellung und Erprobung einer Kunststoffplatte für Senkrechtstarteinsatz, EWR-Bericht 11/66

57 H.M. Dathe, Über die Verteilung der induzierten Geschwindigkeiten in der Umlaufebene einer schräg angeströmten Luftschraube und ihren Einfluß auf die Luftkraftverteilungen und Blattbeanspruchungen, ZfW. 11, Heft 5, 1963

58 G. Löbert, Experimentelle Untersuchung zum Problem der Strömungsablösung bei Kippflügelflugzeugen in der Landetransition, DGLR-Mitt 69-06, pp. 191–236, 1969

59 G. Löbert, Berechnung der Aerodynamik des von mehreren Strahlen beaufschlagten Tragflügels, DGLR-Ber. 70-057, 1970

60 B. Sträter, Ein Beitrag zum Problem der Propeller-Flügelinterferenz, Diss. am FB Maschinenbau der TH Darmstadt, 1976

61 F. Kowalke, Flugphysikalische Aspekte der Entwicklung von Kippflügel-Tandem-Konfigurationen, DLR-Mitt. 69-06, p. 70

62 G. Richter, Probleme der Luftversorgung von Strahltriebwerken für Senkrechtstartflugzeuge, Jb. WGL, 1959, p. 163

63 H. Langfelder, Die Probleme der Lufteinläufe bei Senkrechtstartflugzeugen mit TL-Triebwerken, Jb. WGLR, 1963, p. 233

64 S. Günter, Regelbarer Lufteinlauf, insbesondere für Strahltriebwerke, Auslegeschrift 1078375 des Dt. Patentamtes vom 24. März 1960, Anmeldetag 8. August 1958

65 X. Hafer, Windkanaluntersuchungen zur Entwicklung eines Triebwerkeinlaufs für starke Schräganströmung, Vortrag auf Tagung des Scientific Advisory Board der USAF, Bonn, 1960

66 K. Knauer, Untersuchungen an einem Überschalleinlauf für Doppeltriebwerke, JB. WGLR 1963, p. 219

67 A. Ehrismann, WK-Messungen im Unterschall am Verdichtereintritt eines Gondeleinlaufmodells, EWR-Report Nr. 17/1961

68 F. Aulehla, K. Lotter, Nozzle/Airframe Interference and Integration, AGARD Lecture Nr. 53.

69 J. F. Hart, Diskussion zum Vortrag K. Schwärzler, (7), Jb. WGLR, 1963, p. 43

70 S. Günter, Vorrichtung für ein Gasturbinen-Strahltriebwerk zur Steuerung der Schubstärke bei einem Senkrechtstartflugzeug mit zu beiden Seiten der Flugzeuglängsachse angeordneten Strahltriebwerken unter Anwendung der Schubmodulation, Patent Nr. 1283601 der B.R. Deutschland, 1965

71 AGARD, V/STOL Handling Criteria and Discussions, Rep 577, 1970

72 X. Hafer, H. M. Franke, A Contribution to the Problem of Flight-Control by Thrust-Modulation, AIAA Simulation-Conf. Columbus, Ohio, Conf.Proc., p. 70, 1963

73 J. Schäffler, H. Alscher, G. Steinmetz, J. B. Sinacori, Control-Power Usage for Manoeuvering in Hover of the VJ 101C Aircraft, J. of. Aircr. Bd.4, p.445, 1967

74 G. Schweizer, H. Seelmann, The Control-Moment Distribution for the Do 31 Hovering Rig, AGARD-Rep. Nr. 522, 1965

75 W. Möller, Die Stabilisierung des Flugzeugs um den Schwerpunkt, Jb. WGL 1959, p. 190

76 G. Ernst, Über eine störungsfreie Fluglagenregelung für strahlgetragene Flugzeuge, Jb. WGLR, 1960, p. 221

77 G. Ernst, Lotrechtflug und Stabilisierung, Luftfahrttechnik 5, 1959, p. 17

78 R. Staufenbiel, S. Girlatschek, Automatische Stabilisierung und Sicherheit von VTOL-Flugzeugen, Jb. WGLR 1962, p. 275

79 W. Burkhardt, Optimierung des Schwebeflugverhaltens durch geeignete Kenngrößen an dem Blockschaltbild des Regelkreises. Jb. WGLR, 1963, p. 58

80 G. Kissel, Present-day Role of Control and Stability Augmentation Systems and Future Trends, MBB-Schrift 1973

81 H. M. Franke, G. Döpner, Regeltechnische Untersuchungen über den Einfluß des Stellgliedes im Schwebeflug, Jb. WGLR, 1963, p. 67

82 G. Schweizer, H. Seelmann, H. Weischedel, Die Regelung von VTOL-Flugzeugen im Schwebeflug, Jb. WGLR, 1963, p. 75

83 K. Kracheel, W. Burkhardt, G. Kissel, H. von Diringshofen, Höhenregler für VTOL-Flugzeuge, Jb. WGLR 1965, p. 141

84 M. Rodewald, Die Leistungsfähigkeit der Konzeptionen von Reglern für Schwebegestelle beziehungsweise Senkrechtstarter, Jb. WGLR 1967, p. 162

85 W. Kubbat, Vergleichende Gegenüberstellung der Vermehrfachstruktur und anderer in diesem Zusammenhang getroffenen Maßnahmen der Regler für die VTOL-Flugzeuge Do 31, VAK 191B und VJ 101C-X2. DFL Braunschweig, Dezember 1968

86 G. Schmidt, H. Friedel, G. Schweizer, Regelungs- und Flugführungsprobleme bei zukünftigen V/STOL-Landungen unter Berücksichtigung der Erfahrungen mit der Do 31. Vortrag auf der DGLR-Jahrestagung, Bremen, 1969

87 H. G. Schumann, Flugsimulation als Hilfsmittel der Entwicklung und die deutschen Arbeiten auf diesem Gebiet, Jb. DGLR, 1970, p. 40

88 H. Schmidtlein, Simulation der VAK 191B, VFW-Focker, Bremen, VFW/Epd – Reg. Nr. 189, 1970

89 Gesetz zum Schutz gegen den Fluglärm, Bundesgesetzblatt Nr. 28 (1971), Teil I, p. 282

90 P. Hamel, Noise-Abatement Flight Profils for CTOL and V/STOL Aircraft, Deutsche Luft- und Raumfahrt, FB 71-10, 1971

91 V. Nitsche, Die digitale Simulation von Lärmschutzbereichen in der Umgebung von Verkehrsflugplätzen, Diss. am FB. Maschinenbau der TH Darmstadt, 1978

92 P. Bartels, Schallabstrahlung im Nah- und Fernfeld des V/STOLTransporters Do 31 und Lärmverminderungsmöglichkeiten bei zukünftigen V/STOL-Entwicklungen, BMVg-FBWT 72-23, 1972

93 F.H. Schmitz, W. Z. Stepniewski, The Reduction of VTOL Operational Noise through Flight Trajectory Management, NASATN-D-6246, 1971

94 A. Schneider, Notwendigkeit und Methodik der Lärmminderung bei neuzeitlichen Flugzeugentwicklungen, Luftfahrttechnik/Raumfahrttechnik 11, 1965, p. 6

95 R. Stüssel, Über die Möglichkeit des zivilen Einsatzes von Senkrechtstartflugzeugen, Jb. WGLR, 1962, p. 246

96 D. Stüssel, Der Flugbetrieb von VTOL-Flugzeugen in Flughafennähe, Jb. WGLR, 1965, p. 93

97 A. Schneider, Spezielle Probleme eines Vertikal-Strahl-Luftverkehrs, Jb. WGLR, 1963, p. 146

98 K. Thalau, Vergleichende Beurteilung von Entwürfen der deutschen Luftfahrtindustrie für ein V/STOL-Transport-beziehungsweise Verkehrsflugzeug, Abschlußbericht v. 26. Juni 1970, Anlage 1 (Ergebnisvermerk vom 13. Januar 1969 im BMWI)

99 K. Thalau, s.o., Anlage 2 (Vermerk über Besprechung zwischen DLH, BMwF, BMV, BMWi und BMVg am 28. Januar 1969)

100 K. Thalau, s.o., Anlage 5, (Rahmenforderungen an ein V/STOLKurzstrecken-Transportflugzeug für miltärische und zivile Verwendung v. 30. Mai 1969)

101 K. Thalau, s.o. Seite 1–41

102 K. Thalau, Brief an X. Hafer vom 12. Juli 1971

103 K.-H. Hartmann (Hrsg.), Die deutschen Senkrechtstartflugzeuge EWR-VJ 101C, Dornier Do 31 und VFW VAK 191B, DGLR-Symposium, 31.3.2000, Flugweft Schleißheim des Deutschen Museums, DGLR-Bericht 2000-01

Aerodynamics – from Near-Sonic to Hypersonic Flight

Ernst Heinrich Hirschel

Winning Scientific Parity through National and International Cooperations

The year 1945 represented also a caesura for aerodynamics. Teaching and research, as far as not prohibited by the occupation forces, commenced slowly again at the end of the forties/beginning of the fifties. Many of the leading scientists and engineers were no longer working in Germany or had not yet returned, test facilities were dismantled or brutally destroyed like the wind tunnels in Brunswick in 1948 and institutes had been dissolved.

While the time up to 1945 was the era where the fundamentals of aerodynamics had to a large extend been established in Germany, commenced after 1945, initially under the exclusion of Germany, the full realization of many new findings and knowledge, first in military aircraft engineering then within its civil pendant. In the process, the new main directions were high-speed flight at transonic and supersonic speeds and vertical take-off. These could be accomplished through new aerodynamic forms and the simultaneous engine development. In the case of military aircraft, main problems were the design of slender wings (swept wing, variable-sweep wing) for high-speed flight, which, at the same time, had to have very good low-speed characteristics, and propulsion integration, especially for fuselage-integrated engines: air intake with boundary-layer diverter, diffuser within the channel leading to the engine, and finally nozzle and after-body of the flight vehicle. The requirements, which also hold for a supersonic aircraft, have essentially not changed today, 50 years later. In the eighties and nineties and dependent on the aircraft mission, the unstable flight as a means of performance enhancement, high maneuverability up to post-stall fight, and signature reduction (radar) by shaping of the entire aircraft, including the air intake (radar) and the nozzle/after-body region (radar, infrared) became important areas of design efforts. Here, also thrust vectoring, not only with regard to high maneuverability, has to be mentioned.

For transport aircraft, especially passenger aircraft, the main problems were initially the design of wings for subsonic cruise at a Mach number as high as possible (swept wing, supercritical or transonic airfoil), the integration of jet engines, whose diameters increased with increasing bypass ratio, and the design of high-lift systems for take-off and landing, which were also to ensure low noise emissions in the vicinity of airports. After the first "oil shock", drag reduction by passive and active laminarization and turbulence management became an important topic. Although the fundamentals of laminarization are known since the thirties, the problems related to aerodynamics (three-dimensional boundary layers on swept wings, high Reynolds numbers) and to system technology (integration, combination with high-lift systems, contamination and icing protection, overall efficiency) are so substantial that today considerable research effort related to the development and proof-of-viability is still required. As possible new topics one has to name unstable flight and thrust-vector control, also for passenger aircraft, as well as the further improvement of high-lift systems, for noise-protection reasons, the reduction of airframe noise (approach to land), wake control (in view of the staggering of aircraft during approach to quicker dissipate free vortices trailing an aircraft, which may endanger smaller follow-on planes), the aerodynamics of unconventional configurations, new wing concepts, and the hydrogen aircraft. In addition, there is, especially for wide-body aircraft, the increasing need for highly integrated multi-disciplinary (aerodynamical, structural and flight-mechanical/control-configured) design of air vehicles (for a long time already true for military aircraft). This means a widening of the role of aerodynamics, and has large consequences for the industrial design processes and tools, but also for research and teaching.

Two further types of aircraft shall, within this introduction, briefly be mentioned: sailplanes and space transport vehicles. For the former, the aerodynamic quality (ratio of lift to drag) was developed to high values by the combination of high-aspect ratio wings (ratios of wing span to chord of up to 40) and laminar airfoils (lift/drag up to 60). This was accomplished by the simultaneous introduction of fiber-reinforced composite structures with extremely accurate and smooth surfaces and large stiffness (here, Germany has a monopoly). In case of space-transport vehicles (space plane), aerodynamic/aerothermodynamic problems are treated within system and technology studies for future completely or partly reusable space transport systems. The main types of such systems are firstly pure aero-assisted re-entry vehicles (like the space shuttle of the USA), which are characterized by blunt forms to reduce, via the drag, the reentry velocity up to landing. Secondly they are air-breathing acceleration and cruise vehicles (e.g., NASP, Sänger first stage), which must achieve by their slender form a drag as low as possible. They put quite different demands on the aerodynamic design and, in the case of air-breathing vehicles, on propulsion integration. Here too, is the highly integrated multi-disciplinary design a decisive element.

The research landscape has fundamentally changed after 1945 and 1955. As in aircraft engineering, which is carried out today in Europe on the civil as well as the military side mainly in international consortia one finds, since the seventies, in research at universities and aeronautical research establishments increasingly interlocking European aeronautical research activities. Here, an important role has been played by *Dietrich Küchemann*, who transferred in the autumn of 1946 to the Royal Aeronautical Establishment in England where he soon became a leader in the research on slender wings which later led to the "Concorde", the first and up to now only supersonic passenger transport aircraft [1, 2]. He established the EUROMECH colloquia and the European cooperations EUROVISC and EUROHYP, cooperations in viscous and hypersonic flow, respectively.

GAMM is the "Society of Applied Mathematics and Mechanics" ("Gesellschaft für angewandte Mathematik und Mechanik"). The GAMM "Committee on Numerical Methods in Fluid Mechanics" took up its activities in 1974 with the biennial GAMM "Conferences on Numerical Methods in Fluid Mechanics". These were in 1991, after nine conferences, superceded by the biennial "European Fluid Dynamics Conferences", organized by the "European Committee on Computational Methods in Applied Sciences, ECCOMAS", a union of the European scientific establishments in this field. The GAMM workshops, which were concerned with a systematic comparison and an evaluation of the newly developed numerical computational methods, turned out to be very effective. They were continued later by corresponding AGARD and GARTEUR working groups.

Also important for aerodynamics became the AGARD (since 1951) and GARTEUR (since 1970) activities already mentioned in Part 2. Impetus for joint European aerodynamic research was provided since the eighties by the European Union through the support of pre-competitive research and development (BRITE-EURAM, ESPRIT). This also led to the initiation of further European cooperations comprised of research and industry, such as the "European Research Community on Flow, Turbulence and Combustion ERCOFTAC", and the European database for the validation of aerodynamic computational methods. Of great importance to aerodynamics are, in addition, jointly developed and operated large-scale ground-simulation facilities such as the "Dutch-German Wind Tunnel, DNW" in the Netherlands and the "European Transonic Wind Tunnel, ETW" in Cologne.

The international European research is also so effective since the individual nations continue to sustain, in addition, a strong national research effort. In Germany, aerodynamic problems are for years increasingly being investigated, and led to application, in joint projects between universities, the "German Aerospace Research Center, DLR" and industry. Here, especially the "Priority Programs", the "Research Units" and the "Collaborative Research Centers" ("Sonderforschungsbereiche, SFBs"), financed by the "German Research Association" ("Deutsche Forschungsgemeinschaft, DFG") must be named. There are, furthermore, common activities within the framework of the national "German Aeronautical Research Program", which had various predecessor programs also supported by Federal Ministries. At the end of the seventies, German aerodynamics had created a very effective professional forum, the "Working Association Flows with Separation, STAB" with yearly alternating work shops and symposia [3].

Below, the development in aerodynamics after 1945 is described for four subject areas: "Aerodynamics of near-sonic flight", "Boundary-layer physics and boundary-layer control (laminar wing)", "Supersonic and hypersonic aerodynamics" and "Numerical aerodynamics and new experimental methods".

Aerodynamics of Near-Sonic Flight

Within this section, important developments after 1945 related to the problems of near-sonic flight are presented. This speed range plays for today's transport aircraft an important role since they have to cruise at (subsonic) Mach numbers as high as possible in order to keep the operational costs as low as possible. Starting with the first jet-propelled transport aircraft, the swept wing and the supercritical (transonic) airfoil became the main elements of aerodynamic wing design.

Due to the cost, time and risks associated with the development of aircraft which are getting continuously faster and larger, and due to the increasingly market-driven business of transport aircraft engineering, the design processes and tools are changing: especially the wing design relies stronger on computational methods, while the design verification and the establishment of aerodynamic data sets occurs in more powerful wind tunnels with a steadily improving, often simultaneous, Mach number and Reynolds number simulation [4].

At the beginning of the forties, *J. Weissinger* developed the method for the computation of swept-wing flow on the basis of the extended lifting-line theory. The lifting-wing methods were established at the beginning of the fifties independently by *H. Multhopp* and *E. Truckenbrodt*. These methods determine, for an arbitrary wing form in incompressible flow, the lift distribution in the spanwise as well as in the chordwise direction thus allowing to compute, for the first time, all interesting aerodynamic coefficients of a wing [5, 6]. With the aid of the compressibility rules (*Prandtl-Glauert, Göthert, von Kármán-Tsien* and others) one may compute the results for any Mach number below the critical Mach number, which could, however, at that time not be determined theoretically. A very good survey of the status of aircraft aerodynamics, including gasdynamics and boundary layer theory, is provided by the monograph [7] of *Hermann Schlichting* and *Erich Truckenbrodt*, which is, naturally, partly obliged to the developments, especially in Germany, prior to 1945.

Busemann's swept-wing concept allowed to shift the critical Mach number, also for a thick wing, to higher (subsonic) Mach numbers. The concept of the supercritical or transonic airfoil, respectively wing allowed, furthermore, to fly at Mach numbers higher than the critical one. For modern supercritical airfoil shapes, the drag-divergence Mach number, indicative of a steep drag rise, and the lift-divergence Mach number, indicative of a drop in lift – and due to both effects also indicative of a loss in aerodynamic quality (ratio lift to drag) – are located clearly above the critical Mach number [8]. The swept-wing concept also allows an increase in wing thickness and volume, with the former permitting a larger wingspan (aspect ratio) and hence a higher aerodynamic quality, the latter increasing the (wing) tank volume hence fuel capacity.

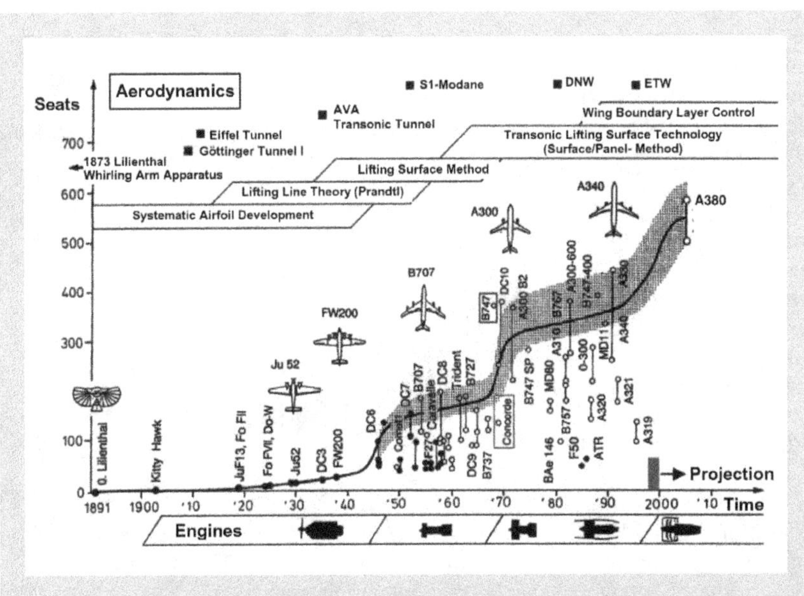

The development of passenger transport aircraft in view of advances in aerodynamic computational methods, in wind tunnel technology and engines.

The swept wing, the supercritical airfoil and the area-rule, in concept all based on pre-1945 German research, exhibit an interesting aerodynamic potential for today's transport jet aircraft. The average flight efficiency in form of the burned fuel (K) per flight distance (F) is dependent on the flight Mach number M as well as on the aerodynamic quality lift/drag (L/D) of the overall aircraft:

$$\frac{K}{F} \sim \frac{b_s}{M} \frac{G}{L/D} .$$

The relation (i.e., the reciprocal of the specific range in Breguet's range equation) also demonstrates the necessity of a low specific fuel consumption (b_S) of the engines and a low aircraft gross-weight (G), synonymous of a high payload share.

The realization of the concept "Supercritical Wing" required long-term research at DLR, at German Universities and the aeronautical industry, as described in the section "Aerodynamics – the Key to Successful Flight". Friction effects, coupled with shock boundary-layer interactions, play an important role. In wing design, beside the design point, off-design points and constraints (e.g., low-speed flight, fuel volume and wing structural weight) have to be considered [9]. This is schematically depicted in the upper left of the figure in the colored plate I. Furthermore shown there is the development of the aerodynamic quality during the last twenty years for various Airbus wings, demonstrating the large contribution of aerodynamics to the increase in the economy of modern civil transport aircraft [10]. The figures also demonstrate that modern computational methods and wind tunnels are necessary tools in aerodynamic research and the industrial realization of its results, which is not the least demanded by competitive pressure.

Shock/boundary-layer interaction effects, mentioned above, were since the first half of the eighties research topics at various universities, at DLR and in industry. The question of measures to reduce the detrimental effects of such interactions, which had initially priority, was extended in view of the design of laminar wings. In the case of the latter, accelerated flow must prevail over as large a part of the wing as possible, which inherently leads to a relatively strong shock wave at the downstream end of the acceleration. During the time period of 1993–1995, the control of this shock became the topic of the EU-project "EUROSHOCK – Drag Reduction by Passive Shock Control", which was coordinated by DLR, and treated and clarified essential aspects of this problem by experimental and numerical investigations [11].

Further research and development topics, which were and still are in essence carried out by German institutions within the overall subject of "Aerodynamics of near-sonic flight", are, besides the laminar-wing technology to be considered in the next section, high-lift systems and variable-camber wings (for high-performance sailplanes already realized). The latter are within the national "German Aeronautical Research Program" [9] subject to a concentrated and continuous effort. The figure on page 391 finally shows schematically the foreseeable drag reduction potential, which may be realized by aerodynamic research and development [12]. The ever closer bidding for the technological boundaries – supercritical cruise Mach numbers as high as possible, very effective propulsion integration, high-performance high-lift systems, all in conjunction with light-weight construction – leads, however, to highly integrated design processes at a high modeling level since the effective aerodynamic shape of the wing and the high-lift elements "under load" become ever more important.

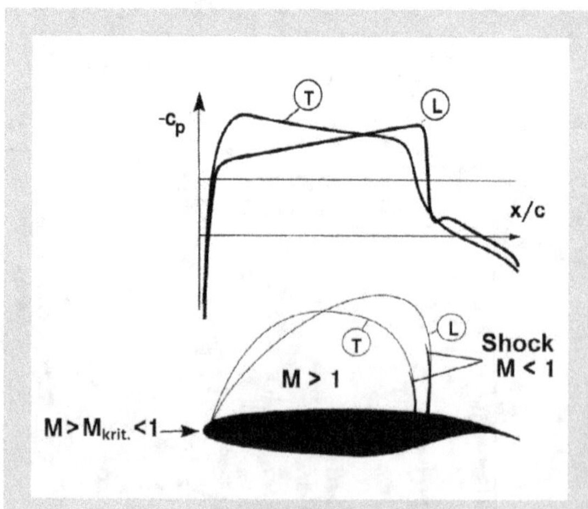

Typical pressure distributions (top), sonic lines and shock pattern (bottom) for a turbulent (T) and a naturally laminar (L) transonic airfoil at equal lift.

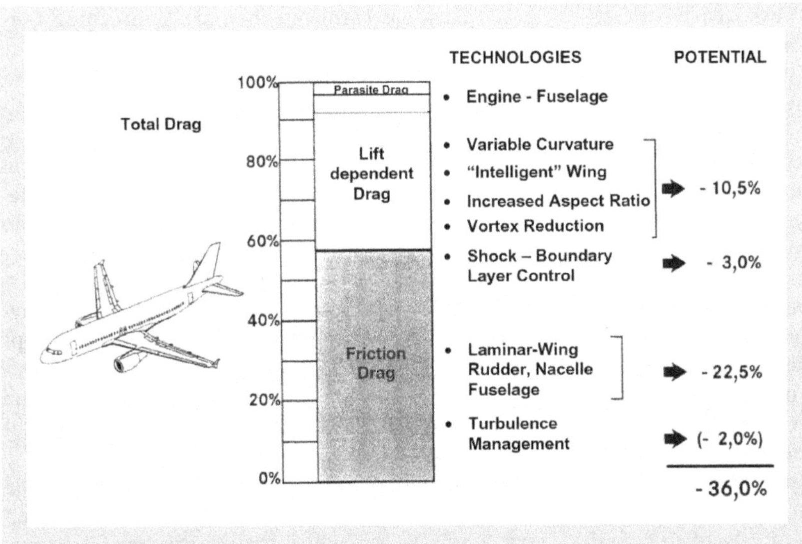

Aerodynamic potential, 20-year realization period.

Boundary-Layer Physics and Boundary-Layer Control (Laminar Wing)

Boundary-layer theory has, up to the early forties, asked all important questions and answered many of them. *Hermann Schlichting's* monograph [13], published in 1951 and going back to lectures during 1941 and 1942, remained for many years, with a total of nine editions, the standard book in the field. Computational methods for the aerodynamic design have been developed from boundary layer theory. Practical methods to compute arbitrary two-dimensional boundary layers were proposed, for instance, by *E. Truckenbrodt* (1952), *R. Eppler* (1963) and *J.C. Rotta* (1971). *Alfred Walz* published in 1966 a monograph on flow and temperature boundary layers [14].

Three-dimensional boundary layers aroused the interest of research with the appearance of the swept wing, at the latest. *Ludwig Prandtl* presented a survey on the status of the knowledge at that time in the *Betz* commemorative publication of 1945 [15]. Problems related to the stall of swept wings, which occurred on the outer wing (tip), hence in the area of the aileron, due to a strong thickening of the boundary layer in this area caused by three-dimensional effects, were counteracted by boundary layer fences, which were tested for the first time in 1938 by *W. Liebe* [16] on the (unswept) wing of the Me 109. The first German publications concerning the solution of the three-dimensional boundary layer equations by finite-difference methods appeared in 1968 at the DFVLR (now DLR) in Cologne-Porz, 1975 at the DFVLR in Göttingen and 1977 at the University of Karlsruhe. An integral method for the solution of the three-dimensional boundary layer equations was worked out in 1978 at Dornier. At the beginning of the eighties, complete computations of the attached three-dimensional boundary layer on any arbitrary configuration were possible [17], however, with the known restriction in predicting boundary layer transition.

The developing capability of computing three-dimensional boundary layers gave new impulses to research regarding the generation of turbulence and turbulence itself within such boundary layers. Initially, mostly only English researchers considered the transition mechanisms (leading from laminar to turbulent flow) in three-dimensional boundary layers on swept wings. These mechanisms are, beside the classical Tollmien-Schlichting instability, the cross-flow instability and the propagation of turbulence on the attachment line (leading edge) of swept wings (attachment-line contamination). Both mechanisms are of great importance for the laminar wing as will be addressed later. Following first activities of *H. Görtler* and *G. Hämmerlin* in the mid-fifties concerning the instability of two-dimensional stagnation-point flows, started in the early seventies extensive investigations related to these mechanisms at the Institute of Applied Gasdynamics of the then DLR in Cologne-Porz, which were continued after 1980, theoretically and experimentally, at the DLR in Göttingen and here build up to a full-fledged research subject [18].

At the beginning of the seventies, work on the "Direct Numerical Simulation, DNS" of two-dimensional boundary-layer instabilities by solving the Navier-Stokes equations started at the University of Stuttgart. The DNS of transition processes, but also of turbulence structures, as well as the numerical simulation of turbulent flow with "Large Eddy Simulation, LES", has, in the meantime, joined the classical wind-tunnel experiment as a full "numerical experiment" (University of Stuttgart, Technical University Munich, DLR in Göttingen and Oberpfaffenhofen, University of the Armed Services in Munich, and others).

A very important development related to stability theory commenced in the early nineties. The so-called non-local stability theory was independently developed at the DLR in Göttingen and at the University of Arizona at Tucson; the theory largely overcomes the inherent limitations of the Tollmien-Schlichting theory [18]. The development at DLR, in conjunction with the Swedish aeronautical research organization FFA and the Royal Institute of Technology (TH) Stockholm, has the objective of the non-empirical prediction with nonlinear approaches of the location where the transition of the boundary layer from the laminar to the turbulent state has been completed. The non-local, non-linear stability and transition theory has also the potential to assess the effect of surface roughness et cetera. A survey of the joined focused research program "Transition", sponsored by the "German Research Association, DFG", which integrates these activities, is provided by *S. Wagner* [19]. Within this research environment, one also has to see the DFG-University-Research Group (Berlin, Aachen, Erlangen-Nürnberg, Stuttgart), which carries out flight experiments on laminar-turbulent transition.

German research into turbulence has, after 1945, received considerable impulses by *J.C. Rotta* in Göttingen [20] (among others, the transport equation for the length scale, 1972 and the anisotropy of turbulence in three-dimensional boundary layers, 1979). His co-workers established in the late seventies for the first time an extensive database on three-dimensional turbulent boundary layers by measurements on a rotational ellipsoid (prolate spheroid). Concerning German universities, focused research on turbulence is mainly conducted at the University of Erlangen-Nürnberg, the Technical-University Berlin, the University of Karlsruhe and the Technical University Munich. At DLR Berlin, researchers deal with coherent structures and especially with turbulence management, where, for instance, by fine longitudinal grooves in the surface (riblets) friction drag is reduced.

Since the end of the seventies, the working association STAB [3] is treating the problem of boundary layer separation and separated flows in research and application. The DFG priority program "Physics of Separated Flows" has provided during the years 1984 and 1990 important results for various classes of separation [21].

In concluding, here are some detailed remarks on the topic of laminar wings and especially of swept laminar wings. Sweep, which is necessary for near-sonic flight, leads, as mentioned, to the additional instability and transition mechanisms "cross-flow instability" and "attachment-line contamination". This impedes natural laminarization, as occurs, for instance, on modern gliders, or renders it impossible. The challenges associated with the laminar wing, which are embedded within the physics of the flow, are schematically sketched in the figure on page 393 [22]. The sweep of the wing leading edge is about proportional to the cruise Mach number and is, therefore, larger for long-range aircraft than for short- or medium-range airplanes. The Reynolds number, decisive for the physics of the boundary layer, is about proportional to the wing chord (that means to the "size" of the aircraft) and to the cruise speed. For small sweep angles, the Tollmien-Schlichting instability of two-dimensional boundary layers is dominant, while the cross-flow instability of three-dimensional boundary layers becomes effective at sweep angles between 10° and 15°; attachment-line contamination is of main importance at sweep angles > 20°. Natural (or passive) laminarization by means of a suitable pressure distribution, which may, as depicted above, be achieved by proper shaping of the airfoil, is only possible on the left of the curve labeled NLF-boundary, and is hence for large-scale aircraft not attainable. For the latter, a combination of shaping and boundary layer suction, i.e., hybrid laminarization (HLF), is needed, which, however, also has its limits; to the right of the HLF-boundary (figure), active disturbance damping, e.g.,

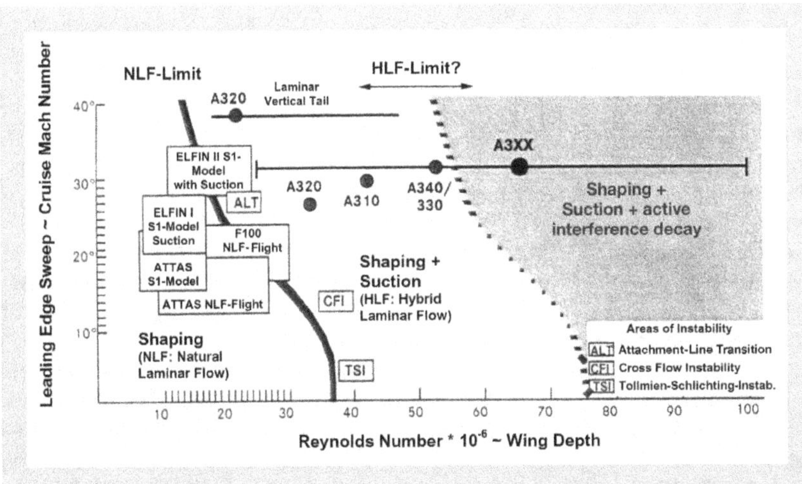

Physical challenges associated with the laminar wing. ELFIN I and II are EU research programs; S1-modell with suction is a wind-tunnel model of the VFW 614-ATTAS with boundary layer suction; F 100 denotes the Fokker F 100 flight tests.

A laminar vertical tail with boundary layer suction through a perforated surface was tested in September of 1998 on an Airbus A 320. Objective is a reduction in drag and, in turn, a reduction in fuel consumption.

by micro-electro-mechanical systems (MEMS) may be possible.

The figure shows the swept-wing dilemma of large-scale aircraft: the higher the cruise Mach number the larger is the wing sweep, the more difficult and elaborate is laminarization. Here, the problems are concentrated in the vicinity of the wing leading edge: due to surface disturbances (edges, steps, insect carcasses) attachment-line contamination is triggered, but the leading edge must house high-lift and anti-icing systems. In the case of hybrid laminarization, the suction system must be integrated into this area. Hybrid laminarization is thus a multi-disciplinary technology [23]. Led by the German industry, systematic research and development work has been and is being performed within the framework of European-Union-sponsored programs as well as nationally within the German Aeronautical Research Program [9]. The introduction of the laminar wing into large-scale aircraft will only occur if one can show that an overall benefit can be achieved. Estimates show that by laminarization fuel consumption can be reduced by up to 10 per cent [24], which will in the future also be important for ecological reasons. The order of magnitude of the reduction in the direct operating cost of an aircraft by laminarization depends on the mission profile [24], but also on other factors such as capital cost, fuel cost, et cetera, and finally also on (exhaust) emission taxes [25].

Supersonic and Hypersonic Aerodynamics

The results of German research in gasdynamics during the time up to 1945 are summarized in the 1952 monograph of *Klaus Oswatitsch* [26]. Further monographs were then the ones of *Ernst Becker* [27] (1966), *Jürgen Zierep* [28] (1966) and *Friedrich Keune* and *Klemens Burg* [29] (1975). The great research merit of this topic after the war is demonstrated by the fact that, after the rebirth of the German aeronautical research establishments at the "German Test Establishment of Aeronautics, DVL" in Aachen the two institutes of Applied and Theoretical Gas Dynamics, respectively, existed. At the "Aerodynamic Test Establishment, AVA" in Göttingen, the departments of Gas Dynamics and Astronautical (Space) Aerodynamics, later the Institute of Rarefied Gas Dynamics were installed. The consequence was a partial duplication of research activities and test facilities, which was only corrected by the foundation of the unified organization DFVLR in 1969. In addition to the predecessor organizations of the DFVLR, existed, since 1959, the "French-German Research Institute Saint-Louis, ISL", France, which had its roots in a working group affiliated with *H. Schardin* in 1945; it also considers gasdynamical problems. Furthermore involved was the "Ernst-Mach Institute" in Freiburg, which is part of the Fraunhofer-Society.

Research topics comprised initially near-sonic flight, supersonic flight, supersonic air intakes, projectile and guided weapons problems (spikes and spoiler control) and, since the end of the sixties together with industry, configurative studies of space re-entry vehicles. At the end of the sixties, *Hans-Georg Knoche* founded the "Missile Working Association" ("Flugkörper-Arbeitsgemeinschaft"), where industry and DFVLR considered directional stability of slender missiles. Theoretical work by *Klaus Oswatitsch*, which was continued in Sweden after the war, resulted in the hypersonic independence principle [30], which is of considerable importance to hypersonic flight. A further important result of theoretical investigations of *Friedrich Keune* and *Klaus Oswatitsch* was the equivalence theorem [31], which contains the area rule of *O. Frenzl* [32] and *R.T. Whitcomb* [33].

Due to the large demands on equipment, experimental research was essentially restricted to the DVL Institute of Applied Gasdynamics, since 1962 in Cologne-Porz, and the Institute of Rarefied Gasdynamics in Göttingen. An exception is the Shock-tube Laboratory of the RWTH Aachen, which was build up by *H. Grönig*.

At the beginning of the seventies, DFVLR, the "Space Research Society" ("Gesellschaft für Weltraumforschung, GFW"), industry and some university institutes, forming the "Working Association Reentry Technology" ("Arbeitsgemeinschaft Rückkehrtechnologie, ART"), created for space aerodynamics the ART-program [34], which was to establish, on a medium-term basis, the technological foundation for future space-transport systems. Through the ART-program, many of the research topics of the time were focused and supplemented as needed: rarefied gas flow, wall accommodation processes, viscous hypersonic interactions, shock/boundary-layer interaction, corner flow, high-temperature effects, heat loads, et cetera. In 1975, the objectives of the German space program changed, and the ART-program was terminated, and

"Tornado" model, scale 1 : 11, with outside loads, including two "Taurus" missiles, in the transonic wind tunnel of ARA Bedford.

with it almost all relevant research work except for the activities related to rarefied gasdynamics, which were of continued importance to satellite technology.

During the second half of the eighties, hypersonic aerodynamics became again a topic due to the "Hermes"-project of ESA (European "space plane" as a purely aerodynamic re-entry vehicle) and a new German hypersonic-technology program. With only a few remaining experts in research and industry, it was possible to reach in a surprisingly short time period a high scientific/technical level. Within the "Hermes"-project, three German industrial aerodynamics groups developed, together with their European partners as "alternative aerothermodynamics group", for the first time computational methods for high-enthalpy hypersonic flows about the "Hermes" configuration. With dissimilar approaches, and as an alternative to the French Dassault Aviation, which was responsible for aerothermodynamics, critical aerothermodynamic design problems were to be treated: heat loads (heat protection system), pitching moment problems, trim and control surface effectiveness. Jointly, Dassault and MBB worked out a proposal for the experimental flight vehicle "Maja", a smaller version of "Hermes", to study these problems – which cannot be investigated with sufficient accuracy in wind tunnels – in flight and to validate the new computational methods. DLR as well as several universities considered fundamentals but were also concerned with configuration-related problems (DLR). In Göttingen, a high-enthalpy tunnel (HEG) was erected, which could, however, no more be utilized for the "Hermes"-project. It must be mentioned that (after the American experiences with the Space Shuttle known by now) within the framework of "Hermes" a pronounced shift in fundamental research occurred from the problems related to rarefied gases, treated in the sixties and seventies, to problems related to heat loads, high-enthalpy effects, trim- and control-surface effectiveness and shock/shock and shock/boundary/layer interactions; this mainly for the development and validation of numerical methods for aerothermodynamics.

The German hypersonics technology program, which, starting in 1988, treated the technological problems of the air-breathing lower stage of the two-stage reference concept "Sänger" (fully reusable "space plane"), gave aerothermodynamic research and development a new direction. The air-breathing (ramjet propulsion) lower stage is, contrary to a reentry vehicle (US Space Shuttle, "Hermes" but also the "Sänger" upper stage), which only performs a decelerating flight from orbit down to earth, a propelled acceleration/cruise flight vehicle, which has to be of a high aerodynamic quality and possesses a highly-integrated lift and propulsion system. Problems, such as, for instance, the aerothermodynamics of radiation-cooled surfaces [35], the aerothermodynamic propulsion integration (hypersonic air intake, nozzle/afterbody), are correspondingly of great importance; they are strongly affected by friction effects, especially the transition from laminar to turbulent flow [36]. A further important problem for a two-stage space transport system is the separation of the upper stage from the lower stage (in case of "Sänger" at a Mach number of $M \approx 6.8$ at an altitude of about 33 km or 21 mi).

The above topics were investigated by industry, DLR and four universities (Aachen, Brunswick, Munich, Stuttgart) within the framework of collaborative research centers of the DFG. Within the SFB 253, for instance, the TH Aachen developed the lower-stage configuration "Elac" as reference concept, the DLR considered, in addition, wave-rider configurations.

Wind-tunnel model of titanium, scale 1 : 160, of "Sänger" from the German hypersonics technology program.

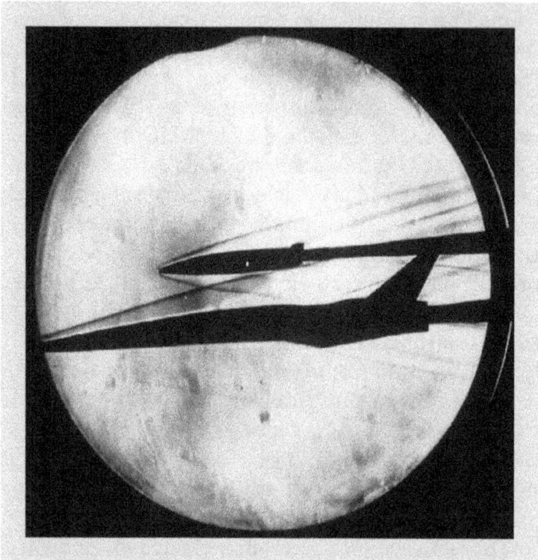

Stage separation of a "Sänger" model simulated in the hypersonic wind tunnel of the DLR in Cologne.

With the termination of the "Hermes"-project in 1994 and the end of the hypersonic technology program in late 1995, both due to the restrictions in the European, respectively German space program, the situation was similar to the one at the end of the ART-program in 1975. Industry and somewhat later the DLR had to suddenly shut down their research and development programs. The fact was accepted that with such a procedure the technological competence, needed to face the American and, in the future, also the Japanese competition, can not be established within Europe in the long run. An exception was here only the DFG, which continued to sponsor three of the original four collaborative research centers. The "Future European Space Transportation Investigations Program" (FESTIP) of ESA, headed by German industry (without the participation of France), which is supposed to work out the technological basis for a future space-transport system, can be considered as a continuation of the German hypersonic technology program. However, it still has to enter its main phase, a "European Launcher Technology Program".

Numerical Aerodynamics and New Measurement Techniques

In the business report [37] of the VDI concerning the years 1919/20 one can find the topics "Numerical integration of differential equations" and "Finite-difference computations and their application in technology". *Ludwig Prandtl* formulated in 1945 in the Betz commemorative publication [15]:

"Already for a long time is it a central idea of the Göttingen Fluid Research to render, by suitable experiment-supported theories, tests superfluous".

The quotations shall demonstrate that the topic "Numerical Aerodynamics" is an old one. The design of modern aircraft is today no longer possible without numerical aerodynamics. The development of its capabilities as analysis, but increasingly as design and optimization tool, "restricts" the classical wind tunnel more and more to the necessary design verification and the generation of data sets, i.e., to the generation of the large amount of aerodynamic coefficients, which characterize the aircraft within its entire flight range. There are three main reasons for this development [38]. The first one is that numerical aerodynamics delivers the aerodynamic information, needed in the aircraft definition and development processes for the design of the aircraft, practically without delays due to model fabrication and wind-tunnel occupancy; in the process, modern inverse and optimization methods purposefully support the design. The second reason is that especially for transport aircraft friction effects become ever more important. These effects could, up to now, not be determined in conventional wind tunnels since the simultaneous Mach Reynolds number simulation (actually duplication) was not possible. The now cryogenic European Transonic Wind Tunnel (ETW) now allows the simultaneous simulation. However, extensive design work is for cost reasons not possible in the ETW. Its role is the one of a high-quality validation tunnel and provider of similarly high-quality data sets. The third reason is finally that modern aircraft engineering requires an increasingly more integrated aerodynamics/structural-dynamics/flight-mechanical/control configured design, which presupposes high-performance disciplinary, and especially multi-disciplinary computational methods able to treat flow/structure interactions and going far beyond the classical aeroelastic methods.

This all does not mean that the wind tunnel will become superfluous. It will, for a long time, remain the main tool for the aerodynamic aircraft design. It delivers furthermore validation data for the methods of numerical aerodynamics, and it is becoming ever more important as tool in the exploration of fluid-physical phenomena. This type of research is very important since fluid-physical modeling (transition, turbulence, separation) is still the weak point of today's numerical aerodynamics. The wind tunnels as well as numerical aerodynamics are, in addition, important means of flow diagnostics, that is the assessment of causes for possible aerodynamic failures of aircraft and their components.

When in the sixties the still continuing increase in digital-computer performance set in (the digital computer was, at that time, still competing with the analogue computer), also the era of numerical fluid mechanics and aerodynamics began. The potential of this new computational tool for research and industry was initially only recognized by a few people. Generally, research disapproved, at best tolerated. In the process, *Ludwig Prandtl* was quoted, obviously not knowing the above statement in the *Betz* commemorative publication. On the other hand, *Ludwig Prandtl* and *Adolf Busemann* had, in 1929, introduced the method of characteristics for supersonic flows, which may be considered as the prototype of numerical methods in aerodynamics; basic elements are still being applied today in many numerical methods [39]. *Ludwig Prandtl* also discussed already in 1938 a "continuation method" [40] for the solution of the boundary layer equations, which contains basic elements of today's difference methods for the integration of the boundary layer equations, and, moreover, of all numerical methods.

Numerical methods are generally mathematical methods, which transform systems of coupled partial differential equations, which describe the flow of a fluid – the equations of motion –, into difference equations, which are then solved algebraically within a discretised space. *G.F.B. Riemann* first considered the solution of *Euler's* equations of motion in 1860. The Frenchman *Jacques Salomon Hadamar* made in 1903 and the Italian *Tullio Levi-Civita* in 1932 important contributions. The Courant-Friedrichs-Lewy criterion, an important theorem, which states that the numerical domain of dependence must include the analytical one (marked by the characteristic features of the equations to be solved), was formulated in 1928 [41]; this shows that the roots of the numerical methods go far back. Numerical methods are of such great interest because the equations of motion can, contrary to the analytical approach, in principle be solved exactly, i.e., without linearization and solution trials, and for arbitrary boundaries and boundary conditions, as far as they are physically sensible. This renders them increasingly an important tool for the investigation of flow phenomena but also for the aerodynamic design of aircraft [38].

The first numerical boundary layer methods were published in the late sixties and in the seventies. The first German panel method, a numerical method for the computation of potential flow (linearized potential equation) about wings and aircraft with body volume and lift, the MBB panel method, was developed during the years 1969 and 1971. In 1971, *E. Krause* and the author of the present section founded the "Working Group on Numerical Methods in Fluid Mechanics" (simultaneously a EUROVISC-working group), which had in 1972 about 30 members from research and industry, and which considered the following topics: "Three-dimensional flow about blunt bodies at supersonic speeds", "Application of multi-point methods for the solution of the boundary layer equations", "Flows with chemical reactions", "Three-dimensional (turbulent) boundary layers", "Subsonic flow about two- and three-dimensional configurations with separation", "Shock/boundary layer interaction". Part of these topics were directly coupled with the ART program. The "Working Group" was in 1974, with strong support from *Henry Görtler*, established as "GAMM Committee on Numerical Methods in Fluid Mechanics", which operated from the beginning on a European basis. The attempt to link the working group with the "German Aerospace Society" ("Deutsche Gesellschaft für Luft- und Raumfahrt, DGLR") failed.

Increasing computer power through the introduction of the vector processor during the second half of the seventies provided strong thrust to the development of numerical aerodynamics. The "GAMM Committee on Numerical Methods in Fluid Mechanics" marked an epoch with its workshop on "Computational methods for inviscid transonic flows with shock waves" [42]. In 1984, MBB initiated the company-wide development project "Numerical Aerodynamics at MBB". Within this project, one of the worldwide first solutions of the Euler equations for a complete fighter configuration was worked out in 1985. At the end of the eighties, MBB and Dornier established, within the framework of the "Hermes" project, equilibrium and non-equilibrium aerothermodynamic solutions of the Euler equations, coupled with a three-dimensional second-order boundary-layer method. Also established were Navier-Stokes solutions, which were then further improved within the German hypersonic-technology program. The application of Euler and Navier-Stokes solutions to complex transport aircraft configurations also commenced at MBB, Dornier and DLR at the end of the eighties. In the mid-nineties, Euler solutions were a rather established tool of numerical aerodynamics at industry and research [38, 43]; Navier-Stokes solutions were already being applied in the development of wings for large-scale aircraft (Airbus A 340-500, A 380).

The above developments had a scientific background in the DFG focused research programs "Finite Approximations in Fluid Mechanics" (1983–1989), and "Flow Simulation with High-performance Computers", which were, starting in 1996, joined by the common German-French program "Numerical Flow Simulation – Simulation Numérique d'Ecoulement" of the German Research Association and the French Centre National de la Recherche Scientifique (CNRS).

With the maturity of the parallel-computing technology in the mid-nineties, a further strong impulse was given to numerical aerodynamics. Within the Aeronautical Research Program, one responded by combining the up to now rather independent developments of the methods in numerical aerodynamics in industry, at the DLR and at universities. Headed by DLR, the MEGAFLOW-system is developed [44], which opens, as the national computational system of numerical aerodynamic, the road to an efficient and complete description around complex aircraft configurations [45] (see colour plate I and II).

In the case of the experimental techniques, one may observe two directions in the development, which are a result of the different requirements associated with the experimental simulation for research purposes on the one hand, and the simulation for industrial development purposes on the other hand. Both are strongly intertwined and both reflect the development of wind tunnel and measurement techniques with the extremes "cryogenic wind tunnels", simultaneously providing Mach number and Reynolds number similarity at transonic speeds, and "high-enthalpy tunnels" with extremely short run times for re-entry aerodynamics.

The measurement of aerodynamic and thermodynamic quantities on the model surface and in the vicinity of the model within the flow field is for research purposes a very important task. In the process, a very fine "granularity" (resolution) in time and space is, dependent on the problem, attempted. Experimental aircraft aerodynamics is interested in and measures mainly integral forces and moments, but also pressure distributions (loads on aircraft elements).

The classical approach to determining aerodynamic and thermodynamic quantities are pressure and temperature/heat flux measurement sensors directly on the body surface and hot-wire measurements in the flow field. To visualize flows with sufficiently strong density changes, Schlieren and interferometer techniques have been and are still being used. Today, piezoelectric arrays and hot-film techniques allow an almost surface-covering measurement of the highly dynamic forces acting on the surface. With pressure and temperature sensitive surface coatings (paints), one may obtain surface-covering time-dependent information, which can immediately be converted to digital signals and stored. This is very important for industrial pressure-distribution measurements since by applying such techniques model instrumentation is rather simplified, faster to be applied and less expensive. Surface-covering temperature measurements with a very high quantitative resolution are made possible by the infrared technique, while liquid-crystal techniques provide more qualitative answers. The measurement of translatory, rotational and vibrational temperatures as well as of gas densities was already possible at the end of the sixties with the aid of electron-beam techniques as long as low-density flows were concerned. Laser-Doppler-Anemometry (LDA) became in the seventies, with the availability of high performance lasers, a standard technique for velocity measurements. Important contributions concerning this technique were made by the German-French Research Institute Saint-Louis, the University of Erlangen-Nürnberg and the "Federal Institute of Physics and Technology" ("Physikalisch-Technische Bundesanstalt") in Brunswick. Laser light-sheet methods allow the visualization and the quantitative determination of the location of vortex structures. Of considerable importance is the Particle Image Velocimetry (PIV), which allows the determination and digital conservation of steady and unsteady velocity fields. Here, important pioneering work is performed at DLR Göttingen. From the measured unsteady velocity fields, coherent structures may be identified with the aid of decomposition techniques. This is also of great importance for the diagnostic of flow problems associated with industrial configurations. For all opto-electronic measurement methods, the development of information technologies plays a big role. The high-performance computer is as important to the (digital) data acquisition and storage as it is for the data reduction, analysis and visualization.

Remark
Concerning this chapter, some figures are presented in color:

Color plate I
- Progress due to developments in aerodynamics (1970–1995).
- Numerical aerodynamics related to the Airbus, development and requirements.

Color plate II
- Examples of numerical aerodynamics: Pressure distributions on a model of the Eurofighter.
- Pressure distribution on an Airbus model.

Advanced Wind-Tunnel Technology

HANS ULRICH MEIER

The Importance of Wind Tunnels to Aeronautical Research

Also at the age of ever increasing computer power is the wind tunnel an indispensable tool for the development of air vehicles of all kind. Besides the wind tunnel, other means are being employed for fluid and flight mechanical investigations of aircraft or their components, such as water tunnels, centrifuges (rotating-arm facilities), shock tubes, rocket sleds and free-flight models and, of course, flight tests with prototypes. The development of wind tunnels was decisively formed by two factors: the changes in the technical demands of research and development and the socio-political conditions such as World War One and World War Two.

An important requirement concerning the test facilities is a simulation of the flow past a model, which exactly represents full-scale conditions. This leads to drive performances and constructive wind tunnel sizes, which would, employing conventional solutions, soon reach their natural limits. How these boundaries may be overcome by physical considerations and the contributions of German aeronautical research to possible solutions shall be addressed below.

Already during the first wind-tunnel tests was the determination of the forces and moments acting on components of the aircraft and their interference effects on the complete model the main objective of the experimental work. The corresponding technical requirements regarding the measurement and testing techniques have led, in the meantime, to investigative methods in wind tunnels, which allow simultaneously with force, moment and pressure distribution measurements also the simulation of the influence of engine and ground effects. Thus, flight performances of an aircraft project can already be accurately predicted during the early development phases and necessary performance-improving changes may be initiated. This is one of the essential prerequisites for a reduction of the development risks associated with civil and military aircraft.

Besides concerns about the performance improvement of new aircraft, which is, due to drag reductions, automatically coupled with a reduction in fuel consumption, and reduced exhaust emissions, aircraft manufacturers are forced by internationally laws to keep noise emissions within prescribed limits. Since the compliance with these limits in the case of subsonic transport aircraft can be tightened by national laws, the accurate prediction of the noise emissions already during the development phase of new wide-body aircraft has become of utmost importance. Similar conditions hold for the helicopter development, especially in the case of civil applications. Here, today's wind tunnel technology with its new aero-acoustic measurement technology may provide important information.

Similarly urgent is the provision of a solution in the area of structural mechanics/aeroelasticity regarding the prediction of the flutter boundaries of new wings and empennages. Here too, wind tunnel tests to determine unsteady forces acting on an aircraft or its components play an important role.

The wind tunnel has thus become a tool for performance predictions. At a stage, where 30 per cent of the development cost have been spent (roughly at the authorization to offer), cost-effective corrections on the configuration can still be carried out. The determination of the corrections needed to go from the wind tunnel model results to the full-scale air vehicle characteristics is becoming more and more decisive. At the same time, wind tunnel energy costs, hence facility costs, must be kept within acceptable limits.

Flow Quality – the Problem of Low-Turbulence Air Ducting

Investigations related to the flow quality in wind tunnels were performed by experimentally determining the transition location from laminar to turbulent flow on a 1:6 rotational ellipsoid (prolate spheroid) [46]. The 2.4 m (7.9 ft) long prolate spheroid was in ten axial locations equipped with surface hot-films, which allow the determination of the magnitude and the oscillations of the local wall shear stresses. Boundary-layer transition measurements were carried out in two DLR wind tunnels and the German-Dutch wind tunnel DNW in open and closed test sections. It is shown in the figure below that the locations of the laminar-turbulent transition regions are quite different in the three wind tunnels. Interesting was the new finding that not the turbulence level measured over the entire frequency spectrum of the turbulent fluctuations was a measure for the location of the boundary layer transition regions. All data can be perfectly correlated if the spectral level is rather taken only over the representative range of the Tollmien-Schlichting waves (in view of boundary-layer instability) to determine the turbulence level. The wind-tunnel-specific correction factors determined from these results are essential for the transfer of the wind-tunnel data to full-scale.

Energy Requirements of Wind Tunnels

One denotes the amount of energy passing through the nozzle cross-section per second with

$$N = \frac{\varrho}{2} U^3 A \left[W\right],$$

where A is the nozzle exit area and ϱ the air density. The drive performance N is hence proportional to the third power of the flow velocity U in the test section.

The performance factor λ is introduced to account for the losses and to be able to determine the actual drive performance $N_A = \lambda\, N$ needed to sustain the flow. *L. Prandtl* determined for the Göttingen wind tunnel with an open test section a performance factor of $\lambda = 0.68$. This relatively high value was essentially caused by the poor efficiency η of the fan ($\eta \approx 0.8$).

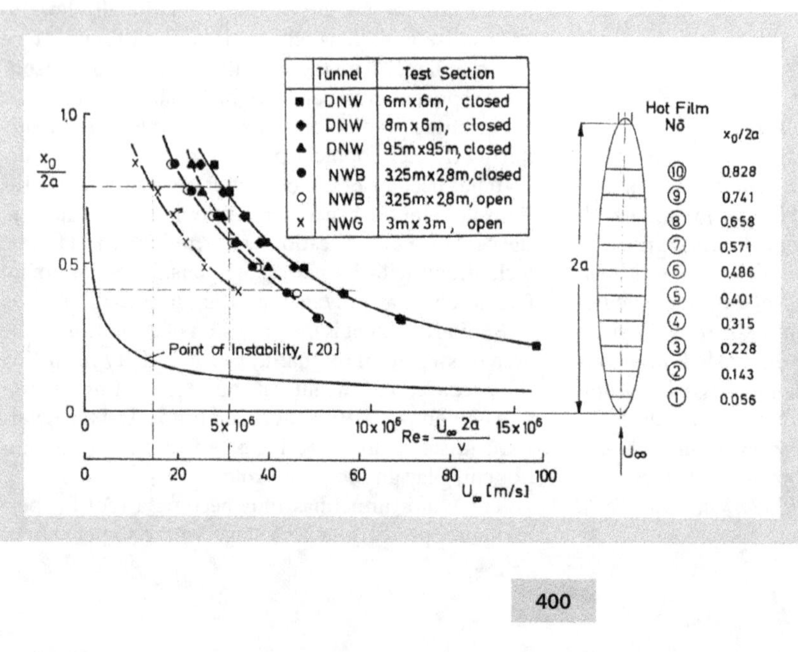

Comparison of transition locations on the DLR prolate spheroid (a/b = 6, a = 1.2 m (3.9 ft)) measured in three different wind tunnels with open and closed test sections. The measurements of transition were performed with surface hot-films in axis-symmetric flow [46].

Reference values for the performance factors of Göttingen-type modern wind tunnels at low speeds are

$\lambda \approx 0.5$ for open-jet test sections,
$\lambda \approx 0.15$ for closed test sections.

Here, it becomes obvious that for large-scale wind tunnels economic operation is only possible with a closed test section. The open test section of the DNW (jet cross-section at the nozzle exit 6 m×8 m (19.7 ft×26 ft)) – compare table on pages 194/195, No. 18 – is, consequently, mainly utilized for special tasks such aero-acoustic investigations on helicopter rotors and engine models. The gigantic increase in drive power can be comprehended by looking at the project of the 8-meter (26 ft) high-speed wind tunnel of the "Aeronautical Research Establishment Munich" ("Luftfahrtforschungsanstalt München, LFM") of 1942 (above cited table, No. 12). This wind tunnel was being constructed at the Ötztal (Austria) but not completed before the end of the war. It was in 1946 disassembled, shipped to Modane, France, and in 1952 as ONERA S1-MA wind tunnel put into operation. The propulsion of the tunnel occurred directly via a water turbine. It is hence the only wind tunnel in the world where not the electricity but the water consumption must be paid for! The maximum drive power is 88,000 kW (118,000 hp), which corresponds to the output of a medium-size power station. At the AEDC (Arnold Engineering Development Center) in the USA, the largest high-speed wind tunnel was put into operation in 1952 having a test section cross-section of 4.9 m×4.9 m (16 ft×16 ft) and a Mach number range of M = 1.5 – 4.7; the maximum drive power is 160,000 kW (215,000 hp). Similarly gigantic is the drive power of the "Full Scale" subsonic wind tunnel at the NASA Ames Research Center, California, USA, with 100,000 kW (134,100 hp). With these facilities, both limits, the one related to the energy supply and the one concerning the structural size, were reached (see above figure on page 402). However, Reynolds numbers were still essentially one order of magnitude below flight Reynolds numbers. Therefore, alternatives had to be found to allow investigations at flight Reynolds numbers at reasonable construction and operational costs. To demonstrate possible solutions, the relation for the drive performance is expressed in terms of the similarity parameters Reynolds number and Mach number and known relations of gasdynamics:

$$N_A = \lambda \frac{1}{2} \frac{A}{l^2} \mathrm{Re}^2\, M \frac{\alpha^2 \mu^2}{\kappa p} \left[W\right]$$

With the aid of this relation, one may easily consider possibilities of energy savings in the operation of wind tunnels.

a) Doubling the Reynolds number by a corresponding increase in the velocity means that the drive performance increases by 8-times the velocity cubed. The Mach number is changed at the same time.
b) Doubling the Reynolds number by doubling the linear model dimensions increases the drive performance by a factor of four since the test section cross-section increases with the square of the model dimension.
c) A doubling of the Reynolds number may be achieved by doubling the pressure p inside the (Göttingen-type) wind tunnel with a closed test section; here, the model dimensions, the freestream velocity and the temperature remain constant. The drive performance thus increases linearly with pressure. The advantage is that the change in Reynolds number can be accomplished at a constant Mach number.
d) Decreasing flow temperature T reduces the drive performance by $T^{3.3}$. Looking at the total energy consumption, one has to account for the energy needed to reduce the tunnel temperature, for instance, by injecting liquid nitrogen. The so-called cryogenic operation of a wind tunnel allows investigations at a constant Reynolds number and variable Mach numbers.

At the beginning of the seventies, an AGARD Working Group considered the technical optimization of a wind-tunnel concept for investigations at very high Reynolds numbers and transonic speeds. The German proposal was made by *Hubert Ludwieg* of the DFVLR in Göttingen. *Ludwieg* had earlier proposed for the supersonic and hypersonic speed ranges a cost-effective test principle, the tube wind tunnel (in the Anglo-Saxon language domain called the Ludwieg-tube) [47, 48]. Based on this principle, he worked out a proposal for a transonic tube wind tunnel (Transonic Ludwieg-tube). At the time, the decision fell on a wind-tunnel principle, which combined the advantages of an increase in pressure with a decrease in temperature in a continuous wind tunnel, a cryogenic pressurized wind tunnel.

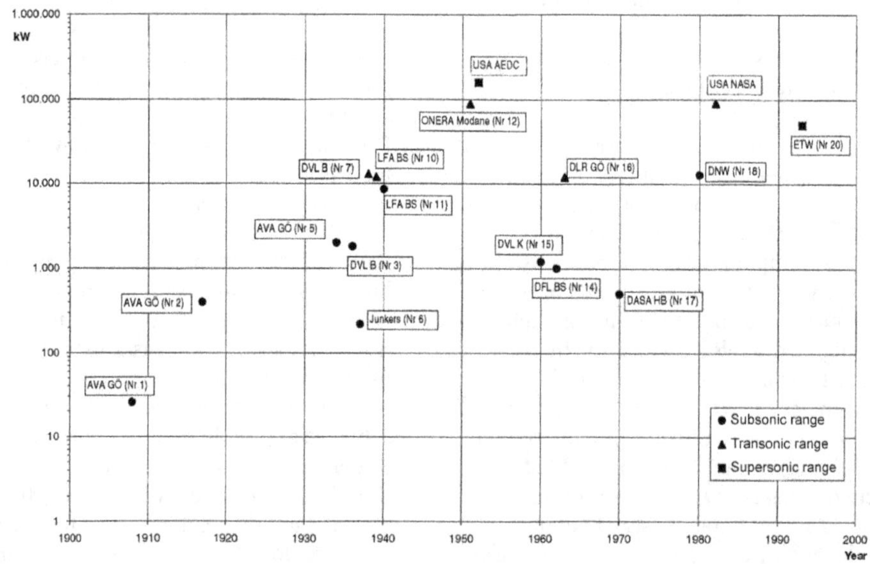

Increase in drive performance of large-scale wind tunnels.

The first country to realize a cryogenic pressurized wind tunnel, the so-called NTF (National Transonic Facility), was the USA. The tunnel is located on the premises of the NASA-Langley Research Center and has a test section diameter of 2.5 m (8.2 ft). In Europe, it was decided in 1985 to build a European Transonic Wind Tunnel (ETW), which went operational in 1993 at its location in Cologne. During the design and construction of this facility, experiences gained with the NTF were taken into account; special attention was paid to an industrial utilization with high productivity. The ETW provides the European aerospace industry with the world's most modern test facility. Especially for the application in research and for industrial predevelopment tasks was in Göttingen the principle of the Ludwieg tube combined with the principle of cryogenic operation. Here, Reynolds numbers of $Re \leq 70 \times 10^6$ can be realized in an adaptive-wall test

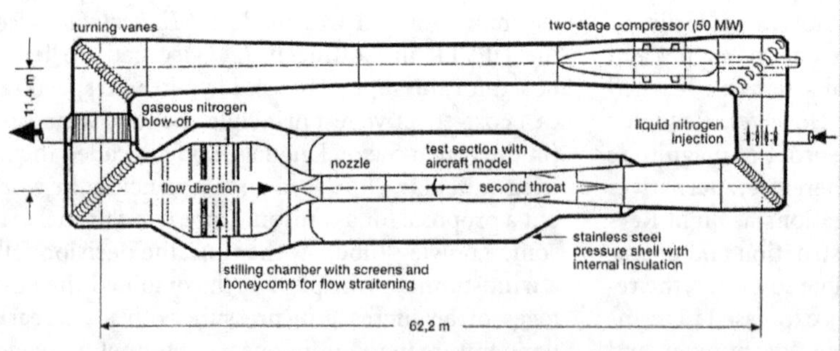

Aerodynamic circuit of the European Transonic Wind Tunnel (ETW) with gaseous nitrogen as test gas. Drive power of the compressor 50 MW (67,000 hp), cross-section of the test section 2.4 m × 2.0 m (7.9 ft × 6.6 ft), commissioning 1993.

section of 0.4 m × 0.35 m (1.3 ft × 1.1 ft) cross-section and a length of 2 m (6.6 ft) on a 200 mm (7.9 in) chord model at transonic speeds. A complete Reynolds number simulation for airfoils of large-scale aircraft is thus possible in a wind tunnel facility at reasonable costs; by means of a special control valve, a Mach number rage of M = 0.2 – 1.0 can be covered.

The Transonic Cryogenic Ludwieg-Tube in Göttingen (KRG)

The tube wind tunnel is in principle a pressure reservoir tunnel (blow-down wind tunnel) for the simulation of subsonic, transonic and supersonic flows with the pressure vessel being a long tube. When starting the tunnel (activating a quick-opening valve), an expansion wave propagates into the pressure tube at the local speed of sound. Behind the wave, conditions of the flow towards the nozzle stay constant until the wave, having been reflected at the closed end of the tube, again reaches the nozzle entrance. The duration of a test period is thus primarily limited by the length of the storage tube. This shows the big advantage of *Ludwieg's* principle compared to a conventional pressure-vessel tunnel, for instance, with a spherical storage tank. Within such a tank, pressure and total temperature steadily decrease due to the outflow from the tank; complicated control mechanisms are then needed to secure constant flow conditions in the test section. Of disadvantage in case of the Ludwieg-tube may be the boundary layer development within the storage tube since it may have an effect on the flow quality in the test section.

The AVA (now part of DLR) had already in 1967/1968 introduced the tube wind tunnels into its test operations with the construction of three supersonic and hypersonic tunnels, respectively (Mach number range 3–12). With the inauguration of the KRG, schematically shown below, in 1994 and its equipment with an adaptive-wall test section in 1994, DLR owns today the most modern test facility for transonic speeds at extremely high Reynolds numbers. The facility is due to its low operational costs especially well suited for research and industrial predevelopment studies.

As in the case of the supersonic and hypersonic facilities, an expansion wave propagates in the case of the KRG, after opening the quick-opening valve, with the local speed of sound into the storage tube and accelerates the gas (nitrogen) to a Mach number dependent on the ratio of the tube cross-section area to the smallest cross-section area within the control valve. The Mach number in the test section can, at the same time, be set by means of the control valve by choosing the corresponding ratio test-section-area to control-valve area. The effective run time (test time) ends when the expansion wave, reflected at the end of the storage tube, reaches again the test section. At that time, the quick-opening valve is being closed. During the passing of the expansion wave, total pressure

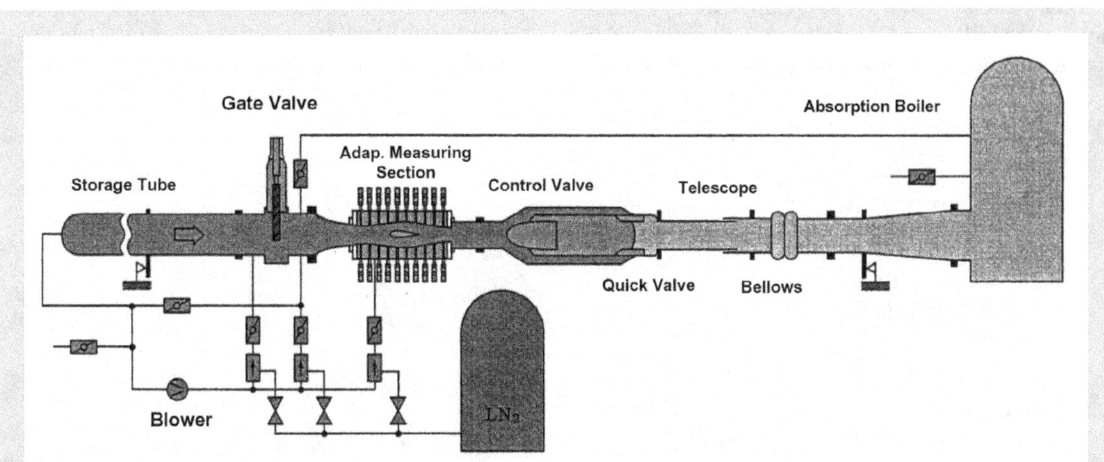

The cryogenic tube wind tunnel (KRG), also denoted Ludwieg-tube, according to *H. Ludwieg*.

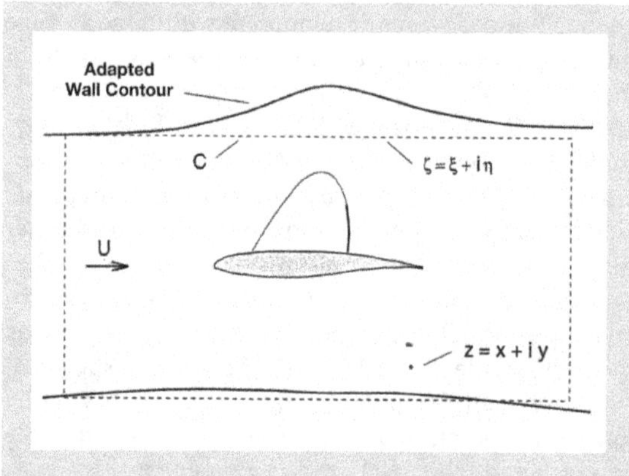

Wind-tunnel wall adaptation based on Cauchy's integral equation.

as well as total temperature drop below the original storage values dependent on the selected test section Mach number. All components, such as the gate valve, the adaptive-wall test section, the combined control valve/quick-opening valve, as well as the insulation of the storage tube and the entire facility, put the highest demands on design and fabrication. Details are described in references [49] and [50]. Here, we shall only address the adaptive-wall test section.

As indicated in the schematic, a specific contouring of the walls can minimize wall effects on the flow about an airfoil. Since run times are too short for a mechanical contour change during a run, the velocity components at the wall, U and V, are determined during a first run by measuring wall pressures and local wall inclinations. Then the wall-induced (disturbance) velocities are determined with the aid of Cauchy's integral equation. The wall contours for nearly interference-free flow may now be determined and set by the previously computed induced velocities. The next run should now, due to the newly adjusted walls, provide nearly interference-free results. The reshaping of the walls is achieved by deforming the flexible walls made of glass and carbon fiber reinforced epoxy resins, which have a similar expansion coefficient as steel. Stepping motors, installed in 19 axial cross-sections, perform the exact positioning. Multiplexing is used to provide electrical energy to each individual stepping motor and to measure the signals of the wall-position sensors. The results of a wind tunnel wall adaptation is schematically shown for pressure distribution measurements at a freestream Mach number of $M = 0.74$. The airfoil and wall pressure distributions for the straight wall and the first iteration in the wall adaptation process show large differences, while the difference between the first and second iterations is very small. The difference in the lift coefficients, the latter obtained by integrating the chord-wise pressure distributions, between the first and the second iteration is correspondingly only minor ($\Delta c_L = 0.0001$). A third iteration would only result in wall shifts of the order of 0.2 mm (0.008 in).

The two-dimensional wall-adaptation method is based on experiences made during the development of a three-dimensional elastic (rubber-tube) adaptive-wall test section for the High-speed Wind Tunnel (HWG) Göttingen. Also for the 1-meter Transonic Wind Tunnel Göttingen (TWG) of the DLR, a two-dimensional adaptive-wall test section is available;

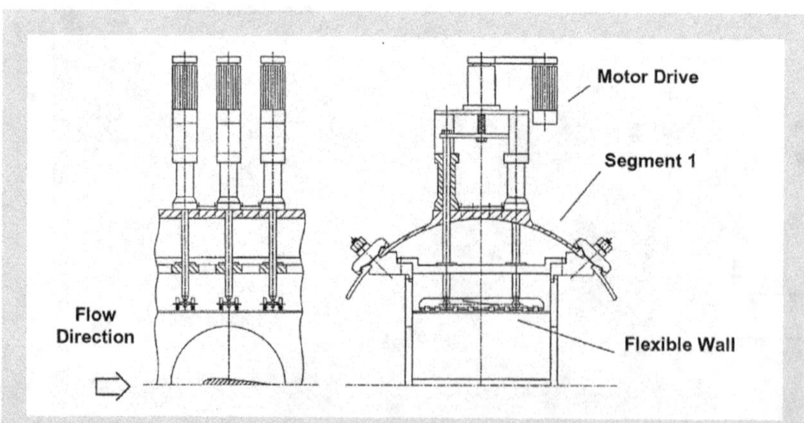

Schematic of the mechanical adjustment mechanism of the KRG adaptive wall in one axial cross-section.

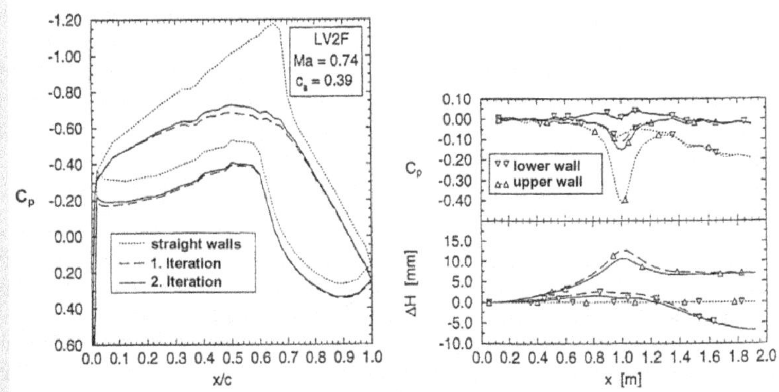

Pressure distribution measurements on an airfoil in the cryogenic Ludwieg-tube (KRG) with the walls straight and after the fist and second iteration in the wall adaptation process [50].

methods are also available to utilize this test section for the two-dimensional adaptation of three-dimensional wind-tunnel model configurations thus also reducing wall interferences here [51].

Test section of the 1-meter Transonic Wind Tunnel Göttingen (TWG) with two-dimensional adaptive walls [51].

The technique of adaptive wind tunnel walls – as developed to maturity at the DLR Göttingen – is an example of how, by intelligent means, wall interferences can be reduced to a degree that the linear model dimensions and hence the achievable Reynolds number can be doubled. The quality and reliability of the measurements are, at the same time, considerably improved.

In the cryogenic Ludwieg-tube (KRG), cost-effective tests on components at transonic speeds may be carried out before elaborate half-model and complete-model tests at the correct Reynolds numbers are being performed in the KKK respectively the ETW. In order to improve the extrapolation of flight performances, determined in conventional wind tunnels, from the model to full scale, it is necessary, besides improving the simulation conditions, to calibrate the computational methods with wind-tunnel results. These aerodynamic and aero-acoustic performance measurements on models are described in the chapter "Ground Test Facilities".

References

1 D. Küchemann, Die aerodynamische Entwicklung von schlanken Flügeln für den Überschallflug, Jhb. 1962 der DGLR, pp. 66–77
2 D. Küchemann, J. Weber, Aerodynamics of Propulsion, McGraw-Hill Book Company, New York/London 1953
3 H. Körner, R. Hilbig (ED.), New Results in Numerical and Experimental Fluid Mechanics, Contributions to the 10th

AG STAB/DGLR Symposium, Brunswick 1996, Notes on Numerical Fluid Mechanics, Vol. 60, Vieweg, Brunswick/Wiesbaden, 1997
4 R. Hilbig, Evolutionen der Verkehrsflugzeugentwicklung, DGLR-Jahrestagung, Berlin 1991
5 H. Multhopp, Methods for Calculating the Lift Distributions of Wings (Subsonic Lifting-Surface Theory), ARC Report and Mem. 2884, 1955
6 E. Truckenbrodt, Tragflächentheorie bei inkompressibler Strömung, WGL Jahrbuch, 1953, pp. 40–65
7 H. Schlichting, E. Truckenbrodt, Aerodynamik des Flugzeuges (I und II), Springer, Berlin/Göttingen/Heidelberg, 1959/1960
8 U. Ganzer, Gasdynamik, Springer, Berlin/Heidelberg/New York/Tokio, 1988
9 J. Mertens, Reduction of Aerodynamic Drag (RaWid)-Status after the First Year of the Program, in Ref. 3, pp. 7–14
10 K. Becker, Numerische Aerodynamik in der Flugzeugentwicklung, MEGAFLOW-Workshop, Braunschweig, 1996
11 E. Stanewsky, J. Délery, J. Fulker, W. Geissler (Eds.), EUROSHOCK-Drag Reduction by Passive Shock Control, Notes on Numerical Fluid Mechanics, Vol. 56, Vieweg, Brunswick/Wiesbaden, 1997, and E. Stanewsky, J. Délery, J. Fulker, P. de Matteis (Eds.), Drag Reduction by Shock and Boundary Layer Control, Notes on Numerical Fluid Mechanics and Multidisciplinary Design, Vol. 80, Springer, Berlin/Heidelberg/New York, 2002
12 R. Hilbig, Adaptiver Flügel/Innovationen im Verkehrsflugzeugbau, Symposium Neue Werkstoffe in Industrie und Forschung, Kaiserslautern, 1995
13 H. Schlichting, Grenzschichttheorie, 1. Auflage, 1951, 9. Auflage, G. Braun, Karlsruhe 1997 (English edition: H. Schlichting, Boundary Layer Theory, 7th edition, McGraw-Hill, New York, 1979)
14 A. Walz, Strömungs- und Temperaturgrenzschichten, G. Braun, Karlsruhe, 1966
15 L. Prandtl, Über Reibungsschichten bei dreidimensionalen Strömungen, Betz-Festschrift 1945, pp. 134–141
16 W. Liebe, Der Grenzschichtzaun, Interavia, 7, 1952, pp. 215–217
17 E.H. Hirschel, Evaluation of Results of Boundary-Layer Calculations with Regard to Design Aerodynamics, AGARD R-741, 1986, pp. 5-1 to 5-29
18 U. Ch. Dallmann, Laminar-Turbulent Transition Research at DLR, in Ref. 3, pp. 24–35
19 S. Wagner, Das Verbund-Schwerpunktprogramm "Transition", DGLR-JT-46-102, Jahrbuch der DGLR II, 1996, pp. 697–705, also S. Wagner, M. Kloker, U. Rist, Recent Results in Laminar-Turbulent Transition, Notes on Numerical Fluid Mechanics and Multidisciplinary Design, Vol. 86, Springer, Berlin/Heidelberg/New York, 2003
20 J.C. Rotta, Turbulente Strömungen, Teubner, Stuttgart, 1972
21 K. Gersten (Herausgeber), Physics of Separated Flows–Numerical, Experimental and Theoretical Aspects, Notes on Numerical Fluid Mechanics, Vol. 40, Vieweg, Brunswick/Wiesbaden, 1993
22 R. Henke, F.X. Münch, Laminar Flow Experiments with Large Half Model in Transonic Flows, ICAS-Paper 90-3.6.4, 1990
23 R. Henke, P. Capbern, A.J. Davies, R. Hinsinger, J.L. Santana, The A320 HLF-Fin Program, Objectives and Challenges, 2nd European Forum on Laminar Flow Technology, Bordeaux, Conference Proceedings, 1996.
24 J.-P. Robert, Hybrid Laminar Flow Control, a Challenge for a Manufacturer, 1st European Forum on Laminar Flow Technology, Hamburg, 1992, DGLR-Bericht 92-06, pp. 294–308
25 J. Szodruch, The Costs and Benefits of Aerodynamic Design: an Industrial View, Nouvelle Revue d' Aeronautique et d'Astronautique, 2, 1997, pp. 88–94
26 K. Oswatitsch, Gasdynamik, Springer, Wien, 1952 (english edition: K. Oswatitsch, Gas Dynamics, Academic Press, New York, 1956)
27 E. Becker, Gasdynamik, Teubner, Stuttgart, 1966
28 J. Zierep, Theorie der schallnahen und der Hyperschallströmungen, Braun, Karlsruhe, 1966
29 F. Keune, K. Burg, Singularitätenverfahren der Strömungslehre, Braun, Karlsruhe, 1975
30 K. Oswatitsch, Ähnlichkeitsgesetze für Hyperschallströmung, Zeitschrift für Angewandte Mathematik und Physik, 2, 1951, pp. 249–264
31 F. Keune, K. Oswatitsch, Nicht angestellte Körper kleiner Spannweite in Unterschall- und Überschallströmung, Zeitschrift für Flugwissenschaften, 1, 1953, pp. 137–145
32 O. Frenzl, Motorgondel-Flügel-Interferenz bei hohen Geschwindigkeiten (die Flächenregel), Zeitschrift für Flugwissenschaften, 5, 1957, pp. 181–183
33 R.T. Whitcomb, A Study of Zero-Lift Drag Rise Characteristics of Wing-Body Combinations near the Speed of Sound, NACA Research Mem. L 52 H108, 1952
34 N. Treinies, Schmidt, C.D. Dickmanns, H. Grallert, I. Widjaja, Themenheft ART-Programm, DGLR, J. Raumfahrtforschung, 19, Heft 4, 1975
35 E.H. Hirschel, Basics of Aerothermodynamics, Springer, Berlin/Heidelberg/New York, 2004

36 N.N., Aerothermodynamics and Propulsion Integration for Hypersonic Vehicles, AGARD-R-813, 1996.
37 H. Gericke, 50 Jahre GAMM, Ingenieur-Archiv, 41, 1972
38 J. B. Vos, A. Rizzi, D. Darracq, E. H. Hirschel, Navier-Stokes Solvers in European Aircraft Design, Progress in Aerospace Sciences, Vol. 38, 2002, pp. 60–697
39 A. Eberle, A. Rizzi, E.H. Hirschel, Numerical Solutions of the Euler Equations for Steady Flow Problems, Notes on Numerical Fluid Mechanics, Vol. 34, Vieweg, Brunswick/Wiesbaden, 1992
40 L. Prandtl, Zur Berechnung der Grenzschichten, Zeitschrift für Angewandte Mathematik und Mechanik, 18, 1938
41 R. Courant, K.O. Friedrichs, H. Lewy, Über die partiellen Differenzengleichungen der mathematischen Physik, Mathematische Annalen, 100, 1928, p. 32
42 A. Rizzi, H. Viviand (eds.), Numerical Methods for the Computation of Inviscid Transonic Flows with Shock Waves, Notes on Numerical Fluid Mechanics, Vol. 3, Vieweg, Brunswick/Wiesbaden, 1981
43 J.W. Slooff, W. Schmidt, Computational Aerodynamics Based on the Euler Equations, AGARD-AG-325, 1994
44 N. Kroll, National CFD Projekt MEGAFLOW – Status Report, in Ref. 3, pp. 15–23
45 K. Becker, POPINDA – ein Musterbeispiel für effiziente industriebezogene Verbundkooperation, Bericht Statustagung des BMBF HPSC97, 1997, pp. 275–284
46 H.U. Meier, U. Michel, H.-P. Kreplin, The Influence of Wind Tunnel Turbulence on the Boundary Layer Transition, in: Perspectives in Turbulence Studies, Ed. H.U. Meier, P. Bradshaw, Springer Verlag, Berlin, Heidelberg, New York, 1987
47 H. Ludwieg, Der Rohrwindkanal, ZfW, Vol. 3, pp. 206–216, 1955
48 D. Küchemann (Ed), The Need for Large Windtunnels in Europe, AGARD-AR-60, Dec. 1972
49 G. Hefer, The Cryogenic Ludwieg Tube at Göttingen, AGARD-R-774, Paper No. 8, 1989
50 H. Rosemann, The Cryogenic Ludwieg Tube at Göttingen, in: Special Course on Advances in Cryogenic Wind Tunnel Technology, AGARD-R-812, Cologne, Germany, 20-24 May 1996
51 H. Holst, K.-W. Bock, W. Lorenz-Meyer, F. Oberdiek, M. Hermes, Operational Experience and Test Results in the Adaptive Test Section of the DLR Transsonic Tunnel, The Aeronautical Journ. of the Royal Aeronautical Society, Nov. 1997, Paper No. 2281

Propulsion Technology – Economic and Ecological Challenge

The Evolution to Modern Jet Engines

HEINZ HOHEISEL

Already in the early phase of the first axial flow compressors for jet engines some aspects of this turbomachinery have been of importance in having exerted a substantial influence on the whole development of jet engines. To these aspects belong the off-design conditions and the aerodynamics of the blading of the axial flow compressor as well as the necessary performance improvement of the compressors in order to produce larger propulsive forces (thrust). These influences will be dealt with here, where also the findings of the early phase will be addressed because these topics refer to typical contributions made in Germany. From that it follows that the subject is not new.

A further aspect of the development of the jet engine was the introduction of the so-called bypass engine as powerplant for civil transport aircraft at the beginning of the sixties and the subsequent increase of the bypass ratio. This led to various research and development tasks of the past two decades and will be dealt with in the following in more detail. Bypass engines were, however, already the subject of investigations at the beginning of the jet engine development and were realized at the company Heinkel as engine He S 10 and particularly in 1943 tested in the successful test-bed runs of the Daimler-Benz bypass engine DB 109-007 with a thrust of 11.3 kN (2,543 lbf) and a bypass ratio of 2.42 [1]. The He S 10 was perhaps the most practical bypass engine that has been designed (1940) in the early days of the jet engine [2].

The Axial Flow Compressor at Off-Design Conditions

Compressors of jet engines do not only operate at their design point but also at the required rotational speed for a given flight condition (altitude and flight velocity), i.e., at off-design conditions. The working range or overall performance of an axial flow compressor depends essentially on the partial load behaviour of the blade cascades employed in the individual stages and their performance data [3]. By partial load behaviour thereby is understood the change of the pressure rise for a change of the rotational speed and of the engine mass flow, which can be represented by a so-called compressor performance map [4]. The most reliable method for the performance map determination is the test on a compressor test bed. Such facilities with the necessary high operating power were not yet available at the beginning of the jet engine development. Therefore, the prediction of the overall performance had to take place by calculation. In the normal case, starting with the blade geometry, that are the blade pitch t, the blade angle β_S, the profile camber and the profile thickness, the velocity triangles associated with a deviation from the design point are to be determined. Owing to the status of axial flow compressor theory at the beginning of the jet engine development, the task could not be solved analytically. It was therefore necessary to use a different approach, which arises if the characteristic of a certain wheel or stage type is known and this is employed for all stages of a multi-stage compressor. The stage characteristic line should also be independent of the Mach number and have a fixed stall line. The application of only one stage characteristic for multi-stage compressors was for the at that time employed pressure ratios quite safe. Also for the first Junkers compressor the performance map was determined in that manner. However, the reliability of such performance map must in each case be investigated experimentally.

Regarding the stall phenomenon or the determination of the stall line respectively, the following is to be mentioned: This phenomenon is known from aerofoil flow and always takes place when, after exceeding a certain angle of incidence, the flow detaches from the aerofoil or blade surface, respectively. Various types of stall exist. Based on the performance of the blading of compressor stages, the stall characteristics can be estimated with reservations, using cascade data [5]. For compressors, the stall condition arises by decreasing the mass flow. Apart from the local stall of the blades of a stage, there arises also the so-called surge of the complete compressor, caused by a strong throttling due to the engine components existing downstream, namely the combustion chamber, turbine and propulsive nozzle. Flow distortion of the air intake can also be responsible for compressor surging of aeroengines. This condition must be avoided under all cir-

First bypass engine Daimler-Benz DB 109-007, test-bed testing 1943.

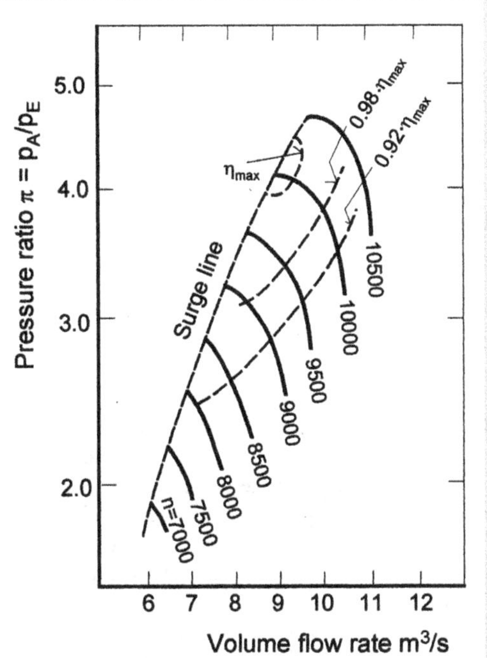

Performance map of an eight stage axial flow compressor determined from performance lines of single stage rotors tested at AVA. Unfavorable design, since η_{max} on the surge line, due to the single stage characteristics. A similar behaviour was found for the BMW 003 and the Jumo 004 compressors.

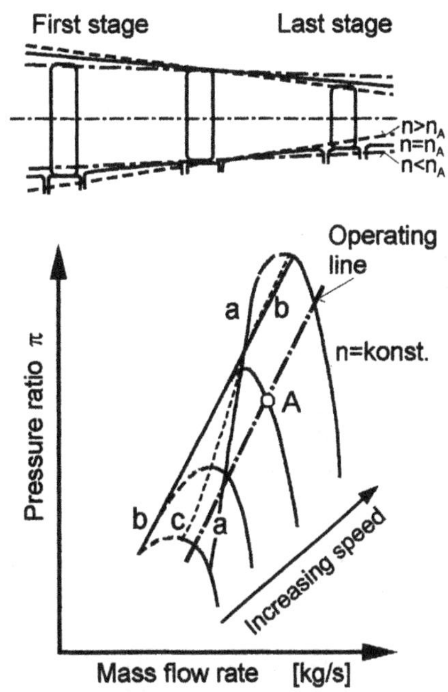

Operating states of axial flow compressors associated with rotational speed change. Top: Flow channel of an axial flow compressor corresponding to the density distribution. Bottom: Compressor performance map (schematic) a) Stall limit of the first stage b) Stall limit of the last stage, c) Surge line, A = design point.

cumstances. Blade stall and surge of the compressor are always associated with changes in the blade loading and lead to blade vibrations and as result of this very often to blade failure.

For the control of the problem of blade stall or the surge, respectively, a governor was developed for the Jumo 004 compressor, which limited during the acceleration phase of the engine the fuel mass flow in order to guarantee a sufficient margin from the stall or surge line, respectively. Furthermore, in this control circuit also a change of the propulsive nozzle exit area was incorporated. This governor was also adopted by BMW and Heinkel for their engines, because surge is a general compressor problem.

This "simple" surge line control becomes immediately more complex if the pressure ratio is to be increased. Already for the more powerful engines Jumo 012 and BMW 018 the compression was raised to pressure ratios 5.5 and 7.0 respectively. The larger compression leads to a more pronounced reduction of the flow cross-section from the front to the rear of the compressor. This compressibility effect involves that actually the flow cross-section for the given flight condition should either be increased or decreased, which means that a "rubber duct" would be necessary as called by *Hans-Georg Münzberg* [2]. Because this is not possible, for the individual compressor stages shifts of the operating points occur, the cooperation of the individual stages is disturbed. For a decreasing rotational speed, this leads to the front stages operating in the stall regime, for an excessive rotational speed the rear stages enter the stall regime. This fundamental problem of the axial flow compressor leads to difficulties of various kinds. To overcome these difficulties, the following means are available:

- Air bleed vents in the middle stages, thereby the front stages operate unthrottled at rotational speeds below the design rotational speed. For the design rotational speed these vents are, however, closed (applied to BMW 018, Jumo 012, Pirna 014 and the E 300).
- Adjustable control of the stator blades. This method is more costly because a special governor is required (applied for the first time to the "Starfighter" engine J 79 built in licence by MTU Munich, today applied in the first stage of the HP-compressor of the EJ 200-"Eurofighter" engine).
- Multi-shaft engine, that is the compressor is distributed on two or three shafts. One thus obtains two or three separate compressors (first time applied to the Rolls-Royce engine RB 211 and in the "Tornado"-engine RB 199).

In addition to these methods casing treatment is used to improve the compressor stall margin. Casing treatment consists of grooves or slots of about 60–70 per cent open area machined into the casing over the rotor tip blades. Through suitably designing the contour of the casing surrounding the rotor wheels the secondary losses can be reduced. Also hub treatment is applied. A disadvantage of the casing treatment can be a small reduction of the efficiency at design point operation of the compressor. Numerical investigations can help to explore various configurations to avoid efficiency reduction [6].

Finally, concerning the off-design behavior, it must be pointed out that an engine must also be started. The starter must be very powerful because compressors possess at very low rotational speeds poor efficiencies. At the starting process of an engine a control system has to take care that during the ignition of the engine combustion chamber the compressor is working stable [7]. This topic was already a subject for research of the first axial flow compressors [8]. Today so-called APUs (Auxiliary Power Units) are applied as starter unit and simultaneously used as secondary power system for the aircraft [9]. The company KHD (Klöckner-Humboldt Deutz, Oberursel, now Rolls-Royce Deutschland), has developed and fabricated different types of APUs.

Systematic Research on Cascade Flow

At all places of development of axial flow compressors and turbines, whether research institutes or industry, it was already recognized very early that it is necessary to obtain data which make possible a reliable prediction of the compressor blading. This required both the theoretical as well as the experimental investigation of the flow processes in detail, as was stated by *Heinz Marcinowski* [8] in a contribution at the meeting of the Lilienthal-Gesellschaft (Lilienthal Society) in Heidelberg in October 1943. The in-depth investigation of the flow process for the blading of axial turbomachinery is facilitated by the development of a coaxial cylindrical surface section in a plane to obtain a two-dimensional cascade. This plane cascade is characterized by the regular arrangement of blade profiles of the same geometry, the same spacing t and the same blade angle β_S. The theoretical treatment of such cascades took place either using wing theory [10] taking into account the cascade influence by means of cascade influence coefficients [11], or applying, according to stream-tube theory, the so-called "angle exaggeration" [12]. The method of angle exaggeration (incidence and deviation angle rules) proves to be very useful for establishing empirical correlations regarding the deflection properties, that is the determination of the outlet flow direction for a given inlet flow direction [13]. These theoretical procedures, however, only produce results for inviscid flow and allow no prediction of the losses which occur in cascades. For the wing it is known that also for viscous flow the lift coefficient is a linear function of the angle of incidence as long as this is small. For compressor cascades this relation is no longer valid. Therefore, in this respect requirements were formulated for systematic measurements on blade cascades because measurements performed at that time were focussed very strongly on other special problems.

Reported on measurements was also at a cascade meeting which took place in Brunswick in March 1944, where topics were dealt with like the definition of loss coefficients, boundary-layer questions for turbomachinery, investigations at high Mach numbers as well as the transfer of plane cascade results to the real machinery. A fundamental goal of this meeting was also to establish "the organization for the collaboration in the field of cascade measurements" [14]. Experimental investigations were carried out up to 1945

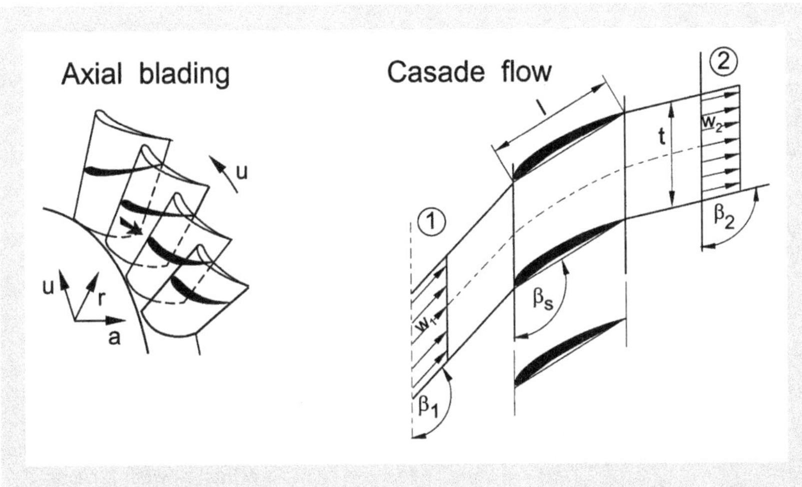

By developing a coaxial cylindrical surface section in a plane a two-dimensional cascade is created.

at many research institutes and firms, where the test stands were still rather simple facilities. A sophisticated cascade test stand, an AVA project for high subsonic speeds, could not be put in operation until 1945. The impetus for a systematic investigation of the plane cascade flow that was initiated at the addressed meeting could only be implemented after 1945. *Hermann Schlichting* of the TH Brunswick started extensive theoretical and experimental cascade work [15], also because his original field of work, aeroplane aerodynamics, was not allowed any longer to be a research topic. A similar situation existed also after 1918, which permitted *Albert Betz* to intensify research work on axial flow compressors. The theoretical work was focussed on the determination of the losses which are of importance for the flow through cascades of airfoils. For a validation of the theoretical results and especially for an understanding of compressible viscous cascade flow experiments are necessary. Thus, under the leadership of *Schlichting* the Hochgeschwindigkeits-Gitterwindkanal (HGK), a high speed cascade wind tunnel, was erected in Brunswick. This tunnel has been put into operation at the Institut für Aerodynamik of the DFL (today DLR) in 1956 [16]. The HGK, designed by *Norbert Scholz*, is characterized by the installation of the cascade test section and the air supply, a six-stage axial flow compressor of the Voith company, in a vacuum chamber. By lowering the pressure in the vacuum chamber the density of the flow in the test section can be changed, whereby it is possible to vary the Mach number and the Reynolds number independently of each other. The cascade testing can therefore take place at the Mach numbers and Reynolds numbers which are experienced by axial flow compressors of aeroengines in practice. A multitude of investigations has been performed in the HGK [17] on compressor [18] and turbine cascades [19].

Since 1985 the HGK is incorporated in the institute of *Leonhard Fottner* at the Universität der Bundeswehr Munich (University of the Armed Forces) and has maintained there until now its importance for cascade investigations [20, 21]. After 1945 also at other institutions cascade test facilities were established, delivering basic data of compressible flow through cascades [22, 23]. In the middle of the sixties *Wolfgang Heilmann* of the Institute of Propulsion Technology of the DVL in Cologne recommissioned facilities closed by NASA [24]. Based on these facilities for subsonic and supersonic cascade investigations research work was carried out in particular for the development of advanced compressor cascades [25, 26]. For turbine cascade research work a special annular cascade with a rotating test wheel was developed and is in operation at the DLR Göttingen [27, 28].

In the past cascade investigations were carried out in the case of compressor blades using profile families like the NACA 65 series or the English C-profiles [29]. In the case of turbine blades additional conditions have to be taken into account, e. g., blade cooling. Within the last two decades the availability of numerical methods allowed to develop new concepts for the blading of compressor and turbine cascades,

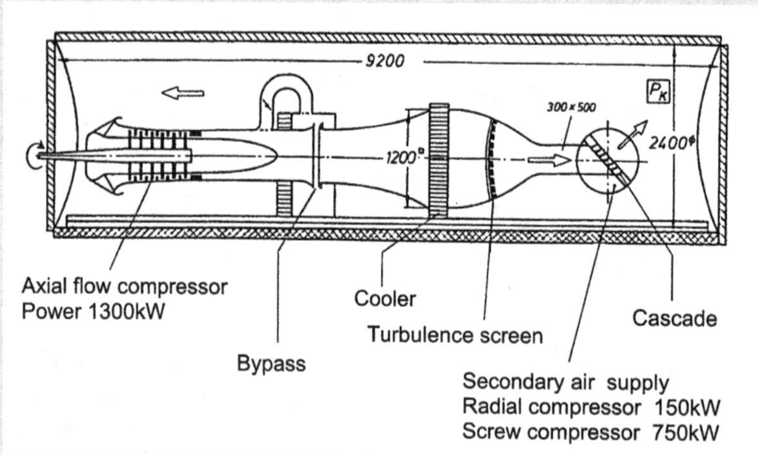

High speed cascade wind tunnel (HGK) of the "Deutsche Forschungsanstalt für Luftfahrt, DFL", Brunswick, 1956.

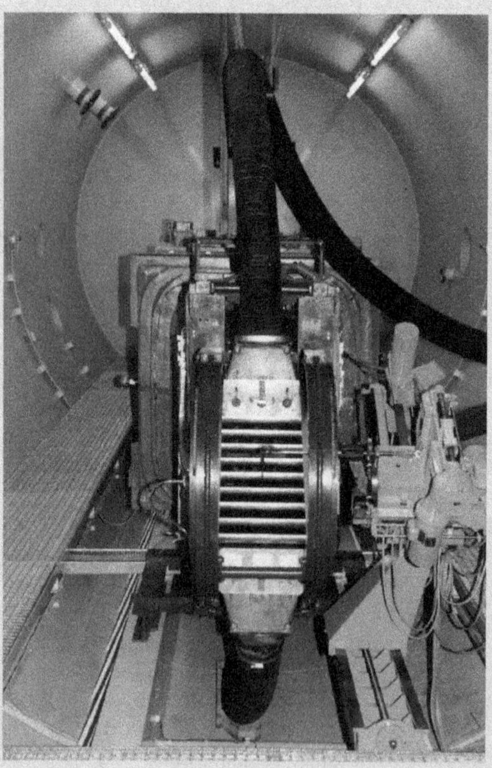

View of the test section of the HGK at the re-commissioning in Munich on November 29, 1985.

based on given pressure distributions on the blade surfaces. In the case of compressor cascades this leads to the so-called PVD (Prescribed Velocity Distribution) – profiles and to the CD (Controlled Diffusion) – blades [30]. The numerical calculation methods have meanwhile reached a high state of development, but the criteria used in the numerical methods have to be verified by means of cascade and turbomachinery test cases [31]. The treatment of cascade flow experimentally as well as with numerical calculation methods has led to fundamental findings and has influenced very favorably the progress in the field of axial flow turbomachinery [32].

Thrust Increase of Jet Engines

On the basis of the initial successes in the development of jet engines with a moderate thrust of 6 to 8 kN (1350–1800 lbf), (BMW 003 and Jumo 004), soon the requirement by the RLM was forthcoming for powerplants with a thrust of approx. 30 kN (6750 lbf). This resulted in the design of the BMW 018 and of the Jumo 012. Both possessed axial flow compressors with the data shown in the survey. Added are data of powerplants that were to follow and also because due to the fact that all these powerplants constitute a kind of "family concept". To speak of a "powerplant family" is justified since in nearly all these projects experts were involved who came from the original BMW or Junkers permanent staff [33], independent of whether they were later temporarily active in the Soviet Union, in France, in Pirna (GDR) or in Kairo-Helwan (Egypt).

The project and production work on the BMW 018 and Jumo 012 was discontinued for the time being at the end of April 1945. In the following weeks the development offices of BMW and Junkers were searched thoroughly by the Western Allies and many important documents were brought to the West. Also a certain number of leading experts were recruited to carry on working in the Western Zones. After the change of the demarcation line in Germany on 1 July, 1945, also the Soviet side got a look at the state of the art and subsequently allowed the continuation of aircraft gas turbine development. This led in October 1946 to test-bed testing of the BMW 018 on a provisional test bed at Staßfurt. The Jumo 012 engine came into operation only in the Soviet Union. The knowledge of the German jet engine development shifted to the area of the victorious powers, especially to France, the Soviet Union (SU) and the USA [34, 35]. The German development teams working there at that time, partially voluntary, partially compulsory, performed outstanding work, leading in the Soviet Union under the leadership of *Ferdinand Brandner* to the world's most powerful turbo-propeller engine (NK 12) [36]. This engine is still in service to this day. All these experts could develop their knowledge steadily, for them there was no interruption of work. The acquisition of foreign publications was, however, not possible for the experts working in the SU.

In order to achieve thrust increases of jet engines, it is necessary to raise the sucked-in air mass flow of the compressor and the compressor pressure ratio. Thereby the aerodynamic conditions in the cascade become more complex and require for the de-

Axial flow compressors with higher thrust (turbojet = TJ) and power (turbopropeller = TP) respectively.

Engine		BMW 018	Jumo 012	Jumo 022	SU NK12	Pirna 014	Pirna 018	Helwan E300	DB 721
Design year		1944	1944	1944	1950	1953	1955	1961	1962
Thrust	(kN)	33.3	29.9	–	–	30,9	–	32,4	–
	(lbf)	7,493	6,728	–	–	6,953	–	7,290	–
Shaft power	(kW)	–	–	4,418	8,820	–	3,680	–	1,618
	(hp)	–	–	6,000	12,000[a]	–	5,000	–	2,201
Mass flow rate	(kg/s)	44	60	30	62	50	31.3	53	10
	(lb/s)	97	132	66	137	110	69	117	22.1
Pressure ratio		7.0	5.5	7.5	13	7.0	10.2	6.1	6.0
Stage number		12	12	14	14	12	13	9	8
Rotational speed (RPM)		5,000	6,100	8,250	8,250	8,000	13,100	10,300	21,000
Efficiency	(per cent)	–	83	87	88	87	86	89	–
Degree of reaction	(per cent)	70	89	70	62	62	50	62	–
Circumferential	(m/s)	–	243	273	324	297	375	405	400
velocity (ft/s)		–	797	895	1,063	974	1,230	1,328	1,312
External diameter	(mm)	–	760	632	750	710	547	751	–
	(in)	–	29.9	24.9	29.5	28	21.5	29.6	–
Hub to tip ratio		0.5	0.5	0.6	0.45	0.53	0.5	0.52	–
Entry guide vane		with fixed	without	with fixed	with adjust.	with adjust.	with adjust.	with fixed	–
Number produced		–	8	18	series	66	3	18	3
		TJ	TJ	TP	TP	TJ	TP	TJ	TP

The project Jumo 022 became in the Soviet Union the basis of the TP TW2 and was put to service as NK4 with a power of 5,000hp on the series aircraft JL 18 for example
[a] Final version 15,000 hp.

sign of the compressor and the turbine more sophisticated computation methods. *Hans-Joachim Schröder* already developed in the SU a method for the determination of the longitudinal flow in the ring chamber of an axial flow compressor [37]. The motivation for this he got from the analysis of measurements of the air mass-flow rates of compressors, which showed that the compressor blading caused a disturbance of the flow which made its influence felt far upstream into the intake duct. Based on the force balance that must exist in a compressor stage in the radial direction between centrifugal force, inertia force and pressure force, the solution of the problem yielded a differential equation which describes the movement of a volume element far in front and far behind of a blade row. Since, however, in a multi-stage compressor the blade rows are arranged closely one behind the other, in *Schröder's* theory a stator or rotor ring is replaced by a disc of continuously distributed vortex filaments (actuator disk theory) with corresponding damping coefficients. The longitudinal flow and the cascade flow were not coupled. This conceptual model found its application for all compressors shown in the survey and also for later compressor designs [38].

The testing of the compressor 014 (already designed in the SU) in the East German VEB Entwicklungsbau Pirna verified the progressive design method. The compressor, however, operated too near the surge line [39]. By a redesign of the first four stages by *Werner Held* this drawback was eliminated, whereby for the determination of the velocity triangles, which resulted from the ring chamber calculation, a correction was applied for the side-wall boundary layers. Especially the estimate of the side-wall boundary layer effects is a basic element for an accurate axial flow compressor design [40]. Today the consideration of secondary flow effects is a fundamental basis of the design process of turbomachinery using 3D numerical methods [41]. Furthermore, at the time of the 014 redesign the compressibility effects on blade cascades were better understood [42]. All this knowledge was applied in the design of the compressors Pirna 018 and Helwan E 300 [36] with considerably increased circumferential velocities. This increase was among other factors also possible because in the front stages profiles were employed which possessed better high speed properties. These better high speed properties were obtained with profiles having their maximum thickness at 60–65 per cent of blade chord. In the meantime controlled diffusion airfoils have been applied in compressor blades as mentioned before. A further increase of the circumferential velocity to a remarkable 470 m/s (1,542 ft/s) has been achieved in the front fan of the "Tornado" engine RB 199, in which case a number of new research results has been incorporated in the design [43].

With this a new generation of compressors was launched, in particular based on the development of the "Transonic Compressor". This type of compressor was more or less the result of the research activities on the supersonic compressor [44] in the fifties in different countries, also in Germany [45, 46]. To achieve an increase in stage loading in the stages of a transonic compressor, very high inlet Mach numbers have to be realized. Typically, between hub and tip there exist Mach numbers in the range of $M = 0.8 - 1.5$. This high Mach number level leads to a complicated compression shock system within the blade passage. The de-

Axial flow compressor of the turbojet Pirna 014 with bleed behind stage 6.

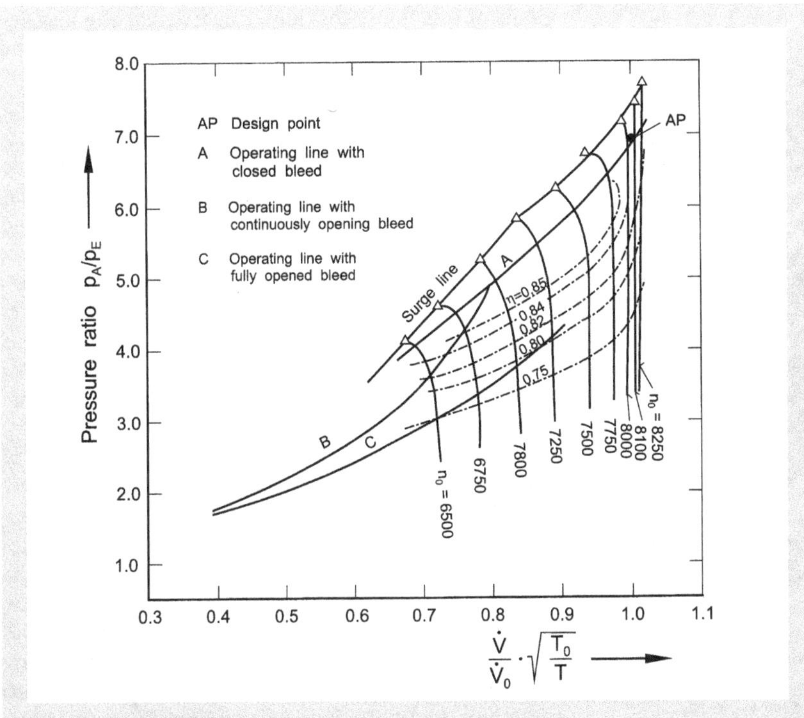

Performance map of the axial flow compressor Pirna 014/A1. Influence of air bleed behind stage 6 on the operating lines of the engine.

Front stages of the rotor of the TP axial flow compressor Pirna 018 featuring transonic blades.

Rotor of the nine stage axial flow compressor of the turbojet E 300 developed, manufactured and tested in Kairo-Helwan. First run of the engine July 1963, first flight (V11) in June 1966 on the AN12 aircraft.

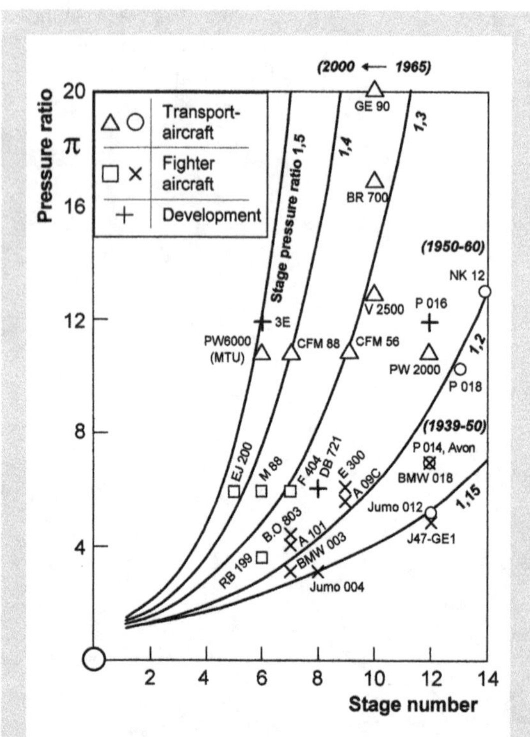

◄ Comparison of compressor data of military and civil high pressure compressors. The abbreviations mean: A = ATAR, B.O. = Bristol Orpheus, BR = Rolls-Royce, CFM = General Electric/SNECMA, DB = Daimler Benz, 3E = MTU, EJ = Eurofighter, F, GE = General Electric, M = SNECMA, NK = Nikolei Kusnezow, P = Pirna, PW = Pratt & Whitney, RB = Rolls-Royce, V = Pratt & Whitney, Rolls-Royce, MTU.

tails of this flow behaviour, in particular the loss development, are the objective of blade design with the aim to minimize the losses [47].

The addressed compressors incorporate all the experience that had been accumulated in the time interval of approximately three decades, being based on a substantial contribution of experts that have been associated with the BMW-Junkers beginnings. If one plots the compressor data regarding pressure ratio, number of stages and stage pressure ratio in a diagram, one finds that the German engine compressors are on the international level. All data plotted in this diagram for later compressors refer to developments to be found in engines associated with international cooperations, such as also the high pressure compressor of the RB 199 engine, based on the 3 shaft concept of Rolls-Royce, designed by *Hubert Grieb*, the first compressor designed in Germany after 1945 that came into service in a quantity production engine. The decision for the development of the RB 199 engine as the power plant for the "Tornado" fighter had been made in 1969. This task was assigned to the companies Rolls-Royce, MTU and Fiat. At that time the company MTU (Motoren-Turbinen-Union), today MTU Aero Engines, had been founded, originating from MAN-Turbo GmbH and the Daimler-Benz AG.

Regarding the engine RB 199, MTU was also involved in the development and production of the intermediate pressure compressor and the intermediate pressure turbine. The development of the RB 199 engine with a thrust of 37–43 kN (8,325–9,675 lbf) without reheat, which started in 1969, could be completed in 1978 with a 150 h acceptance run. The successful three-nation cooperation (D, GB, I) for the RB 199 was continued with the development of the engine EJ 200 as powerplant for the "Eurofighter". Apart from Rolls-Royce, MTU and Fiat also the Spanish company IPT was involved in the development. The German contribution was here again mainly the de-

Three stage transonic compressor of the front fan of the turbojet engine RB 199, MTU alternative design.

velopment and production of the low (LP) and high pressure (HP) compressor. In the field of engine control electronics an international first-rate performance was achieved by MTU [48]. The progress in compressor aerodynamics compared with the beginnings of the jet engine development becomes clearly visible in the increase of the stage pressure ratios (see diagram), in the five high pressure stages of the EJ 200 a pressure ratio of 6 is reached. The continuous, over many years performed fundamental investigations of cascade flow, also at various institutes of the DLR, thus found the confirmation of their usefulness.

While the development work in the area of military engines was mainly directed towards the compressor as a component within the framework of European cooperation, the development activities for civil turbofan engines shifted to the turbine as component and this then in a transatlantic collaboration. For the General Electric engine CF6-50 or 80, respectively, the cooperation (start 1971) was limited to the production of the sophisticated components of the cooled high-pressure turbine. This cooperation ceased in 1994 and is in the meantime restricted to the fabrication of spare parts for existing CF6 engines.

Similarly, at the beginning of the seventies the start of a cooperation between the company Pratt & Whitney and MTU took place and led in 1980 to the design of a five stage low pressure (LP) turbine for the engine PW 2037. With this turbine outstanding efficiencies have been obtained. This is the first turbine in Germany developed after 1945 for a civil quantity production engine. The approx. 1,800 engines of the 2000 series in the thrust class of 165–183 kN (37,125–41,175 lbf) produced up to now are in service on the aircraft Boeing 757, the MD transport aircraft C17, and some engines on the Russian IL96.

The advanced technologies of the PW 2037 also found application on a large scale regarding the five nation (USA, GB, D, I, J) engine IAE-V2500. The MTU contribution was the development and production of the five stage low pressure turbine. This engine came into service as powerplant for the Airbus family A 320 since 1989 and later also for the MD 90. More than 2000 engines are in service, about 4,300 engines are ordered. The successful collaboration of MTU with Pratt & Whitney led in 1991 to the participation in the development and production of the large engine PW 4084 with a take-off thrust of 391 kN (87,975 lbf). The 7-stage low pressure turbine was the MTU contribution. Also for the engines PW 4090 and PW 4098 (the most powerful engine of Pratt & Whitney, 436 kN, (98,100 lbf), MTU is responsible for the low pressure turbine with the world-wide for the first time air-cooled first stage. With a fan diameter of 2,867 mm (9.4 ft), this engine appears to have reached for the time being the maximum engine diameter, especially with respect to the under-wing installation of the engines.

The largest aircraft of Airbus, the new A 380, will be equipped with the Trent 900 of Rolls-Royce and with the GP 7000, a joint development of General Electric and Pratt & Whitney. The GP 7000 engine, delivering a thrust of 311–340 kN (70,000–76,500 lbf) having a bypass ratio of about 9, will have its first run on the test bed in spring 2004. MTU is developing and fabricating the 6 stage low pressure turbine and fabricating the two stage high pressure turbine of the GP 7000.

In the lower thrust regime (<100 kN, 22,500 lbf) too German engine technology could establish itself successfully with the foundation of the company BMW Rolls-Royce. At the beginning of the nineties it came to the development of a complete engine family BR 700 [49]. Already in 1996 the BR 710 (63 kN resp. 14,175 lbf thrust) received the certification as quantity production engine for the long range business aircraft Gulfstream and Global Express. In the year 2000 the BR 710 received the certification also for the maritime patrol aircraft Nimrod MRA4. The more powerful

BR 715 (82 kN resp. 18,450 lbf) entered service on the 100-seat B717-200 of the company Boeing. More than 1,400 BR 710/715 engines have been ordered. The two types of the BR700 engines are two-shaft turbofans with the technology derived from the IAE-V2500 and the Rolls-Royce Trent experience. The BR 700 family comprises engines having the lowest noise level in their class. The pollution emission data of the BR 710/715 engines are clearly under present limits. With the start of the year 2000 BMW Rolls-Royce (today Rolls-Royce Deutschland) became a wholly owned subsidiary of Rolls-Royce and obtained more competence in the product scope of 2-shaft turbofans [50].

For the engines PW 300 and PW 500 in the thrust class of 20–30 kN (4,500–6,750 lbf) of the Canadian subsidiary company Pratt & Whitney, MTU was again involved in the development and production of the low pressure turbine. These are engines for business aircraft of medium range. More than 2,000 of these two types of engine are in service.

Detailed particulars of the engines addressed here and some others can be found in the third edition of "Flugmotoren und Strahltriebwerke [1] (Aeroengines and Jet Engines)".

The thrust increase from approx. 6 kN to about 500 kN (1,350 to about 112,500 lbf) attained in the past six decades since the introduction of the gas turbine as aircraft powerplant are based on a very high research and development effort of all companies. The developments discussed also show that German engine technology has steadily improved and can until now hold on its own as very well accepted partner. This acceptance is also a result of the collaboration between the German engine industry and institutes of the DLR in Cologne, Göttingen and Brunswick. In this collaboration institutes of the University Stuttgart, the University of the Bundeswehr Munich, the TU Munich, Karlsruhe, Darmstadt, Berlin, Brunswick and the RWTH Aachen have participated. Of special importance in this context is the altitude test bed of the University Stuttgart.

Increase of the Bypass Ratio

Since the introduction of turbo-jet engines into civil air traffic in the middle of the fifties, the specific fuel consumption was in several steps reduced by nearly 50 per cent. The first large step was the change from the turbojet (jet engine without bypass flow) to the turbofan (bypass engine), which was preceded already with considerations for the engine DB 109-007 and test-bed runs in 1943. In these turbofan engines only a portion of the total sucked-in airflow flows through the "hot" core of the engine. The bypass ratio (BPR), that is the ratio of the air mass in the cold fan flow to that in the hot core flow, ranged for the first generation of turbofan engines from $\mu = 1–2$. The next step took place towards the end of the sixties with the introduction of the so-called high bypass engines for which the bypass ratio lies between 4 and 6. These are the engines that are still in operation to this day. By improving single components of the engines, this engine concept could be developed to yield the current values of specific fuel consumption. All this progress is the result of a steady process which took place up to the present without particularly spectacular leaps and which will also continue in the future. Ecological aspects, like low pollution emissions and noise levels, are going to influence the further development of engines however more strongly [51].

The large increase of the world air traffic with an annual growth of >5 per cent since the middle of the eighties makes it necessary to aim at further fuel consumption reductions, especially for engines in service with medium and large civil transport aircraft. This also out of consideration for the associated environmental pollution because of the higher fuel consumption owing to the growing passenger miles. In order to achieve this, aircraft aerodynamicists undertake large efforts to reduce the drag of aircraft for example by laminarization of the wing boundary layers [52, 53]. For the engine manufacturers exists the possibility to lower the fuel consumption by increasing the total efficiency of the engine. The total efficiency is the result of the product of the propulsive efficiency, a measure of the efficiency of thrust production, and the thermal efficiency, a measure of the degree of conversion of the available energy to the usable energy. Since the introduction of the turbo-jet engines, the combustion chamber exit temperatures and the total pressure ratios for raising the thermal efficiency have been steadily increased. Further increases are possible, the potential within the framework of conventional cycles is however limited.

The second contribution, the propulsive efficiency, offers a larger improvement potential and is directly coupled with the principle of the thrust produc-

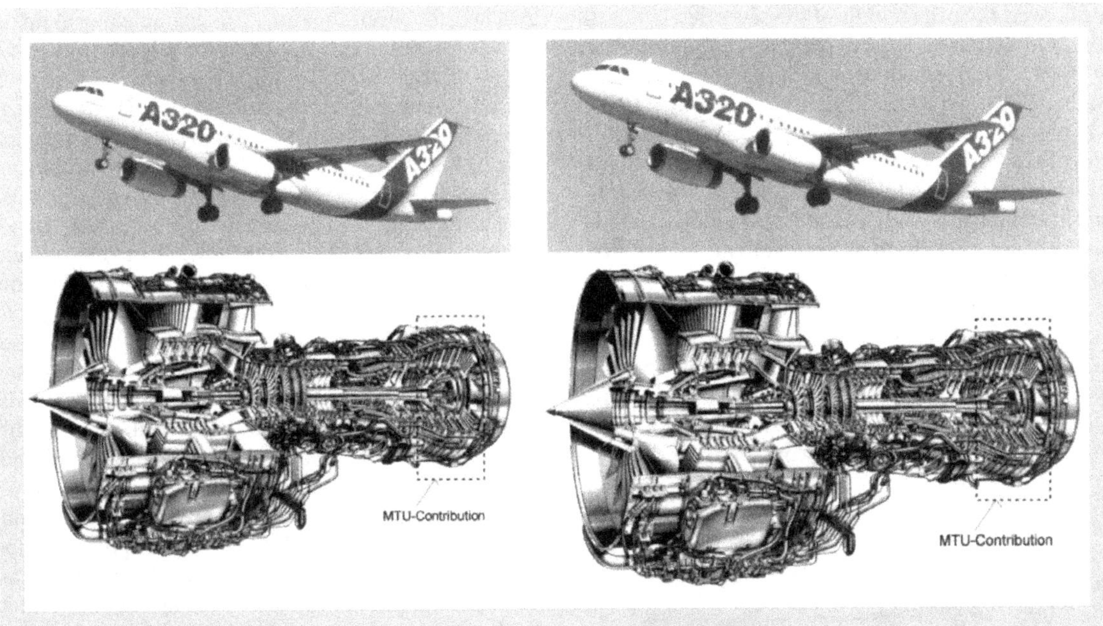

The five nations V2500 bypass engine applied on the Airbus types A 320 and A 321. Indication of the use of this engine type on the aircraft is always the long nacelle.

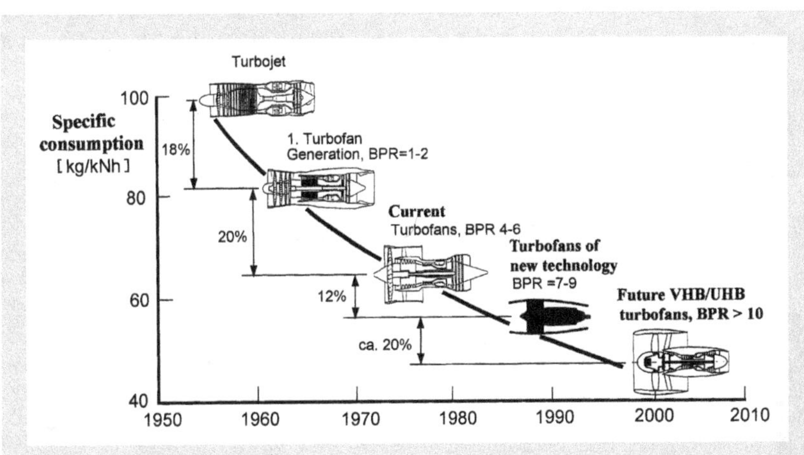

Development of the specific fuel consumption of civil engines in dependence on time.

tion. A turbojet engine achieves its propulsive thrust in the way that at a flight velocity w_0 a relatively small mass flow is exhausted from the propulsion nozzle at a high jet velocity w. A propeller, in contrast, accelerates a relatively large mass flow to an only slightly higher jet velocity compared with the flight velocity. Between these two propulsion units lies the bypass engine and its propulsive efficiency becomes the better the smaller the velocity difference $w - w_0$ is. A smaller velocity difference is obtained by a correspondingly small pressure increase in the fan of a bypass engine or also in a propeller. Two paths have been intensively followed since the middle of the seventies. One path consists of the development of bypass engines further to bypass ratio values of 10–20, the so-called Very-High (VHBR) or Ultra-High Bypass Ratio (UHBR) engines. The other path entails the development of propellers for high flight velocities (called propfans) with bypass ratios of 60–80. As studies of *Peter Schimmig* have shown, with these concepts reductions of the fuel consumption up to 30 per cent can be achieved compared to current bypass engines [54].

The development of a propeller for high flight velocities was the aim of developments in the USA, initiated by the oil crisis in the middle of the seventies [55]. Although the development of the propfan progressed very far, up to flight testing, and also a considerable reduction of the fuel consumption could be reached, for several reasons (for example noise) up to now the propfan has not been introduced into service on aircraft.

In Germany the work concentrated on the further development of the bypass engine concept with the aim to realize a bypass ratio as high as possible. Because of the associated reduction of the jet velocities, this is the decisive measure for noise reduction. Larger changes of the bypass ratio, however, do not remain without influence on the total engine concept. In the turbofan, the fan is directly driven by the low pressure turbine. However, an increase of the bypass ratio requires lower fan rotational speeds because with the bypass ratio also the diameter of the fan grows. This leads to higher circumferential velocities and thus also to higher Mach numbers in the fan blading and as consequence of this to a reduction of the fan efficiency. A decrease of the fan rotational speed would lead to larger stage numbers of the low pressure turbine. By introduction of a gear it is possible to separate the fan and the turbine rotational speed so that both components can be operated at their respective efficiency optimum. With this, the possible reduction of the turbine stage number can compensate the weight increase due to the gear.

An alternative to the single stage fan with outlet stator is a contra rotating fan which also produces a

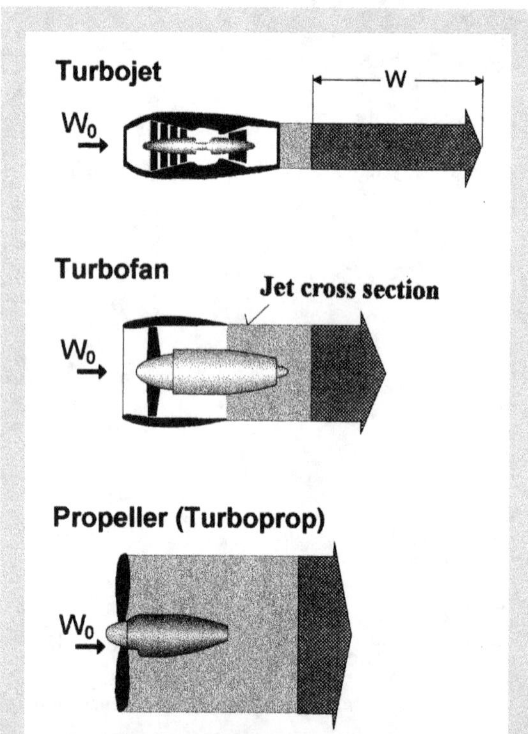

Thrust production with different engine concepts.

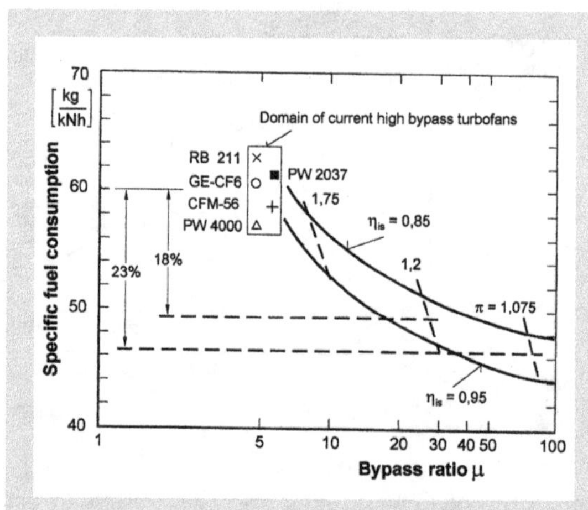

Influence of the bypass ratio μ on the specific fuel consumption for a fixed core engine technology, π = fan pressure ratio, η = fan efficiency, M = 0.8, H = 10,688 m (35,057 ft).

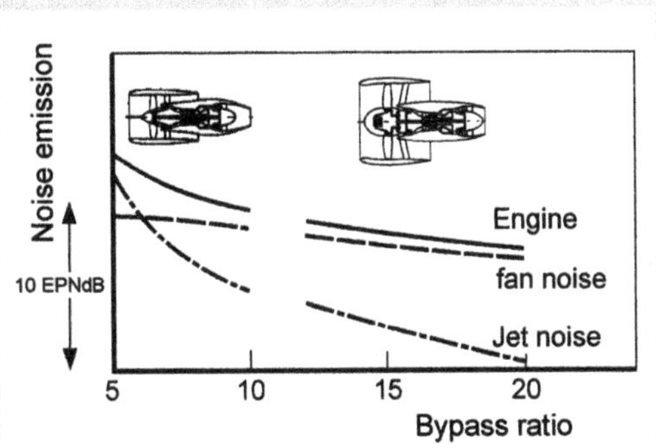

Influence of the bypass ratio on engine noise.

flow free of swirl. Free of swirl means that at the outlet of the fan no velocity component exists in the circumferential direction. The loss producing stator can be dispensed with. Such a concept, described as Counter Rotating Integrated Shrouded Propfan (CRISP), was presented by *Hubert Grieb* and *Dietrich Eckardt* at the beginning of the eighties and led to an extensive technology programme [56]. Involved in this programme were, apart from the Motoren Turbinen Union (MTU), the EADS Airbus, the DLR and some universities [57]. The CRISP engine concept has been conceived for a bypass ratio of 20 and extensive model investigations were performed for an airliner with 150 seats (A 320). This concept also comprises a thrust reverser system in which both contra rotating rotors can be brought to a reverse thrust position.

The CRISP Technology Programme

The research programme supported by the Federal Ministry of Research and Technology focussed since 1985 mainly on investigations of the fan and nacelle aerodynamics, acoustics, new materials as well as of the airframe-engine interference. At the beginning of the technology programme was the aerodynamic design of the fan for the CRISP engine concept. This fan, because of the contra rotating rotors, the swept blades and the wide blade pitch (distance of blade to blade), leaves already for the design condition the area of proven knowledge of blade aerodynamics. Already in 1987 for this purpose investigations of the isolated rotor 1 took place at the Institute of Propulsion Technology of the DLR in Cologne [58]. During these in-

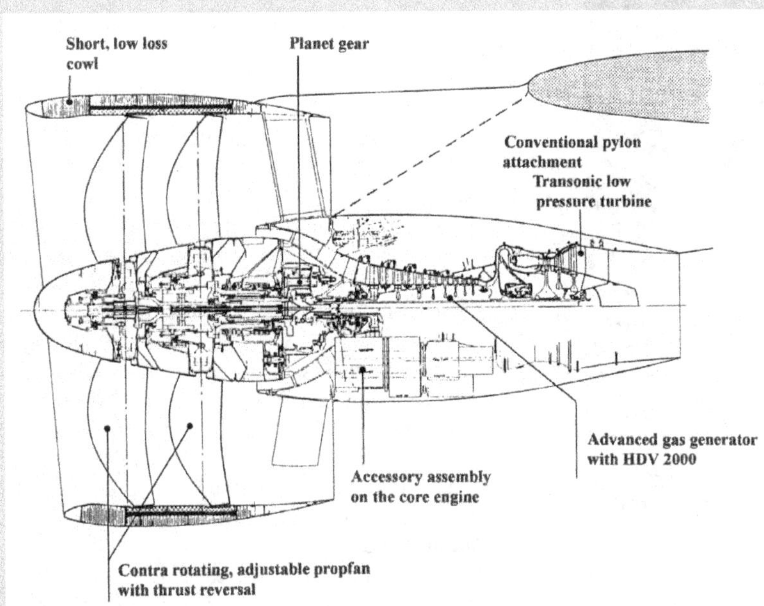

CRISP (Counter Rotating Integrated Shrouded Propfan) engine concept with counter rotating fan driven via a planet gear.

vestigations significant three-dimensional effects in the blade flow became evident which were taken into account in a later new design of the fan.

Focus of the experimental investigations were measurements on a so-called national model of the contra rotating fan with a scale of 1 : 6.25, which is driven by a reverse gear by means of a compressed air turbine. This model with a fan diameter of 0.4 m (1.31 ft) was used in the low speed (NWG) and the transonic (TWG) wind tunnel of the DLR in Göttingen for aerodynamic investigations. Acoustic investigations were performed with the same model in the DNW (German-Dutch Wind Tunnel).

The extensive aerodynamic investigations and the complete verification of the CRISP concept have been reported by *Jörg Sieber* [58]. According to these, the fan shows conspicuously wide performance lines which in no case are limited by reaching the aerodynamic surge line. The most important flight operating points lie within the fan performance map. Furthermore, it was found that only one blade setting of the fan is necessary if in the partial load regime optimum fan efficiencies are to be obtained. A limitation to only few fixed blade settings (take-off, cruise, reverse thrust) becomes possible. Because of the low blade blockage (low number of blades), high axial Mach numbers > 0.74 are achievable upstream of the fan. Consequently, the CRISP can for a given diameter be designed for a higher bypass ratio than other concepts so that a lower fuel consumption is obtained. The contra rotating fan can in addition cope without any problem with large intake disturbances such as those that can arise at take-off.

CRISP model in the low speed wind tunnel of the DLR in Göttingen. (Fan mass flow rate 25 kg/s (55 lb/s), fan pressure ratio 1.24 : 1).

From the acoustic investigations, as dominant noise source the aerodynamic interaction between both rotors and the broad-band noise of the rotors could be identified [59]. By increasing the distance of separation between the rotors, there exists a potential for noise reduction. The investigation of the complex flow field at the fan and the core engine intake for reverse thrust was the goal of investigations in the low speed wind tunnel of the DLR Brunswick [57]. The remarkable reverse thrust potential of the CRISP concept was confirmed. Also a fast rotating low pressure turbine was the object of investigations. Findings of transonic cascade investigations have been used in the design of a two stage turbine. This turbine was tested in the form of a cold rig (turbine operates with cold compressed air) on the altitude test bed of the University Stuttgart. The results are promising for future developments.

The larger fan diameters necessary and the desired blade adjustment for the introduction of engines with higher bypass ratios require blades of light-weight construction. A multitude of possible constructions were investigated. As a possible version the carbon fibre solid blade was found to be suitable [60]. The concept developed by MTU is based on a carbon fibre epoxy resin (CFC) solid blade that is fitted into a fork-like titanium root, bonded and bolted. With the extremely light-weight construction of the blades, the centrifugal loading on the bearings is reduced to the just allowable degree.

Results of additional extended investigations concerning aerodynamics and aeroacoustics of a contra rotating fan are contained in a status report edited

CRISP 1 m (3.28 ft) model during installation on the twin shaft compressor test bed of the DLR in Cologne.

by *Heinrich Weyer* of the DLR [61]. For this, a second CRISP model with a rotor diameter of 1 m (3,28 ft) was manufactured by MTU and put at the disposal of the DLR. This model has been in service on the two-shaft compressor test bed of the DLR. Both motors (with a power of 5 MW each) of this test bed drive the rotors independently of each other so that any arbitrary combination of rotational speeds can be selected and used. The goal of these investigations are flow field measurements in the blade passages of the contra rotating rotors including unsteady phenomena [62]. The data of the model are:

Rotor diameter: 1 m (3.28 ft),
Mass flow rate: 166.2 kg/s (366,5 lbs/s),
Total pressure ratio: 1.24 : 1,
Blade number rotor 1/rotor 2 = 10/12.

For the aeroelastic investigations (in the responsibility of the Institute of Aeroelastics of the DLR Göttingen) the rotor blades were fitted with dynamic pressure sensors. Parallel to the experimental work, the development of a numerical method TRACE (Turbomachinery Research Aerodynamic Computational Environment) for the calculation of the spatial, steady and unsteady flow in contra rotating rotors took place. For such calculations large computer capacities are necessary, because for the flow through both rotors with a blade ratio of 10–12 for reasons of periodicity the calculations must extend over half the circumference, yielding a volume of data of about 800 Gbyte. The result of such calculations are the instantaneous distribution of the static pressure on the rotor blades or in the blade passages. The numerical calculation method has in the meantime been developed to a very high degree [61, 63]. This was accomplished in a long-term development and validation phase so that now the code has reached the application stage and a high reliability. Within the framework of an intensive cooperation with MTU, industrial requirements for the implementation and efficiency have been incorporated in the development of this simulation methodology. The key algorithm of this complex procedure constitutes a structured multi-block flow solver on the basis of the compressible Reynolds-averaged Navier-Stokes equations. Viscous effects are taken into account by relevant turbulence models which however at first had to be adjusted for internal flows. The code represents an essential tool for the aerodynamic design of turbomachinery components and has found its application in research and industry.

A further important element of an engine with increased bypass ratio is the cowling of the fan, the fan cowl. The fan cowls of current transport aircraft possess already an airwashed surface of nearly 20 per cent of the wing surface. With growing bypass ratio, however, the fan cowl diameter increases and with this also the viscous drag producing airwashed surface of the fan cowl. Thus, applying laminar flow technology also to this aircraft component seems to be promising for drag reduction purposes [64]. Extensive investigations of the fan cowl aerodynamics taking into account the different operating conditions, like take-off, cruise, engine failure and reverse thrust, were subject of the technology programme [57]. Included in this program was a flight test program with a laminar flow nacelle. This program was part of a European collaborative flight demonstrator program managed by *Brian Barry* of Rolls-Royce and *Hansgeorg Riedel* of DLR [65, 66].

A substantial reduction of the viscous drag of the nacelle is possible if one succeeds in maintaining the boundary layer on the surface of the fan cowl to a large extent in a laminar state. One way of achieving this is by special shaping of the fan cowl. Since the influence of the realistic flight conditions on the behavior of the boundary layer is difficult to estimate analytically, for the DLR research aircraft ATTAS (VFW 614) a fan cowl was designed and manufactured in CFC

DLR research aircraft ATTAS in flight with laminar flow nacelle and camera pod on the left wing.

construction in collaboration with MTU, Rolls-Royce and the Institute of Design Aerodynamics of the DLR. Furthermore, also a datum fan cowl (ATTAS conventional fan cowl aeroline) was manufactured in CFC construction for an aerodynamic comparison with the laminar flow fan cowl. The laminar flow fan cowl was modified locally after the first successful flight program and fitted with a boundary layer suction system. With the aid of a very sophisticated measurement system, employed by the DLR and the University of Oxford, the validation could be obtained that the laminarization of the boundary layer is possible both by shaping of the fan cowl contour alone as well as by a combination of boundary layer suction and fan cowl shaping so that the laminar boundary layer remained stable up to 60 per cent of the fan cowl length, the design target [66]. The former concept of boundary layer control based on fan cowl shaping is called Natural Laminar Flow (NLF) control, while the latter concept, that is by applying suction over a limited streamwise extent of the fan cowl followed rearwards by fan cowl shaping, is the concept of Hybrid Laminar Flow (HLF) control. The HLF control yields a slimmer fan cowl and thereby a reduced frontal area of the nacelle, which is beneficial for achieving high cruise speeds. The laminar boundary layer, however, reacts sensitively to contamination, for example by insects, and lack of aero-smoothness of the fan cowl surface. The complete subject is being continued in other research programmes [67]. The measurement techniques and also the theoretical studies for laminar flow investigations are very costly in terms of complexity and effort [68].

The Engine 3E Technology Program

The knowledge relating to an engine with higher bypass ratio, which has been acquired within the framework of the CRISP program, has been demonstrated regarding its feasibility in 1992 for a ducted propfan, the so-called Advanced Ducted Propfan (ADP), by Pratt & Whitney (PW), FIAT and MTU and has thus gained a concrete orientation. This engine concept, tested successfully on a test bed, features a single-stage fan yielding a very high bypass ratio > 10. The fan is driven via a gear by the low pressure turbine designed by MTU. For a dedicated continuation of the research and development work therefore the focused research concept Engine 3E (Environment, Efficiency, Economy) with the following content was defined [69]:

- Engine concept
  - bypass ratio > 12,
  - gear with variable-pitch fan.
- Low pressure system
  - fast rotating low pressure compressor,
  - fast rotating low pressure turbine.
- Core engine
  - high combustion chamber exit temperature and high total pressure ratio,
  - low emission combustion chamber.

The following ambitious aims govern this program:
- Decrease of the fuel consumption up to 20 per cent compared with the conventional turbofan with a bypass ratio of 5.
- Reduction of the noise emission up to 10 per cent.
- Decrease of the NOX emission up to 85 per cent compared with the ICAO limit of 1995.

The work is concentrated mainly on the components compressor, combustion chamber and turbine. The necessary knowledge is to be gained and demonstrated within the framework of a long term technology program, with the responsibilities for some key tasks being divided between industry and DLR. Accompanying basic investigations are undertaken by some universities. The complete program was part of the aeronautics research program 1995–1998 supported by the Federal Government (BMBF/BMWi) with the costs of the program shared between the Federal Government and industry in about equal parts. Some ongoing work will be described briefly.

The fan technology, which is important for an engine with increased bypass ratio, is pursued by DLR mainly by investigations on the 1-meter model of the CRISP program. In accordance with the international commitments with its industrial partner, the work of the German side of industry is primarily concentrated on the components of the core engine, which are compressor, combustion chamber and turbine.

Since the efficiencies of the components can only be improved insignificantly, the research and development work is concentrated especially on a weight reduction. In the area of the compressor this becomes possible by the application of new lighter materials for discs and blades as well as new methods of construction and fabrication [70]. The so-called BLISK- (Blade Integrated Disk) manufacturing method can reduce the weight of a compressor up to 30 per cent. Also the experience from the CRISP program concerning carbon fibre composites (CFC) is applied for the low pressure compressor. Thus, already for the 1 meter CRISP model the rotor blading has been designed and manufactured in CFRP and also the rotor blading of the transonic compressor for investigations at the TH Darmstadt [71]. A further means of weight reduction is the decrease of the number of stages. For a reduction of the stage number, however, a prerequisite is a further improvement of the flow quality in the blading. In support of this, for the continuing development of the necessary calculation methods, aerodynamic measurements are carried out, for example, on a single-stage transonic compressor (CFRP blading) of the TH Darmstadt with a laser two-focus method [72]. Detailed data on three-dimensional flow phenomena are obtained with the new 4-stage low speed axial flow compressor of the TU Dresden.

In the area of high pressure compressor development – one of the most demanding components of an aeroengine – use can be made of the experience gained from already built compressors. The high pressure compressor of the EJ 200, which is aerodynamically and mechanically very highly stressed, represents a good basis for this. With a pressure ratio of 11.9, achieved with six stages, the Engine 3E high pressure compressor surpasses, regarding the stage pressure ratio, all up to now known developments on a very high efficiency level. This MTU advanced compressor technology will be applied first time in the histo-

The MTU high pressure transonic axial flow compressor, based on government funded technologies, for the PW 6000 bypass engine. First time use of the BLISK-technology by MTU on the first three stages of a civil engine compressor (not shown in the figure).

ry of MTU in a civil bypass engine, the Pratt & Whitney PW 6000, the propulsion unit of the smallest Airbus A 318. The aerodynamics of the blade design is based on 3D advanced prediction methods. The total aerodynamic design of this transonic axial flow compressor is a complex procedure because of boundary layers, secondary flows, tip leakages, wake interactions, etc. within the flow field. Different requirements have to be fulfilled by the compressor: the desired pressure ratio, the mass flow rate, a high efficiency and an adequate stall margin. No single solution of the equations of motion can at the present fully represent the total flow field. Therefore, the solution of these equations is based on numerical methods. The total design procedure starts defining velocity triangles, blockage influences and losses based on different parameters like geometry and work distributions. A streamline curvature method in the meridional (S2) plane is employed to optimize radial distributions and define boundary conditions for the blade sections on S1 (blade to blade) surfaces. To get an optimum blade design the solution of the task is based on the use of the viscous three-dimensional Navier-Stokes code TRACE [63], developed in cooperation between the Institute of Propulsive Technology of DLR Cologne and MTU. The blading design will be an iterative procedure [73], considering structural and thermal influences. The advantage of the numerical methods is to enhance the understanding of the complex flow phenomena [74] and to improve the models used and correlations in the early design steps. On the other hand the use of the numerical methods can also help to avoid costly measurements. For the realistic compressor design, viscous effects are taken into account in the case of MTU by applying semi-empirical correlations [75] and considering all the extensive experience in HP-compressor development in the past for the military engines RB 199 and EJ 200.

The reduction of the fuel consumption aimed at in the guiding concept Engine 3E also leads to a decreased pollution emission [50, 69]. An exception to this is the NO_X emission which rises with higher pressure ratios and temperatures. The development of low emission combustion chamber concepts is therefore necessary because there is evidence that the nitrogen oxides can contribute to the disturbance of the ozone balance in the stratosphere. The complex flow and combustion processes can no longer be analyzed unambiguously with current methods. In order to better

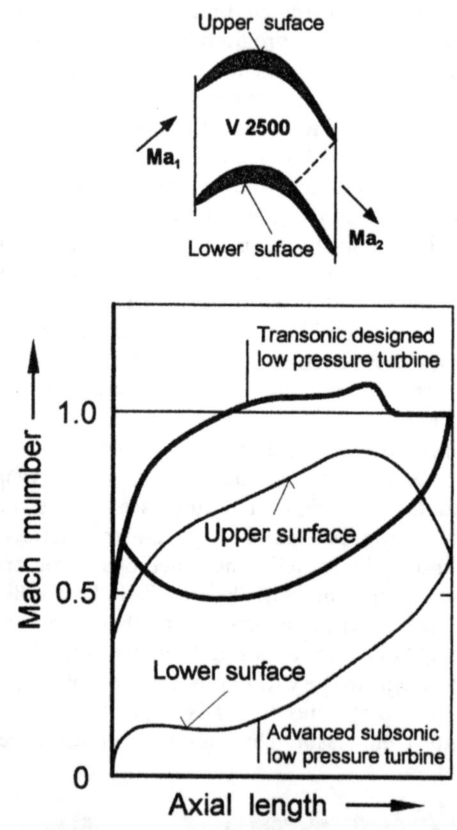

Mach number distribution on the blade surface of different low pressure turbines (*bottom*), LP-turbine cascade (*top*).

understand the creation of pollution emissions, the partners undertook the design of a test stand having a glass combustion chamber and its construction at the DLR Cologne. This test stand allows the use of optical, that means non-intrusive, measurement techniques in realistic combustion chamber geometries [61]. Two fundamental concepts are topic of investigations, the so-called rich-lean combustion (MTU) [69] with a conventional combustion chamber geometry and the more complex axially staged combustion chamber geometry (RR-Deutschland) [76]. The aim of this work is the computer simulation of complex combustion chamber flows on the basis of an adequate modeling supported by experiments. Details of this topic are described in the contribution "Low Emission Combustion Chambers" (page 438).

The Rolls-Royce Deutschland combustor concept was demonstrated successfully under engine conditions in a BR 715 core engine run at the altitude test facility of the university of Stuttgart. The core engine also incorporated other Engine 3E technologies like blisks and an advanced high pressure turbine air system using brush seal technology for increased efficiency and temperature capability.

The fast rotating low pressure system becomes a key technology for the feasibility of an engine with bypass ratio >12. This is especially valid for the low pressure turbine. The high efficiencies of current low pressure turbines of over 93 per cent are due in first instance to the moderate Mach number level on the blading and the application of advanced computational methods [77]. For this, the Mach number distributions on the blade surfaces are selected in such a way that low-loss laminar boundary layers are possible [78]. These blade profiles are developments which were started in the seventies [79] by *Leonhard Fottner* and *Hans-Jürgen Lichtfuß* and have significantly contributed that the low pressure turbine of the engine Pratt & Whitney 2037 could be designed and manufactured in quantity by MTU [80]. For the same specific work as current low pressure turbines, future turbines are going to feature fewer stages, which leads to the transonic turbine. The higher Mach number level requires extensive optimizations of the blade profiling. For a validation of the knowledge gained already within the framework of the current technology programs since about the middle of 1995 the fast rotating transonic low pressure turbine of the ADP demonstrator engine is used as technology tool for continuing aerodynamic work. The results with this 3-stage turbine, which were obtained from measurements on the test bed of the "Institut für Luftfahrtantriebe" "Institute of Aviation Powerplants" of the University Stuttgart, yielded already an efficiency level of markedly over 90 per cent, which corresponds to the level of conventional low pressure turbines [81].

Although the complete technology program is intended as a long term one, already now sufficient knowledge has been gained which could be applied for the design of an engine with a bypass ratio >12. This engine will feature a single stage fan driven via a gear by a fast rotating low pressure turbine. Since this will take place in an international cooperation, a very good prerequisite for a participation in an engine program has been created, for example in the PW 8000 program based on the comprehensive research and development work carried out in partnership by Pratt & Whitney (US), MTU (Germany) and Fiat Avio (Italy). This engine has to transmit a take-off power of about 32,000 hp from the LP-turbine to the fan via a reduction gear. The LP-turbine developed by MTU *"The Specialist"* for such type of turbines, has three stages with 3D aerofoils for high work extraction at high rotational speed. The diameter is significantly less than that for a direct fan drive turbine. Another application of a transonic LP-turbine of MTU is the PW 800-ATFI (Advanced Technology Fan Integrator) turbofan of Pratt & Whitney Canada. This engine is for the application in the 50–90 passenger regional market. Long established examples of geared turbofans are the Honeywell ALF 502 (originally Lycoming [35]) and the TFE 31 engines. Maybe in the near future *Frank Whittle's* vision can be realized: *"A turbofan has a core to drive the fan and the fan provides the thrust."*

Airframe-Engine Interference

With the rise of the bypass ratio the diameters of the engines also increase. The installation of such VHBR or UHBR engines on an aircraft – mostly under the wing – intensifies the problem of the interference between aircraft and engine because of the short coupling with the wing necessitated by the limitations of the ground clearance [82]. Experience shows that this problem is even for a conventional under-wing installation inadequately mastered. The interference effects reduce the benefits of the high bypass ratio and it is therefore necessary to investigate these effects in good time [83]. Already in the CRISP technology program investigations into this topic were undertaken by MTU/Dasa/PW/British Aerospace Airbus [57]. In these investigations, however, a parameter of significant influence, the engine jet, was missing. Merely investigations with so-called through-flow nacelles were carried out.

Also by the Institute of Design Aerodynamics of the DLR the problem was taken up in 1986 and dedicated investigations into propulsion integration were initiated. As first step the design and acquisition of a UHBR engine simulator were undertaken [84] under the project leadership of *Heinz Hoheisel*. At EADS in

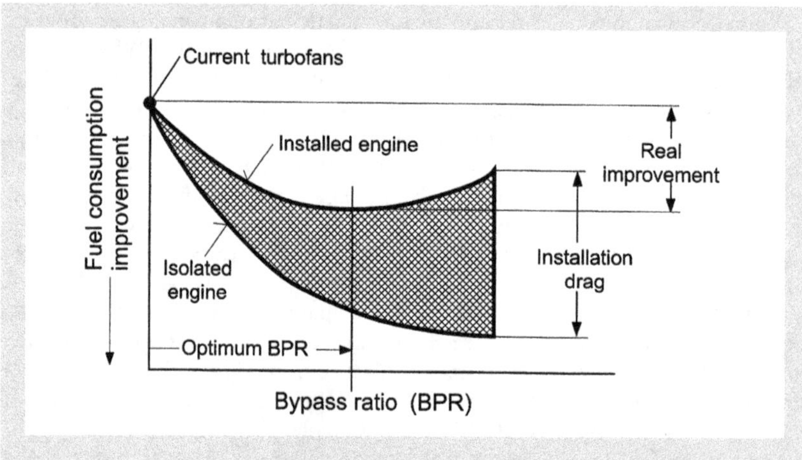

Improvement of the specific fuel consumption dependent on isolated and installed engine. The optimum bypass ratio is reached between 12 and 15 for the installed engine.

Bremen at about the same time the development and manufacture of a conventional turbofan simulator – to the same scale as the DLR simulator – was started for the first time, for such simulators were in the past always acquired from the USA. These prerequisites became finally an essential basis for the start of the task "Ducted Propfan Investigation, DUPRIN" which was taken up by the European Commission (EC) under Area 5 (Aeronautics) in the BRITE/EURAM supported programme [85]. Under the overall direction of *Wolfgang Burgsmüller* of EADS Bremen, this program was launched in 1990 with 14 partners from Europe and was continued in a second and third phase, now designated ENIFAIR (Engine Integration on Future Transport Aircraft). On the German side EADS, DLR, MTU and the TU Berlin were involved from the start of the program. MTU withdrew after Phase 1, instead Rolls-Royce Deutschland is participating since Phase 2. The content of the program is:

- The determination of the installation drag in the low speed and high speed regime with bypass ratios up to about 20 in comparison to the conventional turbofan.
- For the above, development, manufacturing and testing of wind tunnel models and engine simulators for experimental investigations.
- Creation of an experimental data base for the validation of numerical calculation methods
- Dedicated further development and application of numerical methods for an analysis of the interference phenomena.

For the experimental investigations the largest wind tunnel model ALVAST built by the DLR until now has been in use [85]. This model possesses an Airbus like geometry, comparable with the Airbus 320. It is available in the form of a half-span model mainly for high speed tests performed in September 1998 in the wind tunnel S1MA of the ONERA (France) and as a full-span model with a span of 3.43 m equipped with high lift devices. The high lift measurements took place in the DNW (1992–1995). The wind tunnel model was an important constituent element in the programme because the geometry is public. For the determination of the interference effects between airframe and engine, the so-called "Turbo-Powered Simulator (TPS) technology" is applied in the experiment, to this day the best simulation technology, but a very costly technology.

- For the simulation of conventional turbofans (bypass ratio 5) two so-called Turbo Powered Simulators manufactured by EADS are employed [86]. Such TPS features a compressed air turbine which drives a single stage fan of 162 mm (0.53 ft) diameter. The temperature of the turbine entry air is only slightly increased compared to that of the ambient air.
- The simulation of VHBR engines with a bypass ratio of 9.2 is accomplished with two simulators which were manufactured by Airtechnology (Toulouse) in the ENIFAIR programme, the fan diameter of each being 198 mm (0.65 ft). The associated design work was done by Rolls-Royce, BAe and EADS in Phase 2 of DUPRIN.

- The simulation of an Ultra High Bypass engine was performed with a completely new type of simulator, that is a contra rotating fan with designation CRUF, meaning "Counter Rotating Ultra High Bypass Fan". This new type of simulator is based on the MTU-CRISP engine concept and is to simulate a total thrust of 1,150 N (259 lbf) with a bypass ratio of about 16. The 4-stage turbine driven by compressed air produces a power of 162 kW (220 hp) which is transferred by a reverse gear (with the size of an orange) to both rotors, each having 8 blades and a diameter of 254 mm (0.83 ft). The development of this simulator was started in 1987 jointly by DLR and the company Dynamic Engineering Inc. in Newport News, USA [84]. In order to determine and monitor the vibrational behaviour of the fan blades, for rotor 2 a telemetry system had to be installed, which influenced the gear bearing concept since the space available between turbine and fan is very limited. Because of that, some difficulties arose during the first wind tunnel test which could be eliminated with the support of *Gilbert Grasmuck* (Airtechnology, Toulouse). A second simulator based on the same concept was manufactured in Phase 1 of DUPRIN in 1991 by the company Technofan in Toulouse (France).

The operation of such engine simulators puts highest demands on the wind tunnel testing technology. Because the thrust cannot be measured directly in the wind tunnel test, such simulators must be equipped with sophisticated instrumentation which allows the determination of the thrust from pressure and temperature measurements in the fan and in the turbine. For this, prior to the wind tunnel measurements, the simulators have to be calibrated in special test facilities [87].

For the numerical treatment of the integration problem for complex realistic geometries the generation of a computation grid in the complete flow field is necessary. A grid generator developed at the DLR for this task is structured in such a way that geometric changes of the configuration can be taken into account within a reasonable space of time [88]. Using the DLR MEGAFLOW code [89] for the solution of the fluid dynamic equations of motion, as result the lift distribution on the wing subject to the influence of the engine is obtained. The engine leads to a decrease of the lift over the total wing span y/s, with the lift reduction rising with increasing bypass ratio in the region of the engine [90]. This result is qualitatively in good agreement with measurements for this configuration [91].

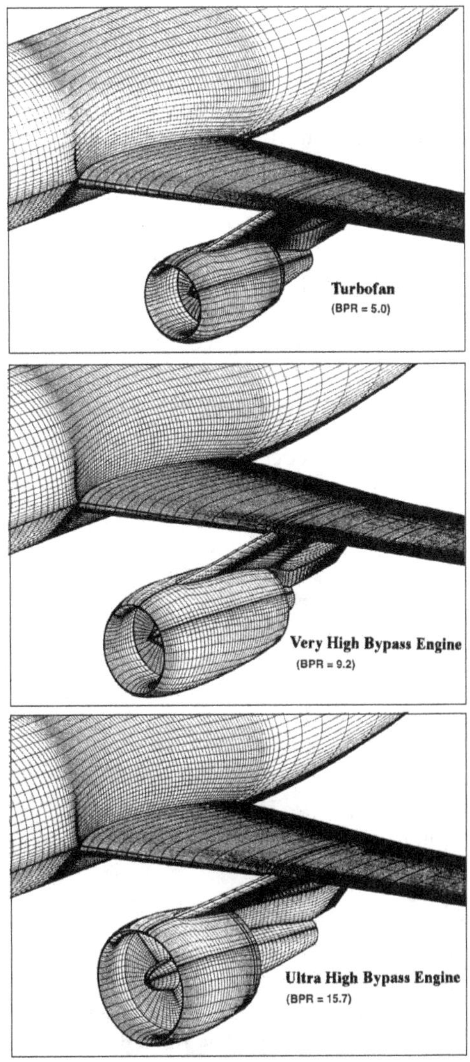

Computation grids for the computation of the flow on the body surface for the ALVAST fuselage/wing/pylon/engine configuration with engines of different bypass ratios. Clearly visible is the marked increase of the engine diameter with growing bypass ratio.

Insight into the drag behavior of an aircraft-engine configuration is available from experiments with the DLR ALVAST model within the framework of the DUPRIN/ ENIFAIR programs. Compared with a turbofan engine, there exists for the VHB and the UHB engine a considerable drag increase, with the jet effect being more marked for the UHB engine than for a conventional turbofan [92, 93]. Owing to the engine size, the engine jet of a UHB engine is closer to the wing and disturbs the wing flow field much stronger than a conventional turbofan jet [83]. In the case of the installed real engine on an aircraft it has to be noted that asymmetrical flow behavior at the fan exit exists [94]. Additional losses are the result of this phenomenon. With the aid of the numerical calculation methods developed in the DUPRIN/ENIFAIR programs it will be possible to gain a deeper detailed knowledge of the flow properties in the wing/pylon/ engine regions in order to discover possibilities of drag reduction for engines having bypass ratios >10. This is valid in particular for the jet effect. Taking into account the viscous effects in the calculation methods for the complete configuration fuselage/wing/pylon/ engine is absolutely necessary. The path leading to this is still very exacting but indispensable, as the activities of the company Boeing for the B 777 have impressively shown [95].

Altogether up to now the work in Germany on the topic "Engines with very high bypass ratio" has gained international recognition in important technological fields and can contribute to future realizations.

Remark

Concerning this chapter, some plates in colour are shown in:

Color plate III

- Airbus models with engine simulation at landing and cruise.

Color plate IV

- Engine simulators with various fan diameters
- Computed shape of the engine jet in the region of the wing
- Complete simulation with engine and wing

Combustion Chamber Development Trends

Dieter Rist

For the aeroengine types, which were developed after 1945 to readiness for quantity production and built in Germany within the framework of international cooperations, the responsibility for the particularly demanding "hot" part of the engine, the combustion chamber – which is to be regarded as technological key field – was never placed on the German partner. This, of course, had amongst other things repercussions on the scope and the intensity of the research and development activities concerning the combustion chamber as engine component. Nevertheless, experts, development teams and departments in German research institutions and in industry worked in this field, for which considerable governmental aid was provided.

After founding the company BMW Rolls-Royce Aero Engines with head office in Dahlewitz near Berlin in the year 1990, the situation changed in this respect. With the system leadership for the development and the manufacture of the engine family BR 700, in this case also the responsibility for the combustion chamber field did lie in the German hands. On January 1, 2000 BMW withdrew from this business and the company has since been continued as Rolls-Royce Deutschland.

Combustion Chamber Research at BMW-Triebwerksbau

The time for the continuation of research and development work on combustion chambers for aero-engines in Germany (Federal Republic) after World War Two can be dated January 22, 1954, when the "BMW Studiengesellschaft für Triebwerksbau GmbH" was founded in Munich. In December 1954 work started with market research and project studies and in March 1956 with the development of the small gas turbine BMW 6002 for 35–55 kW (48–75 hp) shaft power. From this was later – at the "BMW Triebwerksbau GmbH" in Munich – the small gas turbine BMW 6012 with shaft power between 52 and 80 kW (71 and 109 hp) and a thrust of 440 N (99 bf) respectively developed [48]. The 6002 as well as the 6012 were equipped with a novel space saving, flat disc combustion chamber. The fuel was supplied via the hollow rotor shaft and sprayed through a centrifugal injection

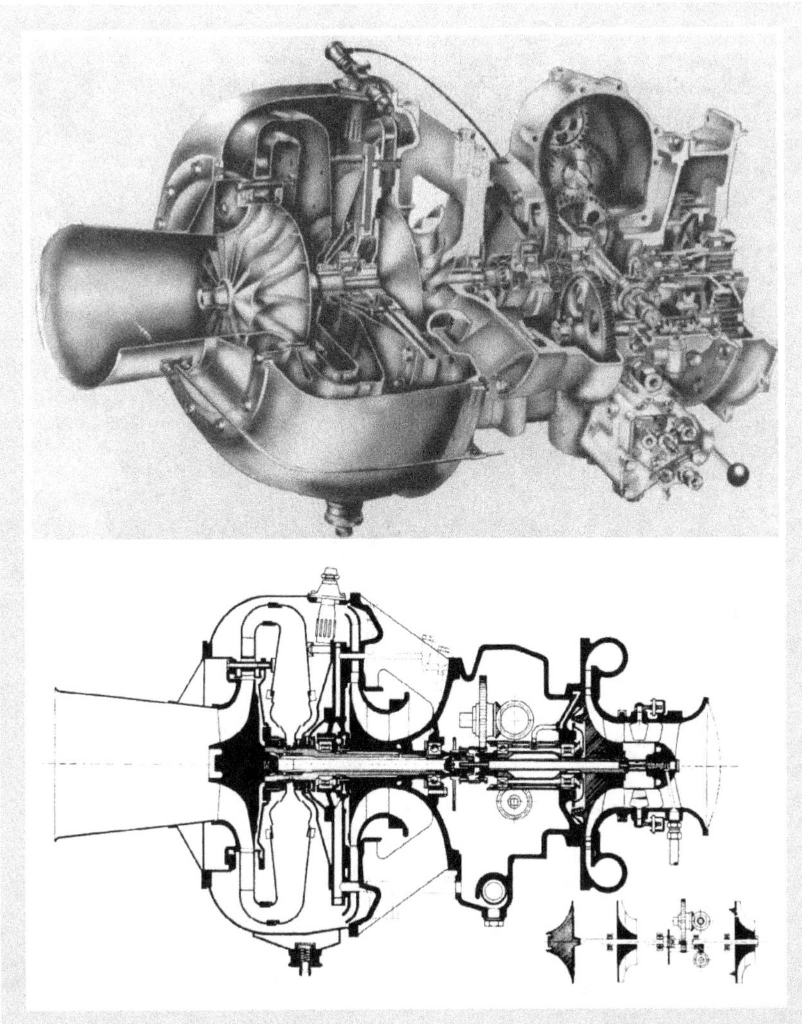

Small gas turbine BMW 6012 with disc combustion chamber. Below: BMW 6012 L as air suplier for the Dornier helicopter with blade tip propulsion.

ring. (This principle of fuel injection into the combustion chamber was at that time also used for the "Marboré" engine family of the French firm Turboméca and the Russian turbostarter TS-1). To gain knowledge regarding the combustion process in the novel disc combustion chamber, already at the beginning a combustion chamber test stand for open combustion was set up, for which a former rocket test bed of BMW in Munich-Allach was converted. At the combustion chamber exit of the BMW 6012 the mean temperature reached a value of 1,040 K at a maximum pressure of about 3 bar (45 b/in^2) [1].

Combustion Chamber Research at MTU Munich

As a continuation of the work on these small gas turbines the development of a new helicopter engine must be regarded. The "Motoren- und Turbinen-Union, MTU" Munich, into which the "BMW Triebwerksbau GmbH" was absorbed, was substantially involved and still is with the components combustion chamber, gas-generator turbine and power turbine, which are the hot parts of the gas turbine. Initially this turboshaft engine was designated MTM 380 (MTU/Turboméca), later MTM 385, and the present very advanced version with approx. 960 kW (1,306 hp)

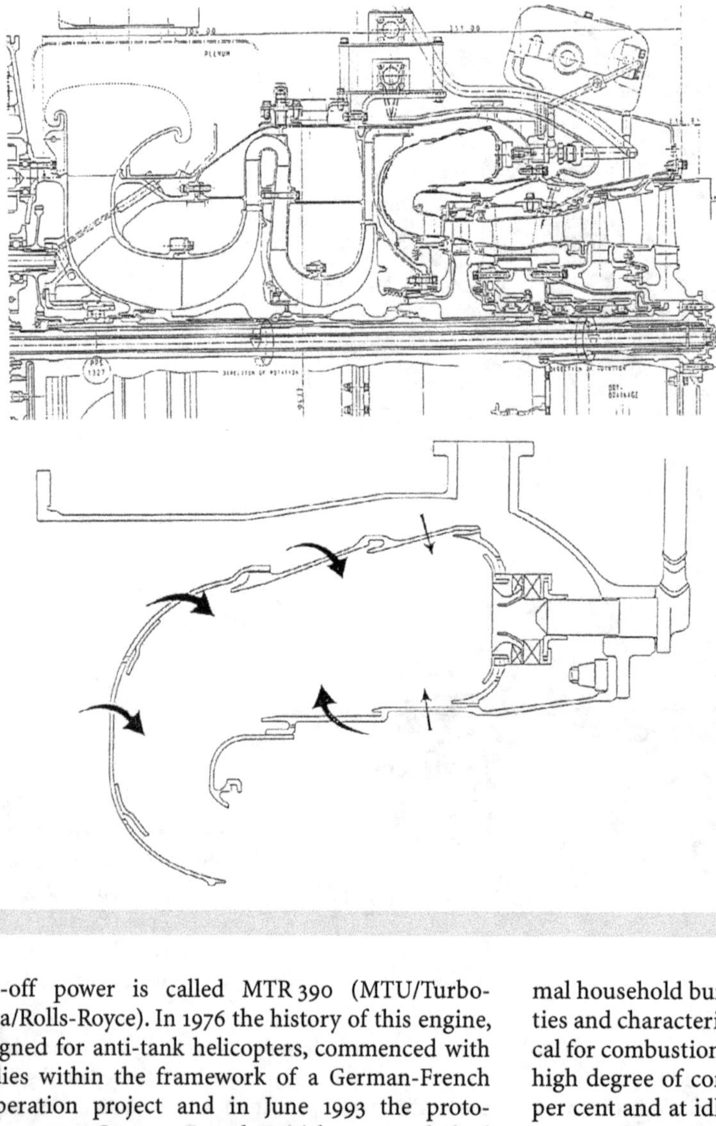

Longitudinal section of the modern helicopter engine MTR 390 equipped with a return-flow annular combustion chamber.

take-off power is called MTR 390 (MTU/Turboméca/Rolls-Royce). In 1976 the history of this engine, designed for anti-tank helicopters, commenced with studies within the framework of a German-French cooperation project and in June 1993 the prototypes, now a German-French-British teamwork, had passed all important certification tests [96–98].

As also for the other components, for the combustion chamber diverse and high requirements were specified, which could only be fulfilled by extensive research and development work. An overall measure for this is the power and loading factor of the combustion chamber referenced to the combustion chamber volume, which is 200 times as large as that of a normal household burner. In detail the following properties and characteristics were required, which are typical for combustion chambers of modern aeroengines: high degree of combustion, that is at full load 99.99 per cent and at idling over 99.5 per cent; low relative pressure losses; good ignition and re-ignition properties; low smoke and pollution emissions; for the high pressure turbine suitable radial and total temperature distributions at the combustion chamber exit; avoidance of burner nozzle choking; long life and high reliability; compact design with a length as short as possible, and low weight.

To meet these demands, the concept of a return-flow annular combustion chamber was selected, which

Return-flow annular combustion chamber with air-atomiser fuel nozzle for small gas turbines.

initially received a novel injection system with pre-vaporised fuel [99]. Later for the fuel injection air-atomiser nozzles were used with which a very fine atomisation, a very good fuel-air mixture, a short combustion zone as well as the desired flow and temperature field in the primary zone could be achieved. In addition this burner nozzle type alters its performance characteristic for changing load conditions only little. To a successful combustion belongs the testing under realistic engine conditions, that is at combustion-chamber entry temperatures and pressures representative for the engine. Such test facilities for small and large mass flow as well as higher pressures are very costly. They are available for some time now at the DLR in Cologne-Porz for research and development work. There a high-pressure test bed was built in cooperation with MTU and used for the development of the MTR 390 combustion chamber, which is represented in the adjacent figure [48].

In 1974/75 various research and pre-development programmes with special test articles for advanced components of aero-engines were commenced by MTU – partly together with other German companies and institutes – with the support of the BMFT. As test articles single components, component assemblies and complete engines, so-called demonstrators, were designed and built. With them knowledge in such different fields as aerodynamics, thermodynamics, structural mechanics as well as construction were to be gained and new test as well as measurement techniques to be experimentally validated. The programmes were conceived from the beginning in such a way that with them many results for all possible future engine programmes could be gained. In this framework since 1974 research was carried out for the development of ceramic components for combustion chambers, turbines and other engine hot parts. Within other programmes the MTU test articles VT 1A and VT 1B in the engine power classes 450 kW (612 hp) and 800 to 1800 kW (1088 to 2449 hp) were built and in April 1977 and December 1979 respectively put into operation. Both test articles possessed a return-flow

The high-pressure combustion-chamber test bed which was erected in collaboration with MTU at DLR in Cologne-Porz.

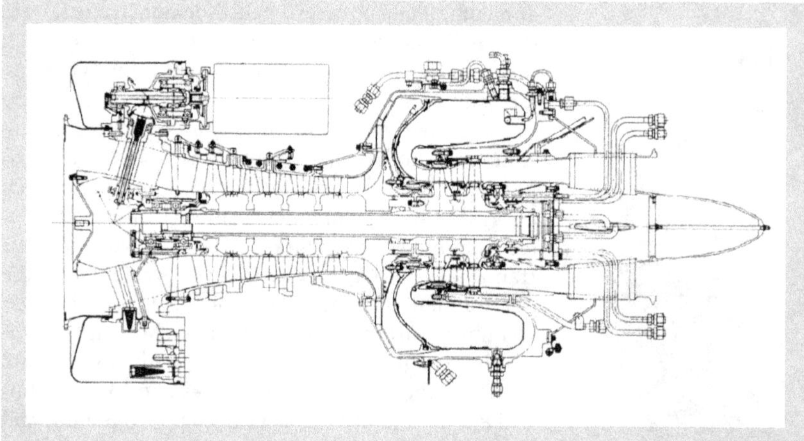

Longitudinal section of the MTU test engine VT 1B with return-flow annular combustion chamber.

annular combustion chamber with high combustion space loading but, however, with different fuel injection systems. The longitudinal section of the VT 1B is shown above.

The combustion-chamber exit temperature of this test unit was in the range of 1,450–1,550 K (1,177–1,277 °C). Return-flow combustion chambers in cylindrical and conical form require special measures of flame-tube cooling for the reduction of the wall temperatures. On this as well as on the fuel preparation and flame stabilisation when using unconventional fuels, MTU worked successfully, for example, in the BMFT supported project [100] "Development of New Combustion Chamber Technologies for Future Alternative Fuels". The fuel injection systems used in the test articles are also – though in different form – to be found in two military engine types in whose development MTU was involved and still is – however without responsibility for the combustion chamber. The multinational "Tornado" bypass engine Turbo-Union RB 199-34R, built in quantity production since 1979, was initially equipped with vaporiser burner nozzles and later with atomiser burner nozzles. In the Eurojet EJ 200 bypass engine of the fighter aircraft "Eurofighter EF 2000", which is presently put into service in the air forces of several countries, this part is also an atomiser burner nozzle.

Combustion Chamber Research at the "Entwicklungsbau Pirna" in the GDR

About at the same time as in the western part of Germany, that is in the year 1954, also in the eastern part, in the "German Democratic Republic, GDR", the rebuilding of an aviation industry [33] and of corresponding research facilities was started [101]. For this experienced German and Austrian aircraft engineering specialists were available, who had been abducted to the Soviet Union in October 1946, were allowed to return home in groups since August 1950 and for the time being went to the GDR. The development of the first civil German turbo-jet engine Pirna 014 and also of the small turboshaft gas turbine Pirna 017 and Pirna 027 respectively was based on projects which had previously been started and realised respectively by the same people in the Soviet Union. Both units were developed and built in the engine development work in Pirna near Dresden since September 1954 under the overall management of the director and chief designer *Rudolf Scheinost*. The official start as "VEB Entwicklungsbau Pirna" was May 1, 1955.

The Pirna 014 was a simple, robust and economic single-shaft engine in the thrust class of 30 kN (6,750 lbf) designed for medium haul operation. It contained a combustion chamber consisting of single combustion elements combined with an annular combustion chamber, as documented in the historical photograph. The in the external casing mounted entry part comprised 12 burner heads air-washed by the primary air and a following annular part, which both were externally airwashed by the cooler compressor air. Via the mixing pockets projecting from the annular part, the rest of the compressor air was mixed with the hot gas and in this way the temperature level toler-

Combined tubo-annular combustion chamber of the turbo-jet engine Pirna 014. The combination of single combustion chamber (Junkers principle) and annular combustion chamber (BMW principle) was already successfully applied in the Soviet Union.

able by the turbine adjusted. This combustion chamber configuration resulted from many years of experience of the BMW and Junkers experts before 1945 as well as the extensive experimental research and development work which these experts had carried out afterwards in the Soviet Union for the design of the combustion chamber of the first Soviet gas turbine aeroengine types. Over a relatively wide operating range a degree of complete combustion of 98 per cent was achieved. After a five year period of development and testing, the Pirna 014 successfully gained in August 1959 the state type certification based on ICAO requirements.

Since the end of 1955, apart from the jet engine Pirna 014, at the Entwicklungsbau Pirna a small turboshaft gas turbine was developed for a power output of 95 and 172 kW (129 and 234 hp) under the leadership of *Paul Stich*, the basis of which was the engine turbostarter TS-1, and which had previously been developed in the Soviet Union by nearly the same small German team [101]. The small gas turbine Pirna 017 and the Pirna 027, subsequently derived from the former, were among other things intended for the not realised propulsion of motor cars and were practically used to a large extent for emergen-

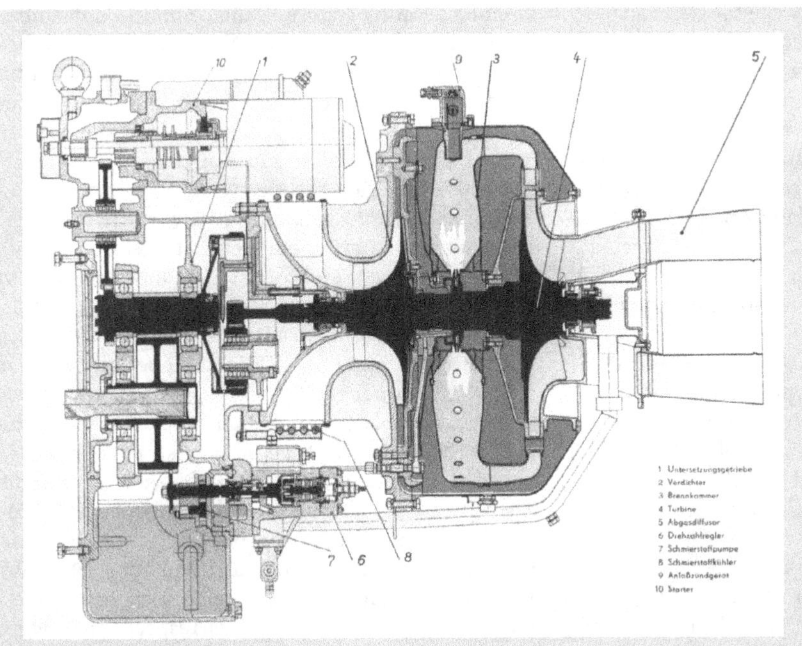

Longitudinal section of the small turboshaft gas turbine Pirna 027, a further development of the Pirna 17 with disc combustion chamber.

cy electric power supplies and fire-fighting pumps. This small gas turbine type possessed – just as the somewhat smaller and very similar BMW 6002 and BMW 6012 which were developed at the same time – a disc combustion chamber to which the fuel was supplied via the hollow shaft and injected with rotation.

In March 1961 the liquidation of the aviation industry in the GDR was ordered by the SED leadership and thus the aviation research and development on the East German territory was brought to an end. Some experts and also young engineers left at that time the GDR in order to continue work in Western Germany or in western countries.

Aeroengine Combustion Chamber Research in the Reunited Germany

It came to a resumption of the aviation research activities in the east of Germany only after the reunification in October 1990. Today also a number of East German university institutes are involved in several programs on topics of combustion and engine combustion-chamber research. In these programs industry, research establishments like the DLR as well as various institutes and chairs of the technical universities cooperate. In the national framework the financing is shared proportionally between the German federal government, the governments of the states and the participating companies. Programs in the European framework are supported proportionally by the EU. For example, within the present national research program "Engine 3E (Environment, Economy, Efficiency)", in which the engine companies MTU Munich and Rolls-Royce Deutschland, the DLR as well as the technical universities in Chemnitz, Darmstadt, Dresden, Karlsruhe and Munich cooperate, exists the project "Technological Fundamentals of Low Pollution Emission Combustion".

At the same time exists with the European BRITE/EURAM program "Low NO_X Combustion" a program with a related topic, in which on the German side both engine companies, the DLR as well as the technical universities in Karlsruhe and Munich are involved. In the research program of the German "Study Group High Temperature Gas Turbine, AG Turbo I", which is now continued with the program "Gas and Steam Turbine for the Power Station of the 21st Century, AG Turbo II", existed the field "Turboflam" in which problems of modern combustion chamber technologies are investigated. This research, sponsored by the BMBF and by the participating companies, does in fact receive no aeronautical research support, however the knowledge gained can naturally be used for stationary gas turbines as well as for aeroengines. In this field of combustion-chamber research, the manufacturers of stationary gas turbines ABB (Asea Brown Boveri) and Siemens KWU (Kraftwerk Union), the aeroengine companies MTU Munich and Rolls-Royce Deutschland, the DLR institute of propulsion technology as well as the chairs of the technical universities in Bochum, Darmstadt, Dresden, Karlsruhe and Munich are active.

Combustion Products and Pollutants

During the combustion of 1 kg (2.2 lb) kerosene in the combustion chamber of a gas turbine or aeroengine the following combustion main products are produced: 1.25 kg (2.75 lb) water vapour (H_2O), 3.15 kg (6.9 lb) carbon dioxide (CO_2), as well as the following, as pollutants defined by-products: 6–20 g nitrogen oxide (NO_X), 0.7–3.7 g carbon monoxide (CO), 0.1–1.3 g unburned hydrocarbons ((U)HC), 0.01–.04 g soot and about 1 g sulphur dioxide and/or other sulphur oxides (SO_2, SO_X, if the fuel contains remnants of sulphur). Depending on the kind of kerosene, the engine operating state, and the combustion chamber layout, these figures can vary. Nevertheless, on this basis it can be calculated what quantities of combustion products are introduced into the atmosphere, if it can be assumed that annually and globally about 5–6 per cent of all mineral oil products, that are 130–180 million tons of fuel, are used in aviation.

The part due to aviation relative to the CO_2 total emission caused by human beings therefore amounts to 1.6–2.2 per cent and for the NO_X total emission to about 1.8 per cent. If the aviation contributions are related to the emissions caused by the motor vehicles only, then the result for CO_2 is 10–14 per cent and for NO_X about six per cent [102].

The world air traffic primarily takes place at the border between troposphere and stratosphere, i.e., between 9 and 12 km (29,500 and 39,360 ft) altitude. There not only the addressed pollutants, especially the nitrogen oxides NO_X, but also the combustion main products carbon dioxide, and (possibly) also water vapour (in the stratosphere) have harmful effects. Sound knowledge in this respect was among other things gained by measurement campaigns and model calculations in the research program "Pollutants in Aviation", which, since the beginning of the nineties, was sponsored by the BMBF. It encompassed 25 individual projects of various research institutions under the leadership of the DLR Institute of Physics of the Atmosphere in Oberpfaffenhofen (Director: *Ulrich Schumann*). As general result it was found that the aviation emissions until 1995 have mainly caused a radiation increase due to different changes of the ozone layer at various altitudes, corresponding to a climatic contribution of aviation to the greenhouse effect of about three to four per cent. These, like other figures relating to possible climatic changes are, however, affected with a considerable element of uncertainty [103].

First guidelines for the permissible pollution emissions of aeroengines were established in the year 1977 by the "International Civil Aviation Organisation, ICAO" and in the following period tightened step by step. As measure for the permissible values, the emission indexes "EI" in g pollutant per kg fuel were specified for NO_X, CO and (U)HC, as well as the smoke number "SN" for soot, a quotient based on the light reflection of clean and soot covered filter paper. In addition the parameter "D_p/F_{oo}" was defined, with which the quantities, which are emitted by the engine in the LTO cycle when referenced to the take-off thrust F_{oo}, are represented in g/kN for the three first pollutants. The LTO cycle (L = landing, TO = take-off) encompasses a certain taxiing – take-off – ascent flight (up to about 1,000 m (3,280 ft) – landing approach – taxiing procedure, with in each case various engine power conditions (thrust per cent) during different periods of time. The pollution emissions for the ascent to cruise altitude, at cruise and for descent to the altitude of the LTO cycle are at present legally not limited. Not as ICAO measure, but otherwise in the gas turbine field customary, the pollutant quantities are also defined as concentrations in ppm (parts per million) referenced to the total exhaust gas volume. In the ICAO limiting value formulas for NO_X the physical-chemical dependence on the increased nitrogen oxide formation at higher pressures is taken into account. In this way the formal disadvantage of modern civil bypass engines with their increased compressor pressure ratios – some new types reach today at full load values slightly below or even above 40 – is excluded. On the other hand with these apparent "concessions" – the engines with higher combustion chamber pressures are allowed to possess higher EI-No_X and D_p/F_{oo} (NO_X) values – it is indirectly taken into account that the thermal efficiency of gas turbines increases with higher pressure ratios, that is the specific fuel consumption, in other words the fuel mass-flow rate per unit of thrust or power, is decreased.

Based on newer findings of climate research, the influence of carbon dioxide and of the pollutants NO_X, soot and SO_X – emitted from aeroengines – on the state of the atmosphere is more serious than those of the other combustion products and pollutants, respectively [103]. Thus, the measures for a reduction of the effects must in first instance be directed towards these species. Various possibilities exist for the propulsion system in that respect, which are related to the engine concept, the thermodynamic basic data of the cycle, the component design, and the fuel of the engines. The specific fuel consumption can be reduced by increasing the thermal efficiency and the propulsive efficiency of the engine [51], whereby for kerosene combustion mainly the emission of carbon dioxide and water is decreased. Improvements of the thermal efficiency can be achieved by increasing the compressor pressure ratio and the turbine entry temperature, which are both related to the combustion chamber operation. Improvements are further possible by increase of all component efficiencies and reduction of the pressure losses as well as the installation of a compressor intercooler and of a heat exchanger in the gas turbine engine core part, the so-called gas generator. Up to the present the installation of the additional components intercooler and heat exchanger has not yet been realised for aeroengines, however, there exist project studies. For the nitrogen oxide formation, the increase of the pressure and temperature level in the combustion chamber, which is beneficial for the thermal efficiency, has a negative effect. This must be counteracted by other measures which will be discussed later.

The propulsion efficiency η_V of aeroengines is defined by the ratio of the propulsive power produced (thrust multiplied by the flight velocity) to the available usable power. The thrust is essentially determined from the engine air mass-flow rate and the difference between engine exhaust velocity and flight velocity. The value of η_V is the better the lower the engine exhaust velocity lies above the flight velocity. Favourable η_V values are achieved with the optimisation of the bypass ratio (fan mass-flow rate to core mass-flow rate) of bypass engines (turbofans) and the corresponding design of the fan.

Since the beginning of the civil air traffic with jet engines, that are the transatlantic flights of the British de Havilland "Comet 4" with Rolls-Royce "Avon" turbo-jet engines in October 1958, the mean specific fuel consumption of turbo-fan engines for a 1,852 km (1,000 nm), flight has been reduced in comparison to the "Avon" by 60 per cent. The consumed fuel quantity per passenger-seat for the Airbus A 330 and A 340 or the Boeing 777 compared with the "Comet 4" has been decreased to about 30 per cent. With regard to the bypass ratios of future engines, possibly being in the range of 6.5 and 13.5, the results of a Rolls-Royce study [104] show that the engines with higher bypass ratios exhibit in the LTO cycle actually lower NO_X emissions, but at cruise flight higher ones for the assumed state of combustion-chamber technology.

Cryogenic fuels

A further possibility to reduce carbon dioxide and other pollutant emissions, investigated in studies, but until now not introduced into civil aviation, is the use of hydrogen or methane as fuel. Either of them must be carried on the aircraft in a deeply cooled liquid state, otherwise they would need too much tank volume. Therefore they are called cryogenic fuels (low temperature fuels). Liquid hydrogen LH_2 has the same energy content as kerosene at only a third of the kerosene mass but nearly the fourfold volume. EADS Airbus has made a proposal how on an A 310 as test vehicle the cryogenic tanks can be placed on the fuselage top because of their volume. In the USA already in the late fifties test flights were carried out with hydrogen engines and considerably later a TU-154 was modified in Russia for such purposes.

In hydrogen combustion no carbon dioxide is produced and also no unburned hydrocarbons like soot, but on the other hand two and a half times as much water vapour as in kerosene combustion. The nitrogen oxide formation, compared to that of kerosene combustion, is about two orders of magnitude weaker. The possible influence on the climate due to the larger water-vapour quantity in the jet exhaust increases probably with flight altitude. For the airframe as well as for the engines with the complete fuel storage and supply system, the use of hydrogen poses many new research and development tasks.

The use of methane and natural gas as cryogenic fuel requires quantitatively lower but also considerable technological modifications on the aircraft and the complete fuel system. For the same energy content only about 85 per cent of the kerosene mass needs to be carried onboard because the mass-specific calorific value of methane is nearly 15 per cent higher than that of kerosene. The tank volume must be larger by about 60 per cent. In comparison with the kerosene combustion about 1.5 times as much water vapour is produced, but however 20 per cent less carbon dioxide.

Low Emission Combustion Chambers

In order to minimise the pollutants, which arise besides the main products during combustion, for some time now fundamentals for novel combustion-chamber concepts are being explored and appropriate technological developments undertaken. Since the early seventies engine pollution emissions are taken into account as criterion in the specifications of research objectives. Today in Germany, at both engine companies MTU Munich [57] and Rolls-Royce Deutschland [50], at the DLR [61] as well as at various institutes of technical universities [76], knowledge concerning the formation of pollutants, especially of nitrogen oxides, and their influencing is obtained by experimental investigations as well as theoretical studies and numerical simulations. The combustion-chamber section of modern aeroengines is designed from the begin in such a way that optional other combustion-chamber types can be installed in order to be able to introduce tested improvements without delay into quantity-production engines.

One of the main reasons for the formation of nitrogen oxides are temperature peaks in the primary zone of the combustion chamber, which exceed 1,500 °C (1,773 K). Therefore, the current combustion-chamber concepts aim for not too high flame and hot-gas temperatures during combustion, to avoid stronger temperature peaks by homogenising the hot gas mixture, or to keep the hot gas portion as small as possible. Another conceptual measure is the purposeful configuring of an as compact and short as possible flame tube, because the nitrogen oxide formation depends also on the residence time of the hot gas in the combustion chamber. At the present time work is mainly proceeding on three concepts of low-emission combustion chambers:

- *"Rich Quench Lean Combustion Chamber (RQL Combustor)"*
  In "rich quench lean combustion" exists in the primary zone of the combustion chamber an oxygen or air, respectively, deficiency (rich combustion). Subsequently in the secondary combustion zone more air is added than is necessary for the burning of the rest of the fuel (lean combustion). In both zones the flame temperature is therefore lower and the nitrogen oxide formation considerably smaller than for a stoichiometric combustion, for which in a limited zone exactly as much oxygen or air, respectively, is added to the fuel, as is required for a complete combustion. Main focus of research and development is the fuel injection and the homogenisation of the fuel-air mixture as well as the design of the air apportioning module for the so-called quenching to achieve a as fast as possible transition from the rich to the lean combustion. Because the configuration of such combustion chambers – which one also calls "stoichiometric optimised combustion chambers" – hardly differs from that of conventional combustion chambers, an engine conversion regarding the combustion chamber section is relatively unproblematic. Based on technology target information of MTU Munich, for the specific heat loading of these combustion chambers values of 60 MW/(m^3bar) = 600 s^{-1} for civil and 80 MW/(m^3bar) = 800 s^{-1} for military engines, are aimed at. On the test bed until now a nitrogen oxide reduction of 57 per cent compared with ICAO 1995 standard has been proven, with the soot emissions and the degree of complete combustion corresponding to the state-of-the art technology [105].
- *"Lean Premixed Prevaporized Combustion Chamber (LPP Combustor)"*
  If the problems associated with rich combustion are to be avoided, the "pure lean combustion" offers itself as a further possibility for a low nitrogen oxide and soot emission operation. For idle and full-load operation with equivalence ratios (the actual fuel-air ratio referenced to the stoichiometric ratio) between 50 and 70 per cent in the primary zone, the temperature level existing there is by about 600–250 °C lower than at stoichiometric combustion. The lean extinguishing limit lies at an equivalence ratio of <40 per cent. To guarantee a still stable combustion with lean mixtures, the liquid kerosene must already be vaporised and uniformly mixed with the air before entry into the combustion chamber. MTU Munich has been working on the development of such combustion chamber concept since 1989, with the focus in the first phase on the fuel preparation, vaporisation length and stability of the lean combustion. In the second phase an extended system encompassing the actual combustion chamber was tested under pressures up to 20 bar (300 lb/in^2) and entry temperatures up to about 550 °C (823 K). In this way it could be shown that a safe operation of the combustion chamber without flame reversal or self-ignition, both the greatest dangers of this concept, is possible. Measurements of the nitrogen oxide emissions have shown that lower emission values indeed did materialize [106].
- *"Staged Combustion Chamber (Staged Combustor: Double Annular Combustor, DAC)"*
  The concept of the radial, axial or axial/radial staged combustion chamber has, regarding the control of the pollutants as top priority, the aim of reducing the nitrogen oxide emission. In Germany, apart from single-stage combustion chambers, especially the concept of the axial/radial staged combustion chamber is object of research and development work at both the research institutes and the engine manufacturers.

The figure on page 440 shows the longitudinal section and a schematic of the staged annular combustion chamber which is being developed at Rolls-Royce Deutschland as alternative component for the BR 700 engine family. Its flame tube contains an inner pilot stage and a shorter outer main stage. The pilot stage is, with its relatively large volume, designed for the low air and fuel mass flow rates at engine partial load and is at present comparable with a conventional diffusion combustion chamber. It ensures the required stability of the combustion and a good degree of complete combustion, which is equivalent to low emissions of carbon dioxide and unburned hydrocarbons for all load conditions of the engine. For an improvement of the pilot stage concerning the nitrogen oxide formation, the application of a rich quench lean concept is possible. At higher load conditions of the engine, in addition the main stage is activated and at full load the largest part of fuel is injected. The for the time being rather conventional burner nozzle is soon going

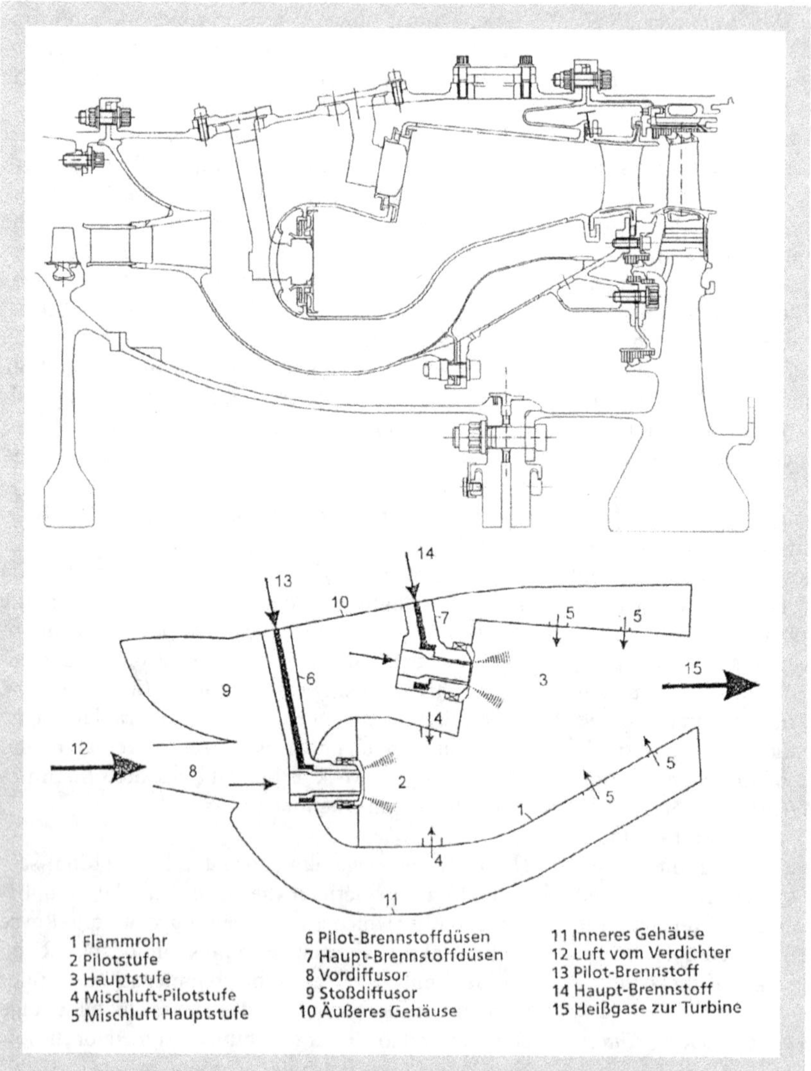

Longitudinal section and schematic of the axial/radial staged annular combustion chamber of BMW Rolls-Royce for the BR 700 engine family as possible combustion chamber alternative (original figure).

to be exchanged by an advanced air atomiser nozzle and in the course of the development to be replaced by a special premixing section for lean combustion in which the liquid fuel is atomised, vaporised and mixed with the metered combustion air before the mixture enters the actual combustion chamber. In this way the nitrogen oxide emissions are to be reduced drastically.

The pollutant formation in combustion chambers is also influenced by the air apportioning in the individual flame tube zones. A considerable portion is used for the film cooling of the walls, whereby temperature inhomogenities and lean gas mixtures in the vicinity of the wall arise, resulting in increased portions of carbon dioxide and unburned hydrocarbons. Because the flame tube surface of staged combustion chambers is larger than that of single-stage combustion chambers, and an as high as possible portion of the air delivered by the compressor is to take part in the combustion process, a relatively low specific cooling-air consumption is the essential criterion. In a technology project in this regard, which was financially supported partly by the BMBF, DLR institutes in Berlin, Cologne and Stuttgart as well as institutes of the technical universities in Chemnitz-Zwickau, Darmstadt, Dresden, Karlsruhe and Munich are involved.

Rolls-Royce Deutschland plans to employ combustion chambers of this type in quantity-production engines in some years after the successful conclusion of the technology work and expects as result in the future a nitrogen oxide reduction of up to 80 per cent compared with the ICAO 1995 limits [107, 108]. On the staged annular combustion chambers of Rolls-Royce Deutschland in the meantime numerous component tests concerning the fluid dynamic and combustion-specific optimisation have been performed. In addition, in core-engine tests of the BR 715 engine on the altitude test bed of the TU Stuttgart, investigations into the emissions behaviour of the staged combustion chamber were carried out, in the course of which nitrogen oxide emissions were measured, which amounted to half the values of the currently valid limiting values. These values correspond to a 25 per cent reduction of the nitrogen oxide emissions compared with the already very low pollution emissions of annular combustion chambers. This decrease of the nitrogen oxide emission did not lead to an increase of carbon monoxide or unburned hydrogen carbons in comparison with the not-staged combustion.

Since a short time ago engines of the companies General Electric (GE 90) and CFM International (General Electric, SNECMA: Special type of the CFM 56 engine family) with radial-staged double-annular combustion chambers (DAC) are in practical service. With regard to the nitrogen oxide emissions of these engines, clear progress is discernible, however, to a certain extent at the cost of increased carbon monoxide emission. In comparison, modern engines with combustion chambers based on other technologies exhibit nitrogen oxide emissions of a similar order of magnitude but possess clearly lower carbon monoxide emissions. The fact that progress must be conquered again and again is shown by the message that aircraft with the first CMF 56-5B-DAC engines had to be taken out of service for a short time in order to rectify structure-mechanical problems (crack formation) of the combustion chambers.

Catalytic Combustion Chambers

A further possibility to operate combustion chambers without or very low nitrogen oxide emissions offers the catalytic combustion. After previous research predominantly undertaken in the USA since about the mid seventies regarding the application of this technology in gas turbines, this topic was taken up in Germany around the early eighties. For example, basic research was performed since 1984 at the chair of aero-propulsion of the TU Munich, supported by the DFG as well as within the framework of the collaborative project "Turboflam" of the "Arbeitsgemeinschaft Hochtemperatur-Gasturbine, AG Turbo" [109, 110]. During the oxidation of a, by necessity, very homogeneous fuel–air mixture, when flowing through a catalyst block having numerous single small channels, the mixture temperature increases along the catalyst and across the cross-section rather uniformly. The catalyst is followed by an afterreactor in which the mixture temperature rises further by self-ignition. By suitable configuring, the temperature field does possess no significant peaks so that no thermal nitrogen oxide can arise as long as the highest temperature level remains $<$1,400–1,500 °C (1,673–1,773 K). Main obstacle for the application of this kind of combustion for aero gas turbines is the too low temperature limit of the currently available catalysts with an acceptable life time.

In addition the space requirements of the component layout consisting of fuel injection system, inlet section, mixer, catalyst and afterreactor as well as the relative pressure losses for an as complete as possible and homogeneous mixing of the fuel-air mixture flowing to and through the catalyst, are too large.

Non-Intrusive Measurement Techniques

Existing and newly specified requirements concerning the propulsive system of aircraft can only be fulfilled with the improvement of the existing technology and the creation of new concepts, if basic and detail knowledge is available. For example, for a reduction or avoidance of the pollution emissions from aeroengines it is important to know where and how these are produced in the combustion chamber. Knowledge from measurements of the total system and in individual regions, during which the actual operation is to be only slightly or not at all disturbed, provides the basis for the optimisation of the combustion chamber as well as the fuel-injection system. In the ideal case models and thermogasdynamic conditions can be simulated in such a way in tests, that they approach as far as possible the real combustion chamber size and configuration as well as the real operation conditions. Conventional measurement techniques employ probes and gauges with sensors for the measurement of pressure, thermo-elements for the approximate determination of temperature, and exhaust gas probes with connected physical-chemical analysers to analyse the gas composition.

Since about more than 10 years increasingly laser measurement techniques and equipment for a practically non-intrusive establishment of velocity and temperature fields, as well as distributions and fluctuations of the gas species in combustion chambers, are being employed. In order to apply such techniques test combustion chambers must be optically accessible and it is advantageous and sometimes necessary to use for the tests fuels with better suited optical, but very similar chemical-thermodynamic properties, than the original fuel, for example, n-heptane instead of kerosene. Because of such requirements and other technical problems, laser measurements have been employed only in very few tests with combustion chambers actually installed in engines.

A simplified schematic of a test set-up for laser-spectroscopic measurements of the distribution of certain species of the hot gas in a test combustion chamber is shown in Colour Plate V. This example is a set-up for photographing the distributions of the combustion intermediate product OH (hydroxide radical) and of the pollutant NO (nitrogen monoxide) after the laser-induced fluorescence (LIF) excitation on a combustion chamber test stand of the chair of aero propulsion (LFA) of the TU Munich [111]. The key piece of the test set-up is the laser unit adjustable to different wavelengths for the excitation of fluorescence signals of species of interest. The laser beam is routed via deflection mirrors, optical lenses and slit apertures to the combustion chamber fitted with quartz side windows. A "Charge Coupled Device, CCD" camera samples the fluorescence signals via filters and suppresses disturbing signals of other wavelengths. With the aid of a PC the scatter, synchronisation and operation of the system as well as the data acquisition and evaluation are carried out. Furthermore, the colour chart shows a result of laser-spectroscopic measurements at the LFA in a staged model combustion chamber. Based on the concentration distribution of the OH radical, which is made visible by the "Laser Induced Pre-dissociation Fluorescence, LIPF" method, it can be shown which form and extent the combustion zone has, as well as which influence the lateral air apportioning jets exercise on the combustion process. The OH radical arises as unstable intermediate product in the decomposition of hydrocarbon combinations in the reaction zone and is thus an indicator of the local combustion intensity.

A number of other methods exists for the optical measurement of concentrations and local thermogasdynamic conditions in combustion [112]. Atomic and molecular concentrations cannot only be measured with fluorescence spectroscopy (LIF, LIPF), but also with absorption spectroscopy and Raman spectroscopy. The light scattering due to the Raman effect results from induced transitions between the rotation and the vibration levels of the molecules. One from the Raman measurement technique derived method is based on the "Coherent Anti Stokes Raman Scattering, CARS" with its relatively strong light-scattering signals. Temperatures and temperature variations can also be measured by means of the fluorescence spectroscopy and the Raman spectroscopy. The two-lines spectroscopy (LIF, LIPF) at a molecule, for example O_2 (oxygen) or OH, is to be carried out with a single or even better with parallel operating laser systems. If only one system is available for the determination of the temperature in turbulent flames, the mean species distributions must be employed. In this context it is a problem that species concentrations as well as the concentration density in the basic state in general do not correlate linearly with the temperature. The in other cases also employed Rayleigh thermometry is very difficult to realise in model combustion chambers, because of their metal or glass walls.

Flow fields of the "cold" and the chemically reacting gas in combustion chambers with the resulting velocities, fluctuations and turbulence can be measured with methods based on the Mie scattering. The Mie scattering is an elastic scattering of particles in transparent gases (and liquids), as, for example, dust and soot particles having dimensions at least as large as the wavelength of the impacting light. The resulting scattered light possesses characteristic properties regarding phase, amplitude, polarisation and distribution, which allow conclusions to be drawn concerning the structure and dimensions of the particles. One method, in which Mie scattering can be used for the determination of the flow velocity, is the laser Doppler anemometry (LDA). Basis for this is the phenomenon of the frequency shift of the light scattering due to a scattering body moving relative to a light source (Doppler effect). Another, relatively new method for the visualisation of flows, which is also based on the Mie scattering, is the so-called particle image velocimetry (PIV). In this case a very narrow plane laser beam is transmitted with two each other shortly succeeding light pulses through a plane of the flow field. By photographing the particle scattering within the illuminated area at both moments of time with an electronic camera and the known space of time between both light pulses, the velocity of the particles can be determined. A survey of individual results of laser-spectroscopic measurements in tests with aero-engine model combustion chambers is, for example, contained in [113].

Supersonic Combustion

In the field of high-speed aircraft engineering completely different combustion processes exist as those described up to now, but also related processes. Up to altitudes of about 40 km (131,200 ft) the atmospheric oxygen can be used for the operation of airbreathing aeroengines. They offer for the propulsion of, for example, reusable space-transportation systems or military flight vehicles, in comparison with conventional rocket propulsion, the advantage that the oxygen must not be carried onboard. Such vehicles fly at high altitudes with supersonic and hypersonic speeds. This governs the requirements on the concept and the design of the propulsion system. For atmospheric flight at very high speeds ramjet engines with supersonic combustion, in short scramjets (supersonic combustion ramjets), possess a very good and competitive potential for propulsion. Since the fifties in the USA and since the sixties in Germany studies and detailed investigations concerning the design and operation of scramjets have been carried out. The research in the field of supersonic combustion since then has been and is being carried out world-wide, especially in the USA, Russia, Japan, Western Europe and also in Australia. Recently the American National Research Council (NRC) has presented recommendations regarding research and developments for hypersonic technology in the USA, applicable in practice with a target date of 2015 [114]. As a matter of priority, airbreathing powerplants of airborne weapon systems with hydrocarbon and hydrogen combustion for flight Mach numbers from 6–8 are first to be developed and the so-called spin-off is to be used to realise new propulsion systems for high-speed flight vehicles.

Although up to now a rather large number of reports on the behaviour of supersonic combustion chambers in the design point, and on problems concerning details, as well also some on partial load behaviour has been published, many important physical and technological questions are still open. Single topics of current research are "Efficiency Improvement over a Broad Operational Envelope", "Operational Stability and Safety", "Material, Structure and Life Span under Extreme Operational Conditions", "Improvement of the Measurement Technology to be Used" as well as "Fundamentals for System Optimisation". In Germany recently work concerning supersonic combustion has been performed and is also continued at the DLR Institute of Space Propulsion in Lampoldshausen near Stuttgart, as well as at the technical universities in Aachen, Munich and Stuttgart and at MTU Munich.

In ramjet engines with supersonic combustion the air passing through is considerably less decelerated as for ramjet engines with subsonic combustion. The combustion chamber entry-velocity can range from 900 to slightly more than 1,800 m/s (2,952–5,904 ft/s). The advantages of this concept are mainly based on three aspects. Firstly, because of the lower static temperature level in the combustion chamber the combustion products dissociate not at all or only slightly, whereby a higher amount of energy is available for the thrust. Secondly, owing to the higher internal Mach number level the flow deceleration in the supersonic intake of the engine and consequently the associated total pressure losses due to shock waves are considerably lower. Thirdly, on account of the lower static pressure and temperatures in the combustion chamber, the structural loading and the heat transfer rates are smaller.

On the other hand with the high velocity of the air flow through the combustion chamber, in a few milli-seconds difficult and complex phenomena occur. They concern the fuel injection and the associated pressure losses in the main flow, the fuel-air mixture, the degree of complete combustion as well as the ignition behaviour and the combustion stability. Experimental investigations of the large and small scale processes and phenomena of supersonic combustion are to the largest extent carried out in model-combustion chambers on stationary test beds. They are supplied with pressurised air from an air supply facility via an air heater located upstream and a Laval nozzle for the simulation of the combustion chamber entry conditions.

In the Coloured Plate VI a longitudinal section of the basic concept of a supersonic model combustion chamber with hydrogen injection is shown. The Laval nozzle in front of the entry can be exchanged for the variation of the inflow Mach number and a wall portion of the divergent combustion space is adjustable for the determination of the most favourable cross-section distribution [115]. In order to make visible the phenomena and the reaction processes, several optical measurement techniques are employed, as already reported above.

Knowledge concerning the fuel-air mixture and the degree of complete combustion can be derived among other things from the spatial flame structure. It can be depicted from the distribution of the hydroxide radicals, which have been exited to fluorescence by the LIPF method. For illustration in the figure the flame of hydrogen combustion in a supersonic model combustion chamber (entry Mach number $M \approx 2$), having an ramp injector, is reproduced [116]. The dominating influence of secondary vortices emanating from the ramp on the mixing is visible and one can classify this combustion-chamber geometry qualitatively as an intermediate step for the determination of a useful engine configuration. Supplementary to the experimental work and for support, numerical simulation methods are being developed and used in order to reduce the time and expense needed.

Remark

Concerning this chapter some figures are presented in colour

Color plate V

- Test set-up for the measurement of combustion intermediate products in a model combustion chamber.
- Influence of mixing air apportioning on the combustion process in a staged test combustion chamber.

Color plate VI

- Concept of a supersonic model combustion chamber.
- Spatial flame structure in a supersonic model combustion chamber.

Progress in Turbine Blade Cooling Since 1945

HELMUT SCHUBERT

From the Convection to the Effusion Cooling

As can be seen from the adjacent figure, the turbine entry temperature alone could be increased since the year 1950 by 5–6 degrees per annum owing to the materials development. Until the year 1960 the leading American and English engine companies refused the application of the turbine-blade cooling because of the associated costly design and also because of the unavoidable additional losses owing to the cooling air. Only when the materials development especially regarding the heat-fatigue strength of the high-temperature resisting materials reached its limits – in about the year 1960 for civil engines – in particular Rolls-Royce in England occupied itself again with the blade cooling. As coolant compressor air was employed. In this context it must be mentioned that *Ernst Schmidt* already during World War Two at the TH Brunswick and also later at his thermodynamics institute of the TU Munich explored water cooling of gas-turbine blades. However, this kind of cooling was in practice applied only occasionally in heavy gas turbines [117, 118]. In the course of time different kinds of turbine blade cooling evolved which led to a considerably increase of the turbine-entry temperature. The different possibilities of cooling-air guidance in turbine blades are schematically represented in the figure.

Convection Cooling

For the so-called internal cooling, or convection cooling, respectively, the cooling air entering at the blade root flows along inside the blade, being heated by convection during this process. This air is in general discharged at the blade tip and/or at the blade trailing edge where it mixes with the main gas flow. The gas temperature achievable with this cooling method is about 1,400–1,500 K (1,127–1,227 °C) depending on the design of the blade internal contours.

Impingement Cooling

An improvement of the convection cooling is the impingement cooling. This implies that the cooling-air flow entering the blade is guided in such a way that it impinges in distributed single jets normal to the in-

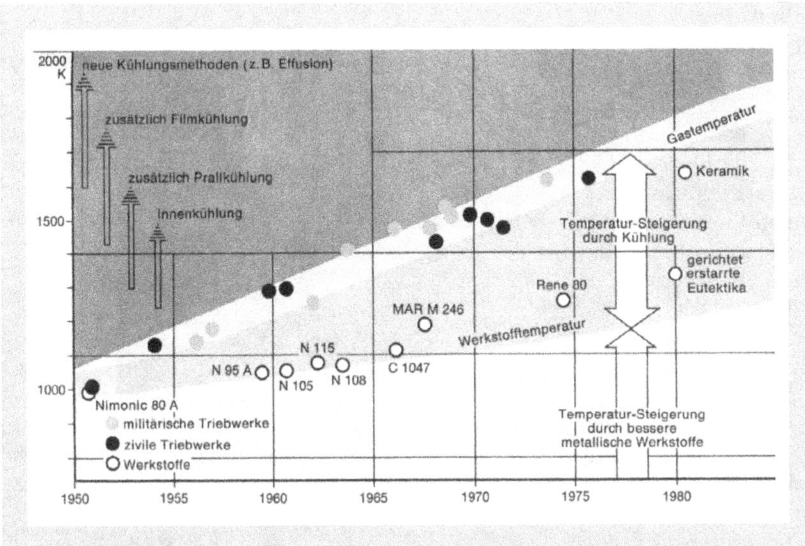

Increase of gas temperature before entry into the high-pressure turbine by better metallic materials and by improved cooling of turbine rotor blades (Werkstofftemperatur = materials temperature, Gastemperatur = gas temperature, Kühlung = Cooling, militärische Triebwerke = military engines, zivile Triebwerke = civil engine, Werkstoffe = materials).

ner surface of the blade leading edge and thus causes a good heat transfer. Turbine blades with pure internal cooling, and combined with impingement cooling, respectively, are however limited regarding their cooling potential.

Film Cooling
A further increase of the turbine entry temperature is achieved by the so-called film cooling. In this cooling method, cooling air exudes from a row of small orifices and slots, respectively, on the external blade surface and forms a protective cooling film between the hot gas and the blade wall. The film is gradually dissolved by mixing with the hot gas so that the cooling effectiveness decreases in the flow direction. At the location, where on the blade the cooling effect falls below a critical value, a further cooling film must be discharged. The current highly loaded turbine blades mostly possess combined internal and film cooling.

Effusion Cooling
The effusion cooling, or transpiration cooling, respectively, implies that the turbine blade mostly consists of a load-bearing core which is surrounded by a porous blade wall through which the cooling air is discharged into the hot gas. In this way the cooling film on the hot-gas side of the turbine blade is continuously renewed, which allows to achieve a uniformly high cooling effect over the complete blade surface.

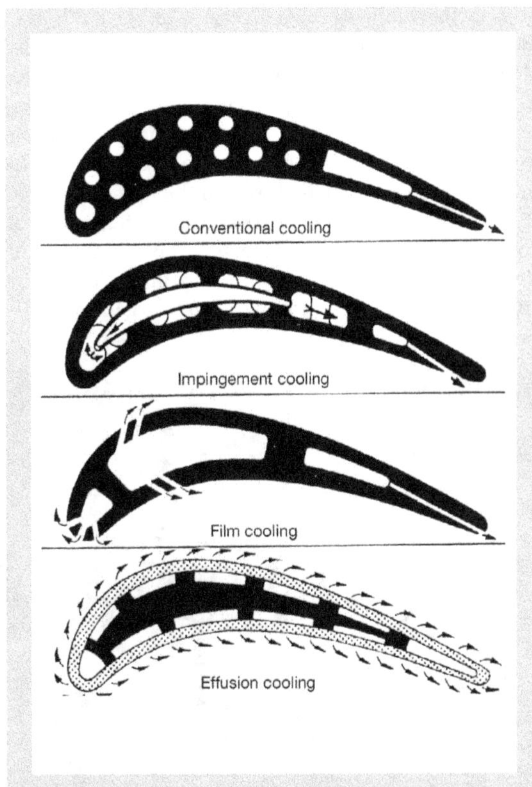

Various kinds of cooling of turbine rotor blades.

Determination of the Blade Cooling Effect

The prediction of the cooling effect, that is the blade temperature for every operating condition, together with the blade loading is possible today. For defined flight programmes or life cycles, by taking into consideration the blade layout as well as the blade profile, the cooling ducts and the life time are optimised. For this it is at first necessary to calculate the heat transfer along the blade contour and to deduce from this the temperature distribution across the blade cross-section by taking into account the internal heat transfer.

The cooling system of a typical highly loaded turbine guide vane consists of a combination of convectively cooled regions as well as of impingement cooling systems and of flow ducts with installations like webs, fins and pins for surface enlargement. Forced turbulence can also be employed to increase the heat transfer. For the cooling of the turbine rotor blade, the cooling air is routed by several flow reversals inside the blade from the blade root to the blade tip. At especially thermally stressed locations on the blade surface, and at the trailing edge, the cooling air is discharged via orifices and slots, respectively, into the main flow. Such for the rotor blade typical cooling systems are also called multi-pass systems. An example for the cooling air guidance of a highly loaded high-pressure turbine stage is shown in the figure above.

When in 1960 at Rolls-Royce the manufacture of cooled turbine blades was started, the calculation of the cooling air requirements was still difficult. One succeeded however to provide blades forged from circular blanks with ducts by drilling, which had already been incorporated in the blades before the forging. Although the forging of these blades was costly, the cooling effect initially did not satisfy. An example of such turbine guide vanes is presented in the figure above.

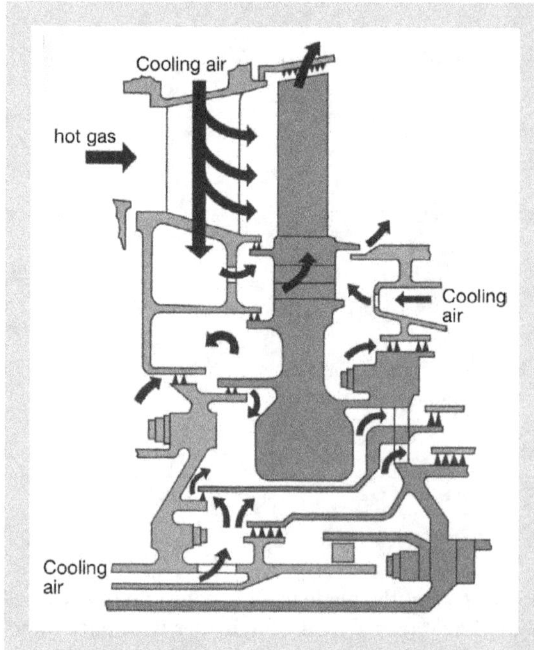

Cooling air guidance for a high-pressure turbine rotor disc as well as for the guide vanes and rotor blades.

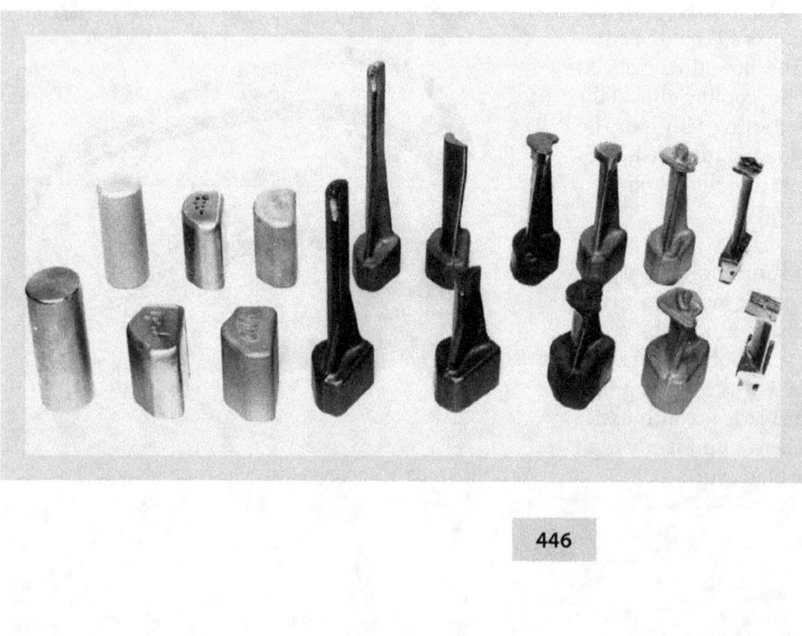

Fabrication stages of cooled turbine rotor blades of the turbo-jet engine RB 153-61 (1963).

The blades were meant for the engine RB 153-61R jointly developed by Rolls-Royce and MAN-Turbo.

The Cast and Cooled Turbine Blade

In the USA soon afterwards several companies succeeded to manufacture turbine blades with cooling ducts cast into them. The material was mostly Inconel 713C. About the year 1970 these cast turbine blades gained acceptance in the quantity production of several engines. Very slowly the technology improved so that it became possible to cast now nearly every desired layout of the cooling ducts. A cooling concept of General Electric with convection, film and impingement cooling is shown in the figure below.

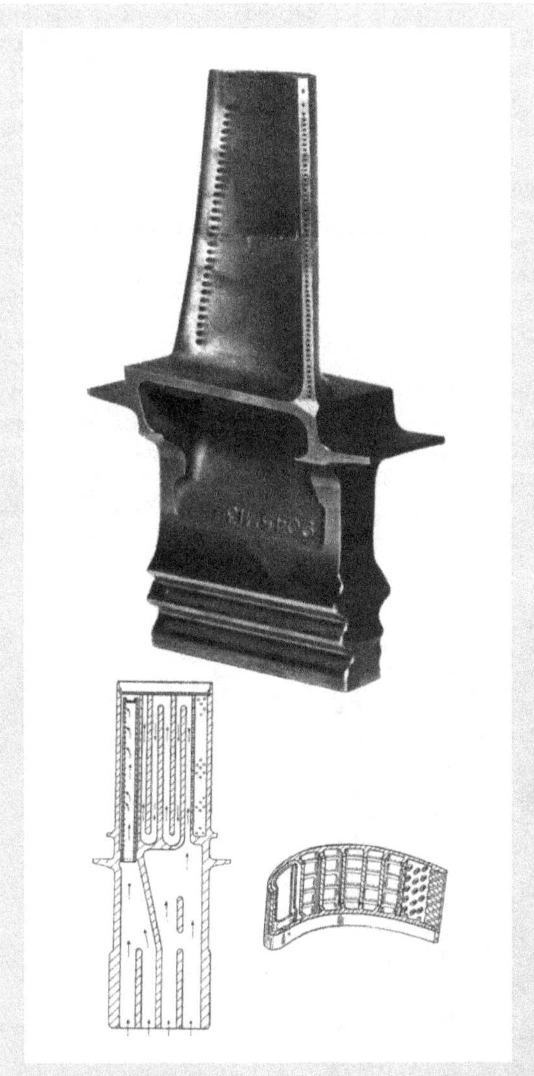

Cooling of the turbine rotor blade of a modern turbo-jet engine. Below the principle of the convection, film and impingement cooling is represented.

By eroding the cast blank, that is the electro-chemical precision drilling of small holes at every desired location and direction from the outside into the cast blade, one succeeds, for example, to deliberately allow cooling air to exude at the blade leading edge. Where the heat transfer from the exterior is especially high, one is able to establish a cooling film around the blade. Important for these turbine blades was, apart from the casting, the mechanical or electro-erosive and electro-chemical processing respectively [119]. *K. Gradl* and *M. Baumgärtner* of the company Leistritz present a comprehensive survey [120] of the developed and employed manufacturing methods for turbine blades in Germany.

The technological development has only led by large effort to fail-safe blades and only few manufacturers have achieved this objective. Since 1980 the materials development of single-crystal blades then came to assistance. If one considers the progress at different companies, then one gains the impression that the developments for turbine-blade cooling and especially for the cooling of the entire turbine are concluded. Remarkable is the penetration of the costly blade cooling into the field of industrial gas turbines.

Research Work for Improvement of the Cooling Methods

Within the framework of supporting the development of technologies for high-temperature gas turbines in the program "TURBOTECH" on the part of the BMBF, the blade-cooling technologies were also supported. After the in the years 1994/1995 discontinued project phase I and an interim phase of about one year, "TURBOTECH II" was started at the beginning of 1996. With this joint program of German university and industry partners in the field of turbomachinery, the continuity of research at the involved universities and research institutions in the field "Turbo-Components for High-Temperature Gas Turbines" was ensured [121].

In "TURBOTECH II", 57 projects with a total budget of about 23.8 million Euro were sponsored. Emphasised must be the by more than 10 per cent increased project volume compared with "TURBOTECH ". Not the least this financial engagement demonstrates the high status which was attached to this research work by the BMBF and industry. Involved in this work, which lasted until the end of 1999/beginning of 2000, were 16 German universities, the DLR and five partners from industry.

In a special project group investigations concerning the optimum use of cooling measures were concentrated. The in total 13 work packages were subdivided into six thematic fields. While for the work in "TURBOTECH I" main emphasis was put on studies of basic phenomena for 2D geometries, in "TURBOTECH II" 3D effects and the influence of high cooling air rates were investigated. Alone four projects dealt with the improvement of the S2 calculation methods in order to be able to make a reliable allowance for the apportioning also of high cooling air quantities already in the design phase. Detailed measurements permitted to set up reliable correlations that made use of the addressed S2 methods. In a further subgroup work on the 3D optimisation of film-cooled blades was concentrated. Based on the aerodynamic measurements carried out in the previous phase I, subsequently on the same blade experimental investigations were performed for the determination of the heat transfer in the presence of film cooling.

The obtained experimental data served for the validation of the calculation methods developed in the accompanying projects. Special work for producing a 3D optimisation system, which allows the layout of a cooled turbine blade along stream-surface sections taking into consideration the strength reserves and life-time aspects, was carried out. In close connection with the addressed tasks were the investigations for the optimum layout and shaping of the duct geometry of cooled turbine blades. Results of detailed measurements are available. Basic work using a plane test cascade is serving the understanding of the flow phenomena in the border zones of cooled turbine guide vane cascades. Extensive aerodynamic, L2F and heat transfer measurements allow the exact analysis of the complex flow phenomena. The results of the aerodynamic investigations into the secondary flow field close to the wall are also available.

The strong increase of the combustion chamber outlet temperatures would not have been possible without an effective cooling. One of the foundations for its realisation was the development of very fine and precise processing methods for materials difficult to process conventionally. Current methods, as, for ex-

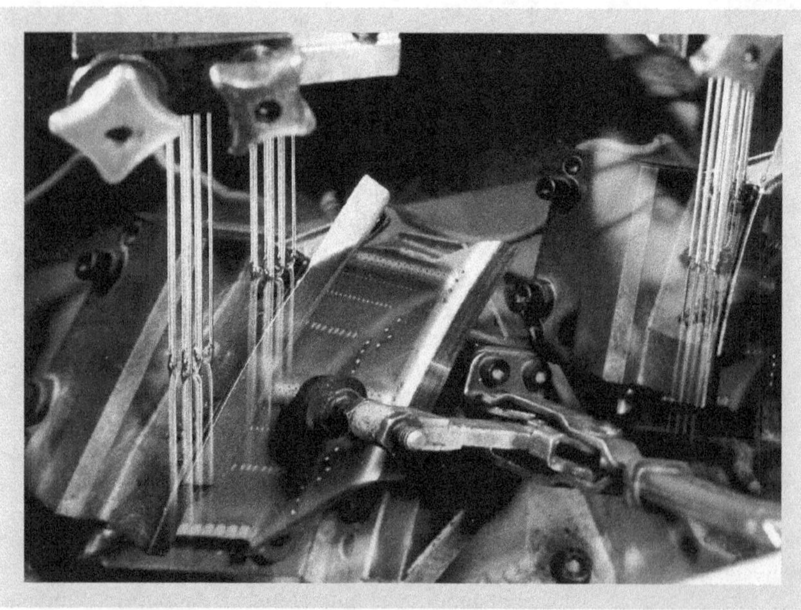

Turbine blade fabrication with the "Electro-Stream Drilling, ESD" method) (1978).

ample, electro-chemical and spark-erosive processing and laser-beam drilling find successful application for many different engine components, especially however for the delicate turbine blades. Protective coatings for hot parts are being used for some time. Added to the eroding and penetration diffusion methods were the spattering and spraying methods. The field of application of the latter extended also to colder engine regions. For example, by application of grindable, not inflammable coatings on casing walls the radial gaps of compressor rotor cascades can be reduced.

By improvement and new development, respectively, of manufacturing processes the associated direct decrease of the manufacturing effort for engines regarding construction volume and engine mass also plays an important role. Formerly frequently used screw-flange joints are replaced in the modern integral construction in many cases by resistance or electron beam-welding joints. A mass saving is also achieved by the use of hollow construction components which are fabricated as thin-walled halfbodies and which obtain their final form by diffusion joining. An example for this are hollow fan blades which are employed as standard in large engines in the USA. Combined fabrication and heat-treatment processes allow to improve considerably the strength and other technological properties of construction components. An example for this is the precision casting of turbine blades fabricated with grain orientation. With the development of other precision-casting methods, for which a post-processing is almost no more required, one moves in the direction of further manufacturing cost savings.

Turbine Cooling for Modern Bypass Engines

The aerodynamic loading of the turbines in modern twin-spool bypass engines continues to grow. At the same time the number of stages is to be as small as possible in order to save mass and to keep the number of construction components (turbine blades) low. Thus, the cooled high-pressure (HP) turbine and the cooled low-pressure (LP) turbine of the EJ 200 have each one stage only. The HP turbine entry temperature is about 1,800 K (1,527 °C). The cooling air requirement of the HP turbine guide vane assembly amounts to about 10 per cent and that of the rotor blade assembly to about 4.5 per cent of the core air mass-flow rate. For the LP turbine, the cooling air requirement is for the turbine guide vane assembly three per cent and for the rotor blade assembly two per cent of the core air mass-flow rate. The cast LP turbine blades consist in the case of the EJ 200 engine of a single-crystal high-temperature material and for the high-pressure turbine of metal powder nickel material. A further increase of the high pressure turbine entry temperature could be achieved for the BMW Rolls-Royce BR 700 engine by using the nickel material U720Li [50].

If one compares the spectrum of the materials employed in engine manufacture today, an interesting development is observed. Steel materials are essentially still used for engine bearings and shafts only. The portion of fibre composites and inter-metal materials grows slightly. The materials development is international. The German portion of the development of high heat-resistant materials was after 1945 no longer very large. However, substantial contributions were made in the field of fibre materials.

Within the framework of the CRISP technology programs of MTU Munich supported in Germany with funds of the BMFT and BMWi, CFC fan blades made of fibre reinforced materials were developed for low mass, favourable production costs and high vibration strength. This took place in collaboration with the DLR research centres Stuttgart and Göttingen. Work was also performed on SiC and C-fibre reinforced metallic and ceramic materials with SiC and glass as matrix for components in hot gas regions. The maximum allowable construction component temperature lies today at 1,100 °C (1,373 K) [122]. The development

High pressure turbine guide vane of the General Electric CF6-50 engine after the ESD fabrication.

Two-part cooled turbine rotor blade with diffusion joint, a development of MTU Munich (1980).

of high temperature resistant and durable CMC structures for gas turbines is the subject of international research and is still in an early stage.

The DLR works with its institutes in Cologne (institute of materials research) and Stuttgart (institute of structural research and design development) at the head of this development. In the case of success these materials can, however, not only be applied in the combustion chamber but also in the entire hot-gas part (turbine, nozzle) and would represent a development leap for the whole propulsion and power-station technology. Also within the framework of the German national aeronautics research programme supported by the BMFT in collaboration with industry novel materials for high-temperature application have been and are being developed. Main research focus are in particular also ceramic heat-insulation layers and ceramic composite materials (CMC) [123]. The maximum controllable turbine entry temperature with cooled turbine blades in today's engines amounts to about 1,800 K (1,527 °C). In special cases already temperatures of 2,100 K (1,827 °C) are reached. The progress in the temperature increase by material improvements resulted only in about 300 K in 50 years, that means an increase of 6 K per annum despite large investments and efforts in the material development. The in the early sixties again introduced blade cooling yielded 400 K temperature increase, that is on the average 12 K per annum. The table next page shows the development stages of the cooled turbine rotor blades from 1939 until 2000.

Development of cooled turbine blades.

	Period of time	Main characteristic	Engine example	Turbine entry temperature [K]	Material temperature [K]	Material	Temperature reduction [K]	Air requirement of hot gas flow [%]	Technology progress	Life [h]
German development	1939–1945	Metal sheet hollow blade without cooling insert	Junkers 004 BMW 003	~1.040	~920	Austenic steel C~0.1 Mn~0.9 Si 1,0 Cr17Mo2 Ni 15 Ta+Ni~1,1	50–100	0,8	Cooling insert	50
English development	1950–1960	Forged cooling ducts	RR Tyne 12 RR Conway 12 RR Conway	1.242 1.310 1.390	~1.050	N 105 N 105	40 120	0,8 1,4	High temperature materials	500 1.300
English development	1955–1970	Forged cooling borings	RR Spey RB 153-61R RR Tyne 21/22	1.360		N 108	220	2	Cooling borings forged	8.000
US and European Development	1960–1970	Cast continuous cooling duct	Allison T56 (1964)	1.580	max. 1.255	INCO 713 C	~180	2	Cast alloys, "lost wax" technique	
US and European Development	1965–1975	Cast branched cooling ducts	RB Olympus General Electric T64 Allison 501-(1960)M62B	1.490 1.369 1.480	1.164	MAR M 246	~150	1,8	Gast borings	> 2.400 5.000
US and European Development	1970–1976	Vacuum fine casting, combined branched cooling ducts and film cooling borings	RB 211-22B CF 6-50 C F 101-GE-100 RB 199-34R	1.440...1.570 1.594 1.650 1.680	~1.150 ~1.165 ~1.170	RENE 80	~300 ~300 ~300	2,5...3,5 2,5...3,5 2,5...3,5	Ceramic core and electrochemical drilling, wall thicknesses 1 mm minimum, cooling ducts 0.6 mm diameter minimum	> 3.500
US and European Development	1976–2000	Aligned solidified eutectica	V 2500 PW 2037 PW 4000 BR700 EJ200	1.750... 1.800	~1.175 ~1.220	MA 600 CMSX4 U720 MARM002 RR2000		max. 36 22		12.000 up to 25.000 1.000

Engines for the German Vertical Take-off Aircraft of the Sixties

Kyrill von Gersdorff, Helmut Schubert

In the fifties the development of light turbo-jet engines in England, in the USA and also in France made such rapid progress that a realisation of military aircraft with vertical take-off and landing (VTOL) capabilities came within the realms of the technical possibility. For the numerous projects of vertical take-off and landing aircraft first in England at Rolls-Royce special VTOL engines were developed which possessed a favourable thrust/mass ratio. There existed three different kinds of VTOL engine. Firstly the ultra-light lift engines which were only in operation for a relatively short time of the take-off and the landing as well during the transition phase. Secondly the lift-cruise engines which by jet deflection also served for the production of the forward thrust. Thirdly the engines for the so-called tail-sitters, that means aircraft which take-off and land vertically but for the forward flight change their attitude. To these are counted aircraft with completely swivelling engines, for example, at the wing tips. In ad-

dition there existed a number of engine combinations, where the aim was mostly to have as much as possible of the installed engine thrust available for the vertical take-off phase.

The Beginning of the German VTOL Engine Development

Soon after the lifting of the ban and the resumption of industrial aviation activities, in the year 1957 the resurrected German aircraft companies, that means Dornier, Heinkel and Messerschmitt, were requested by the German Federal Government to think about vertical take-off and landing aircraft. For the in the time sequence first task – a VTOL interceptor – the flight performance should as far as possible correspond to that of the most modern fighter aircraft – in those days that was the Lockheed F-104 "Starfighter". In accordance with the requirement concerning the design of military vertical take-off aircraft, also the engine company MAN-Turbomotoren GmbH, founded in October 1958 and with headquarters in Munich-Allach, started to address special problems of the VTOL powerplants, based on a contract by the "Federal Ministry of Defence" ("Bundesverteidigungsministerium, BMVg"). Because of the more than ten years lasting interruption of the German engine development, one was dependent on a co-operation with foreign partners.

In March 1960 it then came to a first co-operation contract with the English engine company Rolls-Royce in Derby. Aim of the ten-year contract was to acquire in Germany missing knowledge in the field of aeroengines with the aid of an experienced and forward-looking engine company. In the contract it was also laid down that for the joint development one would seek agreement with each other, share the work and avoid excess capacities for the quantity production of aeroengines. In March 1960 the BMVg placed a development order with the MAN-Turbomotoren GmbH for a light single-spool turbo-jet engine with afterburner which received in the spirit of the co-operation contract the designation RR/MAN RB.153-17. Rolls-Royce became subcontractor to the MAN-Turbomotoren GmbH.

Rolls-Royce/MAN Jet Engine RB.153-17

The engine RB.153-17 was initially only a scaled version of the Rolls-Royce turbo-jet engine RB.108 with a modification for supersonic flight up to M = 2.3. With the start of the development in spring 1960 the consideration of the requirements for vertical take-off and nacelle installation took place. The engine RB.153-17 [1] was intended for installation in pairs in swivelling nacelles of the EWR aircraft project VJ 101 C [124]. The take-off thrust was to be 17.5 kN (3,934 lbf) without afterburner and 24.3 kN (5,463 lbf) with afterburner. A lift-engine version with the designation RB.153-25 was projected in parallel. Several MAN engineers had been assigned to the English develop-

Within the framework of a cooperation contract between the MAN-Turbomotoren company and the English engine company Rolls-Royce, MAN was involved in the development of the turbo-jet engine RB.153-17 which was intended for the installation in the swivelling nacelles of the EWR vertical take-off project VJ 101 C.

ment team in Derby. They soon could take over individual development subtasks.

In December 1961 the development was discontinued after fundamental changes in the VJ 101 aircraft program based on a change of the military mission profile had taken place, for which another engine combination had to be chosen. Instead for a military vertical take-off aircraft, the BMVg only placed an order for two light experimental aircraft VJ 101 C X1 and X2 in accordance with the original interceptor requirement, for which another engine combination had to be selected. Up to the discontinuation of the RB.153-17 development only few construction components had been manufactured by Rolls-Royce and MAN. Test-bed trials had not been undertaken. In retrospective one can consider the years 1960/61 as the apprenticeship of the team of the MAN-Turbomotoren and with this as a prerequisite for a further close co-operation between both companies – in the meantime continued up to the engine of the "Eurofighter" [48].

Rolls-Royce/MAN Jet Engine RB.145

With the experimental aircraft of the VJ 101 C project the requirement arose for a lighter jet engine in the 12.3 kN (2,765 lbf) class. Rolls-Royce proposed a constructional modification of the already proven lift engine RB.108 to yield a lift-cruise engine. The new variant received the designation Rolls-Royce/MAN RB.145 [1] and was developed in two versions. In one it was an in the fuselage vertically installed fixed lift engine with a short propulsive nozzle and in the other a swivelling lift-cruise engine with afterburner for the twin-nacelle installation. The lift engine RB.108 had already been proven in the English VTOL test aircraft short SC1. It distinguished itself by a very favourable thrust/mass ratio of 8.7 : 1. For the RB.145 engine one compressor stage with adjustable guide vanes was added and the control system was designed by the English company Lucas in such a way that rotational speed and degree of afterburning could be varied directly. In the VJ 101 C [124] all six engines two in the fuselage and two each in the swivelling nacelles at the wing tips – were controlled with a single thrust lever. For the control of the rolling and pitching movement in hovering flight by "thrust modulation", the adjustment of the individual thrust levers took place via the flight control system. With the relatively complex control system new ground had to be broken, that is industrial research work was performed and this in close collaboration between the two engine companies, the control-system company and the aircraft developer.

In cooperation also the problems of the air supply to the engines during the transition phase had to be solved. The air intakes of the lift engines were covered up with flaps during fast flight, which were designed in such a way that the engines could be started in flight before the landing transition and set at idling rotational speed. The nacelles had closeable annular slots which allowed the engines to be supplied with sufficient air with nearly no losses at all flight attitudes.

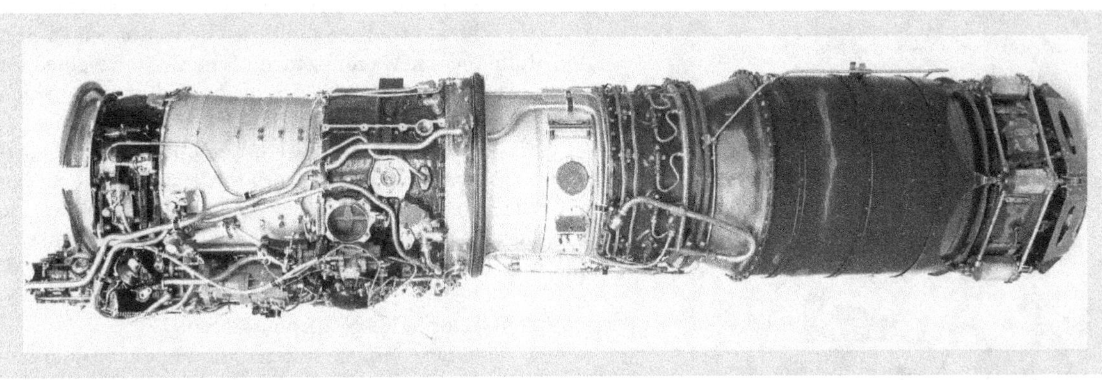

For the EWR vertical take-off test aircraft VJ 101 C X1 and X2 the Rolls-Royce engine RB.108 was modified and received the designation RB.145. It was then used as an in the fuselage vertically installed lift engine and in the swivelling nacelles as lift-cruise engine with afterburning.

The first flights of the EWR vertical take-off aircraft VJ 101 C took place in April 1963 (X1) and June 1965 (X2). Installed in each were six RB.145 jet engines, two in the fuselage and two each in the nacelles.

In the extensive recirculation tests, which were carried out by EWR in Manching in 1962, also MAN engineers were involved [48]. Investigated in the course of this was especially the thrust reduction when stationary on the ground and in hovering flight by re-ingestion of the hot exhaust gases mixed with the sucked-in air. The development of the RB.145 engine started in 1960, the first test run was in March 1961 and two years later the first 50-hour run took place as prerequisite for the flight clearance. Altogether 22 development and flight test engines were built. Many construction components were fabricated at MAN-Turbomotoren in Allach, the final assembly and the acceptance runs took place partly in Derby and partly in Allach. During the flight tests of both VJ 101 C experimental aircraft (first flights: X1 on April 10, 1963 and X2 on June 12, 1965) the engines were serviced by technicians of MAN-Turbomotoren GmbH and since 1969 – after renaming of the company – of the Motoren- und Turbinen Union Munich (MTU). The whole program came to an end in June 1971.

The VJ 101 C X2 on the tripod during an afterburner test run. The exhaust gas jets are deflected via channels.

Rolls-Royce/MAN Jet Engine RB.153-61

When, owing to the changed military requirements, the flight mission of the vertical take-off aircraft VC 101 C was changed from interceptor to fighter aircraft, longer flight periods at low altitude and a top speed of $M = 2.2$ at higher altitude needed to be covered. The subsequent optimisation calculations yielded the result that a bypass engine with low bypass ratio, with afterburner and thrust deflection would best fulfil these new requirements. The EWR presented for this a new aircraft design at the end of 1960, which was given the designation VJ 101 D. This aircraft was to receive two cruise-lift engines installed horizontally in the fuselage with thrust deflector between the core engine and the afterburner as well as five vertically arranged lift engines Rolls-Royce RB.162-31. For the cruise-lift engine RB.153.61 [1] the development began in March 1962. Rolls-Royce was again subcontractor to MAN-Turbomotoren.

The new RB.153 differed however considerably from the earlier RB.153 engines of 1960, despite of nearly the same designation. The engine developed for the VJ 101 D [124] was actually a completely new development. Only very few sub-systems, as for ex-

For the vertical take-off fighter project EWR VJ 101 D, for which longer flight durations at low altitudes and a top speed of over M = 2 at high altitudes was required, two two-spool bypass engines RB.153-61 with thrust deflector and five lift engines in the fuselage were intended. MAN-Turbomotoren was involved to a larger extent in the development work.

ample the combustion chamber, could be taken over from the Rolls-Royce "Spey" engine. The development work was increasingly carried out in Munich-Allach. It encompassed especially the aerodynamic and mechanical strength-specific design and construction of the components as well as model tests for the gas routing and thrust deflection. A large part of the engine tests could be conducted since 1963 on the two full-engine test beds, which had in the mean time been erected in Allach, and which were instrumented with modern measurement equipment. On one of these test beds with vertical thrust measurement capability also the unsteady measurements associated with the VTOL operation could be performed.

The test program, lasting for two years, came to a conclusion at the end of 1963 with a 50-hour qualification run. Six development engines had accumulated altogether 1,400 hours on the test beds in Derby and Allach, of these 50 hours on an altitude test bed. Because the mission specification for the VJ 101 was changed once again, the development was discontinued in September 1965 on instruction by the BMVg. The RB.153-61 engine represented, however, the basis engine for further designs of vertical take-off aircraft such as also the German-American "Advanced Vertical Strike, AVS" project and the EWR project VJ 101 E for which a novel 90° swivelling afterburner [48] was designed, manufactured and tested. The tests with this were carried out at the testing site of the Air Force in Manching. For technical reasons in these tests on the open test bed the jet had to be deflected upwards. The interest in this configuration was that great that in summer 1967 further tests were conducted on order placed by the American engine company General Electric. It is thus not surprising that in some American projects of VTOL aircraft like the

For the German-American project of a vertical take-off fighter AVS and for the EWR project VJ 101 E MTU developed for the engine RB.153-61 a swivel afterburner whose thrust direction could be changed by 90°. The test-bed runs took place in Manching.

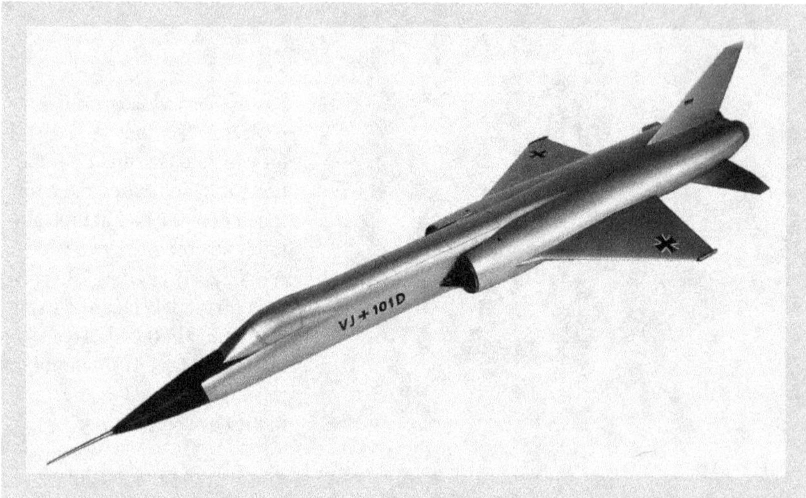

Vertical take-off project EWR VJ 101 D of the year 1960 with two RB.153-61 and five RB.162 turbo-jet engines.

"Joint Strike Fighter, JSF" also swivelling afterburners are being proposed based on the MAN principle.

With the discontinuation of the AVS project at the beginning of 1968 in favour of the short take-off fighter aircraft "Tornado", this promising VTOL engine development had been concluded.

Rolls-Royce/MAN Jet Engine RB.193-12

The history of the aircraft project "Vertical Reconnaissance and Fighter Aircraft" VAK 191 ("Vertikal-Aufklärungs- und Kampfflugzeug") – considered as the successor of the Italian Fiat G 91 – dates back to the year 1961. Basis for this was the "NATO Basic Military Requirement No. 3 (NBMR 3)" of August 1961. The aircraft company Focke-Wulf, Bremen had at that time elaborated a follow-on project of the G 91 with the designation Fw 1162. It followed to a degree the English vertical take-off aircraft Hawker Siddeley P 1127 "Harrier" with the difference that, apart from the combined lift-cruise engine Bristol "Pegasus", additional lift engines in the fuselage were intended. One wanted in this way to improve the range and the payload. The "Pegasus" engine built in quantity production, however, was too large. During cruise flight – for example for the reconnaissance mission – the ensuing engine throttling would have entailed a too high fuel consumption.

In the middle of the year 1965 the Vereinigte Flugtechnische Werke (VFW) in Bremen obtained the development order for the VAK 191 B [124]. In advance one already had contacted the engine companies Rolls-Royce Bristol and MAN-Turbomotoren (later MTU). The choice fell on the quantity production lift-engine Rolls-Royce RB.162-81 with 26.5 kN (5,957 lbf) thrust and a new to be developed lift-cruise engine Rolls-Royce/MAN RB.193-12 [1] with 45.2 kN (10,161 lbf) take-off thrust. This was to be a twin-spool bypass engine with four swivelling propulsion nozzles of which two were "cold" ones behind the compressor and the other two were "hot" ones behind the turbine. All four nozzles should have been able to rotate between −5° and 100° in 1 sec.

After the definition phase, Italy with the company Fiat withdraw from the development project VAK 191. The program was reduced to three single-seat test aircraft and one crash airframe. In contrast to the until now jointly performed engine development with Rolls-Royce, for the RB.193-12 engine the MAN development team [48] was fully integrated, that is *Norbert Scholz* was responsible for the calculations, *Heinrich Leibach* for the project and construction and *Karl Prestel* for the testing. The German development and production portion included among others the front distributor casing of fibre reinforced plastics, the rear trousers pipe, the swivelling nozzles with bearing and actuating mechanism as well as further casings and their internal parts. From the MAN team were designed, constructed and tested the low-pressure compressor and a test version of the high-pressure compressor. A Rolls-Royce "Avon" engine served

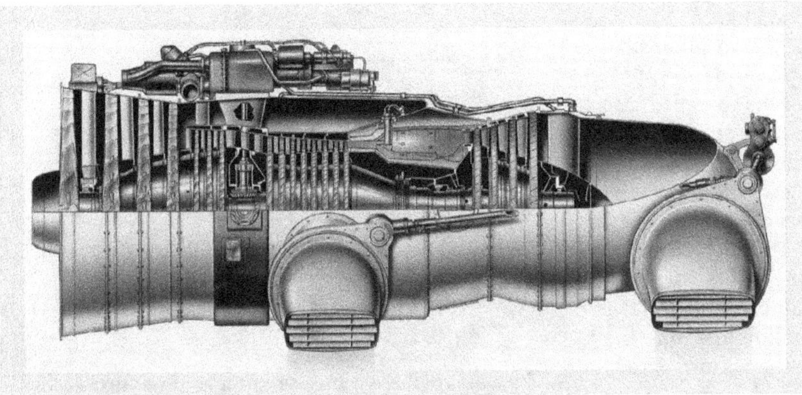

For the vertical take-off reconnaissance and fighter aircraft VAK 191 of the "Vereinigte Flugtechnische Werke, VFW" in Bremen, MAN-Turbomotoren and Rolls-Royce developed the two-spool bypass lift-cruise engine RB.193-12 with four swivelling propulsion nozzles – a derivative of the "Pegasus" engine of the English vertical take-off aircraft "Harrier".

in Allach as hot gas supplier for testing the novel swivelling nozzles and their actuating elements. The test-bed runs were performed to one half at Rolls-Royce in Derby and to the other half at MAN in Allach.

Overlapping of work was avoided with the exception of some comparative tests on the somewhat different test beds. At the end of the testing a 50-hour endurance run with nearly 500 swivelling operations of the nozzles took place. At the beginning of 1967 at MAN-Turbomotoren the proportionate manufacture of the development engines started, with high requirements concerning the forming and welding of thin high-temperature resistant metal sheets – partly of titanium alloy – having to be complied with. For this, modern facilities were introduced for soldering in a high vacuum and for electron-beam welding. Altogether six development and seven flight engines were manufactured, of which the last was delivered in 1971. The first flight of the VAK 191 B took place in Bremen with a hovering flight on September 10, 1971. The subsequent testing was then moved to Manching. The whole program was discontinued in March 1973. It had brought MAN-Turbomotoren (since 1969 "Motoren- und Turbinen Union Munich, MTU") a considerable increase in experience for the fabrication of highly loaded engine construction components, especially for low pressure turbines, which later was of high value for the development of the "Tornado" engine RB.199. Project work on engine versions of the RB.193-12 was undertaken thereafter still in connection with other German military aircraft projects like Dornier D 32 E and the "New Fighter Aircraft" ("Neues Kampfflugzeug, NKF").

Of the vertical take-off reconnaissance and fighter aircraft VFW VAK 191 B three prototypes were built, of which the first began with the hover flight testing in September 1971. Installed were one RB.193-12 and two RB.16 as lift engines in the fuselage.

University Project of a VTOL "Flat Engine"

The BMVg was in the years 1965–1970 interested in new technological concepts for powerplants of VTOL aircraft. Within the framework of the so-called "Mack program", research and study tasks were placed with universities and research establishments. Several interesting and novel propulsion technologies were investigated by *Hans Georg Münzberg* and one of his co-workers, *Dieter Rist*, at the Technical University Munich. Especially a "flat engine" ("Flachtriebwerk, FLT") [125] was extensively examined. The "flat engine" was to have a low construction height to make the installation in the wing or in the bottom of the fuselage of aircraft possible.

A FLT single-circuit engine achieves a very flat configuration by use of a return-flow combustion chamber. For the power transmission the rotors of the compressor and of the turbine are not, as is customary, connected by a shaft but by an integral construction. Air intake and propulsive nozzle lie at the engine lower side. This construction allows with pro-rotating and counter-rotating rotors in single-circuit and two-circuit design for various concepts to realise different versions depending on the application and the installation constraints. Thus, for example, a projected two-circuit (bypass) version (FLT fan) had a diameter of 1,8 m (5.9 ft) and a construction height of 0.8 m (2.6 ft) for a take-off thrust of about 70–80 kN (15,736–17,984 lbf). By *Dieter Rist* a version of a FLT drop engine [126] was investigated which after the take-off of the aircraft could be separated from this and was to land autonomously.

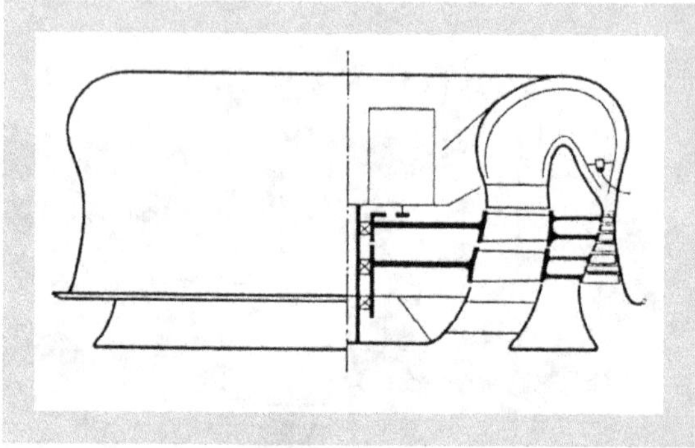

For a flat engine version also a drop unit was investigated, which, after the vertical take-off of the aircraft, would have been dropped. It was then to land on short landing legs on the ground and was to be used again.

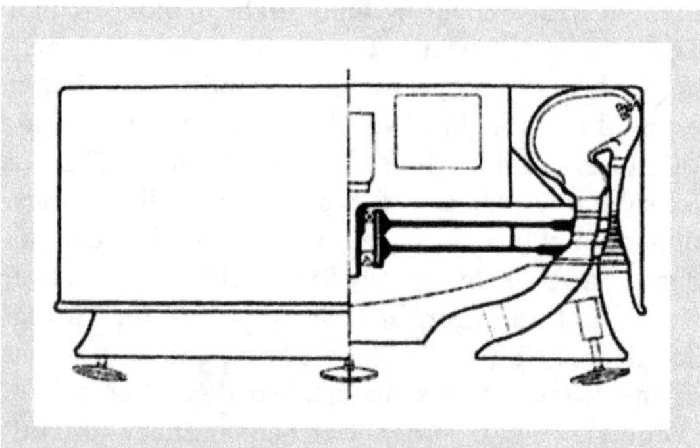

At the Technical University Munich new technological concepts of vertical take-off aircraft engines were investigated in research and study work including also "flat engines" with return-flow combustion chambers in single-circuit and two-circuit design.

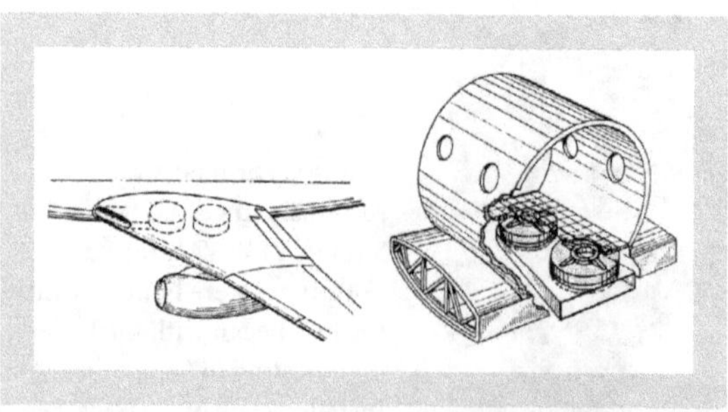

The investigated flat engines could have been installed in wing roots or in the fuselage floor.

Fuel and Lubricant Research

EILHARD JANTZEN

The Period of 1954 until 1990

In April 1954 the work concerning the fuel and lubricant research could be resumed after the Adlershof institute of the DVL, now in Munich, was founded again with the new name "Institut für Flugtreib- und Schmierstoffe" ("Institute for Aviation Fuels and Lubricants"). The leadership had been assumed by the known mineral oil scientist *Günter Spengler*. After equipping the laboratories and test stands, first of all the main task was to catch up with the actual developments in research in foreign countries. In this connection the general reform compared with the Adlershof institute was the thematic change of fuels and lubricants for piston engines to those for turbine engines. The under this aspect commenced fields of work were initially aviation fuels and engine oils. Later added were aviation hydraulic fluids and questions concerning the condition monitoring of aeroengines based on the oils. After the death of *Spengler* in 1973, the work at the DVL was first of all continued under the leadership of *Hans Gemperlein* and since 1979 of *Eilhard Jantzen*. At the end of 1978 the transfer of this work field, with the exception of the fuels, to the Stuttgart research centre of the DFVLR took place.

Since 1970 also further institutions in Germany, which were concentrated in the Munich area, began to occupy themselves with research in partial fields of fuels and lubricants. To be mentioned are the firm MTU (*Kurt Maier*), the "Hochschule der Bundeswehr" in Neubiberg (*Alfred Zeman*) and the "Wehrwissenschaftliche Institut der Bundeswehr" in Erding (*Peter Bartl*).

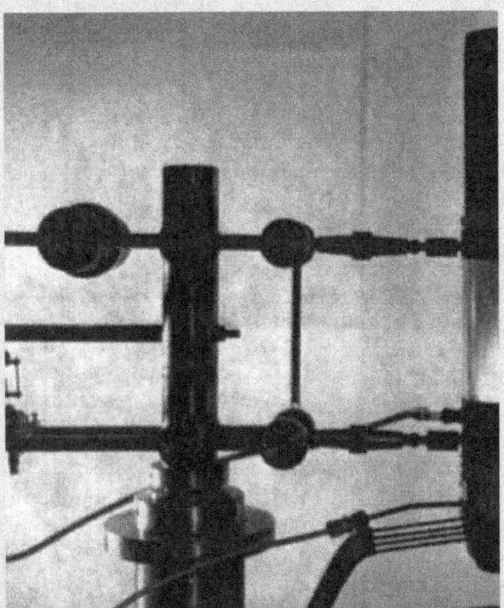

Small combustion chamber of the DVL for fuel and combustion research.

Günter Spengler (1912–1973).

Aviation Fuels

In this field as first research focus at the DVL institute in Munich the combustion behaviour in dependence on the composition and chemical constitution was selected. Since for such formulation of questions no equipment was on the market, first of all a suitable test stand on the basis of a through flow combustion chamber was developed. Important for this was a small sizing with regard to the in part small quantities of fuels for synthesis. The first facility for standard conditions could be erected in 1955/56 [127]. A short time later this was extended by a fuel and air heater. Since 1960 then a facility for below atmospheric pressure conditions was built. To be able to include the behaviour of the fuels under supersonic conditions in the research, in 1966 a facility for pressures up to 20 bar (300 lb/in^2) and temperatures up to 700 °C (973 K) was developed.

With these facilities in the course of years fuels and synthesis products were investigated concerning their combustion behaviour in dependence of composition in wide temperature and pressure ranges [128, 129]. Furthermore, with the aid of the emissions spectroscopy, research work on the combustion process and the in this involved chemical reactions were carried out [130].

Apart from the above work, at the end of the sixties also the field of the supersonic fuels was tackled and here the special topics of required properties, of the synthesis of possible substances and of their op-

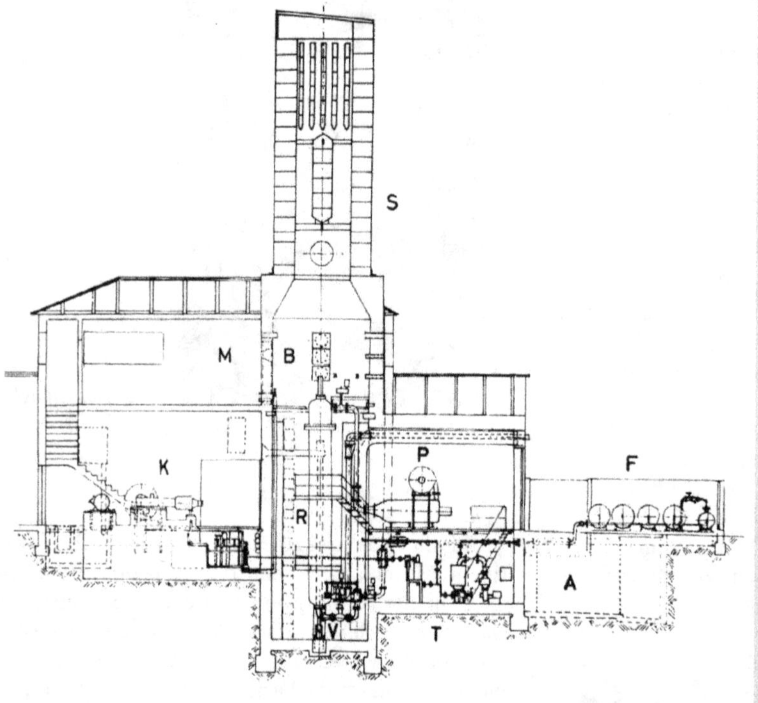

Section of the high pressure – high temperature combustion chamber test facility: (Excess pressure combustion section based on the blow-down principle developed by the "DVL-Institut für Angewandte Gasdynamik" for intermittent operation).
A = Inflow silencer
B = Combustion section
F = Storage bottle row
K = Compressor
M = Observation and control room
P = Propane burner
R = Regenerator
S = Exhaust gas silencer
T = Air drying facility
V = Quick-acting control valve

timisation elaborated [131]. New work in the eighties was then on the influence and possible side effects of additions as well as the friction and wear behaviour of fuels in fuel pumps, their causes and influencing. Apart from this, the adaptation and the use of modern analytic methods for the elucidation of the chemical structure of the substances contained in the new fuels and synthesis products played a growing role.

After the dissolution of the DFVLR Institute of Fuels and Lubricants in the year 1978 and of the transfer of the Department of Fuels to the "Wehrwissenschaftliches Institut für Materialforschung der Bundeswehr" a part of the addressed work was continued there.

Engine Oils

Also for these in the re-established DVL institute the fields of work had to be adjusted to the increasingly employed turbine powerplants and essentially synthetic oils. With this background initially two research foci were regarded as being urgent. In this connection on the one hand it concerned the oil ageing, its causes, simulation and analytics and on the other hand the formation of deposits in the oil system of aeroengines, their emergence causes and the influence of different oils. In both fields there existed, especially for the temperature range under discussion, neither experience nor equipment so that first of all relevant apparatus had to be developed and tested. As especially important requirements were considered: Small quantities of oil, a wide temperature range, addition of defined, moisture controlled air quantities as well as different metals. Also considerable effort had to be expended in order to achieve the high temperature constancy and repeatability for long duration tests. Thus, a first facility for static oil conditions was developed [132]. With a second facility for the first time and for small oil quantities extensively the conditions of the oil circulation in the aeroengine were to be simulated [133]. With these facilities as basis the behaviour of customary turbine oils and later of a large number

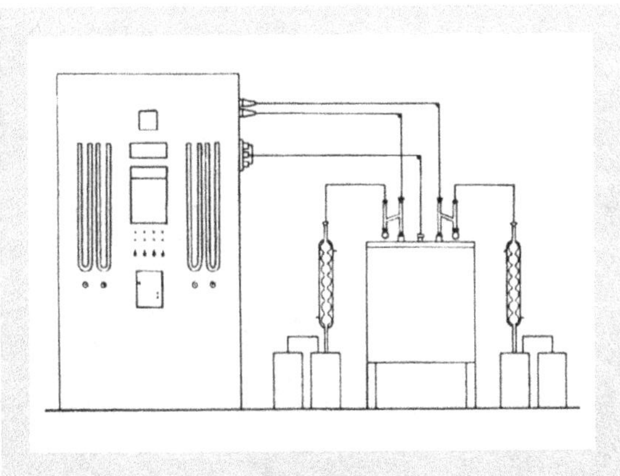

Facility for oils ageing tests under static conditions.

of synthetic basis oils of different structure and the influence of ageing inhibitors were tested [134]. For this also to a large extent new substances had been synthesised [135, 136]. In the eighties then work was started on the high temperature behaviour of oils on the basis of perfluoralcylethers [137]. Also here topics of the influence of the composition and the chemical structure stood in the foreground. This class of substances is of interest for high temperature engines, as, for example, for those that are in the planning for the extended supersonic regime.

Oil circuit simulation facility of the DVL-Institute for Aviation Fuels and Lubricants.

Also for the topic of deposits on hot surfaces initially relevant test equipment had to be developed and tested. To make matters worse, the phenomena and conditions occurring in the engine were only inadequately known. Thus, at the DVL in the course of the years three pieces of equipment were developed which made allowance for the different conditions in aeroengines: Firstly, the exploration of deposit formation owing to droplet impingement, secondly, for a stagnant oil film and, thirdly, for the oil vapour phase. Apart from general questions concerning the prerequisites for the deposit formation, in the course of time a multitude of aircraft turbine oils, novel basis oils and additives were investigated regarding their behaviour [138, 139]. Especially in this field the MTU performed own work orientated on practical aspects and had in this connection an exchange of experience and cooperation lasting over many years with the relevant field of work of the DVL [140].

A further focus was the friction and wear behaviour (tribology) in dependence on the chemical structure of the synthetic compounds as well as materials [141]. Apart from newly developed equipment, for this at the DVL in Munich also an isotope laboratory was equipped in order to be able to measure the smallest quantities of wear on roller bearings that had been made radioactive [142]. In the eighties the in this connection important topic of the reaction layer formation in the friction contact was then included. With this it was possible to gain important knowledge concerning the build-up as well as removal of such protective layers due to measurements of the electric transition resistance in the friction contact as well as with the aid of the XPS surface spectroscopy [143].

Since 1975 the above mentioned laboratories of the Bundeswehr dealt above all with the exploration of the structure and composition of aircraft turbine oils on the basis of ester making use of the most modern methods like IR and mass spectroscopy, chromatography and differential thermoanalysis [144, 145]. Within this framework also falls work which had at its aim the clarification and control of the ageing

Equipment for investigating the formation of deposits from the droplet phase of aircraft turbine oils (DVL-Institute Munich).

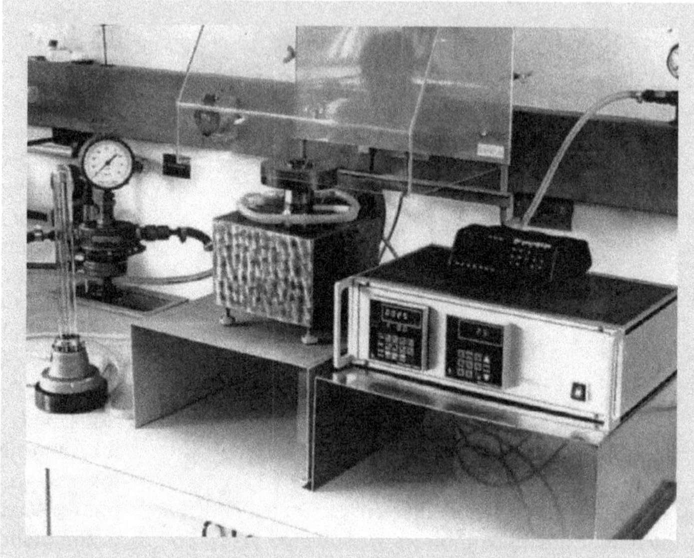

Equipment for investigating the deposit formation under static conditions for oil films (DVL-Institute Munich).

and polymer conditions respectively of aviation oils [146, 147].

Hydraulic Fluids

Cause for dealing with this field was work in the USA on synthetic hydrocarbons as possible future, fire-resistant aviation hydraulic oils. To gather also in Germany experience in this field, the DVL took up partial fields of special interest like temperature stability, friction and wear behaviour, the self-ignition behaviour on hot surfaces as well as fire spreading tendency. For these new tasks a number of new test equipment had to be developed and built. Because in tribology the adjustment to the conditions existing in practice is especially important, for this a facility was developed in which the conditions occurring in an aircraft hydraulic pump could be simulated.

For the exploration of the self-ignition initially an apparatus was developed in which the ignition start in dependence on time and thus the question of the ignition delay could be investigated. A second apparatus was developed in order to simulate the ignition of a fine liquid jet when impinging on a hot surface. A number of fluids of different chemical structures could be examined regarding their behaviour using this equipment [148]. Furthermore, a piece of equipment was built with which questions of the continuing burning behaviour of hydraulic fluids on cable insulation could be elaborated.

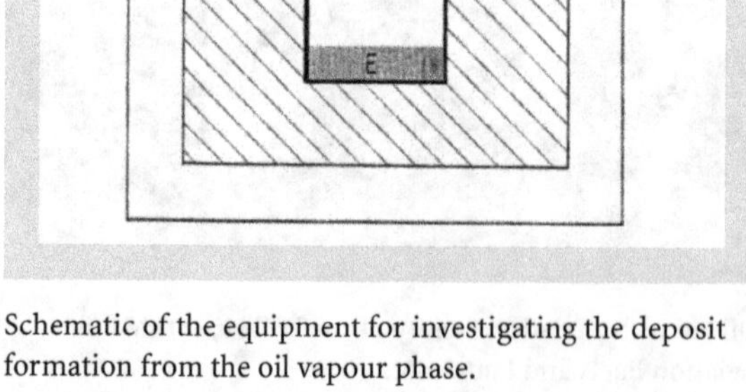

Schematic of the equipment for investigating the deposit formation from the oil vapour phase.

Particles from roller bearings, wear of an aeroengine.

Oils and condition monitoring

At the beginning of the seventies the DVL institute in Munich also took up first work in this field after already in the USA experience was available for some years. It could also be expected that relevant knowledge and methods could become of use in Germany for the engine monitoring. First of all one dealt with research work for the quantitative determination of smallest amounts of particles due to wear by making use of emission and atom absorption spectroscopy. With this, such important questions like that of the influence of the oil ageing condition and the size of the wear particles on the result could be clarified [149].

As additional field, complementary to the spectroscopy, for the condition monitoring of moving elements in engine oil systems at the DVL the particle analytics was taken up [150]. For this, work was initiated in order to record and examine the particles present in the engine oil in the laboratory. In this way detailed knowledge could be obtained concerning the size, the size distribution and the composition. Furthermore, till into the nineties tests for the recording and analytics of solid foreign material in oil was performed. In this connection the determination of size, shape, composition and amount was of special interest. This work had been proposed by roller bearing and engine manufacturers who had discovered a partly serious shortening of the operating life of roller bearings owing to solid foreign materials.

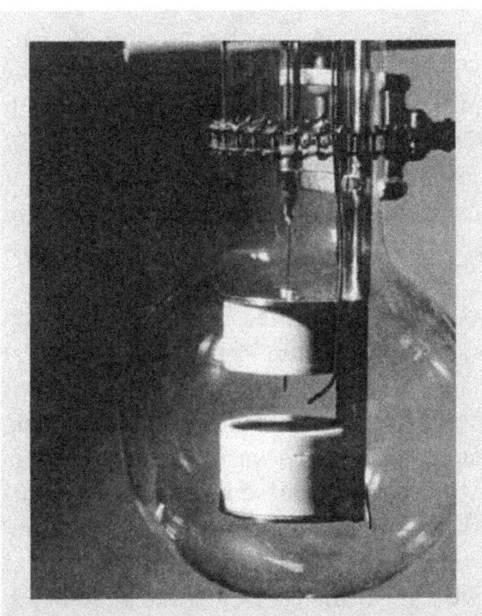

Equipment for droplet ignition (Institute Stuttgart).

Equipment for investigating the ignition of liquid jets on hot surfaces (Institute Stuttgart).

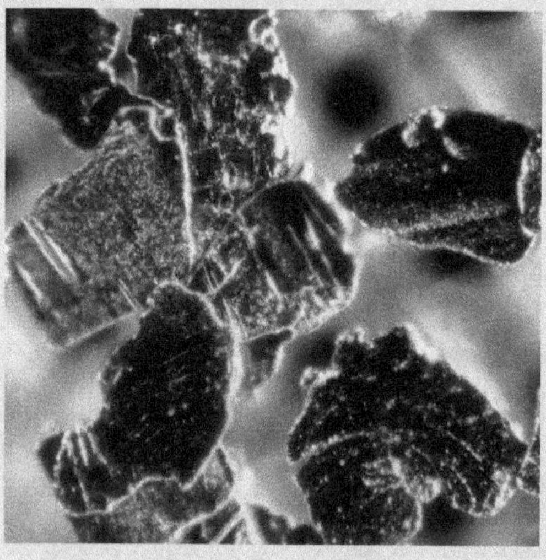

Particles from gear damage.

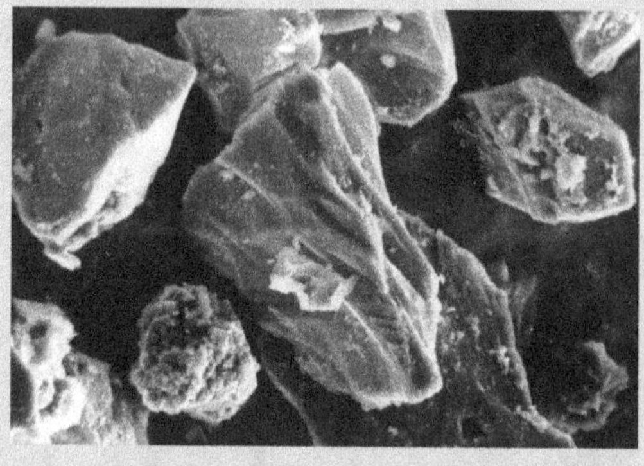

Solid mineral foreign materials in used engine oil after laboratory separation and screen electron microscopy.

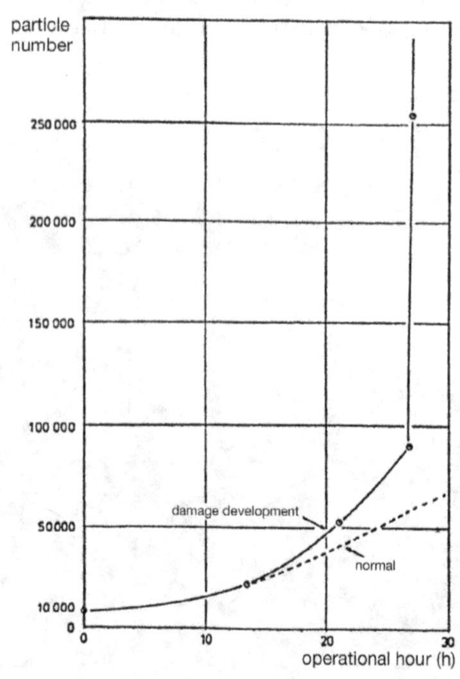

Increase of particle numbers in the range 10–15 µm.

The Stuttgart Altitude Test Bed

WOLFGANG BRAIG

When in 1955 the ban of activities in the field of aviation in Germany was lifted, also at the Technological University Stuttgart the teaching in this field began. Lectures on turbo-jet engines were offered by *Ulrich Senger*, who was since 1948 Professor of "Thermal Turbomachinery" and who knew the jet engines from his former activity as Director of the BBC Turbine Company in Mannheim in connection with the construction of the "Herbitus" altitude test-bed facility. *Ulrich Senger* founded parallel to the "Institute of Thermal Turbomachinery" the "Institute of Aircraft Turbo-Jet Engines" and showed a large commitment for good experimental equipment for the new institute.

His special commitment aimed at the erection of an altitude test bed for a maximum of 80 kg/s (176 lb/s) air mass-flow rate. The test bed, from which *Ulrich Senger* hoped to get a special stimulus for teaching and research at the university, was to be erected parallel to the at that time discussed large national altitude test bed. Favourable for the project was that the State Government of Baden-Württemberg at that time showed great interest in aviation research and the aviation industry, and supported generously the strengthening of the aviation facilities at the TH Stuttgart. An important role concerning the aeronautical activities of the state played the interministry committee which had been set up under the leadership of *Karl Hipp* for the support of aeronautics and the co-ordination of the extension of the aeronautical institutes at the new university site Stuttgart-Pfaffenwald. Members of the committee were the Ministries of Finance, of the Interior, of Education and Arts as well as of Trade and Commerce.

For the planning of the Stuttgart altitude test bed *Ulrich Senger* in 1959 set up a team under the leadership of *Carlotto Martin*, which designed and determined the detailed implementation of the new facility with the support of industry. The fact that numerous characteristic features of the "Herbitus" facility were adopted is not surprising in view of the participation of *Senger* in the erection of that facility.

The design of the Stuttgart altitude test bed took place with the priority for the testing of turbo-jet engines up to an altitude of 16,000 m (52,480 ft) at moderate flight Mach numbers for which the total pressure in front of the compressor is less than 1 bar (15 lb/in^2). On the air entry side the air initially flows through a pre-cooler with cooling brine and is then routed either directly or via cooling turbines, or an air heater with hot water, to the test chamber. By equal setting of the control valves on the entry side the pressure upstream of the engine can be adjusted, while by a different valve setting for the various parallel running entry ducts the temperature can be changed. Via an in the partition of the test chamber axially slideable and to the engine fixed bellmouth entry, the air is routed to the engine mounted on the thrust balance. As in the "Herbitus" test bed, the thrust balance with the engine can be wheeled out of the chamber after dismantling of the exhaust gas duct and detaching the rear test chamber flange.

The exhaust gases are routed through a double-walled, water-cooled pipe to the exhaust gas cooler and from there to the three exhaust gas compressors. These allow in series connection a minimum suction pressure of 0.1 bar (1.5 lb/in^2) corresponding to an altitude of 16,000 m (52,480 ft). By additional switching possibilities two of the three compressors can alternatively be used to increase the pressure of the entry air, in which case also flight conditions at higher flight Mach numbers at medium altitudes can be simulated. Further set-up and switching possibilities allow in addition the testing of propeller-turbine engines and engine components, especially of compressors, in which case a 3.5 MW (4,762 hp) pendulum engine can be used.

The general aim of the design of the Stuttgart test bed was for a cost-effective facility. For this served the application of the storage principle for the cooling water and the cooling brine as well as the use of the exhaust gas heat of the test engine for heating the entry air. In addition the nearby power station of the university could be used for supplying the electrical power, the re-cooling of the cooling water and the provision of the steam for the hot water generation.

The Stuttgart altitude test bed was erected from 1961–1965 on the basis of the original concept, with the compressor as key unit being supplied by the company BBC [151,152] as for the earlier altitude test beds. The financing of this facility was done exclusively with means of the State of Baden-Württemberg. Characteristic for the Stuttgart altitude test bed in comparison with other facilities with the same task is the compact layout in a closed building. Only in this way the operation of such a facility in a closely populated area is at all possible. The compact layout imposed special requirements regarding the closely sized implementation of the machinery and apparatus as well as the layout of the connecting pipelines. On the other hand the installation in a closed heated hall offers substantial advantages concerning corrosion and maintenance effort, while the small volumes of the facility components are advantageous for a simulation of transient processes.

The commissioning of the facility took place in 1965 and 1966 with a turbo-jet engine Bristol "Orpheus". Further tests took place in 1967 with the propeller-turbine engine DB 720 of the Daimler-Benz engine development. In the year 1971 began the co-operation with the company MTU in Munich-Allach. This was decisive for the operation of the Stuttgart altitude test bed and its further extension. First engine tests were performed with the "Orpheus" engine for testing of a new electronic control system. This was followed until 1976 by compressor tests with the intermediate and high-pressure compressor of the "Tornado" engine RB.199 in which case especially the investigation of the influences of Reynolds number and radial gap as well as of intake disturbances played an important role.

In 1977 for the first time low-pressure turbine tests were carried out, initially for the engine Pratt & Whitney JT 10 D, a predecessor version of the engine

Test bed hall of the Stuttgart altitude test bed in the original state (1966) with the compressor rows in the foreground, the fresh-air cooler and fresh-air pipe in the middle top of the figure as well as the test chamber at the top left.

PW 2037, in which MTU was involved. For this project nearly to 2.8 bar (42 lb/in²) compressed entry air was needed. In addition at the turbine exit a low pressure of 0.5 bar (7.5 lb/in²) was required. For this a General Electric engine J-79 was installed and used as suction machinery. Already since 1979 on the Stuttgart altitude test bed trials were conducted with the complete engine RB.199. With the at that time available test bed equipment, however, only a small part of the flight envelope of this engine could be covered. Especially the test bed mass-flow rates at ram condition were too small so that only flight cases with total pressures in front of the engine compressor lying below 1 bar (15 lb/in²) could be explored.

For the for 1982 planned investigations of the low pressure turbine of the engine PW 2037 neither the ram nor the suction capacity of the existing facility components sufficed. With financing by the MTU therefore, in order to increase the ram capacity, a new air heater as well as a new powerplant with a power raised to 9 MW (12,245 hp) for the largest available compressor at that time was installed, while the suction capacity was doubled by two J-79 engines operating in parallel. With this layout especially extensive performance map measurements as well as detailed flow measurements on individual blade cascades were performed. A disadvantage, however, was that, with the two J-79 engines as suction sources, the pressure level was fixed and could not be varied for the investigation of the Reynolds number influence. With the increased ram capacity, also for the RB.199 tests a larger domain of the flight envelope could be covered. Inadequate for the tests with this engine however remained the suction capacity. It was hardly sufficient for the critical propulsion nozzle flow and was completely inadequate for producing the actual altitude pressure for performance tests. Furthermore, tests with afterburning were not possible with the existing exhaust gas system.

With this background it came to a large extension of the Stuttgart altitude test bed in the years 1984–1987 [153]. For this the realisation was decisive that, also for the current development programs undertaken in international co-operation, the German industry must possess testing possibilities in the own country. Aim of the extension was to create for the current and foreseeable testing tasks an adequate test-bed capacity in Germany after the originally planned national

Test bed hall in the first extension stage (1982) for the tests with a PW 2037 low pressure turbine. The air flows from bottom right to the turbine, then into the finned exhaust air collector and from there towards the left in order to be distributed to the supply pipelines of the two J-79 engines serving as suction sources (middle of figure top).

altitude test bed had not been realised. Different from the erection, the financing for the extension was now covered by the Federal Government and the State with a significant contribution by MTU.

Two for parallel operation designed axial compressors with a pressure ratio of 3 as well as a suction volume-flow rate of 80 m³/s (2,823 ft³/s) each can now be used for increasing the ram mass-flow rate but also the suction volume-flow rate and the altitude. Engines with a mass flow rate of 80 kg/s (176.4 lb/s) stationary on the ground can thus be tested on up to M = 1.2 and up to M = 2.2 at an altitude of 11,000 m (36,080 ft),

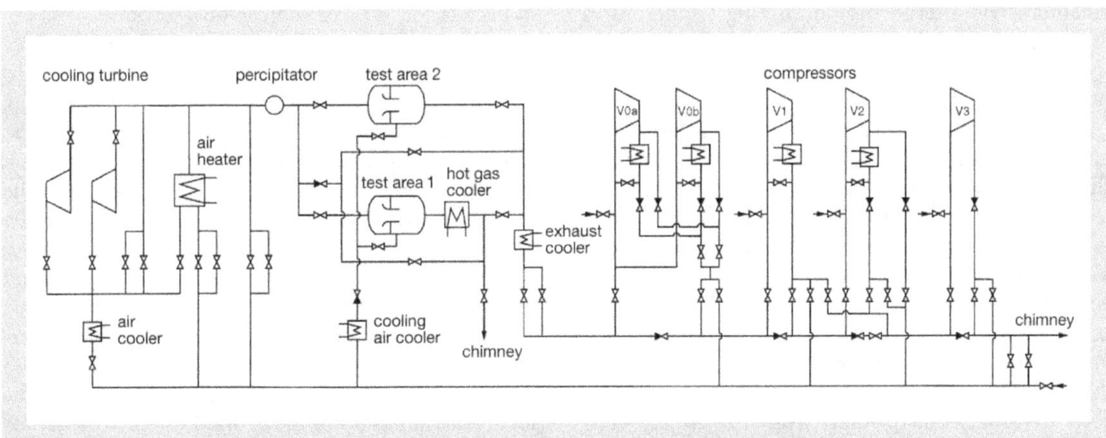

Circuit schematic after the large modification in the years 1984–1987. There are now two test chambers available, which allow a test operation with fewer interruptions. Test Chamber 1 is the in 1987 newly built chamber with large laterally arranged and hydraulically operated opening flaps, while Test Chamber 2 is the original test chamber with opening by unscrewing the exit flange.

status	test items	time	area of investigation
1965	Bristol Orpheus	1965–1966	start of operations 1965
	DB 720	1967	power measurements
	Bristol Orpheus	1971	govenor tests
	RB 199 HD-Compressor	1971–1976	performance map, Reynolds number, radial gap
	RB 199 MD-Compressor	1973–1975	performance map
	JT10D ND-Turbine	1977–1978	performance map
	RB 199	since 1979	system tests
1982	PW 2037 ND-Turbine	since 1982	performance map, blade optimization
	RB 199	since 1982	systems performance in flight range tests
1987	RB 199	since 1987	power and systems performance including after burner operation
	EJ 200	since 1989	power and systems performance including after burner operation
	PW 300	1989–1992	power, emission tests
	PW 4084 ND-Turbine	since 1994	performance map
	BR 700 ND-Turbine	since 1995	performance map
	BR 700 Kerntriebwerk	since 1997	stationary and transient performance

Essential tests carried out in the Stuttgart altitude test bed facility between 1965 and 1997.

while the altitude which can be simulated increases to 20,000 m (65,600 ft). For turbine tests the new compressors replaced the J-79 engines as suction source and allow for the operation in a closed circuit a variable pressure level and thereby a variation of the Reynolds number. A directly behind the new test chamber situated cooler is designed for the at stoichiometric afterburning occurring temperatures and a maximum to be drawn-off heat flow of 320 MW (435,374 hp) without requiring a reduction of the heat loads by water injection.

With the extended facility since 1987 tests were conducted with the RB.199 engine for the complete flight envelope under simulation of extreme temperatures. Since 1989 in Stuttgart the "Eurofighter" engine EJ 200 was tested, which was there subjected to its first altitude tests and since then has passed through an extensive test program. Furthermore, tests were carried out with the engine PW 300, for which emission measurements within the framework of the BMFT program "Pollutants in Aviation" were performed. Apart from the engine testing, with the extended facility test programs with low pressure turbines for the engines PW 2037, V 2500 and PW 300 were conducted, for which, making use of the new possibilities, especially the influence of the Reynolds number and thus the different flow behaviour on the ground and at altitude were determined. While the Stuttgart facility is ruled out for the full testing of large bypass engines with their high mass-flow rates in the fan, tests with core components of large bypass engines can be performed. Thus, the low-pressure turbine of the large engine PW 4084 was tested, for which however the last stage could not be investigated because of its large volume-flow rates.

Apart from the tests by contract of the MTU, since 1995 at first low-pressure turbines and later also high-pressure turbines of Rolls-Royce Germany were tested. Novel requirements were presented by tests with the core engine of the BR 700 engine family [50], for which the test bed must not produce the ram condition at the entry of the turbo-engine but the condition behind the low-pressure compressor and the booster, respectively. Especially for the simulation of the acceleration and deceleration operations, the test bed must change the pressure and the temperature in the same short interval of time as is the case for the full engine. Owing to the compact design of the Stuttgart test bed and the short reaction times of the hydraulic actuators, the for these projects installed test-bed control allows the required quick pressure changes to be safely run.

The core engine BR 700 of the BMW Rolls-Royce turbo-jet engine in the Test Chamber 1.

References

1 K. von Gersdorff, K. Grasmann, H. Schubert, Flugmotoren und Strahltriebwerke. Reihe "Die deutsche Luftfahrt", Vol. 2, 3. Ed., Bernard & Graefe, Bonn, 1995

2 50 Jahre Turbostrahlflug - DGLR-Symposium 1989 Munich, DGLR-Report 89-05 and 92-05

3 B. Eckert, E. Schnell, Axialkompressoren und Radialkompressoren. Springer Verlag, 1. Ed. 1953

4 K. Grahl, Teillastberechnung für Axialverdichterstufen. Zeitschrift für Flugwissenschaften 20, Heft 1/2, 1972

5 H. Hoheisel, Über den Einfluss von Belastung, Lieferzahl und Reaktionsgrad auf das Abreißverhalten von Axialverdichterstufen. DLR FB 69-17, 1969

6 J. Wilke, H.P. Kau, A Numerical Investigation of the Influence of Casing Treatment on the Leakage Flow in an HPC Front Stage. ASME Turbo Expo Amsterdam, Paper GT-2002-30642, 2002

7 R. Müller, Luftstrahltriebwerke, Vieweg Verlag Braunschweig/Wiesbaden, 1997

8 B. Eckert, W. Encke, H. Hagen, H. Marcinowski, F. Weinig, Berichte der Sitzung „Verdichter für L-Triebwerke" am 7./8.10.1943, Bericht 171 der Lilienthal-Gesellschaft

9 W.J.G. Bräunling, Flugzeugtriebwerke, Springer Verlag Berlin, 2001

10 C. Keller, Axialgebläse vom Standpunkt der Tragflügeltheorie - Dissertation ETH Zürich, 1934

11 A. Betz, Diagramme zur Berechnung von Flügelgittern. Ing.-Archiv No. 2, p. 359, 1931

12 F. Weinig, Die Strömung um die Schaufeln von Turbomaschinen, J.A. Barth Leipzig, 1935

13 H. Hausenblas, Axialverdichterberechnung in Großbritannien und den USA. Konstruktion 4, Issue 6, 1952

14 E. Schmidt, Berichte der Gittertagung in Braunschweig am 27./28.3.1944

15 H. Schlichting, Ergebnisse und Probleme von Gitteruntersuchungen. Zeitschrift für Flugwissenschaften, Issue 5, 1953

16 N. Scholz, U. Hopkes, Der Hochgeschwindigkeits-Gitterwindkanal der DFL Braunschweig. Forschung-Ingenieur-Wesen, Vol. 25, 1959

17 H. Hoheisel, R. Kiock, Zwanzig Jahre Hochgeschwindigkeits-Gitterwindkanal des Instituts für Aerodynamik der DFVLR Braunschweig, Zeitschrift Flugwissenschaften Weltraumforschung, Issue 1, 1977

18 U. Stark, H. Hoheisel, The Combined Effect of Axial Velocity Density Ratio and Aspect Ratio on Compressor Cascade Performance, Journal of Engineering for Power, Trans. of the ASME, Vol. 103, January 1981

19 R. Kiock, H. Hoheisel, H.J. Dietrichs, A.T. Holmes, The Boundary Layer Behaviour of an Advanced Gas Turbine Rotor Blade under the Influence of Simulated Film Cooling. AGARDOgraph CP 390, 1985

20 W. Sturm, L. Fottner, The High-Speed Cascade Wind Tunnel of the German Armed Forces University. Munich, 8th Symposium on Measuring Techniques for Transonic and Supersonic Flows in Cascades and Turbomachines, Genua, October 1985

21 L. Fottner, P. Müller, St. Brunner, Turbomachinery Research at the Jet Propulsion Institute of the University of the Federal Armed Forces, Hochschulkurier, Universität der Bundeswehr, No. 7, January 2000

22 E. Lindner, Hochgeschwindigkeitskanal für Gitteruntersuchungen. Maschinenbautechnik 8, Issue 10, 1959

23 E. Lindner, L. Langer, A. Bahr, Die Strömungseigenschaften von geraden Schaufelgittern in kompressibler Strömung. Die Technik, Issue 6/7/8, 1969

24 W. Heilmann, Experimentelle und grenzschichttheoretische Untersuchungen an ebenen Verzögerungsgittern bei kompressibler Strömung, insbesondere bei Änderung des axialen Stromdichteverhältnis und der Zuströmturbulenz. DLR-FB 67-88, 1967

25 H.J. Lichtfuss, H. Starken, Supersonic Cascade Flow, Progress in Aerospace Science, Vol. 5, Pergamon Press, New York, 1974

26 R. Dunker, H. Rechter, H. Starken, H. Weyer, Redesign and Performance Analysis of a Transonic Axial Compressor Stator and Equivalent Plane Cascade with Subsonic Controlled Diffusion Blades. Journ. of Eng. for Gas Turbines and Power, Trans. ASME 106, 1984

27 H.J. Heinemann, The Test Facility for Rotating Annular Cascades of the DFVLR and its Measurement and Evaluation Method. ICIASF-79 Record, IEEE publication, New York, 1979

28 F. Lehthaus, O. Lawaczek, Untersuchungen über die transsonische Strömung durch Turbinenschaufelgitter, VDI-Forschungsheft 586, VDI-Verlag 1978

29 N. Scholz, Aerodynamik der Schaufelgitter. Verlag G. Braun, 1965

30 H. Starken, Design Criteria for Optimal Blading Design. In: Blading Design for Axial Turbomachines, AGARD Lecture Series No. 167, 1989

31 L. Fottner (Ed), Test Cases for Computation on Internal Flows in Aero Engine Components, AGARD-AR 275, 1989

32 L. Fottner, Review on Turbomachinery Blading Design Problems, AGARD Lecture Series No. 167, 1989

33 J. Michels, J. Werner, Die deutsche Luftfahrt, Vol. 22, Luftfahrt Ost 1945–1990, Bernard & Graefe Verlag, Bonn, 1994

34 H. Oestrich, Aus der Entwicklung des Strahltriebwerkes ATAR 101, Jahrbuch der WGL, 1954

35 A. Franz, Zur Geschichte des Strahltriebwerkes – Die Entwicklung bei Junkers und seither erzielter Fortschritt. DGLR Yearbook Vol. 4, 1981

36 F. Brandner, Ein Leben zwischen Fronten, Verlag Welsermühl, München, 3. Ed., 1987, see also "Meine Tätigkeit in Ägypten", Fliegerheft 12, 1969

37 H.J. Schröder, Entwicklung eines Näherungsverfahrens zur Berechnung von dreidimensionalen Gitterströmungen. Jahrbuch der WGL, 1955

38 H. Grieb, Zur Berechnung der inkompressiblen, reibungsfreien Strömungen in axialen Turbomaschinen unter Berücksichtigung der Zirkulationsverteilung und der Verdrängungswirkung der Schaufelprofile. Dissertation Technische Universität Stuttgart, 1974

39 K.H. Mewes, Pirna 014, Flugtriebwerke der DDR. AVIATIC Verlag, Oberhaching, 1997

40 H. Grieb, A. Schäffler, Seitenwandgrenzschichten in Axialverdichtern. Vortrag DGLR-Fachausschußsitzung Luftatmende Triebwerke, Friedrichshafen, 7 December, 1972

41 K. Broichhausen, Aero Design of Turbomachinery Components – CFD in Complex Systems. AGARD-LS-195, 1994

42 W. Held, Über die Kennfeldberechnung mehrstufiger Axialverdichter bei hohen Unterschallgeschwindigkeiten und großen Druckverhältnissen. Wiss. Zeitschrift der TH Dresden, 8.years , issue 2, 1958/59

43 H. Volkmann, L. Fottner, N. Scholz, Aerodynamische Entwicklung eines dreistufigen Transsonikgebläses. Zeitschr. Flugwissenschaften 22, Issue 4, 1974
44 A. Schäffler, Methoden zur Auslegung und Entwicklung transsonischer Axialverdichter. MTU-Berichte 82/36, 1982
45 E. Schnell, Probleme des Überschallverdichters. MTZ, Jahrg. 16, No. 2, 1955
46 G. Cordes, Einfluß des Überschall-Axialverdichters auf die TL-Entwicklung. Berichtsband I. Polytechnische Tagung der TH Dresden, 1956
47 M. Hoeger, K.D. Broichhausen, Prediction of 2D Viscous Transonic Flow in Compressor Cascades Using a Semi-Empirical Shock/Boundary Layer Interaction Method. ASME-Paper 92-GT-277, 1992
48 H. Schubert, Deutsche Triebwerke, Flugmotoren und Strahltriebwerke von den Anfängen bis 1999. AVIATIC Verlag Oberhaching, 1999
49 G. Kappler, R. Moor, J. Hourmouziadis, Hochleistungs-Turbo-Fan Triebwerke. Ein Familienkonzept. DGLR-Paper 93-03-043, 1992
50 H. Richter, Die Bedeutung der Phasen I und II des Luftfahrtforschungsprogramms der Bundesregierung für die Rolle der Rolls-Royce Deutschland GmbH in der Rolls-Royce Gruppe und für die Wettbewerbsfähigkeit ihrer Produkte. Lecture, DGLR Annual Meeting 2000
51 K. Rüd, H.J. Lichtfuss, Trends in Aero-Engine Development. In: Aspects of Engine Airframe Integration for Transport Aircraft. Proceedings of the DLR Workshop Brunswick, DLR-Mitteilung (Note) 96-01, 1996
52 H. Körner, Natural Laminar Flow Research for Subsonic Transport Aircraft in the FRG. Zeitschrift Flugwissenschaften Weltraumforschung, 14, 1990
53 K.H. Horstmann, G. Redeker, A. Quast, U. Dressler, H. Bieler, Flight Tests with a Natural Laminar Flow Glove on a Transport Aircraft. AIAA-Paper No. 90-3044, 1990
54 P. Schimming, Der Propfan: Initialzündung für eine neue Triebwerksgeneration. DGLR-Jahrbuch 1985
55 R.D. Hager, D. Vrabel, Advanced Turboprop Project, NASA SP-495, 1988
56 H. Grieb, D. Eckardt, Turbofan and Propfan as Basis for Future Economic Propulsion Concepts. 22nd Joint Propulsion Conference Huntsville, AIAA 86-1474, 1986
57 J. Sieber, Technologieentwicklung für zukünftige ummantelte Propfanantriebe. DGLR-93-03-045, DGLR-Jahrestagung 1993
58 J. Sieber, Aerodynamic Design and Experimental Verification of an Advanced Counter-Rotating Fan for UHB Engines. Contribution to the Third European Propulsion Forum, Paris, France, 13–15 November, 1991
59 K. Heinig, F. Kennepohl, P. Traub, Acoustics of a Counter-Rotating Shrouded Propfan: Prediction and Data. Journal of Aircraft, Vol. 32, No. 5, Sept./Oct. 1995
60 J. Frischbier, S. Sikorsky, CRISP Modellversuch-Entwicklung der CFK/Titan-Hybridschaufeln. DGLR-Annual Meeting 1991
61 H.B. Weyer (Ed), Statusbericht 1993 - 1997 Institut für Antriebstechnik der DLR, August 1997
62 W. Neise, G. Schewe, L. Wallscheid et. al, Experimentelle Untersuchung zur Aeroakustik, Aeroelastik und Aerodynamik am CRISP 1-m Model. Abschlußbericht 1996-99, Förderzeichen LFT-9601, 1999
63 D. Nürnberger, F. Eulitz, S. Schmitt, A. Zachial, Recent Progress in the Numerical Simulation of Unsteady Viscous Multistage Turbomachinery Flow. 15th ISABE Symposium, Bangalore, India, 2001
64 R. Radespiel, K.H. Horstmann, G. Redeker, Feasibility Study on the Design of a Laminar Flow Nacelle, AIAA Paper 89-0640, 1989
65 P.P. Shipley, N.T. Birch; H. Riedel, K.H. Horstmann, P. Lücking, A European Collaborative NLF Nacelle Flight Demonstrator. First European Forum on Laminar Technology, Hamburg 1992
66 H. Riedel, K.H. Horstmann, A. Ronzheimer, M. Sitzmann, Aerodynamic Design of a Natural Laminar Flow Nacelle and the Design Validation by Flight Testing. Aerospace Science and Technology Vol. 2, No. 1, 1998
67 A.J. Mullender, J.L. Lecordix, E. Lecocceris, J.L. Godard, M. Hepperle, The ELFIN 2 and LARA HLF Nacelles: Design, Manufacture and Test. 2nd European Forum on Laminar Flow Technology, Bordeaux 1996
68 K.H. Horstmann, H. Riedel, D.F. Fisher, Flight Test Measurement Techniques for Laminar Flow. AGARD AG-300 Vol. Y, 2003
69 J. Sieber, Das Engine-3E-Technologieprogramm der MTU. DGLR Annual Meeting 1996
70 R. Walter, G. Heider, J. Henne, W. Kiening, W. Schuette, Neue Wege in der Verdichterentwicklung, DGLR Annual Meeting 1997
71 G. Schulze, C. Blaha, D.K. Hennecke, J.M. Henne, The Performance of a new Axial Stage Transonic Compressor. 12th ISABE 1995
72 C. Blaha, D.K. Hennecke, G. Fritsch, M. Hoeger, M. Beversdorff, Laser-2-Focus Measurements in a Transonic Compressor Blisk-Rotor and Comparison with 3D Numerical Simulation. ISABE Paper 97-7069, 1997

73 G. Fritsch, M. Hoeger, C. Blaha, D. Bauer, Viscous 3D Compressor Simulation on Parallel Architectures. AIAA Paper 97-2876, 1997

74 H.-J. Dietrichs, K. Engel, Modellierung von 3D-Strömungen in Axialverdichtern am Beispiel des Triebwerks PW6000. Lecture 13, DGLR Expert Syposium der STAB, TU Munich, 12–14 November, 2002

75 M. Hoeger, M. Lahmer, M. Dupslaff, G. Fritsch, A Correlation for Tip Leakage Blockage in Compressor Blade Passages. ASME Paper 99-GT-388, 1999

76 H. Klinger, Das Engine 3E Technologieprogramm bei Rolls-Royce Deutschland und begleitende Kooperationen mit Forschungseinrichtungen. Lecture, DGLR Annual Meeting 2002

77 J. Hourmouziadis, Aerodynamic Design of Low Pressure Turbines. AGARD Lecture Series No. 167, 1989

78 H. Hoheisel, R. Kiock, H.J. Lichtfuss, L. Fottner, Influence of Freestream Turbulence and Blade Pressure Gradient on Loss Behaviour of Turbine Cascade. Trans. of ASME, Journal of Turbomachinery, Vol. 109, April 1987

79 L. Fottner, H.J. Lichtfuss, Anwendung neuer Entwurfskonzepte auf Profile für axiale Turbomaschinen. Teil B: Entwicklung und Erprobung optimaler Profilformen für Verdichter- und Turbinengitter. MTU-München. Technischer Bericht 77/045, 1977

80 D. Eckardt, K. Trappmann, PW 2037 Entwicklung bei MTU, Auftrieb für die Boeing 757. MTU heute, 1/82, 1982

81 R. Niehus, K. Rüd, Entwicklung schnellaufender Niederdruckturbinen für künftige wirtschaftliche und umweltschonende ADP Triebwerke. DGLR Jahrestagung 1997

82 H. Hoheisel, Aerodynamic Aspects of Engine-Aircraft Integration of Transport Aircraft. Aerospace Science and Technology, Vol. 1, No. 7, 1997

83 H. Hoheisel (Ed.), Aspects of Engine-Airframe Integration of Transport Aircraft. Proceedings of the DLR Workshop Brunswick, DLR-Mitteilung (Note) 96-01, 1996

84 H. Hoheisel, The Design of a Counter Rotating Ultra-High Bypass Fan Simulator for Windtunnel Investigations. DLR-FB 93-20, 1993

85 W. Burgsmüller, H. Hoheisel, J. Kooi, Engine/Airframe Interference on Transport Aircraft with Ducted Propfans – The European Research Program DUPRIN, ICAS-Paper 94-6.2.1, 1994

86 R. John, TPS Units to meet Future Requirements. Simulation of Very High Bypass Ratio Engines. Proceedings of the International Forum of Turbo Powered Simulation. DNW, NL, 16/17 May, 1995

87 H. Hoheisel, Performance Data of a Counter Rotating Ultra High Bypass Engine Simulator. DLR-FB 98-02, 1998

88 A. Ronzheimer, O. Brodersen, R. Rudnik, A. Findling, C.-C. Rossow, A new Interactive Took for the Management of Grid Generation Process around Arbitrary Configurations. In: Proceedings 4th International Grid Generation in Computational Fluid Dynamics. Swansea, UK, 6–8 April, 1994

89 N. Kroll, C.-C. Rossow, K. Becker, F. Thiele, The MEGAFLOW-Project. Aerospace Science and Technology. Vol. 4, 2000

90 C.-C. Rossow, H. Hoheisel, Numerical Study of Interference Effects of Wing Mounted Advanced Engine Concepts. ICAS-Paper 94-6.4.1, 1994

91 W. Burgsmüller, H. Hoheisel, ENIFAIR – EU Research into Engine Integration on Future Transport Aircraft. Air & Space Europe, Vol. 2, No. 2, 2000

92 H. Frhr. Von Geyr, H. Hoheisel, C.-C. Rossow, Numerical and Experimental Study of Aerodynamic Engine/Airframe Interference Effects for Engines with Different Bypass Ratios. Contribution to the CEAS Aerospace Aerodynamics Research Conference, Cambridge, UK, 10–12 June, 2002

93 H. Hoheisel, Investigations on the Aerodynamic Interference Effects of Wing-Mounted Advanced Engines – Analysis of High Speed Test Results. DLR-FB 2002-20, 2002

94 H. Hoheisel, K.A. Bütefisch, B. Lehmann, R. Henke, H.J. Roscher, U. Seelhorst, The Jet Behaviour on an Actual High-Bypass Engine as Determined by LDA-Measurements in Ground Tests. AGARD CP 498, 1991

95 D.L. Berry, The Boeing 777 Engine/Aircraft Integration Aerodynamic Design Process. ICAS-Paper 94-6.4.4, 1994

96 A. Spirkl, MTR 390, the New Generation Turboshaft Engine. Carl-Cranz-Gesellschaft, Kurs A1, 1995, S. 1–8

97 R. Grichtmaier, MTR 390, MTU-Schulungsunterlagen, Mai 1993

98 – : MTR 390 – die Muskeln des Tigers. AeroSpace 2/96, S. 34 ff.

99 G. Kirschey, A. Fehler, G. Kappler, W. Krockow, Results of Further Investigations of a New Concept of Fuel Prevaporization. 3rd Int. Symp. on Air Breathing Engines, Munich, Germany, March 1976. DGLR-Fachbuchreihe Band 6, 1976, S. 409 ff.

100 B. Simon, Entwicklung einer Umkehrringbrennkammer für den Gasgenerator neuer Technologie (GNT 1)

4. BMFT-Statusseminar Luftfahrtforschung und Luftfahrttechnologie, München, 28.–30. April 1986
101 H. Hartlepp, Entwicklung des Turbostrahltriebwerks Pirna 014 und des Verkehrsflugzeugs 152. Eine Chronik über 15 Jahre Luftfahrtentwicklung im Osten Deutschlands. Deutsches Museum, München, 1. Februar 1991, 37 S.
102 C. Hess, Meßbare Größe – Schadstoffe in der Luftfahrt., Flug Revue, Juni 1998, S.188–121
103 U. Schumann, Atmosphärenforschung für eine umweltfreundliche Luftfahrt. Vortrag im Seminar des Lehrstuhls A für Thermodynamik der TU München, Garching, 11.12.1997
104 A. Owen, H. Low, J. Lewis, Fuel Efficiency versus NOx Control. Nouvelle revue d'aeronautique et d'astronautique, No 2-1998, p. 100–105, 3rd Aero Days Post-Conference Proceedings
105 Zarzalis, NOx-Minderung bei Flugtriebwerksbrennkammern nach dem Konzept der Fett-Mager-Verbrennung. Von der Grundlagenforschung bis zum industriellen Einsatz. Vortrag im Seminar der Lehrstühle B für Thermodynamik sowie Energie- und Umwelt-Verfahrenstechnik der TU München, Garching, 12. 2. 1998
106 – : Deutlich weniger NOx durch neue Brennkammer, MTU-Report 1/97, 1997, S. 42
107 :–: Brennkammer-Technologien für schadstoffarme Triebwerke. Luft- und Raumfahrt 2/1998, S. 22–24
108 N. Brehm, T. Schilling, Technologische Grundlagen schadstoffarmer Verbrennung. Übersichtsvortrag beim Engine3E-Partner-Status-Workshop, Dahlewitz, 4./5. 12. 1997
109 G. Kappler, Theoretische und experimentelle Untersuchungen zur katalytischen Verbrennung in Gasturbinen-Brennkammern. DFGVorhaben Ka654/2, 1987 bis 1991
110 G. Kappler, Katalytisch unterstützte Verbrennung für Mehrstoff-Brennerbetrieb in Hochtemperatur-Gasturbinen. Teilverbundprojekt Turboflam, Teilprojekt 3.2.1.4, 1989 bis 1993
111 R. Fink, R. Lachner, B. Steidle, D. Theisen, D. Rist, Mischung und Verbrennung in schadstoffarmen Brennkammern. Untersuchungen an Brennkammern mit Stufung sowie mit Vorvermischung/Vorverdampfung. Bericht bei der DGLR-Fachausschußtagung, Dresden, 6./7. November 1997
112 F. Mayinger, et al., Optische Meßmethoden in der Wärme- und Stoffübertragung. Handbuch für den Hochschulkurs, Technische Universität München, Lehrstuhl A für Thermodynamik, 1991
113 R. Lachner, Laserspektroskopische Untersuchungen im Primärteil einer Flugtriebwerks-Brennkammer. Dissertation, Technische Universität München, 1998
114 D.A. Fulghum, Mach 8 Missiles Need More Work. Aviation Week & Space Technology, August 24, 1998, p. 24
115 H.P. Kau, D. Rist, Teillastverhalten einer Überschallbrennkammer mit Wasserstoffverbrennung in Über- und Hyperschallantrieben, Teilprojekt B12, Sonderforschungsbereich 255, Lehrstuhl für Flugantriebe, TU München, 1988
116 R. Hönig, Konzentrations- und Temperaturbestimmung in Brennkammern luftatmender Antriebe mit Hilfe laserspektroskopischer Meßverfahren, Dissertation, TU München, 1995
117 E. Schmidt, Gasturbinen hoher Gastemperatur, VDI-Z, 90, 1948, S. 350
118 R. Fridrich, Zur Flüssigkeitskühlung der Schaufeln von Gasturbinen, MTZ 18, 1957, S. 367-368
119 W. Niefer, Fertigung von Turbinenschaufeln für moderne Flugtriebwerke, Interavia, 1976, 3, S. 238–240
120 K. Gradl, M. Baumgärtner: Triebwerkschaufeln im Wandel der Zeit – Bauart, Werkstoffe, Fertigungsmethoden, DGLR-Vortrag 94-G1-109, 1994
121 BMBF (Hrsg.): Tagungsband Sechstes Statusseminar, Arbeitsgemeinschaft Hochtemperatur-Gasturbine, Tagung: 03.-04.12.1998, DLR, Köln-Porz, 1998
122 R. Kochendörfer, Neue Werkstoffe und Strukturkonzepte für die Antriebstechnik, DLR-Nachrichten, 1998, Nr. 91, S. 19–21
123 W. Krenkel, Entwicklung eines kostengünstigen Verfahrens zur Herstellung von Bauteilen aus keramischen Verbundwerkstoffen, DLR-Forschungsbericht 2000-04, 2000
124 O.E. Pabst, Kurzstarter und Senkrechtstarter, Bd. 6, Bernard & Graefe Verlag, Koblenz, 1984
125 H.G. Münzberg, Flugantriebe, Springer Verlag, Berlin/Heidelberg/New-York, 1972
126 D. Rist, Zur Konzeption und zum Flugverhalten eines strahlgetriebenen Abwurftriebwerkes, DGLR-Jhb. 1979
127 G. Spengler, H. Gemperlein, Über einen Kleinstprüfstand zur Untersuchung von Kraftstoffen. Erdöl & Kohle 393–396, 1959
128 H. Gemperlein, The Possibilities of Actually Testing the Combustion Characteristics of Aviation Fuels with Appriate Equipment. AGARD Conference Preprint No 84, 1971
129 G. Spengler, H. Wolff, H. Gemperlein, Über einige Zusammenhänge zwischen Brennverhalten und che-

mischer Konstitution von Kraftstoffen für luftatmende Triebwerke., DLR-Forschungsbericht 64-36, 1964
130 G. Spengler, J. Kern, H. Gemperlein, Untersuchungen zur Kraftstoffverbrennung in Durchströmtriebwerken mit Hilfe der Emissionsspektroskopie. DLR FB 65-45, 1965
131 R. Erlmeier, Optimierung von Überschallkraftstoffen aus Oligomerisaten von C 3- und C 4-Olefinen durch n-Paraffine. DLR FB 73-32, 1973
132 G. Spengler, E. Jantzen, Entwicklung einer Anlage zur Alterung von Flugmotorenölen bei hohen Temperaturen. DVL-Bericht Nr. 287 (1963)
133 G. Spengler, E. Jantzen, Die Simulierung des Ölkreislaufes von Turbinentriebwerken und seine Bedeutung für die praxisnahe Untersuchung von Flugturbinenölen. Zeitschrift f. Flugw. 14, 1966, 2, pp 99–102
134 B. Koch, E. Jantzen, Thermal Oxydation Behavior of Synthetic Oils: Basic Studies in the Polymerisation of Polylol Ester Oils. Journ. Syn. Lubric. 4, 1987, 321–336
135 G. Spengler, M. Koll, Über einige Organozinn-Copolymerisate und ihre Eignung als Schmiermittel beziehungsweise Schmiermittelzusatz. Erdöl u. Kohle 17, 1964, 107–117
136 G. Spengler, M. Koll, Über die Synthese ausgewählter Polyphenyläther und deren Eignung als Schmieröle. DVL Forschungsbericht 68-86, 1968
137 B. Koch, E. Jantzen, Thermo-oxydatives Verhalten von Perfluoralkyläthern. Proc. Tribology 2000. Techn. Akademie Esslingen, 1992 und J. Synth.Lubric, 1993, 191–203
138 E. Jantzen, K. Maier, Zur Entstehung und Simulation von Ablagerungen aus Flugturbinenölen. Trib. & Schmierungstechnik 33, 1986, 84–89
139 E. Jantzen, B. Koch, Eine neue Methode und Untersuchungen zur Ablagerungsbildung aus der Öldampfphase. Erdöl und Kohle 106, 1990, 509–513
140 E. Maier, E. Jantzen, Ablagerungen in Flugtriebwerken, Z. Flugwiss. Weltraumforsch. Teil 1: 9, 1985 pp 211–217, Teil 11-12, 1988, pp 45–50, Teil 111 15, 1991, pp 165–170
141 G. Spengler, H. Klein, Über die Schmierfähigkeit niederviskoser Öle unter den Bedingungen der Misch- und Grenzreibung. Erdöl u. Kohle 21 (1968) pp 217–224
142 G. Spengler, G. Braun, Eine Methode zur Verschleißmessung mit Hilfe radioaktiver Lager, Erdöl u. Kohle 18, 1965, 710–712
143 H. Schröder, E. Jantzen, Untersuchungen zur Tribologie von Flugturbinenölen unter Mischreibungsbedingungen, Proc. Tribology 2000. Techn. Akademie Esslingen 1992, 19.10.1–19.10.16
144 A. Zeman, Analyse synthetischer Flugturbinenöle durch Gaschromatographie-Massenspektroskopie. Fresenius Z. Anal. Chem. 290, 1978, pp 21–28
145 A. Zeman, Charakterisierung von Schmierstoffen mittels thermoanalytischer Verfahren. Proc. Tribology 2000. Techn. Akademie Esslingen 1992, 17.2.1–17.2.9
146 H.-J. Fuchs, A. Zeman, Polymerbildung in thermisch-oxidativ belasteten Flugturbinenölen. Proc. Tribology 2000. Techn. Akademie Esslingen 1992
147 P. Bartl, K. Zuber, M. Leipold, A. Zeman, Quality Control of Used Synthetic Aviation Turbine Oils by Analytical Methods. Fresenius Z. Anal. Chemie 314, 1983, 25–28
148 E. Jantzen, Feuerentstehung und -ausbreitung durch Öle in der Luftfahrt. DFVLR-Nachrichten, 1986, 13–15
149 S.H. Kägler, E. Jantzen, Schadensfrüherkennung mit spektroskopischen Methoden. In: Frühdiagnose von Schäden an Maschinen und Maschinenelementen, expert Verlag, 1988, 94–119
150 E. Jantzen, Abriebpartikel im 01 – Gewinnung und Nutzung zur Schadensfrüherkennung. In: Frühdiagnose von Schäden an Maschinen und Maschinenelementen. expert verlag, 1988, 23
151 Der neue Höhenprüfstand für Torboflugtriebwerke an der Technischen Hochschule in Stuttgart, MTZ 27, 1966, Nr. 7, S. 291/292
152 A. Güth, Eine automatische Vielfach-Druckmeßeinrichtung, ATM J. 132/6, S. 209-212 und 132/7, S. 227–232, 1971
153 W. Braig, Erweiterung des Stuttgarter Höhenprüfstandes, 4. BMFT-Statusseminar Luftfahrtforschung und Luftfahrttechnologie 1986, Band III, S. 598–667

Flying Qualities, Unstable Flight, Avionics, Cockpit, Sensors

Until 1945 Germany was one of the forerunners concerning flying qualities together with USA, France and UK. The break of 1945 terminated all activities on this subject and the leading people moved to USA or Great Britain. The developments in these countries are therefore briefly highlighted prior to the again starting activities since 1956 in Germany. The move of technology into the direction of electronics within the airplane happened at the same time. This technology step was a prerequisite for the unstable flight of high performance combat aircraft. As a typical experimental development the design of highly unstable airplane as well as the flight beyond maximum lift in separated flow is discribed. Following that, the development activities with respect to avionics and cockpits, also for helicopters are briefly summarized.

Flying Quality Requirements After 1945

Conceptual Approaches to the Man-Machine-System

ANDREAS HAFER, GOTTFRIED SACHS

The development of the modern flying quality requirements started rather corresponding in Germany, Great Britain and the USA in the middle of the 1930's. Almost at the same time comprehensive reports were published in the USA and Germany in 1942 and in 1943 [1, 2] resuming the state of art. This simultaneity reveals objective challenges, well-founded in the technical development and the growing importance of aircraft in the civil as well as in the military field. Efforts of developing flying quality requirements were directed towards improving the design of aircraft. Thus the research activities on the flying quality requirements are leading towards a new approach insofar as the handling qualities are respected in an earlier state of the design process of an aircraft. This basic approach went distinctly beyond the former activities of air-traffic-security in the 1920's following the international convention from 1919, which should assure a minimal security standard of flying [3]. Regulations for security standards represent only one aspect of flying quality requirements. The other one is within the context of standardising and optimising the handling qualities.

The work done in the 1930's and in the early 1940's were reflecting the transition of aircraft technology. They started with aircraft with performances and characteristics of design of the late 1920's. In the end of the period their flight performances had changed dramatically: the average speed had nearly doubled, the enlargement of the aircraft led to radical new technical dimensions creating completely new problems for the flying qualities. The problems of adapting the growing stick forces to the ability of pilots are only one example for this development [4]. Finally with the jet-engine a new propulsion-system was available which promised a further increase of speed. Moreover first in Germany shortly later in Great Britain and the USA aircraft with swept wings were developed facing the problems of high speed flight. These wings offered completely new problems particularly regarding the stability behaviour.

Of course the military interests played a major role in developing the flying quality requirements of the 1940's [5]. However, in the beginning of the American regulations of flying qualities civil needs drove the development. This situation was changed radically by World War Two. Thus the approaches towards the flying quality requirements were now discussed primarily regarding military needs, particularly in the USA. The results of the work now were published as military reports by the US Navy and the US Air Force [6,7].

It is important to recognize that, especially in the USA further quantitative research on flying qualities was concentrated within the military sector. The civil requirements remained widely limited to qualitative criteria. To some extent this development is also due to the different needs in civil and military aeronautics. Commercial planes fly within a flight envelope being much smaller than that of military planes, which have to be designed for manoeuvres needing more precise requirements than it is necessary for civil aircraft. Moreover for commercial planes criteria of performance, above all range and economical issues, are essential for the evaluation by the customer. Finally the role of the state is different regarding civil or military aircraft. In the civil case the state is a neutral institution, taking care of security standards, in the military case the state is the customer, responsible for the demanded qualities.

Further Aspects

The different versions of the flying quality requirements reflect both: the essential developments of aircraft technology and the changes in the basic scientific approach toward the research of flying qualities itself [8–10].

The early research was grounded essentially on flying experiences with test aircraft and standard aircraft as well as the interpretation of pilots' evaluation. After the war, new technologies were introduced, enlarging the experimental possibilities. Important for these developments were "variable stability aircraft", based on the use of the more and more refined electronic computers. Thus it became possible to inves-tigate not only the characteristics of one specific airplane, but to change certain aircraft-characteristics during a series of experiments, improving the chance to estimate the flying qualities of aircraft still in planning. The new version of the requirements in 1954 was essentially based on experiences and values won by the "variable stability aircraft" [11]. Further possibilities for the analysis of flying qualities were developed by using simulation technology [12–14], enabled by the fast evolution of electronics. Particularly flight manoeuvres containing a high risk could be tested by simulation during the state of planning. According to MIL-F-8785, published in 1980, as proof for flying qualities a combination of analytical approaches, results of simulation and aircraft testing is required, with the latter particularly related to those flying qualities which were especially relevant for safety [15].

As far as the progress of aircraft technology is concerned, the increasing speeds and flight-altitudes and, connected with them, the changing design configurations of aircraft led to problems which had to be taken into account for the new versions of criteria in the 1950's. These developments of the handling qualities due to the new scope of flight performance corresponded with the introduction of a new control technology, concerning hydraulic and electric control and artificial systems for damping. Beside of the immediate effects due to the change of technological systems this development influenced heavily the role of the pilot as a part of the man-machine-system. The introduction of automatic or semi-automatic systems for flight control contributed to a tendency that the pilot's perception became uncoupled from the direct effects of the dynamic characteristics of the aircraft. The representation of the perception of the pilot on which the flying quality requirements of the late 1940's and early 1950's were based turned out to be insufficient. Thus the problem of the role of the pilot within the man-machine-system became a major subject of research [16, 17]. Particularly by means of the simulation technology investigations regarding the pilot's perception and the pilot's behaviour during flight were carried out [18]. As a result, there were two fields of research related to the problem of handling qualities requirements. The first one was dealing with the systematisation of the context of the pilots' perception and their rating and evaluation. The second one concerned the investigation of the pilot's role as a part of the aircraft system, regarded as a feedback system.

In the 1950's the "ISO-Opinion Plots" were developed, for which the term "Thumb Prints" became familiar because of their specific shape [19]. They represented a useful instrument for the description of pilot's perceptions of the behaviour of the aircraft during flight regarding two connected parameters (for example the frequency of pitch oscillations and the damping ratio). They led to a systematic and far reaching understanding of the characteristics of the evaluation of flying qualities by the pilots and influenced the flying quality requirements in 1969 (MIL-F-8785 B).

A further, mainly complementary way for systemizing the pilots' perception was followed by *G.E. Cooper* developing an evaluation matrix for pilots, gradually improved by him and *R.P. Harper* in the 1960's [20]. This matrix, well known as Cooper-Harper-Scale, was also used for the flying quality requirements of 1969 [21, 22].

A more theoretical approach led to the development of a mathematical model, treating the pilot as part of a feedback system. This approach was started from the assumption that the behaviour of the pilot (likely to other elements of the machine-man-system) can be described by means of a transfer-function, based on the "servo-mechanics theory". Apart from earlier approaches, for example by *A. Tustin* already published in 1947 [23], the work by *Ashkenas* and *McRuer* became fundamental for these models [24, 25]. This theoretical concept led above all in the 1970's to further systematic flight-testing procedures as for example SIFT ("System Identification From Tracking") [26].

The described developments were, as already mentioned, essentially influential for the 1969-version of the flying quality requirements, showing the significance of the MIL-F-8785 B within this context. Thus it can be agreed when *R.P. Harper* spoke of an "almost revolutionary version" [27].

On the other hand the requirements of 1980 revealed no basically new approach. This new version had become necessary because in the 1969-version the development of the "highly augmented aircraft", starting in the 1960's and rapidly progressing in the 1970's, had not been taken into consideration. However this experienced discrepancy between the technological development and the process of formulating requirements led to a more principal discussion of the question, whether the practised form of requirements still meets the needs of aircraft-development at all [28].

There is an essential new aspect of the 1980-version of the requirements, using "equivalent systems". They enable to reduce the great number of parameters appearing at highly augmented aircraft and influencing their flight behaviour to only some parameters. Thus the transfer functions, basics for the formulations of requirements in the version of 1969, could be adapted to highly augmented aircraft [29]. However it is significant that the basic approach of the 1969-version was not changed by using equivalent systems. This mathematical method was introduced for continuing the applicability of the criteria of the MIL-F-8785B, keeping its fundamental sets of data [15].

The mentioned discussion led in the end to a new concept regarding handling qualities requirements, abandoning the demand of obligatory standards. In the new MIL-standards it is only prescribed, what has to be evaluated. The manual itself contains extensive proposals, being adaptable to the different scopes of evaluation. This new strategy reflects the technological development of aircraft in two ways. First active control systems allow to create relatively free the flying qualities according to an optimisation of mission tasks. Secondly the flying qualities have become dependent on such a complex matrix of parameters that it was no longer possible to evaluate them according to criteria related directly to basic aircraft characteristics. Corresponding to this development the investigation of flying qualities led to concepts of criteria with different targets and competences, among them criteria based on the use of equivalent systems, bandwidth criteria [30], the Neal-Smith-criterion, etc. [31].

Abandoning quantifications in the requirements yielded a shift from the specification of distinct criteria for a specific aircraft towards specific tasks and missions. Corresponding to this broader development concerning an aircraft mission, in the MIL-standards the flying qualities were specified as those characteristics of aircraft, which determine the ease and precision allowing the pilot to fulfil his mission [32, 33].

In the United Kingdom the development of flying quality requirements after World War Two has been influenced not only by the American and British research in the 1930's and 1940's but to some extent also by the requirements published in 1943 by the DVL in Germany. The DVL report was translated into English in 1946 and circulated as a RAE publication [34]. Moreover, German engineers who were responsible for the work done on this topic at the DVL, particularly *Karl Doetsch* and *Werner Pinsker*, worked after the war in England in this field of research.

At the end of the 1950's a revised version of the prewar report AvP 970 was published. It contained revised flying quality requirements, and in 1964 this report was slightly modified again [35]. Since the beginning of the 1970's a fundamental revision of these requirements was started [36], leading to a complete version in 1983 [37, 38].

There are some differences between the British and the American approach. Particularly the British requirements were a part of more general airworthiness regulations, currently updated by appropriate additional sheets [39, 40]. However even this updating seemed not to be sufficient for meeting the fundamental technological changes arising in the 1960's. Thus test pilots as well as engineers in the aircraft industry began to consider the requirements to be outdated. The subsequent process of revising the requirements nevertheless turned out to be slow, mainly due to the fact that the American requirements already represented a useful and available fundament for the British aircraft industry. In 1983 the revised British requirements were published as DEF-STAN 00-970, largely based on American research and requirements development.

Thus, American standards became widely used in western aircraft technology. However there existed a remarkable exception to this, regarding the development of the supersonic airliner "Concorde", a British-French joint-venture. For this airplane since 1959 special requirements were worked out, a revised version published in 1968 (TSS-5-2) [41], playing a decisive role in the following development of supersonic transport aircraft in general. Since there was no comparable realised American aircraft project these British-French requirements turned out to become an original European contribution to the development of flying quality requirements. With regard to the contents, there were no basic differences to the regulations formulated in the MIL-F-8785B. However, according to the function of the "Concorde" as civil airliner, the TSS-5 was more orientated towards the questions of flight safety [42, 43].

Special Development: V/STOL-Requirements

In the beginning of the development of VTOL aircraft no requirements were available which were for example suitable for the design of the controllability during hover flight [44]. At that time, about 1953, there existed experience in this field only with helicopters. Since there are fundamental differences in application of the vertical take-off capabilities between helicopters and VTOL aircraft (in principle helicopters use the hover flight for longer lasting flight tasks whereas a VTOL aircraft needs this capability only for the transition from take-off to the conventional flight phase and vice versa) a direct transfer is impossible.

Starting from various research work, chiefly American [45, 46], and utilizing diverse requirements for helicopters [47] and conventional airplanes [48], the Flight Mechanics Panel of AGARD dealt with this problem since 1960, publishing a preliminary report in 1962 [49]. In 1964 this was supplemented by a revised edition [50], containing commentaries essentially based on the experience made in Germany with the handling qualities of the VTOL aircraft VJ 101 X1 [51]. In 1966 an international working group of the AGARD Flight Mechanics Panel, led by *Seth B. Anderson* from NASA, started to work out recommendations for new requirements. As a result the AGARD Report 577 was presented in 1970 [52]. It contained in a first part proposed criteria and comments, in a second part (according to a procedure the first time applied in the MIL-F-8785B of 1969) a reviewing survey on the results of research and flight tests in this field with an appendix containing an extensive list of references [53].

At the same time the Cornell Aeronautical Laboratory began to work on the development of flying quality requirements for VTOL aircraft, within the scope of an US Air Force Program on VTOL integrated flight control systems. These works in which industry, research institutions and state agencies took part led from 1967 step by step to the publication of a final version in 1970 [54].

With regard to the content, there were barely differences between the specifications worked out by AGARD and that by the US Air Force. Apart from that, in the Air Force version the statements were formulated as demands, not as criterions as in the AGARD version. It is remarkable that the military agency within the USA did not renounce on flying quality specifications of their own, although adequate specifications were already developed by AGARD experts since the end of the 1950's. Besides it seems that the scientists working on the Air Force requirements refused a direct cooperation with the AGARD experts.

In Germany, whose aircraft research was interrupted after World War Two, the American standards were taken over almost unchanged after 1955. However the scientific cooperation within the AGARD Flight Mechanics Panel offered German engineers the possibility for a participation in this field of research. As described above this was particularly true for the development of VTOL aircraft flying quality requirements [55, 56], thus influencing the further military requirements for VTOL aircraft within in the NATO.

German Developments

Holger Friehmelt

In the process of establishing the MIL Standards in the eighties, also a German research institution, the DLR in Brunswick, participated [57].

At the successor organization of the "Test Center Rechlin" ("Erprobungsstelle Rechlin"), the "Flight Test Center of the Federal Armed Forces" ("Flugerprobungsstelle der Bundeswehr") in Manching, founded in the early sixties, an independent "Department of Flying Characteristics" was established in the early seventies. A first attempt of the German officials to determine the flying characteristics (qualities) within an aircraft project occurred with the planning of the EFA, for which, under decisive participation of the test center in cooperation with the DLR Brunswick, specifications were established, which were as a contractual basis binding for the participating companies [58]. The corresponding document is to a large degree based on the MIL-F-8785C specs, but contains further reaching requirements.

The task, which was here taken on by a team consisting of specialists from the WTD 61, the IABG and the DLR, represented a novelty in dealing with the flying characteristics specifications MIL-F-8785. Up to now, the last time during the "Tornado"-development, the "8785" was, as far as applicable, utilized as guideline for the assessment and classification of flying characteristics data during testing and certification; now, completely new independent specific aeronautical requirements for an aircraft had to be provided, an aircraft representing the new generation of fighters. When the work started, it was long known that the American colleagues were already working on a new flying characteristics handbook, the MIL standard, which was to replace the old "8785" in total. It became rapidly clear to the German team that the MIL standard was not able to use a fundamentally new flying characteristics database, but employed as base exactly the same, partly containing parameters of motion with rather old pilot assessments attached. In addition, there existed yet nowhere experience in dealing with the MIL standard. It only became known later that the new transport aircraft of the USAF, the C-17, was to be developed and tested according to the requirements of the "old 8785". It was therefore not hard to decide to suggest basing the establishment of the flying characteristics specifications for the EF 2000 on the long-standing "8785C" of 1980. Rather surprising was, however, that all partner countries, their industry and government agencies, charged the purely German team with this task. The task was completed in 1989 with the submission of the documents to the Industry Consortium "Eurofighter". One and a half years of intensive work was required to form, based on the meaningful frame which the "8785" provided, the "Handling Qualities Definition Document, HQDD" for the new fighter aircraft, in the meantime renamed EF 2000. The HQDD is not only part of the performance requirements underlying the development contract but is also the basis for the military admission and certification.

That the "8785" could, indeed, only be a useful frame for the HQDD is due to a number of novel requirements related to the EF 2000. One of the latter is the concept of "Carefree Handling". It was primarily envisaged to relieve the pilot of the task to supervise flight-range limits in the form of critical angles of attack or load factors. Finally, the following important protective functions were defined, which were to provide indeed a "Carefree Handling" of the aircraft: Prevention of a loss in controllability or the transition into a spin, limitation of the maximum load factor and its rapid growth, which may lead to a temporary loss of the pilot's consciousness, aggressive but secure maneuvering utilizing the entire angle of attack envelope, unrestricted operation of thrust levers and dive brakes (air brakes) at optimum response, automatic adjustment of the range to be protected dependent on the configuration, payload, mass, and the center of gravity location. As a consequence of "Carefree Handling", a completely new definition of the classical stall and spin test methodology resulted. While in case of the latter, the conditions and the behaviour during stall and the termination of this condition were to be investigated, is it with a fully functional flight control system not possible to maneuver the aircraft into such a condition. The angle of attack limit for attached flow may, on the other hand, deliberately be exceeded, for instance if decelerating flights at steep-up-to-vertical-pitch attitudes are being carried out. The demand was to overcome these flight conditions by means of the flight control system no matter what the control-stick inputs of the pilot looked like. Emphasis was especially placed on designing the flying characteristics air combat pilot-conform. It was not tried to define here numerical criteria; rather the demand was formulated that the pilot's opinion should be decisive in such a way that, in applying the "Cooper-Harper Rating Scale", at least the Level 1 (that is the lowest rating 3) is certified by the pilot, if he performs air-combat typical maneuvers in the common aggressive way.

Concerning the numerical (quantitative) requirements related to the longitudinal and roll/yaw motions, newest findings in the closed-loop control technology were realized. The team had familiarized itself during extensive preliminary studies with the current status in research on criteria concerning fighter aircraft. MBB, BAe, and Dornier supported the team. The description of the aircraft movement with the aid of the concept of equivalent systems with dead time was adopted from the "8785". The buildup of the equivalent models in the pitch and roll/yaw axes was, however, fundamentally revised and provided with the procedural rules necessary for its application. A novel method, which describes the precise control of the pitch attitude within the tracking range, was adopted and integrated by BAe. The corresponding graphical display of the criterion has become known as "Spider Diagram". It considers between the pilot's manual force input and the response of the aircraft in regard to the pitch axis a number of processes of higher than 2nd order, which result from the control-system, the actuator system, and the unsteady aerodynamics of the control surfaces.

Flying Characteristics Specifications for Helicopters

Heinz-Jürgen Pausder, Carl Ockier

The flying characteristics specifications (flying qualities specifications) for helicopters show in their fundamental development pronounced parallels to the ones of winged aircraft. However, the special operational role of the helicopter has been accounted for in the flying characteristics demands for helicopters. The helicopter distinguishes itself from the winged aircraft most of all in hovering and low-speed flight, in ground-proximity flight close to obstacles, and in the operation from unprepared take-off and landing fields.

As in the case of the winged aircraft were the first flying characteristics specifications issued by the civil flight safety authorities. The first specifications for licensing civil helicopters were produced in 1943. But these contained only general flying quality requirements. The first quantitative flying characteristics specifications for helicopters were published in 1951 with the American Civil Air Regulations Part 6 (CAR-6). These regulations, which are based on NACA research work, were a combination of flying quality criteria, classical stability criteria and specific helicopter criteria. Since also the military application of the helicopter rapidly developed, the first military specifications for helicopter flying characteristics (MIL-H-6501) were published in 1952. A revision of these specifications was carried out in 1961. The specifications (regulations) in MIL-H-8501 most of all specify the static and dynamic stability as well as control forces and the behaviour during autorotation. In 1968, a second revision of MIL-H-8501 was suggested but never realized. A V/STOL specification (MIL-F-83300), developed during that time, was also to cover the helicopter requirements. However, the US Army and Navy never accepted these specifications. Within the civil sector, the CAR-6 regulations were newly established after the foundation of the FAA and issued as FAR Part 27 and 29. In 1973, a new attempt was started in the USA to revise the MIL-H-8501 on the basis of MIL-F-83300. However, this attempt failed due the lack of a database of helicopter flying characteristics.

The US Army started in 1957, together with NASA, a comprehensive research program with the objective of establishing a systematic database, from which the military flying characteristics specifications could be derived. This task was strongly supported by the cooperation of various NATO countries (in Germany by the DFVLR, in England by DERA). *Heinz-Jürgen Pausder* and *Klaus Sanders* of the DFVLR "Institute of Flight Mechanics" ("Institut für Flugmechanik") carried out comparative flying characteristics (qualities) investigations with the helicopters BO 105 and UH-1D [59].

This led to the development of flight-test maneuvers, which were later incorporated into the specifications. Concerning the future understanding of flying characteristics and the interpretation of relevant requirements, the theory of the human controller was developed under the supervision of *Duane McRuer* and *Irving Ashkenas*. This theory was also applied at the DFVLR Institute of Flight Mechanics to the theoretical and practical analysis of flying characteristics criteria and here further developed [60]. First important findings concerning the relation between the pilot stress factor and flying characteristics parameters resulted.

In 1982, the work in the USA had progressed to a stage where the first version of new flying characteristics specifications could be established. During various discussions, with participation of the DFVLR, the document was refined, until finally, in 1988, the new flying characteristics specifications for military helicopters, the Aeronautical Design Standard 33 (ADS-33), were published. This specification represented in many regards a revolution in the assessment of flying characteristics:

- The assessment was no longer carried out exclusively according to prescribed quantitative individual criteria, but also considered the qualitative judgment of the pilot. For that purpose, flight maneuvers, representative of the different helicopter missions, were prescribed, and their fulfillment had to be judged by the pilot with the aid of the Cooper-Harper scale.
- The quantitative as well as the qualitative criteria were divided according to the mission, with the mission element being the smallest unit of a mission, and according to the visibility conditions. In case of the latter, the basis was no longer the classical VMC/IMC division but a special scale was defined, which considers the visibility quality with respect to the (vehicle) control by the pilot.
- The hierarchical structure of the requirements, related to the mission performance and the visibility and the closed-loop control quality, also included the consideration of system degradation.
- There was a detachment from the classical consideration of the static and dynamic stability. Instead, criteria within the frequency domain were defined, which were thus also relevant to highly closed-loop controlled helicopters - where the static and dynamic stabilities are of minor importance. This development was a direct consequence of the development of the theory of the human controller.
- ADS-33 required new means of control for flights with high control aggressiveness at reduced visibility conditions; these are only achievable with closed-loop control systems of high authority ("Attitude Command", "Translational Rate Command").
- All ADS-33 criteria were defined in a hierarchic dependence on control type, mission element, visibility condition, and awareness level.

Despite the fact that the ADS-33 was rather late for the establishment of flying characteristics specifications for the "Tiger" helicopter, various elements for the qualitative assessment were still considered. For the cargo helicopter NH-90, which was equipped with a full-authority digital control, the ADS-33 was adopted to the transport mission and defined as flying characteristics specifications [61]. The DLR was involved in a leading capacity in the establishment of both flying characteristics specifications.

After the publication of the first version of ADS-33 in 1982, interest in the new criteria and hence also the research efforts increased. Especially in the USA, in Germany and in Canada were, with the aid of ground and in-flight (airborne) simulators, the existing criteria refined and new criteria developed. The development of a "model-following control" for the BO 105 "ATTHeS" flight-test helicopter of the DLR Brunswick by *Gerhard Bouwer* allowed the systematic exploration of new criteria [62]. A 1984 test campaign with the "ATTHeS" provided major contributions to the development of the "Attitude Quickness" criterion [63]. Due to the advancement of the "ATTHeS" closed-loop control, it became possible to carry out tests [64], which required a precise high-frequency model following. In 1991, tests were performed at the DLR, in cooperation with the US Army, to investigate the effect of time delays and system bandwidth. These tests under the supervision of *Pausder* revealed considerable deficits in the bandwidth and phase-delay criteria of the ADS-33 [65]. An adaptation of the ADS-33 specifications followed in 1994 (ADS-33D). A further test series in 1992/1993 concentrated on the development of improved criteria for the pitch-roll coupling (decoupling) [66]. Here too, were deficits of the ADS-33 criterion based on the time domain recognized. The completeness of the database obtained in flights with the "ATTHeS" also allowed the definition of a completely new criterion in the frequency domain. This criterion too was incorporated into the ADS-33 specification (ADS-33E published in 2000). In 1995, after the tragic loss of the "ATTHeS" and its crew, the research work in this area had, for the time being, to be suspended. The free capacities were invested into the development of an "ATTHeS" successor, the ACT/FHS ("Active Control Technology Demonstrator and Flying Helicopter Simulator") [67].

Carl Ockier of the DLR Institute of Flight Mechanics provided important contributions to the flight-test technique related to the ADS-33. During the time period from 1992 to 1998, an important flight-test program was carried out with the BO 105 S helicopter of the DLR. This test series provided a unique systematic database for the ADS-33 [68]. One of the consequences of this test series was a corresponding adaptation of several criteria within the ADS-33 specifications (ADS-33D and ADS-33E). This test series is also the basis for a flight-test handbook for ADS-33 to be prepared together with the US Army.

Within the coming years, the vehement development of the flying characteristics specifications will be continued as a result of the technological developments and the extended use of the helicopter. Digital control with graded redundancy, active control organs, carefree handling, artificial vision in ground proximity, as well as active blade control will require further adjustments of the flying qualities criteria. The accepted necessity to test integrated technologies and to account for them in the flying qualities specifications is also reflected in the intense activities related to the development of new test vehicles. These will provide a test potential adapted to the requirements. The German development of the flying simulator ACT/FHS on the basis of the EC 135 helicopter at DLR must, in that regard, especially be emphasized. The development of tilt-rotor aircraft in the USA, also commencing in Europe, has, furthermore, to be considered in the development and improvement of future flying characteristics specifications.

From the Stable towards the Unstable Configuration – Drag Optimization*

HELMUT JOHN

Basic Principles of Flight Mechanical Stability

Within a flight-physical consideration of stability, controllability and flight characteristics, forces and moments acting on the flight vehicle are always dealt with in coefficient form. Here, the forces are divided by the dynamic pressure and the reference wing area, and depicted in non-dimensional form as lift coefficient C_L, drag coefficient C_D and side force coefficient C_Y. The moments about the three axes are additionally being divided by characteristic lengths; the pitching moment divided by the wing mean-aerodynamic chord "c_μ" results in the coefficient C_m; rolling and yawing moments are rendered non-dimensional by the wing semi-span "s" with the corresponding coefficients C_l and C_n. The representation in coefficient form has the advantage for the analytical work in flight physics that coefficients measured in the wind tunnel on different-scale models can directly be transferred to the full-scale configuration. For instance is the plot of the lift coefficient versus the drag coefficient designated a polar. The division of the pitching moment coefficient by, e.g., the lift coefficient, results in the center of pressure location, x_{cp}/c_μ, in per cent of the wing mean aerodynamic chord relative to the origin of the overall coordinate system; it represents the point where the total-lift vector acts on the aircraft. By the deflection of a trim flap on the flight vehicle, mostly an elevator flap, the center of pressure is shifted to exactly coincide with the center of gravity of the flight vehicle, which results in a trimmed (equilibrium) flight attitude.

Stability then denotes the characteristic of a flight vehicle to return to the equilibrium attitude after this condition has been disturbed, for instance, by a gust. So as soon as the angle of attack of the flight vehicle is changed by a certain amount, a small lift force is added to the previous force, which generally no longer acts in the center of gravity – which only corresponds

* The drag optimization by means of unstable flight is essentially a question of the longitudinal movement. The problems related to the lateral movement in the presence of separated flow will be considered in the next chapter.

to the center of pressure at trimmed conditions – but at a different point, called the neutral point. If the neutral point is located downstream of the center of gravity, the effect of the disturbance will be reversed without interference from the pilot, i.e., the flight vehicle is (statically) longitudinally stable. This means, expressed in the form of coefficients, that the neutral point X_{NP} is the change of the pitching moment coefficient divided by the change of the lift coefficient, $dC_m/dC_L = X_{NP}/c_\mu$, expressed in per cent of the mean aerodynamic chord. It describes the location, where the vector of disturbance acts.

During the transition from subsonic to supersonic flight speeds, the position of the neutral point shifts downstream by about 20–30 per cent relative to the subsonic neutral point location due to the occurrence of shock waves on the wing and stabilizers, which means a strong increase in the static longitudinal stability. Since the center of pressure of the untrimmed air vehicle moves proportionally with the neutral point when going from subsonic to supersonic speeds, trimming at supersonic speeds requires substantial deflections of the trim flaps to restore the equilibrium flight attitude. Large deflections result in high trim drag, which negatively affects flight characteristics and flight performance. It is, therefore, obvious to place the wing during the design phase such that the center of gravity lies between the subsonic and the supersonic neutral point location. At subsonic speeds, an unstable basic configuration prevails in that case, which has to be stabilized artificially.

It is today taken for granted that during the design of a new flight vehicle stability and flight characteristics have to meet a large number of criteria to receive a flight certification. *Andreas Hafer* and *Gottfried Sachs* give a very good description [69, 70] and a detailed appreciation of the development of flight-characteristic regulations from the beginnings until today.

The flight characteristics affect the agility of a flight vehicle; they are therefore closely coupled to the flight performance. The essential parameters describing flight performance are lift, drag, thrust, and specific fuel consumption associated with the range and weight. Already during the thirties it was attempted, for instance in the case of long-range reconnaissance aircraft, to reduce drag by reducing the demands on stability and thus improve climb characteristics up to high altitudes, which are essentially determined by the parameter "thrust minus drag".

The subsequent description of the performance improvements due to the consequent utilization of the new artificial-stability technology, while at the same time meeting the flight-characteristics regulations, would, however, not be complete if one would not explain the basic principles, which were already established during the pioneer era of the development of flight vehicles.

The Era of the Pioneers

Science and research as institutional establishments have only at a very late point in time accepted the task of a systematic investigation of aerodynamic problems and flight characteristics. Only when the national governments and authorities realized the possibilities of a military utilization of aircraft and, within the civil sector, the possibility of fast mail transportation by air, were the necessary resources for the buildup of "royal" and governmental research and test establishments provided.

For that reason, designated experiments – to a large degree supported by individuals with a great engagement – formed the forefront in the development of flight models and flight vehicles. The first approaches were based on the careful observation of gliding or soaring birds. But despite of all the efforts, it was not possible to determine whether this is done in an inherently stable condition, or whether the bird carries out non-perceptible control movements via its equilibrium organs to overcome unstable flight phases. The generation of lift to overcome gravity and the knowledge concerning the achievement of inherent longitudinal stability, both necessary prerequisites for a successful flight, was therefore initially obtained purely by experiment with the aid of aircraft models.

A treatise [71] on the early days of flight technology in Germany is given in *Werner Schwipps'* "Heavier than Air". Also *Andreas Hafer, Gottfried Sachs* and *Xaver Hafer* have given in their contribution a vivid description of the early findings. If one collects the individual contributions [72], one finds that *George Cayley* and *Alphonse Pénaud* had, with flight models which were already equipped with adjustable horizontal and vertical stabilizers, successfully researched the physics of free naturally stable flight with results that were interpreted by *Lanchester* and *Zahm*. From these results, the first basic principle of stable flight can be derived, which says that at the trim point a negative slope of the pitching moment coefficient, $dC_m/d\alpha < 0$, has to prevail, which is only possible for a positive zero-lift pitching moment as outlined in the figure. The zero-pitching moment denotes in aerodynamics the pitching moment at zero lift, which results from the camber of the overall configuration. This means that a negative setting of the horizontal stabilizer with respect to the wing must exist.

Accordingly were, as one can easily convince oneself today, *Lilienthal's* flight vehicles statically stable. It is the more remarkable that neither *Otto Lilienthal* himself nor the *Wright Brothers* paid much attention to the problem of longitudinal stability. *Otto Lilienthal* wrote in a paper [73] in 1895: "*The stability during forward flight is a matter of training ... that's why flight tests can only be educational if a human participates in the flight and achieves by an arbitrary control (of the center of gravity) a permanently stable flight*". From this statement we can derive the second basic principle of flight technology, namely, that man and flight vehicle have to be subjected to a common basic consideration in determining flight characteristics. Today this is called "man-machine problem" and undergoes extensive studies. Scientific research did only decades after *Otto Lilienthal* broadly approach this problem.

An explanation of the stability of a flat wing was only given, as already briefly mentioned, by *F. W. Lanchester* in his book published in 1909 [74]. It is his merit to have, with his phugoid theory, given an insight into the dynamic motion of a flight vehicle. His computational check with his method of Lilienthal's flight predicted an unstable dynamic behaviour. His statement concerning the situation was: ..."*it is deeply to be regretted that Lilienthal had to pay for his courage with his life.*"

Otto Lilienthal crashed on August 9, 1896 with his flight vehicle headfirst to the ground and broke his spine. Whether a sudden gust was the reason for this tragedy or the fracture of the wing or some other reason is not clear up to today.

Contrary to *Lilienthal's* flight vehicle was the *Wright Brothers* flying machine of 1903 unstable due to the design with the elevator in the forward position (canard configuration). Wind tunnel checks of *Jex* and *Culick* [75] have shown that the center of gravity was located up to 20 per cent of the wing chord downstream of the neutral point. The front elevator position was obviously chosen for reasons of a good controllability, especially under difficult circumstances,

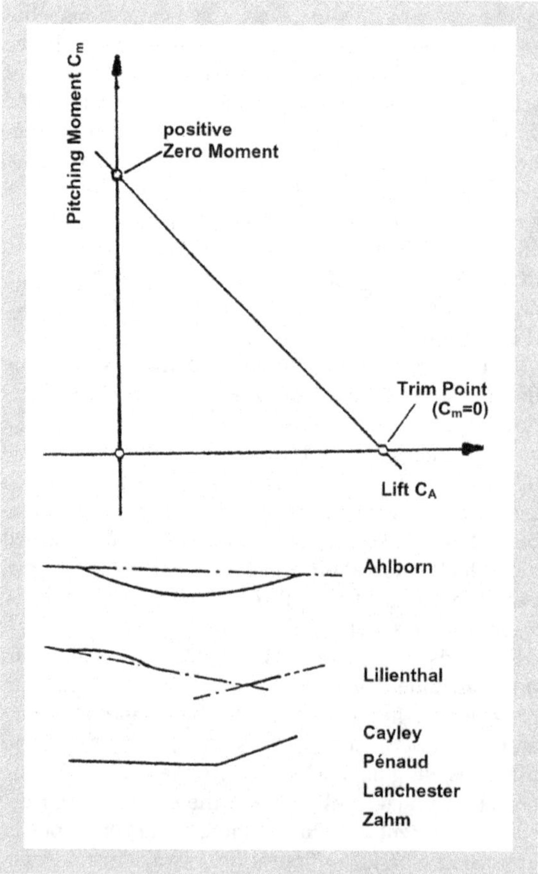

Static stability at the trim point. Longitudinally stable configuration.

such as turbulences, and not for reasons of stability. Stabilizers in an upfront location are exposed to an undisturbed freestream, whereas rear stabilizers in the case of this configuration are located in the wake of the free-seating pilot and are exposed to the downwash of the wings and the spin of the propeller slipstream, all parameters whose effects can only be predicted in detailed wind tunnel tests. There were no test institutions with appropriate wind tunnels, although the *Wright Brothers* employed a simple fan for lift measurements.

The aircraft of the *Wright Brothers* was flyable due to the low flight speeds, the good control-surface effectiveness and the fast response of the forward control surface, but most of all due to the excellent capabilities of the pilots, which were gained by untiring and systematic training.

This leads one to conclude that the second basic principle, expressed above, was also applied and used by the *Wright Brothers* to produce positive flight characteristics. They established, however, a third basic principle which states that also a, to a certain degree unstable, configuration can be flown if the missing inherent stability can be compensated by a suitable control-surface effectiveness and high control surface setting speeds.

Prior to and during World War One, there existed a number of successful aircraft, which were, with the knowledge and by design of the manufacturers, statically unstable. The training of a large number of pilots on such aircraft might have been rather difficult, which soon led to the acceptance of the military's demand of a small inherent stability. Since the rise of civil aviation after World War One, the requirement of sufficient longitudinal stability was made into law.

During the stormy development of aircraft engineering, essentially all designs were, therefore, based on the first basic principle, i.e., a sufficient inherent longitudinal stability had to be guaranteed.

Certain difficulties arose however, when the engine performances rapidly multiplied and the higher velocities of the propeller jet worsened the longitudinal stability. Further difficulties arose with the flight speed approaching the compressible range. The increase in downwash and in the unstable fuselage moment with increasing compressibility also reduces the static longitudinal stability. In studies and in presentations [76], *Heinrich Helmbold* and *Eduard Fischel* already considered, under the supervision of *Ernst Heinkel*, the newly arising problems in 1940. However, the third basic principle, cited above, was not used since the prerequisites for an artificial stabilization were not available to be considered already during the design.

Artificial Stability Through CCV Technology

The realization of the third basic principle within an unstable configuration design is today accomplished by replacing the "well-trained" pilot by a fast and reliable computer respectively by the chain "sensor-computer-actuator". Here, each individual control surface can directly be addressed by the computer, which opens up entirely new, flight-mechanically realizable possibilities. Here, one talks about the so-called "Control Configured Vehicle" technology, for short CCV technology. The CCV technology comprises, among other things, new types of control, such as direct lift control and direct lateral force control, which are, for instance, required in the case of glide path optimization or in the precision approach to the tanker during aerial refueling. There is, furthermore, the so-called fuselage aiming in gun operations during military flight – a decoupling by $\Delta\alpha$ and $\Delta\beta$ of the fuselage attitude from the flight path, which are during normal flight coupled by α – and the maneuver-load optimization. Also local instabilities, as, for instance, the missing roll damping within a limited incidence range, may be compensated by additionally induced aileron deflections. But this has to be done in a way that ensures the loss in roll agility to be as low as possible.

The most important computer-supported kind of control discussed here is related to the possibility of compensating lacking inherent stability by artificial stabilization. The great importance of this element of the CCV technology lies in the fact that a consequent integration into a new design leads to considerable performance improvements, respectively pronounced weight reductions.

Within the present deliberations, the connotation "instability measure" therefore always stands for the value at small angles of attack, i.e., within the linear range, and at subsonic Mach numbers. The denotation "CCV technology" is, furthermore, exclusively used for the application of the computer-supported artificial stabilization. The questions, which have to be answered for such a (artificial stability) design, are:

- Which performance gain can be achieved?
- Which maximum degree of instability can be managed?
- How has the flight-control system to be configured and what degree of redundancy is needed?

The answer to the question of the performance gain due to the CCV technology is closely associated with the configuration to be considered. For a new design, the wing is placed in such a way that the center of gravity is located between the neutral point for sub- and for supersonic speeds. At subsonic speeds, there exists then an unstable basic configuration, whose degree of instability has to be adjusted in a way that it may be artificially stabilized by the technological means (state-of-the-art) available. Since the location of the center of pressure of the untrimmed configuration is proportional to the neutral point location, the negative trim angles are, in this design, at supersonic speeds being reduced by a $\Delta\eta$-deflection and at subsonic speeds equilibrium can be achieved by a positive deflection, which shifts the center of pressure to coincide with the center of gravity. The procedure is explained in detail by corresponding figures in conjunction with the description of the configuration-finding process for the "Tactical Fighter Aircraft, TKF".

Individual contributions are, in any case, to be determined directly in the wind tunnel. For an existing configuration, for which the CCV technology shall be utilized in retrospect, the improvement of performance can be realized in form of an increased range since the fuel consumption goes down due to the drag reduction. However, if this potential is already incorporated into a new design, the improvement to be expected is much larger. For a given design mission, the fuel consumption is also reduced. Since now the tank volume can be reduced, a smaller and lighter aircraft evolves, which, in turn, also consumes less fuel due to the smaller surface area exposed to the flow and less weight. A contribution similar to this aerodynamic "snow-ball effect", is also provided by the engine, something that also has to be mentioned here. A new engine with an improved thrust to weight ratio and a lower fuel consumption reduces, in a similar way, the required tank volume for a given design mission.

The degree of instability that can be coped with assumes a larger "time-to-double" for the pitch-up motion due to the disturbance than is necessary for the buildup of the counter-moment. Or simpler: one has to react faster than the disturbance amplification proceeds.

The reaction time is essentially determined by dead times due to the response and/or computational times of air data sensors, flight control computers and actuators.

The initially posed questions may hence be led back to questions related to the system technology. Relevant are here the configuring of the system for the artificial stabilization, the individual hardware as well as software components, the processing time, the safety architecture and also the reliability of the overall system.

Since the full benefit of the CCV technology is only reached for high degrees of instability, and since such a highly unstable aircraft had not been developed in Europe up to 1972, the BMVg charged in 1974 MBB to execute the corresponding demonstrator program.

The CCV-F-104 G demonstrator program shall be presented here in order to show the relatively large effort necessary to guarantee redundancy and hence the safety of a CCV aircraft. This demonstrator program is exemplary for industrial research; the results are applied today to the EF 2000.

The CCV-F-104 G Demonstrator

Within the framework of the "Experimental Components Program Air" ("Komponenten Experimentalprogramm Luft, KEL") a CCV experimental program employing a F-104 G as test aircraft was established. It was the objective of the investigations to develop the new CCV technology of the control-system-supported aircraft up to maturity (see Color Plate VII).

The experimental program proceeded in two phases. During Phase I, the flight-control guidance system was defined, developed and tested, the test aircraft F-104 G modified according to the task, and in-flight sensor tests were performed. Phase II comprised ground and flight-testing of the control system and flight tests to determine the flight characteristics of the unstable demonstrator. Further insight into the development [104] and trials [105] of the CCV-F-104 G flight-control system is given in the contribution "Control, Navigation, Avionics, Cockpit".

The proof of performance due to the application of the CCV technology was not part of the test program since the test aircraft could only be sufficiently destabilized by considerable external changes to the configuration so that a comparison with the original version was no longer possible. On the other hand evidence of performance increases could be provided with sufficient accuracy on the basis of wind-tunnel tests.

Destabilization Concept

Due to the very large initial stability of the F-104 G test aircraft of more than 20 per cent of the mean aerodynamic wing chord c_μ, the desired degree of instability of about 20 per cent c_μ could only be achieved by the combination of two measures:

Adding a rear trim mass of about 600 kg (1,350 lbs) to shift the center of gravity to 22 per cent c_μ, the downstream limit given by the position of the main landing gear. In a second step, the neutral point is

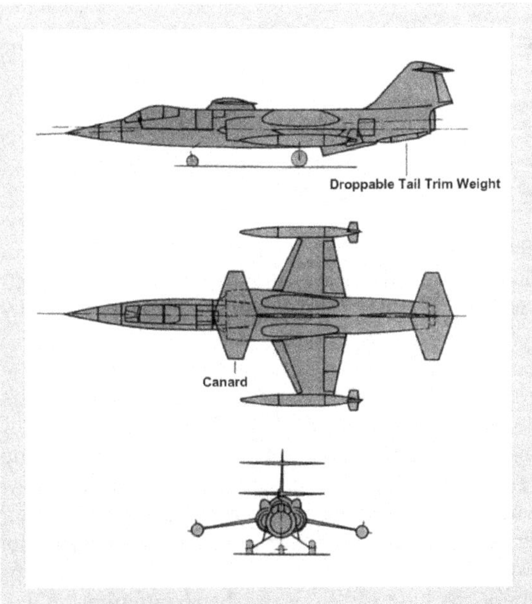

Three view drawing of the CCV-F-104 with the modifications related to the destabilization concept.

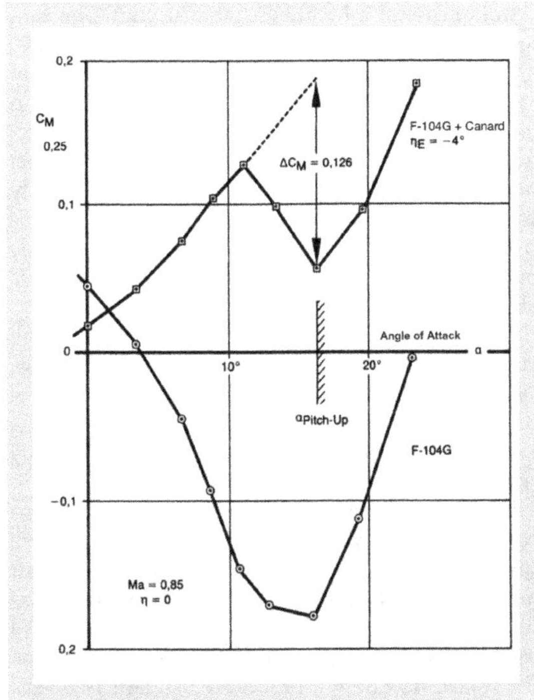

Aerodynamic effect of the canard on the longitudinal motion.

shifted upstream by about 35 per cent by placing a canard on the upper side of the fuselage immediately behind the ammunition compartment.

The effect of the canard on the pitching moment characteristic of the CCV test aircraft is shown in the sketch above. One can observe that the F-104 G with canard shows in the lower angle of attack range the desired unstable behavior with $dC_m/d\alpha > 0$. At $\alpha = 11°$, a strong natural re-stabilization occurs, which prevails up to the pitch-up boundary at $\alpha = 16°$. The instability occurring above this angle of attack is less pronounced than the one of the basis aircraft.

Safety System

The original control system of the F-104 G test aircraft remains as backup for the CCV-flight control system completely in tact. Only immediately in front of the control surface actuators, where the switch from the electrical to the mechanical mode of operation occurs, and inside the cockpit, where the pilot's commands are taken up via inductive motion sensors, the existing system was modified.

The CCV test aircraft was, in addition, equipped with an in-flight droppable afterbody weight (lead-scrap), which allowed to re-stabilize the unstable canard configuration at any time.

A further safety factor is given by the fact that the switch-mechanisms located in front of the control-surface actuators allow to test the CCV flight guidance system during flight first in an open-loop mode (i.e., with the basic control with operable but idle actuators), prior to the full activation of the CCV-control system.

Flight Guidance System

The CCV-F-104 G was equipped with an advanced flight guidance system. For flight safety reasons this system is based on the demand to cover two critical failures (fail-operate, fail-operate). This leads to the requirement to quadruple the individual components of the control system (quadruplex-system).

The failure demand cannot be met by the hydraulic part of the control system since here one has to rely for cost reasons on the hydraulic supply and control-surface actuators of the test aircraft.

In the case of a highly unstable aircraft, one cannot waive the redundant measurement of the angles of attack and yaw. This would mean that in the case of a quadruplex system, one had to provide both, four α and four β transducers. However, by careful arrangement of the angular probes, one may be able to extract the angle of attack and the angle of yaw, respectively, from any arbitrary pair of angular probes. In this case, the double-failure demand may be met by only four obliquely arranged (flow) directional probes (skewed-sensor principle).

The advantages of digital signal processing – high adaptability to configurational and flight-range changes, possibility of software development parallel to hardware manufacturing, and high flexibility in the case of control-law modifications – are sufficiently well known.

In the case of multi-parameter closed-loop control ideally all parameters of state – these are those time-dependent variables, which describe the aircraft motion in the form of first-order differential equations – are proportionally locked on to all control organs thus achieving an optimum control behavior. The control-law design procedure can largely be automated. In classical control technology are, on the contrary, as few measurement parameters as possible being fed back. The lack of feedback for the remaining parameters of state is compensated in that case by suitable filtering of the sensor signals.

The essential additional components of the CCV flight guidance system are:

Four flow probes were located in the CCV-F-104 G fuselage nose area close to the radar compartment.

- flow sensors,
- inertia sensors, consisting of gyroscopes measuring turn-rates and acceleration transducers,
- digital processor,
- CCV actuator motors,
- four additional rectifier units and batteries,
- hydraulic accumulator.

Together with the corresponding cables these devices had a weight of about 220 kg (495 lbs). For a new operational aircraft this number can be reduced by roughly 30 per cent.

Test Concept and Configurational Conditions

Starting from the basis of the F-104 G, the degree of instability of about 20 per cent c_{μ} aimed at is being reached, for safety reasons, in a number of destabilization steps. First, the newly developed CCV control system is thoroughly tested on the basic version of the F-104 G denoted B1. Then, the afterbody trim mass of about 600 kg (1,350 lbs) is attached (configuration B2), which reduces the longitudinal stability to about 10 per cent c_{μ} (see the figure). This test step is based on a special fuel consumption program. Instead of first drawing from and emptying the wing-tip tanks, as is common, the fuel is drawn, at full wing-tip tanks, from the main tank. This enables one to exceed the rear permissible center of gravity location of 22 per cent c_{μ} by 5 per cent c_{μ}, at least during flight. Prior to landing, the fuel is pumped from the wing-tip tanks into the (main) fuselage tanks. The center of gravity shifts as a consequence to a position upstream of the rear limit.

During the third test step, the rear ballast is again removed and at the same time the canard added. This configuration, denoted E1, can be flown within the entire flight envelope, if necessary, manually. It represents the back-up configuration for the unstable canard configurations E2 and E3. The natural stability of the configuration E1 is guaranteed by the installation of a nose ballast of about 320 kg (719 lbs) inside the forward fuselage.

The configuration E2 results from E1 by adding the rear trim mass of about 400 kg (899 lbs). By removing part of the nose ballast and increasing the afterbody ballast by another 200 kg (450 lbs), one finally obtains the configuration E3 where also the special fuel consumption program described above is being applied.

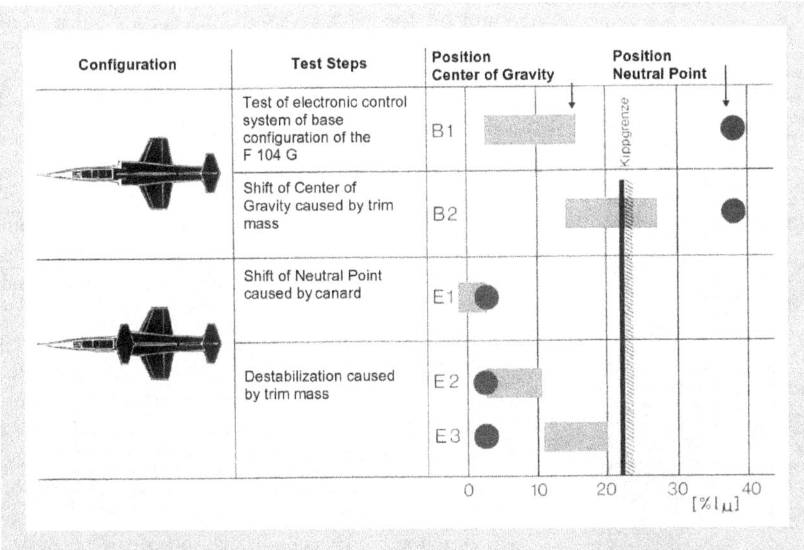

Test concept and configurational conditions.

Flight Test Results

The assessment of the flight characteristics of all configurations flown in CCV operations is similar; therefore, only the most unstable configuration E3 is considered here. A summary of the pilot's assessment gives the following picture:

The movement about the pitch axis was in general at all speeds well damped. The response time was at low speeds relatively good and at medium speeds adequate. The damping about the yaw axis was in general very satisfactory without any problems whatsoever occurring during flight. The roll damping was weak and coupled with a certain rollback tendency. This tendency is typical for aircraft with an extremely low moment of inertia about the roll axis compared to the other two axes.

The weakest assessment was given to the flight control system during landing approach, however, without judging the situation critical. Here, further improvements would have been necessary but they were, for time-reasons, left undone.

In retrospect, it can be said that during the entire test phase no critical flight situations occurred caused by a failure of a CCV component. All occurring failures were securely mastered by the "fail operate-fail operate" design of the system.

Summarizing it can be stated that proof was provided that artificial stabilization is possible while, at the same time, meeting the flight mechanical criteria and that sufficient experience and knowledge could be gathered to specify a CCV flight control system for a new modern aircraft design.

Application of the CCV-Technology within an Integrated Overall Design (TFA/TKF)

The performance improvement due to the introduction of the CCV-supported stabilization concept into the integrated overall design is, naturally, to a large extent dependent on the chosen configuration. Therefore, procedures and results shall here be explained taking as example the configuration-finding process for the "Tactical Fighter Aircraft, TFA" ("Taktisches Kampfflugzeug, TKF"), which, as national predecessor of the joint European project "Eurofighter", was introduced to the consortium by the German side.

For the design and configuration finding of a new aircraft one generally proceeds from a set of demands that is given to industry by the customer in form of "Tactical Requirements, TaR" ("Taktische Forderung, TaF"). However, in the present case of industrial research and optimization studies, financed by the BMVg out of the budget for research and technology, no TaR existed yet. The catalogue of demands was gradually developed in a dialogue between a study group of the Air Force and MBB, to crystallize the potential of a new configuration for the operational concept of the Air Force. The demands, which are im-

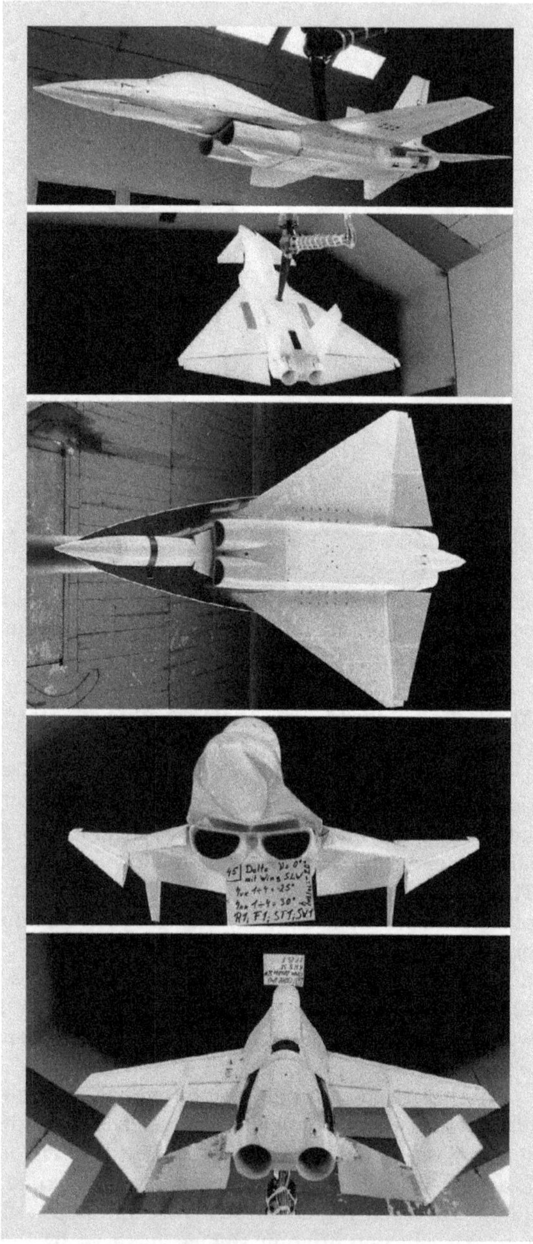

Configurational variations on wind-tunnel models. From top to bottom: Conventional strake-trapezoidal-wing configuration, Conventional delta-canard configuration, Delta wing with an extreme strake, Delta wing with rotatable outer wings, Strake-trapezoidal-wing with a rudder originating on the wing.

portant for the aerodynamic design, may be divided into "point" performances, such as, for instance, unsteady turn rates, steady turn rates, acceleration times from one flight condition to another of higher energy, decelerations and climb performances, and in take-off and landing performance, respectively after-landing roll distances, as well as in mission performances with different mission profiles and external store arrangements. The design mission is defined without external fuel tanks and hence determines the internal fuel capacity, which has a significant influence on the size of the aircraft fuselage. In a first preliminary design loop, a number of possible configurational designs, as well as the main parameters, such as wing load and thrust to weight ratios, are narrowed down to the design with the smallest take-off mass. Here, an electronic data processing design tool in the form of a computer-supported scaling program is being utilized, which scans the neighborhood of the basic design by varying sweep and aspect ratio for favorable trends, and scales the corresponding configurations to the demands. The basic designs must be represented by highly accurate aerodynamic coefficients. For that purpose, a modular wind-tunnel model, as described by *Helmut John* and *Werner Kraus* [77], was extensively tested in the wind tunnel; the coefficients were analyzed and processed into corresponding performance data sets.

The configurational range included two rather conventional designs, viz., a strake-trapezoidal-wing-elevator configuration and a delta-canard configuration. In addition, three rather "exotic" configurations were investigated, a pure delta configuration with an extreme-sized strake, a configuration with rotatable outer wings and outer rudders, which were used as endplates for the rotating surfaces, and a trapezoidal-wing configuration where the vertical stabilizers reached from the wing down to the elevator in order to expose the stabilizers to sufficient air flow also at high angles of attack.

Only after the basic configuration has roughly been determined, components, such as the wing and horizontal and vertical stabilizers, are being aerodynamically optimized and a basic wind-tunnel model is being build, which is then further improved with regard to the longitudinal and lateral motion.

Due to the special features of the TFA/TKF configuration, namely the requirement of maneuverability beyond stall in the presence of separated flow up to angles of attack of $\alpha = 70°$ and under inclusion of thrust vector control, details of the configurational design shall not be discussed here; the reader is instead referred to the Chapter "Controlled Flight in the Presence of Flow Separation". For a better understanding of the approach taken in determining the CCV-improvement potential, the wing optimization shall, however, be addressed here.

From the investigations of configurational variations and optimizations, decisive parameters have been extracted as depicted below. These were: the aspect ratio, Λ, to stay within the limits of the induced drag dictated by the design mission ($\Lambda = 2.2–2.3$); maximum lift, to define the wing area, from the unsteady turn rate with $C_{Lmax} = 1.4–1.45$; wave drag from the requirement of a steady turn rate at supersonic speeds, $WD/q \leq 0.2$ m^2, which points from the start to a delta wing with a leading-edge sweep of $\phi_{LE} > 57°$, since only above this sweep angle no additional drag increase occurs with Mach number, and a new inherent demand that became necessary due to the unstable design: a minimum restoring moment of $\Delta C_m \geq 0{,}03$ at angles of attack around $\alpha_{max} \pm 3°$ dependent on where exactly the minimum is located. Also a certain tolerance is needed since the pilot cannot center C_{Lmax} exactly.

The requirement is related to the fact that for an unstable configuration the movement towards a change in angle of attack is easily initiated, but the restoring moment is needed to stop the motion or to bring the vehicle back from the high angles of attack.

The analysis of the wing planforms with different trailing edge sweeps ϕ_{LE} showed that a minimum in the center of pressure movement with angle of attack lies at $\phi_{LE} = 60°$ (see figure below), which is important for the flight at high angles of attack; this is further explained in the Chapter "Controlled Flight in the Presence of Flow Separation".

For the further optimization of the wing planform geometry, a simple model was therefore build, where the wing consisted of a flat plate with a rounded leading edge and a wedge-like cusped trailing edge. Trailing and leading edge flap deflections were realized by interchangeable inserts. The basic wing had a leading edge sweep of $\phi_{LE} = 45°$, a taper ratio of $\lambda = 0.175$ and a trailing edge sweep of $\phi_{TE} = 0°$. By mounting various wing parts ahead of the leading edge of the basic wing, a whole series of double-delta wings with a variable leading-edge kink location and a variable inner-leading-edge sweep of $\phi_{LE} = 2.5°$, 57° and 65° could be realized. The analysis provided the optimum combination of geometric parameters:

Aspect ratio:	2.3
Inner-wing sweep:	57°
Outer-wing sweep:	45°
Leading-edge kink at 66 per cent semi-span:	
Trailing-edge sweep:	0°
Wing area:	54 m^2 (581 ft^2)

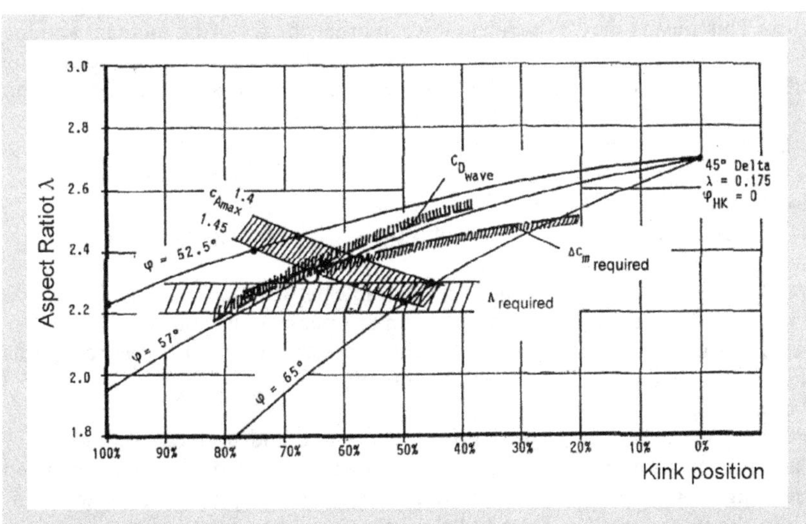

Wing planform optimization.

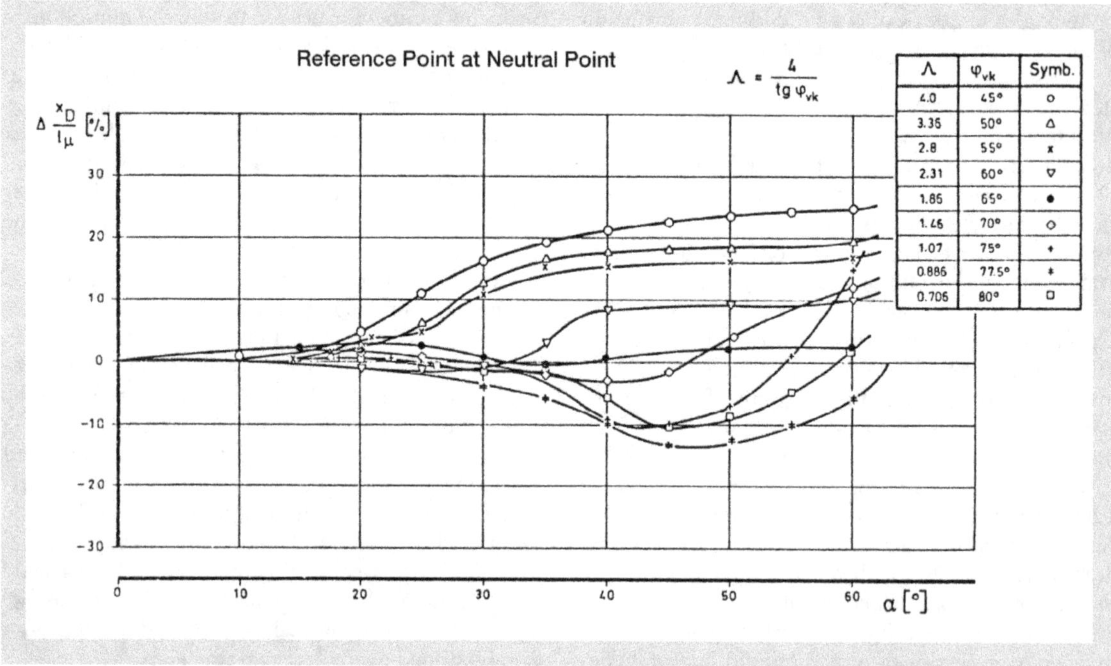

Relative movement of the center of pressure (x_{cp}) of delta wings at angles of attack between 0° and 60°.

The exact analysis of the optimized wing planform shows, however, that for a wing-fuselage-only combination with a degree of instability of 8 per cent c_μ neither can C_{Lmax} be trimmed nor can the requirement of a remaining restoring moment be met, as is also depicted in the next figure. This means first of all that the demand for an unstable design and the achievement of angles of attack or the recovery from angles of attack around C_{Lmax} is contradictory. To meet the latter requirement, the configuration has to be designed neutrally stable to stable, as is also shown in the figure. The solution to the problem of the contradictory requirements is the addition of a surface upstream of the wing – i.e., the addition of the so-called canard. This way, the wing-fuselage combination can be designed neutrally stable and the necessary restoring (recovery) moment can be ensured by the deflection of the trailing edge flaps (bottom curve of the top diagram). The desired optimum in the degree of instability is then achieved by adapting the canard surface and the moment arm (see top curve in the figure). In the range of C_{Lmax}, the canard has to be turned back into the undisturbed flow (freestream) so that it does not provide a contribution to the pitch-up moment, but instead supports the restoring moment.

By adding the canard, one wins an additional degree of freedom, which enables one to adjust independent of each other the restoring moment and the degree of instability. The canard may, in addition, be employed for control and for a gust-load reduction system to be developed later.

Considerations concerning canard geometry and location

The preconditions for the canard layout are:

- Fully controllable between η_c = +20° and –60°
- contribution of 8 per cent instability,
- impairment of the visibility out of the cockpit as small as possible.

To optimize the canard and its location, a series of wind-tunnel tests was conducted, which encompassed forward sweep, medium sweep up to large sweep angles and close coupling up to a strong decoupling.

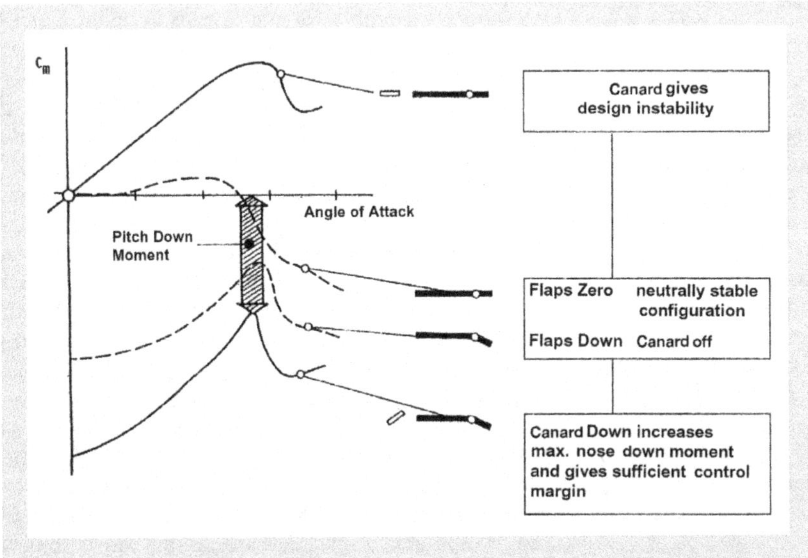

Canard contribution to stability and controllability at different flap deflections.

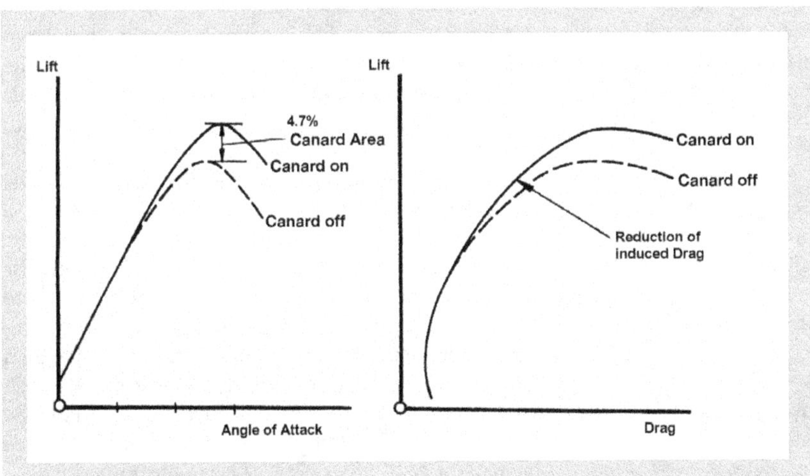

Positive contribution of the canard to lift and drag.

Positive interference of the canard delays the wing vortex bursting at α_{max}, which leads to a more favorable C_{Lmax} and to a better polar. In order to achieve this, the canard has to have a small high-position (z-distance) with regard the plane of the wing.

In the case of a configuration similar to the TFA/TKF with a corresponding canard-wing coupling, and a destabilization of 8 per cent the relative volume coefficient of the canard assumes a value of $V \cong 0.05$ (V represents the exposed canard surface multiplied by the moment arm and divided by the ideal wing area times c_μ).

Despite the fact that the interference canard - wing has a favorable effect on the C_{Lmax} behaviour of the overall configuration, the largest possible decoupling of both surfaces was strived for. There are several reasons for this:

1. The smallest possible canard, which provides an 8 per cent stability contribution, is gained by the

largest possible decoupling. This reduces the loads, which have to be introduced in the area of the cockpit (open U). Also being reduced are flutter problems on the canard.

2. The canard generates by interference a wing pitching moment, which is opposite to the deflection of the canard and reduces the controllability of the canard. Only by decoupling can the loss in control moment be reduced.
3. The cockpit visibility problem requires a negative dihedral of the canard. By a high position of the canard it can be ensured that the canard is on average located above the plane of the wing and that the positive interference effect is thus sustained.
4. The most forward position of the canard is limited by the minimum distance of 200 mm (7,87 in) to the angle of attack and yaw probes, for whom, in turn, a fixed distance to the radar must be maintained.
5. The leading-edge sweep angle of the canard shall not exceed 45°, so that the canard does not generate nonlinear vortex-induced lift thereby reinforcing the instability with the angle of attack.
6. Decoupling of the canard has the further advantage that in respect to the lateral motion the effect of the canard vortex on roll instability ($C_{l\beta}$-dip) can be minimized.
7. A controllable canard, as needed to pitch-down the configuration, allows through the possibility of generating additional moments new control means, such as direct lift generation or fuselage aiming, and the possibility to realize an effective gust alleviation system.

As non-dimensional geometry were defined:

ϕ_{LE}	= 45°	forward sweep angle,
Λ	= 3,43	aspect ratio,
λ	= 0.2454	taper.

X- and z-positions were not fixed to provide to the designer with a certain degree of freedom to find, within the limits described above, the best configurational and structural position.

This optimized wing-fuselage-canard configuration was the starting point for the study concerning drag reduction and the definition of the CCV-potential.

In the subsequent diagram, the improvement of the polar at different trim settings is again shown as a descriptive example. "Stable" means here: Pure delta wing 2 per cent stable, trimmed according to the first basic principle with a negative trailing edge (flap) deflection. "Indifferent" means: Flap without deflection. "Unstable" requires that trimming be carried out only with an additional surface such as provided by the elevator or the canard. For an unstable configuration, which may be trimmed by a positive deflection of the wing trailing edge flap, the variable camber causes an additional improvement of the polar as shown by the enveloping curve. This means that the chosen configuration has to be trimmed first by the trailing edge flaps. However, the drag polar can be further improved in that by the addition of leading-edge flaps optimum deflection schedules can be established for trailing-edge flaps, nose flaps and the canard for any Mach number to trim the configuration at the respective Mach number and respective angle of attack. If one generates in this way trimmed polars for different instability margins, one is able to determine the effect on the size and weight of the configuration with CCV technology with the aid of the computer-aided design method, which scales the configurations considered according to the demands of the (design) point performance and the design mission.

The result is subsequently being displayed. The reduction in take-off weight was here chosen as a measure of the improvement of the configuration. Also included is the so-called snow-ball effect: the fuel reduction in case of the design mission due to the lower drag also results, due to the decrease in tank volume, in a smaller and lighter configuration, which has, in turn, less frictional drag and less mass, which again contributes to further fuel savings. For the chosen TFA/TKF example, an optimum in the degree of instability of 8 per cent c_μ was determined with a maximum of 10 per cent c_μ being required to cover the effect of extreme external stores. This coincides with the system-technological maximum time-to-double, which can still be controlled; this was confirmed by the CCV-F-104 G. The improvement (reduction) in the take-off weight is 8 per cent which constitutes a considerable saving; for instance at a take-off weight of 15 t without CCV technology this corresponds to a saving of 1.2 t due to the CCV technology. The figure shows, furthermore, that above a degree of instability of about 10 per cent the system weight strongly increases in the case of the TFA/TKF design, at least

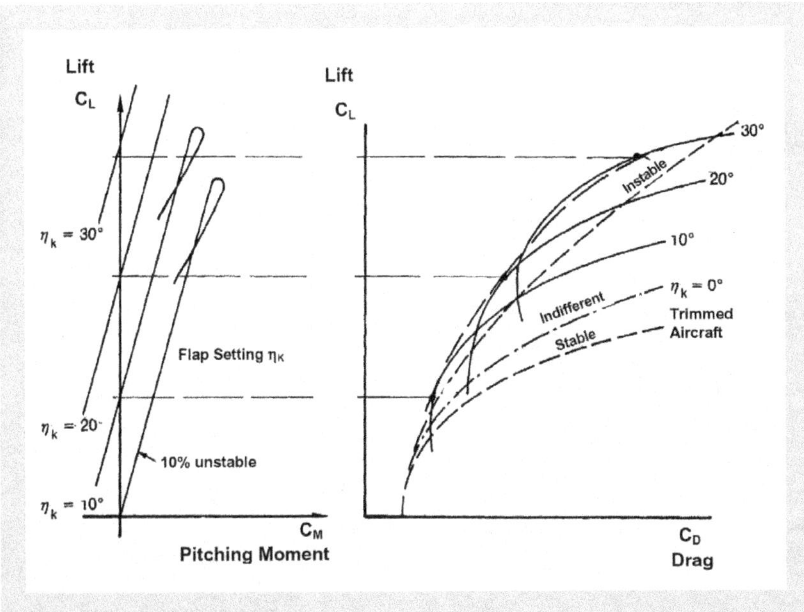

The concept of variable camber for an unstable aircraft design.

based on today's equipment and systems technology (see the CCV-F-104 G treatise above).

For the actual aircraft, certain curtailments may become necessary, if, for instance, in the case of extreme external stores additional local non-linearities occur in the pitching moment.

The expected weight increase in the case of a conversion of the design into a real aircraft is of little importance as is the additional drag due to the antennae, infrared sensors, air refueling probe, et cetera since one can assume that also the configuration without CCV technology is subjected to the same buildup of weight.

Summarizing, one may state that the configuration thus optimized constitutes a very good design for the conventional flight range and an excellent basis for all

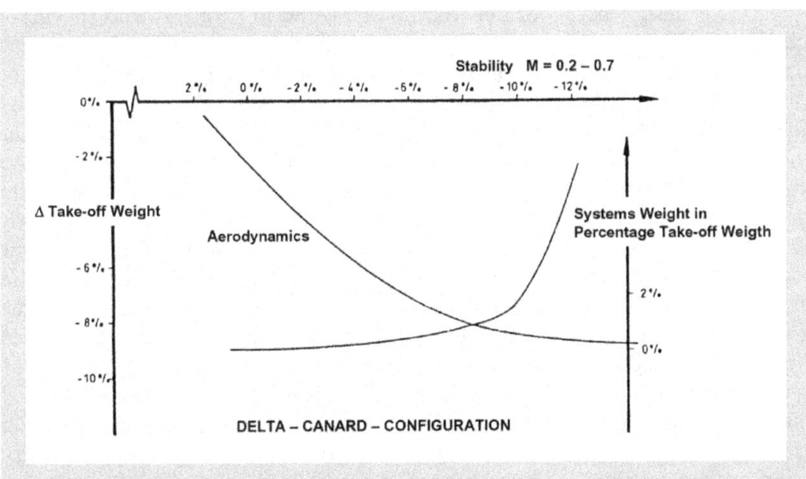

Optimum degree of instability of the delta-canard configuration.

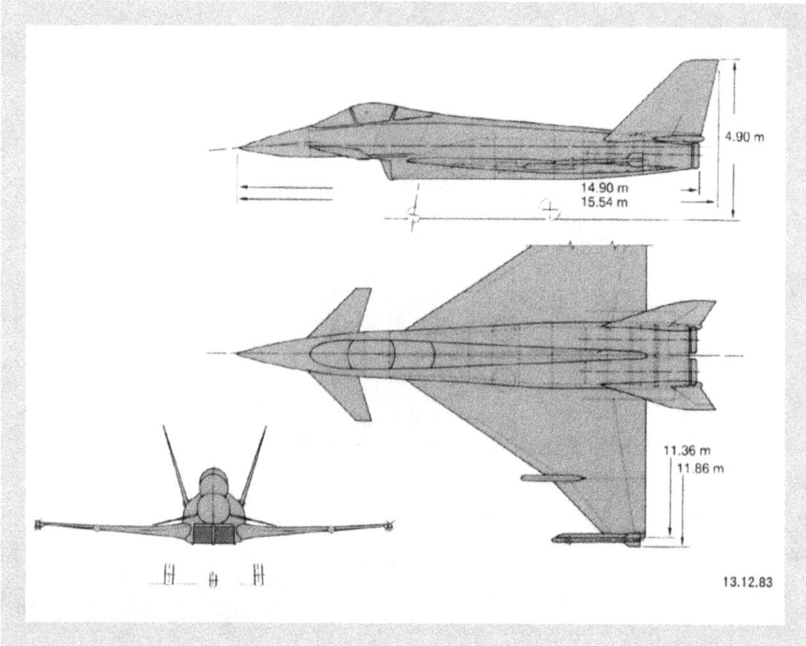

TFA/TKF delta-canard basic configuration, final configuration as a result of the optimization.

further demonstrator derivatives, such as the ACA, EAP and X31, and the "Eurofighter" EF 2000 which is now manufactured.

In April of 1982, an industry initiative for the design and construction of a demonstrator for a new fighter aircraft designated "Advanced Combat Aircraft, ACA" was started by the presidents of the companies AIT, BAe and MBB. The here-developed configuration corresponds in essence to the TFA/TKF design introduced by MBB. The slightly changed requirements resulted in a wing area of $A = 50\ m^2$ ($538{,}2\ ft^2$) instead of the $52\ m^2$ ($559{,}7\ ft^2$) of the TFA/TKF. For cost reasons, also the twin vertical stabilizers of the TFA were foregone, and one fell back to the existing "Tornado" single-stabilizer design. Unfortunately, MBB had to withdraw from the demonstrator program in late 1983 at the request of the BMVg. AIT and BAe completed the aircraft within the now called "Experimental Aircraft Program, EAP". The first flight took place on August 8, 1986 in Warton (see Color Plate VII). Flight-testing was also performed and very successfully completed in Warton. Flight-performance tests were not carried out in detail since the flyable load factors were limited for cost and risk reasons. A long-distance flight similar to the design mission was also not performed since the number of flight hours was limited. Other interesting results are only partly available, which is due to the fact that MBB was not able to contribute financially to the trials.

The results of the industrial research described up to now concern the longitudinal motion of the aircraft. It was the objective to clearly identify the potential, which results from the integration of a number of new technologies, and to turn this into specifications that would enable members of the "Armament Department" ("Rüstungsabteilung, Rü") and the Study Group of the "Command Staff of the Air Force" ("Führungsstab Luftwaffe, FüL") to put down the "Tactical Requirements" ("Taktische Forderung, TaF"). The pre-liminary work, performed with regard to the TFA/TKF, aimed, therefore, only at a relative optimum. Only during the actual development phase of a new fighter aircraft will an absolute optimum be established, based on the exact data of the TaF-catalogue, and supported by comparative design studies under consideration of all experiences and results of the pre-development phase.

The development work for the "Eurofighter" EF 2000 is not being described here, although all preliminary work described has been included. This becomes al-

Exotic, swept-forward canard.

ready obvious by considering the external similarity of the Eurofighter and the TFA/TKF configuration (see Color Plate VII). Why, nevertheless, differences can be observed in some details of the configurations is due to the fact that, within the framework of the four-nationality negotiations of the Armament Directors, TaF parameters were changed and new ones added. Germany always demanded a pure air-to-air aircraft whereas the other nations introduced an additional air-to-ground operational requirement, which naturally led to certain compromises. The post-stall technology (flight at very high angles of attack) was deferred as retrofit potential.

The Armament Directors of the four participating nations prescribed, in addition, the wing area, thrust, and net-weight in absolute numbers, which led to a change in the optimization parameters.

The consequent realization of the CCV technology, as described here, also on large aircraft such as "Airbus" and "Concorde"-like flight vehicles is difficult but possible within limits. The problem in the case of "Airbus" relates to the attachment of the landing gear on the wing. When shifting the wing forward to position the neutral point upstream of the center of gravity, the landing gear moves along and quickly exceeds the balance limit for loading on the ground. A decoupling of landing gear and wing leads to an unacceptable weight increase, which strongly reduces the advantage of the CCV-technology.

This restriction is not given in the case of a military high-wing transport aircraft where the landing gear is not attached to the wing.

It was investigated by Airbus Industries, ONERA and DLR, whether the application of a canard downstream of the cockpit is a viable solution. In order to achieve an optimum degree of instability of about 20 per cent c_μ, one would, however, considering the large wing area, also need a relatively large canard area, which, in turn, might generate undesirable interferences, a matter that has to be explored. There also might arise difficulties to sell such a configuration to the esthetes among the aircraft manufacturers and aircraft operators.

To utilize a certain percentage of the improvement potential, one employs so-called trim tanks to pump, during flight, fuel, for instance, into the horizontal stabilizer thus manipulating the center of gravity location. This is successfully being practiced. Over and above that the further development in the area of large-scale aircraft remains to be seen.

Remark

Concerning this chapter, some figures are presented in color:

Color Plate VII

- Modified F-104 with canard.
- First flight of the EAP-demonstrator.
- Prototype JP 001 of the "Eurofighter".

Controlled Flight in the Presence of Flow Separation

HELMUT JOHN

In order to describe the tasks to be solved within the framework of industrial research more precisely, the basic flight-physics problems of the lateral motion at separated flow conditions and beyond maximal lift shall be discussed first. Especially roll stability and roll damping show deficits, which are symptomatic of slender configurations with a low moment of inertia about the roll axis. To obtain a flyable configuration, the complex flight control laws of the elastic aircraft are partly verified in wind-tunnel tests. The successful realization of the configuration finding process is described on the basis of the X-31 and documented by results of flight tests.

Flight Physics in the Presence of Separated Flow in the Post-Stall Domain

In the case of low-aspect ratio wings with large leading-edge sweep angles, the flow already separates at moderate angles of attack in an orderly and stable fashion at the wing leading edge and rolls up into two vortices, which trail downstream over the wing. Due to the inductive effect on the wing flow, an additional "non-linear" lift is generated which, together with the fact that a thin wing with a high sweep angle also exhibits a smaller wave drag, renders this type of wing planform especially attractive for supersonic flight. *D. Küchemann* already performed a theoretical modeling of this type of flow during the development of the "Concorde" [78].

The roll-up process is associated at the core of the spiral vortex with high velocities relative to the freestream velocity and a corresponding reduction in static pressure. With increasing angle of attack, a total pressure loss occurs due to the dissipation of vortical energy, which renders the vortex unable to negotiate the pressure rise back to the freestream pressure behind the wing. The vortex bursts, forming a bell-like shape with a radius many times larger than the original one, with a reduction in stagnation pressure to 10 per cent of the freestream value, and correspondingly strongly decreased velocities in the burst area. It is possible only today to determine, by means of numerical simulations, the vortical flow development and the vortex bursting. The surface pressure distribution in the form of isobars and the vortex breakdown at the mid-wing position of a delta wing are depicted in Color Plate VIII. For clarity reasons, the flow is shown on one side of the wing only.

If the break-down occurs at the trailing edge of the wing, a series of phenomena occurs due to the unsteady character of the separated flow, such as, for instance, buffeting, wing rock, nose slice, et cetera. They are initially only affecting the pilot's comfort, but exert already a noticeable influence on the flight movement due to the change in the increase of the aerodynamic coefficients. If at a further increase of the angle of attack the breakdown point has moved so far upstream that lift is no longer increased, the maximum lift, C_{Lmax}, has been reached. In the range prior to and beyond C_{Lmax}, a series of unmanageable flight conditions may occur for the uncontrolled aircraft, due to the effect of vortex breakdown on all coefficients, such as, for instance, pitch-up at symmetric breakdown conditions at a yaw angle of $\beta = 0$.

At finite yaw angles, an asymmetric vortex breakdown originates and the windward wing rolls downwards due to a larger loss in lift. This is called wing drop ($C_{l\beta}$ – instability) and may lead to spin. A loss of roll damping may occur during roll due to the asymmetric location of the burst point on the two wing halves and due to an asymmetric vortex intensity. If such phenomena occur prior to reaching maximum lift C_{Lmax}, it is the task of the aerodynamicist to improve the coefficients by appropriate means, such as boundary layer fences, saw teeth, notches, strakes et cetera, so that C_{Lmax} can be reached. In general, maneuvering is then no longer possible – except for pitching down – and should, therefore, be prevented by the flight control system. Exceptions are here aerobatic and trainer aircraft, which are expected to go easily into and out of a spin.

Concerning spin investigations, sufficient treatises with aerodynamic data sets up to angles of attack of $\alpha = 90°$ can be found in the literature. Also for the variable-sweep fighter aircraft "Tornado" various spin data sets are available for wing leading-edge sweep angles of $\varphi_{LE} = 25°$, 45° and 68°, which were established for incidences of up to 120 degrees. An extensive study of the data shows that especially in the range $\alpha_{max} < \alpha \lessapprox 50°$ larger problems are to be expected since, in the presence of partly separated and partly attached flow, at normal and yawing flight conditions, large local instabilities occur. Above $\alpha = 50°$ at completely separated flow conditions, i.e., in the post-stall regime, the coefficients again follow simple and predictable laws.

For the range of separated flow, three statements are possible:

1. Surfaces, which are facing the dynamic pressure, experience no suction on the upper side and it only remains the surface impact effect, i.e., only the windward side of the surface generates forces with the pressure point being located approximately in the center of gravity of the surface.
2. Surfaces, which lie in separated flow (wake flow), as, e.g., vertical stabilizer and rudder, generate no forces and moments due to the lack of dynamic pressure ($q \rightarrow 0$).
3. If a single surfaces can be turned out of the wake flow, as, e.g., a canard, rotatable wing tips, or a rudder on a cantilever away from the fuselage, then the standard aerodynamic laws and calculation methods are valid again.

It had to be found out within the framework of research programs which wing planform would be the most suitable and which measures were necessary to come up with a design, which was not only able to penetrate the region of separated flow but also to fly and maneuver in that region. This requires that sufficient maneuverability is provided in that region, which is, per se, not given by aerodynamic means. Furthermore, especially the air intake has to be considered, since the engine should, at these flight conditions or under any other circumstances, not have a flameout. Beyond C_{Lmax}, engine thrust has to replace the part of the lift that cannot be produced aerodynamically. In the limiting case of steady flight at $\alpha = 90°$, the aircraft is solely carried by engine thrust.

Benefits of Post-stall Capabilities

Before solutions are considered, the question shall first be answered what benefits a pilot has from flying an aircraft beyond maximal lift:

- An aircraft, which is fully controllable beyond C_{Lmax} and up to very high angles of attack, is advantageous for the pilot since he can fully utilize his aircraft without fearing uncontrollable flight conditions (spin). This eliminates a possible crash cause and contributes to a "carefree handling". However, the pilot has to be given support so that he can center C_{Lmax} with the control stick to achieve the maximal turning rate at a given Mach number.
- The new capability leads to new maneuvers and flight conditions, such as, e.g., slice maneuvers, the Wolfgang-Herbst-maneuver, a return flight with an extremely small turn radius, and steady equilibrium flight at high angles of attack. An optimized trajectory [79] of the Wolfgang-Herbst-maneuver is depicted in the artist impression.
- There are tactical advantages in close air combat (close-in-combat and dogfights, respectively), which, however, first has to be proven in digital or manned simulations or in demonstrator programs.

During the years 1977 and 1985, detailed simulations were carried out at MBB and the IABG to prove the "tactical advantages" of the post-stall technology and of the new CCV-control methods.

In all cases, a base-line aircraft, without the new capabilities but with identical conventional performance, equipment and armament, was flown against an aircraft with post-stall and CCV-fuselage-aiming capabilities for the utilization of a cannon and a short-range missile.

If one only considers the close-in-combat results, one would (averaged over all simulations) register a huge superiority of an aircraft with the new capabilities compared to an aircraft without these capabilities. In case of an armament with only one cannon, a statistical average of 13,000 air combat events resulted in a ratio of 5 : 1. Five won and one lost encounter in a total of six encounters can be seen as a definite dominance.

The analysis shows that this advantage is being gained by a series of individual capabilities, viz., the ability to fly at high angles of attack, high agility in roll and pitch, the possibility to reach, at higher angles of attack, significantly smaller turning radii, and the good deceleration capability by using the entire aircraft as air brake. It is to be mentioned furthermore, that in the majority of the encounters frontal attacks with the cannon as well as with the SR-missile were carried out. Such simulations are exemplary of investigations concerning the effectiveness of aircraft under operational conditions; they are necessary in the civil as well as in the military area before developments commence.

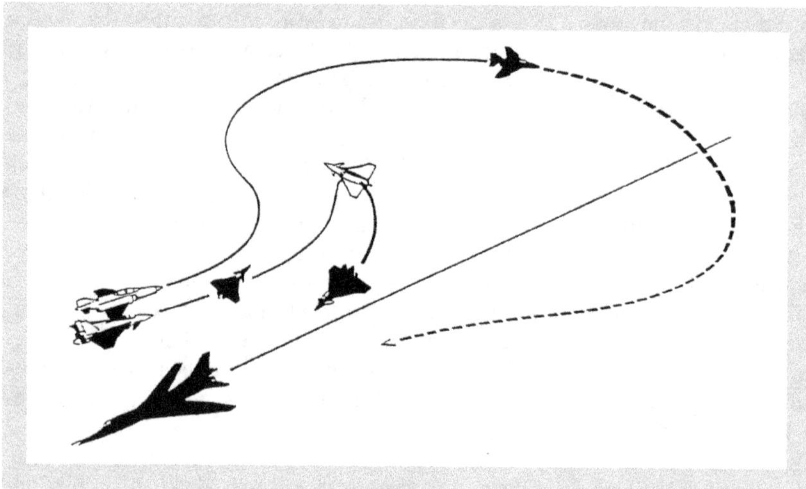

"Wolfgang-Herbst-maneuver": chase of a penetrating enemy.

Optimization of Configurations – The Tactical Fighter Aircraft (TKF)

On the basis of this example it shall be demonstrated how today, prior to a new aircraft design, the configuration is being varied in the computer before wind-tunnel models are being built. It is also an example of the optimization of a "virtual product", even if all design tools have not yet been combined to one mighty tool. The demonstration is done employing as example the "Tactical Fighter Aircraft" ("Taktisches Kampfflugzeug, TKF").

The process of configuration finding in reality is based on numerous configuration studies, wind-tunnel tests, detail optimizations, and re-iterations.

Demands on the configuration were in the considered case:

1. Full application of the new possibilities provided by the active control technology (CCV = Control Configured Vehicle), among them:
   - unstable design and variable camber for drag reduction,
   - new control techniques, especially fuselage aiming capability and drag modulation.
2. Post-stall capability, which means flights up to high angles of attack and maneuverability beyond C_{Lmax}. To support the maneuverability in this range, a thrust vectoring nozzle/ thrust vectorization is envisaged.
3. Good supersonic performance/lift to drag ratios.
4. Application of new materials, e.g., CFRP. This allows, for instance, a twofold curvature of the wing surface and hence uncompromising designs with regard to optimum twist and camber.
5. Good take-off and landing characteristics. Landing roll distances of about 500 m (1,649 ft).
6. Satisfactory gust characteristics (ride qualities).
7. Use of RB 199 engines (the decision to develop a new engine was made only much later).
8. The system optimization to meet the point and mission performances resulted in the following characteristic data:
   Maneuver wing loading: G/F = 220–230 kg/m^2 (46–48 lbs/ft^2),
   thrust to weight ratio S/G = 1.2–1.3,
   and an internal fuel volume of GF = 4.0–4.3 t.

In summarizing, the following reasons for the selection of the concept can be given.

Delta Wing

The delta wing profits proportionally more than the trapezoidal wing from the application of the CCV-technique (unstable flight). The large root chord of the wing allows smaller relative wing thickness, which results, together with the favorable cross-section distribution of delta wings, in lower wave drag.

The lower wing load, which is a design characteristic of delta wings, also leads to small landing roll distances due to a lower approach speed.

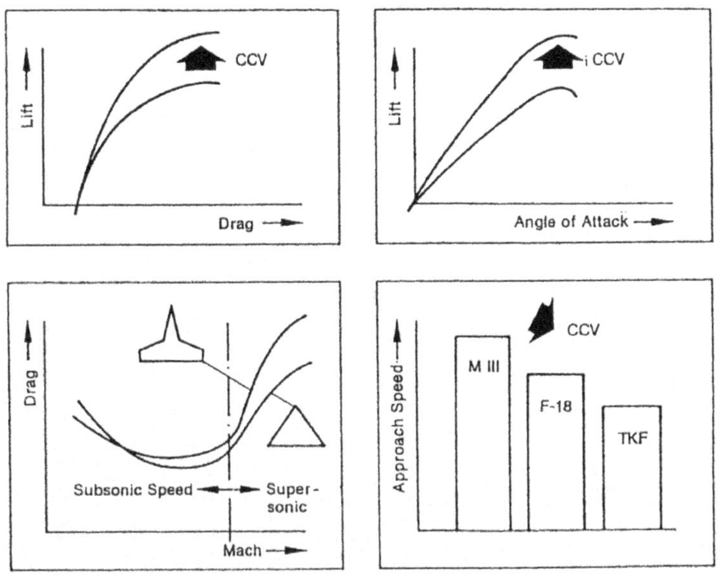

Justification of the delta wing.

The multifaceted requirements of external stores always make their configurative accommodation problematic. Here too, the delta wing performs well since it is aerodynamically comparatively insensitive to external stores.

The choice of the configuration, the optimization of the wing planform, the reason for a canard, and the improvement potential due to the full utilization of the CCV-technique has been comprehensively justified in the preceding chapter.

Wing-design denotes the optimization of the wing with respect to the twist and camber distribution along the span. The work required here was carried out and documented by *Werner Kraus* [80]. Wind-tunnel measurements were conducted to define the leading edge (L. E.) flaps. As the optimization process on the low-speed model showed, L. E. flaps were not required in the area downstream of the canard. Outside of the canard span down to the wing tip, flap deflections normal to the hinge line of up to 30° are required to optimize drag. The L. E. flaps showed, furthermore, that they contribute, up to a Mach number of $M = 1.1$ and in the range of medium to high angles of attack, very effectively to the improvement of the polars. If one takes the turning-rate demand at $M = 0.9$ and an altitude of 11 km (6.8 mi) as a base-line, a comparative design study shows that the improvement in the polar compensates, via the interchange rate, 1.5 times the additional weight due to the L. E. flap, which certainly argues in favor of the introduction of L. E. flaps.

Further deliberations concerning the longitudinal motion – the controllability will be treated later – shall not be made at this point; rather the lateral motion and the selection of the vertical stabilizer and the rudder shall be considered here.

Directional Stability

For the assessment of the aircraft in lateral motion, the "Dynamic-Directional-Stability Criterion", as defined by *R. Weissmann* [81], is employed. This criterion establishes a defined relationship between the static directional stability $C_{n\beta}$ and the static roll stability $C_{l\beta}$ together with the ratio of the moments of inertia of the aircraft about the vertical axis I_{zz} and the roll axis I_{xx}. This allows an assessment of deficits. Since the ratio I_{zz}/I_{xx} can assume values between 5 and 7, roll stability ($C_{l\beta} \leq 0$) becomes of great importance. Delta wings generally encounter in the incidence range between $\alpha = 30°$ and 50° a roll instability ($C_{l\beta} > 0$) due to the asymmetric vortex breakdown, whose elimination requires large efforts in research and basic investigation. It can be influenced by the configuration elements rudder, L. E. flaps, small local surfaces, also called strakes – which locally fix the vortex formation – and similar means.

Single or Twin Vertical Stabilizer

Helmut John and *Werner Kraus* [82] and *Helmut John* and *Ulrich Haux* [83] extensively dealt with the vertical stabilizer design for the TKF. The most important question to be answered is, whether, for the selected configuration and the given requirements, a single or a twin vertical stabilizer is more advantageous. Here, it is first being investigated whether, at the design point, there is a fin-volume advantage for one of the two possible vertical stabilizer configurations since this could initially mean a lower friction drag and possibly also a lower wave drag.

The effectiveness of the vertical stabilizer, depicted in the next figure, is, on the one hand, characterized by the strong decrease with Mach number (low aspect ratio wings) and, on the other hand, by the strong influence of the vortices originating at the canopy, which diminish the effectiveness with increasing angle of attack, dependent on the size of the canopy, more or less rapidly. It can be shown that for the selected design and the wing loading the contribution of the vertical stabilizer has to be considered at

$$M = 2.0, \qquad V_{CAL} = 800 \text{ kn}, \qquad \alpha = 6°–8°.$$

Results of measurements [84] conducted by *Keith McKay* for different Mach numbers and canopy configurations are shown in the subsequent figures. It can be seen that for a "standard" canopy or a "high" canopy configuration at M = 2.0 the twin vertical stabilizer has no disadvantages in the range around $\alpha = 5°$ in comparison to a single vertical stabilizer, but a significant advantage at higher angles of attack. This means that the wake of the canopy influences the single stabilizer much stronger since the gradient in the α dependence is steeper. In comparison to the configuration without canopy, the twin vertical stabilizer becomes increasingly more advantageous with increasing canopy height. The biplane effect becomes stronger with decreasing Mach number; the contribution to the directional stability decreases for a twin vertical stabilizer, however, it is always larger than necessary.

This is an advantage for the twin vertical stabilizer since the aeroelastic effects are smaller in the transonic and subsonic speed ranges and therefore a high rigid directional stability is not required. The reduced directional stability favors the controllability about the vertical axis, which is required for the mechanization of fuselage aiming.

The second area which was investigated in detail, was the low-speed range at medium and high angles of attack to investigate the influence of the various vertical stabilizer configurations on C_{Lmax}, $C_{l\beta}$ and $C_{n\beta}$.

The twin vertical stabilizer in a V-position (tail dihedral) of 18°–20° shows a pronounced stabilizing effect on $C_{l\beta}$. No single-stabilizer configuration displayed this characteristic. *Stephen Ransom* had car-

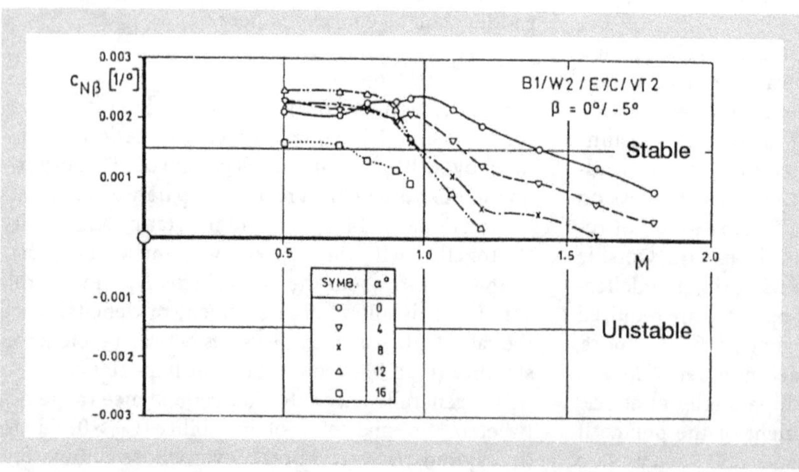

Vertical stabilizer effectiveness as function of Mach number and angle of attack.

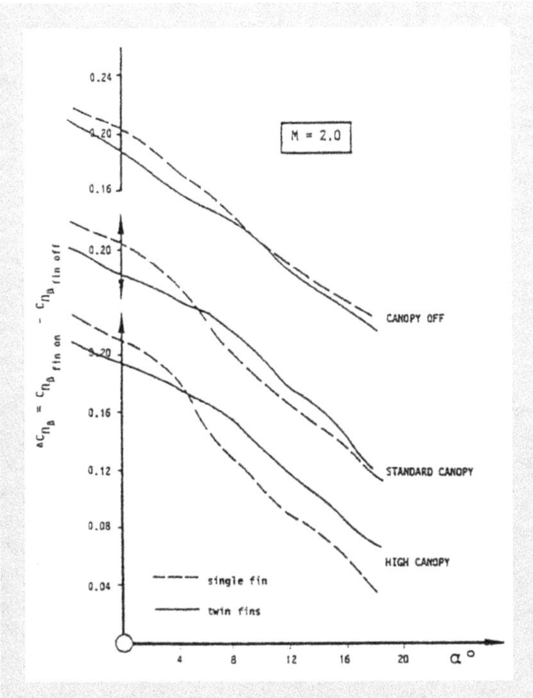

Comparison of single vertical stabilizer and twin vertical stabilizer for various canopy configurations at M = 2.0.

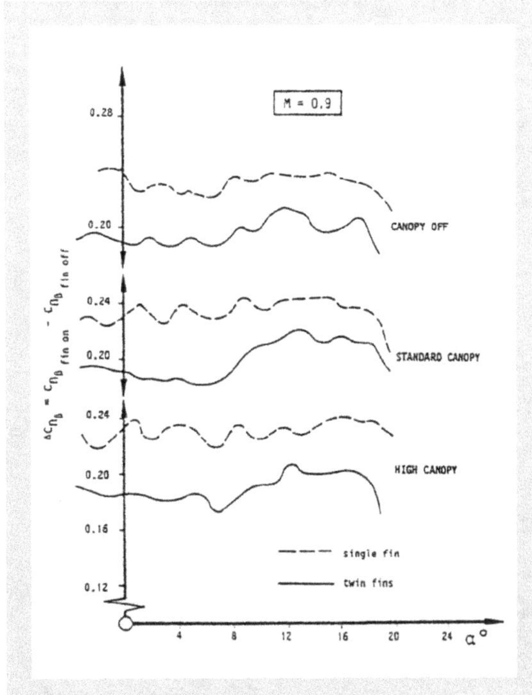

Comparison of single vertical stabilizer and twin vertical stabilizer for various canopy configurations at M = 0.9.

ried out in 1980 at VFW-Bremen investigations [85] on an alternative to the MBB configuration with the objective of showing, how a configuration would look like, that remains restricted to the post-stall envelope. Results are shown in the diagram of the stability characteristics; they are largely in agreement with the findings of MBB.

The $C_{n\beta}$-behavior beyond C_{Lmax} indicates for the single vertical stabilizer an additional strong destabilization over and above the case without stabilizer, as shown in the figure, which lets one conclude that an additional cross flow occurs downstream of the partly separated wing flow, which is caught by the single vertical stabilizer. This was also confirmed by the studies of *Stephen Ransom*.

The twin vertical stabilizer, on the contrary, never shows values in the range beyond C_{Lmax} that exceed the value without vertical stabilizer. This means that the cross flow is already cut off at the very beginning which also results in favorable $C_{l\beta}$-values.

This shows that, only judging aerodynamics, the twin vertical stabilizer is superior to the single one and should therefore be given preference within the conventional flight domain.

The main argument against a twin vertical stabilizer lies in the negative experience encountered on the F-18, namely, so-called "fin vibrations", which led to the local loss of skin structure (panel flutter). The reason for this phenomenon is that the wing vortices move directly towards the vertical stabilizers. This represents a certain risk, especially at flights beyond C_{Lmax}, which could not be reduced despite of considerable research efforts. This later affected the decision against a twin vertical stabilizer. The risk also lies, of course, in the fact that problems most often only are detected during flight-testing, when configuration changes are nearly impossible to make and "repair solutions" are very expensive.

As a consequence of the decision in favor of a single vertical stabilizer, a new design element had to be

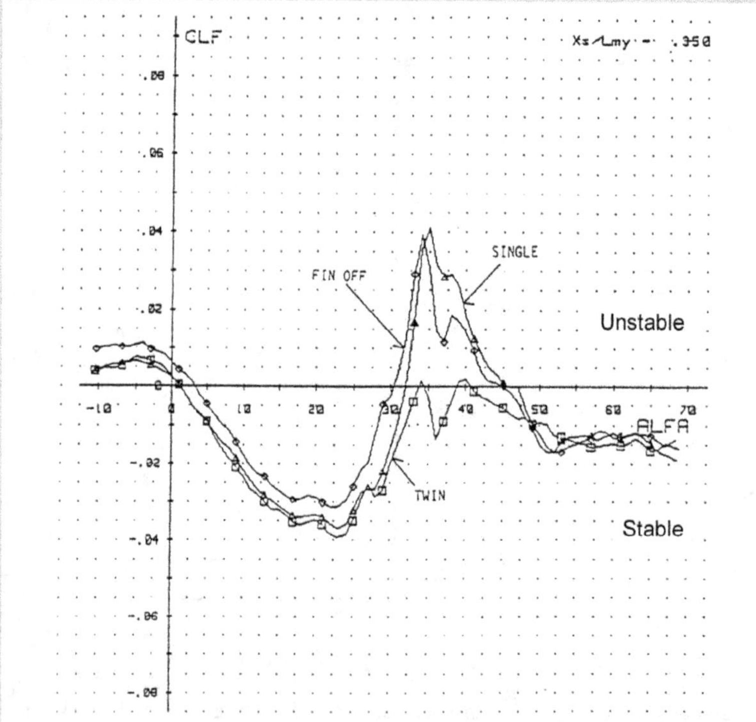

Effect of single and twin vertical stabilizer on roll stability.

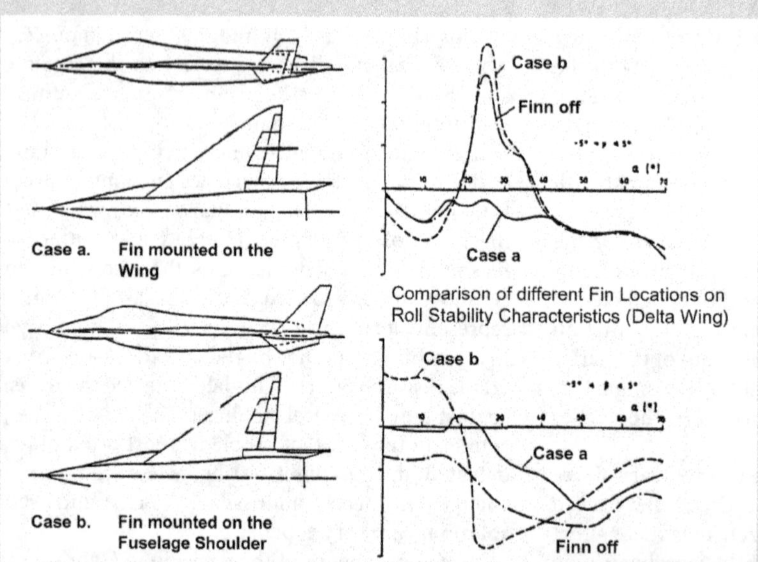

Vertical stabilizer variations to improve roll and yaw stability.

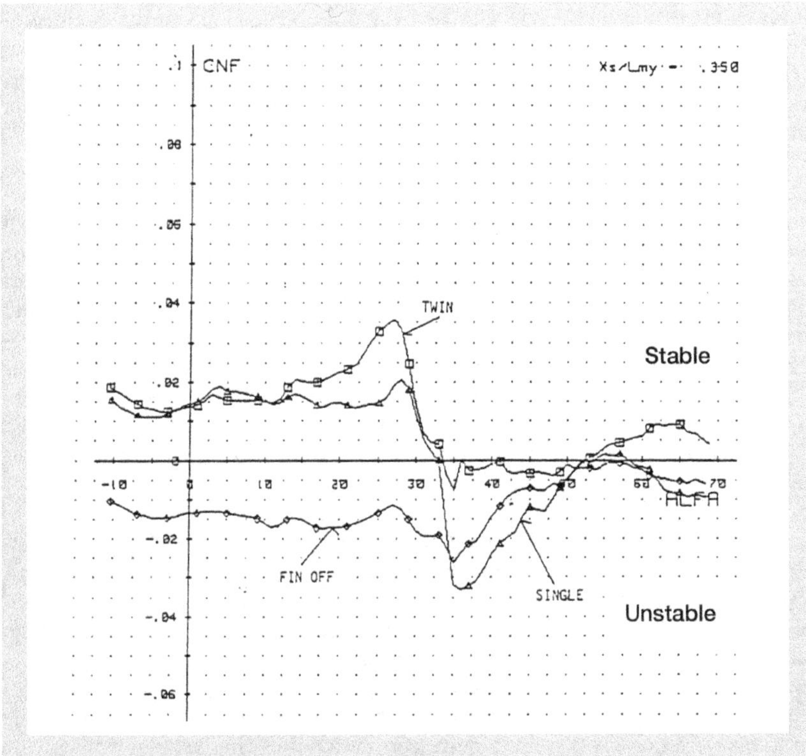

Influence of single and twin vertical stabilizer on the yaw stability.

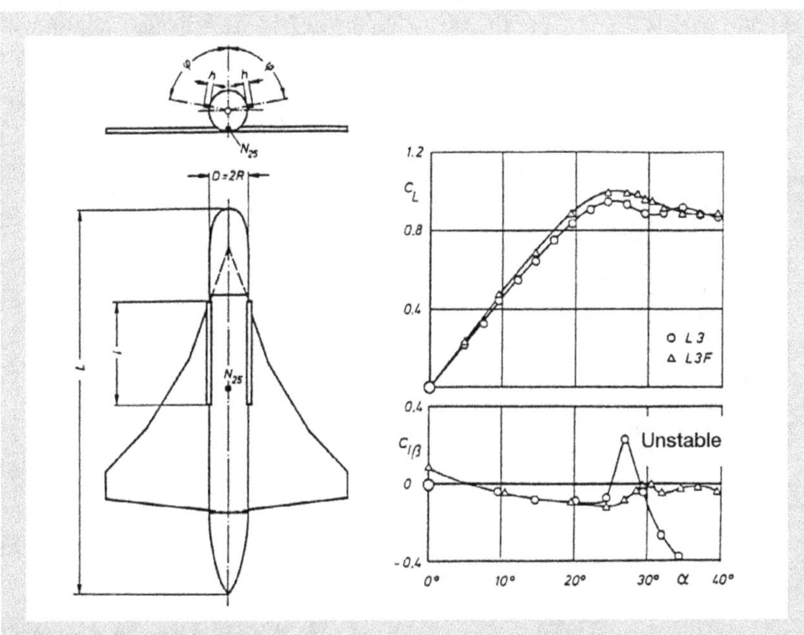

Effect of fuselage strakes on the roll stability of a low-wing aircraft configuration (L3 without strakes).

provided, which resulted in a similarly positive influence on $C_{l\beta}$ and $C_{n\beta}$, in order to remedy the deficits of the lateral motion. *Dietrich Hummel* of the Technical University (TU) Brunswick carried out systematic investigations. He arrived at the conclusion [86], as shown in the figure, that the $C_{l\beta}$-instability can be eliminated by attaching strakes to the fuselage upstream of the wing, which gives, in addition, a small increase in maximum lift. These strakes were transferred to the base-line configuration.

After it had been proven that L. E. flaps also contribute to the improvement of the polars in the case of highly-swept delta wings (see above), it was tried to achieve an improvement in the $C_{l\beta}$-characteristics in the range above C_{Lmax} by increasing the L. E.-flap deflection up to $\eta_{L\,E} = 40°$. The L. E.-flap control law was then extended up into the region of separated flow.

Controllability

Intensive studies have shown that sufficient aerodynamic controllability cannot be guaranteed for post-stall flight. This holds, most of all, for the control about the vertical axis since the vertical stabilizer, and hence also the rudder, are, at angles of attack of $\alpha \geq 35°$, fully embedded in the wake of the separated wing flow. Other vertical stabilizer configurations were not possible for the delta wing, whose trailing edge lies at nearly the same position as the engine nozzles.

The solution for the controllability about the pitch and yaw axes is the utilization of a thrust-vector nozzle, attached via an universal joint to the engine, instead of the fixed convergent nozzle. Very large control moments, which are relatively independent of the angle of attack, can be attained during afterburner operation by thrust deflection.

Thrust vectorization is not an entirely new invention. It was already in the fifties and sixties a key element of vertical take-off and landing (VTOL) aircraft design. However, there it was only required during the take-off and landing phase, with deflection angles of 90° and more, and with maximum thrust. Another characteristic of VTOL-configurations is that the total thrust vector must always pass through the center of gravity to avoid unwanted thrust contributions to the control moments. Necessary control is produced by rotating the nozzles ("Harrier"), by thrust modulation (VJ 101) or by deflecting vanes or baffles (XV-5A).

It has to be stated here that in case of the VTOL configurations the installation of the relatively voluminous propulsion system led to very unfavorable cross-section distributions, resulting in a high wave drag, which excluded supersonic flight in most instances. A new idea, originating in the mid seventies, was the deflection of the thrust by 10–20° against the thrust axis for a quite conventional engine arrangement in the rear fuselage. Designs for a possible mechanization of the thrust vector nozzle are schematically shown in the figure. The maneuverability of the aircraft can now be significantly increased by thrust vector con-

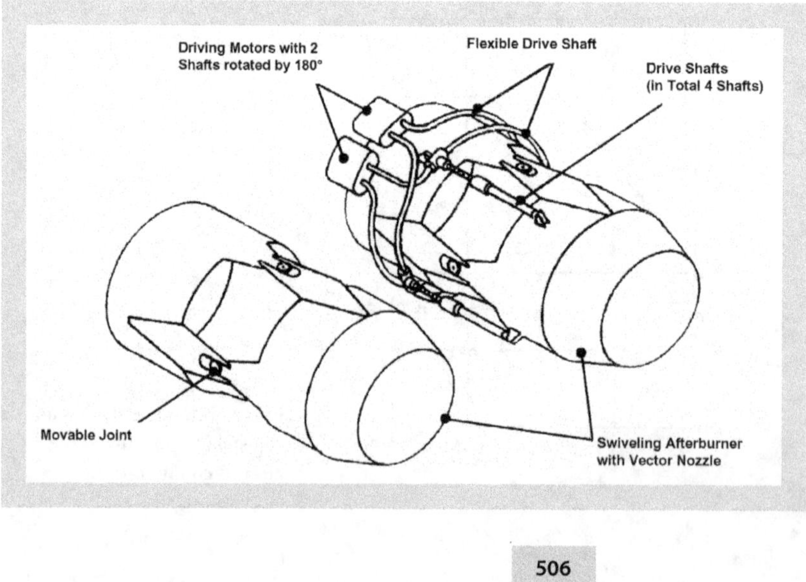

Concept of a vector nozzle.

trol about the pitch and yaw axes and the flight range can be extended into the post-stall area, which strongly improves the agility in close-range air combat. The take-off and landing roll distances can, in addition, be reduced by supporting the rotation by thrust vectorization during take-off and landing.

Several comparative studies were conducted to secure the controllability about the roll axis. As alternatives were investigated:

- utilization of bleed air from the engine at the wing tips (reaction control),
- installation of rotatable outer-wing sections,
- enlargement of the deflection range of conventional ailerons to take advantage of the so-called impingement-surface effect by totally neutralizing the aileron on one side, and optimally positioning it on the other side. This solution was chosen since it was easiest to mechanize.

It has to be mentioned here that rolling about the fuselage axis is no longer permitted at high angles of attack. This can easily be understood since, e.g., at $\alpha = 60°$, a roll about the fuselage axis of 90° leads to a yaw angle of 60°. The consequence is that the air intake experiences such high losses that the engine stalls. Engine flameout has to be avoided under all circumstances since the control by thrust vectorization will no longer function.

At high angles of attack, a roll about the velocity vector – the so-called "barrel roll" – is therefore flown. The roll about the velocity vector is one of the basic maneuvers of post-stall flight and is often called rolling about the aerodynamic axis ("stability roll"). The barrel roll has to be mechanized in a way that it can normally be flown only with the roll stick and that the yaw angle is automatically being adjusted to $\beta = 0$ by the flight control system (FCS). If the pilot uses in addition the rudder or activates the yaw-vector nozzle, a maneuver ensues, which the pilots call "slice-maneuver", during which the yaw angle must, however, stay below values of $\beta \approx 10°$.

For the configuration developed the following control-moment generators are available with regard to controlling and trimming:

pitch axis:	canard, trailing edge flaps and thrust-vector nozzle,
yaw axis:	rudder and thrust-vector nozzle,
roll axis:	aileron on the outer wing section.

Air-Intake Developments for High Angles of Attack

It was the objective of the studies to find an air-intake type and a position of the intake on the airframe that result in small losses, respectively large pressure recoveries for the engine so that sufficient thrust is available for the control by the thrust vector nozzle and for lift support beyond C_{Lmax}. It also has to be guaranteed that intake-engine compatibility is given. The corresponding parameters for the RB.199 engine were known from the "Tornado" development and were, in their limits, not to be exceeded at high angles of attack.

From the configuration-related investigations, an air-intake arrangement resulted where two intake channels were located directly next to each other underneath the fuselage. To determine the type of intake, a wind-tunnel test series was started where the versions shown in the figure were investigated.

The assessment of the different intakes is carried out on the basis of

- pressure recovery (η),
- steady pressure non-uniformity (DC60),
- turbulence (TRMS at the compressor interface).

Summarizing, *Kurt Lotter* and *Jakob Malefakis* [87] made the following statements:

- The intake with variable capture area (B1/γ1 = 40°) and without auxiliary intakes shows throughout the entire investigated range of all investigated configurations the lowest pressure losses and the lowest turbulence values. The steady pressure non-uniformities though exceed the values of the reference intake (B0), but are, however, within the range of the ones of other fighter aircraft (e.g., F-16, "Tornado").
- A reduction of the length of the intake cowl by half (B30/γ) and a simultaneous optimization of the rotation angle γ results only at high angles of attack in a deterioration of the pressure recovery in comparison to the B1/40°-intakes.

All other variations showed disadvantages in comparison to the rotatable cowl design.

The results of the measurements confirmed that the demands of the stationary case and of post-stall operations could be met by an intake with a variable capture area. The underside of the fuselage together

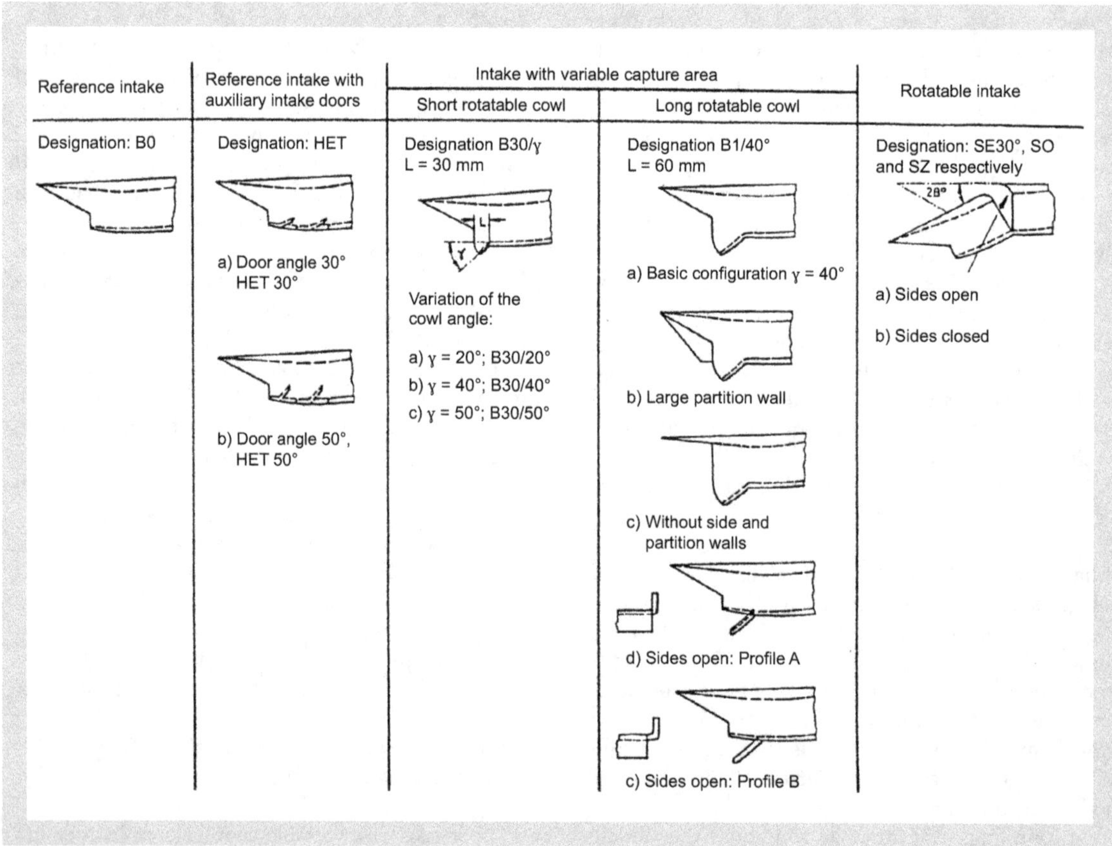

High angle of attack air-intake concepts investigated.

with the deflected cowl causes an effective deflection of the flow into the intake so that a very good pressure recovery can, as shown in the figure, be achieved up to high angles of attack. This type of intake was, therefore, adopted for the basic configuration.

Demonstrator Investigations and X-31 Results

Already in 1982, *Hannes Ross* conducted a study [88] at MBB concerning a one-engine derivative of the TKF, equipped with a Pratt & Whitney F-100 engine. There were also intensive contacts to the Saab Company during that time. The "Viggen" was, for instance, used as demonstrator for fuselage aiming.

With this extensive background, MBB made first contact in 1983 with the US Company Rockwell International, and Rockwell was won as a partner. Rockwell posted in 1983 a first offer to the "Defense Advanced Research Program Agency, DARPA, now ARPA" and, at the same time, established its own research program. The result led to the "Super Normal Attitude Kinetic Enhancement, SNAKE" configuration. This configuration is based on work, which was subcontracted in 1980 by MBB to VFW-Bremen, and whose results [85] had been presented in a lecture by *Stephen Ransom*. Rockwell had adopted this configuration and had built a purely subsonic model, which was tested in quasi-free flight in the 60 ft × 30 ft wind tunnel at the NASA "Langley Research Center" to determine its flight characteristics. The model was supplied via a pipe with compressed air, which was used for driving the actuators as well as for the generation of thrust. Thrust vectorization was integrated into the model. Four paddles were installed on the nozzle, in lieu of a vector nozzle, for thrust deflection, two for

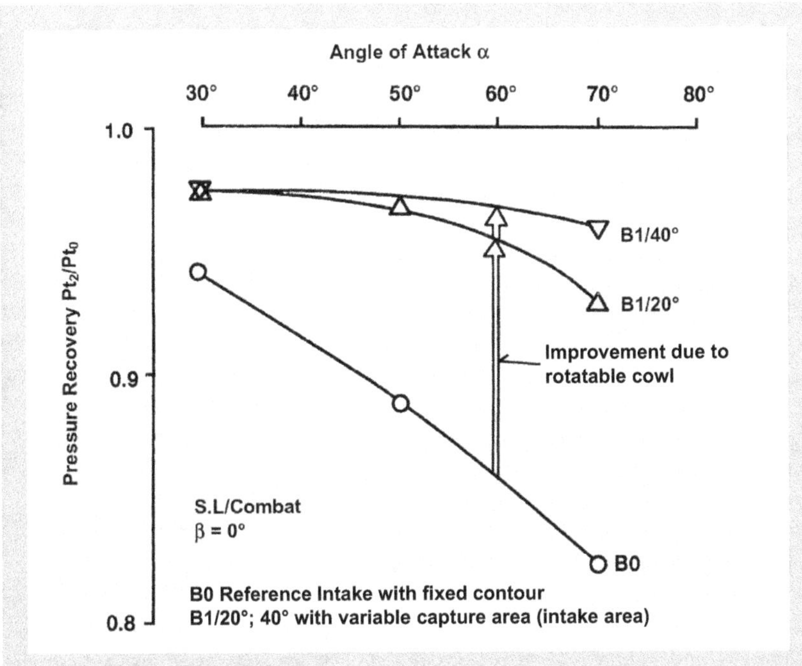

Pressure recovery for the air intake with variable capture area up to high angles of attack at a Mach number of $M = 0.5$.

the pitch axis and two for the yaw axis, to generate the necessary control moments. The picture on Color Plate VIII shows the model in the wind tunnel at an angle of attack of $\alpha = 45°$. With the aid of a computer, set up outside of the model, one was able to stabilize the model in any attitude and to fly all axes by means of a control stick. Here the model is specifically being pointed out since it was part of one of the most important public relations actions preceding the joint X-31 acquisition. Via this PR-presentation, the entire top management of DARPA, NASA and Rockwell, as well as of the Air Force and the Navy, could convince themselves of the flying abilities of a "Post-Stall-Configured" aircraft. Who ever wanted, could fly the configuration in the wind tunnel, which greatly contributed to the elimination of prejudices. Until then, many experienced aircraft designers and pilots considered the stall boundary insurmountable, respectively thought that the penetration of the range beyond maximal lift was undesirable since it was associated with problems of energy loss and uncontrollability.

As a consequence of this successful demonstration, MBB and Rockwell joined and intensified their respective studies. On the basis of the preliminary work of MBB and the single-engine configuration P30, tested in Emmen, the X-31 configuration was developed. The difference of the latter to the TKF configuration lay in the single-engine design (GE-F404) and the smaller scale of the aircraft, which was just able to accommodate the pilot together with an ejector seat. In the process, Rockwell succeeded in enforcing that the X-31 meets certain US-requirements, for instance, supersonic-flight capability and a wing design with regard to twist and camber at a Mach number of $M = 0.9$ and an altitude of 11 km (6.8 mi) to meet the requirement of a steady turn rate. With that, the X-31 acquired a certain "fighter-look". The multiple side view in the next figure shows the size comparison with respect to the EF 2000. Within the framework of the Rockwell International and MBB X-31 cooperation also the expression "Enhanced Fighter Maneuverability, EFM" was coined. This is a merger of the original designations "SNAKE", CCV-technology, new CCV-control methods, such as direct lift, direct side force, fuselage aiming et cetera, together with the post-stall capability, all abbreviated to EFM-technology, respectively even designated

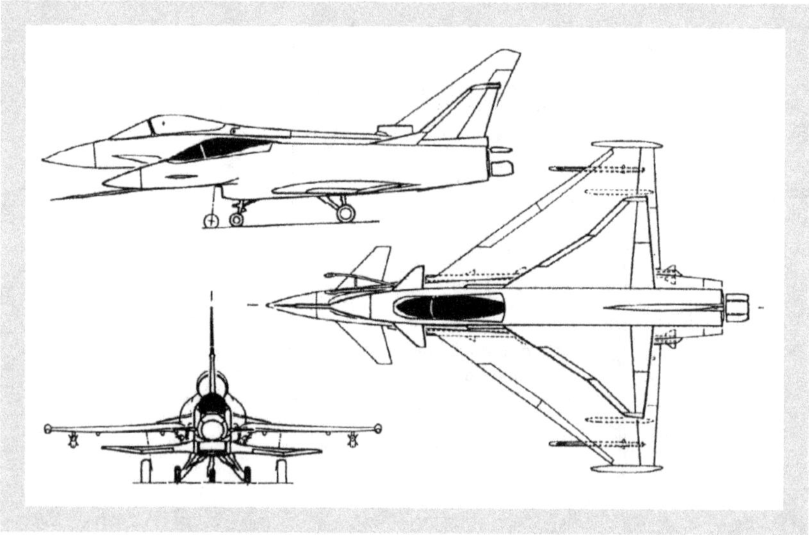

Size comparison of X-31 versus EF 2000.

as "EFM-program". Publications concerning the programmatics exist by *Michael Robinson* and *Hannes Ross* [89], and concerning technology and cost by *Sidney Powers* and *Harvey Schellenger* [90].

The X-31 development phase started officially in 1986 with the signing of a Memorandum of Agreement (MoA) between DARPA and the German Ministry of Defense (BMVg), represented by the Armament Director. DARPA took > 80 per cent of the cost of the overall program and charged the US Naval Air Command to accompany the project and to individually subcontract Rockwell International. The BMVg took over the remaining 20 per cent of the program and gave the contract directly to MBB.

Rockwell International and MBB Military Aircraft signed, in addition, an "Associate Contractor Agreement" in order to control the detail investigations. Four main objectives of the program were defined:

- demonstration of the EFM-technology,
- demonstration of tactical advantages of EFM; the MoA was extended in 1990 to conduct a tactical assessment,
- generation of a database to transfer results and findings of the EFM-technology to current and future aircraft, respectively,
- the proof of a demonstrator concept with a low-cost approach.

The overall responsibility for the X-31 program was with Rockwell. The work sharing envisaged for the US:

- development, construction and maintenance of the fuselage,
- load assumptions, stress analyses and buffet computations,
- basic systems and integration of off-the-shelf equipment,
- data sets and simulation,
- redundancy-management concept,
- flight-control system (FCS) hardware and integration,
- participation in technical and tactical flight tests,
- participation in the assessment of the tactical benefits.

The German work share comprised:

- development, construction and maintenance of the wing and the thrust vector paddles,
- intake definition,
- development of the control laws including thrust vectorization,
- analysis of the flight characteristics,
- parameter identification from the flight tests (DLR),

German-American demonstrator X-31 in flight at high angle of attack.

- design of the "Helmet Mounted Display, HMD" symbology,
- participation in technical and tactical flight tests,
- participation in the tactical assessment of the flight-test results and of the benefits of the EFM-technology.

The MBB-team was supported on location in Palmdale, California, and later NASA Dryden, by the DLR Institute of Flight Mechanics and the Test Center WTD 61 of the German Air Force. For the parameter identification, the DLR Institute of Flight Mechanics developed a test procedure, the so-called single surface excitation [91], which allowed, for the first time, to determine all essential aerodynamic parameters and the control effectiveness of a highly control-supported (control configured) aircraft from flight tests. One focus was here the determination of the thrust-vector effectiveness [92]. The results from test flights with single-surface excitation proved to be considerably more meaningful than from flights with standard maneuvers. The experience gained in this area is already utilized in the EF 2000 program.

The limited budget available for the development and production of two demonstrators forced both companies to a low-cost program. Thus a flight vehicle of minimum geometric size was defined, where, whenever possible, one went back to available components and subsystems such as, e.g., the F-18 ejector seat, the GE-F-404 engine or landing-gear elements of the F-16. Furthermore the cockpit, the display, which projects information onto the windscreen (head-up-display), and the cockpit instrumentation were taken from the F-18, the brakes from the "Cessna Citation", the FCS-computer from Honeywell and a series of subsystems from other aircraft up to the F-20.

A large cost reduction was achieved by not building the ground-test simulator (iron bird) for the integration and testing of the soft- and hardware of the flight control system. In addition, further cost reduction potentials were used. For example, the new thrust-vector nozzle, which was estimated to cost about 60 Million US $ for the GE-F-404, was not built but the mentioned inexpensive solution realized.

MBB had, as German share, taken over the development of paddles made of carbon/carbon. In the figure of the X-31 in flight one can observe that, contrary to the wind-tunnel model that comprised two pitch and two yaw paddles, the design was changed to only three paddles for the X-31. The reason was that three paddles were easier to integrate into the afterbody structure than four. But here, numerous calibrations had to be performed before the contribution of each individual paddle to pitch and yaw control could be hard-wired within the FCS. Initially, the maximum paddle deflection was limited to 26°. But after realizing that this deflection was not sufficient, the paddle control was changed and the calibration extended. The maximal deflection is 35° resulting in a thrust vector deflection of 16°. A further design responsibility of MBB was the definition of the air intake. It was designed according to the optimized geometrical dimensions defined in the wind tunnel. With the aid of an actuator, the cowl can be rotated downwards, in dependence of the angle of attack, to a maximum deflection of 30° to yield a flow deflection into the intake with minimum losses.

The total program required, all in all, roughly 400 million German Mark, of which the German side finally contributed about 25 per cent.

A very good overview [93] and summary concerning the development, the tests and the tactical trials is given *Peter Huber* and shall here not be repeated in detail. The future potential of a thrust-vector nozzle, which is from the very beginning integrated into a new design, is discussed in a presentation [94] by *Peter Huber* and *Hannes Ross*. The X-31 flight tests showed that aircraft design and control laws, under inclusion of the thrust-vectorization system, were successfully tested and that all objectives could be reached. Aircraft No. 1 performed 292 flights and No. 2 an additional 288 flights, i.e., together 580 take-offs with 412 accumulated flight hours. The flight-range envelope, where the X-31 was tested, is depicted in the figure. In the post-stall domain, angles of attack of $\alpha = -5°$ to $70°$ were reached.

The post-stall entrance speed of 265 kn ($M = 0.7$) enables one to exceed the stall boundary up to a load factor of $n_z = 6$ g, which corresponds to the structural limit of the X-31 at typical close-combat weights. The aircraft can be flown at subsonic speeds absolutely "carefree", which means that the aircraft can be recovered to a normal position from any extreme flight attitude. Flight tests to assess the tactical benefit of the EFM-technology were originally not subject of the MoA between DARPA and the BMVg, but were added in 1990 as additional objective. The tactical tests required a certain air-combat training of the pilots. Later, also operational pilots of the US Air Force and the Navy were invited to fly the X-31 against the F-18 and vice versa to achieve an unbiased result. The X-31 flew, moreover, also against the F-14 and the F-18 of the US-Navy as well as against the F-15 and F-16 of the US Air Force in close air combat. The impressive and expectation-exceeding result was that the X-31 reached with the complete EFM-technology employed a superiority of 9.6 : 1 against an F-18. When the EFM capability was omitted by limiting the angle of attack to $\alpha = 30°$, the X-31 lost with 1 : 2.6 against the F-18. In total, the EFM-technology can thus be assessed of having a superiority of 12:1, which exceeded the original simulation results (see above).

After the tactical test phase, a series of further activities was conducted with the X-31. These included:

- pilot support by a HMD,
- increased utilization of thrust vectorization to reduce the vertical stabilizer (rudder) area,
- post-stall relevant activities such as, e.g., investigation of strakes located at the nose of the fuselage,
- determination of aerodynamic and performance data,
- changes on the air-data system.

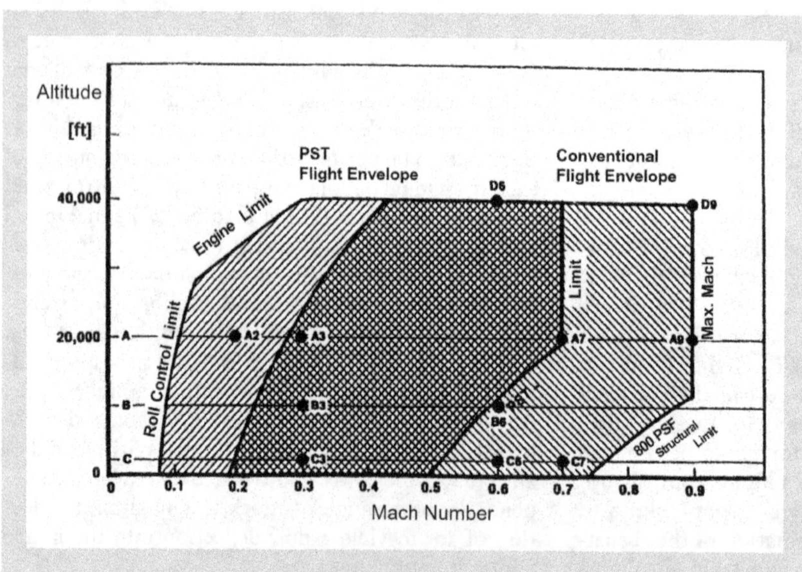

Flight-testing envelope of the X-31.

Here, a total of 55 flights were performed with both aircraft.

Studies concerning further activities, which could lead to a continuation of the program beyond 1997, commenced to show how the X-31 could be used as test vehicle. Possible activities included:

- flights with a reduced vertical stabilizer surface,
- development of a thrust-vector nozzle,
- development of an operational air-data system and,
- studies concerning the reduction of the take-off and landing roll distances.

In the autumn of 1998, the Boeing Company (Take-over of Rockwell International), Dasa (Take-over of MBB) and Saab agreed with their respective government agencies and research institutions to continue the X-31 program under the designation "VECTOR".

However in early 1999 a Swedish defense review resulted in a change of national development priorities and forced the withdrawal of Sweden as a third partner in the VECTOR program. This fact immediately reduced the program funding to two thirds, which in the end caused the cancellation of the development of a thrust-vector nozzle. In August of 1999 the statement of work of the already signed German/U.S. MoU was redefined and the following objectives remained:

- "Extremely-Short-Take-Off-and-Landing, ESTOL" by use of high angles of attack and automated flight-path control (especially the Navy became interested in ESTOL since this feature is very important for a carrier landing, where each knot of speed is of significance.)
- "Flush Air Data System, FADS" incorporated in the aircraft nose tip using mini-sensors to establish α, β, M and altitude information and which can cover an extremely broad flight envelope in terms of α and M.
- "Tailless Aircraft"-investigation of the unstable lateral motion by omitting/reducing the vertical stabilizer and by this contributing substantially to the reduction of radar signature. The stability augmentation is transferred to the thrust vector nozzle by use of a proper control-law design.

Due to budget constraints the reduced tail/tailless study became limited to control-law development, wind-tunnel tests and simulation on a ground-based simulator.

Since the NAVY was to be the U.S. Government sponsor, the decision was made to conduct flight tests at the U.S. Naval Air System Command (NAVAIR) at Patuxent River, Maryland.

Only a few engineers of the first campaign were left. Therefore new teams had to be provided from all partners.

The initial task was to get the aircraft flying again. This activity became the familiarization phase for the many new team members. In parallel changes required to soft- and hardware were being defined, developed and prepared for installation. Details of all of this work can be taken from reference [94a], a paper given by *T. Grohs* and *O. Brieger*.

From the very beginning the ESTOLanding was judged to be "time critical" since in the high angle of attack attitude the aft end of the aircraft is closer to the ground than the main wheels. Therefore the de-rotation of the aircraft had to be automated in dependence of the ground proximity. Numerous trails were performed by using a virtual runway at 5000 ft "on top of the clouds" and by doing touch and go landings.

The flight tests ended with the historical final flight on April 29, 2003. The highlight was a fully automated ESTOLanding with a real touch-down at an angle of attack of 24°, a landing speed of 121 kts and a landing roll-distance of 520 m (1,706 ft). Comparable data for a normal attitude landing of the X-31 are: $\alpha = 12°$, speed 175 kts and 2,400 m (7,872 ft) rolling distance.

The German X-31 partners EADS and BWB on top will profit enormously from the tests of the "Flush Air Data System", which can be used for future aircraft designs. The miniature sensors did show excellent results and do allow to establish α, β, M and altitude with sufficient accuracy, so that a nose boom or angular sensors can be replaced on future designs, which will help to reduce radar signature considerably. The ground test simulation for reduced tail/tailless test objectives showed that the envisaged target can be achieved and that this new aircraft feature will also be incorporated in future designs. Still work is left to be done, namely the in-depth investigation of the records taken regarding the FADSystem to come to final numbers and accuracy.

Concluding it can be stated that the tests with the X-31 aircraft have shown the feasibility of all objectives, the validity of the control-law design and the impacts on landing speed and distance as well as the potential of the FAD System. Thus proven technologies are available to be incorporated in future designs of aircraft. The excellent achievement of objectives was made possible by the high engagement of all of the VECTOR-Team members and the ambition to fulfill their mission.

Remark
Concerning this chapter, some figures are presented in color:
Color Plate VIII
- Flow past a delta wing
- The free-flight model "SNAKE" in the NASA Langley Wind Tunnel
- German-American flight demonstrator X-31

Control, Navigation, Avionics, Cockpit

Ulrich Butter

Electronic Flight Control

In the mid-fifties, when the aeronautical industry in Germany started anew after a ten-year interruption, it was the state of the art that aircraft were equipped with a mechanical flight control system. Here, the controls of the pilot were either directly connected via cables or rods to the control surfaces or with the control valves of the hydraulic actuators, which move the control surfaces.

There were electric controllers with a damping function to improve the dynamic stability of the aircraft. The position commands of the controller were coupled via servomotors into the mechanical control system. The control authority of these controllers was low so that in the case of a failure of a controller, the flight safety of the aircraft was not endangered. There were electric autopilots whose position commands were again coupled via servomotors or clutches into the mechanical control system. Also the autopilots had a low control authority and could fail without endangering the aircraft. An autopilot could, furthermore, be overruled by manual control inputs. Starting in 1959, a new control technology was introduced, electronic flight control, with the realization of the vertical take-off concept. Special characteristics of this technology are electronic signal processing and electrical signal transmission. Electronic flight control is known as "fly-by-wire". It was realized on the vertical take-off aircraft VJ 101 C, DO 31E, and VAK 191 B.

The company "Entwicklungsring Süd, EWR" developed, as already mentioned, since 1959 the vertical take-off aircraft VJ 101 C [95–98]. After the military requirements had changed and the VJ 101 C was no longer seen as the successor of the F-104, it was employed as a test vehicle for flight control concepts.

The aircraft configuration has already been introduced. For the transition from hovering flight into aerodynamic flight, the engine nacelles could be rotated from a vertical into a horizontal position. Control moments during hovering and transitional flight could be produced about the roll axis and the pitch axis by thrust modulation, i.e., by differential control of the engines, and about the yaw axis by opposing rotations of the engine nacelles.

When controlling the aircraft during hovering flight by thrust modulation, the general difficulty exists that the engines respond to the fuel controller with a certain delay and that there exist no stabilizing or damping moments. The controllability during hovering flight was investigated on a seesaw and on a hovering frame. Here, various control concepts were tested, viz., angular acceleration control by direct control of the lift engines without stability augmentation, as well as angular rate control and angular attitude control, each with stability augmentation. The tests on the hovering frame led to the following result: Acceleration control is rather difficult to command since the attitude the pilot wants to control has a phase lag of 180° with regard the control input. The rate and attitude control are, however, well flyable since the angular rates, respectively the angular attitudes are proportional to the control input. It was established that the attitude control is best suited for hovering flight.

Honeywell and "Bodenseewerk Gerätetechnik, BGT" developed control systems for the aircraft VJ 101 C. The control systems had the following characteristics: Attitude control during hovering flight and transition from attitude control to rate control during the transition from hovering flight to aerodynamic flight. This resulted in an optimum adaptation of the control characteristics to the respective flight range. For safety reasons all control systems were of a two-channel design.

As a result of the tests with the two-channel system it was established that hardover failures can only be controlled by the pilot if they do not exceed 40 per cent of the total control authority. Consequently, the control system authority would have had to be limited to 40 per cent of the control authority. However, the control system has to be given full authority to cover engine failures. For this reason it was decided to change to a three-channel fly-by-wire system.

For further tests with the VJ 101 C by Messerschmitt-Bölkow-Blohm (MBB), BGT developed under the supervision of *Waldemar Möller*, and in cooperation with MBB, the three-channel flight control system FRG32. This system was the first real fly-by-wire flight control system for all flight ranges with thrust vector control, i.e., take-off, landing, hovering flight, transitional flight and flights with engine failures within the aerodynamic flight domain. The control of the aerodynamic control surfaces occurred via mechanical control rods. The control system acted as a damper during aerodynamic flight.

During the further utilization of the VJ 101 C as test vehicle, the four-channel fly-by-wire flight control system for automatic vertical landing, FRG33, was developed and tested on the VJ 101 C by BGT in cooperation with MBB.

In the development of the VJ 101 C, valuable experience concerning electronic flight control technology could be gathered. The utilization of electronic signal processing provides to the engineer far-reaching freedom in the design of the control system and allows an optimum adaptation of the control laws to the aircraft and the respective flight condition. The disadvantage, compared to mechanical control, i.e., failures of electronic components, can be compensated by redundancy and monitoring of the individual control channels. The electric signal transmission allows the use of automatic test equipment.

The experience gained could immediately be applied and utilized in subsequent development projects. MBB and BGT became partners in the tri-national development programme concerning the multi-role combat aircraft Panavia "Tornado", the first operational fighter aircraft with a primary flight control system with fly-by-wire technology. MBB had the system de-

Waldemar Möller, born in Preetz (Holstein) on March 11, 1895.

1918 he begins his studies at the TH Berlin, which he concludes in 1924.

1924 during his studies, he is an active glider pilot and a designer of sailplanes. On the recommendation of *August von Parseval*, he joins the Askania Company as head of the Department of Aeronautics. The main focus of his work is already at this time the development of automatic flight controls.

1935 he moves to the "Test Center of the Air Force" in Rechlin. He develops a totally electric three-axes flight control system, which is introduced into the German Air Force as "Standard-Three-Axes Control" ("Einheits-Dreiachsen-Steuerung, EDS").

1940 he goes back to the Askania Company and supervises the quantity production of the EDS. During that time, he also solves some control problems related to the V2, why *Wernher von Braun* wanted him to move Peenemünde.

1945 on Christmas eve, he is being arrested by the Soviets at his house in the East-Sector of Berlin, passed through various camps, before he is transported in

1946 on October 22, together with other Askania-employees and a total of 900 specialists of Siemens, Patin, BMW, and Junkers and 1,500 family members, to the Soviet Union and settled near Kuybischew at the Wolga.

1949 after a desperate move, where he suffered from severe frostbite and had all ten fingers amputated, he is given the supervision of the Askania-group in the Soviet Union, which further developed the Dynuktiv-control.

1950 Mrs. Moeller receives in spring permission to visit her husband; in September, the Askania-group begins its participation in a missile programme in Moscow.

1955 the group is moved to Suchumi at the Black See. The majority of the Suchumi-people is, in

1956 brought to the GDR (German Democratic Republic). Moeller insists on being permitted to move to the Federal Republic, which happened in February of

1958 he becomes, a few weeks later, head of the Aeronautical Department of the "Bodensee-Geräte-Werke, BGT" in Überlingen. He quickly trained qualified neophytes and consulted, overflowing with ideas, up to a high age.

1962 he receives a honorary PhD (Dr.-Ing. Ehren halber) from the TH Darmstadt.

1977 he dies in Überlingen on March 8.

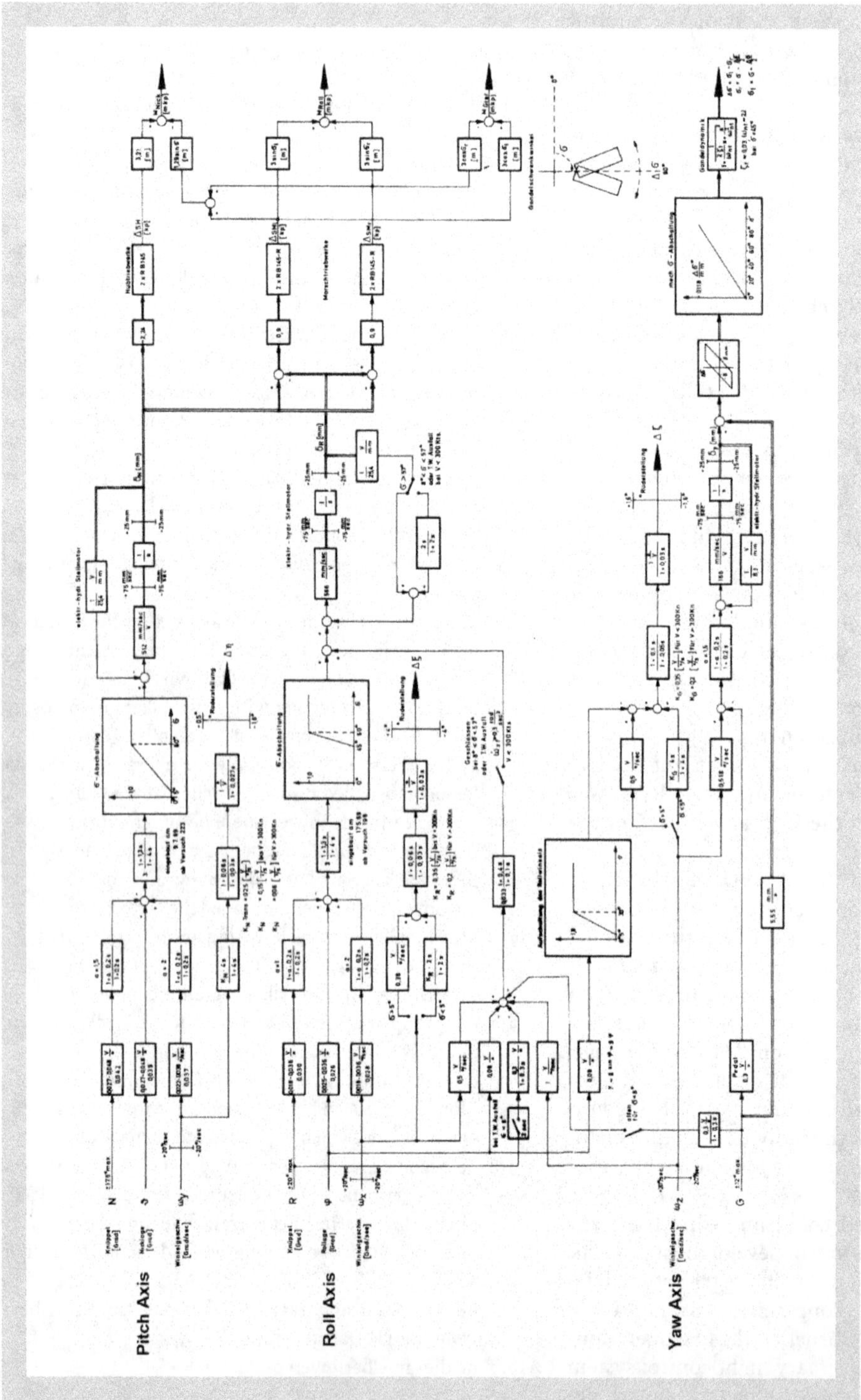

Block diagram of the flight control system of the VJ 101 C.

sign responsibility for the development of the flight control system. The first flight took place in 1974.

In the case of the DO 31E [96, 98–101] in hovering flight, control moments about the roll axis and the yaw axis were generated by thrust modulation of the lift engines, respectively by operating jet-deflection flaps on the lift engines. To produce control moments about the pitch axis, bleed-air drawn from the cruise engines of the aircraft was blown out upwards or downwards.

In order to determine the requirements on the flight control system during hovering flight, tests were performed on two different seesaws, on a control system test frame and on a hovering frame. The tests showed that the pilot is generally able to stabilize the aircraft manually in the roll axis as well as in the pitch axis. The reason was the relatively large moment of inertia in both axes and the undelayed bleed-air control in the pitch axis.

Based on the results of the pre-tests on seesaws, control system test frame and hovering frame, the flight control system FRG10-203/206 was developed by BGT in cooperation with Dornier. For hovering flight, this control system was designed as attitude control system in roll and pitch and as rate control system in yaw. An additional damper was available for the roll axis. The control system was of one-channel design. The mechanical output signals of the control system were coupled into the mechanical control system. If the control system failed in one axis, the pilot had to take over the stabilization.

In the further course of the experimental programme, a triplex flight control system was developed and built by BGT. This control system was, however, never put into operation.

The fighter and reconnaissance aircraft VAK 191 B [95, 98, 102, 103] had two lift engines, one in the forward and one in the rear part of the fuselage, for hovering flight and transition and one engine with thrust deflection in the center section of the fuselage for hovering flight, transition and aerodynamic flight. During hovering flight, control moments in all three axes could be generated by blowing out bleed air off all three engines. The control moments in the pitch axis were, in addition, supported by thrust modulation of the two lift engines.

After extensive pre-investigations on a hovering frame, the flight control system FRG30-302 for the VAK 191 B was jointly developed by BGT and the Vereinigte Flugtechnische Werke (VFW). This control

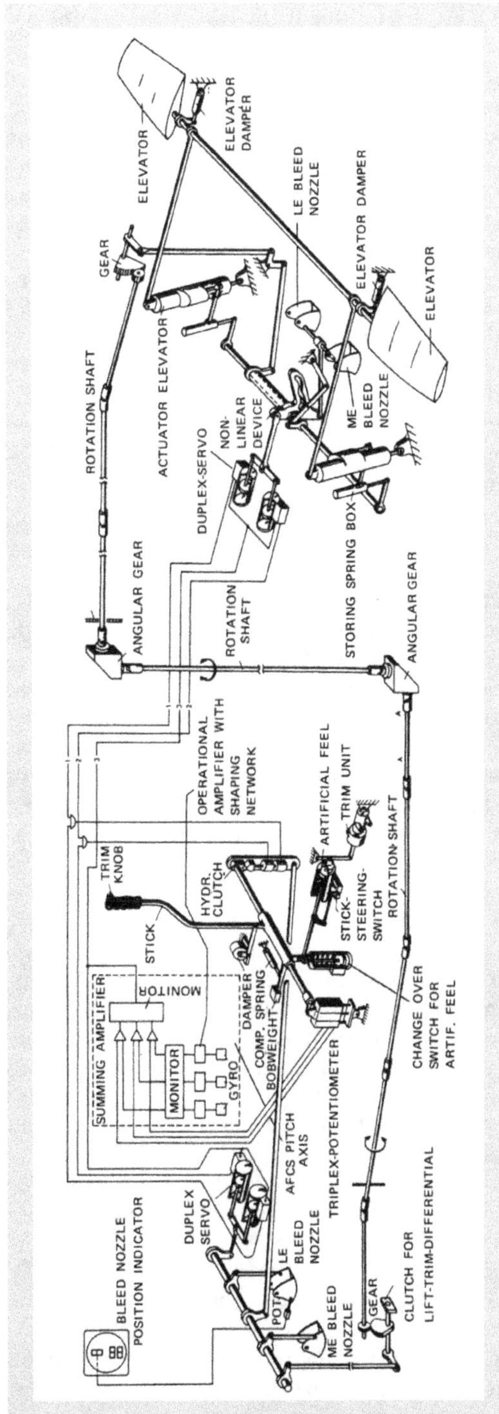

Flight control system of the VAK 191 B (pitch axis).

system was part of the fly-by-wire flight control system of the VAK 191 B. The control system was of three-channel design and had three different operational ranges: in the range up to 120 kts pure attitude command/attitude hold in the roll and pitch axis and rate command in yaw axis; in the range between 120 and 180 kts, transition to rate command; at velocities above 180 kts, operation as damper. In case of a complete loss of the fly-by-wire system, a mechanical backup system could be coupled in. The VAK 191 B flight control concept was no longer pursued after the termination of the aircraft programme in September of 1975.

It has to be kept in mind that the vertical take-off programmes have performed decisive technological groundwork for the realization of the fly-by-wire technology in the aircraft. The vertical take-off programmes paved the way for the German companies to take on responsibility for the flight control systems of upcoming European programmes.

Transition from Analogue to Digital Flight Control

The electronic flight control systems with feedback, developed in the sixties, were designed in analogue technology and formed the basis for the modern flight control systems. Here, one had to allow for the fact that the representation of complex control functions by analogue circuits was only possible within certain limits. The digital technology offers, on the contrary, a much greater flexibility in the design of complex functions. At the beginning of the seventies, there was only limited trust in the reliability of digital computers. On the other hand, one wanted to take advantage of the power of digital technology. This led to the fact that the "Tornado" obtained a hybrid flight control system. The autopilot was designed in digital technology, the command and stability augmentation system in analogue technology. The long-term objective was, however, the general introduction of digital flight control in form of integrated digital flight control systems, which had to be able to cover all functions of flight control.

The introduction of fail-safe electronic flight guidance and control systems provided the possibility to configure aircraft exclusively with respect to performance and to generate static stability artificially by electronically controlled control surfaces. Giving up the requirement for aerodynamic longitudinal stability can be used for improvements such as reduction of the size of the stabilizers, reduction in drag and increase in lift.

In 1974, the Ministry of Defense (BMVg) charged MBB with the development of an experimental aircraft as control configured vehicle (CCV) [104–106]. The objectives were:

- Development and test of an advanced electro-hydraulic flight guidance and control system including the associated safety system.
- Investigation of the flying qualities of an aerodynamically highly unstable supersonic fighter aircraft throughout the entire flight domain.

A modified Lockheed F-104 G was utilized as test aircraft. The aircraft was equipped with a quadruplex redundant digital flight control system. This system was designed in the longitudinal and the lateral axis as an integrated multi-variable control system. The first flight took place in 1977.

The CCV-F-104 G was the first aircraft worldwide, which was equipped with a quadruplex digital flight control system. Here the full integration of the primary control system, the autopilot, and the navigation system was executed for the first time. It was also the first aircraft where a strap-down-system, i.e., an inertial measurement system with body-fixed rate gyros and acceleration sensors, was utilized for navigational tasks and as signal source for the fly-by-wire system.

A detailed description of the experimental programme of the CCV-F-104 G is given in the chapter "From the Stable to the Unstable Configuration (Drag Optimization)".

The experimental aircraft X-31 A [106, 107] was developed in a joint German-American programme by MBB (today EADS Germany) and Rockwell International. The objective of this programme was the demonstration of increased maneuverability by employing thrust vector control for controlled flight in separated flow.

The aircraft X-31 A is a, in its longitudinal axis, highly unstable aircraft with delta wings and canards. The primary aerodynamic control surfaces are symmetrically controllable trailing edge flaps and canards for pitch control, and differentially controllable trailing edge flaps and rudder for roll/yaw control. There is in addition thrust-vector control to support the aerodynamic control in the pitch and yaw axes in the high angle of attack range.

The flight control system is a quadruplex redundant digital fly-by-wire system. The control laws, de-

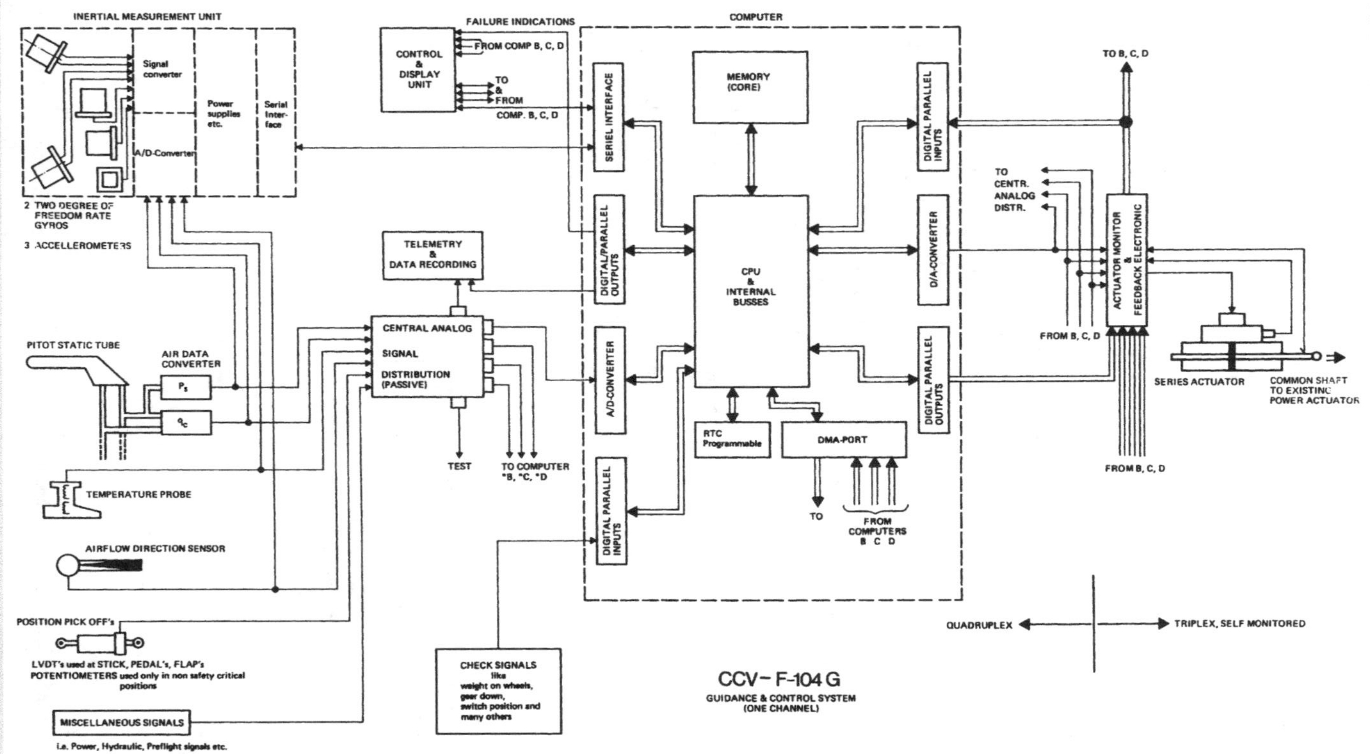

Equipment setup of the flight control system of the CCV- F-104 G.

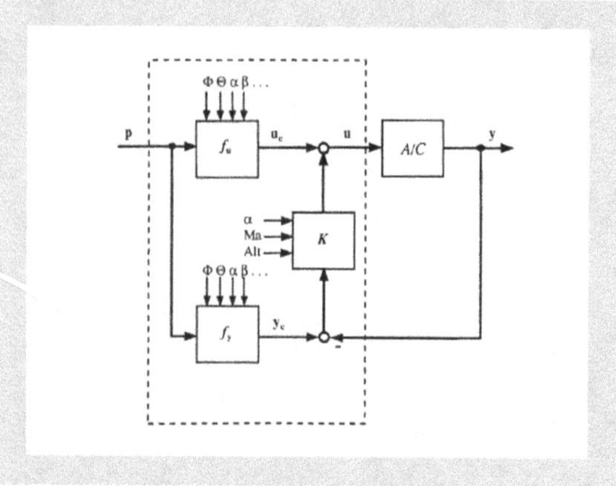

Structure of the control system of the X-31 A.

veloped by MBB (today EADS Germany), are based on the principle of command following control: The pilot commands with his control stick and pedal inputs the roll rate about the flight-velocity vector, the angle of attack and the sideslip angle. In flight domains with high dynamic pressure, the load factor is commanded instead of the angle of attack. From the control commands of the pilot, pre-control values of the control variables and the flight state variables are generated. The control system was designed with a Riccati-optimization according to the method of dynamic programming. The control system contains, in addition, elements to compensate inertial coupling. The first flight of the X-31 A took place in 1990.

For the design of control systems for aircraft, especially for aircraft with reduced or neutral static stability, efficient design procedures are needed. The control system design according to classical procedures is performed in the following steps: First the structure of the control system is established, thereafter suitable control gains are determined for selected flight conditions; finally a curve fitting for the gains is performed to obtain control laws, which are valid throughout the entire flight regime. This procedure frequently leads to control systems of high complexity. Alternatively, the method of robust control can be used.

During the years 1981 and 1982, the "Deutsche Forschungs- und Versuchsanstalt für Luft- und Raumfahrt, DFVLR" developed, together with the Swedish Saab Company, a back-up flight control system for the multi-role combat aircraft JAS39 [108, 109]. The JAS39 is a, in the longitudinal axis, unstable aircraft, which has to be artificially stabilized. The aircraft is equipped with a digital flight control system in fly-by-wire technology. The back-up control system shall stabilize the aircraft after a failure of the digital flight control system. To achieve high reliability, the back-up system was to be as simple and robust as possible.

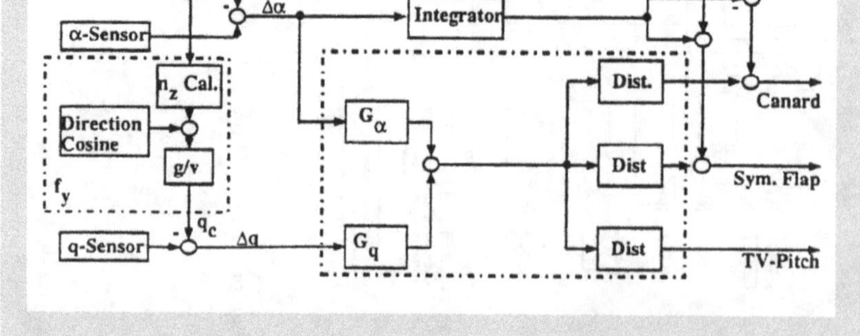

Block diagram of the X-31 A flight control system for the longitudinal axis.

The requirements were: utilization of the pitch rate as the only feedback parameter, constant control gains, and design in analogue technology. It was possible to design a robust control system which provided good flying qualities within the entire flight domain.

Key Technologies of Modern Flight Control Systems

Optical Signal Transmission
A particular challenge in the development of electronic flight control systems is the protection against electromagnetic interference. In the case of systems with electric signal transmission, electromagnetic interference is coupled into the system most of all via the signal lines. Although the interference susceptibility can be reduced by shielding and by filters, it cannot be removed entirely. Contrary to that is the optical signal transmission, i.e., the use of optical fibres as transmission medium, immune against electromagnetic interference. Further advantages of optical fibres are large transmission bandwidth and low weight.

Helicopters are especially exposed to electromagnetic interference since they operate at low altitudes. In 1985, MBB and "Liebherr Aero-Technik, LAT" and the "Deutsche Forschungsanstalt für Luft- und Raumfahrt, DLR" were charged by the BMVg to develop and test a helicopter yaw control system with optical signal transmission [110–113]. Flight control systems with optical signal transmission are known as "fly-by-light" systems. It was the objective of the project to prove the operational maturity of the fly-by-light technology.

The resistance of the system against electromagnetic interference was investigated in laboratory tests. Here, it was shown that the system still worked disturbance-free in a frequency range between 14 kHz and 2 GHz at field strengths of 220 V/m.

The optical control system (OPST1) was installed in the research helicopter BO/105-S3 operated by the DLR. The first flight of the BO/105-S3 with the fly-by-light control system took place in 1988. In about 50 flight hours, the optical control system was tested and assessed. The system was judged positively by the pilots.

The experience gained within the OPST1-programme formed the basis for a later research programme where a helicopter was equipped in all four axes with fly-by-light control (see chapter In-flight Demonstrators "Test vehicle ACT/FHS").

Smart Actuator
The analogue fly-by-wire systems had an architecture where the individual sub-systems were grouped around a central flight control computer. This computer filtered and consolidated the sensor signals, closed the control loops of the actuators and monitored the actuator functions. This architecture was taken over by the first digital fly-by-wire systems since computing power was heavy and expensive. The situation changed with the introduction of microprocessors. The rapid increase in computing power and a progressive miniaturization at sinking costs made it possible to install computing power where it was needed, and to conduct the signal processing and signal monitoring locally within the individual subsystems.

The principle of distributed intelligence was realized and tested as part of the optical tail-rotor control system, described above. LAT developed for the optical yaw axis control system an actuator with digital control electronics for signal consolidation, loop-closure, and self-monitoring of the actuator.

The Active Control Stick
The control stick of a purely mechanical control system does not only serve as input device for control commands but also provides the pilot with a feedback of the flight condition via the stick force. This feedback is not present in an electronic flight control system. If the control forces are simulated by spring force, the control stick acts as a passive input element. The controllability can be considerably improved by feedback of the aircraft response to the control stick as additional information for the pilot.

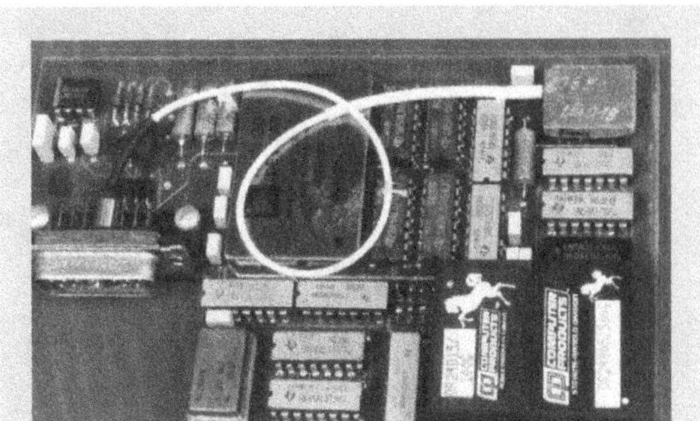

Circuit element with optical interface.

An active control stick [114] for helicopters was developed at Eurocopter Deutschland (ECD) in the form of a two-axes side stick. The stick was equipped with force sensors on the grip and servomotors to move the stick. The stick forces generated by the pilot served as input signals to the flight control computer. The response of the helicopter in the form of attitudes and rates was fed back and used as command input for position control of the servomotors. The control characteristic was designed and optimized by tests on a flight simulator. Here, it could be proven that the active control stick considerably improved the accuracy related to target tracking tasks. The stick was installed in 1995 on the BO/105-S3 of the DLR and tested in flight.

The Technical University of Brunswick developed, in cooperation with the DLR, a two-axes active side stick, whose restoring moments were generated by electromagnets. The stick with the designation MAGSI [115] (Electromagnetic Side-stick) excelled due to its compact and simple design, and large bandwidth.

Flight Guidance

Besides controlling and stabilizing the aircraft about its three axes, it is necessary to guide the aircraft on a suitable flight path through the airspace.

Research in the field of flight-path control was conducted at the Technical University of Brunswick between 1969 and 1994.

An integrated flight-path control system [116] was developed within the framework of the collaborative research center SFB 58 "Flight Guidance" and investigated in simulator and flight tests. The system enabled the pilot to interfere actively at different hierarchy levels with the control process. The pilot could thus either fly via a command following control system on a given flight path, or determine the flight path himself according to a given flight task and engage automatic control for the control task.

Within the SFB 212 "Safety in Air Traffic", the interaction between aircraft, atmosphere and flight task was investigated and new concepts were derived for intelligent flight guidance control [117]. The main characteristic of the intelligent guidance control algorithm is the continuous computation of the optimum demand state and the associated pre-commands under consideration of the operational limits of the aircraft.

Satellite Navigation

For flight guidance, mainly ground based radio-navigation aids are used, such as non-directional radio beacons (NDB), Very High Frequency Omni-Directional Range/Distance Measuring Equipment (VOR/DME), and the instrument landing system (ILS). These facilities work precisely and reliably, they are, however, due to their high investment and operational costs only available in highly developed countries. With the introduction of the Satellite Navigation System GPS (Global Positioning System), designed for military purposes, by the USA and GLONASS (Global Navigation Satellite System) by the former Soviet Union, new possibilities for navigation arose.

At the Technical University of Brunswick, a method was developed for the civil utilization of the military satellite system GPS [118]. One of the difficulties encountered related to the fact that the resolution of the system was reduced to 100–200 m (328–656 ft) for civil users. This problem could be solved by a differential correction procedure. A ground-based receiv-

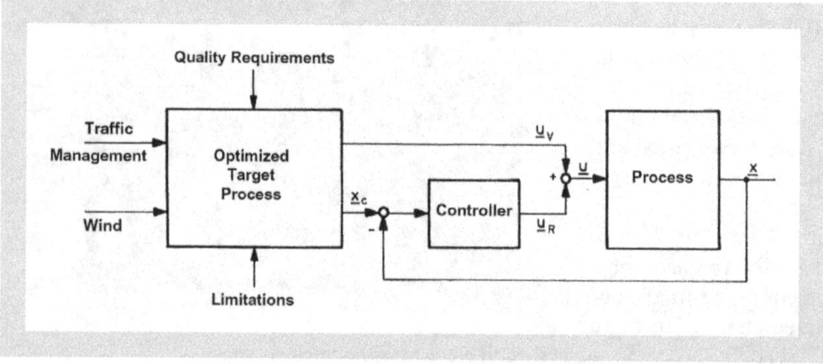

Intelligent flight guidance control.

er, which knows its own position, is used as reference. Deviations in position measured by this receiver can be considered as errors. If these errors are transmitted to an aircraft, the position signals measured on board can be corrected. The accuracy achieved here lays in the decimeter (10-inch) range. This procedure became known as Differential GPS (DGPS). A further problem was the low reliability of the GPS. The error rate was in the order of 10^{-2} errors per hour. To solve this problem, the accurate but unreliable signals of the satellite system were combined with the reliable but not very precise signals of an inertial navigation system. In 1989, the worldwide first automatic landing with a GPS-supported flight guidance system was conducted in Brunswick.

Within the framework of the research project ISAN (Integrity of Satellite Navigation), a satellite navigation system, denoted Integrated Navigation and Landing System (INLS) [119, 120], was developed by the NFS (Navigation and Flight Guidance System) Company. With this system, several hundred precision approaches and automatic landings were carried out during the years 1993 and 996.

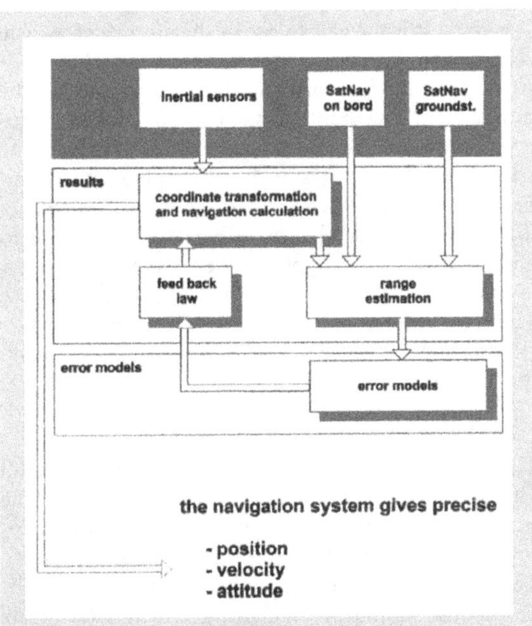

Integrated navigation system.

Flight Guidance with Artificial Visibility

Good visibility or suitable visibility aids are a prerequisite for the safe guidance of aircraft in close proximity to the ground. At the Technical University of Munich, a flight guidance system for poor visibility conditions [121, 122] was developed together with the Technical University of Brunswick. A computer generates a three-dimensional image of the outside world using a data base with information on the topography of and the prominent objects within the area. A flight guidance corridor is projected into the artificially created image of the terrain as a flight director for the pilot. The image is displayed to the pilot in a helmet-mounted display. The sight visibility simulation computer receives the required position and attitude data from a GPS-supported navigation system. In 1994/95, test flights were successfully carried out with this system under difficult flight conditions, amongst them precision approaches and landings, low-level flights in a narrow river gorge, steep curved approaches and terrain-following flights in mountainous terrain.

Avionics

The term "avionics" stands for aviation-electronics and is used as collective a name for the electronic on-board systems. The development of the avionics technology parallels the development of electronics. In the sixties, mainly electro-mechanical equipment was utilized for sensors, signal processing, and information presentation. In the seventies, electronically controlled on-board systems were developed with the introduction of semi-conductor electronics and integrated circuits. The microprocessor technology, introduced in the eighties, led to software-controlled systems.

Integrated Modular Avionics

The complexity of aircraft systems continuously increases. For example, the amount of software and installed computer power doubled from one Airbus generation to the next, i.e., within every five years. Due to the permanently growing performance and the progressive miniaturization of the microprocessors, the volume of electronics nearly stayed the same.

The need to increase the power of the onboard equipment at a simultaneous reduction in the cost for installation, operation, and maintenance led in the early nineties to considerations concerning a new system architecture with the designation "Integrated

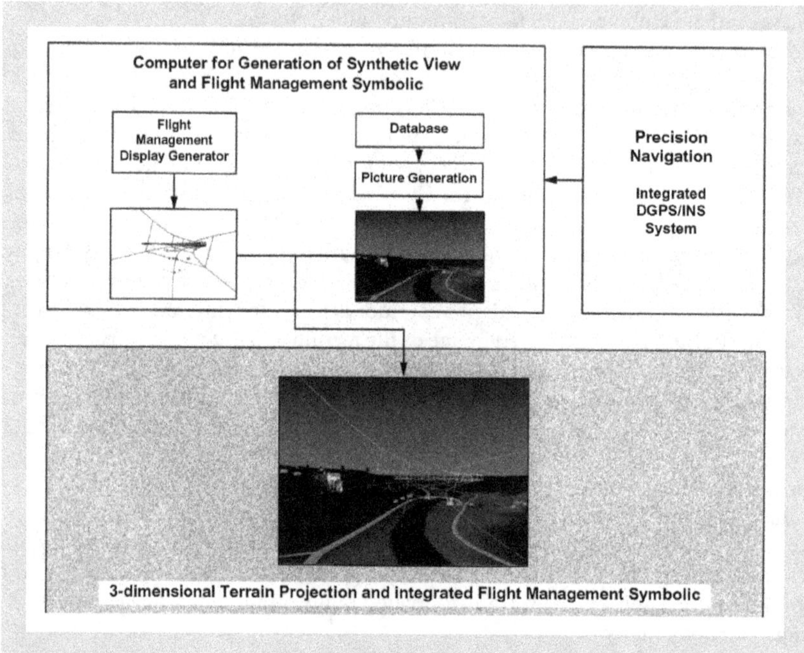

Flight guidance system with artificial sight.

Modular Avionics". This means to combine avionics functions in a system set up according to a modular-design principle. The concept envisages the use of standardized components: hardware modules for input/output control and signal processing, for information processing and power supply, and software modules for the operating system and for the interface to the application software. The hardware modules are installed in a common cabinet. The internal data transmission is carried out via a so-called backplane-bus. The advantages of this concept in comparison to conventional solutions are the reduction in the number of different types of equipment and hence a reduction in the cost for spare parts and increased system reliability and, consequently, a reduction in maintenance cost.

EADS Airbus established within the framework of the Aeronautical Research Programme 1995–1998 the requirements for Integrated Modular Avionics with respect to the architecture, hardware and software elements, specification procedures, integration and certification as well as development tools.

Cockpit

The cockpit is the workplace of the aircraft crew. The control and display elements within the cockpit constitute the man-machine-interface of the aircraft. The design of the cockpit must, therefore, follow ergonomic aspects. The crew must be able to perform the control of the aircraft, as well as the operation and supervision of the aircraft systems, with as little workload as possible so that handling errors are avoided and a maximum in flight safety is achieved. The crew is charged with 60–70 per cent of all flight accidents. The development of the cockpit is, therefore, of particular importance when considering the improvement of flight safety.

Cockpit for a Small Commercial Aircraft

The design of the cockpit and the avionics for a small aircraft is, with regard to cost, a special challenge. The price of the aircraft is, at comparable standards, approximately proportional to the number of seats. The cost for the avionics equipment are, on the other hand, at similar performance characteristics nearly independent of the aircraft size. However, since a small aircraft offers less installation space than a large one, the demands concerning weight, installation space and power supply are particularly high. MBB started in 1986 project work for a small commercial aircraft [123].

The development of the cockpit is an iterative process of design, simulation, integration and assessment. For each phase of this process, specific tools were set up and employed. For the design of the cockpit the design tool CATIA was used. The adjustment range of seats and pedals, the accessibility of the control elements, and the arrangement of instruments was determined with the aid of a three-dimensional model of a human. The spatial models of the cockpit elements and of equipment components were joined to an electronic cockpit mock-up. For the ergonomic and aesthetic design of the cockpit, a full-scale cockpit mock-up was built. Then shape, colour, lighting, interior decorations, and materials in the contact areas were determined. Aided by graphical touch-screen monitors, a virtual cockpit was set up. With this setup, operational processes and the interaction of cockpit systems were investigated. Within the framework of this project, a development simulator for design, integration and testing of the aircraft systems was realized. An essential part of this simulator was a cockpit with all display and control components and with sight and noise simulation.

Cockpit for General Aviation Aircraft

The man-machine interface in general aviation aircraft has remained for decades at the development status of the fifties. Digital computers and electronic display systems were, due to their high cost, initially limited to cockpit systems of commercial and military aircraft. Sinking cost in the area of microelectronics opened up the opportunity to also introduce these technologies for general aviation aircraft.

Within the Aeronautical Research Programme 1995–1998, the Technical University of Brunswick developed in cooperation with the supplier industry a cockpit system [124] for this type of aircraft. The system, consisting of computers, colour monitors and input elements, shall replace the conventional instrumentation board and support the pilot in his tasks of aircraft supervision, flight planning and navigation.

Cockpit Instrumentation

The development of the cockpit instrumentation since the end of the sixties can well be traced by considering the Airbus family. The first model, the A 300, still had the classic electromechanical instrumentation. With the introduction of the Models A 310 and A 320 early and late in the eighties, respectively, the transition to Cathode Ray Tubes (CRTs) occurred. In the nineties, the transition to Liquid Crystal Displays (LCD) followed. The number of instruments within the cockpit could be drastically reduced with the introduction of electronic displays. At the same time, the display technology opened up vast possibilities for the presentation of information.

An analysis of flight accidents shows that the cause of many accidents is a disorientation of the pilot with regard to the terrain and the situation of the aircraft within the airspace. An improvement of the awareness situation at a simultaneous reduction of pilot workload can be reached if the pilot is offered a three-dimensional image of the outside terrain combined with flight guidance information. Prerequisite is the availability of a powerful graphics generator.

The "VDO Luftfahrtgeräte-Werk, VDO-L" developed, therefore, in cooperation with the Technical University of Darmstadt, within the Aeronautical Research Programme 1995–1998 a high performance graphics generator.

Presentation of Air Traffic Control (ATC) Information

The permanently increasing civil air traffic requires that the methods and systems of air traffic control be further developed and improved. Within the framework of a research project funded by the "Deutsche Forschungsgemeinschaft, DFG", the Technical University of Berlin developed a concept for the visual presentation of ATC information [125, 126] inside the cockpit. Via a digital data link between the Air Traffic Control on ground and the flight management system on board of passenger aircraft, it is possible to display ATC instructions and weather information and to negotiate flight plans with the control station on the ground. The system was installed on the A 340 flight simulator at the Technical University of Berlin and assessed by experienced cockpit crews in comparison to conventional radio communication.

Pilot Assistance Systems

Pilots have the task to register a large amount of information and quickly derive corresponding decisions. By applying technologies of knowledge processing

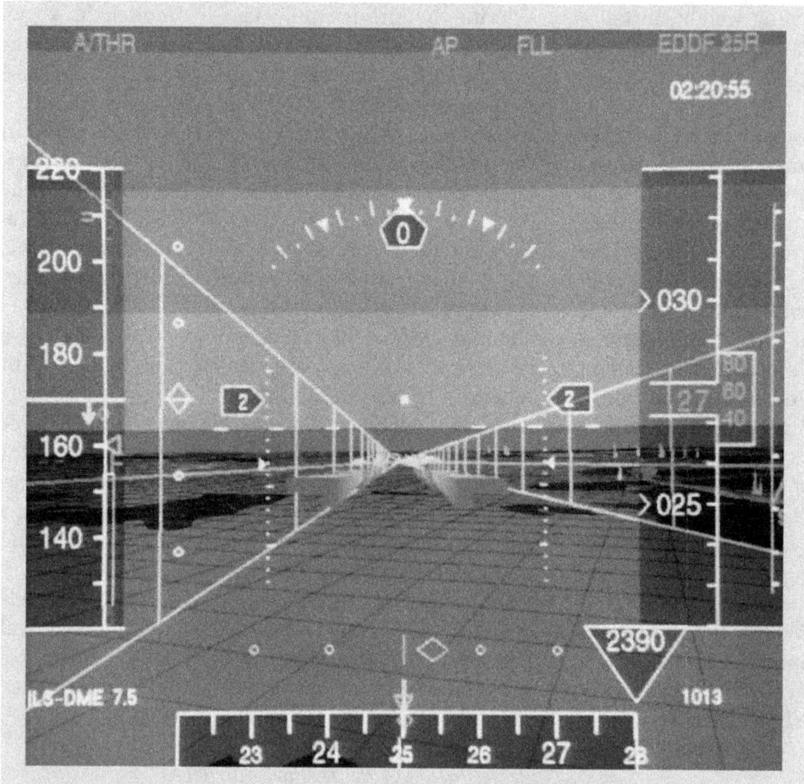

Three-dimensional presentation of flight guidance information.

within the cockpit, the so-called pilot-assistance systems were developed. It is the objective of these systems to improve the interaction between the aircraft crew and the aircraft systems.

At the end of the eighties, the University of the German Armed Forces Munich commenced with the development of knowledge-based pilot assistance systems [127, 128]. The systems ASPIO (Assistant for Single-Pilot IFR Operation) and CASSY (Cockpit Assistant System) were established for civil aircraft, and the system CAMA (Crew Assistant Military Aircraft), developed in a cooperation with the DLR, EADS, and the "Elektroniksystem- und Logistik-GmbH, ESG", for military aircraft. The basic setup of the three systems is the same. The situation of the aircraft, of the environment and of the crew is registered. The data are analyzed and assessed utilizing a data and knowledge base. The result is the basis for a flight plan. The dialogue with the crew is carried out via monitor and voice input and output. The system CASSY was tested and its efficiency was proven in flight tests on the test vehicle "ATTAS" (Advanced Technology and Testing Aircraft System) of the DLR.

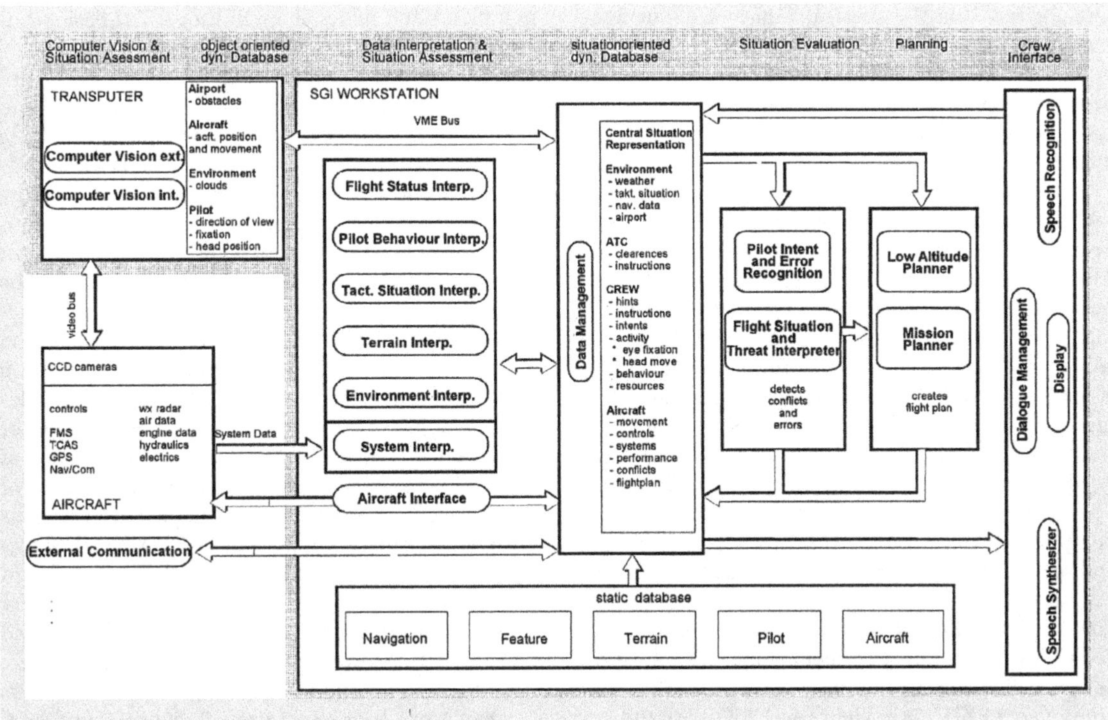

Block diagram of the pilot assistance system CAMA.

Sensors – Night Vision and Warning Systems for Helicopters

GÜNTER BRAUN

The Special Case of the Helicopter

The main advantage of helicopters over fixed-wing aircraft is their ability to take-off from and land vertically on nearly any place of sufficient size. If a landing on rugged terrain or over water is not possible, people or loads can be dropped off or taken onboard in hovering flight. These capabilities predestine helicopters for duty in air rescues, catastrophe prevention, border control and in crime fighting, for numerous transport tasks as well as for a number of military tasks. Considering speed, comfort and cost, helicopters are distinctly inferior to fixed-wing aircraft on longer distances between airports. Flights according to instrument flight rules (IFR) on defined flight paths between landing sites, generally airports, correspondingly equipped, do not fit the typical helicopter operational profile.

Helicopter flights and especially landings are almost exclusively being carried out by sight and according to Visual Flight Rules (VFR). Here exist, on the one hand, legal restrictions that do not allow, for security reasons, visual flights at visions <800 m (2,620 ft) (in the future 1,500 m (4,900 ft)) and a cloud base of 600 ft. On the other hand, also the enormous power of the human eye has its physical limitations, which cannot be exceeded without the use of suitable aids.

For example, the visual power of the human eye is considerably degraded during the night [129]. The adaptation of the eye to darkness can, in addition, be impaired due to the illumination of the cockpit instrumentation. The utilization of headlights is limited to the last phase of the landing approach due to their small range.

Light waves in the spectral range of about 400–720 nm (1 Nanometer = 10^{-9} m) employed by the human sight are reflected or scattered by rain and fog droplets, snow crystals, dust and smoke particles and cannot penetrate bad weather areas or smoke clouds – neither by day nor by night.

The operational possibilities of helicopters can therefore, up to now, not at all or only insufficiently be utilized during bad weather conditions or at night. Also many fatal helicopter flight accidents could have been avoided if suitable sensors had been available. A flight guidance and obstacle warning system, which compensates the deficits of the human perceptibility is long the desire of many helicopter operators. During the last two decades, a number of different sensors were hence developed, which more or (mostly) less closely approach this objective. The most important sensors are "low light-level amplifiers" ("Restlichtverstärker"), infrared cameras, radar, laser-radar, and, as the latest development, HeliRadar.

Low Light-Level Amplifiers

Today, low light-level amplifiers (night vision devices) are avaiable in various designs on the market. Examples are LLLTV (Low Light Level TV), with the image displayed on a monitor, or image amplifier goggles (BiV-goggles or Night Vision Goggles, NVGs, respectively), which are mostly mounted on the pilot's helmet. Low light-level amplifiers (night vision devices) are being manufactured in large numbers and are used by hundreds of thousands of people in terrestrial military and national applications. The basic principle is demonstrated in the figure below.

The photons, coming from the scene to be observed, are mapped via an objective on the photo cathode of the image amplifier tube. Here, they release photo electrons, which are being accelerated towards the anode and which generate secondary electrons by collision processes on their way. Finally, all electrons hit a phosphorous screen exciting it to emit light. This way, light amplification factors of up to 50,000 can be achieved.

The production of the secondary electrons occurs within a so-called Micro-channel Plate (MCP), a thin glass plate consisting of about 2.8 million individual small tubes. The tubes have an inner diameter of about 10 µm and are covered with a film capable of secondary emissions. A voltage applied to the ends of the tubes accelerates the incoming primary electrons until they impinge obliquely upon the walls. The secondary electrons knocked out by the impingement are also accelerated and produce further secondary electrons – this until the end of the channel is reached and all electrons impinge on the phosphorous screen.

A third generation of low light-level amplifiers (night vision devices) is available since about 1980. It distinguishes itself from the second generation by a different coating of the photo cathode and an associated change of its spectral and absolute sensitivity. The normalized spectral sensitivity of both generations, the relative sensitivity of the human eye during daylight as well as the spectral composition of star-

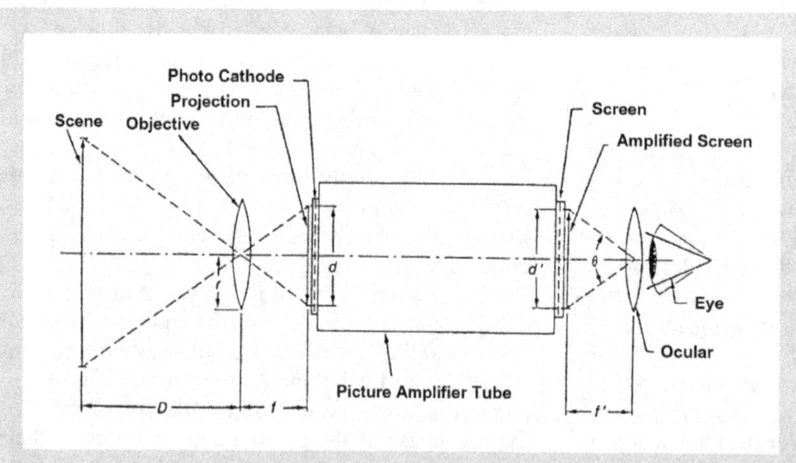

Principle of the low light-level amplifier (night vision device).

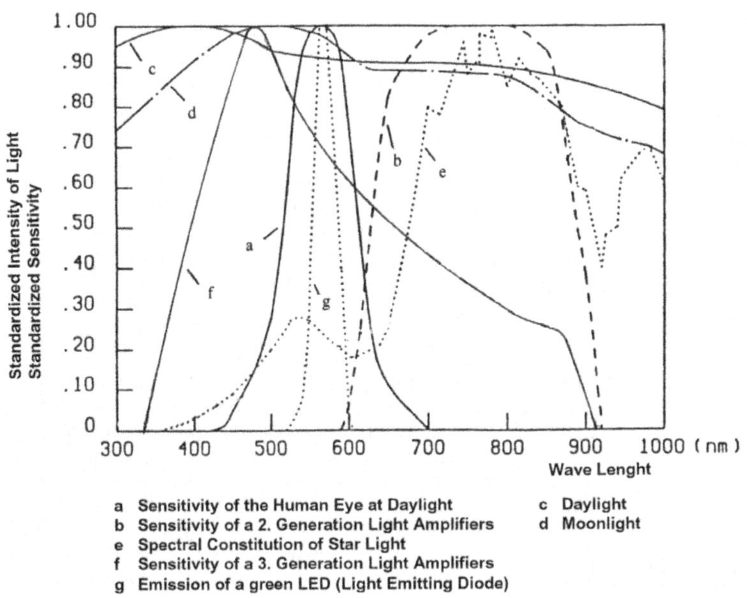

Normalized spectral sensitivity of low light-level amplifiers (night vision devices), spectral distribution of daylight, moonlight and starlight.

light is depicted above. It can clearly be observed that the spectral sensitivity of the third generation is optimally adjusted to the illumination by starlight so that, contrary to previous systems, a sufficient image quality can be achieved without an illumination by the moon. Its amplification below 600 nm is, furthermore, low and can be further reduced by filtering. This guarantees that the instrumentation illumination by green LEDs, easily recognizable by the bare eyes, does not interfere with the function of the night vision goggles. The compatibility of the latter with the helicopter cockpit was investigated in detail in several studies [130, 131].

Low light-level amplifiers (night vision devices) improve the image contrast at night and at good weather conditions. Their field of vision typically includes an angle of about 40°. However, small but dangerous obstacles, such as high-voltage lines, can often not be recognized. Their power degrades, furthermore, at bad weather conditions just as in the case of the human eye so that the latter's deficits in the reliable recognition of obstacles cannot be compensated by low light-level amplifiers. Night vision goggles weigh about 1.5 kg (3.3 bs), but may, in other designs, weigh as much as 5.5 kg (12.1 lbs).

Infrared Camera

Infrared cameras are utilized on helicopters under the designation FLIR (Forward Looking Infrared). They operate in some cases in a spectral range of 3–5 µm, but most often at 8–13 µm, i.e., close to infrared. There are two reasons for the selection of these wavelengths. On the one hand exists at these wavelengths a local minimum in the fundamental atmospheric damping caused by absorption lines of gaseous components of the atmosphere [132, 133]. On the other hand is it just this range, which allows an optimum observation of thermal contrasts between about 250 K and 400 K (−23 °C and 127 °C), temperatures occurring most frequently in our natural environment.

Infrared cameras supply a thermal image of the environment and are able to easily detect hidden heat sources like vehicles and living beings due to their good thermal resolution of about 0.1°. High-voltage lines can, as a rule, be detected, however, mainly after sunset and sunrise there are times where the temperature of the cables, but also of the masts, just correspond to the background temperature so that they are invisible to the FLIR [134]. A reliable obstacle recognition is thus, even in good weather, not possible.

Typical angles of vision lie for FLIR-systems, dependent on the telescope, between a maximum of 60°×40° (azimuth×elevation) and a minimum of 7°×3,25°. Often different telescopes are provided for a FLIR-system. The FLIR can be mounted on a spherical platform, controllable in elevation and azimuth, below the helicopter nose.

The infrared rays proceed through the objective of the camera and impinge on semiconductors, which are arranged individually or in a group in the focal plane and which scan the observed scene, for instance, line by line. The most frequently employed infrared detector is a mercury/cadmium/tellurium compound, which covers wavelengths between 8 and 12 µm and has its maximum sensitivity at 10,6 µm. In order to guarantee a satisfactory functioning, the detector has to be cooled to 77 K, the temperature of liquid nitrogen. Finally, the thermal image is being displayed on a black and white monitor. Monitor and control units are installed inside the passenger cabin of the helicopter and handled by an assistant.

Dependent on outfit, FLIR systems weight between 30 kg and about 80 kg (66 lbs and about 177 lbs). If one combines, for redundancy-reasons, a FLIR system with another night vision sensor, e.g., an LLLTV-system, which is, due to earlier accidents, mandatory for night flights of the German Federal Armed Forces, a weight of more than 90 kg (199 bs) may be reached. Although the additional atmospheric damping by fog or haze is at a wavelength of 10 µm smaller than in the spectral range up to 2 µm (19 dB/km (30.6 dB/mi) instead of 32 dB/km (51 dB/mi) at a range of vision of 500 m (1,640 ft)) [132, 133], the damping is still so high that also infrared sensors do not possess a useful range at bad weather conditions or merely at limited visibility.

Radar Systems

Modern aeronautics is without the application of Radar (Radio Detection and Ranging) inconceivable. Ground based systems monitor the use of air routes and allow approach and landing at airports at any weather condition. Onboard radar systems warn of bad weather and help to avoid collisions – if they are assembled.

Already in 1903, the feasibility and functionality of radar was demonstrated as collision warning system onboard ships. A radar transmitter scans the area with radio waves, which are emitted by a focusing antenna. The same antenna generally receives the reflected signals. They provide information concerning the distance and possibly the kind of the reflecting object by means of the run time and the intensity of the reflected signal. The object position is given by the alignment of the antenna; in many instances only the azimuth position is of interest. Examples are here radar systems for the monitoring of air and waterways; a smaller version for the detection of ships is also applied on marine (navy) helicopters.

The frequency spectrum used by radar systems has shifted over the decades towards higher and higher frequencies, i.e., shorter wavelengths, due to the rapid development of the electronics technology. This allows, due to the linear relation between the antenna diameter (aperture) and the wavelength at the same focus, more and more compact designs at higher power. A natural frequency limit for radar systems exists, however, at about 35 GHz. This is due to the increase in atmospheric damping with frequency with especially the damping due to rain having a negative effect on the range. The graph on the next page shows the atmospheric damping as function of the frequency between 1 GHz and 10 THz (wavelengths between 30 cm and 30 µm) on a logarithmic scale. Curve C represents the basic atmospheric damping due to atmospheric gases, mainly oxygen and water vapor. Curve B shows the additional damping by fog and the curves A the additional damping by rain of different intensity [133, 135]. The curves clearly show that the lowest damping occurs at low frequencies, that the rain damping always plays a role at frequencies above 10 GHz, and that at 33 GHz and 90 GHz damping minima exist (the so-called atmospheric windows).

It also becomes clear from the graph that an all-weather visual system with a reliable obstacle warning can only be realized within the microwave range up to frequencies of about 35 GHz, i.e., in the form of radar. Since helicopters do not operate on defined airways but mostly close to the ground, landing in unknown areas, obstacle-warning devices for helicopters have to meet different and rather significantly higher requirements than fixed-wing aircraft. Demanded are graphic sights of the flight path and the landing area with sufficient angular vision of, for in-

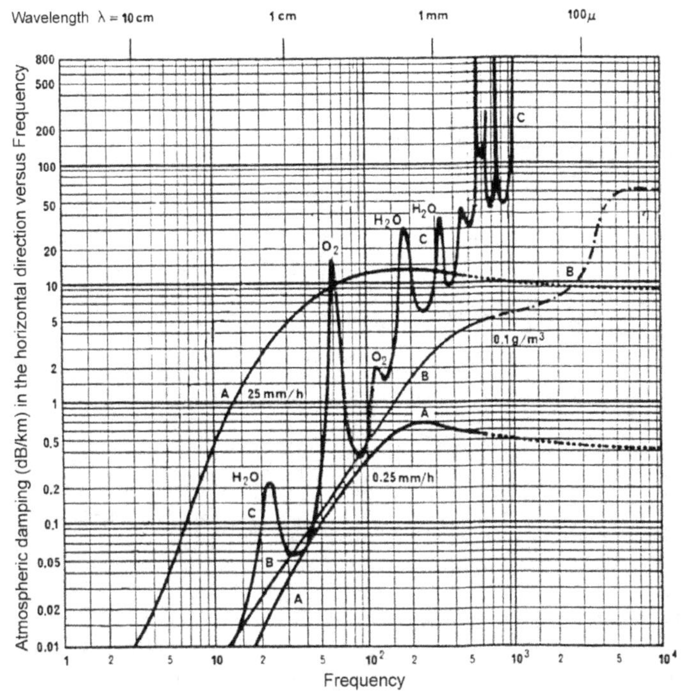

Atmospheric damping in the horizontal direction as function of the frequency.

stance, 70° × 40°, with a good image resolution of, for instance, 0.2°. This corresponds to about 70,000 image dots. The scan rate must finally be high enough so that the distance covered between scans does not become too large. Here, two images per second constitute the minimum. These requirements cannot be met by conventional radar since each image dot must be separately addressed and scanned by the antenna. The antenna would, in addition, at 33 GHz and a resolution of 0.2°, have a diameter of 2.5 m (8.2 ft). This explains why conventional radar systems have, up to now, not been employed on helicopters for flight guidance and obstacle warning.

In the mid-eighties, Telefunken, Ulm (today EADS) carried out a development at 60 GHz. The system had a mechanical scanning arrangement via rotational/folding mirrors. The angular range of vision was 180° × 30° (azimuth × elevation) and the image resolution was 2°. The operational frequency of the system is located at the atmospheric damping maximum. The system exhibits, therefore, only a limited range. It was developed and tested as prototype but was never put into operation. A system simultaneously developed by Thomson CSF scans an angular range of 90° × 30° (azimuth × elevation) at 94 GHz in about 1.5 sec. It also did not prevail.

Laser-Radar

Strictly speaking is laser-radar not radar since it does not employ radio waves for scanning an image but a laser. For that reason, it is also often called Ladar (Laser Detection and Ranging). But its functional principle corresponds essentially to the one of conventional radar. Laser-radar systems are available since 1993/94 as prototypes of Northrop and Fibertek (USA).

Laser-radar systems operate predominantly at a wavelength of 1,54 μm, i.e., slightly above the optical spectrum of the near-infrared range. Due to the short wavelength, a rather good image resolution can be achieved even with a very small aperture. The laser technology allowed, moreover, at 1.54 μm the realization of a pulsed laser with 10 kW peak power for the

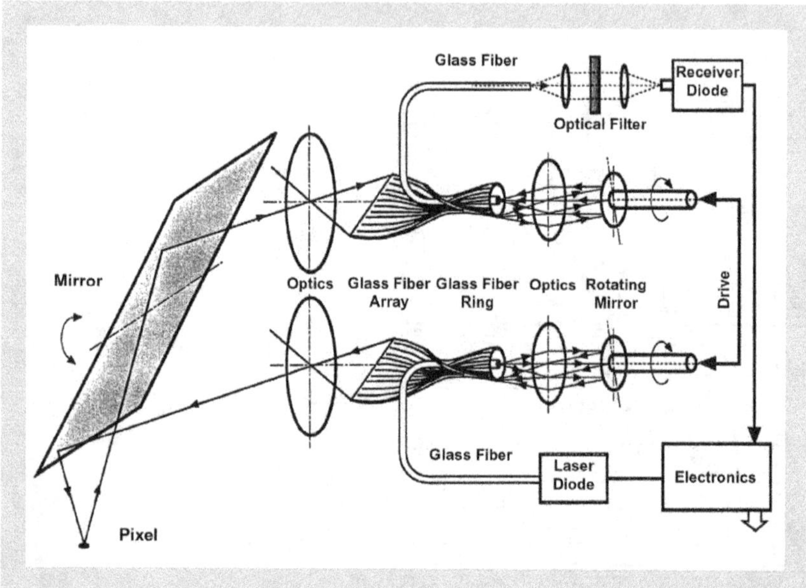

Principle of the Dornier laser-radar.

Dornier-laser-radar, which is required for the detection of high-voltage lines in 500 m (1,640 ft) distance at good visibility. The laser is, according to the manufacturer, also safe for the eyes. However, its average lifespan is only 1,000 hours.

As is already known from conventional radar systems, the difficulty also associated with the laser-radar is to scan as large a number of image points as possible at least twice per second. Various laser-radar manufacturers solve this scanning problem differently. The scanning procedure developed by Dornier is depicted above.

The pulse power of the laser is coupled by a light conduction fiber into a fiber bundle via a mirror that is slightly tilted against its axis of rotation; the fiber bundle is responsible for the beam deflection in the azimuth. A folding mirror realizes the deflection in elevation. The reflected rays received impact, via the folding mirror, on a fiber optical system, which corresponds to the one on the emitting side. For each image-point-associated received signal, the run time since the emission of the laser pulse is determined and the corresponding distance of the reflecting object computed.

With the Dornier laser-radar an angular range of 32° × 32° is scanned twice per second. The image resolution is 0.35° in azimuth and 0.2° in elevation. As range, 1,000 m (3,280 ft) are envisaged for large objects and 500 m (1,640 ft) for high-voltage lines, each at good visibility. The weight of the system shall not exceed a maximum of 30 kg (66 lbs).

At bad weather conditions, the performance of all laser-radar systems deteriorates, analogous to the human eye, since the laser beams, just as light within the visible spectrum, cannot penetrate haze, fog, snow and heavy rain. Investigations of Fibertek show that the range of this system at bad weather conditions always corresponds to only half the range of visibility of the human eye. Laser-radar systems are, therefore, not in the position to compensate the deficits of human vision at bad weather conditions. They improve, however, the detection of hard to recognize obstacles at good visibility.

HeliRadar

As can be seen from the preceding sections that optical and infrared obstacle warning devices fail at bad weather conditions, while conventional radar systems are not able to satisfy the demands on image resolution and image-repetition rates with acceptable antenna apertures. Demanded are for primary flight guidance instruments and obstacle warning systems an image of the flight path and the landing field with a sufficient angular visibility range (for instance 70° × 40°), good image resolution (for instance 0.2°) and a good image repetition rate (> 2 Hz), a reliable recognition of obstacles, including high-voltage lines, as well a sufficient range also at most difficult weather conditions.

The frequency dependence of atmospheric damping clearly shows that the lowest possible operational frequency must be chosen for an all-weather obstacle warning system. It shall, on the other hand, be possible to still recognize horizontally or vertically running wire obstacles of 10 mm diameter, which requires a frequency of at least 30 GHz. However, since the frequency band of 33 GHz not only lies in the minimum of atmospheric damping, but is also available worldwide for radio navigation, it virtually presents itself for the realization of all-weather sensors.

The SAR (Synthetic Aperture Radar) concept was discovered in 1951. SAR-systems were since used in air-supported reconnaissance and, since more than two decades, also utilized on earth-observation satellites. They provide high-resolution images at any weather.

The high azimuth resolution is achieved in that an antenna of a comparatively small aperture looks perpendicular to the flight direction, i.e., laterally, and takes and stores about 1,000 consecutive images. The individual images are taken from different positions due to the linear motion occurring in the intermediate time. By comparison of the individual images, a new image can be computed, whose resolution corresponds to a synthetic antenna aperture, i.e., the flight distance covered during the time the individual images were taken. The distance resolution can be selected by the choice of the bandwidth of the signal to be emitted such that for the image display in elevation a comparable image resolution is achieved.

The applicability of a SAR-system to the flight guidance of helicopters is, however, for two reasons not directly given. On the one hand is the line of vision of the SAR always normal to the direction of the motion of the antennae. Consequently, a standard SAR-system can never look in the direction of flight but only laterally. On the other hand requires the computation of the SAR-images a high performance computer. This holds even more if the computation is to be performed onboard and in real time and not later at a computer center.

The problem of the line of sight of the SAR was solved for helicopter applications in 1989, following a suggestion of *Horst Kaltschmidt*, by the invention of the ROSAR-principle [136]. The outward looking antenna rotates synchronously with the rotor and has at each position of the circular path a velocity component perpendicular to its line of vision. The pilot can select the line of vision of the SAR-system. The size of the spread-out synthetic aperture, and hence the SAR-image resolution, is determined by the length of the antenna arms. At 33 GHz and a length of the antenna arms of 1.5 m (4.9 ft), an image resolution of about 0.2° is achieved.

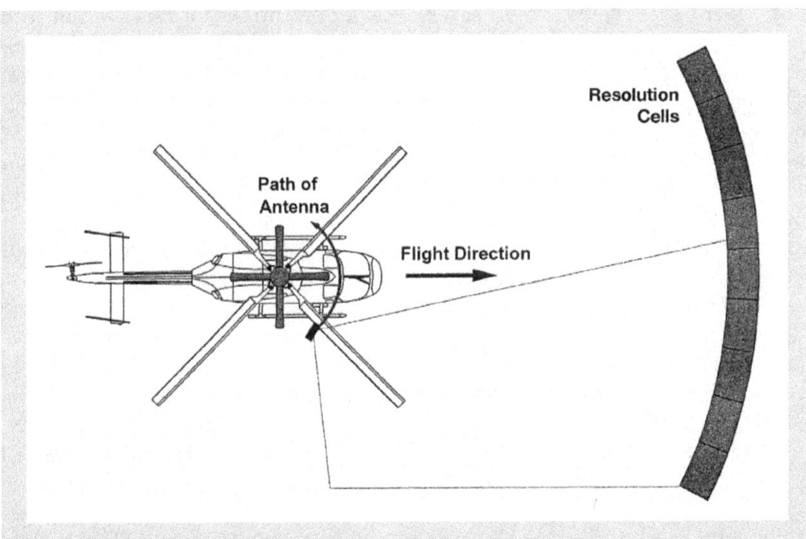

Principle of the Synthetic Aperture Radar (SAR) effect for a rotating SAR (ROSAR).

The system concept of HeliRadar.

The problem of the real-time computation of SAR-images could not be solved at the time of the ROSAR invention. For real time computations, a Cray 2-supercomputer was barely sufficient, which could, however, just for its size, not be utilized as an onboard flight guidance computer. In a joint development program, carried out since 1994 by Dornier Satellite Systems and Eurocopter Germany and having the product name "HeliRadar", a flight guidance computer could be realized, which, for the first time, was able to perform 14 billion floating point operations per second having only the format of a pilot's board case. This way, the real-time computation and display of a SAR-image of 70° × 40° (azimuth × elevation) and a high resolution at a repetition rate of 6 images per second became possible on board of a helicopter [133].

The system concept of HeliRadar is depicted above. Four antenna arms, each equipped with 8 small antennae, scan, according to the ROSAR principle, the environment with a FMCW radar signal. The emitting/reception unit of the device is mounted directly above the rotor head. From there, intermediate-frequency signals are transmitted to the helicopter cabin, transformed into digital signals and processed by the HeliRadar-computer and the subsequent image processing computer for the presentation on a wide-screen monitor inside the cockpit. It is envisaged to superimpose flight guidance information on the screen, such as an artificial horizon, distance lines, obstacles present on the flight paths marked in color, and other information adaptable to the customer's wishes.

HeliRadar produces six times per second an image of the environment with 70° × 40° (azimuth × elevation), an azimuth resolution of 0.2° and a distance resolution of 1.5 m (4.9 ft), which corresponds to an elevation resolution of 0.2° at a distance of about 300 m (984 ft). By fanning out the antennae, an additional discrimination (bias) in elevation of 2.5° is achieved. The system has a range of at least 1.3 km (0.81 mi) for all obstacles, including high voltage lines, at the most severe weather conditions for radar – rain with 12.5 mm (5 in) water per hour. The line of vision can in azimuth be chosen > 360°. HeliRadar has a weight of about 50 kg (110 lbs).

HeliRadar (ROSAR) was awarded the Science Prize of the ITG ("Information Technology Society within the VDE" ("Informationstechnische Gesellschaft im VDE")) in 1990 and in 1995 the Karl-Heinz-Beckurts-Prize; it was nominated 1997 for the "German Prize of the Future" ("Deutscher Zukunftspreis").

References

1 R.R. Gilruth, Requirements for Satisfactory Flying Qualities of Airplanes, NACA-Report 755 (1943), erstmals veröffentlicht als NACA Advanced Confidential Report (April 1941)

2 K.H. Doetsch et al.: Neuentwurf von Flugeigenschaftsrichtlinien, DVL, Institut für Flugmechanik (1943)

3 W. Tye, Civil Airworthiness, Journal of the Royal Aeronautical Society 70 (1966), pp. 253–257
4 Bericht über die Tagung über Rudermomentenausgleich (Herabsetzung der Steuerkräfte), Rostock-Marienehe, September 1940
5 Denkschrift zum Tag der Luftwaffe im Kriegsjahr 1942 (Hg. Von der Deutschen Akademie der Luftfahrtforschung), Denkschrift 1/42g
6 NAVIER SR-119, Specification for Flying Qualities of Piloted Aircraft (1942)
7 AAF-C-1815, Stability and Control Requirements for Airplanes (1943)
8 W.H. Phillips, Flying Qualities From Early Airplanes to the Space Shuttle, AIAA-Paper 88-751 (1988)
9 M.L. Spearman, Historical Trend in the Research and Development of Aircraft, AIAA-Paper 81-1613 (1981)
10 F.C. Haus, The Evolution of Handling Qualities Requirements, 7th Congress of ICAS, Rom 1970
11 W.M. Kaufmann et al.: Flight Tests of an Apparatus for Varying Dihedral Effect in Flight, NACA-TN-1788 (1948)
12 W. Pinsker, Analog-Rechengeräte und Simulatoren in der Luftfahrt, Jahrbuch 1957 der WGL, pp. 339–349
13 X. Hafer, Simulationstechnik in Luft-und Raumfahrt, Luftfahrttechnik-Raumfahrttechnik, 11 (1965), pp. 255-264
14 A.G. Barnes, The Role of Simulation in Flying Qualities and Flight Control System Related Development, AGARD-LS-157 (1988), 8-1–8-21
15 D. Hanke, Anforderungen an Flugeigenschaften, CCG-Kurs Flugregelung, Fly by Wire und CCV-Technology, Teil 1, Brunswick 1988, Manuskript 3
16 S. Baron, Pilot Control, in: E.L. Wiener, D.C. Nagel (Hg.), Human Factors in Aviation, San Diego etc. Academic Press) 1988, pp. 347–385
17 E.D. Dickmanns, Human Factor Aspects, Bericht über Tagungspunkt 7 des AGARD Flight Mechanics Specialists Meeting on "Stability and Control", Cambridge, England, September 1966, DVL 29/66 (1966)
18 R.C.A'Harrah, Flight-Simulation-A significant Aid in Aircraft Design, AGARD-CP-119, 1972, 12-1–12-16
19 C.B. Westbrook, D.T. McRuer, Handling Qualities and Pilot Dynamics, Aero/Space Engineering, Mai 1959, pp. 26–32
20 G.E. Cooper, Understanding and Interpreting Pilot Opinion, Aeronautical Engineering Review 16 (1957), 3
21 R.P. Harper Jr., G.E. Cooper, A Revised Pilot Rating Scale for the Evaluation of Handling Qualities, CAL-R-153, 1966
22 G.E. Cooper, R.P. Harper, The Use of Pilot Rating in the Evaluation of Aircraft Handling Qualities, NASA-TN-D-5153 (1969)
23 A. Tustin, The Nature of the Operator's Response in Manual Control and its Implications for Controller's Design, Journal of the IEEE 94 (1947), Teil II A
24 I.L. Ashkenas, D.T. McRuer, A theory of handling qualities derived from pilot-vehicle system consideration, IAS Paper, 62–39, 1962
25 W. Pinsker, The Role of Theory and Calculations in the Refinement of Flying Qualities, AGARD-CP-119 (1972), 10-1–10-12
26 B.L. Schofield et al., Development of Handling Testing in the 70's, A New Direction, AGARD-CP-33 (1982), 23-1–23-16
27 R.P. Harper, G.E. Cooper, Handling Qualities and Pilot Evaluation, J. Guidance and Control 9 (1986), pp. 515–529
28 R.C. A'Harrah, W.J. Lamanna, J Hodgkinson, Are Today's Specifications Appropriate for Tomorrow's Airplanes, AGARD-CP-260 (1978)
29 J. Hodgkinson, Equivalent Systems Criteria for Handling Qualities of Military Aircraft, AGARD-CP-333 (1982), 3-1–3-11
30 R. Hoh, Advances in Flying Qualities, Concepts and Criteria for a Mission Oriented Flying Qualities Specification, AGARD-LS-157 (1988), 5-1–5-28
31 D.J. Moorhouse, R.J. Woodcock, Present Status of Flying Qualities Criteria for Conventional Aircraft, AGARD-CP-333 (1982), 1-1–1-14
32 R.H. Hoh, D.G. Mitchell, I.L. Ashkenas, Proposed MIL Standard and Handbook-Flying Qualities of Air Vehicles, Vol. I: Proposed MIL Standard, AFWAL-TR-82-3081 I (1982)
33 R.H. Hoh, Advances in Flying Qualities: Concepts and Criteria for a Mission Oriented Flying Qualities Specification, AGARD-LS-157 (1988) 5-1–5-28
34 K.H. Doetsch et al., New Standards for Desirable Handling Qualities of Aircraft, Royal Aircraft Establishment (Aero No 7), Februar 1946
35 AvP 970, Vol.1, Design Requirements for Aircraft for the RAF and RN -Part 6 (Aerodynamics and Flying Qualities), Juli 1964
36 S.J. Andrews, The Nature and Use of the Rules for Judging the Acceptability of the Flying Qualities of Fixed Wing Aircraft, AGARD-CP-106 (1972)
37 G. Bennett, Revised Requirements for the Flying Qualities of Service Airplanes, RAE, Tech. Memo Structures 863, 1975

38 DEF-STAN 00-970, Issue 1, Design and Airworthiness Requirements for Service Aircraft, Vol 1-Aeroplanes, Book 1, December 1983
39 H.A. Mettam, Fifty Years of Military Airworthiness, Journal of the Royal Aeronautical Society 70 (1966), pp. 248–250
40 A. Day, Design Requirements for Military Aircraft, Journal of the Royal Aeronautical Society 70 (1966), pp. 250–252
41 TSS Standard 5-2, Supersonic Transport Aircraft Flying Qualities, 22.3.1968
42 J.C. Wanner, J.W. Carlson, Comparison of French and United States Flying Qualities Requirements, AGARD-CP-106, 1-1–1-10
43 W.E. Lamar, T.L. Neighbor, Handling Qualities Criteria and Requirements, AGARD-CP-119 (1972), 4-1–4-19
44 X. Hafer, G. Sachs, Senkrechtstarttechnik, Springer Verlag, Berlin etc., 1982, pp. 334–359
45 S.B. Anderson, An Examination of Handling Qualities Criteria for V/STOL Aircraft, NACA-RD-331, 1960
46 S.B. Anderson, Handling Qualities Criteria for V/STOL Aircraft, AGARDograph 46.2 (1960), pp. 511–532
47 MIL-H-8501, Military Specification, Helicopter Flying Qualities, Requirements for Military, 5. November 1952 H. Phillips, Flying Qualities
48 MIL-F-8575 (ASG), Military Specification, Flying Qualities of Piloted Airplanes, 12. November 1957
49 AGARD-R-408, Recommendation for V/STOL Handling Qualities, October 1962
50 AGARD-R-408A, Recommendations for V/STOL Handling Qualities with an Addendum Containing Comments and Recommendations, 1964
51 X. Hafer, VTOL Control Systems, Especially Control by Thrust Modulation, AGARDograph 89, 1964, pp. 449–473
52 AGARD-R-577, V/STOL Handling I-Criteria and Discussion, 1970
53 AGARD-R-577, V/STOL Handling II-Documentation, 1973
54 MIL-F-83300, Military Specification, Flying Qualities of Piloted V/STOL Aircraft, 1970
55 AGARD-R-577, V/STOL Handling I-Criteria and Discussion, 1970
56 AGARD-R-577, V/STOL Handling II-Documentation 1973
57 C.R. Chalk, G. Sounders, J. Kroll, F. Eckhart, R. Smith, V/STOL Flying Qualities Criteria Development, Vol.I, A Proposed Military Specification for V/STOL Flying Qualities, 1968
58 Handling Qualities Definition Document for EFA, 4th Draft (Mai 1988)
59 H.-J. Pausder und K. Sanders, DFVLR Flying Quality Research Using Operational Helicopters. Vertica, Vol 10, No 1, 1986
60 H.-J. Pausder und R.M. Gerdes, The effects of pilot stress factors on handling qualities assessments during U.S./ German helicopter agility flight tests. Vertica, Vol 8, No 15, 1984
61 Ph. Benquet, H.-J. Pausder, Ph. Rollet und V. Gollnick, Tailoring of ADS-33 for the NH90 program. American Helicopter Society 52nd Annual Forum, Washington D.C., 1996
62 G. Bouwer und K.B. Hilbert, The design of a model-following control system for helicopters. AIAA Guidance and Control Conference, Seattle, 1984
63 H.-J. Pausder, A study of roll response required in a low altitude slalom task. 11th European Rotorcraft Forum, London, 1985
64 W. von Grünhagen, G. Bouwer, H.-J. Pausder, F. Henschel und J. Kaletka, A high bandwidth control system for a helicopter in-flight simulator. International Journal of Control, Vol. 59, No 1, 1994
65 H.-J. Pausder und C.L. Blanken, Investigation of the effects of bandwidth and time delay on helicopter roll-axis handling qualities. 18th European Rotorcraft Forum, Avignon, 1992
66 C.J. Ockier, H.-J. Pausder und C.L. Blanken, An investigation of the effects of pitch-roll (de)-coupling on helicopter handling qualities. DLR Forschungsbericht 95-08, 1995
67 H.-J. Pausder, U. Butter und V. Gollnick, ACT Demonstrator/Flying Helicopter Simulator – An airborne testbed development project. American Helicopter Society 52nd Annual Forum, Washington D.C., 1996
68 C.J. Ockier, Evaluation of the ADS-33D handling qualities criteria using the Bo 105 helicopter. DLR Forschungsbericht 98-07, 1998
69 A. Hafer, G. Sachs, Zur Entwicklung der Flugeigenschaftsrichtlinien, DGLR Jahrbuch 1988 I, pp. 659–669
70 A. Hafer, Gottfried Sachs, Zur Entwicklung der Flugeigenschaftsrichtlinien nach 1945, DGLR Jahrbuch 1990 II, pp. 983–992
71 W. Schwipps, Schwerer als Luft. Frühzeit der Flugtechnik in Deutschland, Bernard & Graefe Verlag, Bonn, Band 8
72 A. Hafer, X. Hafer, G. Sachs, Die Anfänge der Stabilitätsbetrachtung in der Flugmechanik, ZFW 12 (1988), pp. 27–44

73 O. Lilienthal, Der persönliche Kunstflug, Moldebecks Taschenbuch zum praktischen Gebrauch für Flugtechniker und Luftschiffer pp. 562–600, Berlin: Krayn (Neuauflage 1911)

74 F.W. Lanchester, Aerodynetics, London, 1909 (Dt. Übersetzung von C. und A. Runge, Leipzig 1911)

75 H.J. Jex, F.E.C. Culick, Flight Control Dynamics of the 1903 Wright-flyer, AIAA-Paper 85-1804, 1985

76 H. Helmbold, Eduard Fischel, Vorträge gehalten in der 7. Wissenschaftssitzung der ordentlichen Mitglieder am 20. Sept. 1940

77 H. John, W. Kraus, High Angle of Attack Characteristics of different Fighter Configurations, AGARD CP 247, pp. 2.1–2.15

78 A.B. Haines, Computer und Windkanäle: Ergänzende Hilfen für den Flugzeugentwurf, Dietrich Küchemann-Vorlesung, ZFW 1 (1977), Heft 5

79 D. Kraft, G.C. Shau, K.H. Well, E. Berger, Optimale dreidimensionale Steuerungen von Hochleistungsflugzeugen mit Post-Stall Eigenschaften, DFVLR Bericht A552-78/2, Institut für Dynamik der Flugsysteme, DFVLR, Oberpfaffenhofen, 1978

80 W. Kraus, Wing Design for Delta-Canard Configuration, MBB/UFE/TKF/TN-173/83

81 R. Weissmann, Preliminary Criteria for Predicting Departure Charakteristics/Spin Susceptibility of Fighter Type Aircraft, Journal of Aircraft, Vol. 10, No. 4, pp. 214–219 April 1973

82 H. John, W. Kraus, High Angle of Attack Charakteristics of different Fighter Configurations, AGARD CP 247, pp. 2.1–2.15

83 H. John, U. Haux, Grundlagen der TKF-Konfigurationsfindung, MBB-interner Bericht: M/LKE12/TN/J90/236/84 vom 25.09.84

84 K. McKay, Fin Investigation in the 4 × 4 ft.Tunnel Warton for Trans- and Supersonic Speeds, BAe interner Bericht, BAe 8255/KMck/443/BAL

85 S. Ransom, Configuration Development of a Highly Manoeuverable Research Aircraft with forward Swept Surfaces. International Conference Forward Swept Wing Aircraft, 24–26.3.1982 University of Bristol

86 D. Hummel, H. John und W. Staudacher, Aerodynamic Characteristics of Wing Body Combinations at High Angles of Attack, Paper 84-2.7.1, 14th ICAS Congress, Sept. 9-14, 1984,Toulouse

87 K.W. Lotter, J. Malefakis, Intake Design and Intake Airframe Integration for Post-Stall Fighter Aircraft Concept, AGARD CP 247, Paper No. 32, October 1978

88 H. Ross, Delta Canard F100 – Einmotorige Variante, MBB/TKF/STY/162/20.7.81

89 M.R. Robinson und H. Ross, Joint International Technical Development–The X-31 Program, AIAA Paper 89-2121, AIAA/AMS/ASEE Aircraft Design and Operations Meeting, Seattle WA, Aug. 1989

90 S. Powers, H. Schellenger RI, The X31: High Performance at Low Cost, AIAA/AHS/ASEE Aircraft Design, Operations Meeting Seattle WA, Aug.1989

91 H. Friehmelt, D. Rohlf, E. Plaetschke, S. Weiß, X 31 A Flugversuche mit unkorrelierten Steuerflächenausschlägen zur Bestimmung der aerodynamischen Derivative, DGLR-Jahrbuch 1995, II, pp. 555–564

92 E. Plaetschke, S. Weiß, Identification of Thrust Vector Effectiveness from X-31 A Flight Test Data, ZfW, Vol. 17, 1993, pp. 235–238

93 P. Huber, X-31 Enhanced Fighter Manoeuverability Final Report 22.12.1995, Dasa/LME1/X31A/R/00901

94 H. Ross und P. Huber, Application of Thrust Vectoring for Future Aircraft Design, AGARD Propulsion and Energetics Symposium 25–29 Sept 1995, Seattle USA

94a T. Grohs, O. Brieger, "X-31" VECTOR - Taking ESTOL to the Ground. SFTE-Paper, 14th Symp., June 10–12, 2003, Toulouse

95 G. Kissel, H. Schmidtlein, Operational Proving of Automatic Flight Control Systems for V/STOL Fighter Aircraft, AGARD Conference, Baden-Baden, 1970

96 G. Kissel, H. Wünnenberg, V/STOL Handling Qualities Criteria Compared with Flight Test Results of the V/STOL Supersonic Fighter VJ 101 C and the V/STOL Transport Aircraft DO 31E, AGARD Flight Mechanics Specialists' Meeting, Ottawa, 1971

97 F. Aulehla, G. Kissel, Experience with High Performance V/STOL Fighter Projects at MBB, AIAA/NASA Ames V/STOL Conference, Palo Alto, 1981

98 K. Kracheel, Flugführungssysteme – Blindfluginstrumente, Autopiloten, Flugsteuerungen, Bernard & Graefe Verlag, Bonn, 1993

99 K. Daser, Erläuterung der Probleme einer integrierten Flugsteuerungs- und Regelungsanlage für Hubbläser-V/STOL – Flugzeuge am Beispiel des DO 31-Vertikalgeschwindigkeitsreglers, Gemeinsame Sitzung der DGLR-Fachausschüsse Flugeigenschaften und Flugregelung, Immenstaad, 1972

100 H. Wünnenberg, Stabilität und Steuerbarkeit von V/STOL-Flugzeugen nach Verfahren und Ergebnissen aus der DO 31-Flugerprobung, Dokumentationszentrum der Bundeswehr, 1972

101 R. Draganow, H. Max, Programme Experimentelle DO 31, AGARD Conference, Brüssel, 1972
102 R. Riccius, W. Sobotta, VAK 191 B Experimental Program for a V/STOL Strike/Recce Aircraft, AGARD Conference, Brüssel, 1972
103 H.-H. von Salzen, W. Sobotta, Das Regelungssystem der VAK 191B und daraus resultierende neue Entwicklungen für die nächste Generation von V/STOL Kampfflugzeugen, Gemeinsame Sitzung der DGLR-Fachausschüsse Flugeigenschaften und Flugregelung, Immenstaad, 1972
104 H. Beh, U. Korte, G. Löbert, Stability and Control Aspects of the CCV-F-104 G, AGARD Conference Proceedings No. 260, 1979
105 U. Korte, Flight Test Experience with a Digital Integrated Guidance and Control System in a CCV Fighter Aircraft, AGARD Guidance and Control Panel 35th Symposium, Lissabon, 1982
106 R. Brockhaus, Flugregelung, Springer-Verlag, Berlin Heidelberg New York, 1994
107 H. Beh, G. Hofinger, Control Law Design of the Experimental Aircraft X-31 A, ICAS-94, 1994
108 G. Grübel, D. Joos, D. Kaesbauer, R. Hillgren, Robust Backup Stabilization for Artificial Stability Aircraft, 14th ICAS Congress, Toulouse, 1984
109 G. Grübel, Control Design Engineering, DFVLR, Scientific Report 1993–1996
110 B. Formica, H. König, K. Bender, G. Mansfeld, A Digital Optical Tail Rotor Control System, 14th European Rotorcraft Forum, Milano, 1988
111 H. König, M. Stock, S. Zeller, OPST1-An Optical Yaw Control System for High Performance Helicopters, 45th AHS Annual Forum, Boston/Mass., 1989
112 G. Kissel, A. Faulkner, I'm All "Light" Jack, 16th European Rotorcraft Forum, Glasgow, 1991
113 H. Huber, P. Hamel, Helicopter Flight Control – State of the Art and Future Directions, 19th European Rotorcraft Forum, Cernobbio (Como), 1993
114 R. Knorr, C. Melz, A. Faulkner, M. Obermayer, The Application of Active Side Arm Controllers in Helicopters, Conference on Piloting Vertical Flight Aircraft, San Francisco, 1993
115 J. Meins, C. Deeg, D. Hanke, D. Münscher, Advanced Sidestick Controller with Electro-Magnetic Loading System, CEAS Symposium, Delft, 1995
116 R. Brockhaus, Integriertes Flugbahnführungssystem, Kolloquium Sonderforschungsbereich 58 "Flugführung", Braunschweig, 1981
117 R. Brockhaus, Flugzeug, Atmosphäre und Flugaufgabe, Deutsche Forschungsgemeinschaft, Ergebnisse aus dem Sonderforschungsbereich 212 "Sicherheit im Luftverkehr", Braunschweig, 1994
118 G. Schänzer, Aspects of Safety Critical Applications of Satellite Navigation Systems, GNSS 97, Munich, 1997
119 H. Blomenhofer, T. Jacob, GPS Availability and Integrity bei Präzisionsanflügen und automatischen Landungen, DGON SATNAV 95, Freising/Weihenstephan, 1995
120 H. Blomenhofer, Accuracy, Integrity, and Availability of GPS Based Autopilot Coupled Aircraft Landings, Journal of the Institute of Navigation 1996/1997
121 G. Sachs, K. Dobler, J. Speckels, J. Röper, Precision Control of Air Vehicles with Satellite Navigation and Synthetic Vision. EURNAV 1996, Vehicle Navigation, Location and Control, London, 1996
122 G. Sachs, K. Dobler, G. Schänzer, M. Dieroff, Precision Navigation and Synthetic Vision for Poor Visibility Guidance, AGARD Flight Vehicle Integration Panel, Lissabon, 1996
123 H. Seidel, Konzeption des Cockpits und der Avionik für ein kleines Verkehrsflugzeug, DGLR-Jahrestagung, Friedrichshafen, 1990
124 G. Ardey, R. Blume, G. Schänzer, Man-Machine-Interface Cockpit, A Concept to Integrate New Technologies into General Aviation Aircraft, DGLR-Jahrestagung, Munich, 1997
125 G. Hüttig, G. Anders, A. Tautz, ATC Data-Link in an Airbus A340 Cockpit, 8th International Symposium on Aviation Psychology, Columbus, Ohio, 1997
126 G. Hüttig, G. Anders, A. Tautz, Integration und Evaluation eines ATC Data-Links im Cockpit eines Airbus A340 Simulators, DGLR-Jahrestagung, Munich, 1997
127 Universität der Bundeswehr München, 20 Jahre Institut für Systemdynamik und Flugmechanik, 1995
128 A. Walsdorf, R. Onken, H. Eibl, H. Helmke, R. Suikat, A. Schulte, The Crew Assistant Military Aircraft, Workshop "Human Electronic Crew: The Right Stuff?", Kreuth, 1997
129 H.-J. Hentschel, "Licht und Beleuchtung", Dr. Alfred Hüthig Verlag, Heidelberg, 1987
130 H.-D. Böhm, "The Night Vision Goggle Compatible Helicopter Cockpit", Tenth European Rotorcraft Forum, Den Haag, August 1984
131 S. Haisch, "Auflösungs- und Störgrenzenbestimmung von Bildverstärker-Brillen", Diplomarbeit TU München, Oktober 1992
132 International Telecommunications Union, CCIR-Report 883-1

133 G. Braun, "Operationelle Einsatzfähigkeit von Hubschraubern unter allen Wetterbedingungen, bei Tag und Nacht, eine Grundvoraussetzung für die Zukunft?" 20. Internationales Hubschrauberforum, Bückeburg, September 1995

134 J. Frank, "BK 117 with a Forward Looking Infrared System (FLIR)", MBB Helicopter News Vol. 12, 29th Issue, Januar 1991

135 International Telecommunications Union, CCIR-Report 719-2

136 H. Klausing, "Realisierbarkeit eines Radars mit synthetischer Apertur durch rotierende Antennen", Dissertation Universität Karlsruhe, April 1989

Ultra-Light-Weight Construction – From Metal to Fibre Structures

The Development of Fibre Composite Technology

Heiner Dörner, Horst Prem

All metallic materials have the characteristic of isotropy, which means their physical and mechanical properties are independent of direction. Within light-weight-design of structures, discrete reinforcements in the direction of the main loads, certain directional stiffness characteristics, were already very early realized to reduce the weight of structural elements. The introduction of monocoque constructions into aircraft engineering – and hence the consequent load-conformal design of load-bearing structures by means of beam, rod and plate elements – means a large step towards ultra light-weight structural design. The discrete stiffening of the skin by strips and beams (integral construction) with pronounced orthotrop stiffness directions, i. e., stiffness directions normal to each other, became a design principle of modern light-weight structural design.

A new, large innovative step occurred in the fifties with the consequent utilization of directional strength properties of fibre reinforced plastics (FRP). Here, the excellent unidirectional mechanical properties of certain fibres (e.g., glass and carbon fibres), embedded as construction elements load-oriented within plastics (e.g., polyester and epoxy resins) are advantageously applied. The fibre composite material hence originates during the manufacturing process of the structural element; it can optimally be tailored to the stresses and minimized with respect to its mass. This leads to large specific strengths, which means a high load-capacity at low weight, which is the desirable objective of light-weight construction. It was therefore a logical consequence that aircraft manufacturers, with the usual lead of sailplane engineering, showed great interest in this new fibre composite technology and consequently influenced its further development. It should be mentioned that also the original material of aircraft construction, namely wood, is a fibre material with anisotropic strength characteristics. Even in the unfair comparison with artificial fibres, which constitute only one building block within the anisotropic materials, wood does relatively well with regard to the load-capacity/mass ratio (specific strength).

In Germany, *Ulrich Hütter* fabricated a 17-meter long cantilever rotor blade for a wind energy installation of glass-fibre reinforced plastics (GFRP). *Heiner Dörner* describes in his Hütter-biography [1] "Three Worlds – One Life": *Generally, man tries, in the case of fibre composite materials, to understand, imitate and copy nature. But nature has, during millions of years of evolution, created optimum light-weight structures in plants and animals, which cannot yet be comprehended and copied by man. Up to now, man is incapable of reconstructing the splendid structure of a straw in an artificial building considering the ratio of effort (mass) to load capacity. The ratio of strength to lightness and stiffness is realized in the stem of a plant in a most magnificent and unattainable way. And the straw additionally possesses growth and repair capabilities, which is impressively demonstrated by a wheat field straightening up after it has been flattened by rainfall.*"

Ulrich W. Hütter (1910–1990)

Ulrich W. Hütter, born in Pilsen, Austria, on December 18, 1910

1921 classical high school in Salzburg.

1930 studies of Mechanical Engineering and Naval Engineering at the Institute of Technology in Vienna.

1934 design of the glider-trainer H 17.

1936 studies and graduation in Aeronautical Engineering at the Institute of Technology in Stuttgart.

1938 design of the glider H 28 together with his brother Wolfgang Hütter. Designer at Schempp-Hirth: SH-Hütter dive brake, design of the motor glider MoSe Hi-20.

1939 lecturer at the Engineering School Weimar with the topics being aircraft engineering, aircraft statics, fluid mechanics, flight mechanics, advanced mathematics, machine elements, kinematics; development, design, construction and testing of wind-energy installations with performances between 5 and 50 kW (6.7 and 67 hp) and rotor diameters of 5–18 m (16–59 ft).

1942 PhD from the Institute of Technology in Vienna under Professor *Feifel* and the thesis: "Contribution to the establishment of design fundamentals for wind-energy installations".

1943 drafted to the "Research Institute Count Zeppelin" ("Forschungsinstitut Graf Zeppelin, FGZ); head of the Design Department: aircraft catapult, ribbon parachute, wing bodies (twin bodies), manned missiles, underwater towing facility, long-range bomber Hü 211.

1944 lectureship in Fluid Mechanics and Flight Mechanics at the Institute of Technology in Stuttgart.

1946 Head of Design at the Allgaier Company, Uhingen, Württemberg, in charge of the 10-kW (13.4 hp) wind energy installation, 3-blade fast-rotor.

1952 renewal of the lectureship at the Institute of Technology in Stuttgart.

1957 habilitation (right to lecture) treatise: Talon take-off procedure with damping of the starting shock by a chute.

1957 development of the wind energy installation StGW-34: 34-meter (112 ft) rotor diameter, performance 100 kW (134 hp) at 8 m/s (26 ft/s) wind, 2 rotor blades, first large cantilevered structural elements made of glass-fibre reinforced plastics, GFRP.

1959 associate professor at the Institute of Aircraft Engineering at the Institute of Technology in Stuttgart (IFB).

1965 appointment to full Professor-ship and head of a chair.

1974 Enoch-Thilin-Medal of the Aeronautical Society of Sweden for merits related to sailplane engineering.

1977 award: "Aachen and Munich Award of Technology and Applied Natural Sciences".

1980 retirement, resigning from State Services.

1990 he dies in Kirchheim/Teck on August 12.

And he continues elsewhere: *"This 17-meter long rotor blade could also, certainly somewhat pathetically, be called the primeval mother or father, or the primeval model, of today's modern German sailplanes. And Germany is leading worldwide in this field."*

The first flight of the high-performance glider fs 24 "Phoenix" of *Eppler/Naegele* took already place in 1957; here, a laminar wing in GFRP-Balsa-Sandwich construction was employed and also the fuselage, made of plastics, with a streamlined pilot's canopy, had a laminar-profile shape. While the glide ratios of sailplanes during the time period 1921–1940 were in the range of 16–38, they have reached today, with the "Nimbus 4" of *Holighaus* and the ASW 22 of *Waibel*, in a preliminary culmination glide ratios of up to 60.

Germany is the world's market leader in the area of glider constructions using plastics. The steadily increasing spans of these aircraft made a transition to

The Airbus vertical stabilizer in CFRP-design is in quantity production since 1985.

still stiffer and stronger fibres necessary. Here, carbon fibres, invented in England, offered themselves as material, which is to day, embedded in epoxy resin, in the form of pre-impregnated unidirectional CFRP-mats or CFRP-fabrics mostly mechanically processed. They are also utilized for highly loaded rotor blades and transport aircraft.

Germany has realized the CFRP-technology for the stabilizers of the "Airbus" program. In combination with the trim tank inside the elevator, an increase in the range of the "Airbus" aircraft was achieved. This topic was addressed in the Chapter "From the Stable to the Unstable Configuration". In the meantime, about 1,400 CFRP vertical stabilizers and elevators are in use on wide-body as well as narrow-body aircraft; but the next step is still missing, namely, the design of the aeroelastically critical wing of a Super-jumbo in CFRP-design. For solving the aeroelastic problems of a super-jumbo wing aeroelastic tailoring of CFRP-material seem to be a suitable approach to fulfil the secondary condition of minimum weight and lower operational costs, relative to competing products.

In fighter aircraft engineering, this step has already largely been performed. 70 per cent of the surface of the European fighter EF 2000 consist of carbon-fibre composite materials (CFRP), 12 per cent consist of glass-fibre composite materials (GFRP) and only 15 per cent are made of metallic materials such as Al-Li, cast aluminum, and titanium. The fibre-composite material share amounts to 42 per cent based on the structural weight and is thus much higher than in the case of conservative US-developments, such as the F-22 of Lockheed-Martin, with 28 per cent.

The actual triumphant application of fibre reinforced plastic (FRP) began in Germany during the second half of the sixties with the manufacturing of the BO 105 rotor blades out of GFRP. Hütter had drawn out the essential design elements, even if design simplifications became necessary, especially in the blade junction to the rotor hub [2], to reduce manufacturing costs. In the meantime the bearings of the BO 105 rotor blades have been replaced by aeroelastically dimensioned fibre-composite parts on the EC 135 rotor. The so-called "sandwich construction" was also advantageously applied to rotor blades; here, two cover (surface) layers of metal or fibre-reinforced material are connected with a supporting core (light metal honeycomb, plastics honeycomb, GFRP-fabric layers and hard foams) in between. Today, curved fuselage and partial wing monocoque designs, complete control surfaces and base plates are manufactured in FRP sandwich-construction.

There were still doubts concerning the characteristics of these materials under the influence of humidity, which prevented a wide range of application of these materials. These basic problems could, however, be dispelled during the research projects CFRP-tailer-on for the "Tornado", fibre-composite elevator for the

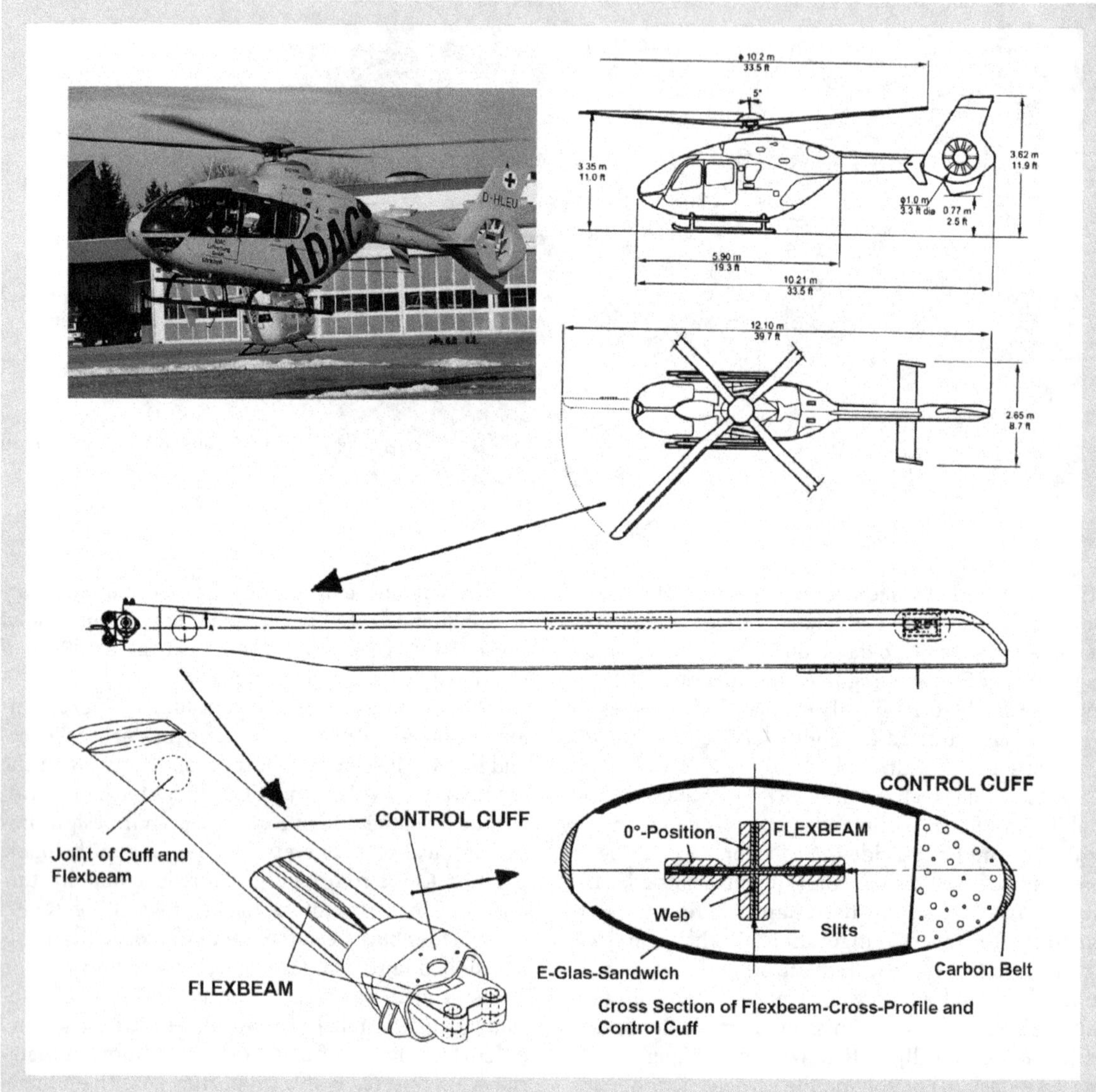

The helicopter rotor blade as aeroelastic ultra lightweight structural element (example: Eurocopter EC 135).

"Alpha-Jet", the CFRP vertical stabilizer of the "Airbus" [3, 4] and in the X-31 wing program.

It was already well known that the specific strength and E-moduli of fibre-composite materials were significantly higher [5] comparing to metal and the fibre material wood, but there were concerns that these values would be degraded under the influence of humidity.

Since the end of the seventies, the so-called softening temperature of various fibre-resin combinations of CFRP-systems, with hardening (tempering) temperatures of 120 °C and 180 °C, were systematically investigated. The softening temperature is here determined by a torsional vibration test; it indicates at which temperature the shear modulus of the material starts decreasing (glass transition temperature). The

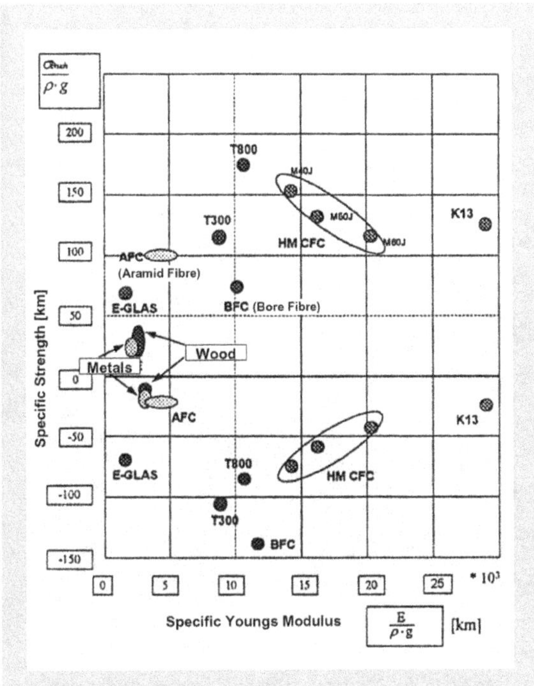

Specific E-moduli and strengths of metals, wood and unidirectional fibre composites with a 60 per cent fibre-volume share.

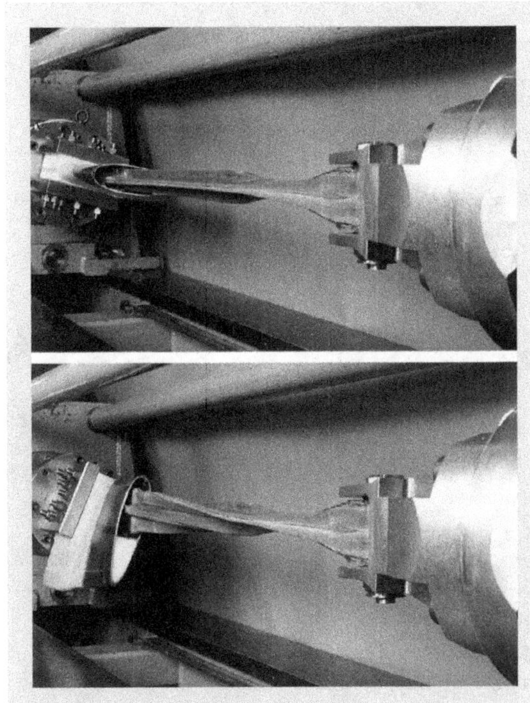

The bending- and torsion-elastic beam of the bearingless, hingeless EC 135 rotor system during tests.

practical operational limits of CFRP materials, for instance also for supersonic transports, are thus described with sufficient accuracy.

A helpful surprise resulted during the construction of the X-31 wing, made of the 180°C hardening resin system Narmco 5245 with the fibre Torayca T800, which made clear that the humidity contents of the resin at the beginning of the hardening process is decisive for the material characteristics to be reached later. Too high a humidity at the beginning of the hardening process leads to delamination and hence to an irreversible damage of the material. The absorption of humidity by a hardened structural element has, on the other hand, for different fibre-resin systems differently strong effects on the softening temperature. The humidity absorption of the hardened structural elements does, however, no longer lead to a material damage.

After these basic relations became known, the widely spread application of fibre-reinforced plastics in aircraft engineering was feasible. In 1990, EADS Germany had a consumption in fibre-reinforced plastics (FRP material) for Airbus of 65 t/year, for the Do 328 of 20 t/year and for helicopters of 11 t/year with a projected 600 t/year for the EF 2000.

Within civil aircraft engineering, the step to manufacture wings and fuselage structures of FRP was only executed in the "niche" of height-altitude research aircraft. This structural weight saving programme is one important step to achieve higher primary-energy efficiencies [6] with respect to transportation performance. Also with regard to the weakness of FRP materials, namely the delamination due to the present build-up of the layered structures, additional research work is needed. Three-dimensional fibre reinforced plastics for aircraft and automotive applications [7] will force methods and procedures of the textile industry upon the aircraft industry and will revolutionize the manufacturing methods. Aircraft engineering will return to fabrics covered wing/fuselage structures of CFRP

At EADS Airbus in Hamburg, the strength tests of a wing of carbon-fibre-composite construction (CFRP) of 15 m (49 ft) length were completed early 2000. The program, concerning the design of a CFRP-wing, was started in 1994 under participation of the Institutes of Technology in Brunswick and in Stuttgart, the "Deutsches Zentrum für Luft- und Raumfahrt, DLR", the Fraunhofer Society in Halle and the "Institut für Materialforschung und Anwendungstechnik" in Dresden. In addition to this "Technology Program CFRP-Wing", designed for a 100-seater, the fundamentals for the computing and manufacturing of a CFRP outer wing of a mega-liner were worked out. In 90,000 simulated flights, including 30,000 with an additional damage, all expectations were met. The maximum deflection of the wing tip amounted to 2.5 m (8.2 ft) under maximum load.

quality. The company, which masters these manufacturing methods first, will gain considerable competitive advantages.

The most essential savings resulting from the use of FRP material shall be demonstrated employing the example of rotor blades. Not only the weight reduction leads to higher flight performance, most of all it is the reduction of the number of parts, which leads to cost savings despite the still higher material cost. The smaller number of parts results in lower operational and maintenance cost. This could be demonstrated in all areas of aircraft engineering where FRP has been utilized.

Design Principles for Ultra-Light-Weight Construction

The objective of the strength-oriented design [8] of an aircraft concentrates on developing load-bearing structures, which meet the required performances with weight as low as possible and with minimum cost at a sufficiently high safety. The minimum safety factors are prescribed in design and construction regulations, for example:

- Safety factors: for loads, this factor is 1.25 limit load and 1.5 for ultimate load, respectively, for lifespan it is 2–4.
- Failure probability is, e.g., 10^{-9}.

A justifiable failure probability is already implicitly accounted for in the safety factors. The demanded performances are very different and cover various areas. Typical demands are:

- Minimum weight at a prescribed payload and a fixed mission profile
- Service life, lifetime
- Resistance to thermal loads and corrosive media
- Repairability
- Surface quality.

Typical costs, which have to be considered during the structural design, are:

- Development cost
- Material cost
- Manufacturing cost
- Operational cost

The interrelation of this Triad: safety-performance-costs can be placed in the corners of an equilateral triangle. Many development activities are not only directed towards one of the three goals. A good protection against corrosion can increase safety; however, large-scale tests are still performed without the simulation of corrosive media. But a good protection against corrosion increases performance due to longer service life time and it may, despite increased manufacturing cost, considerably decrease operational cost (inspections and maintenance), hence increase the economic efficiency.

Design Philosophies and Safety Concepts

There are different design philosophies to guarantee the safety of the structure, which are applied during the development process: Safe Life (no longer commonly applied), Fail Safe and Damage Tolerance.

The proof of safety with regard to static and fatigue strength is provided in the case of the Safe-Life philosophy only for the healthy structure without any damage. For statically defined load paths, this philosophy does not consider damage or crack propagation, respectively; for statically undefined load paths, it does not consider possible load redistribution.

The Fail-Safe philosophy assumes that after failure of a structural component, the maximum operational load, or 80 per cent of it, can be carried by the damaged (overall) structure. Here, statically undefined load paths are prerequisite. But it is also very important that the entire structure can be inspected.

In the case of the Damage-Tolerance philosophy, a structure with damages and cracks is assumed. The damage growth or the crack propagation, respectively, will be investigated, also in large-scale tests, and based on the results inspection intervals will be fixed. If a damage is not detectable, it has to be proven that it does not grow under operational loads (e.g., delamination in CFRP structures).

All principles are not mutually exclusive but can jointly be applied.

Example: A frame consisting of three rods, with an individual attachment of the rods, is subjected to a tensile force at the node common to the rods; the center rod lies in the direction of the force and the outer rods run under 60-degrees towards the node. One can show that the stress ratio of the rods (assuming the same cross section and material) is 4 to 1. If a crack is present or occurs due to fatigue, it will only grow in the center rod. The inspections, necessary due to the damage-tolerance requirement, may then essentially be limited to the center rod. For the given geometry, the rods are nearly equally used to capacity when subjected to the ultimate (breaking) force, which corresponds to the principles of light-weight construction. If during operation the central rod fails, the fail-safe demand is still essentially met.

The milestones in the development of safety concepts in aircraft engineering related to the advancements in fatigue strength after 1945 are being roughly outlined later.

Principles of Light-Weight Design

Concerning the lightweight design principles one distinguishes between:

- Light-weight design related to shape
- Light-weight design related to material
- Light-weight design related to systems.

In the case of a shape-related light-weight construction one may, as a typical example, refer to a thin-walled circular cross section subjected to torsion, which can be realized closed (tube) or slotted (profile). The differences in strength are of the order of magnitude of 10 and in stiffness of the order of 100. An addition of material may, however, have the consequence of a reduction in the moment of resistance and hence in the load capacity.

For the material-related light-weight construction, the situation is more complex. Which material builds the lightest does not only depend on the kind of load but also on the size of the load and on geometric restrictions. The classical light-weight-construction parameters are:

Tensile force: R_m/γ (breaking length)
Compression (pressure)
force (stability): Rod $(E/\gamma)^{1/2}$, Plate $(E/\gamma)^{3/2}$
Bending: $(R_m/\gamma)^{1/2}$
Torsion: R_m/γ.

System-related light-weight construction is strongly interdisciplinary. A classical example is the integrated fuel tank. The wing becomes slightly heavier (densely-spaced riveting, sealing systems, hydrostatic pressures at maneuvers and crashes), but the overall aircraft becomes lighter.

In the case of the so-called "Aeroelastic Tailoring", the possibility of the anisotropic design of reinforced plastic structural elements is directly being utilized to positively influence the deformation of the wing or the empennage and to improve aeroelastic behaviour. For a swept-back wing, a reduction in angle of attack in the flight direction, and hence, aeroelastically, a reduction in aileron effectiveness occurs due to a wing deflection. At higher Mach numbers, the outer ailerons at the wing tips may become entirely ineffective or it may even occur an aileron reversal. For that reason, ailerons on large-scale aircraft with metal wings are rendered inoperable at near-sonic flight. The problem of aileron reversal on swept wings can only be solved under minimum weight by aeroelastic tailoring of a CFRP wing.

The question of cost is very important early in the development process. During the layout of the structural frame, large parts of the global assembly concept are being fixed. Material and the type of manufacturing also have a large effect on the cost. Concerning the latter, there exists a great dependence on the manufacturer since the individual companies have different experiences and hourly machine rates. In astronautics, a large drop in cost cannot be expected due the small numbers built, although the gradient is largest at the beginning. In aeronautics, where today quantity production of 1,000 or more aircraft is possible, learning during manufacturing may, indeed, lead to cost reductions (learning curve).

In the case of the design of structural components, an integration (integral design) of a large number of elements is strived for since the assembly of elements is hard to mold into an automatism. A measure for the degree of integration is the number of attachment elements related to the number of components.

Structural Elements

Any aircraft structure consists of basic tensile, compression, bending and shear elements, which are realized as rod or flat-plate elements, as well as of torsion elements and force-introduction or transmission elements, respectively, which together form the different structural components. These components form structural groups, which are finally assembled to the overall aircraft structure.

In designing and in computational checks, one has to pay attention to the fact that often superimposed loads occur where also non-linear interactions have to be accounted for. For instance, a wing monocoque structure is subject to compression in the spanwise direction due to wing bending but also perpendicular to this direction by the introduction of flap loads; it is also subject to shear due to torsion and cross forces.

Additionally, there is a bending load on the surfaces between the longerons and the ribs resulting from air loads and tank pressures.

During the structural design, however, only the primary loads will initially be considered.

Tension Elements

Some examples of tension and compression elements, respectively, which are subject to tensile forces are:

- Wing lower surface
- Fuselage longerons
- Drag struts of landing gears
- Clevises for wing suspension
- Truss rods
- Elements of mechanisms
- Ropes (not resistant to compression).

One has to observe that tension elements are frequently also subjected to compression (e.g., truss rods of vertical stabilizers or drag struts of landing gears). For statically defined tension elements, an increase in load capacity cannot be achieved by allowing a plastic deformation (ductility) since the stress in tension elements is constant. It is important that the force acts in the elastic center, which does not have to agree with the areal center for FRP designs. Already relatively small eccentricities of the normal force lead here, however, to considerable bending stresses.

Bending Elements
Bending elements exist as beams or as bending surfaces (plates), which may be flat or curved. Some examples are:

- Wing longerons (spars)
- Deflection levers
- Cross beams within fuselages
- Skin panels
- Plane pressure frames and tank end ribs
- Locking mechanisms and arresting hooks.

Since the stress is not constant over the cross-section, an increased load capacity can be achieved by allowing for plastic deformation (ductility), provided the material laws comply to this (CFRP, CFC have limited ductility prior to fraction!).

In order to avoid a coupling of bending and torsion, cross forces must be introduced at the shear center, which is for u-profiles not easy to realize.

For a (bending) beam made of orthotropic (orthorhombic) material, a coupling of bending and torsion can be constructed. The classical shear center looses here its importance. This possibility is being utilized in the case of so-called Aeroelastic Tailoring.

In the case of the curved (bending) beam, three effects occur, which have to be considered:

1. The stress distribution is nonlinear. On the inner side, excessive stresses occur in comparison to the stresses predicted by the classical bending theory of the straight beam.
2. For thin-walled profiles, for instance I- or U-profiles, the flanges evade and one has, therefore, to determine the moments of inertia and resistance with a reduced flange width.
3. The evasion of the flanges causes a bending stress at the foot of the flanges normal to the direction of the primary bending stresses. For designs comprised of plates, it lies in the ST-direction (short traverse, normal to roll direction!). Also in the case of FRP designs, this effect has to be considered especially.

In design and computational checks of plate-like bending elements one must pay attention to the class of elements being considered. As a distinguishing criterion, the ratio of the degree of bending "w" to the plate thickness "h" is being used. It holds:

$w/h < 0.5$	"thick" plate with linear behaviour,
$0.5 < w/h < 5.0$	"thin" plate with nonlinear behaviour,
$w/h > 5.0$	diaphragm.

Concerning the material selection, one must consider whether the construction height is independently selectable by the designer – then the aforementioned classical characteristic numbers for light-weight constructions hold – or whether geometric restrictions are present, in which case the selection of the best material depends on the size of the load.

Compression Elements
Often tension elements are, due to the symmetry of the loads, also compression elements (e.g., the vertical stabilizer). Also primary compression elements are frequently subjected, by other load cases, to tensile forces (for instance, the wing upper surface, wing suspensions clevises, control rods). Typical compression elements are:

- Truss rods,
- Control rods,
- Longerons,
- Upper-wing skin panels.

In dimensioning compression elements, stability criteria as well as strength have to be accounted for. For the compression strength of metals, the Rp_{02}-limit is employed. Since FRP does not have a limit of ductility, the compression strength is used here. In most cases the wrinkling limit is lower than the compression strength. The crease limit is an empirical criterion and constitutes a blend of stability criterion and strength. In the case of FRP, this phenomenon has not yet been sufficiently investigated.

If the force flux is not very large, the stability of the compression elements will be the dimensioning parameter.

When designing a compression element, one has to consider three effects:

- Various forms of instability interact. For instance, the critical bending-torsional-buckling stress is always lower than the lowest Eulerean-stress or the torsional buckling stress. Also the wrinkling stress and the pure instability stresses interact. This is accounted for by the Euler-Johnson relations. There exist no reliable investigations in that regard for FRP.

- High loads or forces cause the critical stresses to be located within the nonlinear range of the stress-strain relation, which means that one has to calculate with a reduced Young's modulus. For the instabilities of rods, the tangent modulus is being used, for a plane load-bearing structure mixed moduli are employed. In order to avoid dimensioning errors, one should consider critical stresses and not critical forces or force effects.
- For FRP components, a bending-torsion coupling is designed on purpose or inherently given (e.g., plates). This may lead to a large reduction in critical stresses if they are calculated by the classical method without accounting for coupling effects. This effect has particularly to be accounted for in the case of FRP plates that are build-up of only a few individual layers (directional angles (0°, 90°)).

The reduction in Young's modulus has the consequence that the choice of the best material depends on the size of the load.

Torsion Elements

Pure torsion elements are rare and exist mainly in mechanisms like the mechanical control system of aircraft. Within structural groups, such as a wing, fuselage or landing gear, torsion plays a large role in dimensioning the elements. The demand for a global stiffness, for instance to stay within the flutter boundaries, is the dimensioning factor.

The principle difference between open and closed torsion elements or structures was already pointed out. Any cutout in a structure with a closed cross section causes a sequence of closed-open-closed cross sections, which results in considerable additional stresses (warp stresses, notch stresses).

This effect is very similar to the effects that occur for warp-resistant fixing of open profiles. In the English language, the torsion due to a warping-force is very appropriately called "differential bending", which can best be understood by considering the torsion of an I-profile (I-beam) due to a warping force.

The "notch-effect" of large cutouts can only be largely avoided by the application of oblique frame structures. In all other designs, an abrupt transition from an open to a closed (Bredt) cross section occurs, which is associated with considerable warping forces.

In designing a structural element, one should take care that over the entire length, i.e., also in the area of global cutouts (for instance within the payload area) Bredt surfaces are present. So-called load-bearing covers and doors are structurally soft in comparison with the primary structures, which has to be accounted for in an FEM (finite element method) analysis.

Elements subject to torsion are relatively yielding since the shear modulus is considerably smaller than Young's modulus. Care should therefore be taken during the design of mechanisms that torsion elements, which cannot be avoided, are as short as possible.

Shear Elements

Shear elements exist in the case of (bending) beams (beam webs), torsion-subjected closed structures (Bredt surfaces enclosing sheet-metal panels), and in the case of force introductions. The example of bonding shows very well that the force introduction by shear is better than the one by tension.

The classical shear element is the sheet metal panel. Idealizing the structure results in the known "shear-panel scheme (Wagner web theory)". Here, a constant shear flux is assumed within the sheet-metal panels (webs) and the tension and compression elements, respectively, have, according to this idealization, a linearly changing normal force. The shear-panel scheme (Wagner web theory) was derived in the thirties from the computational methods for trusses (*Wagner, Ebner, Kölle*).

Most often, shear elements have to be dimensioned for a sufficiently high stability limit. Strength rarely plays a role. Although one achieves for FRP the highest strength with a 45°-composit, the highest stability limit will be reached with different composite structures and other layer directions.

To be considered here is the percentage share of the layers with a certain direction and the sequence of the layers per se. Since the sequence of layers, especially in the case of a small number of layers, has a large effect on the elements of the stiffness-matrix, the strength diagrams given in the literature only hold for very large numbers of layers (theoretically going to infinity).

The transfer of shear to elements primarily transmitting normal forces, is accomplished by riveting, bonding or welding, respectively, but it can also be carried out integrally (milled elements and FRP parts).

For FRP-parts (e.g., I-profiles), shear is normally transmitted by the matrix, i.e., the "bedding mass". Since the "belts" have a large 0-degree share, this often leads to problems. A possibility to improve the design is the "interleaving" of the beam-web layers into the belt layers. This enlarges the shear-transmitting surface considerably.

Modern Computational Methods – Finite-Elements-Method (FEM)

With the appearance of the first large-scale computers during the sixties static and dynamic computing of large aircraft structures became possible, something that was urgently needed. The fundamentals needed here had already been established with the matrix methods of elasto-mechanics at the end of the thirties. They now became the basis of the structural analysis via the application of the "Finite Element Method, FEM".

The method of the finite elements can generally be considered as a numerical computational method for the solution of problems within the area of continuum mechanics. The term continuum mechanics shall emphasize the general validity of this method. In the application to problems of structural mechanics is the basic idea of FEM to divide the entire structure (load-carrying frame) into a large number of small, equal elements whose mechanical or physical behaviour (e.g., material) is either approximately or exactly known. The unknowns of the system are, instead of being described by continuous functions, defined at numerous discrete points within the structure. This results mostly in systems of linear equations, whose setup and solution can be left up to the computer. Here, modern program systems automatically present to the engineer discretization proposals and automate the setup and solution methods.

Today, the notion of the FE-methods is used, somewhat imprecisely, within two areas of continuum mechanics:

- Problems in structural mechanics (statics and dynamics), where structure has to be understood as an elasto-mechanic system in the widest sense, i.e., not only as special load-carrying systems, such as trusses and frames, plane or spatial load-carrying structures, but also as complex structures consisting of different types of load-carrying systems.
- Steady and unsteady field problems related to the theories of heat conduction and fluid mechanics, and to the electromagnetic and acoustic wave theory.

The latter problems refer to the solution of the partial differential equations of fluid mechanics, which are directly being solved by means of finite difference methods. The objective of the computations is here the description of the aerodynamic condition (velocity, pressure) on the surface of a body imbedded within a flow field. According to local considerations, computational grids are generated, which, in the case of three-dimensional problems, divide the outer flow field in discrete volume elements. The grid points or discrete volume elements are assigned functions and their derivatives. Today's FE-methods of structural mechanics and the finite-volume methods of fluid mechanics are, however, physically and numerically completely different. Between the two methods, certain analogies exist but not more. The historical development of the FE-methods commenced already during the last century with the then developed theory of rod and frame load-carrying systems. Names like *Maxwell*, *Castigliano* and *Mohr* have to be spelled out. For modern structural mechanics, this theory represents the foundation of the matrix method of rod statics, which became, in turn, the starting point of the FE-method. In aircraft engineering, it was *Herbert Wagner* who defined structural elements, which had to meet the equilibrium conditions with the deflection parameters as secondary conditions; here, the predecessor of the force-method of the FEM originated.

Initially, practical structural computations concentrated only on the force-method. Here, the force is the unknown. During the twenties, also the deformation parameter method, also called deformation method, had already theoretically been developed. *Osterfeld* published in 1926 his textbook "The Deformation Method" [9]. The deformation parameters are here the unknowns. The deformation method for rods and beams is to be considered the predecessor of the widely used FE-method on the basis of the deformation parameters. The computation of complicated load-carrying structures was limited for a long time due to the computational effort needed to solve large systems of linear equations. When in the early fifties the first practically useful digital computers became available, the previous hand-operated calculation schemes could be replaced. The enormous possibilities of the computers could also be used by consequently applying the matrix computation already to the formulation of the problem and not only to the solution of the equations system.

Two publications can be considered important starting points for the development of the FE-method, publications associated with the names *Argyris* and *Kelsey* [10] and the names *Turner, Clough, Martin* and *Topp* [11]. In these publications, matrix computations are employed for the computer-conform discretization not only of rod systems but also of continuum problems. Especially *John H. Argyris* recognized that the large-scale computer was going to revolutionize the entire theoretical and practical art of engineering.

Argyris, born in 1913 in Greece, descended from an old Byzantine family, who created statesmen, poets and scientists. He studied in Athens until 1934, and graduated in engineering in 1936 from the Institute of Technology (TH) in Munich; he was, until 1939, structural and project engineer at Gollnow, a large steel-construction company in Stettin; subsequently, he studied at the Institutes of Technology in Berlin and in Zurich to consolidate his knowledge in aerodynamics, mechanics, mathematics, and physics. Starting in 1943, he worked at the "Structures and Flutter Department" of the Royal Aeronautical Society; starting in 1949, he gave lectures concerning aeronautical structures at the Imperial College of Science and Technology in London, Great Britain; in 1955, he was appointed to the only chair of Aeronautical Structures within the entire Commonwealth at the University of London; in 1957, he received the D.Sc. (Eng) of this University. In 1959, he followed a call to the Institute of Technology (TH) in Stuttgart to become full professor and head of the newly founded "Institute of Statics and Dynamics of Aerospace Designs" ("Institut für Statik und Dynamik der Luft- und Raumfahrtkonstruktion, ISD"). This appointment was a real lucky break for the setup of today's Faculty of Aerospace Technology of the Institute of Technology in Stuttgart. During his initial time in Stuttgart, *Argyris* organized the procurement of several, increasingly more modern computers: Pegasus, UNIVAC 1107, CDC 6600.

In the beginning, the utilization of computers in the engineering sciences proceeded only slowly, and this not only due to the very limited computational capacities but also for lack of understanding the intelligent and effective use of the computer. For a long time, user programs and intelligent manageable tools were missing.

The book of *Argyris* [10], published in 1960, is still considered the standard textbook on FE-methods. *Theodore von Kármán* called this book the beginning of a scientific revolution.

The subsequent development of the FE-method was characterized by work dealing with the formulation of different element types. The name *Argyris* stands here, for instance, for the elements of the TUBA-family, i.e., plate elements that were later, conceptually adapted, applied within the SHEBA-family to monocoque designs. SHEBA is the only fully compatible monocoque element for general and irregularly curved surfaces.

After the application of the FE-method to stress and deformation analyses within the linear elastic range at static loads, came later the extensions of the method to problems related to linear elasto-dynamics and stability theory, and to thermo-mechanics. The advancement in the direction of material-related nonlinear problems (plasticity, creeping) and nonlinear geometric problems (large deformations no longer small in comparison to the geometrical dimension of the structure) was a similarly logical consequence.

In the following years, the FE-method was further improved at the ISD in Stuttgart with regard to static and dynamic computations in aircraft engineering; with the computer program system ASKA, a software was developed, which was then applied, not only by the aircraft industry, worldwide. In the seventies, such an FE-program system, known as COSA, was also developed at the Dornier Company. In practical aircraft engineering, the FE-method is today universally applied as a standard method for the computation of static deformations and stresses as well as for the determination of the natural-vibration parameters of air and space vehicles. The method is the result of worldwide systematic research and development efforts, going on over decades, and its decisive breakthrough occurred with the appearance of large-scale computers. In aircraft engineering, FE-calculations with 50,000 and more elements are routine and the computational accuracy has naturally been improved. With the FE-method as standard method, practically all static deformation parameters (stiffness respectively influence matrices) and natural vibration parameters of complete aircraft and their substructures, needed within the framework of aeroelastic investigations, are being computed today.

In the meantime, far more than 200 publications constitute the life's work of *Argyris*.

Also chaos theory profited from the fundamental work of the scientist *Argyris* [12]. He is in his high eighties still active in this demanding field.

At this point, also critical remarks concerning FE-methods shall not be withheld. Young engineers too often rely, almost blindly, on the results of FE-computations. However, they forget that the results can only be as good as the initial discretization allows. This means that in the case of an unfavorable element buildup, the results may miss reality by far. A simple, rough calculation by the conventional tools of technical mechanics may very often help in the assessment of the FEM results.

Remark

Concerning this chapter, some figures are presented in color:

Color Plate IX

- Stiffening of the external skin in case of the CFRP rudder of the Airbus
- Manufacturing rig for the central rudder box
- Manufacturing of the GFRP rotor blade of the EC 135 helicopter

High-Altitude Research Aircraft as Further Examples of the Application of the Fiber Composite Technology

Hans Galleithner

The Grob Company manufactured in the early seventies "Cirrus" sailplanes under license and came out about three years later (1974) with its own design, the "Astir". The know-how gathered here could also be utilized in other, technologically more demanding areas of aircraft developments, viz., the design and construction of high-altitude aircraft for military tasks and for atmospheric research.

The altitude world record for sailplanes was held up to 1995 by an "Astir" single-seat glider with noteworthy 15,088 m (49,500 ft). But only the military project of an "airborne distance-keeping primary reconnaissance system" ("Luftgestütztes, abstandsfähiges Primäraufklärungssystem, LAPAS"), with the requirement of long term reconnaisance at an altitude of 15,000 m (49,200 ft) along the eastern borders of the Federal Republic provided the needed financial security to start high-altitude aircraft development at Grob. The first flight of the aircraft "Egrett-1" on June 24, 1987 (proof of concept) at the test center in Manching under contract of the US company E-systems was a visible result of the new task. E-systems was the main contractor of the BMVg (Federal Ministry of Defense) concerning the overall system responsibility and had intentionally selected the Grob Company as subcontractor for the development of the flying platform as carrier of the reconnaissance systems, which were to be integrated in the USA. Substantial was here the unique design and construction experience concerning fiber-composite materials, which was demonstrated by the fabrication of about 3,000 sailplanes, motor-gliders and sports planes in composite design. Between 1987 and 1993, Grob built five single-seat and one double-seat, single-engine "motor gliders" with turboprop propulsion with the in-house designation G 520 "Egrett" and the military designation D- respectively TD-500. During the initial phase of the "Egrett" development the former NASA test pilot *Einar Enevoldson*, now working for Grob, had set an altitude world record for propeller aircraft with 16,330 m (54,000 ft). The aircraft types attained civil and military certifications. The maximum operational altitude was 15,240 m (50,000 ft) at flight durations of between six and nine hours. The maximum take-off mass was 4,700 kg (10,600 lbs) at a payload of 1,000 kg (2,250 lbs). However, the intention of deploying about 16 "Egrett" aircraft for reconnaissance, with the equipment envisaged by E-systems, had to be scrapped due to the changed political situation in Germany. The Federal Ministry of Defense cancelled the order in March 1993. The aircraft produced up to that time were sold except one, which remained as "Strato 1" in the possession of Grob.

Since 1991, while the development of the "Egrett" was in full swing, but probably also inspired by this development, atmospheric and space research circles increasingly demanded high-flying measurement platforms for ozone-research and earth-observation tasks. Corresponding manned aircraft should be able to operate at altitudes between 20 and 26 km (65.600 and 85.301 ft). The Federal Minister of Research and Technology (BMFT) provided the necessary funding out of the department "Environment". The requirements document, presented by the DLR, demanded the operation with an 800-kg (1,800-lbs) payload at an altitude of 24 km (15 mi) for 48 h (high-altitude mission), and for 48 h and a payload of 1,000 kg (2,250 lbs) at an altitude of 18 km (11 mi) (observation mission, surveillance mission); these requirements could be met with a new design. It was again Grob together with the IABG, which presented a proposal, which convinced DLR as project representative of the BMBF (Federal Ministry of Education and Science, Research and Technology). The project was named "Strato 2C" and officially presented to the contract partners at the end of April 1992. After several changes, the following configuration was realized for "Proof-of-Concept" purposes: A wing of 56.5-meter (185-ft) span, with two upper-surface engine nacelles, was as a high wing arranged on a cylindrical, 24-meter (79-ft) long pressurized fuselage with T-tail. All structural components had to be made of carbon fiber-composite-material (CFRP). Computations showed that the required performance could be achieved with a maximum take-off mass of 13,350 kg (30,000 lbs) with a useful fuel load of 5,700 kg (12,800 lbs). The necessary propulsion was accomplished by two 300-kW (402-hp) piston engines via slow-turning five-blade pusher propellers. It was, furthermore, known that a lift/drag ratio of 30 was required to safely reach an operational altitude of 24 km (78.740 ft). For the wing, the DLR laminar airfoil LH37 with a favorable high-lift behaviour was, therefore, selected; here, especially the airfoil characteristics in case of operational contaminations were decisive. A low structural mass, laminar wing flow, a high aspect ratio of 21.3 in conjunction with an optimized wing contour should provide minimum drag.

New ground had to be entered in the manufacturing of the wing and the fuselage with empennage. Besides the size of the structural elements, a long-duration at high altitudes had to be considered, resulting in extreme low temperatures, large pressure differences and UV-radiation. Wing and elevator with box spars were manufactured of upper and lower shells, each of which was embedded in a negative mould, equipped with longerons and rips, and subsequently glued together (just as in the case of plastic models). The center parts of the sealed box spars served as fuel tanks. All primary, load-transmitting structures were built in a sandwich-type design. Highly stressed parts, such as the pressurized fuselage, box spar and empennage, were built up from "Nomex"-honeycombs with inner and outer layers of carbon fiber fabrics. Less loaded structural parts, such as wing noses (leading edges), control surfaces, landing-gear pods, fairings, engine nacelles, and wing-fuselage transitions received instead of the carbon fiber layers, glass- or Aramid-fiber mats as inner or outer layers. The fuselage was as well placed, divided into two vertical halves, into negative moulds glued together shaft-like. All parts were after their completion annealed. Tests at a load factor of $n=1$ as well as pressurization tests of the pressurized cabin were carried out without objections.

The IABG envisaged as propulsion unit the water-cooled 300 kW (402 hp) six-cylinder piston engine "Continental TSIOL-550" with a three-stage exhaust gas turbocharger and interim cooling downstream of each stage. Charging was to insure that also at an altitude of 24 km (78.740 ft), where the ambient static pressure is only about three per cent of the pressure on the ground, ground conditions prevailed at the fresh-gas intake of the piston engine so that also at that altitude nearly the full shaft performance could be provided. The rpm of the engine (max. 3,100 rpm) was reduced by a factor of 4.48 to the desired propeller rotational speed by a gear especially developed by the Zahnradfabrik Friedrichshafen. The pusher propeller, a GFRP-coated wooden construction with a constant-speed hub, was provided by the Mühlbauer Company. Here too, the development entered new grounds with regard to aerodynamics, size and operational range. The components charger (compressor) engine and propeller were qualified in separate test series. Tests in the high-altitude chambers of the "Institute of Aeronautical Propulsion" ("Institut für Luftfahrtantriebe, ILA, University Stuttgart") and of the CEPr Saclay (France) proved that with the selected turbo-charging nearly 300 kW (402 hp) shaft performance could be realized up to altitudes of about 23 km (75.459 ft).

After an intensive development of nearly three years, the "Strato 2C" took off on March 31, 1995 to its successful first flight with *Enevolson/Meyer* (NASA/DLR) at the controls. Only four months later, on August 4, 1995, an altitude of 18,550 m (60,900 ft) was reached on the 29th flight. Unfortunately, this was, at the same time and up to today the last flight of the "Strato 2C", which had to be parked at Grob due to a lack of follow-on funding.

This does, however, in no way diminish the pioneering performance of the Grob Company in the development of high-altitude aircraft in Germany. The "Strato 2C" up to that time was the largest composite airplane ever manufactured, whose primary structure consisted almost exclusively of fiber-composite material. Grob contributed decisively to the development of future wings in composite design. The pressurized cabin is the first, the largest and the only one of this design which flew successfully also at high altitudes. The "Strato 2C" set on its last flight an absolute altitude record for manned, piston-engine powered aircraft which has since then not been surpassed.

Remark
Concerning this chapter, some figures are presented in color:
Color-plate X:
- GFRP-wing of the Grob "Strato 2C"
- Nature as model for the fiber-composite construction
- High-altitude research aircraft Grob "Strato 2C"

Progress in Fatigue Strength After 1945

HANS FÖRSCHING

The great importance of vibration strength and the related structural fatigue and finally the operational safety – especially of aircraft – was at the end of the World War Two well known, and many trend-setting fundamental results in this field had been worked out in Germany. The research activities were enormously increased after 1945 in all countries concerned with aeronautics. In the ensuing civil aircraft engineering, whose development still was frequently based on military aircraft with single-spar wings, a longer lifespan than the 3,000 flight hours of a bomber aircraft was demanded. 30,000 flight hours at sufficient fatigue strength had to be reached. Two spectacular crashes of the British De Havilland "Comet" 1954 near Naples and near Elba, respectively, added to the problem. Also the US Air Force encountered towards the end of the fifties accidents due to fatigue fractures. In 1959, six bombers B-47 crashed within months due to fatigue fractures. These accidents led especially in England and the USA to intensive research in the area of fatigue strength, which led to new safety concepts and lifespan (fatigue life) investigations on complete aircraft.

All these advances in the area of structural fatigue bypassed Germany until the mid-sixties. The up to 1945 leading role of Germany in the field of aircraft fatigue strength had been lost. However, in motor-vehicle engineering, where fatigue fractures increasingly occurred on car and truck parts due to higher speeds, exemplary research and development work was performed. This was most of all due to *Ernst Gaßner* who founded in 1946 in Kempten the "Physical-Technical Laboratory" ("Physikalisch-Technisches Laboratorium"), which moved in 1948 to Darmstadt where it merged with the "Bautz-Bergman Material Consulting, Ltd." ("Bautz-Bergmann-Werkstoffberatungs-GmbH") to form the "Laboratory of Fatigue Strength" ("Laboratorium für Betriebsfestigkeit, LBF"), today's "Fraunhofer Institute of Fatigue Strength". *Gaßner* convinced the entire German automobile industry of the advantages of a fatigue-strength-related design of their automobile components in view of saving weight and cost at a sufficient lifespan. At the LBF, individual investigations were then carried out on numerous motor-vehicle parts before the German automobile companies built up their own fatigue-strength-related laboratories. The experience gathered and the transfer of know-how from the aeronautical industry prior to 1945 as well as new findings thereafter, secured the German automobile industry up to today a certain lead in the area of fatigue strength over their competitors within Europe and overseas.

After the "Comet" accidents due to the rapture of the pressurized fuselage at the window cutouts, flight-load simulations with complete aircraft became in England the rule with the pressurized fuselage being loaded inside a water tank (later with pressurized air) and the wings and empennage being testet by servo hydraulic vibration excitations corresponding to take-off and landing load alterations and gust load encounters. Acceleration and stress collectives were measured over hundred thousands of flight hours, especially in England and the USA by NACA, with newly developed direct analyzing instruments (*Vickers "Strain-Range-Counter and RAE Fatigue Counter")*; the Royal Aircraft Establishment (RAE) rapidly developed into a center for vibration-strength research in aircraft engineering.

Since 1955, a discussion of new structural concepts, based on the newly gained know-how, commenced which were termed "Safe-life" and "Fail-safe". "Safe-life" means that all structural elements must be fatigue-fracture-proof up to the prescribed lifespan. "Fail-safe" demands that in the case of a failure of a load-carrying element due to fatigue or other causes, the flight safety, within the demanded failure probability, must not be jeopardized. Fail-safe design and construction was already, to a certain degree, applied to the first post-war aircraft with piston engines, such as the DC-6, DC-7 or the Lockheed "Super Constellation", but then, to a larger extent, to the jet aircraft B-707 and DC-8, which went into service in 1960. Experiences taught, that fatal, fatigue related damages could not be avoided with certainty as the crashes of the Lockheed "Elektra" in 1959 showed. The high-strength AlZnMgCu-alloys, repeatedly utilized within the Anglo-Saxon sphere, with their unfavorable vibration strength over and over again were the cause for fatigue damages; *Gassner* already warned in 1941 not to use this special Al-alloy. The fail-safe design, later developed into the "damage-tolerant" design, requires, in any case, a careful maintenance and inspection procedure, which was not performed with the necessary intensity by all airlines.

Since 1960, essential progress was made, especially in the USA with large research programs, in the new area of fracture (crack) mechanics, i.e., in crack propagation. On the basis of the respective findings, the US Air Force introduced in the mid-seventies the "Damage Tolerance, DT" design and construction regulations [13]. They assume at critical locations cracks or crack-like defects already present, caused due to manufacturing imperfections. Aircraft manufacturers have to prove that despite the presence of cracks the necessary lifespan is reached. The immediate cause for the introduction of this DT-regulation was the crash of an F-111 after 100 flight hours (!) due to a crack-like material fault within the central wing box, manufactured of ultra high-strength steel D6AC, whose ultimate toughness turned out to react to the smallest deviation in the heat treatment. These DT-regulations were steadily further developed and constitute today an essential part of the design and construction regulations for military aircraft, but also for engines, helicopters and aircraft accessories. The further development of the DT-regulations, which proved generally very valuable for military aircraft engineering, were discussed at regular conferences convened by the US Air Force.

For civil aircraft, DT-regulations were also issued, but more generally phrased. The damage growth of emerging cracks (crack propagation) is investigated at all known critical positions (also on the original structure in large-scale tests) and based on the results, inspection intervals are fixed. For the inspection, all critical areas must be accessible and open to inspection. The fail-safe concept in conjunction with the DT-regulations – put down in aeronautics-related rules – has, in civil aircraft engineering, turned into a design philosophy. The routine inspections are naturally expensive, but have considerably contributed to the safety in air traffic.

Essential progress in research on fatigue strength was provided by the introduction of servo-hydraulic computer-controlled test stands beginning in 1960. They provided the possibility to test structures at a sufficiently high frequency at an arbitrary deterministic load-time sequence. This also allowed an exact check of damage-accumulation hypotheses, crack-propagation hypotheses and of the Gassner block-program tests. Investigations at NASA [14], at the NLR in the Netherlands [15], and at the LBF in Darmstadt [16] showed that the block-program test predicted a longer lifespan than did the tests with a realistic mixture of load cycles, which means that it was no longer up-to-date. Numerous standardized load cycles for fatigue-strength investigations were then proposed and applied for fatigue tests on civil transport aircraft, military aircraft, helicopters and engine components [17].

With the introduction of jet aircraft, also acoustic fatigue due to engine noise became a problem. However, the noise reduction due to environmental concerns constituted here an effective remedy. A new phenomenon in material fatigue arose with the application of fiber reinforced (fiber-composite) constructions, i.e., delamination, which required new extensive fatigue-strength tests. Since the strength of fiber composite materials is also subject to temperature and humidity effects, fatigue strength tests became more difficult. In the case of fighter aircraft with fiber composite material, the problem of repair in the case of damage by bullets (Composite Damage Repair) must finally be added. Cutting, respectively damaging a fiber means a corresponding reduction in strength, hence an increased danger with respect to fatigue, so that in the case of primary structures a complete substitution of the respective part can in most instances not be avoided.

After World War Two, activities in the area of fatigue strength of aircraft commenced in Germany only in 1960 with the restart of independent development work. When the "Organization for the Operation of Industrial Facilities" ("Industrieanlagen Betriebsgesellschaft, IABG") in Ottobrunn was founded, all needed large-scale test facilities were centrally installed there, as shall be addressed in detail in the later chapter "Large-scale Facilities to Test Safety, Reliability and Environmental Compatibility". The conduct of fatigue strength investigations on aircraft components and complete aircraft was performed according to test regulations of American (Federal Aviation Authority, FAA) and European (Joint Aviation Authority, JAA) aeronautical authorities; in the case of military aircraft also according to specifications of the military customer. The structural test facilities at the IABG for the conduct of safe life predictions (fatigue life investigations), static and dynamic strength investigations, thermo-mechanical tests, material tests, were and still are continuously further developed and adhere to international standards. The activities of the aforementioned "Fraunhofer Institute of Fatigue Strength" ("Institut für Betriebsfestigkeit") in Darmstadt essentially concentrated on problems related to mechanical engineering in general, to surface engineering and to vehicle technology. Of the several textbooks related to fatigue strength published in Germany since 1970, the reader is especially referred to the one by *O. Buxbaum* [18].

The fatigue strength investigations on large fiber composite structures of aircraft, as they became relevant since the late seventies with the manufacture of Airbus CFRP-vertical-stabilizers (CFRP = Carbon Fiber Reinforced Plastics), are being performed at the Airbus Test Center in Hamburg. But valuable research work was also carried out in many fatigue tests at a variety of test conditions at the DLR in Brunswick and Stuttgart. Still, an absolutely reliable scientific and practical prediction of the lifespan of structural elements or complete structures subject to vibration loads, at sufficiently low failure probabilities, remains a dream rather unlikely to become true. Many problems, such as fatigue due to corrosion, the interplay of temperature and load cycles, fatigue strength associated with multi-axial loads without a time-wise correlation of the force directions, and hence the lifespan (fatigue life) prediction, are still unsolved. Here, further research over a longer time period is required.

References

1 H. Dörner, Drei Welten – ein Leben, Professor Dr. Ulrich Hütter, Hochschullehrer, Konstrukteur, Künstler, Heilbronn, 1995
2 K. von Gersdorff, K. Knobling, Hubschrauber und Tragschrauber, Bernard & Graefe Verlag, Bonn, 3. Aufl. 1999
3 H. Prem, Faserverstärkte Kunststoffe in der Luft- und Raumfahrt, Kunststoffe 68 (1978) 10, pp. 691–695
4 H. Prem, Faserverstärkte Kunststoffe in der Luft- und Raumfahrt, Kunststoffe 72 (1982) 10, pp. 622–626
5 H. Bansemir, Einfluß von Hubschrauberentwicklungen auf die Faserverbundstrukturen unterschiedlicher Industrieanwendungen, Eurocopter Deutschland, ECD-0058-96-PUB
6 H.A. Niedzballa, D. Schmitt, Vergleich des spezifischen Energiebedarfs von Flugzeugen und anderen Verkehrsmitteln, DGLR-JT97-108
7 F.J. Arendts, J. Brandt, K. Drechsler, J. Filsinger, Dreidimensionalverstärkte Faserverbundwerkstoffe im Flugzeug- und Fahrzeugbau, DGLR-Fachausschußsitzung, 27./28.11.1996
8 F.J. Arendts, Vorlesung Leichtbau I und II, TU Stuttgart, Institut für Flugzeugbau, 1981
9 H. Osterfeld, Die Deformationsmethode, Springer Verlag, Berlin, 1926
10 J.H. Argyris, N.N. Kelsey, Energy Theorems and Structural Analysis, Aircraft Engineering, Vols. 26 and 27, 1955, Butterworth, 1960
11 M. J. Turner, R. W. Clough, H. C. Martin, L. J. Topp, Stiffness and Deflection Analysis of Complex Structures, J. Aero. Sci., 23, P805–823, 1956
12 J.H. Argyris, H.-P. Mlejnek, Die Methode der Finiten Elemente, 3 Bände, Vieweg, Braunschweig/Wiesbaden, 1986
13 USAF, Military Specification, Airplane Damage Tolerance Requirements, MIL-A-83444, July 1974

14 E. C. Naumann, Fatigue under Random and Programmed Loads, NASA TN D-2629, 1989
15 J. Schijve, F.A. Jacobs, Effect of Load Sequences on Crack Propagation under Random and Program Loading, Eng. Fracture Mechanics, Vol. 5, 1974, pp 245–252
16 E. Gassner, G. Jacoby, Betriebsfestigkeit zur Ermittlung zulässiger Entwurfsspannungen für die Flügelunterseite eines Transportflugzeuges, Luftfahrttechnik-Raumfahrttechnik 9 (1964) Nr.1, pp 6–19
17 W. Schütz, Standardized Stress Time Histories – An Overview, in: Development of Fatigue Loading Spectra, ATSM STP 1006, 1989
18 O. Buxbaum, Betriebsfestigkeit, Sichere und wirtschaftliche Bemessung schwingbruchgefährdeter Bauteile, Verlag Stahl und Eisen, Düsseldorf, 1986, 2. Auflage 1992

Aeroelasticity – New Challenges and Problems

Hans Försching

New Beginning of Aeroelastic Research After 1945

The interruption of activities in the area of aeronautical engineering after World War Two, issued by the Allies, also had the consequence that almost all Germans concerned with aeroelastics emigrated to the West, like USA, England, and France. Russia recruited by force the rest of the engineering personnel, or turned it to other occupations within Germany. The DVL in Berlin-Adlershof, located in the Eastern Occupation Zone, disappeared completely, and activities at the AVA in Göttingen ceased on August 21, 1945. *Hans Georg Küssner* and a small team still worked there until the end of 1946 for the British "Ministry of Supply" on the "Göttingen Monographs" [1], edited by *Albert Betz*, where the results of the aeroelastic research between 1939 and 1945 were summarized. The impressive research work "1939-1946" related to unsteady aerodynamics presented there was again as summary published by *Hans Georg Küssner* and *Hans Billing* [2] in 1953. Starting in 1947 and lasting until 1953, all scientific work in the field of aeroelasticity in Germany ceased.

After the Allies lifted the general ban on aeronautics in the Federal Republic of Germany, the AVA in Göttingen began in 1953 her reconstruction. Decisive for the revival of aeroelasticity was then a meeting on September 25, 1953 in Munich, called for by the "Preliminary Federal Office for Aeronautical Equipment and Accident Investigations" ("Vorläufige Bundesstelle für Luftfahrtgerät und Unfalluntersuchung"), to define the objectives of research, and the establishment of design and construction regulations to command self-induced vibrations (flutter). All participants thought it necessary to establish for that purpose a central research institution. Aeroelastic problems gained even more importance with the revolutionary technological progress in aircraft engineering since the end of World War Two and the developments in space-flight technology; especially in the USA and in England, aeroelastic research activities were tremendously increased within governmental research installations (NASA, RAE) and also in industry. *Hans Georg Küssner* was charged with working out suggestions for such a "research institution" in Germany, which was soon established under his supervision at the AVA in Göttingen, becoming in 1957 the department of "Aeroelasticity". This remained until early 1960 the only location of aeroelastic activities in Germany before the arising aeronautical industry (EWR South, VFW and Dornier) established working groups concerned with aeroelasticity.

The restart of scientific work at the AVA progressed only slowly. It was also difficult to find, for the little known scientific field of aeroelasticity, suitable employees, who had first to acquire experience and therefore needed a long familiarization period. The reason for success was then a young, highly motivated staff, to catch-up rather quickly with international standards and the possibility to fall back on the wealth of experience of the former AVA "Institute of Unsteady Proc-esses", represented by *Hans Georg Küssner*. Here, *Küssner's* membership in the AGARD "Structures and Materials Panel" was of considerable value since thus the international research status could best be evaluated. *Hans Georg Küssner* and his new staff confronted the scientific community for the first time with an aeroelastic colloquim [3], organized in Göttingen in April 1957, where well-known foreign scientists participated.

Considering the large scientific progress in aeroelastics, achieved worldwide, and the newly added challenges and problems associated with the rapid technological progress in aerospace engineering, it had become clear that further research within the AVA Department of "Aeroelasticity" had to concentrate, within the limits of its possibilities, on those sub-areas where traditionally the best conditions existed, but where also the largest practical necessity prevailed. These were:

- Further development of computational methods for unsteady aerodynamics (e.g., wing flutter).
- Development of experimental methods for the measurement of unsteady aerodynamic pressure distributions on oscillating wind-tunnel models to validate theoretical results.
- Further development of numerical methods and procedures for flutter computations.
- Further development of ground vibration tests to determine the natural vibration parameters of air and space vehicles.
- Development of methods and procedures for wind tunnel flutter investigations with dynamically similar models.

These topics formed the focused research subjects of further aeroelastic research in Göttingen when, on May 1, 1972, the AVA Department "Aeroelasticity" was turned into the "Institute of Aeroelasticity" within the newly founded DFVLR. Its director became *Hans Försching*, who had already on October 1, 1968 succeeded *Hans Georg Küssner* as head of the AVA department. The aforementioned activities in aeroelastics within the German aeronautical industry concentrated most of all on the establishment of the necessary soft and hardware to cope with the routine problems within the framework of type certifications, as well as on new aeroelastic problems related to the development of the "Tornado" fighter (variable-sweep wing with external stores), on the introduction of new flight mechanical control and stabilization systems (aero-servo-elastics and active control technology), and on the problem of in-flight vibration tests. All these activities between the DLR Institute of Aeroelasticity in Göttingen and industry were performed in close cooperation, partly within government sponsored joint research projects.

Advances in Computing Methods of Unsteady Aerodynamic for Oscillating Wings

Two- and Three-Dimensional Wing Methods for Incompressible Flow

The continuation of the traditional work in the area of the unsteady aerodynamics of (harmonically) oscillating wings commenced at the AVA in Göttingen at the end of the fifties when a small electronic computing system, the IBM 650, became available. *Hans Georg Küssner* and *Heinz Göllnitz* [4] initially calculated extensive tables of unsteady aerodynamic derivatives for the oscillating "infinite airfoil" ("Streckenprofil") with two kinks and steps in incompressible flow. *Hans Georg Küssner* and *Friedrich Wille* [5] established similar tables in 1968 for compressible subsonic flow up to a Mach number of 0.95. For harmonic oscillations, the motion-induced unsteady aerodynamic forces and the corresponding coefficients may be divided into a component in phase (or in opposite phase) with the oscillatory movement (real part) and a component with a ± 90-degree phase shift (imaginary part); this considerably simplifies the analytical treatment of the flutter problem. For large aspect-ratio wings in incompressible flow, *Hans Georg Küssner* and *Horst Hertrich* [6] published in 1965 an unsteady theory, which transforms for steady flow into Prandtl's lifting line theory. On the basis of this work, flutter computations were carried out in Germany until the mid seventies – for sailplanes still today. It was finally also possible, with the aid of the now available electronic computer system, to successfully complete basic work related to the computation of pressure distributions on harmonically oscillating two-dimensional wings of finite thickness, of several symmetrical Joukowsky and Mises airfoils. Work started already in 1944 by *Ludwig Schwarz* and continued since 1955 by *Guntram von Gorup* [7].

The arising large-scale computer systems now provided the possibility of a numerical solution of the integral equation [8] for oscillating three-dimensional wings and wing systems at arbitrary Mach numbers in subsonic flow, derived by *Küssner* in 1940. Here, mathematical groundwork was already performed in the mid-fifties in the USA, and numerical solutions of this integral equation, singular in its kernel function, were provided with the aid of so-called collocation methods (kernel function methods), or the division of the wing into a large number of elemental panels (Doublet-Lattice-Method and Panel Method). In the sixties, several of these solution procedures were developed for the subsonic and supersonic speed ranges, among others in Germany at MBB by *Boris Laschka* [9]; he also provided internationally an important contribution with the application of his collocation method to interfering multi-surface systems [10] in connection with the "Tornado" development and the extensive and partly basic investigations of the "wing-with-external-loads" flutter problem [11] carried out at MBB. Alternatively to these computational methods developed the Göttingen Institute of Aeroelasticity in the seventies and eighties, on the basis of the velocity potential, several panel methods for oscillating three-dimensional wings with control surfaces [12, 13], for fuselages [14] and for cylindrical wings (flow-through engine nacelles) [15].

With these methods, based on linearized theory, there existed no chance for a satisfactory solution to the problematic of the oscillating wing in transonic flow and its inherent aeroelastic problems. But such a solution was urgently needed since flutter-free flight had to be proven for aircraft operating at near-sonic speeds far inside the transonic range. Here, the flow exhibits discontinuities (shock waves) and the relevant fluid-mechanical basic equations are therefore inherently nonlinear which renders their solution more difficult by far. Only the arrival of a new generation of large computer systems at the early eighties offered possibilities for the numerical solution – steady and unsteady – of these nonlinear fluid-mechanical equations and hence for the computation of the unsteady aerodynamic pressure distributions on oscil-

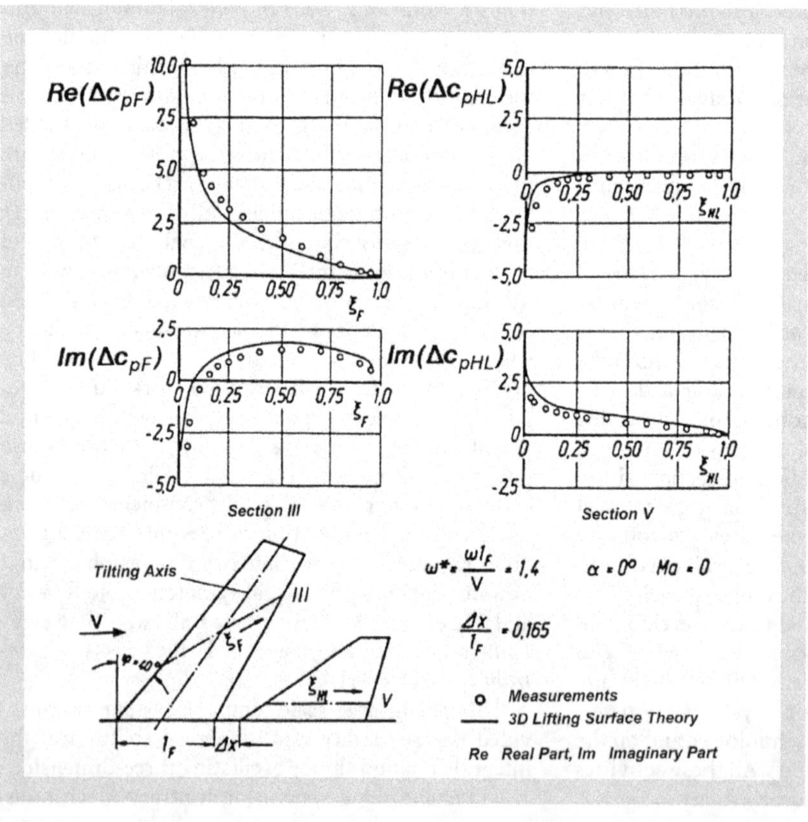

Real and imaginary parts of the unsteady aerodynamic pressure-difference coefficients, Δc_p, of a wing-elevator configuration performing harmonic wing torsional oscillations in compressible flow.

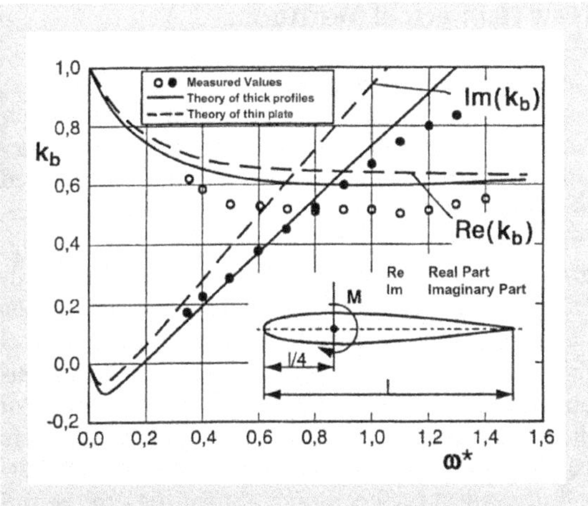

Real and imaginary parts of the derivative k_b of the unsteady aerodynamic moment of a harmonically flapping Mises-airfoil of 8.8 percent thickness as function of the reduced frequency ω^*.

lating wings in conjunction with aeroelastic investigations within the transonic speed range.

Numerical Computational Methods (CFD)

The development of numerical methods for the computation of aerodynamic pressure distributions on oscillating wings in transonic flow started in the USA already at the end of the seventies, where especially NASA performed pioneering work. In order to account for the shock waves occurring in transonic flow, the so-called Transonic Small Disturbance (TSD) Potential Equation was derived from the Euler equations under the assumption of small disturbances (e.g., oscillation amplitudes), which may be considered as the initial stage in the nonlinear modeling of unsteady aerodynamics. On the basis of the TSD equation, *Ralph Voss* [16] developed in the mid-eighties at the Göttingen Institute of Aeroelasticity a transonic panel method for three-dimensional wings, which was later developed into a transonic doublet-lattice method [17]. With the latter, numerous flutter computations were performed within the framework of the Airbus development. Such TSD methods, partly coupled with unsteady boundary layer methods to account for viscous effects, are today routinely applied as design methods in aeroelastic applications.

With the availability of large-scale computers commenced in about 1980 within fluid mechanical research also efforts to solve the complete (inviscid) Euler equations and the (viscous) Navier-Stokes equations as the utmost stage in aerodynamic modeling. The basic pioneering work concerning the solution of these nonlinear partial differential equations with the aid of finite-difference (FD) methods, especially the derivation of the necessary mathematics for the generation of the corresponding computational grids, was also already being performed since the mid-seventies in the USA. Here, progress in the area of unsteady aerodynamics was trailing the advances in steady aerodynamics by several years since unsteady solutions are by far more difficult and elaborate to achieve. For unsteady flow processes, one needs real-time solutions by means of time-stepping methods as well as contour- and solution-adaptive computational grids, whose consequence is a drastic increase in computer time, which turned out to be the limiting factor.

In Germany, at the Göttingen Institute of Aeroelasticity, several CFD-methods for the numerical computation of unsteady pressure distributions on (harmonically) oscillating 2D and 3D wings were developed since 1980: as an advanced stage of the aforementioned TSD panel method, at first a nonlinear full-potential method [18], at the next level, a 2D and 3D, respectively, Euler method [19] and, at the same

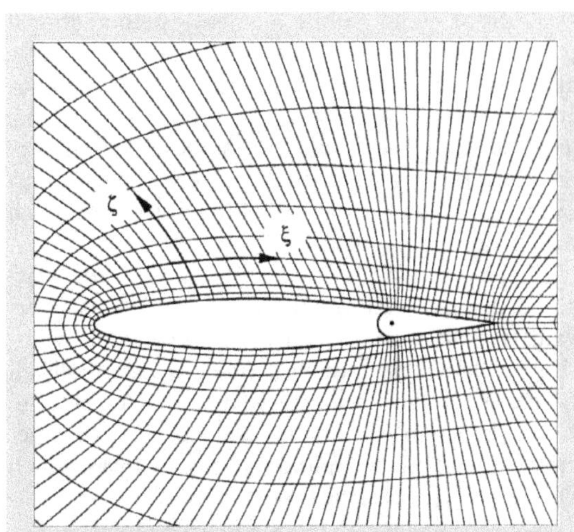

Two-dimensional CFD grid for an airfoil with control surface (flap).

level, a method for the computation of the unsteady subsonic and transonic flow about oscillating engine nacelles (cylindrical wing with jet simulation) [20]. All these methods have by now achieved complete maturity with regard to their industrial application.

To reduce computer time when also considering viscous effects, the coupling of a low-level inviscid (potential) CFD-method with an unsteady boundary layer method has turned out to be attractive. Such unsteady aerodynamic interaction (coupled) methods were developed at the DLR Institute of Aeroelasticity in Göttingen, in cooperation with EADS Airbus in Bremen, on the basis of the TSD and full-potential (FP) models and applied within the Airbus development. At the time, a further development on the basis of the Euler equations is being carried out.

Computation of the Natural Vibration Parameters of Aircraft

New perspectives concerning determination of the static deformation and natural vibration characteristics of wings and finally of entire aircraft arose. The prerequisite for the analytical treatment of aeroelastic problems is the knowledge of the unsteady aerodynamic forces. Within the framework of an AGARD research program, initiated by *Hans Georg Küssner*, a "natural vibration computation" for the entire F-104 G aircraft was performed for the first time at the end of the sixties at the ISD of *John H. Argyris* in Stuttgart. By means of the FE (Finite Element)-method almost 8,000 degrees of freedom were considered [21], a tremendous performance bearing in mind the computer power available at that time. A comparison of the theoretical results with the natural vibration parameters measured in ground vibration tests was quite encouraging but also showed that there was still much work to do.

The FE-method for the determination of the static and dynamic deformation characteristics of aircraft was subsequently further developed worldwide, with excellent contributions by the ISD. Essentially all static deformation parameters (stiffness-, respectively influence-matrices) and natural vibration parameters of an entire aircraft and its substructures, which are needed in aeroelastic investigations, are practically computed today with this standard method.

New Numerical Methods

General Formulation of the Aeroelastic Equations

Any arbitrary time-dependent deformation of a linear elasto-mechanical system, subject to arbitrary time-dependent external forces, can be developed as series (superposition) of the characteristic eigenfunctions (natural vibration modes as modal degrees of freedom) and the associated time-dependent free coefficients, the so-called generalized coordinates (modal normal-coordinates). These express, as the unknowns of the respective problem, the contributions of the eigen-functions involved in the form of the dynamic response, which is then also known. On this mathematical-physical basis, *Hans Georg Küssner* derived, at the beginning of the fifties, the aeroelastic equations of motion in generalized form. These generalized basic equations of aeroelasticity are, according to Lagrange's principle, energy equations and contain as coefficients integral terms of the respective kinetic, potential (elastic) and dissipative energies in the form of the generalized masses, natural frequencies and the global "damping loss-angle" of the natural vibration modes involved; they also contain the energy provided by the aerodynamic forces for the shift in these modal degrees of freedom in the form of the generalized aerodynamic forces (force integrals). To compute the latter, arbitrary aerodynamic models can, corresponding to the task, be utilized. The theoretical treatment of dynamic aeroelastic problems on the basis of these generalized aeroelastic integral-differential equations has in several regards proven to be advantageous:

- Due to the orthogonality of the natural vibration modes, the generalized mass and stiffness matrices become diagonal matrices, which simplify the numerical solution of the equations, and reduce the complex eigenvalue problem in the case of flutter computations to a special eigenvalue problem.
- With the modal degrees of freedom as initial functions in *Galerkin's* sense, the solutions become more transparent and easier to interpret.
- The modal natural vibration parameters (natural frequencies, natural vibration modes, generalized mass and damping loss-angle) may also be determined experimentally with high accuracy in ground vibration tests and directly introduced into the aeroelastic equations.

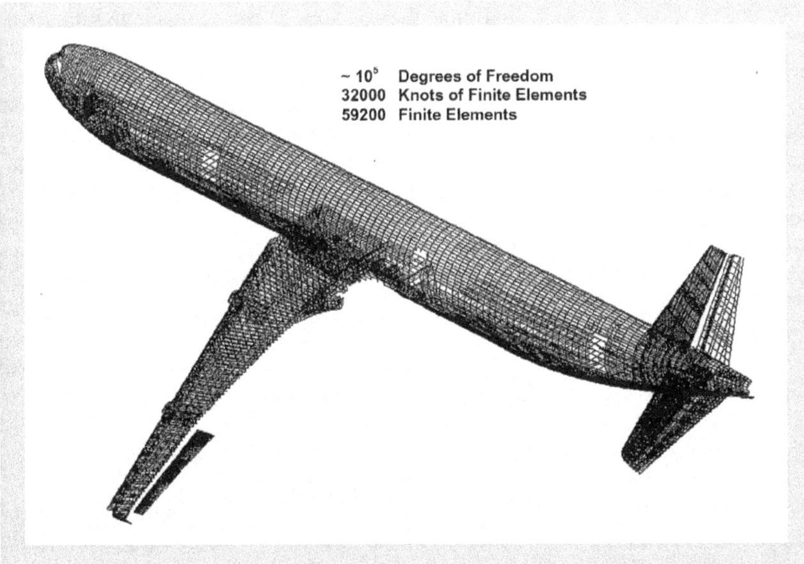

Finite element (FE) modeling of the Airbus A 340-300 (half model).

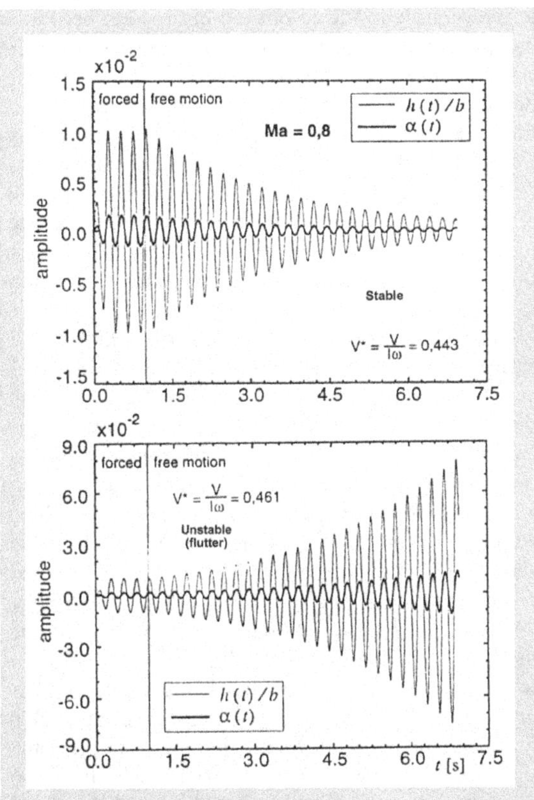

These generalized aeroelastic equations, formulated by *Küssner* during the early fifties, very soon found worldwide acceptance. Practically all dynamic aeroelastic computations in modern aircraft engineering are being carried out today with these basic aeroelastic equations. With today's computers, complete aircraft flutter and dynamic-response computations, with twenty and more modal degrees of freedom, have become routine [22].

The analytical solution of dynamic aeroelastic problems, especially the flutter problem, was from the very beginning based on the assumption of a linear dependence of the system forces on the elastic deformation (shift) within the so-called frequency range with the reduced frequency as a fundamental parameter. In classical flutter computations, with harmonic oscillations as limit of the dynamic stability in inviscid homogeneous, i.e., shock-free, flow, the problem reduces to the solution of a system of linear (complex) algebraic equations. Such classical solutions within the frequency range will also in the future preferably be applied, whenever possible.

◂ Numerical real-time solution of the flutter-stability behaviour of a 2D-airfoil with the degrees of freedom flapping h and torsion α.

Numerical Real-Time Solutions – Computational Aeroelasticity

With the arrival of super-computers and the development of advanced unsteady CFD-methods on the basis of the fundamental nonlinear aerodynamic equations (Euler and Navier-Stokes equations), there emerged the new possibility of a real-time numerical solution of the nonlinear aeroelastic equations, hence of the numerical simulation of dynamic aeroelastic processes. Such real-time solutions are especially required in dynamic aeroelastic investigations within the transonic speed range, but also when accounting for certain structural non-linearities in the aeroelastic equations. This led to a new field within future aeroelastic research, which is, within the Anglo-Saxon language domain, called Computational Aeroelasticity (CAe).

The numerical solution of nonlinear aeroelastic equations in the time domain requires a simultaneous step-wise integration in time of the fluid mechanical and structure dynamical equations, coupled via the oscillatory motion by the aerodynamic force terms, applying time-stepping methods and efficient solution algorithms. The integration in time of the corresponding differential equations in the case of such an aeroelastic simulation is here essentially determined by the Finite-Difference-Time-Stepping method of the CFD solution and the highest natural system frequencies. The dynamic aeroelastic stability and response characteristics are obtained from the damping behaviour of the modal degrees of freedom after an artificial excitation, similar to the wind-tunnel flutter and in-flight vibration tests [23].

In the development of such CAe-methods on the basis of an elasto-mechanical FE modeling on the one hand and modern CFD-models on the other hand, dominates the problem of fluid-structure-coupling statically as well as dynamically. In case of the fluid mechanical discretization, the outer flow field is being divided into finite volumes and the flow condition is described by a local Eulerian analysis. The discretization of the elastic structure (on the inside) is accomplished by finite elements; the description of the deformation condition is based on a *Lagrange* material consideration. The computational grids of the two different discretizations are entirely different and have, in general, no common point on the body surface. The central task of the CAe fluid-structure-coupling hence consists in the acquisition (capture) and functional description of the equivalent numerical data transfer between the aerodynamic forces (pressures) – computed at discrete grid points of the aerodynamic CFD computational model – and the static or dynamic deformations (natural vibration modes) at the discrete nodal points of the elasto-mechanical FE computational model. Here, corresponding interpolation strategies and procedures are needed, whose establishment is presently the subject of intensive research work also at the Göttingen DLR Institute of Aeroelasticity [24].

Aero-Servo-Elasticity

Aeroelastic Interactions with Flight-Mechanical Control and Stability Systems

The concept of aerodynamically balanced control surfaces with auxiliary flaps was nearly completely abandoned when, in the fifties, servo-hydraulic control-surface drives were introduced. Due to the nearly irreversible mode of operation of these hydraulic mechanisms, the natural torque frequencies of the control surfaces could be considerably increased, decisively influencing the danger of control-surface flutter. Servo-hydraulically activated control surfaces were then utilized later within automatic flight-mechanical control systems (AFCS) for numerous tasks, e.g., as autopilot and to artificially increase aerodynamic damping with respect to yaw, roll, and pitch, thus improving flight performance and flight comfort.

Such automatic (active) flight control and stabilization systems may have a large effect on the dynamic aeroelastic characteristics of an aircraft. The interaction between the dynamics of the active control-surface control system, structural dynamics, and oscillation-induced unsteady aerodynamic forces leads to a number of new, highly interdisciplinary problems, which were summarized under the term Aero-servo-elasticity and can be clearly presented by the aero-servo-elasticity triangle. The left side of this triangle describes the classical dynamic aeroelastic problems, the lower side the classical flight-mechanical aero-servo-dynamic problems related to the design of the flight-control system for the ideally rigid aircraft in the classical flight-mechanical sense, and the right side describes the important servo-mechanical coupling between the natural oscillations of the elastic

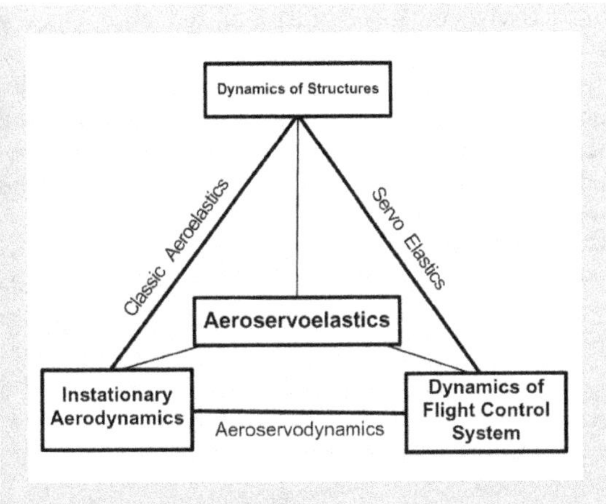

Aero-servo-elasticity triangle.

aircraft and the dynamics of the flight control system. This coupling together with the aerodynamic feedback of the servo-activated control surfaces, results in aero-servo-elastic interactions. These may have an undesirable energy exchange between the structural natural oscillations of the aircraft and the AFCS since the sensor signals contain not only the movement of the rigid-body degrees of freedom but also of the natural vibration modes. This may lead to serious consequences with regard to the aeroelastic and flight-mechanical stability behavior.

The aeroelastically malfunction-proof design of active flight control systems has become in today's aircraft engineering, especially for large-scale wide-body aircraft, in several respects an important design aspect [25].

Aero-Servo-Elastic Active Control Technology

The interaction between the automatic flight-control system and the dynamic characteristics of the elastic aircraft, with its actively controlled and servo-hydraulically activated control surfaces, may, with regard to aeroelasticity, also be utilized advantageously within the framework of the so-called Active-Control Technology, viz., for the

- Increase in the critical flutter speed up to flutter suppression,
- Reduction of the structural gust-loading,
- Improvement of flight comfort by a reduction of vibrations.

The principal concept of controlled artificial aerodynamic damping by means of an automatically controlled active control surface (flap) is depicted in the figure below. The servo-activated flap is being controlled in amplitude and phase via a closed-loop control circuit, following a given control law (σ, ϕ), and the signal (σ) of a sensor embedded in the wing in such a way that the unsteady aerodynamic forces (A), induced

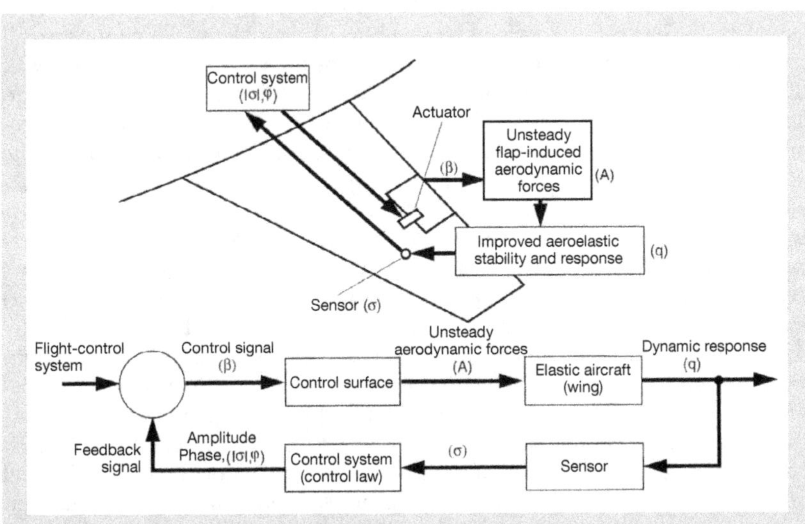

Functional scheme and block diagram of an actively controlled flap to improve the aeroelastic stability and to reduce the structural dynamic response.

Actively controlled GARTEUR wind-tunnel model in the 3-m Low-Speed Wind Tunnel in Göttingen.

by the flap movement (β), generate an aerodynamic damping hence counteracting the oscillatory motion of the wing. This results in a reduction of the oscillation amplitude and hence in improved aeroelastic stability, respectively a reduction in the structural loads (q→0).

Since the eighties, intensive research work was performed in this promising field of aeroelasticity in almost all countries active in aeronautics. In Germany, extensive measurements were carried out in 1985/86, within the framework of a GARTEUR research program, by DLR and MBB, together with ONERA, RAE and NLR, with a dynamically similar (complete) model with five active control surfaces (2 ailerons, 2 spoilers and the taileron); the program was concerned with comparative flutter and dynamic-response investigations; the tests were performed in the Göttingen 3-meter low-speed wind tunnel equipped with a gust generator [26]. In the process, many important practical results were attained, and the application potential of the aero-servo-elastic active-control technology impressively demonstrated. Applying different control laws, the critical flutter speed could be increased by 60 per cent, and the dynamic response to discrete and stochastic gusts could be noticeably reduced. On the basis of the results, an aero-servo-elastic actively controlled gust-load-reduction system was later realized on the Airbus A 340; here, the (aerodynamically only weakly dampened) lowest fuselage-side bending vibration is additionally damped by the actively controlled rudder [25], which results in a considerable improvement of the flight comfort. The aero-servo-elastic active-control technology was already, with various objectives, earlier applied in military aircraft engineering. Its integration into an existing AFCS requires corresponding multi-disciplinary design and development strategies, which will become of increased importance in the foreseeable development of a new generation of (highly-flexible) super transport aircraft (Mega-liner) [27].

Current Aeroelastic Challenges and Problems

Aeroelasticity in Transonic and Separated Flows

While the solution of the classical aeroelastic problems in inviscid, attached flow – especially the required proof of flutter-free conditions – is today routine due to the availability of the necessary aerodynamic and elasto-mechanic computational methods, developed after World War Two in worldwide, decade-long research work, there still remain many important aeroelastic problems, whose satisfactory solution still needs long-term intensive research work. Aerodynamic non-linearity is the essential characteristic of these aeroelastic problems, which most of all occur at transonic speeds as well as at high angles of attack in the presence of flow separation. The transonic range remained in the advancement of aeroelastics over decades an unexplored island, and, in the case of flutter computations, it was simply extrapolated from high subsonic to low supersonic speeds.

The typical characteristics of attached and separated flows, respectively, and the identification of the associated traditional (classical) and modern (current) work areas of aeroelasticity may be graphically depicted in an angle of attack/Mach number diagram also showing the division into three typical flow regimes. In Regime I, typical of flight at low angles of attack at subsonic and supersonic speeds, the flow is essentially attached. This is the domain of the classical (linear) aeroelasticity. Flying at transonic speeds and/ or at increasingly higher incidences, a transitional

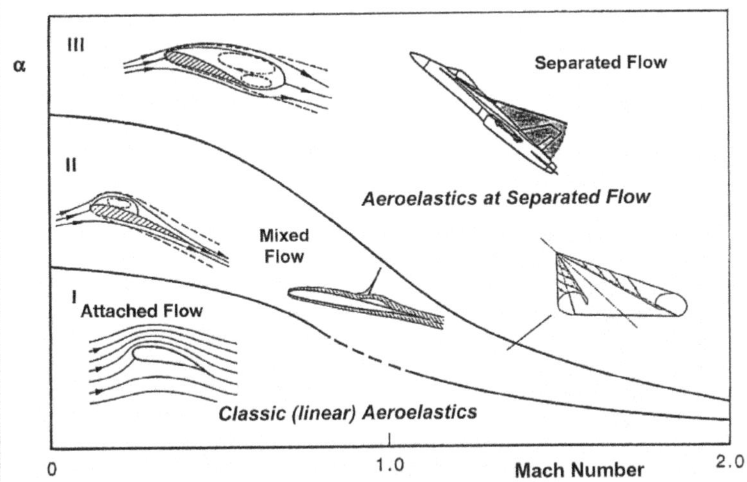

Characteristics of attached and separated flows and identification of the associated aerodynamic/aeroelastic work areas.

range with mixed flow is entered in Region II, characterized by areas of local separations at subsonic Mach numbers, embedded local supersonic pockets with unsteady flow behavior (shock waves) at transonic speeds, and the typical vortex roll-up at the leading edge of delta-wings at supersonic Mach numbers. These non-linear aerodynamic processes, already occurring on the wing in undisturbed flight, may, together with the unsteady aerodynamic reactions induced by wing oscillations, in their aeroelastic interaction lead to undesirable and dangerous (non-classical) dynamic aeroelastic problems; especially within the transonic range they may lead to

- transonic wing bending and torsion flutter with primarily one degree of freedom and, for reasons of the aerodynamic non-linearity, final amplitudes,
- transonic control-surface buss, equivalent to control-surface flutter, with one degree of freedom and also finite amplitude, and to
- transonic buffeting and here to resonance-like interactions with intense vibrations of the entire aircraft in its elastic natural vibration modes.

These typical transonic dynamic aeroelastic phenomena may considerably narrow-down the flight envelope. The buffeting boundary thus defines the upper limit of the flight speed of today's subsonic transport aircraft. Considering aeroelastics, the transonic range is for still another reason of importance. The stagnation pressure, at which flutter occurs, is here, as experience shows, lowest. The critical stagnation pressure is characterized by a gradual decrease to a minimum value and a subsequent rapid increase for $M \rightarrow 1$. The precise prediction of this typical flutter-stability characteristic, in the Anglo-Saxon language domain called Transonic (Flutter) Dip, is of great importance to aircraft operating at subsonic speeds close to the boundary to transonic flow. Here, advanced unsteady CFD methods, based on viscous flow models (Navier-Stokes equations) are needed.

Region III, finally, is characterized by a completely separated flow on the wing or the entire aircraft (e.g., in case of a fighter during maneuvers under so-called post-stall conditions). Within the separated regions, the flow is highly turbulent and unsteady, which can lead to considerable structural responses and to resonance-like excited buffet oscillations of tail surfaces located in the wake of the wing. Flow separation at high angles of attack (stall) may also cause stall-flutter, i.e., aeroelastically self-induced torsional oscillations of lifting surfaces operating at angles of attack close to the stall angle. Especially in the case of propellers, helicopter rotors and blades of turbo-machines may stall-flutter become a serious problem. The entire problematic of aeroelasticity in separat-

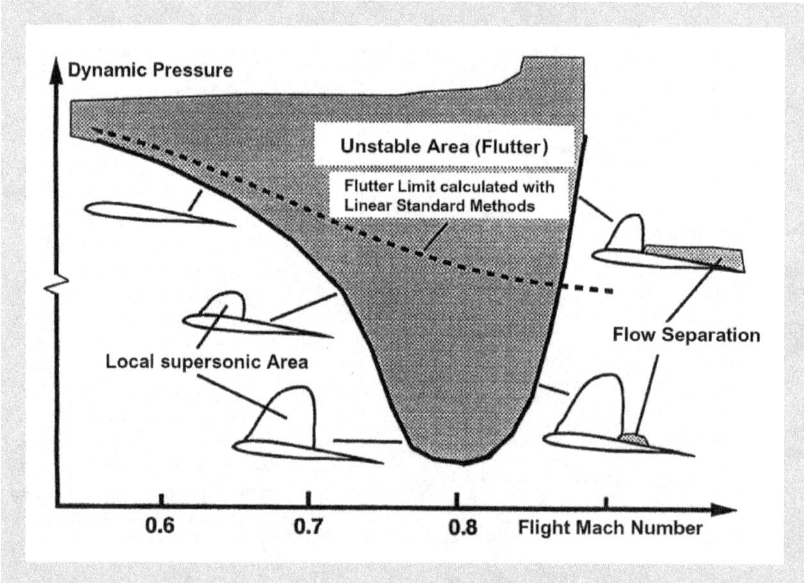

Typical transonic flutter-stability characteristic (Transonic dip).

ed flow (Region III) constitutes a big challenge and represents an area of long-term future aeroelastic research.

During the last 10 years, considerable progress was made worldwide, but most of all in the USA, in the theoretical treatment of the specific aeroelastic problems related to the transonic speed range. This research work, performed with great effort and supercomputers as well as advanced unsteady CFD-methods, has considerably contributed to the general understanding of these extremely complex interactions. Some contributions to the problematic of the Transonic Dip [17], stall-flutter of wings and rotors [28], and the unsteady aerodynamic forces on oscillating wings with separated flow [29] were in Germany also provided by the DLR in Göttingen.

Blade Flutter in Axial Turbo-Machinery

The development towards higher efficiency, coupled with the demand of a reduction in the primary energy consumption at a simultaneous reduction in noise and pollutant emissions, led in modern turbo-machinery engineering to a steady increase of the pressure ratios, mass-flow rates and flow velocities. In the case of aircraft engines, this development led to ever increasing bypass ratios and, in turn, to compressors (fans) with ever increasing blade lengths and aspect ratios. This caused the blades to become, due to the aeroelastic interaction, more susceptible to vibrations with the problem of the flutter stability being most important. Since the sixties, intensive research was performed, most of all in the USA and England, concerning the physical understanding and the analytical treatment of aeroelastically excited oscillations in turbo-machinery cascades; a new field of aeroelastics arose: Aeroelasticity in Turbo-machinery.

The flutter problem concerning the blading of axial turbo-machines differs in many ways from the one of the isolated wing and is, due to the existing parameter variety, also by far more complex. A fan or compressor rotor, with a number of equidistant, thin and twisted blades under load from centrifugal and aerodynamic forces, is an extremely complex aeromechanical system. The cascade geometry (pitch/chord ratio, stagger angle, blade twist) and, in the case of oscillations, the aeroelastically adjusting phase angle between blades, are important additional parameters, which considerably aggravate the analytical treatment of the problem. A further peculiarity exists in that flutter of turbo-machinery cascades mostly occurs with only one elasto-mechanical degree of freedom (natural vibration modes) of the blades and one characteristic phase-shift angle, i.e., the flutter oscillation form consists of a wave circulating within the blade row.

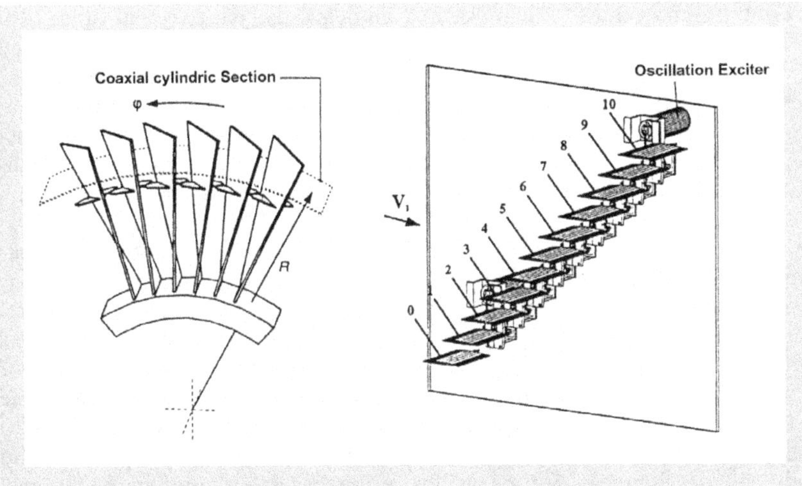

Idealization of an annular cascade by a two-dimensional straight cascade and its installation as a finite cascade with elastically arranged blades for aeroelastic investigations in the Göttingen Cascade Wind Tunnel.

A basic prerequisite for the theoretical investigation and determination of the aeroelastic stability of the blades within turbo-machinery is again the knowledge of the natural vibration characteristics of the blades on the one hand, and of the motion-induced unsteady aerodynamic pressure distributions on the other hand. Most of all the computation of the unsteady aerodynamic forces on the oscillating cascade turned out to be an extremely difficult problem, which could only be attacked satisfactorily during the last twenty years when large-scale computer systems became available. In successfully treating this problematic, considerable research work was also performed in Germany since that time, especially at the "Institute of Jet Propulsion and Turbo-machinery" ("Institut für Strahlantriebe und Turboarbeitsmaschinen") of the Institute of Technology (TH) Aachen (*Heinz Gallus*) and the Göttingen Institute of Aeroelasticity. Here, work concentrated in recent times on:

- Development of methods and procedures for the computation of the unsteady aerodynamic forces on harmonically oscillating 2D and 3D cascades in compressible subsonic and transonic flow on the basis of the Euler equations (TH Aachen and DLR Göttingen) [30, 31].
- Aeroelastic experimental investigations in stationary cascade test facilities (TH Aachen) [32] and in a dedicated cascade wind tunnel (DLR Göttingen) [33].
- Theoretical parametric investigation of the flutter stability of 2D turbine and compressor cascades and further development of the flutter-stability theory (DLR Göttingen) [34].

This research, conducted in coordination with the German turbo-machinery and aircraft-engine industry, led to many basic results and to computer programs valuable to industry. In spite of the considerable progress made during recent years remains the field of turbo-machinery aeroelastics a focus of the research work in aeroelastics.

Advances in the Aeroelastic Experimental Technology

Ground Vibration Test Technology

The "Ground Vibration Test" technology for the measurement of the natural vibration parameters (natural frequencies, natural vibration modes, generalized masses and damping) of aircraft was in Germany since 1953 at the focus of further activities. Here, most of all the Göttingen Institute of Aeroelasticity performed trend-setting work. In conducting a ground vibration test according to the classical phase-resonance method, the following main tasks have to be performed:

- Elastic suspension or support of the aircraft must be as soft as possible.
- Identification of the natural vibrations existing in the frequency range of interest; separation and individual excitation of the corresponding natural vibration modes.
- Measurement of the respective vibration parameters.

For that purpose, numerous measurement devices (among others, frequency generators, vector-component gages, vibration transducers), electro-dynamic vibration exciters, identification methods (localization and indicator function methods), as well as suspensions and supports to simulate free-floating flight conditions, were developed. As a result of decade-long systematic research and development work regarding ground vibration tests, a mobile facility was designed and constructed at the DLR Institute of Aeroelasticity, which is, in its kind, exemplary; it has been utilized to test, in Germany within the framework of flutter-proofing, all prototypes of newly developed aircraft, and also to carry out for ESA basic experimental research on space structures. This mobile facility has also been utilized in numerous ground vibration tests abroad. At the present time, the facility possesses 700 data channels (vibration transducers and data processing) for the measurement of natural vibration modes at as many positions, therewith also satisfying the internationally most modern demands [35]. In recent times, also so-called phase-separation methods were developed as alternative to the classical phase-resonance method, where it is tried to determine the natural vibration parameters of the test structure from measured time series of the structural response, at modally non-matching harmonic excitations, by means of a numerical-analytical separation of the natural vibration modes. Such semi-experimental ground vibration methods were developed [36] at the DLR Göttingen and also at VFW in Bremen. The test setup and effort is here practically the same as in the case of the classical phase-resonance method. Since in case of the phase-separation procedure the relatively time consuming adjustment of the exciter configuration to each natural vibration mode can be dropped, certain time savings during the test execution can be realized. One has to accept the disadvantage that the natural vibration parameters can no longer be directly obtained but only later be determined from the numerical computer analysis. Possible test insufficiencies can no longer be directly corrected on side but are most likely only being detected during the analysis. For a reliable application of the phase-separation method is, in addition, an absolute linearity of the test structure a necessary prerequisite, which is not given in the case of modern aircraft with servo-hydraulic control surface actuation.

The ground vibration test has today become an indispensable means for the measurement and valida-

Ground vibration test on an Airbus A 321 at EADS-Hamburg.

tion of the natural vibration parameters of aircraft needed for aeroelastic investigations. It is also the basis for aeroelastic wind-tunnel model and in-flight vibration tests; it also provides, in addition, the possibility for investigations and research concerning important structural-dynamic characteristics, which are hard or impossible to determine theoretically, such as, for instance, the structural damping behavior, structural non-linearities and stiffnesses, and transfer functions of servo-hydraulic control-surface actuators.

Wind-Tunnel Testing Technology

Experimental investigations on wind-tunnel models have played an important role in aeroelastics from the very beginning. The tasks and objectives of such aeroelastic model investigations in the wind tunnel are manifold:

- Measurement of unsteady aerodynamic pressure distributions and of global coefficients on (fixed or semi-fixed) oscillating and/or rotating models of wings (e.g., wings with control surfaces, empennages, cascades of fluid machinery, and rotors) with the objective of validating theoretical results and researching basic unsteady aerodynamic processes.
- Aeroelastic investigations on dynamically similar models, especially flutter investigations to validate theoretical methods, and investigation of specific dynamic aeroelastic phenomena, for instance, at transonic speeds or in the presence of separated flow (stall and buffeting).

The experimental determination of unsteady pressure distributions on oscillating wings was for a long time trailing the expeditious development of computational methods. The first known unsteady aerodynamic pressure measurements on a harmonically force-excited wing model with control surface were already carried out by *Hans Drescher* in 1936, and in 1944 at the AVA in a water tunnel (!) [37]. Here, the pressures were measured, via small orifices in the model surface and tubes, by external multi-tube manometers. This indirect method, which requires the knowledge of the unsteady acoustic transfer function of each tube, is also today still frequently used; it was until the seventies the standard method. With this method, many measurements were carried out since about 1960 in various wind tunnels on harmonically oscillating wings with and without control surfaces up to supersonic speeds by the DLR Institute of Aeroelasticity in Göttingen.

Direct pressure measurement, without transfer corrections, became possible with the availability of miniature pressure sensors, which were installed inside the model in the immediate vicinity of the pressure orifices. This direct method is the basis for a mobile system to measure unsteady aerodynamic pressure distributions on oscillating lifting configurations (AMIS I), developed during the eighties at the DLR Institute of Aeroelasticity. The systems consists of 720 data channels and is based on data acquisition via multiplexing and an analog to digital (AD) converter, which is connected to a parallel computer by way of a programmable interface. The measurement range covers frequencies up to about 2 kHz. The system AMIS II was then developed during the early nineties, especially for unsteady pressure distribution measurements on helicopter rotors, within the framework of a large DLR research project (CP-ROT/Helinoice), with a required frequency resolution per data channel of 20 kHz [38]. The concept of the system consists in the combination of the many possibilities of modern computer-supported workplaces (work stations) with a multi-channel modular and highly accurate signal processing and data acquisition system. It consists today of more than 360 simultaneous (parallel) channels for time-dependent sensor signals (acquisition system TEDAS on transputer basis) and as many channels for steady signals (on the basis of multiplexing). Connected to a standard IEC-bus interface are a generator, providing the desired vibration-excitation signals for the hydraulic actuators, and the sensors for measuring the wind-tunnel parameters; a parallel input/output interface controls the sensor amplifiers. Specifically developed extensive equipment-specific software for the various computer systems allows a parallel real-time processing, presentation and visualization of the measured results. With the mobile systems AMIS I and AMIS II, the DLR Institute of Aeroelastics conducted in various wind tunnels in Göttingen, in the DNW and at ONERA in Modane, within the framework of national and international research programs, numerous steady and unsteady pressure distribution measurements on oscillating and rotating wings (wings, turbo-machinery cascades and helicopter rotors). The AMIS II system is unique in Eu-

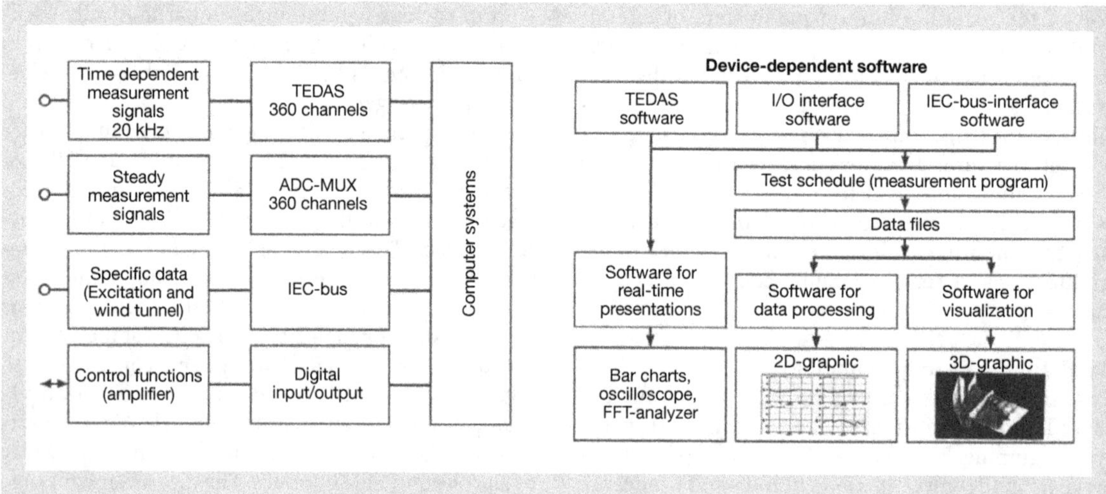

Components of the data acquisition system (left) and software structure (right) of the mobile AMIS II system for the measurement of steady and unsteady aerodynamic quantities in wind-tunnel tests.

rope and satisfies the highest international demands. It is an important tool of future research.

Flutter-model investigations on dynamically similar complete- and half-models of aircraft in the wind tunnel are since many years routine and constitute another tool of aeroelastic research. For instance, MBB conducted numerous wind-tunnel-model investigations in Emmen/Switzerland during the "Tornado" development to study the wing-with-external-load flutter problem [11]. Many valuable results were also obtained in flutter-model investigations in the transonic wind tunnel S2-Modane during the Airbus A 340 development within the so-called "Aeroelastic Model Program", a cooperation between MBB/Bremen and Ottobrunn, ONERA and DLR. Many other wind-tunnel flutter investigations were also performed as part of the scientific work at the Göttingen DLR Institute of Aeroelasticity.

In-Flight Vibration Test Technology

Similar to the wind-tunnel flutter test is also the in-flight vibration test, due to the steady advances of the electronic measurement and data-processing technology, today routine. It is the last member in the long chain of aeroelastic investigations related to the development of modern aircraft. The problematic of in-flight vibration tests is related to the solution of the following individual tasks:

- Realization of an in-flight oscillation (vibration) exciter.
- Measurement of the dynamic response of the excited natural vibrations at a sufficiently large number of locations on the aircraft, and the subsequent determination of the natural frequencies and damping characteristics.
- Establishment of flutter-stability diagrams from the measured data.

Here, the vibration excitation is the biggest problem; several methods are relevant:

- A harmonic excitation with electro-dynamic vibration exciters, similar to ground vibration tests (if the installation is possible, as is the case for large-scale aircraft), with artificially generated control-surface oscillations (for servo-hydraulically controlled aircraft), and with additional vibration-excited wings (winglets).
- An impulse-like excitation with especially developed rocket pulsars or by a jerky movement of the control surfaces by the pilot (preferred in the case of fighter aircraft).
- A natural excitation by gusts.

All these means of exciting in-flight vibrations need different methods of data processing and analysis;

Aerodynamical vibration excitation by winglets at the wing tip of an Airbus A 300.

they are also applied in the same way in wind-tunnel flutter investigations.

Generally, in-flight vibration tests are being conducted on prototypes of new aircraft by the manufacturer, where also the specific soft- and hardware is being developed. Many tests of MBB with the "Tornado" fighter aircraft were thus carried out in Manching with the vibration excitations being generated by rocket-pulsars and winglets mounted on the external wing loads. These types of excitation were also applied in case of the Airbus A 300 in the early seventies and on successor developments. The excitation by winglets at the wing tips (tip vanes) and the empennage surfaces and the corresponding data acquisition and processing was developed by VFW in Bremen [39].

Concluding Remarks

The necessity of aeroelastic research in aircraft engineering was an inevitable consequence of the technological development. The current state of knowledge in aeroelasticity was always trailing the overall technological development and was hence faced with increasing challenges. New problems had to be added continuously and new solutions had to be worked out. That the status of knowledge in aeroelasticity was at all able to keep up the pace with the rapid technological development in aircraft engineering is due to the similarly stormy development in computer technology, which, in its beginnings, received remarkable impetus from the specific requirements of aeroelasticity.

Although aeroelastic investigations in conjunction with new aircraft developments are today routine, there still exists a wide field for future research. Especially the numerical real-time simulation of aeroelastic processes (CAe-Computational Aeroelasticity), with the application of advanced CFD-methods, opens promising perspectives; but progress depends here too on the future development of computer technology. In the long run, it can be expected that these CAe-activities will finally culminate in the capability of numerically simulating and visualizing the total aeroelastic and flight dynamic behaviour of aircraft at all flight conditions. CAe is the great challenge of "Aeroelasticity 2000 – and beyond" with far-reaching technological and economic consequences for the future of aerospace engineering.

References

1 A. Betz (Hrsg.). Instationäre Vorgänge. Britische Monographien Göttingen, Band G (1946)
2 H. G. Küssner, H. Billing. Instationäre Strömungen. Naturforschung und Medizin in Deutschland 1939–1946. Weinheim 1953, Band 11, S. 127–180
3 Aeroelastisches Kolloquium in Göttingen am 16. und 17. April 1957. Mitt. MPI für Strömungsforschung und AVA Nr.18, Göttingen 1958
4 H.G. Küssner, H. Göllnitz. Tabellen der aerodynamischen Derivativa des schwingenden Streckenprofils mit Knicken und Stufen. DLR-FB 64-05 (1964)
5 H.G. Küssner, F. Wille. Theorie des schwingenden Streckenprofils bei kompressibler Unterschallströmung, Teil IV. DLR-FB 68-76 (1968)
6 H.G. Küssner, H. Hertrich. Theorie der schwingenden Tragfläche großer Streckung im Unterschallbereich. DLR-FB 65-19 (1965)
7 H.G. Küssner, G. von Gorup. Instationäre linearisierte Theorie der Flügelprofile endlicher Dicke in inkompressibler Strömung. Mitt. MPI für Strömungsforschung und AVA Nr. 26, Göttingen 1960
8 H.G.Küssner.AllgemeineTragflächentheorie.Luftfahrtforschung 17 (1940), S. 370–378
9 B. Laschka. Zur Theorie der harmonisch schwingenden tragenden Fläche bei Unterschallanströmung. Zeitschr. Flugwiss., Band 11 (1963), S. 265–292

References

10 B. Laschka. Interfering Lifting Surfaces in Subsonic Flow. Zeitschr. Flugwiss., Band 18 (1970), S. 359–368

11 O. Sensburg, A. Lotze, G. Haidl. Wing With Stores Flutter On Variable Sweep Aircraft. AGARD-CP-1 62 (1975), S. 6-1 bis 6-19

12 W. Geissler, Ein numerisches Verfahren zur Berechnung der instationären aerodynamischen Druckverteilung an der harmonisch schwingenden Tragfläche mit Ruder in Unterschallströmung. Teil 1: Inkompressible Strömung, DLR-FB 75-37 (1975). Teil 2: Kompressible Strömung, DLR-FB 77-15 (1977)

13 W. Geissler, Nonlinear Unsteady Potential Flow Calculations for Three-Dimensional Oscillating Wings. AIAA-Jounal, Vol. 16, No. 11 (1978), S. 1168–1174

14 W. Geissler, Der harmonisch schwingende Rumpf in Unterschallströmung. DLR-FB 78-24 (1978)

15 E. Katzer, Stationäre und instationäre Potentialströmung um dünne Ringflügel und Triebwerke mit Simulation des Triebwerkstrahls. DFVLR-FB 89-19 (1989)

16 R. Voß, Calculation of 3D Unsteady Transonic Potential Flow by a Field Panel Method. 2nd Intern. Symposium on Aeroelasticity and Structural Dynamics, Aachen 1985. DGLR-Bericht 85-02 (1985)

17 S. Lu, R. Voß, TDLM-A Transonic Doublet-Lattice Method for 3D Potential Unsteady Transonic Flow Calculations. DLR-FB 92-25 (1992)

18 R. Voß, L. Zhiliang, DLR-FP+VII-Code Improvements and Computation of Unsteady Transonic Flows, DLR-1B 232-96J03 (1996)

19 W. Wegner, Prediction of Unsteady Aerodynamic Forces for Elastically Oscillating Wings Using CFD-Codes. Proceedings of the Intern. Symposium on Aeroelasticity and Structural Dynamics, Manchester, Juli 1995

20 A. Knipfer, Numerische Untersuchung der transsonischen Strömung an schwingenden Triebwerksgondeln. DLR-FB 95-39 (1995)

21 H.P. Mleinek, M.M. Mai, H.J. Bühlmeier. Theoretische Standschwingungsanalyse des Starfighter F 104 G. ISD-Bericht Nr. 68, Juli 1969

22 H. Försching, Grundlagen der Aeroelastik. Springer-Verlag Berlin/Heidelberg/New York, 1974

23 S. Schulze. Transonic Aeroelastic Simulation of a Flexible Wing. AGARD-CP-822 (1998)

24 A. Beckert. Ein Beitrag zur Strömungs-Struktur-Kopplung für die Berechnung des aeroelastischen Gleichgewichtszustandes. DLR-FB 97-42 (1997)

25 H. Molzow, H. Zimmermann. Influence of Active Controls on the Design Process of a Large Transport Aircraft. AGARD-R-794 (1993), S. 10–1 bis 10–12

26 H. Försching (Hrsg.), Active Control Application for Flutter Suppression and Gust Load Alleviation, Results of Phases 1 to 3, GARTEUR-TP-022 (1985), Addendum: Results of Phase 4 (1986)

27 H. Försching, New Ultra High Capacity Aircraft (UHCA)-Challenges and problems from an aeroelastic point of view, Zeitschr. Flugwissenschaften, Band 18, Nr. 4 (1994), S. 219–231

28 W. Geißler, Numerical Investigation of Dynamic Stall on Oscillating Airfoils. Proceedings of the Intern. Symposium on Aeroelasticity and Structural Dynamics, Aachen 1991, S. 85–93

29 H. Försching, H. Zingel, Motion-Induced Unsteady Airloads on an Oscillating Low-Aspect-Ratio Trapezoidal Half-Wing in Separated Flow, Journ. Fluids and Structures (1988)2, S. 515–539

30 D. Peitsch, Berechnung der dreidimensionalen, instationären Durchströmung schwingender Turbomaschinengitter mit einem expliziten Zeitschrittverfahren. Dissertation RWTH Aachen, 1995

31 V. Carstens, Computation of Unsteady Transsonic 3D Flow in Oscillating Turbomachinery Bladings by an Euler-Algorithm with Deforming Grids. Proceedings of the 7th Intern. Symposium on Unsteady Aerodynamics and Aeroelasticity of Turbomachines, Fukuoka/Japan, Sept. 1994

32 H. Gallus, H. P. Kau, C, A. Poensgen, Schaufelschwingungen, FVV-Hefte 467-1 und 467-2, 1990

33 H. Hennings, Windkanal-Flatteruntersuchungen an einem endlichen, geraden 2d-Verdichtergitter in inkompressibler Strömung. Dissertation, RWTH Aachen,1996

34 H. Försching, Aeroelastic Stability of Cascades in Turbomachinery. Progr.Aerosp.Sci., Vol.30, S. 213–266 (1994)

35 M. Degener, U. Füllekrug, M. Sinapius, E. Achenbach, W. Peschel, Standschwingungsversuche am neuen Airbus A 321. DGLR-Jahrbuch 1993, S. 537-546

36 H. G. Natke, Einführung in Theorie und Praxis der Zeitreihen- und Modalanalyse. Vieweg-Verlag Braunschweig, 3. Aufl., 1992

37 H. Drescher, Messungen an einem Tragflügel mit harmonisch bewegtem Ruder in ebener Strömung. AVA-Bericht B 44/J/10 (1944)

38 J. Wagener, W. Wagner, B. Weitemeier, M. Winter, AMIS II-Anlage zur Messung zeitveränderlicher mechanischer Größen in Windkanalversuchen. DLR-Nachrichten, Heft 76 (1994)

39 K. König, Experience From Flight Flutter Testing With Tip Vanes On Airbus. Proceedings of the International Symposium on Aeroelasticity, Nürnberg 5–7. Oktober 1981, S. 130–142

Advances in Helicopter Rotor Technology

VALENTIN KLÖPPEL

Development of Hingeless Helicopter Rotors

Main Rotors

Since the beginning of the helicopter development (see "First Helicopters and Rotor Systems") there was a need to place hinges and bearings as close to the rotor center as possible to guarantee low rotor stiffness and hence good control characteristics at low cabin vibrations. This design concept is characterised by the distance of the flapping hinge (delta hinge) from the rotor center and demonstrated in the subsequent figure, which covers the time period from the beginnings until today.

With time, the flapping-hinge distance increased, which improved maneuverability but detrimentally affected vibrational loads and hence comfort. Today, the tendency is again towards smaller flapping-hinge distances, which is also a result of the current technological possibilities.

Contrary to the winged aircraft, helicopters experience almost always asymmetrical freestream conditions. (Addition of circumferential and flight speed in case of the advancing blade and subtraction for the retreating blade.) The blade-tip plane has, in addition, to be tilted (pitched) to generate controlling moments for trimming the helicopter about the pitch and roll axes, to generate positive and negative accelerations, and to change directions. *Walter Just* has decisively contributed to the formulation of the fundamentals during the initial post-World-War-Two phase.

Both requirements are met by the rotor blades via a cyclic blade control and the flapping blade movement. The latter generates, in turn, via the Coriolis acceleration the in-plane motion of the blade. To con-

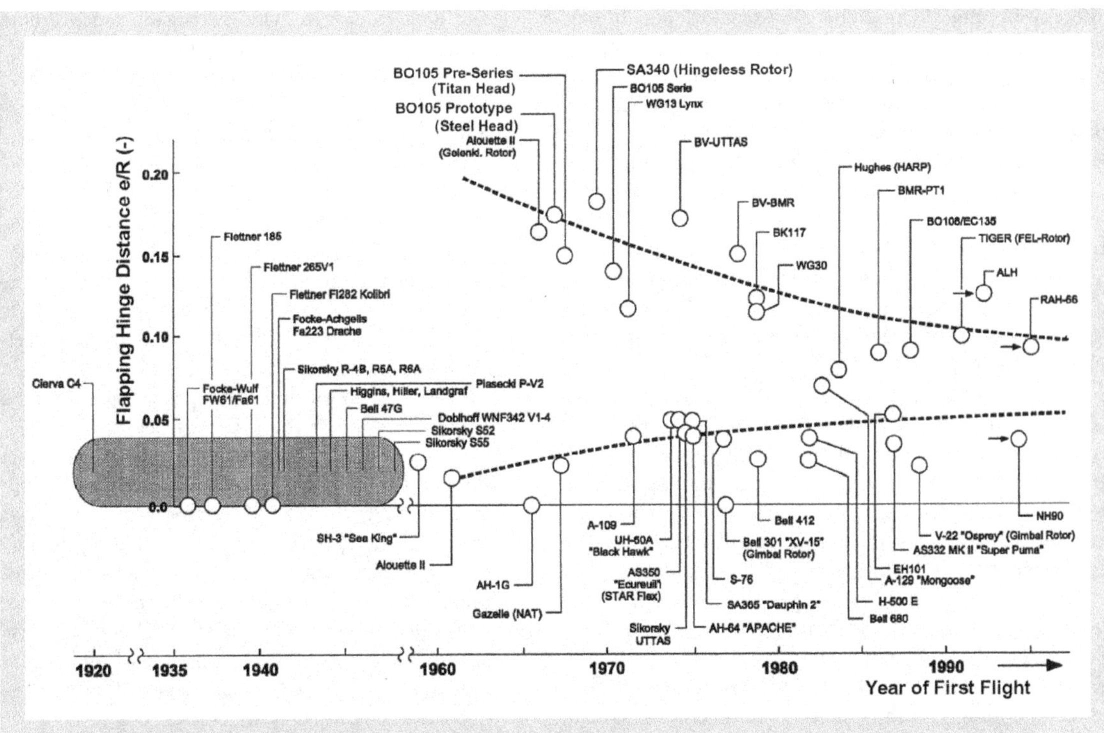

Development with time of the flapping-hinge (delta hinge) distance.

trol these degrees of freedom, three hinges are generally being employed in classical rotor technology for the blade adjustment, the blade flapping motion, and the blade in-plane motion (lead lag), respectively. Exceptions are the two-blade "pendulum-hub" or "constant-velocity-joint" rotor heads, where a lag (drag) hinge is not needed.

The standard antifriction bearings are increasingly replaced by Elastomer bearings – by combining the three hinges in one spherical Elastomer-joint (Sikorsky, Agusta, Eurocopter-Spheriflex) or in two Elastomer-joints (Eurocopter "Tiger", Sikorsky UH 60 A, Bell-FEL, Westland-Agusta-EH 101).

The so-called hingeless (rigid) rotor systems replace flapping and lag bearings by flexible composite/metal designs (Westland-Lynx, Eurocopter-BO 105, BK 117 [1], Eurocopter "Tiger", ALH), require, however, larger flapping-hinge distances.

The replacement of the bearings by composite structures allows, on the other hand, the reduction of the flapping-hinge distance (Eurocopter-EC 135, McDonell-Douglas "Explorer", Bell-412, Boeing-BMR, Sikorsky-Boeing "Comanche", Kawasaki). For both systems, the hingeless as well as the hinge- and bearingless system, hereafter denoted bearingless, the flapping-hinge distance e/R developed towards a range of $0.05 \leq e/R \leq 0.1$.

A comparison of the principles of the different rotor-head systems is given in the figure on page 573. The hingeless BO 105 rotor system, developed by MBB (today's Eurocopter Germany, ECD), was subjected to severe tests within the high-speed program HGH to prove its reliability and performance capacity [1]. With the system, a speed of 404 km/h (218 kts) was reached during a dive.

ECD developed as an experimental version of the "Tiger" rotor system [1] the hingeless FEL-system [2], a combination between a hingeless and an Elastomer-rotor system. All together, important impulses were provided by the rotor systems of ECD in conjunction with the development of the BO 105 rotor system and subsequently with the bearingless EC 135 system [3].

Blade adjustment is accomplished in case of a bearingless rotor by a system consisting of a flexible element and a control cuff. The flexible element transfers the centrifugal force and allows blade incidence-, flapping-, and in-plane movements. The control cuff bypasses the flexible element and introduces at its end the blade-control angle. The flexible ele-

Walter Just, born in Niederhermsdorf in Silesia on November 13, 1909.

1934 he finishes his studies at the University Breslau with an examination related to higher-educational teaching in mathematics, physics and chemistry.

1935 he graduates with a master's degree (Diplom-Ingenieur) from the Institute of Technology (TH) Breslau; as a manager, he participates in establishing the subject experimental aircraft design and construction at the TH. He receives his PhD with a thesis on new wind motors with a vertical axis and obtains the lecturing qualification (habilitation) in

1937 with a treatise on the computation of the Voith-Schneider propeller. During the year, he joins the AGO Aircraft Company as department head of aerodynamics and in

1939 he joins the Helicopter Company Focke-Achgelis in Delmenhorst. As group supervisor of aerodynamics, he participated decisively in the development of helicopters and autogyros. Shortly before the end of the war he was, as department head at Messerschmitt and Junkers, charged with investigations concerning fighter aircraft with rocket propulsion.

1946 *Just* went to Turbomeca and SNCASE in France where he made essential contributions to the design of jet aircraft such as, for instance, the "Caravelle". After a short stay with the Messerschmitt-team in Egypt, he took over in

1954 as head of the "Study Association Helicopter" ("Studiengemeinschaft Hubschrauber, DSH") in Stuttgart.

1968 he becomes a professor at the TH Stuttgart. In the meantime, he had become director of the "Institute of Rotational-wing Aircraft" ("Institut für Drehflügelflugzeuge") within the DVL, since

1969 the DFVLR, into which the DSH had merged. After the DFVLR moved her helicopter activities in

1972 to Brunswick, he continued his teaching obligations at the TH Stuttgart. Indispensable tools for the development of helicopters and autogyros became the DSH reports No. 2 and 3 "Aerodynamics of helicopters and autogyros, Part 1: Introduction and performance computations" and "Part 2: Rotor computations" by *W. Just* and *K. Jaeckel*. Also the standard textbooks "Control and stability of rotational-wing aircraft" and "Flight mechanics, control and stability of aircraft" are results of his extensive teaching activities.

1980 he dies on September 3 in Nürtingen.

ment must withstand torsions of > 40°. In the case of a preceding experimental version, the torsional control moment was still introduced into the blade by a torque tube [1], but the control cuff has greater advantages due to a better attachment, lower drag, and the possibility to integrate in-plane damping and kinematically coupled damping.

A special development concerning the rotor-head mast integration is the "integrated dynamical system, IDS", which contains the rotating control rods inside the mast. This system was realized on the joint German-Indian project ALH, which is depicted in Color Plate XI.

Tail Rotors

Since tail rotors essentially deliver thrust to compensate the propulsive moment of the main rotor and to control the movement about the vertical axis and the yaw angle, cyclic control may be omitted. The number of degrees of freedom is, however, the same as for the main rotor. Tailrotors contain correspondingly, in the case of more than two blades in conventional design, three bearings for the flapping and in-plane movements and blade adjustment, respectively. The two-blade rotor exhibits, as in the case of the main rotor, a special design. Since only thrust but no controlling moment is transferred to the helicopter fuselage, the mast torque can be minimized. This happens in case of the two-blade rotor by means of a central "seesaw" bearing and a rigid connection of the two blades in the flapping direction (seesaw or teetering system). Bearings related to the in-plane movements can here be omitted; the $1/\Omega$ in-plane mode manifests itself as rpm variation. In case of more than two blades, one tries to place the bearing for the flapping as close as possible to the rotor hub.

Similar to the main rotors, the classical bearings were in the past replaced by Elastomer bearings or flexible fiber structures (bearingless systems). In case of the latter, one distinguishes between systems soft or stiff with respect to in-plane motions dependent on whether the first natural in-plane-motion frequency is lower or higher than the rotational frequency. The former exhibit a lower load level, but require a specific damping of the in-plane motion, especially since they may enter into resonance with the natural modes of the tail boom, a subject that will be addressed later.

ECD developed and tested all aforementioned systems – conventional as well as advanced [4] – starting with the classical two-blade tail rotor (BO 105, BK 117 [1]) and going on to the two-blade rotor with Elastomer bearings as well as with a bearingless blade adjustment, respectively (both flown on the BO 108, a preliminary version of the EC 135), and advancing to the in-plane-motion stiff three-blade rotor [1] and the bearingless four-blade rotor as in-plane-motion soft [5] and in-plane-motion stiff design, respectively [1, 4]. When going from the BO 108 to the EC 135, the

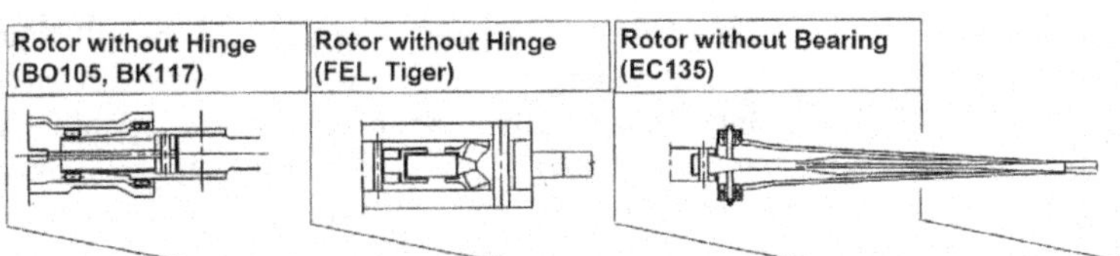

Hub	Titan Star	FRC-Plates	Metal
Blade Adjustment	2 Rolling Contact Bearing, oil lubricated	2 Elastomer Bearings, maintenance-free	Torsion flexible Element (Drill Element)
Flapping / Lead lag	Angular flexible Neck (FRC)	Angular flexible Neck (FRC)	Angular flexible Neck (FRC)
Centrifugal Force Transmission	Bendix-Element	Elastomer Taper Bearing and Blade Bolt	Drill Element and Blade Bolt
Controlling	By Control Lever	By Control Lever	By Control Cuff

Comparison of principles of rotor-head systems.

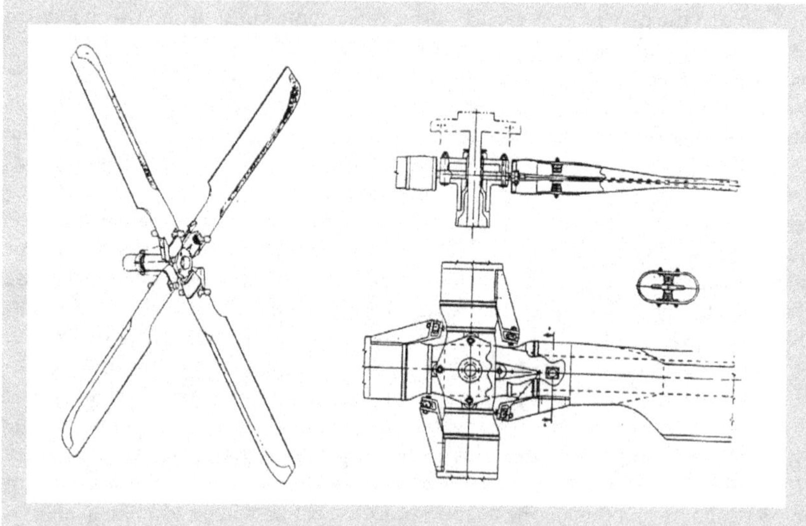

In-plane-motion stiff four-blade tail rotor system of the German Indian light helicopter ALH.

principle of the two-blade rotor with central lag hinge was abandoned in favor of the "Fenestron" – a product of the French Eurocopter partner [6]. For larger helicopters, this system, protecting the rotor as well as ground personnel, would require too large rudders. Thus for the joint German-Indian project ALH, the in-plane-motion stiff four-blade rotor is utilized, which proved to be an efficient, modern tail rotor system. The preceding figures clarify these interrelations.

Fiber Composite Design and Construction

Demands on Rotor Blades

The replacement of the flapping and lag bearings in case of the hingeless BO 105 main rotor system required in the blade-neck area a blade material of high elasticity. MBB used as first manufacturer glass fiber for the entire blade. This trend was followed by all other manufacturers since glass and soon also carbon proved to be considerably saver and more cost effective than the regular aluminum. In order to reduce manufacturing costs, MBB soon switched over to designs that could be mechanized. For instance, the hand-inserted loops of the blade attachment in case of the EC 135 and "Tiger" blades were replaced by pre-impregnated fabric (prepreg). The blade structure in fiber-composite design is depicted in the figure on page 577.

The fiber structure of such blades can no longer be checked for possible faults, damages or abnormalities by conventional non-intrusive means, such as ultrasound and X-rays. Here, computer tomography proved well suited. It is, since 1979, routinely utilized [7]. With it, defects, such as broken fibers, delamination between fiber layers, wavy lamination and the quality of bonding, can be detected.

Blade-Geometry Design

The blades of the BO 105 and BK 117 are still characterized by a rectangular planform and NACA airfoils. But already for the enhanced-combat-capability version (KWS [8]) of the BO 105-PAH1 and then also for the "Tiger" and the EC 135, pointed and swept blade tips were investigated. Here, the DLR Brunswick developed the DMH3/DMH4 blade profiles (airfoils) with an increased maximum lift and a low drag and control moment. After positive tests of the new developments in blade planform and airfoils, both were utilized for the mass production of the aforementioned types. Here, engine power could be saved and a thrust increase and a noise reduction could be realized. This development was continued within the joint German-French program HLR/ORPHEE [9]. Within the program, four different blade geometries were investigated in the German-Dutch Wind Tunnel (DNW)

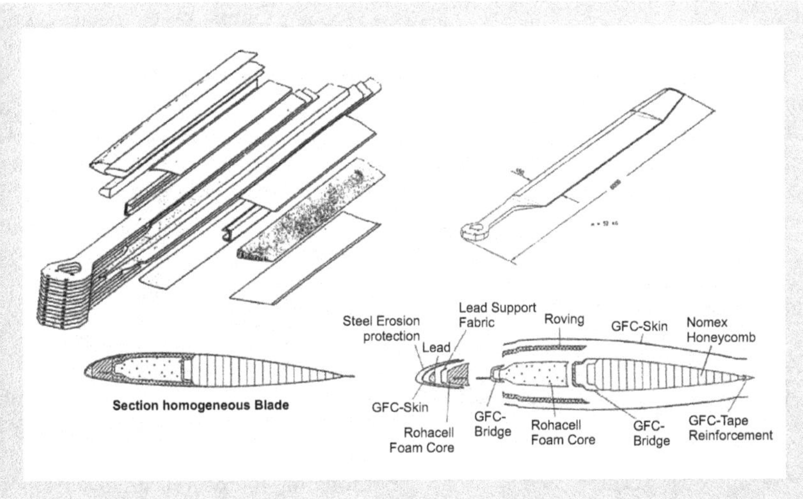

Rotor-blade designs suitable for mechanization.

and the ONERA Wind Tunnel S1-Modane. The experiences gained are presently being processed within the project "Advanced Rotor Technology, ART", a part of the Research Program of the Federal Government. Technical characteristics of the ART are negative taper of the main part of the blade, interchangeable blade tips to investigate their effect on power requirements and noise, OA-airfoils of ONERA, and nonlinear twist. After flight tests with a hingeless BK 117 rotor, a bearingless BK 117 five-blade rotor with the same blade geometry shall be developed. The rotor (ART) with interchangeable blade tips (hingeless rotor system) is depicted below.

Rotor Power Requirements

The computer programs for the determination of the rotor power requirements are continuously being improved. They are initially based on the blade-element theory under utilization of aerodynamic coefficients, in form of tables, and on rotor mass flow computation by means of the momentum equation. DLR and ECD tried to validate existing theories by rotor mass flow measurements in the wind tunnel. At the time, Navier-Stokes methods are being established (DLR Brunswick, University Stuttgart) which will increasingly replace the aforementioned methods.

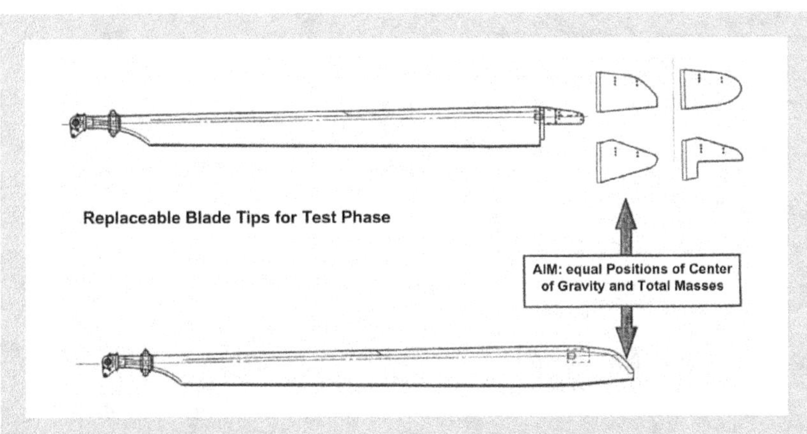

Blade tips for ART-A.

Blade Dynamics/Blade Loads

The complexity of the dynamics of helicopter rotors requires extensive investigations [10] prior to the introduction of a new rotor into quantitiy production. To control blade dynamics and hence also blade loads, research considers, since the beginnings, the phenomena:

- Placement of the blades natural modes relative to the rotor harmonics (Example: FEL-rotor) to minimize load excitations.
- Selection of flapping and in-plane motion stiffnesses to define in-plane damping requirements (control of vibrations as well as ground and air resonance) and agility.
- Placement of center of gravity, tensile-stress center and elastic blade axis to govern control forces and flutter stability.

The natural frequencies of the FEL-rotor (non-coupled computation) are presented in the figure below. By means of a dedicated arrangement of the material during construction of the "fiber-composite-Elastomer" rotor (FEL-rotor), the natural frequencies of the blade may be shifted to the desired frequency range. This is a typical example of "Aeroelastic Tailoring".

Methods for the computational determination of loads were investigated within European research programs. Partners were Eurocopter in France, Agusta in Italy, Westland in England, the DLR and other national research institutions as well as universities. The programs aimed at the experimental validation of the following methods:

DACRO:	Full-potential and Euler methods.
SCIA:	Panel method for the computation of aerodynamic interferences.
HELISHAPE:	Navier-Stokes methods for airfoil stall, Euler methods for the hovering helicopter, full-potential and boundary-element methods for the overall flight envelope [11].
EROS:	Euler method for the overall flight envelope.
HELIFUSE:	Navier-Stokes method for fuselages.

Within the current project HELIFLOW, load-specific, dynamic and flight-mechanical phenomena, arising from coupling and interference mechanisms, are thoroughly investigated. This shall most of all vali-

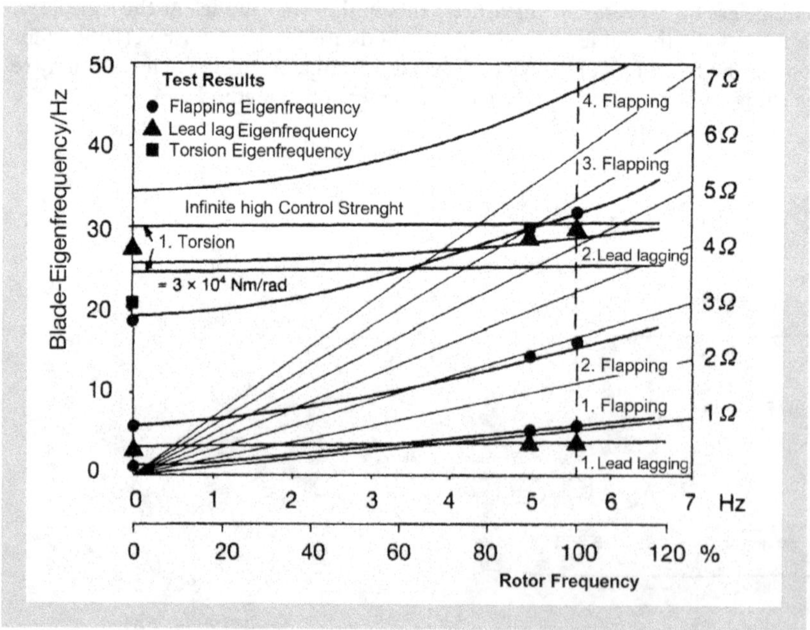

Eigen frequencies of the FEL rotor.

date the research tool "wind tunnel" with regard to this complex helicopter-related field.

Specific Helicopter-related Problems

Rotor-Fuselage-Stability

A typical and problematic helicopter phenomenon is the ground/air resonance. Here, the regressive blade in-plane-motion natural mode (excited by the rotation of the overall rotor-center-of-gravity against the sense of rotation) couples with a fuselage natural mode (on the ground mostly pitch, in the air roll). In order to avoid this as far as possible, one has initially always tried, by employing prediction models, to separate the corresponding natural frequencies. Considerable research effort was, in addition, invested to increase the in-plane motion damping by means of a suitable coupling of the in-plane motion and the torsional motion, as well as by viscous-elastic dampers ("Tiger") or Elastomer dampers (EC 135). The possibility of damping by recovering (reducing) the in-plane-motion velocity is being described below.

Separation of the Blade Flow/Aeroelasticity

The separation of the flow on the retreating blade in the case of high-speed forward flight essentially determines the limits of the flight regime of a helicopter. ECD has analyzed this phenomenon for various helicopter/rotor systems in flight tests and, together with the DLR, in wind-tunnel tests. An improvement of the prediction methods is presently being pursued within a European cooperation (GARTEUR). The aeroelastic blade characteristics, associated with flow separation, are primarily being investigated with CFD methods at the DLR in Göttingen.

Testing Helicopter Technology in Wind Tunnels

For some time now, the wind tunnel is utilized for the investigation of very different helicopter-related phenomena. Initially, MBB used, besides university wind tunnels, facilities of the automobile industry and the tunnel of the "Swiss Aircraft Company Emmen" ("Eidgenössische Flugzeugwerke Emmen"). Increasingly, the German-Dutch Wind Tunnel (DNW) is developing into a very effective tool. Here, ECD and DLR work closely together operating two test rigs, the DLR rig and the modular DLR-ECD wind-tunnel model MWM [12]. With both, dynamic and Mach-number-scaled model rotors of four-meter diameter were tested in closed or open test sections having cross sections of 8 m × 6 m and 9.5 m × 9.5 m (26 ft × 19.6 ft and 31 ft × 31 ft), respectively. The testing technology related to complex interference and coupling phenomena is, as already mentioned, presently validated within the European research program HELIFLOW.

Special Developments as Alternatives to the Conventional Helicopter

Besides the development of the BO 105 with a hingeless rotor system, MBB investigated special rotational-wing forms [1] such as:

- the one-blade helicopter,
- the tilt rotor for high speeds,
- the 31-meter (102-ft) low-pressure reaction rotor (tip-driven rotor),
- the transformable aircraft "Rotor-jet".

Except for the BO 105, none of these systems was so convincing that a quantity production would have been justified.

Active Rotor Control

Conventional rotor control can, over and above the trim and control tasks, not contribute further to the exploitation of the technical potential of helicopters. Here, additional degrees of freedom within the fixed and rotational areas of control must be superimposed by the pilot. In the case of the former (fixed areas), the hydraulic actuators below the swash plate may be utilized to move the swash plate vertically or tilt it at low frequencies according to the input of the pilot or the autopilot.

If one extends the actuator dynamics such that in the case of a four-bladed rotor the swash plate can be moved with the blade-number harmonic $f_n = \Omega$ (Ω = rotational frequency), one may realize on the rotating blade a higher harmonic control (HHC) with the harmonics 3-, 4-, and 5-Ω.

If one intervenes with the rotation-related control system, for instance, by means of actuators instead of the rotating control rods, or by flaps attached to the blades, one may prescribe to each blade an arbitrary control signal ("Individual Blade Control, IBC").

Higher Harmonic Control (HHC)
Together with the DLR, as well as Eurocopter France, ONERA, the US-Army and NASA, ECD carried out wind-tunnel tests in the DNW with the Mach number scaled 4-meter-diameter (3.3-ft) model rotor. The objective was the reduction of the vibration excitation and the blade-vortex-interference noise (BVI-noise) during descent (sinking). The blade pressure distribution (Δc_p) and the noise contour of the BO 105 model rotor in the presence of blade-vortex interference during descend (DNW measurements by DLR) are presented in the Color Plate XII.

It could be proven that the BVI-noise can be largely suppressed [13]. Also the rotor forces and moments, mainly responsible for cabin (frame) vibrations, could be markedly reduced. Both effect are caused by different interference phenomena. The simultaneous reduction of noise and vibrations by higher-harmonic control is, therefore, problematic. The reduction of one effect, keeping the consequences of the other effect constant, could, nonetheless, be shown.

Individual Blade Control (IBC)
The individual blade control has compared to the higher harmonic blade control a noticeably higher technical potential. Application possibilities are a simultaneous reduction in vibrations and noise, stall delay, rotor stabilization and automatic track correction. To demonstrate the concept ECD, DLR and ZFL (formerly "Henschel Aircraft Company") carried out BO 105 flight tests and – together with NASA – full-scale wind-tunnel tests in the Ames 40 ft × 80 ft wind tunnel [14]. ECD, DLR and ZFL continued these investigations within the framework of the Aeronautical Research Program of the Federal Government of Germany [15].

The test vehicle BO 105 S1 for the investigation of individual blade control is shown in Color Plate XII. Here, the objective of ZFL was the improvement of the actuator technique, of ECD and DLR to find suitable control algorithms and to test trailing-edge-flap control in lieu of blade-root control [16]. The flap acts basically as servo-flap. It is driven by a piezoelectric-stack actuator with a mechanical amplitude enlargement.

The piezoelectrically actuated servo-flap, developed by Daimler-Benz Research, is depicted in the colored Plate XII in its wind-tunnel configuration [17].

By imbedding the corresponding research activities within a common reference concept "Adaptive Rotor Systems, AROSYS", carried by ECD, Daimler-Benz and DLR, it is intended to reach the risk-prone objectives faster and more securely.

Vibration Isolation and Noise Reduction

If there is no possibility to fight the vibration excitation in the domain of rotor-blade aerodynamics – as in the case of the aforementioned higher harmonic or individual blade control – the vibrations, originating there, must be damped by active or passive systems at the blade root, the gearbox suspension or within the airframe (cabin) structure.

Passive/Semi-Active Systems
ECD has already early, as have other manufacturers, investigated pendulums at the blade root, which reduce in the case of four-blade rotors the 3 Ω-excitation moments, which are transmitted to the frame as 4 Ω-vibrations. As a next step, the replacement of the gearbox suspension – normally a rigid connection with the fuselage roof – by spring-mass systems was tested, which counteract the 4-Ω exciter loads of the four-blade rotor (System ARIS [18]).

Even further away from the location of the excitation, but closer to the seats, act cabin dampers (erasers), also spring-mass systems, which attack the 4 Ω structural vibrations directly. Due to the tuning of the aforementioned systems to the frequency 3 Ω or 4 Ω, respectively, changes in the rotor rpm, as they are common with modern rotors, act as detuning mechanisms. In case of the cabin damper, ECD investigated therefore, together with Eurocopter France, a system with a variable moment of inertia to adjust the natural frequency of the system. Here, a controllable motor moves masses along a spring lever. Such a system can optionally be utilized on the EC 135.

An alternative system with a constant moment of inertia – developed in cooperation by DB and ECD – varies the spring stiffness via an active coating of the spring lever with piezoelectric ceramics.

Active Systems
A further reaching adjustment with respect to the variable rotor rpm can be accomplished by active gearbox suspension systems. Thus ECD investigated ear-

ly, together with the DLR, on a test rig a hydraulic digitally controlled system (ASIS [19]), which moves the gear (box) in such a way that the resulting inertia (mass) forces largely compensate the excitation forces coming from the rotor.

ECD together with Liebherr (LLI) investigated within the Aeronautical Research Program of the Federal Government active gear (box) struts, which were, contrary to the aforementioned ones, driven by a piezoceramic actuator. The amplitude was hydraulically enlarged.

Noise Reduction

Noise is an essential obstacle to the utilization of helicopters close to inhabited areas. Most of all the impulse noise, generated by blade-vortex interference (BVI) during landing approach ("carpet beater" noise), is very disturbing. To reduce BVI noise the following possibilities exist:

a) the avoidance of BVI flight paths,
b) a vortex-affecting blade-tip design, and
c) a higher-harmonic blade control (HHC or IBC, respectively).

ECD has tested during various flight-test campaigns approach procedures, which allow, most of all by a steeper approach, to bypass the critical BVI flight conditions [20]. The DLR and universities have in the past conducted several theoretical/experimental investigations to explore the interference phenomenon and the effect of specific blade-geometry designs. ECD currently investigates on the BK 117 different blade tips with respect to their suitability to influence the blade-tip vortex; this is being carried out within the framework of the "Advanced Rotor Technology, ART" program.

An effective influence on rotor noise can be achieved by a reduction of the compressibility effects in the blade-tip area by means of a reduction of the circumferential velocity. For that reason, ECD tested a coupling of the flight altitude with the rotor rpm on the EC 135, which was then permanently installed on the production vehicles. The system causes an automatic rpm reduction, and thus a two-dBA noise reduction, when approaching the ground.

In order to advance and validate the different existing noise-prediction models, the three European helicopter manufacturers, together with national research institutions and universities, carried out several EU research projects. In the projects HELINOISE and HELISHAPE, Mach-number scaled 4-meter-diameter (3.3-ft) rotors of different geometries were investigated in the German-Dutch Wind Tunnel (open test section) with the aid of traversable microphone arrays. The sound-source strength in form of the blade surface pressure was determined by more than 100 electronic absolute-pressure gages on one instrumented blade. The test setup for HELISHAPE is shown in Color Plate XII.

Within the present project HELIFLOW, especially tail-rotor noise is being investigated. Since the latter is essentially determined by the interference of the tail rotor with the wake of the main rotor, both rotors were equipped with pressure sensors.

The results of the aforementioned projects contribute decisively to a better noise prediction and to the analysis of the acoustic consequences of the noise-reduction measures initially mentioned.

The total of the research activities introduced shows a manifold picture of the utilization of innovative technologies meant to steadily improve the system "Helicopter".

Remarks

Concerning this chapter, some figures are presented in color:

Color Plate XI

- EC-135 helicopter with hingeless and bearingless rotor system
- Indian light helicopter ALH with integrated dynamic system

Color Plate XII

- External noise measurements on a tilt-rotor-aircraft model
- External noise measurements on a modular helicopter model
- BO-105 helicopter with individual blade control
- Individual blade control by means of trailing-edge flaps

References

1 K. von Gersdorff, K. Knobling, Hubschrauber und Tragschrauber, Bernard & Graefe Verlag, Bonn, 3. erweiterte Auflage, 1999

2 D. Braun, H. Frommlet, A. Schwarz, FEL-New Main Rotor System, 12th European Rotorcraft Forum, Garmisch-Partenkirchen, 1986

3 H. Huber, Will Rotor Hubs loose their Bearings? A Survey of Bearingless Main Rotor Development, 18th European Rotorcraft Forum, Avignon, 1992

4 V. Klöppel, H. Huber, B. Enenkl, Development of Bearingless Tail Rotors, Conference on Helicopter Yaw Control Concepts, The Royal Aeronautical Society, London, 1990

5 B. Enenkl, V. Klöppel, Design Verification and Flight Testing of a Bearingless Soft Inplane Tail Rotor, 11th European Rotorcraft Forum

6 S. Attlfellner, Eurocopter EC135 Qualification for the Market, 22nd European Rotorcraft Forum, Brighton, 1996

7 R. Oster, Computertomografie als zerstörungsfreie Prüfmethode an faserverstärkten Hubschrauberrotorblättern in Entwicklung, Serie und Betrieb, DGLR-Tagung, Faserverbundwerkstoffe und -bauweisen in der Luft-und Raumfahrt, 1996, München

8 G. Polz, D. Schimke, New Aerodynamic Rotorblade Design at MBB, 13th European Rotorcraft Forum, Arles, 1987

9 P. Stahl, G. Niesl, Messungen der Lärmabstrahlung und der Rotor/Rumpf-Interferenz von Rotoren für den hohen Geschwindigkeitsbereich im Deutsch-Niederländischen Windkanal, DGLR-Jahrestagung 1994, Erlangen

10 H. Strehlow, Applied Helicopter Aeroelastics-Modelling and Testing, 22nd European Rotorcraft Forum, Brighton, 1996

11 V. Klöppel et al., The BRITE/EURAM Programme HELISHAPE – a Successful Step to Better Helicopter Aerodynamic and Aeroacoustic Description, 23rd European Rotorcraft Forum, Dresden, 1997

12 M. Stephan, V. Klöppel, H.-J. Langer, A New Wind Tunnel Rig for Helicopter Testing, 14th European Rotorcraft Forum, Milano, 1988

13 W. Splettstößer et al., A Higher Harmonic Control Test in the DNW to Reduce Impulsive BVI Noise, Journal of the American Helicopter Society, Vol 39, No 4, 1994

14 S.M. Swanson, S.J. Jacklin, A. Blaas, G. Niesl, R. Kube, Acoustic Results from a Full-Scale Wind Tunnel Test Evaluating Individual Blade Control, 51st Forum of the American Helicopter Society, Fort Worth, 1995

15 D. Schimke, U.T.P. Arnold, R, Kube, Individual Blade Root Control Demonstration-Evaluation of Recent Flight Tests, 54th Forum of the American Helicopter Society, Washington, 1998

16 V. Klöppel, Adaptive Rotorsysteme, Adaptronic Congress, Berlin 1997

17 P. Jänker et al., Development of High-Performing Piezoelectric Actuators for Transport Systems, Actuator 98, 6th Int. Conf. On New Actuators, 1998

18 D. Braun, Ground and Flight Tests of a Passive Rotor Isolation System for Helicopter Vibration Reduction, Vertica, Vol. 8, No. 1, 1984

19 H. Strehlow, R. Mehlhose, M. Obermayer, Active Helicopter Rotor Isolation with Application of Multi-Variable Feedback Control, 3rd European Rotorcraft Forum, Aix-en-Provence, 1977

20 S.A. Yoshikami, C.R. Cox, FAA/HAI Helicopter Flight Operations–Noise Tests and Initial Results, 41st Forum of the American Helicopter Society, Fort Worth, 1985

Large-scale Facilities to Test Safety, Reliability and Environmental Compatibility

Safety, Reliability and Environmental Compatibility within the Global Air Transport System

Jürgen W. Bergmann

During the first half of the 20th century, German aeronautics succeeded in establishing concepts and setting up trend-setting large-scale test facilities. Especially in the years immediately prior to and during World War Two, politics and government support led to a facility technology that represents peak performances. Examples are, for instances, the former engine high-altitude test facility of the "Bavarian Motor Company" ("Bayerische Motoren Werke, BMW") in Munich-Milbertshofen, which is still in operation at the American Arnold Engineering Development Center (AEDC), located near the city of Tullahoma in the State of Tennessee [1], *and the large wind tunnel erected in the alpine Ötztal for energy-supply reasons in the immediate vicinity of a power station; Americans and the French fought bitterly after the war over this wind tunnel, which is still today in Modane (France), as wind tunnel S1, one of the wind tunnels used by the French "Office Nationale des Etudes et Recherches Aéronautiques, ONERA" within the Airbus program* [2]. *Both facilities were, like many other German test facilities, after the war disassembled and as reparations installed and operated by the Victorious Powers.*

The reestablishment of the test facilities after the war could only commence in the fifties with the new beginnings of aeronautics. The aeronautical industry had lost almost everything due to the effect of the war. On the one hand was most of the equipment destroyed or dismantled and the assets confiscated; on the other hand remained the debts, also devalued in account with the currency reform. Personnel were, in addition, on the initiative of the Victorious Powers (utilization of the German know-how and prohibition of employment in Germany), largely employed abroad. The aeronautical industry depended, therefore, more than ever on governmental support. Non-university research was – except for basic aerodynamic research – completely shattered [3]. *However, since the research establishments were able to sustain their civil-law status during the Third Reich, they were protected from the complete dissolution otherwise imposed by the Allies. Already in the mid-fifties did the research institutions again receive research contracts. The government tried from the very beginning to enforce its political principles in the reestablishment of aeronautics in research and industry by only providing conditional support. This had an especially lasting effect with regard to the reestablishment or establishment, respectively, of large-scale research facilities* [3].

The Origin of the IABG

In the autumn of 1957, the companies Bölkow-Developments, Ltd., Ernst Heinkel Aircraft Engineering, Ltd., Messerschmitt AG, and Siebel ATG (Bölkow-Entwicklungen KG, Ernst Heinkel Flugzeugbau GmbH, Messerschmitt AG, Siebelwerke ATG) discussed the foundation of a working association for the acquisition of larger development contracts. The "Federal Ministry of Defense, BMVg" pressed the negotiating partners to merge in a suitable way [3a] by delaying the issuance of the order for the, up to that time, largest development contract, viz., for the vertical take-off and landing fighter VJ 101. The merger to form the "Development Group South" ("Entwicklungsring Süd, EWR") occurred in 1958. Munich was selected as seat (inside the Deutsches Museum) to be able to also include the engine industry (MAN and BMW) as well as the electronic and equipment industry. Further aspects were the closeness to the University and the Institute of Technology (TH) Munich, to libraries and the German Patent Office. In the north of Germany, the "Development Group North, ERNO" was formed.

The VJ 101 development required to a large extent research and test equipment. *Ludwig Bölkow* had, on February 19, 1959, invited the companies of the EWR, Dornier Ltd., the two engine manufacturers MAN and BMW, and a representative of the Bavarian State Government to discuss ways of obtaining the required equipment. The "Southern German planning committee for development-related facilities for the aeronautical industry, PEL" was founded. It had initially the task to determine and specify the need for test equipment within the individual companies. Short-term inquiries were carried out in the following areas:

1. Computer center
2. Simulation facilities
3. Supersonic wind tunnel
4. Ground test rig for helicopter rotors
5. Reliability laboratory
6. Electronics and (measurement) instrument laboratory
7. Antennae test facility
8. Free-fight and supersonic-sled measurement system
9. Strength laboratory, Design and construction laboratory.

The engine industry did, during this phase, not express any needs. Also the electronics industry wanted to specify their requirements at a later date. On May 12, 1959, the PEL submitted the results of the inquiry to the BMVg. The cover letter expresses in form of a memorandum (see next page) impressively the basic idea of a future joint central-facility utilization by industry.

The last meeting of the PEL took place on July 10, 1959. Here, also the electronics industry and the equipment (supply) industry participated. The creation of a "legal person" as contractor for the conduct of the definitive technical planning was suggested, as was the form of a "Study Society" and as location Ottobrunn near Munich (following the "Aeronautical Research Establishment Munich, LFM", established during the war) [3]. On July 14, 1959 followed the foundation of the "Society of Flight Technology" ("Gesellschaft für Flugtechnik e.V., GfF") as a registered association. Founding members were:

- the Free-state of Bavaria, represented by the State Ministry of Economics and Transportation,
- the "Apparatus Construction Gauting, Ltd." ("Apperatebau Gauting GmbH"),
- the BMW Engine Engineering, Ltd.,
- the Bölkow Developments, KG,
- the Dornier Company, Ltd.,
- the Ernst Heinkel Aircraft Engineering, Ltd.,
- the MAN Turbo-motor, Ltd.,
- the Messerschmitt AG,
- the Siebel Company ATG.

The organizations:

- Daimler-Benz AG (today DaimlerCrysler),
- Siemens & Halske AG,
- Telefunken, Ltd.,
- Standard Electric Lorenz AG

had the possibility to join GfF within a month, which also happened.

The GfF, in the meantime extended by the companies of ERNO, submitted to the BMVg a planning application, unanimously agreed upon by the members of the GfF, for the first construction phase of the new facilities, which were considered particularly urgent. The application comprised the following facilities at an overall financial volume of about 17.8 Million German Marks:

- Building for the computer center (computer leasing envisaged)
- Rotor test rig south (2,000 hp)
- Solid rocket fuel test stands
- Strength laboratory (implementation 1. Step)
- Reliability laboratory
- Building for strength and reliability laboratories
- Analog computer for Focke-Wulf
- Water tank for "Aircraft Engineering Hamburg" ("Hamburger Flugzeugbau").

Southern German Planning Committee for development-related facilities for the aeronautical industry

Memorandum

To be presented to the Federal Minister of Defense

The development of modern air vehicles requires today, compared to the status during the last war, an effort in engineering hours, which has increased by orders of magnitude. Coupled with the increase in the demand for qualified personnel, also the requirements concerning the technical facilities needed for the performance of the developmental work have unimaginably increased. The establishment of such facilities has correspondingly become very expensive.

One should have therefore attached more and more importance to this general knowledge and accounted for it during the last decade. Especially in the USA, in England and in France, industrial research and development facilities were with the aid of government funds established and centralized at suitable locations and jointly provided to the aeronautical enterprises.

With the reconstruction of the development-performing aeronautical industry, a decision for the construction of government-owned development facilities and for a corresponding suitable concentration of the aeronautical companies was in Germany urgently needed since non of the enterprises were able to come up with the necessary investments.

For two years now, mainly Messrs. (Ministerialdirigent) Dr. Benecke and (Ministerialrat) Wahl, ministerial-department officials, have repeatedly approached the aeronautical industry in that sense, which is gratefully acknowledged, and have thereby also in Southern Germany initiated the concentration of the respective companies.

Around the turn of the years 1958/1959, the association of the Bölkow-Developments KG, the Ernst Heinkel Aircraft Engineering, Ltd. and the Messerschmitt AG within the "Development Group South" was realized. Here, the point in time had arrived to attack jointly the overall planning of the development facilities.

For this task, the "Southern German Planning Committee for development-related facilities for the aeronautical industry" was founded on February 19, 1959.

The committee presents subsequently a survey of such development-facilities, which are considered urgent and which are in their size, technical realization, performance, et cetera today already characterized. It is intended to establish a corporate body by the foundation of a non-profit institution.

The named members of the "Southern German Planning Committee for development-related facilities for the aeronautical industry" hereby permit themselves the request to be charged by the Federal Minister of Defense with the planning of the aeronautical development facilities in Ottobrunn.

Munich on the 12th of May 1959

Museum Island 1 (on the premises of the Development Group South)

Bavarian State Ministry of Economics and Transportation
Bölkow-Developments KG
Dornier Company Ltd.
Ernst Heinkel Aircraft Engineering Ltd.
Messerschmidt AG
BMW Engine Engineering Ltd.
MAN-Turbo Motors Ltd.

"Southern German Planning Committee
for development-related facilities
for the aeronautical industry"
signed: Dr.-Ing. Kopfermann

The engine industry handed in a jointly established rough plan later, which essentially comprised facilities to be erected on their own premises (MAN-Turbo and BMW) for

- the supply of air,
- energy generation, and the
- functional and endurance testing of components,

having a total financial volume of 31 million German Mark. These facilities were, however, not recognized as GfF facilities. They were later successively setup within research and development contracts or production orders, respectively.

On September 7, 1960, the BMVg agreed with the GfF planning in principle, provided a rough overall plan would still be submitted and the financial scale would be restricted to 16.4 million German Marks. At the same time BMVg recognized the necessity of a second construction stage.

Industry tried to secure from the BMVg, beyond the planning of the test facilities, also the guarantee for the expert personnel required for their operation. However, the BMVg had already charged on August 5, 1960 the "Industry Administration Organization, Ltd." ("Industrieverwaltungsgesellschaft mbH, IVG") in Bad Godesberg with the setup and operation of the Development Center Ottobrunn. For that purpose, the "Credit Bank for Reconstruction" ("Kreditanstalt für Wiederaufbau") provided a credit over 16.4 million German Mark. The part of the operational cost not covered by yields was taken over by the BMVg wholesale.

This development, alarming for the industry, led to discussions of an operational committee, founded by the GfF, with BMVg and IVG. Results of these discussions were the formation of a Preparatory Committee (VA) for the planned "Organization for the Operation of Facilities" ("Anlagenbetriebsgesellschaft, ABG", and a participation of the GfF in the ABG with a blocking minority of 26 per cent, for which the IVG should, however, provide the entire working capital. This solution was for the BMVg only justified under the following considerations (minutes of March 7, 1961): "The tasks of the development center, to be established in Ottobrunn and whose setup and operation is to be headed by the ABG, are predominantly of a technical nature and shall therefore be solved in closest cooperation with industry". The IVG was, furthermore, authorized by the BMVg to transfer the rights and duties, acquired by the contract of August 5, to a subsidiary organization.

The industry also gained the agreement "...to participate decisively in the work of the supervisory board by appointing a corresponding number of members..." and "...to propose a managing or the managing directors..."; furthermore, the addition of a technical-scientific advisory board to the supervisory board and the management "...which was to act on an advisory basis without organization-related legal functions".

On order of the BMVg, the IVG established the ABG on September 12, 1961, utilizing the umbrella of the "Industrial Property Administration Organization, Ltd." ("Verwaltungsgesellschaft für Industriegrundstücke mbH, VfI"), and at the same time changed the status of the company, the seat and the business object of the organization. The VfI was, at the time, a no longer operational but still registered subsidiary of the IVG located in Munich. The name ABG was later extended by the letter "I" of the IVG and obtained its final name "Organization for the Operation of Industrial Facilities, Ltd." ("Industrieanlagenbetriebsgesellschaft mbH, IABG". Contrary to earlier intentions of the industry, was the IABG to have not only the control over its own budget but also its own personnel; industrial delegations were to be only temporary.

Of the facilities originally envisaged by the PEL only

- the computer center,
- the simulation facilities,
- the strength and reliability laboratory,

as well as later

- the thermal-strength laboratory,
- the vibrational-strength laboratory, and
- the gyro test laboratory

were installed and subsequently operated at and by the IABG. The other test facilities were erected at the industries and independently operated by industry.

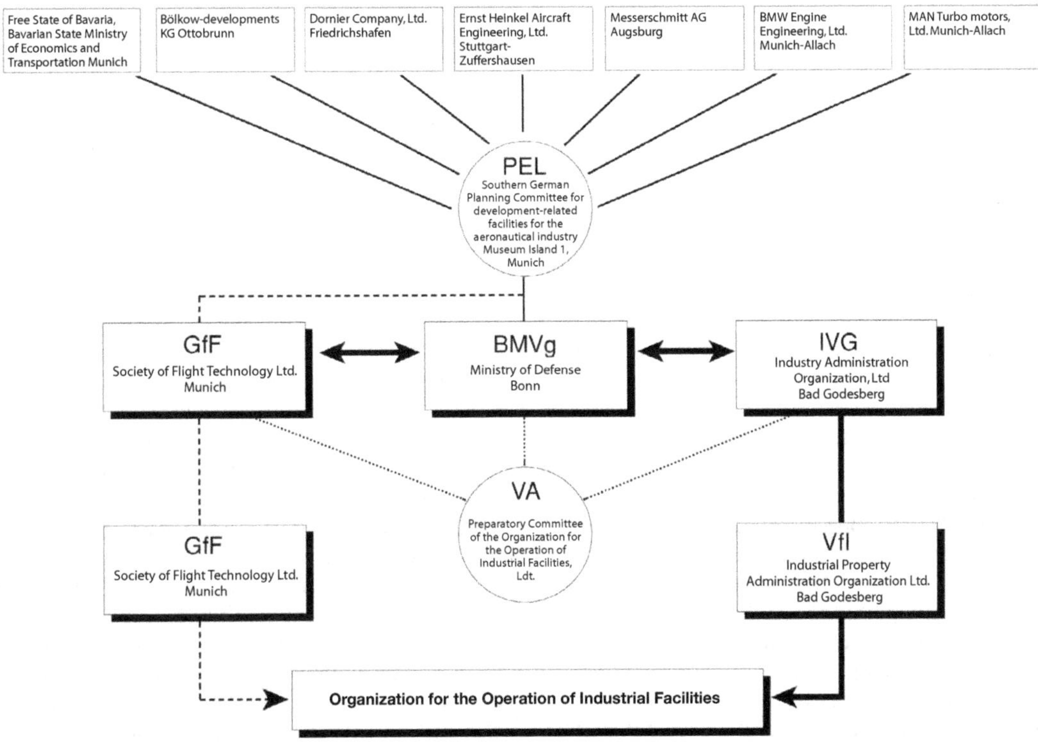

19.2.1959 The flight-technical industry in the Munich area establishes an honorary planning committee for the industry-demand on development-related facilities (PEL).

12.5.1959 The PEL presents to the Federal Ministry of Defense a first, comprehensively justified preplan for the common development-related facilities to be located in Ottobrunn.

14.7.1959 The group of the participating companies is being extended to include the Southern German area by founding the GfF e.V. During the summer of 1959, more companies, now from the entire Federal Republic, join.

7.2.1961 FOUNDATION of the IABG: Prominent representatives of the BMVg and the IVG with the member companies of the GfF decide at the KRUPP-House in Bonn the foundation of the "Organization for the Operation of Facilities, ABG" with seat in Munich.

17.2.1961 A preparatory committee (VA) under the participation of representatives of the BMVg, the IVG (chairmanship) and the GfF works out in 13 meetings, during the time February 17, 1961 and July 27, 1962, the legal and organizational forms.

1.3.1961 The VA asks the envisaged managing directors – Messrs. Obermüller (GFK) and Dr. Kopfermann (GFT) – to commence their managing activities.

1.6.1961 The "Organization for the Operation of Facilities" starts its work.

12.9.1961 On order of the BMVg, the IVG establishes the ABG utilizing the company share of the "Industrial Property Administration Organization, Ltd., Bad Godesberg, VfI".
At the same time it is decided to rename the organization into "Organization for the Operation of Industrial Facilities, IABG".

9.3.1962 The "Society of Flight Technology, Ltd, GfF" with seat in Munich is founded for the purpose of a participation in the IABG. Distribution of the company shares: GfF 26 per cent/IVG 74 per cent.

10.3.1962 Starting time of the Studies Teams: The "OR-group Ottobrunn", comprised of a Bölkow-EWR team with some Northrop people, commences its activities in the Building 7a of the IABG.

1.8.1962 Starting time of the Technical Teams: Commencement of operations in the strength laboratory by personnel delegated by the Dornier Company, Ltd., Friedrichshafen (2 Engineers, 3 Technicians).

1.9.1962 Take-over of the so-called construction office Dr. Kopfermann from the Bölkow-development KG, Ottobrunn, and of IVG personnel for the business administration.

25.7.1967 The general assembly decides to move the seat of the IABG from Munich to Ottobrunn.

Extended Tasks of the IABG

The technical requirements related to aeronautical equipment increased enormously with the growing air traffic demands during the second half of the 20th century. The drastic increase in the number of passengers, following the improvements in economic efficiency and comfort due to the introduction of turbojet engines, the growing globalization of the economy, and the increasing wealth of the industrialized nations caused safety, economic efficiency, reliability and finally environmental compatibility to become more and more important.

The demand of the air lines for economic aircraft efficiency to increase aircraft productivity, such as an increase in pay-load capacity and an extension of flight times compared to the times on the ground, required an extension in material utilization, and hence an increase in material stresses, as well as longer maintenance intervals and hence longer operational times for all systems. The larger economic efficiency, in turn, made the aircraft to a global transportation means of ever increasing importance.

This required a new dimension in safety and reliability. Operational (fatigue) strength, damage accumulation models, damage tolerances, and lifespan predictions based on these characteristics, became decisive design criteria for air vehicles. But the tremendous efforts to clear up damage and failure mechanisms did not make, up to today, a theoretical lifespan prediction possible.

The considerable widening of the world air traffic made, in addition, a reduction in pollution by exhaust emissions and noise necessary. The guarantee of the aerodynamic and economic performance data to the airlines as well as governmental conditions led to considerable quality demands on the manufacturer. The public supervisory authorities, the federal aviation authorities – most of all the American Federal Aviation Authority (FFA) – became pacemakers in air safety by establishing qualification standards. As a consequence, many test facilities are required today to qualify an aircraft. They serve, on the one hand, to clear up damage mechanisms and they allow, on the other hand, experimental performance predictions to guarantee, in the interest of the public, safety, reliability and environmental compatibility of the global air-transportation systems. Noise reduction, one component of environmental compatibility, is always coupled with performance improvements since noise is synonymous with the dissipation of performance in the aero-acoustic interaction with the environment. Here, essential improvements have been achieved, especially with respect to helicopters but also in the case of winged aircraft during the landing approach, as will subsequently be discussed.

Ground Test Facilities

Aerodynamic and Aero-acoustic Performance Measurements on Models

Hans Ulrich Meier

The successful post-war products were created within European cooperations. The technological precursor (pre-competition) programs for these products were, nevertheless, frequently still performed within national competitions. It was known to all participants in this process that a competitive technological lead also meant a division of the competence on a European level. Crystallization point for this European technology partnership became the large-scale test facilities, which could only have been competitively built within a European cooperation. For this, wind tunnels are but one example.

The German-Dutch (Netherlands) Wind Tunnel (DNW)

The DNW, the largest European wind tunnel, went into operation in 1980. It arose from the two planned national predecessor projects, the 8 m × 6 m (26 ft × 19.7 ft) LST (Low-speed Tunnel) of the NLR and the "Large Subsonic Tunnel" ("Großer Unterschallkanal, GUK") of today's DLR, whose combination was possible due to similar design criteria. Seat of the organization, conceived as a Dutch foundation, is Emmelord located within the Northeast Polder, The Netherlands.

DLR and NLR had similar concepts for their national wind-tunnel projects since, in both cases, wind tunnels of Göttingen design at ambient pressure conditions were envisaged. The aerodynamic design of the wind-tunnel circuit (see the figure below) was determined by high demands on the flow quality and good aero-acoustic test conditions. Characteristic features of such a design are separation-proof diffusers, low-loss deflections of the flow in the corners, optimized installations in the settling chamber (flow straighteners and screens) as well as an extremely high efficiency of the fan. In order to meet the national requirements concerning the operational spectrum, the DNW includes three closed, interchangeable test sections, viz.,

9.5 m × 9.5 m (31 ft × 31 ft): $U \approx 60$ m/s (197 ft/s),
8.0 m × 6.0 m (26 ft × 19.7 ft): $U \approx 110$ m/s (360 ft/s),
6.0 m × 6.0 m (19.7 ft × 19.7 ft): $U \approx 145$ m/s (476 ft/s),

and a free-jet (open-jet) configuration. Here, the free jet from the 8 m x 6 m (26 ft × 19.7 ft) nozzle is being captured by the re-entry scoop of the 9.5 m × 9.5 m (31 ft × 31 ft) test section. The free-jet test section is enclosed by a building, which exhibits good anechoic characteristics due the sound absorbing material of its walls. The schematic sketch below shows the open test section with model and model support in a three-dimensional view.

For force and moment measurements, the closed test section – at least in the case of large-scale facilities – prevail since wall interferences can reliably be corrected by means of suitable correction methods.

Another reason is due to the fact that for fluid-physical reasons the open-jet velocity is in the case of large-scale facilities limited since especially low frequency disturbances may generate dangerous resonance oscillation of the entire facility. For the large-scale open-jet test section of the DNW, *Hermann Holthusen* [4] has analyzed the existing flow and oscillation phenomena.

In the preceding years of operation, the DNW played an important role within the entire aeronautical technology development and has secured itself a permanent position in the respective test programs. This holds for the research programs financially supported by the "Federal Ministry of Education and Research" ("Bundesministerium für Bildung und Forschung, BMBF"), the European BRITE/EURAM projects but also for pure (industrial) project work. Here, most of all excellent cooperations between research and industry proved a success, which already commenced with the design of the facility. The basic idea of the partners was to develop a wind tunnel for investigations related to the take-off and landing phases of new aircraft. Here, two boundary conditions were to be met:

- Sufficient model size for the integration of new measurement and testing techniques,
- an open-jet test section for aero-acoustic investigations.

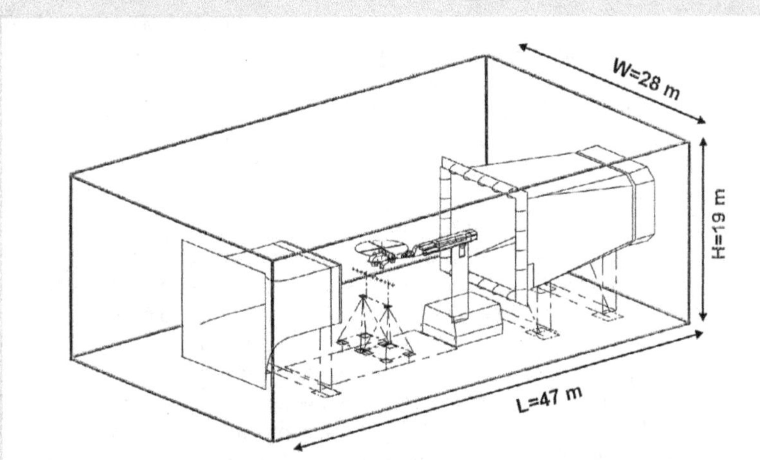

The open test section of the German-Dutch Wind Tunnel (DNW-LLF) with a model set-up for acoustic measurements. The dimensions given refer to the inside of the wind-tunnel building.

The aero-acoustic measurements represent today a large part of the overall occupancy of the DNW. Here, the DNW has developed into the most important instrument in reducing noise emissions, for instance, of helicopter rotors, through the investigation of fluid mechanical and acoustic phenomena. Examples are the programs HART (Higher Harmonic Control Aero-acoustic Rotor Test) and the BRITE/EURAM project HELINOISE. Basic knowledge concerning the complex aero-acoustic interactions could only have been obtained by the intensive cooperation of the national partners, i.e., the DLR institutes, universities and industry, with their international counterparts.

The results show that the active blade control changes the vortex structure within the wake of the blade. Especially a minute shift of the blade-tip vortices from the plane of rotation leads to an essential reduction in the noise and vibration problems. With the aid of the measured vortex structure, ideas concerning the corresponding computer models could be adapted to reality and a better prediction could be realized. In the process, the following measurement and test techniques were applied:

- Simultaneous unsteady pressure measurements with more than 100 pressure sensors on the rotor blade.
- On-line data reduction during tests with a data acquisition and processing system based on transputer technology, especially developed for that purpose.
- Measurement of the deformation of the rotor blades.
- Localization of the vortex core position relative to the currant blade location.
- Velocity distributions in vortical flows interacting with the rotor blade. These measurements were performed by means of the 3-D-Laser-Doppler-Anemometry and the PIV method (Particle Image Velocimetry).
- Measurement of the current noise-emission distributions with the aid of "inflow" microphones, which are traversed below the rotor model.

The complexity of the aerodynamic and aeroelastic relations described shows clearly the demands on model and measurement techniques. It is, therefore, not surprising that NASA, the US Army and ONERA joined these tests.

Noise-source determination on an original Airbus A 320 landing gear in the open-jet test section of the DNW.

The noise emission of large-scale wide-body aircraft confronts the aerodynamicist and the aeroacoustician with completely new problems if reliable predictions are to be made during the project phase. While ten years ago engine noise predominated, may today, during the take-off and landing phases, the noise generated by the flow about the high lift system and the landing gear be dominant. In the future, two new measurement techniques will be implemented in the DNW for aero-acoustic investigations: the Acoustical Mirror Telescope and the Microphone Array Technique.

The principle of the acoustic mirror telescope, which was developed by *Friedrich-Reinhard Grosche* (DLR Göttingen) already 30 years ago [5–7], is depicted in the sketch next page. The elliptical concave mirror forms the canopy of a rotational ellipsoid, which is crosshatched in the figure. The ellipsoid has two foci. The sound waves, which are emitted by a sound source located in the object focus (here the right focus), are focused by the concave mirror at the image focus (in front of the mirror) where the "mirror microphone" is located. The intensity of the sound waves focused by the mirror at the image focus, i.e., at the mirror microphone, is 20–40 dB higher than the intensity of

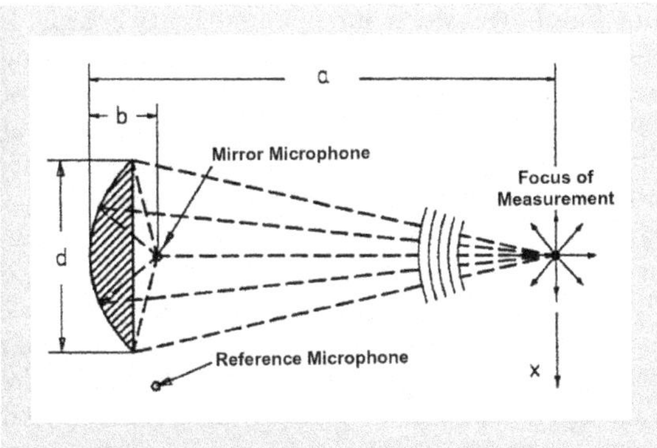

Acoustical (concave) mirror for locating noise sources (according to *F.R. Grosche* [7]).

the direct sound waves. The magnitude of the amplification factor as well as the spatial resolution of the acoustical mirror are determined by the diffraction of the sound waves at the edge of the mirror. The resolution is the better the higher the sound frequency and the ratio mirror diameter/sound-source distance is. The amplification factor G also increases with the sound frequency and the mirror diameter. First measurements with the aid of an acoustic mirror telescope made clear that this method was excellently suited to detect local sound sources.

A second important field of work of the DNW within the Airbus development is the high-lift optimization related to the take-off and landing phases. There, especially the engine integration plays an important role. Here, 50 per cent of the German funds were, for instance, spent for wind-tunnel measurements in the DNW within the framework of the Airbus A 330/A 340 programs. Within the scope of a typical development process, studies are being carried out on half-model, high-lift configurations with and without engine simulation in the wind tunnel LSWT (Low-speed Wind Tunnel, compare wind-tunnel Table No. 17, and the figure on page 592) during the preliminary technology and concept phase. Reynolds number effects were, up to now, determined in the pressurized wind tunnel F1 of the ONERA in Fauga. This task is envisaged to be taken over in the future by the Cryogenic Wind Tunnel Cologne [8] (KKK, compare wind tunnel Table No. 15). The above figure on page 593, shows the test section of the KKK with a six-component balance and a half-model ready for cryogenic testing. The balance, developed at the TH Darmstadt under the supervision of *Bernd Ewald* [9], already performed extremely well during the first tests. The com-

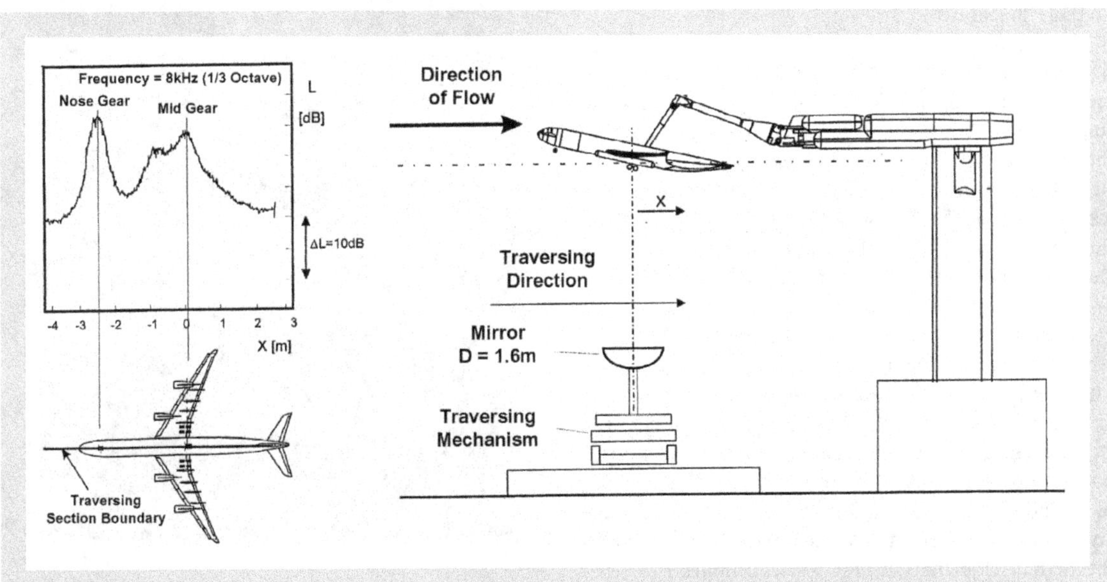

Sound level measurements of the noise generated by the flow past a 1:10.6-scale A 340 Airbus model in the DNW.

plete-model tests with engine simulations, applying Turbine Powered Simulators (TPS) fabricated by DA, are at low speeds exclusively being carried out in the DNW. The photograph of the Airbus A 340 model (Scale 1:10.6) in the 8 m × 6 m (26 ft × 19.7 ft) test section documents the extreme technical effort involved in six-component measurements with engine simulations and a rolling floor for ground-effect simulations (see photograph on page 593).

Within the DNW foundation, the Low-speed Wind Tunnel Brunswick (NWB) plays an important role especially in the investigation of new measurement techniques. For example, the Doppler Global Velocimetry (DGV) method was, in 1997, for the first time applied to engine air-intake measurements for a fighter aircraft project of DASA, supervised by the DLR Cologne [10]. As in the case of the Laser Doppler Anemometry (LDA) or the Particle Image Velocimetry (PIV), measures the Doppler Global Velocimetry (DGV) system the velocity of extreme small particles added to the flow. By means of an analysis of the scattered light, this technique provides all three components of the local velocity vector in the area, which is illuminated by a laser-light sheet. This measurement method constitutes an important supplement to the other two non-intrusive methods (LDA, PIV) and will, at a later date, also be available at the DNW (see figure on page 594, above).

The occupancy days for the wind-tunnel investigations of the Airbus A 320 and A 330/A 340 in the low- and high-speed domains are plotted on page 594 in form of a bar chart. Here, the large share of the low-speed investigations, which is essentially a result of the optimization of the high-lift aerodynamics, becomes again obvious. Within the framework of the joint project "Technological Fundamentals of the Design of Novel High-lift Configurations for Future Transport Aircraft", sponsored by the BMBF, flow investigations were performed in the DNW, on flap systems, within the flow field as well as on the surface. The objective was the improvement of computational methods for high-lift wings by means of detailed experimental information about, for instance, laminar-turbulent boundary layer transition, the location of boundary layer separation, the wake, the boundary layer confluence, and the vortex/boundary layer and the (exhaust) jet/flap interactions, compare *Hans-Peter Kreplin, Gebhard Höhler* [11]. As an important result of these investigations, the position of the stagnation line, of transition and of boundary layer separation and reattachment on the slat, on the main wing and on the flap could be determined with the aid of the surface hot-film technique for an entire angle of attack sweep (see figure on page 595). The design aerodynamicist has, for the first time, instantly flow parameters at his disposal, which can not yet be made available by today's computational methods, since these results, for corresponding model modifications, can be determined simultaneously with the force and pressure distribution measurements. It will be the challenge of the future to supply these experimental results, such as transition and separation data, to the computational methods in an "on-line" like procedure quasi simultaneously with the wind-tunnel measurements in order to improve the prediction capabilities with the experimental "input" so that part of the routine measurements can be omitted.

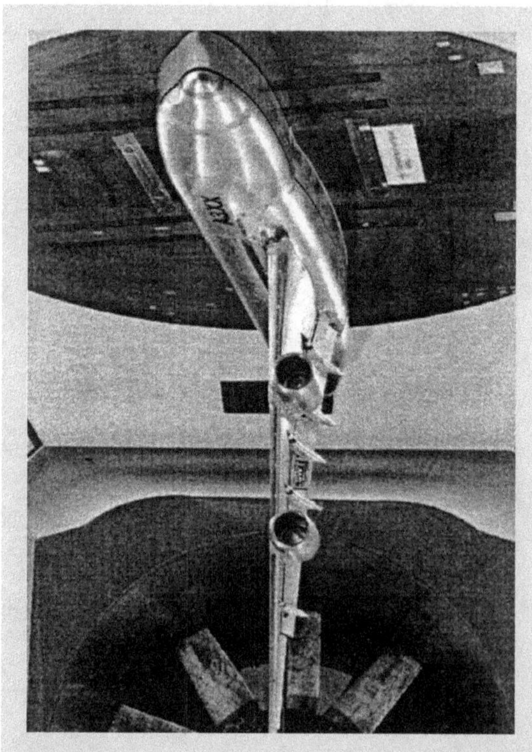

The Airbus A 340 half-model (Scale 1:22.5) in the DA-LSWT (Low-speed Wind Tunnel) in Bremen.

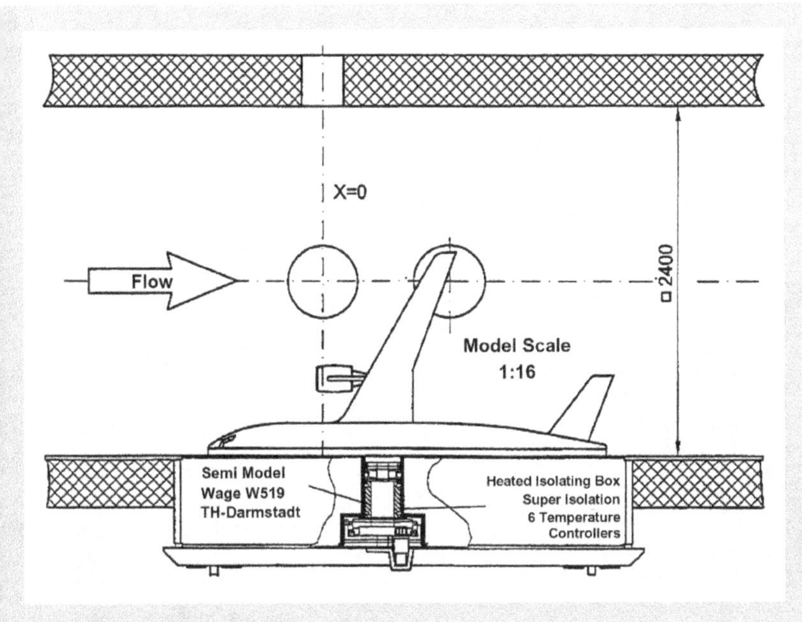

Half-model test setup of an Airbus model in the Cryogenic Wind Tunnel in Cologne (DNW-KKK).

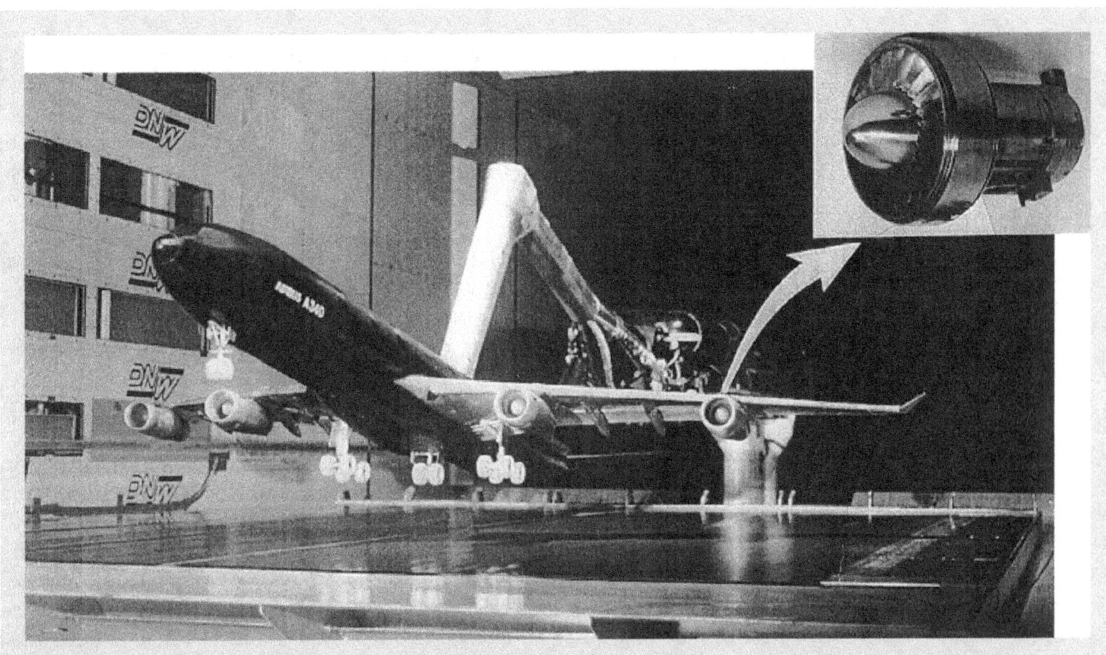

Investigation of an Airbus A 340 model (Scale 1:10.6) in the DNW-LLF Wind Tunnel with engine and ground-effect simulations.

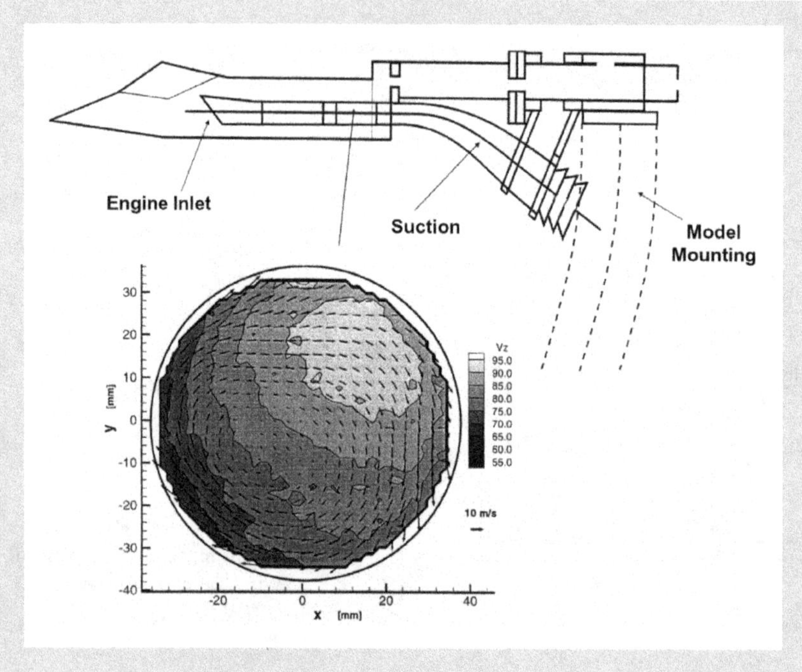

Axial velocities in the air intake of a fighter configuration. The measurements were carried out in the Low-speed Wind Tunnel in Brunswick (NWB) with a Doppler-Global-Velocimetry system developed by the DLR in Cologne [10].

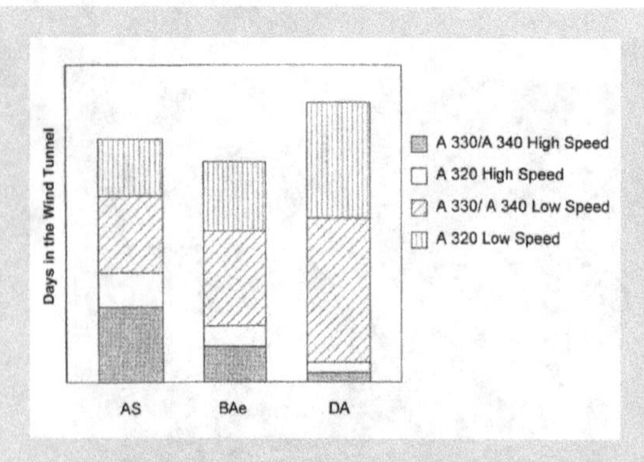

Occupancy time in days for the low- and high-speed wind-tunnel investigations of the Airbus Models A 320/A 330/ A 340 (AS: Aérosaptiale, BAe: British Aerospace, DA: Daimler-Benz Aerospace Airbus).

Especially for the certification of the next generation of large-scale (wide-body) aircraft, but also for the effective utilization of today's central airports like Frankfurt and Munich, is research concerning aircraft wakes of essential importance. An aircraft may, during its landing approach, encounter the wake of a preceding aircraft; here, the magnitude of "turbulence" and the vortex strength are basically dependent on the weight, and hence on the lift to drag ratio, of the preceding aircraft. Tasks of corresponding research activities are:

- The experimental determination of the vortical flow generated by the aircraft with respect to vortex strength and interactions (e.g., between the wing-tip and the flap-generated vortices) downstream of the aircraft wind-tunnel model.
- The development of computational models, which are able to determine the position of the vortices and the reduction in vortex strength by dissipation at a location several kilometers downstream of the aircraft.
- The design and realization of means to change the wake of large-scale aircraft so that landing frequencies may be increased. (The safety distance between a Boeing 747 and an Airbus A 320 amounts, for instance, at the present ICAO regulations, to about 5 nm).

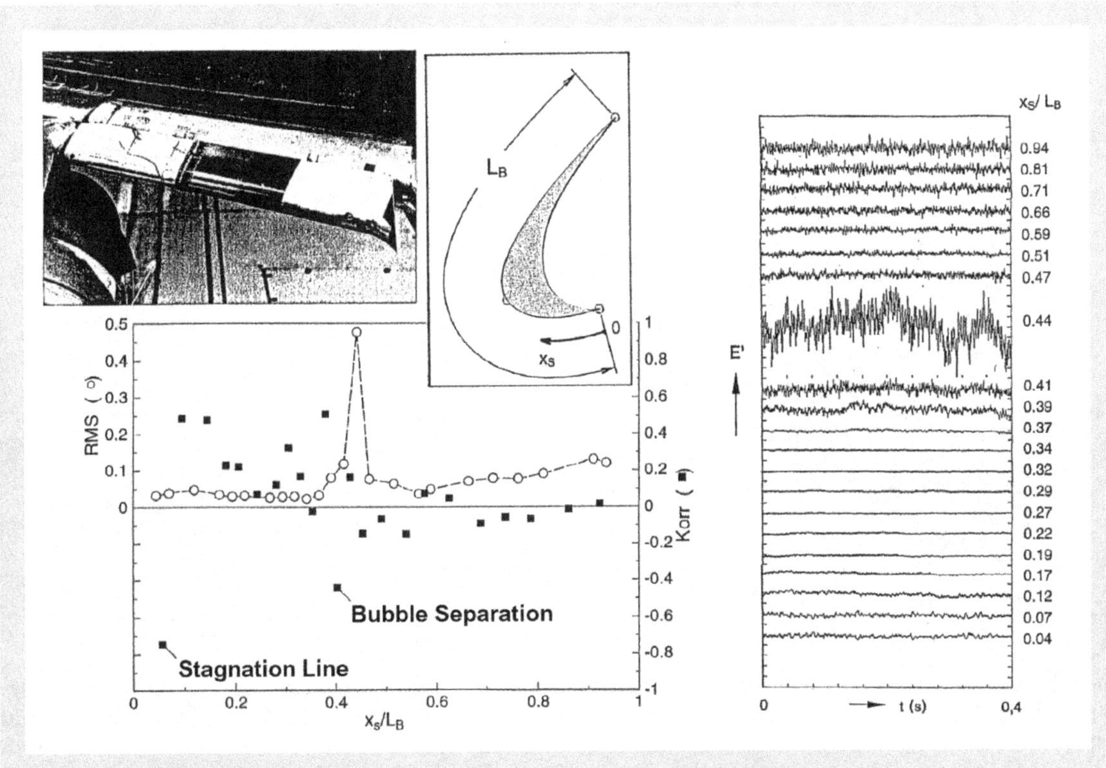

Hot-film measurements on a slat of an Airbus A 321 model in the DNW. Evaluated were the average fluctuation amplitudes (RMS) and the cross correlation coefficients (korr) of the signals of neighboring sensors at M = 0.2 and an angle of attack of 16°. Together with the time signals one is able to clearly identify a separation bubble at $x_s/L_B \approx 0.4$ (according to *H.P. Kreplin, H. Höhler* [11]).

In Europe, aircraft industry, airport organizations and the certification authorities work jointly on this problem. The DNW has an especially low-turbulence flow and a long test section, which allows detailed flow-field measurements downstream of a model. Besides conventional multi-hole probes, a non-intrusive measurement method was utilized for the first time with the application of the PIV-method, which was readied for the use in large-scale facilities like DNW by the DLR-Göttingen experts *Jürgen Kompenhans, Markus Raffel* and *Christian Willert* [12]. Principle results are depicted in the sketch on the next page showing streamlines of the vortical flow downstream of a 1:13.6-scale Airbus A 321 high-lift model computed from measured pressures. The merging of the vortices generated by the flaps and the tip vortices is clearly demonstrated by the PIV measurements (upper left part of the figure).

The European Transonic Wind Tunnel in Cologne (ETW)

The new European Transonic Wind Tunnel in Cologne closes a gap, which existed up to now in the wind-tunnel simulation of cruise conditions of large-scale transport aircraft, of military transports, but also in the case of fighter aircraft and spacecraft, at high speeds. Through the utilization of an extremely-low-temperature gaseous nitrogen flow and an additional pressurization of the wind-tunnel circuit flow physics are adjusted to the model size such that decisive aerodynamic model laws are exactly met; the results of the model measurements are directly transferable to

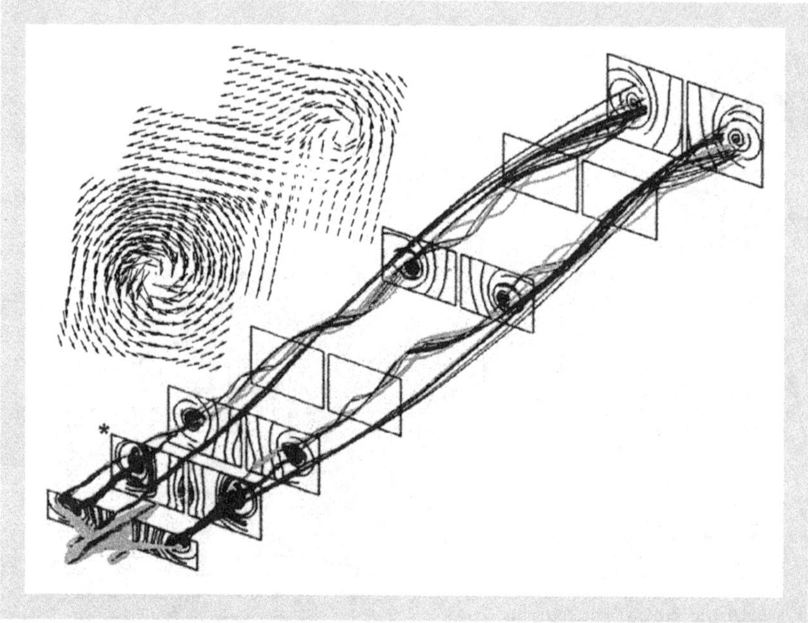

Wake measurements in the DNW-LLF wind tunnel downstream of a high-lift configuration of an Airbus A 321 model at a freestream velocity of 60 m/s (197 ft/s); the measurements covered a range downstream of the wing corresponding to seven spans. Compared are computed streamlines (from fife-hole probe measurements) with cross-flow velocity vectors in several sections (PIV measurements).

the full-scale aircraft (compare increase in drive performance of large-scale wind tunnels, page 402). The cryogenic technology employed to increase the Reynolds number has already been mentioned when considering the KKK (Cryogenic Wind Tunnel Cologne) and the KRG (Cryogenic Ludwieg Tube Göttingen). The effect of the temperature reduction on the power requirement is demonstrated in the right figure where, at a constant Mach number and constant tunnel pressure, the Reynolds number is increased by reducing the temperature. If the temperature is reduced from 310 K to 100 K by the injection of liquid nitrogen into the tunnel circuit, the Reynolds number increases by a factor of five. Here, the fan drive power is about 50 per cent lower, but the total energy consumption increases considerably due to the use of liquid nitrogen. Total energy balance sheets for the NASA-Langley Cryogenic High Reynolds Number Wind Tunnel were worked out by *E.C. Polhamus et al.* [14].

Proof of the indispensability of such a facility was already provided by measurements with the ETW reference model (DLR F4 wing). Here, it was shown that the extrapolation of the measured drag at two lift coefficients and free transition to higher Reynolds numbers is unsure and therefore not to be justified (see the left figure on the next page).

If one plots the operational range of the three cryogenic wind tunnels (ETW, KRG, KKK) available in Germany in the form of a Reynolds number-Mach number diagram, it becomes clear that these three facilities cover the entire flight domain of modern transport aircraft enabling a corresponding full-scale Reynolds number simulation. However, one has to see clearly that the cryogenic Ludwieg tube (KRG) is employed within project predevelopments and re-

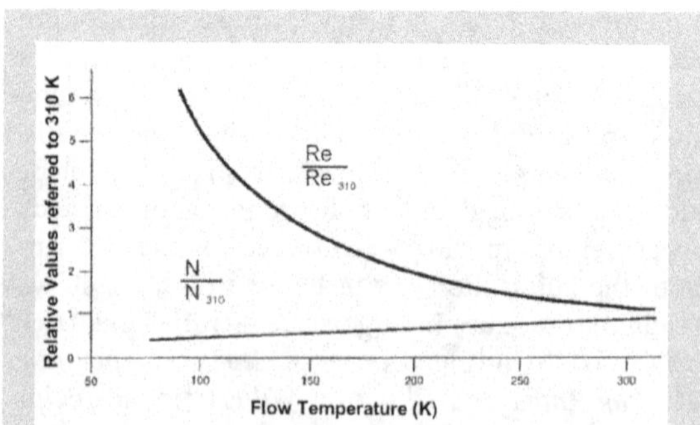

Effect of the wind-tunnel temperature, referred to the reference temperature of 310 K, on the Reynolds number (Re) and the drive power (N).

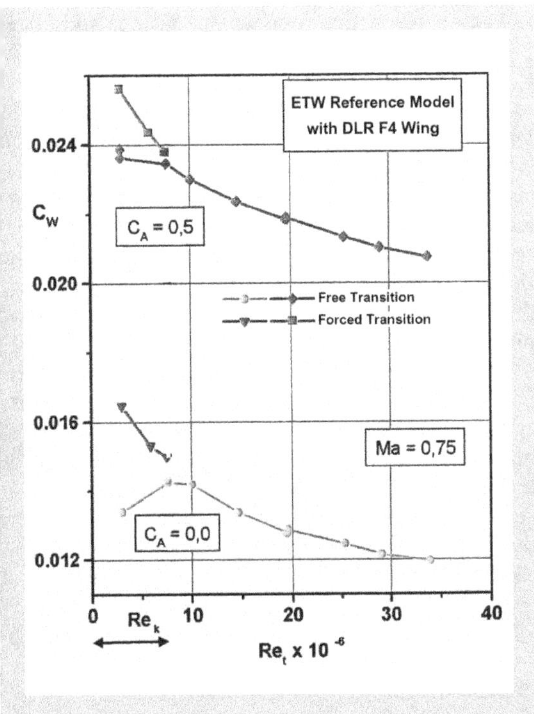

Variation of the drag coefficient (C_W) with Reynolds number (Re) for the ETW reference model at two lift coefficients C_A.

Re_t is the Reynolds number based on the wing reference chord "t",

Re_k is the Reynolds number range available in conventional European wind tunnels up to now.

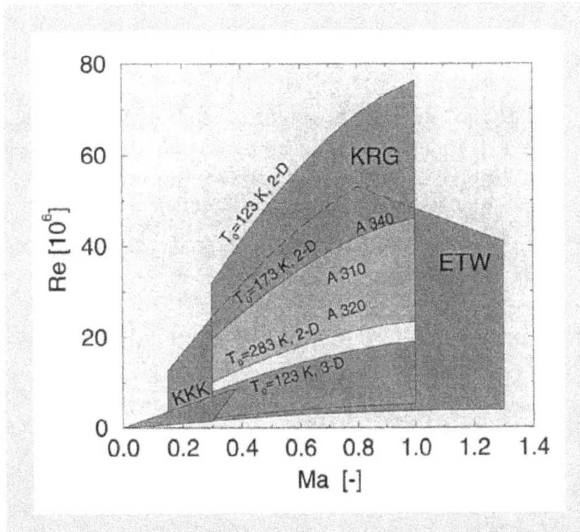

Reynolds number-Mach number diagram of the KRG related to airfoil measurements, the ETW for complete-model measurements and the KKK for half-model measurements with the Airbus derivatives A 320, A310 and A 340 at cruise conditions.

search concerned with airfoil design. Within the concept and definition phases of the predevelopment, the cryogenic wind tunnel in Cologne (KKK) is employed for Reynolds number investigations on half-models and the European transonic wind tunnel (ETW) for project investigations with complete models. These European wind tunnels, supervised multi-nationally, provide for the German and European aeronautical industry design and qualification tools, which are imperative for the guarantee of safety, reliability and environmental compatibility of the air transportation systems.

Test Facilities for Airframes and for Dynamic Systems of Rotary Wings

JÜRGEN W. BERGMANN

Static and Dynamic Tests on Airframes

Research in materials and structural-construction-elements mainly serves the selection of materials and of construction elements for airframe and engine. Many of the test facilities can be found at universities, research establishments and at material test centers. Of special importance are here the facilities available at research establishments and among them especially the ones related to fibre reinforced plastics (FRP). Trend setting are here the test facilities of the DLR, for instance, at the "Institute of Structural Mechanics" ("Institut für Strukturmechanik") in Brunswick. At the institute, research is being carried out with regard to the fabrication and testing of advanced fibre-reinforced plastics and construction elements. The following facilities are here in operation:

Facilities for the fabrication of material and construction elements:

- Universal winding facility for the fabrication of wound fibre-composite structures with up to ten axial alignments.
- Facility for the fabrication of fibre-reinforced structures from fabrics utilizing a new "Resin-Transfer-Molding, RTM" method (DLR-patent for highest fibre share at a uniform wetting by resin).
- Laboratory autoclave for thermal tempering of fibre-reinforced plastic elements.

Facilities to test fibre-composite materials and construction elements:

- Ultrasound-, X-ray- and sound-emission-facilities for non-intrusive material/construction-element tests.
- Servo-hydraulic test rig to investigate the fatigue strength.
- Buckling facility for testing cylindrical monocoque designs and panels.
- Facility for testing monocoque designs and plates under combined mechanical and thermal loads.

At the former test center Trauen near the town Celle DLR still today has the possibility to test the ignition respectively fire sensitivity or resistance of materials and construction elements based on fibre-reinforced resins (plastics). This was and is for a material selection, for instance, for the interior lining or cable sheaths of modern passenger aircraft, decisive since these materials are only allowed to gas-out non-toxic products in case of fire.

In order to determine the strength of structural parts of the airframe or the strength of the complete airframe, numerous tests and examinations are required. For large civil aircraft, but similarly for general aviation aircraft, the proof of the conductance of certain prescribed test procedures is demanded by the respective aeronautical authorities in the USA (Federal Aviation Authority, FAA) and in Europe (Joint

Project	Test structure	Type of test	Time period
VJ 101	Complete frame	Static	1963 - 64
DO 31	Complete frame	Static	1965 - 66
F 104	Wing and fuselage	Fatigue	1969 - 73
Tornado	Lag-bearing area, Wing and fuselage center part	Fatigue	1971 - 74
	Prototype	Fatigue	1974 - 76
	Series	Static/Fatigue	1978 - 82
	Wing (special version)	Static/Fatigue	1983 - 85
	Complete frame	Fatigue	1981 - 96
Fiat G-91	Wing and fuselage center part	Fatigue	1972 - 74
Alpha Jet	Wing and fuselage center part	Fatigue	1975 - 76
Airbus A300	Wing and fuselage center part	Fatigue	1973 - 80
A310	Wing and fuselage center part	Fatigue	1982 - 85
A320	Wing and fuselage center part	Fatigue	1986 - 88
A330/340	Wing and fuselage center part	Fatigue	1990 - 96
DO 228	Wing and fuselage center part	Fatigue	1983 - 84
Pilatus PC-7	Complete frame	Static	1983
Speed Canard	Wing	Static/Fatigue	1986 - 87
Seastar	Wing Fuselage Empennage	Static/Fatigue	1984 - 90
Grob G115	Wing	Static/Fatigue	1987 - 88
Helikopter BK117	Fuselage	Static	1986 - 88
G520 Egrett	Wing Fuselage Empennage	Static/Fatigue	1989 - 90
EFA	Complete frame	Fatigue	1990 - 96
Pilatus PC-12	Complete frame	Fatigue	1994 - 95
	Wing	Static	1994
Extra 400	Complete frame	Static/Fatigue	1998

Structural tests on complete aircraft frames and large structures, respectively, at the IABG since 1963.

Aviation Authority, JAA). In the case of military aircraft, the user or customer generally sets the test requirements to be followed. The test facilities, operated on behalf of aeronautics, essentially aim at meeting these design and construction requirements. The aviation authorities have, for this reason, lastingly affected the test facilities and the facility technology, especially the FAA during the time of burgeoning aviation after the last war and particularly since the introduction of the turbojet engine into civil aeronautics. The corresponding facilities in Germany are essentially installed at the IABG. Here, investigations are being carried out within the qualification (certification) procedure of aircraft with respect to life cycle, static and dynamic strength, and functionality; also thermo-mechanical tests and material investigations on components are performed (also see "Progress in Fatigue Strength after 1945").

Since 1963, IABG has advanced the testing technology and the facilities to test aircraft structures. This development can be divided into four consecutive periods. In Period I (1963–1970), structural tests related to the development of the vertical take-off aircraft VJ 101 and DO 31 were carried out. Here, within the bounds of airframe certification, investigations are being performed as to life cycle, static and dynamic strength, and functionality as well as thermo-mechanical tests and investigations on component materials.

The induction of the forces was effected by a mechanically controlled hydraulic system.

In Period II (1970–1973), conduction of the structural tests on the F 104 G; the force induction was already effected by means of electronically controlled servo-hydraulic actuation systems.

The Period III (1973–1992) introduced innovations in force induction by employing computer-controlled servo-hydraulic actuation systems (for structural tests on the G 91, "Alpha Jet", and "Tornado" as well as on the Airbus models A 300, A 320 and A 330/340).

In Period IV (since 1992), the force induction (for the structural tests on the "Eurofighter") was accomplished by digitally controlled servo-hydraulics with digital closed-loop control technologies.

The figure shows impressively the development in the experimental complexity of aircraft structural tests; the aircraft structural components investigated

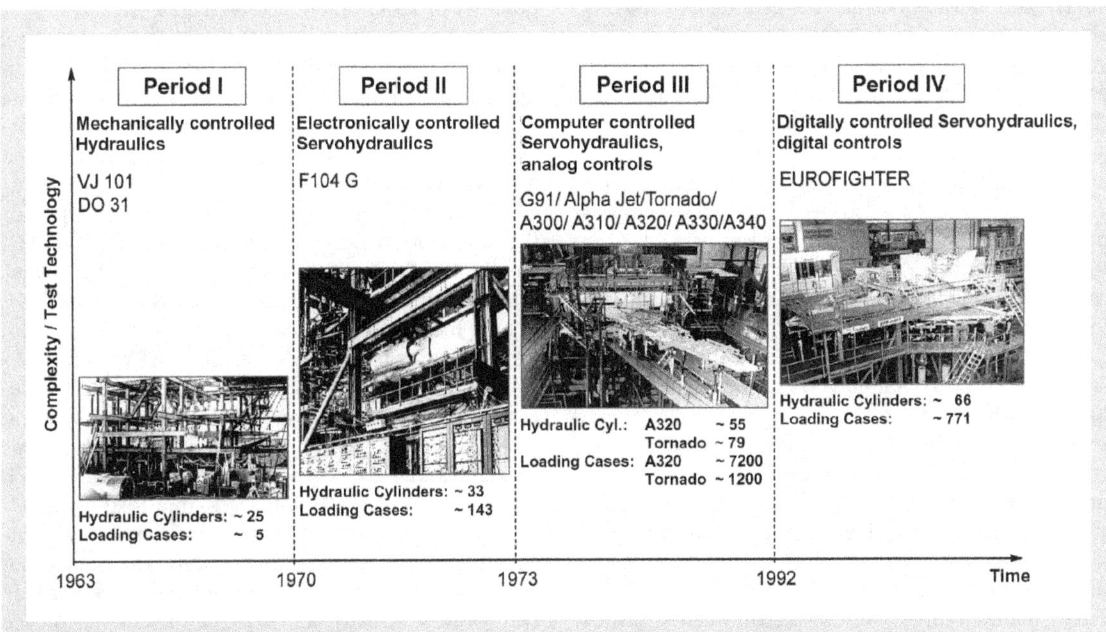

Development of the testing technology and the complexity of aircraft structural tests at the IABG in Ottobrunn during the time period 1963–1998.

between 1963 and 1998 in structural tests at the IABG are presented in the table (aircraft complete frames/ large-scale structural tests of the IABG). Load traverses and load chains in operational (fatigue) strength tests on the Airbus A 340 are shown in Color Plate XIII.

As an example for the investigation of fibre-composite structures, the static structural test on the helicopter frame of the BK 117, a joint project of MBB and Kawasaki, shall be cited. In the tests, flight and landing load cases were investigated. The objective of the tests was the verification of the chosen design of the frame, the manufacturing process and the static strength of the composite structure.

To simulate mechanical loads, temperature and humidity conditioning in static and fatigue tests on fibre-reinforced plastic structures, as well as on other structures and equipment components from various technology areas, the IABG has developed a test facility for fibre-reinforced structural elements and built it under contract of the Federal Ministry of Research and Technology in 1986.

In the eighties, carbon-fibre composite material was introduced for the construction of vertical stabilizers and elevators of all Airbus aircraft (starting with the A 310-300) in order to save weight hence increase range. The test facilities needed for the strength tests of the vertical stabilizer were installed at EADS-Airbus in Hamburg. Since that time, strength testing of the carbon-fibre vertical stabilizers of the Airbus-family, manufactured at the Plastic Component Center in Stade, is being carried out in these test facilities in Hamburg. The last, and simultaneously the largest product of this fabrication series is the vertical stabilizer of the A 330-200 shown within the test facility in Color Plate XIV.

Test Facilities for Dynamic Systems of Rotational-Wing Aircraft

A specialty in the area of aircraft are the helicopters since their rotor systems (dynamic systems) are simultaneously propulsion and lifting systems. Rotor systems therefore need investigations and testing of their individual blades as well as of the overall rotor arrangement. Rotor blades are today almost exclusively manufactured of fibre-reinforced plastics (FRP), glas or carbon fibre (GFRP or CFRP). In Germany, the development of main rotors of glass-fibre reinforced plastics was started by *Ludwig Bölkow* already in the sixties. By utilizing these materials, it was possible to omit, at the rotor head, the discrete flapping and lag hinges necessary up to that time. Needed during the development of rotor blades are benders for the dynamic loading of the blade juncture as well as ultrasound and X-ray facilities up to computer tomography to verify the blade build-up and the quality of the laminate.

In the course of their development, rotor blades took on more and more functions, which had to be integrated into the blade structure. As a consequence, the structural design became more and more elaborate and required, in turn, more and more accurate test methods. While for testing the initial FRP rotor blades standard non-intrusive test methods, such as ultrasound and conventional X-ray methods, would be sufficient, new display-capable test methods, such as the X-ray-computer-tomography (CT), are needed for the new monolithic FRP laminates. This method is, with regard to optimizing manufacturing, today indispensable in the development of FRP rotor blades due to the complexity and increased wall thickness of the rotor-blade constructions. The method can, at the same time, be utilized for quality control during quantity production and for damage analysis.

In order to investigate the overall dynamic system, rotor test rigs are required as test facilities; they are in operation at all developers of rotor systems. With the aid of these test facilities, propulsive and lifting characteristics as well as vibrational and control-related characteristics of rotor systems are being investigated. Eurocopter Germany has three rotor test rigs. For the cooperation project between the Technical University Brunswick, DLR Brunswick, and the "Henschel Flugzeugwerke, HFW" in Kassel for investigating the higher harmonic control of dynamic systems, a special rotor test rig was erected at HFW in Kassel.

In order to be able to conduct aerodynamic and aeroelastic investigations on rotary-wing rotor systems, DLR Brunswick designed and built rotor test rigs (ROTOS and ROTEST) [15] for the "German-Dutch Wind Tunnel, DNW", which are, due to their excellent quality, also being utilized together with American and French research institutions. The pioneering results achieved in applying higher-harmonic blade control were already addressed in the chapter on the DNW.

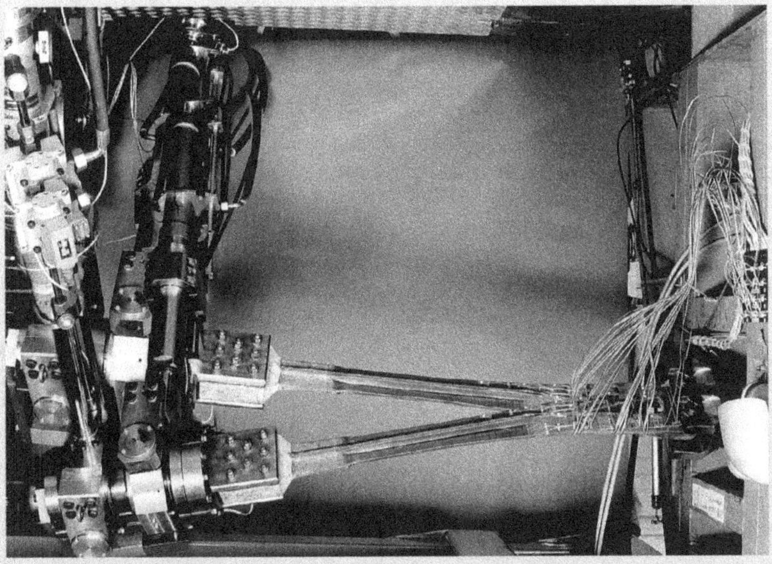

"Flapping and lag hinges" in the form of a flexible GFRP beam shown in the respective extreme test positions (EC 135 rotor).

Engine Test Stands

Jürgen W. Bergmann

Today, the engine industry has many test facilities for complete engines, which may serve for acceptance tests as well as developmental work. The test facilities for large scale research installed at the research establishment DLR serve applied research. Concerning universities, there exists a test rig for complete jet engines each at the University of the Armed Forces in Hamburg (HSBwH) and in Munich (HSBwM). A test rig for jet engines/turboshaft engines exists each at the Institute of Technology (TH) Aachen (thrust, t = 70 kN (15,700 lbs)) and at the Technological High-school at Aachen (HTL) (t = 50 kN (111,200 lbs), power = 1 MW (1,340 hp)). A test facility for small jet engines is located at the Institute of Technology of Berlin (t = 4.8 kN (1,080 lbs)). The only high-altitude test facility – jointly financed by the Federal Ministry for Research and Technology (BMFT), the Federal State of Baden-Württemberg, and the MTU Company – was erected in Stuttgart at the Institute of Technology (TH). A survey of the engine test facilities is given in the table "Engine Test Facilities for Research and Development".

The test facilities for engine components operating in Germany are also installed at the engine industries, at the establishments for large scale research and at universities. The utilization of these facilities is generally not limited to turbojet engines but includes all turbo-machinery. They consist essentially of test facilities for the investigation of (blade) cascade elements, i.e., so-called cascade wind tunnels. The DLR operates four of these cascade wind tunnels. Two are located at the Research Center Göttingen: a wind tunnel for two-dimensional (plane) cascades (EGG) and a wind tunnel for rotational cascades (RGG). The other two are installed at the DLR Research Center Cologne-Porz: a transonic cascade wind tunnel (TGK) and a supersonic cascade wind tunnel (SGK). The Institute of Technology (TH) Aachen also operates two turbine-cascade test facilities: a transonic one (G1) and a hot-gas test rig (G2). Two more are being operated at the Institute of Technology of Brunswick: a small cascade tunnel (KGK) and a large one (GGK). Also the Institute of Technology of Hanover owns two cascade wind tunnels for the investigation of two-dimensional cascades, one for blow-down and one for suck-down operation. The Universities of the Armed Forces also operate each a cascade wind tunnel: the HSBwH a tunnel for axial (blade) cascades, the HS-BwM a high-speed cascade wind tunnel.

In addition, a low-speed airfoil tunnel with unsteady freestream flow is being operated at the Institute of Technology (TH) Darmstadt (THD). The industry (MTU), finally, owns a cascade test rig for the investigation of cooled blades. An overall survey of the cascade test installations is provided in the Table "Cascade Test Facilities".

Especially important for engine research are the test facilities of the DLR "Institute of Propulsion Technology" ("Institut für Antriebstechnik") in Cologne-Porz. Most impressive of the facilities is the "Multistage Two-shaft Compressor Test Facility, M2VP" for compressor research with a performance of 2 MW × 5 MW (2 hp × 6,700 hp). Coupled to the facility is a suction (extraction) fan for mass-flow rates of 185 kg/s (408 lbs/s). The test facility serves the investigation of two-shaft multi-stage compressors as well as counter-rotating (ducted) fans of turbojet engines and fan-compressor combinations. It was sponsored by the BMFT and is in operation since 1987. In this facility, various components of the "Eurofighter" engine EJ 200 were investigated. Also for the MTU-design of a counter-rotating ducted prop-fan ("Counter-rotating Integrated Shrouded Prop-fan, CRISP") investigations were carried out here.

There exist, furthermore, in Cologne-Porz test facilities for combustion-chamber research with the high-pressure combustion chamber 3 (HBK 3) especially designed for combustion-chamber investigations of aircraft engines at realistic cruise conditions. Of particular interest is here the all-around access, which allows the application of the various Laser-optical methods. The test facility, also utilized for nozzle investigation, allows the study of two-phase flows by optical means.

Concerning propulsion technology, the DLR maintains at her "Institute of Chemical Propulsion and Processing Technology" ("Institut für Chemische Antriebe und Verfahrenstechnik") in Lampoldshausen

Engine test facilities for research and development (from the "Aeronautical Technology Handbook" ("Luftfahrttechnisches Handbuch, LTH") Engine Technology).

Operator/Location	Designation		Performance	Range of operation
BMW-RR, Oberursel (KHD)	Jet engine test rig	A1	Thrust = 75 kN, Mass flow rate = 120 kg/s	Development and series acceptance
BMW-RR, Oberursel (KHD)	Jet engine test rig	A2	Thrust = 75 kN, Mass flow rate = 120 kg/s	Development and series acceptance
BMW-RR, Oberursel (KHD)	Turboshaft engine test rig	B1	Performance = 3,500 kW	Development and series acceptance
BMW-RR, Oberursel (KHD)	Turboshaft engine test rig	B2	Performance = 3,500 kW	Development and series acceptance
BMW-RR, Oberursel (KHD)	Turboshaft engine test rig	6	Performance = 700 kW, M. fl. r. =10 kg/s	Development and acceptance
BMW-RR, Oberursel (KHD)	Thrust engine test rig	8	Thrust = 2 kN, Mass flow rate = 10 kg/s	Development and acceptance
BMW-RR, Oberursel (KHD)	Turboshaft engine test rig	9A	Performance = 132 kW, M. fl. r. = 5 kg/s	Development and acceptance
BMW-RR, Oberursel (KHD)	Turboshaft engine test rig	9 B	Performance = 132 kW, M. fl. r. = 5 kg/s	Development and acceptance
DLR, Braunschweig	Jet engine ground test rig	SSB	Thrust = 40 kN	Research, Perform.-/functional tests
FH, Aachen	Flight engine test stand		Thrust = 50 kN, Mass flow rate = 50 kg/s	Teaching, applied research
HSBwH, Hamburg	Gas turbine test rig		Performance = 1,700 Kw, M. fl. r. = 8 kg/s	Research and development
HSBwM, Neubiberg	Jet and turbo shaft test rig		Thrust = 30 kN, Mass flow rate = 60 kg/s	Teaching and research
MTU-München, Manching	Free/rotatable test rig		Thrust = 80 kN	Development-related test rig
MTU-München, München	Development-related test rig	I	Thrust = 80 kN, Mass flow rate = 100 kg/s	Development-related test rig
MTU-München, München	Development-related test rig	II	Thrust = 80 kN, Mass flow rate = 100 kg/s	Development-related test rig
MTU-München, München	Develop.-related small-engine test rig	3	Performance = 660 kW	Develop.-related performance test rig
MTU-München, München	Small development-related test rig	4	Performance = 400 kW	Development-/series-related test rig
MTU-München, München	Small development-related test rig	6	Performance = 660 kW	Develop.-related performance test rig
MTU-München, München	Small development-related test rig	9	Gas generator for mass-flow rate = 5 kg/s	Develop.-related performance test rig
RWTH, Aachen	Jet engine test rig		Thrust = 70 kN	Teaching and research
TU-Berlin, Berlin	Small test rig Turbomeca Marboré	VI	Thrust = 4.8 kN, Mass flow rate = 9.6 kg/s	Research, student exercises
Universität Stuttgart (TH)	High-altitude test rig		Engine gas-mass-flow rate = 70 kg/s	Research a. teaching, Contract res.

Cascade test facilities for research and development (from the "Aeronautical Technology Handbook" ("Luftfahrttechnisches Handbuch, LTH") Engine Technology).

Operator/Location	Designation		Performance	Range of operation
DLR, Göttingen	Two-dimensional cas. w. tun.	EGG	Mach Numb. (Ma) < 1; Reynolds Numb. (Re) <10^6	Research on turbine cascades
DLR, Göttingen	Rotational cascade wind tun.	RGG	Ma < 1; Re < 1,2 x 10^7	Research on turbine cascades
DLR, Köln-Porz	Transonic cascade wind tunnel	TGK	0,3 < Ma < 1,4; 0,15 x 10^6 < Re < 1,5 x 10^6	Research on straight cascades
DLR, Köln-Porz	Supersonic cascade wind tun.	SGK	1,3 < Ma < 2,4; 5 x 10^5 < Re < 3 x 10^6	Research on straight cascades
HSBwH, Hamburg	Cascade wind tunnel		0,1 < Ma < 1;	Res. on axial compressor/turbine cascades
HSBwM, Neubiberg	High-speed cascade wind tun.	HGK	0,2 < Ma < 1,05; 106 < Re < 1,5 x 107	Res. on two-dimen. compressor/turb. casc.
MTU-München, München	Cascade test rig f. cooled blades		Ma depend. on cascade; 1,5 x 10^5 < Re < 7 x 10^5	Research on concepts of blade cooling
RWTH, Aachen	Turbine cascade test rig	G1	0,2 < Ma < 0,5; 10^6 < Re < 3 x 10^6	Research on turbine blade airfoils
RWTH, Aachen	Turbine cascade test rig	G2	0,2 < Ma < 0,4; 5 x 10^5 < Re < 3 x 10^6; Hot gas	High-temperature-low-load-cycle fatigue
TH, Darmstadt	Low-speed tunnel		40 m/s Re 10^6	Res. concerning unsteady cascade flow
TU, Braunschweig	Large cascade tunnel	GGK	40 - 60 m/s; 2 x 10^5 < Re < 8 x 10^5	Research on compressor/turbine cascades
TU, Braunschweig	Small cascade tunnel	KGK	40 - 80 m/s; 2 x 10^5 < Re < 4 x 10^5	Research on compressor/turbine cascades
Universität Hannover	Blow-down cascade wind tunnel		0,2 < Ma < 0,7;	Research on two-dimensional cascades
Universität Hannover	Suck-down cascade wind tunnel		Re < 10^6	Res. on acceleration/deceleration cascades

near Heilbronn a number of test facilities for missile-related ramjet engines with excellent measurement-technological equipment.

Within the Hypersonic Technology Program of the Federal Ministry for Research and Technology, some very powerful test facilities for research in the field of ramjet engines were installed at industry. Especially the combustion chamber test installation, designed, built and operated by MBB, comprises a particularly powerful air heater facility for investigations up to flight Mach numbers of M = 7. A corresponding nozzle test facility for the investigations of supersonic/hypersonic variable-geometry nozzles was built and is being operated at the MTU Company.

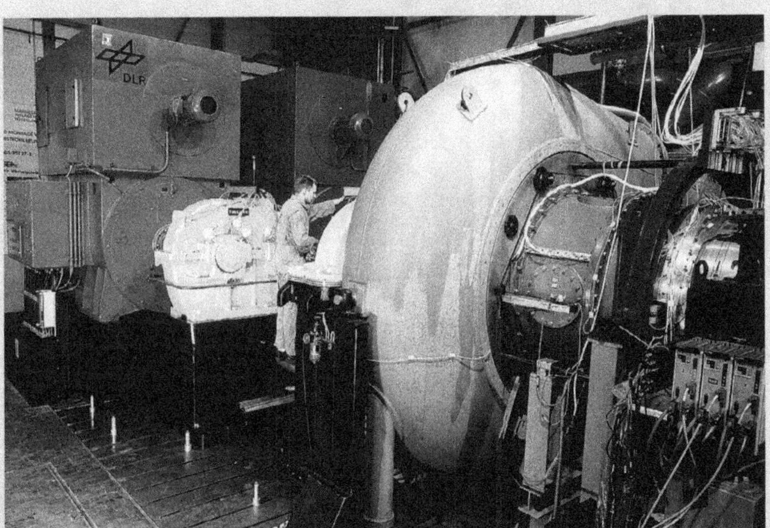

Multi-stage two-shaft compressor test installation (MZVP) of the DLR in Cologne-Porz.

Test setup for ramjet and combination propulsion engines at the test stand M 11.4 for air-breathing propulsion of the DFVLR (now DLR) in Hardthausen-Lampoldshausen (from the DFVLR-News No. 49, Nov. 1986).

A hydrogen-ramjet engine, designed for a flight speed corresponding to M = 7 and envisaged for the first stage of the "Sänger" project, was tested by MBB in 1989 on a test rig in Ottobrunn-South.

Test Facilities for Aircraft Equipment

Jürgen W. Bergmann

The test facilities for aircraft equipment are probably the most differentiated and the most extensive. They exist for all drives (actuators) of control surfaces and flaps related control, lift increase and trimming, for hydraulic systems, electrical systems as well as for flight-guidance and navigation systems; they exist, furthermore, for the other cockpit equipment, fuel systems, communication systems, air conditioning, and, last not least, the landing gear. In a wider sense, also the electronic systems, which the ground stations provide for air-traffic surveillance and control as well as for aircraft navigation, are part of this equipment.

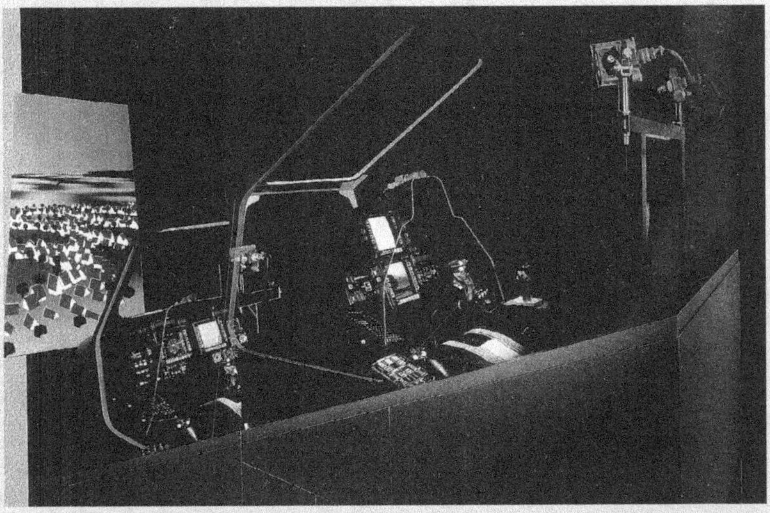

Avionics test facility of the ESG for the integration of helicopter equipment.

One of the most interesting test facilities in the area of aircraft equipment is the drum test facility for aircraft landing gears at the IABG. With this facility investigations in addition to pure rolling of landing gears, the overrun of obstacles at take-off and landing speeds, as well as hard touchdowns of landing gears on the runway, for research and qualification purposes (see Color Plate XIV) are possible.

Many equipment systems run through the entire aircraft frame since they have to reach, everywhere on the periphery, for instance control surfaces or flaps, respectively their actuators, as well as entrance or exit openings (hydraulic or electric systems, air conditioning systems). For each of these systems, complete engineering mock-ups (iron birds) are being erected, which have to contain all subsystems arranged as required on the aircraft. Control surface and flap systems need a coordination of their drive systems (shafts, which run along the entire rear wing longeron connecting all individual drives). It holds for all facilities that they serve research as well as development and qualification and acceptance testing, respectively. They are, therefore, as a rule, being established at, and operated by the industry responsible. They are located at the equipment companies' (e.g., concerning hydraulics, air conditioning and flap drives at Liebherr-Aero-Technology in Lindenberg in the Allgäu Mountains). The test facilities for the cockpits (equipment) or other electronic equipment are to be found at the respective equipment companies (e.g., gyro systems: LITEF, Ltd. in Freiburg, Nord-Micro Elektronik Feinmechanik GmbH in Frankfurt; Navigation instruments: VDO Luftfahrtgeräte Werk Adolf Schindling GmbH; Communication systems: Becker Flugfunkwerk GmbH in Rastatt near Baden-Baden.

The equipment industry had, as all of the aeronautical industry, during reconstruction the handicap of a strong fragmentation. However, no merging could be acomplished comparable to the situation of the airframe industry. Also with the aid of research sponsorships there was no comparable political success in promoting a similar consolidation. An enterprise, comparable to the IABG in the area of aircraft airframe investigations, is in the area of electronic aircraft equipment the "Elektroniksystem- und Logistik GmbH, ESG" in Munich. In 1967, the predecessor company was founded as "Elektronik-System-Gesellschaft mbH" with the objective to take over the system supervision for the planning and development of electronic systems for modern weapons systems. The ESG merged in 1970 with the "Elektronik-Gesellschaft mbH, FEG" founded already in 1963 by the electronics companies participating in the German "Starfighter" program. The final merger of the companies occurred in 1992. The activities were, at the same time, divided into business fields, each acting today largely in-

Helicopter-cockpit design rig of the ESG.

dependently. Today, the ESG plans, develops and integrates mainly complex electronic-, guidance- and information systems. The company furthermore develops and operates logistics systems with the focus in the area of system-engineering and software development. Special test facilities of the ESG are in that regard an avionics-integration rig and several avionics test facilities.

Remarks

Concerning this chapter, some figures are presented in color:

Color Plate XIII

- Structural tests on an Airbus wing at the IABG.
- Structural tests on the plastic frame of the helicopter BK 117.
- Fatigue tests of the frame of the "Eurofighter" at the IABG.

Color Plate XIV

- Static and dynamic tests on the CFRP vertical stabilizer of the Airbus.
- Drum test stand of the IABG for the test of aircraft land ing gears.
- Test rig of Liebherr-Aerospace for the Airbus wing-flap adjustment.

Development and Flight Simulators

Ulrich Fligge

Simulation facilities (simulators) are test facilities for the support of development work and training. Here, development simulators support the complete development of aircraft and subsystems such as, for instance, engines. Therefore aircraft and subsystems as well as construction assemblies or components, including their equipment, can be investigated and tested on the ground during the development phase in order to save cost. The simulation structure is generally arranged according to the system components, becoming during the course of the development successively more realistic. The characteristics of individual newly formed components can thus directly be determined and, by a constant connection to large model units, their effect on the overall system be tested. Real and simulated (i.e., in their consequences electronically reproduced) components are connected to each other according to the status of the respective development. Besides the obvious cost savings together with a decisive reduction in the development time, simulation facilities may considerably contribute to a reduction in the development risks and to an increase in certainty during the development stages. Many of the new technologies (flight control system, unstable flight, thrust-vector control, et cetera) cannot be flight-tested without prior simulation. Flight simulators shall, for the personnel to be trained, provide training as realistic as possible (dangerous flight, take-off or landing situations and maneuvers for flight or air traffic services personnel) without a loss of flight vehicles or human lives at cost as low as possible. Aeronautical simulators belong, in the widest sense, to the oldest aeronautical equipment since the risk of flying proper was always judged higher than the one of moving on the ground.

Modern simulation facilities for aircraft and airfield simulations are build up of many integrated individual components (external view simulation, noise

A modern training simulator for the Airbus A 330/A 340 is installed at the Center of Flight Simulation in Berlin. It is also being used for research purposes.

and movement simulations, et cetera). The sight system provides here the most important human sense with information. The development of digital sight systems – made possible by the high-speed computer technology with high storage capacities – leads to pictures comparable to video photographs of the real world. Here, day times and seasons as well as random weather situations or other events, such as natural catastrophes and war influence, can arbitrarily be simulated. Sight systems require, therefore, the highest amounts of data within the simulation technology. The sense of hearing has, besides the sense of sight, the second highest information density. Flight noise (boundary-layer noise of fuselage and wings, engine and environmental noise, noise from actuators and valves, et cetera) and cockpit noise (warning signals, switches, communication systems, et cetera) are, therefore, simulated. The movement system imitates the accelerations occurring at the simulated maneuvers in accordance with the other simulation systems in order to avoid the so-called "Simulator Thickness" (indisposition up to nausea caused by the time difference between sight and aircraft reaction). Of special importance is the simulation of the flight behavior, respectively the flight characteristics, of the aircraft and the engine performance (power) simulation. Here, the control computers may lead to considerable differences between the perception and the reality. Responsible are here, on the one hand, the strength limits of

The flight simulator, installed at the "Institute of Aeronautics and Astronautics" ("Institut für Luft- und Raumfahrt") of the Technical University Berlin, is especially equipped for the transport aircraft models Airbus A 330-200 and A 340-200/300. It can be rearranged within 30 min to accommodate other engine (aircraft) models.

the aircraft and, on the other hand, the performance boundaries of the engines and, in addition, possible limits concerning comfort-detrimental maneuvers known to the control computer as limits of the controllability. Concerning control, avionics simulation plays a particularly important role. The overall control of the simulation and the networking with other simulators, respectively, is finally of great importance. At the center of flight simulation in Berlin [16], a training simulator, identical to a Lufthansa A 330/A 340 simulator, was extended for research purposes by a parallel computer. In this configuration, the research simulator, which can also incorporate a ground station of the air traffic control and other type-related simulators, is able to realistically simulate operations within the cockpit interacting with air traffic control and other aircraft.

Forecasts point to an enormous future air traffic growth. Already today, the course has to be set to be able to command the future traffic scenarios. One focal point is, for instance, the rationalization of the data exchange between cockpit and ground stations. Where today an extensive voice communication prevails, an automatic data exchange via data link will exist in the future, i.e., a transmission of information in the form of pictures and text. Here, changes to the navigation monitor are required to come to solutions, which do not need additional monitors inside the cockpit. The said research simulator can also be utilized as avionics and software test installation and for the solution of flight-medical tasks, which require an optimization of the man-machine interface thus contributing to a reduction in the workload of the pilot.

In-Flight Simulation in Germany

Precursors of In-Flight Simulation

PETER HAMEL

During the 7th scientific meeting of the German Academy of Aeronautical Research on September 20, 1940, the earliest recognition of the need for stability augmentation was given by *Ernst Heinkel* [17]: "*The transition to control configured stability, which will be created by an automatic control system, will, therefore, become an unavoidable necessity in near future*", while introducing two lectures on the subject presented by Heinrich Helmbold [18]: "*With increasing flight speed the introduction of artificial stability appears to be unavoidable,* and Eduard Fischel [19, 20]: "*Future aircraft with high flight speed can no more be designed fully natural stable. Therefore the use of automatic control will more and more become a necessity*". *Heinkel* briefly remarked that longitudinal stability of very fast airplanes will diminish, and that the transition to *artificial stability* produced by an automatic pilot will become an inevitable necessity in the near future. Contrary to the natural stability (eigenstability) of a flight vehicles the authors defined artificial stability as the total stability of a flight vehicle in combination with an automatic flight control system. In the early 1940's increased piston engine performance resulted in increased propeller slipstream interference effects yielding poorer natural longitudinal flight stability. Similar negative effects on longitudinal stability were due to newly developed wing profiles for improved laminar boundary layer flow.

At the same time with increasing flight speeds compressibility effects on wing and tail components had an detrimental influence on the flight stability, and at the same scientific meeting of the German Academy of Aeronautical Research the well-known aerodynamicist *Albert Betz* notified possible configurational changes required for increased *Mach*-numbers such as canard and variable sweep wing configurations.

One of the lectures on this subject was presented by *E. Fischel* [20] who discussed for the first time control-theory approaches for artificial stability to improve existing eigenstability or to generate missing stability "*by taking hold of automatic control systems with the objective to increase eigenstability or to create missing stability*". The analytical approach was part of *Fischel's* doctoral thesis. With this theory the fundamentals became available to design variable stability flight control systems.

A further problem in the area of control and stabilization was associated with ever increasing control surface forces and hinge moments of very large flight vehicles. Whereas the aircraft weight increase from

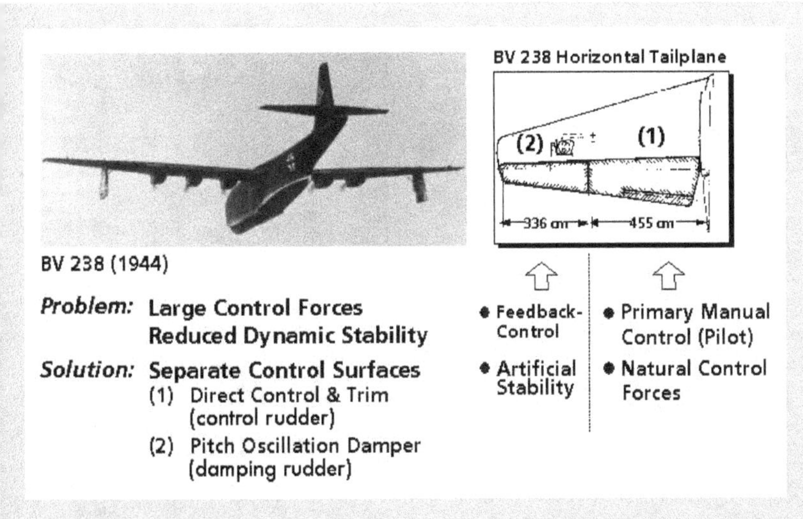

Separate elevator control surfaces in combination with an electromechanical pitch damper resulted in acceptable pilot control forces and improved longitudinal dynamics (BV 238).

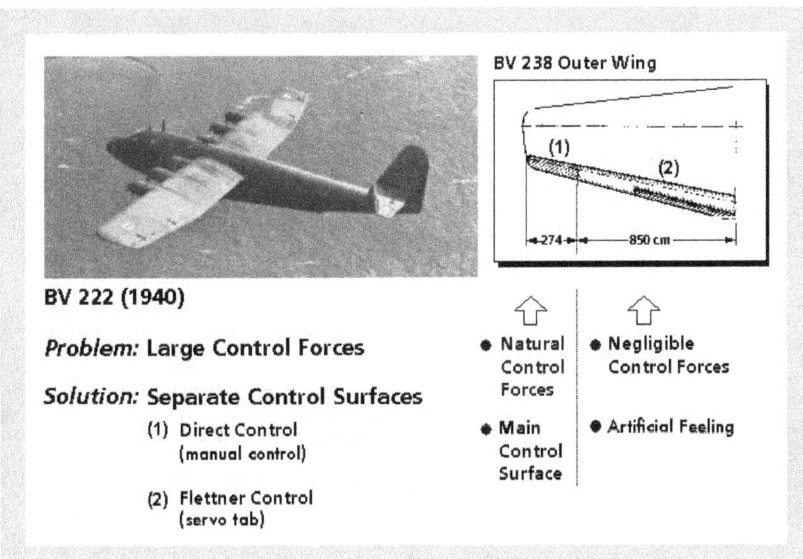

Separate aileron control surfaces in combination with a Flettner-tab resulted in satisfactory lateral flying qualities and acceptable manual control forces (BV 222/238).

a FW 90 to the giant flying boat BV-238 of Blohm & Voss was in the order of 30 the rise in control hinge moments was of the order of 117. Blohm & Voss chief design engineer *Richard Vogt* [21] mastered the colossal aerodynamic control forces of the BV 222 (1940) and BV 238 (1944) by splitting the elevator into a (1) manually controlled element with limited natural control forces generated by a linked tab (Flettner-tab), in which the pilot effort is shared between the tab and the elevator control surface element and (2) a feedback-controlled surface element for artificially improving flight stability. A slightly different approach was provided for the lateral axis: the aileron control surface was separated into (1) a smaller outer element which is pilot-controlled via natural control column forces and (2) a larger inner aileron surface with neg-

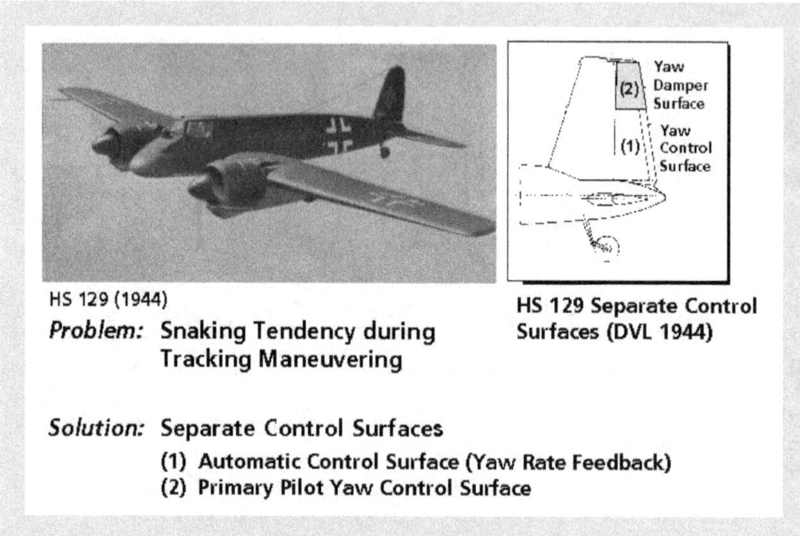

Separate rudder control surfaces in combination with an electromechanical yaw damper resulted in improved directional flying qualities for target tracking.

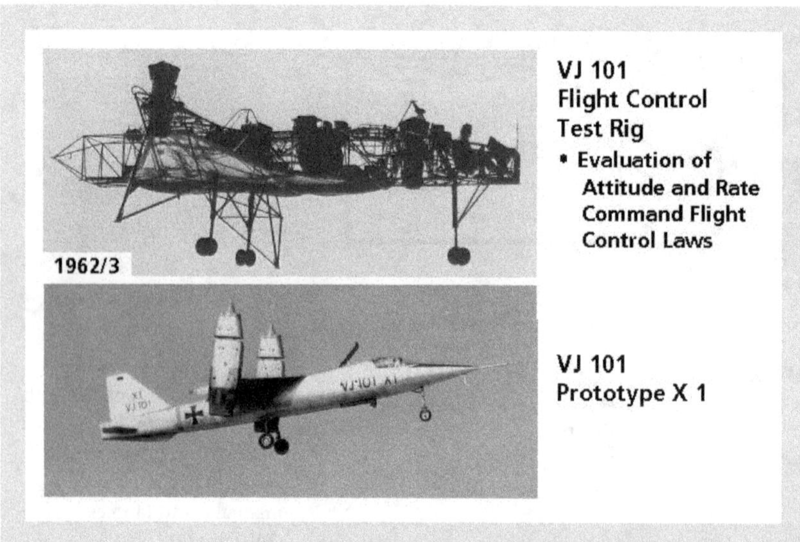

EWR VJ 101 flight control law testing on a hovering rig reduced development risks and cost.

ligible additional control forces due to a linked *Flettner*-tab which is directly attached to the outer manual control surface. *R. Vogt* achieved excellent flight test results with this separate surface control methodology in order to artificially improve the controllability and dynamic stability of large seaplanes. This technology was directly implemented into the quantity production of the BV 222 [22].

Finally, the flight test experiments by „Flugbaumeister" *Karl Doetsch* and his associate *E. G. Friedrichs* at the Deutschen Versuchsanstalt für Luftfahrt (DVL) in 1944 became one of the first real applications of artificial stability. The objective was to actively reduce directional snaking oscillations which were aircraft configuration dependent and extremely annoying for the pilots during anti-tank target tracking tasks.

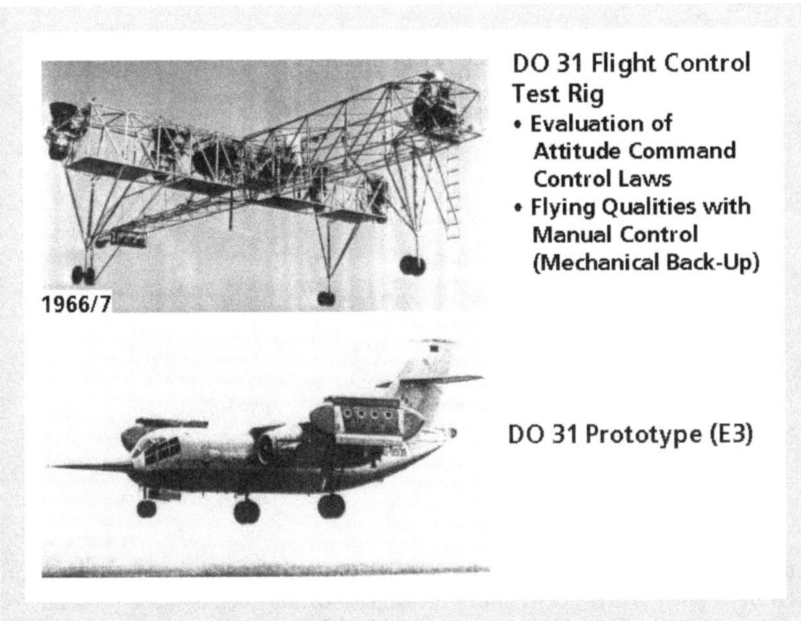

Dornier Do 31 flight control law and manual back-up mode testing on a hovering rig.

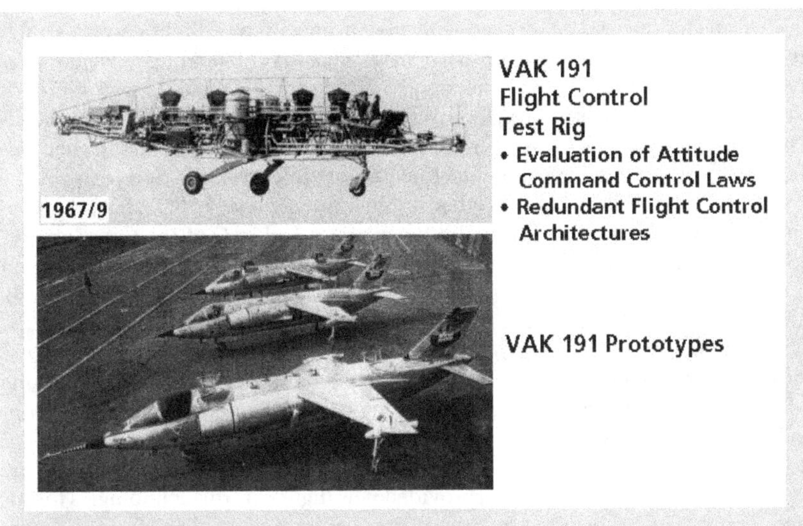

VFW VAK 191B flight control law and hardware system architecture (redundancy) testing on a multipurpose hovering rig (SG 1262).

Doetsch and *Friedrichs* replaced the rudder of a Henschel HS 129 with two separate yaw control surfaces, one for pilot manual control and the other for rate gyro controlled yaw damping [22, 23]. Similar investigations were accomplished at the end of WW II for the first German operational turbojet combat aircraft Me 262. In the United States the first yaw dampers were designed in the late 1940s for the Northrop YB-39 "Flying Wing" and the already discussed Boeing XB-47 swept wing aircraft with considerable success. Since then, the notions "stability augmentation" or "augmented aircraft" are in aeronautical colloquial usage.

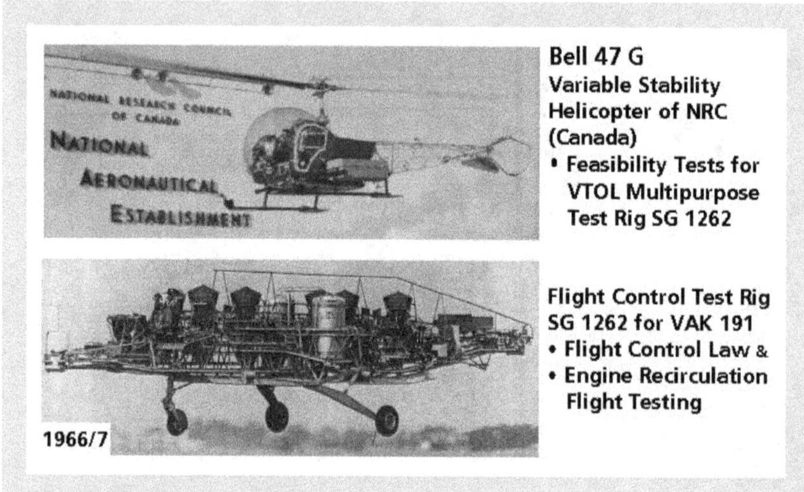

VFW VAK 191B multi-axes flight control law testing was complemented by a variable stability helicopter.

From a today's viewpoint the afore-mentioned flight experiments by *Richard Vogt* and *Karl Doetsch* to artificially change the control and stability characteristics were the first investigations towards the design of aircraft with variable flying qualities. This technique was used very early after World War Two in Germany as an important tool to investigate and evaluate the control and stability characteristics of new aircraft which were still in the design stage. The research tool "airborne simulation" using variable stability and control aircraft was originally initiated at DVL Oberpfaffenhofen in the early 1960's and consequently developed at the DFVLR-Institute of Flight Mechanics (later: DLR-Institute of Flight Research and since 2002: DLR-Institute of Flight Systems) in the early 1970s. At the latter institute the diligent flight test methodology of airborne (in-flight) simulation has advanced to modern operational fly-by-wire and fly-by-light flight test demonstrators such as HFB 320 FLISI, VFW 614 ATTAS, BO 105 ATTHeS and EC 135 FHS [23, 24] (see also the following two sections).

In-flight simulation gained special importance during the industrial development of three German vertical take-off and landing (VTOL) aircraft systems, e.g. VJ 101, Do 31 und VAK 191.

The application of in-flight simulation was concentrated in the early 1960's on the development of VTOL hovering test rigs (flying test beds) generally equipped with the original propulsion systems of the later VTOL prototypes. The utilization of these flying hovering rigs was not only limited to control and stability investigations but included the preliminary evaluation of flight control subsystem hardware of the future VTOL aircraft projects. Various questions concerned with the optimization of flight control strategies during VTOL take-off and landing as well as during the aerodynamic transition phase gave sufficient rise to also investigate with hovering test rigs the effects of the propulsion configurations on the phenomena of engine exhaust/inlet recirculation [25].

Further to the hovering test rig SG 1262; the variable stability helicopter Bell 47 G of the Canadian Research Council was utilized during the development phase of the VTOL combat aircraft project VAK 191. The flight tests had clearly demonstrated that the safety pilot of the variable stability Bell 47 G had excellent monitoring and control capabilities during safety-critical flight conditions of the flying helicopter simulator [25, 26]. Finally, the industrial F 104 CCV demonstrator technology program of MBB in the 1970s, provided new flight control technologies for future aircraft projects with unstable airframes. Again, artificial stability and the potential of variable flying qualities of this technology demonstrator provided sufficient insight to flight evaluate key control technologies in time for the up-coming multinational EF 2000 development program. More details about this program have already been discussed in the section "From the Stable towards the Unstable Configuration – Drag Optimization".

In-Flight Simulation with Fixed Wing Aircraft

Dietrich Hanke

In the mid-sixties, scientists at the DVL Institute of Flight Mechanics (*Gerhard Brüning*) in Oberpfaffenhofen started to think about the use of a variable stability aircraft as a research tool for the in-flight investigation of handling qualities. The idea was born during a 1964 visit of DVL scientists to several research institutions and flight-test installations in the United States.

After the visit, a variable stability aircraft was designed based on the piston-engine powered Piaggio P 149 D of the German Air Force (*Alfred Pietraß*). Initially, the stability concept was only set up for the elevator control channel. The on-board computer system was based on an analogue computer; the control surfaces were actuated by electrically driven servo motors [27–30]. The total system, including the data acquisition system, was integrated into the Piaggio and flight tested in 1967 for the first time. An explicit model-following concept was applied and eight different model dynamics could be chosen in flight. When the DVL obtained an HFB 320 Hansa Jet in 1968, it was decided to convert the aircraft into an in-flight simulator for research purposes (German Ministry of Defense Contract 1969).

At the DFL Institute of Flight Guidance in Brunswick (*Karl-H. Doetsch*), the development of digital-electronic flight control (fly-by-wire systems) commenced in 1965. A Dornier Do 27 aircraft was equipped with a digital fly-by-wire control system in combination with an advanced side stick (*Ludwig Tacke*) as control element for the pilot. First flights with the freely programmable flight control system were carried out in 1967. In 1968, the digital fly-by-wire system (Honeywell H 516 computer [31]) was also implemented on a larger aircraft, the Percival "Pembroke" of the DFL, which was also equipped with a side stick. First flights took place in 1969 (*Walter Metzdorff*). On June 3, 1971, the fly-by-wire system of the "Pembroke" was flown by Neil Armstrong, the first man on the moon, on the occasion of a visit to DFVLR as a NASA representative.

Due to the merger of the three German research establishments DFL, DVL and AVA to DFVLR (1969), it was decided by the board of directors of the DFVLR and desired by the Ministry of Defense (*Heinz v. Halem*) to transfer the work on in-flight simulations with the HFB 320 "Hansa" jet to the Institute of Flight Mechanics (*Peter Hamel*) in Brunswick. *Dietrich Hanke* [32] was appointed as project head for the development of the airborne simulator HFB 320. The development of the electro-mechanical landing-flap control system for the direct lift control (the basic aircraft landing flaps were used) was carried out by MBB-UF in Ottobrunn. The electro-mechanical control of both engines was realized by MBB-UH in Hamburg. The primary flight control electro-hydraulic actuators were implemented by the Institute of Flight Guidance (*Ludwig Tacke, Johannes Tersteegen*). Already in December 1972, first flight tests were conducted to evaluate direct lift control concepts (*Dietrich Hanke*); for flight-speed control an in-house developed analog computer was used. In 1973, a digital computer, type Honeywell H 316, was integrated into the HFB 320 (*Rainer Dombrowsky*) and first flight tests concerning "fly-by-wire" control were carried out (*Reiner Onken, Volkmar Adam*). The development of the HFB 320 to an in-flight simulator was consequently advanced during the next years (*Dietrich Hanke, Hans-Heinz Lange, Frohmut Henschel, Horst Meyer, Bernhard Gartung*). Finally, in 1977, the first in-flight simulation of a complete aircraft an A 310 configuration (B10), could be carried out and evaluated. A digital control system and an explicit model-following control with a multi-parameter control system, optimized according to Ricatti, was employed (*Frohmut Henschel*). The HFB 320 was used in a number of research programs of both DFVLR institutes. A summary of the most important research programs is presented in the table below.

The HFB 20 was a very successful research tool playing an important role especially within the German "Civil Component Programs I and II" of the German Ministry of Research and Technology (BMFT). The objective of these nationally funded programs, jointly carried out by research (DFVLR) and the German aircraft and aeronautical-supply industry, was the development and testing of new technologies and procedures.

In addition, important contributions were provided by the HFB 320 in-flight simulator concerning flying-characteristics databases and criteria, direct lift control, dead-times in flight control systems, as well as reduced static stability; the findings were adopted by the military flying-characteristics requirements

HFB 320 Flight-test Programs.

Flight-test programs	Cooperation	Period
• Integration and flight testing of a digital fly-by-wire system for elevator, aileron, rudder, direct lift flaps and engines [32–34]		1972–1973
• Landing approaches with direct lift control [35, 36]		1972–1973
• Steep noise abatement 2-segment approaches with and without direct lift control (DLC) [37]		1972–1973
• Evaluation of the effect of signal quantification in elevator control for a manual altitude tracking task		1973
• Dual side grip configuration as control element for fly-by-wire control in pitch, roll and yaw axes [38]		1973
• Specific approach to measure the vertical movement of an aircraft [39, 40]		1973
• Flight testing of a Doppler-radar-system		1974
• Landing approaches with decoupled flight path control by direct lift control [41–43]		1975
• Hybrid navigation by using a combination of DLS, INS and pressure altitude measurements		1975–1976
• Flight path tracking with direct lift control (DLC), Pilot modeling [44]	NLR	1975–1977
• Decelerated approaches with auto-flap and auto-throttle [45]	NLR	1976
• Digital flight control with command modes and automatic flight guidance in longitudinal and lateral motion • Flight demonstration of a fully integrated flight control system with command modes, autopilot and auto-throttle (DFVLR) • Flight testing of a digital AFCS mode control unit developed by BGT [46–48]	ZKP I, MBB, VFW, BGT, DLH	1976–1978
• Steep noise abatement 2-segment approaches with and without direct lift control (DLC)		1974–1976
• First in-flight simulation with digital explicit model-following control of a 6-DOF model, • Multi-axes-controller with Ricatti optimization [49]		1977
• A 310 (B10) in-flight simulation in landing approach with direct lift control [50–52]	ZKP I	1977–1978
• Flight testing of a hybrid navigation system based on INS and Multi-DME		1979
• Flight testing of a digital AFCS computer (VFW) • Flight testing of an integrated flight control system (DFVLR) • Flight testing of automatic flight guidance for decelerated approaches (MBB)	ZKP I	1979
• Flight testing of a rate command/attitude hold system with DLC and side grip controller [53, 54]	USAF-MoU	1979–1980
• Integration and test of a digital air data computer [55] • Automatic control of aerodynamic states (angle of attack, speed) [56] • Fully automatic 4D approach guidance from metering fix (TMA entry point) to the approach gate (VFW)	ZKP II MBB, VFW, BGT, DLH	1980
• Flight testing of a DFVLR developed strap down system [57]		1979–1982
• Handling qualities evaluation of aircraft characteristics with reduced static stability [58]	USAF-MoU	1981–1983
• Integration and flight testing of an INS (Carousel IVB) • Failure monitoring and failure handling of duplex sensors in a flight control system under consideration of analytical redundancy [59–61]		1982
• Fully automatic 4D approach guidance from metering fix (TMA entry point) to the approach gate with on-board wind estimation and 4D idle descent (DFVLR) [62, 63]		1982–1983
• Handling qualities investigation on the effect of time delays in digital flight control systems [64]	USAF-MoU	1983

VFW-614 "ATTAS" Flight-test Programs.

Flight-test programs	Cooperation	Period
• 1. FBW/Light-flight, Flight testing of the ATTAS fly-by-wire system with electro-optical computer data link [65]		1986
• In-flight measurements on a laminar glove	DA Bremen	1987
• Flight testing of a model-following system for in-flight simulation purposes [66, 67]		1988
• In-flight simulation of a large transport aircraft model (A 300 type) [68]		1988–1989
• Flight testing of a nonlinear model-following control concept [69]		1989
• Wind estimation by using on-board sensors [70]		1990
• RACOSS, Flight testing of a rate command/attitude hold system for the pitch and roll axes		1990
• Integration and flight testing of an experimental cockpit located in the cabin (aft cockpit)		1990–1997
• LARS, Load alleviation and ride smoothing system by using DLC flaps [71, 72]		1990–1992
• In-flight simulation of an IPTN N250 aircraft	IPTN	1991
• In-flight simulation of the Hermes space plane [73]	CNES	1991
• Flight testing of an AFCS with command control, autopilot and auto-throttle		1991–1993
• Integrated thrust and EGT control		1992–1994
• AIDA, Airborne identification and development of simulation fidelity by using ATTAS		1993–1999
• SAFIR I, II, III, Small airliner flight control law investigation and refinement, Flight testing of flight control laws and flight envelope protection functions for a small transport aircraft [74]	DA Hamburg	1993–1996
• SCARLET, Saturated commanded rate limited elevator time delay, Flight testing of a phase compensator for rate-limited actuators [75]	NASA, USAF-MoU	1993–1994
• ARCORE, Artificial redundancy concept for reconfiguration, Flight testing of a flight control system with automatic reconfiguration after system failures [76]		1993–1996
• In-flight simulation of an A3XX design configuration [77]	AI	1994–1996
• CASSY, Cockpit assistant system, Flight testing of an intelligence based pilot assist system	University of BW, Dornier	1994
• Total energy control for a 4D-descent flight path without thrust		1994
• Flight management system with 4D trajectory planning, trajectory negotiation via data link, and 4D guidance		1994–1997
• Duplex configuration of the ATTAS FBW/light system with failure monitoring and failure handling		1997
• MIROR, In-flight measurements of engine exhaust gases with the MIROR spectrometer in the case of low- sulfur fuel		1994–1998
• HGG, Flight testing of a high performance graphics generator, synthetic vision with 3-D terrain visualization and advanced symbology on a flat-panel display	TU-D, VDO-L, DASA L, Honeywell	1995–1999
• SATCOM, Flight testing of a satellite communication system	SOFREAVIA, PHARE-Partner, ISTEFE, Sextant, Thomson-CFS	1996–1998

VFW-614 "ATTAS" Flight-test Programs – Continued.

Flight-test programs	Cooperation	Period
• HUD (Head Up Display), Flight testing of a HUD with new symbology for low level flight guidance	DASA-L ,TU-D	1996–1999
• PHARE Demo PD/2P, Program for harmonized air traffic management research in EUROCONTROL, Demonstration of an advanced air traffic management system with air/ground negotiation of the flight trajectory and with fully automatic 4D guidance [78, 79]	EUROCONTROL, CENA, DFS, DRA, NLR	1997
• GRACE, Genetic algorithm for control system enhancement, Evaluation of an genetic algorithm for reconfiguring the flight control system after system failures		1997
• Flight testing of a simple phase compensator for rate limitation elements in flight-control systems [80]		1998
• Advanced GPWS, Flight-testing of an advanced ground-proximity warning system	DASA NFS	1998
• OLOP (Open Loop Onset Point), Flight tests for the OLOP-PIO criterion verification for rate-limited actuator systems [81]	NASA	1998
• Advanced flight-management system with interactive graphical user interface for the pilot	EPNER	1999
• TESTPACK, Flight control characteristics demonstration for test pilots and engineers	NLR, Sextant	1998
• AWARD, All-weather arrival and departure system, Flight-testing of an enhanced-vision system by using microwave radar data and infrared data displayed on head-down and head-up displays.	ONERA, AS, DASA, NLR	1999–2000
• REAL, Robust and efficient autopilot control laws, • Modifications to allow touch down and landing in the experimental fly-by-wire mode	DA Bremen, LLI	1999–2000 1994–1999
• Touch down in the electrical mode		1994–1999
• CAMA, Cockpit assistant military aircraft • Pilot assistant at low level flight • Fly-by-wire landing validation • Flight controller evaluation for ETPS • AFCS with aft cockpit flight guidance • MIROR		2000
• WAKE, Wing vortex measurements by ground based LIDAR system • S-WAKE, in-flight wing vortex measurements		2001
• FD 728 Jet In-flight simulation, Flight control system evaluation • ATTAS s UAV, Technologies demonstration flights	FD	2002

documents of the US Air Force (German-USAF Memorandum "Flight Control Concepts").

For the first time, also an integrated digital flight control and flight guidance system with fully automated 4-D navigation capabilities was realized and flight-tested.

With the HFB 320 Hansa Jet, an in-flight (airborne) simulator was created that was, equipped with most modern digital multi-axes control methods, operationally utilized for the in-flight simulation of very different aircraft dynamics and flight control characteristics. The HFB 320 research aircraft helped DFVLR to gain an international reputation in airborne simulation and flight-control research.

DFVLR started in 1977 with the planning of a successor for the HFB 320 since the HFB 320 was too noisy and the cabin space and payload was limited. Finally, a VFW 614 aircraft was selected, the first passenger jet developed in Germany. This aircraft was converted between 1981 and 1986 into a flying simulator and technologies demonstrator ("ATTAS, Advanced Technologies Testing Aircraft System"). The modifications were sponsored by the German Ministry of Research and Technology (BMFT) and jointly carried out by MBB-Bremen and DFVLR (Project supervisor at DFVLR: *Heinz Krüger*, Project supervisor at MBB: *Hartmut Griem*, Systems and concept supervisor at DFVLR: *Dietrich Hanke*, Flight operations and test pilot at DFVLR: *Hans-Ludwig Meyer*); the aircraft was, after its first flight with digital electro-optical control, available for research starting in December 1986.

Similar to the HFB 320, "ATTAS" was designed as a flexible and "freely programmable" in-flight simulator and technologies demonstrator mainly to be used for flight-control and flight-guidance research investigations. The "fly-by-wire" flight control was designed by DFVLR as a dual-redundant system with optical data bus communication between nine on-board computers (ROLM MSE 14, Hawk 32); one of the computers was as freely programmable computer available to the user.

A total of 15 electro-hydraulic actuators with digital self-monitoring electronics and a MIL-Bus 1553B interface (Liebherr Aerotechnik) were implemented to actuate the control surfaces, the engines, and the direct-lift control system, consisting of six individual flaps located at the trailing edge of the landing flaps (MBB, Bremen).

Parallel to the aircraft modifications, a complete systems simulator was developed at DFVLR, which contained the on-board computers as hardware-in-the-loop. This system was employed to develop the flight control software for "ATTAS" and to test all functions under real time conditions. The simulator was simultaneously used for flight test preparations, flight crew training and user software validation.

The flight-control software development was headed by *Hans-Heinz Lange*, the nonlinear in-flight simulation software was further developed by *Michael Bauschat* and *Henning Heutger*, and the control law design was carried out by *Shyam Chetty* and *Frohmut Henschel*.

Important in-flight simulation applications that must be mentioned were the simulation of the "Hermes" space-plane characteristics under contract of ESA/CNES (1991), and the simulation of the preliminary configuration of a very large transport aircraft (A 3XX type) under contract of Airbus Industries.

"ATTAS" was, since 1986, successfully employed in numerous flight test programs as indicated in the table. During the time period 1994–1999, the flight-test equipment of "ATTAS" was improved and the flight-control duplex system was made fully operational. Safety features were, furthermore, implemented to expand the fly-by-wire flight envelope to provide experimental operation all the way to touch down. With "ATTAS", the DLR owns a powerful flying test-bed, which is with regard to its concept and wide range of capabilities unique in Europe and worldwide competitive.

Remarks

Concerning this chapter, some figures are presented in color:

Color Plate XV

- HFB 320 "HansaJet", Flying Simulator for digital fly-by-wire control
- Dornier Do 27 as test aircraft side-grip control
- BO 105 "ATTHeS" in-flight simulator in fully automatic hovering flight

Color Plate XVI

- The experimental variable stability helicopter EC 135-ACT-FHS in flight
- VFW 614, Flying simulator and technology demonstrator ("ATTAS")

In-Flight Simulation with Helicopters

Gerd Bouwer, Heinz-Jürgen Pausder

Simulating aircraft with vertical or short take-off and landing (V/STOL) capabilities needs an air vehicle that can fly these manoeuvers. The only proven flight vehicles having these characteristics are helicopters. It was thus investigated theoretically in 1975 how far helicopters were suitable to simulate V/STOL aircraft [82]. It was found that a helicopter with a modern hingeless rotor system, like the BO 105 helicopter, was particularly well suited for this task.

At the same time, the BO 105 3 (3rd quantity production model) was at MBB equipped with an experimental simplex "Fly-by-wire" actuator system. The airframe was modified in a way that in the center forward part of the cockpit the seat for the simulation pilot was arranged, while on the left rearward side the seat for the safety pilot was installed. When activating the "fly-by-wire" system, the actuators were airframe-fixed via magnetic brakes. Their movements, controlled by the simulation pilot or by computer, altered the position of the safety pilot's controls so that he was able to permanently supervise the overall air vehicle control. In case of an error, the safety pilot was able to override the "fly-by-wire" actuators and switch off the system. This arrangement guaranteed high safety in case of a failure. A minimum safety altitude of 100 ft was additionally requested. Results of first flight tests with this helicopter, including remotely controlled flights, were published in 1975 [83].

One of the possible control methods in in-flight simulations is the explicit model-following control: the control commands of the simulation pilot are input to a simulation model of the air vehicle to be simulated; the model computes the flight state changes resulting from the pilot's commands. These are then processed as set values to a control system. The latter controls via the actuators the "actual" helicopter so that it follows the flight conditions of the simulation model. The simulation pilot has, in this case, the impression to fly the simulation model rather than the "actual" helicopter. First investigations concerning the realization of a model-following control as prerequisite for the in-flight simulation of another helicopter by means of the BO 105 commenced in Brunswick at the Institute of Flight Mechanics of the DFVLR already in 1980 [84]. Here, the model of the air vehicle to be simulated and the (closed-loop) control system were treated as separate blocks. This method of explicit model-following control was consistently advanced and tested.

On the initiative of the Institute of Flight Mechanics, DFVLR purchased in 1982 the BO 105 S3. The helicopter was equipped with the necessary sensors, such as acceleration sensors, rate gyros, attitude gyros, and a digital computer to control the actuators. It had its first flight at DFVLR with a digital control system on March 4, 1983. Already in 1983/1984, first flight tests were performed concerning the effect of certain control characteristics on the pilot workload [85]. During these tests, the cyclic longitudinal and lateral control inputs of the simulation pilot were dynamically deformed and added via analog filters. This resulted for the pilot in different couplings between the control about the pitch and the roll axes.

Within the framework of the German-American Memorandum of Understanding on Aircraft Flight Control Concepts [86] between the US-Army and the BMVg, the model-following control was jointly investigated by scientists of the Institute of Flight Mechanics, the US-Army and the NASA Ames Research Center (RC) in 1983/1984 on a simulator at the Ames RC. The control system was determined and implemented for the helicopters BO 105 and UH-1 D. On the simulator, several pilots performed the flight tasks "Dolphin" and "Slalom", i.e., flying over and around obstacles with the approach and departure altitudes being 75 ft. The results [87] showed that with the control system the flying characteristics and the task realization could be considerably improved for both helicopters at a simultaneous reduction of the pilot's workload. The system was realized for the first time in 1985 on the tandem-rotor helicopter CH-47 in investigations related to the roll axis in hovering flight [88]. These simulation and flight-test results represent a milestone in the joint German-American development of explicit model-following systems.

The helicopter was equipped at the end of the eighties with a "fly-by-light" yaw control system [89, 90]. The "fly-by-wire" yaw actuator was replaced by a duo-duplex actuator. This actuator was controlled by a triplex "fly-by-light" computer via optical fibers. In 1990/91, the system was tested in conjunction with a yaw control system for heading hold [91]. Especially the on-board computer system was further developed into a fault tolerant system and tested in 1993 [92].

An essential improvement of the in-flight simulation quality was achieved in 1990 by implementing a dynamic pre-control [93] within the onboard computer system of the BO 105 S3. The dynamic pre-control generates actuator commands, which depend on a change in the set values rather than on the flight condition of the helicopter proper. Theoretically, this already achieves a perfect in-flight simulation. In reality, it is necessary to have state feedback or a flight control system. These are, however, only responsible for controlling gust disturbances and modeling errors.

At this time, the helicopter also received its name "Advanced Technologies Testing Helicopter System, ATTHeS". The onboard computer systems were improved and the computer software was newly designed especially with regard to computing time. All essential systems of the "ATTHeS" and the flight-test results regarding the highly dynamic explicit model-following control system were in conclusion described in the book "Advances in Flight Control" [94].

In July 1991, a symposium on in-flight simulation took place at the DLR in Brunswick where the experts exchanged their respective knowledge and experience [95]. Various airborne simulators and their applications and benefits were presented and discussed. The Journal "Aviation Week & Space Technology" reported in its issue of October 7, 1991 extensively on the meeting [96].

Since 1992, flying characteristics were successfully investigated complementary with the partners in the USA: at the NASA Ames research center tests were conducted together with the US-Army in the "Vertical Motion Simulator, VMS" and in Germany the corresponding flight tests were performed with the "ATTHeS" [97]. The results constitute important basic data for the new "Aeronautical Design Standard" for helicopters of the US-Army [98].

The "ATTHeS" was, in addition, since 1992 regularly chartered once a year by the two large test pilot schools "Ecole du Personel Navigant d'Essais et de Reception, EPNER" in Istres, France, and "Empire Test Pilot School, ETPS" in Boscombe Down, England. Each prospective helicopter test pilot had to fly the "ATTHeS" for some hours in various configurations. Even the various very dangerous flight situations known as "Aircraft Pilot Coupling, APC" in the case of fixed-wing aircraft, which have led to the total loss of modern aircraft prototypes (Saab JAS-37 "Gripen", Lockheed YF-22A), could be practiced with the pilots.

The first series helicopter, which was tested with the aforementioned control-technology for the in-flight simulation of helicopters, was the RAH-66 "Comanche" of Sikorsky/Boeing [99]. Here, the explicit model-following control was, for instance, utilized to impose different flying characteristics on the helicopter in low-speed and high-speed flight, respectively.

To develop and test modern autopilot systems, flight tests were carried out in 1993 with the ATTHeS regarding fully automatic navigation [100], and in 1994 with regard to the automatic hovering-flight stabilization above a moving object [101]. During the automatic navigation exercise, the onboard computer programmed way-points and the autopilot directed the helicopter at fixed altitude and fixed speed along these way-points. The only thing the pilot had to do was switch on the system.

To stabilize automatic hovering flight, a downward looking camera was attached to the helicopter. A special computer was able to follow defined objects in the pictures provided by the cameras. The computer transferred the position data of an object to the control-system computer, which controlled the helicopter so that it hovered above the object. It was shown that the system was able to move the helicopter automatically, partly flying backwards, in unison with and above a moving vehicle without the pilot being forced to use the controls.

Also in 1994, flight tests with the "ATTHeS", equipped with an active side-stick, were performed within the European project "Active Control Technology, ACT". Different flying characteristics were investigated during defined flight tasks.

Within the framework of a cooperation with the ONERA, France, flight tests with the A"ATTHeS" were carried out in 1995 to determine helicopter flying characteristics during hover [102]. The task consisted of following a vehicle in a lateral flight as precisely as possible. The test pilots had to meet this task for different pre-programmed flying characteristics.

On May 14, 1995, the "ATTHeS" crashed during a routine training flight due to a technical defect. The pilot *K. Sanders* and the flight engineer *J. Zimmer* died in the crash.

Already prior to the tragic crash of the "ATTHeS", the Institute of Flight Mechanics initiated the project "Active Control Technologies-Flying-Helicopter-Simulator, ACT-FHS" as a joint project of the BMVg, DLR and industry (Eurocopter Germany, Liebherr Aerospace). Since November 7, 2002 this helicopter, based on the EC 135 and equipped with a quadruplex "fly-by-light" system and several onboard computers, is operated at DLR in Brunswick as a variable stability helicopter also for in-flight simulations. With this helicopter, the pioneering work carried out at the DLR with the "ATTHeS" can be continued and European research and industry provided with a unique flight vehicle for technology testing and demonstration [103–107].

References

1 F.E. Peñaranda, M. S. Freda, Aeronautical Facilities: Catalogue, Vol.2, NASA RP 1133, Washington D.C., 1985

References

2 F.E. Peñaranda, M.S. Freda, Aeronautical Facilities: Catalogue, Vol.1, NASA RP 1132, Washington D.C., 1985

3 H. Trischler, Luft- und Raumfahrtforschung in Deutschland 1900–1970, Frankfurt M./New York, 1992

3a K. Kopfermann, Aufzeichnungen zur Entstehung der IABG. IABG-Archiv unveröffentlicht

4 H. Holthusen, Model and Full Scale Investigations of Frequency Phenomena of the DNW Open Jet. FDP Symposium on "Aerodynamics of Wind Tunnel Circuits and Their Components", AGARD-CP 585

5 F.-R. Grosche, Schallmessungen an Luftfreistrahlen aus Schlitzdüsen mit und ohne Abschirmung, Jhb. DGLR 1968, pp. 213–225

6 F.-R. Grosche, J. H. Jones, and G. A. Wilhold, Measurements of the Distribution of Sound Source Intensities in Turbulent Jets, Progress in Astronautics, V l.37 (AIAA), pp. 79–92, 1975

7 F.-R. Grosche, H. Stiewitt, and B. Binder, Acoustic Wind Tunnel Measurements with Highly Directional Microphones, AIAA-Journal 15, pp. 1590-1596, Nov. 1977

8 G. Viehweger, R. Rebstock, B. Stahl, K. Wichmann, W. Becker, R. Kronen, D. Distelrath, Der Kryo-Kanal Köln (KKK) der DLR (Stand 1992), DLR-Mitt. 93-10, 1993

9 B. Ewald, Windkanal-Handbuch, Technische Hochschule Darmstadt

10 I. Roehle, C. Willert, R. Schodl, Recent Applications of Three-Dimensional Doppler Global Velocimetry in Turbo-Machinery, 9th Int. Symp. On Applications of Laser Techniques to Fluid Mechanics, 13.–16. July 1998, Lisboa, Portugal, Paper No.34-2

11 H.P. Kreplin, H. Höhler, Multi-Sensor Hot Film Array on an Airbus High-Lift Wing, DNW Annual Report 1994, Emmeloord, NL 1995

12 M. Raffel, C. Willert, J. Kompenhans, Particle Image Velocimetry, Springer-Verlag, Berlin, Heidelberg, New York, 1998

13 K. Hünecke, European Investigation on Wake Vortex Phenomena, Nouvelle Revue d'Aeronautique et d'Astronautique, No. 2, 1998

14 E.C. Polhamus, R.A. Kilgore, J.B. Adcock, E.J. Ray, The Langley Cryogenic High Reynolds Number Wind-Tunnel Program, Aeronautics & Astronautics, Vol. 12, No. 10, 1974

15 DLR Institut für Flugmechanik, A Historical Account and Technical Guide, Brunswick, April 1999

16 G. Hüttig, Im Cockpit der Zukunft, Aus der Arbeit am Berliner Forschungs-Flugsimulator, Fliegerrevue 11/1998

17 E. Heinkel, Natürliche und künstliche Stabilität, 7. Wissenschaftssitzung der Ordentlichen Mitglieder am 20. September 1940, Jahrbuch der Deutschen Akademie der Luftfahrtforschung 1940/41, Berlin 1941, pp. 283–284

18 H. Helmbold, Statische Längsstabilität bei hohen Fluggeschwindigkeiten, 7. Wissenschaftssitzung der Ordentlichen Mitglieder am 20. September 1940, Jahrbuch der Deutschen Akademie der Luftfahrtforschung 1940/41, Berlin 1941, pp. 285–293

19 E. Fischel, Die vollautomatische Flugsteuerung, Ringbuch der Luftfahrttechnik, VE4, 31. Oktober 1940, pp. 1–47

20 E. Fischel, Künstliche Stabilität bei hohen Fluggeschwindigkeiten, 7. Wissenschaftssitzung der Ordentlichen Mitglieder am 20. September 1940, Jahrbuch der Deutschen Akademie der Luftfahrtforschung 1940/41, Berlin 1941, pp. 296–303

21 R. Vogt, Steuerung von Großflugzeugen im Hinblick auf die bei Blohm & Voss beschrittenen Wege, 3. Wissenschaftssitzung der Ordentlichen Mitglieder am 9. Mai 1944, Schriften der Deutschen Akademie der Luftfahrtforschung, Heft 1080/44g, 1944, pp. 1–24

22 K.-H. Doetsch, Probleme der automatischen Flugsteuerung von Flugzeugen, Jahrbuch der Wissenschaftlichen Gesellschaft für Luftahrt e.V., 1955, pp. 72–78

23 P. Hamel, The Evolution of In-Flight Simulation in Germany, Contribution to Vistas in Aerospace Engineering, University of Maryland, Vortrag, 9. Februar 1996

24 P. Hamel, Systemintegration und Fliegende Simulation zur Minderung des Entwicklungsrisikos zukünftiger Flugzeuge, DLR Jahresbericht 1996, pp. 28–33

25 H.-G. Schumann, Flugsimulation als Hilfsmittel der Entwicklung und die deutschen Arbeiten auf diesem Gebiet, Jahrbuch der DGLR 1970, pp. 40–58

26 A. Pietraß, Flugeigenschaftsuntersuchungen mittels eines Flugzeugs mit modellgesteuertem Übertragungsverhalten, 1. Zwischenbericht DVL TV, 13/65, 8.10.1965, 2. Zwischenbericht DVL TV, 1/66, 20.01.1966, 3. Zwischenbericht DVL TV, 5/66. 20.04.1966, 4. Zwischenbericht DVL TV, 14/66, 21.07.1966

27 B. Uhrmeister, Untersuchung des Regelkreises der umgerüsteten Piaggio P 149 D mittels Analog- und Digitalrechner. Zum Forschungsauftrag T 01-529-I-205 des BMVg. DVL TV 15/67, 1967

28 A. Brünner, H.-J. Kuper, Umrüstung einer Piaggio P 149 zu einem Flugzeug mit modellgesteuertem Übertragungsverhalten. DVL TV 11/67, 1967

29 A. Pietraß, Bestimmung von Trägheits- und Reibungsmomenten des Piaggio-Landeklappensystems aus Einschwingvorgängen. DVL TV 6/68, 1998

30 A. Pietraß, B. Uhrmeister, Umrüstung eines Flugzeugs zu einem fliegenden Simulator, DVL-Nachrichten, Heft 38, Januar 1969

31 W. Metzdorff, Neuere Arbeiten zur Flugzeug-Primärsteuerung am Institut für Flugführung der DFL, DFL-Mitteilungen, Heft 8/1968

32 D. Hanke, Testflugzeug mit variablen Flugeigenschaften. Umschau in Wissenschaft und Technik, Umschau 4 (1974), Heft 3

33 DFVLR-Jahresbericht 1973

34 R. Onken, H.-P. Joenck, L. Tacke, M. Gottschlich, Digital Fly-By-Wire Control System with Selfdiagnosing Failure Detection., 17th AGARD Symposium, Geilo/Norway, 1973

35 D. Hanke, Erprobung der direkten Auftriebssteuerung beim Versuchsflugzeug DFVLR-HFB 320, DFVLR Nachrichten, Heft 7/1972

36 D. Hanke, H.-H. Lange, Flugmechanische Probleme beim Landeanflug mit direkter Auftriebssteuerung am Beispiel der HFB 320 Hansa. DGLR/DGON Symposium Neue Anflugverfahren, Düsseldorf, 2.–4. Mai 1973, DGLR 73-024

37 P.G. Hamel, K. Wilhelm, D. Hanke, H.-H. Lange, Steep Approach Flight Test Results of a Business-Type Aircraft with Direct Lift Control. AGARD CP-160, Paper No.21, 1974

38 L. Tacke, Beitrag zu Übersichtsbericht der Gemeinschaftsaufgabe "Digitales Flugsteuer- und Regelsystem mit Fail-Safe-Eigenschaften". DFVLR/Dornier, ZTL 1974

39 H. Winter, Stützung des Vertikalkanals einer Trägheitsplattform mit einem barometrischen Höhenmesser. DGON-Symposium "Kreiseltechnik", Heidelberg, April 1974

40 H. Winter, H.P. Zenz, J. Jenschek, Messung der Vertikalbewegungen eines Flugzeugs, DLR-FB 75-41, Braunschweig, 1975

41 D. Hanke, H.-H. Lange, Fliegbarkeitsuntersuchungen zur Anwendung der direkten Kraftsteuerung in der Längsbewegung. Vortrag DGLR-Symposium CCV-Technologien, 17.9.1976, München, DGLR 76-238

42 D. Hanke, In-Flight Investigation of Direct Lift Control in Landing Approach, DFVLR IB 154-77/25, August 1977

43 D. Hanke, H.-H. Lange, In-Flight Handling Qualities Investigation of various longitudinal short term dynamics and direct lift control combinations for flight path tracking using DFVLR HFB 320 581 variable stability aircraft. AGARD Conference Proceedings Nr. 260, Stability and Control, Ottawa, Canada, 25-28 Sept.1978

44 M.F.C. v. Gool, D. Hanke, H.-H. Lange, Flight Path Angle Tracking Experiments in the DFVLR HFB 320 Equipped with Direct Lift Control. DFVLR IB-154-75/32, Dezember 1975

45 C.F.G. Hofmann, H.-H. Lange, D. Hanke, In-Flight Investigation of Decelerating Approaches, DFVLR IB-154-76/10, Juni 1976

46 V. Adam, R. Onken, Evaluation of a new Flight Path Command Control Concept, 10th ICAS Congress, Ottawa/Canada, 1976

47 V. Adam, H. Leyendecker, Erhöhung der Führungsgenauigkeit durch den Einsatz eines integrierten digitalen Flugführungssystems. DGLR-DGON Symposium, Hamburg, April 1979

48 V. Adam, H. Leyendecker, Control Law Design for Transport Aircraft Flight Tasks, Agardograph 251, 1981

49 F. Henschel, The determination of flight-tracking controllers using Solheim's pole shifting method., Z.Flugwiss. 1 (1977) h.5, P. 375–377

50 D. Hanke, Fliegbarkeitsuntersuchungen von DLC-Konfigurationen für den Airbus A310 im Landeanflug mit dem In-Flight Simulator DFVLR-HFB 320. DFVLR-FB 79–18, Februar 1979, ESA-TT-630, November 1980

51 H.-H. Lange, Flugerprobung des Modellfolgereglers für die HFB 320 zur Simulation des Airbus A 310 im Fluge. DFVLR-Mitt.79-13, September1979, ESA-TT-660, 1981

52 H.-H. Lange, D. Hanke, Use of DFVLR In-Flight Simulator HFB 320 Hansa for Handling Qualities Investigations. Z.Flugwiss., Band 6, Juli/August 1982, Heft 4, pp. 225–234

53 D. Hanke, Flying Qualities Experiments of Rate Command/Attitude Hold Systems in the HFB 320 In-Flight Simulator, DFVLR-FB 83-25, Juli 1983

54 D. Altenkirch, Untersuchungen des Pilotenverhaltens bei Flugversuchen mit einem Rate-command/attitude-hold Steuerungssystem. DFVLR-FB 84-25, 1984

55 V. Adam, R. Dombrowsky, Integration eines Luftwerterechners in das Versuchsflugzeug HFB 320, DFVLR-Mitt. 81-09, Braunschweig, 1981

56 V. Adam, Flight Control Modes for Control of Aerodynamic State Parameters, 12th ICAS Congress, Munich, October 1980

57 H.-J. Hotop, B. Stieler, Modulares Strapdown System (MSS)

58 K. Wilhelm, D. Schafranek, In-Flight Investigation of Landing Approach Flying Qualities of Transport Aircraft with Relaxed Static Stability. DFVLR-FB 84-11, März 1984

59 N. Stuckenberg, Ein Beitrag zur Erkennung und Isolation von Sensorfehlern in Flugregelsystemen unter Verwendung von Beobachtern. DFLR-FB 81-26, Braunschweig, 1981
60 N. Stuckenberg, A diagnosis scheme for sensors of a flight control system using analytic redundancy, 36th AGARD-Symposium of the GCP, Toulouse/France, May 1983
61 N. Stuckenberg, Sensor Failure Detection in Flight Control Systems using Deterministic Observers, IFAC/IFORS Symposium, York/UK, July 1985
62 V. Adam, W. Lechner, A Concept for 4D-Guidance of Transport Aircraft in the TMA, 13th Congress, USA, August 1982
63 V. Adam, W. Lechner, Investigations on Four-Dimensional Guidance in the TMA. AGARD Special Session, Lisbon/Portugal, October 1982
64 K. Wilhelm, D. Altenkirch, In-Flight Investigation of the Effects of Time Delay in Control System on Flying Qualities in Landing Approach. DFVLR-FB 84-35, Oktober1984
65 D. Hanke, G. Bouwer, DFVLR In-Flight Simulators ATTAS and ATTHeS for Flying Qualities and Flight Control Re-search. AGARD Conference Proceedings Nr. 408, Flight Simulation, Cambridge, UK, 30 Sept.–3. Oct. 1985
66 F. Henschel, Über Regelungskonzepte zur In-Flight Simulation unter Berücksichtigung von Nichtlinearitäten und Totzeiten in den Stellsystemen. DFVLR-FB 85-24 (1985)
67 F. Henschel, S. Chetty, Design Technique for model-following control system in in-flight simulation, DLR-FB 87-09, 1987
68 D. Hanke, J.-M. Bauschat, H.-H. Lange, H. Heutger, In-Flight Simulator ATTAS-System Design and Capabilities. International Symposium "In-Flight Simulation for the 90's", DLR Brunswick, 1–3 July 1991, Conference Proceedings DGLR-91-05, Paper No. 91-05-02
69 J.-M. Bauschat, On the Application of a Nonlinear Inflight Simulation Technique, Proceedings of the European Conference ECC 91, Grenoble, 2–5 July 1991
70 Guofeng Wu, Wind Measuring and Filtering on Board the ATTAS Aircraft, DLR-IB 112-90/27, Brunswick 1990
71 R. König, K.-U. Hahn, Load Alleviation and Ride Smoothing Investigations Using ATTAS. Proceedings of the 17th ICAS Congress, Stockholm, 9-14 Sept. 1990
72 K.-U. Hahn, R. König, ATTAS Flight Test and Simulation Results of the Advanced Gust Management System LARS. Proceedings of the AIAA Atmospheric Flight Conference, Hilton Head, SC, USA, 10–12 August, 1992
73 W. Mönnich, ATTAS – Flight Test Evaluation. HERMES In-Flight Simulation Test. No.203, DLR-IB 111-92/34
74 T. Heintsch, R. Luckner, K.-U. Hahn, Flight Testing of Manual Flight Control Lawa for Small Transport Aircraft. AGARD Flight Vehicle Integration Panel Symposium on "Advances in Flight Testing", Lissabon, Portugal, 23–26 September 1996
75 J.R. Martin, J.-J. Buchholz, Scarlet: DLR Rate Saturation Flight Experiment. AGARD-AR-335, FVP Workshop on Pilot Induced Oscillation, May 1994, Turin, Italy
76 J.-J. Buchholz, W. Heine, G. Baumgarten, Genetic Algorithms for Reconfiguration. ZFW, Vol. 20, No. 3, 1996. 77 A3XX IFS
78 V. Adam, U. Teegen, 4D-Flying with EFMS in PD/2. DLR-Mitt. 97-03, Brunswick, 1997
79 V. Adam, U. Teegen, Airborne Aspects of PD/2. PD/2 Final Report, Annex F, EUROCONTROL DOC 97.70-13, Brussels/Belgium, 1998
80 D. Hanke, Phase Compensation: a means of preventing aircraft-pilot coupling caused by rate limitation. DLR-FB 98-15, June 1998
81 H. Duda, G. Duus, Recent Results of APC Testing with ATTAS, SAE Aerospace Control & Guidance Systems Committee Meeting, Salt Lake City, UT (US), 12 March 1999
82 M. Marchand, K. Wilhelm: "Flugmechanische Untersuchungen über die Einsetzbarkeit eines Hubschraubers zur Simulation von V/STOL-Flugzeugen", DLR-FB 75-59, Braunschweig 1975
83 S. Attlfellner, M. Rade, "Bo 105 In-Flight Simulator for Flight Control and Guidance Systems" First European Rotorcraft and Powered Lift Aircraft Forum, Southampton, England, September 1975
84 G. Bouwer, "Realisierbarkeit einer Modellfolgeregelung für Hubschrauber mittels digitaler Simulation a. Beispiel der BO 105", DFVLR IB 154-80/20, Braunschweig, 1980
85 H.J. Pausder, "A Study of Roll Response Required in a Low Altitude Slalom Task", Proceedings of the 11th European Rotorcraft Forum, London, UK, Vol.2, 1985, pp. 79.1–79.15
86 D.L. Key, B. Gmelin, "A Decade of Cooperation", Vertiflite, An American Helicopter Society Publication, Sept./Oct. 1989
87 G. Bouwer, K.B. Hilbert, "A Piloted Investigation of Decoupling Helicopters by Using a Model Following Control System", AHS Paper No A-84-40-07-4000, Arlington, Virginia, USA, 1984
88 K.B. Hilbert, J.V. Lebacqz, W.S. Hindson, "Flight Investigation of a Model-Following Control System for Rotor-

craft", AIAA 3rd Flight Testing Conference, Las Vegas, NE, April 1986
89 G. Ried, "Spezifikation LWL-Stellantrieb", LAT 0-0558, Lindenberg 1988
90 J. Tersteegen, "Geräteintegration für LWL-Giersteuerung (Zulassungsunterlagen für MPL)", BS 003-88-01.500, Braunschweig 1989
91 K. Bender et. al, "Flugerprobung des FbL-Giersteuerungssystems mit der Bo 105 S3", DLR IB 112-91/02
92 K. Bender, H. Leyendecker, "Untersuchungen zu einem fehlertoleranten Hubschrauber-Flugregelungssystem mit abgestufter Redundanz", DLR-Mitteilung 97-01, Braunschweig, 1997
93 W. von Grünhagen, J. Kaletka, "Application of System Identification for Rotorcraft Control System Design", 4. Braunschweiger Aerospace Symposium "Sicherheit im Luftverkehr", Technische Universität Braunschweig, September 1994
94 M.B. Tischler (ed), "Advances in Aircraft Flight Control", Taylor & Francis, ISBN 07484 0479 1, 1996
95 P. Hamel (ed), In-Flight Simulation for the 90's, International Symposium, DLR Brunswick, 1–3. July 1991, Conference Proceedings DGLR 91-05 8 (1991)
96 M. Mecham, "Airborne Simulation is a Driving Force in Advanced Control System Research", Aviation Week & Space Technology, October 7, 1991
97 H.J. Pausder, C.L. Blanken, "Investigation of the effects of bandwidth and time delay on helicopter roll-axis handling qualities", 18th European Rotorcraft Forum, Avignon, France, September 1992
98 N.N, "ADS-33-PRF Aeronautical Design Standard", United States Army Aviation and Troop Command, 13 May 1996
99 P.J. Gold, J.B. Dryfoos, "Design and Pilot Evaluation of the RAH-66 Comanche selectable Control Modes", AHS Conference on Flying Qualities and Human Factors, San Francisco, California, January 1993
100 M. Hamers, "Vollautomatische Navigation mit ATTHeS", Deutsche Forschungsanstalt für Luft- und Raumfahrt, DLR-IB 111-94/16, Braunschweig 1994
101 G. Bouwer, C-H. Oertel, W. von Grünhagen, "Autonomous helicopter hover positioning by optical tracking", The Aeronautical Journal, Volume 99, Number 984, April 1995
102 G. Bouwer, A. Taghizad, H. Mödden, "ONERA & DLR Cooperation on the "Smart Helicopter Concept"–Handling Qualities Data Base for Hover and Low Speed Flight", AGARD CP-592, Ottawa, Canada, May 1996
103 H.J. Pausder, U. Butter, V. Dion, "ACT Demonstrator und Fliegender Simulator–Stand des Projekts und zukünftige Nutzungsprogramme", Proceedings of the 21st International Helicopter Forum, Bückeburg, Germany, March 1998
104 H.J. Pausder, U. Butter, V. Gollnick, ACT Demonstrator and Flying Helicopter Simulator, 52th Annual AHS Forum, Washington DC, USA, June 1996
105 H.J. Pausder, U. Butter, F. Steinmaier, ACT – FHS for the Next Generation Technologies Evaluation and Demonstration, 25th European Rotorcraft Forum, Rome, Italy, September 1999
106 U. Butter, H.J. Pausder, F. Steinmaier, Safety Concept of the Technology Demonstrator ACT – FHS, 56th Annual AHS Forum, Virginia Beach, USA, May 2000
107 J. Kaletka, H. Kewscheck, U. Butter, FHS, the New Research Helicopter Ready for Use, 29th European Rotorcraft Forum, Friedrichshafen, Germany, September 2003

Part 4

International and Economical Relevance and Perspectives

The Significance of Aeronautical Research in Germany – Nationally and Internationally

GERO MADELUNG

The following considerations are fundamentally looking upon aeronautical research as a subdivision of the greater spectrum of research in the natural and engineering sciences. Aeronautical research, however, has the peculiar characteristic that it involves a multitude of disciplines – as already exposed in Part 1. Thus, the content of aeronautical research – involving aerodynamics as well as materials and structures, and propulsion as well as avionics – encompasses an extraordinarily wide scope.

The point of view for these considerations is primarily that of product-related research, following the spirit of the preceding volumes of the series "German Aeronautics" ("Die deutsche Luftfahrt") of the Bernard and Graefe publisher, on airplane-manufacturers and airplane-operators in Germany – to which the present volume is contributing particularly the aspect of research.

Some Remarks on the Evaluation of Research

For a consideration of the significance of aeronautical research in Germany it is perhaps appropriate to first review the traditional practice of research evaluation. A fundamental measure is the attention which a relevant publication on a subject of research is gaining among the experts, which in turn is reflected in the respective scientific journals. Important may also be the judgement of the responsible official research administration, not at least in terms of the financial support given to the particular subject of research. This holds also for the attention of the respective national professional society – within Germany today the DGLR – evidenced by inviting a presentation at a scientific congress. The societies also give recognition through awards which often refer to specific research contributions [1]. Such recognition of course is also given internationally.

A union of national societies founded 1958 on the initiative of *Theodore von Kármán* is the global professional society on aeronautical research, the "International Council of Aeronautical Sciences, ICAS". It recognizes significant multinational aeronautical developments such as, for example, the "Airbus" or the "MRCA/Tornado". In both cases the technical and scientific contributions also from Germany have been specifically highlighted.

Transatlantic exchange of aeronautical research had been facilitated since the 50's by NATO's "Advisory Group of Aeronautical Research and Development, AGARD" [2], including recognition of outstanding contributions. Scientists working in Germany have from time to time been awarded also by this multinational organization.

In the realm of applied research and experimental development, positive contributions are directly recognized through their successful application to airplanes, engines or items of equipment. "Success" can be officially recognized records or rankings in contests (such as for example the famous "Kaiserpreis-Contest" for aircraft engines or the "European Flying Contests" of the 20's and 30's). Finally, commercial success of a product can be taken as a measure, even if this may often not be attributable to a single item of technology.

Sometimes an item of key-research is accomplished independently by more than one researcher or research group. The "priority" is then difficult to allocate. Thus, for example, the revolutionary technology of propulsion by turbo-jet-engines has been researched and developed (under national security restrictions in each case) simultaneously in Britain and Germany up to the first experimental demonstration in flight.

Further down in this chapter there will be – supplementing the "highlights" presented in Part 3 – an attempt of an evaluation of significant contributions of aeronautics research in Germany along the above line of thought.

The Environment of Aeronautical Research – Compared Internationally

A comparative view of the environment in which research has been established and performed in Germany may provide some insight into its relative significance. The expositions of Part 2 may serve as a guide.

The most basic environmental element of research is probably the general motivation to utilize the scientific approach of exploration rather than the empirical approach. While the urge of exploration seems to be universal to all humans, the rigorous scientific approach is not so. In aeronautics – as a subject of applied research – man's ancient dream of flight provided also a basic impetus. Only later arose the more rational motive of developing an "ideal mode" of travel which would readily overcome natural barriers such as the sea and the mountains. And not at least was there the motive of gaining sovereignty and prestige relative to ones fellow humans. One or more elements of this spectrum of motives is moving – to a greater or lesser degree – probably all individuals and groups participating in aeronautical research.

Within Germany and – now – its federal states such a spectrum of motives to perform aeronautical research has been – and still is – prevailing, just like in the other developed nations of this world.

The effectiveness of research depends to a fair degree on the freedom to exchange information between individual researchers as well as groups of researchers. In Germany a society for the advancement of the aeronautical sciences is active in this sense already since 1912.

Aeronautics and especially aeronautical research however also require an industrial environment of the kind highly industrialized nations have. Such a background is essential for most of the individual researchers in aeronautics. This applied already to the early research towards unpowered flight and even more so towards powered flight. Germany belonged at the turn to the 20th century to the small group of highly industrialized nations, providing an environment comparable with that of countries like Britain, France, Italy, Japan and the USA.

With respect to the critical technology of air-vehicle propulsion the circumstances in Germany were particularly favorable, since there the research towards the high-rpm internal combustion engine had achieved a breakthrough as early as 1883. This technology has indeed been a deciding prerequisite for a successful aero-engine.

However, the early (1900) enthusiasm for the airship, which had been generated by the *Count Zeppelin,* obscured in the early pioneering days of powered flight the larger potential of the airplane. Looking back in time this is not really surprising, comparing the impressive sight of an airship with the flimsy appearance of contemporary airplanes. But nevertheless, the early research in Germany towards the construction of airships contributed later also to the research towards the development of airplanes. An example is the development of light-weight structures for airships including the first application of aluminum – since 1909 already as a heat-treated high-strength alloy. Another example is the development of light-weight engines of low fuel consumption and with propellers for the use on airships. Though these propulsion systems were to big for the use on airplanes they provided useful technology. Still another subsystem of later airplanes, which had first (1909) been developed for the airship, is wireless communication. Finally, the first research establishment for aerodynamics, the AVA at Göttingen, had initially been founded for the purpose of doing research for airships. Actually its work then immediately supported the development of the airplane. Thus, the to a degree delayed interest towards powered airplanes in Germany had been compensated in a way by the fallout of technology from the airship.

An important basis of research is of course an appropriate education system. A comparatively high developed standard of the faculties of the natural sciences at the universities as well as at the engineering colleges could be found in the German states already at the turn of the century. These institutions were and still are state-run, and until the 30's they were in the highest regard also of foreign students. The education in the engineering sciences in Germany demands still a background of industrial shop practice. Also worth mentioning in this context is the pervading concept of combining "research and teaching" ("Forschung und Lehre") at the German universities and institutes of technology.

This "practical attitude", for example, brought about that professors at the university of Göttingen convinced their governmental administrators as well as sponsors from industry to establish the first aeronautical research institute in the country, the AVA. It also led already in 1909 to the first lectureship for "Aeronautics" at universities, followed in 1911 in Stuttgart by the installation of the first chair of aeronautics. The subsequent establishment of such chairs at a number of other technical universities also brought into being in each case an officially funded facility for aeronautical research.

This level of professional education provided a most positive environment for aeronautical research in Germany. Thus, the pioneer *Otto Lilienthal* had been a graduate of the "Trade Academy" (later "College of Engineering" and finally "Institute of Technology (TH)") of Berlin, – however, before a course in aeronautics had been offered. Also the renowned *Hugo Junkers* has been a professional engineer and even professor of mechanical engineering before engaging in aeronautical research – and like *Lilienthal* he had earlier founded his own enterprise.

However, also the non-academic educational system in Germany facilitated and still facilitates successful research in aeronautics through skilled technicians and craftsmen, who contribute in an important way to experimental work and prototyping. Finally, the training of apprentices in handicraft and industry in Germany is fairly unique.

Subsequent to the achievements of the early pioneers, the official sponsoring of research at institutions outside of the universities became an important contribution to the development of aeronautics in some countries. In Germany such sponsoring of research became most prominent after the turn of the century with the establishment of the "Kaiser-Wilhelm-Society for the Advancement of Sciences, KWG", founded in 1911 (and renamed in 1948 in "Max-Plank-Society, MPG"). Through this society a network of institutes and research establishments is sponsored, including the segments of chemistry and physics. The official sponsoring of aeronautical research in Germany nevertheless followed a somewhat different – but mostly also highly productive – scheme, as outlined below and in Part 2.

In the context of illustrating the general environment of research in Germany, the "Deutsche Forschungsgemeinschaft, DFG" ("German Research Association") should not be overlooked. The DFG had been founded in 1920 in the aftermath of World War One as "Emergency Society for German Sciences" to support research projects through official as well as private sponsorship. Grants by trade and industry where – and still are – channeled through the "Stifterverband für die Deutsche Wissenschaft" ("Union of Sponsors of German Research").

And finally, in the general context of today's scientific research, the non-profit "Fraunhofer-Gesellschaft" performing tasks under contract by industry and the "Helmholtz-Union" of the 16 major state-run research establishments in Germany should be mentioned, as well as the official "Wissenschaftsrat" ("Research Counsel"). In summary – so far – it may be stated that aeronautical research in Germany always enjoyed a comparatively positive environment.

Consideration is now given to the aspect of "application" of aeronautical research, since this segment of research in general belongs to the category of "applied research" – in contrast to "basic research", such as for example in astronomy or in the arts. While the latter is traditionally being sponsored by the sovereign or by foundations as an element of the cultural development of the country, the sponsoring of applied research needs ultimately justification through its "practical return", be it in economic terms or through other usefulness.

In the upstart-phase of a new technology (see also Part 1), – either by enterprising "investment" or through official interest and sponsoring – it is generally expected that, at least in the long run, the expenses will be recovered in a successful application on the "market". An element of risk nevertheless is tacitly accepted. The "market" may in fact also be a governmental organization, such as for example the mail service. Official sponsoring of applied research may ultimately seek a "return" in terms of the potential development of a branch of national economy and the related creation of jobs.

This expectation of "return" may not be evident to an individual researcher, working for example in an academic laboratory, since he will have hardly any control over the application of the results of his research (such as for example a particular tool for structural analysis). But the expectation will in the long run apply to the "system" of aeronautical research as an element of the national environment. Thus, countries without a segment of aircraft manufacturing – i.e. without a "useful" application – will hardly host aeronautical research.

The environment for an useful application of results of aeronautical research has in Germany often not been as favorable as in most of the other highly industrialized nations. A brief account of developments in Germany and elsewhere may contribute to illustrate this point. *Otto Lilienthal* – who financed his research from the returns of his small manufacturing business – already had in mind the application of his research in short term for aero-sports. In fact, he executed orders for a small number of gliders to other experimenters. But his fatal flying accident (after about 2000 flights!) delayed the development of a "gliding-movement" at that time. The *Wrights* in contrast succeeded a little later not only in proceeding with their also privately funded research to the demonstration of powered flight, but also in the development of their airplane to perform useful missions. Thus, they were able to win a competition of the US Army (1909) for reconnaissance aircraft and also orders for deliveries of aircraft.

The market for military aircraft subsequently provided over several decades the major financial source for aeronautical research and development in many countries– in France since 1909 and in Germany as of 1911. The reason for this was, that the military users of aircraft where mostly prepared and able to fund aeronautical research – either contracted for or within their own establishments – ahead of specific orders for new equipment. The often private aircraft manufacturers could, with the financial returns of military orders, also fund their own "free" research and development projects. In this way, for example, *Ernst Heinkel* could undertake his research – together with *Pabst von Ohain* – to do the first demonstration of the feasibility of turbo-jet propulsion. On the other hand, however, the market for military aircraft was subject to fluctuations – depending on national budget allocations – and the funding of research was affected accordingly.

In Germany these fluctuations were more extreme than elsewhere due to the political conditions imposed from 1919–1933 and from 1945–1955. In fact, no military contracts at all were permitted in these periods. Even after 1955 until the present the military market in Germany is subject to extraordinary conditions, which shall be outlined later [3].

Researchers of course strive also in aeronautics for regular funding – apart from the varying official budgets – of their projects, the scientific nature of which may actually be largely independent of the source of funding.

Aero-sports provided very early such a potential source, and air shows as well as flying contests raised public interest, also in Germany. Thus, the development of aero-sports had been a positive environmental element for the young aeronautics research. Sponsored flying contests provided valuable impulses to research until the 30's and again since the 60's, in the latter phase to the research towards high performance gliders.

Nevertheless the ultimate aim of aeronautical research became the above mentioned "ideal mode" of travel which would contribute to bring the people of the world together. Visionary researchers quite early had in mind a market-based system like ocean shipping.

Just how long it would take to achieve this goal had, however, been beyond the imagination of most researchers. Actually, the seemingly commercial market of air travel or postal services was government-controlled over an extended period of time. Most of the worlds airlines depended well beyond 1945 on subsidies of their respective governments or were government owned [4]. This was motivated by the interest of the governments in the development of national ties to overseas possessions as well as in foreign trade. The latter had been true also for Germany after World War One, and provided the government with a reasonable argument to continue funding of aeronautical research, now without military needs and through the "Ministry of Transport". This mode of official support was at that time essential for the survival of aeronautical research in the country. In the background however there was also the aim of maintaining an option for the later reestablishment of military aviation.

One of the positive elements of the environment of aeronautical research had been – as mentioned above – official initiatives to foster and maintain aeronautical research establishments. The purpose was to scientifically support the development, procurement and operation of airplanes mainly for the governments own needs – e.g., for the military forces. And later also to act as the official authority for the airworthiness certification of airplanes, engines and other components, further to provide technical advice to the manufacturers.

Such research institutes had been founded in Germany already since 1910 and 1912, and they proved to be as highly qualified as the best comparable establishments in other countries. In some areas of research they were, at least temporarily, leading establishments. After World War Two the facilities had to be largely dismantled and for about 10 years no aeronautical research was permitted. As of the mid 80's there was a phase of directing the official funding of these establishments towards space research at the expense of aeronautical research, perhaps under the assumption that aeronautics was becoming commercially viable enough to fund its required research without official support. Nevertheless, there were to remain official functions related to airworthiness, even if these are being increasingly harmonized on a European level.

Last but not least the airplane manufacturers are contributing to the environment of aeronautical research. This applies to most countries with such an industry, and certainly to Germany. The contribution may be one of stimulation, or provision of research contracts or conducting research in own facilities. The research of the manufacturers is generally related to their products, and the demarcation between general applied research and product development is often rather fluid (see also Part 1). The more the modeling of a product departs from established standards, the more related research will be called for. The road to success in airplane manufacturing has often been one of innovation, however sometimes at serious commercial risk.

Gaining and retaining leadership in the market for a certain segment of products requires in aeronautics comparably heavy research expenditures in terms of business turnover. Expensive facilities such as a structural laboratory, wind tunnels and a flight test center are called for. In Germany *Hugo Junkers* had been probably the first airplane manufacturer with his own wind-tunnel facility, which he initially funded from the earnings of his factory for self invented gas heated geysers. In other countries some airplane manufacturers were government owned, not at least to enable such investment. Also a government sponsored establishment to render laboratory services to manufacturers – as in Germany after 1955 the IABG – may relieve manufacturing companies from heavy investment.

Most branches of manufacturing are of course required to provide for their funding of research and product development, in the long run through earnings accruing from sales and services of their products. In the interim it is done by funding through the financial resources of the enterprise. The airplane manufacturers however often lacked financial strength relative to the business risks. But official operators of airplanes such as the military services were often prepared – not only in Germany – to fund agreed programmes of research and development. Private use of the results of such research then required official consent, which was generally granted. Thus, for example, *Hugo Junkers* could launch his commercial transport type F 13 – the worldwide first dedicated design for this use – in a revolutionary all metal design in 1919 at relatively modest technical risk, building on the experience gained in the previous development, manufacturing and operation of military aircraft (several hundred airplanes) of this new technology. The dominating role of the Boeing Company (and its engine suppliers) over several decades in the rise of the jet-age of commercial transport grew in an analogous way, albeit on a much bigger scale.

The level of earnings from sales and services – and thus the potential of a manufacturer to fund proprietary research – depends also to a degree upon strength on the export market, since the national market for airplanes, engines and equipment may be relatively small. This is the case for airplane manufacturers in most countries, and also in Germany. Actually, only the manufacturers in the USA do enjoy a relatively large home market. Therefore most countries with an aircraft industry do officially support the export of both civil and military airplanes, engines and equipment. Germany nevertheless has been, for extended periods, an exception in this respect due to political restraints after the two World Wars. Thus, while industry in general in Germany is relatively strong on the export markets, the airplane manufacturers did – apart from soaring planes – not fit into this positive pattern. On the other hand, official support has been given in Germany to transnational projects, especially on a European level. While such projects have often been advantageous in terms of strengthening the manufacturers, they have however generally not been as beneficial as regular export business.

The above mentioned official interest in the development of air-transport has of course not only been a positive environmental element for aeronautical research as such, but also specifically for the aeroplane manufacturers. In many countries – and also in certain periods in Germany – actually the respective government itself established airlines and even funded the research at the manufacturers towards new or improved commercial airplanes, engines and equipment. This, because the manufacturers often did not have the financial strength to carry the entire burden of such programs on their own. A spectacular example is the Anglo-French development program of the supersonic airliner "Concorde" in the 60's.

The manufacturers in Germany however enjoyed this kind of civil market only in the 20's and 30's. After 1955 a less advantageous kind of official support of research for non-military use was instituted instead: support was provided to the national industry under the heading of alleviating its dependence upon military contracts. Actually, these military programs, which were due to Germany's participation in NATO, had been the basis to reestablish this industry after World War Two. However, already in the early 60's it became apparent that the related favorable employment could not be maintained without the reentry also into the civil airplane market. Nevertheless, this new kind of governmental support – where the official side itself was not the "user" of the research results – remained rather halfhearted and not really effective. Probably the competing trade interests in the aeronautical market of the major NATO allies played a role, too. This is perhaps also one of the reasons why official support in Germany was often politically justified in terms of the valuable, but secondary "spin-off" of aeronautical research into other branches of industry rather than with outright (primary) economic interests.

In the late 60's an additional political purpose of official support of aeronautical research towards non-military application came into the foreground: The contribution of transnational research to the political unification of Europe. The research towards the ultimately highly successful Airbus-Program was supported on the side of Germany largely under this heading, and this programme finally also opened the door for large scale export sales.

Thus European unification has been most beneficial to the airplane manufacturers in Germany, and to the research activities. At the same time it should not be overlooked that some of the manufacturers had developed good transatlantic business ties, not only in Germany, but also in France, Britain and other countries of Europe.

Finally some of the more recent environmental factors of particular relevance to aeronautical research in Germany – such as the international trade agreements, trade balance, political and environmental trends, defense and the influence of the European Union – shall be assessed.

The environment for applied research and product development in aeronautics has been influenced since the 90's by new international trade agreements (GATT/WTO), which particularly concern and restrain the official support of research outside the realm of national defense. In those nations where the defense policy gives high priority to aeronautical research and development – such as traditionally the USA – the non-military research may however benefit from considerable "spill-over". In Germany, with a rather low priority of aeronautical research within the defense policy (since the reunification in 1990), this valuable "indirect official support" is no longer so substantial. This – together with the restrictions arising from the trade agreements – is a disadvantage for aeronautical research in Germany relative to that in other countries.

However, there is the possibility that the political movement towards strengthening the European Union with a "defense-identity" will alleviate this disadvantage. The transnational amalgamation of the aeronautical manufacturing industry in Europe alone though will probably not be entirely beneficial to aeronautical research in Germany, because the envisioned "centers of excellence" of this industry in the various countries will have to rely – to a degree – on ocal support. In addition and contrary to some countries in Europe, influential political forces in Germany are no longer really "airminded". A positive element of the European Union is nevertheless the introduction of the "Euro", since this new currency is alleviating former disadvantages in the export of airplanes, engines and equipment and thus is potentially strengthening the funding of research.

In summary, the environment of aeronautical research has been in some aspects quite favorable in Germany. Otherwise, it is subject to more fluctuations than in comparable countries. Relative to the very positive environment of other research in the country, it has often been hampered by political circumstances.

The Significance for the Global Development of Aeronautics

Aeronautical research is strongly engaged in international exchange, and thus the researchers in Germany have often benefited from "imported" results. This despite of periodic restrictions in the transfer of industrial know-how or of restrictions due to considerations of national defense. On the other hand, researchers from Germany and the results of their work have often been beneficial to research in other countries. This shall be exemplified with a selection of highlights from Part 3. The sequence of the presentation shall not signify their relative importance.

The influence of the research of *Otto Lilienthal* and his publications on the general airplane development can be assessed as significant. The pioneers of powered flight – the *Wright Brothers* – documented this influence on their own work. *Ludwig Prandtl* has been recognized also outside of Germany for his basic contributions to aerodynamics as well as those of his school. *Theodore von Kármán*, his doctoral student (1906) and later member of his research staff and subsequently colleague, advanced aeronautical research in the USA significantly beginning in, say, 1929. After World War Two he contributed to the reestablishment of aeronautical research in continental Europe through the foundation and direction of AGARD.

The concept and development of the airplane with a cantilever "thick", internally braced wing of all metal construction was pioneered by *Hugo Junkers*. He also pioneered the quantity production of such revolutionary airplanes, which *von Kármán* judged to be one of the fundamental contributions towards the global development of aviation. Junkers airplanes established important records in terms of payload-range, speed and altitude at the time.

The "Sailplane Movement" – beginning 1920 in Germany – contributed "*more to the development of the aeronautical sciences than most of the powered flight during World War One*", so *von Kármán* [5]. While this assessment may be exaggerated, he also points to specific contributions: the high aerodynamic efficiency of the wing of high aspect ratio; the optimized lightweight structure; the research on meteorological phenomena. Research results of the sailplane move-ment have such contributed indirectly as well as directly to the global development of efficient transport airplanes of good aerodynamic efficiency and high payload. Even earlier, the "Light Airplanes" for aerosports took advantage of this technology.

The research towards modern high-speed flight with propulsion by propellers had been advanced in the USA since 1930. But already in 1932 the "aerodynamically smooth" design of the He 70 demonstrated the corresponding level of applied research also in Germany. For the first time an FAI speed record – over a distance of 1,000 km (504 nm) with a payload of 500 kg (1,102 lb) – had been achieved in the country. The famous British engine maker Rolls Royce bought a He 70 for the testing of his newest engine developments. By 1937, a racing version of the fighter Me 109 powered by an engine from Daimler Benz achieved the FAI speed record for land-planes, and subsequently a Me 209 the absolute FAI speed record, which stood (for propellers aircraft) for the next 30 years.

The research in Germany towards high-speed flight with turbo-jet propulsion in the period from the mid 30's until ist termination in 1945 was of pioneering significance. This applies also to the research towards the new propulsion system, with the axial flow compressor being the key technology as well as to the aerodynamic configuration – especially of the wing. The swept wing, the distinguishing mark of all modern transport airplanes, has been a result of research in Germany. The first prototype airplanes with this technology have actually been built there. Finally, in 1944, the introduction of the jet-technology into military service in Germany demonstrated convincingly its advantage in flight performance and thus provided a global impetus to its quick further development. The technology of the "thin wing" with a high lift system and the podded engine installation of the first jet airplane in large scale production also paved the way for future technical developments.

Research on aeroelastics and unsteady aerodynamics became important with increasing speed of flight, and the relevant work in Germany was of global significance. Also the technology of deceleration of airplanes in flight ("aerodynamic braking") became a widespread operational requirement, and the research on effective parachute systems for this purpose – and subsequently also for the reentry of space vehicles – was pioneered in Germany.

Research on stability and control of airplanes including helicopters achieved international recognition. More lately research on controlled flight at angles of attack beyond stall („post-stall flight") had been sparked in Germany and successfully explored in a joint effort with researchers in the USA in the X-31 program. Controlled flight in this regime had previously not been possible.

The important technology of light-weight structures and materials benefited significantly from research in Germany. The early work on the light-alloy airframes with the advantage of ruggedness, suitability for high speed flight and crash worthiness has been mentioned above. The technology of manufacturing airframes and components with fiber re-enforced plastics has been pioneered in Germany first for sailplanes and subsequently also for large transport airplanes like the "Airbus", but also for the rotary components of helicopters like the blades and hubs. Such structures are very light and rigid and are not prone to corrosion and fatigue.

With regard to avionics, research in Germany has also found international recognition, for example with the early introduction of airborne RADAR on night fighters including jet airplanes.

The contribution of aeronautical research in Germany to the success of multinational programs of airplanes, aero-engines and equipment became effective since the late 50's. An example is the product line of Airbus, as well as the military programs of the "Transall", the "Tornado" and lately the "Eurofighter-Typhoon", and also the family of modern helicopters of Eurocopter.

This brief summary of contributions of aeronautical research in Germany should of course not be taken as an all inclusive listing – it is limited to certain highlights. In many other areas original research has been performed, sometimes without finding exiting applications and sometimes also in parallel to research in other countries. But this is in the nature of all work in research. At last, the numerous non-spectacular contributions in the wide spectrum of refinement of the engineering sciences – which are indispensable for the safety and progress of global aviation – also merit recognition.

The Significance for the German Economy

Evident is the influence of the science of aerodynamics with the concept of "streamlining" of the shapes of fast ground vehicles such as automobiles or high speed trains. All manufacturers of automobiles have invested in their own wind tunnels and are testing their products during development. The major tool of aerodynamics, the wind tunnel, is also used to predict wind loads on large buildings and structures, as well as to clarify aspects of ventilation of buildings.

The aeronautical concept of the "light-weight structure" has gained a major influence on the automotive industry as well as on the construction of modern railroad cars. Aeronautic research has been extended towards these applications, which have often additional or differing requirements as to materials, production techniques and shapes than aircraft structures. Aeronautical research also continues to influence analytical methods of design for such light-weight structures. An other area where the "light-weight concept" has been taken on is sporting goods, such as for example tennis rackets made of carbon fiber. The research on modern wind turbines using fiber reinforced plastics for the rotors was successfully initiated by an aircraft engineer.

The research on internal combustion engines has been positively influenced by the special aeronautical requirements of low weight and low fuel consumption at high levels of power and at high altitudes. Already in 1916 research on the supercharged engine came to fruition and later that on the turbo-supercharger. Both features where subsequently applied to automotive engines. Research on fuel injection had already in the 30's been successful for application on aircraft engines and has later been adapted to many automotive engines.

The jet engine and the related research has been successfully adapted for gas-generators of naval propulsion systems, and also of powerplants, for pipelines and for electric generation of peak power, lately also as an element of optimizing the efficiency of steam power-plants.

Aeronautical research has also provided an important impulse to the research on guided missiles including the launchers for space-vehicles. Ist significance for the ongoing and most challenging research towards reusable space launchers is evident.

Last, but not least, aeronautics and its related research has always been significant as a pioneering user of other developments of high technology such as for example electronic data processing. *Konrad Zuse* found here the very first practical application of his revolutionary developments, and the requirements of aeronautical research are still providing an impetus to the development of the most powerful computers. This is also true for the research and development on highly integrated circuits ("computer chips") as well as on displays.

The discipline of "Stability and Control" of airplanes called for the pioneering development of electronic simulation as a tool for research and training. Subsequently this tool has been successfully applied to "man-machine-interface" problems. The concept of "Artificial Stability" has first been researched in aeronautics and elements thereof, such as anti-skid-control are now being applied also to cars.

The requirements of airplane manufacturing have often provided significant impulses to the technology of tool machinery and the related automation of multi-axis control, of tape laying of fiber reinforced structural components or electron-beam welding as well as laser cutting and welding.

It can be expected that the potential economic benefits from aeronautical research will continue to be of significance also to a spectrum of branches of trade and industry in Germany and elsewhere.

References

1 DGLR-Awards, http://www.dglr.de
2 J. van der Bliek (Editor), AGARD –The History 1952–1997, The NATO Research and Technology Organization, RTO, Paris, 1999
3 G. Madelung, 40 Jahre Flugzeugbau in der BRD, DGLR Jahrbuch 1955, Band III
4 K.-D. Seifert, Der deutsche Luftverkehr 1926–1945 – auf dem Weg zum Weltverkehr, Bernard & Graefe Verlag, Bonn 1999
5 Th. von Kármán (with Lee Edson), The Wind and Beyond: Theodore von Kármán, Pioneer in Aviation and Pathfinder in Space, Little, Brown and Company, Boston, Mass., USA, 1967

Conclusions and Perspectives

Inventors, Researchers, Entrepreneurs and the State in their Various Roles

Horst Prem

Throughout the history of every field of research, thus also research in aeronautics, major accomplishments have been achieved by extraordinary individuals. However, when discoverers or inventors create such achievements, their environment such as society, universities and state are virtually absent and may even try to show, that this new idea or concept is unrealistic and not usable. This was also the case in the initial phase of flying with apparatuses heavier than air.

In *Otto Lilienthal*, the "inventor", "scientist", "pilot" and "entrepreneur" were combined in one man. He had the idea to imitate the flight of birds, invented an apparatus, researched the basic principles together with his brother *Gustav*, published these principles [1] and no one on home soil was interested in it! Nevertheless *Lilienthal* succeeded in crossing the threshold of inventor and scientist to becoming an entrepreneur by selling, even exporting, his "Standard Glider" ("Normal-Segelapparat"). *Lilienthal* was a pilot, a test pilot and a flight instructor all at the same time, which was essential for his success. It allowed him to convince other aviation enthusiasts to follow his example.

Following *Lilienthal's* deadly accident on August 9, 1896, his idea was largely dismissed in Germany for the time being. The idea of the airship dominated the scene. In Göttingen on July 31, 1906, the "Society for the Study of Motor-Airships" ("Motorluftschiff-Studiengesellschaft") was founded by the visionary entrepreneur *Walther Rathenau* and the Prussian administration of education, which *Ludwig Prandtl* soon thereafter formed into a society, primarily concerned with the problem of flying with machines heavier than air. It became thereby a forerunner of the aerodynamic laboratory of the "Kaiser Wilhelm Society", in which the basics of aerodynamic were researched.

In every country, also in Germany, the state has played an important role in the development of traffic infrastructure through its military institutions: whether it is the channel system of navigation, road construction, or the railroad. There was no exception with aeronautics in Germany. In 1910 the military flight school in *Doeberitz* was founded, which served at the same time as the "Teaching and Laboratory Establishment for Military Aviation" ("Lehr- und Versuchsanstalt für Militärflugwesen"). Finally, twenty years after the first successful flights of *Lilienthal* (1891), the German Reich began to actively appear as promoter and buyer of aviation equipment.

It was one of the first tasks of the "German Test Establishment of Aeronautics" ("Deutsche Versuchsanstalt für Luftfahrt, DVL"), created on June 3, 1912 to organize the tendering procedure for the "Competition for the best German flight engine" ("Kaiserpreis") and to function as the technical-scientific judge in the evaluation process. The legal and responsible body of the DVL was a registered association, which combined the interests of the "German Office of the Interior" ("Reichsamt des Inneren") and of business (Daimler, Benz and Bosch). Additionally the DVL took on responsibilities as the publisher of construction regulations and as the type-testing authority, the so-called "Flugzeugmeisterei".

The history of German aeronautical research is characterized by the time gaps during and after the two world wars, which also allowed opportunities for a new beginning. With the prohibition of the air force imposed in Versailles in 1919, the German government was voided of its important role as "procuring party and operator" of airplanes. A national airline as a potential customer did not yet exist. But politicians and civil servants were of the opinion that this restriction should be no permanent condition and that a basic capacity should be maintained. This task was transferred to the "State Ministry of Transportation" ("Reichs-Verkehrsministerium"), which promoted aeronautical research at universities and research institutes. It also supported competitions in aircraft design and in air shows. The national promotion of aviation technology in the 20's was so effective that a high-technology air force could quickly be set up after the unilateral lifting of the Versailles restrictions by Germany in 1935 and despite the lack of technical support from other countries. With the creation of the "State Ministry of Aviation" ("Reichsluftfahrtministerium, RLM") in 1933, the state became very influential and promoted technically outstanding achievements, whose principles had mainly, however, been previously achieved.

The number of personnel involved to produce an airplane in the USA and in Great Britain, 11–18 workers per airplane, was already in the year 1937 lower than in Germany or France with approximately 21 workers. The German aircraft industry produced various models and even the forced coordination of the industry ("Gleichschaltung") by the state did not result in a set of standard types or in a more productive cooperation between the political leadership and the industry. *Lutz Budraß* documents this situation in his book concerning the aircraft industry and air armament in Germany [2].

However on a competitive market it is crucial to create an atmosphere between scientist, inventor, manufacturer and state which in addition to basic research also permits creative applied research. It is important to speed up the transition from researching the fundamentals to developing and verifying technologies. In every established industrial sector, scientists, inventors, entrepreneurs and administrative officials are embedded in institutions such as universities, laboratories, industry and governmental departments, which today is often referred to as a "System of innovation". Despite the embedding of different institutions, an effective system determines the competitive ability by minimizing the time-delays between the different institutions. Nevertheless the question must be permitted, whether today's nearly-zero fluctuation of personnel between these institutions in Germany and the splitting of aviation competencies into four governmental departments, encourages effectivness: BMV for air-traffic control and certification, BMBF for civilian research and technology, BMVg for the military sector and BMWi for the industrial policy. In order to focus on the problem, let us only concentrate on universities, research establishments and the industry.

The present cooperation between universities, research establishments and industries is not only a consequence of the restructuring process at the end of the fifties, when the "German Society of Flight Sciences" ("Deutsche Gesellschaft für Flugwissenschaften, DGF") was assigned the task to reorganize German aeronautical research establishments on May 29, 1959 [3]. This cooperation also takes into account the experience of the industry made during the restoration period: consistently shared responsibilities between technological development and technological verification result in more cost-effective industrial structures. Accordingly testing facilities are not duplicated and hence can be utilized more effectively. After 1945 it was realized that research in aviation, which had conducted its technological verification mainly through ground-facility experiments since *Otto Lilienthal's* first attempts, now had to establish a larger simulation sector and that this investment-intensive expansion could not be financed by the industry alone. This trend, especially in view of the strong enhancement by the evolving information technologies, is outlined in the following section "On the Way to the Virtual Product".

Let us return to the chronological order of events and to the essential innovation steps after *Lilienthal*. While *Hugo Junkers* introduced the all-metal aircraft and the cantilever wing [4] as his personal contributions to aircraft design, *Ernst Heinkel* focused on high-speed aircraft by utilizing theoretical results of research in order to develop aerodynamically "clean" aircraft shapes [5]. Fueled by the necessity to speed up the transition from the results of fundamental research to their industrial application, both *Junkers* and *Heinkel* pursued the philosophy of maintaining testing facilities such as wind tunnels or material testing installations on their company sites or of establishing construction standards themselves.

Claude Dornier is considered the pioneer of the flying boat, and at the same time *Willy Messerschmitt* contributed to innovation with the extreme lightweight design [6] and simplified structures for efficient manufacturing of high-performance aircraft.

Not until the middle of the thirties, did research establishments such as the "Aerodynamic Test Establishment" ("Aerodynamische Versuchsanstalt, AVA") in Göttingen under *Prandtl* and *Betz*, the "German Test Establishment of Aeronautics" ("Deutsche Versuchsanstalt für Luftfahrt, DVL") in Berlin Adlershof and the "Air Radio Research Institute Oberpfaffenhofen" ("Flugfunk-Forschungsinstitut Oberpfaffenhofen, FFO") near Munich become more effective partners of the industry parallel to the foundation of the "German Research Establishment for Aeronautics" ("Deutsche Forschungsanstalt für Luftfahrt, DFL/LFA") in Brunswick [7].

Before this time they were occupied with expanding and training their staff and researching the later essential fundamentals in their respective fields: aerodynamics, propulsion, flight characteristics, construction specifications, and aeronautical radio communication and navigation. Thus it is not surprising that it was important to *Messerschmitt* to find a complementary cooperation between industry and AVA/DFL in aerodynamics. The crucial swept-wing measurements (*H. Ludwieg*) could be completed in 1939 and beginning in 1941 the BF 109 E could undergo aerodynamic tests in original size in the speed range of up to 95 m/s (312 ft/s) in the Brunswick wind tunnel [8]. In cooperation with the industry, swept wings were measured in 1944 in the Brunswick high-speed wind tunnel on a Ju 287 model with four jet engines installed under the wings. Another exampel in this time is the design and special manufacturing of compressor and turbine blades for axial flow jet engines. Thus a cooperation between the industry and large-scale research establishments was generated, also caused by the state of war. After World War Two this cooperation continued through the division of labor between "Messerschmitt-Bölkow-Blohm, MBB" and the "Organization for the Operation of Industrial Facilities, Ltd." ("Industrieanlagenbetriebsgesellschaft, IABG") in Ottobrunn. Thereby the IABG took over more the role as a "technical monitoring board", supervising structural fatigue tests for the industry. In the north with the *Focke Wulf* and the *Weserflug* aircraft manufacturing companies, which later merged to become the "Vereinigte Flugtechnische Werke, VFW", the old *Junkers/Heinke* philosophy was continued. There were and still are today test facilities in industrial possession, which shoulc have been closed following the consequent division of basic research and technology development as well as large scale testing, even if there may be advantages in obtaining first experimental results without delay.

This development continues on a European level, and there is no one today who would seriously propose that industry builds, maintains and runs ground-test facilities such as the "German-Dutch Wind Tunnel, DNW" or the "European Transonic Wind Tunnel, ETW". However the lack of investing capital calls for the need for the different research institutions in Europe to seek coordination and concentration. Similar to the industry, they have to be involved in the process of European concentration in order to be ready to do civil as well as military business in the future. European "reference concepts" in the sector of technology demonstration should function as guides to facilitate this concentration and the assignment of specific competencies. The table, published in the January 1994 memorandum [9] gives an overview of the seven aviation research establishments in Europe. It also shows existing parallels, which can only be adjusted following a forward strategy.

Since the European aerospace industry must keep up in global competition [10], in particular with the USA, an effective organization of cooperation between universities, research establishments with their testing facilities and the industry is crucial. Here, the competitive ability of this technology-intensive industry is decided. It must subordinate prestige as a secondary value under the primary value of its long-term competitiveness.

The complexity of cooperation between the three partners is sketched in the following regarding the temporal succession of the phases of airplane development. Basic research, technological development and technological verification/demonstration describe thereby the field of technology and research, in which university research, large-scale research and the industry must pass the baton, in this order, however in a flexible way. The responsibility for the final phase of the technological development must be exercised by the industry, since the product phases, "definition", "development", "manufacturing" and "support" in the utilization phase up to the disposal of the product are the responsibility of the industry. Breaks can only be avoided

European aeronautic research establishments – an overview.

1. Organisation	CIRA	DLR (DVL)	DRA	FFA	INTA	NLR	ONERA
2. Nation	Italy	Germany	United Kingdom	Sweden	Spain	Netherlands	France
3. Headquarters	Capua	Cologne-Porz	Farnborough	Bromma	Torrejo de Ardoz	Amsterdam	Chatillon
4. Other sites (aeronautical)	Roma (Representative Office)	Stuttgart, Brunswick, Göttingen, Oberpfaffenhofen, Berlin	Bedford, Pyestock,	Stockholm Malvern		Northeast-polder	Chalais, Palaiseau, Modane, Le Fagua, Toulouse, Lille, Salon
5. History	1984	1969 (1912)	1918	1940	1942	1919	1946
6. Legal Status	Private Research Establishment	Private Research Establishment	Government Agency	Government Agency	Public Research Establishment	Private Research Establishment	Public Research Establishment under MoD Supervision
7. Owner	Campania Region (33%) Italian Aeronautical Industry (66%)	Government authorities, private cooperations and entities	Secretary of State for Defense	Government (MoD)	Ministry of Defense	NLR (funds partly provided by government)	Government
8. Composition of supervisory board	Government (M.U.R.S.T.)[1] Employees Repres.	Government Industry, Academia,	Government Repres	Government Industry, Employees scientists	Government members, Relevant independent members, Employees Rep.	Government Industry, Academia Repres.	Government Industry, Employees
9. Pricing (Facilities)	true costs oriented	market oriented	market oriented true costs	market oriented	market oriented	market oriented true costs	market oriented true costs
10. Profit/non Profit	non profit	non profit	non profit	non profit	non profit	non profit	non profit
11. Funding sources	institutional funding, contracts	institutional funding (grants), third party funding	100% contracts	20% gov. grants, 80% contracts (gov.) and industry	public funds (gov. budget), contracts	65% contracts, 35% public	subsidy of MoD, contracts
12. Total Funding (MECU) Operational Investment	125 MECU[2] aeron.: 81 MECUS 16% 84%	318 MECU aeron.: 91 MECU 87% 13%	1000 MECU aeron.: 150 MECU 90% 10%	13 MECU aeron.: 11 MECU 100% 0%	136 MECU aeron.: 68 MECU 30% 70%	71 MECU aeron.: 61 89% 11%	257 MECU aeron.: 116 MECU 86% 14%
13. Contract work (versus operational)	10%	10%	100%	80%	35%	70%	65%

European aeronautic research establishments (continued).

14. Staff (research /total)	350/500	2440/4200	3551/10760	140/190	514/1350	700/920	1285/2380
15. Aeron./space others % (of funding)	65/30/5	30/60/10	15/3/82	83/8/9	50/30/20	86/10/4	45/15/40
16. Civil/military (%) (orig. funding aeron.)	100/0	70/30	25/75	40/60	40/60	67/33	12/88

1 M.U.R.S.T. = Ministry of University, Scientific Research & Technology
2 1993 only
3 About 10% of total funding are used for investments but not earmarked

by maintaining continuity of responsibility across all phases. In addition the aircraft industry cannot afford major design changes in order to correct design flaws because of the small number of units produced and the longevity of the products (of over 40 years).

The decision of go-ahead with product development cannot be made before completing the technology verification. However the history of European aircraft projects shows that technological verification often had to be part of the development program. This sometimes resulted in high financial support by the involved governments, bringing Europeans under the allegation of subsidiary policies contradicting international (GATT/WTO) trade rules.

Future projects will have to clearly separate technological verification from the phase of development. At the same time such a proceeding means that the members of the European Union with an aviation industry must agree on technology demonstrators similar to the USA, which would have to be tested and assessed by a "European NASA". It is the question how fast Europe can carry out this philosophy, which will be the deciding factor for the competitive edge of its aircraft industry.

The technology demonstrators in the USA show the networking between the military and the civilian domain. In the context of a common European foreign and security policy, Germany should exercise a well-adjusted technology policy on the military sector, which can then eventually meet civilian needs. According to international standards, merely civilian technological programs are always interpreted as non-permitted subsidies as the example of the "Institute of World Economy" ("Institut für Weltwirtschaft") shows. In its Kieler study [11] it follows the US argumentation of inadmissible subsidies and has not taken into account until today subsidies, which have gone to the US industry via cross financing from military programmes. More realistic financial figures of this problem were collected recently by *P. Lawrence* and *D. Braddon* [10]. Yet in order to come to a long-term and stable cooperation between universities, research establishments, and the industry, also Germany should turn more to the international standards and should try to argue against the subsidy reproach by emphasizing the ambivalence of technology programs for military and civilian applications.

An analysis from the year 1992 shows that the USA invested 850 million US $ of public funds per year more than Europe in research and technology of the aircraft industry. Today this is done very skillfully through military programs, for example the cockpit design for the Boeing 777 evolved "purely coincidentally" from the military HIDEC program and did not need permission from the Pentagon as with the tanker KC 135, when it was transformed into the civilian Boeing 707. The continuing concentration process on the industrial side leads to structures of the industry which are suited to optimally organize the transfer from a military to a civilian technology and vice versa. The merger "Boeing" and "McDonald Douglas" is only one of many examples. More examples for the technology transfer from the military to the civilian side go far beyond the range of aviation and are summarized in the table on page 642.

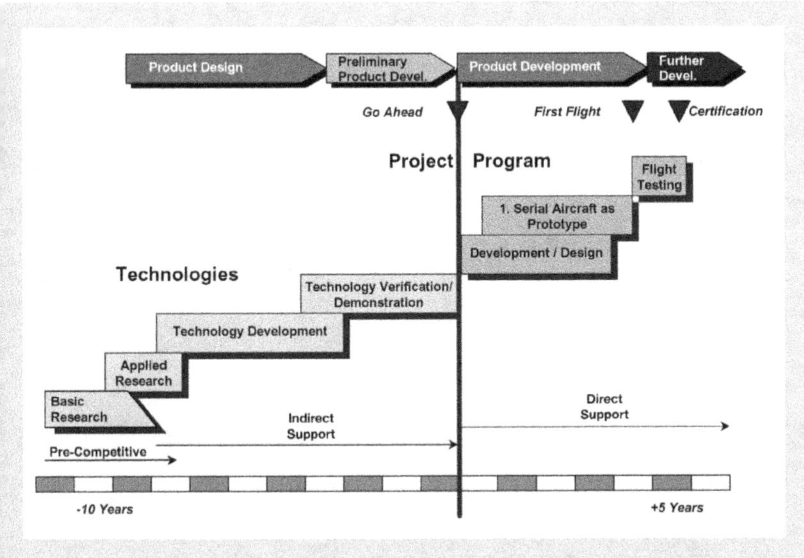

Phases of aircraft development [10].

Experimental Demonstrators USA (1992)		Experimental Demonstrators Europe (1992)	
B-757	Hybrid laminar flow testbed (subsonic)	F-100	Technology Demonstrator for laminar flow research (ELFIN)
B-737	High lift flap technology		
F-18	High angle-of-attack research vehicle (HARV) and thrust vectoring control system (TVCS)		
SR-71	Flight research testbed for high-speed civil transport (HSCT)		?
V22	Osprey — Noise testing		*National Demonstrators, which are not comparable with US Experimental Demonstrators*
XV-15	Certification profiles for noise abatement, failure modes, and handling qualities		
F-15	Highly integrated digital electronics control (HIDEC)		
VAV-8B	Vertical/short take off and landing research aircraft (VSRA)		
2 F16-XL	Hybrid laminar flow control (supersonic), for the highly swept wings of a high-speed civil transport!		
X-31	Enhanced fighter maneuverability (EFM)	X-31	Enhanced fighter maneuverability (EFM)
X-30	National Aero-Space Plane		

Technology demonstrators in the USA and in Europe [10].

Europe has achieved great successes in aviation research. The "Airbus" characterizes one of these successes, which today together with the supplying industry secures 40,000 jobs alone in Germany and which has brought trade balance advantages to the national economies in Europe by the selling of airplanes on a global market.

Approximately 2,000 sold aircraft alone of the "family" A 319/A 320/A 321 in the time period from 1987 until May 1999 speak for the success of Europe's commitment to the Airbus under technical terms and its philosophy of flight-safety. Now it is a commercial success. In view of an increasing air traffic market, it nevertheless is important to continue this success and to bring the market share of "Boeing" from today's 74

Military	*Transfer to*	Civil	Benefit
KC 135 (tanker)	➔	Boeing 707	90 % of technology
C-5A (large transport)	➔	Boeing 747 DC 10	Structural & aerodyn. configuration e.g. wing design and nacelles
TF-39 (large engine)	➔	CF 6 (DC 10 /Airbusengine)	Same technologies as for the TF-39
NASP (Nat. Aerospace Plane)	➔	HSCT (High Speed Civ. Transp.)	Similar technical requirements
Internet for Research Center	➔	Internet for public domaine	Priority of US-SW-industry worldwide; use of European development WWW
GPS	➔	GPS	US-onboard equipment worldwide; full use of technologies

Examples of military projects with civil application in the USA [10].

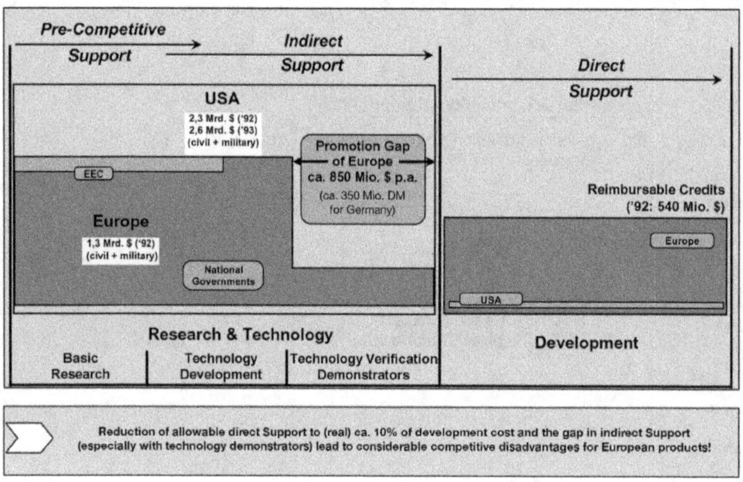

Sponsorship of research and development in the field of aeronautics. (Mio = million, Mrd. = billion).

per cent (in the year 2000) of the flying fleet to a more balanced level.

Germany's contribution to securing the European position can only consist of opening its national technology concepts to a cooperation within Europe and by convincing other European members through its competency. A key role has been assigned in Germany to the general agreement of cooperation regarding technology concepts, which was agreed on by the industry, the research establishments and the institutes of technology, respectively the technical universities. This does not mean a rejection of transatlantic cooperation but only an enforcement of the European position for more cooperation. Only after reaching an agreement on a realistic set of goals in the civilian aircraft industry concerning the "Airbus", could the European aerospace industry become an almost equivalent vendor in comparison to US-American companies in the market. Boeing alone reached a revenue of 45,8 billion US $ in 1997: a total revenue higher than the three prominent European companies "Aerospatiale", "BAe" and "Daimler Benz Aerospace, Dasa" together. Here the magnitude of the problem becomes evident.

Like the USA, "Europe" should privatize its aerospace industry. The "EADS" is the important first step. The heads of governments and the foreign and defense secretaries of the seven aviation nations will play a crucial role, since they are responsible for a common foreign and security policy, which cannot be enforced if technologically unfounded. A pre-competitive technological policy in a European framework is not sufficient. The European aerospace industry alone will not be able to adjust its competitive disadvantages regarding the USA. This means that also a European responsibility exists in order to secure jobs in this high-technology field.

On the Way to the Virtual Product

Ernst Heinrich Hirschel

The triumphant advancement of information and communication technologies is greatly changing aviation (which on the other hand has helped to further this advancement). This is not only present in product technologies from avionics to flight control, and from navigation to the "Infotainment" for the passengers of airliners. It also refers to process technologies, particularly those applied during the definition and development phases of airplanes. The potential and the development of information and communication technologies for these process technologies are presented and discussed briefly in the following section since they are a challenge to several areas of today's and future aeronautical research.

Aircraft design was already early an intensive user of electronic data processing [12]. The first main field of its application was in structural mechanics. With the introduction of the so-called vector processor at the beginning of the eighties, the computation-intensive numerical aerodynamics began to find its place within aircraft design. In the beginning of the nineties, extraordinary developments began with chip, storage and network technologies, which opened the actual age of information technologies. Consequently, the performance of personal computers increased dramatically, and new computer architectures such as the parallel computer became operational. Teraflops computers are now available [13], and in a few years Petaflops computers ("flops": "floating point operations per second", Tera $= 10^{12}$, Peta $= 10^{15}$). Also graphic interfaces for users, visualization systems, database systems and broadband high-speed communication are being developed to a previously unimaginable degree.

In recent years, the aircraft industry, as other industries, began to systematically utilize the potentials of the new information technologies. Meanwhile the term "Virtual Product" has acquired various definitions, which in the long run however will only refer to different aspects of one issue: the holistic representation of the product with the aid of linked data sets, which represent with high fidelity its physical and functional characteristics numerically or graphically. In the strict sense of the word there was always a "virtual product", even if it was "only" the concept of a product in the mind of an individual. The crucial point is now the qualitative and quantitative power of the virtual product, which can become possible by the employment of information technologies. Different aspects of the virtual product are the "Computer Aided Design, CAD", the "Computer Aided Engineering, CAE", and the still or animated graphic representation of the physical and functional product properties with the help of "Virtual Reality", which are supplemented by the possibilities of "networking" and of "distributed computing" beyond spatial limits, with the aid of broad-band network techniques.

Business Peculiarities of the Aircraft Industry

Within the last decades, the aircraft industry has strongly changed due to a change of paradigm in product philosophy: from the "technology-driven design" to the "design to cost" philosophy [14]. At the same time the competitive situation has not only changed worldwide, but also within Europe [11]. The concentration of the classic aircraft manufacturing companies in the USA into a few major corporations, the growth of the Asian aviation industry and the potential of Russian aircraft manufacturers represent enormous competition problems for the German aircraft industry, which must increasingly be seen within a European framework.

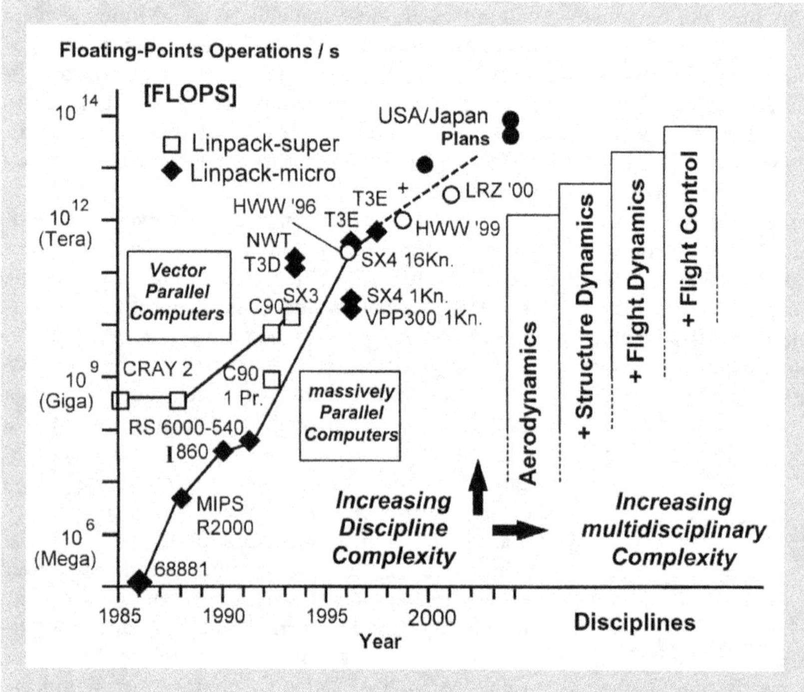

Increase of computer performance as well as potential disciplinary and interdisciplinary complexity in view of aircraft design. (HWW: "High-performance Computing Center for Science and Industry" (Stuttgart), LRZ: "Leibniz Computing Center" (Munich)).

In addition the peculiarities of aircraft manufacturing come into play, which represent a very large challenge for management and creditors in today's general business atmosphere (net yield expectations, currency rate of exchange risks). Product characteristics such as performance, cost of operation, risks of operation, life span, life-cycle cost, environment issues as well as product development and manufacturing (risks, cost and time to the market) are for every industry the quintessential points of business. The aircraft industry, however, has special peculiarities, such as large and long technology development efforts at a high cost and with great risks, long product-definition and development phases, small production series (a few hundred to a maximum of a few thousand aircraft) with high cost per unit and a restricted correction potential of product deficits, due to the small series. On the creditors' side it creates on the other hand a large and long capital tie-up, a very late net output (at least ten and more years up to the "break even" point) and a long product life, which lead to the typical long-term cycles of this business.

This discussion is greatly simplified. If for example a new airbus is derived from an existing model, the characteristic development time and the cost are much lower than for a new development (for instance the A 380). Military aircraft development has its own characteristics in this regard. The particularities of aircraft design are in general strongly coupled technical disciplines, systems/subsystems and functions, which go far beyond those of other industries, and which therefore make aircraft design, also with regard to cost issues, a technology driver par excellence, which no highly developed national economy can do without.

Particularities of aircraft design, such as the so-called (aerodynamic) flow path and extreme light-weight construction, are especially relevant to information technology and in particular to high-performance computation. Aircraft and their engines are extremely light-weight, flow-dominated entities, whose functions "lift" et cetera, and "propulsion" are "supported" and controlled by very flexible structures. The special aerodynamic phenomena in the outer (airframe) and the inner (propulsion) flow path encompass laminar-turbulent transition, turbulent and diffusive transport and mixing, separation, combustion and surface/flow interactions. Special structure phenomena include nonlinear structure behavior (material properties, deflections, contact problems), joint behavior (non-linearities, damping) and post buckling. The interdisciplinary issues concern flow/structure interactions (aeroelastics, flutter, buffeting), the static and dynamic airframe behavior with regard to flight control, and finally signature characteristics (noise, radar, infrared).

The problems with the definition and the development of the product lie in the description and quantification of these phenomena, testing on a reduced scale and system engineering. The physical restrictions of ground-facility simulation, and time and cost considerations continue to require more effective product-definition as well as development processes and tools.

The Prospects of the Virtual Product

In order to correctly comprehend and evaluate the potential of information technologies and in particular of high-performance computing, it is necessary to consider the processes of project definition and development, as shown in the following schematic and "linear" illustration. The product definition phase, which may require approximately two to four years, usually consists of three sub-phases: the concept phase, the pre-design phase and the design phase. At the end of the pre-design phase the concept is "frozen", meaning that its feasibility has been proven. Drastic changes are no longer made at this point, unless large problems, which arise in the subsequent phases, make it necessary. At the end of the pre-design phase "sensitivities" and the technological needs have also been identified.

During the design phase, details of the design, data sets and so on are achieved. At the end of this phase the design is "frozen". Once the appropriate decision has been made, the development phase begins, which can likewise take several years. In this phase the detailed construction (engineering) and the assembling of the first airframe take place. Finally, after one or more prototypes have been built and tested, the actual manufacturing begins. In each definition and development phase, the necessary functions, components, and subsystems are defined, examined, et cetera, and tested in ground-test facilities.

In the definition phase however not only shape and performance of the aircraft are specified, but also, implicitly, to a very large extent production, operation and maintenance cost [14–16]. These aspects must all

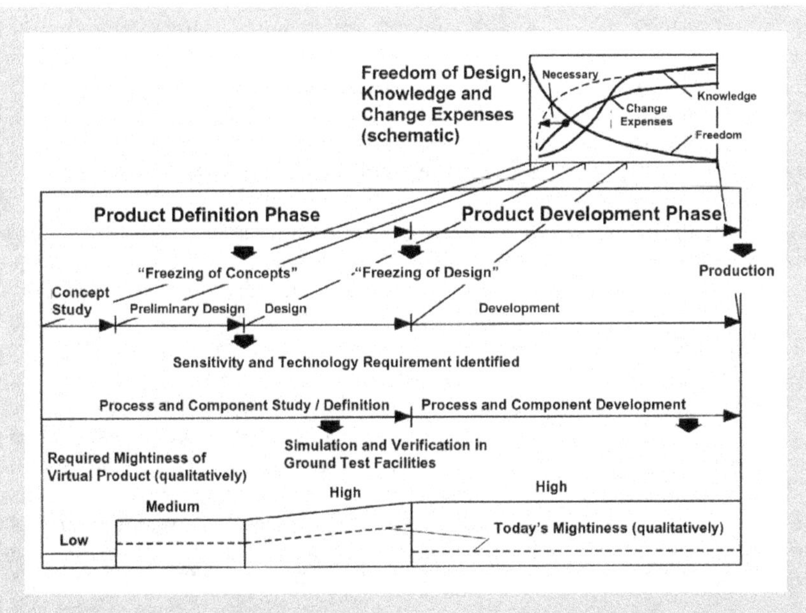

Schematic representation of product definition and development processes of aircraft design.

be sufficiently considered. Errors in this phase, above all if they are discovered too late, can affect the other phases catastrophically. This problem is schematically represented in the upper right corner of the illustration. As a general rule, the "knowledge" about the future product (it's characteristics, functions and so on) increases throughout the phases. The "freedom" to make changes decreases however, since the "cost of change" increase the more the product is already defined or developed. Therefore it is necessary and desirable to gain as much knowledge as possible about the product already during the early product definition phases.*

As a consequence, product-definition processes in particular, but also other processes including their tools must be permanently expanded and improved. Large progress in information technologies now allows the formalization and computerization of the product-definition and development processes in order to create the "virtual product". All of this is an approach of "concurrent engineering", which has a large potential in the aircraft industry, as well as in other industries.

Current – and past – product definition and development relies on approximate and linear methods and the wind tunnel in aerodynamics, on finite element methods and ground-facility structural tests in structure mechanics as well as on approximate and linear interdisciplinary linkages of subsystems and their functions. Linear "bookkeeping" methods, scaling methods et cetera, are used sequentially in the design cycles. This approach is well proven and largely successful. The products of the aircraft industry are safe and accepted, however the potential of this approach is to a large extent exhausted [16].

Developments in information technologies and in particular high-performance computing in principle now open the following two perspectives:

- high-fidelity description and optimization of the functions and the physical properties of the aircraft,
- holistic treatment of the aircraft in formalized computer-aided product-definition and development processes, the "virtual product",

which make an innovative and evolutionary development of the current approach possible with justifiable risks. The objectives of the "virtual product" are summarized as:

- holistic consideration of the aircraft,
- improved description of the functions and the physical properties of the aircraft,
- more effective pursuit of sensitivities including growth factors, et cetera,
- early recognition of problems,
- sensitivity-driven use of design tools (from simple tools to high-performance computing),
- improved feasibility statements,
- improved baseline for disciplinary and multidisciplinary optimization,
- improved control of life-cycle costs,
- improved product quality (performance guarantees, system efficiency, and operational risks).

The expected effects on the business are:

- reduction of the number of process cycles and of process duration,
- avoidance of "unnecessary" process iterations,
- avoidance/reduction of possible repair solutions [16, 14] in later product phases (cost, time),
- reduction of process and change cost,
- reduction of the number of flight-tests,
- reduction of development risks.

In the background of this development is a change of the classical design paradigm, i.e., Cayley's design paradigm, which however already began in the 60's and 70's and is now increasingly supported by new information and communication technologies.

The classical approach in the aircraft industry, as well as in other technical industries, has been to assign, in the case of the aircraft, functions like lift generation or propulsion et cetera to clearly corresponding subsystems such as the wing or the propulsion system et cetera, while these subsystems and their functions are to be only weakly and linearly coupled ("Cayley's design-paradigm" [16, 14]). If this is the case, each subsystem in itself can be optimized,

* Between 6 and 10 per cent of the development cost are spent at the end of the product-definition phase. The relation between development cost, manufacturing cost and operation/maintenance cost amounts to an approx. 1:3:6 ratio.

which at the same time optimizes the whole product, the aircraft.

Yet in modern of aircraft design, an increasingly strong and nonlinear coupling of subsystems takes place in order to improve flight performance, to lower fuel consumption, to reduce the impact on the environmental et cetera. Examples are airliners where high-by-pass engines must be effectively "coupled" with the wing, or the unstable flying (CCV) combat aircraft (and possibly in foreseeable time also the airliner), which is characterized by a strong coupling of aerodynamics, the flexible structure and the flight-control system. Combat aircraft additionally have a highly integrated propulsion system and a large number of various configurations of external stores. All of the above interfere with signature problems (radar, infrared), which can influence the aerodynamic shape of the airplane and the integration of the propulsion system (inlet, nozzle).

In order to limit cost and risks of development to an acceptable amount, these developments require more and more the outlined high-fidelity holistic design and holistic optimization.

High-Performance Computing as a Requirement of the Virtual Product

High-performance computing in aircraft design is essential when it comes to the definition and optimization of the physical properties and the functions of the aircraft; in other words, the aerodynamic characteristics of flight and maneuver performances and the structural mass as well as the database of the flight control system. According to the requirements of the respective product process the concept may be provisional or as complete as possible [14]. In this respect the disciplines of aerodynamics, structure mechanics and flight control play key roles, each having the highest computational need.

The roles of aerodynamics are

- the definition and optimization of the aerodynamic shape of the aircraft,
- the aerodynamic propulsion integration and optimization,
- the determination of aerodynamic structural loads,
- the specification of surface properties in regard to the external and internal (propulsion) flow paths.

The roles of structure mechanics are on the other hand:

- the definition and optimization of the structural topology of the airframe,
- the structural layout,
- the determination and optimization of the aero-servo-elastic properties of the airframe,
- the dimensioning of the construction parts.

All of these roles result in very close interdisciplinary connections, which have a high potential in cost reduction and quality increase through high-performance computing. Since the interdisciplinary links go far beyond these "core connections" high-performance computing in aircraft development can be seen in an extended perspective. This is the surrounding environment of general developments in information technologies and broad-band high-speed communication, which has the potential to achieve large progress in system engineering, process organization and process management, data storage ("product data management"), analysis and visualization techniques et cetera. This potential exists also with regard to disciplinary and multidisciplinary numerical simulation and optimization methods and regarding flow-physics and structure-physics modeling [14, 16].

The development of these elements of the "virtual product" has been initiated through focused technology programs in industry and in research (DLR, universities), with promotion by several federal ministries, by the DFG, and on a European level by the European Union.

The program system "MEGAFLOW" of numerical airframe aerodynamics (CFD) is currently in an advanced state of development [17]. The next steps have been introduced, considering the presented perspective of the "virtual product" and the fact that high-performance computing can and must be made available to the industry because of the large innovation and cost-saving potential and for competitive reasons. This will lead from the computational problems of the individual disciplines to an all-encompassing computational description of the entire aircraft, in the sense of "concurrent engineering" as an approach to the "virtual product".

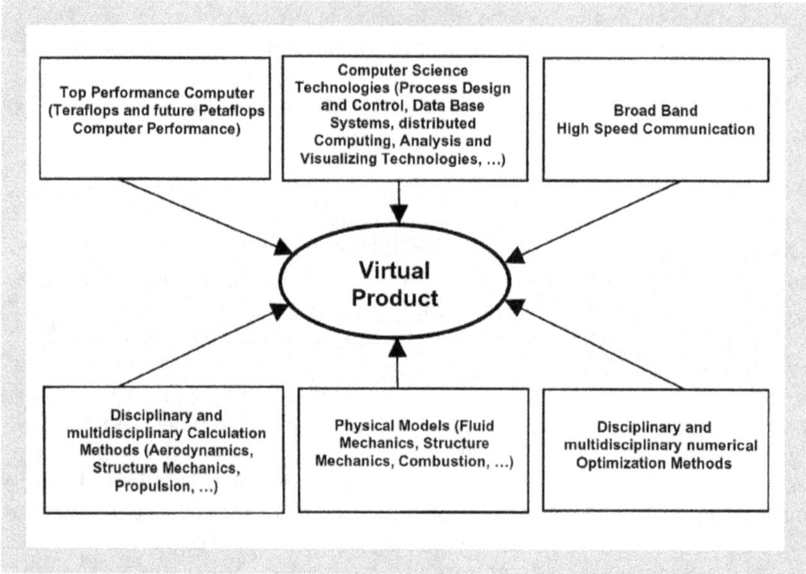

The basic elements of High-Performance-Computing as components of the virtual product.

Developments of Concurrent Engineering

Up to now the product definition and development processes in aircraft development have been strongly discipline-oriented. This orientation was driven by the worldwide requirements of high-technology standards [16]. Due to the mentioned reasons a change is being registered also here, essentially caused by modern information technologies. Today the concerned disciplines are beginning to reduce sequential work wherever possible in favor of parallel work, i.e., of "concurrent engineering", which requires the common use of software languages, data formats, computing systems, data bases and visualization systems. With the Airbus partners' "Airbus Concurrent Engineering, ACE" project the appropriate prerequisites have been established.

If the involved departments work in different locations, wide-band high-speed communication is important, in order to make "multi-site concurrent engineering" possible. The "distributed work" takes place through modern videoconferences, by which the engineer can directly communicate audio-visually with his or her partner over the screen.

A further important element for concurrent engineering is three-dimensional visualization of construction parts on the screen. This concerns on the one hand the design process with "CAD", on the other hand installation, accessibility and maintenance problems, which could so far be worked on only with the help of an engineering mock-up. The "virtual mock up", even with "virtual mannequins", permits cost and time saving in design work, particularly in the product-development phase.

The "simulation" of the product for verification and also for training purposes, is a further discipline, which profits enormously from new information technologies. Extremely high data rates permit a very realistic visualization, of, e.g., movement procedures in virtual scenarios (airport with environment, etc.).

The information and communication technologies thus are extensively changing many product processes also in aviation. Large efforts in research are being undertaken in order to make their potential usable. In addition, they are demanding new requirements in the education and qualification of staff and management [16].

How will it Continue?

Horst Prem, Ernst Heinrich Hirschel

Why Predictions are Possible

The computerized representation of the "virtual product" shows a trend, which will gain special importance in aircraft manufacturing because of the relatively small number of aircraft of a kind produced. This trend is obvious. Trends can be observed and predicted with the help of different means. In Part 1 the S-curve approach was mentioned. It is sketched now as a somewhat different and unconventional approach to predict trends as a function of time.

The S-curve approach can be traced back to *Vito Volterra*. He knew about *D'Ancona's* observation of fishing in the Adria [18], specifically the relative increase of the number of predatory fish in comparison to the number of their prey. Since the large fish eat the small and therefore their survival depends on the small fish, one should expect an interrelation of the population sizes. First the number of predator fish rises, until the prey becomes rarer. Now the large fish begin to suffer from hunger, and their decreasing population gives the smaller survivors the chance to increase. *Volterra* described this procedure mathematically. The time dependence of the population N(t) presents itself in the form of an S-curve:

$$N(t) = \frac{N_0}{\left[1 + e^{-a(t-T)}\right]}.$$

This "growth-relation" can also be used to describe the implementation of new products on the market, whereby 'N_0' indicates the saturation value and the half-life time 'T' marks the inflection point of the S-curve, while the parameter 'a' is a measure of the "slope" of the curve. The first derivative of this growth curve describes the change of a population, and the second the rate of this change.

With this rather simple mathematical formulation, *A. Grubler* and *Th. Modis* examined the infrastructure of the American system of transportation and found a well arranged parade of S-curves with 55–56 years distance from inflection point to inflection point (remember the length of the Kondratieff cycles of about 40–50 years (Part 1)). Although the saturation values are very different, they can be represented in a standardized form:

Traffic system	Saturation rate
channels	4,000 miles
railways	300,000 miles
roads	$3.4 * 10^6$ miles
air traffic routes	$3.2 * 10^6$ miles

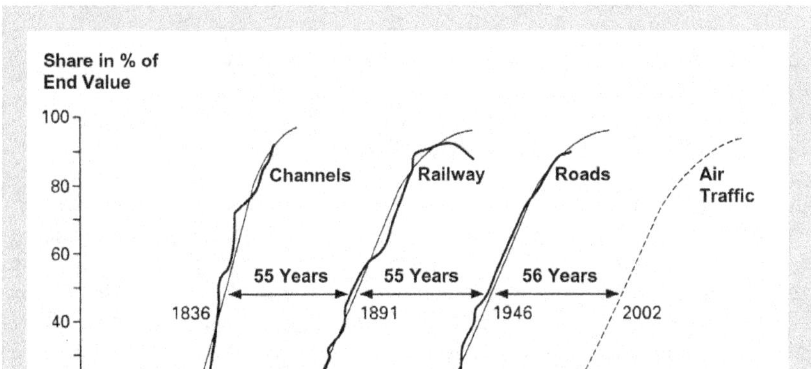

Parade of S-curves representing the growth of traffic throughout the history of the American transportation systems in per cent with variable total distances in miles. The magnetically levitated vehicle "Maglev" could be introduced as a transportation system sometime around the turn of the century, but the time at which it will have reached half of its capacity, will probably be in the year 2058 (after *Arnulf Gruber*).

If the trend observed with the ground-transportation systems is accepted for aeronautics, too, the largest increase in aircraft transportation capacity was present around the year 2002, i.e., the commercial airplane fleet had reached then half of its transportation potential. We can simultaneously see that the saturation value for the air-transport system can be expected around the year 2060. In order to tackle this increase, we certainly need an increased transportation capacity per airplane, instead of an enlargement of the fleets, since airports are pushed to their capacity limits. This might also help to decrease the environmental impact of NO_X (increase of the ozone concentration in the upper troposphere) and CO_2 (greenhouse effect, amplified by NO_X).

Increase of Transportation Capacity

The transportation capacity of an airplane can be increased by increasing either the airspeed, or increasing the payload, or by increasing both. Today's airliners fly in the transonic range (flight Mach number $M < 1$). For different reasons an increase of the airspeed leads immediately into the supersonic range of $M = 2–2.5$, in which the "Concorde" operated at $M = 2.2$, and a further increase into the low hypersonic range with $M = 5–6$.

Today's studies of supersonic airliners project flight Mach numbers somewhat higher than those of the "Concorde", with almost twice the number of passengers. Technological progress is viewed as necessary with regard to propulsion (increase of efficiency, emission reduction), aerodynamics (increase of aerodynamic quality in cruise, decrease of the sonic boom, improvement of the high-lift system for take-off and landing), materials and structures (structural weight) as well as to flight control/avionics.

Today it is expected that the payload and not the airspeed will continue to increase. Hence the airspeed will remain in the transonic range ($M = 0.7–0.86$), however with projected passenger numbers of 600–1,000 per airplane. In terms of aircraft configurations the classical form (now the Airbus A 380 with two passenger decks) will be kept while other "unconventional" configurations are being studied. As in supersonic flight, technological challenges lie in all disciplines, however with other emphasis.

Ground operations (airport infrastructure, passenger boarding and de-boarding, passenger evacuation in an emergency, et cetera) will cause difficult problems with such large aircraft. Another problem is the far-field vortex wakes of these airplanes, which are proportional in their strength to the gross flying weight. Since vortex wakes can become dangerous to following smaller airplanes especially during the landing approach, the benefit of very large aircraft regarding the transportation capacity is reduced by the necessary spacing of aircraft and hence a limited take-off and landing frequency. Separating aircraft according to their flight weight on the runways is generally no solution, because of the bottlenecks in the capacity of the airports. Today's criteria in determining the spacing of aircraft are based on the empirical values of the 70's. In the meantime a scientific basis was developed, which analyzes and considers also the interactions between the different vortex sheets and vortices behind the aircraft (main vortices, flap vortices, pylon wakes). A decrease or "management" of the strength of vortex wakes by means of classical aerodynamic approaches is only marginally possible because they reflect the laws of physics, such as the sonic boom. Nevertheless if traffic continues to rise, larger airplanes will surely have positive effects. The wing and flap architecture must however evolve in a way that will not demand larger separation distances than today. Configuring the wing with minimum wake intensity during landing is becoming an additional criterion for optimizing the transportation capacity. New unconventional configurations, maybe delta configurations, could possibly be an advantage in this regard.

Ecological Compatibility

Noise and pollutants are substantial factors of the environmental impact of airplanes. Different technological concepts are being pursued in order to decrease noise during take-off and landing: increasing the bypass ratio of the engines and thus decreasing the excessive speeds of the jet-engine exhaust, reduction of the airframe noise (open landing-gear bays, extended landing flaps and control surfaces, et cetera are sources of this noise) and in connection with the improvement of high-lift systems, introduction of special flight-path control measures (steeper climb after lift-off and landing approaches).

The efforts to decrease the amount of pollutants begin first with the engine, then with the aerodynamics of the airframe (drag reduction by refined aerodynamic shaping and propulsion integration, boundary-layer management (laminar wing, turbulence management)), the structural weight, and finally with the fuel, where today the substitution of kerosene by hydrogen is regarded as most promising. Hydrogen as a fuel requires a larger tank volume, which again affects the configuration and the drag problem (infra-structure problems pose further questions). The technological developments concerning these concepts must continuously be advanced in long-term research, which is becoming less popular with today's shareholder-value-thinking. Here a stable cooperation between politics and economics must help to overcome the purely financial viewpoint and create a technology-friendly atmosphere for provident research. In the Federal Republic of Germany, Bavaria has been a pioneer in this field of the aircraft industry for a long time.

Increasing environmental impacts are going to lead to emission taxes for air traffic in some countries and regions in the near future [15], which will then change the whole perspective on this issue. A similar effect regarding technical advancement was caused by the increased price of crude oil in the energy crisis of 1972/73 and in 1983. This can however be dealt with only through internationally coordinated actions in the context of fiscal legislation.

Another thought is nevertheless important in the reduction of the environmental impact, taking into account the well arranged parade of the transportation system's infrastructure. *Modis* writes [18]:

"Every transport system is very strongly connected to a certain primary source of energy: channels with draft animals (and thus animal fodder), railways with coal and cars with oil. The emergence of a new energy source precedes the saturation phase of the transport system in use at a specific time. Following this discussion the airplane fuel of the future will not rely on oil anyhow! The use of kerosene with airplanes is directly comparable with the use of wood in locomotives in the early period of the railway (at least until the 1870s), although coal had already gained a strong significance as the primary energy source. The most modern primary source of energy today is nuclear energy, which speaks for the development of an engine system with hydrogen from nuclear energy, which could greatly advance the air traffic cycle".

Even if the appropriateness of nuclear energy must be doubted today because of its potential accidents and unresolved disposal problems, it is out of the question that hydrogen will gain even more significance as a secondary source of energy. Whether the burning or the electro-chemical transformation of hydrogen in electrical driving power will become the predominant source of energy certainly depends on the ability to build light-weight electrical engines in the future which are based on electrostatics and not on electromagnetics.

Minimization of Infrastructure Cost by Intelligent Traffic Management

The attractiveness of air traffic compared to traffic over longer distances on tracks, will remain consistent, due to cost reasons alone. In this sense the same also applies to shipping. Air traffic is characterized by minimum infrastructure cost, since an effective air-traffic control makes it unnecessary to maintain networks such as railroad tracks or roads. The "intelligence" of the on-board systems of airplanes must however still be further developed. The non-availability of anti-collision devices has pushed air traffic throughout its historical development into air-route thinking, which is actually unnecessary and only causes cost. Intelligent flight routing in principle can be implemented on airplanes with today's computer and sensor technology involving satellite-based communication and navigation systems. In that case the flight crew would become almost autonomous regarding navigation decisions, so that infrastructure cost could finally be limited to airports and satellites.

Again military developments are setting the pace in the development of the "autonomous" airplane on a long-term basis. This terrain has been well prepared in the USA. Low infrastructure cost will result in inexpensive transportation and low transport times in air traffic with distances more than approximately 800 km (500 miles). On the other hand, the advantages of track-based systems, i.e., the railroad, lie in their spatial distribution, the access to the city centers, as well as the connection to airports. Connecting airports to efficient track-based systems for passengers and cargo would create a network of traffic systems, which so far has often been limited to the building of new roads when extending or creating new airports. Since only a few airports today are fully linked to the conventional railroad system, the "*Maglev*" should become interesting for the planners of traffic infrastructure. This could apply also to the development of a container system "aircraft-Maglev".

What kind of new transport system will be developed "following" air traffic once it has reached its saturation around the year 2060 is an interesting question, which will not be further pursued here. In any case there are sufficient incentives for young people to participate in the development of an efficient, environmental friendly, and economical air traffic system in the future, which will look different than today. Holistic and network thinking are essential in order to create the traffic infrastructure of the future, which will require more than the mere optimization of only one means of transportation.

Quality Management, Bypassing Human Inadequacies

The aircraft industry was already in the past a pioneer in quality management. The term suggests that this procedure exceeds quality control, which is usually practiced in the industry. Quality management means that during the design of a product, product quality and security are taken into account from the beginning rather than only their later examination.

Redundancy principles, like the ones used for air-traffic control systems, where they eliminate catastrophic errors before they can happen at all, are also typical measures of quality management. They include the mathematical analysis of the overall system in order to gain reliable data on the product as well as the prediction of life expectancies on the basis of such data. Today these considerations also include the problem of human inadequacies in the piloting of airplanes. Piloting an airplane no longer means that the pilot's input is translated at a 1 : 1 ratio during a maneuver. The flight-control system actually examines the context of this input, and may even annul it.

The methods and procedures of quality management in the aircraft industry can be transferred to all other branches of industrial activity. Therefore young people, who are active in the area of quality management, have chances in every field of the industry.

The Space Plane

The space plane is definitely not an aircraft which can increase the transportation capacity of aviation. It is addressed briefly here, because it presents the greatest challenge in aeronautics today in interaction with astronautics. Under space planes we understand, on the one hand winged reentry devices like the American "Space Shuttle", which are brought into orbit via rockets and which glide back into the earth's atmosphere ("space glider"). On the other hand space planes resemble airplane-like devices, which ascend into orbit by means of airbreathing jet engines (ramjet ("ram") and supersonic ramjet ("scram") engines) and rocket propulsion.

Although the latter have been in the technological preparation stage [19] for some time, it will take a few decades before they can replace the purely rocket-propelled space gliders. The first ideas for such a space plane came from *Eugen Sänger*, who proposed the antipodal bomber during World War Two. In 1964 he projected a space-transportation system consisting of two stages, including air-breathing propulsion of the first stage.

Large technology development projects for air-breathing space-transportation systems began in the middle of the eighties in the USA ("National Aerospace Plane, NASP" program), and in Germany ("National Hypersonics Technology Program" with the lower stage "Sänger" of a two-stage-to-orbit system as reference concept) followed by English, French ("PREPHA" program), Russian and Japanese technology programs.

Besides airbreathing space-transportation systems, possible hypersonic airliners were investigated, particularly in the USA, in Germany and France, which would be in a way natural by-products of space-transportation systems (also applying to its military counterpart of course). Hypersonic airliners, which fly at a Mach numbers of between $M = 4\text{–}6$ at an altitude of approximately 30 km (100,000 ft), are a potentially very interesting means of air traffic for long distances, but not for large-scale transportation. They would also be propelled by methane or hydrogen, making them "clean" aircrafts, although the problem of water exhaust into the atmosphere (stratosphere) would have to be considered.

The airbreathing space-transportation system, as well as the hypersonic airliner, has extraordinary technology requirements, regardless whether it is realized as a two-stage-to-orbit system or as a single-stage-to-orbit system. The difficulties are the very large speed span between take-off (and landing) and orbital flight in the extreme case of the single-stage-to-orbit system (combined then with a rocket engine), the large operational range of the airbreathing propulsion system, the high surface temperatures of the airframe, and guidance, control and navigation at extremely high speeds. In addition these aircraft are confronted with the problem of an extremely high nonlinear coupling of the lift and the propulsion system (Cayley's design paradigm is completely annulled here), the problem of insufficient ground testing facilities, et cetera.

Space planes, whether space gliders, airbreathing systems or hypersonic airliners, put very high demands on basic and applied research, as well as on experimental development in the industry. Although presently only second-generation space gliders, after the "Space Shuttle", are under consideration and airbreathing systems are expected to evolve only much later, the influence and effects of the corresponding technology developments on other industries are immense [20]. Germany's aeronautical research and industry are devoted to this topic and are developing these very promising and challenging technologies in national and European (ESA) technology programs.

References

1. O. Lilienthal, Der Vogelflug als Grundlage der Fliegekunst – Ein Beitrag zur Systematik der Flugtechnik, R. Gaertners Verlagsbuchhandlung, Berlin, 1889. (English edition: Otto Lilienthal, Birdflight as the Basis of Aviation: A Contribution toward a System of Aviation, Motorbooks Intl. /MBI Publishing Company, St. Paul, MN, USA, 2001)
2. L. Budraß, Flugzeugindustrie und Luftrüstung in Deutschland 1918–1945, Droste Verlag, Düsseldorf, 1998
3. H. Trischler, Luft- und Raumfahrtforschung in Deutschland 1900–1970, Campus Verlag, Frankfurt/M, 1992, p. 371
4. W. Wagner, Hugo Junkers Pionier der Luftfahrt – seine Flugzeuge, Bernard & Graefe Verlag, Bonn, 1996
5. H.D. Köhler, Ernst Heinkel – Pionier der Schnellflugzeuge, Bernard & Graefe Verlag, Bonn, 1983
6. H.J. Ebert, J. B. Kaiser, K. Peters, Willy Messerschmitt – Pionier der Luftfahrt und des Leichtbaues, Bernard & Graefe Verlag, Bonn, 1992. (English edition: H.J. Ebert, J.B. Kaiser, K. Peters, Willy Messerschmitt – Pioneer of Aviation Design, Schiffer Publishing Ltd., Atglen, PA, USA, 2000)
7. A. Betz, J. Brämer, W. Georgii, M. Dieckmann, H. Blenk, Beiträge zur Geschichte der Deutschen Luftfahrtwissenschaft und -technik, Deutsche Akademie der Luftfahrtforschung, 1941
8. H. Schlichting, Die Entwicklung der Windkanäle in den letzten beiden Jahrzehnten, Jahrbuch 1952 der WGL
9. W. Kröll et al., Joint Position on the Future Role of the Aeronautical Research Establishments in Europe, German Aerospace Establishment, Cologne, 1994
10. P. Lawrence, D. Braddon, (Eds.), Strategic Issues in European Aerospace, Ashgate, London, 1999
11. G. Bletschacher, H. Klodt, Strategische Handels- und Industriepolitik, J.C.B. Mohr, Tübingen, 1992, p. 69
12. R. Eppler, Rechner und Rechenmethoden, In: Ein Jahrhundert Flugzeuge (L. Bölkow, Editor), VDI Verlag, Düsseldorf, 1990, pp. 268–285
13. E.H. Hirschel, H. Hönlinger (Eds.), Höchstleistungsrechnen in der Luftfahrtindustrie und -forschung, Memorandum, Dasa und DLR, München und Göttingen, 1996
14. J. Vos, A. Rizzi, D. Darracq, E.H. Hirschel, Navier-Stokes Solvers in European Aircraft Design, Progress in Aerospace Sciences, Vol. 38, 2002, pp. 601–697
15. J. Szodruch, The Costs and Benefits of Aerodynamic Design: an Industrial View, Nouvelle Revue d'Aeronautique et d'Astronautique, 2, 1997, pp. 88–94
16. E.H. Hirschel, Towards the Virtual Product in Aircraft Design?, Proceedings of NFD 2000 – Towards a New Fluid Dynamics with its Challenges in Aerospace Engineering, CNRS, Paris, November 2000. In: M. Champion, J. Periaux, J. Pironneau, P. Thomas (eds.), NFD 2000, Towards a New Fluid Dynamics, CIMNE Handbooks on Theory and Engineering Applications of Computational Methods, Barcelona, 2003
17. N. Kroll, National CFD Projekt MEGAFLOW – Status Report, in : New Results in Numerical and Experimental Fluid Mechanics (H. Körner, R. Hilbig (Eds.9), Notes on Numerical Fluid Mechanics, Vol. 60, Vieweg Verlag, Brunswick/Wiesbaden, 1997, pp. 15–23
18. Th. Modis, Predictions. Societies' Tell-tale Signature Reveals the Past and Forecasts the Future, Simon & Schuster, New York, 1992

References

19. D.E. Koelle, Beiträge der Raumfahrt zur Luftfahrt, in: Ein Jahrhundert Flugzeuge, Hrsg.: L. Bölkow, VDI Verlag, Düsseldorf, 1990, pp. 576–597

20. E.H. Hirschel, P. Krammer, W. Uhse, F. Hormann, H. Kuczera, P.W. Sacher, Technologietransferpotentiale de BMFT-Förderkonzeptes Hyperschalltechnologie, DGL 92-03-087, 1992

Appendix

Milestones*

Events Related to German Aeronautical Research until 1945

1881 the "German Association for the Promotion of (Airship) Aviation" is founded.

1889 *Otto Lilienthal's* book "Bird flight as basis of the art of flying" is being published.

1890 *Lilienthal* starts his flight tests with cambered wings,

since **1891** *Lilienthal*, essentially only performs flight tests.

1895

August 10, 1896 *Otto Lilienthal* has his fatal accident.

February 26, 1898 the "Association for the Promotion of Applied Physics of Göttingen, GV" is founded, *Hugo Hergesell* becomes at the conference of the directors of the meteorological institutes, held in Paris, chairman of the "International Committee for Scientific Airship Aviation".

1900 **July 2, 1900** first flight of the Airship **LZ 1.**

December 17, 1903 first powered flight of the *Wright Brothers.*

1904 *Prandtl's* boundary layer theory, *Prandtl* receives a call to Göttingen.

1905

1906 first powered flights in Europe, *von Kármán* becomes assistant of *Prandtl*, foundation of the "Powered-airship Study Association" in Göttingen.

November 1, 1907 foundation of the "Model Test Establishment of the Powered-airship Study Society", first closed-circuit wind tunnel.

1908 foundation of the "Radiotelegraphic and Air-electricity Test-station Gräfelfing, **DVG**".

1909 *Prandtl* starts lectures in "Aeronautics", **July 10 until October 17, 1909** first "International Airship Aviation Exhibition, **ILA**" in Frankfurt/Main, **July 25, 1909** *Blériot's* flight across the English Channel.

1910 **January 1910** the "Journal of Flight-techniques and Powered-airship Aviation, **ZFM**" is being published, **May 17, 1910** first flight of the "Etrich-pigeon".

January 11, 1911 "Emperor-Wilhelm-Society for the Promotion of the Sciences, **KWG**", **1911** first chair of "Aeronautical Technology" (Stuttgart).

1912 call for a "National Flight Donation", **January 27, 1912** announcement of the "Emporer's Prize (Competition)" for the best German flight engine, **April 3, 1912** foundation of the "Scientific Flight Technical Society, **WGF**" and **April 20, 1912** foundation of the "German Test Establishment of Aeronautics, **DVL**".

August 1, 1914 begin of World War One, the **WGF** becomes in **1914** the "Scientific Aeronautical Society, **WGL**".

---

*Also see "Selected dates of German aeronautical history" on Page 137

1915 **1914/1915** commencement of metal aircraft design and construction. **June 1915** the Göttingen "Model Test Establishment" becomes the "Model Test Establishment for Aerodynamics, **MVA**".

1917 "Scientific Consultation Agency for Aviation, **WAF**".

1918 *Prandtl's* lifting-wing theory, **November 11, 1918** World War One ends.

June 28, 1919 Versailles Treaty, ban on aircraft manufacturing in Germany.

1920 **May 11, 1920:** the **MVA** becomes the "Aerodynamic Test Establishment in Göttingen, **AVA**", initially within the **KWG, 1920** "Emergency Association of the German Sciences", first glider competition at the Wasserkuppe (Rhön mountains)

1922 "Definitions" for German aeronautical engineering with a partial lifting of the manufacturing ban.

Late 1923 the inflation in Germany has been overcome.

1924 foundation of the "Rhön-Rossitten Society, **RRG**", since **1924**, secret flight tests in Lipezk, Russia.

1925 **1924/25** "Emperor-Wilhelm-Institute of Fluid Research" in Göttingen.

1928 "German Aeronautical Research Council".

October 24, 1929 begin of the World Economic Crisis.

1930 since **1930** the "Emergency Association of German Sciences" is (until 1945) called "German Research Association, **DFG**".

January 30, 1933 *Hitler's* power seizure in Germany, *A. Baeumker* starts the reform of German aeronautical research. **March 23, 1933** "Center of Scientific Publications, **ZWB**" and "Aeronautical Research Association, **VLF**", **April 29, 1933** "State (Reichs) Ministry of Aviation, **RLM**", **May 1933** the "German Research Institute of Gliding (Soaring), **DFS**" develops out of the **RRG**. Since **1933** "Faculties of Aeronautical Technology" at various universities, career as "Aircraft Master Builder" possible.

1935 **March 16, 1935** unilateral cancellation by Germany of the restrictions on German aircraft industry imposed by the Versailles Treaty.

February 10, 1936 foundation of the "German Research Establishment of Aeronautics **DFL**". **Mai 1936** the **WGL** dissolves, **June 12 1936:** the **VLF** becomes the "Lilienthal-Society of Aeronautical Research". **July 24, 1936** foundation of the "Academy of Aeronautical Research".

1937 out of the **DVG** evolves the "Air Radio Research Institute Oberpfaffenhofen, **FFO**".

1938 the **DFL** is renamed into "Aeronautical Research Establishment Herrmann Göring, **LFA**".

September 1, 1939 begin of World War Two, restrictions on aeronautical research.

1940 **1940** new fife-year plan for aeronautical research, the "Aeronautical Research Establishment Munich, **LFM**" is being founded, in **autumn 1940** air-battle over England.

1941 planning of the new form of research control and sponsorship, foundation of the "Research Establishment Count Zeppelin, **FGZ**".

1942 creation of the "Aviation Research Command, **FoFü**", reformation of the "State (Reichs) Research Council".

1945 **May 7, 1945** end of World War Two, end of aeronautical research activities in Germany.

since **May 1945, CIOS/FIAT** assessment and inventory at German aeronautical research establishments and industry, **1945/46** compilation of the Göttingen and Brunswick monographs, in the Soviet Occupation Zone experimental developments concerning German jet aircraft are secretly being continued, exodus of German aeronautical scientists and engineers.

Events Related to German Aeronautical Research after 1945

October 1946 transfer of **SBZ** aeronautical personnel with families to the **USSR**.

1948 the **KWG** is renamed "Max-Planck-Society, **MPG**".

1949 foundation of **NATO**, the "Federal Republic of Germany, **FRG**" and the "German Democratic Republic, **GDR**" are being established.

1950 **1950–1954** return of the aeronautical researchers and engineers from the **USSR**.

1951 FRG: establishment of the "German Research Association, **DFG**" out of the "German Aeronautical Research Council" and the "Emergency Association of the German Sciences". **DVL** Working Groups resume their research, the ban on gliding is lifted.

1952 DFG "Committee on Aeronautical Research", re-founding of the **WGL**, "Journal of Flight Sciences, **ZFW**", **1952/54** buildup of the aeronautical industry in the **GDR**.

1953 DFL and **AVA** (within the **MPG**) resume their work, the "German Study Association Helicopter, **DSH**" is founded.

1954 foundation of the "Research Establishment of Jet Propulsion Physics, **FPS**", the **FRG** joins **NATO**.

1955 **May 1955: FRG** becomes essentially sovereign, **1955/56** rebuilding of the aeronautical industry, the **FRG** joins the "Advisory Group for Aeronautical Research and Development, **AGARD**", the **FFO** merges into **DVL**. **GDR** "Research Center of the Aeronautical Industry, **FZL**".

1956 Wind-tunnel program of the **FRG**, "Presidential Council of the Aeronautical Research Establishments", foundation of the international "Von Kármán Institute, **VKI**" in Belgium.

1958/59 start of the West European industrial cooperation with the "Transall" (C 160) project

1959 "German Society of Flight Sciences, **DGF**", commencement of the development of military vertical take-off aircraft in the **FRG.**

1960 **1960** the **DFS** becomes the "Research Establishment of Flight Sciences Munich, **FFM**" and merges in **1963** into **DVL**, the **DSH** becomes the "German Research Establishment of Helicopters and Vertical-flight Techniques, **DFH**" and merges in **1962** with the **DFL**.

February 28, 1961 Termination of all aeronautical R&D activities in the **GDR**.

1963 the **FPS** is incorporated into **DVL**, since **1964** "**EUROMECH**-colloquia".

1965

December 1967 the **WGL** becomes the "German Aerospace Society, **DGLR**".

1967/68 begin of a new phase in technology support for military aircraft design with "Future Technologies Air, **ZTL**".

1968 first "Collaborative Research Center, **SFB**" of the **DFG** for aeronautical research.

1969 "Messerschmitt-Bölkow-Blohm, **MBB**" unites most of the aeronautical companies of the **FRG**, on **April 1, 1969, DVL, DFL,** and **AVA** join to become the "German Aerospace Research and Test Establishment, **DFVLR**". **Mai 1969** start of the Airbus **A 300** program.

1970 **1970** the first yearly report of the Federal Government on aeronautical research is issued, **1970-1974** basis program of the German aerospace industry, **DFVLR** introduces Research Planning, since **1971** "European Program on Viscous Flows, **EUROVISC**".

1973 foundation of the "Group for Aeronautical Research and Technology in Europe, **GARTEUR**".

1975 **1974/75** first overall civil technology support program "Aeronautical Research and Aeronautical Technologies" with the "Civil Components Program, **ZKP**".

1980 **1980** start of operations of the "German-Dutch Wind Tunnel, **DNW**".

1984 harmonization of the research plans of **DFVLR** and **MBB.**

1985

1987 "European Cooperative Measures for Aeronautical Research and Technology, "**EUROMART**" study for the **EU**,

since **1989 EU** sponsorship of aeronautical research and technologies, the **DFVLR** becomes the "German Aerospace Research Establishment, **DLR**", **May 1989** "Deutsche Aerospace AG, **Dasa**".

1990 on **October 3, 1990,** the **GDR** ceases to exist.

1990/92 the Airbus loans conflict with the **USA** leads to a bilateral agreement Europe-USA.

1992/93 "Confederation of European Aerospace Societies, **CEAS**".

1994 the "European Transonic Wind Tunnel, **ETW**" starts its commissioning phase.

1995 **1995-1998** first national "Aeronautical Research Program".

1996 "Reference Concepts" and "Focused Technology Areas" of **DLR, Daimler-Benz** und **Dasa.**

1997 DLR becomes the "German Aerospace Center", foundation of the "Association of

European Research Establishments in Aeronautics, **AEREA/EREA**", the journals **ZFW** und **La Recherche Aérospatiale** unite to become the journal "Aerospace Science and Technology, **AST**", **AGARD** becomes the "**NATO** Research and Technology Organization, **RTO**".

1999 continuation of the national "Aeronautical Research Program".

2000 on **July 10, 2000** the "European Aeronauti Defense and Space Company, **EADS**" is estab lished by merger of "DaimlerChrysler Aero space, **Dasa**", the French "Aerospatiale Matra **ASM**" and the Spanish "Construcciones Aer onáuticas, **CASA**".

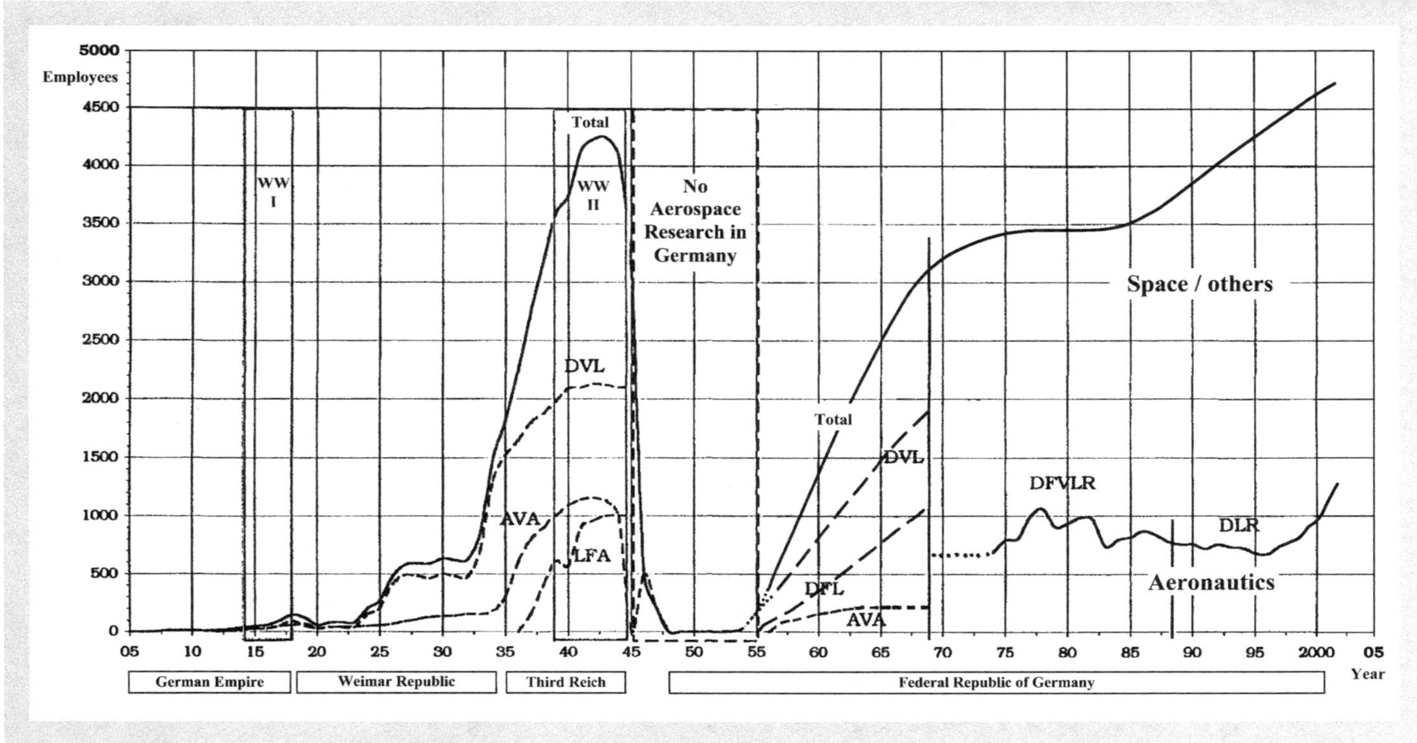

Development of number of staff at the major German research establishments (AVA, DFL, DVL, since 1969 DFVLR/DLR) from 1907 until today (after H.U. Meier, E. Meyer, 1998).
1955 – 1968: total number of staff including space and other activities.
1968 – 1973: exact numbers are not available.
1973 – today, lower curve: only aeronautics.

Conversion of Units

Length
- 1 mm = 0.0394 in
- 1 cm = 0.394 in
- 1 dm = 3.94 in
- 1 m = 3.28 ft
- 1 km = 0.54 nm (nautical mile)
- 1 km = 0,621 mile (statute mile)

Surface
- 1 cm^2 = 0.155 sq in
- 1 m^2 = 10.76 sq ft

Volume
- 1 l (Liter) = 1 dm^3 = 0.264 gal (US)
- 1 l (Liter) = 1 dm^3 = 0.220 gal (UK)
- 1 l (Liter) = 1 dm^3 = 61.02 cu inch

Weight/mass
- 1 kg = 2.205 lb
- 1 t (Tonne) = 1.102 ton (US)
- 1 t (Tonne) = 0.984 ton (UK)

Specific weight
- 1 g/cm^3 = 1 kg/dm^3 = 0.0336 lb/cu in

Mass flow
- 1 kg/s = 2.205 lb/sec

Pressure
- 1 bar = 14.5 lb/sq in
- 1 kg/cm^2 = 14.69 lb/sq in
- 1 kg/m^2 = 0.2048 lb/sq ft

Power
- 1 PS = 0.736 kW
- 1 PS = 0.986 hp
- 1 kW = 1.341 hp

Thrust
- 1 kN = 225 lbf
- 1 kg (kp) = 0.0098 kN

Specific fuel consumption
- 1 g/kNs = 0.0353 lb/lbf h
- 1 kg/kN h = 0.00981 lb/lbf h
- 1 kg/kp h = 1 lb/lbf h
- 1 g/kW h = 0.00164 lb/lbf h
- 1 g/PS h = 0.00224 lb/hp h

Speed
- 1 m/s = 3.28 ft/sec
- 1 km/h = 0.54 kn
- 1 km/h = 0.621 mph

Temperature
- K = °C + 273,16
- °C = (°F – 32) 5/9

Abbreviations

ACT	Active Control Technologies
AECMA	Association Européenne des Constructeurs de Matériel Aérospatiale
AEDC	Arnold Engineering Development Center
AFCS	Aircraft Flight Control System
AGARD	Advisory Group for Aerospace Research and Development
AIAA	American Institute of Aeronautics and Astronautics
Akaflieg	Akademische Fliegergruppe (Academic Flying Group)
ART	Arbeitsgemeinschaft Rückkehrtechnologie (Working Association Reentry Technology)
ART	Advanced Rotor Technology
AST	Aerospace Science and Technology
ATAR	Atelier Aéronautique de Rickenbach
ATTAS	Advanced Technologies Testing Aircraft System
ATTHeS	Advanced Technologies Testing Helicopter System
ALA	Allgemeine Luftfahrt-Ausstellung (General Aeronautical Exhibition)
AVS	Advanced Vertical Take-off and Landing System
AVA	Aerodynamische Versuchsanstalt Göttingen (Aerodynamic Test Establishment Göttingen)
BDLI	Bundesverband der Deutschen Luft- und Raumfahrtindustrie (Federal Association of the German Aerospace Industry)
BLISK	Blade Integrated Disk
BMFT	Bundesministerium für Forschung und Technologie (Federal Ministry of Research and Technology)
BMV	Bundesministerium für Verkehr (Federal Ministry of Transportation)
BMVg	Bundesministerium der Verteidigung (Federal Ministry of Defence)
BMWi	Bundeswirtschaftsministerium (Federal Ministry of Economics)
BMW	Bayerische Motoren Werke (Bavarian Motor Company)
BRITE	Basic Research in Industrial Technologies for Europe
BTZ	Bureau Technique Zborowski
BWB	Bundesamt für Wehrtechnik und Beschaffung (Federal Office of Defense Technology and Procurement)
CALTECH	California Institute of Technology
CCV	Control Configured Vehicle
CEAS	Confederation of European Aerospace Societies
CFD	Computational Fluid Dynamics
CFK	Carbonfaser verstärkter Kunststoff (Carbon Fiber Reinforced Plastics)
CFRP	Carbon Fiber Reinforced Plastics
CIOS	Combined Intelligence Objectives Subcommittee
CIRA	Centro Italiano Ricerce Aerospaziali, Italy
CNRS	Centre National de la Recherche Scientifique
CRISP	Counter-Rotation Integrated Shrouded Propfan
CT	Computer Tomography
DARA	Deutsche Agentur für Raumfahrtangelegenheiten (German Agency of Aeronautical Affairs)
Dasa	Deutsche Aerospace AG (German Aerospace Corporation)
DERA	Defense Evaluation Research Agency, UK
DFG	Deutsche Forschungsgemeinschaft (German Research Association)
DFH	Deutsche Forschungsanstalt für Hubschrauber und Vertikalflugtechnik (German Establishment of Helicopters and Vertical-Flight Technique)
DFL	Deutsche Forschungsanstalt für Luftfahrt (German Research Establishment of Aeronautics)
DFS	Deutsche Forschungsanstalt für Segelflug (German Research Institute for Gliding)

DFVLR	Deutsche Forschungs- und Versuchsanstalt für Luft- und Raumfahrt (German Aerospace Research and Test Establishment)
DFW	Deutsche Flugzeugwerke (German Aircraft Company)
DGF	Deutsche Gesellschaft für Flugwissenschaften (German Society of Flight Sciences)
DGLR	Deutsche Gesellschaft für Luft- und Raumfahrt (German Aerospace Society)
DLH	Deutsche Lufthansa
DLA	Deutscher Luftfahrzeug-Ausschuß (German Aircraft Committee)
DLR	Deutsche Luftfahrtreederei (German Airship Company)
DLR	Deutsches Zentrum für Luft- und Raumfahrt (German Aerospace Center)
DNW	Deutsch-Niederländischer Windkanal (German Dutch Wind Tunnel)
DRA	Defense Research Agency, UK
DSH	Deutsche Studiengemeinschaft Hubschrauber (German Study Association Helicopters)
DVG	Drahtlos-Telegraphie und Luftelektrische Versuchsstation Gräfelfing (Radiotelegraphic and Air-Electricity Test-Station Gräfelfing)
DVL	Deutsche Versuchsanstalt für Luftfahrt (German Test Establishment of Aeronautics)
EADS	European Aeronautic Defense and Space Company
ECD	Eurocopter Deutschland
E-center	Erprobungsstelle (Flight-test Center)
EFA	European Fighter Aircraft
EFM	Enhanced Fighter Maneuverability
ELDO	European Launcher Development Organisation
EPNER	Ecole du Personnel Navigant d'Essais et de Réception, France
EREA	European Research Establishments of Aeronautics
ERNO	Entwicklungsring Nord (Development Group North)
ESG	Elektroniksystem- und Logistik GmbH (Electronic System and Logistic Company)
ESPRIT	European Strategic Program for Research and Development in Information Technologies
ESTOL	Extremely-Short-Take-off and Landing
ETH	Eidgenössische Technische Hochschule (Swiss Institute of Technology)
ETPS	Empire Test Pilot School
ETW	European Transonic Windtunnel
EUCLID	European Cooperation for Long Term Defense
EURAM	European Research in Advanced Materials
EUREKA	European Research Coordination Agency
EUROMART	European Cooperation Measures for Aeronautical Research and Technology
EWR	Entwicklungsring Süd (Development Group South)
FAA	Federal Aviation Authority
FAI	Fédération Aéronautique Internationale
FAG	Flugtechnische Arbeitsgemeinschaft (Flight Technical Expert Group)
FCS	Flight Control System
FDP	Fluid Dynamics Panel
FEM	Finite Element Method
FFA	Flygtekniska Försöksanstalten
FFE	Freie Forschung und Entwicklung (Free Research and Development)
FFG	Flugtechnische Fachgruppe (Flight Technical Working Group)
FFG	Flugzeugbau Friedrichshafen (Aircraft-Factory Friedrichshafen)
FFM	Flugwissenschaftliche Forschungsanstalt München (Research Establishment of Flight Sciences Munich)
FFO	Flugfunk-Forschungsinstitut Oberpfaffenhofen (Air Radio Research Institute Oberpfaffenhofen)
FGZ	Forschungsanstalt Graf Zeppelin (Research Establishment Count Zeppelin)
FIAT	Field Information Agency Technical
FIST	Flugtechnisches Institut Stuttgart (Flight-Technical Institute Stuttgart)
FKFS	Forschungsinstitut für Kraftfahrwesen und Fahrzeugmotoren (Research Institute of Motor Vehicles and Motor-vehicle Engines)

FLZ	Flugzeugmeisterei des Heeres (Aircraft Establishment of the Army)
FoFü	Forschungsführung (Aviation Research Command)
FPS	Forschungsinstitut für Physik der Strahlantriebe (Research Establishment of Jet Propulsion Physics)
FRG	Federal Republic of Germany
FRP	Fiber Reinforced Plastics
FSV	Flugsport-Vereinigung Darmstadt (Flight-Sports Association Darmstadt)
FVA	Flugwissenschaftliche Vereinigung Aachen (Flight Sciences Association Aachen)
FVP	Flugtechnische Versuchsanstalt Prag (Flight Technical Research Establishment Prag)
GAMM	Gesellschaft für Angewandte Mathematik und Mechanik (Society of Applied Mathematics and Mechanics)
GARTEUR	Group for Aeronautical Research and Technology in Europe
GDR	German Democratic Republik
GFW	Gesellschaft für Weltraumforschung (Space Research Society)
GFK	Glasfaserverstärkter Kunststoff (Glass-Fiber Reinforced Plastics)
GFRP	Glass-Fiber Reinforced Plastics
GV	Göttinger Vereinigung zur Förderung der angewandten Physik (Association for the Promotion of Applied Physics of Göttingen)
GWR	Gesellschaft für Weltraumforschung und Raumfahrt (Society of Space Research and Astronautics)
HFB	Hamburger Flugzeugbau (Hamburger Aircraft Company)
HOG	Hermann-Oberth Gesellschaft (Hermann-Oberth Society)
HVA	Heeresversuchsanstalt Peenemünde (Test Establishment of the Army Peenemünde).
HWK	Hochgeschwindigkeits-Windkanal (High-Speed Wind tunnel)
IABG	Industrieanlagenbetriebsgesellschaft (Organization for the Operation of Industrial Facilities)
ICAS	International Council of Aeronautical Sciences
ILA	Internationale Luftschiffahrt Ausstellung (International Airship Aviation Exhibition)
ILÜK	Interallierte Luftfahrt-Überwachungs Kommission (Inter-allied Aviation Control Commission)
INTA	Instituto Nacional de Técnica Aerospacial, Spain
IRA	Institut für Radiotechnik und Flugfunkwesen (Institute of Radio Technology and Air-Radio Transmission)
ISL	Deutsch-Französisches Forschungsinstitut Saint-Louis (German-French Research Institute Saint-Louis)
ISTUS	Internationale Studienkommission für den motorlosen Flug (International Study Committee for Unpowered Flight)
IVG	Industrieanlagenverwaltungsgesellschaft (Industry Administration Organization)
JAA	Joint Aviation Authority
JPL	Jet Propulsion Laboratory
KEL	Komponenten-Experimental Programm Luft (Component-related Experimental Program Air)
KKK	Kryo Kanal Köln (Cryogenic Wind Tunnel Cologne)
KRG	Kryogener Rohrwindkanal Göttingen (Cryogenic Ludwieg-Tube Göttingen)
KWG	Kaiser-Wilhelm-Gesellschaft (Emperor-Wilhelm Society)
KWI	Kaiser-Wilhelm Institut (Emperor-Wilhelm Institute)
LBF	Laboratorium für Betriebsfestigkeit (Institute of Operational Strength)
LFA	Luftfahrtforschungsanstalt Hermann Göring (Aeronautical Research Establishment Hermann Göring)
LFM	Luftfahrtforschungsanstalt München (Aeronautical Research Establishment Munich)

LRBA	Laboratoire de Recherches Balistiques et Aérodynamiques, France
LST	Low-Speed Tunnel
LSWT	Low-Speed Windtunnel
LZ	Luftschiffbau Zeppelin (Airship Construction Company Zeppelin)
MBB	Messerschmitt-Bölkow-Blohm
MPG	Max-Planck Gesellschaft zur Förderung der Wissenschaften (Max-Planck-Society for the Promotion of Science)
MRCA	Multi Role Combat Aircraft
MStG	Motorluftschiff-Studiengesellschaft (Powered Airship Study Association)
MTU	Motoren- und Turbinen Union (Engine and Turbomachinery Association)
MVA	Modell-Versuchsanstalt für Aerodynamik (Model Test Establishment for Aerodynamics)
NACA	National Advisory Committee for Aeronautics
NASA	National Aeronautics and Space Administration
NKF	Neues Kampfflugzeug (New Fighter Aircraft)
NLR	National Lucht- en Ruimtevaarlaboratorium, The Netherlands
NWB	Niedergeschwindigkeitswindkanal Braunschweig (Low-Speed Windtunnel Brunswick)
OKB	Opytnoje Konstruktordkoje Bjuro, USSR
ONERA	Office National d'Etudes et de Recherches Aéronautique, France
OSTIV	Organisation Scientifique et Technique International de Vol à Voile
R&D	Research and Development
RAE	Royal Aircraft Establishment, UK
RFR	Reichsforschungsrat (State Research Council)
RLA	Reichsluftamt (State Aviation Office)
RLM	Reichsluftfahrtministerium (State Ministry of Aeronautics/German Air Ministry)
ROSAR	Rotating Synthetic Aperture Radar
RRG	Rhön-Rossitten Gesellschaft (Rhön-Rossitten Society)
R&T	Research and Technology
RTO	Research and Technology Organisation
RVM	Reichsverkehrsministerium (State Ministry of Transportation)
RWTH	Rheinisch-Westfälische Technische Hochschule Aachen (Rhine-Westphalia Institute of Technology Aachen)
SAR	Synthetic Aperture Radar
SFB	Sonderforschungsbereich (Collaborative Research Center)
SEPR	Société d'Etudes de la Propulsion par Réaction
SNCASE	Société Nationale de Construction Aéronautique de Sud-Est
SNCASO	Société Nationale de Construction Aéronautique du Sud-Ouest
SNECMA	Société Nationale d'Etudes et de Construction des Moteurs Aéronautiques
TaF	Taktische Forderung (Tactical Requirement)
TKF	Taktisches Kampfflugzeug (Tactical Fighter Aircraft)
TH	Technische Hochschule (Institute of Technology)
TU	Technische Universität (Technical University)
UHBR	Ultra-High Bypass Ratio
VDI	Verein Deutscher Ingenieure (Association of German Engineers)
VEB	Volkseigener Betrieb (A People's Enterprise)
VFW	Vereinigte Flugtechnische Werke Bremen (United Flight-Technical Company Bremen)
VHBR	Very-High Bypass Ratio
VKI	Von Kármán Institute, Belgium
VLF	Vereinigung für Luftfahrtforschung (Aeronautical Research Association)
VTOL	Vertical Take-off and Landing
VVB	Vereinigung volkseigener Betriebe (Association of the People's Enterprises)

WAF Wissenschaftliche Auskunftei für Flugwesen (Scientific Consulting Agency for Aviation)

WGF Wissenschaftliche Gesellschaft für Flugtechnik (Scientific Flight-Technical Society)

WGL Wissenschaftliche Gesellschaft für Luftfahrt (Scientific Society of Aeronautics)

WGLR Wissenschaftliche Gesellschaft für Luft- und Raumfahrt (Scientific Aerospace Society)

WNF Wiener-Neustädter Flugzeugwerke (Wiener-Neustädter Aircraft Company)

WTD Wehrtechnische Dienststelle (Test Center of the German Airforce)

WVA Wasserbau-Versuchsanstalt Kochelsee (Hydraulic Engineering Test Organisation Kochelsee)

ZAGI Central Aero-Hydrodynamics Institute (Moscow)

ZFM Zeitschrift für Flugtechnik und Motorluftschiffahrt (Journal of Flight-techniques and Powered-airship Aviation)

ZFW Zeitschrift für Flugwissenschaften (Journal of Flight Sciences)

ZLDI Zentralstelle für Luftfahrtdokumentation und -information (Central Office of Aeronautical Documentation and Information)

ZTL Zukunftstechnologie Luft (Future Technologies Air)

ZWB Zentrale für wissenschaftliches Berichtswesen über Luftfahrtforschung (Center of Scientific Publications concerning Aeronautical Research)

Sources of Illustrations

AGARD (2), Akaflieg Braunschweig (5), Akaflieg Darmstadt (1), Akaflieg Stuttgart (2), Archiv Rietz (2), Autoflug (4), AVA (19), H. Baer (1), H. Bansemir (2), BMW (6), BMW-Rolls Royce (2), Daimler-Benz (1), Dasa (28), Dasa-Airbus (10), Deutsches Museum (206), DFL (1), DFVLR (2), DGLR (12), K. H. Doetsch (4), Dornier (5), DLR (14), DNW (8), Drei-Brunnen-Verlag (6), DV L (49), EADS (3), Elbe-Flugzeugwer-ke (2), Elze-Jacobs (1), ESG (2), ETW (4), Eurocopter (23), R. Eyb (2), EWR (3), R. Flentje (1), Flettner (3), H. Försching (21), Focke-Achgelis (4), Focke-Wulf (2), Flugzeug-Union Süd (1), K. Graichen (1), Grob (2), X. Hafer (11), X. Hafer/G. Sachs (3), H. Hagen (1), P. Hamel (10), Heinkel (6), W. Heinzerling (1), E. H. Hirschel (25), H. Hoheisel (14), Hubschrauber Museum Bückeburg (1), HSBw-München (1), IABG (3), Junkers (2), Jumo (2), K. Kimm (2), LFA (6), LTH (2), Lycoming (1), G. Madelung (3), MAN (3), MAN-Turbo (3), MBB (39), H.U. Meier (38), K. H. Mewes (5), MTU (19), MTZ (1), NASA (1), Oldenbourg-Verlag (1), ONERA (1), Physika-Verlag (1), H. Pflügel (1), H. Prem (8), E. Rabau (1), J. Roeder (2), D. Schmitt (1), SNECMA (5), K. Soestmeyer (3), TU Berlin (2), TU München (7), Universität Stuttgart (3), VFW (2), VFW-Fokker (1), H. Wagner (2), H. Weingarten (2), B. Williams (1), WNF (2), H. Zacher (1), ZFW (1), H. Zacher (1), ZFW (1)

Bibliographicaly Sketches of Outstanding Persons

Main Authors*

Hirschel, Ernst Heinrich, Prof. Dr.-Ing., Dr.-Ing. habil. (*8.10.1934). 1958 to 1963: Study of Aerospace Engineering at the RWTH Aachen. 1966 to 1967: NASA University fellowship at the University of Alabama, Huntsville, Ala., USA. 1963 to 1979 research work at the DFVLR Institute of Applied Gas Dynamics, Köln. Since 1973 Chief of "Theory and Numerics". 1970 Dr.-Ing. at the RWTH Aachen. 1975 to 1982 Lecturer (Dr.-Ing. habil.) and Professor Extraordinarius at the RWTH Aachen, from 1982 to 1991 at the Technical University Munich, since 1992 at the University Stuttgart. 1974 to 1977 first Chairman of the GAMM committee for Numerical Methods in Fluid Mechanics. Since 1980 with MBB (now EADS) Military Aircraft Division. 1988 to 1990 chief of "Theoretical Aerodynamics". 1988 to 1995 Technical Manager of "Alternate Aerothermodynamics" in the Hermes project and of "Aerothermodynamics and Propulsion Integration" in the German Hypersonics Technology Programme ("Sänger"). 1990 to 1994: Manager Technology Air Vehicle Engineering, and 1994 to 1999 Flight Physics. 1995 to 1999 Technical Manager of the Technology Development and Verification Plan of the Future European Space Transportation Investigations Programme (FESTIP) of ESA. October 1999 retired from EADS. Since 1984 General Editor of the "Notes on Numerical Fluid Mechanics" of Vieweg publishing house, now of Springer. Scientific and technical expertise in Fluid Mechanics, Aerodynamics, Aerothermodynamics, Flow-Physics Modelling, Numerical Methods, Multidisciplinary Design. About 100 journal and conference publications. Member GAMM, senior member AIAA.

* E.H. Hirschel also is editor of Part 1 and Part 2, H. Prem of Part 3. All authors are co-editors of Part 4.

Prem, Horst, Dipl.-Ing. (*25.11.1940). After studies of Mechanical Engineering and Aerospace Engineering at TH Brunswick from 1966 to 1970 Assistant Professor. Since 1971 with Messerschmitt-Bölkow-Blohm (MBB). Substantial participation at the German-US Helios solar probe programme. 1977 Head Technology within the Corporate Staff Development. Setting up of the development planning and initiation/planning of the technology programmes Airbus CFRP vertical stabilizer, Infrared Technologies, Micro-electronic capabilities; preparation in establishing the European Euromart and Euclid technology programmes. 1992 Dasa, Corporate Staff Research and Technology. Competition Analysis USA/Europe on the aerospace sector led to the implementation of the national aerospace research programme and enhanced promotion of the European Union (EU) aerospace technology programmes. Since 1997 author on aerospace research matters. President of a company for renewable energies. Member DGLR, AIAA.

Madelung, Gero, Prof. Dipl.-Ing. (*2.2.1928). Professional education: TH Stuttgart, Clarkson College of Technology at Potsdam, N.Y., and TU Munich. Employment: General Electric Co. (USA), Test Engineer; Messerschmitt AG (Germany), Design Engineer and subsequently Project Engineer at the Messerschmitt design office in Sevilla (Spain) with La Hispano Aviacion. As of 1958 Head of Preliminary Development at the Messerschmitt Team of Entwicklungsring Süd (EWR) for the VTOL development VJ-101. 1964 Vice President of Engineering and Board Member at Messerschmitt and as of 1969 CEO of Panavia Aircraft GmbH (trinational MRCA (Multi-Role Combat Aircraft/"Tornado" project management). 1978 to 1983 CEO of Messerschmitt-Bölkow-Blohm GmbH and subsequently Member of its Supervisory Board. Since 1969 lecturing at the TU Munich on aircraft design. 1989 to 1995 Chair of aircraft design at TU Munich. 1978 to 1997 Member of national delegates board of AGARD, including chairmanship from 1986 to 1988. Recipient of von Kármán Medal. Member of DGLR since 1953, recipient of Ludwig-Prandtl-Ring.

Advisors, Contributors, Coordinators

Bergmann, Jürgen W., Dr.-Ing. (*5. 4. 1939). Study of mechanical engineering/aerospace technology at the TH Darmstadt, University Stuttgart. 1980 PhD thesis on flow and combustion phenomena in aerospace propulsion. 1979 to 1983 Head of DFVLR Missile Propulsion. Scientific and development-related work on air-breathing and autogenous engines. 1984 to 1989 Head of Project Monitoring in Aeronautical Research at the IABG. Participation in the establishment of the basic concept of the Hypersonic Technology Program of the BMFT. Participation in the contents-formulation of the first EU Aeronautical Research Program within the framework of BRITE/EURAM as member of various expert teams. Member of the EU Steering Committee for Aeronautical Research within BRITE/EURAM as expert adviser of the BMFT. Setup and head of a contact and advisory office concerning the EU Aeronautical Research Program on order of the BMFT. Since 1985 establishment of Space Projects Monitoring at the IABG. 1990 to 1997 employee at Dasa Corporate Development with main responsibilities in the areas of astronautics (space), propulsion, flow phenomena, and new technologies.

Bouwer, Gerhard, Dr.-Ing. (*21. 7. 1951). 1981 completion of study in electrical engineering at the TU Brunswick. 1982 to 1999 Associate Scientist at the Institute of Flight-systems Technology of the DLR Brunswick. 1983 to 1984 research work at NASA-Ames Research Center. 1987 PhD from the TU Brunswick. Since 2000 Associate Scientist at CargoLifter, Yard Briesen-Brand.

Braig, Wolfgang, Prof. Dr.-Ing. (*9. 1. 1935). 1959 completion of study of mechanical engineering at the TH Stuttgart. 1959 to 1963 Associate Scientist at the Institute of Turbo Flight Engines and at the Material Test Establishment of the TH Stuttgart. 1964 PhD from the TH Stuttgart (Thesis "Metal bonding"). 1964 to 1968 MAN-Turbo Munich, engine-performance computations. 1968 to 1973 AEG Kanis Company Essen, Head of Development and Design. 1973 Professor of Aeronautical Propulsion at the University Stuttgart. Head of the Institute of Aeroengines.

Braun, Günter, Dr.-Ing. (*16. 6. 1949). Study at the TU Munich and the University of Southampton. Scholar of the Study Foundation of the German People (Studienstiftung des deutschen Volkes). 1974 graduation from the TU Munich. 1985 PhD from the University of the Armed Forces Munich (High-frequency Engineering). 1974 to 1982 Siemens, Business Units concerned with construction elements, communication, and central technology business unit. Research and development concerning microprocessors, film-circuit technology, construction elements (building blocks) for directional-radio systems. Product management, sales. 1983 to 1986 University of the Armed Forces Munich. Research and teaching in high frequency and microwave technology. 1986 to 1997 MBB/Dasa/Dornier Satellite Systems. System analyses and head of development projects concerned with antennae and microwave systems for communication and earth-observation satellites and planetary probes. Innovation management, strategies, sales. 1997 up to today, Siemens Business Units concerned with information and communication, enterprise networks, responsible for business strategy, strategic planning, instruments of business control, strategic controlling. Two patents, 35 publications.

Butter, Ulrich, Prof. Dr.-Ing. (*9. 8. 1939). 1965 graduation from the TH Darmstadt (mechanical engineering). 1965 to 1970 Assistant at the TU Berlin. 1970 PhD. 1970 to 1973 Development Engineer at VFW-Fokker in Bremen. 1973 to 1986 MBB Military Aircraft Business Unit, last position was head of the Main-department Flight-control/Flight closed-loop control/Flight-guidance. 1987 Business Unit Rotary Wings (today Eurocopter). 1990 Chief Engineer flight-guidance systems. Since 1986 teaching at the TU Berlin, since 1995 as Honorary Professor of Flight Control. Since 1995 teaching at the University Stuttgart, since 2000 as Honorary Professor.

Dörner, Heiner, Dipl.-Ing. (*24. 4. 1940). Training as mechanic, study of aeronautical technology at the TH Stuttgart. 1967 graduation (MS). Since 1968 Associate Scientist and teaching assignment at the Institute of Aircraft Engineering of the TU Stuttgart as well as Head of the Trainee Office at the Faculty of Aerospace Technology. 1991 to 1996 teaching assignment at the Technical College Esslingen. Supervision of more than 100 study and diploma theses. Lectures on wind energy in Egypt, Argentine, Brasilia, India, Sweden, USA, Venezuela. Co-author of several text and reference books.

Fabian, Hans, Dr.-Ing. (*28. 3. 1928). Study of physics at the TU Berlin. 1959 PhD from the Faculty of Mechanical Engineering of the TU Berlin in the area of fluid technology. 1956-1961 employee of DVL in Berlin, Aachen and Porz-Wahn. 1961-1963 buildup of the Office of Astronautics (Space) at the BDLI. 1963-1988 at MBB with activities in astronautics and within the Central Development Department. 1979-1982 Program Supervisor of the German-Japanese development program of the helicopter BK 117. Until 1988 work at the MBB Central Business Unit Sales.

Fligge, Ulrich Ewald, Dr.-Ing. (*28. 11. 1947). Study at the Wuppertal Engineering School and the Ruhr University Bochum. PhD in 1983 from the University Wuppertal in the area of traffic safety. Large spectrum of experience in the areas of electrical engineering, electronics, closed-loop control technology, traffic-safety technology and real-time simulation technology including Virtual Reality. Since 1985 system engineer and project head at MBB/Dasa/EADS for training and simulation projects; responsible for the training-technology team Education and Training Systems within the EADS Business Unit Military Aircraft.

Försching, Hans, Prof. Dr.-Ing. (*15. 4. 1930). 1949 apprenticeship as machinery mechanic. 1951 study of mechanical engineering at the TH Karlsruhe, 1955 graduation (MS) and subsequently test engineer at Daimler-Benz in Gaggenau (strength and vibration investigations, design of test rigs/test facilities, vehicle acoustics, cooler design). 1958 Associate Scientist at the mechanics department of the Physikalisch-Technische Bundesanstalt (Federal Physical-Technical Establishment). 1958 to 1963 Associate Scientist at the Department of Aeroelasticity of the AVA. Since 1962 Deputy Department Head. 1962 PhD from the TU Brunswick. 1963 Chief Engineer for structural dynamics and aeroelasticity within the Messerschmitt development team for the HA 300 in Heluan, Egypt. 1965 AVA, 1968 Head of the Department of Aeroelasticity. 1968 Habilitation (qualification as university lecturer) at the TU Brunswick. 1972 Director of the Institute of Aeroelasticity of the DFVLR at the Research Center (RC) Göttingen. 1980 to 1982 Acting Head of the RC. 1975 extra-curricular professor at the TU Brunswick. More than 100 books and publications. Several scientific awards and prizes, among others in 1999 the Aachen and Munich Prize for Technology and Applied Natural Sciences.

Friehmelt, Holger, Dr.-Ing. (*12. 4. 1966). Study of aerospace technology at the TU Brunswick. 1993 to 1998 Institute of Flight-systems Technology of the DLR. During that time, two years flight testing of the X-31A at the NASA Dryden Flight Research Center, 1.5 years flight testing of the "Eurofighter" at Dasa in Manching. 1998 advanced design and technology, Dasa Business Unit Military Aircraft. 2000 Chief Engineer of the experimental program "VECTOR" (as successor program of the X-31). 1999 PhD from the TU Brunswick with the topic "Contribution to the modeling and identification of a thrust-vector controlled aircraft". 2001 Member of the Board of DGLR.

Galleithner, Hans, Dipl.-Ing. (*17. 4. 1940). Study of aeronautical technology at the TU Munich. Since 1967 employment at Flight-test Dornier Oberpfaffenhofen (Do 31 program). 1972 until today DLR, Head of the Oberpfaffenhofen Branch. 1973 graduate of EBNER (INE). Since 1975 responsible for the cooperation with WTD 61 in Manching within the defense-technological flight-testing. Working areas: overall-system flight-testing, flying qualities, flight performance, departure/stall/spin. Programs: meteorological test aircraft "Falcon", "Alpha-Jet", "Tornado", F-4, F-15, F-16, "Do-Seastar", X-31, "Transall", "Strato 2C", "Egrett", "Eurofighter".

Gersdorff, Kyrill von, Ing. (*4. 1. 1916). Study in Weimar. 1941 German Air Ministry (RLM), Senior Flight Engineer – Office of Foreign Engines. 1946 to 1958 participation in the development of the "ATAR" turbojet engine at SNECMA in France. 1958 Bölkow-Development KG Stuttgart and Ottobrunn. Head of the Business Unit Helicopters and Light Aircraft. Preparation of the quantity production of the Bölkow "Junior" and the helicopter BO 105. As head of the business unit responsible for the extension of the MBB plant Ottobrunn. 1973 Advisory Board Deutsches Museum in Munich, 1977 adviser concerning the planning of the extension of the new aerospace exhibition hall. Since the same year coordination of the book series "Die deutsche Luftfahrt" ("German Aeronautics"). Co-author of the volumes "Piston and Jet Engines" and " Helicopters and Autogyros", author of "Ludwig Bölkow and his Work – Ottobrunn Innovations". 1976 Oskar-von-Miller Medal of the Deutsches Museum, 1986 Honorary Membership of the German Aerospace Society (DGLR).

Hafer, Andreas, Dr. (*21. 3. 1951). Study of history, mathematics and philosophy in Darmstadt, Frankfurt, Buffalo and Cambridge. 1987 to 1991 participation in a research project concerning the history of flight mechanics at the Chair of Flight Mechanics and Flight Control of the TU Munich. Book and journal publications concerning the history of the city, the history of mathematics and the history of technical sciences.

Hafer, Xaver, Prof. Dr.-Ing. (*6. 10. 1915). Study at the TH Hanover. 1957 PhD from the TH Brunswick. 1939 to 1945 Design Engineer at the Heinkel Company in Rostock-Marienehe and Vienna-Schwechat. 1951 to 1954 at the Körting Company in Hanover-Linden. 1954 to 1959 again working in aircraft engineering at the Heinkel Company in Stuttgart-Zuffenhausen. 1959 to 1964 Head of the Main Department Aerodynamics and Flight Mechanics at the Entwicklungsring Süd. 1961 to 1965 teaching assignment in flight mechanics at the TH Munich. 1964 to 1966 Technical Director at the Vereinigte Flugtechnische Werke Munich. 1966 call to the Chair of Flight Technology at the TH, later TU Darmstadt. 1982 Retirement. 1964 to 1974 Member of the Flight Mechanics Panel of AGARD. More than 40 scientific publications and contributions to encyclopedias and documentations. Three reference books with translations into Russian and Chinese. 1984 Honorary Member of the DGLR. 1993 recipient of the Ludwig-Prandtl-Ring.

Hamel, Peter, Prof. Dr.-Ing. (*12. 7. 1936). Study at the TU Brunswick. 1968 PhD in the aeronautical-technology area. SM MIT. 1963 to 1970 work in the area of flight mechanics and aeronautical systems for the TU Brunswick, 1970/1971 at Messerschmitt-Bölkow-Blohm (MBB) in Hamburg. Since 1971 Director of the Institute of Flight-systems Technology of the DFVLR/DLR. Research foci: rescue and recovery systems (until 1980), system identification, dynamic helicopter model-test techniques, flying qualities, as well as in-flight simulations, flight-test technology and flight-testing. Since 1988 lectures at the TU Brunswick on experimental flight mechanics, since 1993 honorary professor. More than 50 publications concerning flight mechanics. 1993 AGARD Scientific Achievement Award, 1995 AHS Gruopo Agusta Award, 1997 AIAA Survey Paper Citation Award, 1998 von Kármán-Medal of the NATO Research & Technology Organization, 2001 AIAA Fellow and AHS Dr. Alexander A. Klemin Award.

Hanke, Dietrich, Dipl.-Ing. (*9. 9. 1942). Study of aerospace technology at the TU Brunswick. Since 1970 Associate Scientist at the Institute of Flight-systems Technology of the DLR in Brunswick. 1972 Poject Supervisor for the "Flying Simulator" HFB 320 "Hansa Jet". Basic work concerning in-flight simulation, direct lift control and handling qualities. 1981 Head of the department In-flight Simulation and technical supervisor for the airborne simulator and technology carrier "Advanced Technologies Testing Aircraft System, ATTAS" on behalf of the DFVLR. Essential work concerning the system design and technical specifications of "ATTAS". 1987 concept and specification for a "Hermes Training Aircraft, HTA" under contract of ESA/CNES. Since 1993 work related to pilot-induced oscillations (PIO) on rate limitations, extension of the ATTAS flight envelope to include landing in the "experimental mode" and development of an active side-stick with electromagnetic force generation.

Hansen, Holger, Dipl.-Ing. (*24. 9. 1946). Study of mechanical engineering in Hamburg (MS). Since 1971 Program Manager at the Autoflug Company. Development and design of aeronautical safety and rescue systems.

Heinzerling, Werner, Dipl.-Ing. (*16. 8. 1939). Study at the TH Munich (MS). 1964 to 1966 Assistant. 1967 to 1989 Development Engineer at MBB. Participation in planning and advisory committees on aerodynamic large-scale test facilities. Until 1990 President of the Permanent International Wind-tunnel Committee within the Association of the European Aerospace Industry. Since 1989 Head of the Aerospace Section of the Deutsches Museum in Munich. Establishment of the new museum branch for aeronautics in Oberschleißheim (opening 1992). 1984 Oskar-von-Miller-Badge.

Hoheisel, Heinz, Prof. Dipl.-Ing. (*24. 8. 1932). Study of mechanical engineering at the Engineering School Mittweida. 1954 to 1956 Test Engineer at the Gas-turbine Department of the Design and Development Office for Fluid Machinery in Dresden. 1956 to 1960 group supervisor within the engine pre-design of the VEB Development Engineering Pirna (VEB Entwicklungsbau Pirna). 1960 to 1965 project engineer for the design and testing of jet-engine compressors at the Engines-Turbines-Pumps AG in Cairo/Egypt. 1966 to 1995 Associate Scientist at the Institute of Aerodynamics (in 1981 renamed Design Aerodynamics) of the DFL/DLR. 1976 Head of the Department Experimental Aerodynamics, 1985 Head of the Department Propulsion Integration. Since 1996 advisor to DLR especially concerning EU projects (Workshop-organization). More than 40 publications on the topics cascade-aerodynamics and aerodynamics of airframe – engine integration. Since 1983 teaching assignments at the Faculty of Mechanical Engineering of the FH (Engineering School) Brunswick/Wolfenbüttel. Honorary Professorship at the FH.

Jantzen, Eilhard, Dr. (*19. 5. 1927). 1950 study of natural sciences at the TH Stuttgart and TH Munich with the main subject chemistry. 1960 PhD from TH Munich in the area of mineral oils; subsequently at the Institute of Flight-fuel and Lubricants of the DVL in Munich. Establishment and supervision of the work areas Flight-engine Oils, Hydraulic Fluids for Aircraft, Monitoring of Movable Elements, and Tribology in Aerospace with the Practical Application in Satellites. Since 1982 additional activities related to the topic Aeronautical History. More than 100 publications.

John, Helmut, Dr.-Ing. (*22. 3. 1935). Study at the TH Darmstadt of mechanical engineering and aerodynamics. 1962 to 1966 Assistant at the Institute of Flight Technology and Aerodynamics of the TU Darmstadt. 1966 PhD thesis on "Ground Effects in Aerodynamics". 1966 to 1971 at VFW in Munich. 1968 to 1972 as member of the tri-national design team responsible for "MRCA/Tornado" aerodynamics, wing design, and aeroelasticity. 1972 to 1977 at MBB Department Head Aerodynamics and Performance concerned with the Airbus A 300 under contract of German Airbus. Basic design work for the A 310. 1977 to 1980 as supervisor of the TKF project aerodynamics responsible, among other things, for post-stall aerodynamics. 1980 to 1982 Head of Technology for the TKF. 1982 to 1984 Head of Technology for the EAP/EFA. 1984 to 1990 Head of the Main-department Aerodynamics and Propulsion of the Business Unit Military Aircraft. 1990-1998 Head of Flight Physics at Dasa-M.

Junkers, Bernd, Dipl.-Wirtsch.-Ing. (*29. 5. 1940). Study of economic engineering at the TU Berlin. 1974 to 1976 Assistant to the Chairman of the Board of Directors of the Thyssen North See Company ("Thyssen Nordseewerke") in Emden. 1976 to 1997 Managing Director of the Junkers Company in Munich and Gernlinden (manufacturing of free-piston compressors and hydraulic rescue equipment.)

Klöppel, Valentin, Dr.-Ing. (*28. 10. 1943). Study of aeronautical engineering at the RWTH Aachen. 1971 Scientific Assistant at the Aerospace Institute. 1977 PhD thesis on "Sound emission due to acoustic feedback of air jets deflected at right angles". 1977 MBB, now Eurocopter Germany. 1981 Head of the Department Fundamentals of New Systems. 1993 Supervisor of the programs Aeromechanics and Flight-guidance.

Knobling, Kurt, Prof. Dr.-Ing. (*12. 4. 1928). Study of mechanical engineering at the TH Munich. 1953 Scientific Assistant at the Laboratory of Thermal Engines. 1959 Lecturer in aerodynamics, later also in hydraulics and helicopter technology at the Higher Technical School of the Air Force in Neubiberg/Munich. 1962 PhD from the TH Munich. In addition to teaching, establishment of a fluid-technical laboratory with a wind tunnel. 1974 Advisory Board of the Deutsches Museum, Munich. 1979 Professor at the University of the Armed Forces Munich. 1981 Vice President, Dean of the Faculty of Mechanical Engineering. After retirement in 1994, teaching assignment at the University of the Armed Forces in Munich. Publications on fluid technical problems and helicopters.

Meier, Hans Ulrich, Prof. Dr.-Ing. habil. (*22. 1. 1938). Study at the TH Brunswick. 1964 AVA Göttingen. 1970 PhD from the TU Brunswick. 1973 Visiting Assistant Professor at the University of Maryland, Visiting Consultant at the Naval Ordnance Laboratory. 1974 to 1997 Scientific Officer of the US-FRG Data Exchange Agreement "Viscous and Interacting Flow Field Effects". 1977 to 1988 Head of the Department Boundary Layers at the Institute of Experimental Fluid Mechanics of the DFVLR. 1983 to 1988 Head of the DFVLR-MBB Working Association Missile Aerodynamics. 1986 Habilitation (right to lecture) in the area of Experimental Fluid Mechanics at the TU Clausthal. 1988 to 1998 Director of the German-Dutch Wind Tunnel (DNW). 1990 Extra-curricular Professor at the TU Clausthal. 1991 Honorary Professor at the Nanjing Aeronautical Institute, VR China. 1997 member of the Board of Directors of the von Kármán Institute for Fluid Dynamics, Brussels. More than 50 publications.

Ockier, Carl, Dipl.-Ing., MA (*14. 10. 1965). Study at the TU Delft. MA at the University of Maryland. Education as flight-test engineer at the Empire Test Pilots School in England. 1991 Associate Scientist at the Institute of Flight-systems Technology of the DLR in Brunswick. Main work areas: helicopter flying qualities and airborne simulators. 2001 flight-test engineer at Eurocopter Germany in Donauwörth.

Pausder, Heinz-Jürgen, Dipl.-Ing. (*26. 11. 1944). Study of mechanical engineering/aeronautical technology at the TU Brunswick. Graduation (MS) 1973. 1973 up to today Associate Scientist at the DLR Institute of Flight-systems Technology. Main tasks: flight dynamics, airborne simulation and flying qualities of helicopters. 1995-1999 Technical Project Supervisor for the development of the "Airborne Helicopter Simulator, ACT/FHS". Since 1999 Head of the Department Helicopters at the DLR Institute of Flight-systems Technology.

Peters, Klaus, Dipl.-Ing. (*28. 12. 1938). Study of mechanical engineering at the RWTH Aachen, MS of aeronautical engineering/rocket technology. 1966 technical and general editor of the "deutscher aerokurier" aerospace part "Soldier and Technology", editing of "Avicentra-Nachrichten Ausrüstung", "Space Flight/Rocket Technology". 1971 Technical Press Publications Messerschmitt-Bölkow-Blohm (MBB), later Deputy Head Public Relations. 1983 Head of the Protocol, 1989 Acting Head of Public Relations. Since 1990 until 2003 Dasa/EADS Head of Corporate Protocol. Coordinator of the Book Series "Die deutsche Luftfahrt" ("German Aeronautics"). Co-author of Volume 17 "Messerschmitt". Advisory Board Deutsches Museum. Publisher of the planned Aeronautical Dictionary. Member of the Board and Senate DGLR, member LPC, cofounder and committee member of the RAeS, Munich Branch. Member of the Board of the Society for the Preservation of Sites of German Aeronautical History (GBSL).

Pletschacher, Peter (*1. 5. 1941). 1964 activities as freelance collaborator of several journals, among others, "Flugwelt" and "Flugrevue". 1966-1974 Press Department Dornier. 1974 to 1980 Editor in Chief "Flugrevue". 1981 freelance journalist working, among others, for Interavia Journal, Flight, and Aviation Week. 1985 foundation of the Aviatic Publishing House for aeronautically oriented titles. Since 1990 Editor of the DGLR journal Aerospace (Luft- und Raumfahrt). Since 2000 editorially responsible for the aeronautical-history-related Journal "Flugzeug Classic". Since 1982 President of the German Aeronautical Press Club (LPC).

Rist, Dieter, Prof. Dr.-Ing. habil. (*14. 6. 1936). Study of mechanical engineering and flight technology at the TU Berlin. 1971 PhD, 1982 Habilitation (right to lecture) at the TU Munich with investigations in the area of flight propulsion. Teaching in mechanical engineering in the areas of gas turbines and flight propulsion. Supervision of the research groups Engine Technology, Optimization and Combustion chambers at the Chair of Flight Propulsion of the TU Munich. Since 1979 development of high-performance fans, turbo compressors and flow channel systems for aircraft and railroads together with, among others, aircraft and equipment companies. Since 1983 cooperation with companies of the gas industry and government agencies in the areas of gas transportation, gas drying, combustion, pollutant emissions, safety systems and economic efficiency investigations. 1996 author of the Springer-Book "Dynamik Realer Gase" (Dynamics of Real Gases).

Sachs, Gottfried, Prof. Dr. Ing. (*16. 1. 1939). 1960-1965 study of mechanical engineering at the TH Darmstadt. 1965 to 1977 activities at the University (TH Darmstadt), Research (German Aerospace Test Establishment DVL, Oberpfaffenhofen) and Industry (Dornier Ltd, Friedrichshafen). 1977 to 1983 Professor of Flight Mechanics and Flight Guidance, University of the Armed Forces Munich. Since 1983 Professor of Flight Mechanics and Flight Guidance, TU Munich. More than 260 publications, four reference books, 12 scientific movies/videos.

Schmitt, Dieter, Prof. Dr.-Ing. (*22. 6. 1942). Study of mechanical engineering at the TH Darmstadt. 1968 to 1975 Associate Scientist at the Institute of Flight Technology. 1976 PhD with a topic from flight mechanics. Employed at MBB in Hamburg in the area of flight mechanics. 1980 at Airbus Industries in Toulouse as General Manager of Aerodynamics, subsequently setup of the Department Research and Technologies. 1990 at German Airbus in Hamburg Head of the Department Future Projects, 1994 Chief Engineer New Projects at Airbus Industries in Toulouse. 1996 Professor of Aeronautical Technology at the TU Munich. Numerous presentations and publications. Member of the Board and the Senate of DGLR, Head of the Study Group Aerospace of the VDI in Bavaria.

Schubert, Helmut, Dipl.-Ing., Dipl. Wirtschafts.-Ing. (*17. 9. 1935). Study mechanical engineering at the TU Berlin and TU Munich. Post-graduate study related to Labor- and Economic Sciences at the TU Munich. 1964 MAN-Turbo-motors as Computation Engineer within the Engineering Sciences Department. Participation in the development of the engines RB.153-61, RB.193-12 and RB.199-34R. 1973 Head of the Staff Office Technical Information of the MTU. Author of several reference books on aeronautical propulsion and aeronautical history. Member of the Advisory Board Deutsches Museum. Long-time head of the DGLR-section on Aerospace History.

Sedlmayr, Gerhard, Dr. (*19. 11. 1924). Son of the flight pioneer Gerhard Sedlmayr, founder of the Autoflug Company. Study in Göttingen and Hamburg. 1954 to 1955 Assistant at the University Hamburg, PhD. Since 1956 rebuilding of the paternal Autoflug Company. Specialization on aerospace safety and rescue equipment. 1972 Federal Distinguished Service Cross (Bundesverdienstkreuz), 1984 Oskar-von-Miller-Medal.

Trischler, Helmuth, Prof. Dr. (*29. 4. 1958). Study of history and German philology in Munich. 1986 PhD, 1991 Habilitation (right to lecture) with a treatise on the history of aerospace research in Germany between 1970 and 1990. Since 1992 Research Director of the Deutsches Museum. 1998 Extra-curricular Professor of Newer History and Technological History at the LMU Munich. Since 2000 Executive Board Member of the Münchner Zentrum für Wissenschafts- und Technikgeschichte (Munich Center of the History of Science and Technology). Numerous publications on the history of science and technology in the 19th and 20th century.

Zacher, Hans, Dipl.-Ing. (22. 11. 1912 – 5. 8. 2003). After graduation from highschool and training as mechanic, study of aircraft engineering at the TH Darmstadt. 1933 Academic Flying Group (Akaflieg) Darmstadt, later its head. Development of aircraft. Powered-flight training at the DVL in Adlershof. 1940 to 1945 Department Head of Flight Mechanics at the DFS. After 1951 renewal of the licenses. 1952 type tester for sailplanes and light aircraft. 1958 DFS, later DFVLR, then DLR. Until 1977 Head of the Department Sailplanes and Light Aircraft, responsible for, among other things, flight measurements, Idaflieg Comparative Flights, flight courses for engineers, construction regulations with corresponding presentations and publications. Advisory Board Deutsches Museum, Corresponding Member of the DGLR. Golden Daidalos-Medal of the DAeC, Federal Distinguished Serice Cross (Bundesverdienstkreuz). Co-author of Vol. 19 "Evolution of Sailplanes", contributions to Vol. 23 "Sports and Light Aircraft" of the book series "Die deutsche Luftfahrt" ("German Aeronautics").

Name Index

Subject Index

Subject Index

Printing and Binding: Stürtz AG, Würzburg

www.ingramcontent.com/pod-product-compliance
Lightning Source LLC
Chambersburg PA
CBHW081435060526
44446CB00047B/243
* 9 7 8 3 5 4 0 4 0 6 4 5 7 *